2nd International Conference on Innovation in Engineering and Vocational Education (ICIEVE 2017)

IOP Conference Series: Materials Science and Engineering
Volume 306

Manado, Indonesia
25 - 26 October 2017

Editors:

Ade Gafar Abdullah
Asep Bayu Dani Nandiyanto
Isma Widiaty
Verry Palilingan

ISBN: 978-1-5108-5887-9
ISSN: 1757-8981

Printed from e-media with permission by:

Curran Associates, Inc.
57 Morehouse Lane
Red Hook, NY 12571

proceedings.com

Some format issues inherent in the e-media version may also appear in this print version.

Printed by Curran Associates, Inc. (2018)

Institute of Physics
Dirac House, Temple Back
Bristol BS1 6BE UK

Phone: 44 1 17 929 7481
Fax: 44 1 17 920 0979

techtracking@iop.org

Additional copies of this publication are available from:

Curran Associates, Inc.
57 Morehouse Lane
Red Hook, NY 12571 USA
Phone: 845-758-0400
Fax: 845-758-2633
Email: curran@proceedings.com
Web: www.proceedings.com

TABLE OF CONTENTS

IOP Conf. Series: Materials Science and Engineering **306** (2018) 011001 doi:10.1088/1757-899X/306/1/011001

Preface

The 2nd ICIEVE 2017, the International Conference on Innovation in Engineering and Vocational Education, held on October 25-26, 2017 at Aryaduta Hotel, Manado, North, Sulawesi, Indonesia, is hosted by Universitas Pendidikan Indonesia (Indonesia), Universitas Negeri Manado (Indonesia), and Rajamangala University of Technology Thanyaburi (Thailand).

The conference was a platform for scientists, scholars, engineers, industrial professionals, and researchers to exchange, share, and discuss their innovation, experiences, research works and problem solving techniques in all issues in engineering and vocational education.

The participants of ICIEVE 2017 were from around the world with a variety of background, including academics, industry, and even well-known enterprise. In general, there were 140 papers discussing such various topics as engineering and technology innovation (mechanical engineering, chemical engineering, civil engineering, etc.), engineering education (basic science in engineering education, engineering education reforms, new technologies in education, etc.), and vocational education and training (industry-driven training programs and collaborations, lifelong learning – reskilling and upskilling, government and policy, etc.).

We would like to thank all of those who helped and supported ICIEVE 2017. Each individual and institution's support was very important for the success of this conference. Specifically, we would like to acknowledge the advisory board, scientific committee, and organizing committee for their valuable advice, help, suggestions, and support in the organization and helpful peer-reviewing process of the papers. This year, we would like to express our deepest gratitude for all the co-hosts of ICIEVE 2017, UNIMA, Indonesia, and Rajamangala University of Technology Thanyaburi, Thailand for the collaboration. We would also extend our best gratitude to keynote speakers for their valuable contribution for sharing ideas and knowledge in the ICIEVE 2017.

We sincerely hope that ICIEVE 2017 will be a forum for excellent discussions for improving the quality of research and development in relation to innovation in engineering and vocational education. We also hope that this forum will put forward new ideas and promote collaborative researches among participants. We believe that the proceedings can serve as an important research source of reference and the knowledge. Indeed, the proceedings will lead to not only scientific and engineering progress but also other new products and processes for better science and technology in vocational education.

The Editors
Dr. Ade Gafar Abdullah
Dr. Eng. Asep Bayu Dani Nandiyanto
Dr. Isma Widiaty
Dr. Verry Palilingan, M. Eng

IOP Conf. Series: Materials Science and Engineering **306** (2018) 011001 doi:10.1088/1757-899X/306/1/011001

LIST OF COMMITTEE

Conference Chair:
Dr. Kamin Sumardi, M.Pd.

Co- Conference Chair:
Dr. Rita Patriasih, M.Si.

International Advisory Boards:

1. Dr. Michael Grosch – Karlsruhe Institute of Technology, Germany
2. Dr. Tugba Ozturk – Ankara University, Turkey
3. Prof. Paryono, SEAMEO VOCHTECH, Brunai Darussalam
4. Erica Smith (Federation University, Australia)
5. Prof. Tetsu Kubota (Hiroshima-University, Japan)
6. Luisa Brotas (London Metropolitan University, UK)
7. Prof. Sirilak Hanvatananukul (RAVTE, Rajamangala University of Technology Thanyaburi, Thailand)
8. Asnul Dahar Mingat (UTM, Malaysia)
9. Prof. Muhammad Sukri Saud (UTM, Malaysia)
10. Ramlee bin Mustapha(UPSI, Malaysia)
11. Mohd. Sattar bin Rasul (UKM, Malaysia)
12. Mohd. Nazeri (IPGK Perlis, Malaysia)
13. Phil. Dadang Kurnia (GIZ-Germany)
14. Joachim Dittrich (ITB-Germany)
15. Sigit Dwiananto Arifwidodo (KASETSART Univ-Thailand)
16. Henri DOU – Aix Marseille University, France
17. Henri GOUIN – Aix Marseille University, France
18. Guido BALKEMA – Han University, Netherland

Scientific Committee

1. Prof. M. Syaom Barliana – Universitas Pendidikan Indonesia
2. Prof. Asep Kadarohman – Universitas Pendidikan Indonesia
3. Prof. Dr. Revolson Mege, M.S – Universitas Negeri Manado
4. Prof. Dr. Harry Sumual – Universitas Negeri Manado
5. Dr. Eng. Agus Setiawan – Universitas Pendidikan Indonesia
6. Dr. Ida Hamidah – Universitas Pendidikan Indonesia
7. Dr. Eng. Asep Bayu Dani Nandiyanto – Universitas Pendidikan Indonesia
8. Dr. Ade Gafar Abdullah – Universitas Pendidikan Indonesia
9. Dr. Budi Mulyanti, M.Si – Universitas Pendidikan Indonesia
10. Prof. Dr. Ratih Hurriyati, M.Si – Universitas Pendidikan Indonesia
11. Prof. Khairurrijal – Institut Teknologi Bandung
12. Dr. Ika Amalia Kartika – Institut Pertanian Bogor
13. Prof. Luthfiyah Nurlaela – Universitas Negeri Surabaya
14. Prof. Ivan Hanafi – Universitas Negeri Jakarta
15. Dr. Wagiran – Universitas Negeri Yogyakarta
16. Dr. Putu Sudira – Universitas Negeri Yogyakarta
17. Prof. Jasruddin Daud Malago – Universitas Negeri Makasar
18. Prof. Dian Fiantis – Universitas Andalas

IOP Conf. Series: Materials Science and Engineering **306** (2018) 011001 doi:10.1088/1757-899X/306/1/011001

19. Prof. Dr. Heindrich Taunaumang – Universitas Negeri Manado
20. Dr. Jimmy Waworuntu – Universitas Negeri Manado
21. Dr. Hendro Sumual – Universitas Negeri Manado
22. Dr. Sutthiporn Boonsong – Rajamangala University of Technology Thanyaburi (RMUTT), Thailand
23. Dr. Thosporn Sangsawang – Rajamangala University of Technology Thanyaburi (RMUTT), Thailand

Organizing Committee

1. Dr. Kamin Sumardi, M.Pd – Conference Chairman (Universitas Pendidikan Indonesia)
2. Dr. Verry Palilingan, M. Eng – Technical Chairman (Universitas Negeri Manado)
3. Dr. Isma Widiaty, M.Pd (Universitas Pendidikan Indonesia)
4. Dr. Ana, M.Pd (Universitas Pendidikan Indonesia)
5. Dr. Ing. Drs. Parabelem Rompas, MT (Universitas Negeri Manado)
6. Dr. Jane Sumarouw, M. Si (Universitas Negeri Manado)
7. Dr. Ferol Warou, M. Eng (Universitas Negeri Manado)
8. Dr. Debbie A. J. Harimu, S.T., M.T. (Universitas Negeri Manado)
9. Made Krisnanda, S.T, M.T (Universitas Negeri Manado)

IOP Conf. Series: Materials Science and Engineering **306** (2018) 011001 doi:10.1088/1757-899X/306/1/011001

CONFERENCE PHOTOGRAPH

IOP Conf. Series: Materials Science and Engineering **306** (2018) 011001 doi:10.1088/1757-899X/306/1/011001

IOP Conf. Series: Materials Science and Engineering **306** (2018) 011001 doi:10.1088/1757-899X/306/1/011001

IOP Conf. Series: Materials Science and Engineering **306** (2018) 011002 doi:10.1088/1757-899X/306/1/011002

Peer review statement

All papers published in this volume of *IOP Conference Series: Materials Science and Engineering* have been peer reviewed through processes administered by the proceedings Editors. Reviews were conducted by expert referees to the professional and scientific standards expected of a proceedings journal published by IOP Publishing.

IOP Conf. Series: Materials Science and Engineering **306** (2018) 012001 doi:10.1088/1757-899X/306/1/012001

New Model of Information Technology Governance in the Government of Gorontalo City using Framework COBIT 4.1

A A Bouty*, M H Koniyo, and D Novian

Department of Informatics Engineering, Faculty of Engineering, State University of Gorontalo, Indonesia

*abd.azizbouty@ung.ac.id

Abstract. This study aims to determine the level of maturity of information technology governance in Gorontalo city government by applying the COBIT framework 4.1. The research method is the case study method, by conducting surveys and data collection at 25 institution in Gorontalo City. The results of this study is the analysis of information technology needs based on the measurement of maturity level. The results of the measurement of the maturity level of information technology governance shows that there are still many business processes running at lower level, from 9 existing business processes there are 4 processes at level 2 (repetitive but intuitive) and 3 processes at level 1 (Initial/Ad hoc). With these results, is expected that the government of Gorontalo city immediately make improvements to the governance of information technology so that it can run more effectively and efficiently.

1. Introduction

Information technology (IT) is one of the important aspects of governance organizations as a whole. The success of the organization will be greatly influenced by its ability in utilizing IT optimally [1]. IT governance is a procedure directive implementation of organizational arrangements to support the management of IT integrally and follow up the goals and strategies of organizations that have responsibility [2]. Another definition of IT governance is the decision making that ensures the allocation of IT usage in the organization's strategies [3].

IT governance at government institutions is critical to ensure the successful use of IT in supporting the performance of government institutions, as mandated by Presidential Instruction No. 33 of 2003 on "National policies and strategies for e-government development" [4].

To create a progressive and competitive government, Gorontalo City has a great vision "The implementation of the excellent service of Gorontalo city government to realize the prosperous, progressive, active, religious and educated of Gorontalo". In line with that vision, one of the mission related to IT field is to increase the availability of IT infrastructure in urban area by opening cyber city and techno park area, as well as organizing IT-based government service program [5].

At the implementation level, the use of IT in governance needs to take into account the limitations of resources such as data, technology, facilities, and human resources including the fact that IT is relatively costly in its utilization. Limited resources is then an important factor the need for IT governance governing the use of IT in government organizations. IT governance in the process of data management is less good will cause some problems that are weaknesses that will cause threats such as the occurrence of loss, destruction, theft and tapping of important data company or organization.

IOP Conf. Series: Materials Science and Engineering **306** (2018) 012001 doi:10.1088/1757-899X/306/1/012001

Based on that, it is necessary to do an analysis of IT needs for the government of Gorontalo city through the measurement of maturity level of IT governance so that it can be known the level of IT governance maturity that runs in Gorontalo city government institutions. In measuring the maturity level of IT governance it will be used COBIT (Control Objective for Information and Related Technology) which is a framework that provides a standard within the framework of a domain consisting of a set of IT processes that represent controllable and structured activities [6] [7] [8].

COBIT is an integrated framework for alignment between IT resource management and organizational objectives with the direction of integration and risk associated with IT and is an IT governance management concept addressed to management, IT service staff, departmental control, audit functions and business process owners for confidentiality, integrity and availability of sensitive and critical data and information [9] [10].

2. Method

This research method using case study method by conducting survey, collecting data through questionnaire questionnaire and interview related support of information technology service at government institutions of Gorontalo City. There are four stages in this research that is 1). Surveying and data collection, 2). Preparation of Information Technology Governance Framework, 3). Measurement of Information Technology Governance, 4). Information Technology Governance Analysis.

3. Results and Discussion

3.1. The Result of Compilation of IT Governance Framework

Referring to the COBIT standard 4.1 the scope of information technology governance is divided into several stages of the process: Identification of Business Goals, IT Goals Identification, IT Process Identification and Identification of Control Objectives.

3.1.1. Identification of Business Goals

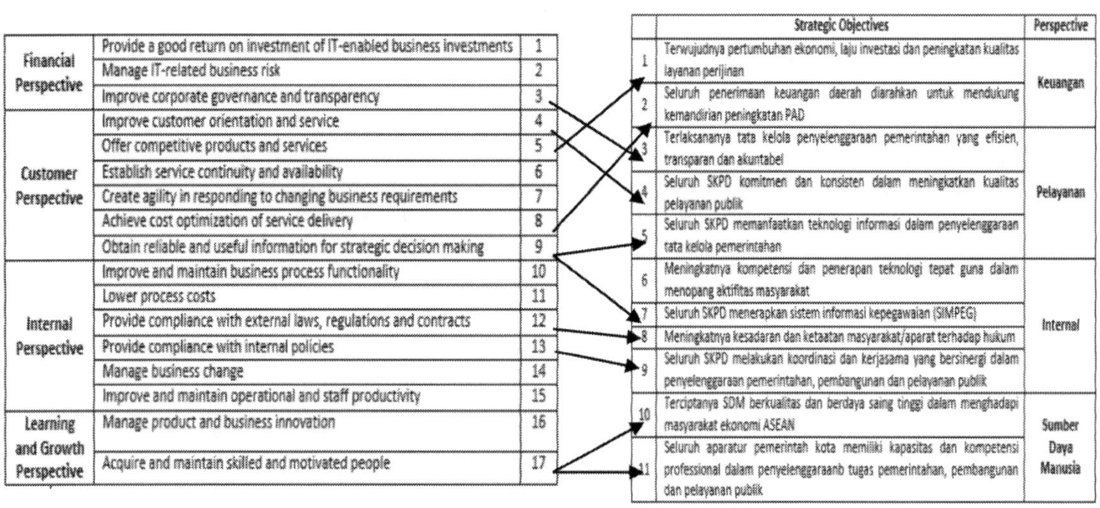

Figure 1. COBIT Business Goals Linkage with strategic goals the government of Gorontalo City

Referring to figure 1 on the linkage between COBIT business goals and strategic goals the government of Gorontalo City can be seen from four perspectives that exist, that the priority of business target relationship is in the perspective of service and human resource. This indicates that the government of Gorontalo City wants to improve the service to the society and want to increase the capacity and competence of the government employs to create good and quality governance.

3.1.2. IT Goals Identification

Table 1. Mapping business goals with IT goals of Gorontalo City Government

Perspective		Business Goals COBIT							
Financial	3	Improve corporate governance and transparency	2	18					
	4	Improve customer orientation and service	3	23					
	5	Offer competitive products and services	5	24					
Customer	8	Achieve cost optimization of service delivery	7	8	10	24			
	9	Obtain reliable and useful information for strategic decision making	2	4	12	20	26		
Internal	12	Provide compliance with external laws, regulations and contracts	2	19	20	21	22	26	27
	13	Provide compliance with internal policies	2	13					
Learning and Growth	17	Acquire and maintain skilled and motivated people	9						

Based on the results of the mapping in Table 1, the COBIT IT goals identified in the Gorontalo City Government as described in table 2. the following :

Table 2. IT goals identified in the government of Gorontalo City

2	Respond to governance requirements in line with board direction
3	Ensure satisfaction of end users with service offerings and service levels
4	Optimize the use of information
5	Create IT agility
7	Acquire and maintain integrated and standardized application systems
8	Acquire and maintain an integrated and standardized IT infrastructure
9	Acquire and maintain IT skills that respond to the IT strategy
10	Ensure mutual satisfaction of third-party relationships
12	Ensure transparency and understanding of IT cost, benefits, strategy, policies and service levels
13	Ensure proper use and performance of the applications and technology solutions
18	Establish clarity of business impact of risk to IT objectives and resources
19	Ensure that critical and confidential information is withheld from those who should not have access to it
20	Ensure that automated business transactions and information exchanges can be trusted
21	Ensure that IT services and infrastructure can properly resist and recover from failures due to error, deliberate attack or disaster
22	Ensure minimum business impact in the event of an IT service disruption or change.
23	Make sure that IT services are available as required
24	Improve IT's cost efficiency and its contribution to business profitability
26	Maintain the integrity of information and processing infrastructure
27	Ensure IT compliance with laws, regulations and contracts

3.1.3. IT Process Identification. At this stage, the results of the identification of IT processes selected in accordance with the IT Goals applicable in the Government of Gorontalo City.

Table 3. Result of Identification IT Process

IT Process	IT Domain
PO1, PO4, PO7	Plan and Organize
AI3, AI4	Acquire and Implementation
DS1, DS4, DS7	Deliver and Support
ME1	Monitor and Evaluation

3.1.4. Identification of Control Objectives. Based on the research that has been done there are 9 processes and 62 detailed control objectives

3.2. The Result of Measurement of Maturity Level from Selected IT Process

The calculation of the maturity level of the selected IT process is done gradually for each IT process by using COBIT 4.1-Maturity Tools. Table 4. Below shows the weight of each level of maturity

Table 4. Weight of the level maturity of each process

No	Value of Maturity	Description of Maturity
0	0 – 0.50	Non-Existent
1	0.51 – 1.50	Initial/Ad Hoc
2	1.51 – 2.50	Repeatable but Intuitive
3	2.51 – 3.50	Define Process
4	3.51 – 4.50	Managed and Measurable
5	4.51 – 5.00	Optimised

Furthermore, for each selected IT process will be presented in Table 5 which is the result of maturity level calculation on IT governance that runs on the government of Gorontalo City.

Table 5. The Value of the maturity level of the selected IT Process

No	IT Process	IT Process Description	Value of Maturity (existing condition)	Level of Maturity (existing condition)	Description of Maturity Level
1	PO1	Define a strategi IT Plan	3.30	3	Define Process
2	PO4	Define the IT process, Organisation and Relationships	2.35	2	Repeatable but Intuitive
3	PO7	Manage IT Human Resource	2.48	2	Repeatable but Intuitive
4	AI3	Acquire and Maintain IT Technology Architecture	2.15	2	Repeatable but Intuitive
5	AI4	Enable Operation and Use	2.80	3	Define Process
6	DS1	Define and Manage Service Levels	1.20	1	Initial
7	DS4	Ensure Continuous Service	1.15	1	Initial
8	DS7	Educate and Train Users	2.40	2	Repeatable but Intuitive
9	ME1	Monitor and Evaluate IT Performance	0.85	1	Initial

Based on table 5 it can be seen that the value of maturity level of each selected IT process is mostly at level 2 i.e. there are 4 processes (PO4, PO7, AI3 and DS7). Which get the highest level (level 3) that is on process of PO1 and AI4. While there are 3 processes at the lowest level (level 1) i.e. the process DS1, DS4 and ME1.

Table 6. The result of the compilation of the IT Process findings

	IT Process Code	Level of Maturity	Compilation of Findings
Domain PO	PO1. Define a strategic IT	Define Process	1. The institution has formal and written standard procedures in defining IT strategy and has been socialized to all parts to be obeyed and worked on. 2. It still allows the occurrence of many irregularities, because there is no supervision in running the procedure.
	PO4. Define the IT process, Organisation & Relationships	Repeatable but Intuitive	1. The institution already has an IT organizational structure but has not yet elaborated on duties and responsibilities 2. There is no mapping of staff needs and skills
	PO7. Manage IT Human Resource	Repeatable but Intuitive	1. The percentage of certified IT personnel is still lacking 2. There is no list of competence and training needs on a regular basis in each institution

IOP Conf. Series: Materials Science and Engineering **306** (2018) 012001 doi:10.1088/1757-899X/306/1/012001

Table 6. Cont.

Domain AI	AI3. Acquire Automated Solutions	Repeatable but Intuitive	1. Institutions have planning in acquiring technology in accordance with the infrastructure plan 2. The institution has not implemented internal controls related to security and auditing measures
	AI4. Enable Operation and Use	Define Process	1. Institutions already have operational documents, user manuals, technical and administration 2. Many applications are stand-alone and not yet integrated in various business processes
Domain DS	DS1. Define and Manage Service Levels	*Initial*	1. Institutions already has formal procedures for managing service levels 2. The service level management process is still manuall 3. Undefined responsibility and accountability of service management
	DS4. Ensure Continuous Service	*Initial*	1. There is no document of contingency test result and backup plan document and storage protection 2. No service requirement document in the event of an incident
	DS7. Educate and Train Users	Repeatable but Intuitive	1. Institutions have been aware of the need for education and training programs and are being developed 2. There is no standard and documentation related to the training curriculum 3. Lack of monitoring and reporting on the effectiveness of training
Domain ME	ME1. Monitor and Evaluate IT Performance	*Initial*	1. There is no document and standard of IT performance planning in every institution 2. There is no measurement of IT performance on each institution

3.3. Gap Analysis / Gap Maturity Level of Selected IT Process

Table 7. The result of the compilation of the IT Process findings

No	IT Process	IT Process Description	Assessment of maturity level			
			Index	Level Exist	Target	Gap
1	PO1	Define a strategi IT Plan	3.30	3	4	0.7
2	PO4	Define the IT process, Organisation and Relationships	2.35	2	4	1.65
3	PO7	Manage IT Human Resource	2.48	2	4	1.52
4	AI3	Acquire and Maintain IT Technology Architecture	2.15	2	4	1.85
5	AI4	Enable Operation and Use	2.80	3	4	1.2
6	DS1	Define and Manage Service Levels	1.20	1	4	2.8
7	DS4	Ensure Continuous Service	1.15	1	4	2.85
8	DS7	Educate and Train Users	2.40	2	4	1.6
9	ME1	Monitor and Evaluate IT Performance	0.85	1	4	3.15

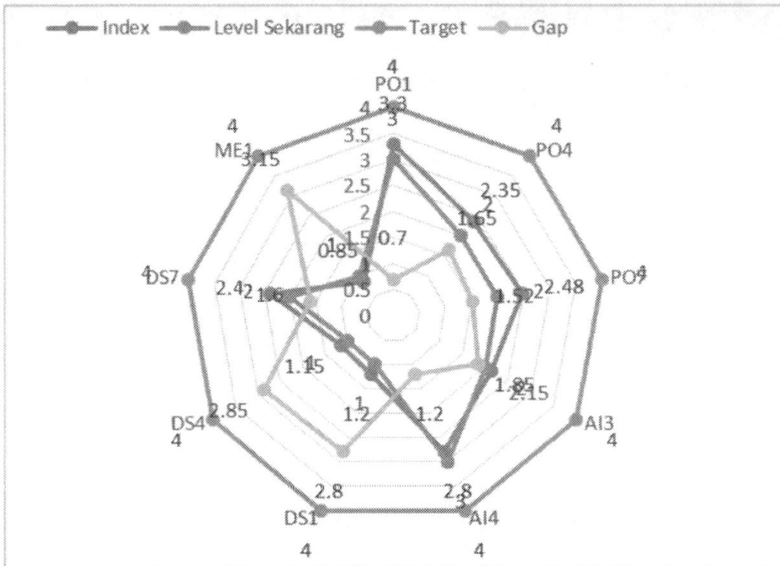

Figure 2. Gap Level Diagram Maturity Level

3.4. Discussion

In this research produce a measurement value of maturity level of IT governance at government of Gorontalo City. Acquisition of the value of this maturity level refers to the application of the COBIT 4.1 framework. Referring to Gap maturity level produced, the government of Gorontalo City should be willing to cover the gap of the existing gap so that all IT resources owned can be managed optimally, and all business processes in the government of Gorontalo City more effective and efficient, accordingly with strategic objectives that have been established through the vision and mission the government of Gorontalo City. The following recommendations are given from this study to be improved by the government of Gorontalo city so as to increase the value of maturity.

4. Conclusion

Based on the result of this research, can be described the conclusion as follows:

1. IT management in the government of Gorontalo city has not been run in accordance with the government's strategic plan, although the government has standard procedure which defines IT implementation strategy
2. There is still a lack of supervision in carrying out the procedure causing frequent occurrence of various deviations in planning and management of IT
3. To measure the maturity level of IT governance in the government of Gorontalo city, the results of the evaluation show that there are still many business processes running at the bottom level, where from 9 processes there are 4 processes at level 2 (Repeatable but Intuitive) and 3 processes are at level 1 (Initial / Ad Hoc)
4. It needs a comprehensive improvement effort from the government of Gorontalo City to the policies, management, and IT resources to produce a good and quality IT governance

References

[1] Weber, Ron 1999 *Information System Control Audit* (New Jersey: Prentice Hall)

[2] Haes, Steven De, and Wim Van Grembergen 2004 *IT Governance and Its Mechanisms." Information Systems Control Journal 1*

[3] Jeanna W Ross, Peter Weill, David C Robertson 2004 *IT Governance, How Top Performer Manage IT Decision Rights for Superior Result* (Harvard Business School Press)

[4] Instruksi Presiden Republik Indonesia No. 3 Tahun 2003. *Kebijakan dan Strategi Nasional Pengembangan E-Government. 2003*
[5] Peraturan Daerah Kota Gorontalo No. 03. 2014 *Rencana Pembangunan Jangka Menengah Daerah Kota Gorontalo Tahun 2014 – 2019*
[6] Budi Yuwono, Muhammad Nasri, Rein Nusa Triputra 2009 Measuring the Effectiveness of a Simplified COBIT based IT Process Maturity Assesment Method *Jurnal Sistem Informasi. Fakultas Ilmu Komputer* (Universitas Indonesia)
[7] ITGI 2001 *Board Briefing on IT Governance*. From.http://www.itgi.org.
[8] ITGI 2003 *Board Briefing on IT Governance*.
[9] IT Governance Institute 2007 *COBIT 4.1*. (IT Governance Institute, Illinois)
[10] Triputra Group 2010 *Buku Pedoman Triputra Management Information System* (Corporate Information Technology Triputra Group)

Photodecomposition Profile of Curcumin in the Existence of Tungsten Trioxide Particles

A B D Nandiyanto[1,*], R Zaen[1], R Oktiani[1], and A G Abdullah[2]

[1]Departemen Kimia, Universitas Pendidikan Indonesia, Jl. Dr. Setiabudi no 229, Bandung 40154, Jawa Barat, Indonesia
[2]Departemen Elektro, Universitas Pendidikan Indonesia, Jl. Dr. Setiabudi no 229, Bandung 40154, Jawa Barat, Indonesia

*nandiyanto@upi.edu

Abstract. The purpose of this study was to investigate the stability of curcumin solution in the existence of tungsten trioxide (WO_3) particles under light illumination. In the experimental method, curcumin extracted from Indonesian local turmeric was added with WO_3 microparticles and put into the photoreactor system. The photostability performance of curcumin was conducted for 22 hours using 100 W of Neon Lamp. The results showed that the curcumin solution was relatively stable. When curcumin without existence of WO_3 was irradiated, no change in the curcumin concentration was found. However, when curcumin solution was mixed with WO_3 particles, decreases in the concentration of curcumin was found. The concentration of curcumin with WO_3 after light irradiation was about 73.58%. Based on the results, we concluded that the curcumin is relatively stable against light. However, its light-irradiation stability decreases with additional inorganic material.

1. Introduction

Curcumin (diferuloylmethane) is a polyphenol derived from the plant Curcuma longa, commonly called turmeric.[1] Since curcumin has beneficial health effects on againsting many diseases and modulating immune system, this material is often applied in many different types of food products, including bread, instant rice, and noodles.[2] This substance has also pharmacological potentials as anticancer, antitumor, antioxidant, activities.[3; 4; 5] Although curcumin has widely applied, this material has been less stability. [6] Curcumin is sensitive to bases condition, light, existence of metal ions, enzymes, and temperature.[7] Therefore, the stability of curcumin is considered as one of the attractive challenges in the recent years.

The main decomposition products have previously been identified as feruloyl methane, ferulic acid, and vanillin.[6] A number of studies have reported several factors involving the curcumin decomposition process, such as ultraviolet and visible light radiation [7], alkaline condition [8], solvent [9], oxygen [10], ionic and non-ionic cyclodextrins [6], and the existence of inorganic material such as titanium dioxide.[11] However, investigation of photodecomposition of curcumin in the existence of tungsten trioxide (WO_3) is typically rare.

Here, the purpose of this study was to investigate the stability of curcumin solution by adding the amount of WO_3 particles as a photocatalyst. In the experimental method, we compared the stability of curcumin when interacting with WO_3 microparticles under light irradiation for 22 hours. The results showed that no change in the curcumin concentration without existence of WO_3 was irradiated. However, additional WO_3 into curcumin solution had significant impact. It cause decreases in the concentration of curcumin was about 73.58%. We conclude that the curcumin is relatively stable against light. However, its light-irradiation stability decreases with additional inorganic material.

2. Experimental Method

Curcumin was extracted from turmeric, which was collected from Bandung, Indonesia. In short of the process, turmeric was washed, cut into small pieces (size about 1 x 1 mm), then dried at 100°C to remove the water content in turmeric. Next, the dried turmeric was then grinded, dissolved in aqueous solution, and filtered. The filtered solution is then used as the curcumin solution.

To investigate the photodecomposition process, the concentration of curcumin was set at 100 ppm. Then, 0.411 gram of tungsten trioxide (WO_3) was added into 300 mL of solution of curcumin. In addition, WO_3 was prepared based on our previous work.[12]

The mixed solution was then placed into a home-made photocatalytic reactor to confirm the stability of curcumin under light irradiation. The photocatalytic reactor consisted of a batch-glass reactor (250 mL), a magnetic stirrer (600 rpm), and a bubbler. The mixed material (containing curcumin and WO_3 in the aqueous solution) then irradiated by 100 W of Neon Lamp. The air (1.0 L/min) was bubbled into the reactor during the photodecomposition process in order to keep the oxygen concentration constant. During the photodecomposition process, the concentration was measured using a turbidity analysis based on the light exposed into the mixed solution (containing curcumin and WO_3) using a home-made portable equipment. In short, the home made portable spectrophotometer consisted of light source and an arduino system equipped with a light sensor (lux meter; BH1750FVI, Rohm Co. Ltd., Japan). The spectrophotometer is integrated to computer to analyze the real-time concentration measurement. In addition, several analyses were conducted to support the present study, such as a Scanning Electron Microscope (SEM;JSM-6360LA; JEOL Ltd., Japan), an X-Ray Diffraction (XRD; PANalytical X'PertPRO; Philips Corp., Netherland) and a Fourier Transform Infra Red (FTIR; FTIR-8400 dan Prestige 21, Shimadzu Corp., Japan).

3. Results and Discussion

Figure 1a shows the FTIR spectra of WO_3 particles prepared with synthesis temperatures of about 800°C. There is a broad band in the range 500-1000 cm-1.[13] This range is characteristic of the different O-W-O stretching vibrations in the WO_3 crystal lattice.[14]Figure 1b depicts the XRD patterns of WO_3 particles. The detection of diffraction peaks at 2θ peaks of 23.57; 30.00; 35.71; 50.00; and 58.26 suggested that the prepared particles had monoclinic crystal structures type (JCPDS no. 72–1465).[15]

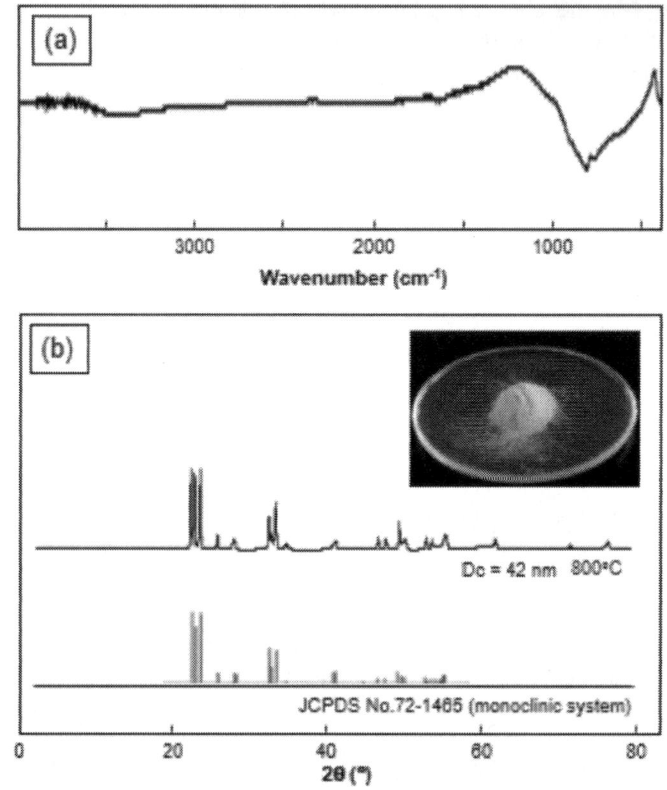

Figure 1. The FTIR (a) and the XRD (b) analysis results of WO_3 particles.

Figure 2 shows the SEM images of WO_3 particles produced using synthesis temperature of 800^0C. Particles with cubical shape are found (Figure 2a). Particle size range of prepared sample was about 20-70 micrometers. The high-magnified SEM image (Figure 2b) shows asymmetric surface particle. [16]

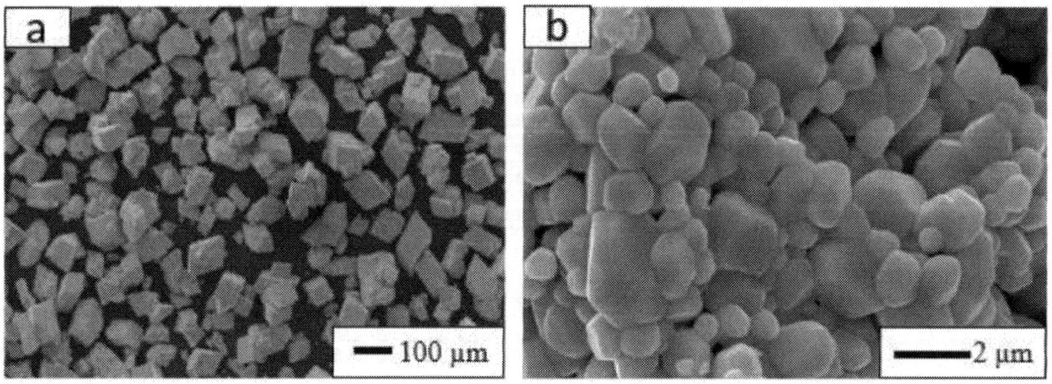

Figure 2. SEM images of WO_3 particles. Figure 2b is the high magnification of SEM images.

Figure 3 shows the photodecomposition profile of curcumin in the existence of WO_3 particles. As a standard comparison, the curcumin only solution was also tested. The result showed that curcumin only solution was relatively stable, while the concentration of curcumin with additional WO_3 gradually decreased along with increasing time. The concentration of curcumin with additional WO_3 particles was near to 20% after 22 hours of light illumination. This result confirmed that WO_3 is the main reason for the decrease of 73.58% of curcumin concentration.

To ensure the photodecomposition profile, we analyzed the concentration in every hour. We found that the reaction was done in two steps. The first step is the reaction before 13 hours of process, in which the step has an insignificant reaction rate (the slight change in the curcumin concentration). Then, the next step is after 13 hours of the process, in which the curcumin concentration gradually decrease.

The fundamental reason is explained in **Figure 4**. In the common process, curcumin is degraded into ferulic and vanilic (see dashed line in the reaction route). However, for some cases, the reaction must through the formation of intermediate compound (see solid line in the reaction route).[17] This intermediate compound is not identical with the isolated degradation product. The postulated structure was probably formed by ring closure within molecule. Indeed, the additional route in the formation of intermediate makes the reaction route longer, resulting the reaction rate decreases

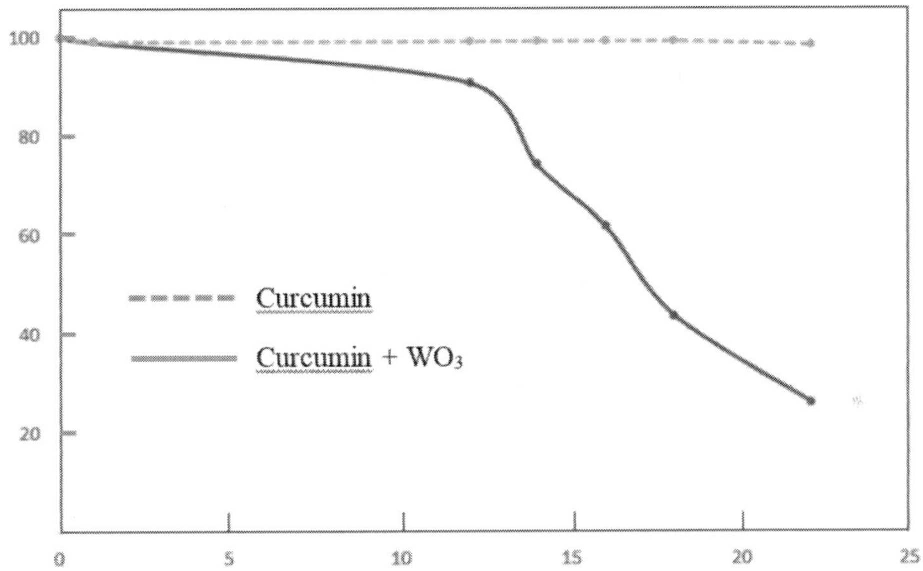

Figure 3. Photodecomposition Profile of Curcumin in the existence of WO_3 particles.

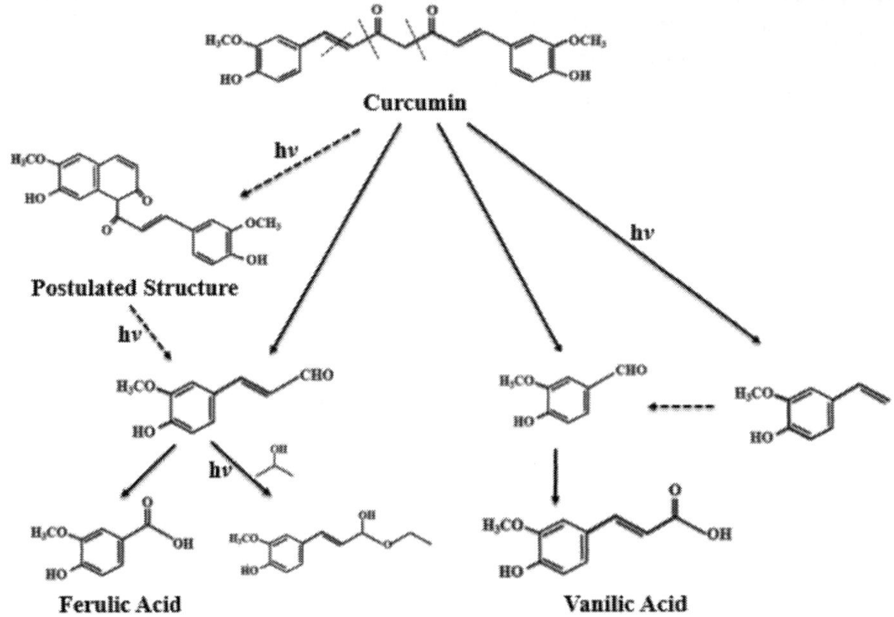

Figure 4. Particles Photochemical degradation of curcumin.

4. Conclusl̇ions

This study investigated the photodecomposition behavior of curcumin in the existence of WO_3 particles. In the experiments, curcumin solution was mixed with WO_3 particles and iluminated under light radiation. These materials were placed into the reactor that was integrated to the computer system. To confirm the influence of WO_3 particles on the photodecomposition of curcumin, several analyses were conducted such as a SEM, XRD, and FTIR. The results showed that the concentration of curcumin gradually decreased along with increasing time, while curcumin was stable relatively in the absence of WO_3 particles. The lost curcumin concentration in the existence of WO_3 particles after 22h photodecomposition process were 73.58%. The reaction rate in early 13 hours was not significant. This is probably caused by the forming of the other molecule (intermediate), which is called as postulated structure since this structure was not identical with the isolated degradation product. Based on the results, we can conclude that avoiding contact with WO_3 particles can minimized decomposition of curcumin can be minimized.

References

[1] Nagpal M & Sood S. 2013 Role of curcumin in systemic and oral health: An overview *Journal of natural science, biology, and medicine* **4** 1 3

[2] Basniwal, R. K., Buttar, H. S., Jain, V. K., & Jain, N. 2011 Curcumin nanoparticles: Preparation, characterization, and antimicrobial study. *J. Agric. Food Chem* **59** 5 2056-2061

[3] Barclay, L. R. C., Vinqvist, M. R., Mukai, K., Goto, H., Hashimoto, Y., Tokunaga, A., & Uno, H. 2000 On the antioxidant mechanism of curcumin: Classical methods are needed to determine antioxidant mechanism and activity *Organic letters* **2** 18 2841-2843

[4] Aggarwal, B. B., Kumar, A., & Bharti, A. C. 2003 Anticancer potential of curcumin: Preclinical and clinical studies *Anticancer research* **23** 1/A 363-398

[5] Kunnumakkara, A. B., Guha, S., Krishnan, S., Diagaradjane, P., Gelovani, J., & Aggarwal, B. B. 2007 Curcumin potentiates antitumor activity of gemcitabine in an orthotopic model of pancreatic cancer through suppression of proliferation, angiogenesis, and inhibition of nuclear factor-κB–regulated gene products *Cancer research* **67** 8 3853-3861

[6] Tønnesen, H. H., Másson, M., & Loftsson, T. 2002 Studies of curcumin and curcuminoids.

XXVII. Cyclodextrin complexation: solubility, chemical and photochemical stability *International Journal of Pharmaceutics* **244** 1 127-135

[7] Lee, B. H., Choi, H. A., Kim, M. R., & Hong, J. 2013 Changes in chemical stability and bioactivities of curcumin by ultraviolet radiation *Food Science and Biotechnology* **22** 1 279-282

[8] Chen, X., Zou, L. Q., Niu, J., Liu, W., Peng, S. F., & Liu, C. M. 2015 The stability, sustained release and cellular antioxidant activity of curcumin nanoliposomes *Molecules* **20** 8 14293-14311.

[9] Khopde, S. M., Indira Priyadarsini, K., Palit, D. K., & Mukherjee, T. 2000 Effect of solvent on the excited-state photophysical properties of curcumin *Photochemistry and photobiology* **72** 5 625-631

[10] Price, L. C., & Buescher, R. W. 1996 Decomposition of turmeric curcuminoids as affected by light, solvent and oxygen *Journal of food biochemistry* **20** 5 125-133

[11] Nandiyanto, A. B. D., Wiryani, A. S., Rusli, A., Purnamasari, A., Abdullah, A. G., & Riza, L. S. 2017 Decomposition behavior of curcumin during solar irradiation when contact with inorganic particles. *IOP Conference Series: Materials Science and Engineering* **180** 1 012135

[12] Nandiyanto, A. B. D., Sofiani, D., Permatasari, N., Sucahya, T. N., Wiryani, A. S., Purnamasari, A., Rusli, A. & Prima, E. C. 2016 Photodecomposition profile of organic material during the partial solar eclipse of 9 march 2016 and its correlation with organic material concentration and photocatalyst amount *Indonesian Journal of Science and Technology* **1** 2 132-155

[13] Díaz-Reyes, J., Dorantes-García, V., Pérez-Benítez, A., & Balderas-López, J. A. 2008 Obtaining of films of tungsten trioxide (WO_3) by resistive heating of a tungsten filament. *Superficies y vacío* **21** 2 12-17

[14] Bai, S., Zhang, K., Wang, L., Sun, J., Luo, R., Li, D., & Chen, A. 2014 Synthesis mechanism and gas-sensing application of nanosheet-assembled tungsten oxide microspheres *Journal of Materials Chemistry A* **2** 21 7927-7934

[15] Aksel, E., Forrester, J. S., Jones, J. L., Thomas, P. A., Page, K., & Suchomel, M. R. 2011 Monoclinic crystal structure of polycrystalline Na 0.5 Bi 0.5 TiO 3 *Applied Physics Letters* **98** 15 152901

[16] Nandiyanto, A. B. D., Munawaroh, H. S. H., Kurniawan, T., & Mudzakir, A. 2016 Influences of temperature on the conversion of ammonium tungstate pentahydrate to tungsten oxide particles with controllable sizes, crystallinities, and physical properties *Indonesian journal of chemistry* **16** 2 124-129

[17] Tønnesen, H. H., Karlsen, J., & van Henegouwen, G. B. 1986 Studies on curcumin and curcuminoids VIII. Photochemical stability of curcumin *Zeitschrift für Lebensmitteluntersuchung und-Forschung A*, **183** 2 116-122

Acknowledgement
We acknowledged RISTEK DIKTI (Grant: Program Unggulan PerguruanTinggi (PUPT).

IOP Conf. Series: Materials Science and Engineering **306** (2018) 012003 doi:10.1088/1757-899X/306/1/012003

E-Portfolio Web-based for Students' Internship Program Activities

A Juhana*, A G Abdullah, M Somantri, S Aryadi, D Zakaria, N Amelia and W Arasid

Electrical Engineering Education Department, Faculty of Electrical Engineering Education, Universitas Pendidikan Indonesia, Bandung, Indonesia

*agus.juhana@student.upi.edu

Abstract. Internship program is an important part in vocational education process to improve the quality of competent graduates. The complete work documentation process in electronic portfolio (e-Portfolio) platform will facilitate students in reporting the results of their work to both university and industry supervisor. The purpose of this research is to create a more easily accessed e-Portfolio which is appropriate for students and supervisors' need in documenting their work and monitoring process. The method used in this research is fundamental research. This research is focused on the implementation of internship e-Portfolio features by demonstrating them to students who have conducted internship program. The result of this research is to create a proper web-based e-Portfolio which can be used to facilitate students in documenting the results of their work and aid supervisors in monitoring process during internship.

1. Introduction

Internship program is a kind of experimental learning in which students are required to apply the knowledge they have learned in the industry. Internship program is important for students as they will be able to integrate intellectual skills with live actions in real life, enhance critical thinking ability, communication, and better business capability [1][2].

Undoubtedly, internship program should be observed and assessed. Advisors are expected to the extent of students' development in the internship. Conversely, in Universitas Pendidikan Indonesia (UPI), the monitoring process is not entirely carried out. In fact, not all of the advisors visits internship program location and directly recognizes students' routine work achievement in the internship [3]. The restraints of distance, time and cost turn out to be the most common reasons. In the end, internship program is assessed based on students' final report only. However, the assessment during internship program is the most fundamental thing [4].

One of attempts considered effective to monitor students' tasks in internship program is to observe their portfolio during the internship. This portfolio consists of a collections of students' invention and work report throughout the internship program [5]. Every student is obliged to produce portfolio as a means of advising and monitoring to improve students' work in the internship [6].

For the reasons mentioned above, the authors create a platform of website-based electronic portfolio (e-portfolio) which is expected to offer solutions to these problems. This platform has been frequently used in advising process to directly measure students' work outcome during internship program [7]. E-Portfolio designed in this research acts not only as a platform to store and observe the result of daily

IOP Conf. Series: Materials Science and Engineering **306** (2018) 012003 doi:10.1088/1757-899X/306/1/012003

work reports, but also as a means for students so they can directly communicate with internship advisor and internship supervisor. The outcome of e-Portfolio planning will be tested and demonstrated towards students who have accomplished the internship program in UPI. This research aims to acquire information related to students' responses towards the system of documentation, as well as its operation, and accessibility of website-based e- Portfolio to be implemented in internship program activities.

2. System Design

This section will explain the system requirements and design as a whole that becomes the target of the research. The features provided in e-Portfolio are classified according to the needs of students, lecturers, and internship. Several standard criteria mandatory for e-Portfolio features intend to aid the three sides. E-Portfolio features should meet some criteria including it being website-based, effective to help students in documenting e-portfolio, easily accessed, and most importantly, user-friendly, and especially for students [8].

2.1. Entity Relationship Diagram (ERD)

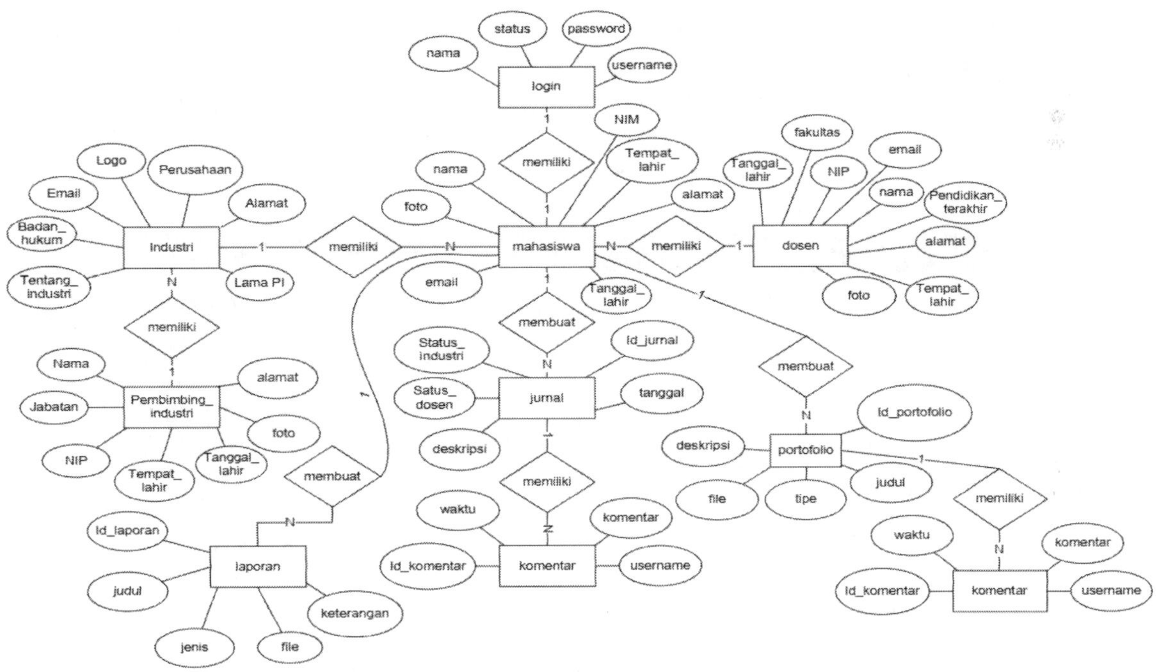

Figure 1. Entity Relationship Diagram (ERD)

Figure 1 describes the relation among the system users. An administrator manages the information system of internship program, every user and the entity of internship program system owns details of incoming portfolio. Table 1 defines the users, entity and attributes of the diagram.

Table 1. *Attributes of Entity Relationship Diagram* (ERD)

No	User/entity	Attribute
1	Lecturer	Nip, nama, tempat_lahir, tanggal_lahir, fakultas, pendidikan_terakhir, email, foto, alamat
2	Student	Nim, nama, tempat_lahir, tgl_lahir, alamat, foto, nip_dosen_pembimbing, email, id_perusahaan
3	Company supervisor	Nip, id_perusahaan, nama, tempat_lahir, tanggal_lahir, foto, posisi, email, alamat
4	Login	*User*name, password, status, nama
5	Journal	Id_jurnal, nim, tanggal, sts_dosen, sts_perusahaan, deskripsi
3	Portfolio	Id_portofolio, nama, nim, type, file, deskripsi
6	Comment	Id_komentar, id_jurnal, *user*name, komentar, waktu
7	Report	Id_laporan, judul, nim, jenis, file, keterangan
8	Company	Id_perusahaan, nip_pembimbing, perusahaan, badan_hukum, email, alamat, pembimbing, logo, lama_prakerin, tentang

2.2. Interface

When the users access e-Portfolio, they will be welcomed with several pages showing how to-steps and facilities to operate the system and communicate among the users (student, lecturer, and internship supervisor). Each page of e-Portfolio is explained below:

a) Main Page
This page provides much information related to internship program guide, news/announcement related to the program, advisor, schedule, department contacts, and help. Every user can register and login to access his/her account in this main page.

Figure 2. Interface of Main Page Internship Program Portal

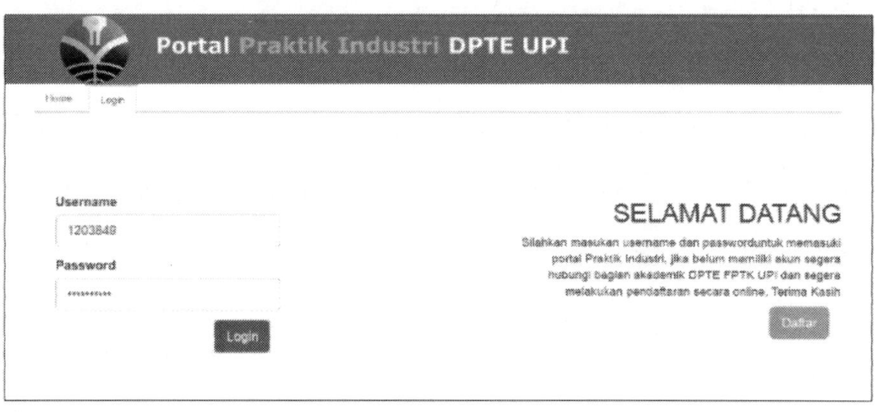

Figure 3. Interface of Login Page of Internship Program Portal

Figure 2 is the most front page of internship program website, while figure 3 shows the starting page when users are about to login.

b) e-Portfolio Field

The students' page has an e-Portfolio field, in which students can upload files and outcome of students' work in the internship. The files uploaded to e-portfolio can come in various types; the extension types acceptable to upload are: .jpg, .jpeg, .doc, .xls, .ppt, .pdf, .mp4, etc. There are parts in e-Portfolio that should be comprehended by the users, students in particular, include types of buttons to upload files, e-Portfolio page to display the uploaded files, and comment field as a means of communication among students, lecturers, and internship supervisors.

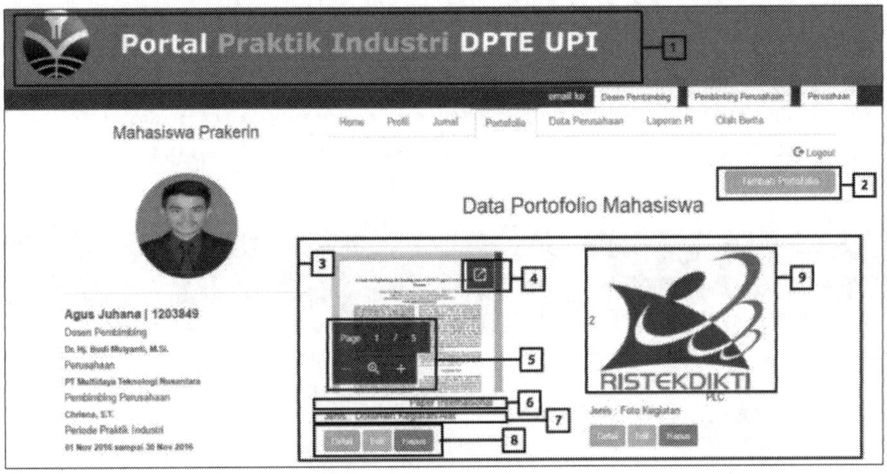

Figure 4. Interface of e- Portfolio Page

Figure 4 can be described as follows:

1) Main Header of Internship Program Portal.
2) Upload File button, this button opens folders in computer to choose the file which will be inserted into e-Portfolio. When the button is clicked, a new page will pop up as shown in figure 5.

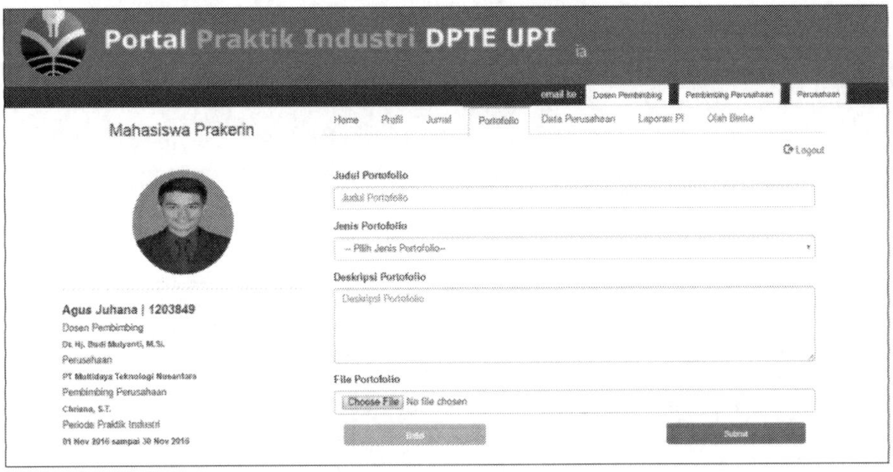

Figure 5. Interface of e-Portfolio File Upload Page

The display of figure 5 is the page in which the users can upload a file which will be included in e-Portfolio, firstly, the users should write the title of the file that will be uploaded, pick the file type, add a description, upload the file from computer, and then click finish.

3) Main Body of e-Portfolio, in this part, the successfully uploaded file will be displayed in the form of e-Portfolio.
4) Preview Pop-out button, when users click the preview button on one of the uploaded files, a new page displaying the file will show up.
5) Regular Preview button, this button can be used by the users to directly see the content of a file from the page without opening a new one.
6) The Title of the File, the file title which is shown in the same way as it is uploaded by the user in the e-Portfolio.
7) The Type of the File, this part describes the type of the file uploaded by the users. The file types provided by this system include tools/activities document, activities pictures, and activities videos.
8) Detail, Edit, and Erase button. The button detail functions to open the complete information about the file uploaded by students, and on this page, a comment field is also provided for them to have a discussion with internship advisor and internship supervisor. The detail page view is displayed on figure 6.

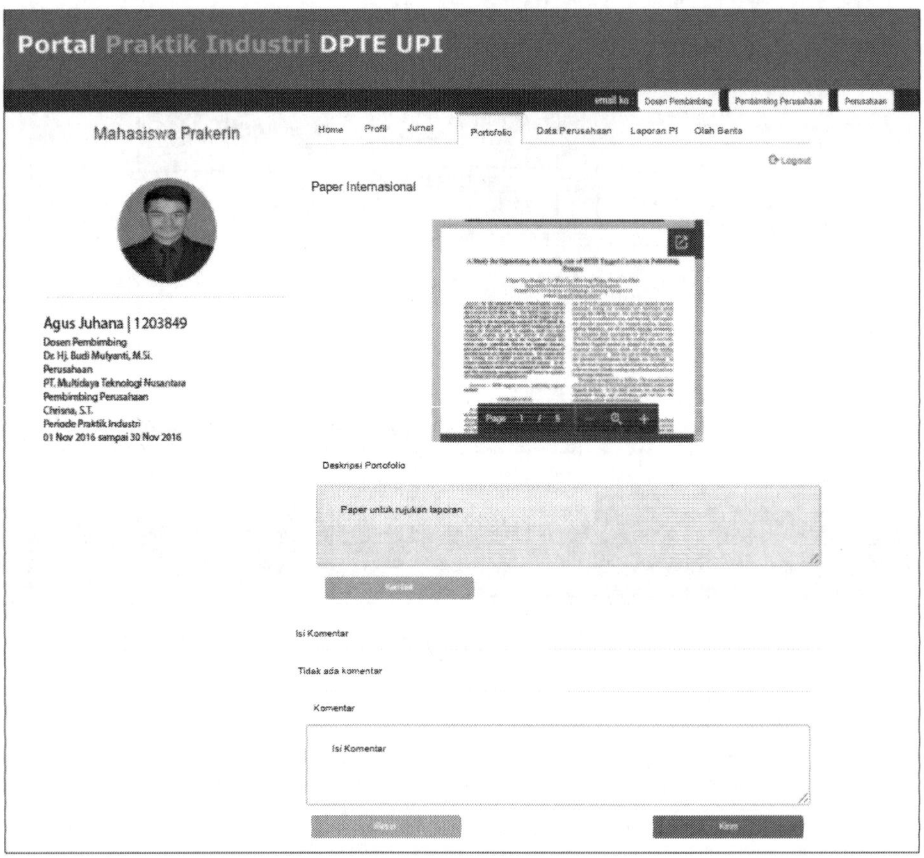

Figure 6. Interface of File Details Page on e-Portfolio

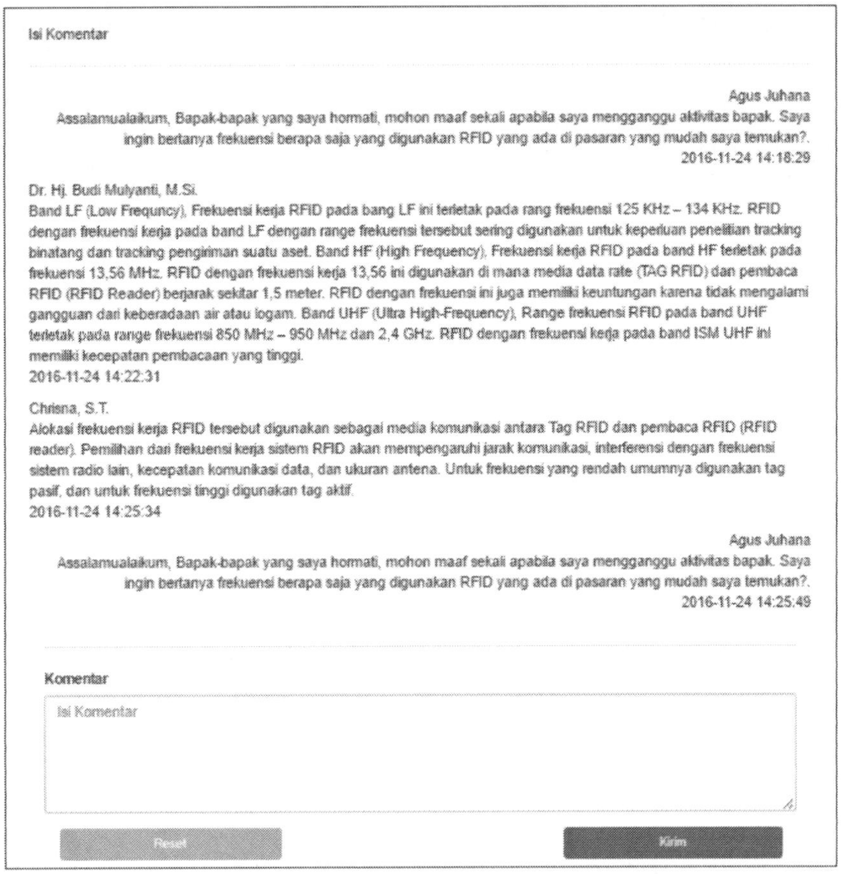

Figure 7. Discussion among Student, Lecturer, and Internship Supervisor

The comment field on this page is beneficial for students as they can ask questions to, coordinate, and evaluate the results of their work with lecturer and internship supervisor. Figure 7 shows an example of discussion among student, lecturer, and internship supervisor. The edit button functions to alter data, description, and file details, while the erase button can be used to erase the unwanted data from being shown or stored.

9) The view of file first page. This part presents the first page or picture of the uploaded file, to allow the users to easily view the content of the file.

3. Methods

This research is conducted using fundamental research method started with literature study [9]. In this research, students are invited to discuss the information system of internship practice and to demonstrate/operate e-Portfolio. Based on the results of discussion and trial, various responses are collected regarding respondents' experience in using website-based e-Portfolio.

The data are collected from 25 students as respondents of Departemen Pendidikan Teknik Elektro Fakultas Pendidikan Teknik Elektro (DPTE FPTK) UPI from batches of 2012-2015 who have conducted internship program. The research instrument employed in this research is questionnaire. The questionnaire used is a closed-questionnaire, utilizing Likert scale. Instrument in this research is adapted from previous scientific journal references dealing with information system of internship program [10][11]. Next, the presentation of product quality predicate is designed based on Rating Scale measurement. Rating scale is the alteration of the collected data outcome from a quantitative one to a qualitative one [12]. Finally, the last process is to interpret the data in the form of discussion, findings and conclusion.

4. Results and Discussion

4.1. Result of System Analysis

Below is the results of planning and testing of internship program e-Portfolio at DPTE FPTK UPI. According to the system testing that has been conducted, the system is functionally ready to use and produce produces outcome which corresponds with what is expected. If the outcome of the system is compatible with the plan, then the system can be considered good [13].

4.2 Result of Data Analysis

4.2.1 Documenting e-Portfolio. Based on the data collected from 25 respondents, the results of questionnaire points regarding documenting internship program e-Portfolio at DPTE FPTK UPI are illustrated in table 2.

Table 2. Data of Students' Responses towards Documentation Process in Internship Program e-Portfolio at DPTE FPTK UPI

No	Question	Score					Grade	Mean	SD
		1	2	3	4	5	(%)		
1	E-Portfolio helps me collect the outcome of the works during internship program.	0	0	4	19	2	78,4	3,92	0,493
2	E-Portfolio can reflect what I have learned in internship program.	0	1	9	11	4	74,4	3,72	0,792
3	Using e-Portfolio, I can integrate things I have learned (in internship program as well as university classes)	0	0	12	7	6	75,2	3,76	0,831
4	Using e-Portfolio, I can remember and revisit what I have learned during internship program.	0	0	2	18	5	82,4	4,12	0,526
5	Using e-Portfolio, I can summarize my learning contents gradually.	0	1	7	13	4	76	3,8	0,764
6	Using e-Portfolio, I can evaluate what I have done during internship program.	0	0	5	14	6	80,8	4,04	4,04
7	I can see mu development during internship practice and my personal development through e-Portfolio system.	0	0	6	13	6	80	4	0,676

Table 2 presents the average percentage of students' responses towards e-Portfolio documentation 78.17%. Regarding the percentage scored, it can be implied that e-Portfolio documentation process can be said as considerably good [12].

4.2.2 Operating e-Portfolio. The data collected on questionnaire points related to operating internship program e-Portfolio are displayed in table 3.

Table 3. Students' Responses in Operating Internship Program e-Portfolio at DPTE FPTK UPI

No	Question	Score					Grade	Mean	SD
		1	2	3	4	5	(%)		
1	Generally, this e-Portfolio is easy to use/operate	0	0	5	16	4	79,2	3,96	0,73
2	I feel helped by the tools/facilities provided in this e-Portfolio	0	0	5	16	4	79,2	3,96	0,61
3	If there are problems in e-Portfolio then I will easily recognize the problems and how to solve them	0	1	16	6	2	67,2	3,36	0,7
4	The login process into e-Portfolio system is easy.	0	0	6	11	8	81,6	4,08	0,76
5	E-Portfolio system provides considerably required functions to document work during internship program.	0	0	5	17	3	78,4	3,92	0,57
6	The appearance of e-Portfolio is good and organized.	0	0	11	12	2	72,8	3,64	0,64
7	The file uploading process in e-Portfolio in internship program website can be performed effortlessly.	0	0	5	16	4	79,2	3,96	0,61
8	I can easily review my previous e-Portfolio in internship program website.	0	0	4	15	6	81,6	4,08	0,64
9	The logout process can be performed easily.	0	0	6	13	6	80	4	0,71
10	The appearance of e-Portfolio can simply be understood and operated.	0	0	9	13	3	75,2	3,76	0,66
11	The settings of e-Portfolio system can be configured according to my needs.	0	1	7	15	2	74,4	3,72	0,68

Table 3 shows the average percentage of students' responses in operating e-Portfolio is 77.16%. This result implies that the process of operating e-portfolio can be considered applicable [12].

4.2.3 The Accessibility of e-Portfolio. The data processing of questionnaire points to discover the accessibility of internship program e-Portfolio at DPTE FPTK UPI is shown in table 4.

Table 4. Students' Responses towards the Accessibility of Internship Program e-portfolio at DPTE FPTK UPI

No	Question	Score 1	2	3	4	5	Grade (%)	Mean	SD
1	It allows me to interact or consult online with lecturers and internship supervisors during the internship program	0	0	7	11	7	80	4	0,76
2	In this website, the reason and the aims of e-Portfolio is explicated	0	0	10	13	2	73,6	3,68	0,63
3	The contents of e-portfolio is orderly arranged and is helpful in the process of submitting work documentation	0	0	6	15	4	78,4	3,92	0,64
4	Using e-Portfolio, I can develop my potential and skills I have learned during the internship program, since my work documents can be reviewed and re-learned anytime	0	0	12	9	4	73,6	3,68	0,75
5	In e-Portfolio, my work documents can be observed anytime by the advisor	0	0	8	14	3	76	3,8	0,65
6	I can ask for feedback from university supervisors as well as internship supervisors regarding the work documents I uploaded in e-Portfolio	0	0	8	16	1	74,4	3,72	0,54
7	This e-Portfolio is an innovation which can help and benefit every side (students, university supervisors, and internship supervisors)	0	0	5	12	8	82,4	4,12	0,73

Based on table 4, it can be seen that the average percentage of students' responses towards the accessibility of e-Portfolio reaches the percentage of 76.91%. According to this result, it can be -implied that the accessibility level of this e-Portfolio is feasible [12].

4.3 Discussion

The final result of this research shows that e-Portfolio features can be used according to the users' needs. Since the result of the system is in line with the plan, this system can be considered good [13]. The operation system of e-Portfolio is deemed easily accessed. Starting from the login process, operating the tools found in e-Portfolio, downloading and uploading process, filling comments, until logout process have been assessed in this e-Portfolio operation.

The last assessment from students concerns with the accessibility of e-Portfolio. Students testify that e-Portfolio is easily accessed; in these features, the organized content arrangement helps students in processing the data and the observation from lecturers. Basically, e-Portfolio features should be easily utilized for documentation process, easily accessed, operated, and developed [8].

5. Conclusions

According to the research which has been carried out, it can be concluded that an easily accessed information system poses as an important aspect when space and time hinder and limit the access of observation. According to the results of this research, e-Portfolio is highly recommended to overcome the problems. E-Portfolio system built in this website-based Internship Program Portal has been deemed applicable in documenting students' work outcome during internship program and useful in providing an effective means of consulting. Moreover, students' responses towards the operation of e-Portfolio are good and these features e-portfolio are all set to be applied in internship program activities. The accessibility of website-based e-Portfolio is considered to be user-friendly for students while operating the system as well as during interacting with university or internship supervisors.

References

[1] Lam T and Ching L 2007 An exploratory study of an internship program: The case of Hong Kong students *Int. J. Hosp. Manag.*, vol. **26**, no. 2, pp. 336–351
[2] Sarlan A B, Ahmad W F B W and Bismo D 2008 Student Industrial Internship Web Portal *Proc.*

- *Int. Symp. Inf. Technol. 2008, ITSim*, vol. **1**
[3] Zakaria D 2016 *Evaluasi Keterlaksanaan Praktik Industri Mahasiswa di Departemen Pendidikan Teknik Elektro FPTK UPI* (Bandung: Universitas Pendidikan Indonesia)
[4] Salamonson Y *et al.* 2016 Measuring final-year nursing students' satisfaction with the viva assessment *Nurse Educ. Pract.*, vol. **16**, no. 1, pp. 91–96.
[5] Neades B L 2003 Professional portfolios: All you need to know and were afraid to ask *Accid. Emerg. Nurs*, vol. **11**, no. 1, pp. 49–55.
[6] Kuter S 2009 Examination of the merits and pitfalls of the internship portfolio *Procedia - Soc. Behav. Sci.*, vol. **1**, no. 1, pp. 84–90.
[7] Halada G P and Wozniak N M 2014 Enhancing Assessment of Experiential Learning in Engineering Education through Electronic Portfolios in *2014 Zone 1 Conference of the American Society for Engineering Education (ASEE Zone 1)*.
[8] Zaini R and Quqandi E 2015 A Multi-criteria Decision Making Model for Software Selection to Build E-Portfolio *2015 Fifth Int. Conf. e-Learning*, pp. 131–134
[9] Liviu N M and Ana-Andreea M 2013 Issues on Improving Internships in Technical Universities in Romania *Procedia - Soc. Behav. Sci.*, vol. **84**, pp. 1757–1762
[10] McNeill M and Cram A 2011 Evaluating E-portfolios for University Learning: Challenges and Opportunities in *Changing Demands, Changing Directions. Proceedings ascilite Hobart 2011*, pp. 862–873
[11] ChanLin L J and Hung W H 2015 Evaluation of An Online Internship Journal System for Interns *Procedia - Soc. Behav. Sci.*, vol. **191**, pp. 1024–1027
[12] Riduwan 2011 *Dasar-dasar Statistika* (Bandung: Alfabeta)
[13] Noviyanto F 2008 Membangun Sistem Pembelajaran Pengenalan Bentuk untuk Anak Berbasis Multimedia dan Game Interaktif *J. Inform.*, vol. **2**, no. 1, pp. 858–869.

IOP Conf. Series: Materials Science and Engineering **306** (2018) 012004 doi:10.1088/1757-899X/306/1/012004

Designing on-Board Data Handling for EDF (Electric Ducted Fan) Rocket

A Mulyana* and L A A Faiz

Computer Engineering Departement, Universitas Komputer Indonesia, Dipati Ukur Street, 112-116, Bandung, West Java, Indonesia

*agus.mulyana@email.unikom.ac.id

Abstract. The EDF (Electric Ducted Fan) rocket to launch requires a system of monitoring, tracking and controlling to allow the rocket to glide properly. One of the important components in the rocket is OBDH (On-Board Data Handling) which serves as a medium to perform commands and data processing. However, TTC (Telemetry, Tracking, and Command) are required to communicate between GCS (Ground Control Station) and OBDH on EDF rockets. So the design control system of EDF rockets and GCS for telemetry and telecommand needs to be made. In the design of integrated OBDH controller uses a lot of electronics modules, to know the behavior of rocket used IMU sensor (Inertial Measurement Unit) in which consist of 3-axis gyroscope sensor and Accelerometer 3-axis. To do tracking using GPS, compass sensor as a determinant of the direction of the rocket as well as a reference point on the z-axis of gyroscope sensor processing and used barometer sensors to measure the height of the rocket at the time of glide. The data can be known in real-time by sending data through radio modules at 2.4 GHz frequency using XBee-Pro S2B to GCS. By using windows filter, noises can be reduced, and it used to guarantee monitoring and controlling system can work properly.

1. Introduction

Rocket and space technology is one of the leading technology targets for countries in the world to get the title as a developed country. No doubt, a country capable of mastering this technology will be respected in the world political arena. Indonesia as a large and vast archipelago country should have independence in the mastery of rocket and space technology. Therefore it is necessary to continue efforts to realize this independence, one of them through the effort to develop a love of aerospace technology, especially the technology of rocking from an early age. The importance of the development of rocket technology, then we must have the will and ability to examine the field of rocket technology, both rocket and rocket technology itself. Starting from designing, creating, functional testing to carry out flight test. Through understanding the behavior of rockets, both RUM rockets, and EDF rockets. Meanwhile, in this research will focus on making monitoring system, data processing of rocket behavior and Control of EDF Rocket by OBDH (On-Board Data Handling). OBDH has a function to perform commands and processing existing data on a rocket via TTC (Telemetry, Tracking, and Command) [1-4].

The purpose of making this EDF rocket is as follows, designing an OBDH system that contains a data measurement system (sensor), data processing (microprocessor), data communication (radio frequency), and power supply (battery).

With OBDH intended to conduct telemetry, telecommand, and tracking [2-4]. Building Ground Control Station Software (GCS) using Microsoft Visual Studio application where should be able to:

- Communicate with OBDH system on EDF rocket.
- Display data of OBDH measurement results in numeric or graphic formats.

2. OBDH (On-Board Data Handling) and IMU Sensor

2.1. OBDH

OBDH is a microcontroller that has input, output, and sensors that are used to support OBDH function. OBDH also serves as an interface or link between other subsystems. Such as communication, and sensors that have been integrated with the microcontroller. The OBDH function has two main functions, the first function is to receive, validate, decode, and route commands to other rocket sub-systems. The second function is to collects, processes and manages the work and status of OBDH. In addition, OBDH generally has additional functions, such as timeliness, monitoring (watchdog), and security interfaces [1-3].

The size of the OBDH subsystem is usually directly proportional to the complexity and functionality of the system to be created. The more systems and functions you want to use, the more components you need in OBDH. In addition, to make the OBDH sub-system more reliable, OBDH sizes can be enlarged and added some peripherals that can support OBDH stability. To collect data, we use a lot sensor such as accelerometer, gyroscope, and magnetometer [4,5,7].

2.2. IMU Sensor

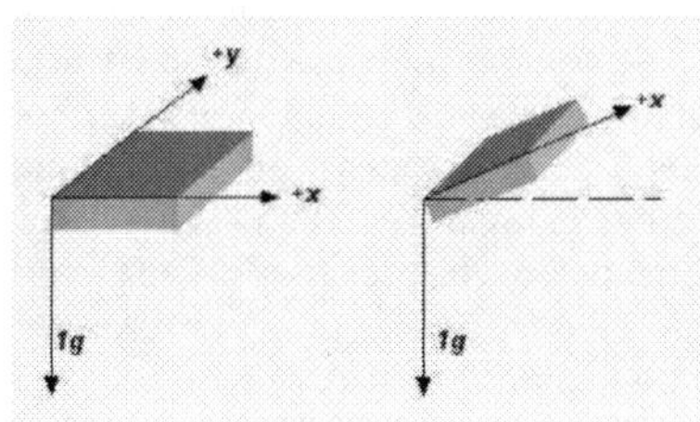

Figure 1. Illustration of accelerometer sensor

Captions of Figure 1:

- + 1g, the unidirectional position of the sensor in the vertical direction of the earth and facing upwards.
- 0g, the unidirectional position of the sensor in the direction of the horizontal earth. Accelerometer sensors can also be used to measure the angle of an object when in a static state [4-7]. The slope angle can be calculated by the acceleration using the following equation:

$$x = tan^{-1}(\frac{akselerometer\ X}{akselerometer\ Z}) \quad (1)$$

$$\theta_y = \tan^{-1}(\frac{akselerometer\ Y}{akselerometer\ Z}) \quad (2)$$

The gyroscope is a device that serves to measure angular velocity or maintain rotational motion. The unit of angular velocity measured in degrees per second (° / s). By utilizing the angular velocity data can be known angle (roll, pitch, yaw) an object. To find the angle value, angular velocity data must be integrated [4-7]. The equation is the following:

$$\theta_{(x,y,z)} \int gyro(x, y, z)dt \quad (3)$$

$$dt = T = t_n - t_{n-1} \quad (4)$$

where:

T	= time sample
n	= {1,2,3...} values
gyro (x,y,z)	= Gyroscope data (°/s)

IOP Conf. Series: Materials Science and Engineering **306** (2018) 012004 doi:10.1088/1757-899X/306/1/012004

For rocket attitude, we need to find it's position, included pitch, yaw, and roll. The mathematical equations applied to the microcontroller are as follows [4-7]:

$$\theta_{(x,y,z)t_n} = gyro(x,y,z)\ x\ T + \theta_{(x,y,z)t_{n-1}} \quad (5)$$

A magnetometer is a useful electronic device for measuring the strength of a magnetic field. The existence of the earth's magnetic field originating from the north makes the magnetometer sensor can be used to measure the angle to the northern directions of the earth [5-7]. The equation to know the value of the compass is as follows:

$$Angle\ (\theta) = arc\tan\left(\frac{Mag\ Y}{Mag\ X}\right) \quad (6)$$

Where:
MagX = output value form magnetometer for X axis
MagY = output value form magnetometer for Y axis
Angle(θ) = value which can be derived from magnetometer (angle)

A barometer is a sensor to measure air pressure with an output value of Pa (Pascal). By utilizing the air pressure, then this sensor can also measure the altitude (Altimeter) [5-7].
The equation measuring the altitude using a barometer sensor is as follows:

$$altitude = 44330\ x\left[1 - \left(\frac{p}{p_0}\right)^{\frac{1}{5.255}}\right] \quad (7)$$

Where:
P = pressure(Pa)
p_0 = *sea level standard atmospheric pressure* (101325Pa)
altitude = Altitude value (Meter above sea level)

3. OBDH Design

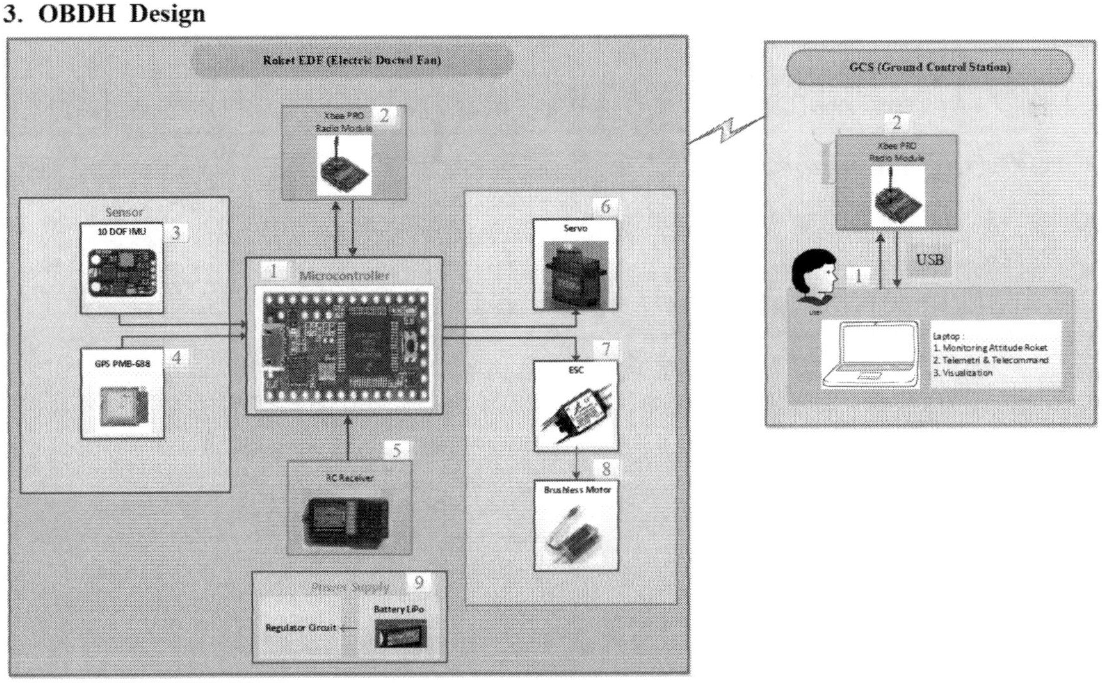

Figure 2. OBDH block diagram

IOP Conf. Series: Materials Science and Engineering **306** (2018) 012004 doi:10.1088/1757-899X/306/1/012004

Here is a list of a lot of component for OBDH:

- Teensy 3.1
- A microcontroller to process a data from IMU sensor and also communicate with Ground Control Station
- XBee-PRO S2B
- It is a radio modem for transmitting and receive data
- 10 DOF IMU
- This a 10 DOF inertial measurement unit sensor for measure acceleration, angle, pressure, altitude of EDF Rocket
- GPS Ublox NEO-6
- A GPS module for *tracking*.
- RC (*Remote Control*) receiver
- Receiving signal and convert it into PPM (*Pulse Position Modulation).*
- Servo Motor
- For trigger to open/close parachute
- ESC (*Electronic Speed Control*)
- ESC (*Elektronic Speed Control)*to control a speed of *brushless motor.*
- Brushless Motor
- It is used as an actuator for EDF Rocket
- Power Supply
- Based on DC power by using Lithium Polymer (LiPo) with 11,1 Volt and 1300 mAh.
- Based on those part, here is a result of OBDH system for EDF Rocket

Figure 3. OBDH board top

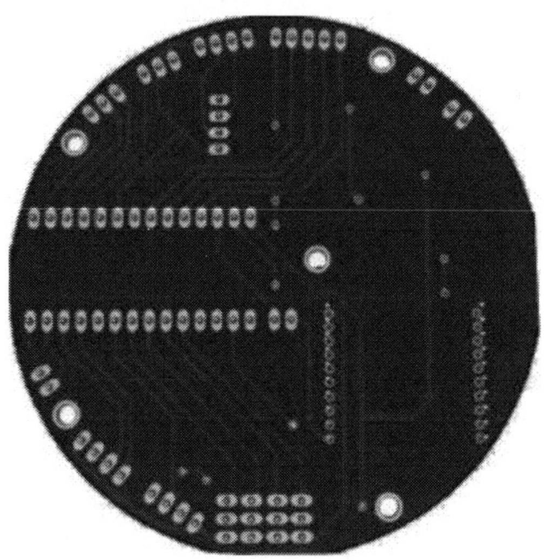

Figure 4. OBDH board bottom

4. Ground Control Station

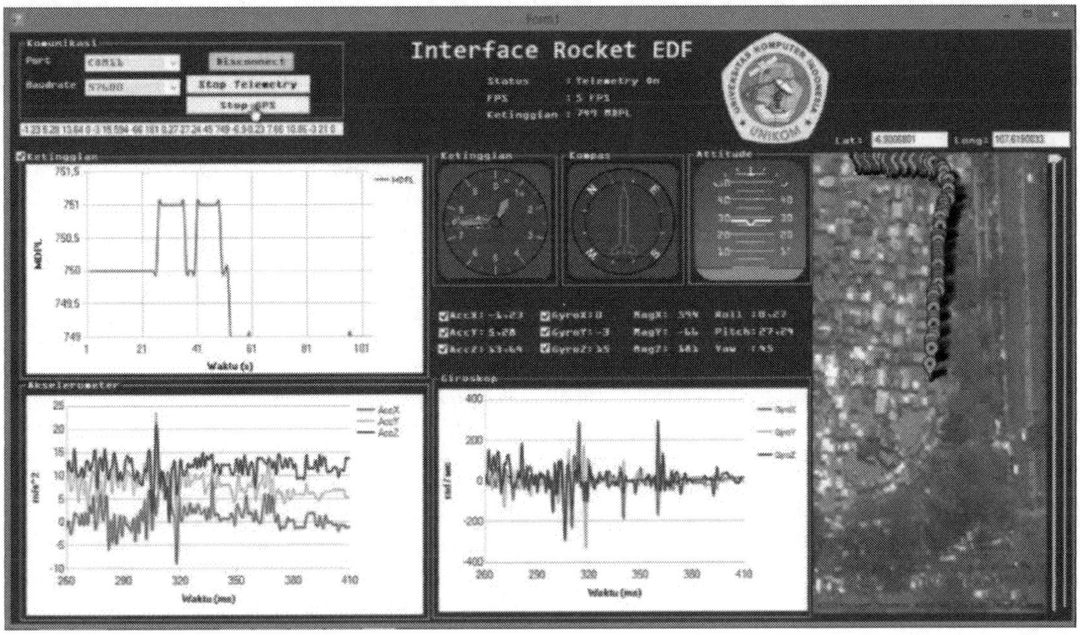

Figure 5. Ground Control Station (GCS)

Ground Control Station (GCS) is used to telecommand, telemetry, tracking GPS, enable EDF motor and display sensor data in the form of numbers and graphics. All of the data as stored in databases system and it can be used for the further purpose [5-8].

5. Results and Discussion

Mechanical Filtering This window is applied to data acceleration and angular velocity only. The goal is that the noise from the acceleration data and the angular velocity that appears when the stationary device can be filtered [6-8].

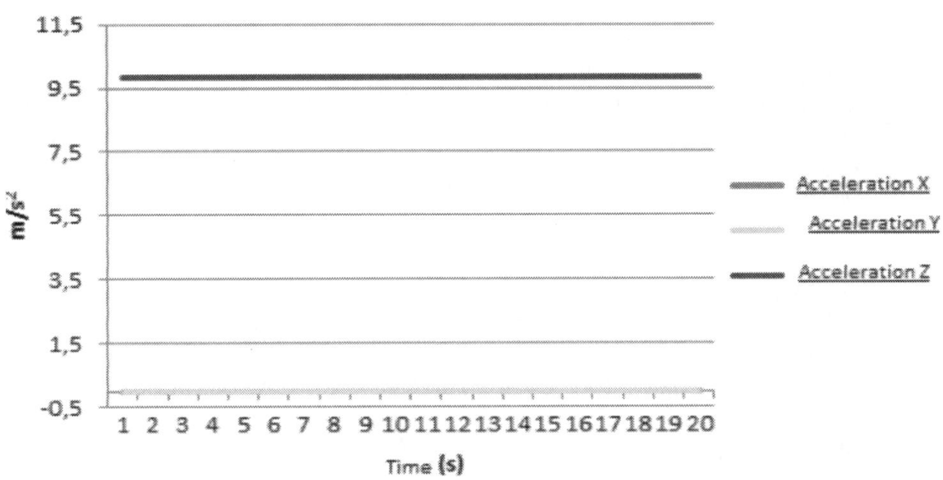

Figure 6. Windows filter for accelerometer

Based on Figure 6 can be seen that by using the Mechanical Filtering Window can overcome the error data on the OBDH device in a state of silence (immobile) caused by very small vibration, mechanical noise or less precisely in determining the offset value [6-8].

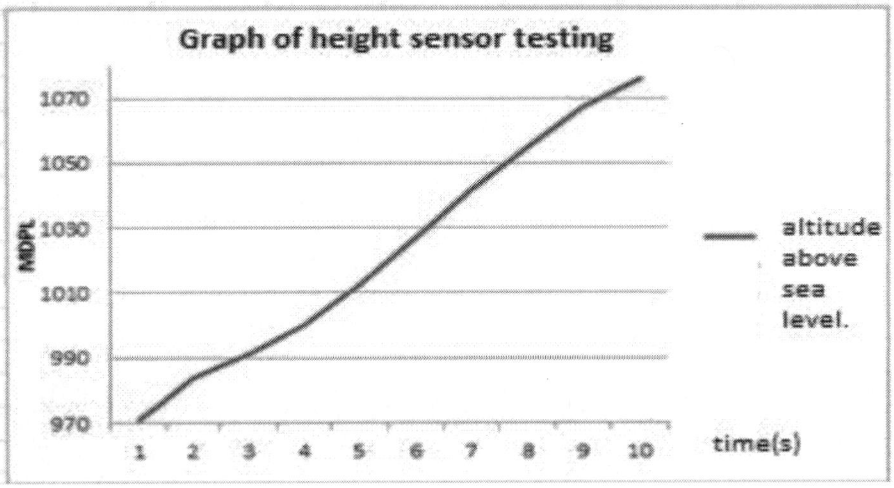

Figure 7. Altitude graph

The launch time of the rocket is 10 seconds and at 10 seconds the height of the rocket reaches 1076 meters above sea level or about 106 meters. Thus the altitude sensor can function properly.

6. Conclusion

On-Board Data Handling (OBDH) design results in power supply systems, sensors, microcontrollers and communications (radio frequency) have been successfully designed. And OBDH Hardware has successfully done telemetry, telecommand, and GPS tracking. This is based on the launch test results where data can be well received on GCS applications. Using the Mechanical Filtering Window can discriminate the small error value that arises when the object is idle. The error that appears when OBDH hardware is silent 99% successfully removed.

References

[1] Haryono, On-Board Data Handling (OBDH) based on pc104, international journal of advanced studies in computer science and engineering ijascse volume 5, issue 2, 2016

[2] Dinesha H. A, H. N Bhagavan, Dr. V. K Agrawal (2013), Design Of Component-Based Software For Telemetry, Tracking And Commanding (TTC) Operations Of Nano Satellite, International Journal Of Technology Enhancements And Emerging Engineering Research, Vol 1, Issue 5 29 Issn 2347-4289

[3] Dinesha H A, Dr.V.K Agrawal (2013), Design of Telemetry in Student Imaging Satellite, International Journal of Innovative Research in Computer and Communication Engineering, India

[4] Jie Liu, Hao Zhang, Wei Zheng, Liuguo Yin (2011), Design of telemetry and telecommand subsystem of a microsatellite-based on industrial-level chips, International Journal of Circuits, System and Signal Processing

[5] Cahyono, D., Nugroho, A., & Hidayat, I. (2012). Design and realization On-Board Data Handling for remote sensing payload of nanosatellite FPGA base. *19th ITS Biennial Conference*. Bangkok: ECONSTOR.

[6] Daniel Rodríguez-Martín, C. P.-L. (2013). A Wearable Inertial Measurement Unit for Long-Term. *Open Access - Sensor*.

[7] Heroe Wijayanto, B. S. (2016). A preliminary design and testing of the On-Board Data Handling (OBDH) for nano-satellite using an atmospheric balloon. *2015 1st International Conference on Wireless and Telematics (ICWT)*.

[8] Kaspars Laizans, I. S. (2014). Design of the fault tolerant command and data handling subsystem. *Proceedings of the Estonian Academy of Sciences.* Tartu, Estonia: Eap.ee.

Estimation of Compaction Parameters Based on Soil Classification

A S Lubis*, **Z A Muis, I P Hastuty and I M Siregar**

Civil Engineering Department, Faculty of Engineering. Universitas Sumatera Utara, Medan, Indonesia

*adinasarilubis@usu.ac.id

Abstract. Factors that must be considered in compaction of the soil works were the type of soil material, field control, maintenance and availability of funds. Those problems then raised the idea of how to estimate the density of the soil with a proper implementation system, fast, and economical. This study aims to estimate the compaction parameter i.e. the maximum dry unit weight (γ_{dmax}) and optimum water content (Wopt) based on soil classification. Each of 30 samples were being tested for its properties index and compaction test. All of the data's from the laboratory test results, were used to estimate the compaction parameter values by using linear regression and Goswami Model. From the research result, the soil types were A4, A-6, and A-7 according to AASHTO and SC, SC-SM, and CL based on USCS. By linear regression, the equation for estimation of the maximum dry unit weight (γd_{max}*)=1,862-0,005*FINES-0,003*LL and estimation of the optimum water content (w_{opt}*)=-0,607+0,362*FINES+0,161*LL. By Goswami Model (with equation Y=mLogG+k), for estimation of the maximum dry unit weight (γd_{max}*) with m=-0,376 and k=2,482, for estimation of the optimum water content (w_{opt}*) with m=21,265 and k=-32,421. For both of these equations a 95% confidence interval was obtained.

1. Introduction

Subgrade strength on the pavement construction is highly dependent on the value of the density of the subgrade. Field density is obtained by either Sand Cone Test or Dynamic Cone Penetrometer Test which yielded the Maximum Dry Maximum Content (γd_{max}) value in the field [1]. Whereas laboratory density is determined by performing Proctor Compaction Test on several soil samples with varying water content. Determining process for the Maximum Dry Unit Weight (γd_{max}) and Optimum Water Content (w_{opt}) in the laboratory required a quite considerable material, expert laboratory operators as well as time consuming. If the results of this classification can be used to estimate Maximum Dry Unit Weight (γd_{max}) and Optimum Water Content (w_{opt}) subgrade material then it would save time, effort and cost on the execution of this work. This is also a clarification (cross check) of the work that has been done by technicians in the laboratory [2].

Several studies in predicting the value of compaction parameters (maximum dry unit weight and optimum moisture content) have been widely developed. The first study to find out the relationship between compaction parameters was performed by Johnson and Sallberg, where the compaction values were related by means of linear regression based on the value of the index properties [3]. Then a study by Metacalf and Romanoschi [4], predicting the value of maximum dry unit weight and optimum water content also by using the linear regression equation method based on plastic limit, modulus plasticity, gradient coefficient and gradient ratio. A research by Blotz, et.al and then by Al-Khafaji were done and formulated the relationship between the value of compaction with the Atterberg limit into an equation for several types of soils [5]. Another research by Ugbe proposed equations in predicting the maximum dry unit weight and optimum water content by using the value of index properties: fine grain percent, liquid limit, and specific gravity [6,7]. Isik F and Ozden G used artificial neural network (ANN) prediction models for estimating the compaction parameters of both coarse- and fine-grained soils [8]. Z A Muis determinated the maximum dry unit weight of base material on a road project based on soil classification data [2] and estimated the compaction parameter values of pavement subgrade based on soil properties [3]. A S Lubis et al. estimated the compaction parameter values based on soil properties for pavement subgrade stabilized with portland cement [9] and for pavement subgrade stabilized with lime [10]. This study aims to estimate the compaction parameter i.e. the maximum dry unit weight (γ_{dmax}) and the optimum water content (Wopt) based on soil classification. Each of 30 samples were being tested for its properties index and compaction test. All of the data's from the laboratory test results, were used to estimate the compaction parameter values by using linear regression and Goswami Model.

2. Literature Review

Soil compaction is a process where air in the pores of the soil is removed by mechanical means. The mechanical way used to solidify the soil can be done in various ways. In field, grinding is widely used, whereas in the laboratory hitting with proctor is more common. There are two methods for Compaction testing in the laboratory, known as: Standardized Testing Test (Standard Proctor Test) and Modified Proctor Test (Modified Proctor Test) [1].

The index properties show soil properties that indicate the type (classification) and soil conditions which relate to the mechanical properties such as strength and compression or tendency to expand and permeability. The physical properties of the soil (index properties) are water content (w), specific gravity (SG), grain gradation (Fine=F), Atterberg consistency (LL, PL, PI), and many others.

The prediction for the value of maximum dry unit weight (γd_{max}) and optimum water content (w_{opt}) can also be calculated from the model suggested by Goswami with the following equation [2]:

$$Y = m \,\mathrm{Log}\, G + k \tag{1}$$

Y = maximum dry unit weight (γd_{max}) and optimum water content (w_{opt})
m = the slope of the curve
k = constant
G = constant gradation (1 + F) (AX1 + BX2 + CX3)
X1 = % weight retained shieve 4.75 mm
X2 = % weight retained between shieve 4.75 mm and 0.075 mm
X3 = % weight through shieve 0.075 mm
A, B, C = constants for shieve number
F = % fine grain

The m and k constants are obtained from the graph between Log G with the value of the maximum dry unit weight and the optimum water content from the experimental results in the laboratory. Whereas, F is fine grain percent that is determined based on percent through shieve 0.075 mm and Plasticity Index (IP) value (Table 1).

Table 1. Determination of F value [6]

% weight through shieve 0.075 mm	F value	
	IP < 10%	IP > 10%
0 – 25	0,0	0,0
26 – 40	0,2	0,2
41 – 60	1,0	1,0
61 – 85	1,0	0,0
86 – 100	1,0	1,0

3. Research Methods

Before testing, the 30 samples as subgrade material were dried to get dry air condition. Laboratory testing [1] consists of the index properties testing that were Water Content Test, Specific Gravity Test, Atterberg Limit Test, Shieve Analysis Test and Soil Classification Test (USCS and AASHTO). After that, the samples got The Compaction Test i.e. Standard Proctor Test.

All of the data's from the laboratory test results, were used to estimate the compaction parameter values. The estimation of the relationship between parameter compaction values and index properties was done in two ways: firstly, by linear regression and secondly, by using Goswami model. In the estimation by using linear regression, the required data were value of index properties, while in the estimation by using Goswami model, the required data was only fine grain percentage.

From the estimation result by using linear regression and Goswami model, the relationship between compaction parameter values and index properties was analyzed. The compaction parameter values estimated by using the Goswami model then were compared with the compaction parameter values from the laboratory. The compaction parameter values estimated were also analyzed based on the soil classification. Afterward, the level of accountability was observed by the validation method, to get a positive correlation level of accountability.

4. Results and discussion

Tests results in the laboratory can be seen in Table 2.

Table 2. Tests Results in Laboratory

No.	Sample	w	SG	LL	PL	PI	FINES	γ_{max}	w_{opt}
1.	**PTB-1**	30.71	2.606	29.67	17.21	12.46	45.28	1.519	20.79
2.	**PTB-2**	36.95	2.604	39.56	23.73	15.82	50.98	1.488	24.21
3.	**PTB-3**	32.09	2.617	29.83	17.37	12.46	50.11	1.533	21.67
4.	**PTB-4**	30.40	2.632	29.08	17.21	11.87	50.03	1.538	21.45
5.	**PTB-5**	35.43	2.598	32.64	21.10	11.54	50.15	1.506	22.15
6.	**PTB-6**	32.25	2.603	33.09	21.63	11.46	50.07	1.519	22.17
7.	**PTB-7**	33.15	2.637	32.78	20.97	11.82	43.37	1.521	20.66
8.	**PTB-8**	33.24	2.603	29.71	17..22	12.49	50.63	1.527	21.18
9.	**PTB-9**	34.46	2.603	39.11	24.65	14.46	50.59	1.466	24.01
10.	**PTB-10**	33.27	2.605	29.08	17.33	11.75	50.67	1.522	21.58
11.	**PTB-11**	29.60	2.603	38.05	25.00	13.05	51.87	1.481	24.08
12.	**PTB-12**	33.00	2.625	33.17	20.81	12.36	48.84	1.522	22.16
13.	**PTB-13**	33.06	2.637	29.77	13.79	12.38	44.43	1.556	20.34
14.	**PTB-14**	30.74	2.604	29.75	17.52	12.23	42.03	1.566	19.81
15.	**PTB-15**	33.95	2.633	29.78	16.56	13.22	50.93	1.530	21.75
16.	**PTB-16**	34.00	2.660	32.29	20.48	11.81	56.94	1.454	25.51
17.	**PTB-17**	32.92	2.670	38.26	20.11	18.15	57.03	1.457	26.62
18.	**PTB-18**	32.94	2.615	23.95	15.96	7.99	53.76	1.490	25.14
19.	**PTB-19**	34.67	2.640	41.90	22.06	19.84	57.86	1.427	26.66
20.	**PTB-20**	31.95	2.699	39.00	18.87	20.13	57.22	1.456	25.08
21.	**PTB-21**	30.80	2.613	40.28	24.44	15.84	55.91	1.442	26.22
22.	**PTB-22**	31.55	2.682	36.98	19.30	17.68	56.34	1.489	25.55
23.	**PTB-23**	30.32	2.657	31.06	19.20	11.86	55.91	1.471	25.94
24.	**PTB-24**	32.92	2.692	20.39	15.97	4.42	52.95	1.502	21.58
25.	**PTB-25**	33.15	2.644	30.96	16.35	14.61	53.09	1.498	23.34
26.	**PTB-26**	32.90	2.650	39.84	22.23	17.61	49.17	1.447	25.78
27.	**PTB-27**	30.99	2.646	36.67	21.18	15.49	55.03	1.453	25.86
28.	**PTB-28**	32.06	2.670	34.00	20.81	13.19	56.37	1.464	25.46
29.	**PTB-29**	32.46	2.671	30.23	12.37	17.86	48.19	1.512	20.18
30.	**PTB-30**	30.86	2.640	31.26	15.89	15.37	50.19	1.472	25.20

Where: w = water content, SG = specific gravity, LL = liquid limit, PL = platic limit, PI = plasticity index, FINES = grain gradation, γ_{max} = maximum dry unit weight and w_{opt} = optimum water content.

The compaction parameter values estimated by using the linear regression are the following equation:

$$\gamma d_{max}^{*} = 1{,}862 - 0{,}005 * FINES - 0{,}003 * LL \quad (2)$$

$$w_{opt}^{*} = -0{,}607 + 0{,}362 * FINES + 0{,}161 * LL \quad (3)$$

From the equation (2) and (3), the maximum dry unit weight estimated (γd_{max}^{*}) and the optimum water content estimated (w_{opt}^{*}) were obtained as shown as Tabel 3. Equation (2) has R^2 values = 0,75 and equation (3) has R^2 values = 0,80; both have very good accuracy.

Table 3. Compaction Parameter Values Estimated by Using the Linear Regression Model

No.	Sample	AASHTO	USCS	LL	FINES	γd_{max}	γd_{max}^{*}	w_{opt}	w_{opt}^{*}
1	**PTB-1**	A-6 (2)	SC	29.67	45.28	1.519	1.537	20.79	20.57
2	**PTB-2**	A-6 (4)	CL	39.56	50.98	1.488	1.477	24.21	24.23
3	**PTB-3**	A-6 (3)	CL	29.83	50.11	1.533	1.512	21.67	22.34
4	**PTB-4**	A-6 (2)	CL	29.08	50.03	1.538	1.515	21.45	22.19
5	**PTB-5**	A-6 (3)	CL	32.64	50.15	1.506	1.503	22.15	22.81
6	**PTB-6**	A-6(2)	CL	33.09	50.07	1.519	1.502	22.17	22.85
7	**PTB-7**	A-6(3)	CL	32.78	43.37	1.521	1.538	20.66	20.38
8	**PTB-8**	A-4(1)	CL	29.71	50.63	1.527	1.510	21.18	22.51
9	**PTB-9**	A6-(4)	CL	39.11	50.59	1.466	1.481	24.01	24.01
10	**PTB-10**	A6-(2)	CL	29.08	50.67	1.522	1.511	21.58	22.42
11	**PTB-11**	A6-(3)	ML	38.05	51.87	1.481	1.478	24.08	24.31
12	**PTB-12**	A6-(3)	SC	33.17	48.84	1.522	1.508	22.16	22.42
13	**PTB-13**	A6-(5)	SC	29.77	44.43	1.556	1.541	20.34	20.28
14	**PTB-14**	A6-(1)	SC	29.75	42.03	1.566	1.554	19.81	19.41
15	**PTB-15**	A6-(3)	CL	29.78	50.93	1.530	1.508	21.75	22.63
16	**PTB-16**	A-6(4)	CL	32.29	42.67	1.454	1.469	25.51	25.21
17	**PTB-17**	A6-(7)	CL	38.26	41.79	1.457	1.450	26.62	26.21
18	**PTB-18**	A4-(1)	CL	23.95	39.66	1.490	1.511	25.14	22.71
19	**PTB-19**	A7-(7)	CL	41.90	39.70	1.427	1.435	26.66	27.10
20	**PTB-20**	A6-(12)	CL	39.00	39.75	1.456	1.447	25.08	26.40
21	**PTB-21**	A6-(6)	CL	40.28	40.62	1.442	1.450	26.22	26.13
22	**PTB-22**	A6-(7)	CL	36.98	41.25	1.489	1.458	25.55	25.75
23	**PTB-23**	A6-(4)	CL	31.06	40.80	1.471	1.468	25.94	25.39
24	**PTB-24**	A4-(2)	CL	20.39	40.65	1.502	1.526	21.58	21.84
25	**PTB-25**	A6-(1)	CL	30.96	43.12	1.498	1.493	23.34	23.60
26	**PTB-26**	A6-(5)	SC	39.84	42.45	1.447	1.486	25.78	23.62
27	**PTB-27**	A6-(5)	CL	36.67	41.63	1.453	1.466	25.86	25.23
28	**PTB-28**	A6-(4)	CL	34.00	42.17	1.464	1.467	25.46	25.28
29	**PTB-29**	A6-(4)	SC	30.23	41.55	1.512	1.521	20.18	21.71
30	**PTB-30**	A6-(4)	CL	31.26	39.71	1.472	1.507	25.20	22.60

Where: LL = liquid limit, FINES = grain gradation, γ_{max} = maximum dry unit weight, $\gamma dmax^{*}$ = estimated maximum dry unit weight with linear regression, w_{opt} = optimum water content, and w_{opt}^{*} = estimated optimum water content with linear regression.

According to Goswami model (eq. 1), the correlation between maximum dry unit weight (γd_{max}^{*}) and Log G was shown as figure 1, with m=-0,376 dan k=2,482. While the correlation between optimum water content (w_{opt}^{*}) and Log G was shown as figure 2, with m=21,265 dan k=-32,421. Using the m and k constants, maximum dry unit weight estimated ($\gamma d_{max}^{\#}$) and optimum water content estimated ($w_{opt}^{\#}$) were obtained as shown as Tabel 4.

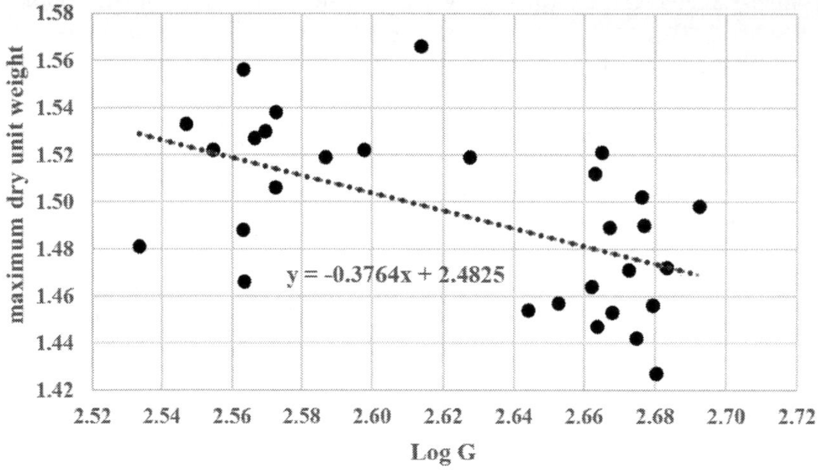

Figure 1. Correlation maximum dry unit weight (γd_{max}^{*}) with Log G

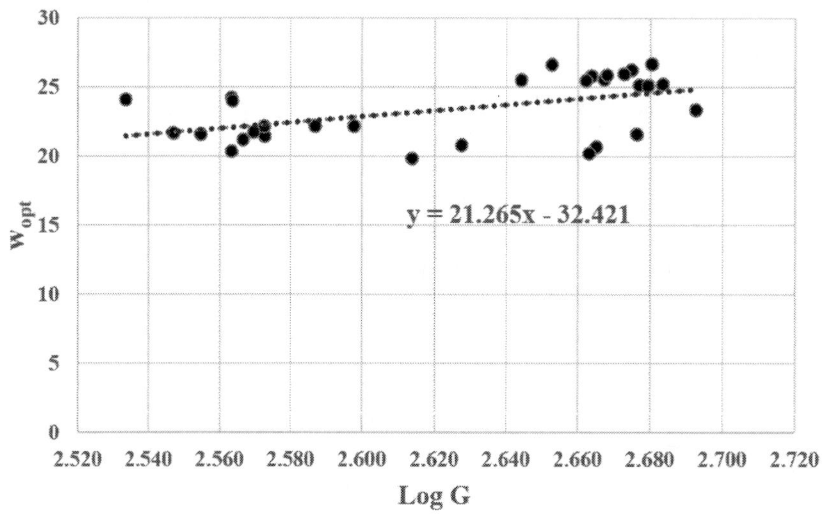

Figure 2. Correlation optimum water content (w_{opt}^{*}) with Log G

Table 4. Compaction Parameter Values Estimated by Using the Goswami Model

No	Sample	AASHTO	USCS	FINES	γd_{max}	$\gamma d_{max}^{\#}$	w_{opt}	$w_{opt}^{\#}$
1	PTB-1	A-6 (2)	SC	45.28	1.519	1.494	20.79	23.452
2	PTB-2	A-6 (4)	CL	50.98	1.488	1.518	24.21	22.085
3	PTB-3	A-6 (3)	CL	50.11	1.533	1.524	21.67	21.743
4	PTB-4	A-6 (2)	CL	50.03	1.538	1.514	21.45	22.286
5	PTB-5	A-6 (3)	CL	50.15	1.506	1.514	22.15	22.283
6	PTB-6	A-6(2)	CL	50.07	1.519	1.509	22.17	22.589
7	PTB-7	A-6(3)	CL	43.37	1.521	1.479	20.66	24.252
8	PTB-8	A-4(1)	CL	50.63	1.527	1.516	21.18	22.155
9	PTB-9	A6-(4)	CL	50.59	1.466	1.518	24.01	22.092

Table 4. Cont.

10	PTB-10	A6-(2)	CL	50.67	1.522	1.521	21.58	21.905
11	PTB-11	A6-(3)	ML	51.87	1.481	1.529	24.08	21.453
12	PTB-12	A6-(3)	SC	48.84	1.522	1.505	22.16	22.821
13	PTB-13	A6-(5)	SC	44.43	1.556	1.518	20.34	22.087
14	PTB-14	A6-(1)	SC	42.03	1.566	1.499	19.81	23.160
15	PTB-15	A6-(3)	CL	50.93	1.530	1.515	21.75	22.221
16	PTB-16	A-6(4)	CL	42.67	1.454	1.487	25.51	23.807
17	PTB-17	A6-(7)	CL	41.79	1.457	1.484	26.62	23.989
18	PTB-18	A4-(1)	CL	39.66	1.490	1.475	25.14	24.506
19	PTB-19	A7-(7)	CL	39.70	1.427	1.474	26.66	24.578
20	PTB-20	A6-(12)	CL	39.75	1.456	1.474	25.08	24.556
21	PTB-21	A6-(6)	CL	40.62	1.442	1.476	26.22	24.459
22	PTB-22	A6-(7)	CL	41.25	1.489	1.479	25.55	24.298
23	PTB-23	A6-(4)	CL	40.80	1.471	1.476	25.94	24.415
24	PTB-24	A4-(2)	CL	40.65	1.502	1.475	21.58	24.492
25	PTB-25	A6-(1)	CL	43.12	1.498	1.469	23.34	24.836
26	PTB-26	A6-(5)	SC	42.45	1.447	1.480	25.78	24.224
27	PTB-27	A6-(5)	CL	41.63	1.453	1.478	25.86	24.315
28	PTB-28	A6-(4)	CL	42.17	1.464	1.480	25.46	24.192
29	PTB-29	A6-(4)	SC	41.55	1.512	1.480	20.18	24.211
30	PTB-30	A6-(4)	CL	39.71	1.472	1.473	25.20	24.640

Where: FINES = grain gradation, γ_{max} = maximum dry unit weight, $\gamma dmax^{\#}$ = estimated maximum dry unit weight with Goswami Model, w_{opt} = optimum water content, and $w_{opt}^{\#}$ = estimated optimum water content with Goswami Model.

By Using linear regression and Goswami model, the relationship between compaction parameter values Estimated and index properties were obtained, included the soil classification. So, the purpose of this research has been achieved.

For both estimation models, correlation between the compaction parameter values estimated with the compaction parameter values laboratory, then were calculated for their level of confidence intervals, as in Table 5.

Table 5. Level of Confidence Intervals

	Linear Regression Model		Goswami Model	
	Maximum dry unit weight (γd_{max}^{*})	Optimum water content (w_{opt}*)	Maximum dry unit weight ($\gamma d_{max}^{\#}$)	Optimum water content ($w_{opt}^{\#}$)
Correlation Coefficient	1,000	1,000	0,988	1,000
T value	3,268	3,674	3,453	3,068
T table (95% of Confidence Intervals)	2,055	2,055	2,052	2,052

5. Conclusion

By linear regression, the equation for estimation of the maximum dry unit weight (γd_{max}*) =1,862-0,005*FINES-0,003*LL and estimation of the optimum water content (w_{opt}*) =-0,607+0,362*FINES+0,161*LL. By Goswami Model (with equation Y=mLogG+k), for estimation of the maximum dry unit weight (γd_{max}*) with m=-0,376 and k=2,482, for estimation of the optimum water content (w_{opt}*) with m=21,265 and k=-32,421. For both of these equations a 95% confidence interval was obtained. Estimation by using Goswami Model is easier and more effective than Linier Regression Model because it only require fine grains (Fines) value. Based on the classification value, the types of soil were A4, A6 and A7 (AASHTO) or SC and CL (USCS).

Acknowledgements
The authors also would like to thank to Sustainable Energy and Biomaterial Center of Excellence (SEBCOE), Universitas Sumatera Utara for the financial aids.

References
[1] J E Bowles 1993 *Sifat-sifat Fisis dan Geoteknis* Edisi Kedua Erlangga Jakarta
[2] Z A Muis 1998 Penentuan Berat Isi Kering Maksimum Bahan Aggregat Base Berdasarkan Data Klasifikasi Tanah Pada Proyek Jalan Raya *Seminar Highway Engineering Medan Academic Commitee (MAC)* Teknik Sipil USU
[3] Z A Muis and D Siagian 2013 Estimasi Nilai Parameter Kompaksi Bahan Subgrade Berdasarkan Nilai Index Properties Pada Proyek Jalan Raya *Skripsi Sarjana Departemen Teknik Sipil USU*
[4] J B Metacalf and S A Romanoschi 2008 Prediction Od Maximum Dry Density And Optimum Moisture Content From Simple Material Properties
[5] A M Nendi 2010 Korelasi Antara Hasil Ujian Mampatan Dengan Had Atterberg, *Skripsi Sarjana*, Fakultas Teknik Sipil, Universitas Teknologi Malaysia
[6] F C Ugbe 2011 Estimating Compaction Characteristics From Fines in A-2 Type Lateritic Soils *Research Journal Of Environmental And Earth Sciences* 3 (4): 433-7, 2011
[7] F C Ugbe 2012 Predicting Compaction Characteristics Of Lateritic Soil Of Western Niger Delta Nigeria, *Research Journal Of Environmental And Earth Sciences* 4 (5): 553-9
[8] Isik F, dan Ozden G 2013 Estimating Compaction Parameters Of Fine- And Coarse- Grained Soils By Means Of Artificial Neural Networks *Environ Earth Sci* 69:2287-2297
[9] A S Lubis et.all 2017 The Estimation Of Compaction Parameter Values Based On Soil Properties Values Stabilized With Portland Cement, *IOP Conf. Ser.: Mater. Sci. Eng.*180012139
[10] A S Lubis et.all 2017 The Estimation of Parameter Compaction Values For Pavement Subgrade Stabilized With Lime (*still waiting for publishing on IOP*)

IOP Conf. Series: Materials Science and Engineering 306 (2018) 012006 doi:10.1088/1757-899X/306/1/012006

Video Tutorial of Continental Food

A S Nurani*, A Juwaedah, and A Mahmudatussa'adah
Universitas Pendidikan Indonesia

*atatsiti@upi.edu

Abstract. This research is motivated by the belief in the importance of media in a learning process. Media as an intermediary serves to focus on the attention of learners. Selection of appropriate learning media is very influential on the success of the delivery of information itself both in terms of cognitive, affective and skills. Continental food is a course that studies food that comes from Europe and is very complex. To reduce verbalism and provide more real learning, then the tutorial media is needed. Media tutorials that are audio visual can provide a more concrete learning experience. The purpose of this research is to develop tutorial media in the form of video. The method used is the development method with the stages of analyzing the learning objectives, creating a story board, validating the story board, revising the story board and making video tutorial media. The results show that the making of storyboards should be very thorough, and detailed in accordance with the learning objectives to reduce errors in video capture so as to save time, cost and effort. In video capturing, lighting, shooting angles, and soundproofing make an excellent contribution to the quality of tutorial video produced. In shooting should focus more on tools, materials, and processing. Video tutorials should be interactive and two-way.

1. Introduction

Continental Food is a course on food processing Erofa began its development from France. In Continental Food courses are discussed from the history, the development of menu order, Handling Knives, Vegetable Pieces, Appetizer (oderve (hot, and cold): pate, balotin, galantine, aspic and Salad, soup (Stock), Canape, Sandwich, Processing broth and sauce, Poultry Processing, Vegetable Processing, Flour and Eggs, Soma Sauce: Sauce and Meat, Poultry Seafood. Competence is so much of course requires the help of learning resources that can be learned at any time such as tutorial media.

Media is derived from Latin which is the plural of the word medium which means intermediary. Media is a medium for channeling learning information or message channeling [1]. Arsyad stated that "learning media have several terms such as hearing instrument, instructional material, audio visual communication, visual education, educational technology, visuals and explanatory tools [2]. The eight criteria of good instructional media that need to be addressed in the media selection process are as follows: (1) Clear and neat, (2) Clean and attractive, (3) Matches the target, (4) Relevant to the topic being taught, (5) (6) Practical, flexible and resistant, (7) Good quality, (8) The size is in accordance with the learning environment [3]. The five principles of media selection are suitability, clarity of presentation, accessibility, affordability, availability, quality, alternative, interactive, organizational, novelty, and student oriented [3]. Good media criteria and have principles in choosing instructional media according to Sudjana [4]: (1) Media accuracy with teaching objectives, (2) Support for content, (3) Ease of access to media, (4) Teacher skills in using it, (5) Available time to use it, (6) In accordance with the child's thinking level. Ariani and Haryanto suggested several benefits of multimedia learning, among others: (1) More interesting, (2) More interactive, (3) Total teaching time (lecture) teachers / educators can be reduced, (4) Student learning quality can be more motivated (5) Attention to the text by including sounds, pictures, music, animation and video, (7) Interesting because it is a combination of sight, sound and movement [5].

The benefits of multimedia are: (1) Students can learn according to their abilities, readiness and desires, (2) Students learn from a patient tutor (computer) that adapts to students' abilities, (3) Students will be motivated to pursue knowledge and acquire (4) Students face an objective evaluation through their participation in the provided practice / test, (5) Students enjoy privacy (6) Learn-while-needs ("just-in-time" learning), (7) Learn whenever they want to be unattached a predetermined time, (8) Overcome weaknesses in group and individual learning.

Learning video media is a medium that presents audio visual learning messages relating to principles, concepts, procedures, theories, and applications to aid understanding of learning materials [6]. According to Susilana and Riyana tutorial model is a learning through computer media that has been programmed with the presentation of materials and exercise questions [7]. Media tutorial is a learning audio visual media that can be operated by using computer facilities that contains learning objectives, materials and exercise questions.

2. Research Methods

Research development aims to produce new products through the development process. Product development begins with survey, product creation and product evaluation. According to Borg and Gall development research is intended to create and then evaluate the resulting product. The steps in this study begin with a survey of student needs, literature studies, composing materials, making scripts, storyboards, story board evaluation, storyboard revision, product creation, product evaluation and product revision.

3. Results and Discussion

The first step is to identify the objectives of learning, learning needs, problems that often arise. This is important to do. Because by knowing and identifying the learning objectives, the researcher can design a program that allows the program user to perform the learning steps consistently and not deviate from the learning objectives to be achieved. Similarly, knowing the needs of learning then the designer can adjust the program to be made in order to meet the learning needs of learners as the main target in learning. As for identifying problems that arise can fix the problem

Preliminary student survey results have failed in Hollandise sauce making. Hollandise sauce is a type of yellow sauce made from egg yolk, lemon and clarified butter un salted made with auben marrie technique. This sauce is perfect to be enjoyed with poached eggs or boiled vegetables. The characteristics of Hollandise sauce are creamy and soft, flowed perfectly and the viscosity is moderate. Failure in the manufacture of hollandaise sauce is characterized by the occurrence of clots due to broken or oily eggs because the emulsinya system is not achieved. Generally this is due to overheating or incomplete clarified butter so that the Hollandise sulphid emulsion fails to form.

Step make Hollandise sauce is by preparing 2 chicken egg yolk, store in a heat-resistant container. Water in a pot is about the size of a container containing heated egg yolk. In another container, liqueze 150 g of unsalted butter and then remove it and leave it to room temperature. White-white and bubbles that exist in melted butter are thrown away. Beat the egg yolks until well blended. Place this bowl on a pot of boiling water. Keep the fire small. Roll the lemon on the table, press, then add 1 tbsp lemon juice and 1 tsp of mustard, shake it flat with a wire shake. Pour the melted butter little by little while shaking it on until the eggs and butter emulsively emulate. Do not add too much liquid butter to avoid the dough. Once emulsified properly, Hollandise sauce will look shiny and creamy. Remove from the pan and stir until slightly cool. Tips for not failing in making Hollandise sauce is when heating or whisking eggs steam should not be too hot, done by pulling out occasional egg whisk, so it is not continuously on the fire. To create a video tutorial making Hollandise sauce begins with the making of story board.

Story board is a media creation plan that consists of scene number, description, visual and audio. The description is a global record and an explanation of the contents of the scenes that are inclined. Visual contains an explanation of the contents of the scene concerned consisting of components of writing, pictures, and colors. In each scene the visual columns must be filled with illustrations to illustrate clearly the scene plan to be created. The audio column is filled with an audio plan that will be displayed in the tutorial video.

Assessment of feasibility by the learning expert through validation of the assessment of the learning expert with the question indicator such as: objectives, usability, quality of content, quality of language and other technical aspects with percentage of assessment result as follows: goal aspect 90%, 80% 77%, linguistic 80% and other technical aspects by 83%. If the average assessment of expert learning by 82%.

4. Conclusions

The results show that the making of storyboards should be very thorough, and detailed in accordance with the learning objectives to reduce errors in video capture so as to save time, cost and effort. In video capturing, lighting, shooting angles, and soundproofing make an excellent contribution to the quality of tutorial video produced. In shooting should focus more on tools, materials, and processing. Video tutorials must be interactive and bi-directional.

Acknowledgements

The authors would like to thank the Directorate General of Higher Education, Ministry of research and technology, Republic of Indonesia, for funding this research through Penelitian Penerapan Teknologi research grant 2017.

References

[1] Djamarah S B and Aswan Z 2010 *Strategi Belajar Mengajar* (Jakarta: Rineka Cipta)
[2] Arsyad A 2007 *Media Pembelajaran* (Jakarta: PT Raja Grafindo Persada)
[3] Asyhar R 2012 *Kreatif Mengembangkan Media Pembelajaran* (Jakarta: Referensi)
[4] Sudjana N and Rivai A 1991 *Media Pengajaran* (Bandung: Sinar Baru)
[5] Ariani N and Haryanto D 2010 *Pembelajaran Multimedia di Sekolah* (Jakarta: PT Prestasi Pustakaraya)
[6] Riyana C 2007 *Pedoman Pengembangan Media Video* (Bandung: Universitas Pendidikan Indonesia)
[7] Susilana R and Riyana C 2009 *Media Pembelajaran: Hakikat, Pengembangan, Pemanfaatan, dan Penilaian* (Bandung: CV Wacana Prima)

Characterization of Co:TiO_2 Thin Film Grown by MOCVD Technique

A Saripudin* and W Purnama

Program Studi Teknik Elektro, Fakultas Pendidikan Teknologi dan Kejuruan, Universitas Pendidikan Indonesia, 3rd floor FPTK Building, Jl. Dr. Setiabudhi 229 Bandung 40154

*aipsaripudin@upi.edu

Abstract. The Co:TiO_2 thin film was grown on n-type Si(100) by metal organic chemical vapor deposition (MOCVD) technique. The film's growth parameters are as follow: substrate temperature of 450°C, bubbler temperature of 70°C, reactor chamber pressure of 2 militorr, and growth time of 2 hours. We characterized the structure of film by X-ray Difractometer (XRD), the morphology was characterized by Scanning Electron Microscope (SEM), and the fraction of Co atoms in TiO_2 was characterized by Energy Dispersive x-ray Spectroscopy (EDS). The XRD result shows that the Co:TiO_2 thin film is an anatase phase crystal dominated by A(213) orientation. Using Warren-Scherrer's formula, the average grain size of Co:TiO_2 is 169 nm. The SEM result shows that the Co:TiO_2 film surface is quite coarse with relatively homogeneous grain shape. the average growth rate of Co:TiO_2 film is 0.78 μm/h. In addition, the EDS result shows that Co atoms have been incorporated into the film replacing a portion of the Ti atoms by 0.085%.

1. Introduction

Titanium dioxide, TiO_2, is one of semiconductor materials that studied for many purposes. Because of its properties, TiO_2 can be applied in various fields [1]. TiO_2 have been applied in optical components [2] and photocatalysis [3] and have a high potential application in as a nonlinear optic material, sensitive solar cell, and dynamic random access memory [4]. TiO_2 also have an interesting properties for spintronic application. TiO_2 doped by Co atoms shows the ferromagnetic properties at room temperature [5].

TiO_2 can be grown on a substrate by various methods such as molecular beam epitaxy (MBE) [5], DC and RF Sputtering [6], atomic layer deposition (ALD) [7], pulsed laser deposition (PLD) [8], and metal organic chemical vapor deposition (MOCVD) [1], [9], [10]. Compared to other methods, MOCVD has many advantages, i. e., film can be grown epitaxial, film can be deposited selectively, and film can be grown by simple growth parameter [10].

In this paper, we focused on a structural a morphology characterization of TiO_2 doped by Co atoms (Co:TiO_2).

2. Experimental Details

The Co:TiO_2 was grown on n-type Si(100) by MOCVD technique. Materials used in this experiment are $Ti(C_3H_7O)_4$ (titanium (IV) isopropoxide or TTIP), $Co(C_{11}H_{19}O_2)_3$ (Cobalt tris 2,2,6,6-tetramethyl -

3,5-heptanedionato or Co(TMHD)$_3$), and C_4H_3O (tetrahydrofuran or THF). TTIP 99,99% from Sigma Aldrich Chemical Co., Inc. was used as TiO_2 precursor, Co(TMHD)$_3$ from Strem Chemical Inc. was used as Co precursor, and THF 99,99% from Sigma Aldrich Chemical Co., Inc., was used as Co solvent.

The process of the film growth was done in the following way. The vaporized tube containing the 20 mL TTIP and 900 mg Co(TMHD)$_2$ (in 20 mL THF) were heated to a constant temperature of 70°C. TTIP and Co(TMHD)$_2$ vapor pressurized 20 kPa was driven by argon gas at a rate of 70 sccm to the reactor chamber through a connecting pipe. In the reactor chamber, n-type Si(100) substrate has been pasted at molybdenum disk for the film growth process at temperature of 450° C. The total reactor pressure was 2 militorr. The growth time was 2 hours.

Characterization of Co:TiO_2 thin films was done using XRD, SEM, and EDS. XRD was used to find out the structure and orientation of the Co:TiO_2 film. In this case, XRD uses Cu K_α radi radiation at a wavelength of 1,54060 angstrom (Philips PW3710). SEM (JEOL type JSM-6510LA) was used to characterize the surface morphology and the thickness of Co:TiO_2 film. Meanwhile, EDS was used to measure the fraction of Co atoms in TiO_2.

3. Results and Discussion

3.1. XRD

Figure 1 shows the XRD result of Co:TiO_2 thin film. There are two diffraction peaks that appear at 2θ = 55.00 degree and 2θ = 61.70 degree. These peaks belongs to anatase phase, i.e., A(211) and A(213), respectively. Intensity of A(213) is more higher than A(211). Therefore, it can be revealed that the Co:TiO_2 thin film is an anatase phase crystal dominated by A(213) orientation.

Referring to the standard diffraction peak for TiO_2, the peak of A(213) is present at 2θ = 62.119° and A(211) at 2θ = 55.060°. This means that there is a diffraction peak shift for Co:TiO_2. Such a shift may occur because the bit doping of Co atoms in TiO_2 films change the film lattice constants.

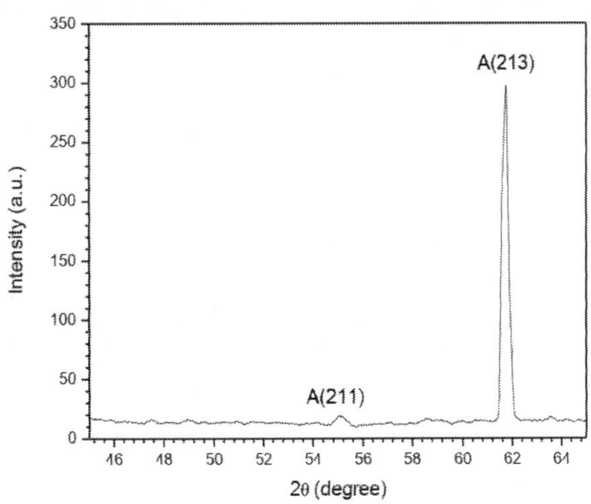

Figure 1. XRD result of Co:TiO_2 thin film

The size (diameter) of the crystals was obtained by using the following Warren-Scherer equations [11]:

$$D \approx \frac{K\lambda}{B\cos\theta}$$

where D is the diameter of the crystal, λ is X-ray wavelength, θ is the angle of Bragg diffraction, *B* is half of FWHM (full-width at half-maximum) of the selected peak, and *K* is the constant. Generally, the value of K chosen is 0.89.

IOP Conf. Series: Materials Science and Engineering **306** (2018) 012007 doi:10.1088/1757-899X/306/1/012007

By substituting the value of $K = 0.89$, $\lambda = 1.5046$ angstrom, $B = 0.000955$ rad (for $2\theta = 61.70$ degree), and $\theta = 0.53804066$ rad into the Warren-Scherrer equation, the obtained crystalline diameter is 169 nm. The size of this crystal diameter indicates that the MOCVD method has successfully grown film by grain size in the order of nanometers.

3.2. SEM

The surface morphology and cross section of the film was characterized by SEM. **Figure 2**(a) shows the SEM image of the Co:TiO_2 film surcae. It can be seen that, in general, all surfaces of the film appear to be quite coarse with relatively homogeneous grain shape.

The SEM cross-sectional image of the film is shown in **Figure 2(b)**. The film is composed of crystal-like grains of crystal. In the substrate-film interface, the crystal grains appear irregular. This is because the growth of crystals is competitive so that the grains formed depend on the diffusion process of the surfaces of the precursor atoms. This competitive crystal growth is due to the diverse orientation of crystal grains that are bound to the surface of the substrate [12].

For 2 hours growth of the Co:TiO_2 film, the average thickness of the film is 1.56 um. This shows that the average growth rate of the Co:TiO_2 film is 0.78 μm/h.

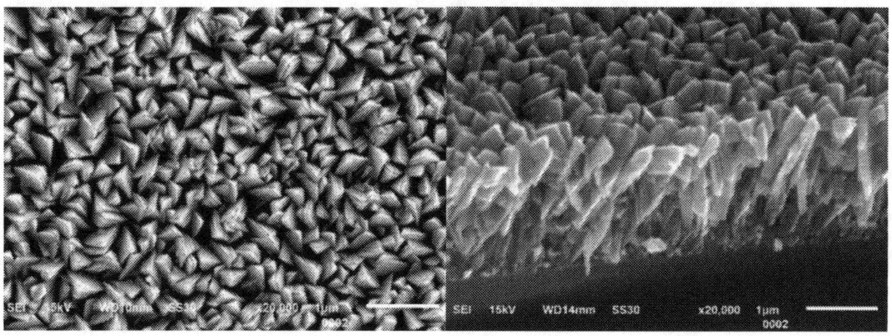

(a) (b)

Figure 2. SEM image of Co:TiO_2 thin film: (a) surface and (b) cross-section.

3.3. EDS

EDS characterization was performed to determine the composition of the atoms of the film. **Figure 3** shows the EDS graph of the Co:TiO_2 film. Based on the tabulation of the percentage of atoms, it is obtained that O: Ti: Co = 64.84: 35.13: 0.03. Referring to this comparison, the fraction of Co atoms to Ti is $0.03 : 35.13 \times 100\% = 0.085\%$. Thus, the Co atoms have been incorporated into the film replacing a portion of the Ti atoms by 0.085%.

4. Conclusion

The Co:TiO_2 thin film have been successfully grown by MOCVD technique. XRD result show that the Co:TiO_2 thin film is an anatase phase crystal dominated by A(213) orientation. From the SEM image, in general, all surfaces of the film appear to be quite coarse with relatively homogeneous grain shape, the film is composed of crystal-like grains of crystal, and the average growth rate of Co:TiO_2 film is 0.78 μm/h. EDS results show that Co atoms have been incorporated into TiO_2 film.

IOP Conf. Series: Materials Science and Engineering **306** (2018) 012007 doi:10.1088/1757-899X/306/1/012007

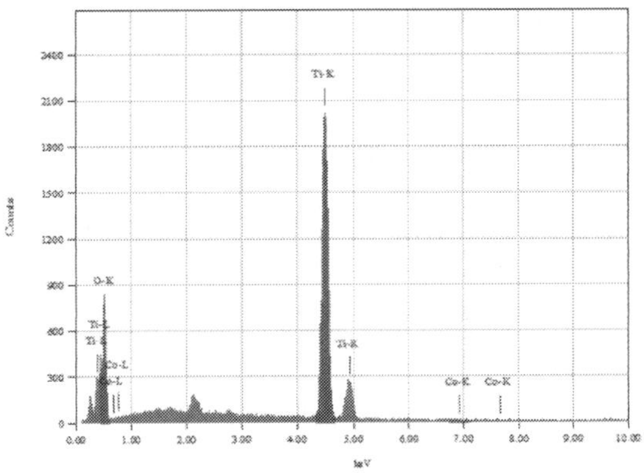

Element	(keV)	Mass%	Error%	Atom%	Compound	Mass%	Cation	K
O K	0.525	38.11	0.45	64.84			16.5866	
Ti K	4.508	61.83	0.37	35.13			83.3370	
Co K	6.924	0.06	1.16	0.03			0.0764	
Total		100.00		100.00				

Figure 3. EDS Result of Co:TiO_2 thin film

References

[1] Saripudin, A., Saragih, H., & Arifin, P. (2014). Effect of Growth Temperature on Cobalt-Doped TiO_2 Thin Films Deposited on Si (100) Substrate by MOCVD Technique. In Advanced Materials Research (Vol. 896, pp. 192-196). Trans Tech Publications.

[2] Asahi, R., Taga, Y., Mannstadt, W., & Freeman, A. J. (2000). Electronic and optical properties of anatase TiO_2. Physical Review B, 61(11), 7459.

[3] Di Fonzo, F., Casari, C. S., Russo, V., Brunella, M. F., Bassi, A. L., & Bottani, C. E. (2008). Hierarchically organized nanostructured TiO_2 for photocatalysis applications. Nanotechnology, 20(1), 015604.

[4] Magyari-Köpe, B., Tendulkar, M., Park, S. G., Lee, H. D., & Nishi, Y. (2011). Resistive switching mechanisms in random access memory devices incorporating transition metal oxides: TiO_2, NiO and Pr0. 7Ca0. 3MnO3. Nanotechnology, 22(25), 254029.

[5] Matsumoto, Y., Murakami, M., Shono, T., Hasegawa, T., Fukumura, T., Kawasaki, M., & Koinuma, H. (2001). Room-temperature ferromagnetism in transparent transition metal-doped titanium dioxide. Science, 291(5505), 854-856.

[6] Tuna, O., Selamet, Y., Aygun, G., & Ozyuzer, L. (2010). High quality ITO thin films grown by dc and RF sputtering without oxygen. Journal of Physics D: Applied Physics, 43(5), 055402.

[7] Shin, H., Jeong, D. K., Lee, J. A. E. G. A. B., Sung, M. M., & Kim, J. I. Y. O. U. N. G. (2004). Formation of TiO_2 and ZrO_2 nanotubes using atomic layer deposition with ultraprecise control of the wall thickness. Advanced Materials, 16(14), 1197-1200.

[8] Hitosugi, T., Ueda, A., Furubayashi, Y., Hirose, Y., Konuma, S., Shimada, T., & Hasegawa, T. (2007). Fabrication of TiO_2-based transparent conducting oxide films on glass by pulsed laser deposition. Japanese journal of applied physics, 46(1L), L86.

[9] Kang, B. C., Lee, S. B., & Boo, J. H. (2000). Growth of TiO_2 thin films on Si (100) substrates using single molecular precursors by metal organic chemical vapor deposition. Surface and Coatings Technology, 131(1), 88-92.

[10] Saragih, H., Arifin, P., & Barmawi, M. (2006). Co-doped TiO_2 rutile thin films deposited by MOCVD method. Nanoscience and Nanotechnology, 87-91.

[11] Abdullah, M., & Okuyama, K. (2004). Zinc oxide nanoparticles prepared by a simple heating: Effect of polymer addition and polymer absence on the morphology. Journal of Engineering and Technological Sciences, 36(2), 141-153.

[12] Barna, P. B., & Adamik, M. (1998). Fundamental structure forming phenomena of polycrystalline films and the structure zone models. Thin solid films, 317(1), 27-33.

Effect of Higher Order Thinking Laboratory on the Improvement of Critical and Creative Thinking Skills

A Setiawan*, A Malik, A Suhandi, and A Permanasari

Science Education Program, School of Postgraduate Studies, Universitas Pendidikan Indonesia, Jl. Dr. Setiabudhi No. 229 Bandung 40154 Indonesia

*agus_setiawan@upi.edu

Abstract. This research was based on the need for improving critical and creative thinking skills of student in the 21-st century. In this research, we have implemented HOT-Lab model for topic of force. The model was characterized by problem solving and higher order thinking development through real laboratory activities. This research used a quasy experiment method with pre-test post-test control group design. Samples of this research were 60 students of Physics Education Program of Teacher Educatuon Institution in Bandung. The samples were divided into 2 classes, experiment class (HOT-lab model) and control class (verification lab model). Research instruments were essay tests for creative and critical thinking skills measurements. The results revealed that both the models have improved student's creative and critical thinking skills. However, the improvement of the experiment class was significantly higher than that of the control class, as indicated by the average of normalized gains (N-gain) for critical thinking skills of 60.18 and 29.30 and for creative thinking skills of 70.71 and 29.40, respectively for the experimental class and the control class. In addition, there is no significant correlation between the improvement of critical thinking skills and creative thinking skills in both the classes.

1. Introduction

The learning achievement described in the National Standards of Higher Education implies that every university needs to equip students with the 21st century skills including critical thinking, creative thinking, problem solving, cooperative, communication as well as ICT-literate. Those skills are well known as 21st century skills or transferable skills. In general, transferable skills are skills developed in one situation to be transfered to other situations, from classroom learning situations to real life situations [1].

Transferable skills can be developed through science learning process in classrooms or practicum in laboratories. Practicum is important to (1) generate motivation for science learning, (2) improve basic skills of experiment, (3) apply scientific approach and (4) support mastery of learning materials [2]. The main purpose of physics practicum is to increase physics knowledges, improve practicum skills, aroush interest, develop creative thinking and problem-solving skills, improve scientific thinking skills as well as apply methods of experiment [3]. In addition, practicum also offers context-rich learning experiences, enhance conceptual understanding, develop practical skills [4] and the best way to learn the nature of science [5, 6].

In fact, purposes of practicum as mentioned above have not fully implemented by teachers in physics teaching and learning as reflected in the results of questionnaires distributed to several senior high schools in West Java. Most physics teachers face difficulties to develop practicum procedures. The current practicums tend to cookbook model and lact of practicum equipments. Most physics teachers only organize practicum twice per semester [7]. A research reported that implementation of physics practicum in senior high school in Bengkulu Province is in the low category [8]. Main constrains experienced by teachers to do practicum are as follow: (1) lack of practicum facilities, (2) time limited to do practicum and (3) difficult to design practicum. We also identified that the existing practicums havenot developed higher order thinking skills. By considering the previous findings, it is urgent to develop a Higher Order Thingking Laboratory (HOT Lab) to facilitate 21st century skills (transferable skills). Furthermore, the HOT Lab developed in this research should be able to solve problem with practicum equipments and limited time for practicum.

Some empirical studies aimed at improving critical and creative thinking skills as part of transferable skills has been widely carried out. Research on strategy and model of learning to enhance critical thinking skills that have been done include guided inquiry learning model [9], problem-based approach [10], collaborative work approach [11], application of writing strategies [12]. Research on improving creative thinking skills using strategies and learning models that have been done include brainstorming [13, 14], lateral thinking [15], creative problem solving [16, 17], project-based learning [18] and creative inquiry learning [19]. While research on the improvement of two high-level thinking skills, critical and creative thinking skills together through practicum in university level is as an innovation. Therefore this research addresses what can the HOT lab improve critical and creative thinking skills and what is the relationship between students' critical and creative thinking skills.

2. Experimental Method

This research employs a quasi-experimental method with control group pretest-posttest design. The experiment class implemented HOT Lab consisting of 11 stages: real world problem, determine and evaluate ideas,experimental questions, materials and equipment, prediction, question method; exploration, measurement, analysis, conclusion and presentation. While the control class implemented verification lab consisting of 9 stages: objectives, basic theory, tools and materials, preliminary assignment, trial procedure, measurement, analysis, conclusion, and final report.

Population in this research is students of Physics Education Program of UIN Sunan Gunung Djati Bandung at semester IV in academic year of 2015/2016. The samples are 60 students divided into 30 students for experimental class and 30 students for control class. The samples were selected by using a simple random sampling technique. The instrument used in this research is an essay test to measure critical and creative thinking skills on the concept of Force. Indicators of critical thinking test refers to Binkley et all framework [20], while indicator of creative thinking skills test refers to Torrance framework.

Data were analyzed by using a normalized gain $<g>$ according to the equation expressed by Hake [21]. The calculated $<g>$ was then interpreted by the Hake criterion [22] ($<g> <0.3$ (low); $0.3 \leq <g> \leq 0.7$ (medium); and $<g>> 0.7$ (high). The difference of the $<g>$ for experiment and control classes was tested by t-test after previously conducting normality test using One-Sample Kolmogorov Smirnov test and homogeneity test by using test of homogenity of variances. Then, correlation between critical thinking skill and creative skills was calculated by corelation bivariate test.

3. Result and Discussion

3.1. Critical Thinking Skills

Figure 1 provides average score of pretest, posttest and N-gain for creative thinking skills. N-gain of critical thinking skills for HOT Lab class is 60.2% (medium category) and verification lab class is 29.30% (low category). Result of statistical calculation is shown in Table 1. It can be concluded that there is a significant difference of students' critical thinking skills skill between HOT lab class compared to lab verification class. This means that HOT lab is better than verification lab to improve critical thingking skills.

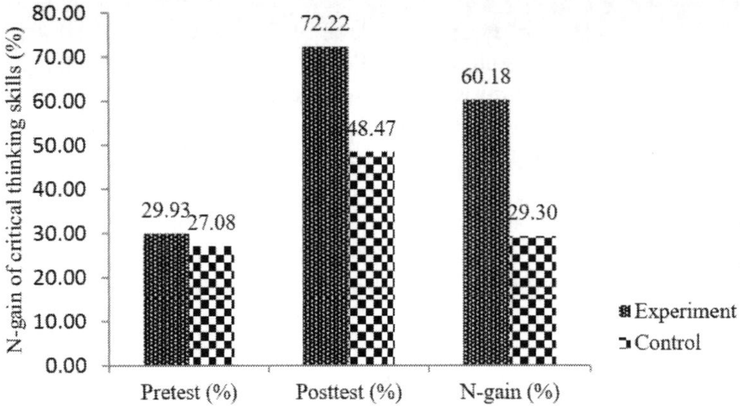

Figure 1. Average N-gain scores of critical thinking skills

Table 1. Statistic calculation of critical thinking skills data

Data type	Normality (α = 0.05)				Homogeinity (α = 0.05)	Uji t (α = 0.05)	
	Level Significance (2-tailed)		Information			Level Significance (2-tailed)	Information
	Experimental	Control	Experimental	Control			
Critical thinking skills	0.782	0.983	Normal	Normal	0.109 Homogeneous	0.000	Significant

Further calculations are performed for each aspect of critical thinking skills in both the groups. The result is shown in Figure 2. The N-gain for each aspect of critical thinking skills in the experimental class is higher than the control class. All aspects of critical thinking skills (CTS1 to CTS6) in the experimental class are in modium category. The N-gain of critical thinking skills aspect in the control class classified into two categories, namely medium category for CTS1 and CTS2 and low category for CTS3 to CTS6.

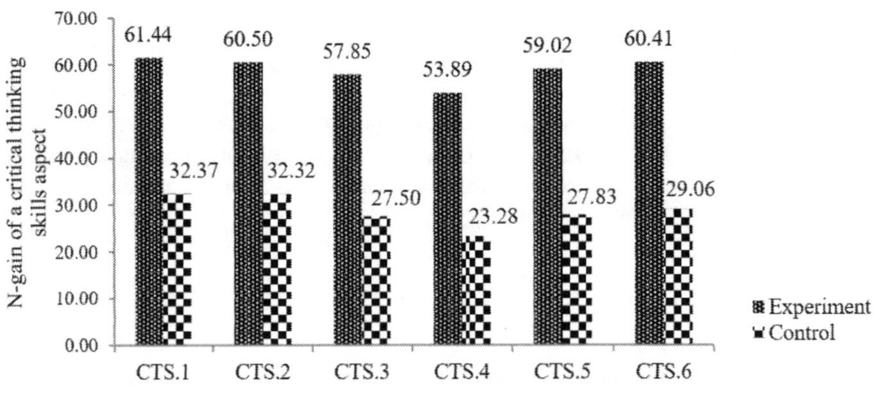

Figure 2. Average N-gain for each aspect of critical thinking skills

Respectively:
1. CTS1. Explain. State results, justify procedures, and present arguments
2. CTS2. Analyze how parts of a whole interact with each other to produce overall outcomes in complex systems
3. CTS3. Interpret information and draw conclusions based on the best analysis
4. CTS4. Synthesize and make connections between information and arguments
5. CTS5. Inference. Query evidence, conjecture alternatives, and draw conclusions
6. CTS6. Evaluate. Assess claims and arguments

The CTS1 aspect shows the greatest N-gain for both the class, in medium category. However, the experimental class shows higher improvement than the control class. The aspect of explanation shows the greatest N-gain in both the classes. Students of the HOT Lab class are capable to present and provide results, procedures and arguments. Those skills have been developed in the stage of exploration and presentation in the HOT Lab design. In the both stages, the students design and check system of work from provided choice of answers to solve problems and present experimental results obtained in the form of verbal and non-verbal. In contrast, students of verification lab class are already informed what to measure, perform the given procedures. They do not present the the obtained experiments. This result agreed with previous study that cookbook experiments less facilitates to know what have been done during the experiment [22].

Aspect of CTS4 shows the smallest N-gain in both the classes. But, the experiment class has higher N-gain (medium category) compated ro the control class (low category). Stages of experimental questioning and questions method in the HOT lab design facilitate much better opportunity for students to connect the provided information with the presented arguments. While the student of of verification lab lass only decide the details of the analysis and emphasize the concept comprehension (quantitative). This is in accordance with the focus of developed competency in the lab verification design that supports what students are learning, teaching experimental techniques but lacking the practice of thinking skills [23].

3.2. Creative thinking skills

Figure 3 provide the average score of pretest, posttest and N-gain of creative thinking skills for the experimental class and control class on the topic of force. Here, N-gain scores are provided in decimal numbers. We noted that creative thinking skills of the experimental class and the control class are increased. However, N-gain of the experimental class is higher (0.7, high category) than that of the control class (0.29, low category). Table 2 provides results of statistical calculation of creative thinking skills in experiment and control class. It can be concluded that there critical thinking skills of the experimentalkclass (HOT lab class) is significantly better than that of the control class (Verification Lab class).

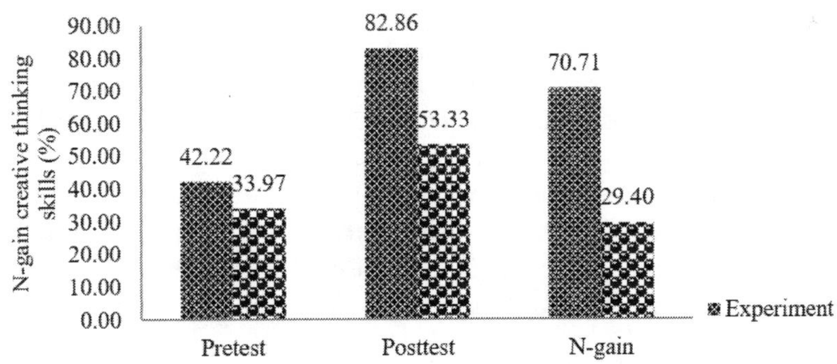

Figure 3. Average N-gain of creative thinking skills

Table 2. Statistic calculation of students' creative thinking skill

Data type	Normality (α = 0.05)				Homogeinity (α = 0.05)	Uji t (α = 0.05)	
	Level Significance		Information			Level Significance	Information
	Experiment	Control	Experiment	Control			
Creative thinking skills	0.528	0.509	Normal	Normal	0.798 Homogeneous	0.000	Significant

Further calculations have been made for each aspect of creative thinking skills in both the classes (Figure 4). The average N-gain for each aspect of creative thinking skills in the experimental class is higher than that of the control class. In the experimental class, three aspects of creative thinking skills (fluency, flexibility and elaboration) showed high category while the aspect of orginalities showed medium category. The N-gain aspects of creative thinking skills in the control class classified into two categories: fluency and elaboration are in the medium category while the flexibility and orginality aspects are in low category.

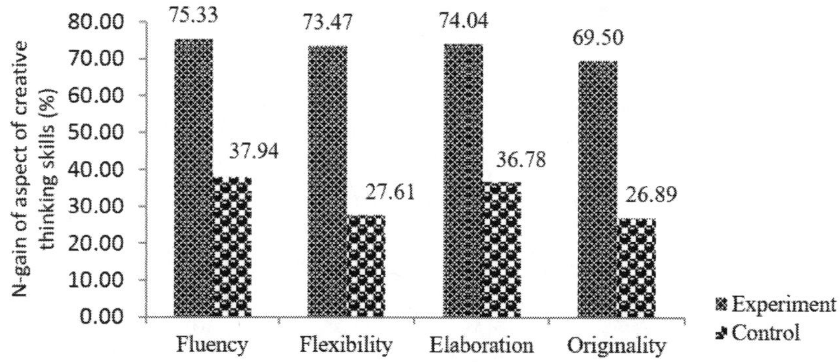

Figure 4. Average N-gain for each aspect of creative thinking skills

Fluency aspect shows the greatest N-gain both in the experimental class and control class. N-gain of the experimental class is higher than the control class. In the stages of question method, activity in the HOT lab design provides a better opportunity for students to think fluently related to asking questions that serve a guide for a problem solving. One can find the most basic principles of physics and its problems by asking questions [24]. The fluency aspect in the control group is less developed, because in the verification lab design students are just doing the procedures already given. This agrees with other report that the implementation of verification lab rarely cacilitate to understand what has been done [25].

The originality aspect shows the smallest N-gain for both the experimental class and control class. N-gain of the experimental class is higher than the control class. The stages of exploration and analysis activities in the HOT lab design give better opportunities for students to think originaly to produce a new and unique phrase. In contrass, students who impelemened verification lab design only decide the details of analysis and emphasize the concept comprehension. The results are similar to other studies reprting that the application of CPS strategy as the basis of HOT Lab design does not significantly improve originality of thinking skills as compared to non-CPS strategies [26].

3.3. Correlation between creative and critical thingking skills

In this study, we also explore the correlation between N-gain of critical thinking skills and creative thinking skills in the experimental class and control class. The result of statistical calculation is provided in Table 3. We found that significantly there is no correlation between N-gain of critical thinking skills and N-gain of creative thinking skills both for experimental class and control class.The result indicate that critical thinking skills and creative thinking skills as high-level thinking skills are associated with other variables such as age, cognitive level and field of study[27]. Other aspect related to the science process is the logic of science as a form of logical thinking. Therefore, the laboratory activities also give a rich context for using logical thinking together between creative thinking skills and critical thinking skills [28]. Every activity in the HOT Lab promotes together convergent thinking (critical thinking) and divergent thinking (creative thinking).

Table 3. Correlation between creative thinking and critical thinking skills of the experimental class and control class

Group	Normality (α = 0.05)				Homogeinity (α = 0.05)	Correlations test			
	Level of Significance		Information			Level of Significance	Information	Pearson corelation	Information
	Critical	Creative	Critical	Creative					
Experiment class	0.782	0.528	Normal	Normal	0.205 Homogeneous	0.149	No corelation	-0.270	Negative corelation
Control class	0.983	0.509	Normal	Normal	0.165 Homogeneous	0.310	No corelation	0.192	Positif corelation

4. Conclusion

We have successfully carried out a research to investigate effect of HOT lab in improving critical thinking skills and creative thinking skills of students. The results showed that HOT lab increased better critical thinking and creative thinking skills of students as competed to verification lab design. This research also found that there is no siginificat correlation between improvement of creating thinking skills and critical thingking skills. However, further researchs are required to elaborate factors that influence no-correlation between critical skills and creative thinking skills.

Acknowledgments

The authors would like to thank to the chairperson, Lab operator and assistant as well as students of the Program Studi Pendidikan Fisika UIN Sunan Gunung Djati Bandung who allowed and contributed in this research. The authors also would like to thank The Minsitry of Research, Technology and Higher Education for financial support of this research through Hibah Pascasarjana research scheme year 2017.

References

[1] Denicolo P and Reevers J 2013 Developing Transferable Skills (London: SAGE Publications Ltd)
[2] Hofstein A and Lunetta V N 2004 The laboratory in science education: Foundation for the 21st century Science Education **88** (1) p 28–54
[3] Deacon C and Hajek A 2011 Student perceptions of the value of physics laboratories International Journal of Science Education **33** (7) p 943–977
[4] Byers W 2002 Promoting active learning through small group laboratory groupes University Chemistry Education, **6** p 28–34
[5] Brewe E., Kramer L., and O'Brein, G., Modeling instruction: positive attitudinal shifts in introductional physics measures with class, *Physics Review Special Topics Physics Educational Resource*, **5** (1): 013102 (2009)
[6] Wenning C J 2006 A framework for teaching the nature of science Journal of Physics Teacher Education Online **3** (3) p 3-10
[7] Malik A and Setiawan A 2016 The Development of Higher Order Thinking Laboratory to Improve Transferable Skills of Students Proceedings 2015 International Conference on Innovation in Engineering and Vocational Education (Bandung), Vol **56** (Amsterdam: Atlantis Press) p 36–40
[8] Sutarno M and Putri D H 2013. Pengembangan laboratorium virtual fisika untuk meningkatkan keterampilan proses sains siswa SMA di wilayah miskin Propinsi Bengkulu. *Laporan Penelitian Hibah Bersaing*. Universitas Bengkulu
[9] Malik A 2010 Model pembelajaran inkuiri dengan menggunakan *virtual laboratory* dan *real laboratory* untuk meningkatkan penguasaan konsep dan keterampilan berpikir kritis siswa SMA pada topik listrik dinamis, Tesis (Bandung: Universitas Pendidikan Indonesia)
[10] Ku K Y L., Ho I T., Hau K T., and Lai E C M 2014 Integrating direct and inquiry-based instruction in the teaching of critical thinking: an intervention study, Instr Sci, **42** p 251–269

[11] Cowden C D and Santiago M F 2016 Interdisciplinary explorations: promoting critical thinking via problem-based learning in an advanced biochemistry class, Journal of Chemical Education, **93** (3) p 464–469
[12] Temel S 2014 The effects of problem-based learning on pre-service teachers' critical thinking dispositions and perceptions of problem-solving ability, South African Journal of Education, **34** (1) p 1-20
[13] Osborn A F 1963 Applied Imagination: Principles and Procedures of Creative Problem Solving (3rd revised ed.) (New York: Charles Scribner's Sons).
[14] Runco M A 2007 Creativity, Theories and Themes: Research, Development and Practice, (Amsterdam: Elsevier)
[15] De Bono E 1971 Lateral Thinking for Management (New York: McGraw-Hill).
[16] Isaksen S G and Treffinger D J 2004 Celebrating 50 years of reflective practice: versions of creative problem solving, *The Journal of Creative Behavior*, **38**, p 75–101
[17] Busyairi A and Sinaga P 2015 Strategi pembelajaran *Creative Problem Solving* (CPS) berbasis eksperimen untuk meningkatkan kemampuan kognitif dan keterampilan berpikir kreatif, *Jurnal Pengajaran MIPA*, **20** (2), p 133-143
[18] Şener N., Türk C., and Tas, E., 2015 Improving science attitude and creative thinking through science education project : a design, implementation and assessment, *Journal of Education and Training Studies*, **3** (4), p 57–67
[19] [Yang K., Lee L., Hong Z., and Lin H., 2016 Investigation of effective strategies for developing creative science thinking, *International Journal of Science Education*, **0693** (September), p 1–19
[20] Binkley M., Erstad O., Herman J., Raizen S., Ripley M., Miller-Ricci M., and Rumble M., 2012 Defining Twenty-First Century Skills (Assessment and Teaching of 21st Century Skills) eds Grifin P McGaw B and Care E (New York: Spinger) chapter 2 p 17–66
[21] Hake R R 1999 Analyzing change/gain scores, American Educational Research Association, [online] di:http://www.physics.indiana.edu/~sdi/AnalyzingChange-Gain, pdf
[22] Bowen G M and Roth W M., 2005 Data and graph interpretation practices among preserve science teachers, J Res Sci Teach, 42 (10) p 1063–1068
[23] Kukliansky I 2009 The characterization of data analysis and understanding of related concepts in the physics laboratory among science and engineering students Unpublished doctoral dissertation (Ben-Gurion University of the Negev)
[24] Stenberg R., 2012 The assesment of creativity: an investmen-based approach. *Creativity Research Journal,* **24** (1), p 3-12
[25] Von Aufschnaiter C and Von Aufschnaiter S 2007 University students' activities, thinking and learning during laboratory work European Journal of Physics **28** (3) S51–S60
[26] Wang C W.and Horng R Y 2002 The effects of creative problem solving training on creativity, cognitive type and R & D performance, *The Journal of Research, Technology and Innovation Management*, **32** (1), p 35-45
[27] Lin W C 2004 The relationship among creative, critical thinking and thinking styles in Taiwan high school students. *Journal of Instructional Psychology,* **31** (1), p 33-45
[28] Koray Ö and Köksal M S 2009 The effect of creative and critical thinking based laboratory applications on creative and logical thinking abilities of prospective teachers, Asia-Pacific Forum on Science Learning and Teaching, Volume 10, Issue 1, Article 2, p.1

IOP Conf. Series: Materials Science and Engineering **306** (2018) 012009 doi:10.1088/1757-899X/306/1/012009

Identification of the Thickness of Nugget on Worksheet Spot Welding Using Non Destructive Test (NDT) – Effect of Pressure

A Sifa[1*], A S Baskoro[2], S Sugeng[2], B Badruzzaman[1], and T Endramawan[1]

[1]Mechanical Engineering Department, Politeknik Negeri Indramayu, Indramayu-Indonesia

[2]Mechanical Engineering Department, Faculty of Engineering, Universitas Indonesia, Depok-Indonesia

*agus.sifa@polindra.ac.id

Abstract. Resistance Spot Welding (RSW) is a process of connecting between two worksheet with thermomechanical loading process, RSW is widely used in automotive industry, the quality of splicing spot welding is influenced by several factors. One of the factors at the time of the welding process is pressure. The quality of welding on the nuggets can be determined by undertaking non-destructive testing by using Non Destructive Test (NDT) - Ultrasonic Test. In the NDT test is done by detecting the thickness of the nugget area, the purpose of research conducted to determine the effect of pressure to welding quality with Nugget thickness gauge measurement with Non Destructive Test method and manual measurement with micrometer, Experimental welding process by entering the welding parameters that have been specified and pressure variables 1-5 bars on the worksheet thickness of 1 mm. The results of testing with NDT show there is addition of thickness in nugget superiority after compare with measurement result of thickness of nugget with micrometer which slightly experience thickness in nugget area, this indicates that the welding results have a connection between worksheet 1 and worksheet 2.

1. Introduction

Resistance Spot Welding (RSW) is a process of connecting between two worksheet and an electrode with thermal-mechanical loading process, RSW is widely used in automotive industry, the quality of splicing spot welding is influenced by several factors, such as welding time, current, electrode voltage, welding cycle, metallurgical properties, diameter tip electrode, force/pressure and others.[1-6] One of the factors at the time of the welding process is pressure, where two plates are combined with a thermal-mechanical load, the mechanical load is in the form of electrode pressure on the surface of the worksheet, and at the same time there is a heating process.

The quality of welding on a nugget can be determined by performing a damage test or undamaged test. The damage test on welding results can be done with shear test worksheet, metallographic test and others, while the test is undamaged can be by using Non Destructive Test (NDT) - Ultrasonic Test (UT). [7] NDT commonly used test quality testing or looking for defects in SMAW welding results. In the NDT test is done by detecting the thickness of the nugget area, detecting the thickness of nugget to assess the quality spot weld [8]. NDT is performed by using a probe to scan the thickness of the

IOP Conf. Series: Materials Science and Engineering **306** (2018) 012009 doi:10.1088/1757-899X/306/1/012009

conjugated material, where the test is performed by scanning the welded material by utilizing the propagation power of the material, NDT is efficiency and realibity method to assess the damage [9].

In this study experimental welding process will be carried out to determine the quality of welding influenced by the mechanical loads of pressure, pressure variations made in the welding process, then measured the quality of spot welding using Non Destructive Test method to determine the joining between the two worksheet [10], and impact of the electrode pressure factor. The purpose of research conducted to determine the effect of pressure on the quality of welding with Nugget thickness gauge measurement with Non Destructive Test method (NDT) and manual measurement with micrometer. Experimental welding process by entering the welding parameters that have been specified and pressure variables 1-5 bar on the worksheet thickness 1 mm with used material Fe.

2. Methods

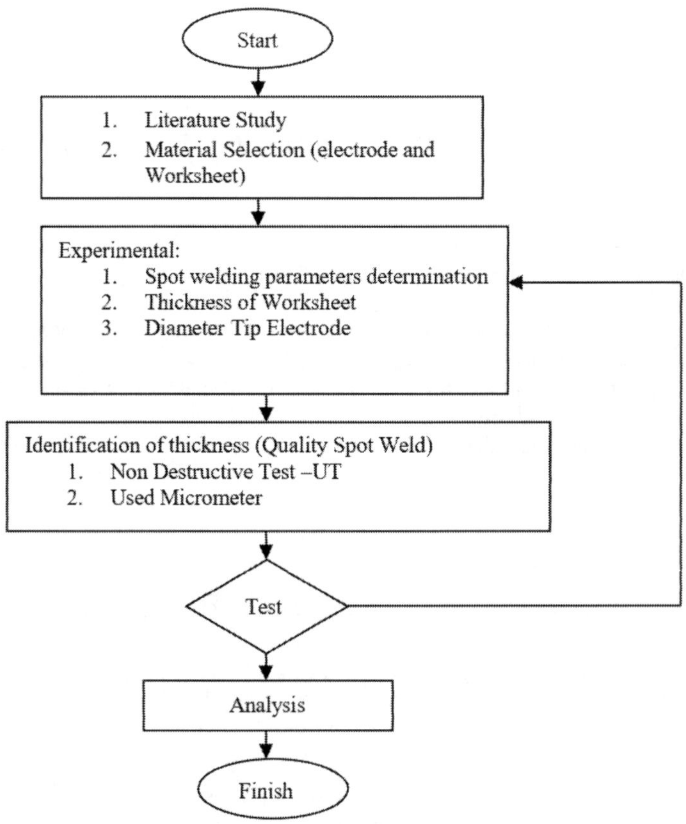

Figure 1. Flow chart Identification of thickness

In the process of identifying the thickness of nuggets to determine the quality of the spot welding results, then performed several stages of the process, the determination of materials and dimensions of the electrodes and worksheets is very important in relation to the welding parameters at the time of the experiment, which is the determination of the parameters to determine the quality of the welding corresponding to the worksheet pairs and electrodes.

2.1. Material

Before doing experimental spot welding, the composition test was performed on the worksheet, the test specimen was tested using ARL 3460 testing machine spectrometry, at room temperature 25° C, with moisture 51% with the result of composition test as follows:

IOP Conf. Series: Materials Science and Engineering **306** (2018) 012009 doi:10.1088/1757-899X/306/1/012009

Table 1. Composition of Worksheet (Fe)

Composition	Percentage (%)
C	0,05
Si	0,01
S	0,05
P	0,01
Mn	0,22
Ni	0,06
Cr	0,02
Cu	0,01
Al	0,04
Fe	99,65

Spot welding process, using Cu electrode with tip diameter ($5\sqrt{t}$), where t = thickness worksheet (mm), in the worksheet welding process using Fe material with thickness (1 + 1) mm [11].

2.2. Experimental

In this study experimental method, the pressure parameters on the spot welding machine vary between 1 Bar – 5 Bar, with 30% current output from Output Current Spot Weld JPC35 (9000 A), cooling used by water cooling.

Table 2. Specification of Machine Spot Weld

Model	JPC 35
Rated Capacity	35 KVA
Rated Primary Voltage	380 V
Rated Frequency	50/60 HZ
Rated Sec. Current (max)	14,000 A
Duty Cycle	8.5/7.5%
Output Current	9000A

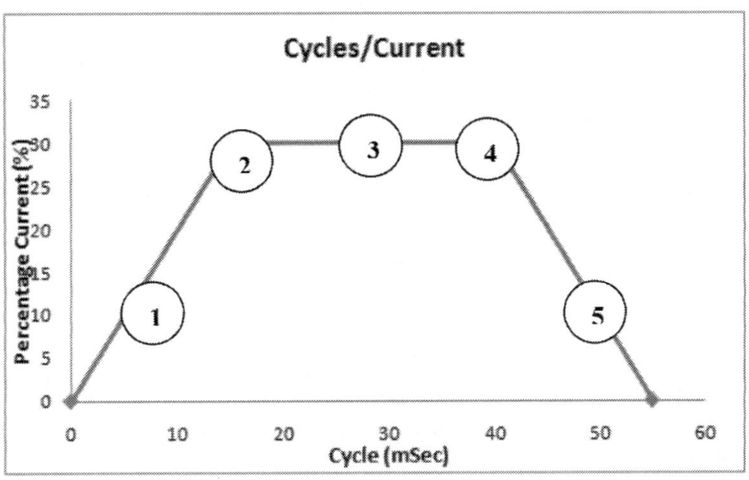

Figure 2. Parameters of Spot Welding

Experimental welding process with input parameters Current voltage 30%, then output current 270 KA and welding time 55 cycle (55 x 20ms) ≈ 1100 ms, size diameter tip of electrode 5 mm, and thickness of worksheet (1 + 1) mm, with experimental process as follows:

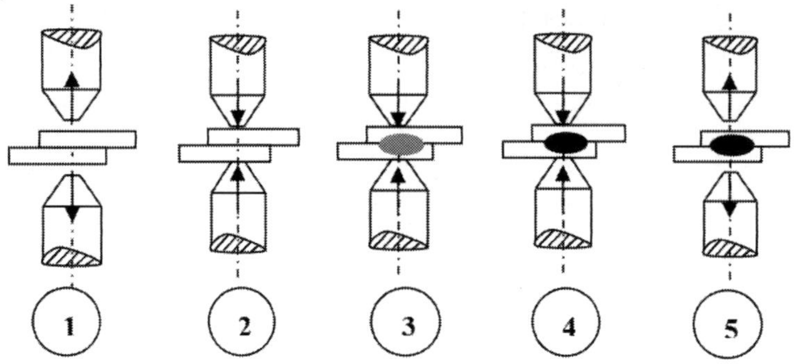

Figure 3. Spot Welding Process

2.3. Identification

The results of welding tested with Non-Destructive Test- Ultrasonic Test using normal TR (Transmitter and Receiver) probes. Frequency 4 MHz, Sound Velocity of 2267 m / s, range 100 and couplant using oil, boundary selection method using 6 dB drop, The tools used Ultrasonic Flaw Detector Karl Deutsch Echograph 1090 [11].

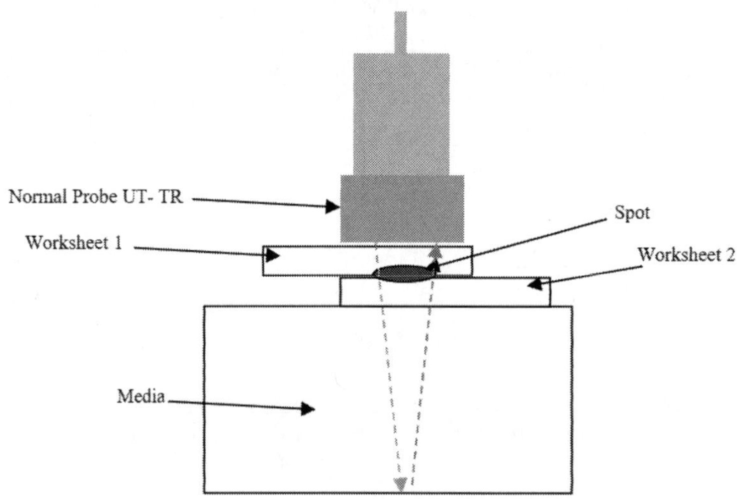

Figure 4. Identification Process

Some steps to do Non Destructive Test (NDT) using normal Transmitter –Receiver (TR) ultrasonic testing (UT) [11]:

1. First step Standard UT probe on surface of reference specimen (Calibration block).
2. The second step Ultrasonic Testing Turn on the device and select the type of menu that is normally used probe.
3. The third step select the menu with a range of 50 mm.
4. Step four Choose two-point calibration on the display menu and make sure the material speed for steel is 2267 m / s.
5. Make sure the five-pulse steps will appear on display scales 2.5, 5.0, 7.5, and 10.0
6. The final step of the final stage probe reading will appear on the display instrument.

3. Results and Discussion

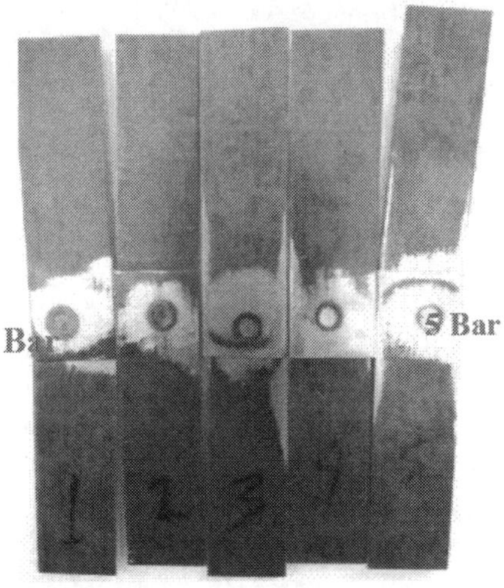

Figure 5. Spot Welding Results

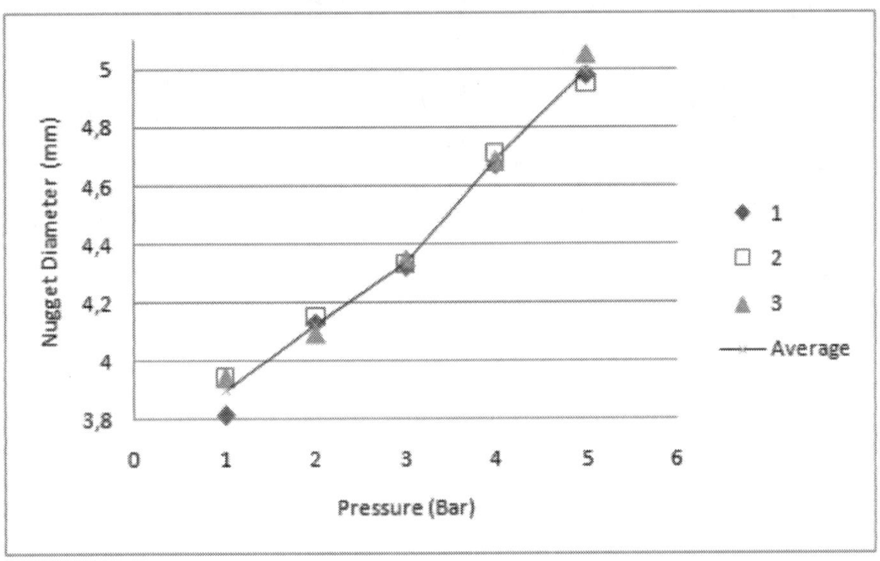

Figure 6. Graph the results of measurement of nugget diameter

Zhao DW. *et al.* has argued that the greater the force applied to the electrode the larger the nugget diameter of the worksheet [12]. The experimental welding process with a pressure of 1-5 bar, spot welding results with different pressure variations can be seen in Figure 5 and 6, where it can be seen the difference of nugget diameter between 1 bar to 5 bar. Heat Affected Zone (HAZ) at 1 bar pressure has bigger area compared to HAZ on welding results with a pressure of 5 bar.

IOP Conf. Series: Materials Science and Engineering **306** (2018) 012009 doi:10.1088/1757-899X/306/1/012009

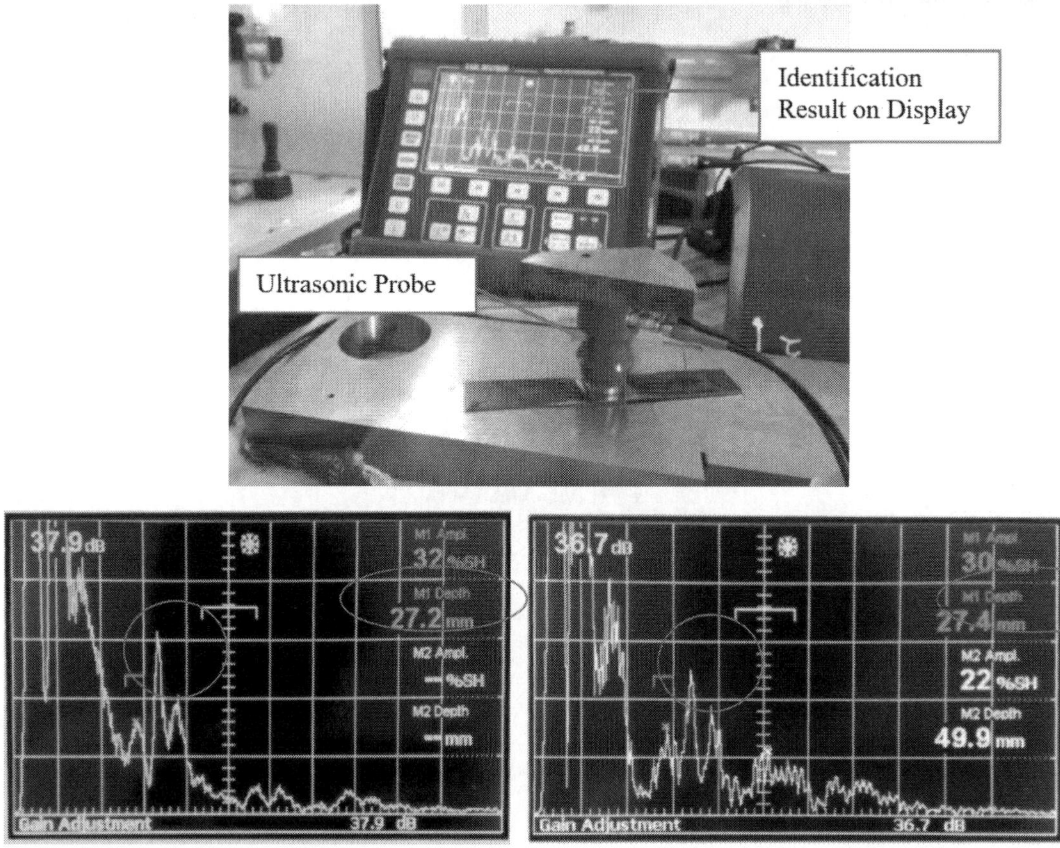

Figure 7. Results Non Destructive Test – Pressure of 1 Bar and 5 Bar

Non Destructive Test Result on welding result, using TR type normal probes, the following test result with different pressure show in Figure 7. The results of the test with NDT, where the amplitude detected by the probe shows the same character but with different results, wherein (a) shows the result of identification at the welding result with a pressure of 1 bar with the result of the identification 27.2 mm (25 mm media + worksheet 1 + worksheet 2), and (b) shows the result of identification on the welding result with pressure of 1 bar with the result of identification 27,4 mm (25 mm media + worksheet 1 + worksheet 2).

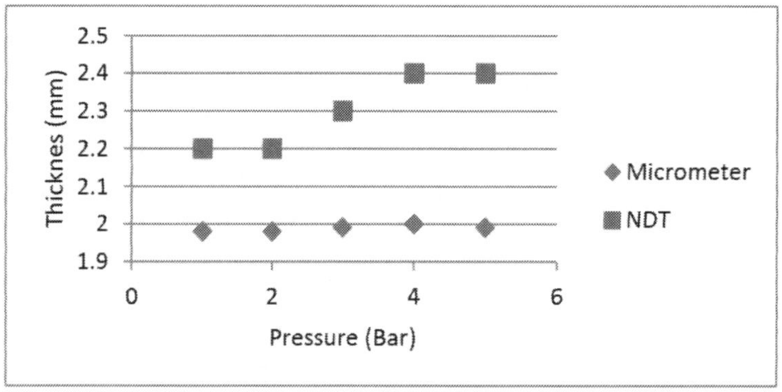

Figure 8. Results Measurement of Thickness NDT and Micrometer

Figure 8 shows the comparison of nugget thickness measurements on the worksheet, after identification with NDT and micrometer has a difference. The blue line on the graph shows identification with NDT, where at the pressure of 1 bar and 2 bar have the same value, seen from the visual (figure) the HAZ area is close to the same, whereas at the pressure of 2 bar has a value higher than the pressure of 1 and 2 bars, whereas at the pressure of 4 and 5 bar has the same thickness value, but greater than the pressure of 1,2 and 3 bar, the pressure shows a greater thickness, this is in comparison with the manual measurement with the micrometer. The average of the measurement with the micrometer is smaller than the measurement using NDT, it shows visually there is a pressure mark on the nugget area, whereas identification with NDT shows the overall impact of the area on the nugget against the thickness, this indicates that the welding results have a connection between worksheet 1 and worksheet 2.

4. Conclusions

In this study, it can be concluded that welding experiments are given 270KA output current parameters, Welding Time 1100 ms or 55 cycles with pressure parameters on the spot welding machine 1-5 Bar, with worksheet thickness each 1 mm and diameter tip electrode 5 mm, where test results are tested with Non Destructive Test (NDT) with 25 mm propagation media , the results of testing on the welding results with Pressure 1 Bar and 2 Bar has a thickness value of 27,2 mm (media + worksheet1 + worksheet2), at the pressure of 3 Bar test results show 27,3 mm. (medium + worksheet1 + Worksheet2), on welding results with 4 bar and 5 bar pressure with same 27,2 mm thickness, NDT test results show the addition of thickness is proportional to the pressure and visually the smaller HAZ is inversely proportional to pressure and re-tested manually with a micrometer showing a reduction in thickness on the surface of the nuggets, identification using NDT identifies the edge of nugget surface area, this indicates that the welding results have a connection between worksheet 1 and worksheet 2.

Acknowledgments

Directorate of Research and Community Service-Ministry of Research, Technology and Higher Education, and To Department of Mechanical Engineering, Faculty of Engineering, University of Indonesia, and Department of Mechanical Engineering-State Polytechnic of Indramayu.

References

[1] Ru-xiong LI. Quality monitoring of resistance spot welding based on process parameters. Energy Procedia. 2012 Jan 1;14:925-30.

[2] Baskoro a. s. and Muzakki h. The Effect of Welding Time and Welding Currents on Weld Nugget and Tensile Properties of Thin Aluminum a1100 by Micro Resistance Spot Welding. 2016 Jan;11:1050-1055.

[3] Wang B, Hua L, Wang X, Song Y, Liu Y. Effects of electrode tip morphology on resistance spot welding quality of DP590 dual-phase steel. The International Journal of Advanced Manufacturing Technology. 2016 Apr 1;83(9-12):1917-26.

[4] Mikno Z, Bartnik Z. Heating of electrodes during spot resistance welding in FEM calculations. Archives of Civil and Mechanical Engineering. 2016 Jan 31;16(1):86-100.

[5] Ma N, Murakawa H. Numerical and experimental study on nugget formation in resistance spot welding for three pieces of high strength steel sheets. Journal of Materials Processing Technology. 2010 Nov 1;210(14):2045-52.

[6] Ho JE, Wei PS, Wu TH. Workpiece property effect on resistance spot welding. IEEE Transactions on Components, Packaging and Manufacturing Technology. 2012 Jun;2(6):925-34.

[7] Thornton M, Han L, Shergold M. Progress in NDT of resistance spot welding of aluminium using ultrasonic C-scan. NDT & E International. 2012 Jun 30;48:30-8.

[8] Zhao D, Wang Y, Lin Z, Sheng S. An effective quality assessment method for small scale resistance spot welding based on process parameters. NDT & E International. 2013 Apr 30;55:36-41.

[9] Kharrat M, Gaillet L. Non-destructive evaluation of anchorage zones by ultrasonics techniques.

Ultrasonics. 2015 Aug 31;61:52-61.
[10] Ulbrich D, Kowalczyk J, Jósko M, Selech J. The analysis of spot welding joints of steel sheets with closed profile by ultrasonic method. Case Studies in Nondestructive Testing and Evaluation. 2015 Nov 30;4:8-14.
[11] Sifa A, Endramawan T. Identification of the Quality Spot Welding used Non Destructive Test-Ultrasonic Testing:(Effect of Welding Time). InIOP Conference Series: Materials Science and Engineering 2017 Mar (Vol. 180, No. 1, p. 012101). IOP Publishing.
[12] Zhao DW, Wang YX, Zhang L, Zhang P. Effects of electrode force on microstructure and mechanical behavior of the resistance spot welded DP600 joint. Materials & Design. 2013 Sep 30;50:72-7.

IOP Conf. Series: Materials Science and Engineering **306** (2018) 012010 doi:10.1088/1757-899X/306/1/012010

Analysis of Quality and Output of Entrepreneurship in the Field of Refractionist Optician

A Wesnita[1] and M Dewi[2]

[1]Akademi Refraksi Optisi YLPTK, Padang, Indonesia
[2] FKIP Universitas Putra Indonesia YPTK, Padang, Indonesia

*awesnita@gmail.com

Abstract. The launching of the Asean Economic Community (AEC) caused a rivalry on the exchange of the work, especially workers who are involved in the sector of specific expertise that increased sharply. The solution offered is through the implementation of entrepreneurship learning, but despite entrepreneurship courses have been given to students since 2007, data from the last three years states only 21% of graduates of Refractionist Optician Academy in Padang, Indonesia, which became entrepreneurs in the fields of optics and eye health. Therefore, there needs to be an assessment whether the 5 variables of learning process that has been done three variables affect the output quality of learning entrepreneurship. This study has a quantitative method to determine the effect on output quality of the learning process of learning through regression analysis, test t, R2 and the percentage of contribution. The population consist of 121 students and the sample is 54 people. The study states that the academic skills of entrepreneurship is influenced by four variables of learning process, personnel skills of entrepreneurship is not influenced by any learning variable process, and social skills of entrepreneurship is influenced by three variables learning process.

1. Introduction

National Education aims to educate the nation and develop fully Indonesian based on Pancasila and the Constitution of National 1945. Educations are expected to produce well educated of faith, pious, noble character, knowledge able, skilled, and responsible personality. According to PP No 19 of 2005 on National Education Standards, stated that the conduct of education in each educational institution must be in accordance with National Education Standards, which is the minimum standard of education systems across Indonesia and serves as a basis by planning, implementation and supervision of the implementation of education in order to achieve quality education. This is necessary particularly in anticipation of the free market, especially in the ASEAN countries such as the launching of the ASEAN Economic Community (AEC) which requires that the Indonesian have special competence in the world to be able to compete internationally. AEC caused to increased competition on the exchange of the work, especially workers who are involved in the sector specific expertise. Problems anticipate are the international competition by improving the quality of Human Resources (HR) through institution education, one through entrepreneurship education because through innovation in the entrepreneurship spirit would be developed the advantage in competitive.

According to Fontana the problem that inhibits the development of entrepreneurship in Indonesia is the weak institutional factors and individual, which include: the ability finding business opportunities, entrepreneurship culture, entrepreneurship leadership, market availability, education level, technology information and communication avaibility, and the ability to innovate (2013) [1]. For that education Entrepreneurship conducted in university is an important thing and even getting more attention from all parties, especially the State through curriculum learning entrepreneurship.

But there were many problems in the process of entrepreneurship education in universities are primarily associated with universities that are the output of unemployment continues to rise each year. As reported by the Central Statistics Agency (CSA) in 2015 Unemployment Rate (UR) in Indonesia in August 2015 retracing the figure of 7.56 million people. This figure is equivalent to 6.18 % of the total 122.4 million work forces. That number has increased compared to CSA in February 2015 amounted to 5.81 % and CSA in August 2014 amounted to 5.94 %. About 600 thousand openly unemployed are college graduates either diploma or Bachelor Digri (webside BPS, 2016).

The data also showed that the unemployment which passed the college is still relatively a lot of the work force in Indonesia, it is indicated that the employment of college graduates tend to be slow so fertilize unemployed college graduates. Danuhadimejo states that one of the causes of unemployment mistake is the education system that only produces power of technical skill, which has not been beneficial to the State [2], while Dewi stated that the factor of incompetence and lack of courage for self-employed job seekers or self-employment is a cause problem of unemployment [3]. Therefore, to solve the problem of unemployment and overcome the competition in the job market International then increased ability to innovate through entrepreneurship education are important, this is done as an alternative to solving the problem didn't absorb college graduates by units of the industry and that students have the ability and courage in entrepreneurship.

One of the deliveries of education Vocational education institutions with specialized expertise competence is the Akademi Refraksi Optisi YLPTK Padang. The Academy is to prepare professional personnel in eye health and optics trained to be skilled and expert on scientific competence. Graduates are also expected to be a filler job opportunities and entrepreneurship in the field of optics after finishing education, therefore entrepreneurship education given to the students as a compulsory subject in the field of supporting special ability on optics. However, based on the fact that there is still the issue of lack of courage of students in entrepreneurship, students tend to show antiphaty towards entrepreneurship, and note also that the interests of entrepreneurship students still low. Data obtained last 3 years that there are 46% of graduates who work with others, 21% were self-employed in the field optics and the remaining 33% still existed unemployment. This indicates that the number of students that entrepreneurship is still small. Therefore, it is necessary to study how the quality of entrepreneurship learning processes and learning outcomes (output) at the Academy of refraction and Optisi Padang that impact of is not achieving educational goals Entrepreneurship.

2. Literature Review

2.1. Entrepreneurship

The term entrepreneur is a translation of words into Indonesian entrepreneur, originally known as the self-employed, which means to stand on his own strength. Suharsono Sagir in Buchari Alma wrote that the entrepreneur is a main capital is the perseverance that is based on an optimistic attitude, creative and doing business as the founder of the first accompanied by the courage to the risk based on a calculation and proper planning [4]. Meredith says entrepreneurs are people who have the ability, to see and assess business opportunities, gather the resources needed to take the profit there of and take appropriate action, in order to ensure success [5]. Zimmerer says entrepreneurship is the result of a disciplined and systematic implementation process of creativity and innovation in meeting the needs and opportunities in the market [6]. That is Entrepreneurs are the subject of conducting Entrepreneurship, while Entrepreneurship is an activity that does.

2.2. Entrepreneurship Education

Educators who plays an important role in the learning process needs to carry out the development of learning through a learning approach that fits the needs of learners related to the purpose of

Entrepreneurship Education. Developments in doing achieved such Entrepreneurship learning in many countries around the world should be a reference for Entrepreneurship at Indonessia education, this is done as benchmarks to conduct quality learning for students in the field of Entrepreneurship. Premand states that entrepreneurship education has the potential to enable learners to gain the skills and create their own jobs, research shows that entrepreneurship education significantly increase the level of self-employment among university graduates about one year after graduation [7]. In the official recommendations, educators are encouraged to adopt innovative pedagogical for entrepreneurship education in order to achieve a positive impact on learners. Therefore McGrath and MacMillan states that the fostering of entrepreneurship as a mindset can be considered as an educational competence, based on instructional learning experience [8].

Research of Jiménez states that the Enterprise has been an indication of economic growth in a country, a positive impact on formal education Entrepreneurship is the ability gained through the education necessary to detect and evaluate business opportunities with better, boost confidence to run the risk perceived , as well as foster care and employment opportunities [9]. Coduras underlines that people tend to acquire knowledge that can be of giving benefits to the ability of skills through education (especially formal) [10]. Based on the study on the subject of entrepreneurship, it can be said that fostering entrepreneurship attitude for students despite the efforts made by the approach through a learning process that is appropriate to the purpose Entrepreneurship courses. Take into weigh the relevance of the educational process with the skills needed factually through a learning approach that fits the needs of learners related to the goal of Education Entrepreneurship should be done.

Entrepreneurship Education in college High to apply the principles and methodologies towards internalization of values in their students through a curriculum that is integrated with developments both at educational institutions and their communities as well as the use of models and learning strategies relevant to the purpose of learning it own. Educational institutions must not only serve spawned many graduates, but much more important is how much the gaduation it can help themselves to face the challenges in the community or in other words the institution must improve life skills graduates.

Today there are many activities in the field of entrepreneurship education in universities and colleges around the world [11]. United States seems to lead in entrepreneurship education. As Kuratko notes in his study, the number of colleges and universities that offer entrepreneurship education has grown from a handful in 1970 and rose to 1,600 in 2003 [12]. Elsewhere, there is a fact that occurred increasing number of Australian universities offer entrepreneurship programs and business small business, entrepreneurship development has also been listed as one of the four strategic objectives for universities in the UK [13]. Thus it can be seen that the Education Entrepreneurship education has become a trend around the world, this is of course due to the importance of studying entrepreneurship so that an increasing number of citizens who were self-economic contracting State to support the country.

2.3. Learning and Output Quality of Entrepreneurship Learning

According to the terms, the word quality means quality, the level of the good and bad things. According to ISO 2000, the quality is the totality character a product (goods and services) that support the ability to satisfy the requirement that is specified or defined. One of the qualities that need to be considered in education is the quality of the learning process. The learning process implies a change in the perception and behavior, that behavior improvement [14]. The results of the learning process that can be viewed from a variety of forms, such as changing the knowledge, understanding, attitude, behavior, skill and some other aspects that exist on the learner. As with learning, teaching is a process of arranging, organizing around the existing environment of learners, so as to foster and encourage learners teaching and learning process. Both of these concepts by Sudjana are two concepts that can not be separated and be integrated in one activity, when there is interaction between educators and learners during the teaching takes place [15] [16]. Basically in the learning process consists of three components, namely teachers, who taught and teaching materials provided by the teacher. Suryo Subroto, effective teaching and learning process can be in the review of two aspects [17]: 1. Taught by educators, which concerns the extent to which the planned teaching and learning activities take place. 2. Study of participants, which concerns the extent to which the desired learning objectives achieved through Teaching Learning. The development of the concept of education assesment that existed at this time shows the broader direction. But specifically targets three main aspects of the assessment include a program of education, teaching and learning processes and learning outcomes [15].

The indicator of the success of the learning process based curriculum perfected at this time are: a. Absorption of the material taught lessons in high achievement, either individually or in groups, and b. Behavior that outlined in the teaching objectives have been achieved by students either individually or in groups. Similarly, the quality of learning on Entrepreneurship subjects. The process and learner outcomes are things that the indicators of learning quality. Presented by Widoyoko outputs a learning program is always associated with the process that has taken place [18]. Therefore, assessment of learning outputs Entrepreneurship is not only in the aspect of students' academic skills, but also to reach an assessment of the personal skills and social skills of students. Learning essence of the process is an interaction between educators and students that doing in Context specific classroom with the support of learning facilities. In this study, which will be the topic of research is how the quality of the learning process with the indicator Learning Facility. Lecturer Performance, Climate Classes, Students Attitude, Motivation. On the Component Output Quality Learning evaluation results known to the academic field of Entrepreneurship Skills, Personal Skills entrepreneurship field and Social Skills in the field of Entrepreneurship [18].

3. Method

This study uses a quantitative approach to descriptive to analyze how quality of facility education. Lecturer performance, climate classes, student attitude, motivation. Quality output of learning on the academic field of entrepreneurship skills, personal skills entrepreneurship field and social skills in the field of entrepreneurship Akademi Refraksi Optisi YLPTK Padang academic year 2015-2016. And analyze the influence of each process variable to each output. Respondents were given a questionnaire instrument learning process is performed to output a lecturer while learning assessed by a lecturer in Entrepreneurship

Population and sample is the entire student class of 2013 and 2014 amounted to 121 people and the sample size is 54 students, with thought being given to that student who studied are students who have followed the course Entrepreneurship. To collect research data conducted by using a questionnaire instrument (questionnaire). Questionnaire was conceived and developed in accordance with the Evaluation Model EKOP [18] and then compiled based on the study of lectures Entrepreneurship. By using a Likert Scale modified in accordance with the measurement indicators with a positive attitude measurement scale up negative.

Data analysis was performed using percentage and categorization by the formula:

$$\frac{\text{Score Result}}{\text{Highest Score}} \times 100\%$$

The results of the analysis based on the above formula and then do the categorization of the results Achievement Level Respondents by scale Sudjana according to Table 1 below [19]:

Table 1. Scale Criteria

Achieved Percentage	Criteria
90 – 100%	Highest
80 – 89%	Hight
65 - 79%	Middle
55 - 64%	Low
0 – 54%	Lowest

Analysis of data using multiple linear regression analysis and t test to see and analyze the effect of each process variable to each output. In this study the quality of the process is the five independent variables (X_1, X_2, X_3, X_4, X_5) quality of the output is three dependent variables (Y_1, Y_2, Y_3). Analyses were performed separately to see the effect of independent variables on the dependent variable Analysis of data using multiple linear regression analysis and t test to see and analyze the effect of each process variable to each output. In this study the quality of the process is the five independent variables (X_1, X_2, X_3, X_4, X_5) quality of the output is three dependent variables (Y_1, Y_2, Y_3). Analyses were performed separately to see the effect of independent variables on the dependent variable.

4. Results and Discussion

4.1. Description of the data statistics of each independent variable can be seen in Table 2 below:

Table 2. Description of Statistic Variables

Variabel	Mean	SD	Min	Max	Sum
X_1	23,37	4.787	11	31	1262
X_2	34,63	5,830	18	44	1870
X_3	27,81	5.978	17	40	1502
X_4	22,35	5.662	11	35	1207
X_5	32,96	7,468	18	48	1780

The above data shows that the average composition of highest to lowest is 34.63 lecturer performance, motivation to learn 32.96, Climate classes 27.81, learning facilities 23.37 and attitude 22.35.

The categorization is based on a formula percentage can be seen in the following figure 1:

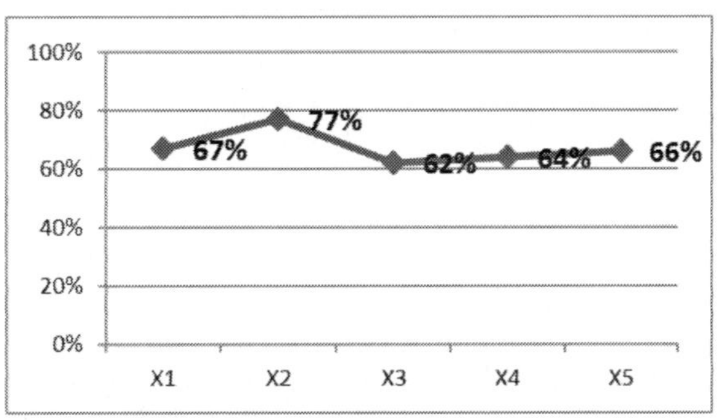

Figure 1. Categorizing Variables

The interpretation of each variable based on percentage score is known that learning facilities, lecturer performance and motivation category were middle, climate class and attitude of students categorized still low.

4.2. Description of the data statistics each dependent variable can be seen in Table 3 below:

Table 3. Description of Statistic Variables Bound

Variable	Mean	SD	Min	Max	Sum
Y_1	32,69	7.797	14	44	1765
Y_2	19,56	5.414	9	30	1056
Y_3	21.80	5.764	11	34	1177

The above data shows that the average composition of highest to lowest are skill academic field of Entrepreneurship 32.69, social skills 21.80, personal skills 19.56. The categorization is based on a formula percentage can be seen in the following figure 2:

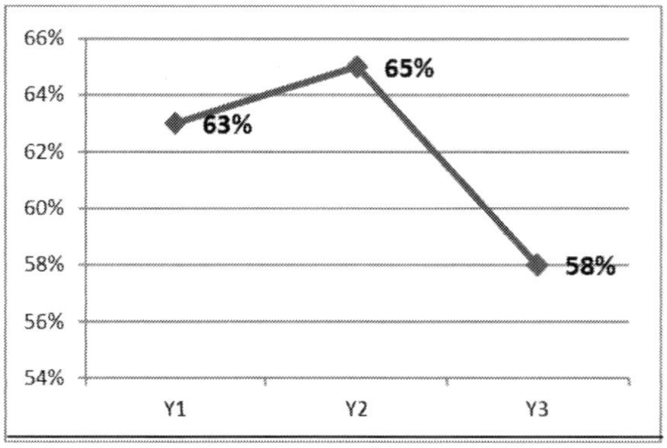

Figure 2. Categorization of Variables

The interpretation of each variable based on the percentage score is 65% personal skills, academic skills 63% and social skills 58%.

4.3. Analysis of the influence of independent variables (X_1, X_2, X_3, X_4, X_5) of the Academic skills in the field of entrepreneurship (Y_1) can be seen in terms of the following analysis:

Table 4. Results of Analysis of the Effect of Y_1

Variable	t score	p.value	Sd. Error
X_1	-3.353	0.000	0.173
X_2	3.033	0.007	0.106
X_3	-3.233	0.002	0.111
X_4	1.745	0.308	0.218
X_5	4.565	0.000	0.151

Based on the above analysis, it is known that there are four variables that significantly affect academic skills that student are entrepreneurship learning facilities, faculty performance, classroom climate, and motivation to learn. The fifth independent variable simultaneously bekontribusi 77.2% (R^2 = 0772) with Score F of 32 559 F p-value 0.000, which means simultaneously fifth variable real impact on the ability academic entrepreneurship or at least there is one free variable influenced. The results of the study in accordance with the statement of Volkmann this time educators need innovation, new solutions, creative approach and new ways to increase entrepreneurship education system the attitudes, motivation in entrepreneurship learning [20]. Because according to research factors that affect learning such as facilities, educators, motivation and learning climate affects academic skills in entrepreneurship, while the attitude factor has no effect on academic ability.

4.4. Analysis of the influence of independent variables (X_1, X_2, X_3, X_4, X_5) on the dependent variable of entrepreneurship academic personal skills in the field of entrepreneurship (Y_2) can be seen in terms of the following analysis:

Table 5. Analysis of the Effect of Y_2

Variable	t score	p.value	Sd. Error
X_1	-1.310	0.196	0.212
X_2	1.186	0.241	0.131
X_3	0.512	0.611	0.137
X_4	0.666	0.509	0.269
X_5	1.769	0.083	0.186

Based on the above analysis, it is known that there are no independent variables that significantly affect personnel entrepreneurship skills of students. These five independent variables simultaneously bekontribusi only 2.8% (R^2 = 0.028) with Score F of 3.812 p-value 0.710, which means that no free variable influence in silmutan the personal ability entrepreneurship field students. Personality or entrepreneurship skills by characteristic of a person is the determining factor the emergence of desire entrepreneurship. Superior character an entrepreneur is inovatif. Innovation is recommended as a behavior characteristic of entrepreneurship and orientation entrepreneurship [21]. That is, the innovation allows someone turns out to be entrepreneurs by being able to look for new opportunities [22]. Drucker, for example, claimed that innovation is a tool main entrepreneurship. Referring to innovation as the systematic search for change as new market opportunities, product, or idea [23]. In line with this in a study by Thomas and Mueller [24], innovation has also been the central feature in defining the profile of entrepreneurship.

4.5. Analysis of the influence of independent variables (X_1, X_2, X_3, X_4, X_5) on the dependent variable of social skills (Y_3) can be seen in terms of the following analysis:

Table 6. Analysis of the Effect of Y_3

Variable	t skor	p.value	Sd. Error
X_1	-0.787	0.435	0.169
X_2	-0.459	0.648	0.104
X_3	4.947	0.000	0.109
X_4	3.116	0.003	0.214
X_5	2.554	0.014	0.148

Based on the above analysis, it is known that there are three variables that significantly affect the social skills of entrepreneurship in the students themselves, classroom climate, attitudes and motivation to learn. The fifth independent variable simultaneously bekontribusi 59.8% (R2 0.599) with score F is 14.343, p-value 0.000, which means simultaneously fifth variable real impact on the ability of social entrepreneurship, or at least there is a free variable that influence. Social skills to become an entrepreneur is a factor that can not be underestimated. Birch and Whittam defines the concept of social entrepreneurship is how someone have the ability to combined motivation and vision in accordance with the aim of bringing together different interests [25]. Many higher education institutions prepare education program to become a social entrepreneur because of the interest and demand for the entrepreneurship education [26]. The profit of social entrepreneurship education is able to increase the entrepreneurship population day by day due to their investment social entrepreneurship uploading one's motivation in trying.

5. Conclusion

This study aimed to analyze the effect of learning on output quality learning for students Entrepreneurship Akademi Refraksi Optisi YLPTK Padang. The first research results indicate that there are four variables that significantly affect academic skills that students entrepreneurship learning facilities, faculty performance, classroom climate, and motivation to learn. The fifth independent variable simultaneously constributed is 77.2%. Thus concluded that sucessed academic achievement in entrepreneurship can be influenced by the quality of the learning process, but the knowledge gained can be an experience of a student's skills in promoted entrepreneurship as though the attitude of no effect, but the attitude can be trained in entrepreneurship through education. Dees argues that we experience in life can support the development of entrepreneurship skills by getting an education is the first step to acquire and develop entrepreneurship skills [27].

The second conclusion results of the study revealed that there is no any independent variables that significantly affect personnel entrepreneurship skills of students. The fifth independent variable simultaneously contributed only 2.8%. The main characteristics of the entrepreneur was innovation, solution for generating inovation characteristics that exist within the student self is through learning method. Many researchers have found a way of learning that can enhance student motivation to innovate, previous research on the effect of learning strategies in entrepreneurship education have implemented the model and learning approaches such as model problem based learning, interactive learning, mentoring, experiential learning and active learning and project based learning.

The results of a third study is there are three variables that significantly affect the social skills of entrepreneurship in the students themselves, classroom climate, attitudes and motivation to learn, the fifth independent variable simultaneously constributed 59.8%. Social skills in the field of entrepreneurship can be improved by investment the mental attitude of entrepreneurship, motivate students in entrepreneurship by bringing them closer to the needs in life. This approach is consistent with constructive learning theory, which proposes authentic learning, such as the use of real life problems, to make learning meaningful for students [28]. Meaningful learning and relevant to the life and social issues in everyday life will increase the social skills of entrepreneurship students.

References

[1] Fontana, Alvanti 2013 *Kewirausahaan Indonesia Menyambut Masyarakat Ekonomi ASEAN 2015* Article Online http://wwwavantifontanacom/blog
[2] Danuhadimedjo 1998 *Kewiraswastaan dan Pembangunan* (Bandung: Alfabeta)
[3] Dewi, Endah Rita Sulistya, Sumarno dan Prasetiyo 2011 Pengembangan Model Pembelajaran Berperspektif Kewirausahaan *Jurnal Bioma* Vol **1**, No 2, Oktober 2011
[4] Buchori Alma 2005 *Kewirausahaan* (Bandung: Alfabeta)
[5] Meredith, Geoffrey G, Nelson, Robert E, and Neck, Phllip A 2002 *Kewirausahaan Teori dan Praktek (The Practice of Entrepreneurship)* (Jakarta: Penerbit PPM)
[6] Zimmerer W Thomas Et al 1996 *Entrepreneurship and The New Venture Formation* (New Jersey: Prentice Hall Inc)
[7] Premand Patrick, Stefanie Brodman, Rita Almeida, Rebekka GRUN and MAHDI BAROUNI 2015 Entrepreneurship Education and Entry into Self-Employment Among University Graduates *World Development Journal* Vol **77**, pp 311–327
[8] McGrath, R G, and MacMillan, I 2000 *The entrepreneurship mindset: Strategies for continuously creating opportunity in an age of uncertainty* (Boston: Harvard Business New Jersey 07458)
[9] Jiménez Alfredo, Carmen Palmero-Cámara, María Josefa González-Santos,Jerónimo González-Bernal, Juan Alfredo Jiménez-Eguizábal 2013 The impact of educational levels on formal and informal entrepreneurship Business *Research Quarterly* Vol **34**, hal 1-9
[10] Coduras A, Levie J, Kelley D J, Saemundsson J R, Schott T 2010 *Global Entrepreneurship Monitor Special Report: A GlobalPerspective on Entre-preneurship Education and Training Global Entrepreneurship Research Association* (Wellesley, MA)

[11] Thompson J L 1999 "The world of the entrepreneur – a new perspective", *Journal of Workplace Learning* Vol **11** No 6, pp 209-24
[12] Kuratko D F 2003 *Entrepreneurship Education: Emerging Trends and Challenges for the 21st*
[13] Kirby D A 2004 Entrepreneurship education: can business schools meet the challenge? *Education and Training* Vol **46** Nos 8/9, pp 510-19
[14] Oemar Hamalik 2002 *Psikologi Belajar Mengajar (*Bandung: Sinar Baru Algesindo)
[15] Nana Sudjana 2010 *Penilaian Hasil Proses Belajar Mengajar* (Bandung: Remaja Rosdakarya)
[16] Sadirman Am 2010 *Interaksi Dan Motifasi Belajar Mengajar* (Jakarta: Rajawali Pers)
[17] Suryo Subroto 1997 *Proses Belajar Mengajar Di Sekolah* (Jakarta: Rineka Cipta)
[18] Widoyoko and Eko Putro 2012 *Evaluasi Program Pembelajaran, Panduan Praktis Bagi Pendidik dan Calon Pendidik* (Yogyakarta: Pustaka Pelajar)
[19] Nana Sudjana 1995 *Dasar-Dasar Proses Belajar Mengajar* (Bandung: Sinar Baru Algesindo)
[20] Volkmann C, Wilson K E, Marlotti S, Rabuzzi D, Vykarnam S, and Sepulvuda A 2009 *Educating the next wave of entrepreneurs: Unlocking entrepreneurship capabilities to meet the global challenges of the 21st Century A report on the global Education Initiative Century* (Coleman Foundation White Paper Series, http://usasbeorg/pdf/CWP-2003)
[21] Entrialgo M, Fernandez E and Vazquez C 2000 Characteristics of managers as determinants of entrepreneurship orientation: some Spanish evidence *Enterprise and Innovation Management Studies* Vol **1** No 2, pp 187-205
[22] Zacharakis A 1997 *Entrepreneurship entry into foreign markets Entrepreneurship: Theory and Practice* Spring, pp 23-40
[23] Cromie S 2000 Assessing entrepreneurship inclinations: some approaches empirical evidence *European Journal of Work and Organizational Psychology* Vol **9** No 1, pp 7-30
[24] Thomas A S and Mueller S L 2000 A case for comparative entrepreneurship: assessing the relevance of culture *Journal of International Business Studies* Vol **31** No 2, pp 287-301
[25] Birch K and Whittam G 2006 *The role of the social economy, social enterprise and social entrepreneurship in sustainable regional development Centre for public policy for regions,* Discussion Paper No 12
[26] Brock D D and Kim M 2011 *Social entrepreneurship education resource handbook Ashoka U, the University Division of Ashoka*
[27] Dees G, Emerson J and Economy P 2001 *Enterprising nonprofits: A toolkit for social entrepreneurs* (John Wiley and Sons, Inc New York)
[28] Woolfolk A 2008 *Educational Psychology 10th Edition* (Boston: Allyn and Bacon)

Economic Evaluation of the Production Magnesium Oxide Nanoparticles via Liquid-Phase Route

A B D Nandiyanto[1,*], R Fariansyah[1], M F Ramadhan[1], A G Abdullah[2], I Widiaty[3]

[1]Departemen Kimia, Universitas Pendidikan Indonesia, Jl. Dr. Setiabudi no 229, Bandung 40154, Jawa Barat, Indonesia
[2]Departemen Pendidikan Teknik Elektro, Universitas Pendidikan Indonesia, Jl. Dr. Setiabudi no 229, Bandung 40154, Indonesia
[3]Departemen Pendidikan Kesejahteraan Keluarga, Universitas Pendidikan Indonesia, Jl. Dr. Setiabudi no 229, Bandung 40154, Indonesia

*nandiyanto@upi.edu

Abstract. The purpose of this study was to evaluate the production of magnesium oxide (MgO) nanoparticles. The evaluation was done in two perspectives: engineering and economic evaluation. The engineering perspective concerned about the analysis of the production rate based on the available apparatuses and raw materials, completed with mass balance calculation. The economic analysis was conducted based on several economic parameters: gross profit margin (GPM), internal return rate (IRR), payback period (PBP), cumulative net present value (CNPV), break even point (BEP), and profit to investment (PI). The engineering perspective showed that the production of MgO is feasibly done in small scale industry. This is verified by the potential production using current available apparatuses and raw materials in the market. Economic analysis obtained that the present project is profitable. But, for some cases, further studies must be done to get the present production process is attractive for investor.

1. Introduction

Magnesium oxide (MgO) is one of the most useful ceramics materials because this material has high melting temperature (about 2800°C) [1]. MgO usually used for catalysis and toxic waste remediation or as additive material on paint [2]. MgO also has good additive properties, which can be applied in high fuel oil [3].

Several synthesis methods for the production of MgO have been reported: liquid-phase, gas-phase, and vapor-phase. But, in this study, we delivered the analysis in the liquid-phase method only [3]. The liquid-phase (known as wet method) is valuable from conservative to conventional process since this method uses simple equipment for the experiment [4].

MgO is needed in industry, environmental, and health product. MgO in industry used as catalyst in many applications include organic carbonates synthesis catalyst [5]. Another application of MgO is used in steel manufacturing because of its highly corrosion resistant [6]. MgO in recent year has been investigated and reported for its effectiveness in absorbing Uranium [7]. In addition, an experiment for MgO has been done for maintain the balance of the environment with green synthesis by using extract of natural source such as Rambutan (*Nephelium lappaceum* L.) [8]. The development of MgO research in health product has been found due to its antibacterial activity. This material is toxic for bacteria at specific concentration [4].

Based on current developments of MgO material, here, the purpose of this study was to analyze the production of MgO material. The evaluation was done in two perspectives: engineering and economic evaluation. The engineering perspective concerned about the analysis of the production rate based on the available apparatuses and raw materials, completed with mass balance calculation. The economic analysis was conducted based on several economic parameters: gross profit margin (GPM), internal return rate (IRR), payback period (PBP), cumulative net present value (CNPV), break even point (BEP), and profit to investment (PI). In this study, the analysis was focused on the production of MgO based on reference [3].

2. Methods

In short of this study, the first analysis is the production of MgO based on literature [3]. In this case, we selected Agrawal et al [3] as the main reference. Then, the analysis was continued by analyzing available apparatuses and raw materials in the online shopping web that are used for the production of MgO. This analysis is important to ensure the possible production of MgO in the small scale industry.

To ensure the engineering perspective, the analysis is followed by mass balance calculation. This mass balance is important for understanding how much mass flow must be done for the production of MgO per year.

The mass flow was then compared to the economic analysis to get gross profit margin (GPM), internal return rate (IRR), payback period (PBP), cumulative net present value (CNPV), break even point BEP), and profit to investment (PI).

3. Results and Discussion

3.1. Engineering analysis

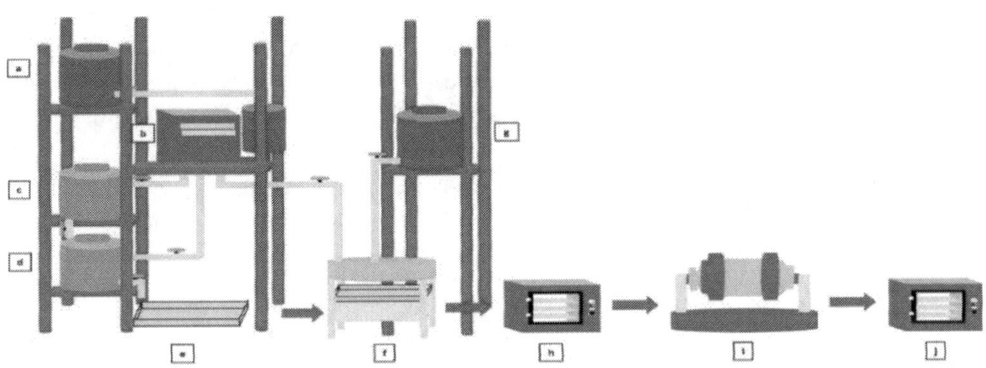

Figure 1. Illustration model for the production of MgO nanoparticles. Apparatuses (a), (b), (c), (d), (e), (f), (g), (h), (i), and (j) are water container, water purification process, basic mixing tank, batch reactor, decantation container, filtration, ethanol container, oven furnace, ball mill, and oven vacuum, respectively.

Figure 1 shows an illustration model for the production of MgO material. The system contains several apparatuses. Figure 1(a) is the water container for storaging that water from natural resource, such as river. Then, prior to using, the water must through water purification process (shown in Figure 1(b)). This purified water is then used for dissolving natrium hydroxide in basic mixing tank (See Figure 1(c)) as well as being used as solvent. The solute natrium hydroxide (as a basic solution) is then put into the Batch Reactor (see Figure 1(d)). In this batch reactor system, the basic solution is reacted with magnesium nitrite (as $Mg(NO_2)_3.6H_2O$). The process is stirred for 2 hours [4]. The reaction occurs can be described as [4]:

$$Mg(NO_3)_2.6H_2O + 2\ NaOH \rightarrow Mg\ (OH)_2 + 2NaNO_3$$

After homogenous magnesium hydroxide ($Mg(OH)_2$) is obtained, the solution is flowed into decantation container (See Figure 1(e)). The decantation process is precipitated for 24 hours [4]. After sediments are formed, the solution is moved to other containers: one is for waste and the other is for the product. The product is then washed and filtrated (See Figure 1(f)). In short of the filtration process, three steps of purification are done [4]. First step is the washing process with water from water purification system (see Figure 1(b)), the second is by ethanol from from ethanol container (see Figure 1(g)), and the last step is by rewashing process by water from water purification system (see Figure 1(b)). All utilized water and ethanol is streamed into waste container. The sediment is then dried using oven furnace (See Figure 1(h)) at 80 °C for 4 hours [3]. In the drying process, oxidation reaction of magnesium hydroxide happens, resulting MgO. The reaction is shown in the following [4]:

$$Mg(OH)_2 \rightarrow MgO + H_2O$$

After the oxidation process, dry solid of MgO is destroyed using ball mill process (See Figure 1(i)) [3]. Ball mill is selected since this method is effective to break down the particle size [9].

The next process is calcination process [3]. MgO, which has been destroyed to nanoparticles, is heated again in oven vacuum (see Figure 1(j)) [8]. This process takes 3 hours at 400°C [3]. After the calcination process, MgO is wrapped and packaged for the market.

To calculate the mass balance, we used the following assumptions:

- As a raw material, we used magnesium nitrate ($Mg(NO_3)_2.6H_2O$) and natrium hydroxide (NaOH)
- The loss of material during the washing process is assumed approximately 5%.
- The efficiency of the reactions is assumed approximately 100%.
- The material lost during moving/transferring process is assumed approximately 5% for each movement. The movement occurs twice: after the decantation process and after the filtration.
- The loss of material mass during heating process is assumed approximately 5%.
- The loss of material mass during ball milling process is assumed approximately 5%.

Based on the ideal calculation, 260.5 g of $Mg(NO_3)_2.6H_2O$ reacts with 40.1 g of NaOH. This reaction can create 116 g of MgO. However, using above assumptions, the mass balance calculate that the process can create 85.3 g of MgO nanoparticles.

3.2. Economic evalution

To ensure the feasibility of the production of MgO nanoparticles, economic evaluation must be done. Several assumptions were made

- The analysis used USD as a main currency, in which 1 USD is equal to 13,000 IDR.

- The basic electricity cost is assumed to be IDR 1.467,28/kWh
- The labor wage is IDR 18.000.000 for the total of 5 workers.
- Selling price for MgO is assumed approximately 139 USD/Kg.
- The discount rate is 15% annually.
- Import and shipping fees for the material or purchased equipment are omitted.
- The project period is 20 years.

Figure 2 shows the relationship between total investment cost (TIC) with CNPV and the year of MgO production. the GPM being created is 2.631,82 USD per year. The result from the mass analysis showed that the PBP for this project is 11,89 years from starting year. From the graph, we concluded that the business still profitable near to 20 years production although in the initial 11 years results in the negative CNPV value.

Based on above analysis, the profit from this project is not attractive. The reason of this analysis is due to several reasons: high-cost equipment, high-cost precursor materials, and the high loss of mass in some processes such as filtration process.

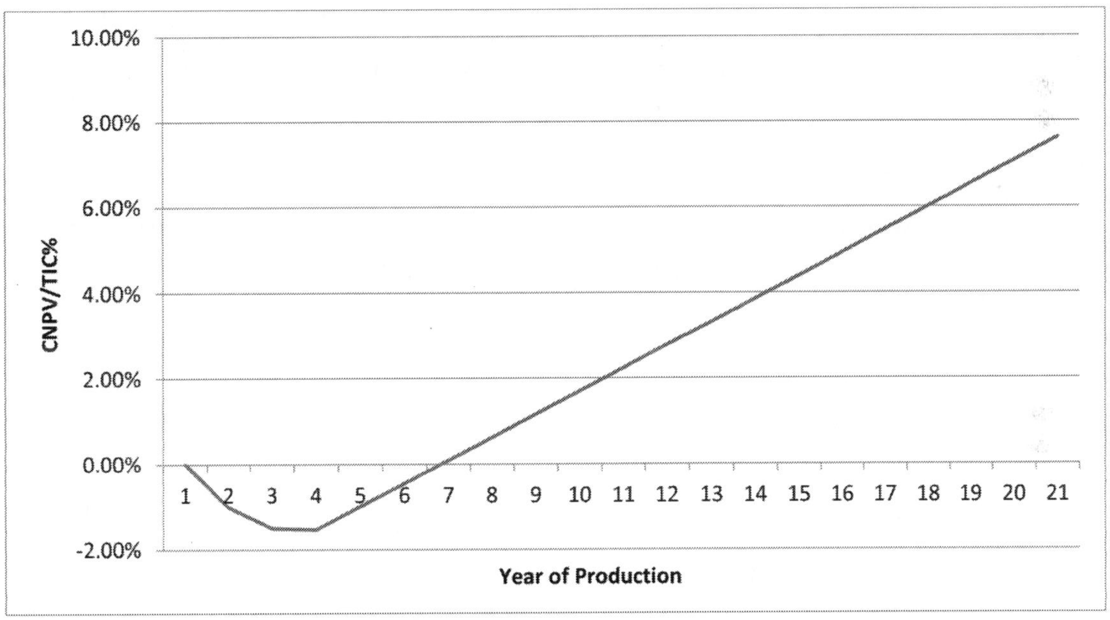

Figure 2. The relationship between CNPV/TIC with the year of production of the production of Magnesium Oxide nanoparticles.

4. Conclusions

The present study shows the economic evaluation and engineering perspective for the production of MgO nanoparticles via liquid-phase method. Engineering analysis showed that the process can be done in small scale industry. However, the result of economic evaluation shows approximately not too attractive and not profitable. This is because several reasons include loss is too high in filtration process, high-cost equipment and precursor materials.

References

[1] C Y Tai, C T Tai, M H Chang, and H S Liu 2007 Synthesis of Magnesium Hydroxide and Oxide nanoparticles using a Spinning Disk Reactor *Industrial and Engineering Chemistry Research* **46** p. 5536-5541.

[2] Y Ding, G Zhang, H Wu, B Hai, L Wang, and Y Qian 2001 Nanoscale magnesium hydroxide and magnesium oxide powders: control over size, shape, and structure via hydrothermal synthesis *Chemical Material* **13** p. 435.

[3] R M Agrawal, S D Charpe, F C Raghuwanshi, and G T Lamdhade 2015 Synthesis and characterization of Magnesium Oxide nanoparticles with 1:1 molar ratio via liquid-phase method *International Journal of Application or Innovation in Engineering & Management* **4** p. 141-145.

[4] J Suresh, R G R Gandhi, S Selvam, and M Sundrarajan 2013 Synthesis of Magnesium Oxide nanoparticles by wet chemical method and its antibacterial activity *Advanced Material Research* **678** p. 297-300.

[5] M L Kantam, U Pal, B Sreedhar, and B M Choudary 2007 An efficient synthesis of organic carbonates using nanocrystalline Magnesium Oxide *Advanced Synthesis and Catalysis* **349** p. 1671-1675.

[6] Z Zhang, Y Zheng, J Chen, Q Zhang, Y Ni, and X Liang 2015 Facile synthesis of Monodisperse Magnesium Oxide microspheres via Seed-induced Precipitation and their applications in High-Performance Liquid Chromatography *Advanced Functional Materials* **17** p. 2447-2554.

[7] Z Camtakan, S Erenturk, and S Yusan 2012 Magnesium Oxide nanoparticles: preparation, characterization, and Uranium sorption properties *Environmental Progress and Sustainable Energy* **00** p. 1-8.

[8] J Suresh, R Yuvakkumar, M Sundrarajan, and S I Hong 2014 Green synthesis of magnesium oxide nanoparticles *Advanced Material Research* **952** p. 141-144.

[9] A B D Nandiyanto, Z A Putra, R Andika, M R Bilad, T Kurniawan, R Zulhijah, and I Hamidah 2017 *Porous Activated Carbon Particles from Rice Straw Waste and Their Adsorption Properties* **12** p. 1-11.

IOP Conf. Series: Materials Science and Engineering **306** (2018) 012012 doi:10.1088/1757-899X/306/1/012012

Engineering Analysis and Economic Evaluation of the Synthesis of Composite $CuO/ZnO/ZrO_2$ Nanocatalyst

A B D Nandiyanto[1]*, W R Hayati[1], T A Aziz[1], R Ragadhita[1], A G Abdullah[2], I Widiaty[3]
[1]Departemen Kimia, Universitas Pendidikan Indonesia, Jl. Dr. Setiabudi no 229, Bandung 40154, Jawa Barat, Indonesia
[2]Departemen Pendidikan Teknik Elektro, Universitas Pendidikan Indonesia, Jl. Dr. Setiabudi no 229, Bandung 40154, Indonesia
[3]Departemen Pendidikan Kesejahteraan Keluarga, Universitas Pendidikan Indonesia, Jl. Dr. Setiabudi no 229, Bandung 40154, Indonesia

*nandiyanto@upi.edu

Abstract. The purpose of this study was to evaluate the synthesis process of composite $CuO/ZnO/ZrO_2$ (CZZ) nanocatalyst. The CZZ nanocatalyst is used as a catalyst in the hydrogenation reaction, which is used to convert harmful CO_2 gas into liquid fuels such as methanol. In this study, CZZ produced using the sol-gel auto-combustion method was selected as a model of the production procedure. The evaluation was conducted into two factors: engineering and economic analysis. Engineering analysis was analyzed based on the available apparatuses and raw materials in online web. Then, these data were used and compared with the mass balance approximation. The economic analysis was done using several economic parameters, including such as gross margin, internal rate return, payback period, cumulative net value, break even point (BEP), profitability index on sales to investment. Engineering analysis showed the present method can be applied using available apparatuses and raw materials in market. Economic analysis result confirmed that the present project is profitable. The analysis concluded that the present project for the production of CZZ nanocatalyst is prospective in small scale industry and profitable (by positive values in all economic parameters).

1. Introduction

Nanocatalysts is a catalyst with nanometer-size, with the aim of producing it is to make the catalyst active center have a larger area between substrate and catalyst [1]. Heterogeneous catalyst is one type of catalyst, in the form of a single metal catalyst, a metal oxide, or a supported catalyst. The component of a supported catalyst is a catalyst with supporting the active metal and having a large surface area. Examples of supported in catalyst are alumina, silica, zirconia, magnesia, zeolite, and titanium. The addition of this support will cause the catalyst active components to be uniformly distributed, increasing the surface area for contract with the reactants, increasing the mechanical strength, leveling the reaction heat and increasing the catalyst stability [2].

Catalyst is indispensable in industry due to the possibility loses during the reaction [2]. The use of catalysts exists in various applications such as conversion of crude oil, coal, and natural gas to fuel and chemical raw materials, the production of various petrochemical and chemical product, and control of CO, hydrocarbon and NO emission [3]. The catalyst is one of the components that play an important role in hydrogenation [4].

The advantages of incorporating catalyst technology and nanotechnology will have an impact on the properties it produces. If in materials such as crystal solids, the process of making nanomaterials will make the size of the structure components decreased, there is a much larger interface area in material, and this can greatly affect the mechanical and electrical properties [3].

The catalyst generally has three components; active, support, and promoter; and CZZ catalyst to be analyzed is Cu-based catalyst because it has the active component of Cu. This active component is the most responsible element for the reaction. The support component serves as a component that expands the surface area of active components. The promoter that serves to reduce the process of sintering on active components [5].

This type of CZZ nanocatalyst is more promising in delivering high methanol yields than conventional methanol catalysts. This is due to the unique benefits of zirconium that has reducing, oxidizing, and high thermos table properties [6]. Besides CZZ nanocatalyst, the highly investigated nanocatalyst is CZA or Cu-ZnO-Al_2O_3, which has high efficiency for methanol catalysis reaction. But, CZZ nanocatalyst claimed to have a more active catalytic system due to better zirconia tolerance than alumina [7]. Nikel nanocatalyst supported on silica and alumina is an alloy with the most commonly used aluminium on commercial reactions. But, the residual nickel after the reaction can cause toxicity to the oil. Another study that examines the use of alternative catalysts is Cu-Cr/diatomea, which has the function of Cu-Cr as an active cock, while diatomaceous is used as a supported. This diatomeceous contains Al_2O_3 and SiO_2 with large surface area [4].

Research on hydrothermal, sol-gel, and reversed microemulsion methods have been reported for the preparation of Cu-based catalyst and have been described differently. Many reports have shown the feasibility for the production of Cu-related material [1-14]. But, most of the studies have several limitations, especially relating to complexity and time-fixing procedures, as well as expensive materials requirements [4]. There is also research by solid state method [8]. Another technique used to synthesize this CZZ nanocatalyst is a reverse co-precipitation technique under ultrasonic irradiation with a catalytic precipitation sequence CZZ. The result showed CZZ with total surface area and high dispersion [9]. Other studies with co-precipitation oxide techniques to enhance the catalyst activity is a citrate method [10], which produces solids with high surface area and at relatively low calcination temperatures [11]. But, from various methods, the combustion method is the best. This method has a simple route and a short time process, and can produce the optimal product [12].

In this paper, the purpose of this study is to evaluate the process of synthesis of CZZ nanocatalyst. The CZZ nanocatalyst is synthesized from precursor of Cu, Zn, ZrO_2 nitric metals, and by the addition of a complexing agent with glycine, with preliminary methods and nitric precipitation. The CZZ nanocatalyst production was analyzed using engineering and economic evaluation. The economic evaluation was done using several economic evaluation parameters, such as gross margin (GPM), internal rate return (IRR), payback period (PBP), cumulative net value (CNPV), break even point (BEP), profitability index (PI) on sales to investment. Optimizing the process to get a good profit for solving environmental problems and outcomes was also done. Research begins with analyzing and the needs of the factory. Then, the study was analyzed financially from the industrial economy evaluation with the source of research.

2. Methods

The method analyzed in this study is a combination of sol-gel with auto-combustion. This method is a very popular for nanomaterial synthesis. This method is started from local ignition techniques, gradual reactions, to the production of desired product oxide. In general, the precursor gel is prepared in a metal nitrate solution with the addition of an organic complexing agent such as citric acid, urea or glycine. Especially for glycine, glycine has a dual role in addition to complexing agents as well as fuel for combustion reactions to be more optimal. The nitrate ion itself acts as a burning oxidizing agent. This procedure has several advantages: use of economical precursor material, short preparation time, easy use of equipment, and good powder quality [1]. The factory model design uses a medium scale, in which the designed process is illustrated by Figure 1.

To support its evaluation and economic analysis, the evaluation data collection process is based on the average price on the online shopping web. All data is calculated by simple mathematical analysis. The nanocatalyst CZZ synthesis is analyzed by several economic evaluation parameters, including GPM, IRR, PBP, CNPV, BEP, and PI sales to investment.

3. Results and Discussion

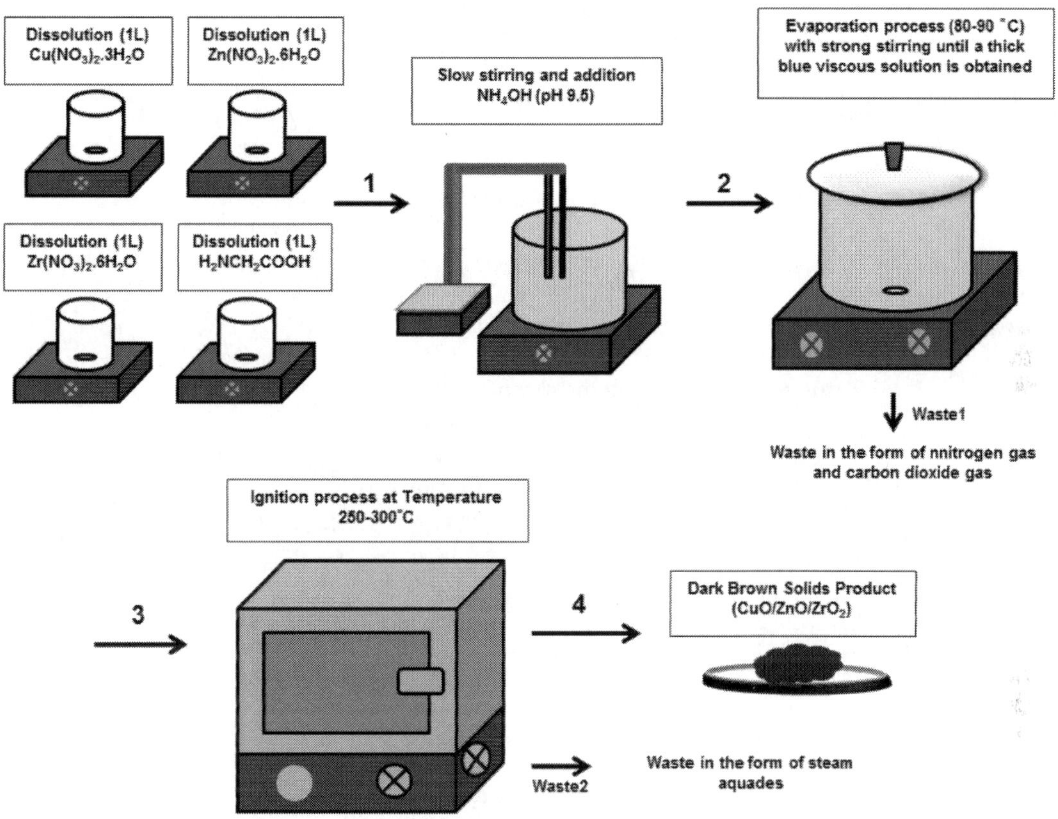

Figure 1. Illustration model for the production of CZZ Nanocatalyst

3.1. The engineering perspective

Figure 1 illustrates the steps of the composite $CuO/ZnO/ZrO_2$ nanocatalyst synthetic process. The synthetic process begins with the dissolution of nitrate metal solids (Cu^{2+}, Zn^{2+}, Zr^{4+}) as precursors with aquadest with constant stirring until completely dissolving. Also, the process is also done dissolution method of glycine (H_2NCH_2COOH) with aquadest. This glycine acts as an adjuvant that serves as an organic complexing agent. Glycine is also compatible with citric acid, carboxylate azides, and urea. After all materials dissolve perfectly in the aquadest, the process is mixed in one container. Then, the solution is added ammonium hydroxide (NH_4OH) dropwise until the pH of the mixture solution reaches 9.5. This addition is accompanied by vigorously stirring to the expected pH value. Then, the solution is evaporated by heating the mixture at temperature of 80-90°C and accompanied by stirring. The addition of oxygen is carried out during oven combustion because the evaporator uses heating with pressure techniques; so it is expected to contact with oxygen.

Assuming a complete combustion, the reaction that takes place between nitrates and glycine. Detailed reacton can be expressed as follows:

$$\tfrac{1}{2}\, Cu(NO_3)_2.3H_2O + \tfrac{1}{5}\, Zn(NO_3)_2.6H_2O + \tfrac{3}{10}\, Zr(NO_3)_4.5H_2O + mH_2NCH_2COOH + (\tfrac{9m}{4} - \tfrac{13}{4})\, O_2$$
$$\rightarrow (CuO)_{0.5}(ZnO)_{0.2}(ZrO_2)_{0.3} + (\tfrac{m}{2} + \tfrac{13}{10})\, N_2 + 2mCO_2 + (\tfrac{5}{2m} + \tfrac{21}{5})\, H_2O$$

After the evaporation process, dark blue viscous liquid is obtained in the form of intermediates of product with water and other wastes. Then, by adding oven combustion process and accompanied by a vacuum process at a temperature of 250-300°C, solution with dark brown dry powder was obtained.

The waste from this synthesis process consists of gas waste in the form of nitrogen gas, carbon dioxide gas, and water vapor. Utilization of CO_2 waste can be used as a gas source for hydrogenation reactions if the catalyst product has been obtained. However, the processing of waste from the process and production of fuel from carbon dioxide gas is not described in this study.

To calculate the mass balance of this process, we use the following assumptions:

- The process using sol-gel auto-combustion method.
- The operating temperature range in temperature 90 and 300°C
- The combustion efficiency is approximately on 100%
- The assumption loss of solvent in this form is aquades as much 80%

Based on the mass balance calculations listed above, with the composition in Table 1. Above generates about 248.16 grams of CZZ nanocatalyst. The price calculation approach is not based by price comparisons because the production of CZZ synthesis is more in the field or research and still sought variables and circumstances that produce optimal results.

Table 1. Composition comparison of reactants and products

Name of molecule	**Phase**	**Molecular Mass (g/mol)**	**Mass (gram)**	**Mol**
Cu(NO3)2.3H2O	Solid	241.6	120.8	0.5
Zn(NO3)2.6H2O	Solid	297.47	59.494	0.2
Zr(NO3)4.5H2O	Solid	447.32	134.196	0.3
$CuO/ZnO/ZrO_2$	Solid	248.16	248.16	1.0

3.2. Economic perspective

GPM analysis showed that the excellent value 10 US/kg product. These results are calculated from the production of CZZ nanocatalyst for one manufacturing cycle. A technical point of view indicates the project is profitable. However, to ensure analysis, additional other economic parameters are required. Therefore, to evaluate the economic evaluation, the following assumptions used:

- All the calculation was used USD with currency rate of 1 USD = Rp. 10,000
- The labor wage is Rp. 3,600,000 /month (near to 360 USD/month).
- The basic electricity cost is assumed to be Rp 1,467 / kWh (close to 0.15 USD / kWh).
- The length of business operation is 20 year.
- Discount rate is 15% /year.
- Income tax is 10% annually.
- Imports and shipping costs of purchased materials or equipment are eliminated in the economic evaluation.

From the relationship between investment and the production year of the CZZ nanocatalyst (Figure 2), the PBP is about 2 years to recover the initial investment fund. From the curve, this business still survives up to 20 years of production. Based on GPM even this is more profitable, because of its price and quality.

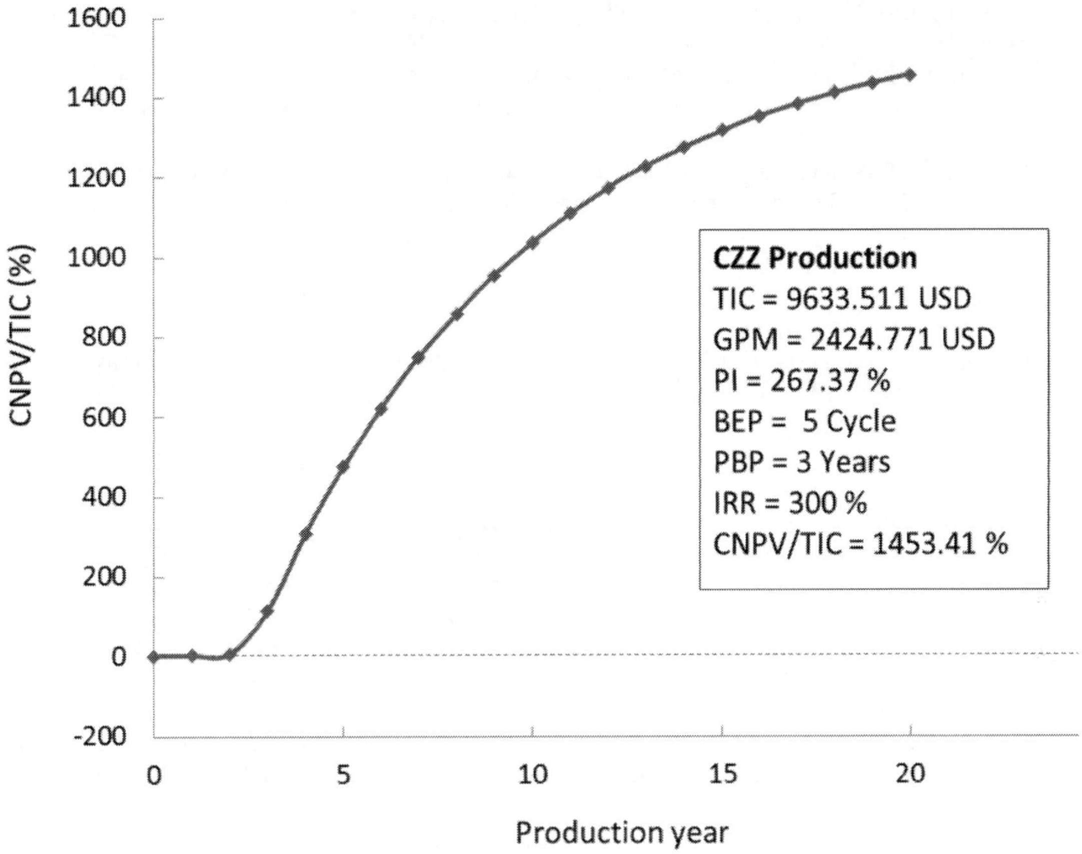

Figure 2. The relationship between the CNPV/TIC with the year of production of CZZ

To support the economic evaluation of CZZ nanocatalyst production in the perspective of PI, IRR, and BEP, insert table in Figure 2 is shown. CNPV last was 1453.41% compared with total investment fund. BEP analysis of this production is profitable after 5 production cycle. In 20 years, the investment profit from this production is USD 2,424,771 per year, meanwhile the total investment cost is USD 9,633,511. From this economic evaluation information, the project is profitable.

4. Conclusions

This study shows the engineering perspective and economic evaluation of the production CZZ nanocatalyst. The engineering perspective suggests this project is potential to be done in small scale industry. Economic analysis shows that the project is profitable. However, for being more profitable, further processing of the plant is required in the form of a hydrogenation reactor, in which this will be used for converting the emission gas into a liquid fuel with CZZ nanocatalyst as the catalyst.

References

[1] Syahroni A 2012 *Sintesis dan Karakterisasi Nanokatalis CuO/SiO_2 dari Sekam Padi* (Skripsi: Jurusan Kimia Universitas Islam Negeri Sunan Gunung Djati).
[2] Lubis, S 2009 Preparasi Katalis Cu/Silikat Gel dari Kristobalit Alam Sabang serta Uji Aktivitasnya pada Reaksi Dehidrogenasi *Jurnal Rekayasa Kimia dan Lingkungan* **7**(1) p 29-35.
[3] Chaturvedi S, Dave P N, and Shah N K 2012 Applications of nano-catalyst in new era *Journal of Saudi Chemical Society* **16**(3) p 307-325.

[4] Arena F, Italiano G, Barbera K, Bordiga S, Bonura G, Spadaro L, and Frusteri F 2008 Solid-state interactions, adsorption sites and functionality of Cu-ZnO/ZrO 2 catalysts in the CO 2 hydrogenation to CH_3OH *Applied Catalysis A: General* **350**(1) p 16-23.
[5] Liherlinar, dkk. 2009 Sintesis Nanokatalis CuO/ZnO/Al2O3 Untuk Mengubah Metanol Menjadi Hidrogen untuk Bahan Bakar Kendaraan Fuel Cell *Jurnal Nanosains & Nanoteknologi* (Edisi Khusus).
[6] Dong X, Li F, Zhao N, Xiao F, Wang J, and Tan Y 2016 CO_2 hydrogenation to methanol over Cu/ZnO/ZrO 2 catalysts prepared by precipitation-reduction method *Applied Catalysis B: Environmental* **191** p 8-17.
[7] Frusteri F, Cordaro M, Cannilla C, and Bonura G 2015 Multifunctionality of Cu–ZnO–ZrO 2/H-ZSM5 catalysts for the one-step CO 2-to-DME hydrogenation reaction *Applied Catalysis B: Environmental* **162** p 57-65.
[8] Guo X, Mao D, Lu G, Wang S, and Wu G 2010 Glycine–nitrate combustion synthesis of CuO–ZnO–ZrO_2 catalysts for methanol synthesis from CO 2 hydrogenation *Journal of Catalysis* **271**(2) p 178-185.
[9] Arena F, Italiano G, Barbera K, Bordiga S, Bonura G, Spadaro L, and Frusteri F 2008 Solid-state interactions, adsorption sites and functionality of Cu-ZnO/ZrO 2 catalysts in the CO 2 hydrogenation to CH 3 OH *Applied Catalysis A: General* **350**(1) p 16-23.
[10] Bonura G, Cannilla C, Frusteri L, Mezzapica A, and Frusteri F 2017 DME production by CO 2 hydrogenation: Key factors affecting the behaviour of CuZnZr/ferrierite catalysts *Catalysis Today* **281** p 337-344.
[11] Karelovic A, and Ruiz P 2012 CO 2 hydrogenation at low temperature over Rh/γ-Al 2 O 3 catalysts: Effect of the metal particle size on catalytic performances and reaction mechanism *Applied Catalysis B: Environmental* **113** p 237-249.
[12] Guo X, Mao D, Wang S, Wu G, and Lu G 2009 Combustion synthesis of CuO–ZnO–ZrO 2 catalysts for the hydrogenation of carbon dioxide to methanol *Catalysis Communications* **10**(13) p 1661-1664.
[13] Marimpul R 2017 Effect of substrate temperature on quality of copper film catalyst substrate: A Molecular Dynamics Study *Indonesian Journal of Science and Technology* **2**(2) p 183-190.
[14] Muslih E Y, Ismail A, and Kim K H 2017 Synthesis CuInSe2 (CISe) Thin Films Prepared from Metal-Ethanolamine Complex Compound *Indonesian Journal of Science and Technology* **2**(2) p 191-196.

IOP Conf. Series: Materials Science and Engineering **306** (2018) 012013 doi:10.1088/1757-899X/306/1/012013

Utilization of Baked-Smashed Sweet Potato and Vegetables on Patisserie Product

A Ana*, S Subekti, S Sudewi, E N Perdani, F Hanum, T Suciani, and V Tania

Home Economics Department, Universitas Pendidikan Indonesia, Indonesia

*ana@upi.edu

Abstract. The research was an experimental study in Green Skill Patisserie Course using Project-Based Learning model. It aims to complete the project development of pie named guramnis rainbow pie. Several experiments were carried out to produce a pie dough crust mixed with baked-smashed sweet potato and added with vegetables extract as the food coloring. The experiment method in order to make a better appearance or an attractive shape and to have more nutrition. In addition, the pie was filled with a mixture of sweet and sour gurame as Indonesian traditional food. By applying an organoleptic test to 10 respondents, the result shows that pie dough recipe using flour substituted by baked-smashed sweet potato with 2:1 of a ratio. Coloring pie dough adding extract vegetables (carrots, beets and celery) as color. We found that pie dough has more interesting pie color (90%) and the texture of the pie with a quite level of crispness (60%). Moreover, the pie taste is fairly (70%) and tasty (70%). Nutritional analysis results show that per size, serving guramnis rainbow pie contains energy as much as 81.72 calories, carbohydrates 12.5 grams, fat 2.32 grams and 2.77 grams of protein. The main findings are the pie appearance and taste was different compared to the previous pies because of the pie was served with gurame asam manis as the filling and had flour and cilembu sweet potato as the basic ingredients. The color of guramnis rainbow pie was resulted not only from food coloring but also from vegetables extract namely carrot (orange), bit (red), and salary (green). Thus, it had many benefits for health and adds the nutrition. The researchers recommend a further study in order to make pie dough with baked sweet potato and vegetables extract having an optimal level of crispness.

1. Introduction

Student who joined the course of patisserie project based learning was required to accomplish a project under the theme of patisserie product development [1]. The study develops pie production using sweet potato and vegetables as additives in making pie dough. The pie filling used Indonesian food namely *gurame asam manis* (fried *gurame* fish with sour-sweet sauce) or *guramnis*.

Sweet potato (*Ipomoea Batatas L*) is an alternative food as a carbohydrate source besides rice, corn, and cassava. Sweet potato has a big potential to be used as an alternative ingredient of many snacks [2]. It grows relatively fast in all seasons and does not need fertile soil. The researchers made a modification of pie dough by using Cilembu sweet potato comes from Sumedang West Java. Nowadays, *cilembu* sweet potato can be found easily there. It has a different characteristic compared to other sweet potatoes. It has a natural sweet taste. Moreover, it consists of carbohydrate, beta carotene (pro vitamin A), minerals (Ca, P, Fe, and K), fibers, and a good color and smell. The smashed *cilembu* sweet potato in this study was used to enrich the pie dough nutrition. To make the pie dough more attractive, it was added with some natural colors from vegetables namely bit, carrot, and salary.

The modification of pie dough by using a mixture of sweet potato and vegetables was preferred to enrich the nutrition. Besides, a pie dough commonly had no variation in its color. The filling was also modified by using Indonesian food namely *gurame asam manis* (fried *gurame* fish with sour-sweet sauce) which is very popular and never been used before. A good pie has a tender, flaky crust that can top a variety of fillings. A pie can be sweet and served as a dessert. With various fruits as filling, a pie is a favorite way to add more fruit to your menus. It can be savory and filled with meat or vegetables [3]. Therefore, the researchers were interested to do an experiment to a patisserie product namely pie by making the dough from a mixture of *cilembu* sweet potato and vegetables and by filling it with an Indonesian food. The experiment included testing the product until the appropriate receipt formula was got, doing an organoleptic test, and measuring the nutrition facts.

2. Material and Method

2.1. Material

In the experiment, the ingredients of pie dough were *cilembu* sweet potato, moderate-protein flour, milk, egg, sugar powder, salt, green extract from salary, yellow extract from carrot, red extract from bit, and natural food coloring. For the filling, the researchers used *gurame* fish, pineapple, onion leaf, Bombay onion, garlic, onion, red paprika, yellow paprika, chili sauce, tomato sauce, salt, pepper powder, sugar, corn powder and cooking oil.

2.2. Method

To get good pie dough, we did the experiment three times by combining taro and *cilembu* sweet potato. The researchers used a pie basic recipe formula as the following:

Table 1. Pie dough basic formula

No	Ingredients	Amount
1	Wheat flour	200 g
2	Shortening	100 g
3	Salt	0.5 g
4	Water	10 ml

Then, the researchers developed the pie dough and the filling.

3. Result and Discussion

In the first experiment, the researchers developed a basic formula recipe of pie dough by adding 100 gr steamed taro. Besides, the amount of flour was lessened 50 g (25%), shortening was added 20 g (20%), egg yolk was added 25 g and water was replaced by milk. Taro was added into the pie dough to make it more nutritious and having a good taste. Moreover, the pie dough was given colors from bit for red, salary for green, and sweet corn for yellow. Table 2 shows the formula of the first experiment:

Table 2. The formula of pie dough receipt in experiment 1

No	Ingredient	Amount	Product Characteristic
1	Flour	125 g	The pie dough was less flaky crust, the color was too soft, the taste was plain
2	Taro	100 g	
3	Margarine	120 g	
4	Milk	10 ml	
5	Egg Yolk	25 g	
6	Salt	0.5 g	

Table 2. Cont.

7	Bit extract	5 ml
8	Sweet corn extract	5 ml
9	Salary extract	5 ml

In making the pie dough, the researchers found many obstacles such as it became sticky, slick, and dropped when baking. For the filling in the first experiment, the researchers used *rendang* Padang but it was too oily and made the dough wet. As the result, when eating the rainbow *rendang* pie, the taste was not balanced for the filling was too dry. Meanwhile, the dough was tough but easily broken for it did not use sugar.

In the second experiment, the researchers tried to develop the pie dough by adding 100 gram baked *cilembu* sweet potato. The flour was lessened 50 g (25%), shortening was added 20 g (20%), egg yolk was added 25 g, and water was replaced by milk. The researchers also added 20 gram sugar powder to make the dough tastier and sturdier rather than the previous experiment. The sugar caramel when baking strengthened and united all dough ingredients; thus, it was not easily broken [5]. The filling used in this experiment was *gurame asam manis* (fried gurame fish with sour-sweet sauce) which was made from *gurame* fillet which was fried crispy with flour and egg and was poured with a sour-sweet sauce on top. The sour sweet sauce was preferred to balance the pretty sweet taste of the pie dough. The table 3 below shows the pie dough recipe formula in the second experiment.

Table 3. The pie dough recipe formula in the second experiment

No	Ingredient	Amount	Product Characteristic
1	Flour	125 g	The dough was pretty flaky crust, the color was too soft, and the taste was plain
2	*Cilembu* sweet potato	100 g	
3	Margarine	120 g	
4	Milk	10 ml	
5	Sugar powder	20 g	
6	Egg Yolk	25 g	
7	Salt	0.5 g	
8	Bit extract	20 ml	
9	Sweet corn extract	20 ml	
10	Salary extract	20 ml	

In the previous experiment, the pie dough color was less attractive. Therefore, the researchers added carrot extract to make it yellow, bit extract to make it red, and salary extract to make it green. Besides, they also added food coloring but the formula recipe of the dough and the filling was still the same. The following table 4 shows the recipe.

Table 4. The recipe formula of pie dough in the third experiment

No	Ingredient	Amount	Product Characteristic
1	Flour	125 g	The pie dough was flaky crust, the color was interesting, and the taste was good.
2	*Cilembu* sweet potato	100 g	
3	Margarine	120 g	
4	Milk	10 ml	
5.	Sugar powder	20 g	
6	Egg Yolk	25 g	
7	Salt	0.5 g	
8	Bit extract	20 ml	
9	Carrot extract	20 ml	
10	Salary extract	20 ml	
11	Food Color (green, red, yellow)	3 ml	

In this experiment, the pie dough which previously used color from sweet corn extract was added with carrot extract for the color of sweet corn was less striking. The color of vegetable extract was soft, which made the dough pale. Therefore, the researchers added a small amount of natural food coloring to get a good color. They tried to perfect the experiment so the result can be maximal. For the technique of dough color fusion, the researchers piled up some dough with different colors and rolled them up. Thus, the dough colors were spread evenly, which made easy the process of molding the pie dough into pan. The researchers named it rainbow pie for the dough color was like rainbow. The scheme 1 shows the process of dough making in the last experiment.

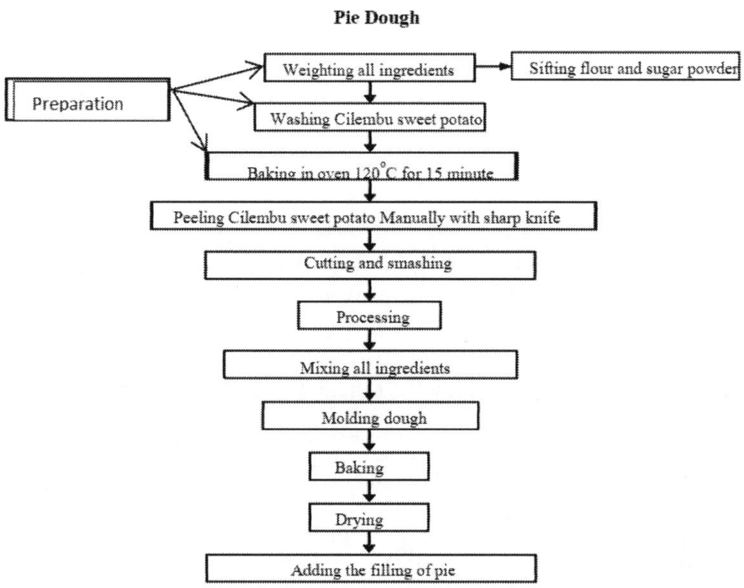

Figure 1. The process of pie dough making

In the last experiment, the researchers found the recipe formula of pie dough, analyzed the nutrition facts, and did an organoleptic test to 10 panelists. The recipe formula of pie filling is shown in table 5 below.

Table 5. The recipe of pie with *gurame asam manis* (fried gurame fish with sour-sweet sauce) filling

No	Ingredients	Amount
Basic Ingredient		
1	*Gurame* fish	500 g
Sauce		
1	Water	400 ml
2	Pineapple	250 g
3	Leek, slice aslant	2
4	Onion, slice round	½ clove
5	Garlic, slice	3 cloves
6	Red onion, slice	2 cloves
7	Red paprika	1
8	Yellow paprika	1
9	Chili sauce	4 spoons
10	Tomato	5 spoons
11	Salt	1 tea spoon
12	Pepper powder	1/2 tea spoon

Table 5. Cont.

13	Sugar	1/2 spoon
14	Cornstarch, dissolve	3 spoons
15	Cooking oil	100 g
Dressing		
1	Garlic	2 cloves
2	Lime	1
3	Pepper powder	1/2 teaspoon
4	Salt	1/2 teaspoon
Coating		
1	Flour	50 gram
2	Cornstarch	2 spoons
3	Pepper	1/2 teaspoon

Based on the recipe formula in table 5, the researchers made the pie filling from *gurame asam manis*. The following is the procedure of cooking the pie filling.

1. Discard fish scales, fillet, and cut into some parts. Wash them and also the head. Smear them with the dressing, keep for 30 minutes.
2. Roll the fish parts on the coating until spreading evenly. Pat them to avoid clumps of flour. Fry the fish in a hot cooking oil with a medium fire until the color is golden. Lift, drain, and set aside.
3. Heat two spoons of cooking oil, fry onion, garlic, red onion, and paprika until the smell is good. Pour water then add salt, sugar, and pepper. Add pineapple chunks and leek.
4. Pour tomato sauce and chili sauce into pan, stir well. Cook all ingredients until boiling then add cornstarch solution. Stir well until the sauce becomes thickened. Turn off the heat. Serve the fish with sour and sweet sauce as the dressing.

The experiment combined pie dough with *gurame asam manis*. For the next, the researchers named it *guramnis* rainbow pie. The result of organoleptic test of the product, which was tested to 10 panelists, involved some aspects namely pie dough color, pie dough flaky crust, taste, and aroma. Organoleptic method is comparing the food products at different variations with standard of the food products [6]. The following is the result of organoleptic test.

Table 6. The result of organoleptic test of the product

No	Aspect	Percentage (%)
1	Pie dough color	90
2	Pie dough flaky crust	60
3	Taste	70
4	Aroma	70

Serving size 100 g

According to the result of organoleptic test in table 6, it was known that the pie dough color, the taste, and the aroma were categorized as good. However, the dough flaky crust still needed improvement. The researchers found the product primacy as follows:

1. The pie appearance was different compared to the previous pies because of the rainbow colors which made the pie more beautiful.
2. The color of *guramnis* rainbow pie was resulted not only from food coloring but also from vegetables extract namely carrot (orange), bit (red), and salary (green). Thus, it had many benefits for health and adds the nutrition.
3. The taste of *guramnis* rainbow pie was different compared to the previous pies. Basically, a pie filling had a sweet taste such as vanila la, strawberry vla and so on. Nevertheless, *guramnis* pie was served with *gurame asam manis* as the filling and had flour and *cilembu* sweet potato as the basic

ingredients. They made a new taste for this pie was never exist before, which may increase the selling price.

The nutrition of per serving size of *guramnis* rainbow pie was shown in table 7 below.

Table 7. The nutrition facts of *guramnis* rainbow pie

Energy	Carbohydrates	Fat	Protein
81.72 calories	12.5 grams	2.32	2.77 grams

Serving size 60 g

Table 7 shows that the first fruit pie carp weighing 60 grams contains 81.72 calories of energy, carbohydrate 12.5 grams, 2.32 grams of fat, protein content by weight of 2.7 grams. Pie *guramnis* is a snack that can be eaten as a snack or a small meal that can be served as a complement to a cup of tea in the morning or afternoon. Presentation *guramnis* pie as a snack is very suitable, due to two pieces of pie *guramnis* is equivalent to two bowls of green bean porridge consisting of 25 grams of green beans + sugar + milk 13 grams and 50 grams = 175 calories can be eaten for 2 times in the morning and afternoon [7].

4. Conclusion

Guramnis rainbow pie is an innovative patisserie product which is resulted from a project-based learning. The innovation can be seen from the use of complementary ingredients namely *cilembu* sweet potato and vegetables extract for the pie dough. The study recommends a further experiment on pie dough flaky crust.

Acknowledgments
The authors would like to thank the Directorate of Higher-Education, Ministry of Education and Culture of the Republic Indonesia, which has provided funding of this research.

References

[1] Ana, Subekti, S, and Hamidah, S 2015 The patisserie project based learning model to enhance vocational students' generic green skills *Proceeding Atlantis Press* ISBN on line 978-94-62520-53-0
[2] Zuraida, N 2003 Sweet potato as an alternative food supplement during rice storage *Journal Lubang Pertanian* **22**(4), pp 150-155
[3] National Food Service Management Institute 2009 *Culinary techniques for healthy school meals (2nd ed) University* (MS: Author)
[4] Sampson, G O,Tetteh, AY, and Oldham, JH 2015 Rheological Behaviour Of Maize B-Glucan And Its Application As A Fat Replacer In Baked Goods *International Journal of Advance Research*, IJOAR org **3**(1), January 2015, Online: ISSN 2320-9186
[5] Bharathiar University 2008 *Food Production And Patisserie Rontline Institute Of Hotel Management Studies 14*, (Vallalar Salai: Pondicherry)
[6] Gita Bisla,Shailza Choudhary, and Vijeta Chaudhary 2014 Evaluation of the Nutritive and Organoleptic Values of Food Products Developed by Incorporated Catharanthus roseus (Sadabahar) Fresh Leaves Explore Their Hypoglycemic Potential *The Scientific World Journal* Volume **2014** (2014), Article ID 304120, 5 pages
[7] Sunita Almatsier 2005 *Penuntun Diet* Instalasi Gizi Perjan RS Dr Ciptomangun Kusumo dan Asosiasi Dietesien Indonesia (Jakarta: PT Gramedia Pustaka Utama)

IOP Conf. Series: Materials Science and Engineering **306** (2018) 012014 doi:10.1088/1757-899X/306/1/012014

Rapid Measurement of Soil Carbon in Rice Paddy Field of Lombok Island Indonesia Using Near Infrared Technology

B H Kusumo*, S Sukartono and B Bustan

Department of Soil Science, Faculty of Agriculture, University of Mataram, JL. Majapahit No. 62 Mataram, Lombok, Indonesia.

*bambanghk@gmail.com

Abstract. Measuring soil organic carbon (C) using conventional analysis is tedious procedure, time consuming and expensive. It is needed simple procedure which is cheap and saves time. Near infrared technology offers rapid procedure as it works based on the soil spectral reflectance and without any chemicals. The aim of this research is to test whether this technology able to rapidly measure soil organic C in rice paddy field. Soil samples were collected from rice paddy field of Lombok Island Indonesia, and the coordinates of the samples were recorded. Parts of the samples were analysed using conventional analysis (Walkley and Black) and some other parts were scanned using near infrared spectroscopy (NIRS) for soil spectral collection. Partial Least Square Regression (PLSR) Models were developed using data of soil C analysed using conventional analysis and data from soil spectral reflectance. The models were moderately successful to measure soil C in rice paddy field of Lombok Island. This shows that the NIR technology can be further used to monitor the C change in rice paddy soil.

1. Introduction

Soil organic matter (SOM) plays important roles to the fertility of soil. SOM can improve the physical, chemical and biological condition of soil [1]. It may improve the ability of soil to retain water, to improve soil structure [2], to release plant nutrients through its decomposition [3] and to increase the ability of soil in buffering contaminants [4]. Due to these roles, the adequate amount of SOM should be retained in soil and its change should be monitored.

So far, measurement of SOM is commonly conducted using conventional procedure i.e. Walkley and Black method. However this method is tedious and time consuming. It becomes expensive procedure if a large number of samples are needed to be analysed, for example, for producing a soil carbon (C) map.

Near infrared (NIR) technology, which is rapid and non-destructive technique and no chemical needed, has nowdays become an extremely important analytical technique [5]. This technology works by relating the vibration of covalent bonds of atoms such as C-H, O-H, and N-H with properties of object interest i.e. soils, minerals, and organic products. It has been reported able to measure soil C and N [6] [7] [8] [9], soil moisture content, organic matter, CEC, total C, ammonium N, nitrate N, and total N [10], root density [11, 12, 9], and biochar properties [13].

However, no information has been found on the accuracy of this technique in measuring soil C in rice paddy field of Indonesian soils. This paper reports the use of this technique to measure soil C collected from rice paddy field of Lombok Island, Indonesia.

2. Materials and Methods

2.1. Soil sample collection, analysis, scanning and spectral pre-processing

Top soil (0-10 cm depth) of 150 soil samples were collected using soil corer (2.54 cm diameter) from rice paddy field in Lombok Island, Indonesia (Figure 1). Coordinates of the samples were recorded, and the samples were shifted to the laboratory for air drying, grinding and sieving (with 0.2 mm diameter sieve). Each of the samples was divided into two parts. One part was anaysed in the laboratory for total C using Walkley and Black method, and another part was scanned using ASD FieldSpec 3 V-NIR Spectrometer (Analytical Spectral Device, Boulder, CO, USA). When collecting the spectral reflectance through scanning, the sapphire window of the soil probe was touched to the soil sample which was put on a small container. The spectral data were then imported to ParLeS [14], a software that was used to transform the spectral data; transformation to log (1/R) - R, wavelet detrending, and Savitzky-Golay smoothing filter. The smoothed data were then transformed into first derivative, and treated by mean centering.

Figure 1. Location of soil sample collection in rice paddy field of Lombok Island (area with yellow colour in the map)

2.2. Developing calibration models and parameter indicating accuracy of the model

Calibration models for soil C prediction were developed using Partial Least Square Regression (PLSR) from two sources of data; pre-processed spectral data of 780 – 2500 nm and the data of soil analysis using Walkley and Black method. In order to avoid over fitting, the model was developed using the number of factors (principal components) that produce the lowest root mean square error (RMSE) [14]. The accuracy of the model was tested internally using leave-one-out cross-validation. The ability of the PLSR models to predict soil C was assessed using the following statistic: (i) RMSE (root mean square error) of measured and predicted soil C, (ii) coefficient determination (R^2), and (iii) RPD (ratio of prediction to deviation). RPD is defined by standard deviation of the reference data divided by root mean square error (SD/RMSE). The best model is shown by the largest RPD and R^2, and the smallest RMSE.

3. Results and Discussion

3.1. Summary of Soil Carbon

Data of laboratory analysis of soil organic C which were analysed using Walkley and Black were present at Table 1. It can be seen that the C concentration on rice paddy field of Lombok Island varies from very low medium. Bogor Soil Research Centre (Pusat Penelitian Tanah Bogor) classifies the C-organic content into 4 categories; very low if the C-organic < 1.0%, low if the C-organic content 1.0-2%, and medium if the C-organic content 2.0-3.0%. Of all the samples, 71% samples contain very low until low C-organic, 26% samples contain medium C-organic, and only 3% samples contain high C-organic. The low concentration of soil C-organic in rice paddy field of Lombok Island is probably due to (i) low return of organic waste to the land and (ii) low input of organic fertilizer. Low C concentration in soil can also be influenced by soil cultivation. Heavily cultivated soils expose soil to the atmospheric oxygen and promote soil organic matter decomposition if soil contains less clay with coarse texture.

Table 1. Soil property

Soil property	Range		Median	Mean	Variance	Standard deviation
	Min.	Max.				
Total C-organic (%)	0.90	2.98	1.75	1.66	0.209	0.46

Soil C-organic status with various soil types of Lombok Island is presented at Table 2. Mediteran and Regosol contain C-organic varied from very low to medium. Both soils contain low average of soil C. Grumusol, Latosol and Alluvial vary from low to medium and these soils in average contain more C-organic than in Mediteran and Regosol.

Table 2. Soil C-organic status with various soil types.

	Soil type	Range		Average	Status
		Min	Max		
C-organic	Grumusol	1.41	2.98	2.06 (medium)	Low - Medium
	Mediteran	0.85	2.73	1.56 (very low)	Very low - Medium
	Regosol	0.90	2.79	1.83 (very low)	Very low - Medium
	Latosol	1.44	2.58	2.04 (medium)	Low - Medium
	Alluvial	2.00	2.20	2.10 (medium)	Low - Medium

3.2. Soil Spectral Shape

Spectral shape of soil reflectance based on soil types is depicted at Figure 2. There are variations of soil color due to soil types which can be seen on the visible band (400-780 nm). Alluvial soil reflects highest visible bands while Regosol soil absorbs more visible bands. Very strong absorption at around 1400 and 1900 nm at Figure 2 is due to water absorption; the first overtones of the O-H bond of water and the combination of the H-O-H bend and O-H stretching, respectively [15] [16] [6]. While strong absorption at around 2200 nm is the combination of metal O-H stretch [15].

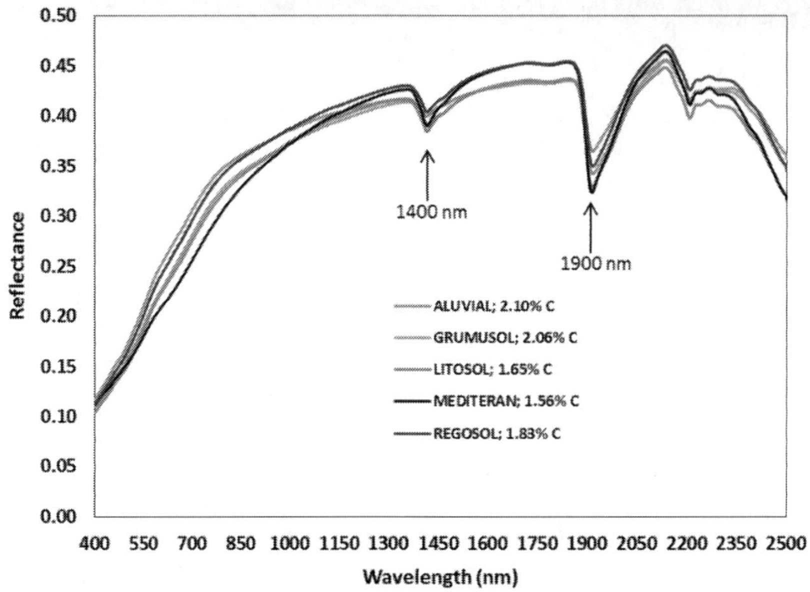

Figure 2. Soil reflectance of each soil type.

3.3. Accuracy Measurement of Soil C using NIRS

Prediction values of soil C-organic using leave-one-out cross-validation are shown at Table 3. It can be seen that NIRS technique was able to moderately predict soil C which was showed by RPD around 2.00 and R^2 0.7. Chang et. al. [17] classify the prediction values of models into moderately successful if R^2 0.5-0.8 and RPD 1.4-2.0. While, Malley et. al. [18] considered models with moderate accuracy if the R^2 0.7-0.8 and RPD 1.75-2.25. Accuracy of a model may be influenced by the accuracy of laboratory analysis and the signal to noise ratio of the soil reflectance. Soil water content, organic matter, clay and non-clay soil minerals, carbonates, iron oxides, and particle size of the soil sample [19] may also influence the model accuracy.

Table 3. Prediction values of soil C using leave-one-out cross-validation

Properties	Prediction values (leave-one-out cross-validation)		
	R^2_{CV}	$RMSE_{CV}$	RPD_{CV}
C Total	0.761	0.217	2.04

3.4. Relationship of laboratory measurement and NIR prediction

The relationship between soil C data of laboratory measurement and NIRS prediction is shown at Figure 3. Moderate accuracy of soil C prediction (R^2_{CV} 0.75; RPD_{CV} 2.04) shows that NIRS technique can be used to measure soil C-organic in rice paddy field of Lombok Island. Moderate accuracy using NIRS technique was also reported by Kusumo et. al. [6] and Chang et. al. [17] when dealing with soil C measurement.

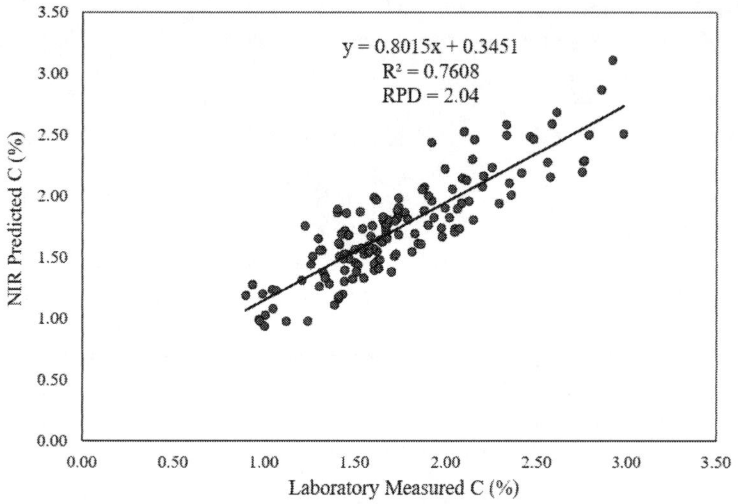

Figure 3. Relationship between soil C measured by laboratory analysis and predicted by NIRS.

3.5. Important Wavelength to the Model

Very important wavelength of soil reflectance contributing to the PLSR model is presented in Figure 4. Wavelengths with high VIP scores play more important role to the model [14]. The wavelengths around 1420 nm and most wavelengths between 2100 – 2500 nm are more important than other wavelengths in predicting soil C.

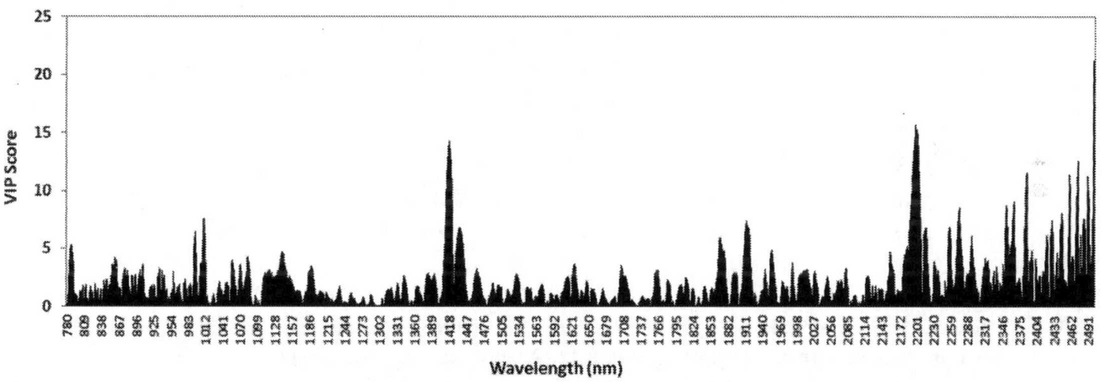

Figure 4. Very important wavelength to the model.

4. Conclusion

Soil C concentrations in rice paddy field of Lombok Island Indonesia are mostly very low to low. Some samples contain medium amount of soil C, and only small samples contain high C. This indicates that the concentration of soil C is in the critical condition. Near infrared technology was moderately successful to rapidly measure soil C in the study area. This successful measurement may give benefit for rapid monitoring of C concentration in soil which may be used as the information of the fertility status of the soil and as the information of soil C change and sequestration.

References

[1] Blume H P, Brümmer GW, Fleige H, Horn R, Kandeler E, Kögel-Knabner I, Kretzschmar R, Stahr K and Wilke B M 2016 *Soil Organic Matter. In Scheffer/Schachtschabel Soil Science* (Berlin Heidelberg: Springer) pp 55-86

[2] Kibet L C, Blanco-Canqui H and Jasa P 2016 Long-term tillage impacts on soil organic matter components and related properties on a Typic Argiudoll *Soil and Tillage Research* **155** pp 78-84
[3] Paul E A 2016 The nature and dynamics of soil organic matter: Plant inputs, microbial transformations, and organic matter stabilization. *Soil Biology and Biochemistry* **98** pp 109-126
[4] Venegas A, Rigol A, Vidal M 2016 Changes in heavy metal extractability from contaminated soils remediated with organic waste or biochar *Geoderma* **279** pp 132-40
[5] Stenberg B, Viscarra Rossel R A, Mouazen A M and Wetterlind J 2010 Chapter Five-Visible and Near Infrared Spectroscopy in Soil Science *Advances in Agronomy* **107** pp 163-215
[6] Kusumo B H, Hedley C B, Hedley M J, Hueni A, Tuohy M P and Arnold G C 2008 The use of diffuse reflectance spectroscopy for in situ carbon and nitrogen analysis of pastoral soils *Australian Journal of Soil Research* **46** pp 623-35
[7] Kusumo B H, Hedley M J, Hedley C B and Tuohy M P 2011 Measuring carbon dynamics in field soils using soil spectral reflectance: prediction of maize root density, soil organic carbon and nitrogen content *Plant and Soil* **338(1-2)** pp 233-45
[8] Minasny B, McBratney A B, Malone B P and Wheeler I 2013 Digital mapping of soil carbon *Advance in Agronomy* **118** pp 1-47
[9] Vågen T G and Winowiecki L A 2013 Mapping of soil organic carbon stocks for spatially explicit assessments of climate change mitigation potential *Environmental Research Letters* **8 (1)** 015011
[10] Kodaira M and Shibusawa S 2013 Using a mobile real-time soil visible-near infrared sensor for high resolution soil property mapping *Geoderma* **199** pp 64-79
[11] Kusumo B H, Hedley M J, Hedley C B, Hueni A, Arnold G C and Tuohy M P 2009 The use of Vis-NIR spectral reflectance for determining root density: evaluation of ryegrass roots in a glasshouse trial *European Journal of Soil Science* **60** pp 22-32
[12] Kusumo B H, Hedley M J, Hedley C B, Arnold G C and Tuohy M P 2010 Predicting pasture root density from soil spectral reflectance: field measurement *European Journal of Soil Science* **61(1)** pp 1-13
[13] Kusumo B H, Camps Arbestain M, Mahmud A, Hedley M J, Hedley C B, Calvelo Pereira R, Wang T and Singh B P 2014 Assessing biochar stability indices using near-infrared (NIR) reflectance spectroscopy *Journal of Near Infrared Spectroscopy* **22** pp 313-28
[14] Viscarra Rossel R A 2008 ParLeS: Software for chemometric analysis of spectroscopic data *Chemometrics and Intelligent Laboratory Systems* **90** pp 72-83
[15] Clark R N 1999 *Spectroscopy of rocks and minerals, and principles of spectroscopy. In Remote sensing for the earth science: Manual of remote sensing* 'ed Rench A N (New York: John Wiley & Sons)
[16] Demattê, J A M, Campos R C, Alves M C, Fiorio P R and Nanni M R 2004 Visible-NIR reflectance: a new approach on soil evaluation *Geoderma* **121(1/2)** pp 95-112
[17] Chang C W, Laird D A, Mausbach M J and Hurburg C R J 2001 Near-infrared reflectance spectroscopy – principal component regression analysis of soil properties *Soil Science Society of America Journal* **65** pp 480-90
[18] Malley D F, Martin P D and Ben-Dor E 2004 *Application in analysis of soils. Near-Infrared Spectroscopy in Agriculture*, Agronomy Monograph no. 44. 'ed Barbarick K A, Roberts C A, Dick W A *et al* (Madison, Wisconsin, USA: American Society of Agronomy, Inc., Crop Science Society of America, Inc., Soil Science Society of America, Inc.) pp 729-84
[19] Ben-Dor E, Irons J R, Epema G 1999 *Soil reflectance. In Remote sensing for the earth science: Manual of remote sensing* 'ed Rench A N (New York: John Wiley & Sons)

IOP Conf. Series: Materials Science and Engineering **306** (2018) 012015 doi:10.1088/1757-899X/306/1/012015

Student's Entrepreneur Model Development in Creative Industry through Utilization of Web Development Software and Educational Game

B Hasan*, H Hasbullah, S Elvyanti, and W Purnama

Departemen Pendididkan Teknik Elektro, FPTK Universitas Pendidikan Indonesia, Jl. Dr. Setiabudhi 229, Bandung 40154, Indonesia

*bachtiarhasan55@yahoo.com

Abstract. The creative industry is the utilization of creativity, skill and talent of individuals to create wealth and jobs by generating and exploiting creativity power of individual. In the field of design, utilization of information technology can spur creative industry, development of creative industry design will accommodate a lot of creative energy that can pour their ideas and creativity without limitations. Open Source software is a trend in the field of information technology has developed since the 1990s. Examples of applications developed by the Open Source approach is the Apache web services, Linux and Android Operating System, the MySQL database. This community service activities based entrepreneurship aims to: 1). give an idea about the profile of the UPI student's knowledge of entrepreneurship about the business based creative industries in software by using web software development and educational game 2) create a model for fostering entrepreneurship based on the creative industries in software by leveraging web development and educational games, 3) conduct training and guidance on UPI students who want to develop business in the field of creative industries engaged in the software industry . PKM-based entrepreneurship activity was attended by about 35 students DPTE FPTK UPI had entrepreneurial high interest and competence in information technology. Outcome generated from PKM entrepreneurship is the emergence of entrepreneurs from the students who are interested in the creative industry in the field of software which is able to open up business opportunities for themselves and others. Another outcome of this entrepreneurship PKM activity is the publication of articles or scientific publications in journals of national/international indexed.

1. Introduction

The development of business and industry with the globalization of the economy has risen sharply. Economic globalization and the information era to encourage industry to use the human resources of college graduates who are competent and have an entrepreneurial spirit. On the other hand, the economic crisis led to the number of jobs is not growing, and even decreases due to bankruptcy. Therefore, in order that the college is able to meet these demands, various innovations is required. One of them is learning innovations in build technopreneurship generation in today's information era [1]. In Indonesia, the creative economy is quite a role in national economic development though not many untouched by government intervention. According to data from the Creative Economy Study Team, the creative industry sub-sectors of Information Technology in 2014 has a Value Added Subsector (ADHB)

Rp. 8.610 billion or on average accounted for 0.11% of the total GDP added Subsector (ADHB) during 2010-2013 [2]. As for the labor force participation rate of the Creative Economy Employment sector by 0.58% during 2010-2013. In addition, this sector has a workforce subsector 67.888 workers with employment growth rate of the national average of 0.06% in 2013 [3].

Creativity and innovation in business become a matter to be considered and this can be obtained from the experience of interacting with the consumer to the product as well as the quality of the product itself. Each of the products or services produced will be qualified if they always learn from the symptoms of needs and desires of consumers who increasingly critical and educated [4]. Therefore to foster the innovation is necessary to create an environment that encourages capacity and ability to create local creative economy. Economic progress associated with the level of development of technical change which means the stage of technological mastery. Technical change vast majority are not codified and built on experience. Also are cumulative. Time mastery of this technology relies on the industrial sector and the accumulation process follows a specific trajectory [5].

The popularity of the Windows operating system made by Microsoft, can be seen riding the rapid sales of the iPhone, iPod, and iPad apple, as well as the spread of social networking sites Facebook and Twitter throughout the world. Of course, the opportunity to establish a business in the field of information technology is very large in an era like today. However, we must know in advance some of the main factors before starting a business in the field of information technology, such as: market feasibility, preferences, skills or familiarity, funds, raw materials, human resources, technology and personality [6].

As an effort to foster entrepreneurship in information technology, especially based creative industries through the use of software web development and educational games from the students, PKM entrepreneurship activities can be a solution to the growth of young entrepreneurs from the students who has a role to boost economic growth on the local scale, regional and international levels.

2. Method

Sustainable development is to create such sustainability into a form of advantages. In this context, Barney and Clark suggests VRIO models (Value, Rareness, Imperfect Imitability, and Organization) [7]. According to him, Sustainable Competitive Advantage can be obtained by performing continuous process of discovery that will continue to innovation. This process will run properly when first build the organization's core competencies, those are resources and capabilities. Business development that rely on nature, culture, and daily life of the community is a very closely in developing a sustainable business. UPI as the college has hosted a variety of entrepreneurial programs in an effort to increase the independence of his students that after graduating from college not be job seekers but can create jobs. Coaching methods compiled here are trying to synergize some coaching methods that already exist in the UPI.

Coaching is done by conducting a needs assessment regarding what training is needed by students in order to become entrepreneurs in the creative industries field of software technology. Then do the simulation in the form of business ideas competition. This competition aims to enhance student creativity in exploring creative and innovative ideas that give solutions to the problems that takes advantage of information technology. Further, training and coaching process is. To produce an entrepreneur, a phased approach that can be mentally stimulating entrepreneurship among the students, which is as follows:

The first stage of assistance, aims to make students understand deeply the business process, from the ability to identify and translate business opportunities into value-added superior, how it transformed the added value in the form of products or services, and can be consumed by the target customers. Analytical skills and creativity is trained in the process of this intervention. This stage we call the understanding entrepreneurial process.

The second stage of assistance, intended to allow students to learn how an entrepreneur running his business begins with making a business plan to convince a third party in order to finance its business; committed to what you want to achieve in the future to always be looking for ways to achieve those goals; partnership with third parties; or sell some or all ownership in the business. Through this simulation, students are expected to feel the bitterness and sweetness of being an entrepreneur.

The third stage of assistance, was felt to be an entrepreneur. Start determining value through market research, translate added value into new product concepts and business concepts, then required to maximize entrepreneurship capital such as the competence of the group, funding access, and social capital among other immediate family, relatives, friends, and parents in order the risks failure when launching the business can be minimized.

The fourth stage of assistance, is the process of consolidation of the business and future business development plans. For the purpose of long-term, sustainability of the program establishment of a coaching program as an innovation house think tank for developing products and services of the creative industry technology entrepreneurs in software web development and educational games.

3. PKM Activities Entrepreneurship

3.1. Implementation Phase

3.1.1. Phase 1: Preparation. At this stage the team offers to audience of students widely in the UPI (especially students of DPTE FPTK UPI) who have business skills. The students were mostly those who have a strong commitment to become an entrepreneur and have a basic competency in developing creative industries business based Web Development and Educational Games. At this stage, selected and netted a total of 35 DPTE students who have the desire to follow the PKM Technology training activities based creative industries in the field of software.

3.1.2. Phase 2: Implementation of Entrepreneurship PKM Workshop. At this stage, all participants who are interested and have the potential of entrepreneurial technology-based creative industries are given training in how to design marketing business by utilizing e-marketing, the website designing of company profile that can be accessed widely by business partners and easier. As for instructors involved are Entrepreneurship PKM team. In addition PKM Team also assisted by some IT experts, Web Designer and Educational Games, Materials provided in this training are web development and educational games in learning through the use of information technology (technopreneur).

3.1.3. Phase 3. Development and Assistance Phase. The participant's continously to be monitored their developments and Entrepreneurship PKM teams and instructors are intensively provide advisory services and technical assistance needed by the participants. In the expansion is positive for all groups because all of them can creatively develop its business profile through the website.

3.1.4. Stage 4: Monitoring and Evaluation. Evaluation is done with direct discussions with students who fostered and also discussions with the PKM faculty team. What are the difficulties in implementing these activities and solutions to overcome them. Students also continues to provide valuable inputs to the faculty team that PKM entrepreneurial activity goes well. Follow-up of this activities are students' capacity enhancement training about the creative industry business field of software-based Web Development and Educational Games.

4. Results and Discussion

4.1. Overview PKMK Development Activity in 2016

Entrepreneurship training activities in IT fields that utilize web development and educational games can run smoothly and in accordance with the target planned. It can be seen from the seriousness and discipline of participants who did their best to be able to develop the training materials provided.

IOP Conf. Series: Materials Science and Engineering **306** (2018) 012015 doi:10.1088/1757-899X/306/1/012015

Product or entrepreneurial achievements of web design and educational games created and developed by participants in general are related to the fashion business, education, services, engineering and online selling (e-commerce). Products developed by some of the participants can be seen at the address specified, as follows :

1. http://electroclub.esy.es
2. http://karisma.pe.hu/
3. http://dadantampan.pe.hu/
4. http://rinilestriani.pe.hu/
5. http://sopyansm.pe.hu/

4.2. The Results of Student Entrepreneurial Development Web Based Development

1. (http://rinilestriani.pe.hu/)

Figure 1. The look of online selling blog in the fields of fashion

2. (http://dadantampan.pe.hu/)

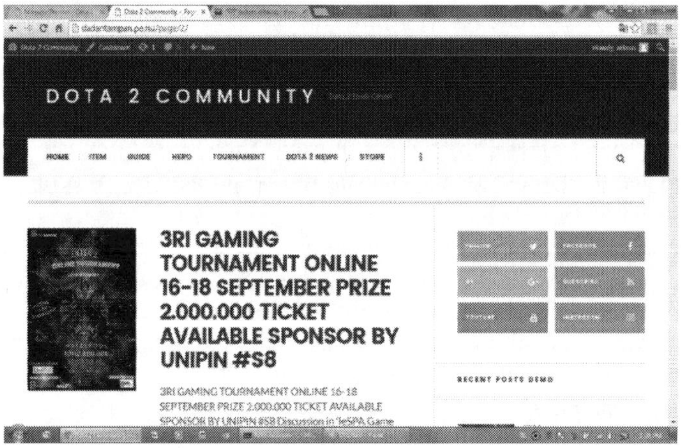

Figure 2. The look of online game selling blog

3. http://sopyansm.pe.hu/

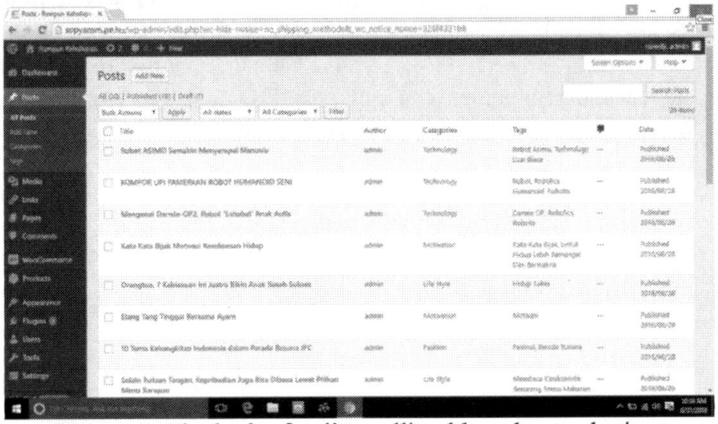

Figure 3. The look of online selling blog about robotic

4. http://karisma.pe.hu/

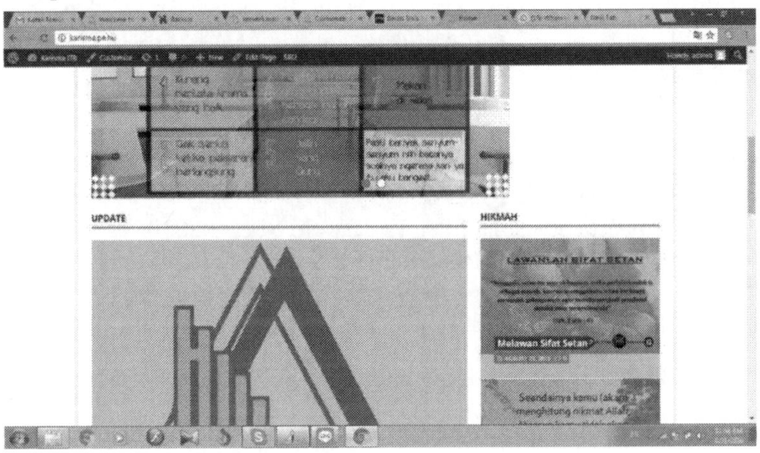

Figure 4. The look of educational blog

5. http://nuurmiftah.pe.hu/

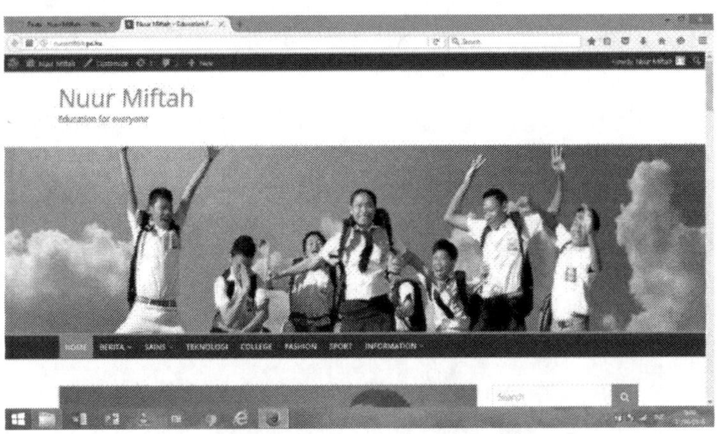

Figure 5. The look of educational blog

5. Conclusions

Based on the results of the implementation of the Community Service Activities (PKM) based entrepreneurship there are some conclusions that can be obtained, as follows:

1. Entrepreneurship fostering in the field of creative industries software to the utilization of web development and educational games can run smoothly and successfully, it is seen from the enthusiasm of those who follow these events and achievements generated by participants who have a high innovation.
2. Based on the student profile and based on the observation and mentoring activities, approximately 70% of students entrepreneurial UPI with various types of businesses, are already using and use the advantages of information and communication technologies, especially web development as one way for the promotion and marketing efforts.

References

[1] Chris Boulton and Patrick Turner 2011 *Mastering Business in Asia: Entrepreneurship* Wiley (MBA Publications Reproduced)
[2] Tim Studi Ekonomi Kreatif 2014 Statistik Ekonomi Kreatif Subsektor Teknologi Informasi Kementerian Pariwisata dan Ekonomi Kreatif
[3] Iqbal, Ahmad 2010 *Pengembangan Kewirausahaan Teknologi Untuk Mendorong Tumbuhnya Industri Kreatif Software Melalui Pemanfaatan FOSS* Pusat Pengkajian Kebijakan Peningkatan Daya Saing Badan Pengkajian dan Penerapan Teknologi
[4] Wening S 2012 *Menjadi Entrepreneur yang Kreatif dan Inovatif berbasis Konsumen Terdidik Prosiding Seminar Nasional* (Malang: Jurusan Teknologi Industi, FT UNM)
[5] Departemen Perdagangan Republik Indonesia 2008 *Pengembangan ekonorni kreatif Indonesia 2025* (Studi industri kreatif Indonesia)
[6] Rahmawaty, Penny Suwarto, Dyna Herlina, Endarwati, M Lies 2010 *Technopreneurship Course Development Program* (Jurusan Manajemen FISE Universitas Negeri Yogyakarta)
[7] Barney J B, Clark D N 2007 Resource-based theory: Creating and sustaining competitive advantage (Oxford University Press on Demand)

IOP Conf. Series: Materials Science and Engineering **306** (2018) 012016 doi:10.1088/1757-899X/306/1/012016

Destination Information System for Bandung City Using Location-Based Services (LBS) on Android

B Kurniawan* and H Pranoto

Universitas Komputer Indonesia (UNIKOM), Jl. Dipatiukur No. 112-116, Bandung, Jawa Barat, Indonesia

*bobi@email.unikom.ac.id

Abstract. Bandung is a city in West Java, Indonesia with many interesting locations to visit. For most favourite destinations, we can easily look for it on Google and we will find some blogs there discussing about related content. The problem is we can not guarantee that the destination is frequented by visitor. In this research, we utilizes an application to help everyone choosing destination frequented by visitor. The use of information technology in the form of picture, maps, and textual on Android application makes it possible for user to have information about destination with its visitor in a period of time. If destination has visit history, selection of proper destination will be given with fresh informations. This application can run well on Android Lollipop (API Level 21) or above with a minimum RAM of 2 GB since it will compare two coordinates for every data. The use of this app make it possible to access information about location with its visitor history and could help choosing proper destinations for the users.

1. Introduction

Bandung is a city in West Java, Indonesia with many interesting locations to visit. Travelling to Bandung can be used as an option to relieve fatigue after a full day activity. The problem is when people go for a travel but do not feel like travelling instead of creating a new burden of thoughts. Selection of proper destination influential in this regard.

Based on the number of foreign and domestic tourists statistics who visit Bandung, since 2012 the number of tourists who come to Bandung showing an enchancement[1] [2]. In 2012 the number of tourist arrivals reached 3,513,705 people and increased by 9.84% to 3,897,429 people in 2013. Then in 2014 increased by 11.79% to 4.418.781 people. However, a decrease occurred in 2015 by 9.37%, bringing the total tourists coming into 4,004,492 people.[2][3]

Table 1. Number of Foreign and Domestic Tourist in Bandung Year 2010-2015[2][3]

Year	Tourist		
	Foreign	Domestic	Total
2010	180,603	3,024,666	3,205,269
2011	194,602	3,882,010	4,076,072
2012	158,848	3,354,857	3,513,705
2013	170,982	3,726,447	3,897,429
2014	176,487	4,242,294	4,418,781
2015	130,039	3,874,453	4,004,492

The table above shows the data arrival of foreign and domestic tourists in 2010-2015. In 2010 there is a significant increase towards the year 2011 compared to other years, it is foreign tourists which increase by 7.19% and for domestic tourists increase by 22.08%. However, entering the year 2012 decrease by 18.373% for foreign tourists and 13.58% for domestic tourists. The same thing happen when entering year 2015 where foreign tourists decrease by 26.31% and 8.67% for the domestic tourists. Nevertheless, in 2013 and 2014 the data shows a fairly steady rise, both in foreign and domestic so that the average number of tourist arrivals in the last six years stands at 3,852,624 people. Seeing the number of tourists scattered in the city of Bandung and its surroundings, it is possible to generate dissemination of tourist information based on the areas frequented in a certain period of time using information technology from an application. Everyone will certainly not have the same destination at the same time and that makes the visitor data can be an information. Although social media applications can already provide user location sharing features but there are still shortcomings that they can not directly display information about what the users share because the social media application does not discuss about profile of a location. Similar research are found like Manav Singhal [4], and Amit Kushwaha [5] which are implementing the use of GPS technology to retrieve users data.

This study aims to improve the information of destination in Bandung and nearest with its visitor using Android applications that can meet these needs so that the parties involved can immediately access the desired information. Android is used because 80% users are using Android in the last five quarters [6].

2. Research Methodology

Research methodology in this research include the identification of problems, formulation of problem identification and problem formulation, formulation of objectives, data collection, system analysis, planning system, building system, testing system, and withdrawal conclusion. The following are described in detail.

a. Problem Identification and Problem Formulation
In this stage is the initial stage of research that begins by identifying problems that may arise when someone tries to select proper destinations. Once found the problem then the next stage is to formulate the problem and determine the appropriate method to implement the system in order to solve the existing problems.

b. Formulation of Objectives
In this stage the researcher determines the purpose and objectives to be achieved to solve the problems that have been described in the previous stage.

c. Data Collection
In this stage the researcher determines how to obtain the data needed to be used on the system as input data. The steps used in data collection are Using questionnaires, interviews, literature studies to find out previous research.

d. System Analysis
In this stage the researcher performs the analysis of the system that is identifying and evaluating the problems, opportunities, constraints that occur and the expected needs, so as to fit the needs. The steps used in the analysis are system analysis, problem analysis, system architecture analysis, non-functional requirements analysis, and functional requirements analysis.

5. Planning System
In this stage is finding the optimal form of application to be built with consideration of the factors problems and needs that exist in the system as has been determined by combining the use of appropriate hardware and software technology so that obtained the optimal and easy to implement. The steps in system planning are designing relation schemes, table structures, menus, interfaces, and error messages.

6. Building System
In this stage is to build a system based on the results of analysis and system design so that the system built will be in accordance with the needs of the results of analysis and system design.

7. Testing System
   In this stage the software testing process is performed to ensure that all statements are tested as well as to uncover errors and ensure that defining inputs will provide output in accordance with the required results.
8. Withdrawal Conclusion
   In this stage is the last stage where will be drawing conclusions on the system that has been built. From these conclusions will be assessed whether the system built is in accordance with the original purpose of research or not.

3. System Design

3.1. System Overview

To make it works these components are necessary; Android smartphone, a server with web services and database server. First of all, application needs tobe installed on Android smartphone. Users can access information on the server by making request to web services using application installed. Furthermore, web services will provide the information requested by client in the form of JSON response. JSON response reprocessed into user friendly information for the user.

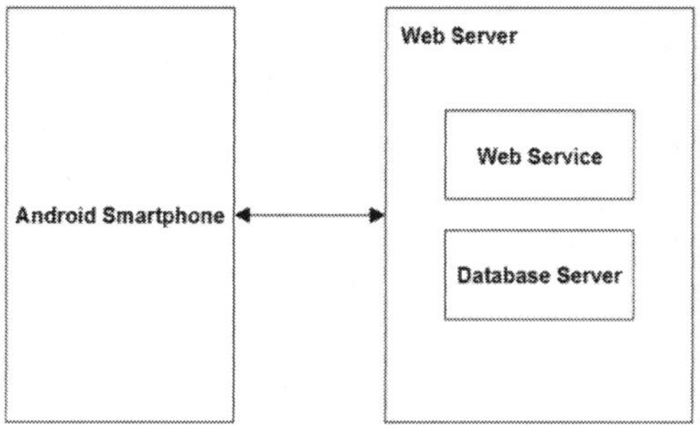

Figure 1. Block Diagram of System

3.2. System Analysis and Planning

With the use of GPS technology it is possible to provide real time location of device even with room-level accuracy like research by Yohan Chon [7]. It is useful if we are going to build a system with location features, but in this research will not talking about accuracy since we can tolerate it in region-level accuracy. The system to be built involves two important things. The first is getting the latest location data from the user. And the second thing is displaying it back as visitor data over a place based on location within specific radius.

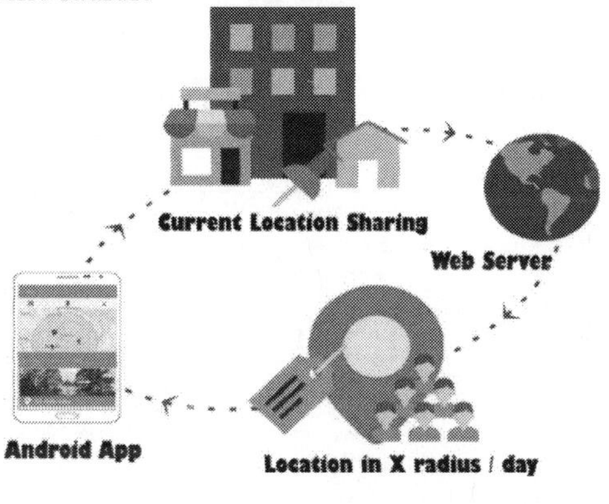

Figure 2. The Concept Of The System To Be Built

a. Updating Current Location
   Location sharing features are used in this study to generate user location data. This data is used to support the research because with the existence of this data then the system can produce information based on the closest coordinates. Just like research which done by Bagreecha Komal [8], a system can be built to provide a facilities for users when user is newer to any place through the application named places directory. We actually could implement this as visitor log data in the future.
b. Selection of proper destination
   The system will assist the user in selecting the destination by displaying the destination profile along with the closest visitor data on the current day. this is possible by calculating the radius of the location between the destination and the previously received visitor data. So the system will download all data today first, then calculated whether each data is near the destination or not.

3.3. Use Case Diagram
Use case diagram of monitoring system consists of 8 use cases that is create an account, help signing in, login, logout, edit profile, find destination, and share my location. The use case diagram can be seen in Figure 3 below.

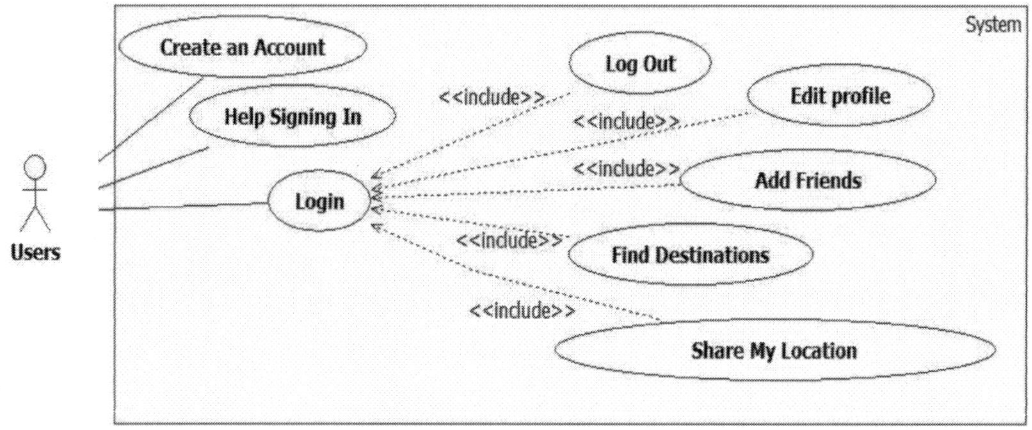

Figure 3. Use Case Diagram

4. Results And Discussion

4.1. Hardware Implementation
The hardware used to test the application is as follows :

Table 2. Hardware Implementation Testers

No.	Hardware	Spesification
1	Processor	Quad-core with 1.2 GHz speed
2	RAM	3GB
3	Layar	5 inch
4	Storage	40MB
5	Kamera	16MP & 3.7MP

4.2. Software Implementation
The software used to test the application is as follows:

Table 3. Software Implementation Testers

No.	Software	Spesification
1	Computer Operating System	Windows 8.1 Pro
2	Android Studio	Versi 2.3.2
3	Android SDK	Level 22 and above
4	Programming language	Java, PHP, JSON
5	Database	MySQL
6	Operating system Smartphone	Android Lollipop 5.1 (API Level 22) and above

4.3. Screenshot Program

Here is an example of screenshot program. The figure 4 describes that user is sharing their location.

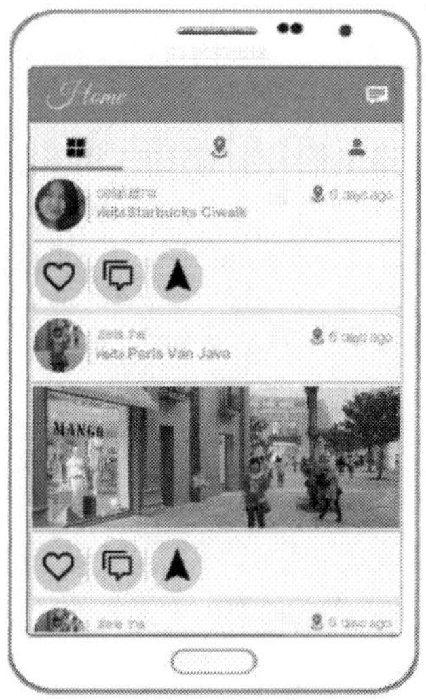

Figure 4. Screenshot Program

Locations will be used for displaying statistic visits of a destination like the figure below.

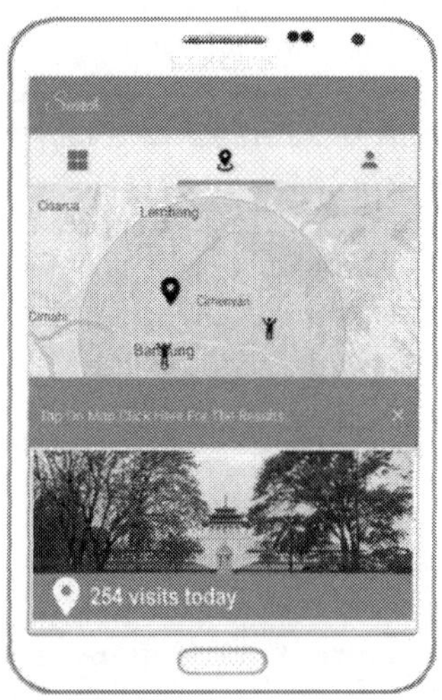

Figure 5. Screenshot Program

This application enables users to find destinations that are frequented based on starting position determined on the map in a certain radius and it is based on data supplied by application.

4.4. System Testing
Testing is done in order to find errors and deficiencies in the software being tested. With the testing we can know whether the software is made already meet the criteria in accordance with the purpose of software design. Testing is done by black box testing techniques that focus more on finding functional fault program.

Black box testing is performed for create an account, help signing in, login, logout, edit profile, find destination, and share my location. The results can be seen in below table.

Table 4. Software Implementation Testers

No.	Test items	Testing Type	Status
1	Create an account	Black Box	Accepted
2	Help signing in	Black Box	Accepted
3	Login	Black Box	Accepted
4	Logout	Black Box	Accepted
5	Edit profile	Black Box	Accepted
6	Find destination	Black Box	Accepted
7	Share my location	Black Box	Accepted

5. Conclusion
Based on the results of research the following conclusion can be drawn, by using this application, users can display destination with visitor history per day. Selection of proper destination can be done with the statistic of visitor

References
[1] Badan Pusat Statistik Kota Bandung 2015 *Statitistik Daerah Kota Bandung* (Bandung: Katalog BPS No. 9312.3237)
[2] Badan Pusat Statistik Kota Bandung 2015 *Kota Bandung Dalam Angka* (Bandung: Katalog BPS: 1102001.3273)
[3] Rekapitulasi Data Kunjungan Wisatawan Yang Datang Ke Kota Bandung Tahun 2010 – 2015. *ppid.bandung.go.id/?media_dl=14563* [Accessed: July 30, 2017]
[4] S Manav and S Anupam 2012 Implementation of Location based Services in Android using GPS and Web Services *IJCSI International Journal of Computer Science Issues* Vol **9**, No 2
[5] K Amit and K Vineet 2011 Location Based Services using Android Mobile Operating System *International Journal of Advances in Engineering & Technology* **2011**
[6] *http://www.idc.com/prodserv/smartphone-os-market-share.jsp* [Accessed: July 30, 2017]
[7] C Yohan 2011 *LifeMap: A Smartphone-Based Context Provider for Location-Based Services IEEE Pervasive Computing,*Vol. **10**
[8] K. Bagrecha 2012 Android Application Using Gps Navigation *International Journal of electronics, Communication & Soft Computing Science & Engineering*

IOP Conf. Series: Materials Science and Engineering **306** (2018) 012017 doi:10.1088/1757-899X/306/1/012017

Teacher Professionalism in Technical and Vocational Education

B L L Tampang[*] and D Wonggo
Universitas Negeri Manado, Tondano 95618, Sulawesi Utara, Indonesia.

[*]benyamintampang@unima.ac.id

Abstract. The advancement of a nation is largely determined by the advancement of education, and the field of education is on the teachers' shoulders. Therefore, professionalism of teachers should be improved in performing their duties as facilitators, initiators, and motivators in achieving the competence of graduates. Teachers should be able to facilitate the learning process and pay attention to the development of learners in its various dimensions, leading to the ownership and development of intelligence, learning skills, attitudes, work skills, and social independence. Vocational education carried by Vocational High School, should not only provide theory, but also be coupled with the application or application of theory. It is time for us to rise up to give birth to professional experts in their fields, in accordance with the needs of the workforce. A professional teacher is required to have a work ethic and a high commitment to his profession, and always through continuous development through professional organizations, the internet, books, seminars, and the like. With the professionalism of teachers, then the future teacher does not appear again only as a teacher (teacher), such a prominent function during this, but switch as coach, counselor, and learning manager. Each Educational Institution of Education Personnel graduate and other college graduates who wish to become teachers are required to have at least the competence of the standard. Thus, teacher professionalism is a life-long and never-ending process. Developing and developing vocational education is an important and absolute thing to do, considering that approximately 80% of the workforce in the field is lower middle-level workforce.

1. Introduction

Education is a very essential thing in the process of humanity in a society. In this era of globalization there is a tremendous change in people's lives. We can not escape the globalization tsunami that has entered every facet of modern human life. In the era of globalization there are jumps or the transformation of life values and therefore there is also a change in the process of humanity or education.

This relatively rapid change also influences the dynamics of policy change in education. National development in the field of education is directed to support efforts to improve the quality of human resources who are faithful and devoted to God Almighty, have noble character, healthy, knowledgeable, capable, creative, independent and become citizens of democratic, and responsible [1].

Among the issues relating to teachers and teachers are usually about inadequate qualifications and teacher competence, lack of teacher welfare, low work ethic and teacher commitment, and lack of community awards in the teaching profession. Although the government, together with its parents and the community, have made various efforts to improve the teaching profession, teachers' issues remain as the main problem of the national education world.

In the context of education, to be able to realize these goals, very many factors that must be prepared. One of the forms of education and training services based on technology and vocational that has a strategic role in national development, including Vocational High school, Polytechnic, Technology and Vocational Education within Educational Institution of Education Personnel, and Vocational Training Institutions both government and private. Educational personnel who handle these educational institutions are teachers, instructors and lecturers who have the characteristics of competence vary widely, in accordance with the areas of expertise required. Teachers in the field of education and training based on technology and vocational, viewed from the background of his expertise, both qualifications and relevance keompetensinya, until now still varied and not relying standard.

Thus, teacher professionalism is a life-long and never-ending process. Developing and developing vocational education is an important and absolute thing to do, considering that approximately 80% of the workforce in the field is lower middle-level workforce.

2. The Essence of Teacher Professionalism

Departing from a question whether job (title) teacher has been referred to as a profession? Teacher work if done correctly, demands seriousness from making a teaching-learning plan, organizing, organizing, controlling, guiding, nurturing, and being responsible for the implementation of teaching and learning process relevant, efficient and effective. Further assess the process and learning outcomes, as well as diagnose the factors that affect the success of the learning process. Thus the process of teaching and learning can be improved (perfected) continuously, which is a professional work that is guaranteed quality. Based on this fact, it is not exaggerating to say that the job or professorship is a professional job, as are other professional positions, such as doctors, engineers, lawyers, etc., because a teacher's work requires systematic and scientific training.

According to Aqib, the teaching profession is not just a vehicle to channel a hobby or as a side job, but it is a job that must be known to realize maximum professional skill. As professionals, teachers play important roles and responsibilities in the implementation of teaching programs [2].

While Budiarso states the performance of professional teachers, among others: (1) the desire to always display behavior approaching the ideal standard, (2) to improve and nurture the profession, (3) the desire to always develop the profession by increasing the knowledge and mastery of technology, (4) pursuing quality and ideals in the profession, and (5) pride in the profession. Similarly, Professional teachers among others have the following characteristics: (1) have mature and developing personality, (2) have the skills to arouse students' interest, (3) strong mastery of knowledge and technology, and (4) have a sustained professionalism attitude [3].

Thus, it can be said that the professional teacher is a teacher who has the expertise, responsibility, and sense of sustainability supported by strong professional ethics. To that end teachers should have adequate competence qualifications that include pedagogic competence, social competence, personality competence and professional competence. Teacher professionalism performance is basically a manifestation of professionalism of teachers consciously and directed to carry out education both at school and outside school.

3. Teacher Qualification

Education will be qualified if it has quality resources. Teachers as one of the educational resources is a strategic component. Any effort to improve educational outcomes will be significant if it involves the teacher as the key and at the same time the central point of any educational reform. This means that qualified teachers must be realized. The teachers at least have: professional skills, professional efforts, time devoted to professional activities, and rewards for the results of his work. Teachers as a professional figure must have work kecapakan harmonious with the demands of the work field in cultivation, so have a clear authority in improving the quality of education results. Teacher competence is highly demanded in carrying out the task professionally. This means that teachers as a profession will determine teacher performance and quality. Qualifications of educators for formal education in Indonesia are required as follows: (a) Kindergarten, Primary and Extraordinary School minimum D3 qualification. (b) Junior and Senior High School (SMP and LB), Vocational and Extraordinary School (SMK and LB) minimum S1. The current condition of teachers in Indonesia is 44% unqualified and 56% are qualified [4].

4. Teacher Competencies

Competence can be defined as knowledge, skill and personal qualities, which include the motive, attitude, value, self image and trait necessary to carry out work activities effectively in line with educational goals. In Kepmendiknas Number 045 / U / 2002, it is stated that what is meant by competence is a set of intelligent and full of responsibility which is owned by a person as a requirement to be deemed capable by the society in carrying out tasks in certain job field (Guidelines for Certification of Educator Competency, 2004) . In conjunction with educators or teachers, the competency refers to performance or actions that are rational and meet certain specifications in the performance of educational tasks. Thus the competence of educators puts special emphasis on the formation of competence by reviewing and testing the relation between the competency task requirements and the learning experience given to the students.

Teachers are required to have academic qualifications, competencies, certification of educators, physically and mentally healthy, and have the ability to realize the goals of national education. The competencies teachers must possess are pedagogical competence, personality competence, social competence, and professional competence gained through professional education [1].

Competence is the characteristic or ability of the knowledge, skills and attitudes of behavior required in accordance with the duties of his position [6]. This limitation can be interpreted that everyone who performs certain functions must be supported with competence. In general, competence is defined as knowledge, skills and basic values that are reflected in the habit of thinking and acting. However, the understanding of this competency should not be interpreted as a knowledge, attitude, ability and skills alone, but competence must be conceptualized as a form of behavior, action, performance of a person after experiencing the learning process.

According to Finch and Cruncilton [7] knowledge can be learned through education, while skills, attitudes, norms can be mastered through work experience, meaning that competent teachers are teachers who have adequate educational qualifications and have adequate teaching experience in the field. In relation to the competency test, reliable and reliable competence tests are not sufficiently tested for a moment, but should include performance tests and personality tests through observation and observation processes over a period of time, involving, superiors, peers and students (head assessment, peer assessment, student assessment).

5. Vocational Education Now and Forever

Vocational education or skill education is an education with combined theory and practice 40:60% percentage with the aim of preparing graduates who can compete in the business and industry. The curriculum in vocational education is concentrated on skill-learning systems. Constantly developing and advancing science and technology and widespread industrialization, demanding vocational education continues to be carried out continuously.

The problem that has been developed in Indonesia in the world of education is the relevance of education to the world of work, especially for academically oriented education programs. The direction of education in Indonesia is less relevant to market needs, because education in Indonesia is more focused on academic education than vocational education that produces skilled labor. "The education that now exists produces arrogant people who feel smart, but lack the skills and competence to take action that should be done to make Indonesia prosper and prosper". From the opinion of Sadino, if we think of it as a generation who has been educated for several years until now sitting on the bench lectures do have a point. The statement, also a picture of Indonesia's education is still in need of improvement.

These factors, which led to the existence of vocational education. Educational program by prioritizing the sustainability of skills competence. Briefly vocational education is an education that includes curriculums that are sensitive to the needs of the necessary experts in a real industry in life. It is expected that vocational education produces quality human resources that are able to optimize the wealth of natural resources owned, capable of prospering the community. Vocational education is expected to provide the competence or ability that is needed in the life of the community and the world of work.

Activities of students conducted in the business world or in the is a concrete manifestation of the function of vocational education to prepare human resources experts. In practice, the vocational school students are expected to be able to develop their skills, because they will be directly in touch with their fields, which will also be directly practiced in the world of work. This is the hallmark of vocational education, with the direct employment of industry industry will shape mentally and provide a valuable experience for students. Very useful and useful when jumping straight into the real world of work competition.

6. Conclusions

Professional teachers are teachers with expertise, responsibility, and a sense of well-being supported by strong professional ethics. To that end teachers should have adequate competence qualifications that include pedagogic competence, social competence, personality competence and professional competence. Teacher professionalism performance is basically a manifestation of professionalism of teachers consciously and directed to carry out education both at school and outside school. Teacher professionalism is a life-long and never-ending process. Developing and developing vocational education is an important and absolute thing to do, considering that approximately 80% of the workforce in the field is lower middle-level workforce. Vocational education carried by SMK, should not only provide theory, but should be coupled with the application or application of theory. It is time for us to rise up to give birth to professional experts in their fields, in accordance with the needs of the workforce.

References

[1] Law of the Republic of Indonesia Number 20 Year 2003 *About National Education System* (Jakarta)
[2] Aqib Z 2002 *Profesionalisme Guru dalam Pembelajaran* (Insan Cendekia: Surabaya)
[3] Budiarso, E 2004 Sertifikasi Lulusan Prodi Kejuruan Tekik (Ex FPTK) dan JPTK, Makalah, Prosiding Konvensi Nasional Aptekindo II Jakarta
[4] Budiarso E 2004 Graduation Certificate of Technical Studies Program (Ex FPTK) and JPTK, Papers, Proceedings of the National Convention Aptekindo II Jakarta
[5] Depdiknas 2004 Standard of Beginner Vocational Teacher Competence Directorate General of Higher Education Directorate of Education Education and Teaching Staff of Higher Education Jakarta
[6] LAN 2003 State Administration System of the Republic of Indonesia State Administration Institute of the Republic of Indonesia Jakarta
[7] Wardaya 2006 Competency Test is a Workshop of Performance Assessment of Educators Increasing Vocational Human Resources in the Framework of Regional Potential Development Potential National Aptekkindo Convention and Temu Karya XIV FP / FPTK / JPTK University of Indonesia Gorontalo, 15-17 February 2006
[8] Mungin W 2003 Opportunities and Challenges Entering the Global Era and Autonomy Paper Seminar Semarang UNNES

IOP Conf. Series: Materials Science and Engineering **306** (2018) 012018 doi:10.1088/1757-899X/306/1/012018

Learning Application of Astronomy Based Augmented Reality using Android Platform

B Maleke*, D Paseru, and R Padang

Teknik Informatika, Universitas De La Salle Manado, Indonesia

*bydeb_raul@yahoo.com

Abstract. Astronomy is a branch of science involving observations of celestial bodies such as stars, planets, nebular comets, star clusters, and galaxies as well as natural phenomena occurring outside the Earth's atmosphere. The way of learning of Astronomy is quite varied, such as by using a book or observe directly with a telescope. But both ways of learning have shortcomings, for example learning through books is only presented in the form of interesting 2D drawings. While learning with a telescope requires a fairly expensive cost to buy the equipment. This study will present a more interesting way of learning from the previous one, namely through Augmented Reality (AR) application using Android platform. Augmented Reality is a combination of virtual world (virtual) and real world (real) made by computer. Virtual objects can be text, animation, 3D models or videos that are combined with the actual environment so that the user feels the virtual object is in his environment. With the use of the Android platform, this application makes the learning method more interesting because it can be used on various Android smartphones so that learning can be done anytime and anywhere. The methodology used in making applications is Multimedia Lifecycle, along with C # language for AR programming and flowchart as a modelling tool. The results of research on some users stated that this application can run well and can be used as an alternative way of learning Astronomy with more interesting.

1. Introduction

Astronomy as a branch of science involving the observation of celestial bodies is very interesting to learn. However the learning media used today is less interesting; just using a book containing 2D astronomical images. While other learning media using telescopes is quite expensive so this research will use Augmented Reality technology as one of technology in introducing astronomy interestingly.

There are several studies on astronomy using Augmented Reality technology. Soga et al. [1] in 2008 already did a research using Augmented Reality in learning Star. The result of this research stated that the learning process is more interesting. Another research also comes from Fleck and Simon [2]. In their research, they also using Augmented Reality to create astronomy environment. They compared the physical mode and the AR environment in learning astronomy for elementary grades. The result concluded that learning astronomy in AR environment significantly improved the students' way of learning astronomy.

In Indonesia, there is a similar research related to AR and astronomy. For example the research of Sagita and Amalia [3] on Learning of Solar System Using Augmented Reality Technology. This study focuses on the planets that exist in the solar system. Planetary objects are displayed in 3D using markers. When a marker is detected, a 3D object will appear along with information about the planet in text, animation and sound. The weakness in this study lies in the sensitivity of the marker when it is not detected by the camera.

A similar study was also found by Sujati et al. [4] that makes apps about "Multimedia Application Development for NASA Astronomy Satellite Learning with Augmented Reality technology based on Android". This research focuses on creating applications with satellite objects that combine multimedia technology and Augmented Reality on the Android operating system so that applications are easy to use anywhere. Just like the previous research, this application also has the same weakness that is light sensitivity to marker detection. Both of these studies also cannot interact with the 3D object.

Based on these two studies, this research also will make a similar research using Augmented Reality technology on the Android operating system with objects of celestial bodies (Milky Way galaxy and constellations). This research is expected to enrich the learning media for astronomy and make learning astronomy more interesting because there is an interaction with the 3D object.

2. Literature Review

2.1. Research Method

The research method used is Literature Review, by studying several applications and/or similar research, Milky Way galaxy theory and constellation, how to make 3D object from various sources and methodology of multimedia life cycle for making the application. The design in this study using flowchart, followed by making 3D galaxy object and constellation using Cinema 4D, designing marker and then making program using Unity.

2.2. Augmented Reality

Cushnan and Habbak [5] explain that Augmented Reality (AR) is a technology that allows the addition of virtual content to the real world. This is usually associated with the addition of 3D content for direct view of the camera. Then Suryawinata [6] found the same definition of Augmented Reality (AR) which is a combination of virtual world (virtual) and real world made by computer. Virtual objects can be text, animation, 3D models or videos that are combined with the actual environment so that the user feels the virtual object is in his environment. AR is a new and fun way in which humans interact with computers, because it can bring virtual objects into the user's environment, providing a natural and fun visualization experience.

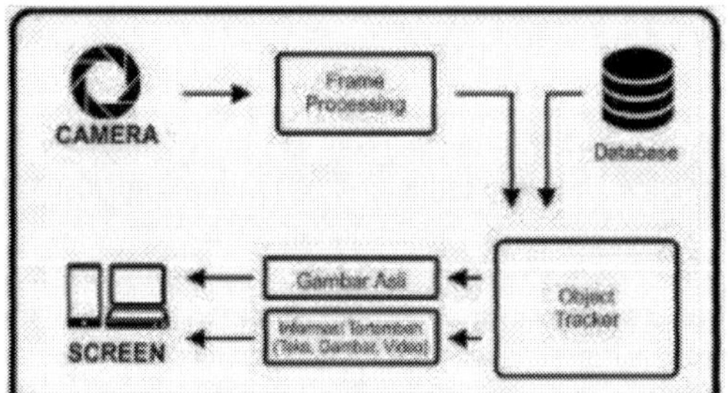

Figure 1. Workflow of AR

Figure 1 explains the workflow of Augmented Reality. It starts from camera to capture the image on the marker, then processes it to the database and if it is valid, the real picture will appear.

The AR application workflow generally begins with the capture of a marker image from the camera or webcam. The marker is recognized by features owned, then tracked by the object tracker provided by the Software Development Kit (SDK). But earlier, the marker has been registered and stored into the database.

The tracker object will track and match the marker in order to display the appropriate information. The result of marker detection is immediately displayed into the screen. The information displayed is attached to the corresponding marker in real time [7].

2.3. Vuforia
According to Fernando [8], Vuforia is a software for augmented reality developed by Qualcomm, which uses a consistent source of computer vision that focuses on image recognition. Vuforia has many features and capabilities, which can help developers to realize their thinking without any technical limitations.

With support for iOS, Android and Unity 3D, the Vuforia platform supports developers to create applications that can be used on virtually all smart phones and tablets.

2.4. Unity
According to Cushnan and Habbak [5], Unity is a cross-platform game engine developed by Unity. This game engine has been built with IDE and the ability to run on multiple platforms. Unity is designed for user convenience and high productivity.

The greatest strength of Unity is its ability to run on a number of large platforms with ease and some changes to the project structure. Unity can be run on Window, OS X, iOS, Android, Web Plugin, Flash, Xbox 360, Playstation 3 and Wii U.

Unity allows three languages to write scripts, available languages are Javascript, C #, and Boo.

2.5. Cinema 4D
Aaron [10] stated that Cinema 4D is a 3D application created by Maxon Computing. This app is relatively easy to learn and use. However, this app also has a number of powerful features that can be used to create dynamic images and animations that are dynamic and interesting quickly and efficiently. Cinema 4D has an intuitive interface and if you have questions about the features provided, the help documentation can be accessed easily and widely.

Cinema 4D has a long list of advanced features, including modeling, texturing, animating, lighting, and rendering. Cinema 4D also matches other applications, both 2D and 3D.

2.6. Related Article
Soga et al. [1] already did a research using Augmented Reality in learning Star. They made an interactive learning using finger pointing and augmented reality. They developed an environment to learn star using magnetic sensors. The environment had three functions. First function was star name. When a learner pointed out a star with the finger sensor, the system told the star name and constellation with voice. Second function was star navigator. If the learner know the name of a star, the system showed the path from current pointing position to the target star for the learner if the leaner did not know where it was in the real sky,. Third function was constellation tutor. By doing this, the learning process is more interesting. The similar research also comes from Fleck and Simon [2]. In their research, they also using Augmented Reality to learn astronomy. They stated that AR environment is particularly suitable for astronomy learning compared to the physical one. Furthermore, they said that using Augmented Reality to learn astronomy significantly improved the students' way of learning astronomy.

3. Results and Discussion
Based on the previous researches, this research also will using Augmented Reality for learning astronomy. This research focus on how to make the application for supporting astronomy learner to learn astronomy more attractive and interesting. As mentioned earlier, AR application creation will use Multimedia Lifecycle method. This method is suitable for projects that have multimedia elements. According to Binanto [11] quoted from Luther (1994), multimedia lifecycle method consists of six stages, namely concept, design, collecting materials, assembly, testing and distribution. These six

stages are not necessarily sequential in practice, they can be exchanged positions. However, the concept stage should be the first thing to do.

a. Concept

This application is made for all users especially those with an interest in astronomy. The materials to be displayed in this application include the Milky Way galaxy and some constellations in the form of 3D objects on the mobile phone screen. In addition, 3D objects can interact through touch to bring up sound and information.

The existence of this application makes learning astronomy interesting and the use of learning media more sophisticated because it uses AR technology and combines multimedia elements.

b. Design

It is the stage of creating a navigation structure, a story board to describe the description of each scene, by listing all the multimedia objects and links to other scenes and flowcharts to illustrate the flow from one scene to another.

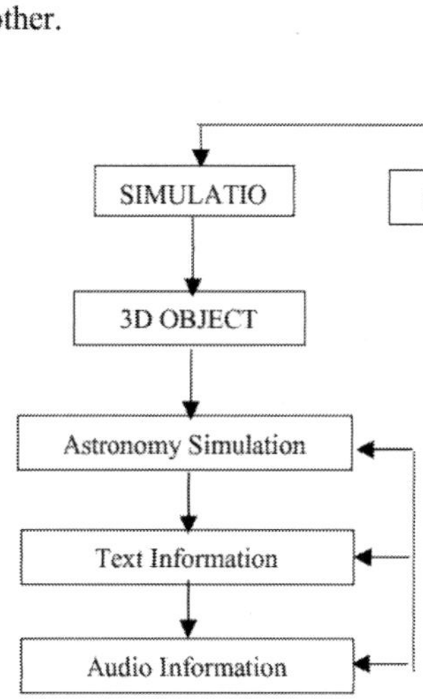

Figure 2. A Navigation Structure

This figure 2 describes the navigation for using the application. When we run the application, there are 3 menus: start the simulation, how to use the application and exit. On the first menu, there is an astronomy object and if you play the object, there will be a simulation, information of the object in text and voice.

Based on this structure, we can draw the flowchart of this application as follow:

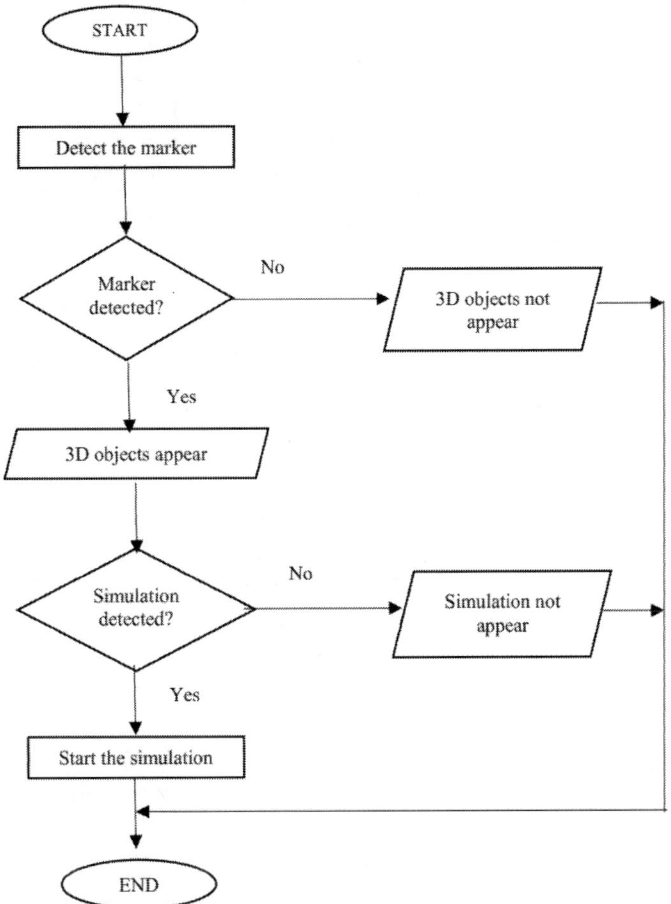

Figure 3. Flowchart

Figure 3 shows the flow of the application when it detect the marker. If the marker detected, then the 3D object will appear directly and also the simulation of the object will be started. Otherwise, the position of the marker should be arranged again.

Story boards describe the interface of the application, which is used as a link between one page and another. Here are some examples of designed interfaces:

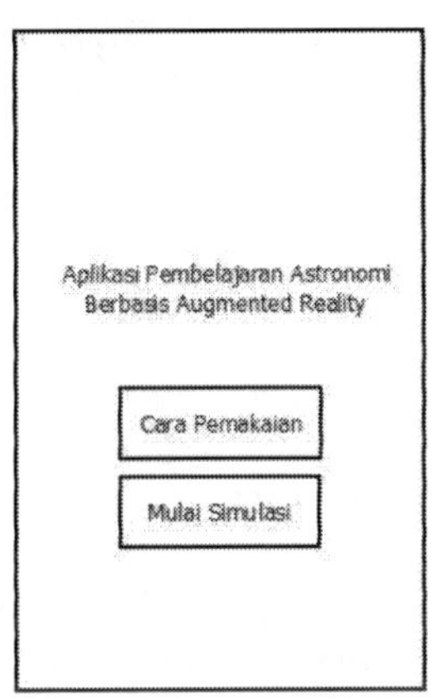

Figure 4. Introduction

This is the initial view of the application (figure 4). In this view there is the title of the application, how to use button to see how the application usage, and start the simulation button to start using the application. Next is the page to set up the camera.

Figure 5. Set up the camera

In this view (figure 5), there is a camera to receive input in order to display 3D objects in augmented reality.

c. Material Collecting

This stage describes the collection of materials / data required in making this application. Required data in the form of images or 3D objects of Milky Way galaxies and constellations and software to be used. The software used to create this Augmented Reality application is Unity, Vuforia SDK and Cinema 4D. The required software can be downloaded from the official website of the software while the images are obtained from various electronic and non electronic sources.

d. Assembly

Tools, materials and software used in making this learning application are as follows:

Table 1. Specification of Smartphone

Name	Specification
CPU	Dual Core 1 GHz
RAM	1 GB
GPU	PowerVR SGX540
Camera	3.15 MP
Operating System	Android 4.0 (*Ice Cream Sandwich*)
Display	1,024 x 600 (WSVGA)
Chipset	TI OMAP 4430
Sound	Stereo Speaker

Table 1 explains the hardware of the mobile phone that used in making the application. The operating system minimum requirement for the mobile phone is Ice Cream Sandwich (Android 4.0) for running the application.

Table 2. Hardware Specification

Name	Specification
CPU	Intel Core i5 3.20 GHz
RAM	4GB
VGA	NVidia GForce GTS 450
Web Cam	Logitec 2 MP
Operating System	*Windows* 7 64bit
Harddisk	1 TB

Table 2 explains the hardware computer that used in building the application. Higher specification requirements for computer should be provided because this application uses several 3D objects.

Table 3. Software Specification

Name	Use as
Unity	Making animation and AR
Cinema 4D	Making 3D object
Adobe Photoshop	Making the marker
Sound Recorder	Recording the sound

Table 3 only explains the software used in designing and making the application. There are Unity to make animation and AR, Cinema 4D to build the 3D object.

e. Testing

This stage aims to test learning applications built and meet all the needs of the application. Testing is done to ensure all functions run properly, markers can be detected and bugs in the application can be resolved. Testing on this application will use black box testing. After testing with black box, there are several factors causing marker not detected, such as camera distance, light intensity, and the marker itself.

Table 4. Testing

Activities	Expected Result	Actual Result
Run the application	Application runs well	As expected
Test the main menu	Main menu can be displayed	As expected
Test the marker	Marker can be detected	As expected
Test the constellations	Camera can read the marker and the 3D objects of the constellations appeared	As expected
Test the Milky Way	Camera can read the marker and the 3D objects of the Milky Way appeared	As expected
Test to the dark location	Marker can be detected	Marker cannot be detected
Test to the dark location using flash	Marker can be detected	As expected
Touch the 3D object	Information and voice appeared	As expected

There are some interfaces that tested as follows:

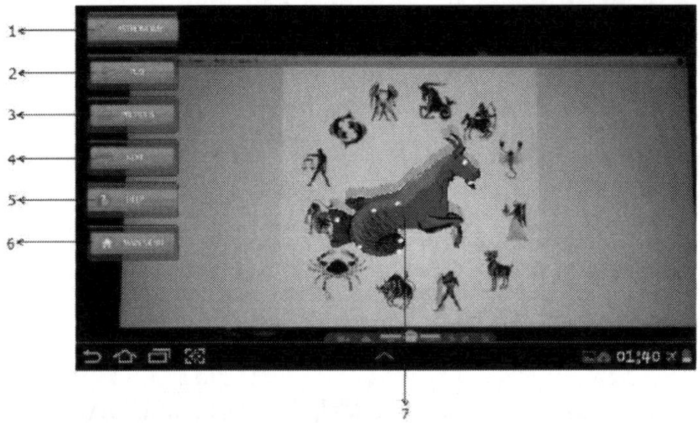

Figure 6. Constellations

Figure 6 describes the constellations object in the application. There are 12 zodiacs in the picture and each zodiac has its 3D object to appear. Information for each zodiac provides in figure 7. This application uses Bahasa Indonesia so the learner will easy to understand the information.

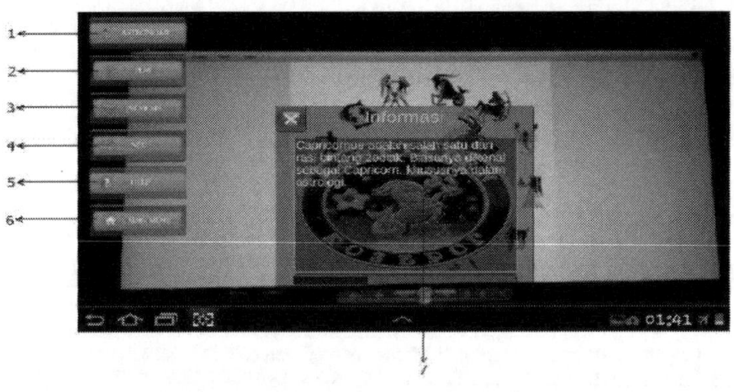

Figure 7. Information of Constellations

4. Conclusions

The making and testing of this learning application yields the following conclusions:

1. Augmented reality-based astronomy learning application using the android platform can draw interest to learn astronomy attractively.
2. This application help the learner to learn astronomy more interesting and attractive.
3. Augmented reality-based astronomy learning applications can complement existing learning media.
4. Light intensity determines marker detection.

References

[1] Soga M, Matsui K, Takaseki K, Tokoi K 2008 Interactive Learning Environment for Astronomy with Finger Pointing and Augmented Reality. Published in Advanced Learning Technologies, 2008. ICALT '08. *Eighth IEEE International Conference* on 1-5 July 2008,

Spain. Available at http://ieeexplore.ieee.org/document/4561760/
[2] Fleck S, Simon G 2013 *An Augmented Reality Environment for Astronomy Learning in Elementary Grades: An Exploratory Study. 25`eme conf'erence francophone sur l'Interaction Homme-Machine,* IHM'13, Nov 2013, Bordeaux, France. ACM
[3] Sagita S M, Amalia R 2014 Pembelajaran tata surya menggunakan teknologi *augmented reality Factor Exacta* **7(3)** 224-235
[4] Sujiati A D, Isnanto R R, and Martono K T 2016 Pengembangan aplikasi multimedia untuk pembelajaran satelit astronomi nasa dengan teknologi augmented reality berbasis android *J. Teknologi dan Sistem Komputer* **4(2)** 249-258
[5] Cushnan D and Habbak H E 2013 Developing AR games for iOS and android Packt: Publishing Brimingham
[6] Suryawinata B A 2010 *Jurnal pemanfaatan augmented reality dalam memvisualisasikan produk perumahan melalui internet* Vol. **I** No.2. Tersedia di: http://library.binus.ac.id/eColls/eJournal/48-bonny-PEMANFAATAN%20AUGMENTED%20REALITY.pdf. [Accessed 03 Feb 2015 : 12:03]
[7] Yudiantika A R 2013 Implementasi augmented reality di museum: studi awal perancangan aplikasi edukasi. *Konferensi Nasional Teknologi Informasi dan Komunikasi* **Vol I 4**
[8] Fernando M 2013 Membuat aplikasi android augmented reality menggunakan Vuforia SDK dan Unity. Manado: Buku AR ONLINE
[9] PTC Inc. 2011 *Vuforia developer guide* (online). Available: https://ui-dev2.vuforia.com/resources/devguide/getting-started
[10] Aaron K 2013 Instant cinema 4D starter. Packt: Publishing Birmingham
[11] Binanto I 2010 *Media Digital*-Dasar Teori dan Pengembangannya. Yogyakarta: Andi.

Study Orientation Ply of Fiberglass on Blade Salt Water Pump Windmill using Abaqus

B Badruzzaman* **and A Sifa**

Mechanical Engineering Department, Politeknik Negeri Indramayu, Indramayu, Indonesia

*badruzzaman@polindra.ac.id

Abstract. Windmill is one tool to generate energy from wind energy is converted into energy motion, salt production process still using traditional process by utilizing windmill to move sea water to salt field With a windmill driven water system, a horizontal axis type windmill with an average windmill height of 3-4 m, with a potential wind speed of 5-9 m / s, the amount of blade used for salt water pumps as much as 4 blades, one of the main factor of the windmill component is a blade, blade designed for the needs of a salt water pump by using fiberglass material. On layer orientation 0°,30°,45°,60° and 90° with layer number 10 and layer thickness 2 mm, the purpose of this study was to determine the strength of fiberglass that was influenced by the orientation of the layer, and to determine the orientation of fiberglass layer before making. This method used Finite Element Analysis method using ABAQUS, with homogenous and heterogeneous layer parameters. The simulation result shows the difference in von misses value at an angle of 0°, 30° ,45°,60° homogeneous value is greater than heterogeneous value, whereas in orientation 90 heterogeneous values have value 1,689e9 Pa, greater than homogenous 90 orientation value of 1,296e9 Pa.

1. Introduction

Windmill is one tool to generate energy from wind energy is converted into energy motion, Indramayu is one of salt producer in west java Indonesia, salt production process still using traditional process by utilizing windmill to move sea water to salt field with a windmill driven water system[1], a horizontal axis type windmill with an average windmill height of 3-4 m, with a potential wind speed of 5-9 m/s [2], the amount of blade used for salt water pumps as much as 4 blades, one of the main factors of the windmill component is a blade, a blade designed for the needs of a salt water pump by using fiberglass material.

Composite materials show different materials with materials in general, composite materials are strong fibers either in the form of continuous or noncontinuous surrounded by weak matrix materials. The matrix serves to distribute the fiber and also transmits the load to the fiber. Composite materials are not new, composite materials have been used since time immemorial. Composites have also been used to optimize the performance of some conventional weapons [3] [4] [5].

The purpose of this study was to determine the strength of fiberglass that was influenced by the orientation of the layer, and to determine the orientation of fiberglass layer before making. This method used Finite Element Analysis method using ABAQUS, with homogenous and heterogeneous layer parameters. In the blade-making process other than the thickness of the blade should be noted the effect of fiberglass layer orientation on the blade that aims to determine the strength of the blade when exposed to drag and lift force style. In this study, it is necessary to study the effect of fiberglass layer ornament on blade.

2. Method

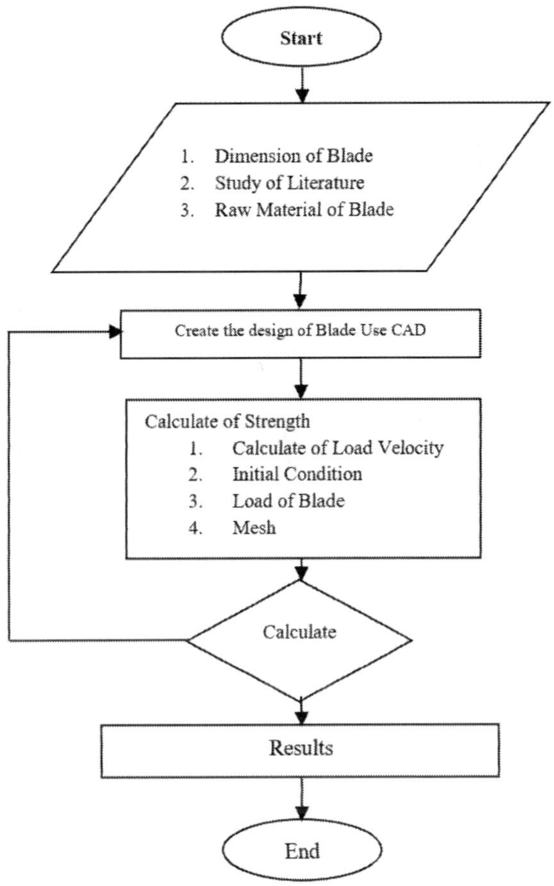

Figure 1. Flow Chart Methodology

The method of study in this paper is done by designing the blade first by using Computer Aied Design (CAD), then calculation of load due to wind load, then calculation simulation by loading and load parameters and material, simulation of blade strength calculation of orientation difference, Simulation of static strength of ply orientation is distinguished by homogeneous or heterogeneous type using ABAQUS software.

Table 1. Property Material [4]

Material	Density $\rho\ (\frac{kg}{m^3})$	Modulus Elasticities E (MPa)	Shear Modulus G (MPa)	Ratio Poison	Tensile strength $\sigma\ (MPa)$
E-Glass	2600	74000	30000	0.25	2500
Epoxy	1200	4500	1600	0.4	130

Table 2. Characteristics Mechanic *Fiberglass* [4]

Modulus Elasticity Longitudinal E_l (MPa)	Modulus Elasticity TranversalE$_t$ (MPa)	Shear Modulus G_{lt}(MPa)	Ratio Poisson (v_{lt})
46200	10309,6	3719	0,31

3. Results and Discussion

a. Dimension of Blade

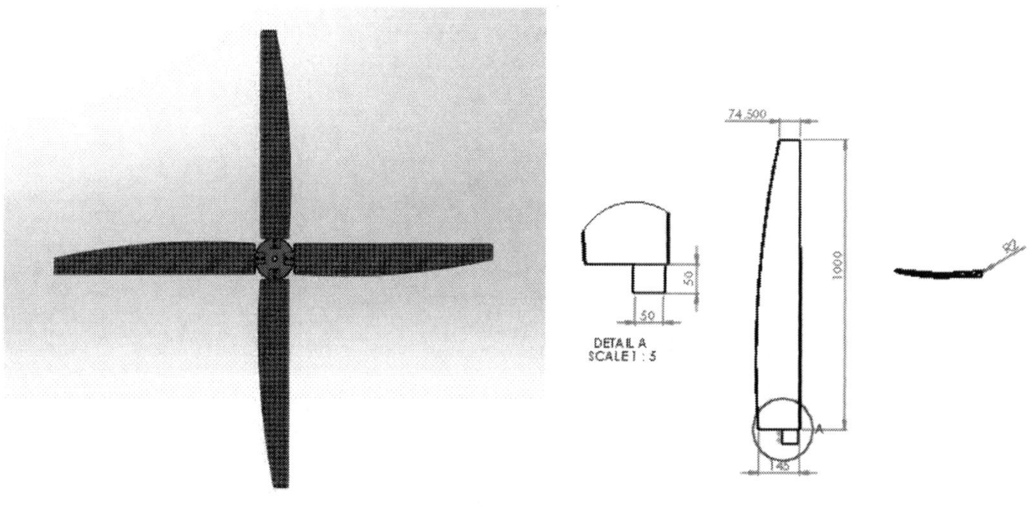

(a) (b)

Figure 2. Design of Blade

Table 3. Dimension [2]

Thickness (mm)	Material	Length (mm)	Wide (mm)	Ply
2	Fiberglass	1000	145	10

b. Initial Condition

Blade design that has been made with CAD surface planar type, can be known big blade area to be made, with blade area (A) 130766,64 mm^2. To find out the magnitude of the force that occurs due to the wind speed and from the blade has a coefficient drag (CD) 2,3 with a wind type of 1,2 kg / m^3,[5] it is possible to find a large drag force that occurs in the blade;

b.1 Minimal Velocity

$$F_{drag} = \frac{1}{2} CD.\rho.A.v^2 \qquad \text{Eq.1 [2]}$$

$$F_{drag} = \frac{1}{2} 2{,}3.1{,}2kg/m^3.\,130257{,}73\ \text{mm}^2.\,5^2 m/s$$

$$F_{drag} = 4{,}49\ kg \approx 5\ kg$$

b.2 Maximum Velocity

$$F_{drag} = \frac{1}{2} CD.\rho.A.v^2$$
$$F_{drag} = \frac{1}{2} 2{,}3.1{,}2kg/m^3.\,130257{,}73\ \text{mm}^2.\,9^2 m/s$$
$$F_{drag} = 11{,}56\ kg \approx 12\ kg$$

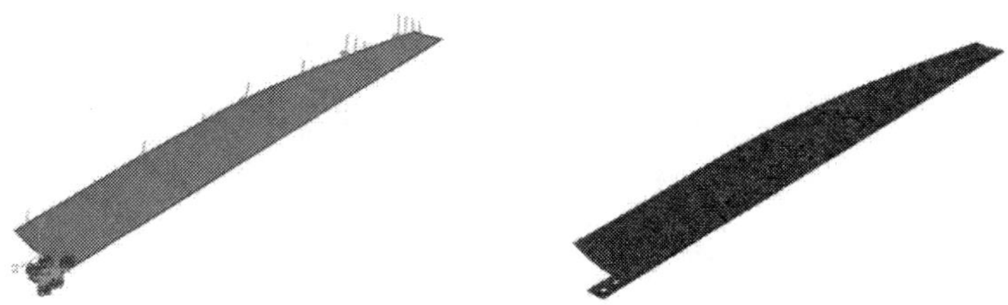

(a) (b)

Figure 3. Initial Condition

In Figure 3 shows the initial condition, (a) shows the load on the blade surface due to the drag wind load, (b) denotes the meshing blade before it is calculated by ABAQUS.

c. Homogenous

In the calculation simulation using ABAQUS software, orientation is made with homogenous, at an angle 0°, with a maximum loading parameter of 12 kg, the following results are obtained:

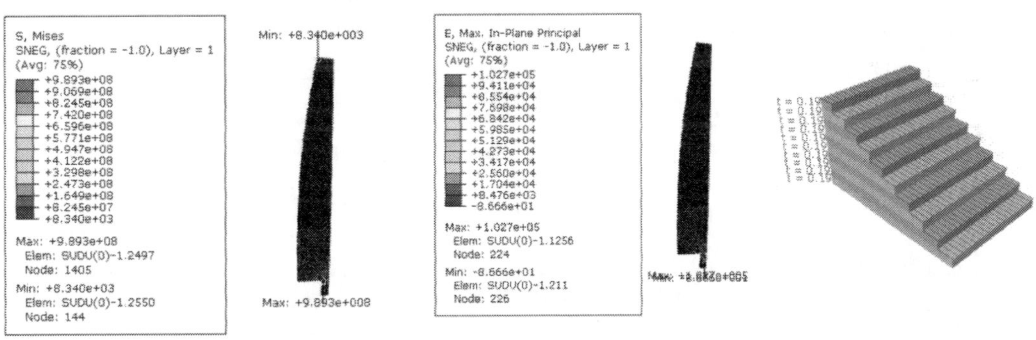

(a) (b) (c)

Figure 4. Results of Orientation 0°

Figure 4 shows the simulation results of the calculations at homogeneous orientation, (a) the von mises value occurs at the maximum conditions at the pedestal area of 9,893e8 Pa, (b) The value of the strain 1,027e5 Pa, (c) the orientation direction of the homogeneous layer on Angle 0°.

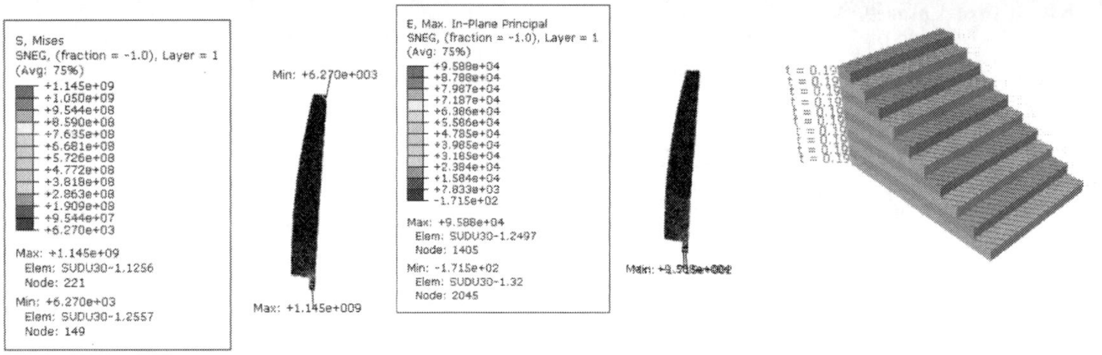

(a) (b) (c)

Figure 5. Results of Orientation 30°

Figure 5 shows the simulation results of the calculations at homogeneous orientation, (a) the von mises value occurs at the maximum conditions at the pedestal area of 1,145e9 Pa, (b) The value of the strain 9,588e4 Pa, (c) the orientation direction of the homogeneous layer on Angle 30°.

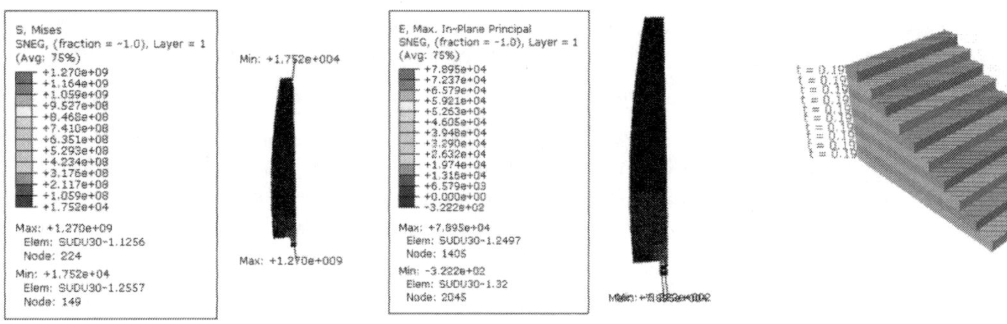

(a) (b) (c)

Figure 6. Results of Orientation 45°

Figure 6 shows the simulation results of the calculations at homogeneous orientation, (a) the von mises value occurs at the maximum conditions at the pedestal area of 1,145e9 Pa, (b) The value of the strain 9,588e4 Pa, (c) the orientation direction of the homogeneous layer on Angle 45°.

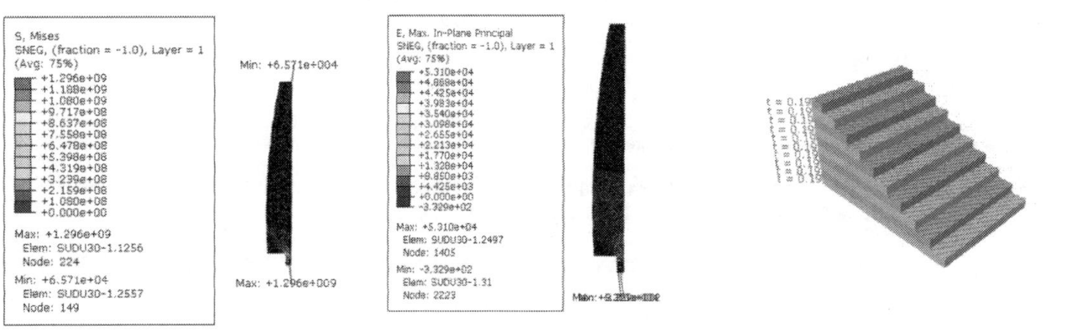

(a) (b) (c)

Figure 7. Results of Orientation 60°

Figure 7 shows the simulation results of the calculations at homogeneous orientation, (a) the von mises value occurs at the maximum conditions at the pedestal area of 1,296e9 Pa, (b) The value of the strain 5,310e4 Pa, (c) the orientation direction of the homogeneous layer on Angle 60°.

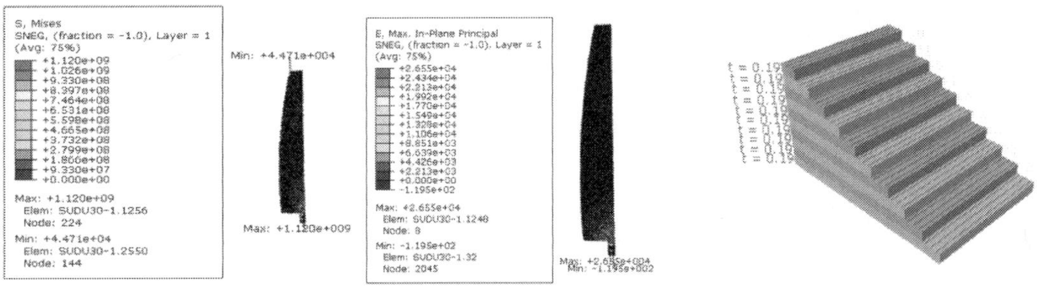

(a) (b) (c)

Figure 8. Results of Orientation 90°

Figure 8 shows the simulation results of the calculations at homogeneous orientation, (a) the von mises value occurs at the maximum conditions at the pedestal area of 1,120e9 Pa, (b) The value of the strain 2,655e4 Pa, (c) the orientation direction of the homogeneous layer on Angle 90°.

d. Heterogeneous

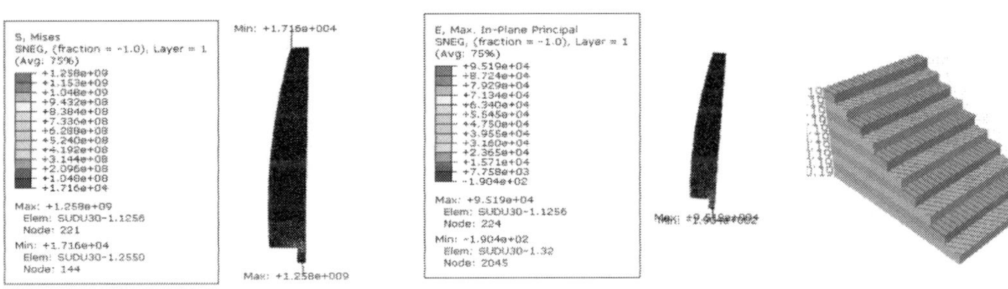

(a) (b) (c)

Figure 9. Results of Orientation 0°,30°

Figure 9 shows the simulation results of the calculations at heterogeneous orientation, (a) the von mises value occurs at the maximum conditions at the pedestal area of 1,120e9 Pa, (b) The value of the strain 2,655e4 Pa, (c) the orientation direction of the homogeneous layer on Angle 0°,30°

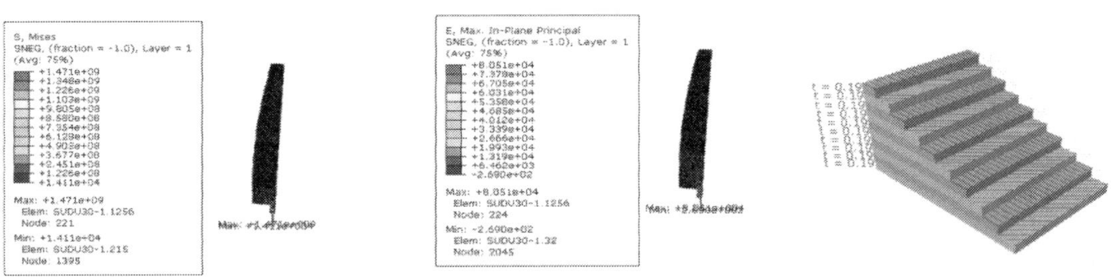

(a) (b) (c)

Figure 10. Results of Orientation 0°,45°

Figure 10 shows the simulation results of the calculations at heterogeneous orientation, (a) the von mises value occurs at the maximum conditions at the pedestal area of 1,471e9 Pa, (b) The value of the strain 8,051e4 Pa, (c) the orientation direction of the homogeneous layer on Angle 0°,45°

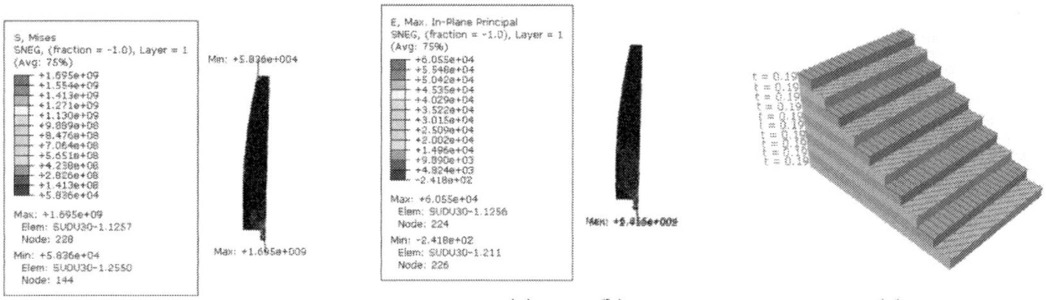

(a) (b) (c)

Figure 11. Results of Orientation 0°,60°

Figure 11 shows the simulation results of the calculations at heterogeneous orientation, (a) the von mises value occurs at the maximum conditions at the pedestal area of 1,695e9 Pa, (b) The value of the strain 6,055e4 Pa, (c) the orientation direction of the homogeneous layer on Angle 0°,60°

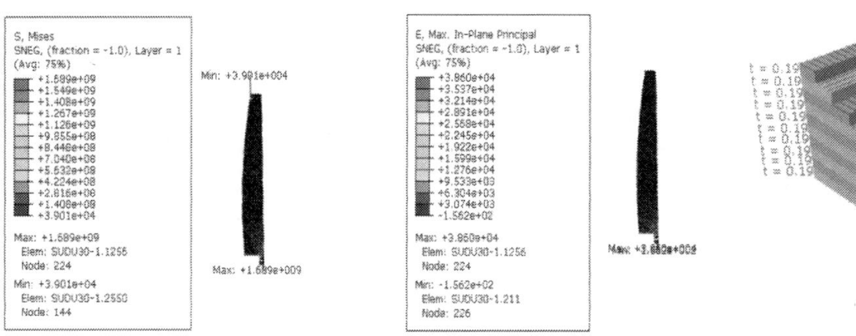

(a) (b) (c)

Figure 12. Results of Orientation 0°,90°

Figure 12 shows the simulation results of the calculations at heterogeneous orientation, (a) the von mises value occurs at the maximum conditions at the pedestal area of 1,689e9 Pa, (b) The value of the strain 3,860e4 Pa, (c) the orientation direction of the homogeneous layer on Angle 0°,90°

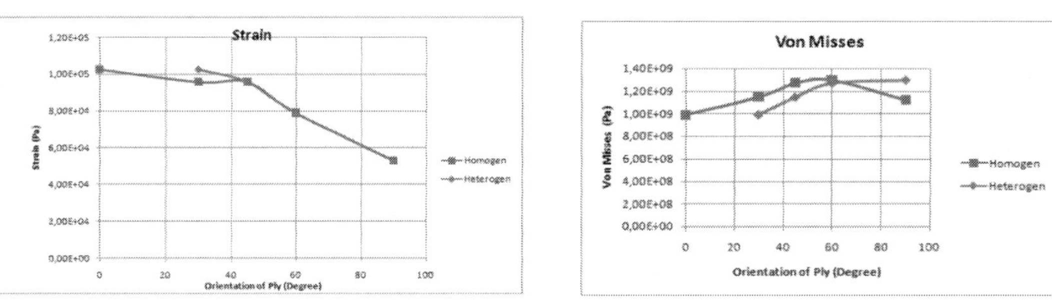

(a) (b)

Figure 13. Results Strain and Von Mises

Comparison of simulation results of strain and von mises calculation on homogenous and heterogeneous, in Figure 13 shows the graph of strain and von mises, (a) the value of strain shows at the angle 45°,60°,90° homogenous and heterogenous condition have the same value, while at 30° has a value different, heterogenous strain values, whereas (b) show the value of von mises at homogenous at oriented angle 0°,30°, 45° has von mises value greater than von mises value in heterogeneous orientation, whereas at the angle of 90°heterogenous von mises value more large compared to the value of homogenous von mises.

4. Conclusion

The result of simulation calculation using ABAQUS, on layer orientation 0,30,45,60 and 90 with layer number 10 and layer thickness 2 mm, in homogenous and heterogeneous condition show the value of strain shows at the angle 45°,60°,90° homogenous and heterogeneous conditions have the same value, while at 30 has a value Different, heterogeneous strain values, whereas (b) show the value of von mises homogeneous at oreinted angle 0°,30°, 45° has von mises value greater than von mises value in heterogeneous orientation, whereas at the angle of 90°heterogenous von mises value more large compared to the value of homogenous von mises.Accordingly in the layer 90° orientation has the largest von mises value under heterogeneous conditions and has the lowest strain value and the same as the homogenous strain value.

Acknowledgments

Directorate of Research and Community Service of the Ministry of Research, Technology and Higher Education and Director of State Polytechnic of Indramayu.

References

[1]. Petunjuk Teknis produksi garam menggunakan teknologi geomembran di kabupaten Indramayu http://diskanla.indramayu kab.go.id/accessed on 10 April 2017.

[2]. Sifa, Agus, et al. Ply Thickness Fiber Glass on Windmill Drive Salt Water Pump. In: IOP Conference Series: Materials Science and Engineering. IOP Publishing, 2016. p. 012024. Available from : http://iopscience.iop.org/article/10.1088/1757-899X/128/1/012024/meta DOI : 10.1088/1757-899X/128/1/012024

[3]. Peter Grand dan Carl. Q Roussean. Composite Structure (USA: Theory and Practice, ASTM International), 2000.

[4]. Daniel Gay. Composite Materials Design and Applications, Editions Hermes (California, USA) 2003.

[5]. The Drag Coeficient, Glenn Research Center, National Aeronautics and Space Administration, accessed on 30 Agustus 2017 www.grc.nasa.gov

ADDIE Model Application Promoting Interactive Multimedia

B Baharuddin*

Universitas Negeri Medan, Indonesia

*baharuddin.bah@gmail.com

Abstract. This paper presents the benefits of interactive learning in a vocational high school, which is developed by Research and Developmet (R&D) method. The questionnaires, documentations, and instrument tests are used to obtain data and it is analyzed by descriptive statistic. The results show the students' competence is generated up to 80.00 %, and the subject matter aspects of the content is up to 90.00 %. The learning outcomes average is 85. This type media fulfils the proposed objective which can enhance the learning outcome.

1. Introduction

Nowadays Indonesia needed increasing the high quality of the performance of the graduates of the vocational high school by improving the learning quality with providing a new learning paradigm [1]. Some problems faced in this case including the limitation media.

All this time the print media is still often used in learning, because of the quite easy and widely known in the world of education, as well as easy to develop and to find the various sources. It becomes the weakness because of the highly dependent on the verbal symbol.

The development of instructional media is very important to overcome these limitations which is can be designed by educators themselves according to the defined goals. It is expected to create an atmosphere of learning and the learning experience more meaningful, and to enrich the student learning experience. Thus the passive and boring of the learning atmosphere can turn into active participation and then the students to be more interactive. The success of the learning in the vocational education determine the successful of them in the work. Therefore it is important to build the media which can be to fulfill its which can be used to deliver messages from the sender to the receiver to stimulate their thoughts, feelings, interests and students' attention such as the learning process occurs [2], and make the learning process efficient and effective manner. So the learning material is accepted by students more quickly with an intact and attract students to learn more. It is as a tool for physical and non-physical form that is deliberately used as an intermediary between teachers and students in understanding the learning materials to more effectively and efficiently [3].

The Instructional design as a system and consider learning is systematically process. This systematic way of working always refer to the general stages of learning development system. The stages are analysis, design, development, implementation, and evaluation [4]. By designing the creative learning, it becomes innovative, interested, more iterative, more effective, and the quality learning which enhance the student learning [5]. The teaching and learning can be done anywhere and anytime, and the student's attitude and interest in learning can be enhanced [6].

The interactive multimedia is a multimedia equipped with a controller that can be operated by the user, so that it can be chosen what they wants to proceed. Its benefits are (1) the introduction of information and communication technology to students, (2) giving out new experiences and fun both teachers themselves and the students, (3) Pursue pace with the knowledge of science and technology in the field of education, (4) the use of multimedia can raise the learners' learning motivation, (5) Multimedia can be used to help learners to form mental models that will make it easier to understand a concept, (6) Following the development of science and technology [7]. It intended to assist educators in the delivery of material and help students involved in the learning process to understand the material being taught [8].

By designing the creative learning, the process is expected to be innovative, interested, more iterative, more effective, quality learning, and which can enhanced student learning, teaching and learning. And also it can be done anywhere and anytime. Furthermore the students' attitude and interest in learning can enhanced student learning [9]. The effectiveness of this media is based on (1) ease of navigation (2) the content of cognition. (3) the presentation of information, (4) the integration of media, (5) aesthetics (6) functions as a whole [10], and the changes in behavior are: (1) knowledge; (2) understanding; (3) habits ; (4) skills; (5) appreciation; (6) emotional; (7) social relations; (8) physical; (9) the ethical or moral character; (10) attitude. These were obtained through a learning process includes the overall change of the behavior on the understanding learning [7] [11]. Furthermore the interactive multimedia learning is develop in order to achieve an effective and efficient of learning in electrical installation on vocational school.

2. Methods

The stages of the development process done are (1) Gathering data from the field related to students needs by observation accompanied by the school had computer and internet, and also student and teacher have been familiar to use it; (2) Designing and manufacturing the media, where the results of the evaluation in each of phase of the development of learning can bring to the earlier phase. The final results from a phase become an initial product for the next phase. It is a bridge between learners, materials, and all forms of media based technology, which is based on the five learning process. Analyzing the process is to define what will be learned by the participants learn. It is done by needs assessment, identify problems, and perform analysis tasks. Therefore, the output will we generate is in the form of a characteristic or profile of potential participants learn, identify gaps, needs identification and analysis of the detailed tasks based on the need. In the design stage also is built blueprints which is said the formulating learning goals, namely specific, measurable, applicable, and realistic. Furthermore devise a test, where the test is based on the learning objectives that have been formulated earlier; (3) Implementing the model such as a concrete step in learning system. In this stage, all of them have been developed and installed or set in such a way appropriate to the role or function that could be implemented; (4) Evaluating is done to see if the system built a successful learning, in line with initial expectations or not. Actually, the evaluation phase can occur at any stage in the top four. Evaluation happens at each of four stages above it is called formative evaluation, because the goal of the revised requirements.

The questionnaires is used to find out data related to the useful and quality of the media, and the students' learning outcomes assessed by the tests. In these case is initiated the criteria in Table 1.

Table 1. Interval Assessment Criteria

No	Interval	Criteria
1	81 % $\leq$ score $\geq$ 100 %	Very Good or Strongly Agree
2	61 % $\leq$ score $\geq$ 80 %	Good or Disagree
3	41 % $\leq$ score $\geq$ 60 %	Enough
4	21 % $\leq$ score $\geq$ 40 %	Less good or less Agree
5	0 % $\leq$ score $\geq$ 20 %	No Good or Disagree

3. Results and Discussion

Based on the need assessment is found (a) the delivery of learning materials using the lecture makes the students less active so that students experiencing rapid saturation; (b) students expressed the need for the use of interactive media which is able to facilitate their students learning process in electrical lighting installation of the simple building, and also the media which can be used on the practice of a virtual so that they can understand theoritically and then reinforced by the practice of the material submitted, (c) students need to know the lighting electrical installation material of the simple buildings because these materials is the basic materials of the next instructional material.

Based on these the media is developed and packaged in the form of an interactive CD by the specifications of (a) the display of interest, (b) easy to use, (c) to be used in computer, (d) equipped with supporting software, (e) material according to the need to learn, and easy to understand. This media is made in simulations, animations, audio, video, and images.

It is considered to (a) the user; to facilitate the learning process and increase the interest and learning outcomes Vocational middle school students; (b) equipment output as an interactive CD which can be used independently by the students.; (c) figure, it is accompanied by a picture as the background and illustrations; (d) audio; to refresh students; (e) animation; students can see the illustrations move from material learned; (f) simulation and video; students can press the play button to run the video; (g) answers form; the answers provided in menu form of exercise in the learning media.

In the design and development of this CD interactive the following factors of major concern, namely: (1) the presentation of information, (2) the use of media, (3) ease of use, (4) expediency. The material contained in the media is still limited to a few sub competencies that remain intertwined, so although only consists of several sub competencies media still has a value of the knowledge needed to achieve the learning objectives in accordance with the curriculum. Limitations of this material due to time constraints in the study.

The material consists of (1) Instructions for use, (2) General purpose, (3) the material, (4) Simulation, (5) Application, (6) test. Once the design and content of the CD interactive has been completed, the next action is a concrete manifestation. In its' operations, the execution starts from choosing the background and putting the buttons in accordance with a predetermined design. This product is validated by media expert and subject matter experts.

The validation results of media experts and matter experts is noted respectively 80% (good), 95,58 (very good). It presents the development of these media have succeeded in achieving the ultimate goal of the research is to see whether the media response experts that can be developed or fit for using as a medium of learning.

Disseminating the media is done in two phase, namely (1) phase 1 involved 10 students of class XI Mechanical Power Installation in obtaining the information about the students' responses using the media in the learning process. Each student used the media and then give the comment to the used of media, by answering the provided questionnaire and take the test. Based on the analysis results of these is conducted the revision of the media. (2) phase 2 involve 29 students. In this stage is found the average of the students' comment is 85. It is made in revising the media. Based on the research done, and the results of a questionnaire distributed to media specialists and subject matter experts, the multimedia ratings experts expressed the assessment of aspects is a good guide , the following discussion are listed in the Table 2.

Table 2. Validation Results of expert Aspects

No.	Statement	Expert			Average	Category
		I	II	III		
1	Instructions for use application program	4	5	5	90%	Very Good
2	Ease of use program	4	5	5	90%	Very Good
3	Structure and practicality of application program	4	4	4	81%	Good

Aspects of assessment in the aspect of the guide is a manual application usage by the average value of 90.00 % with a very good category. The second point is the ease of use assessment program with an average value of 90.00 % with a very good category. The third point is the structure and practicability assessment application program with an average 81.00 % in both categories. Overall the average assessment of aspects of guidance is 88 % or in the excellent category. Expert assessment of the media states that media content is excellent as the Table 3.

Table 3. Results of Validation Experts on Aspects of Content

No.	Statement	Expert			Average	Category
		I	II	III		
1	Clarity of learning objectives statement	5	5	4	90.00%	Very Good
2	The suitability of the content / materials with the aim	4	5	4	87.00%	Very Good
3	The suitability of the content / materials with the curriculum used	4	4	4	80.00%	Good
4	Instructions and application guides the use of the content / materials	4	5	5	90.00%	Very Good
5	Exposure limits on the concept and content of materials	4	5	5	90.00%	Very Good
6	Giving examples and illustrations	3	5	5	87.00%	Very Good
7	Recency of the content / material presented	4	4	4	80.00%	Good
8	Ease understand the language used	4	4	5	87.00%	Very Good

In the aspect of the media content multiple ratings include clarity of statement of purpose with average value of 90.00 % in the excellent category , the suitability of the content with the aim to value 87.00 % in both categories , the suitability of the content with the curriculum used with 80 % in either category , user and application guidelines use the content with a value of 90.00 % in the excellent category , and exposure to concepts and restrictions on the content of the material , giving examples and illustrations as well as the recency of the content 87.00 % in the excellent category , as well as the ease of understanding the language used by the value of 87.00 % in the excellent category . Overall the average ratings aspects of content media was 88.00 % in the excellent category.

Assessment of the aspects of the exposure in the media is very good, as shown in Table 4.

Table 4. Validation Results expert on Aspects of Exposure

No.	Statement	Expert			Averag e	Category
		I	II	III		
1	Quality exposure to content / material presented	4	5	4	87.00%	Very Good
2	Mechanical and systematic presentation of the content / materials	3	4	5	80.00%	Good
3	The use of colours on screen display (screen)	4	4	5	87.00%	Very Good
4	The composition of the text used	4	5	4	87.00%	Very Good
5	The composition of graphics and images presented	3	5	4	80.00%	Good
6	Quality video/audio	4	4	5	87.00%	Very Good
7	Ease of understanding the information presented	4	5	4	87.00%	Very Good
8	The use of directional keypad / controller	4	4	5	87.00%	Very Good

Assessment aspects of exposure at each point is the quality of the exposure of the content/materials are presented as well as techniques and systematic presentation of the content with a value of 87.00% ;80% in both categories, the use of color on screen display (screen) and, composition of the text that is used with a value of 87.00 % in the excellent category, the arrangement of graphics and images are presented with 80 % in both categories, the quality of video/audio, the ease of understanding the information presented, use directional button/controller with a value of 87.00 % in the excellent category. Overall the average assessment of aspects of exposure was 85 % in the excellent category.

Based on the research that has been done, the effective learning media such as a medium of learning in subjects of lighting electrical installations on the simplified building is relatively easy and simple, just insert the CD via DVD or CD drive on the laptop or computer and displays via the projector screen or also via computer, following the practice directly as presented in the learning media. It is in line to the states of Ryan and Bagley (2015). It is also showed by the language which important to avoid the constraints in terms of language. In this case Indonesian language is used to guide using the media.

4. Conclusions

The investigation presents the interactive learning media developed is feasible and effective as a medium of learning, especially in the and also can improve the skills of teachers in making learning more interesting for students. So students more interested in learning, because they learn in a way that is more interesting.

References

[1] Rayandra and Asyhar 2011 *Kreatif Mengembangkan Media Pembelajaran* (Jakarta: GP Press)

[2] Deng, Liping 2015 Exploring university students' use of technologies beyond the formal learning context: A tale of two online platforms *Australasian Journal of Educational Technology,* **31**(3)

[3] Banas, Jennifer R 2014 Authentic learning exercises as a means to influence preservice teachers' technology integration self-efficacy and intentions to integrate technology *Australasian Journal of Educational Technology,* **30**(6)

[4] Dick W, Carey L J O, and Carey 2005 *The Systematic design of intruction Boston* (New York: Walter Dick and Lou Carey)

[5] Liaw et al 2007 Surveying instructor and learner attitudes toward elearning, *Computer and Education,* **49**(4)

[6] Thoutenhoofd et al 2015 From self-regulation to learning to learn: observations on the construction of self and learning, *British Educational Research Journal,* **41**(1), 72-84

[7] Niken Ariani dan Haryanto 2010 *Pembelajaran Multimedia di Sekolah* (Jakarta: Prestasi Pustaka)

[8] Broekman, Pauline van Mourik, *et al* 2014 Reviews *British Journal of Educational Technology,* **46**(1)

[9] Tsai et al 2001 Developing an Internet Attitude Scale for high school students *Computer and Education,* **37**(1), 41-51

[10] Wery, J, and Thomson, M M 2013 Motivational strategies to enhance effective learning in teaching struggling students *Support for Learning,* **28**(3), 103-108

[11] Benny A Priyadi 2009 *Model-Model Desain Sistem Pembelejaran* (Jakarta: PPS-UNJ)

Characteristics from Recycled of Zinc Anode used as a Corrosion Preventing Material on Board Ship

B Barokah[1], S Semin[2*], D D Kaligis[1], J Huwae[1], M Z Fanani[1] and P T D Rompas[3]

[1]Politeknik Kelautan dan Perikanan Bitung 95226, North Sulawesi, Indonesia
[2]Departement of Marine Engineering, Institut Teknologi Sepuluh Nopember Surabaya 60111, Indonesia
[3]Universitas Negeri Manado, Tondano 95618, Sulawesi Utara, Indonesia

*semin@its.ac.id

Abstract. The objective of this research is to obtain the values of chemical composition, electrochemical potential and electrochemical efficiency. Methods used were experiment with physical tests conducted in metallurgical laboratory and DNV-RP-B401 cathode protection design DNV (Det Norske Veritas) standard. The results showed that the composition of chemical as Zinc (Zn), Aluminium, Cadmium, Plumbumb, Copper and Indium is suitable of standard. The values of electrochemical potential and electrochemical efficiency were respectively. However it can be concluded that the normal meaning of recycled zinc anode with increasing melting temperature can produce zinc anode better than original zinc anode and can be used as cathode protection on board ships. This research can assist in the management of used zinc anode waste, the supply of zinc anodes for consumers at relatively low prices, and recommendations of using zinc anodes for the prevention of corrosion on board ship.

1. Introduction

Corrosion is a degradation of a substance or its nature due to its interaction with the environment. The corrosion mechanism is very damaging to materials and equipment in the industrial structure [1]. Corrosion is assumed to be the result of three sequential processes: degradation of the paint layer, pitting point formation, and pitting point progress. Corrosion rate can be identified quantitatively. Modeling probabilistic corrosion by analyzing existing data collected from plate thickness measurements. Data is verified by comparing the approximate behavior of corrosion and dispersion progress with actual data [2].

The corrosion rate of the sacrificial anodes of Al-Zn-In-Mg-Ti in sea water is higher than that of sea mud [3]. Many types of corrosion such as pitting corrosion is a material failure mechanism of a component that occurs due to an aggressive environmental influence, this is indicated by the presence of chloride and sulphide that accelerate the corrosion process above 50-55°C [4].

Steel is a metal that is widely used in marine ship structures that are strongly influenced by moisture and atmospheric ocean temperatures resulting in extremely vulnerable corrosion. So it is necessary to prevent by using sacrificial anode to reduce the corrosion rate. The marine environment is a corrosive environment which is as a factor that greatly affects metal degradation. Corrosion can occur continuously if there is no corrosion precaution on the metal [5]. Treatment needs to be done to minimize corrosion. There are several methods to protect material against corrosion commonly known as cathodic protection method (CP) and cathodic prevention (CPrev) [6]. Voltage generated by the anode can function as an antifouling on the hull of the ship [7].

Generally the conventional anode used as a sacrificial anode on board is the Aluminum-Zinc-Indium (Al-Zn-In) alloy. This type of alloy has high efficiency and high potential protection in the seawater environment [8]. Various types of sacrificial anodes such as Al-Zn-In-Mg-Ti have the same relative dissolution. Several factors affecting corrosion rates are conductivity and dissolved oxygen in seawater [3]. When electrochemical testing is applied, sampling for testing shall be carried out for each heat produced, and has to be standardized. Unless agreed by others, the testing shall be carried out according to the procedure in standard and the following acceptance criteria shall apply for Zinc based anodes electrochemical capacity: minimum 780 Ah/kg, closed circuit potential: ≤ -1.00 V [9].

Zinc is a type of metal widely used as a sacrificial anode to protect iron from rust as well as aluminum and magnesium. The working principle of the anode is metals such as zinc, aluminum, and magnesium, because it is more reactive to the carat element and has a lower oxidation point than the iron so that when mounted on other metals such as iron, it first reacts to the rust and neutralize the rust before attacking the iron. Anode can work because it is made of zinc, aluminium, and magnesium metal which is more active against corrosion than iron so that if installed adjacent to the arrangement based on the correct calculation, then the rust will go to the anode first. Due to its inherent nature, after a certain period the anode should be replaced to ensure maximum protection. The zinc anode work will increase naturally to a maximum level to protect the hull from corrosion caused by the high salinity content of seawater present in the ocean waters [10]. Recycling used products from metal to be used again has been done in many western countries [11]. Recycling allows the recovery of metal parts, reducing the resource requirements of the natural environment. Recycling will save natural resources, reduce pollute, and save energy [12].

2. Methods

The method used in this research is experimental method that goes through several stages: conducting test of making zinc anode by recycling used zinc anode; laboratory testing of recycled zinc anode product to obtain value of chemical composition, electrochemical potential value and electrochemical efficiency value. Tools and materials such as: Furnace, Brander, Cast/Mattress, Thermometer, Scales, Iron strip, Rubber gasket, Zinc anode, Calliper, Camera, ATK, Grinder, Epoxy, and Scissors/cutter.

Data was obtained in the form of quantitative data from laboratory test results include chemical composition, electrochemical potential value (V), and electrochemical efficiency value (Ah/kg) analyzed and compared with DNV standard.

3. Results and Discussion

This research started with the manufacture of zinc anode products using a zinc anode material. There are several stages on how to recycle, among others:

3.1. Preparing materials to be recycled (zinc anode) used

This material is obtained from zinc disposal waste from vessels that is docking. In accordance with the provisions of SOLAS and the provisions of the Classification Bureau of the vessel, which performs the annual dock, must perform zinc anode replacement.

3.2. Preparing the smelting equipment (furnaces, steel pans, fuel, scoop, tongs and other equipment)

This equipment must be available before starting the smelting work to make the smelting work process runs smoothly and safely.

3.3. Set up printing equipment

Mattress used for this research is sand casting, by sprinkling the powder on the mattress and then fill in the sand and solidify it by adjusting the shape of the mat. The compacting is done evenly to get the desired result

3.4. Melting Process
The melting of the material is carried out on a steel skillet above the furnace by directing the combustion directly on the material to be melted. Fuel used by LPG gas. The heated material melts at a temperature of 609 ° C in approximately 30 minutes and then stirs well and separates the dirt that floats on the surface of the already melted material.

3.5. Pouring
Pouring is done with a careful heart, by inserting the liquid material in the sand casting hole. The material liquid is put into the cast until it is full.

3.6. Opening the cast
After the material is put into the cast, the ingredients will be cooled for about 5 minutes, then the cast can be dismantled to retrieve the printed zinc anode results.

3.7. Finishing
The end of the zinc anode recycling casting work is to do the finishing including checking and smoothing.

Zinc anode testing was conducted in corrosion laboratory and material failure of Ten November Institute of Technology Surabaya. The tests were performed using DNV ((Det Norske Verity) standard to test the material characteristics [9].

3.8. Testing the chemical composition that is contained in the zinc anode.
The compositions tested include Zn (Zinc), Al (Aluminium), In (Indium), Cd (Cadmium), Fe (Iron / Ferro), Cu (Copper), Pb (Led / Plumbumb).

From the tests performed, the results are as listed in the table below.

Table 1. Comparison between DNV standard compositions with laboratory test results

Chemical Composition	Result (%Wt) Std DNV	Result (%Wt) Lab
Zn	rem	99.300
Al	0.10-0.50	0.333
In	na	0.002
Cd	≤0.07	0.029
Fe	≤ 0.005	< 0.002
Cu	≤0.005	0.021
Pb	≤0.006	0.015

Table 1 illustrates that the Zn composition of the tested product is 99.300% Wt and this proves the Zinc compound is excellent. Aluminium Composition 0.333% Wt is in the range between 0.10-0.50% Wt indicating under normal conditions. Indium composition is not specified on the standard while the results of the lab test are 0.002% Wt. Standard Composition Cadmium ≤ 0.07% Wt is 0.029% Wt in accordance with the provisions. Copper Composition Standard ≤ 0.005% Wt turns out in the laboratory test 0.021% Wt as well as the standard value of Plumbumb composition ≤ 0.006% Wt on laboratory test 0.015% Wt. Of the eight chemical compositions, Copper has a difference of 0.016% Wt and Plumbumb 0.009% Wt. Since both are the last compositions of relatively small values in Zinc Anode material, they are still categorized as normal.

3.9. Potential Electrochemical Testing and Electrochemical Efficiency
The results of testing the characteristics of Electrochemical Potential and Electrochemical Efficiency are as in Table 6.

Table 2. Comparison of Potential Electrochemical and Electrochemical Efficiency between DNV standard and Laboratory test results

Characteristics	DNV Standard	Test in Lab
Electrochemical Potential (V)	≤ -1.00	-1.07
Electrochemical Efficiency (Ah/kg)	780	842

Referring to the DNV standard that the required potential electrochemical quantity is ≤ -1.00, while the test result shows -1.07, it means that the electrochemical potential in the recycled zinc anode is adequate (or qualified). Similarly, electrochemical efficiency at a minimum standard is 780 whereas in a recycled zinc anode shows 842, it can be defined that it is adequate with the required standard.

4. Conclusions

Process for the zinc anode by recycling zinc anode former takes approximately 30 minutes and a melting temperature of 609 ° C to obtain a higher percentage of remaining zinc 99.30%. The results of laboratory tests on the chemical composition performed in the laboratory of corrosion and material failure Institute of Technology (ITS) Surabaya, is normal, which means that the recycled zinc anode can be used as protection / cathode protection in preventing corrosion on ships.

References

[1] Singh R 2014 *Corrosion control for offshore structures, cathodic protection and high efficiency coating* First Edition (Elsevier BV) 131
[2] Yamamoto N and Ikegami K 1998 A study on the degradation of coating and corrosion of ship's hull based on the probabilistic approach *Eng.ASME* **120** 121
[3] Wan H and Huangfu W 2017 Influence of sea mud state on the anodic behavior of Al-Zn-In-Mg-Ti sacrificial anode **136** 11
[4] Vazdirvanidis A and Pantazopoulos 2017 Corrosion investigation of stainless steel water pump components *Engineering Failure Analysis* **82** 466
[5] Shengping Q and Weicheng C 2003 Effect of corrosion models on the time-dependent reliability of steel plated elements *Marine Structure* **16** 15
[6] Carmona J and Garces P 2015 Efficiency of a conductive cement-based anodic system for the application of cathodic protection, cathodic prevention and electrochemical chloride extraction to control corrosion in reinforced concrete structures *Corrosion Science* **96** 102
[7] Jia M Y and Zhang Z M 2017 PANI-PMMA as cathodic electrode material and its application in cathodic polarization antifouling *Electrochemistry Communication* **84** 57
[8] Ferdian D and Pratesa Y 2017 Development of Al-Zn-Cu Alloy for low voltage aluminum sacrificial *Procedea Engineering* **184** 418
[9] Det Norske Veritas 2010 *Recommended practice Dnv-Rp-B401 cathodic protection design* (Norway - Det Norske Veritas) 27
[10] Fleury P 2014 *Performance Metal Product*. Bechtelsville : Sci Cast Company
[11] Spengler T and Püchert H 1997 Environmental integrated production and recycling management *European Jurnal of Opreational Research* **97** 308
[12] Rosebrock C and Bracke S 2017 Comparison and evaluation of concepts for a future recycling process regarding attrited lead anodes used in the electrowinning process of non-ferrous metals *Procedia Manufacturing* **8** 627-634

IOP Conf. Series: Materials Science and Engineering **306** (2018) 012022 doi:10.1088/1757-899X/306/1/012022

How do the Polytechnic Students Cope with the Difficulties in Composing Abstracts for Their Final Projects?

C Niswatin[1], M A Latief[2] and S Suharyadi[2]

[1]Politeknik Kota Malang, Indonesia
[2]Universitas Negeri Malang, Indonesia

*choirun.niswatin@poltekom.ac.id

Abstract. This research aims to uncover the fact about engineering students in dealing with composing abstracts for their final projects. The research applies a descriptive qualitative quantitative design. The data were collected through questioners involving 104 engineering students, including the alumni at Politeknik Kota Malang, Indonesia. Furthermore, interviews were carried out to explain the details where necessary to support the primary data. It is found that the common problems faced by engineering students include 1) combining words into sentences, 2) identifying the most appropriate technical terms in engineering, and 3) applying grammar in context. To cope with those difficulties they demanded translation application machines, supported by peer-proofreaders. In addition, they considerably engaged personal tutoring with the lectures more than three times.

1. Introduction

English proficiency of Indonesian tends to be credited to low level in general. Empirical studies which were carried out by Indonesian scholars revealed that the higher education outcomes were still poor [1] [2]. Moreover, English Testing Service (ETS) reported that the average of test takers of Indonesian ranked the lowest out of 50 countries [3]. Meanwhile, English First (EF) has listed that Indonesia ranked the 32th of 72 countries all over the world and marked 52.94 in English Proficiency Index Ranking (EPI) [4]. Statistically, test takers performance by IELTS partners reported that the general band score of Indonesia was 6.3 out of 9, but the writing was likely to be lower than six [5].

Those phenomena seemed to have a significant impact to the scarcity of the number of Indonesian scientific publication in the worldwide. However, international publication has been conquered by the English language for the last decades [6].

As a matter of fact, the preliminary research on 100 abstracts written by engineering students of Politeknik Kota Malang found that only 80% containing of aims and 88% of them provided the methods. However, the students seemed to overlook the result of their research because it was only 43%. In addition, more than half of those selected abstracts on purpose (67%) were found 582 errors. The errors were ranging from *admission* type (6.54%), *omission* (11.51%), misordering (12.71%), misinformation (22.16%), and miscellaneous (47.08%).

Therefore, this paper purposes to quote the verity of how college students cope with the difficulties in composing academic writing, particularly in writing abstracts of their final projects. It is expected that their final projects will globally recognized since their abstracts are written in grammatically correct sentences. This could also increase the quantity and quality of Indonesian scholars' publication, as well as the researchers' publication in the international level. This will open a wide partnership in conducting the projects as well.

To disclose the engineering students' experience in writing abstract particularly their ways to cope with the difficulties, these questions are broken down as follows:

1) What are the engineering students' major problems in composing their abstracts?
2) How do the engineering students cope with their difficulties in composing the abstracts?

2. Literature Review

To pursue a certain academic degree, students must submit their scientific report known as a thesis or dissertation or the like referring to final project. Without it, one may fail to achieve his study. It is one of the requirements to pursue the Diploma 3 program at Politeknik Kota Malang (Here after refers to POLTEKOM) which is generally given at the last semester. For some reasons, some departments offer it at the fifth semester. The final project covers three aspects of assessments: engineering products, book of report commonly known as final project, and oral examination to defend their product and report. The engineering products could be in the form of designing tools, creating machines, applying engineering formula or software application, etc. In addition, these products must be well-written into a book of final assignment. It contains five chapters; introduction, review of literature, research method, findings and discussion, and conclusion. These will be presented orally in front of the examiners. Each of the students is obliged to submit the first two requirements. If the individual fails to complete one of those, he fails. It does not matter how good it is. However, if he fails in presentation, he still gets one more chance to do so.

Abstract is a concise short synopsis of scientific works covering the main parts of it, i.e. background, method, result, and conclusion. A Poor abstract neglects the authors' crucial message and usually provides the ambiguity [7]. On the other hand, it must convey the prominent information representing the whole picture of papers.

The abstract is the main part of the paper which is the main target for the general readers before thoroughly reading all information that they need. It is constantly submitted to publication before the whole papers.

Generally, publication requires a certain format such as word count, font type and size, indents and spacing, single paragraph or more, bilingual, etc. The abstract of the final assignment at POLTEKOM must be written in both languages, Indonesian and English. The word count should be not more than 300 words, the fonts and spaces are typed in smaller size than other parts of papers, and the paragraph is indented.

Apart of abstract, the whole paper should be well written, grammatically correct, and show the clear idea. As a matter of fact, the most common problem of rejected paper is the problems with the writing. In other words, writing is the major issue of academic writing and is claimed to be the tremendously discouraging skill. Some research findings showed that many ESL students had got difficulties and stress on their academic writing [8]. The difficulties were identified to several problems with; among of them are; differing between spoken and written language, composing grammatically correct sentences, diction, and so forth. Other Asian ESLs had struggled on academic writing and information synthetizing [18]. They claimed some issues underlying those problems are English proficiency, learning experience, etc.

Despite the various difficulties and the reason, it calls for the appropriate strategies to cope with. Self-monitoring strategies in academic writing is predominant factor to cope the difficulty. At this point, students who acknowledged the additional self-monitoring strategy intervention were better in terms of acquisition of academic writing skills and the quality of their texts than those who did not do so [9]. Another empirical study highlighted that evolving writing practice conceivably benefited to deal with the writing problems [14]. In addition, planning self-study supporting by applying one to one course and getting to know how to study at college would impact on students' achievement in writing [15]. Individuals should be encouraged to harness the critical reading to their own compositions. Hence, this strategy could make them intensively correct and improve what they wrote about. It was contended to have a considerable changes in personal revision [16]. A particular treatment such as self-regulated strategy development (SRSD) instruction for the POW (pick my idea–pay attention to the prompt, organize, write and say more) + TREE (topic sentence, reasons, explanations, ending) + COPS (capitalize, organize, punctuate, sense) strategies could be applied to overcome the difficulties on writing. Those treatments made significant improvement on students' writing achievement [17].

3. Research Method

This research employed the descriptive quantitative qualitative research design. The quantitative aspect is related to the students' responses about how they deal with the difficulties in writing abstract presented in the form of percentage. The qualitative aspect supports the data in meaningful words to make them easier to be understood. Therefore, the research will acquire the description about the fact of the students on how to cope their difficulties in composing the abstract.

The population of this study is diploma engineering students at POLTEKOM majoring Mechatronics, Telecommunication, and Informatics. Whereas, the sample of this study is those engineering students who had finished writing the abstract as a part of their final assignment. There were 104 students who involved in this study. 64.08% of them were alumni, whereas 35.92% of them were the students who were at the last semester of the program and finish writing the abstract for their final assignment. All of the respondents write abstract in their submitted final assignments.

The data were collected through questioner and interview. The questionnaire contained open and closed question type. However, it still gives the option *other* assuming that they have their own answer instead of given options. The former attempts to gain the details of certain cases. While the latter was subjected to obtain the short answer. See the sample of the questions in Table 1.

Table 1. Sample of the Questions

No.	Questions
1.	Did you consult it to your supervisor when you are writing abstract?
2.	What was his feedback?
3.	How many time did you consult it to him?
4.	Actually, how many time to did you need it to make it a good abstract?
5.	What made you difficult to write it?
6.	Did you proofread it?
7.	If the error had been found on your abstract, what might you think the most reasonable cause?
8.	Did you need other assistance to write it instead of the supervisors? Explain briefly.

Besides, the personal interview serving as a secondary source of data had been conducted to support the primary data form questionnaire. It was semi-structured interview which took place in classroom. It was carried out in periods of time when the English course 4 had been given. See the sample of the questions in interview in Table 2.

Table 2. The Questions in Interview

No.	Questions
1.	Did/do you still remember what is abstract?
2.	What do you think about abstract? Was it important? Why?
3.	Did you write the abstract?
4.	How did you find it? Easy/difficult? Why
5.	What made it easy/difficult?
6.	What did you do to finish it? / How did you deal with the difficulties, if there is any?

The collected data had been analyzed using a descriptive statistics. It simply obtains the general picture of students' way to cope with the difficulties in writing abstracts together with the major problems when writing abstract. It identifies the quantitative analysis presented in percentage on maximum as well as minimum. Furthermore, this study required a qualitative analysis process. It manages the collection data from the interview session involving organizing, describing, classifying, and categorizing them. Subsequently, it connects and interrelates within those data and the data from the questionnaire. At the final stage, it interprets to attempt the explanation by providing meaning to be easily understood.

4. Results and Discussion

The findings of this research determines that the main problem faced by the students in this study is combining words into sentences (70%), identifying the most appropriate technical terms in engineering (15.7%), and applying grammar in context (14.3%).

Other findings reveal how they cope with those difficulties. The majority of them looked up dictionary (87%) including printed-dictionary as well as the online dictionary. They also consulted their works to lectures (85.9%) more than three times (56.1%). Some of them asked their peers to be proofreaders (14.1%) and few of them attempted to do by themselves.

In spite of having proofreading and revision more than three times (87.6%), supposing there are errors found they said that the errors occurred because of they did writing abstract carelessly (74.2%), missed typed (12.4%), and some of them did not proof read it (13.4%).

The data analysis showed that the majority of respondents experienced the difficulties in composing the abstract for some reasons. The first reason is that they were struggle to combine word by word into sentences. Indeed, word order is a part of challenging problems of language learners [10]. As they are EFL students, they are resistant to face the difficulty in writing skills [11] and they are non-English department students as well, i.e. engineering students [12]. To overcome this problem, most of the students utilized the application of online translation, such as *translate.google.com* (55.9%), *Bing.com/translator* (2.6%), sederet.com/translate (5.2%), and the rest had easily installed some various translation application provided by the *Play Store* in their gadgets without naming what kind of it. They preferred to have the former translation application since it is trustable, worldwide application, and free of charge, they asserted. A few of them had to pay sum of money for asking the translation agent for translation service.

The second problem was they were hard to opt the particular technical terms in engineering context that are related to their topic. To deal with it, again students tent to utilize the application of translations. Eventually they needed to discuss with the lectures and their classmates and friends whom they thought they were good at English. Few of them look up printed-dictionary and the digital one, including engineering dictionary. However, they preferred the Indonesian-English to English-Indonesian or English-English dictionaries since the former is easier to understand.

The following problem was that applying the grammar in context. They felt afraid of making mistakes in grammar. As a matter of fact, grammatical matters has been major issue among EFL context for ages [13][14]. To solve this, they directly looked into translation application mentioned above. Simply, they only typed the sentences in native language then consider the displayed complete sentences in target language as the result of translation machines. A few of the students intended to check their works comparing to application whether theirs were right or not. However, they still needed the intensively personal tutoring with their lectures especially English lectures. Unfortunately, most of the students were reluctant to look up the dictionary, though this strategy is useful for ESL students and teachers to achieve better result in grammar.

5. Conclusion

This research underlines the findings that the major problems of the engineering students covers combining words into sentences, identifying technical terms in engineering, and applying grammar in context. To tackle those problems, they endeavored some strategies 1) utilizing digital application, 2) having personal tutoring with lectures, 3) having peer-proofreading, 4) looking up the dictionaries, 4) giving a try their best efforts but still looking into digital application at the end.

The technology advances may support students to learn in general. They are useful for students to tackle the writing problems [22]. In the same line, they keep reinforcing the writing process for writers [20][21].

Future research could be focused on the impact using digital applications towards academic writing comparing before and after giving the technology tools. It also compares the control group and experimental one as well to discover the significant of using digital technology advances.

References

[1] Hidayatai, S 2011 Error Analysis on a Short Speech: a Case of an ESL Indonesian Learner *Article published in E-CLUE (Journal of English Culture, Language, Literature, and Education), English Department, Manado State University* **5** 1

[2] Trisyanti, U 2011 *Developing an English Syllabus for D3 Civil Engineering Students of ITS Surabaya* Thesis, Graduate Program in English Language Education

[3] https://www.ets.org/s/toeic/pdf/ww_data_report_unlweb.pdf

[4] http://www.ef.co.uk/epi/

[5] https://www.ielts.org/teaching-and-research/test-performance

[6] Armstrong T 2011 Reader-Friendliness and feedback: German-L1 Scholars' perceptions of Writing for publication in English *Journal of academic writing* **1** 1 pp153-164

[7] Andrade C 2011 How to write a good abstract for a scientific paper or conference presentation *Indian Journal of Psychiatry* **53** 2 pp172175

[8] Fadda H A 2012 Difficulties in Academic Writing: From the Perspective of King Saud University Postgraduate Students *ELT English Language Teaching* **5** 3 pp123

[9] Wischgoll A 2016 Combined Training of One Cognitive and One Metacognitive Strategy Improves Academic Writing Skills *Frontiers in Psychology* **7** 187

[10] Gharabally M 2015 The Writing Difficulties Faced by L2 Learners and How to Minimize them *International journal of English Language, Linguistics Research* **3** 5 pp:42-49

[11] Salma, U 2015 Problems and practical needs of writing skill in EFL context: An analysis of Iranian Students of Aligarh Muslim University *IOSR Journal of Humanities and Social Science (IOSRJHSS)* **20** 11 Ver III Pp 74-76 e-ISSN:2279-0837 P-ISSN:22790845

[12] Xiou G and Chen X 2015 English Academic Writing difficulties of Engineering Students at Tertiary level in China *World Transaction on Engineering and Technology Education* **13** 3 Al-Mekhlafi AM and Nagaratnam RD 2011 Difficulties in Teaching and Learning Grammar in EFL Context *International Journal of Instruction* **4** 2 ISSN:1694-609x e-ISSN:1308-1470

[13] Younes Z B and Albalawi F S 2015 Exploring the most common types of writing problems among English Language and Translation major sophomore Female Students at Tabuk University *Asian Journal of Basic, and Applied Science* **3** 2 ISSN2313-7797

[14] Raija P, Marjatta T and Tarja L 2015 Students in higher education with reading and writing difficulties, Education Inquiry, **6** 1, 24277 DOI: 103402/eduiv624277

[15] Daiute C and Kruidenier J 1985 A self-questioning strategy to increase young writers' revising processes *Applied Psycholinguistics* **6** 3 307-318

[16] Ann BB, Linda HM 2010 *The Effects of Self-Regulated Strategy Development on the Writing of Expository Essays for Adults With Written Expression Difficulties Preparing for the GED* First Published June 23, 2010 33 2 pp: 124-136 Article first published online: June 23, 2010; Issue published: March 1, 2012

[17] Aek P and Lulu L 2011 *General Academic Difficulties and Reading and Writing Difficulties among Asian ESL Postgraduate Students in TESOL* at an Australian University First Published December 14, 2011 **42** 3 pp: 227-264 Article first published online: December 14, 2011; Issue published: December 1, 2011

[18] Julia M 2006 An investigation into the effect of English learners' dictionaries on international students' acquisition of the English article system *International Education Journal* **7** 4 435-445 ISSN 1443-1475 © 2006 Shannon Research Press

[19] Hayes J R 2006 New directions in writing theory In C MacArthur, S Graham, & J Fitzgerald (Eds), Handbook of writing research pp 28–40 (New York, NY: Guilford Press)

[20] Hayes J R 2013 *A new framework for understanding cognition and affect in writing In C M Levy and S Ransdell (Eds) The science of writing: Theories, methods, individual differences, and applications* pp 1–27 (Mahwah, NJ: Erlbaum) (Original work published 1996)

[21] Graham S, Harris K and Larsen L 2001 Prevention and intervention of writing difficulties for students with learning disabilities Learning *Disabilities Research & Practice* **16** 2 74–84

IOP Conf. Series: Materials Science and Engineering **306** (2018) 012023 doi:10.1088/1757-899X/306/1/012023

An Expert System for Diagnosing Eye Diseases using Forward Chaining Method

C P C Munaiseche*, D R Kaparang and P T D Rompas

Universitas Negeri Manado, Tondano 95618, Sulawesi Utara, Indonesia

*cindymunaiseche@unima.ac.id

Abstract. Expert System is a system that seeks to adopt human knowledge to the computer, so that the computer can solve problems which are usually done by experts. The purpose of medical expert system is to support the diagnosis process of physicians. It considers facts and symptoms to provide diagnosis. This implies that a medical expert system uses knowledge about diseases and facts about the patients to suggest diagnosis. The aim of this research is to design an expert system application for diagnosing eye diseases using forward chaining method and to figure out user acceptance to this application through usability testing. Eye is selected because it is one of the five senses which is very sensitive and important. The scope of the work is extended to 16 types of eye diseases with 41 symptoms of the disease, arranged in 16 rules. The computer programming language employed was the PHP programming language and MySQL as the Relational Database Management System (RDBMS). The results obtained showed that the expert system was able to successfully diagnose eye diseases corresponding to the selected symptoms entered as query and the system evaluation through usability testing showed the expert system for diagnosis eye diseases had very good rate of usability, which includes learnability, efficiency, memorability, errors, and satisfaction so that the system can be received in the operational environment.

1. Introduction

One of the medical problems occurring recently is the imbalance between the number of patient and the number doctor. Limitations of an expert (doctor) sometimes become an obstacle for people who will consult to get the best treatment solution associated with the disease suffered. In addition, most people are not trained medically so they do not know what to do when they experience symptoms of illness. It is unfortunate when the symptoms which can actually be dealt with early develop into a more serious disease due to lack of knowledge. People can obtain knowledge about health from books or internet sites. However, it is not easy to learn that way because it takes a long time. In addition, these sources cannot diagnose types of diseases as the doctors do. In this case, expert system is presented as an alternative in solving the problem.

Expert system is software designed specifically based on Artificial Intelligence, where the system seeks to adopt human knowledge to the computer so that the computer can solve a particular problem by imitating the work of the experts. Expert systems development requires knowledge acquisition from people, involving both knowledge engineers and application domain experts in specialist interactions with computing systems. Expert systems may be used to provide support and advice to a user of any complex information system and hence to improve the human-computer interface [1].

There are different areas in medicine where an expert system has been designed and implemented to profers solution to health status stability in human. Among these diverse areas includes an expert system for Eye, Skin, Pregnancy, Blood Disorder and several other human diseases. In previous study, Naser and Zaiter [2] in their research work for eye expert system, they used CLIPS language in their research thereby serving as a tunnel to the inner workings of the body. The scope of the expert system is the following eye diseases: Discharge from the Eye, Bulging Eye, Double Vision, and Drooping Eyelid (four types of eye diseases) [2]. Also, Gudu et.al [3] in their research for expert system to diagnosis and treat hypertension in pregnancy stated that the diagnostic and treatment expert system for hypertension in pregnancy has so far remained at the testing phase of its life cycle and is yet to be implemented [3].

Ayangbekun et. al [4] develops an expert system for diagnosing brain diseases, using the C#.NET programming language and Microsoft SQL Server 2012 served as the RDBMS. From the study, this application serves as a model tool that will enable hospitals to effectively monitor patients medical records without ambiguity [4]. In addition, Ayangbekun et. al [5] also developed an expert system for diagnosis of blood disorder. There were two hospital which was taken as the case study of the research. The information was gathered from the hematology department and the blood department of the two hospitals respectively. The information gotten was analyzed and manipulated based on the symptoms and causes of the blood disorders and then turned into rules for easy programming into the computer [5].

This study aimed to design an expert system application for diagnosis of eye diseases using forward chaining and analyze the level of user acceptance to this application through usability testing. The scope of the study is extended to 16 types of eye diseases with 41 symptoms of the disease, arranged in 16 rules that were called the Rule-Based System.

The main contribution is the experts system for diagnosing eye disease has become an expert knowledge sharing tool to be used by other medical personnel who are not specialists in diagnosis of eye diseases, especially for hospitals that do not have an ophthalmologist. The research's novelty is the expert system based on web and user friendly so that can be accessed by everyone wherever and whenever easily and using PHP programming language and MySQL as the Relational Database Management System (RDBMS). The advantage of this research is as a guide for the patient in taking initial action if they know the possibility of suffering eye disease (early detection).

2. Literature Review

2.1 Basic Concepts of Expert System

Expert system is a piece of software programmed using Artificial Intelligence (AI) techniques. Such systems use databases of expert knowledge to offer advice or make decisions in such areas as medical diagnosis and trading on the stock exchange [6]. An expert system is a system that employs human knowledge captured in a computer to solve problems that ordinarily require human expertise. Expert system seeks and utilizes relevant information from their human users and from available knowledge bases in order to make recommendations. With the expert system, the user can interact with a computer to solve a certain problem. This can occur because the expert system can store heuristic knowledge [2]. Generally to develop an expert system, a rule based method is required to analyze and compute the knowledge base [7].

2.2 System Architecture

Expert system consists of domain expert, designer, inference engine, knowledge base, user interface and user. There is relationship between these subdivisions which makes it expert system. The domain expert is connected to the knowledge base in order to give rules and fact. The domain experts are normally the expert in the body or field. The knowledge base stores the rule and fact collected. The knowledge base is also connected to inference engine in which is used to process the rule to deduce another set of rule or fact. The inference engine is normally designed by the programmer or designer. The inference engine is then connected to the user interface in which is used to collect data from the users. This is also developed by the designer. This trend can also be followed backward. The user interface gives information to the inference engine and the knowledge base for user data to be processed. Also for the knowledge base update, a need to contact the domain expert is needed. All this can be represented below (Figure 1) [4]:

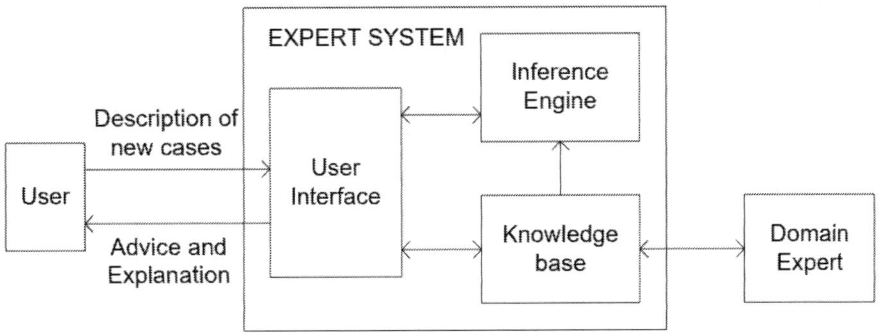

Figure 1. Expert System Architecture

2.3 Medical Knowledge

The medical knowledge of specialized doctor is required for the development of an expert system. This knowledge is collected in two phases. In the first phase, the medical background of eye diseases is recorded through the creation of personal interview with doctors and patients. In the second phase, a set of rules is created where each rule contains in IF part that has the symptoms and in THEN part that has the disease that should be realized. The inference engine (forward chaining) is a mechanism through which rules are selected to be fired. It is based on a pattern matching algorithm whose main purpose is to associate the facts (input data) with applicable rules from the rule base. The search is done by using rules whose premise matches the known facts to gain new facts and continue the process until the goal is reached or until there is no more rules whose premises match the known facts as well as the facts obtained. Finally, the eye diseases are produced by the inference engine [2].

3. Research Method

Research procedures consist of: preliminary study, data collection, data analysis, system design, system implementation and evaluation, and drawing conclusion.

Preliminary Study. At this stage, the authors collected information, study materials and data sources related to expert systems, forward chaining methods, rule-based reasoning, the types of eye diseases in humans, symptoms of eye diseases and treatment or preventive solutions.

Data Collection. Data sources used in expert systems to diagnose eye diseases in humans include data of the type og eye disease, eye disease symptoms, disease information and solutions provided. The data required in this study were obtained from Literature Study and Consultation/interview with experts, in this case ophthalmologist.

Data Analysis. Based on the collected data, the researcher conducted following analysis steps: (i) made a list and coded eye diseases along with symptoms. This expert system software can diagnose 16 types of eye diseases with 41 symptoms of the disease; (ii) Made Rule-Based System. In order to recognize the type of eye disease, rules in expert system tracing are required. There are 16 Rules and a forward chaining hierarchy called the Rule-Based System.

System Design. The design of this system includes design process described by using decision tree, context diagram explaining the relationship between input / output between system with outer world, data flow diagram (DFD), the design of the database and user interface.

System Implementation. The activitiy performed at this stage was the programming or coding. This stage was the translation of the design into the form of computer programming language. This research employed PHP programming language.

System Evaluation. The evaluation of this expert system uses usability testing. This evaluation aims to find out how easy an interface can be used by the user so that the system can be accepted in the operational environment [6].

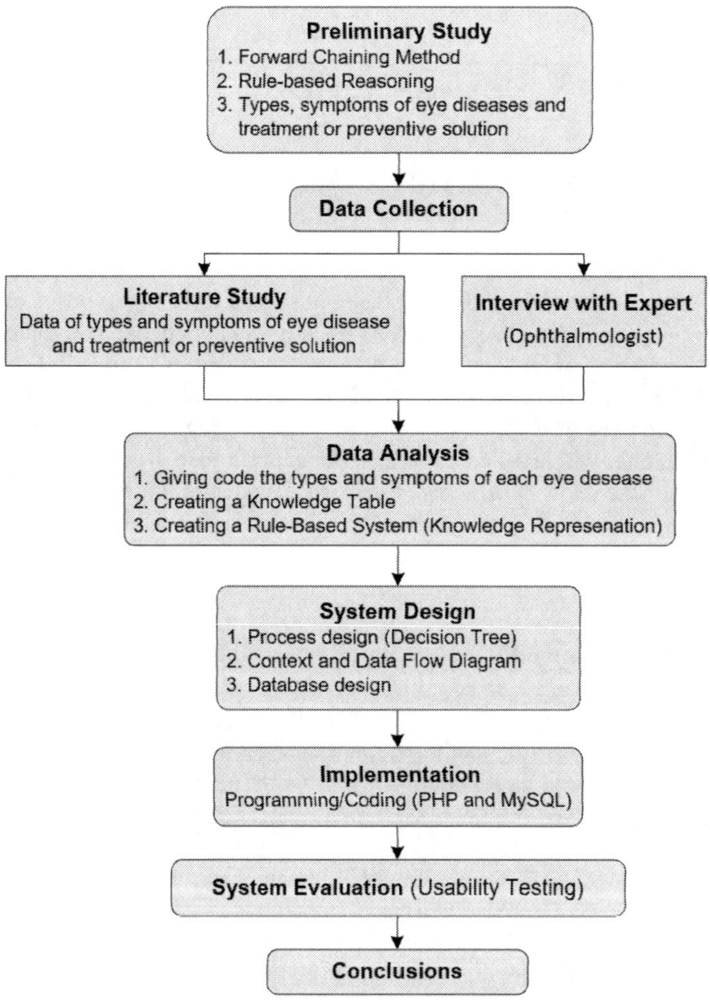

Figure 2. Research Prosedure

IOP Conf. Series: Materials Science and Engineering **306** (2018) 012023 doi:10.1088/1757-899X/306/1/012023

Table 1. Plot of Usability Aspects

No.	Questions	Aspects of Usability				
		Learn	Eff	Mem	Err	Sat
1.	Are you able to recognize from the Home Menu that this application is an expert system to diagnose eye disease?	■				■
2.	Are you able to register as a patient by using Consultation Menu?	■	■	■		■
3.	After registering, are you able to answer questions related to symptoms presented by the applications through Consultation Menu?	■	■			■
4.	Are you able to obtain final diagnosis regarding the type of disease, symptoms, description, prevention and solution offered by the systems?	■	■			■
5.	Are the letters and texts in the web readable?				■	■
6.	Are the symbols easy to understand?				■	■
7.	Is the color design comfortable to see?					■
8.	Are you able to find the login form?		■	■		■
9.	Are you able to access information at each page?		■			■
10.	Are you able to recall menus and view from the web after not using the application for a while?			■		■

Note: Learn = **Learnability**; Eff = **Efficiency**; Mem = **Memorability**; Err = **Errors**; Sat = **Satisfaction**

Usability is the degree of a software's ability to assist users in completing a task. The initial step of this usability testing is to provide some tasks to the user when interacting with the system being tested. These tasks are assigned to 30 respondents who are accustomed to using Web Browser features. These tasks are used as 'means of interaction' in usability measurements [8].
After he users had completed all of the tasks, the researcher distributed the questionnaire containing 10 questions representing the five aspects of usability. Users completed the questionnaires based on their experiences (what they saw and felt) during the completion of the tasks.

Each question of the questionnaire aimed to figure out usability level according to user acceptance, which would be assessed on a scale of 5 [9]. The questions given in this questionnaire can be seen in Table 1. According to Jacob Nielson [10], aspects of usability testing includes five points, namely:

1) Learnability, describes the level of user convenience to complete basic tasks when they first see or deal with existing systems.
2) Efficiency, explains how quickly users can complete the tasks when they first learn the system.
3) Memorability, explain the level of user convenience in using the system after not using the system for a while.
4) Errors, describes the possibility of errors or mistakes made by users and how easily they can overcome them.
5) Satisfaction, describes the level of user satisfaction in using the system that has been made.

The plot result of the five aspect of usability above on the 10 questions of questionnaire can be seen in Table 1.

Conclusion. A conclusion will be drawn from the results of these evaluations based on the results of testing (usability testing) of the expert system.

4 Results and Discussion

4.1 Results

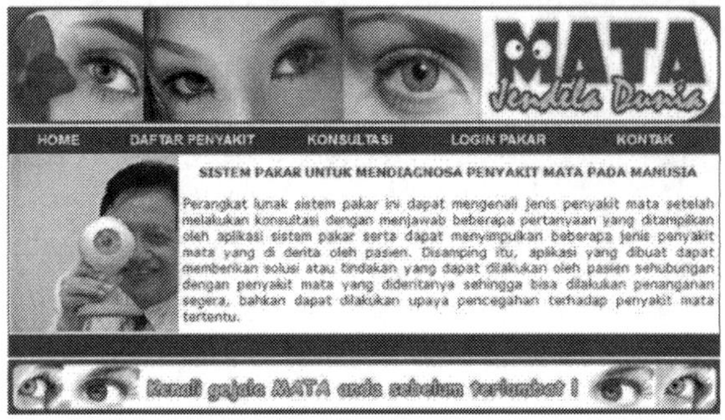

Figure 3. Main Menu Interface (Home Menu)

This research results in a web-based expert system that can recognize the type of eye disease in humans based on the symptoms experienced by patients. This system performs analysis based on the dialogue between the system with the user/patient. The web-based expert system of eye disease diagnosis is designed by using PHP programming language and MySQL for database processing. Figure 3 is the main display image of the eye disease diagnosis expert system.

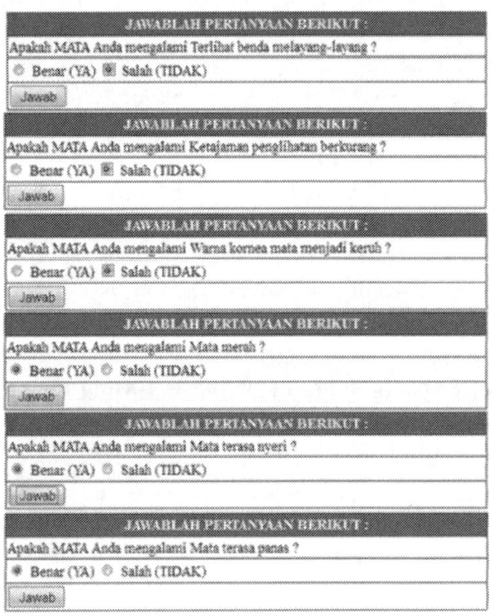

Figure 4. Consultation Application Form

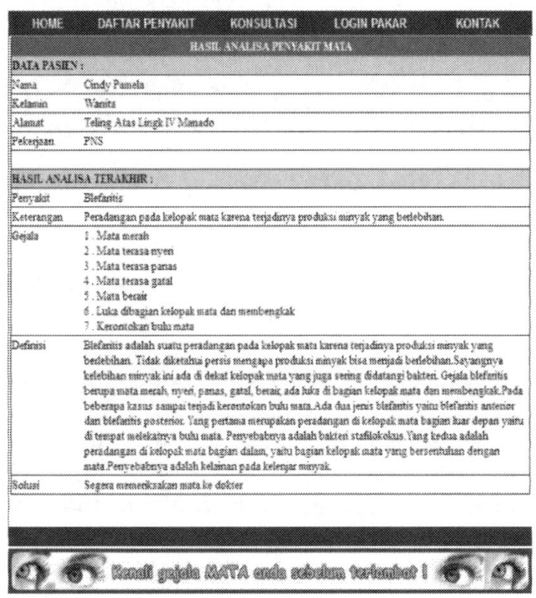

Figure 5. Diagnosis Result

Prior to consultation on the consultation menu, the patient is expected to fill out registration form first so that each patient will be in the database information. The patient cannot enter the consultation application window (Figure 4) before registration. To answer the questions, the patient (user) can directly select True (YES) or False (NO) options as shown in Figure 4.Then, the system will infer the type of illness suffered by the patient on the Result of Eye Disease Analysis showing patient data and results final analysis of the name of the disease, symptoms, the definition of the disease and the solutions offered (Figure 5).

4.2 Discussion

The results of system evaluation through usability testing are managed into the percentage of questionnaire answers from 30 respondents [9].

Table 2. *Usability* Value Recapitulation

No.	Attribute	Usability Value
1.	The interface of the expert system is easy to recognize	**4.32**
2.	The ease to register through Consultation Menu	**3.93**
3.	The ease of the patient to answer the question of the system according to the symptom he / she experienced	**4.38**
4.	The ease of the patient to obtain the final diagnosis	**4.29**
5.	Readable writings on the web page	**4.17**
6.	The picture symbols are easy to understand	**3.78**
7.	The color design is comfortable to see	**3.92**
8.	Ease of patient to find login form	**4.34**
9.	The ease to access information in each webpage	**4.28**
10.	Ease of recall menus after exiting the application	**4.12**

Table 2 shows the values of satisfaction or user acceptance of each attribute [9]. It can be seen that "Easy recognition of Interface" attribute has a user acceptance value of usability of 4.32 (already above 3 or above the middle value) in the scale of 5. It means that the web application is easy to recognize by the user from the web interface home page.

Related to each aspect of usability in Table 2, it can be said that the application software has **excellent usability value**, which includes Learnability, Efficiency, Memorability, Errors, and Satisfaction. It is shown by the value of usability result on the five attributes, which are presented as follows:

- The value of the "Easy Interface recognition" attribute of 4.32 indicates that the web has the value aspect of **Learnability**.
- Attribute "Ease of patient registration" has value of 3.93; Attribute "The ease of the patient to answer the question of the system according to the symptom he / she experienced" has value of 4.38; attribute "The ease of the patient to obtain the final diagnosis" has value of 4.29 and "the ease of accessing information on each page" has value of 4.28. Those values indicate that the web has **Efficiency** aspect value.
- Attribute "Ease of patient to find login form" has value of 4.34; the "Ease of calling back menus after exiting the application" has value of 4.12. Those values indicate that the web already has a **Memorability** aspect.
- The attribute of "Readable writings on the web page" has value of 4.17; attribute "Picture symbols are easy to understand" has value of 3.78. Those value mean that the web has minimized aspects of **Errors**.
- overall attributes have average values above 4, indicating that the system has aspects of **excellence Satisfaction.**

5. Conclusions

The results obtained showed that the expert system was able to successfully diagnose eye diseases corresponding to the selected symptoms entered as query and recapitulation of result usability value indicates that all attributes of usability acceptance value by user have average value above 4. Therefore, it can be said that expert system application software has excellent usability value including learnability, efficiency, memorability, errors , and satisfaction so that the system can be received in the operational environment and it can be applied as a system that will be operated by the user.

References

[1] Brian R, Gaines 2010 Designing Expert System for Usability Available from: http://citeseerx.ist.psu.edu/viewdoc/download?doi=10.1.1.414.583&rep=rep1&type=pdf
[2] Naser S S A, Zaiter A O 2008 An Expert System for Diagnosing Eye Diseases Using Clips *J. of Theorit. and Applied Inform. Technol* pp 923-930 Available from: http://www.jatit.org/volumes/research-papers/Vol4No10/5Vol4No10.pdf
[3] Gudu J, Gichoya D, Nyongesa P, Muumbo A 2012 Development of a Medical Expert System as an Expert Knowledge Sharing Tool on Diagnosis and Treatment of Hypertension in Pregnancy *Int. J. of Bioscience, Biochemisry and Bioinformatics* **2**(5) pp 297-300 Available from: http://www.ijbbb.org/papers/120-CB301.pdf
[4] Ayangbekun O J, Jimoh I A 2015 Expert System for Diagnosis Neurodegenerative Diseases *Int. J. of Comp. and Inf. Technol.* **04** pp 694-698 Available from: https://www.ijcit.com/archives/volume4/issue4/Paper040413.pdf
[5] Ayangbekun O J, Olatunde A I and Bankole F O August 2014 An Expert System for Diagnosis of Blood Disorder *Int. J. of Comp. Applic.* (ISSN: 0975-8887) **100** pp 36-40 Available from: http://research.ijcaonline.org/volume100/number3/pxc3898061.pdf
[6] Munaiseche C P C and Liando O E S 2016 Evaluation of Expert System Application Based On Usability Aspects *IOP Conference Series: Materials Science and Engineering* volume 128 Available from: http://iopscience.iop.org/article/10.1088/1757-899X/128/1/012001/pdf
[7] Patra P S K, Sahu D P and Mandal I 2010 An Expert System for Diagnosis of Human Diseases *Int. J. of Comp. Applications* **1**(13) Available from: http://citeseerx.ist.psu.edu/viewdoc/download?doi=10.1.1.206.4184&rep=rep1&type=pdfA
[8] Sastramihardja et al. 2008 Usability Measurement By Means Task Model In User Center Software Development p 4
[9] Wingnjosoebroto S, Sudiarno A and Harenda D 2009 Designing Web Interface Prototype Based On Usability Aspects (Case Studi: Ergonomic and Work System Design Laboratory) p 7 Available from: http://digilib.its.ac.id/public/ITS-Undergraduate-8877-2504100030-Paper.pdf.
[10] Jakob Nielsen. Usability 101: Introduction to Usability. 2014 Jan 2012. Available from: http://www.useit.com/alertbox/20030825.html

IOP Conf. Series: Materials Science and Engineering **306** (2018) 012024 doi:10.1088/1757-899X/306/1/012024

Green Building Implementation at Schools in North Sulawesi, Indonesia

D A J Harimu*
Universitas Negeri Manado, Tondano 95618, Sulawesi Utara, Indonesia

*dr_debbie_harimu@unima.ac.id

Abstract. This research aims at investigating the green building implementation at schools in North Sulawesi, Indonesia; and to analysis the relationship between implementation of green building concept at school with students' green behaviour. This research is Survey Research with quantitative descriptive method. The analysis unit is taken purposively, that is school that had been implemented the green building concept, Manado's 3rd Public Vocational High School, Lokon High School at Tomohon, Manado Independent School at North Minahasa, and Tondano's 3rd Public Vocational High School. Data collecting is acquired by observation and questionnaire. The Assessment Criteria of green building on Analysis Unit, is taken from Greenship Existing Building ver 1. There are 4 main points that being assessed, which are Energy Conservation and Efficiency; Water Conservation; Indoor Health and Comfort; Waste Managerial. The Analysis technique used in this research is the simple regression analysis. The result of the research shows that there is a significant relation between green building implementation at school and students' green behavior. The result is accordance with the Gesalts Psychologist theories, that architecture can change the user's behaviour.

1. Introduction

Global warming is not only a story comes from a land far away; about gas pollution, huge factory disposal, melting ice, polar bear in the verge of extinction, and descending agricultural land in Africa are truly real. On the other side, most of people assume that the effect of global warming will only happen in tens year ahead. Unaware that the horrible effect from the global warming is coming closer to visit our lives. The impact can be felt now, likes the disorderly seasons, flood and drought happens at the same time, disaster often occurs related to the climate. In Indonesia, since 2003, has occurred 4 times escalation of disaster. In 2003 to 2005, there were 1.429 disasters and approximately about 53.3% among them, related to the climate, hydrology, such as flood, drought, forest fires and typhoon. In North Sulawesi, in recent years has occurred a wretched disaster [1].

IOP Conf. Series: Materials Science and Engineering **306** (2018) 012024 doi:10.1088/1757-899X/306/1/012024

Figure 1. In 2014 and 2015, a great flood happened in Manado

Figure 2. In 2015 forest fires, including the forest in the valley near Mount Lokon

In 2014 and 2015, a great flood happened (**Figure 1**), caused many people lost their lives, damaged home, and landslide followed by a long drought that caused forest fires in 2015 (**see Figure 2**), including the forest in the valley near Mount Lokon [2]. In truth, every of our choices and habits, like in choosing transportation, building, up to the habits of saving energy, water and not to carelessly littering, contributes importantly in the deteriorating global warming, or we can be the solution [3][4][5].

Based on the experiences from the disasters occurred in recent years, seems that the schools, especially in Manado, in Minahasa, and North Sulawesi in general, has started to introduce students about the go green act, and trying to implement the green building concept into their infrastructures at school. Such act can be seen through the slogans hang on the wall and school's magazine board (*madding)*, also the availability of trash can, separating the organic and non-organic waste, and etc. This is interesting and urgent to be investigated and discussed, about the application of the concept of green building at school in Minahasa affecting the go green habits of theirs students. According to the explanation in the background, the research questions of these researches are; what are the green building concepts that have been implemented at schools in Minahasa? And how is the effect of implemented green building concept toward students' go green habits at schools in Minahasa?

2. Research Methods

The analysis techniques used in this research are the simple regression analysis and descriptive analysis. In order to get a representative picture, thus the research location was held in Manado, Tomohon, North Minahasa and Minahasa. The objects of the research from schools can see in **Figure 3**.

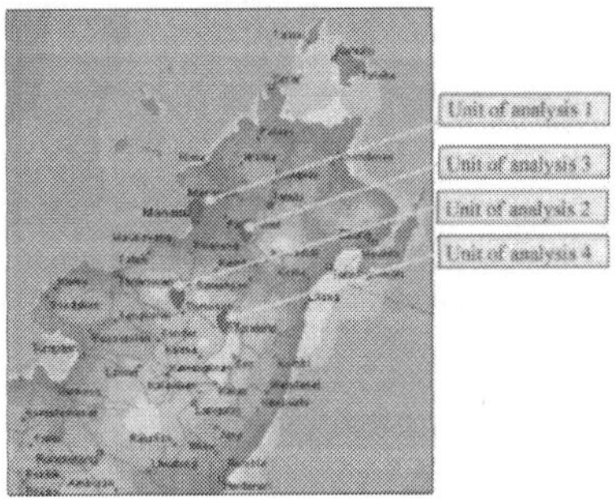

UNIT OF ANALYSIS 1	SMKN 3 Manado JL. POMOROUW Taas Kec. Tikala Kota Manado Prop. Sulawesi Utara.
UNIT OF ANALYSIS 2	Persekolahan Lokon Santa Nikolaus Tomohon, JL. PERLOMBAAN NO. 99, KAKASKASEN DUA, Kakaskasen Dua Kec. Tomohon Utara
UNIT OF ANALYSIS 3	Persekolahan Manado Independent School MIS JL. WALANDA MARAMIS, Kolongan Tetempangan Kec. Kalawat, Kab. Minahasa Utara Prop. Sulawesi Utara
UNIT OF ANALYSIS 4	Persekolahan SMKN 3 Tondano, JL. GUNUNG AGUNG Rinegetan Kec. Tondano Barat Kab. Minahasa Prop. Sulawesi Utara

Figure 3. The object of the research location in Manado, Minahasa, North Minahasa, North Sulawesi, Indonesia.

IOP Conf. Series: Materials Science and Engineering **306** (2018) 012024 doi:10.1088/1757-899X/306/1/012024

3. Results and Discussion

3.1. Implementation of "Green building" concept

Criteria assessment of the implementation of green building concept at analysis unit, conclude from green ship rating tools for established building, version 01. There were 4 main points that was assessed, which were; 1) energy efficiency and conservation; 2) water conservation; 3) indoor health and comfort; 4) waste managerial [6] [7].

The field observation result of the energy efficiency and conservation shows in **Table 1.** There are four unit of analysis schooling in North Sulawesi only applied given electric meter for every building (**Figure 5**), only some of them regularly do the evaluation to calculate the use of energy, and three unit of analysis prioritizing natural lighting (**Figure 4**) and the management of room temperature by air ventilation of each room. Four unit of analysis schooling in North Sulawesi didn't apply OTTV circulation, on site renewable energy, and anticipate climate change impact.

Table 1. Implementation of the energy efficiency and conservation

No	Green Building Criteria	Unit of Analysis			
		1	2	3	4
1	Electrical Sub Metering	•	•	•	•
2	Energy Efficiency Measure	X	•	X	•
3	Natural Lighting	•	•	X	•
4	Ventilation	•	•	X	•

Figure 4. Natural lighting at unit of analysis 2

Figure 5. Electrical lighting for all day at unit of analysis 3

The field observation result for water conservation shows in **Table 2.** The application of water conservation of four unit of analysis at schools in North Sulawesi is the implementation of the total amount of water use, water circulation, rain water catchment (**Figure 6**), and efficiency (**Table 7**) in using the water for the field, and also the availability of water absorption at school environment. Four unit of analysis schooling in North Sulawesi didn't apply water use reduction, water fixtures, and alternative water resources.

IOP Conf. Series: Materials Science and Engineering **306** (2018) 012024 doi:10.1088/1757-899X/306/1/012024

Table 2. Implementation of the water conservation

No	Green Building Criteria	Unit of Analysis			
		1	2	3	4
1	Water metering	•	•	•	•
2	Water circulation	•	•	•	•
3	Water recycling	•	X	X	•
4	Rainwater harvesting	X	•	X	X
5	Water efficiency landscaping	X	•	X	X

The field observation result for indoor health and comfort shows in **Table 3**. The application of indoor health and comfort aspect of four unit of analysis at schools in North Sulawesi is about the application of fresh air circulation from outside to inside the room (**Figure 8**), minimizing the use of machine or motor transportation that produces CO, no smoking area at every corner of the schools, paying attention to the thermal comfort and protecting each classroom from noises.

Figure 6. Rainwater harvesting at unit analysis 2

Figure 7 Water inefficiency landscaping at unit analysis 3

Table 3. Implementation of indoor health and comfort

No	Green Building Criteria	Unit of Analysis			
		1	2	3	4
1	Outdoor air introduction	X	•	X	X
2	CO monitoring	X	•	X	X
3	Environmental tobacco smoke control	X	•	X	•
4	Thermal comfort	X	•	X	•
5	Acoustic level	X	•	X	X

Figure 8. Landscape of unit of analysis 2 application of fresh air circulation from outside to inside the room

Figure 9. Availability of trash bin and waste separation in unit of analysis 1 and unit of analysis 4

The field observation result for waste managerial shows in **Table 4**. The availability of trash can in every corner of school environment; along with the regular schedule to dump the trash are the aspects that have been done at the observed four unit analysis (see on **Figure 9**). Four unit of analysis schooling in North Sulawesi didn't apply composting the waste and didn't use waste incinerator

Table 4. Implementation of waste managerial

No	Green Building Criteria	Unit of Analysis			
		1	2	3	4
1	Availability of trash bin	•	•	•	•
2	Waste separation	X	X	•	•
3	Regular Schedule to dump waste at the landfills	•	•	•	•

In general, the implementation of green building concept at four unit of analysis at schools in North Sulawesi, which are: the first unit of analysis has implemented 29.63%, the second unit of analysis has implemented 55.65 %, the third unit of analysis has implemented 18.52%, and the fourth unit of analysis has implemented 40.74%.

3.2 Students' Green Behavior

Students' green habits toward energy conservation and efficiency aspect can be measured from; students preference to use natural lighting than man-made lighting or electric lamp, students always turn off the lamp after they use the room, students that choose to let the window open and the air can circulate freely, students are automatically can always make sure that they have turned off the electrical tools before they leave the room. Students' green behavior toward the water conservation can be measured from, students are sure to turn off the tap water every time they leave the kitchen, bathroom/ lavatory, students also care to make sure that they have turned off the tap water at outside the room after they finish. Students' green habits toward indoor health and comfort aspect measured by, students prefer to use natural air conditioner than machine one, students do not smoke at school, and students do not make unnecessary noises around the learning environment. Students' green ha-bits toward the waste management aspect can be measured by, students always throw the trash at the bin, students are able to separate organic and non-organic waste, and dangerous waste toward the environment, students with their creative effort use the waste to create a product, and also the students know how to make fertilizer [8][9].

The result of questionnaire data toward students' green behavior shows in **Table 5**.

Table 5. Questionnaire data result toward students' green habits

Unit of Analisys	1	2	3	4
Students' green behavior	52.56%	88.21%	38.46%	70.00%

3.3 The implementation effect of green building concept at school toward students' green habits

The analysis impact of the schools' effort to apply green building concept toward students' green habits using simple regression has shown in **Figure 10**. The simple equation regression output shows: The intercept or constants worth 13.18 % shows that without the implementation of green building concept, students' green behavior are 13.18 %. From the positive equation relationship in which every escalation of variable X; which is the implementation of green building concept, can escalate the variable Y; which is the students' green behavior. The coefficient regression 0.1318 shows every improvement/escalation of the green building implementation concept at every unit of analysis 1% can escalate 13 % students' green behavior. The discussion of simple regression analysis shows the wider/bigger the green building implementation concept can improve the establishment of students' green behavior. This result suits with the behavioral theory argue that every architect can influence those who use the building.

$$Y= 1.3606X + 0.1318 \quad (1)$$

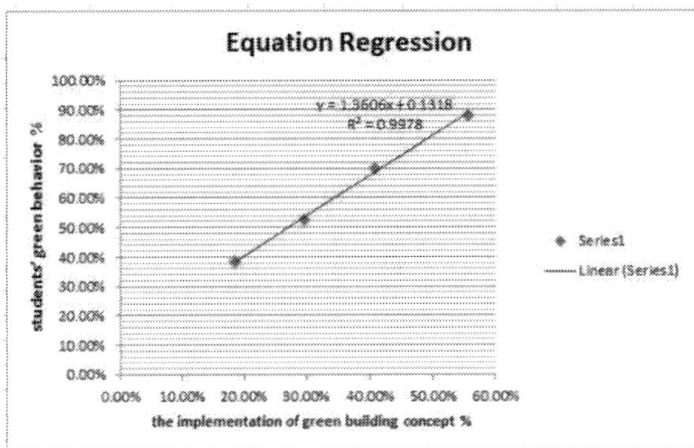

Figure 10. The graphic of equation regression

4. Conclusions

There are 4 Green building concept aspects that has been implemented at schools in North Sulawesi, Indonesia, which are: energy efficiency and conservation, water conservation, indoor health and comfort, and waste managerial. The implementation of green building concept has proven to give significant influence toward students' go green behavior at school in North Sulawesi, Indonesia. The wider/ bigger the implementation of green building will improve/escalate the establishment of students' green behavior. The result is accordance with the Gesalts Psychologist theories, that architecture can change the user's behavior [10].

References

[1] Khishnani N 2012 *Greening Asia Emerging Principles for Sustainable* BCI Asia Construction Information Pte L.td

[2] *http://www.mongabay.co.id/2014/01/17/banjir-bandang-manado-18-warga-tewas-1-000-an-rumah-rusak* Accessed Oct 1, 2017

[3] Esa M R B et al 2011 Obstacles in Implementing Green Building Project in Malaysia *Australian Journal of Basic and Applied Sciences* **5**(12) 1806-1812

[4] Li Y et al 2017 A Review of Studies on Green Building Assessment methods by Comparative Analysis *Energy and Building Journal* **146** 152-159

[5] Huang K T S et al, 2015 Implementation of Green Building Specification Credis for better thermal conditions in Vaturally Ventilated School Buildings, *Building and Environment Journal* **86** 141-150
[6] Wines J 2015 *Green Architecture* Taschen: Hongkong
[7] *Green Building Rating tool for Existing Building ver 1.0* 2012 Rating Development Departemen Green Building Council Indonesia
[8] Elham N A W and Shyan T S 2011 Factors Influencing the Green Purchase, Behavior of Penang Environmental Volunteers *International Business Management* **5** 38-49
[9] Zhao D-X, He B-J, Meng F-Q 2015 The Green School Project: A means of Spending up Sustainable Development? *Geoforum Journal* **56** 310-313
[10] Canter D and Stringer S 1975 *Environment Interaction in Architecture* International Universities Press, New York

Acknowledgments
The authors wish to express appreciation to Lembaga Penelitian dan Pengabdian Masyarakat of Manado State University which had finance all of the research activities, and Rector of Manado State University, Indonesia who has proposed research grant.

Student Learning Strategy and Soft-skill in Clothing Business Management

D Ampera

Universitas Negeri Medan, Medan, Indonesia

*ampera.din@gmail.com

Abstract. Clothing Business Management course is a subject delivering knowledge and skills about how to manage clothing business. This course requires students' ethics, leadership, commitment, toughness, honesty, and the ability to take initiative, argue logically, and work together. However, students are not fully efficient in managing a business during practical task. The purpose of the study is to investigate: (1) the differences of students' learning achievement between those receiving gradual and conventional learning strategy, (2)the effect of interactions between learning strategy and soft skills toward students learning achievement, (3) the differences of learning achievement of high soft skills students who receive gradual and conventional learning strategy, (4) the differences of learning achievement of low soft skills students who receive gradual and conventional learning strategy. The 2x2 treatment-by-level experimental study was carried out in Dressmaking Study Program. The result proved significantly that, overall, differences of students learning achievement exist, except of low soft skills students.

1. Introduction

Education is essential for human progress and involves many participants including faculties and students. Good learning should be managed professionally by people who have special competence in the field of education. Therefore, learning needs a special attention to the development of intellectual potential and skills including soft skills of students.

Clothing Business Management course is a subject that delivers knowledge and skills about how to manage clothing business. This course requires students' ethics, leadership, commitment, toughness, honesty, and the ability to take initiative, argue logically, and work together. However, the students are not efficient in managing a business during practical task. This is caused by low of creativity and imagination, lack of facilities and infrastructure, limited funds for the materials laboratory, difference between learning environment and real life, and unsuitable learning strategy.

The problems of low quality of learning can be classified into the problem coming from (a) learning conditions, including student and material characteristics, and (b) learning methods used by educators [1]. Learning strategy, until now, is still characterized by a classical and a teacher centered learning so that learning objectives are achieved only in cognitive aspect, while affective and psychomotor aspects are not yet achieved. From those problems, the learning strategy should be turned into a student centered learning by using an active approach, multi-method, and multi-source learning.

Based on the above, it is necessary to do research on the effects of student learning strategy and soft skills of in Clothing Business Management course. The purpose of the study is to investigate: (1) The differences of learning achievement in Clothing Business Management course between students who receive gradual learning strategy and of those who receive conventional learning strategy, (2) The effect of interactions between learning strategy and soft skills toward students learning achievement in Clothing Business Management course, (3) The differences of learning achievement in clothing business management course between students with high soft skills who receive gradual learning strategy and of those who receive conventional learning strategy, (4) The differences learning achievement in clothing business management course between students with low soft skills who receive gradual learning strategy and of those who receive conventional learning strategy.

1.1. Clothing Business Management Learning Result

Richey describes that learning is a relatively permanent change in knowledge or behavior due to experience. The experience is not only meant physically but also cognitive and mental experience [1]. It happens because there is an interaction among the students and the learning environment including the school. Learning result is an achievement gained by the student in learning process, written as the numbers and applied on everyday life over competence that the student get.

High or low learning result demonstrates the success of educators in delivering learning materials. The result of one's learning depends on what he/she has known, goal, and motivation that affect the interaction process between one and the material being studied. Djaali says that the factors affecting the success of students in study [3]: 1) Internal factors, (derived from inside themselves) such as: health, intelligence, interest, and motivation, and 2) External factors (derived from outside themselves), such as family, school, community, and environment.

Clothing business management learning consists of production learning including development, characteristic, opportunity and the feasibility of clothing business including analysis of clothing business opportunity, economic analysis of clothing business, and the basics of garment production systems covering clothing production systems and feasibility studies of clothing business clothing business management learning focuses on production planning that is cut to order and cut inventory. The production process that has a smaller risk is cut to stock. It means cutting garment clothing if there is an order.

1.2. Learning Strategy

Learning strategy has five basic elements, namely (1) syntax, the operational steps of learning, (2) social system, the atmosphere and norms in learning, (3) principles of reaction, describes how teachers should look, treat and respond to students, (4) support system, all the facilities, materials, tools, or a learning environment that supports learning, (5) instructional and nurturent effects of learning result obtained directly based on the objectives that were targeted (instructional effects) and the results of learning beyond the target (nurturent effects) [4].

1.3. Gradual Learning Strategy

Molenda suggests a gradual strategy is a learning strategy that was developed to create learning activities effective and efficient, especially on learning activities requiring skills, gradually [5].

According to Arends gradual learning strategy is specifically designed to support the learning process for students with regard to structured procedural knowledge. It can be taught with the pattern of activity gradually or step by step. Additionally, gradual learning strategy is also aimed to help students learn the basic skills and obtain information about material presented and taught by lecturer step by step [6].

The gradual learning strategy, in five stages, is mainly used as a base for activities. The stages according Romiszowski are [7]:

Phase 1: Gaining knowledge about what to do, for what purpose, in what order and in what manner.
Phase 2: Implementing action step by step. Characteristics of this phase is being aware of application to knowledge, information and perception to initiate and control the action.
Phase 3: Transfer understanding from the eye to the other senses.
Phase 4: Movement skills through skill needs and what actions will be carried out.

Phase 5: All skills in various situations is collected, but the physical skills involved with strategy and skills stages of work that will be done.

1.4. Conventional Learning Strategy

Sanjaya writes that the conventional learning strategy is similar to the strategy of direct instructional, because learning material is delivered directly by the teacher. Sanjaya, also, interpret the strategy is a learning process that is more centered to teachers. Teachers centered approach is an approach where the teachers tend to control the learning process actively and as major information providers [8]. Ausebel argues that basically conventional study emphasizes the spoken process, learning materials given directly by educators. The student in this strategy is listening to learning material to be mastered from their teacher [9].

1.5. Soft skills in Learning

Soft skills refer to a set of characteristics of personality, social charm, language ability, personal habits, sensitivity/awareness, as well as optimism. Soft skills is a complement of hard skills. Soft skills cover (a) the personal qualities such as responsibility, selfconfidence, social skills, selfcontrol mana gement, and integrity or honesty (b) interpersonal skills, for instance participates in group, teaching orsharing knowledge, serving customers, leadership, negotiation skills, and work ability in diversity.

Elfrindi explained that soft skills are skills of a person to keep in touch with other. Attributes of soft skills possessed by each person at different levels, influenced by the habits of thinking, talking, acting and having an attitude [10].

To assess soft skills of students, evaluation by convergent answer is not enough, it needs to be complemented by a diverging item with various answers. Observation is needed to determine if the students apply the soft skills in daily life. Therefore, observation sheet is needed to know the performance of the students in doing the task and daily behavior.

2. Methods

The research was conducted in Dressmaking Study Program, Department of Family Welfare Education, Faculty of Engineering, Universitas Negeri Medan. The research method was experimental research. The design experiment was to distinguish the effect of learning strategy to the learning result of students in Clothing Business Management course with a focus on soft skills. The experiment design is the design of treatment by level 2 x 2.

The study population are 82 students of Dressmaking Study Program, who are following Clothing Business Management course. To determine the sample, random sampling was used to determine class A as an experimental class, and class B as a control class. Then, students soft skills scores were calculated, followed by determining the category of low and high of students soft skills score. After the scores were sorted, 27% lowest score and 27% the highest score from the experimental class (class A) and the control class (class B) were taken. The experiment subjects were 11 students who have a high soft skills score and 11 students who have a low soft skills score from class A and B. Therefore there are 44 students as a sample.

The dependent variable is the result of studying Clothing Business Management. It is the form of the result competency assessment based on competency standards specified in the curriculum of Dressmaking Studies Program. The result of student learning is measured by test to measure cognitive aspect and the observations sheets for the result of the practicum students to measure psychomotor aspect.

3. Results and Discussion

For the first hypothesis, the average value of students learning results using a gradual learning strategy is 84.68, while the average value of students learning results using conventional learning strategy is 82.73. Based on the summary result of Anova two lanes treatment by level indicates that $F_{count} = 4.75$ is greater than $F_{table} = 4.06$. So H_o rejected and H_a accepted at the 5% significance level it was concluded that students learning result using a gradual learning strategy is higher than that using conventional learning strategy.

For the second hypothesis, Anova summary shows that the value of $F_{count} = 6.25$ is greater than $F_{table} = 4.06$, it can be decided that H_o rejected and H_a accepted at the 5% significance level. It was concluded that there is an interaction between learning strategy and student's soft skills affecting the students learning results in Clothing Business Management subject.

For the third hypothesis, the average value of students group learning result who have high soft skills by using a gradual learning strategy is 89.27, while using conventional learning strategy is 84.77. Based on the summary results of Anova shows that $F_{count}= 52.65$ is greater than $F_{table}= 4.06$. So it can be decided that H_o rejected and H_a accepted at the 5% significance level. It was concluded that the students group learning result who have high soft skills by using a gradual learning strategy is higher than that the group who use conventional learning strategy.

For the fourth hypothesis, the average value of students groups learning result who have low soft skills by using a gradual learning strategy is 80.41, while using conventional learning strategy is 81.41. It can be concluded that the students group learning result who have low soft skills given the gradual learning strategy is not better than that the group who use conventional learning strategy.

There is an inconsistency between score that students get and their attitudes and behaviour in everyday life in implementing the Clothing Business Management learning. The students can not apply the concept of Clothing Business Management course in the real entrepreneur context yet. It means the learning target is not achieved maximally.

The first hypothesis shows that there is a difference in Clothing Business Management learning result, between the students group who was delivered a gradual learning strategy and the group who was delivered conventional learning strategy. It means, different learning strategy causes different learning result.

The second hypothesis shows that there is interaction between the gradual learning strategy and soft skills of students towards Clothing Business Management learning result. This is an indication that the treatment towards the group of students who have high soft skills is different from the group of students who have low soft skills. It means that one of the two groups will yield better learning using gradual learning strategy, while the other will be better using conventional learning strategy. This is because the students who have high soft skills will further demonstrate the desire to be independent, creative and innovative, and actively look for information about the clothing business. In contrast, the students who have low soft skills always wait for information from the lecturers to improve their learning result and do their task late.

The third hypothesis shows the Clothing Business Management learning result of the high soft skills student groups by delivering gradual learning strategy are higher than the group of those who use conventional learning strategy. It means, there is a difference between the two groups because of different learning strategy.

Student learning result is influenced by various factors, both internal and external factors. In general, the students who have high soft skills have these characteristics: they put more effort to achieve the feat, like challenges, accept personal responsibility, more creative, happy to complete the task, able to overcome failure, are sure that their actions lead to success, not anxious, not easy to feel tired, have more positive mood, more confidence, no hesitation, , do not easily tense encounter a problem, do not get bored quickly, jolly, so as to adapt to the situation and learning conditions. The student group that has high soft skills is more likely to interact optimally in learning activities and they are more enthusiastic in learning achievement based on high confidence in their ability.

4. Conclusions

Overall, there is difference in learning result between the Clothing Business Management student group who were delivered the gradual learning strategy and of that who were delivered the conventional learning strategy. There is an interaction between the learning strategy and soft skills affecting the learning result of Clothing Business Management student. The learning results of Clothing Business

Management students group with high soft skills by delivering gradual learning strategy is better than of those who is delivered conventional learning strategy. The learning results of Clothing Business Management students group who have low soft skills by delivering gradual learning strategy is not better than of those who was delivered conventional learning strategy.

References

[1] Reigeluth C M 2006 Functional contextualism: An ideal Framework for Theory in Instruction Design Techology *Educational Techologi Recearch and development* Vol **54**
[2] Richey R C and Klein J D 2007 *Design and Development Research; Method, Strategy, and Issues* (New Jersey: Lawrence Erlbaum Associates)
[3] Djaali 2011 *Psikologi Pendidikan* (Jakarta: Bumi Aksara)
[4] Joyce B W and Weil M 2007 *Strategi of Teaching* (New Jersey: Printice Hall Inc, Englewood Cliffs)
[5] Molenda M and Januszewski A 2008 *Educational Technology A Definition with Commentary* (New York: Lawrence Erlbaum Associates taylor & Francis Group)
[6] Arends R I 1998 *Learning to Teach* (New York: Mc Graw-Hill book Company)
[7] Romiszowski A J 2000 *The Development of Physical Skills*: *Instruction in the Psychomotor* Domain Volume II Chapter 19 Instructional Design Theories and Strategis: A New Paradigm of Instructional Theory (KOTA: Lawrence Erlbaum Associates)
[8] Sanjaya W 2010 *Kurikulum dan Pembelajaran, Teori dan Praktik Pengembangan Kurikulum Tingkat Satuan Pendidikan (KTSP)* Edisi 1 cetakan 3 (Jakarta: Kencana Predana Media Grup)
[9] Ausebel D et al 1982 *Educational Psycology: A Cognitive View* 2nd Spanish edition (New York: Holt, Rinehart & Winston)
[10] Elfrindi et al 2010 *Soft skills untuk Pendidikan* (Batam: Banuose Media Praninta Offset)

IOP Conf. Series: Materials Science and Engineering **306** (2018) 012026 doi:10.1088/1757-899X/306/1/012026

Blended Learning Implementation in "Guru Pembelajar" Program

D Mahdan*, M Kamaludin, H F Wendi, and M V Simanjuntak

Technology and Vocational Education, Universitas Pendidikan Indonesia, Jl. Dr. Setiabudhi 207 Bandung, 40154 Indonesia

*d2mahdan@gmail.com

Abstract. The rapid development of information and communication technology (ICT), especially the internet, computers and communication devices requires the innovation in learning; one of which is Blended Learning. The concept of Blended Learning is the mixing of face-to-face learning models by learning online. Blended learning used in the learner teacher program organized by the Indonesian department of education and culture that a program to improve the competence of teachers, called "Guru Pembelajar" (GP). Blended learning model is perfect for learning for teachers, due to limited distance and time because online learning can be done anywhere and anytime. but the problems that arise from the implementation of this activity are many teachers who do not follow the activities because teachers, especially the elderly do not want to follow the activities because they cannot use computers and the internet, applications that are difficult to understand by participants, unstable internet connection in the area where the teacher lives and facilities and infrastructure are not adequate.

1. Introduction

Implementation of vocational education is currently entering an important stage, the stage where vocational education graduates will be at stake in the workforce in the Asian region. Efforts must be made is to make the arrangements and improvements as much as possible in the vocational education sector, both arrangement in the pattern of recruitment, development of education and training programs, innovation of education and training process, evaluation and certification development.

Today's rapid technological developments, especially in the field of Information and Integration of ICT-based learning model into one of the new learning in learning in the classroom is accessible [1]. The influence of ICT in the world of education is increasingly felt because with the ICT learning patterns slightly different. From a conventional or regular face-to-face pattern to open and medium education. With the technology that can be remotely, learning in the present also exist through distance learning or e-learning by using the Internet network.

What is Blended Learning? Blended Learning comes from Blended and Learning. Blend means mixed and Learning means learning. "The combination of different training "media" (technologies, activities, and types of events) to create an optimum training program for a specific audience. The term "blended" means that traditional instructor-led training is being supplemented with other electronic formats. In the context of this book, blended learning programs use many different forms of e-learning, perhaps complemented with instructor-led training and other live formats"[2]. It represents an opportunity to integrate the innovative and technological advances offered by online learning with the interaction and participation offered in the best of traditional learning [3] From both elements of the word can be seen that Blended Learning is a mixture of learning patterns. According [4] said that the mixed learning patterns are the two main elements of learning in the classroom with online learning. In this online learning there is learning to use internet network in which there is web-based learning.

In this paper will be discussed the implementation of blended learning in the learner teacher program organized by the Indonesian department of education and culture that a program to improve the competence of teachers, called the Guru Pembelajar (GP).

2. Objective

This research aims to see how implementation of blended learning in Guru Pembelajar program and What are the factors that affect implementation of blended learning in Guru Pembelajar program?

3. Research Methodology

This research data is obtained to using observation, interviews and documentation. This research was conducted to evaluate the implementation blended learning in Guru Pembelajar Program. the activity is carried out in SMKN 2 Cimahi as one of the Learning Center for teachers of SMK Cimahi City and West Bandung Regency is followed by teachers from subjects of Mathematics, Entrepreneurship and engineering machinery, from October to December 2017 where teachers do face-to-face with mentors in designated learning centers and then do online learning to work on worksheets, consult with mentors and video calls for online learning. activities are held for 3 months, face-to-face learning 3 times where the first one-day face-to-face mentors provide materials on how to use an application called Learning Teacher Information System (SIGELAR), how to fill out worksheets and present learning materials in the first learning module.

Then the teachers conduct online learning for 30 days to learn modules, work on worksheets, and consult with mentors without being limited by time and place. Face-to-face two mentors evaluate online learning activities and explain the second learning module material, after which it is conducted again online for 30 days to learn modules, work sheet, consult with mentor. Third face-to-face teachers undertake teacher competency exams to assess overall learning, while the implementation as shown in the following figure:

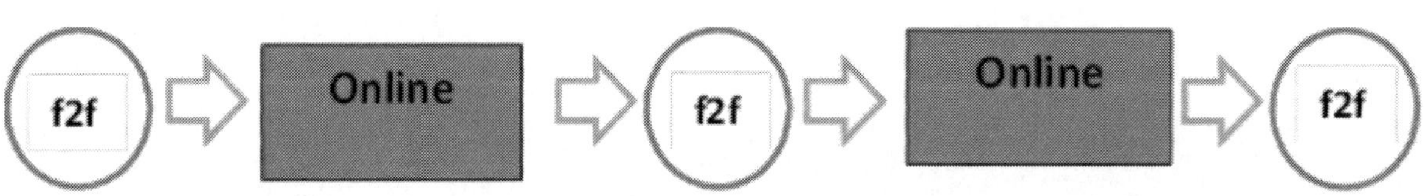

Figure 1. Blended Learning Implementation in Guru Pembelajar Program

4. Results and Discussion

Table 1. Face to face attendance of participants

No	Subject	Participant	f2f 1	%	f2f 2	%	f2f 3	%
1	Kewirausahaan	40	27	68%	17	43%	19	48%
2	Matematika	80	49	61%	54	68%	57	71%
3	Teknik Pemesinan	20	18	90%	11	55%	12	60%

Table 2. Online attendance of participants

No	Subject	Participant	Online 1	%	Online 2	%
1	Kewirausahaan	40	20	50%	25	63%
2	Matematika	80	15	19%	17	21%
3	Teknik Pemesinan	20	10	50%	15	75%

Based on table 1and table 2, problems in this activity is the presence of participants on face to face and online activeness less than should, from the interview and observation to all participant results difficulties in understanding the application 17%, unstable internet connection application 32% programs that still need improvement 19% and facilities and infrastructure are not adequate 32%.

5. Conclusion

Blended learning concept is one of innovation in learning. This innovation involves mixing conventional learning model and online learning model with internet network. Blended Learning will require learners to play a more active role in their learning. Because Blended Learning is a mixed learning model, the theory used also consists of various learning theories of some experts by adjusting the situation and learning conditions of learners. Blended learning model is perfect for learning for teachers, due to limited distance and time because online learning can be done anywhere and anytime. but the problems that arise from the implementation of this activity are many teachers who do not follow the activities because teachers, especially the elderly do not want to follow the activities because they can not use computers and the internet, applications that are difficult to understand by participants, unstable internet connection in the area where the teacher lives and facilities and infrastructure are not adequate.

References

[1] R Schibeci, J Maccallum, W C Potvin, C Durrant, B Kissane, and E J Miller 2008 Teachers' journeys towards critical use of ICT *Learn Media Technol* vol **33**, no November 2014, pp 313–327
[2] Bersin, J 2004 *The blended learning book: best practices, proven methodologies and lessons learned* San Francisco (CA: Pfeiffer Publishing)
[3] Kaye Thorne 2003 *Blended Learning How to Integrate Online and Traditional Learning* (London: Kogan Page Limited)
[4] Rusman dkk 2011 *Pembelajar dan Berbasis Teknologi Informasi dan Komunikasi* (Jakarta: PT Raja Grafindo Persada)

Priority of VHS Development Based in Potential Area using Principal Component Analysis

D Meirawan, A Ana, and S Saripudin

Universitas Pendidikan Indonesia, Bandung, Indonesia

*dmeirawan@gmail.com

Abstract. The current condition of VHS is still inadequate in quality, quantity and relevance. The purpose of this research is to analyse the development of VHS based on the development of regional potential by using principal component analysis (PCA) in Bandung, Indonesia. This study used descriptive qualitative data analysis using the principle of secondary data reduction component. The method used is Principal Component Analysis (PCA) analysis with Minitab Statistics Software tool. The results of this study indicate the value of the lowest requirement is a priority of the construction of development VHS with a program of majors in accordance with the development of regional potential. Based on the PCA score found that the main priority in the development of VHS in Bandung is in Saguling, which has the lowest PCA value of 416.92 in area 1, Cihampelas with the lowest PCA value in region 2 and Padalarang with the lowest PCA value.

1. Introduction

The quality of VHS (vocational school) graduates in Indonesia encountered several problems. In Indonesia, the problem is the lack of trust from the industrial world to VHS which includes aspect of knowledge, skills and attitudes possessed by VHS graduates which is seen to be inadequate to enter the industrial world.

According to data analyzed by West Java BPS in 2017, the number of unemployement in West Java is approximately 1.9 million people. Of that number, about 38.11% are graduates of SMA / VHS with age range between 20-24 years. The high unemployment rate of VHS graduates is due to the incompatible competencies between the VHS graduates and the required by the industry [1].

Government programs to expand the education provision is by increasing the number of schools such as VHS. The problem is how the Office of Education (DISDIK) can accurately locate the school in line with the region potential so that the location of VHS is not centered on a single point. The school location should not be far from the activities of educators, settlements, and transportation access to enable them easily accessed. The location of schools should also be supported with strategic conditions in the hope of supporting the learning so as to generate qualified graduates. Essentially, education planning is very influential on the quality of education [2].

VHS graduates are expected to work within the city/regency by developing the potential of the respective region. Regional potential can be more economic value if there are human resources who are capable to manage it well.

The opening of VHS and competency of expertise is conducted based on the interest of the society instead of the needs [3]. Many vocational graduates are unemployed due to the little availability of jobs and relevance to their competence or expertise. In addition, VHS graduates tend to seek for jobs in other areas.The slow growth of the local economy is due to the lack of skilled relevant workers to the needs of the community besides the VHS graduates tend to work in other areas [4].

The increasing number of students, educators and schools at vocational school level as well as the school potential have not been properly recorded. The data recording are done manually and has not been well integrated so the fast and accurate information need and reporting is still difficult to do. School mapping is an important part of educational planning process because it is dynamic, meaning that it follows the ongoing development of education.

2. Methods

The research approach used is descriptive qualitative analysis, using the principle of this component of analysis is the reduction of secondary data of education. The method used for analysis of potential development area is by using Principal Component Analysis (PCA) analysis. PCA is one form of multivariate analysis with high degree of difficulty but it allows to generate accurate data compared to other methods [5] [6]. PCA in this study is used to measure the variables of educational indicators, i.e the number of population aged 16-18 years, the ratio of school students, the ratio of students per class, the ratio of students per teacher, the ratio of students per new student, APK, ratio of SMA / VHS. Meanwhile, the needs of vocational school is done with PCA analysis with the help of Minitab Release Statistics software that is by considering the educational indicators used. The subject of this research is all the potential of regional and industrial world in West Bandung regency, West Java. The sample of research used is VHS and potential industrial area in West Bandung Regency (KBB). Data collection tool used in this research is interview, observation and sheet for data grouping or recapitulation table. This data collection tool is prepared by researchers by adopting the standards used by the Central Bureau of Statistics (BPS). The research designs developed are as follows. See Figure 1:

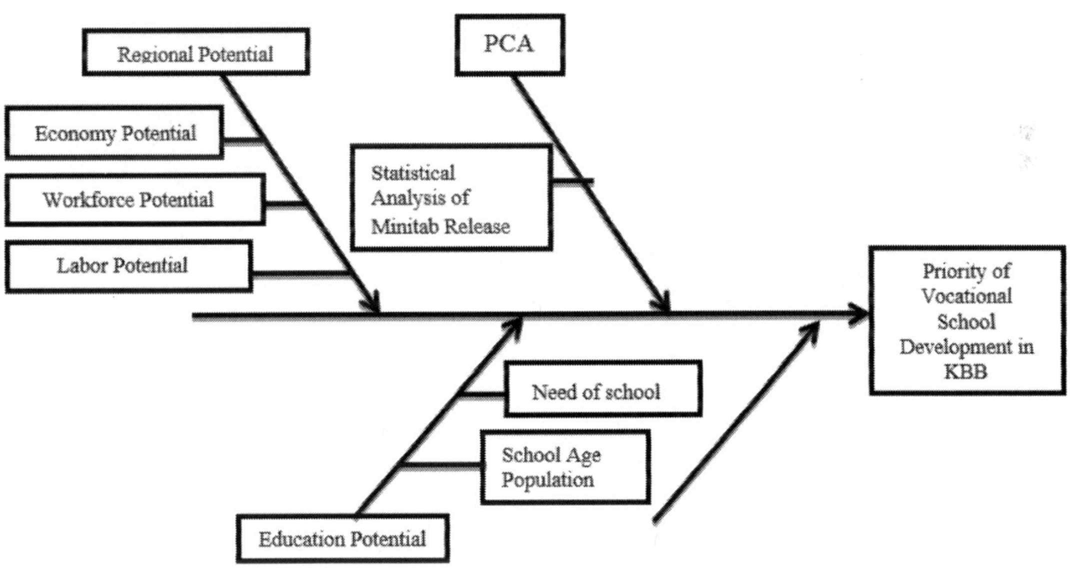

Figure 1. Research Design

3. Results and Discussion

The priority development of West Bandung Regency has 16 Districts with 165 Villages. West Bandung regency consists of 16 districts, namely: Rongga District, Gununghalu District, Sindangkerta

IOP Conf. Series: Materials Science and Engineering **306** (2018) 012027 doi:10.1088/1757-899X/306/1/012027

District, Cililin District, Cihampelas District, Cipongkor District, Batujajar District, Saguling District, Cipatat District, Padalarang District, Ngamprah District, Parongpong District, Lembang District, Cisarua District, Cikalongwetan District and Cipeundeuy District. West Bandung Regency is 1,305.77 km2 wide.

Figure 2 shows the widest district in West Bandung Regency is Gununghalu District and the smallest one is Batujajar District. District of Padalarang, Lembang, Cililin, and Batujajar is the region growth point. District of Ngamprah, Cikalongwetan, and Cipatat serve as the main service center. District of Parongpong, Cisarua, Sindangkerta, and Cihampelas serve as local service centers. Meanwhile, District of Cipeundeuy, Cipongkor, Gununghalu, and Rongga serve as the smallest service centers.

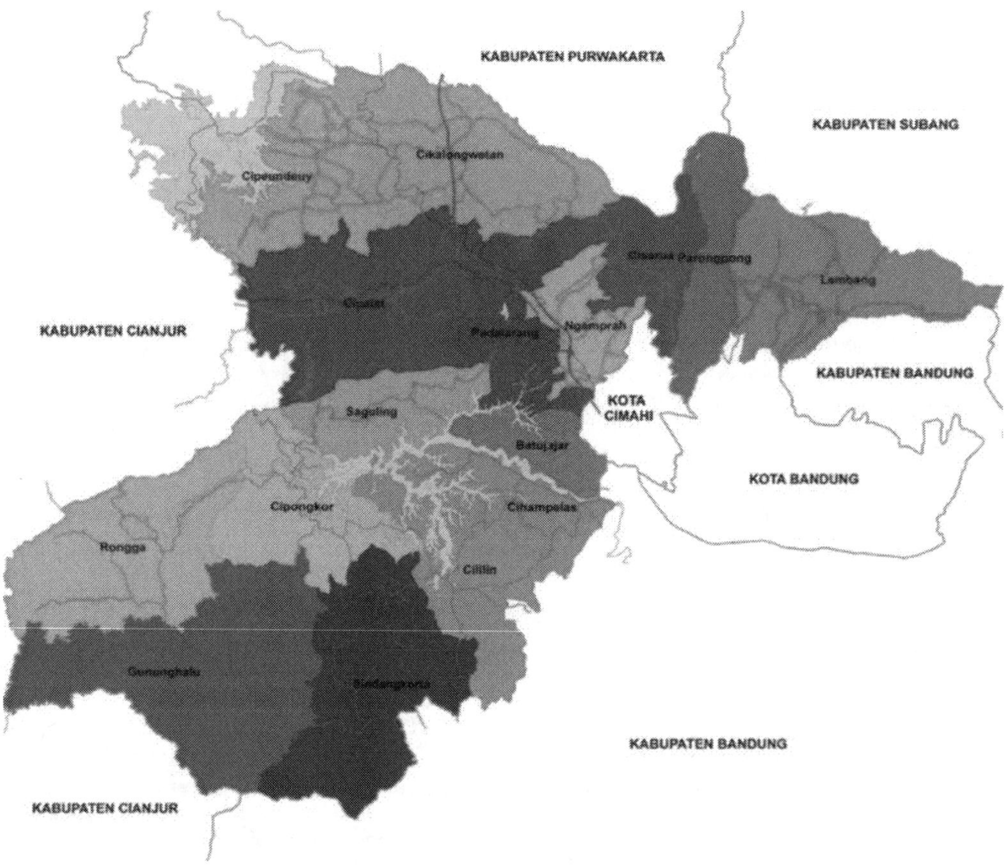

Figure 2. Regency of Bandung Barat
Source: BPS Regency of Bandung Barat 2016

The number of VHS in West Bandung regency is 91 schools with 201 skill packages. The number of Public VHS in KBB is 8 and Private VHS is 83. Education Office of KBB divides them into 6 clusters as follows:

Cluster 1: Lembang, Cisarua, dan Parompong
Cluster 2: Ngamprah, Padalarang dan Cipatat
Cluster 3: Batujajar, Cihampelas, dan Cililin
Cluster 4: Cipendeuy dan Cikalong Wetan
Cluster 5: Saguling dan Sindangkerta
Cluster 6 : Cipongkor, Rongga dan Gunung Halu

(a) Based on survey results and data from NIEP (National Indicators for Education Planning) in 2016 and referring to VHS 2016 spectrum, VHS in KBB includes 6 areas of expertise, 15 Expertise Programs with 201 skill competencies. The technological and engineering expertise is 29.85% (60 skill competencies). The technology and information expertise is 27.36% (55 skill expertise). The health expertise 10.45% (21 skill competencies). The agribusiness and agro-technology expertise is 8.96% (6 skill competencies). The business management expertise is 20.40% (41 skill competencies) and the tourism expertise is 8.96% (18 skill competencies).
(b) The PCA method is used to analyse the VHS development prioritizing by considering the educational indicators. Indicator of VHS education based on SWP in KBB can be seen in table 1 below:

Table 1. Indicator of Education Equality of SMA/MA and VHS According to SWP in Bandung Barat Regency Year 2017

Distribution Area	Number of Student	Population Age of 16-18 (XI)	Student Ratio			APK (X5)
			School (X2)	Class (X3)	Teacher (X4)	
Area I	3335	15722	1693,25	282,98	97,62	134,37
Area II	8233	27098	1372,17	191,45	95,63	158,26
Area III	11703	24356	1567,45	250,55	114,05	236,51

The analysis of the need level of vocational school is done using PCA calculation method. The educational indicators used as the basis for PCA calculation are Age Population 16-18 (X1), Ratio of Students Per School (X2), Ratio of Students Per Class (X3), Student Ratio Per Teacher (X4), and Participation Rate (APK / X5) which can be seen in table 2:

Table 2. Indicator of Education Equality Based on District in Bandung Barat Regency Year 2017

SWP	Name of Distric	Number of Student	Population Age of 16-18 (XI)	Student Ratio Per			APK (X5)	Ratio of SMA/VHS	
				School (X2)	Class (X3)	Teacher (X4)		SMA	VHS
1	Rongga	582	1797	582	34,24	17,64	32,39	2	1
2	Gununghalu	1224	4042	153	28,47	15,9	30,28	1	8
3	Sindangkerta	364	2241	121,33	91	13,48	16,24	4	3
4	Cililin	1173	5042	195,5	33,51	20,22	23,26	7	6
5	Cihampelas	2631	4245	438,5	38,69	20,72	61,98	1	6
6	Cipongkor	1006	4594	167,67	34,69	17,96	21,90	4	6
7	Batujajar	2624	2856	291,56	82	24,75	91,88	3	9
8	Cipatat	2748	7076	249,82	56,08	27,76	38,84	2	11
9	Padalarang	383	2633	348,18	42,56	22,93	14,55	6	11
10	Ngamprah	2111	7572	351,83	37,7	21,11	27,88	3	6
11	Parongpong	611	4029	152,75	47	17,97	15,17	3	4
12	Lembang	4724	7749	524,89	41,44	22,71	60,96	10	9
13	Cisarua	320	2066	160	40	13,91	15,49	2	2
14	Cikalong Wetan	1312	5645	218,67	46,86	15,62	23,24	2	6
15	Cipeundeuy	1147	4547	573,5	31,86	20,48	25,23	4	2
16	Saguling	311	1042	103,67	38,88	14,14	29,85	2	3
	Total	23271,00	67176,00	4632,87	724,98	307,30	529,12	56	93

Based on indicators of educational equality, the lowest percentage of APK for the secondary education is located at District of Padalarang (14.55%), and District of Parongpong 15.17%.

Calculation through PCA method can be done by finding the value of PCA 1 first, as can be seen in table 3:

Table 3. Score of PCA 1 for Each District

SWP	Name of District	Number of Student	Population Age of 16-18 (XI)	Student Ratio Per School (X2)	Student Ratio Per Class (X3)	Student Ratio Per Teacher (X4)	APK (X5)	Score PCA 1	Label
I	Rongga	582	1797	582	34,24	17,64	32,39	620,22	Priority II
	Cisarua	320	2066	160	40	13,91	15,49	857,23	
	Cipeundeuy	1147	4547	573,5	31,86	20,48	25,23	1839,77	
	Sindangkerta	364	2241	121,33	91	13,48	16,24	936,06	
	Saguling	311	1042	103,67	38,88	14,14	29,85	416,92	Priority I
	Parongpong	611	4029	152,75	47	17,97	15,17	1725,21	
II	Cililin	1173	5042	195,5	33,51	20,22	23,26	2161,66	
	Cihampelas	2631	4245	438,5	38,69	20,72	61,98	1733,80	Priority I
	Cipongkor	1006	4594	167,67	34,69	17,96	21,90	1971,62	Priority II
	Ngamprah	2111	7572	351,83	37,7	21,11	27,88	3235,68	
	Cikalong Wetan	1312	5645	218,67	46,86	15,62	23,24	2420,30	
III	Gununghalu	1224	4042	153	28,47	15,9	30,28	1731,18	
	Batujajar	2624	2856	291,56	82	24,75	91,88	1145,02	Priority II
	Lembang	4724	7749	524,89	41,44	22,71	60,96	3258,74	
	Cipatat	2748	7076	249,82	56,08	27,76	38,84	3037,54	
	Padalarang	383	2633	348,18	42,56	22,93	14,55	1055,02	Priority I
	Total	23271	67176	4632,87	724,98	307,3	529,12		

Based on educational indicators, the lowest PCA 1 score for SWP I was found in Saguling District (416.92). This means for SWP I, the priority location for vocational school development is Saguling District. Next is followed by Rongga District which became the priority location for vocational school development. For SWP II, the lowest PCA 1 score was Cihampelas District (1733.80). Therefore, Cihampelas District is a priority location for the development of vocational schools. Then, followed by Cipongkor District (1971.62) which became the second priority of vocational school development. As for SWP III, the vocational priority point for the development of vocational schools is Padalarang District which has the lowest PCA 1 score (1055.02). Then, the next priority is Batujajar District (1145.02).

Table 4. Eigenvalue (Output MINITAB)

Eigenanalysis of the Correlation Matrix					
Eigenvalue	2,1705	1,3346	0,8196	0,3786	0,2968
Proportion	0,434	0,267	0,164	0,076	0,059
Cumulative	0,434	0,701	0,865	0,941	1,000

The first point to note in table 4 above is the eigenvalue value of the five components. The factoring process is terminated in a component with an eigenvalue value below 1. Therefore, the factoring process will stop at the second component where its eigenvalue value is below 1 (0.8196).

IOP Conf. Series: Materials Science and Engineering **306** (2018) 012027 doi:10.1088/1757-899X/306/1/012027

Eigenvalue is the value of the main component variant (principal component, PC). The output shows the eigenvalue for the first main component (PC1) and the second (PC2) is 2.1705 and 1.3346, respectively. Eigenvalue of the two main components represent 43.6% and 26.7% of all variability. When accumulated, the two main components represent 70.1% of the total variability. This means that if the five variables are reduced to 2 variables, then the two new variables can account for 70.1% of the total variability. The following table 5 is the component matrix table of each variable:

Table 5. Component Matrix

	Eigenvectors				
Variable	PC1	PC2	PC3	PC4	PC5
The population aged 16-18 (X1)	0,442	-0,291	-0,690	0,006	-0,493
School (X2)	0,467	-0,272	0,614	-0,499	-0,286
Class (X3)	-0,006	0,800	-0,176	-0,522	-0,238
Teacher (X4)	0,600	0,085	-0,193	-0,155	0,756
APK (X5)	0,476	0,440	0,280	0,674	-0,217

Scores for established components can be calculated by looking at the coefficient values for each variable. For PC1 components, the score can be calculated as follows:

PC1 = 0,442 population age of 16-18 years old + 0.467 schools - 0.006 class + 0.600 teacher + 0.476 APK

The second component (PC2) has an eigenvalue of 1.3346 and can account for 26.7 percent of the diversity. Together with the first component (PC1), both represent 70.1 percent of total diversity. Scores for PC2 are calculated as follows:

PC2 = -0.291 population age of 16-18 years old - 0.272 schools + 0.800 class + 0.085 teacher + 0.440 APK

Determination of the number of components to be used is very subjective. In this case, both PC1 and PC2 components which represen 70.1 percent of total diversity can be judged to have adequately captured the data structure. Or, using the first component is adequate to capture the data structure, when viewed with the eigenvalue value criterion greater than 1. Other components have a small proportion of diversity can be considered unimportant.

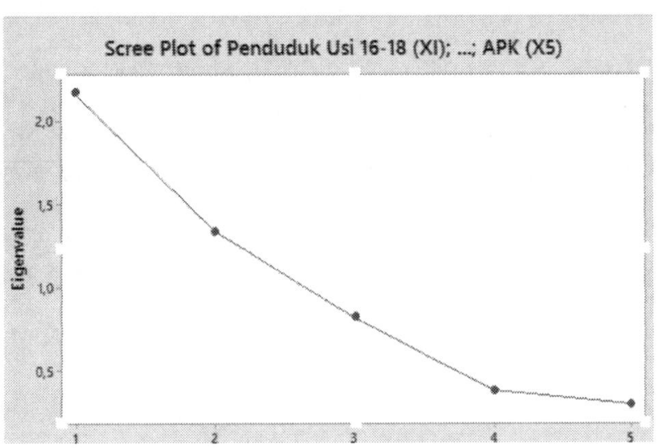

Figure 3. Scree Plot

From the scree plots of Figures 3 and 4 it can be seen that the eigenvalue above 1 is only 2. This proves that there are only two factors can be established.

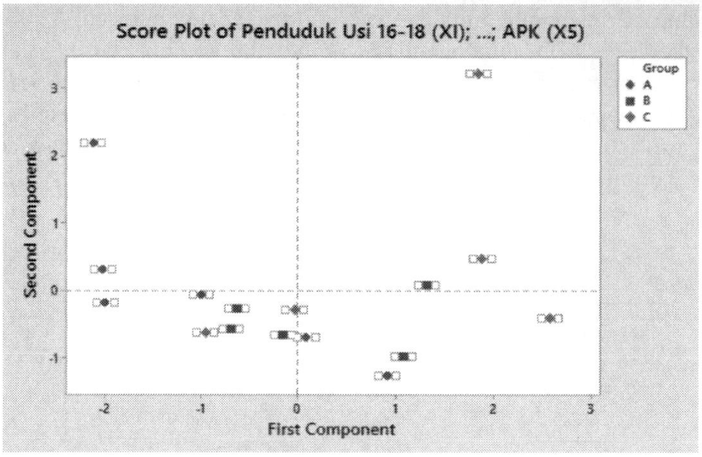

Figure 4. Score Plot

4. Conclusions

The regional potential of Bandung Barat Regency (KBB) is largely agriculture, the prominent potential of a prominent is Agribusiness and Agrotechnology, Livestock, Plantation, Creative Industries, Food and Beverage and Tourism. The main finding of this research is that there is a disparity between the vocational skills program in KBB that is not in accordance with the industry availability in every district.

References

[1] Badan Pusat Statistik 2017 *Keadaan Ketenagakerjaan Agustus 2017*

[2] Enoch and Jusuf 1992 *Dasar-Dasar Perencanaan Pendidikan* (Jakarta: Bumi Aksara)

[3] Pramono Y 2009 *Kajian Kebutuhan dan Penyediaan Sekolah Menengah Kejuruan di Kabupaten Rembang* (Semarang: Universitas Diponegoro)

[4] Jatmoko D 2013 Relevansi Kurikulum VHS Kompetensi Keahlian Teknik Kendaraan Ringan terhadap Kebutuhan Dunia Industri di Kabupaten Sleman *Jurnal Pendidikan Vokasi* **3**(1), 1–13

[5] Bodroža-solarov M, Kljajić P, Andrić G, Filipčev B, Šimurina O, Golić M P, and Adamović M 2011 Application of Principal Component Analysis in Assessment of Relation Between the Parameters of Technological Quality of Wheat Grains Treated with Inert Dusts Against Rice Weevil (Sitophilus oryzae L) *Scientific Journal "Pesticides and Phytomedicine,"* **26**(4), 385–390

[6] Žnidaršič T, Verbič J, and Babnik D 2006 Prediction of Chemical Composition and Energy Value of Grass Silage By Near-Infrared Reflectance Spectroscopy Napovedovanje Kemične Sestave in Energijske Vrednosti Travne Silaže Z Bližnjo Infrardečo Refleksijsko Spektroskopijo *Journal of Central European Agriculture* **7**(1), 127–134

IOP Conf. Series: Materials Science and Engineering **306** (2018) 012028 doi:10.1088/1757-899X/306/1/012028

Teaching Quality and Learning Creativity in Technical and Vocational Schools

D R E Kembuan, P T D Rompas, M Mintjelungan, T Pantondate, and B M H Kilis

Universitas Negeri Manado, Indonesia

*eddy.kembuan@yahoo.co.id

Abstract. The purpose of this study is to obtain information about the teacher quality of teaching and learning creativity with the outcomes of student learning in a vocational high school in Indonesia. This research is a survey research. The sample used in this research is 50 teachers, selected by simple random sampling. Data were analyzed by using correlation analysis. The findings of this study are as follows: (1) There is a significant and positive correlation between teacher quality of teaching with the outcomes of student learning at the vocational high school; (2) There is a significant and positive correlation between learning creativity with the outcomes of student learning at the vocational high school, and (3) there is a significant and positive correlation between the teacher quality of teaching and learning creativity with the outcomes of student learning at the school. That is, if the use of appropriate the teacher quality of teaching and learning creativity, then the outcomes of student learning at the school. Finally it can be concluded that to improve the outcomes of student learning, it has to be followed by an improvement of teacher quality of teaching and learning creativity.

1. Introduction

Education is the process of carrying out a conscious effort to change the knowledge, attitude, and equip one's or group's skills through the interaction of learning to make an individual fit with his or her social life and to help pass on the habits, laws, beliefs, languages and institutions Social institutions from one generation to the next.

Law of the Republic of Indonesia No. 20 of 2003 on the National Education System, namely that the national education aims to develop the potential of learners to become human beings who believe and cautious to God Almighty, noble, healthy knowledgeable, capable, creative, independent and become citizens of a democratic and to be responsible.

The main objective to be achieved by every educational institution, especially formal education is to produce graduates who excel. Achievements in question is the achievements generated through the learning process. The expectation of a good learning achievement is of course supported by the quality of teachers that are teachers who have social competence and professional competence.

At the senior high school level, the problem of learning outcomes is very important to be able to continue in the next education, but in reality there are still vocational graduates who lack the ability in the field of science, consequently adjustments in education level many experiencing problems, especially the Related to the adjustment with the lesson. Increased student learning outcomes is a process that must be continuous in every aspect that shapes the overall service of education.

Improvement of student learning outcomes as a result of educational process can not only be done by one or more aspects only and at one particular period only, but must be done systemically, continuously, and done by all elements.

Based on the researcher's observation of SMK, many of the teachers who pay less attention to the elements of teaching quality, such as the quality of teaching teachers, students' learning creativity, the implementation of learning activities, teaching evaluation, the use of instructional media that less support the smooth learning process, and the use of teaching methods Inefficient, and effective. This results in less satisfactory learning outcomes.

The role of the teacher in the learning process strongly determine the success of students in learning. In the context of academic learning, creativity can be thought of as occurring at both a subjective (creativity as part of the act of learning) and an intersubjective (learning as a creative act) level [1]. At the subjective level, students exercise their creativity by developing new and personally meaningful ideas, insights, and understandings within the context of particular academic constraints. Thus the role of teachers in the learning process is crucial to student learning outcomes.

The elements that affect the quality of teaching teachers according to Miller are the experience of teaching teachers and professionalism [2]. Both of these elements can affect the quality of teachers, according to him that the more experienced teachers in carrying out their duties as teachers, the more professional qualifications are adequate. While the quality of teacher professionalism can be improved through teacher education.

Darling states that in the effort to realize the quality of teaching teachers, it is necessary to pay attention to several factors such as teachers must meet professional standards by selecting prospective teachers, the level of good school accreditation, the knowledge and skills of teachers who continue to be improved [3]. In an effort to improve student learning outcomes required creativity of learning and creativity of student learning.

Both factors are important in the effort to improve student learning outcomes through learning activities. A "value-added" assessment of the degree to which teachers who are already in the classroom contribute to their students' learning, as indicated by higher-than-predicted increases in student achievement scores [4].

Student learning creativity is a process that is reflected in fluency, regularity, and originality in thinking. In learning activities, student learning creativity plays an important role, because thanks to the creativity of learning, step by step to make the world of education more advanced. Creativity is yet another student characteristic that shares a conceptual, albeit equivocal, link with academic achievement. As we have discussed, researchers have reported associations that are relatively strong [5].

In general, the process of thinking trained in schools is limited to cognition, memory, and convergent thinking, whereas divergent thinking and evaluation are overlooked. Divergent thinking is to provide a variety of possible options for solving problems based on the information provided, with an emphasis on diversity and suitability. According to Lampe, that learning creativity is important that support student achievement. Creative students are students who have the ability to work with others in learning [6]. It can support the achievement of learning achievement in accordance with the purpose of learning. Khalid argued that in the learning activities, teachers are required to understand the creativity of student learning, by understanding the creativity of student learning, facilitate teachers to apply appropriate and strategic teaching approach in accordance with student learning needs.

Davies (1981: 246) states, learning outcomes are not the same as performance, learning outcomes involve acquiring knowledge, skills or attitudes while performance concerns the use of demonstrated knowledge, skills and attitudes that can show that a person has indeed mastered them accordingly with predetermined standards as a result of its learning [7]. Learning outcomes are the abilities students have after they have received their learning experience. Learning outcomes is a person's maximum performance in mastering materials or materials that have been taught or have been studied.

The research hypothesis consists of 3 (three) as follows: 1) There is a positive relationship between the quality of teaching teachers with the results of student learning SMK Negeri 2 Manado, 2) There is a positive relationship between student learning creativity with student learning outcomes SMK Negeri 2 Manado, 3) A positive relationship between the quality of teachers teaching, and creativity of learning with student learning outcomes SMK Negeri 2 Manado.

2. Method

This research was conducted in SMK Negeri 2 Manado, and implemented three months, beginning in March 2016 and is expected to be completed in May 2016. This research use survey method with correlation analysis technique. In this study the population is all students of SMK Negeri 2 Manado a number of 98 students, 50 teachers were selected randomly as the sample. Data collection techniques were conducted using questionnaire-shaped research instruments. Testing of the instrument includes validity test (validity) and reliability test (reliability). From the test results obtained valid and invalid instrument items. Invalid instrument removed. Instruments to be used to obtain research data, conducted by using the five scale questionnaire. Instruments are based on the theories used. Questionnaires were distributed to teachers who became respondents in this study. Questionnaires are used to obtain research data as described below.

3. Results

To test the correlation between teacher teaching quality (X_1), and learning creativity (X2) with student learning result of SMK Negeri 2 Manado (Y), either individually or jointly, hypothesis testing with analytical technique: (1) Simple correlation; (2) double correlation. The results of testing the research hypothesis are as follows:

3.1. The correlation of teacher teaching quality (X_1) with student learning outcomes SMK Negeri 2 Manado (Y)

The first hypothesis proposed states that there is a positive correlation between teacher teaching quality variables (X1) with student learning outcomes SMK Negeri 2 Manado (Y). To know the correlation is used simple correlation analysis between X1 with Y or ry1. The result of regression significance test showed that Fh = 9,548> F (0.01; 1/48) = 7,12 and result of regression linearity test showed Fh = 0.848 <Ft (0,05,29/19) = 4.04, Student learning SMK Negeri 2 Manado (Y) for teacher teaching quality (X1) is very significant and their correlation is linear. Thus, the quality of teaching teachers is one predictor of student learning outcomes SMK Negeri 2 Manado.

The strength of the correlation between the qualities of teaching teachers with student learning outcomes SMK Negeri 2 Manado can be shown by the correlation coefficient ry1 of 0.771. The result of significance test of correlation coefficient shows that the correlation between teacher teaching quality (X1) with student learning result variable of SMK Negeri 2 Manado (Y) is positive and very significant, in other words the higher the quality of teaching the student teacher, the better the student learning outcomes. Thus, it receives the first hypothesis which states: "there is a positive correlation between teacher teaching qualities (X1) with the variable of student learning outcomes of SMK Negeri 2 Manado (Y)." The coefficient of determination of 0.594 shows that 59.4% Learning student of SMK Negeri 2 Manado, explained by teacher teaching quality through regression equation line $\hat{y}$ = 22,824 + 0,779X1.

3.2. The correlation of learning creativity (X2) with student learning outcomes SMK Negeri 2 Manado (Y).

The second hypothesis proposed states there is a positive correlation between learning creativity (X_2) with the variable of student learning outcomes SMK Negeri 2 Manado (Y). To know the correlation is used simple regression analysis. The linearity results show that the null hypothesis is accepted and the alternative hypothesis is rejected, since Fh = 0.919 <Ft (0.05; 22/26) = 2.07. Thus it can be concluded that the regression equation Y = 18.720 + 0.405X2 is linear. It means that every increase of one score creativity of learning will lead to increase student learning outcomes SMK Negeri 2 Manado of 0.405 at Constanta 18.720. The strength of the correlation between learning creativity (X_2) and student

learning outcomes SMK Negeri 2 Manado (Y) can be shown by the correlation coefficient ry_2 of 0.407.

The test results show that the correlation between learning creativity (X_2) and student learning outcomes of SMK Negeri 2 Manado (Y) are positive and very significant. That is, the better the quality of teaching teachers the higher the results of student learning SMK Negeri 2 Manado. Coefficient of determination (ry_2) 2 = 0,166 shows that 16,6% variance that happened in student learning result of SMK Negeri 2 Manado can be explained through student's learning creativity. Thus the second hypothesis that states: "there is a positive correlation between learning creativity with student learning outcomes SMK Negeri 2 Manado is accepted".

3.3. The correlation between the quality of teaching teachers, and creativity of learning together with student learning outcomes SMK Negeri 2 Manado.

The third hypothesis proposed is: there is a positive correlation between teacher teaching quality, and creativity of learning with student learning outcomes SMK Negeri 2 Manado. To know the existence of correlation together between the three independent variables with the dependent variable, used multiple regression. From regression and multiple correlation test results obtained by multiple regression equation shown by equation $\hat{Y} = 25{,}412 + 0{,}757X1 + 0{,}045X2$.

To know the degree of significance of the equation of multiple regression lines, then tested F, the result is as follows: The result of regression significance test is Fh = 34,599> F (0.01; 2/47) = 7,271. This shows that regression of student learning result of SMK Negeri 2 Manado (Y) for teacher teaching quality (X1), and creativity of learning (X2) is very significant. This means the quality of teaching teachers, and creativity of learning together are predictors of student learning outcomes SMK Negeri 2 Manado. Regression equation $\hat{Y} = 25{,}412 + 0{,}757X1 + 0{,}045X2$ shows that everyone score increase on teacher quality teaching quality, and quality of teaching teacher together, can cause increase of learning result of student of SMK Negeri 2 Manado. The strength of the correlation between these variables is indicated by the double correlation coefficient of 0.822. Significant test of correlation coefficient, as follows: Duality Test of Coefficient of Double Correlation of teacher teaching quality (X1), and creativity of learning (X2) together with student learning outcomes SMK Negeri 2 Manado (Y)

The test results show that the correlation between teacher teaching quality, and creativity of learning together are positive and very significant, with the coefficient of determination of 0.676. These findings indicate that the higher the quality of teaching teachers, and creativity of learning together, it will also increase student learning outcomes SMK Negeri 2 Manado, 67.6% variance that occurs in the learning outcomes of students SMK Negeri 2 Manado, described together Equals by variable X1 and X2 through regression line equation $\hat{Y} = 25{,}412 + 0{,}757X1 + 0{,}045X2$.

4. Conclusions

Testing of proposed research hypothesis proved that teacher quality teaching quality (X1), creativity of learning (X2), both individually and jointly have positive and significant relation with learning result of SMK Negeri 2 Manado (Y) students.

First, the correlation coefficient between the qualities of teaching teachers with the learning outcomes Students SMK Negeri 2 Manado is equal to 0.771. This shows that the quality of teaching teachers can contribute to improve learning outcomes Students SMK Negeri 2 Manado 59.4%. With a large contribution of teacher quality teaching quality to the learning outcomes Students SMK Negeri 2 Manado is equal to 59.44%.

Second, the correlation coefficient between learning creativity and learning outcomes Student SMK Negeri 2 Manado is equal to 0.407. This shows that the creativity of learning also contributes high enough to increase the learning outcomes of students of SMK Negeri 2 Manado is 16.6%. With a large, variable contribution of learning creativity to learning outcomes Student SMK Negeri 2 Manado is 16.6%.

Third, the result of correlation and multiple regression on self-control, and creativity of learning together with consumer Satisfaction obtained a large correlation coefficient of 0.822 or 67.6% contribution that occurs in the learning outcomes Student SMK Negeri 2 Manado can be explained together Through teacher teaching quality and learning creativity condition through regression equation $\hat{Y} = 25{,}412 + 0{,}757X1 + 0{,}045X2$.

Based on the result of the research, there are some suggestions that can be stated concerning the learning outcomes of SMK Negeri 2 Manado for the quality of teaching teachers, the creativity of learning as follows: 1) School parties need to encourage teachers, especially students to improve competitive and competitive learning achievement and utilize (2) Schools need to work with parents to manage or monitor the activities of students, whether at school or at home, 3) Students need to pay attention to the interests needed in achieving learning achievement to provide convenience when faced with the problem- (4) Shiva needs to change behavior that is ugly or less acceptable to teachers and classmates, 5) Research study can be applied in other learning, 6) The author suggests research on the same topic but the object is the vocational school.

References

[1] Gajda A, Karwowski M, and Beghetto R A 2016 Creativity and Academic achievement: A Meta-Analysis *Journal of Educational Psychology* Accessed Nov 10 2017
[2] Miller and D Cindy Hartzler 2009 Learning to teach teachers in astandars-based program: when experiences is'n enough *Action in teacher education* v **21** no 3
[3] Darling, Hammond and Linda 2006 What matters most: a competent teacher for every child *Phi Delta Kappan* v **78** Nov 2006
[4] Goe Laura, Leslie M and Stickler 2008 *Teacher Quality and Student Achievement: Making The Most of Recent Research* (Washington: National Comprehensive)
[5] Gajda A, Karwowski M, and Beghetto R A 2016 Creativity and Academic Achievement: A Meta-Analysis *Journal of Educational Psychology* https://www.researchgate.net/publication/306296585 [accessed Nov 10 2017].
[6] Lampe, Judith R 2006 *Effect of Cooperativie Learning among Hispanic Students ini Elementary Social Studies, The Journal of Education Research* (Washington DC) v **89** Jan/peb 2006
[7] Davies and Ivor K 1981 *Instrucktional Technique* (New York: McGraw-Hill Book Company)

IOP Conf. Series: Materials Science and Engineering **306** (2018) 012029 doi:10.1088/1757-899X/306/1/012029

The Use of Geometry Learning Media Based on Augmented Reality for Junior High School Students

D Rohendi*, **S Septian, and H Sutarno**

Computer Science Education, Universitas Pendidikan Indonesia, Bandung, Indonesia

*dedir@upi.edu

Abstract. Understanding the geometry especially of three-dimensional space is still considered difficult by some students. Therefore, a learning innovation is required to overcome students' difficulties in learning geometry. In this research, we developed geometry learning media based on augmented reality in android flatform's then it was implemented in teaching three-dimensional objects for some junior high school students to find out: how is the students response in using this new media in geometry and is this media can solve the student's difficulties in understanding geometry concept. The results showed that the use of geometry learning media based on augmented reality in android flatform is able to get positive responses from the students in learning geometry concepts especially three-dimensional objects and students more easy to understand concept of diagonal in geometry than before using this media.

1. Introduction

Geometry is an area of mathematics which is essential to be learned at schools. By learning geometry, students may be able to identify shapes and space around them. Geometry can also help them to represent their surroundings. The comprehension of geometry models and their properties may give students a new perspective [1]. Thus, they are able to analyze and communicate geometry-related things in their daily lives. The geometry comprehension relates to spatial ability.

Basically, geometry lesson has been taught to students since they are enrolled in elementary schools. This competence is one of the fundamental aspects which is important to be acquired by students to understand further lessons regarding geometry. The material for geometry lesson at elementary school itself is designed to introduce students with simple geometric shapes by adjusting it with children's level of thinking.

To achieve the required geometry competence, the geometry material is included in the mathematics curriculum, starting from elementary school to university. Every student shall try to develop their spatial ability and sense which is very useful in solving problems in mathematics and everyday life [2]. The importance of spatial ability which is required in engineering sciences and mathematics, especially geometry [3]. It is in line with [4] opinion which argues that spatial ability has a significant role in developing science, technology, engineering and mathematics skills.

The process of learning geometry is not always running well. Although, geometry instruction as one main factor of improving the students' spatials abilities [5]. Moreover, there is an assumption that geometry or mathematics is one of the most difficult subject and is not preferred by most learners [6]. In learning geometry particularly, many students are facing difficulties [7]. Points out that many geometrical problems require certain visualization in problem solving and students generally find it difficult to construct three-dimensional spaces.

It is emphasized that the learning process in all subjects should apply a certain method which involves students more (student center) and optimizes the use of innovative learning media. However, what actually happen in the current situation is that the requisites have not been properly implemented.

The learning method used is still dominated by teachers and the utilization of innovative learning media is still limited. This particular learning process is not appropriate, because children's memory is limited and they only remember visible things. Learning mathematics with applying rote learning method will only burden the brain memory and make students reluctant to learn mathematics, decrease their motivation and mathematical ability.

To anticipate these problems, a multimedia is tried to be applied to improve student learning outcomes, such as Graphical User Interface (GUI) based applications such as Augmented Reality (AR) application. AR-based multimedia is able to display both two-dimensional and three-dimensional objects by showing every part of the objects in detail. By having this capability, we are able to explain geometric concepts to improve students' spatial ability. However, to date, the use of multimedia learning using AR technology is still very limited, whereas AR technology is very promising and has advantages when it is applied in learning processes.

Therefore, this study is conducted to facilitate the process of learning geometry and attract students' interests, as well as provide them with new experiences in interacting through learning media. The formulation of the research problem is: 1) How to develop augmented reality media that can be used in teaching geometry? 2) How are the studenst responses and learning activities of junior high school students when they are using augmented reality-based geometry learning media?

2. Literature Review

2.1 Augmented Reality

Augmented Reality (AR) is the combination of two concepts which are *Virtual Reality* (VR) and *world reality*. Therefore, two-dimensional or three-dimensional virtual objects are seemingly real and compromise within the real world. The AR is a new technology which is capable in presenting possibilities that are difficult for other technologies to offer and meet. In this research paper, we will give a brief description of what is Augmented Reality and how it will change the way we see the world. Augmented reality (abbreviated as AR) is a new technology that blurs the line between what's real and what is computer generated by enhancing what we see, smell, hear and feel. It is said to change the way we see the world around us. It basically adds a layer of graphics and other sensory enhancements on the natural world as it exists in real time [8].

Augmented Reality (AR) allows the user to see the real world, with virtual objects superimposed upon or composited with the real world. However, the users cannot see the real world that surrounds them. In AR technology, users can see the real world around them by adding virtual objects which are generated by computer [9]. To make the AR 3D objects appear directly on the media, a special device called Head Mounted Display (HMD) is required [10].

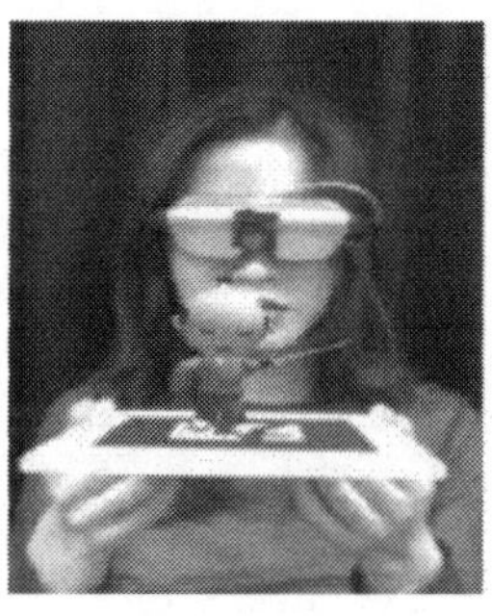

Figure 1. The example of the use of AR.

In general, the work principle of ARToolkit is as follows.

1. The camera captures images from the real world in real time and sends them to the computer.
2. The software in the computer will search for markers on each video frame.
3. If the marker has been found, the computer will mathematically proceed the relative position of the camera to the black box contained in the marker.
4. If the camera position is known, then the model will be depicted in the same position.
5. The 3D object model will be displayed on the marker as if the virtual object is added into the real world.

2.2 Spatial Ability

Spatial ability is children's high level of understanding skill towards three-dimensional objects, what they can see from the objects, and what the objects' names. Spatial ability or the development of geometry ideas can be stated according to these purposes: (a) Include various characteristics or properties of the objects in categorizing and building activities; (b) Use a suitable language to describe geometry (correspond with students level of thinking); (c) Show the evidence of geometrical thinking in solving puzzles, exploring shapes, designing, and analyzing shapes; (d) Recognize shapes in environment; and (e) Solve problems regarding spatial ability

Spatial ability is an essential competence; thus, NTCM explains required geometry abilities which are important to be understood by students [11], which are:

1. Able to analyze the geometry characteristics and properties of both two-dimensional and three-dimensional.
2. Able to construct mathematics arguments concerning geometric relationships and so on.
3. Able to decide the position of a more specific point and spatial relationships by using coordinate geometry and connect it with other systems.
4. Apply the transformation and use it systematically in analyzing mathematics situation.
5. Use visualization, spatial thinking, and geometry model in solving problems.

The standard of geometry and spatial ability is when students are able to identify, classify, compare and analyze the characteristics of properties and relationships on areas and one, two and three-dimensional geometry objects [12]. Students apply spatial reasoning, the properties of geometry objects and transformations into mathematical situation analysis and problem solving. The general indicator of spatial ability in geometry are: (1) Identify and draw geometric shapes; for example, circles, squares, rectangles, and triangles; (2) Identify solid objects; for example, cylinders, cones, spheres, quadrilic prisms; (3) Build three-dimensional objects using blocks; (4) Compare and group two-dimensional shapes according to their attributes, then explain the reasons for their grouping and comparison; (5) Able to describe, and show the relative position of the object, for instance: its upper view, lower view, inside view, outside view, side view, and between view; and (6) Investigate and predict the results of two-dimensional shapes which are placed together and taken separately.

Several experts state the definition of spatial visualization in [13]: 1) Spatial visualization as the mental skills concerned with understanding, manipulating, reorganizing, and interpreting relationship visually 2) Spatial visualization is the process of representing, transforming, generating, and recalling symbolic, non-linguistic information.

Some experts in [14] argue that, two major components of spatial visualization have been identified: spatial relations and spatial visualization. In standardized spatial ability tests, spatial relations tasks involve 2D and 3D rotations and cube comparisons. Spatial visualization is described as the ability to imagine rotations of objects or their parts in 3-D space.

3. Research Methodology

This study is started by developing augmented reality media for teaching geometry. Steps conducted in developing augmented reality media includes: the analysis of software and hardware necessity, designing, development, and testing.

In analyzing step, researchers conducted introductory study in the form of exploratory and literature. Exploratory study was carried out by using browsing technique, i.e. the search for looking information about spatial ability of yunior high school students and the media usage, including learning methods used. While the literature study was conducted by collecting writings both in the form of journals, books, and articles related to interactive multimedia that can be used in learning geometry. Continued with user analysis, software analysis, and hardware analysis for knowing a sufficient hardware is also needed so that augmented reality media will run well in minimum specifications of hardware. Then designing and developing step was also done. A testing toward multimedia is conducted to make sure the functions run well. The testing is conducted by using blackbox testing. After that the AR media is judged to be considered eligible by experts and the repair related to theirs suggestion and recommendation. After that steps continued by investigate the implemented media to know students responses towards the use of geometry learning media based on augmented reality by using questionnaire. This steps was to examining students' responses towards the implementation of geometry learning media based on augmented reality. Because this media is still prototype the implementation is in limited scopes, it was implemented to below some students in Bandung.

4. Results

4.1 AR-based Geometry Learning Media

The learning media that is used in this study is the geometry learning media based on augmented reality for three-dimensional objects lesson. The marker that is used is in the form of a figure which is deliberately made to display three-dimensional objects. The three-dimensional objects which are covered in the lesson include: cubes, prisms, beams, pyramids, and cones. In this lesson, it is discussed how to determine the face diagonal, the space diagonal, the height or altitude, and the slant height on three-dimensional objects. The AR media illustration is shown in figure 2. The AR media illustration used is shown below.

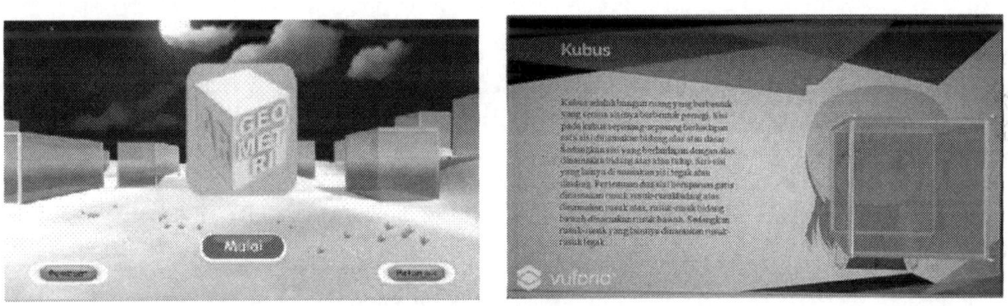

Figure 2. The main display of AR geometry media and geometry shape appears

Based on the implementation of AR-based geometry media, it can be seen that junior high school students, who learn three-dimensional objects, find the media helpful and can quickly understand the concept of space which is discussed. They can see three-dimensional objects from various perspectives: front, side, back, top and bottom part of the objects. All this time, students have difficulty if they are asked to determine the location of the diagonal in a three-dimensional object, especially if the diagonal is located beside or behind the objects. They find it difficult to imagine how the form of the diagonal is. However, by using AR-based geometry media, they can rotate the three-dimensional objects freely, by merely moving the highlighted marker, rotating it, until they can find the part that they want to see.

4.2 Questionnaire Results for Student Response

After AR-based geometry learning media is implemented, students give their responses towards the use of media. Students' responses are collected gradually from the first phase to the third phase. The following results are obtained.

Table 1. Students' responses towards the use of geometry learning media based on AR

Indicators	1st phase	2nd phase	3rd phase
Learning Goal Alignment	63%	73%	75%
Feedback and Adaptation	68%	72%	78%
Motivation	65%	74%	80%
Presentation Design	70%	70%	78%
Interaction Usability	69%	69%	77%
Accessibility	68%	72%	80%
Average	67,1%	71,7%	78,0%

Based on Table 1, there is an increase in all indicators. In the aspect of learning goal alignment, the increase is from 63% to 75%, the percentage of feedback and adaptation increases from 68% to 78%, the percentage of motivation goes up from 65% to 80%, the percentage of presentation design increases from 70% to 78%, the percentage of interaction design usability increases from 69% to 77% and the percentage of accessibility aspects increases from 68% to 80%. Averagely, students' responses towards the use of geometry learning media based on AR increase along the phases from 67,1% to 78,0%.

5. Conclusions

Based on the results, the conclusions are obtained as follows:

1. AR media is used in teaching geometry developed in accordance with the steps appropriate media development, by considering aspects of students' needs and the material being taught.
2. AR-based geometry media facilitate students in understanding the concept of three-dimensional geometry objects. They can rotate the three-dimensional objects freely, making them easier to see the three-dimensional objects from various perspectives.
3. AR-based geometry media gets positive responses from students. Hence, this media is suitable to be applied in the process of learning geometry in junior high schools.

References

[1] Muabuai, Y 2010 Teaching geometry using cooperatif model STAD based on cabri geometry programe II Plus for improving students communication ability. Mater thesis.

[2] National Academy of Science 2006 Learning to Think Spatially. Washington DC: The National Academics Press.

[3] Nemeth, B 2007 Measurement of the Development of Spatial Ability by Mental Cutting Test. Anales Mathematicae et Informaticae 34 pp. 123-128 at: http://www.ektf.hu/tanszek/matematika/ami.

[4] Wai, J. Lubinski, D. & Benbow, C.P 2009 "Spatial Ability for STEM Domains: Aligning Over 50 Years of Cumulative Psychological Knowledge Solidifies Its Importance". Journal of Educational Pschology, Vol. 101, No. 4, 817-835.

[5] Gittler, G & Gluck 1998 Diferential transfer of learning: Effects of istruction in Descritive Geometry on Spatial Test Performance. Journal for Geometry and Graphics, vol. 2, No. 1, 71-84.

[6] Adolphus, T 2011 Problem of Teaching and Learning of Geometry in Secondary Schools in Rivers State, Nigeria. International Journal of Emerging Science 1(2), 143-152, June 2011.

[7] Kariadinata, R 2010 Geometry visualization and spatial abilities student of Madrasah Aliyah Negeri (MAN) grade X using learning individual software *Journal of EDUMAT* vol **1** no. 2.
[8] Angrawal, M., Kulkarni, A., Joshi, S., Tiku, N 2015 Augmented Reality. International Journal of advance Research in Computer Science and Management Studies: India
[9] Azuma 2015 A Survey of Augmented Reality. *In Presence: Teleoperators and Virtual Environtments* **6**, 4 August 1997. Pp. 355-385
[10] Billinghurst, Mark, et. al. (2008), *Tangible Augmented Reality.* International Conference on Computer Graphics and Interactive Techniques. ACM SIGGRAPH ASIA, Singapore.
[11] NCTM 2000 Principles and Standards for School Mathematics. Reston, VA: NCTM.
[12] Diocese of Toledo Mathematics Course of Study 2010 Geometry and Spatial Sense Standard. [Online]. Tersedia: http://www.cyss.org/Schools/CofS/ MathPilotByStandards/Pilot MathGeometry SpatialSense.pdf. [17 Maret 2011] 23
[13] Pitalis, Mousoulides, dan Christou 2006 Developing the 3D Math Dinamic Geometry Software: Theoretical Perspectives on Design, In International Journal for Technology in Mathematics Education. Volume. 13 No.4
[14] Olkun, S 2003 Making Connections: Improving Spatial Abilities with Engineering Drawing Activities. International Journal of Mathematics Teaching and Learning. April 2003

Designing Production Based Learning as a Basic Strategy for Creating Income Generating Units at Universitas Pendidikan Indonesia

D Suryadi* and N Supriatna

Universitas Pendidikan Indonesia

*dedysuryadi@upi.edu

Abstract. The establishment of Universitas Pendidikan Indonesia (later to be referred as UPI) Statute as a State-Owned State University (PTN-BH) has implications for UPI requirements. One of them is the need for UPI to generate an Income Generating Unit (IGU) of at least IDR 100 Billion (one hundred billion rupiah). This requirement is considered difficult since UPI is one of the universities whose focus is on the world of education and not the business and industry. Surely this becomes the thinking of the entire academic community to make a breakthrough by optimizing their potential. This study aims to find the pattern of learning practice that produces economic value products as one indicator of IGU value achievement as an effort to support UPI as PTN-BH. Learning strategy is done by designing and implementing the production base learning (PBL) approach as the basis strategy for the development of production units capable of becoming IGU in UPI. The research method used refers to research and development methods with adjustments taking into account the effectiveness in validating and conducting field model trials. The result of this research is the basic design of PBL model as the development strategy of production unit in the achievement of IGU UPI PTN-BH.

1. Introduction

The Statutes of Universitas Pendidikan Indonesia (UPI) as state-owned universities (PTN-BH) stipulated by Government Regulation No. 15 of 2014 have a major impact on the overall activities of education in UPI [1]. It is characterized by various policies within the context of authority in organize and conducts educational and educational finance processes within the university. One consequence of the funding aspect is the demand from PTN-BH to generate income of 100 (one hundred) billion rupiah in one budget year. Obviously this becomes the thinking of the leadership and academic civitas to programmatically able to direct the resources owned to increase the income generating unit (IGU).

Based on the analysis of potential resources owned by UPI in addition to human resources is the presence of laboratories and workshops owned by UPI in each faculty unit. One of them is in the Faculty of Technology and Vocational Education with 6 (six) departments that have a number of laboratories and workshops. The Department of Civil Engineering Education is one of the departments in FPTK having several laboratories (hydraulics, structure, transportation, soil mechanics, soil and computer sciences) and building workshops (Wood, Concrete, and Steel / Plambing). Of course the potential of this resource can be optimized in addition to special to serve the lecture students can also be developed unit services and production. In the context of technical and vocational education, learning outcomes of learners is the mastery of a set of work competencies. Technical and Vocational

Education and Training (TVET) refers to education and training that prepares persons for gainful employment [2].

Optimalization of its resources certainly does not interfere with the learning process, but synergistically by compiling the lectures, especially the workshop in addition to practicing the skills of students in the production process also produces products of economic value. Moving from this problem, in the study program of Engineering Education Building practicum lectures in building workshops are designed learning practices that bring closer to the situation and conditions in the field or industrial world.

The learning of best practices in technology and vocational education is the one that leads to the pattern of work developed in the industry. Of course, when done in the industry is not possible, but can be done by setting workshop as a miniature of the industry. The learning pattern must follow the concepts done in the industry, so that the practice lectures in the workshop are selected and designed with a product-based learning approach (PBL).

The importance of this PBL approach is to provide meaningful experience for the students of the Building Engineering Education in mastering a set of competence skills in the production process. The implementation of production based learning with entrepreneurship approach using workshop based lectures, qualitatively improved the quality and meaningfulness of the learning [3]. In addition, this research is directed to form a model of development of production unit which is developed based on workshop, where one of them is in wood workshop on lecture Practice of Wood and Practice of Furniture and Finishing Technique. The products produced in this course will refer to the standard product that is selling value so that it can be directed in achieving economic value that impact on the achievement of IGU both in study program and university in general.

2. Methods

The method used in this research refers to research and development methods. In its application make adjustments taking into consideration the effectiveness in validating and conducting field model trials. This method is chosen in accordance with the characteristics and objectives, namely to produce the form of production base learning design. Research and development is a process used to develop and validate educational product [4].

Based on the research objectives and the factual conditions in the field and the consideration of the feasibility of the research, then of course the steps developed by Borg and Gall should be adjusted. Therefore, in this study modification includes the following stages.

2.1. Introduction study stage

In this first phase, it begins first with literature study and field study as a study to know the approach of PBL and wood products in the field. The literature study is intended to understand the subject related to the models, approaches and learning methods used in practical learning in the workshop. In addition it also refers to the search and search references of research results that have been done previously that have a close to the topic of research that we will do.

While the field study was conducted by survey approach and descriptive documentation study that describes the condition of what is about the variables to be studied. The main purpose of field study is to collect information on as many variables as possible. At this stage will be traced all aspects in the implementation of practical learning that has been implemented as a basis in determining the design of PBL as an effort to lay a solid foundation in its implementation in the lecture of wood working practices.

2.2. Planning stage of PBL design

In the planning stages starting from the preliminary study results include findings on the practical methods used, the products that have been produced and the factual condition of practicum learning in the Wood Workshop. Then a study of similar research results that have been done in the application of methods and procedures practical learning. Then a series of planning activities of research were prepared.

Stages of production-based learning design that is implemented schematically in this study as the following table 1:

Table 1. Stages of preparation of PBL design

NO	PHASE OF ACTIVITY
1	Review of the purpose of lecturing on woodworking practices in the curriculum and syllabus
2	Preparation of the scope of teaching materials / teaching materials
3	Preparation of learning tools
4	Preliminary drafting of production-based learning design
5	Prepare the job sheet framework or worksheet for each practice product
6	Preparation of draft observation sheet on PBL process
7	Drafting of process and product assessment sheets
8	Planning of processing techniques and data analysis of research results
9	Planning the scenario stages of applying the PBL approach
10	Determination of the object / respondent for the application of the PBL approach
11	Practice lessons based on PBL design
12	Observations and assessment of PBL processes on individual/ student groups
13	Assessment of product quality practices based on specified product criteria
14	Analysis of the economic value of products and product development opportunities as part of the development framework of production units

2.3. Development stage

The next stage is the development of PBL designer, Job sheet, Observation Sheet and Processing and Product Assessment Sheet by validating in the form of pilot design of PBL approach and consulted to vocational education and training expert group as well as educational instructor until final draft from the concept of the model of the developed model instrument.

The development of the model at this stage of development is done by using judgment technique and expert judgment to validate PBL design concept. Then empirically validate by the education practitioner about the indicator's load and the items of the process and product valuation instrument. At the development stage, also conducted planning process assessment of products. Based on the design made, the researcher will use this application program only by filling out the assessment scale columns on each item of assessment on the process with the marking of Check (V). Systemically, the performance values of each aspect and each indicator will be seen, as well as the existing gap values obtained by comparing product performance with product standards.

3. Results and Discussion

3.1. Design validation model PBL

Validation of PBL model concepts using expert judgment in the study environment of Building Engineering Education includes lecturers, instructors, and laboratory educational institutions (PLP) of building workshops. The result of consideration is the input received to reinforce the concept of PBL model design in the practice lecture directed at the creation of the Income Generating Unit (IGU) unit within the study program environment.

After the improvement and refinement, the concept of PBL model then discussed again to strengthen the concept of the concept of PBL model. Based on the results of discussion and review of the concept of the model that has been improved, the concept of PBL model can be accepted and considered to meet the rules of building the concept and as a reference in making design tools PBL model.

3.2. Empirical validation design build model PBL

The design of production base learning that has been tested, then empirically validated to obtain information about the performance and value of implementation gap implementation of PBL approach, so it will be faster also determine the decision to improve the implementation. The following is a plan for stages of PBL implementation based on validation results that have been done, namely:

- Learning about the selection of wood tools and materials
- Learning about the types and criteria of processed wood products
- Learning about working techniques from the beginning to finishing the product
- Learning about work safety implementation in PBL
- Practical learning begins from material selection to finishing processed products
- Observation and assessment of PBL processes on each individual / group of students
- Assessment of product quality practices generated based on specified product criteria
- Analyze the economic value of products and product development opportunities as part of the development framework of production units

Empirical validation for the design of PBL model is directed to obtain data about the appropriateness of PBL model design with its implementation in the practicum lecture. Based on the validation, it is found that the aspects of the design developed in the PBL syntax are seen in accordance with the characteristics of the purpose of the course, the lecture material and the learning achievement of the learners. is considered to represent the whole process of curriculum implementation. This study reinforces the importance of implementation PBL in vocational education with the syntax of learning include: (1) analysis of curriculum and learners characteristics, (2) identification and product analysis, (3) creating important questions about product, (4) questions mapping, (5) analysis of equipment and material needed of the product which was going to be made, (6) making schedule of implementation of making product, (7) process of making product, (8) evaluated regularly, and (9) creating business plan [5].

Nevertheless, there are some things to consider, namely the validation is still in a limited trial so it has not been used in all lectures in one semester. Thus it does not cover the entire course material taught in one or two courses so that it still needs to be developed and expanded in its implementation as part of competency learning model.

Empirical validation conducted from trial results in practice lectures for four lectures on one competency or one discussion, provides an assessment that all aspects, indicators and items are evaluated both on the planning and implementation dimensions are in accordance with the improvement to be developed again in the lecture practice as a whole. Thus, it is necessary to carry out another large scale trial and practice for one semester so that a PBL model can be constructed which can be used to increase the possibility of developing the existing production unit in the Building Engineering Education study program.

The following are the products resulting from trials on the application of PBL to lecturing on wood working practices (see figure 1).

Figure 1. Some furniture products from practice lectures

3.3. The framework of establishing a production unit as an IGU strategy

The existence of UPI as PTN-BH with one of the demands to produce an IGU value oriented on institutional strengthening and learning becomes something conditional on the role of UPI. Various steps that have been developed either through the BPPU established under the UPI Statute to optimize the resources owned by UPI. One of them is to optimize the resources of infrastructure facilities that can be optimized in addition to lectures can also be used to get the value of profit or input in the form of funding.

Facilities owned by FPTK UPI with various laboratory devices and workshops provide basic capital for the development of IGU UPI. The current practice is optimized for lectures only so that many are not used at certain times. Even though the practice model produces a product but not directed how the product will be used and for whom.

UPI position and funding demands collected apart from funds from the central government and the community, then the opportunity to optimize the infrastructure becomes a necessity. However, it is necessary to integrate a practice lecture model that leads not only to the achievement of lecture objectives and competencies to be achieved, but also how to make efforts to market the product of lectures into financial value added to increase lecture financing. The model developed in PBL design is basically to provide a foundation for developing a special production unit established in the environment of Building Engineering Education study program.

Stages of establishment of wood processing units developed from the results of this PBL modeling is done with the following steps:

- Determination of system and mechanism of establishment and management of production unit based on provision as determined by BPPU UPI.
- Determination of the type of processed wood products for the interior of buildings, furniture, and building component materials (frames, doors / windows, balustrade) that can be produced in wood workshops.
- Determination of quality and price of various processed products produced to see the needs of the community, especially in the campus environment through research of product standards and market price.
- Preparation of the Business Plan by the task force of the production unit as a direction in developing the production unit as part of duty to develop the revenue value of the IGU Building Engineering Education study program.
- Determination of assigned task forces to carry out the management of production units based on legal umbrella developed by BPPU UPI with transparent and accountable principles.

IOP Conf. Series: Materials Science and Engineering **306** (2018) 012030 doi:10.1088/1757-899X/306/1/012030

4. Conclusions

This affirmative study is emphasized as the basis for the development of a production-based learning model. Thus the extent of the validity is limited to constructing a PBL modeling framework that is piloted in one or two practical courses only. Thus, the next stage of the research will be conducted to develop a valid and reliable PBL model for the overall learning of the practice lecture.

Based on preliminary research results and limited trial this obtained the following conclusions:

- Designed forms of production-based learning in lectures at workshop buildings have been limited in validation.
- Stages of implementation of the PBL approach in producing practicable products of economic value have been achieved on a limited product scope.
- The production unit framework in the workshop as part of the strategy of achieving the value of IGU for UPI as PTN-BH is based on the result of the research which is conducted based on the rule of law and mechanism stipulated by BPPU UPI.

References

[1] Government Regulation No. 15 of 2014 on the Statutes of the University of Education of Indonesia as State-Owned College of Legal Entities.

[2] Finch C, Crunkilton J R 1999 *Curriculum Development in Vocational and Technical Education* (Boston, MA: Allyn & Bacon).

[3] Ganefri 2013 The Development of Production-Based Learning Approach to Entrepreunerial Spirit for Engineering Students *Asian Social Science* **9**(12) pp 1911-2025.

[4] Borg W R, Gall M D and Gall J P 1998 *Education research, an introduction* (Boston: Pearson Education, Inc).

[5] Ganefri and Hidayat H 2014 Production Based Learning: An Instructional Design Model in The Context of Vocational Education and Training (VET) *Global Journal of Science Frontier Research: E Interdiciplinary* (Publisher: Global Journal Inc. (USA). Online ISSN:2249-4626&Print ISSN: 0975-5896).

Improvement of Students' Environmental Literacy by Using Integrated Science Teaching Materials

D Suryanti[1*], P Sinaga[2], and W Surakusumah[3]

[1]Postgraduate Science Education, Universitas Pendidikan Indonesia, Jl. Dr. Setiabudhi No. 229, Bandung 40154, Indonesia
[2]Department of Physics, Universitas Pendidikan Indonesia, Jl. Dr. Setiabudhi No. 229, Bandung 40154, Indonesia
[3]Departement of Biology, Universitas Pendidikan Indonesia, Jl. Dr. Setiabudhi No. 229, Bandung 40154, Indonesia

*desisuryanti@student.upi.edu

Abstract. This study aims to determine the improvement of student environmental literacy through the use of integrated science teaching materials on pollution topics. The research is used weak experiment method with the one group pre-test post-test design. The sample of the study were junior high school students in Bandung amounted to 32 people of 7th grade. Data collection in the form of environmental literacy test instrument consist of four components of environmental literacy that is (1) Knowledge, (2) Competencies (Cognitive Skill), (3) Affective and (4) Environmentally Responsible Behavior. The results show that the student's environmental literacy ability is improved after using integrated science teaching materials. An increase in the medium category is occurring in the knowledge (N-gain= 46%) and cognitive skill (N-gain=31%), while the increase in the low category occurs in the affective component (N-gain=25%) and behaviour (N-gain=24%). The conclusions of this study as a whole the improvement of students' environmental literacy by using integrated science teaching material is in the medium category (N-gain=34%).

1. Introduction

Environmental issues are increasingly considered to be an emergency since human domination of the environment, this is exacerbated as technology advances [1]. It is very difficult to address this increasingly acute environmental problem without any public awareness to be involved and responsible. Therefore it becomes priority major countries to seek to raise the awareness of the community, by taking preventive measures against environmental conditions continue to decline through the role of environmental education [1, 2]. The importance of environmental education was first plagiarized by almost all world leaders, in the Tibilisi Declaration in 1977. The declaration contains an important message for the changing world of education and the starting point of the substance mandatory environmental education [2]. Attention and great support towards a sustainable environment through the contribution of environmental education a top priority in a number of countries [1, 2]. Since then environmental education has begun to be integrated into school curricula for each level and level, including in Indonesia. The culmination of environmental education is to make students who are literate in the environment [3].

Environmental literacy is a knowledge of how the natural environment functions and how humans play in preserving and nurturing their environment. Environmental literacy can also be defined as basic education for all people, providing basic knowledge, skills and motivations related to environmental needs and contributing to sustainable development [2]. The environmental literacy is very important for the students in order to prepare the people who understand and can solve the environmental problems, so that it can be prepared by environmental reform agencies that have a positive attitude and action toward the environment [4]. One of the basic consistency of the environmental literacy curriculum approach is integrated into the science lesson [2, 3, 5, 6]. This is in accordance with the current curriculum objectives (curriculum 2013), which develop attitudes, knowledge, and skills and apply them in various situations in schools and communities.

Research that has been done related to environmental literacy such as instructional strategy applied by teacher has a big effect on the development of students' environmental literacy [5, 6, 7, 8]. For example with the application of interactive interactive learning techniques as an alternative solution to the monotony and not contextual learning environment so far. In it integrated the use of different learning methods ranging from field trips, project delivery, and use of ICT media, there are communication exercises, and the implementation of assessments for learning and assessment of learning. Learning is more memorable and understood as they experience directly [5]. Furthermore, the use of concept maps to obtain a description of the mastery of ecological concepts as a prerequisite of environmental literacy [6]. Then another example is the use of the Environmental Education Kit (EE-Kit) in environmental studies in an effort to improve students' environmental literacy in Malaysian high school students. The EE-Kit used is a module that contains information about knowledge and skills in teaching environmental education, so that EE-Kit can help teachers in teaching about the environment effectively. EE-Kit can increase students' level of literacy environment [7]. Another study is the use of Summer Environmental Education Program (SEEP) in the achievement of students' environmental literacy. This research is based on the facts of previous research that one is often involved in environmental activities and engaging in direct learning in nature, will have connections and a sense of belonging to the environment to increase, as well as the tendency to protect the environment will be formed. In the SEEP program the learning was conducted outdoors (field trips) for a week and provided six modules (consisting of different subjects / subjects) associated with environmental literacy [8].

Based on the explanation of the environmental literacy research that has been done, the researchers tested the improvement of students' environmental literacy through the use of integrated science teaching materials for students of grade 7th junior high school. The influence of teaching materials on students is huge because between students and teaching materials have an important interaction in the learning process, meaning that students have frequent interaction frequency with teaching materials, not only in the classroom but also outside the class [9]. Learning by using teaching material allows students to learn individually, they learn actively without maximum help of a teacher. Based on previous research, teaching materials reduce teacher-oriented learning, students are more focused on finding concepts and more active in learning. In addition, teaching material can enhance students' intellectual potential which in this case is a learning goal [10]. Thus, the development of integrated teaching science teaching materials for students needs to pay attention for the characteristics and needs of students in answering 21st century challenges. The author tries to create an integrated science teaching material for students who can trill environmental literacy using multimodal representation on pollution topic. These pollution materials include material contain phenomena and events close to the students' life and the surrounding environment so as to facilitate the students in understanding the concept of environment in depth. The development of this teaching material uses the development flow of Design Representational Approach Learning to Write developed by Sinaga [11]. One of the advantages of teaching materials used is to use multimodus representation. Multimodus representation can help students in understanding the concept described. If students have difficulty in understanding a concept on verbal explanation can be helped by other visual explanations. With so many representations being used, a concept becomes easier to understand. This instructional material is also used to demand students learning activities that involve aspects of knowledge, cognitive skills, affective and behavioral with the aim of training students' environmental literacy. The problem of this research is how to improve of students' environmental literacy which includes the aspect of knowledge , skill, affective and behavior of junior high school students toward the use of integrated science teaching materials on pollution topic?.

2. Experimental Method

2.1 Research method
The method used in this research is quasy experiment with design as one group pretest-posttest without control group design (see Table 1.) [12].

Table 1. One Group Pretest-Posttest Design.

Pre-test	Treatment	Post-test
O	X	O

Note: X: Treatment by using integrated teaching materials

The sample used in this research consists of 32 junior high school students in Bandung class VII include of 15 males and 17 females. The sampling technique in this study using purposive sampling with the purpose of certain purposes [13]. The research design can be seen in Figure 1 as follow:

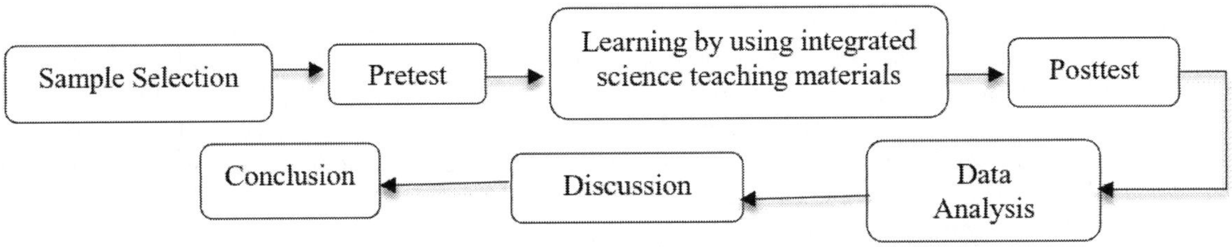

Figure 1. Research Design

2.2 Research instrument
The students' environmental literacy capability was measured using an environmental literacy test adapted from the Middle School Environment Literacy Survey (MSELS) developed by the National Environmental Literacy Assessment [14]. Then modified according to pollution content amounting to 74 questions representing the entire component of environmental literacy include of knowledge, cognitive skill, affective, and environmentally responsible behavior. The environmental literacy test used in this study is tested for its validity and reliability considering the change, addition, and adjustment of the test questions with the pollution content. Grid tests of environmental literacy can be seen in Table 2.

Table 2. Grid about Student's Environmental Literacy Test.

Component of Environmental Literacy	Sub Component	Question Number	Range Score	Score
a. Knowledge (Cognitive)	Ecological knowledge	1-20	0-20	60
b. Competency (Cognitive Skills)	- Identify environmental issues - Analyze environmental issues - Action Plan	21-23 24-29 30-34	0-14	60
c. Affective about the Environment	- Verbal commitment - Sensitivity - Feeling	35-49 50-57 58-59	5-125	60
d. Environmentally Responsible Behavior	A real commitment (pro-environmental action)	60-74	5-75	60
Total		74		240

Notes: For scores on Ecological Knowledge and on Cognitive Skills, Low = 0-20, Moderate = 21-40, and High = 41-60; For scores on Environmental Affect and on Environmental Behavior, Low = 12-27, Moderate = 28-44, and High = 45-60; For Composite Scores, Low = 24-96, Moderate = 97-168, and High = 169-240.

2.3 Analysis data

The improvement of students' environmental literacy in the form of pretest and posttest result is analyzed using Normalized gain $\langle g \rangle$ statistic test with the following formula (1).

$$\langle g \rangle = \frac{\langle posttes \rangle - \langle pretest \rangle}{maximum\ score - \langle pretest \rangle} \quad \dots\dots\dots\dots\dots \quad (1)$$

The category of score acquisition can be seen in the Table 3 [15].

Table 3. Categorization of N-gain Scores.

Range	Category
$\langle g \rangle \geq 0{,}7$	High
$0{,}7 > \langle g \rangle \geq 0{,}3$	Medium
$\langle g \rangle < 0{,}30$	Low

3. Results and Discussion

3.1. Environmental knowledge

Improvement of the literacy environment on overall knowledge aspects can be seen in Table 4.

Table 4. Pretest and posttest comparison for the score of knowledge test

Aspect	Test	N	Mean	Std. Deviation	t	Df	Sig 2-tailed	Result
Knowledge	Pretest	32	31,03	10,32	-11,576	31	0,000	Differ significantly
	Posttest	32	44,25	8,95				

Based on Table 4, overall obtained value of P-value (*Sig 2-tailed*) pretest and posttest (Sig) <0.05. This means that there are significant differences in students' environmental literacy ability in the knowledge aspect between before and after learning using integrated science teaching materials. The negative t value indicates that the posttest result is higher than the pretest which means an increase in posttest result. The average percentage of pretest, posttest and N-gain scores on the knowledge aspect can be seen in Figure 2 below.

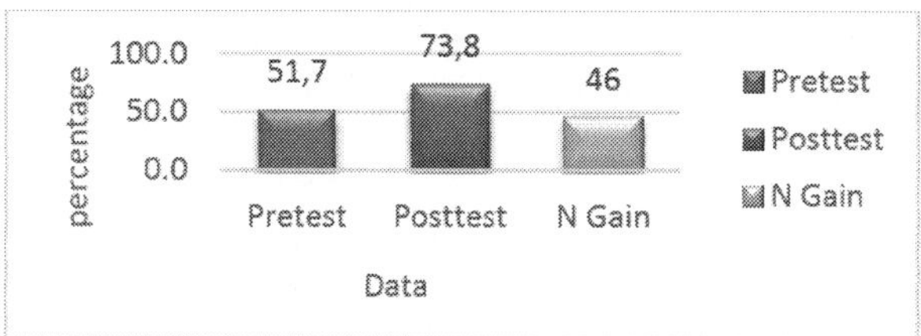

Figure 2. Percentage of pretest, posttest and N-gain scores on the knowledge aspect

From Figure 2 it can be seen that there is an increase of knowledge component with 46% N-gain value is in medium category. This increase in the knowledge component is likely due to the students having little difficulty in understanding the teaching materials used. This is supported by the presentation of materials in teaching materials using phenomena and events that are close to the student's (contextual) life and the surrounding environment so as to facilitate the students in understanding the environmental concept in depth. The use of examples and facts contained in the environment around the students will facilitate students in understanding a concept. Therefore, teachers should be able to associate the concept with the facts contained around the students in accordance with the principle of meaningful learning. Meaningful learning is a learning that directs students to deep understanding and is able to apply concepts to real, new and different conditions [16]. Science learning should be more emphasis on giving environmental issues or problems contextually to develop the competence of students to be able to explore and understand the surrounding natural environment scientifically and directed to find and do, so as to help students to gain a deeper understanding of the surrounding environment especially related to environmental pollution.

3.2. Cognitive skills

Improvement of environmental literacy capability in the aspect of knowledge as a whole can be seen in Table 5.

Table 5. Pretest and posttest comparison for the score of cognitive skills test

Aspect		N	Mean	Std. Deviation	T	df	Sig 2-tailed	Result
Cognitive Skills	Pretest	32	24,67	10,32	-13,867	31	0,000	Differ significantly
	Posttest	32	35,66	8,95				

Based on Table 5. Overall values of P-value pretest and posttest (Sig) <0.05 were obtained. This means that there are significant differences in students' literacy ability on the aspects of cognitive skills between before and after learning using integrated science materials. The negative t value indicates that posttest result is higher than the pretation which means an increase in the posttest result. The average percentage of pretest, posttest and N-gain scores on the cognitive skills aspect can be seen in Figure 3 below.

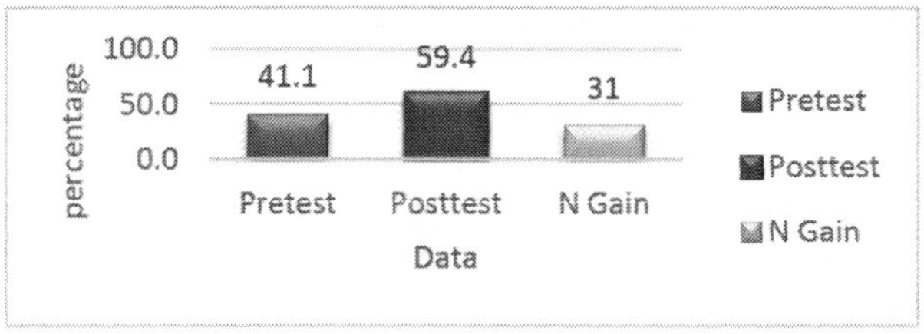

Figure 3. Percentage of pretest, posttest and N-gain scores on the cognitive skills aspect

From Figure 3 it can be seen that there is an increase in the component of cognitive skills with 31% N-gain is in the medium category. Improvement in the component of cognitive skills is inseparable from the experience that students gain during learning as contained in the component of the teaching materials where students are given exercise questions about environmental issue articles to be able to identify, analyze and make action plans to address the issues / issues of the environment. Therefore, students get training on the problem solving ability of the environment, so that students' cognitive skills can be achieved well. In addition to the component of teaching materials there are also Scientific Activities and discussions that invite students to experiment and simple inquiry about environmental issues that occur around the students and make it to be more actively involved in discussion activities in discussing environmental pollution issues. Only individuals with literacy, awareness, and sensitivity will contribute to addressing environmental issues. Therefore, the cultivation of environmental literacy from within students is very important so that they can become environmental reform agents that can overcome environmental problems through their action plan [17].

3.3. Affective

The improvement of environmental literacy capability in the overall affective aspect can be seen in Table 6.

Table 6. Pretest and posttest comparison for the score of affective test

Aspect		N	Mean	Std. Deviation	t	df	Sig 2-tailed	Result
Affective	Pretest	32	41,59	4,14	-7,758	31	0,000	Differ significantly
	Posttest	32	46,24	4,05				

Based on Table 6, overall values of P-value pretest and posttest (Sig) <0.05 were obtained. This means that there are significant differences in the ability of environmental literacy on affective aspects of students between before and after learning using integrated science materials. The negative t value indicates that the posttest result is higher than the pretation which means an increase in the posttest result. The comparison of the average percentage of pretest, posttest and N-gain scores on the affective aspect can be seen in Figure 4 below.

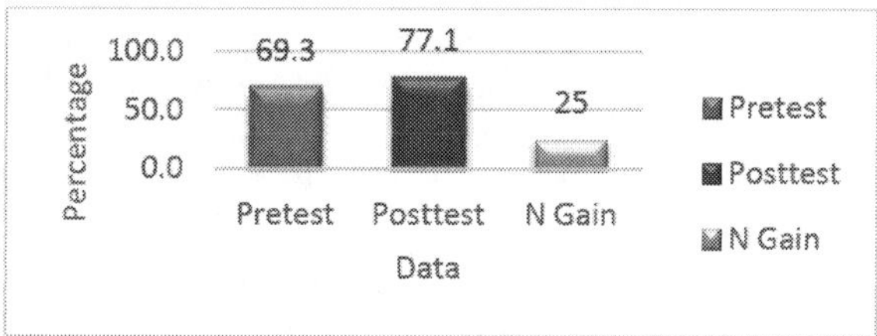

Figure 4. Percentage of pretest, posttest and N-gain scores on the affective aspect

Based on Figure 4 it can be seen that there is an increase in the affective component with a 25% N-gain value being in the low category, although when seen from the comparison of pretest results with posttest seen there is an increase from moderate to high category. The high pretest results make it less likely to be able to raise an ideal posttest result in high categories.

The lack of maximum affective component increase is probably due to limited time allocation for the implementation of teaching materials developed and at the time of implementation is not followed by the use of strategies and teaching methods that support such as outdoor activities, field trips or hands-on activities. Although affective aspects show an increase in the low category, the contribution of the use of teaching materials in growing and improving students' affective to the environment is evident during the learning activities. When students discuss Environmental Issues contained in teaching materials can increase students' knowledge of environmental issues and provide a separate learning experience for students. From this process it is clear that the results of discussion answers most students have a great concern for the environment. The provision of opportunities to students to interact directly with the facts and issues that develop in the environment will be able to train the students' sensitivity to the environment [18]. The knowledge embedded in the target will affect his attitude towards the environment that will be manifested in an action / behavior [19].

3.4. Environmental responsible behavior

The improvement of environmental literacy capability in the overall behavioral aspect can be seen in Table 7.

Table 7. Pre test and post test comparison for the score of behavioral test

Aspect		N	Mean	Std. Deviation	t	df	Sig 2-tailed	Result
Cognitive skills	Pretest	32	40,98	5,06	-6,465	31	0,000	Differ significantly
	Posttest	32	46,05	3,62				

Based on Table 7, overall values of P-value pretest and posttest (Sig) <0.05 were obtained. This means that there are significant differences in environmental literacy capability in the aspects of behavioral of students between before and after learning using integrated science materials. The negative t value indicates that the posttest result is higher than the pretation which means an increase in posttest result. The average percentage comparison of pretest, posttest and N-gain scores on the behavioral aspect can be seen in Figure 5 below.

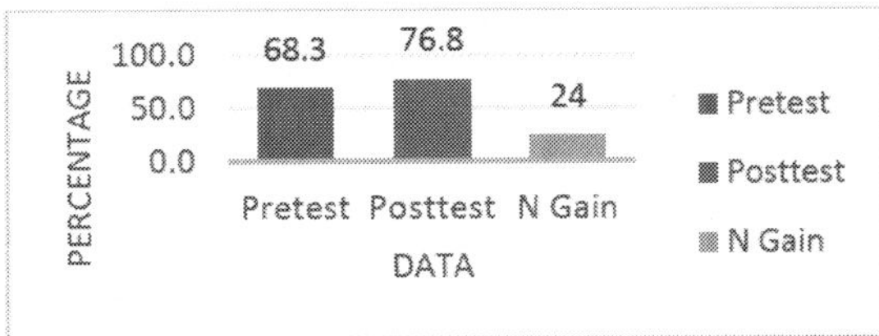

Figure 5. Percentage of pretest, posttest and N-gain scores on the behavioral aspect

Based on Figure 5 it can be seen that overall there is an increase in the component of cognitive skills with a 24% N-gain value in the low category. The low increase in the behavioral component is due to the learning time using this learning material only two weeks only and without accompanied by the use of strategies and learning methods that support such as outdoor activities, field trips or hands-on activities. Expecting large behavioral changes in a short time is difficult because behavior is a component that is influenced by many factors so that behavior can not change instantly. To make students who have good environmental literacy skills is a process that takes a long and continuous time The slightest change in behavior will give meaning to environmental sustainability. That small changes in behavior can have a major impact[20]. Then the contribution of teaching materials in changing the behavior of a person to the environment can be seen on the content of teaching materials through strategies by changing the way students think in advance through the development of students' thinking skills to be able to take decisions and have the competence to act [3].

3.5. Criteria for all students' Environmental literacy

The improvement of students' environmental literacy in overall can be seen in Table 8.

Table 8. Pretest and posttest comparison for the total score of environmental literacy test

Aspect		N	Mean	Std. Deviation	t	df	Sig 2-tailed	Result
Environment literacy	Pretest	32	138,27	16,31	-18,979	31	0,000	Differ significantly
	Posttest	32	172,21	14,95				

Based on Table 8, overall values of P-value pretest and posttest (Sig) <0.05 were obtained. This means that there are significant differences in environmental literacy of students between before and after learning using integrated science materials. The negative t value indicates that the postest result is higher than the pretation which means an increase in the posttest result. The percentage of improvement of students' environmental literacy ability in overall a whole can be seen in Figure 6.

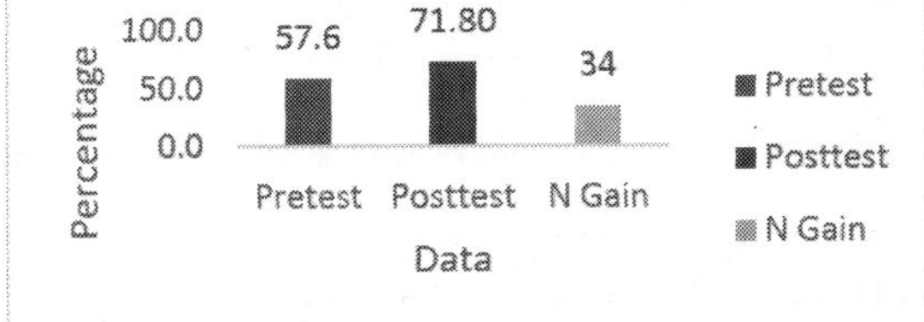

Figure 6. Percentage The improvement of students' environmental literacy

Based on Figure 6, it can be seen that overall there is an increase in overall students' environmental literacy with a 34% N-gain value in the medium category. Thus it can be concluded that the integrated science teaching materials used in learning can increase the students' environmental literacy. Although the improvement of students 'environmental literacy is not yet optimal (being in the medium category), the teaching materials have contributed greatly to the students' environmental literacy. To make students have the environment literacy skills of the real is a long process and cannot be done instantly. Real environmental literacy takes a long time [20]. The use of integrated science teaching materials is one of the efforts to improve of students' environmental literacy. The effort to improve environmental literacy is an effort that must be done continuously given the ability of one's literacy environment greatly affects how a person treats his environment.

4. Conclusions

Based on the results of the use of integrated science teaching materials in learning can improve students' environmental literacy for each component there is an increasing component of knowledge and cognitive skills in the medium category, as well as an increase in affective and behavioral components in the low category. Overall, improvement of students' environmental literacy by use teaching material is in the medium category. It can be concluded the use of Integrated science teaching material can further increase the students' environmental literacy.

Acknowledgments

The authors would like to thank for teacher and students who have provided input and direction on this research.

References

[1] Karatekin K 2012 Environmental Literacy in Turkey Primary School Social Studies Textbooks *Procedia Social and Behavioral Sciences* **46**, p 3519-352

[2] Erdogan M, Kostova Z, Macinkowski T 2009 Components of Environmental Literacy in Elementary Science Education Curriculum in Bulgaria and Turkey *Eurasia journal of Mathematics, Science & Technology Education,* **5** (1), p 15-26

[3] Kernel D and Naglic S 2009 Environmental Literacy Comparison between Eco-School and Ordinary School in Slovenia *International Council of Association for Science Education,* **2** (½), p 5-24

[4] K S Hollweg, J R Taylor, R W Bybee, T J Marcinkowski, W C McBeth, and P Zoido 2011 *Developing a framework for assessing environmental literacy* (Washington, DC North Am Assoc Environ Educ) p 11

[5] Kostova K and Vladimirova E 2010 Development of Environmental Literacy by Didactic Strategies *Chemistry* **19** (3), p 50-70

[6] Meagher T 2009 Looking Inside a Student's Mind: Can an Analysis of Sutend Concept Maps Measure Changes in Environmental Literacy *Electronic Journal of Science Education* **13** (1), p 1-28

[7] Jannah M, Halim L, Meerah T S M, Fairuz M 2013 Impact of Environmental Education Kit on Students' Environmental Literacy *Asian Social Science,* **9** (12)

[8] Erdogan, M 2015 The Effect of Summer Environmental Education Program (SEEP) on Elementary School Students' Environmental Literacy *International Journal of Environmental & Science Education,* **10** (2), p 165-181

[9] M M Chingos and G J R Whitehurst, 2012 "*Hoosing lindly*," no April, p 4

[10] Ifeoma M M 2013 Use of Instructional Materials and Educational Performance of Students in Integrated Science (A Case Study of Unity Schools in Jalingo, Taraba State, Nigeria) *IOSR Journal of Research & Method in Education* (IOSR-JRME), Vol **3**, Issue 4

[11] Sinaga, P, Suhandi, A, and Liliasari, 2015 The effectiveness of scaffolding design in training writing skill physics teaching materials *International journal of instruction* **8** (1), p 9

[12] Fraenkel J R and Wallen N E 2007 *How to Design and Evaluate Research in Education Eight Edition* (New York: McGraw Hill)
[13] Sugiyono 2010 *Metode Penelitian Pendidikan (pendekatan kuantitatif, kualitatif, dan R&D)* (Bandung: Alfabeta)
[14] National Environmental Literacy Assessment (NELA) (2008) *Final Research Report Part 1 & 2* (Washington DC: NELA Project)
[15] Hake R 1998 *Interactive-engagement versus traditional methods: A six-thousand-student survey of mechanics test data for introductory physics courses,* **61** (1), hlm 65
[16] Achyani 2010 *Meningkatkan kepedulian siswa terhadap lingkungan ekosistem persawahan dengan model penulisan buku ajar biologi SMA berwawasan lokal dan ekologi Universitas pendidikan Indonesia (*Bandung)
[17] Mayer R E, and Moreno R 2003 Nine Ways do Reduce Cognitive Load in Multimedia Learning *Educational Psychology* **38**(1), p 43-52
[18] Kose S 2011 Investigation of undergraduate students' environmental attitudes *International Electronic Journal of environmental education* **1** (2) p 85-96
[19] Eggen P and Kauchak D 2012 *Strategi dan Model Pembelajaran Mengajarkan Konten dan Keterampilan Berpikir* (Jakarta: Indeks)
[20] Littledyke 2006 Science education for environmental awareness: approaches to integrating cognitive and affective domains in: Evangelos I Manolas (editor), *Proceedings of the 2006 Naxos International Conference on Sustainable Management and Development of Mountanious and Island Areas* (Thace: Media University of Crete) p 254-268

Internet Literacy of Vocational High School Teachers

D Vernanda*, A G Abdullah, and D Rohendi

Universitas Pendidikan Indonesia, Bandung, Indonesia

*yogurt.nda@gmail.com

Abstract. Internet literacy is needed to know the development of the world in various things quickly and precisely, as well as in the world of education, especially teachers. Seeing the importance of internet literacy, there is an interest to discuss and analyze the level of internet literacy of teachers. The method used in this research is descriptive qualitative. The sample of research is vocational school teacher (SMK) as many as 99 respondents. Data are collected through questionnaires. The theory used is adopted from Suitable Learning Object Type (LRE APV4.7) and Three Elements of Literacy Digital. The result data of the questionnaire is processed and analyzed using Capability Maturity Model theory, the result of the research shows that the level of internet literacy of vocational teachers is at level 2, meaning that SMK teachers have used the internet many times to assist their daily activities and have pattern of repetition in internet utilization. The use of internet by teachers of SMK with various needs that support the process of teaching, communicating, sharing knowledge, but the most dominant is to communicate through social media. Factors affecting internet usage include age, gender, and employment status.

1. Introduction

Recent studies reveal that the use of technology and the Internet will offer many benefits in the development of effective education and learning [1]. With the Internet can broaden the information needed by the teacher as a basis for knowledge development, improve the efficiency of the learning process as well as with the Internet can cultivate lifelong learning [2]. The productivity of internet use in education is reflected in the fact that students can learn easily and they are willing to accept tasks with new technology [3]. The Internet in education is used for the purpose of improving the quality of teachers especially in providing teaching resources, sharing ideas and opinions, working on projects and enabling teacher or group collaboration [4].

In addition, the use of the internet in education is considered an efficient teaching process and is followed by interactions such as presentations, demonstrations, practices, and collaborations [5].

Studies ever conducted on internet use, teachers rarely or never use the internet related applications such as online discussion, voice chat, and video conferencing [6]. Some teachers with low self-esteem do not have sufficient capability in information retrieval resulting in limited use of internet resources [7]. The research ever conducted states that teachers are not ready to use the internet in education [8], as well as in the study [9] states that it is difficult to build relationships between teachers and the internet in the learning process. This problem is one of them occurred in Prabumulih City, South Sumatera Province, most of teachers in Prabumulih have not utilized internet in teaching and learning process and making lesson plan.

Research [10] states that schools use the Internet to design instructional plans, provide individual guidance and how to use the internet, use the internet for teacher professional development, and provide assistance to students through online courses. In an effort to provide creative, innovative, and easy-to-understand education in education and the demands of 21st century education, teachers are encouraged to develop a sense of confidence and competence in using the internet to teach [11].

A large number of studies have been conducted on teachers' perceptions, opinions, and attitudes on issues such as subject matter, teacher professional development, internet usage training, implementation of Internet innovations in education, etc. [12]. Small study that discusses the internet teacher's literacy. Similar studies discussing the internet literacy for the learning process begin by utilizing social media, web-based learning, and creating online discussion groups [13]. The results revealed the teacher's internet literacy is influenced by the intention of a teacher. Therefore this study will evaluate the internet literacy of teachers in the learning process, knowing the intensity of internet usage time and how the internet is used in daily work by a teacher in education. This research was conducted in Prabumulih City, especially in Vocational High School (SMK) teachers.

2. Concepts

2.1. Internet Literacy

The Internet began in August 1962 which was first proposed by J.C.R. Licklider from MIT (Massachutts Institute of Technology). The Internet is a computer network that connects millions of computers spread all over the world, all computers can be connected in the internet network even though located in different places. The Internet can also be interpreted as a global network connecting about one million organizations of computer networks to more than two thousand countries on all continents [14].

An association of "The Association of Colleges and Research Libraries (2010)" defines Internet literacy as people associated with the use of computers, software / applications, databases, and other technologies to achieve certain goals. Internet literacy can also be interpreted as a skill that includes a location to acquire internet and use the internet for knowledge of interpreting and evaluating information [15].

Internet literacy by Shapiro and Hughes (1996) includes seven constructs, including:

1. Tool, refers to the ability to understand and use practical and conceptual information technology tools in their respective professional lives
2. Resources, is the ability to understand the form, location, access method, and format of the information source
3. The social structure of literacy, expressing an understanding of how information is socially located and produced
4. Research, demonstrating the ability to understand and use relevant information technology for research
5. Publishing, reflecting the ability to format and publish research and ideas in text and multimedia
6. Important technology, refers to the awareness and ability to adapt in understanding, evaluating, and using existing information technology
7. Literacy is critical, revealing the ability to critically evaluate the strengths and weaknesses, capabilities and limits of information technology.

2.2. Personal Capability Maturity Model (P-CMM)

Each generation in each country has different ICT / e-literacy literacy levels, as well as individual e-literacy capabilities will have different patterns according to the needs of people's lives and maturity. According to the Personal Capability Maturity Model (P-CMM) [16], a person's e-literacy level can be described more or less as below:

1. Level 0, if an individual is totally ignorant and does not care about the importance of information and technology for everyday life

2. Level 1, if an individual has ever had one or two experiences in which information is an essential component of achieving desire and solving problems and has involved information or communication technology to look for it
3. Level 2, if an individual individual has repeatedly used information and communication technology to assist his daily activities and has had a pattern of recurrence in its use
4. Level 3, if this person has a standard of knowledge and mastery of information or technology required and consistent use of standards as a reference
5. Level 4, if an individual individual has been able to significantly improve the performance of his daily life activities through the utilization of information technology
6. Level 5, if an individual has regarded information and technology as an integral part of everyday activities and directly or indirectly has colored the behavior and culture of his life.

2.3. Technology as a Learning Resource

The growing use of technology in teaching and learning has led to an expansion in the production and dissemination of digital content. This digital content is known as Learning Objects (LOs). LOs are commonly used by teachers and learners in the learning process, LO repositories are specialized software systems used to manage the collection of LOs, a collection of several LOs namely The Learning Resource Exchange (LRE) [17].

3. Research Methods

3.1 Research Design

This research is a descriptive research aimed at giving systematic and factual description which is in accordance with facts about the level of internet literacy of vocational teachers. The data collection used a qualitative approach to the survey through questionnaires [18].

3.2. Development of Research Instruments

Data collection in this study using questionnaires. Indicators measured refer to the internet literacy area such as: Tools (availability of tools for accessing the internet), resources (use of support resources, location, access methods), literacy social structures (understanding and ability to use the internet in education), research Ability to understand and use the internet to support research and improve the professional competence of teachers), publishing (the ability of teachers in using the internet to create or create something like simple media and shared on the internet), important technology (awareness and ability to understand the internet, Education), critical literacy (revealing the ability to critically evaluate the strengths and weaknesses, abilities and limits of the internet) [19].

The questionnaire adopted from the Learning Resource Exchange (LRE) Metadata Application Profile Version 4.7 and Three Elements of Literacy Digital and Model of SCONUL 2006, these three elements are among the first to connect with the Internet including Internet use orientation, surfing the internet, internet focus and location Access the internet. The second is internet interaction including critical thinking, evaluating internet usage [20]. Third is utilizing the internet that includes changing, communicating, and implementing the internet [21] [22]. The instrument grille is shown in Table 1.

Table 1. The instrument grille

Indicator	Item Questions / Statements
Availability of tools to access the internet	The following tools are already available to access the internet A. Computer B. Laptop C. Smartphone D. Tablet E. Internet connection (data / modem package) F. Internet connection (wifi)
Location and amount of internet access time in one day	1. At school using a computer 2. At school using a smartphone or tablet 3. Home using a computer 4. Home using a smartphone or tablet 5. In public use smartphone or tablet
Use of the Internet in education	6. Search for teaching materials from search engines (google, yahoo, bing, amazon) 7. Looking for training / replication questions from search engines (google, yahoo, bing, amazon) 8. Search for information related to prakerin places or street vendors for students from the internet 9. Search for job related information for vocational graduate students from internet 10. Using the glossary to collect specific terms and their meanings in enriching the teaching materials 11. Use of an online dictionary, an encyclopedia to find specific information about a learning topic 12. Downloading electronic books (e-books) 13. Download (download) video, audio, animation from internet / youtube 14. Download Power Point for learning 15. Demonstrate the downloaded video / audio, animation that has been downloaded / downloaded 16. Using the online editor tool (edit photos, change pdf-word files, etc.) 17. Learning using e-learning (edmodo) 18. Doing exams / replicates online 19. Leveraging online educational games 20. Save the file / resource in cloud computing (google drive, drop-box) 21. Using social media for communication in learning 22. Incorporated in an online discussion forum
The type of social media used	23. Facebook 24. Instagram 25. Whatsapp 26. Blackberry Messenger (BBM) 27. Line 28. Google+ 29. Path 30. Twitter
Utilization of the Internet for Research	31. Searching research information from internet 32. Join the forum or research group 33. Active in forum or research group 34. Submit research results in a forum or journal
Utilization of the Internet for sharing knowledge	35. Create an electronic mail address (e-mail) 36. Create a blog 37. Creating a website 38. Actively publish useful things on your blog or website 39. Actively publish material related matters and resources on blog or social media 40. Actively publish things related to prakerin or job vacancy on social media 41. Create and upload learning media on your blog or website 42. Create and upload learning videos on Youtube

3.3. Maturity Model Level Mapping

Once known the score of each category followed by calculating the value of each respondent for each variable. Furthermore, the calculation of index maturity or index maturity with the formula:

$$\text{Index maturity} = \frac{Number\ of Answers}{Number\ of\ Questions}$$

After known index maturity (IM) for each variable then followed by scaling that refers to Personal Capability Maturity Model (P-CMM) [23]. The maximum maturity index is 2 and the minimum maturity index is 0, so the obtained range is 2, after it is known that many classes are 6, then there is a long interval (I) = range / many classes, i.e. 2/6 = 0.33. Index maturity scale for mapping to maturity model level can be seen in table 2. Index maturity scale.

Table 2. Index maturity scale

Indeks Maturity (IM)	Personal Capability Maturity Model (P-CMM)	
	Scale	Information
$0,00 \leq IM \leq 0,33$	0	Someone does not know the internet
$0,34 \leq IM \leq 0,67$	1	Someone has one or two experience using the internet
$0,68 \leq IM \leq 1,00$	2	Someone has repeatedly used and used the internet
$1,01 \leq IM \leq 1,34$	3	Someone has standard and understanding in using internet
$1,35 \leq IM \leq 1,68$	4	Someone has been able to improve performance through the use of the internet
$1,69 \leq IM \leq 2,00$	5	Someone has assumed the internet is partly inseparable in everyday activities

Table 2 shows scaled questionnaires referring to the level of maturity model. Of the total variables, four variables mapped in the level of maturity model including the use of the internet in education, social media used, the use of the Internet for research, and the use of the Internet for sharing knowledge.

3.4 Population and Sample

In this study population is all teachers of SMK in Prabumulih City amounting to 358 people, consisting of 2 SMK Negeri and 8 SMK Private. The number of sample members is often expressed by the size of the sample, the larger the number of samples approaching the population, the less probability of generalization error and the smaller the number of samples away from the population, the greater the generalization error. The sample size is 99 respondents.

4. Results and Discussion

4.1. Availability of Internet Access Tool

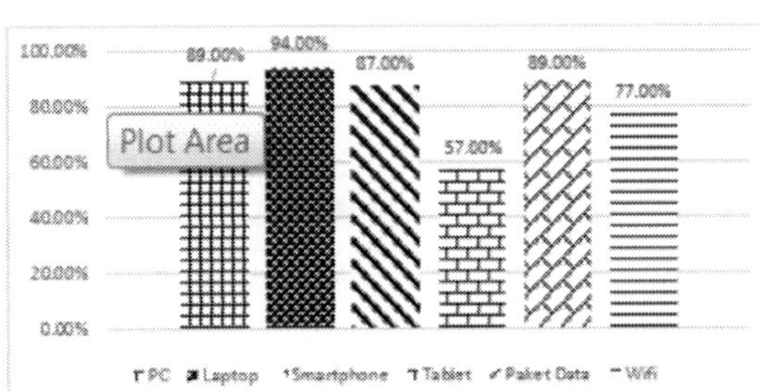

Figure 1. Availability of Internet Access Tool

Figure 1 shows the availability of internet access tools owned by both school and private inventory holders. If averaged ownership of internet access tools SMK teachers have had internet access tools such as computers or personal computers (PCs), laptops, smartphones, and tablets. The dominant internet access tools owned by SMK teachers are laptops. In many countries accessing the Internet via cellular networks and portable equipment such as laptops are experiencing a surge in demand [24]. Research has been conducted on the selection of types or internet access tools that are influenced by gender, age, experience of internet usage. The study revealed a younger age and male respondents preferred mobile access tools [25]. Bottom of Form.

4.2 Location and Number of Times Accessing the Internet

Table 3. Location and number of times accessing the internet

Internet Access Location	Number of Respondents						
Average Time of Day (Hour)	0	<1	1	2	3	4	>5
At school using PC/Laptop	11	19	15	26	13	5	10
At school using smartphone/tablet	10	8	25	24	13	10	9
At home using PC/laptop	16	11	25	21	17	4	5
At home using smartphone/tablet	6	9	16	19	22	15	12
At public area using smartphone/tablet	11	11	27	16	17	10	7

The internet access time of SMK teachers in Prabumulih if averaged in one day approximately 2.6 hours and if in one month approximately 57,2 working hours. Based on the above categories (table 3), the intensity of internet access of SMK teachers in Prabumulih City is included in the category of heavy users or internet addiction. Some researchers have conducted research on the effects of Internet addiction, in this topic of psychological security for Internet addicts [26]. Internet access made by SMK teachers in Prabumulih for some interests such as seeking information, other fun activities such as playing games, communicating, and making transactions such as buying and selling [27].

4.3 Use of the Internet in Education

Figure 2. Use of the Internet in Education

Figure 2 illustrates that the use of internet in education comes in level 3, respondents have standard and understanding of internet usage in education. Most vocational school teachers at Prabumulih have downloaded and demonstrated the video as a teaching support especially on subjects requiring labs and props.

4.4. Use of the Internet for Social Media

Based on the findings on the use of internet for social media entry in level 3, that is, respondents have standards and understanding in the use of social media. The results of data processing stated that SMK teachers in Prabumulih already have standards and understanding in the use of social media, social media most widely used to interact with others locally and globally, in various languages both oral and written. Social media used by SMK teachers in Prabumulih besides whatsapp is facebook. Social media allows teachers and students to have fast and reliable information anywhere in the world. The use of social media for those actively participating will produce accurate new knowledge [28].

4.5. Use of the Internet for Research

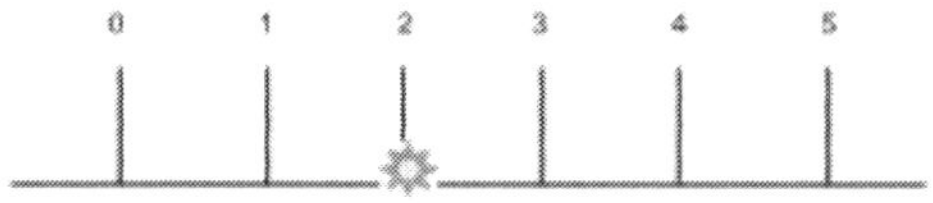

Figure 3. Grafik Maturity Use of the Internet for Research

Figure 3 shows the internet use by SMK teachers in Prabumulih for entry research at level 2, the majority of respondents using the internet only to search for research information. To advance the science of the profession as a teacher in the world of education, teachers are required to conduct research and publish the results [29].

The average teacher of SMK in Prabumulih has never sent or published the results of his research via the internet, respondents use the internet to search information and join research forums.

4.6. Use of the Internet for Sharing of Knowledge
The results of data processing of 8 statements to determine the utilization of the internet for sharing knowledge is at level 1. Level 1 means an individuals never have experience once or twice in the use of the Internet for information sharing. Research reveals the use of social networking sites for sharing information online in teaching and learning, such as downloading and downloading social networking based resources [30].

4.7. Maturity Rate of Internet Usage SMK Teachers
Based on the calculation of maturity model level starting from the variable of internet utilization for education, internet usage in social media, internet utilization for research, and exploiting internet for sharing knowledge obtained average score that is 0,93. The internet literacy of SMK teachers if rounded up on a Personal Capability Maturity Model scale of the internet literacy level of SMK teachers entered in level 2.

The use of internet by teachers of SMK in Prabumulih with various needs such as supporting the teaching process, communicating through social media, and sharing knowledge, but the most dominant is to communicate through social media. Previous research on internet use by math teachers shows that most math teachers use the internet for personal affairs. The results of the analysis suggest the need for technical, pedagogical, and didactic training on the use of digital resources [31].

Research has been conducted by surveying more than 200 teachers, research results suggest teachers have Internet usage patterns but they do not use the internet to explore the wider world and learn about alternative ways of thinking. The study also revealed that the majority of prospective teachers use the Internet only a little to find information on various world perspectives and the search for very minimal supporter information [32]. There are many factors that influence teachers to use the internet in education that is individual perspectives, organizational influences and technology to be used [33].

5. Conclusion
The analysis of internet literacy of SMK teachers using six indicators such as availability of internet access tool, internet access location and internet access intensity, internet utilization in education, internet usage for social media, internet utilization for research, and for sharing science / information. After calculated and analyzed from the six indicators, the level of maturity of internet literacy of vocational teachers has reached level 2, meaning that SMK teachers have been using the internet many times to help the daily activities and has had pattern of repetition in the utilization of internet.

References
[1] Sirakaya M, Armak B A Ş and Baltaci Ö 2015 Analysis of teacher candidates' educational internet self -efficacy beliefs in terms of various variables *Procedia - Social and Behavioral Sciences*, **174**, 3094–3101
[2] Chuang S, Lin F, and Tsai C 2015 An exploration of the relationship between Internet self-efficacy and sources of Internet self-efficacy among Taiwanese university students *Computer in Human Behavior*, **48**, 147–155
[3] Mura G, and Diamantini D 2014 The Use and Perception of ICT Among Educators: The Italian Case *Procedia - Social and Behavioral Sciences*, **141**, 1228–1233

[4] Zhang X, Ordonez De Pablos P, and Zhou Z 2013 Effect of knowledge sharing visibility on incentive-based relationship in Electronic Knowledge Management Systems: An empirical investigation *Computers in Human Behavior*, **29**(2), 307–313

[5] Marinagi C, Skourlas C, and Belsis P 2013 Employing Ubiquitous Computing Devices and Technologies in the Higher Education Classroom of the Future *Procedia - Social and Behavioral Sciences*, **73**, 487–494

[6] Subekti L B, Bandung Y, and Gondokaryono Y S 2015 A Design of Teacher Portal for Supporting Teacher ' s Internet Literacy In *The 5th International Conference on Electrical Engineering and Informatics 2015* (pp 618–623)

[7] Karaseva A 2016 Relationship of Internet self-efficacy and online search performance of secondary school teachers *Procedia - Social and Behavioral Sciences*, **231**(May), 278–285

[8] Stošić L, and Stošić I 2015 Perceptions of teachers regarding the implementation of the internet in education *Computers in Human Behavior*, **53**, 462–468

[9] Machin S, McNally S, and Silva O 2007 New technologies in schools: Is there a payoff? *The Economic Journal*, **117** (2234), 1145–1167

[10] Wells, J, and Lewis, L 2007 Internet Access in US Public Schools and Classrooms: 1994-2005 *National Center for Education Statistics*

[11] Kiran B, Aktas E, and Tuncer I 2013 An analysis of university students internet use in relation to loneliness and social self-efficacy *Procedia - Social and Behavioral Sciences* **84** *(2013) 1504 – 1508 3rd, 84*, 1504–1508

[12] Kurilovas E, Kubilinskiene S, and Dagiene V 2014 Web 30 - Based personalisation of learning objects in virtual learning environments *Computers in Human Behavior*, **30**, 654–662

[13] Acarli D S and Sağlam Y 2015 Investigation of Pre-service Teachers' Intentions to Use of Social Media in Teaching Activities within the Framework of Technology Acceptance Model *Procedia - Social and Behavioral Sciences*, *176*(Social and Behavioral Sciences), 709–713

[14] Rainer R K, Prince B, and Cegielski C G 2014 *Introduction to Information Systems, 5th Edition International Student Version* (Fifth Edit) (International Student Version)

[15] Leung L, and Lee P S N 2012 Impact of Internet Literacy, Internet Addiction Symptoms, and Internet Activities on Academic Performance *Sosial Science COmputer Review - SAGE*, **30**(4), 403–418

[16] Curtis B, Hefley W E, and Miller S A 2001 *People Capability Maturity Carnegie Mellon Software Engineering Institute* (Pittsburgh)

[17] Massart D, Shulman E, and Assche F Van 2010 *Learning Resource Exchange Metadata Application Profile version 45 European Schoolnet* (Fourth Edi) (Belgian) Retrieved from http://lre.eun.org/sites/default/files/docs/LREMAPv4p5w.pdf

[18] Creswell J W 2009 *Research Design: Qualitative, Quantitative and Mixed Approaches (3rd Edition) Research Design: Qualitative, Quantitative, and Mixed Methods Approaches (3rd ed)* (Third) (United States of America: California SAGE)

[19] Rosenthal M 2016 Qualitative research methods: Why, when, and how to conduct interviews and focus groups in pharmacy research *Currents in Pharmacy Teaching and Learning*, **8**(4), 509–516

[20] Bas G, Kubiatko M, and Sunbul A M 2016 Teachers' perceptions towards ICTs in teaching-learning process: Scale validity and reliability study *Computers in Human Behavior*, **61**, 176–185

[21] Markless S and Streatfield D 2001 Three decades of information literacy : redefining the parameters *Change and Challenge: Information Literacy for the 21st Century*, 15–36

[22] Martin, A, and Madigan, D 2007 Literacies for the digital age *Digital Literacies for Learning* (London : Facet Publishing 242) **1**(3), 1–3

[23] Cobit and Govermance, I I 2007 *Cobit 41 Framework Control Objectives Managment Guidelines Maturity Model* Rolling Meadows

[24] Gerpott T J 2011 Attribute perceptions as factors explaining Mobile internet acceptance of cellular customers in Germany-An empirical study comparing actual and potential adopters with distinct categories of access appliances *Expert Systems with Applications*, **38**(3), 2148–2162
[25] Thomas S, Weichert M, and Gerpott T J 2013 Characteristics and mobile Internet use intensity of consumers with computer-centric communication devices: An exploratory empirical study of iPad and laptop users in Germany *Elsevier-Telematics and Informatics*, **37**(4–5), 357–371
[26] Jia J, Li D, Li X, Zhou Y, Wang Y, and Sun W 2017 Psychological security and deviant peer affiliation as mediators between teacher-student relationship and adolescent Internet addiction *Elsevier-Computers in Human Behavior*, **73** (Computers in Human Behavior), 345–352
[27] Buente W and Robbin A 2008 Trends in internet information behavior, 2000-2004 *Journal of the American Society for Information Science and Technology*, **59** (11), 1743–1760
[28] Cevik A A, Aksel G, Akoglu H, Eroglu S E, Dogan N O, and Altunci Y A 2016 Social media, FOAMed in medical education and knowledge sharing: Local experiences with international perspective *Turkish Journal of Emergency Medicine*, **16** (3), 112–117
[29] Mahon P Y 2014 Internet research and ethics: Transformative issues in nursing education research *Journal of Professional Nursing*, **30** (2), 124–129
[30] Asterhan C S C and Bouton E 2017 Teenage peer-to-peer knowledge sharing through social network sites in secondary schools *Computers and Education*, **110**, 16–34
[31] Brahim N, Mohamed B, Abdelwahed N, Ahmed L, Radouane K, Khalid S, and Mohammed T 2014 The Use of the Internet in Moroccan High Schools Mathematics Teaching: State and Perspectives *Procedia - Social and Behavioral Sciences*, **116**, 5175–5179
[32] Redekopp R and Kalanda K 2015 Internet use: A Study of Preservice Education Students in Lesotho and Canada *Procedia - Social and Behavioral Sciences*, **182** (Social and Behavioral Sciences), 529–534
[33] Yassin F, Salim J, and Ashaari N S 2013 The Influence of Organizational Factors on Knowledge Sharing Using ICT among Teachers *Procedia Technology*, **11** (Iceei), 272–280

IOP Conf. Series: Materials Science and Engineering **306** (2018) 012033 doi:10.1088/1757-899X/306/1/012033

Evaluation of an Affordable Wireless Node Sensor (Mote69) Designed for Internet of Thing (IoT) Device

Z F Ruhiyat, Y Somantri, D Wahyudin*, and D L Hakim

Electrical Engineering Education, Universitas Pendidikan Indonesia, Bandung, Indonesia

*deewahyu@upi.edu

Abstract. This research aims to determine the student's response to the implementation of Internet of Things (IoT) device based on RFM69, called Mote69, for practical work of Aircraft Electronic Circuits and Controllers. Participants in this study were students of a vocational high school of Aircraft Electronics which consisted of three groups. The first group is the students who have grades above the average class. The second group is the students who have grade the average class. The third group is the students who have grades below the average class. The research phase consisted of (1). Observation and Assessment of Empirical Issues; (2). Testing of Media Feasibility and Research Instruments; (3). Accumulate and Processing of Field Data; and (4). Results and Data Conclusions. The result of media feasibility showed that Mote69 is appropriate to be used in practical work of Aircraft Electronic Circuits and Controllers subject.

1. Introduction

Media for learning is one of the essential elements of the vocational high school to support the education process. Application of media in learning can raise interest, motivation, and stimulation of learning activities [1]. The results of some research also proved that media for learning could increase the student's interest and progress in learning [2]. Based on the experience in School Experience Program especially in the subject of Aircraft Electronic Circuits and Controllers, the researcher found problems in the learning process. In the practical work, teachers only used conventional learning methods and they did not use any media to learning. Hence, the students unmotivated to study hard. On the other hand, lab equipment's are costly. It impacted to the school committee that could not be able to fund lab equipment. Hence, it is necessary to develop a training kit that is user-friendly, low cost, industry-oriented requirements, and portable [3]. In recent years, Arduino microcontroller is an appropriate device for education. It has been increased the learner's interest efficiently and progresses in learning [4]. Also, the results of research in also showed that the development of technologies such as the Internet of Things (IoT) should be utilized in the field of education. It provides an intelligent and innovative learning media that can be used by many learners [5].

This paper explains the evaluation of an affordable wireless sensor node, named Mote69, designed for IoT device. We used Arduino Pro Mini microcontroller, a wireless module of RFM69, and various sensors as an IoT trainer to be used in the subject of Aircraft Electronic Circuits and Controllers. It has low cost, Arduino Pro Mini also has many advantages, i.e., has a lot I/O pins. Hence learners could use various sensors in the IoT projects. The printed circuit board (PCB) in this trainer had been designed to compatible with Arduino Mega. So, the device would have more space that could be more flexible to add some module or shield

2. Mote69 Design

IoT device based on RFM69 is an IoT trainer that can be used to support students in the practical work of microcontroller programming course. This section describes the design and materials of Mote69 with Arduino Mega PCB layout as shown in Figure 1. However, the block diagram of the system is shown in Figure 2.

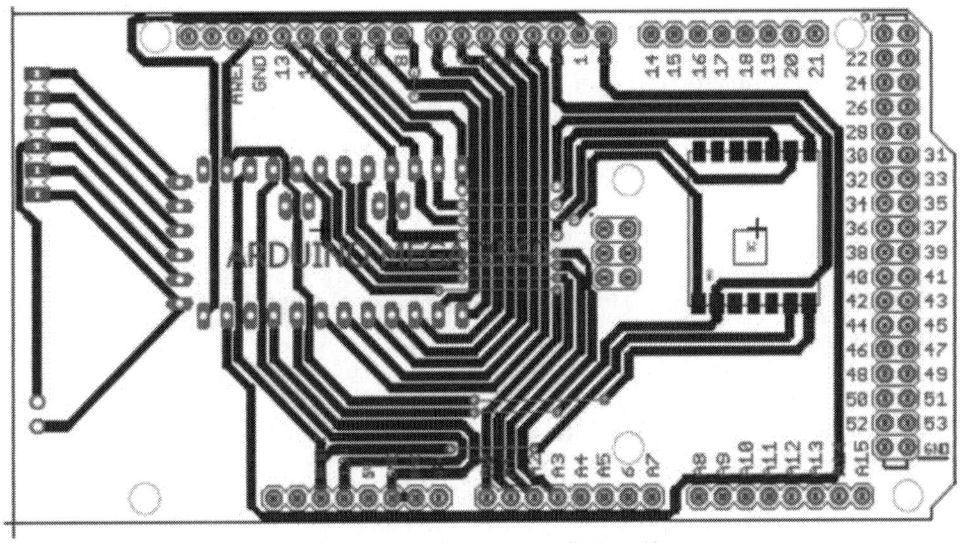

Figure 1. PCB layout of Mote69

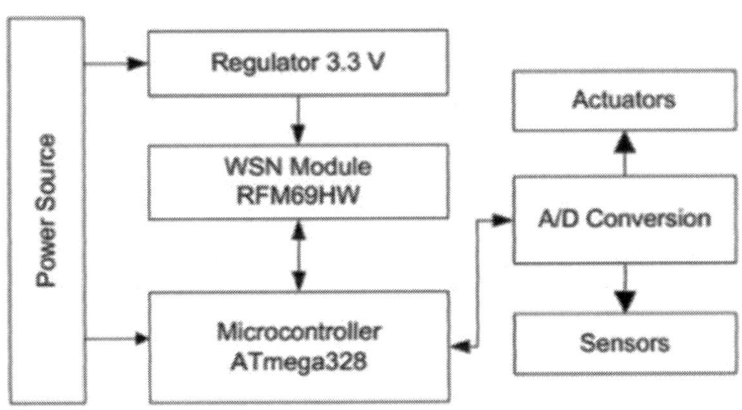

Figure 2. Diagram block of Mote69

Based on the block diagram in Figure 2 several components used in this trainer such as a microcontroller, sensors, and a transceiver module. The following are the explanation of each part.

2.1. Microcontroller Unit (MCU)

The microcontroller is the brain of the trainer kit system. This component serves to process data obtained from inputs such as sensors and then process it before transmitting by the RFM69 module. Most of the microcontrollers can work in real-time. It worked by inserting the program into the microcontroller [6]. We used Arduino Pro Mini version 3.3 V based on ATmega328 microcontroller. The Arduino has 14 digital input/output pins that are six pins as PWM output, six analog inputs, an on-board resonator, reset button, and holes for pin header installation. Six pins can be connected to the FTDI series. Arduino Pro Mini is selected as a microcontroller that used in this trainer because this microcontroller is more efficient than the other type. This microcontroller has all the pins that required for the Trainer kit, such as the SPI pins (MISO, MOSI, and SCK) that the controller needs to communicate with the RFM69, the PWM pin and the ADC pin that required for sensors and actuators. Also, Arduino Pro Mini also has the lowest price compared to other types of Arduino. It has a small size so that this Trainer becomes more portable or easy to carry anywhere.

2.2. Sensors
The sensor is the front end of the IoT systems. These are so-called "sense" of the system. The primary purpose is to gather data from its surrounding (sensors) or provide data to its surrounding (actuators). Sensors may be analog, or digital sensors, for example, Temperature, Humidity, Pressure, Gas, GPS, IR, and LDR [7]. In this IoT device, the various sensor was utilized so that the learners can practice many experiments.

2.3. Wireless Module
Mote69 used RFM69 as the transceiver module that similar to Wireless Sensor Network (WSN) devices. RFM69 is capable of operating over a full frequency in the range 315, 433, 868 and 915 MHz. However, due to reason that Mote69 could be used in the indoor area, hence it uses RFM69 with the frequency of 433 MHz. RFM69 can be optimized that able to use a low power consumption but with high RF output power. Besides that, in Indonesia, the regulation on the use of radio frequency is regulated in the Minister of Communication and Informatics Regulation. Radio frequency band 433-435 MHz is used for the operation of tools and communication devices at close range (Short Range Device). The research about WSN as one of the applications of IoT technology has been done, such as research on data security on WSN [8]. In Mote69 design, wireless module is used to send and receive data between two or more Mote69 as a mesh network.

3. Method
The method used in this study using qualitative methods that intends to understand the phenomenon of subjects experienced such as perception, motivation, and action. Moreover, presented by the description in a natural context and by utilizing various natural methods [9].

There are five approaches to qualitative research: biography, phenomenology, grounded theory research, ethnography, and case studies [10]. The researcher chose case study as an approach to the implementation of this research. The case study implies the researcher do intensive analysis on one unit of analysis examined (case). A case can be one individual, one organization, one event, one decision, one period, or a system that can be studied thoroughly and holistically [11]. This study focused on the analysis of student's perceptions of Mote69 implementation.

The experiment was conducted at the department aircraft electronics in an aircraft vocational high school in Bandung, the capital of Java province, Indonesia. The research activity was focused on the feasibility of Mote69 as a trainer, and examined the student's response to the implementation of this trainer on the subjects of Aircraft Electronic Circuits and Controllers.

3.1. Usability Study of Mote69
The Feasibility of Mote69 is obtained based on an expert judgment by an expert of media and learning, and the usability testing involved some students.

3.2. Students Response to the Implementation of Mote69
Participants in this study were students of grade 12th of Aircraft Electronics study which consisted of three group of students. The first group is the students who have grades above the average class. The second group is the students who have grade the average class. The third class is the students who have

grades below the average class. The participants were asked to do practical work with the scenario as shown in Figure 3. Their practical work activity then monitored and observed by the researcher. They also asked to give opinion about Mote69 usability.

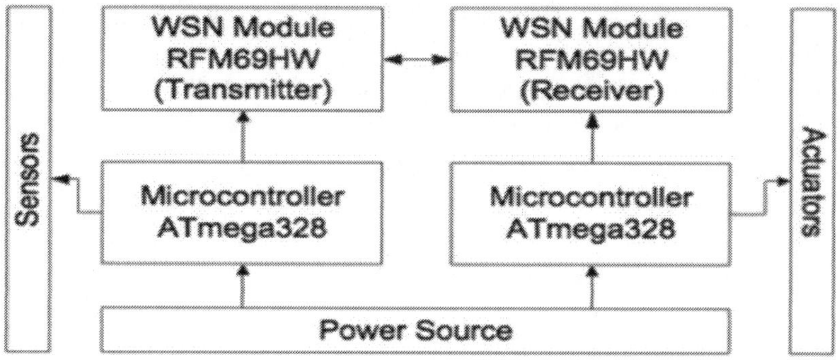

Figure 3. Practical work design using two Mote69 devices

4. Results and Discussion

4.1. Usability Study of Mote69

The result of media feasibility showed that Mote69 is proper to be used in practical work of Aircraft Electronic Circuits and Controllers course. It is capable be used in classrooms with the aim of developing educational programs related to microcontroller programming.

4.2. Students Response to the Implementation of Mote69

The results of evaluation of Mote69 implementation showed the positive response from all participants. This IoT device could promote innovation and motivation for students during the learning process especially on the subject of Aircraft Electronic Circuits and Controllers. The implementation of this mote also helps the teacher to arouse the curiosity of students through experiments that enhance the students' affective and skills.

5. Conclusion

This paper is concerned with two primary objectives: the first one is to approve the feasibility of Mote69 as a trainer used in practical work on the subject of Aircraft Electronic Circuits and Controllers. The evaluation evidence also shows the student's response. It is proven that Mote69 could be used in classrooms with the aim of developing educational programs related to the microcontroller. It promotes innovation and motivation for students during the learning process.

References

[1] Arsyad A 2015 *Media Pembelajaran* (Jakarta: Rajawali Pers)

[2] Plaza P Sancristobal E Fernandez G Castro M Perez C 2016 Collaborative Robotic Educational Tool based on Programmable Logic and Arduino (IEEE Transaction on Education)

[3] Abdullah A G Hakim D L Auliya M A 2014 Preliminary Design of Industrial Automation Training Kit Based Real Mobile Plant *International Conference on Advances in Education TechnoloAgy* pp 133-135

[4] Souza M R S B Tavares J J P Z Ribeiro J F Rocha R R 2015 Design, Manufacture and Construction of a Wireless Robotic Arm for Educational *Purposes IEEE Transaction on Education* pp 376-380

[5] Rahman M Deep V Rahman S 2016 ICT and Internet of Things for Creating Smart Learning Environment for Students at Education Institutes in India *IEEE Transaction on Education* pp

701-704
[6] Bhadoriya R Chattopadhyay M K Dandekar P W 2016 Low Cost IoT for Laboratory Environment *IEEETransaction on Education*
[7] Bhadoriya R Chattopadhyay M K Dandekar P W 2016 Low Cost IoT for Laboratory Environment *IEEETransaction on Education*
[8] Alghamdi A Alshamrani M Alqahtani A Ghamdi S S A A Harrathi R 2016 Secure Data Aggregation Scheme in Wireless Sensor Networks for IoT *IEEE Transaction on Education*
[9] Moleong L J 2015 *Metodologi Penelitian Kualitatif* (Bandung: PT. Remaja Rosdakarya)
[10] Satori D Komariah A 2010 *Metodologi Penelitian Kualitatif* (Bandung: CV. Alfabeta)
[11] Sarosa Samiaji 2012 *Penelitian Kualitatif: Dasar-Dasar* (Jakarta: PT. Indeks)

A Remote PLC Laboratory (RLab) for Distance Practical Work of Industrial Automation

E Haritman, Y Somantri, D Wahyudin*, and E Mulyana

Electrical Engineering Education, Universitas Pendidikan Indonesia, Bandung, Indonesia

*deewahyu@upi.edu

Abstract. A laboratory is an essential equipment for engineering students to do a useful practical work. Therefore, universities should provide an adequate facility for practical work. On the other hand, industrial automation laboratory would offer students beneficial experience by using various educational PLC kits. This paper describes the development of Web-based Programmable Logic Controller (PLC) remote laboratory called RLab. It provides an environment for learners to study PLC application to control the level of the non-interacting tank. The RLab architecture is based on a Moodle and Remote Desktop, which also manages the booking system of the schedule of practical work in the laboratory. The RLab equipped by USB cameras providing a real-time view of PLC environment. To provide a secured system, the RLab combines Moodle and Remote Desktop application for the authentication system and management of remote users. Moodle will send PartnerID and password to connect to TeamViewer. It has been examined that the laboratory requirement, time and flexibility restrictions constitute a significant obstacle facing traditional students desiring to finish the course. A remote access laboratory can be eliminating time and flexibility restrictions. The preliminary study of RLab usability proved that such system is adequate to give the learners a distance practical work environment.

1. Introduction

Laboratories are the most important facilities for engineering students to do a useful practical work. Therefore, universities should provide the appropriate environment for practical work. Hence, an industrial automation laboratory will offer students experience with educational PLC kits. A Programmable Logic Controller (PLC) is a type of industrial controller that could be used to control machinery which industrial productivity is an essential role. The PLC can be controlled using Ladder Diagram, Function Block Diagram, and Structured Text, which is easy to learn and to use. However, this sophisticated facility is high cost. Thus, a university should be eager to provide it.

The invention of internet technology took effect as an accelerator to implement new teaching methods. One of the benefits of the internet on the laboratory is the emergence of the concept of the remote laboratory. The concept of the remote laboratory can be an alternative solution for remote measurement, supervision or control. In the education, this concept can be used as a supporting tool for distance learning. Thus, this concept allows students to operate the laboratory remotely. The concept of the remote laboratory has been used in many universities around the world. The use of a remote laboratory is an urgent need with consideration of the ratio cost, benefits and the time. Remote labs are a smart and useful solution, although local laboratories will always be necessary [1]. One of the most significant contributions of the remote laboratory is the improvement of self-study for learners [2].

Practical work in industrial automation course is often inadequate to get much knowledge of various PLC applications. To answer the problems, an accessible remote-control software is adopted for realizing RLab which would provide the real experiment environment [3]. However, equipment of laboratories and its maintenance require a considerable investment. It was appeared because of the lack of laboratory budget. On the other hand, available resources insufficient some time inappropriate due to the limitation of the flexibility of schedules and time to access the laboratories, and too many learners in the queue to carry out a practical work activity. To cope these problems, virtual instrumentation and the internet capability are being introduced to implement sharing the resources among different institutions. It can be helping to reduce costs and to provide greater flexibility of practical work schedules, thus giving the opportunity to a more significant number of students to perform experiments in their style, at any time and from anywhere [4]. It can be done using the resources among different institution are sharing through remote labs. Therefore, the experiment is designed to stay on and available for 24 hours a day and seven days a week. Thus the learner can use this experiment tool anytime. To be further developed, all experiments are designed in such a way. With RLab, then the teacher can improve the quality of learning without extra time of teaching and learning [5].

2. Model of The Remote Lab

The Remote Laboratory (RLab) utilized the CE111 and CE123 industrial automation kits produced by TQ. It is to provide practical experience of PLC to sense and to control the flow rate and the level of water in a two-tank system. Each tank included in two internal reflection types optical level sensors. It was to sense the maximum and minimum water levels. A variable-speed pump was to transfer water from a reservoir located in the unit base into the upper tank. It was also equipped with an in-line flow sensor providing a pulsed output for flow rate and volumetric monitoring. Solenoid operated valves to be separately closed or opened to control the flow path of the water from the upper tank back to the sump via the lower tank. Diagram of this model is shown in figure 1. The Rlab model can be described into following formulas.

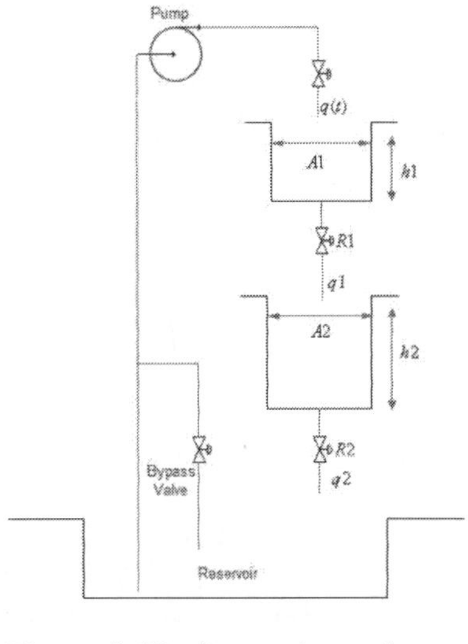

Figure 1: Noninteracting tank system

From figure 1, we can write mass balance at tank1 as:

$$q - q1 = A1\left(\frac{dh1}{dt}\right) \qquad (1)$$

IOP Conf. Series: Materials Science and Engineering **306** (2018) 012034 doi:10.1088/1757-899X/306/1/012034

A mass balance at tank two is given as

$$q1 - q2 = A2\left(\frac{dh2}{dt}\right) \tag{2}$$

The expressions give the flow head relationships for the two linear resistances in the non-interacting system.

$$q1 = \left(\frac{h1}{R1}\right) \tag{3}$$
$$q2 = \left(\frac{h2}{R2}\right) \tag{4}$$

combining (1) and (3)

$q - \left(\frac{h1}{R1}\right) = A1\left(\frac{dh1}{dt}\right) = R1.A1\left(\frac{dh1}{dt}\right) = R1.q$ Taking the Laplace transforms of both sides of Equation, assuming the zero initial condition, we obtain $(R1.A1s + 1)H1(s) = R1.Q(s)$

$$\frac{H1(s)}{Q(s)} = \frac{R1}{R1.A1s+1} = \frac{R1}{\tau_1 s+1} \tag{5}$$
$$\frac{Q1(s)}{Q(s)} = \frac{1}{\tau_1 s+1} \tag{6}$$

combining (2) and (4), using the same steps as above, we obtain:

$$\frac{H2(s)}{Q1(s)} = \frac{R2}{R2.A2s+1} = \frac{R2}{\tau_2 s+1} \tag{7}$$

Therefore the overall transfer function is determined as

$$\frac{H2(s)}{Q(s)} = \frac{1}{(\tau_1 s+1)}\frac{R2}{(\tau_2 s+1)} = \frac{R2}{(\tau_1 s+1)(\tau_2 s+1)} \tag{8}$$

Table 1. The variable for the PLC system

Variable	Address on PLC	Function
Fluid-level, L1	X1	Tank 2 LOW-level switch
Fluid-level, L2	X2	Tank 2 HIGH-level switch
Fluid-level, L3	X3	Tank 1 LOW-level switch
Fluid-level, L4	X4	Tank 1 HIGH-level switch
Fluid-level, L5	X5	Tank 2 LOW-level switch
Valve, V1	Y1	Reservoir LOW level float switch
Valve, V2	Y2	Pump to reservoir valve (by-pass)
Valve, V3	Y3	Tank 1 feed valve
Valve, V4	Y4	Tank 1drain valve
Pump, P1	Y0	Pump
Flowmeter, F1	X0	Flowmeter

The variables of input and output in these kits could be seen in Table 1. The wiring diagram is kept in offering the manual instructions. Remotely, the learners could not change the wiring diagram that has been set to a fixed configuration. However, the learners could do some experiments for this configuration.

A simple set of the sequence for RLab might be as follows:

1. Open V2, and then close V3, and set P1 ON (this is to start filling tank 1).
2. Wait until L4 is set and P1 OFF (this is to stop the process of filling Tank 1 when it is full).
3. Wait for 10 seconds.
4. Open V3, then close V4 (this is to start emptying the Tank 1 and flowing the content into Tank 2).
5. Wait a minute until L3 is set and close V3
6. Repeat the sequence from step 1 indefinitely.

3. Design and Development of The Remote Lab

This section describes the design and development of RLab. The RLab system consists of remote user client installed on notebook or desktop computer. The RLab server provides the ability to access distance learning system using Moodle. On the remote lab, a desktop computer connected to the PLC equipped by a web camera that can provide a real-time view of the lab environment. The construction of RLab could be seen in Figure 2.

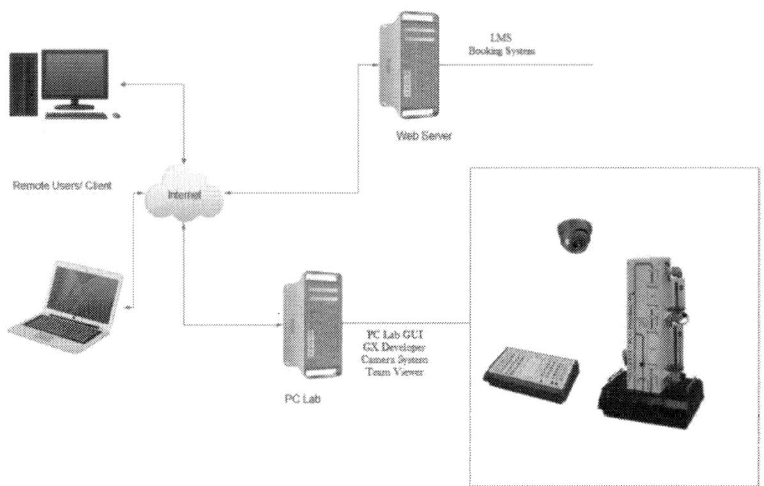

Figure 2: Construction of RLab

In general, three methods for granting remote access to RLab with resources are as follows. (1) the direct access through the standalone authentication system of single laboratory instance. (2) the access provided by the supervisory and management system like LMS or CMS. (3) Access provided by platforms for interconnection and laboratory sharing [6]. RLab using a combination of access provided by the supervisory and management system of Moodle. The learners access provided by platforms for interconnection and laboratory sharing using TeamViewer.

The learners can use a browser installed on their computer or laptop to connect with RLab with a proper internet connection. Also, the learner should install the TeamViewer on his computer. To start the experiment, the learners should log in through Moodle website stored on the web server. Hence, they could also access the available instructions of using RLab, job sheet, and booking facility.

After logging on the Moodle website, learners could book a schedule to use the remote lab. Learners could select the desired day and time slot. Each slot has 30 minutes, and each learner can only use three slots in a day. When the booking time has arrived, the learner will get a partner ID and password from a particular link. The partner ID and password are then used in the TeamViewer application.

At this stage, the learner is already connected with the RLab. However, the security of these protocols can become a problem. While SSH and Remote Desktop protocols are considered secure by design,

VNC is not secure by default [7]. To solve this problem, RLab's GUI is designed to protect the target PC. Hence, the learners could only access specific applications. Furthermore, avoidable applications will not be accessible to prevent the system vulnerability. The RLab's GUI is shown in figure 3.

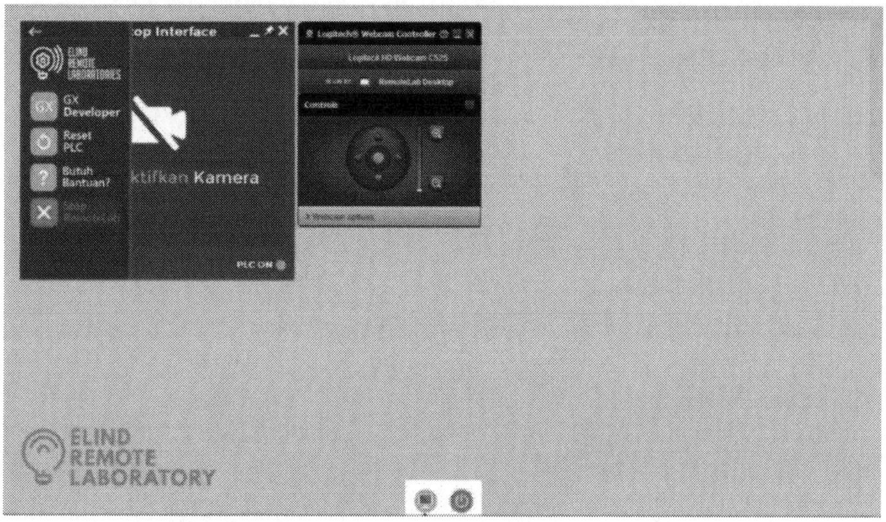

Figure 3: RLab's GUI

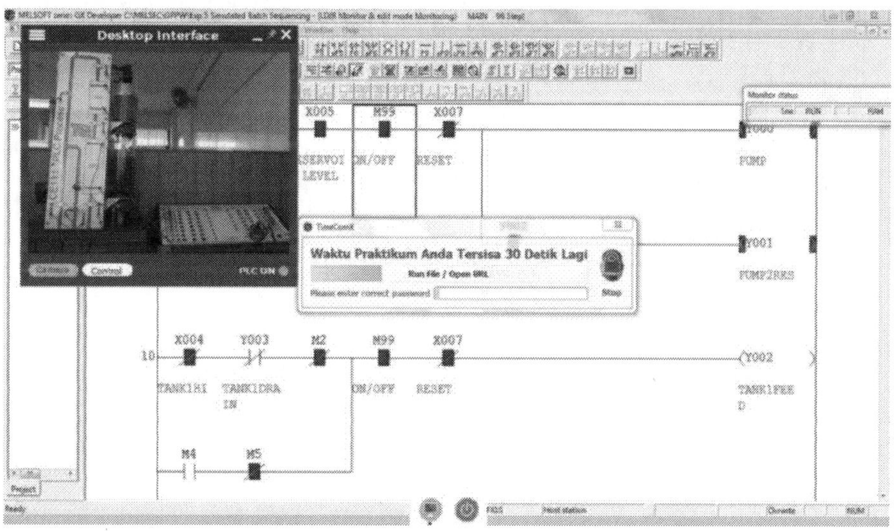

Figure 4: The learner side of the remote lab.

4. Preliminary Examination

To assess the early development of RLab, we conducted the usability examination. The examination involved expert and potential learners, i.e., students of electrical engineering education. The participants asked to log in to the Moodle website study the desired experiment provided on the job sheet. They asked to draw diagram ladder provided by the job sheet and then uploading the compiled program to target PLC. After that, the learner can operate and see directly the experimental tools that are operated. They can monitor the experiment progress in a real-time view through webcam installed in the target PC. The available time for each participant to use RLab for one session is 30 minutes. When the experiment time is almost over, a notice will appear that can be used to store the experiment results in

RLab. The system will automatically close the RLab when the 30-minute time is up. Screenshot of this stage is shown in figure 4.

5. Conclusion and Future Work

Setting up and implementing RLab can be successful and practical. The research success is evidenced by the fact that the learners in the remote client can connect and operate both RLab from a remote location.

Further examination in improving the remote laboratory can deliver of the RLab more flexible, more accessible and increase a rapidity in practical. A pedagogical study and examination also benefit to assess the learner's interest, along with investigating specific equipment that needed in deciding of the scale of application.

References

[1] Luis, et al 2016 Remote Laboratory: Application and usability *IEEE Proceeding 2016 Technologies Applied to Electronics Teaching (TAEE)*

[2] Alfredo, et al 2010 Remote Automation Laboratory Using a Cluster of Virtual Machines *IEEE Transactions on Industrial Electronics*

[3] Xiaomei Chena, Hongyi Gao 2012 A Remote PLC Laboratory Design and Realization Elsevier *Procedia Engineering* ***31*** (2012) 1168 – 1172

[4] Farah Helúe, et al 2013 Electropneumatic system for industrial automation: a remote experiment within a web-based learning environment *Procedia Technology* **7** (2013) p198–207 Elsevier

[5] Petr Sládek, et al 2011 Remote laboratory – new possibility for school experiments *Procedia Social and Behavioral Sciences* **12** (2011) 164–167 Elsevier

[6] Martin Kal′uz, et al 2013 Sharing Control Laboratories by Remote Laboratory Management *System the 10th IFAC Symposium Advances in Control Education The International Federation of Automatic Control*

[7] P Orduña, et al 2011 *Adding New Features to New and Existing Remote Experiments Through Their Integration in WebLab-Deusto International Journal of Online Engineering (iJOE)* Volume **7** special issue 2

IOP Conf. Series: Materials Science and Engineering **306** (2018) 012035 doi:10.1088/1757-899X/306/1/012035

Strategic Planning towards a World-Class University

E J Usoh[1]*, D Ratu[1], A Manongko[1], J Taroreh[1], and G Preston[2]

[1]Universitas Negeri Manado, Kampus Tondano 95618, Indonesia
[2]The University of Newcastle, University Drive, Callaghan, NSW 2308, Australia

*elni.usoh@unima.ac.id

Abstract. Strategic planning with a focus on world-class university status is an option that cannot be avoided by universities today to survive and succeed in competition as a provider of higher education. The objective of this research is to obtain exploratory research results on the strategic plans of universities that are prepared to generate world-class university status. This research utilised exploratory qualitative research method and data was collected by in-depth interviews method. Interview transcripts were analyzed by using thematic content analysis through NVivo software analysis and manual systems. The main finding of interview shows that most interviewees agreed that UNIMA has been engaged in strategic planning. Contribution from faculties and schools are acknowledged and inform the planning process. However, a new model of strategic planning should be adopted by UNIMA due to the shift towards a "corporate university". The finding results from documents, literature review and interview were the addition of world-class university characteristics and features to current strategic planning of UNIMA and how to upgrade by considering to use the characteristics and features towards world-class university.

1. Introduction

The word of world-class university has become a catchy phrase. This status can describe the university's research level as the culmination point of higher education. The paradox of world-class university mentioned as statement "everyone wants one, no one knows what it is, and no one knows how to get one "[1].

However, the existence of a world-class university today is inevitable. The ranking status makes the universities of the world competing to improve the achievements which lead to elitism that brings pros and cons. Building world-class university does require high cost and very expensive investment. To meet the world class university status requires a lot of hard work. Internationalization ranking is a tool for strengthening educational goals and improving the performance of universities.

The need of universities to obtain world-class university status becomes a very important framework for most universities in Indonesia. Higher Education Strategic Planning is an official document that determines policy direction, decision making and institutional strengthening within the university. Strategic planning with a focus on world-class university status is an option that cannot be avoided by universities today to survive and succeed in competition as a provider of higher education.

The Indonesian government has also emphasized the need for internationalization for universities through the University's Long-Term Strategy 2003-2010, which declared that universities in Indonesia are expected to improve their competitiveness and quality significantly to compete globally [2]. It is undeniable that the status of world-class university brings more prestigious benefits and impressions for the university, despite the many challenges to be faced. Most universities in Indonesia cannot carry out high quality research and have below standard quality of teaching [3]. Universities in Indonesia must improve quality, ensure the reformations for key issues such as funding, policy, institutional and academic quality and access to have better positioning at regional and global levels.

World-class university has a different understanding both the target and assessment criteria. The ranking of universities is conducted by several institutions or organizations such as Shanghai Jia Tong University Rankings, Times Higher Education Supplement (THES), Webometrics Ranking of World Universities, International Rankings Experts Group & The Berlin Principles, Global University City Index, World Education News, Academic Ranking of World Universities and QS World University Ranking. Each of organization has different indicators, for example QS World University emphasizes criteria on research quality, teaching quality, graduate employability, and international outlook. It is very important to pay attention to these indicators in designing strategic planning to embody world-class university status.

1.1 Research Background

The main problem faced by universities during the preparation of a strategic plan is the concept of strategic planning which should adapt to the need for world-class university status. This situation is relatively new to the world of higher education institutions in Indonesia, especially Uniiversitas Negeri Manado (UNIMA). In fact, the current ranking of UNIMA is far from world-class university. According to Unirank's website UNIMA's world rankings is 8759 [4]. Data from Webometric UNIMA's national ranking is 232 and world rank is 13582 [5].

Based on the above information, UNIMA's ranking is still far below the standard to achieve world-class university status. Therefore, strategic planning which emphasis on achieving world-class status is an urgency to be implemented. Through the strategic plan UNIMA can prepare the reformation and restructure its organization. To achieve world-class university status the institution requires solid leadership, vision, mission and bold objectives and articulate strategic planning to clarify the vision in the form of concrete targets and programs. Colleges must be able to achieve maximum results, show improvement, set new goals, design and implement new strategic plans to achieve the desired results [6].

This research was conducted to improve the development of strategic planning in Universitas Negeri Manado, Indonesia to upgrade existing strategic planning in preparation for world-class university. The specific purpose of this study is to provide advice and references to strategic planning at UNIMA, by proposing alternative innovative strategies; recognise the various institutional resources that can support the strategic planning process at UNIMA; and recommended strategic planning by considering the criteria and indicators of world-class university that can be used as a preparation, to be able to compete globally which lead to the initiation of world-class university.

2. Research Methods

This research used exploratory qualitative research method and explored the strategic planning of university, obtained from the in-depth interview result and library study result about strategic planning in the higher education institution, collecting the documents that relevant and researchers' observations to gain a deeper understanding of the actual state of strategic planning at UNIMA.

2.1 Data Collection Technique

Data collection techniques are through interviews, documents and research notes from observations (field notes) [7], [8].

1. Interview

   The Interview was conducted to obtain information on the strategic planning process at UNIMA, various resources that can support strategic planning in UNIMA, as well as suggestions to be taken into consideration of the preparation of strategic planning for UNIMA towards world-class university.

2. Documentation
   The relevant UNIMA documentation, such as planning documents, resources, facility development, and other documents, collected after approval.
3. Observation
   Researchers did observations by noting the important and relevant data to the purpose of research.

2.2 Respondents and Selection Criteria
Interviews were conducted for the Rector and Vice Rectors, Deans and Vice Deans, academic and administrative staff who have experience and involved in the preparation of strategic planning. Respondents were selected randomly. The number of respondents targeted in this study were 40 respondents with the duration of the interview for 45 minutes - 1 hour.

Interview conducted with guidance of research instrument containing questions. After the interview, respondents had the opportunity to read the transcript of the interview and respond if necessary before the researcher concludes the interview.

2.3 Data Analysis Technique
In this study, interview results were analysed using thematic content analysis. Documentation and observation results (field notes) analysed and selected based on relevance to the interview results. Interview transcripts are coded and partially processed using NVivo software and some were done manually. The researcher recorded each interview process, issues related to interview answers and other additional information. The transcript was read and the narrative meanings were identified with the codes. The codes were reviewed again and developed in the form of "themes". The codes are then developed with NVivo. Some notes were written manually, and transcript interviews were made in hardcopy. This study has the possibility of inaccurate data collection and interpretation. In addition, researchers have limitations in the presentation of research results in a comprehensive and precise. Recognising these possibilities, researchers applied triangulation protocols, which include triangulation of data sources, research and theories, and methodologies, so the validity and reliability of the research can be maintained.

3. Results and Discussion

3.1 Strategic Planning in UNIMA
The vision in the strategic plan of UNIMA is "Becoming an excellent, innovative, competitive institution in performing education, research and community service programs. The missions are:
- Improve the quality of institutions in UNIMA
- Improve the quality of education, research and community service
- Create the availability of affordable education for community
- Improve the quality of resources
- Prepare the financial management to be a corporate university alike
- Produce the qualified and professional alumni [9]

The findings on the interview show that there are several items that should be considered in designing UNIMA strategic planning. The respondents' comments in the theme of suggestions for UNIMA strategic planning are presented below:
- The strategic planning process required the bottom-up approach; the initial process is started from the lower management to the top management.
- Should review the planning term, persons, model of planning and sufficient finance.
- Human resources should be the great asset in strategic planning.
- The planning should be followed by accountability and well implemented
- Strategic Planning should involve stakeholders and expert and the feedback of the stakeholders should be accommodated.
- Involved every department concerned in strategic planning process.

- Strategic Planning should also concern with the facilities and Information and Technology.
- Team is required for monitoring and evaluating
- Establish a high-quality planning but easy to be implicated.
- Well implemented of strategic planning is required
- Strategic Planning should fit with the condition in UNIMA.
- Involve the local government in strategic planning.
- Develop the established vision and mission.
- Planning should emphasised the academic activities

The respondents agreed that in the process of generating vision and mission of the university, recommendations were collected from all faculty and school and there was assessment of external and internal environment. There is an annual evaluation to the implementation of short and long-term planning. The long-term planning consists of UNIMA's future direction which is achieved by fulfilling short-term goals and objectives. As one of the interviewees expressed as follows;

> "The planning process in university has developed significantly, it was started from the work meeting in faculty which is never done in previous years. The work meeting held beginning from the program in faculty and study program then the result of this was presented into the level of university work meeting. The outcome of the meeting then implemented to the whole faculties in university as strategic plan. At the end of term each of faculty report the accountability and gathered together to construct the next planning." (Interviewee no.3)

All interviewees emphasised the need to develop new model of strategic planning. Two interviewees spoke at length and in depth about a corporate university, primarily because UNIMA should adopt "corporate university" paradigm and to be able to accommodate opinions revealed from annual evaluation. In addition, all interviewees asserted the need to involve all stakeholders in planning processes. In terms of important items to be included in an appropriate planning model for UNIMA, 3 interviewees suggested that the model should be able to figure out systematically operational activities to achieve goals and objectives because there are financial constraints for UNIMA. The interviewee number 2 stated:

> "Unima should develop a new model of strategic planning due to this university will shift to the corporate university, and in this new status it is important for UNIMA to approach the stakeholders and involve them in the process of strategic planning. In the corporate university, the fund from government will be deducted and UNIMA should make the effort to fund the university. In the next future, the process of strategic planning will be supported by stakeholders that consist of community, industry, association and non-government office". (Interviewee no.2)

Regarding the resources in designing strategic plan UNIMA has some academic staff who have returned to the university after finishing their study in overseas university which classified as world-class university. They can provide their expertise to the planning processes in UNIMA. External experts can be hired as consultants and not involved directly in most parts of the processes. Similarly, three other interviewees emphasized the need of support from stakeholders and the importance of gaining advice from management experts, said that adequate finance is a pre-requisite to develop appropriate planning processes. One of the interviewees also stated that one of the most important resources is experienced administrative staff and previous strategic planning documentations. As one of the interviewees articulated below:

> "UNIMA has sufficient human resources which is 80-85% of academic staff possess post graduate and doctorate degree that graduate from internal university in Indonesia or from abroad. Their expertise can bring great contribution for strategic planning process. Another resource is the fund granted from government to facilitate learning process such as books, computers and laboratory instruments. In fact, government funding and university revenue also important for budgeting in constructing strategic planning". (Interviewee no 5)

All interviewees asserted importance of stakeholders' participation in the development of UNIMA generally. When asked about importance of good communication, one of the deans stated that close and mutual relationships that are generated from good communication among all stakeholders are essential in gaining their participation.

The results show that most interviewees stated that UNIMA has been engaged in strategic planning at some level. Contribution from faculties and schools are acknowledged and inform the planning process. However, a new model of strategic planning should be adopted by UNIMA due to shift towards a "corporate university". It was recognised that the involvement of stakeholders is crucial in planning processes. UNIMA may have the human resources from staff who have returned from overseas study, which will be able to contribute their expertise in planning process and the university may hire the external experts as consultants in strategic planning process.

3.2 Strategic Planning towards World-class University

The finding results from documents and literature review regarding the current strategic planning of UNIMA and how to upgrade by considering the characteristics and features towards world-class university can be seen in the table below:

Table 1. UNIMA Strategic Planning and Features of World-class University

No	Strategic Planning of UNIMA	Program Indicators	The essential features and characteristics towards world-class University
1	Improve the quality of institutions in UNIMA	-Refinement of regulations -Improvement of service	-Commitment to be an excellence university -Participation in the latest and most advanced researches and regarding education, powerful and meaningful communication between education and research. -Researchers' freedom and academic freedom. In this case, they have freedom in: 1) creativity and innovation, 2) freedom to get rid of bureaucracy and unnecessary limitation 3) freedom from arbitrary decisions of financing 4) Freedom to choose partners 5) Have wide access to cooperate with the industry through open interdisciplinary of education and research [10], [11]. -Promotion of Internationalisation strategy: 1) Bilingual teaching and learning 2) Summer schools, exchange students and internship 3) Dual degree programs and joint institutes [12].
2	Improve the quality of education, research and community service	-Enhancement of research activities -Strengthening the Research and Community Service Center -Strengthening the Instructional Activities Fostering and Development Agency -Empowerment of Quality Assurance Agency and Internal Audit, and Internal Control Unit	
3	Improve the quality of resources	-Increase the number of professors -Doctorate degree for academic staff -Certified lecturers -Competence Training for academic and administrative staff -Perform linearity between educational background and competence -Performance improvement of technician, administrative, and laboratory staff -Reward and punishment system -Discipline for academic and administrative staff	

IOP Conf. Series: Materials Science and Engineering **306** (2018) 012035 doi:10.1088/1757-899X/306/1/012035

4. Conclusion

Constructing strategic planning towards a world-class university is a challenging, brave, and time consuming effort. In reaching the world-class status, the university such UNIMA may turn the current conditions into opportunities, for example; to transform the research center at international standard and the outdated management style to strategic management. The focus can be shifted from local standard to international standard and from national competition to global competition.

The road to world-class university will have an exorbitant budget and seems impossible to reach. The ambition to reach world-class university status in UNIMA is still very long way to go. However, through strategic planning, which designing with thoughtful and realistic plan, the university can initiate the program by inserting the criteria and features of world-class university. It may not reach the status immediately but as first attempt, it can start to elevate the national ranking and may lead to increase the international ranking. This starting point can lead the university to slowly fulfil the ambition in achieving world-class university status.

References

[1] Altbach P G 2004 The costs and benefits of world-class universities *Academe* (January-February) Available from: https://ejournals.bc.edu/ojs/index.php/ihe/article/viewFile/7381/6578.

[2] Direktorat Jenderal Pergururuan Tinggi (DIKTI) 2004 *Strategi pendidikan tinggi jangka panjang 2003-2010 mewujudkan perguruan tinggi berkualitas* Departemen Pendidikan Nasional Republik Indonesia Available from: http://www.inherent-dikti.net/files/HELTS2003-2010B.pdf.

[3] Hill H and Wie T K 2013 Indonesian universities: Rapid growth, major challenges In: Suryadarma D and Jones G W (eds) *Education in Indonesia* Institute of Southeast Asian Studies, Singapore

[4] UniRank 2017 Available from: http://www.4icu.org/reviews/universities-urls/10169.shtm

[5] Webometric 2017 Webometric Asia. Available from: http://www.webometrics.info/en/Asia/indonesia%20

[6] Salmi J and Liu N C 2011 Paths to a world-class university In Liu N C Wang Q and Cheng Y(eds) *Paths to a world-class university* Rotterdam, The Netherlands, Sense Publishers, p ix

[7] Zhang Y 2008a The Influence of Mental Models of undergraduate students' searching behaviour on the web *Information Processing and Management* **44** (3), 1330-1345

[8] Zhang Y 2008b Undergraduate Students' Mental Models of the web as an information retrieval system *Journal of the American Society and Technology,* **59** (13), 2087-2098

[9] Rencana Strategis UNIMA 2016-2020 2016 UNIMA: Tondano – Indonesia

[10] Li Y Whalley J Zhang S and Zhao X 2008 Higher educational transformation in China and its global implications *The Centre for International Governance Innovation Policy Brief*, **3**

[11] Kerr C 2001 *The uses of the university* Cambridge,MA: Harvard University Press

[12] Wang Q H Wang Q and Liu N C 2011 Building world-class universities in China: Shanghai Jiao Tong University In Altbach P G and Salmi J (eds) *The road of academic excellent,* pp 33-62 Washington, DC: The World Bank

IOP Conf. Series: Materials Science and Engineering **306** (2018) 012036 doi:10.1088/1757-899X/306/1/012036

Design of Integrated Database on Mobile Information System: A Study of Yogyakarta Smart City App

E K Nurnawati* and E Ermawati
Jurusan Teknik Informatika, Institut Sains & Teknologi AKPRIND, Yogyakarta, 55222, Indonesia

*ernakumala@akprind.ac.id

Abstract. An integration database is a database which acts as the data store for multiple applications and thus integrates data across these applications (in contrast to an Application Database). An integration database needs a schema that takes all its client applications into account. The benefit of the schema that sharing data among applications does not require an extra layer of integration services on the applications. Any changes to data made in a single application are made available to all applications at the time of database commit - thus keeping the applications' data use better synchronized. This study aims to design and build an integrated database that can be used by various applications in a mobile device based system platforms with the based on smart city system. The built-in database can be used by various applications, whether used together or separately. The design and development of the database are emphasized on the flexibility, security, and completeness of attributes that can be used together by various applications to be built. The method used in this study is to choice of the appropriate database logical structure (patterns of data) and to build the relational-database models (Design Databases). Test the resulting design with some prototype apps and analyze system performance with test data. The integrated database can be utilized both of the admin and the user in an integral and comprehensive platform. This system can help admin, manager, and operator in managing the application easily and efficiently. This Android-based app is built based on a dynamic client-server where data is extracted from an external database MySQL. So if there is a change of data in the database, then the data on Android applications will also change. This Android app assists users in searching of Yogyakarta (as smart city) related information, especially in culture, government, hotels, and transportation.

1. Introduction

The development of increasingly complex applications requires an increasingly complex database design as well. The database is one of the decisive factors in the power and success of the application. The factor of normalization, nonduplication, and optimization of memory usage become the main factor in database design [1]. If the designed database does not meet the above requirements, it will be fatal; process errors can occur, programming logic errors, output errors generated, and excessive memory usage resulting in economic wastage, especially if the application requires large amounts of data. The database technology develops more rapidly. If the first data is limited to text and images, then the current data already includes more kinds, such as imagery, genomics, heart rate and so forth. An integrated database is a database that acts as a data store for multiple applications, resulting in data integrity between applications [2]. An integrated database requires a scheme that can unify all user needs into a database design. The result of this scheme is a more general (general) or complex database design or both because it must incorporate bounded contexts if each application is built separately. The advantage of sharing data between applications is that it does not require an extra layer of application integration services. Any changes to data performed on an application will automatically change all the data in other applications so that synchronization is easy to do. The integrated database contains information collected from more than one data source. In the process of combining different data sources, individual data elements often need to be altered or synced to represent a consistent or standard context. This transformation must be done without loss of data integrity and interpretation. Combining data from multiple sources increases statistical power to identify important relationships between data elements. An integrated database is used by research institutes, pharmaceutical companies, and regulatory agencies during clinical development and post-marketing surveillance. Integrated databases have been used for years, but in recent years, the use of integrated databases has evolved beyond classical applications to clinical trial data to incorporate applications into imagery, genomics, and other digital data. More advances in data storage, agency software applications, and information standardization efforts have affected the construction and use of integrated databases [3-4].

Information Systems are the key to success in Enterprise systems. Integration can replace functionally oriented, impacting the cost of infrastructure support [5]. Currently, the integration of an information system is a necessity, more important than ever, because in a large institution there are many separate applications, which require a lot of costs, a long time in syncing information between applications [5]. According to Pichler [6], data integration is a combination of multiple heterogeneous data sources that display data through virtual similarities and global schemes. Mapping the (local) data source to the target (global) scheme, so the global scheme as a view and locale is also a view with a strong connection to the data query. Sutanta [7] creates an integrated database model for population systems by utilizing Indonesian Resident's ID Number (NIK). The design is made so that the primary key values that have been previously defined and used in each database can still be used. Gal [8] performs a systematic analysis of process properties and stabilizes the criteria for evaluating the quality of matching algorithms by using mapping attributes between heterogeneous schemas. Nurnawati [9] developed location-based mobile applications to present healthcare location information in Yogyakarta. The information provided is among others the locations of hospitals, clinics, pharmacies, and specific regional locations presented on the digital map feature. The application is also developed based on Android. The app is intended for people and migrants using Android mobile device. This app presents the closest Health Service detection feature of the user's location using the Google Map i.e. location distance, location route along with photos and a short information. Nurnawati [10] succeeded in developing a database integration model for GIS application prototypes that can be applied to mobile device. Applications in this study have been tested on 4 (four) types of information, namely transportation, tourism, hotels, and government offices. This research succeeded in implementing database usage for several applications at once. The objective of this study is to design an integrated database used by the application package i.e. an application that integrates multiple GIS applications on Android-based mobile hardware where it can be accessed and used freely by users.

According to Yang and Hsu [11], Location Based Service (LBS) is a service that reacts actively to position entity changes to be able to detect the location of the object and provide services according to the location of the known object. The LBS works on "Mobile Positioning" technology. This service is an application that depends on a particular location and is also defined as an information service by utilizing technology to know the position of something. The Positioning System technology on LBS facilitates users to obtain location information according to their needs. The LBS has similar technology category as like as Geographic Information System (GIS), and Global Positioning System (GPS) application, known as geospatial technology. This technology consists of devices for collecting, storing, analyzing and distributing data by the needs of users of the earth coordinate system. This service becomes very important for users because it can connect with geographic location information to the location of its users. The LBS application is the result of a merger of three technologies: New Information and Communication Technologies (NICTS), the Internet, and GIS using spatial databases

IOP Conf. Series: Materials Science and Engineering **306** (2018) 012036 doi:10.1088/1757-899X/306/1/012036

[12]. This LBS technology consists of devices used to collect, store, analyze and distribute data and information based on the real-time geographic coordinate system. Identification of user coordinates allows LBS applications to provide services for mobile device users.

2. Experimental Setup

2.1. Domain of study

The scope of the study includes data on tourist objects, transportation, hotels and lodging, government offices, and education in the Daerah Istimewa Yogyakarta. The data is provided by the Office of Dinas Komunikasi dan Informatika (Kominfo) Daerah Istimewa Yogyakarta.

2.2. Methods of study

The studies use survey, observation, and analysis methods with details as follows:

- Preliminary studies. This stage is intended to conduct a preliminary study with the goal of finding gap information needs between users and existing application systems and determining alternative relevant solutions.
- Data collection. Data collection is done through literature review and field survey to obtain supporting data for the needs of database integration of applications that will use the database.
- Needs analysis. At this stage, an analysis is done to find the technical and non-technical requirements for the system. The analysis is done by examining the needs of all applications, looking for patterns of data requirements, making the design normal, not duplicate and meet the needs.
- Design Databases. In this stage consists of two parts, the first database design is realized by building the database with the use case tools, class diagrams, and build the relation. The second part to make the system design logically using diagram-shaped tools consisting of use case diagrams, class diagrams, activity diagrams, and sequence diagrams. In this stage also compiled design display, input, output, and integrated database that allows for use by many GIS applications on mobile devices.
- Build prototypes. The prototypes are used to test the performance of the database. At this stage, the system design is implemented into an integrated application package. The developed application package consists of 2 (two) versions, which is the version of the manager (web-based) and the user version (based on Android mobile).
- Testing. Testing is done to find possible errors in the system. Testing on the reliability of database design is done in 3 ways [13], namely component testing, connectivity testing and system testing. While the prototype testing is performed using an emulator device that allows displaying the results of each module developed in the coding stage. If the testing time found an error then immediately fixed so that in the end obtained the system by the design that has been set [14].
- Implementation. At this stage, the developed application system is implemented in a real Android mobile device. At this stage also tested the compatibility of the application system against several mobile devices and different browsers, so that in the end will be obtained system application as expected.
- Documentation and reporting of research results.

3. Results and Discussion

The main result of this study is the design of an integrated database which can be seen in the relational diagram as shown in Figure 1. In Figure 1, there are 13 tables in the database, consisting of "Provinsi", "Kabupaten", "Kecamatan", "kode_pos", level, user, object, Log_edit, temp_login, sub_category and details_objects.

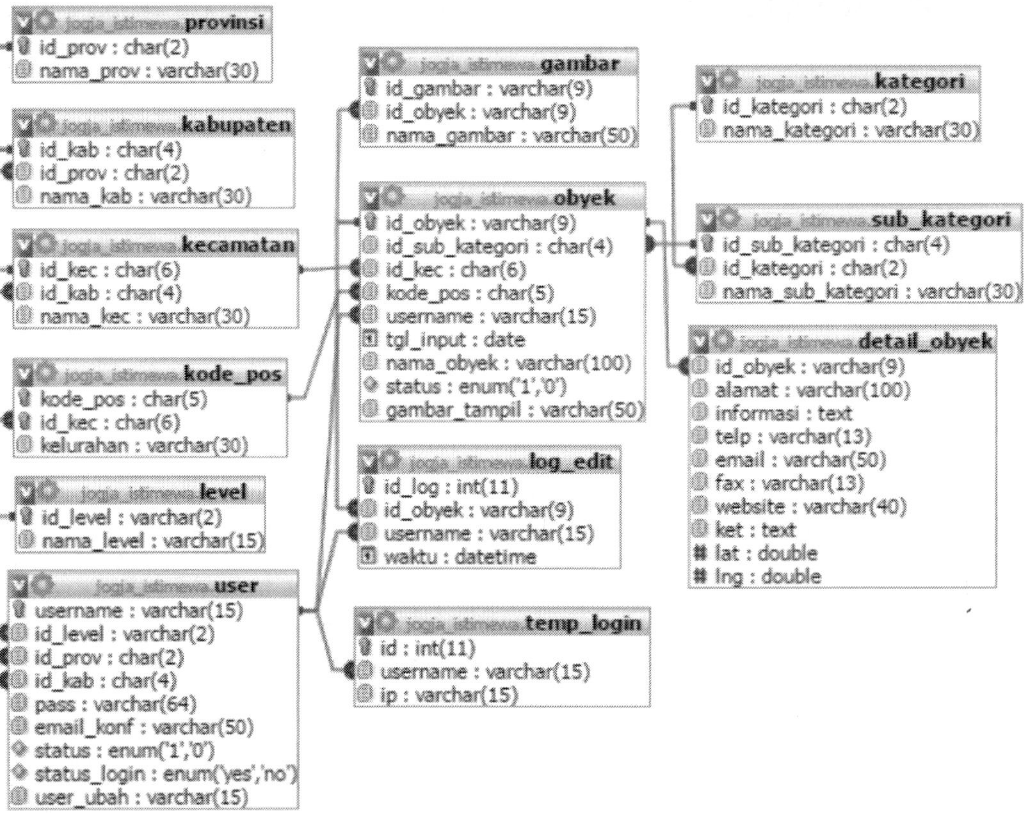

Figure 1. Database design with its clarity.

The design of this database has represented all the objects in this study, whether tourist attractions, schools, cultures, hotels, and inns, government offices with each distinguished by the object code. The object type can be expanded by adding new code. The key to integrity is in the provincial code (if the data increases in addition to the Daerah Istimewa Yogyakarta), as well as other area codes.

The class diagram shows the class relationship formed from the database along with its actions, as shown in Figure 2. The class diagram is a logical representation of the database design. In the class, the diagram has classified the classes to be formed along with methods to be done by the class. Each class has its task, whether it be an interface class, a connecting class or a master class [4].

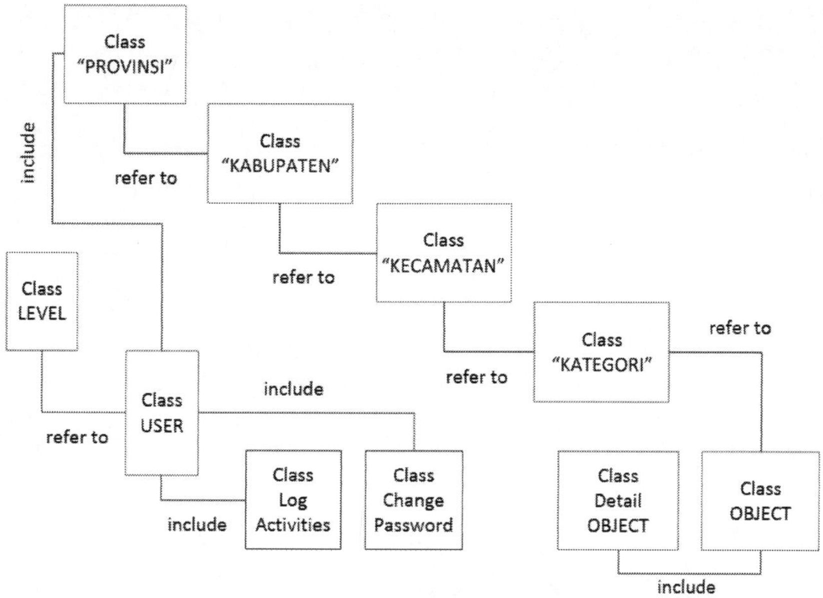

Figure 2. Simplified class diagram.

As a test of database design, the application prototypes were built as support in this research. The system is built with two different systems, the first web-based system used by the admin, the operator and the manager and the second Android-based mobile application used by the user. This system has several actors namely admin, manager, operator, and user. Each actor has different access rights. Admin has access rights to manage level data, user data, change the password, change email and view and print activity log data. Managers have access rights to manage provincial, district, sub-district, postal code, category, sub-categories, change passwords and change emails. The operator has access rights to manage object data, image data, change password and change email. Operators in this system are divided into operators per district. Users have access rights to view and search for information about the intended object. After admin enters the level and user data, then this system can be used. Managers enter data of provinces, districts, sub-districts, zip codes, categories, and subcategories. After that, the operator entered the data object and image. The data will be stored in the integrated database. This Android-based app is built based on a dynamic client-server where data is extracted from an external database MySQL. So if there is a change of data in the database, then the data on Android applications will also change. The application results are described in Figure 3 to11.

The login page is the page used by the user, i.e., Admin, Manager or Operator to enter the system. Users must enter the username and password correctly to enter the system. The display of the login page is shown in Figure 3.

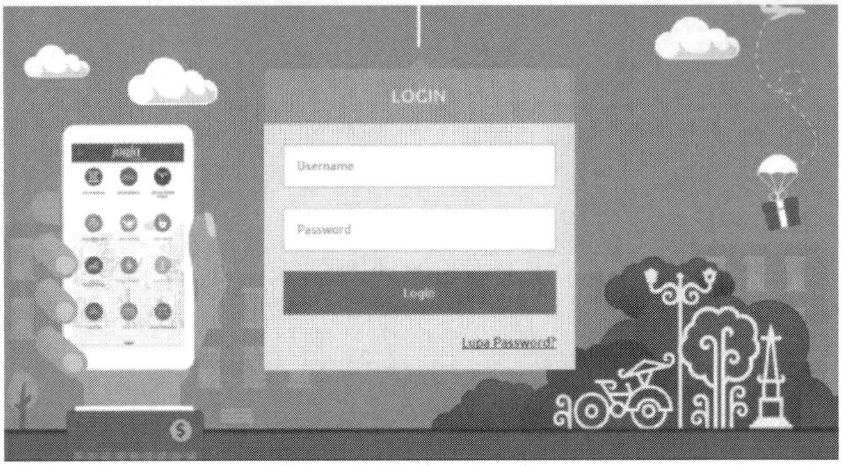

Figure 3. Login page view.

Figure 4. Admin page view.

The main page view of the admin is the view after the admin successfully login. There are five menus in this view, i.e., user menu, level, activity log, change password and change email. The main admin page view is shown in Figure 4. The event log page is the page that displays the activity log when the operator performs the edit activity on an existing object. The activity log page is shown in Figure 5.

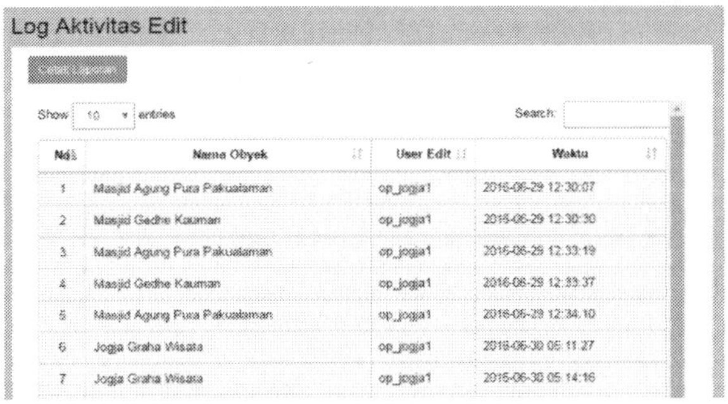

Figure 5. Log page view.

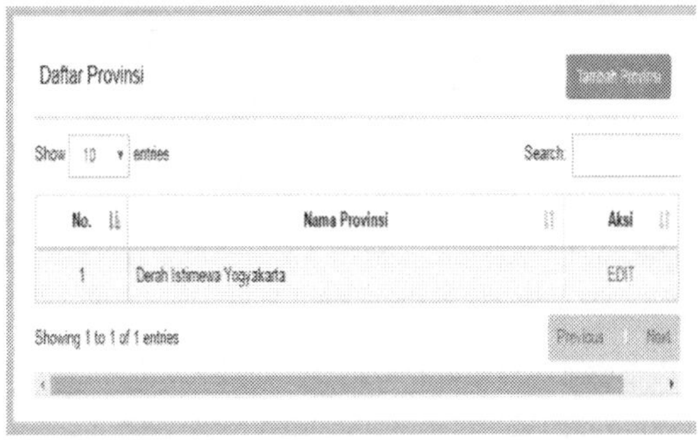

Figure 6. Manage provincial view.

IOP Conf. Series: Materials Science and Engineering **306** (2018) 012036 doi:10.1088/1757-899X/306/1/012036

The provincial governance page is used to manage provincial data. Managers can view, add and edit provincial data. The provincial list of page views is shown in Figure 6. The Objects management page is used to manage object data. The operator can view a list of objects, input data of object, view detail of objects and edit data of an object as shown in Figure 7. The prototype application for the user is a mobile-based app. The app's homepage for users is a caller for other mobile apps. Currently has been developed four applications namely Jogja Budaya, Jogja Wisata and Jogja Transportasi as shown in figure 8.

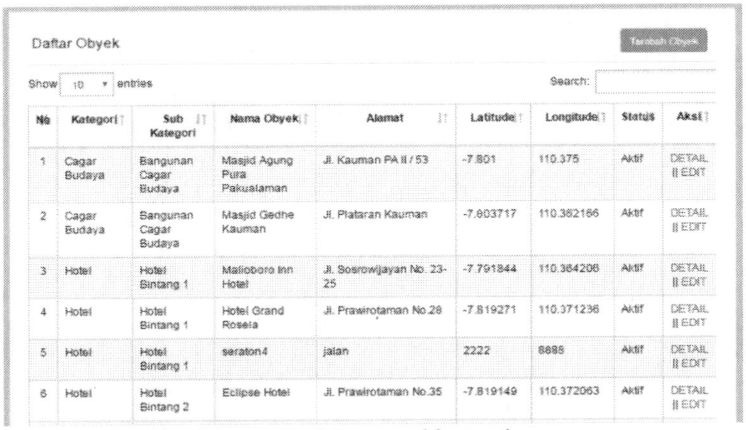
Daftar Obyek

Show 10 entries Search:

No	Kategori	Sub Kategori	Nama Obyek	Alamat	Latitude	Longitude	Status	Aksi
1	Cagar Budaya	Bangunan Cagar Budaya	Masjid Agung Pura Pakualaman	Jl. Kauman PA II / 53	-7.801	110.375	Aktif	DETAIL || EDIT
2	Cagar Budaya	Bangunan Cagar Budaya	Masjid Gedhe Kauman	Jl. Plataran Kauman	-7.803717	110.362166	Aktif	DETAIL || EDIT
3	Hotel	Hotel Bintang 1	Malioboro Inn Hotel	Jl. Sosrowijayan No. 23-25	-7.791844	110.364208	Aktif	DETAIL || EDIT
4	Hotel	Hotel Bintang 1	Hotel Grand Rosela	Jl. Prawirotaman No.28	-7.819271	110.371236	Aktif	DETAIL || EDIT
5	Hotel	Hotel Bintang 1	seraton4	jalan	2222	8888	Aktif	DETAIL || EDIT
6	Hotel	Hotel Bintang 2	Eclipse Hotel	Jl. Prawirotaman No.35	-7.819149	110.372063	Aktif	DETAIL || EDIT

Figure 7. Manage objects view.

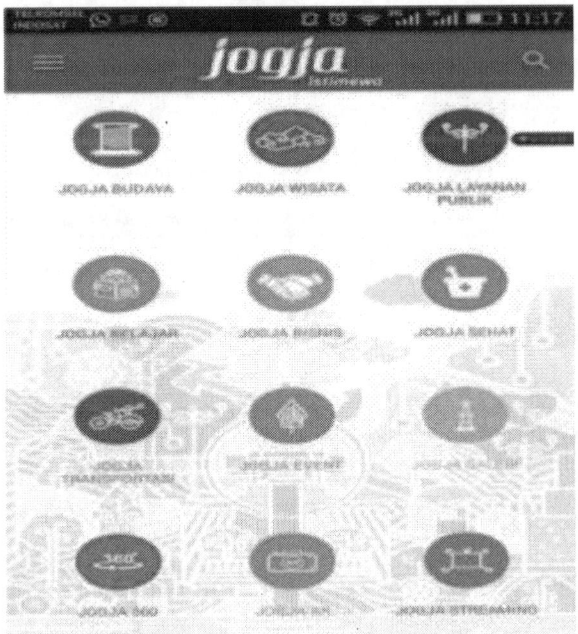

Figure 8. App controller view.

The main page of the application of Jogja Budaya has six main menus namely cultural heritage building menu, cultural heritage objects, cultural heritage strata, cultural heritage area, cultural preservation site and search menu. The search menu is used to search for the object the user wants as shown in Figure 9.

The Heritage List View displays a list of cultural presences with list view based on the subcategory of cultural preservation selected by the user. The data displayed is the object images, object name, object address and object distance from the user location to the location of the destination object. There is a button maps view at the bottom right of this page which is used to display a list of cultural presences through maps as shown in Figure 10. The list of cultural presence is displayed in the form of maps view according to the subcategories selected by the user. The data displayed when the object markers are selected the object name and sub-category name of the object as shown in Figure 11.

Figure 9. App jogja budaya menu.

Figure 10. Heritage list view.

IOP Conf. Series: Materials Science and Engineering **306** (2018) 012036 doi:10.1088/1757-899X/306/1/012036

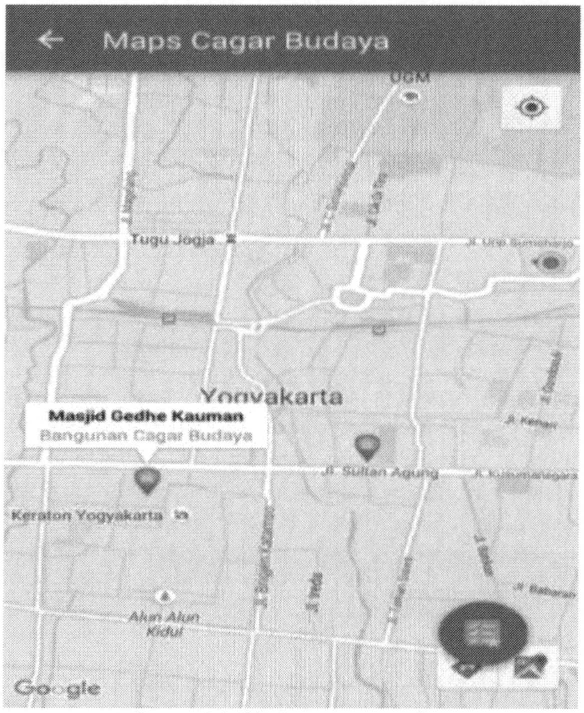

Figure 11. Heritage maps view.

3.1. Testing

Testing includes component testing, integration testing, and system testing [15]. Component testing is a test of system components. In this study, the interface component was tested as subject of the component testing. Component testing for the component interface is a test performed to determine the functionality of the interface that has been made whether it works as expected or not. Testing interface is done in the menu and button on the application. From the results of component interface testing, it can be concluded that all the menu components and buttons that exist in the application have led to the right page or form. The testing has been running as expected, for example, menu, added data has led to the form added data, edit data link has led to the data edit form, and the delete data button has succeeded to delete the data. Integration testing is a test group of components that are integrated to form sub-system or system. In this study, after each component was tested on component testing, tested the integration of the components. Integration testing related to the communication between applications with databases in the form of testing application functions associated with the database such as add, edit, delete and search. System testing is a test of the integration of sub-systems, namely connectivity between sub-systems. In this research, system testing is done to validate data before entering into the database. Each interface already has validation so that the data entered into the table in the database is valid data. The results of each test are shown in Table 1 below:

Table 1. Result of integration and system test.

Integration Test			System Test		
Atribut Test	Table	Result	Type	Atribut test	Result
Read,add,edit, seek	Level	(√) 0.25 sec	login	Username and password unmatch	Alert : 0.1 sec
Read,add,edit, seek	User	(√) 0.25 sec		User status inactive	Alert : 0.1 sec
Read,add,edit, seek	Provinsi	(√) 0.25 sec		User active in same IP address	Alert : 0.1 sec
Read,add,edit, seek	Kabupaten	(√) 0.3 sec		User and password match	Direct to menu 0.1 sec
Read,write,edit, seek,delete	Gambar	(√) 0.35 sec	Data	Add data fail	Alert 0.1 sec
Read,add,edit, seek	Kategori	(√) 0.2 sec		Add data success	Alert 0.1 sec
Read,add,edit, seek	Obyek	(√) 0.2 sec		Edit data fail	Alert 0.1 sec
Add, delete	Temp_login	(√) 0.1 sec		Edit data success	Alert 0.1 sec
Read,add,edit, seek	Log_edit	(√) 0.15 sec		Delete data success	alert 0.1 sec

From the results of Table 1, it can be concluded that integration testing and system testing has been running well with a fairly short time. The result was excellent refer to Mittal et al. [15] have done. The length of time depends on the internet access and test equipment used. In this test used Oppo F1s mobile device and Acer Travelmate 9700 laptop. The final product proved to be a very satisfactory platform for online smart city data management.

4. Conclusions

From the design, development, and testing of the database using prototype application, it can be concluded that the integrated database can be utilized both on the admin and the user in an integral and comprehensive platform. This system can help admin, manager, and operator in managing the application easily and efficiently. This Android-based app is built based on a dynamic client-server where data is extracted from an external database MySQL. So if there is a change of data in the database, then the data on Android applications will also change. This Android app assists users in searching Yogyakarta related tourism information, especially in culture, government, hotels, and transportation.

References

[1] Garcia I, Sebastia L and Onaindia E 2011 On the design of individual and group recommender systems for tourism *Expert Systems with Applications* **38** (6) pp 7683-7692
[2] Filis I V, Sabrakos M, Yialouris C P, Sideridis A B and Mahaman B 2003 GEDAS: an integrated geographical expert database system *Expert Systems with Applications* **24** (1) pp 25-34
[3] Chen Z 2015 Towards Integrated Study of Data Management and Data Mining *Procedia Computer Science* **55** pp 1331-1339
[4] Vavliakis K N, Symeonidis A L, Karagiannis G T and Mitkas P A 2011 An integrated framework for enhancing the semantic transformation, editing and querying of relational database *Expert Systems with Applications* **38** (4) pp 3844-3856
[5] Litan D, Copcea L, Mocanu A M, Surugiu J and Raduta O 2011 Information Systems Integration, a New Trend in Business *Recent Researches in Communication, Electrical and Engineering* ISBN: 978-960-474-286-B
[6] Pichler R and Savenkov V 2010 Towards practical feasibility of core computation data exchange *Theoretical Computer Science* **411** (7-9) pp 935-957
[7] Sutanta E 2011 Model Integrasi Database Penduduk Indonesia Dengan Berbagai Sistem Informasi Berbasis Komputer *Jurnal Informatika* **5** (2) ISSN 1978-0524
[8] Gal A, Anaby-Tabor A, Trombetta A and Montesi D 2003 A Model for Schema in Heterogeneous Databases *Proceedings of the Seventh International Databases Engineering and Application Symposium* (IDEAS'03) IEEE 1098-8068.

IOP Conf. Series: Materials Science and Engineering **306** (2018) 012036 doi:10.1088/1757-899X/306/1/012036

[9] Nurnawati E K 2014 Aplikasi Mobile Berbasis Lokasi Untuk Penyedia Lokasi Layanan Kesehatan di Yogyakarta *Prosiding Seminar Nasional Aplikasi Sains & Teknologi (SNAST) 2014* IST AKPRIND Yogyakarta.

[10] Nurnawati E K 2016 Pemanfaatan Basis Data Terintegrasi Pada Sistem Informasi Perangkat Bergerak *Proceeding of Seminar Nasional Aplikasi Sains & Teknologi (SNAST)* November 2016.

[11] Yang SY and Hsu CL 2016 A location-based service and Google maps-based information master system for tour guiding *Computer & Electrical Engineering* **54** pp 87-105

[12] Ait-Cheik-Bihi W, Bakhouya M, Nait-Sidi-Moh A, Gaber J and Wack M 2011 A Platform for Interactive Location-Based Services *Procedia Computer Science* **5** pp 697-704

[13] Lochau M, Lity S, Lachmann R, Schaefer I and Goltz U 2014 Delta-oriented model-based integration testing of large-scale systems *Journal of Systems and Software* **91** pp 63-84

[14] Assuncao W K G, Colanzi T E, Vergilio S R and Pozo A 2014 A multi-objective optimization approach for the integration and test order problem *Information Sciences* **267** pp 119-139

[15] Mittal M, Sangani R and Srivastava K 2015 Testing Data Integrity in Distributed Systems *Procedia Computer Science* **45** pp 446-52

Acknowledgments

The authors would like to thank profusely to Rector of Institut Sains & Teknologi AKPRIND Yogyakarta, Indonesia that has agreed and funded our research.

IOP Conf. Series: Materials Science and Engineering **306** (2018) 012037 doi:10.1088/1757-899X/306/1/012037

The Relevance of Vocational High School Curriculum with the Requirement of the Heavy Equipment Industries

E P Asfiyanur*, K Sumardi, Y Rahayu and R C Putra

Program Pascasarjana Universitas Pendidikan Indonesia, Jl. Dr. Setiabudhi No. 227, Bandung, 40153, Indonesia

*ekaasfianur@student.upi.edu

Abstract. This study aims to determine the relevance of competencies developed by vocational high schools with the needs of workers in the industrial world. This is to answer the statement from the world of industry that the competencies possessed by vocational secondary education graduates are not in accordance with industry standards. This research is a qualitative research conducted by collecting data, presenting accurate and objective information. Respondents of this research are industrial institutions in the field of heavy equipment, vocational education institutions and government agencies in charge of manpower. Selection of informants / sources of information used is purposive sample technique, which aims to give consideration to select informants who meet the criteria in providing accurate information. The results of this study provide insight and input to vocational secondary education managers about various information in developing vocational secondary education curriculum, major issues include curriculum content relevance, technological advances in the heavy equipment industry, updating of school facilities and collaboration between educational institutions and labor institutions and the industrial world in the development of vocational secondary education.

1. Introduction

One of the most frequently discussed issues is the impact of globalization on the world economy. The globalization process makes the world economy more competitive and challenging. Globalization affects all aspects of people's lives [1] [2]. There are two processes in globalization: compression and awareness, these two processes will create a world without boundaries [3]. In a global context, each country should have a strategy to conduct long-term educational planning [4]. Globalization gives rise to the social and economic order that demands new policies and strategies for the educational process [5]. This change will create new demands on the labor market. The education sector is a key factor in generating skilled labor for the service industry and the economy [3]. Thus the role of vocational education is expected in improving the quality and quantity of human resources [6].

One of the challenges facing Indonesia in the ASEAN Economic Community (MEA) is where the free labor market will occur. This means that the workforce in Indonesia will compete with workers from other ASEAN countries. The main concept of the MEA is to create ASEAN as a single market and unity of production base where free flow of goods, services, factors of production, investment and capital and elimination of tariffs for trade between ASEAN countries are then expected to reduce poverty and economic disparities among countries, member countries through a number of mutually beneficial cooperation [7].

The main purpose of this study is to identify the extent to which the curriculum of vocational education is able to provide the competence of work required by learners to be able to compete in the world of work whether the vocational education curriculum is aligned with the Indonesian National Work Competency Standards issued by the Ministry of Manpower of the Republic of Indonesia and standard of labor requirements issued by the industrial world. The method for obtaining data is by analyzing the structure and content of the curriculum published by the Ministry of Education and Culture of the Republic of Indonesia, then studying the Indonesian National Work Competency Standards and the required labor competency standards according to the industrial world. From the results of the analysis is expected to describe the ideal curriculum to be applied to vocational education in the field of mechanical heavy equipment so as to meet the needs of skilled workers in accordance with its qualifications.

2. Research Methods

This research is a qualitative research conducted by collecting data, presenting accurate and objective information [8]. The form of qualitative research data is the words and actions, the rest are additions such as documents and others. The form of data or respondents of this research is industrial institutions in the field of heavy equipment, vocational education institutions and government agencies in charge of employment. Selection of informants / sources of information used by researchers is a purposive sample technique [9], which aims to provide consideration to select informants who meet the criteria in providing accurate information. Selection of informants is a very important thing that must be done carefully, then the researchers decide the first informant or key informant. Initial data of a number of respondents based on sources of informants who have been selected through interviews continue to roll until they get saturated (Snowball Sampling) [10].

3. Results and Discussion

The purpose of this research is to know how far the relevance of vocational education curriculum in Indonesia in the context of the needs of the workforce in the heavy equipment industry. The results of this study are expected to answer the complaints from the industry about the competence of vocational education graduates who are unable to compete in the world of work, and can be an input for vocational education institutions to make improvements in educational planning [11].

Over the past decade, vocational education in Indonesia has been subjected to criticism due to the lack of skills and knowledge of vocational education graduates that resulted in them not being able to obtain employment opportunities. Central Bureau of Statistics recorded in February 2017, the number of open unemployment reached 7, 02 million people (5.33%). Of these, 9.27% are graduates of vocational secondary schools [12].

In this new reality, vocational high schools face the challenge of continuing to evaluate programs, content, implementation, in updating the curriculum[13]. Vocational secondary schools need to integrate theoretical material or vocational practice with existing competencies in the industrial world [14]. Vocational education should be able to teach competencies that meet the future needs of graduates by looking at workplace realities and emerging technologies.

Vocational education institutions should be able to prepare learners who are able to compete in the world of work, therefore the relationship and relevance of vocational education institutions with the industrial world becomes a very important [15]. Improving the quality of education is done by improving the quality of relevance to the world of work. Relevance in the world of education is the suitability between the process and the material given in education to the needs of the market [16]. The classical problem for vocational education in Indonesia is that the link and match between the output of vocational education and industry as a graduate user of vocational education has not been achieved, one of the problems is on the quality of vocational education graduates who are not in accordance with the competence standards established by the world of work [17].

The program of heavy equipment engineering expertise as one of the existing fields in vocational education is expected to answer the needs of skilled workers as heavy equipment mechanics. The geographical condition of Indonesia as a country with many natural resources has an effect on the number of industries in the field of mining, be it petroleum, gas and other minerals [18]. Infrastructure preparation can not be separated from the construction services, especially to open the land and build roads that connect economic channels in various regions. The realization of the construction services will encourage the utilization of heavy equipment which further brings the consequences of preparing human resources in accordance with their expertise [19]. Consequently, equipment resources are also needed so that this condition increases the number of heavy equipment populations with various brands and types, in the mining, construction, agro-industry and forestry sectors. Data from Indoanalysis shows that the market share of heavy equipment based on industry sector 2000-2016 shows: forestry sector 15%, construction 50%, Agroindustry 15%, and Mining 20% (IndoAnalisis.com, 2016). With so many industries will be directly proportional to the needs of workers in the field of heavy equipment mechanics as one part of the work in exploring the natural resources. In accordance with the purpose of vocational education in Indonesia, which equip students with certain skills to enter the workforce, the development of vocational education should always refer to the needs of the labor market [20]. Similarly, in the field of heavy equipment mechanics, the success of vocational education in the field of heavy equipment mechanics can be reflected from the number of graduates absorbed by the industrial world engaged in mining and heavy equipment. To achieve this, the suitability of graduates' competency from the heavy equipment mechanics program to the needs of the world of work is of paramount importance.

The curriculum is an integral part and plays a central role in the process of implementation of education, the curriculum directs all educational activities for the achievement of educational goals [21]. In accordance with the purpose of vocational education to produce graduates who have the skills according to the needs of the world of work, the necessary curriculum relevant to the needs of the world of work. Therefore, it is necessary to evaluate and improve the sustainable curriculum as a necessity for the national education system to be relevant and competitive [22]. In order to produce competitive graduates who are able to compete in the world of work, vocational education institutions must have the structure and content of the curriculum in harmony and in line with the Indonesian National Work Competency Standards published by the Ministry of Manpower of the Republic of Indonesia as well as the standard of mechanical work competence of heavy equipment based on industry needs.

3.1. Structure and content of vocational high school curriculum
The preparation of vocational high school curriculum in the field of heavy equipment mechanics is done by referring to the curriculum structure issued by the Directorate General of Secondary Education Ministry of Education and Culture of the Republic of Indonesia. Based on the analysis of the structure of the curriculum, the vocational education in the field of heavy equipment mechanics aims to equip learners with the skills, knowledge and attitudes to be competent in 1) Carrying out basic machinery engine work. 2) Carrying out basic electrical work of the machine. 3) Perform basic heavy equipment power train work. 4) Perform basic work under heavy equipment cariage. 5) perform basic heavy equipment hydraulic work. 6) Implement preventive maintenance work [20].

3.2. Indonesian national competency standards

Law No. 18/1999, on: Construction Services and its implementing regulations mandate that workers undertaking construction planning, implementation and supervision must have a certificate of expertise or skill. The requirement to have a certificate of skill and or skill reflects the demands of truly reliable quality of workforce. Such conditions require a concrete step in preparing the device (standard standards) required to measure the quality of construction service labor [23].

Indonesian National Work Competency Standards (SKKNI) is a formulation of work skills that includes aspects of knowledge, skills and / or expertise and work attitude relevant to the implementation of duties and terms of office stipulated by the provisions of applicable legislation. With the compilation and enforcement of the Indonesian National Work Competency Standards in Mechanical Sector of Construction Services for Heavy Equipment Mechanics, all stakeholders can use them to develop the quality of mechanical power of heavy equipment [24]. Figure 1 illustrates the level of competence that a machine mechanic must have at the junior and senior level. General competence and special competence must be mastered to become a machine mechanic.

Job Competence : Engineering of Heavy Equipment Hydraulic			
No	Unit Code	Competency Unit: Heavy Equipment Hydraulic Junior Engineering	Competency Unit: Heavy Equipment Hydraulic Senior Engineering
General Competency			
1	MEK.PW1 2.231.00	Implement safety and health as well as the environment	Implement safety and health as well as the environment
2	MEK.PW1 2.232.00	Implementing communication and teamwork at workplace	Implementing communication and teamwork at workplace
Specific Competency			
	MEK.PW2 2.231.00	Identifying components at the heavy equipment hydraulic system	Identifying components at the heavy equipment hydraulic system
	MEK.PW2 2.232.00	Implementing maintenance at the heavy equipment hydraulic system	Implementing maintenance at the heavy equipment hydraulic system
	MEK.PW2 2.233.00		Repairing components at the heavy equipment hydraulic system
	MEK.PW2 2.234.00		Overcome troubleshooting at the heavy equipment hydraulic system
	MEK.PW2 2.235.00		Created job report

Job Competence : Heavy Equipment Engine Engineering		
No	Unit Code	Competency Unit
General Competency		
1	INA.5220.2 22.x x.01.08	Implement safety and health as well as the environment
2	INA.5220.2 22.x x.02.08	Implementing communication at workplace
	INA.5220.2 22.x x.03.08	Implementing teamwork at workplace
Specific Competency		
	INA.5220.2 22.x x.04.08	Identifying main components of the engine
	INA.5220.2 22.x x.05.08	Implementing the maintenance of engine
	INA.5220.2 22.x x.06.08	Implementing the minor repair of engine
	INA.5220.2 22.x x.07.08	Implementing the major repair of engine
	INA.5220.2 22.x x.08.08	Analyzing and overcome troubleshooting at the engine
	INA.5220.2 22.x x.09.08	Created job report

Figure 1. Units of competence in heavy equipment engineer qualifications or occupations according to SKKNI (Kepmennakertrans RI No. KEP.88/MEN/V/2010).

3.3. Heavy machinery competency standards based on industrial requirements

According to the industry engaged in heavy equipment, teaching materials delivered in training activities for mechanics adapted to the level of the mechanics. For the ideal mechanic according to minimum work requirements is a mechanical level III [19]. Overall, teaching materials that must be mastered by machine mechanics based on mechanical level are the following:

- Mechanical Level: Helper, material to be mastered: Basic Mechanic
- Mechanical Level: I, materials to be mastered: Minimum: a) Basic mechanic, b) Basic engine. Additional: a) Basic hydraulic system, b) Basic powertrain system, c) Basic vehicle control, d) Basic electrical system.

- Mechanical Level: II, the material must be mastered: a) Basic mechanic, b) Basic engine, c) Basic hydraulic system, d) Basic powertrain system, e) Basic vehicle control, f) Basic electrical system.
- Mechanical level : III, the material must be mastered: a) Basic Mechanic, b) Basic engine, c) Basic hydraulic system, d) Basic powertrain system, e) Basic vehicle control, f) Basic electrical system, g) Intermediate engine system, h) Intermediate hydraulic system, i) Intermediate powertrain system.
- Mechanical Level: Specialist, material to be mastered: All Level III materials, plus, a) Product training (at least one type of machine), b) Electronic machine control (engine ecm, transmission ecm, etc.) [19].

Based on the data obtained, the teaching materials that must be submitted to the candidate ideal mechanic of heavy equipment include: safety, engine & machine design, jacking, blocking, lifting; seals, bearings, gaskets; measuring tools, fastener, workshop tools, literature, contamination control, fundamental engine, fundamental powertrain, fundamental hydraulic, fundamental electric, and battery maintenance. However, for senior mechanical personnel, in addition to some materials are also submitted materials: intermediate engine, powertrain, hydraulic, ET Tool, electronic engine, troubleshooting, and AFA 1 (Applied Failure Analysis). When viewed from material required by some industries, most of the material presented has similarities [19].

4. Conclusion

The ultimate goal of this research is to provide information, improve and develop vocational education system to be more relevant in the context of preparing graduates to be competent workforce in accordance with the needs of the industrial world. The development of vocational education programs requires clear justification or justification. Justification for vocational education programs is the real need for manpower in employment or in the business and industry. From the results of the discussion, found that the curriculum of vocational education in the field of mechanical equipment is still not relevant to the needs of the industrial world, there are still many competencies that have not been given to the learning process. So that the work competence of vocational education graduates have not been able to meet the standards of work competence set by the industry. Collaboration between the world of education, government institutions and industrial institutions is very important and indispensable in the process of curriculum preparation of vocational education, so that the resulting curriculum design is appropriate and proportional to the needs of the world of work. The vocational school curriculum should be more flexible so that it can keep up with the world of work.

Reference

[1] N Singh and R Papa 2010 The Impact Of Globalization in Higher Education, *Int. J. Educ Leadersh Prep*, vol **5**, no 2
[2] M Meyer, M Bushney, and W I Ukpere 2011 The impact of globalisation on higher education: achieving a balance between local and global needs and realities, *African J. Bus Manag*, vol **5**, no 15, pp 6569–6578
[3] K Rostam 2009 Economic Globalisation and Change: Implications on Geographical Education in Malaysia, *Eur. J. Soc. Sci.*, vol **9**, no 4, pp 533–543
[4] Y Kamin, A Ahmad, and D Cartledge 2013 Students' Perceptions On The Relevance Of a Diploma In an Automotive Curriculum to The Workplace, *Procedia - Soc. Behav. Sci.*, vol **93**, pp 90–96
[5] G S Mouzakitis 2010 The role of vocational education and training curricula in economic development, vol **2**, no 2, pp 3914–3920
[6] S McGrath 2012 Vocational education and training for development: A policy in need of a theory?, *Int. J. Educ. Dev.*, vol **32**, no 5, pp 623–631
[7] M G Plummer 2006 ASEAN-EU economic relationship: Integration and lessons for the ASEAN economic community, *J. Asian Econ.*, vol **17**, no 3, pp 427–447

[8] M D Myers 1997 Qualitative research in information systems, *Manag Inf. Syst. Q.*, vol **21**, no June, pp 1–18
[9] A R Smith, J M Colombi, and J R Wirthlin 2013 Rapid development: A content analysis comparison of literature and purposive sampling of rapid reaction projects, *Procedia Comput. Sci.*, vol **16**, pp 475–482
[10] K B Jensen 2012 *A Handbook of Media and Communication Research: Qualitative and Quantitative Methodologies*
[11] W Eichhorst, N Rodríguez-planas, R Schmidl, and K F Zimmermann 2015 A Road Map To Vocational Education And Training In Industrialized Countries, *ILR Rev.*, vol **68**, no 2, pp 314–337
[12] B P Statistik 2017 *Keadaan Ketenagakerjaan Indonesia Februari 2017* (Jakarta)
[13] J Schlingensiepen 2014 Competence Driven Methodology For Curriculum Development Based On Requirement Engineering, *Procedia - Soc. Behav. Sci.*, vol **141**, pp 1203–1207
[14] H-C Hsiao, M-N Chen, and H-S Yang 2008 Leadership of vocational high school principals in curriculum reform: A case study in Taiwan, *Int. J. Educ. Dev.*, vol **28**, no 6, pp 669–686
[15] S Sermsuk, C Chianchana, and P Stirayakorn 2014 5 th World Conference on Educational Sciences 2013 A Study of Model of Vocational Curriculum Development Under Vocational Education Commission Using Cross-Impact Analysis, *Procedia - Soc. Behav. Sci.*, vol **116**, pp 1896–1901
[16] P Chatterton and J Goddard 2000 The Response of Higher Education Institutions to Regional Needs, *Eur. J. Educ.*, vol **35**, no 4, pp 475–496
[17] C Natalia S, Maria S Vladislav, and P Paquet 2015 Problems of educational processes development and labor market needs analysis in Russia, *Procedia - Soc. Behav. Sci.*, vol **166**, pp 309–316
[18] H Hidayat 2011 Pengelolaan Sumber Daya Alam Berbasis Kelembagaan Lokal, *J. Sej. CITRA LEKHA*, vol **15**, no 1, pp 19–32
[19] M B Triyono 2016 The Content Curriculum Development Of Education Technic In Heavy Equipment Based On Industrial Needs, vol **6**, no 3, pp 355–363
[20] Mendikbud 2016 *Permendikbud RI Nomor 20 Tahun 2016 tentang Standar Kompetensi Lulusan Pendidikan Dasar dan Menengah* Indonesia
[21] R W Tyler 2013 *Basic principles of curriculum and instruction* Chicago
[22] E Mulyasa 2002 *Kurikulum Berbasis Kompetensi: Konsep, Karakteristik, dan Implementasi* (Bandung: PT Remaja Rosdakarya)
[23] UURI 1999 *Undang - Undang Republik Indonesia Nomor 18 Tahun 1999 tentang Jasa Konstruksi*, no 1 Indonesia, pp 1–49
[24] Kemenakertrans 2010 *Keputusan Menakertrans RI tentang Penetapan Rancangan Standar Kompetensi Kerja Nasional Indonesia* Indonesia

IOP Conf. Series: Materials Science and Engineering **306** (2018) 012038 doi:10.1088/1757-899X/306/1/012038

Simulation and Failure Analysis of Car Bumper Made of Pineapple Leaf Fiber Reinforced Composite

E S Arbintarso[1,*], **M Muslim**[2] **and T Rusianto**[3]

[1]Jurusan Teknik Mesin, Institut Sains & Teknologi AKPRIND, Yogyakarta 55222, Indonesia
[2]Engineering Department, PT. Duraquipt Cemerlang, Depok 16515, Indonesia
[3]Fakultas Teknologi Industri, Institut Sains & Teknologi AKPRIND, Yogyakarta 55222, Indonesia

*ellyawan@akprind.ac.id

Abstract. The bumper car made of the Pineapple Leaf Fiber Reinforced Composite (PLFRC) is possible to be produced with the advantage of easy to get, and cheap. Pineapple leaf fiber has chosen as a natural fiber, which the maximum of the strength of 368 MPa. The objective of this study was to determine the maximum capability of front car bumpers using Pineapple Leaf Fiber Reinforced Composite materials through the process of simulating stress analysis with Solidworks 2014 software. The aim also to know the distribution of loads that occur on the front car bumper and predict the critical point position on the design of the bumper. The result will use to develop the alternative lightweight, cheap and environmentally friendly materials in general and the development of the use of pineapple fiber for automotive purposes in particular. Simulations and failure analysis have been conducted and showed an increased impact speed in line with increased displacement, strain, and stress that occur on the surface of the bumper. The bumper can withstand collisions at a speed of less than 70 kph.

1. Introduction

In line with the intensively using of green technology, the focus of researchers and developers of an alternative material comes back to research on the mechanical properties of natural fiber-reinforced composites. In general, composite is reinforced with fiberglass, which is relatively expensive and harmful to the environment. There are some advantages to using a natural fiber, among which are: a) natural fibers are biodegradable (capable of being broken down by bacteria/microorganisms); b) natural fiber, one of which pineapple leaf fiber has low production cost, low density, and low energy consumption. According to Kaewpirom and Worrarat [1] amidst the rapidly evolving material science, natural fibers seem to be a remarkable material, which can be used as a viable replacement of synthetic fibers as well as abundant resource and inexpensive. Natural fiber reinforced composites have a specific tensile strength of 250-650 MPa smaller than 2850 MPa for carbon fiber reinforced composites [2]. However, natural fibers have advantages, especially cheaper, lighter and better environmentally [3]. Natural fiber reinforced composites form a new class of materials science that has good potential in the future as a substitute for wood-based materials in a wide range of applications [4]. However, resistance to poor water absorption makes the use of natural fiber reinforced composites less attractive.

Car bumpers are made of lightweight metals such as aluminium alloys or thin steel plates. Nowadays, as further technological developments, car bumpers are made of fiber reinforced composites (most use synthetic fibers) [3]. However, in recent years the price of aluminium and plastic materials continues to increase, so the demand for a replacement material with similar properties and low price also increases. Natural fibers are chosen because they have advantages with biodegradable and low prices. Natural fiber reinforced composites in addition to having low-cost, lightweight and eco-friendly advantages also have drawbacks where the specific tensile strength and weight depend on the type of fiber, the type of webbing, and the process of preparing the natural fibers. Arbintarso [5] suggested that pineapple fiber has the highest mechanical properties compared to coconut fiber fibers and Roselle fibers.

The objective of this study was to determine the maximum capability of front car bumpers using natural fiber composite materials through the process of simulating stress analysis with Solidworks 2014 software and to know the distribution of loads that occur on the car bumper and predict the critical point position on the design of the bumper. The result will use to develop the alternative lightweight, cheap and environmentally friendly materials in general and the development of the use of natural fiber (pineapple fiber) for automotive purposes in particular.

1.1. Front car bumper model

The main purpose of front car bumper systems is to protect the engine and radiator, to absorb energy at the start of a collision and to guide the remaining impact forces into the rest of the body structure [6]. The front bumper system protects the vehicle in a low-speed crash (at ≤ 8 km/h (kph) ~ ECE R-42 Europe standard) which indicates without damage of functionally relevant parts and only minimal or no plastic deformation of any other vehicle component. The modern design front car bumper consists of plastic fascia, a propylene foam or a honeycomb energy absorber and a reinforcing bumper beam. The energy absorber is placed between the plastic fascia and reinforcing bumper beam [6]. This modern design also improves pedestrian protection (leg impact) as shown in Figure 1.a. The fascia is a non-structural component that reduces the air resistance, while the energy absorber dissipates part of the kinetic energy during the collision and the bumper beam is a structural component, which absorbs the low impact energy by bending resistance [7]. The traditional design only used bumper beam without fascia and energy absorber is seldom used today. A visible metallic transverse beam was chosen to decorate the front or rear end of the vehicle and will act as the primary energy absorber in a collision [6]. The Isuzu Elf NKR 55 Microbus adopts the traditional bumper design with adding an energy absorber as shown in Figure 1.b. This design is chosen because of simple, compact and inexpensive. According to the speed and the acceleration of the car, the total impact energy at 40 kph is 825 J [8] that the load is distributed throughout the bumper.

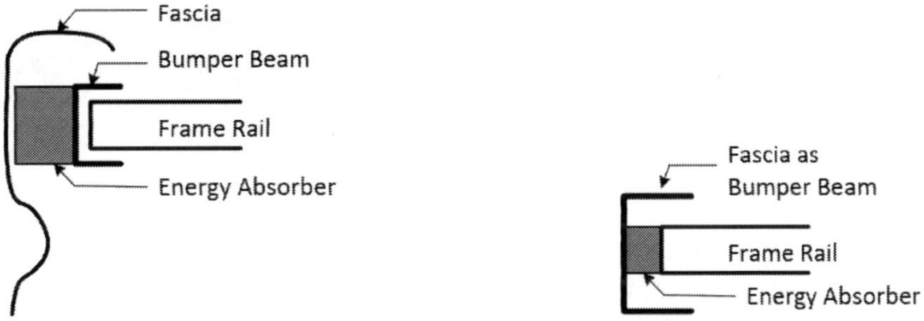

(a) (b)

Figure 1. a) Common modern front car bumper design [6] and b) Front car bumper of Isuzu [9].

1.2. Pineapple leaf fiber reinforced composites

Pineapple is a perennial herbaceous plant with 1-2 m for height and width belongs to family Bromeliaceae [10]. It is chiefly cultivated in coastal and tropical regions, mainly for its fruits purpose. In Indonesia, one of the big pineapple industry is PT. Great Giant Pineapple with produced 200,000 Ton per year in 2015. It is cultivated on about 81,510 acres in Lampung Province of Sumatra, Indonesia [11] and is continuously increasing its production. PALF has high specific strength and stiffness; it is hydrophilic due to high cellulose content [12], as shown in Table 1.

Table 1. Physical and mechanical strength of pineapple leaves fiber [12].

Density (g/cm^3)	Tensile Strength (MPa)	Young's Modulus (GPa)	Elongation At break (%)	Diameter (μm)	Microfibril Angle
1.520	170-1627	6.21-82.51	1.6-3	20-80	-
1.526	170-413	6.26-62.10	1.6-3	30-60	-
1.440	413-1627	34.50-82.51	1.6	1.56-80	8-15
1.070	126.6-1627	4.40-82.50	1.6-2.2	20-80	14
1.523	413	4.20	3.0-4.0	50	14

The disadvantage of PALF as fiber on reinforced composite is the lower degree of compatibility with hydrophobic polymers due to its hygroscopic nature. The existence of natural waxy substance on the surface of fiber layer provides low surface tension, which does not allow a strong bond with polymer matrix [12]. Santosh Kumar et al. [13] obtained the maximum tensile, and flexural strength of pineapple leaf fiber reinforced composites as 65.95 MPa and 121.83 MPa at 30% volume ratio, respectively.

2. Experimental Setup

Solidworks 2014 was used in making car bumper design and the simulations. Using Surfacing feature to make the 3D model and Drop Test and Static Analysis for the simulations. Drop Test is a test of the strength/reliability of an object, where the object is dropped from a certain height. Drop Test simulation is chosen because the condition is similar when a car has a collision with a wall (the target is motionless). The Drop Test simulation uses the velocity of impact that depends on some parameters, such as gravity magnitude, velocity magnitude and target stiffness (in this simulation using target rigid as shown in Figure 2.a). The simulation runs at speeds of 40 kph, 70 kph, 100 kph, and 130 kph.

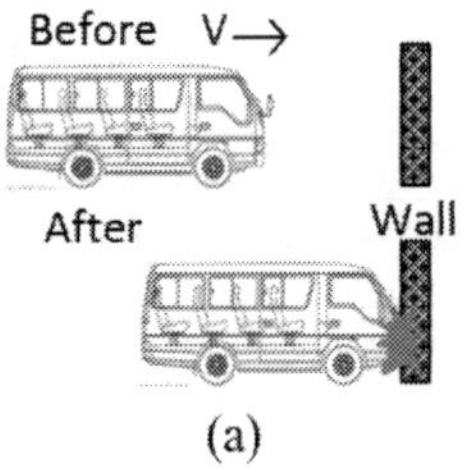

(a)

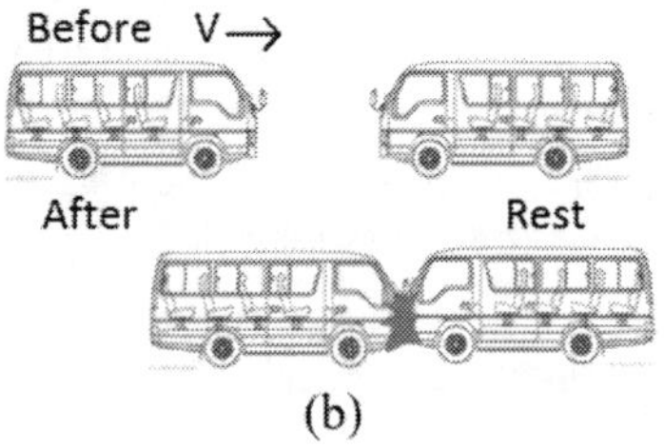

(b)

Figure 2. a) Drop Test illustration, and b) Static Analysis illustration (car figure is property of [9]).

Static Analysis is a testing method to figure out the maximum ability of material to receive a moving load in a state condition. The results of the Static Analysis will be used to review a design of front car bumper and see the possible damage when the bumper -in the rest condition- was hit by another car (Figure 2.b.). The location of fixed points, the location of loading and determine the amount of force are a parameter that will be used to Static Analysis simulation.

Figure 3. Blueprint of Isuzu Elf NKR 55 Microbus [9].

Front car bumper was designed from the Isuzu Elf NKR 55 Microbus car blueprint [9], as shown in Figure 3. The failure that occurs in the bumper is analyzed with variable speed of the car, and the critical loading test is manifested through graphics by using striking colors for critical points where composites fail over material capabilities.

3. Results and Discussion

In the simulated static test, the Isuzu Elf NKR 55 Microbus vehicle scenario is in rest condition and hit by the same vehicle (mass and dimension) in the middle of the front vehicle. The simulation of the drop test is similar to the case of the static test, but in this case, the Isuzu Elf vehicle NKR 55 Microbus accelerates and hit a strong and heavy wall (motionless). The determination of force based on the mass, speed and time, by converting the force value ($F = (mV)/t$ ...(1)) both for static and drop test, time (t) is expected to occur for 1 second. The simulation results for the static test can be seen in Table 2. Emphasis the observed properties are the displacement (mm), the maximum strain that will occur, and the maximum stress (MPa) occurring in the bumper. The results of the simulation drop test can be seen in Table 3.

Table 2. Static test simulation results.

Velocity (V) (kph)	Force (F) (kN)	Displacement (max) (mm)	Strain (max)	Stress (max) (MPa)
40	57,11	8,59	0,018	75,10
70	99,94	15,04	0,031	131,42
100	142,78	21,48	0,045	187,74
130	185,61	27,92	0,058	244,06

Table 3. Drop test simulation results.

Impact Velocity (V) (kph)	Displacement (max) (mm)	Strain (max)	Stress (max) (MPa)
40	6,03	0,007	69,71
70	10,54	0,012	129,09
100	15,10	0,016	189,54
130	19,70	0,021	244,72

Figure 4 shows the location of the critical stress (the area that has the highest potential damage when subjected to force) on the car bumper. The maximum stress value (indicated in red) is 75.10 MPa for speed 40 kph (Figure 4.a), and increase as the speed increases. Figure 5. Shows the location of the highest strain value on the car bumper shown in red, with a value of 0.018 for speed 40 kph at Figure 5.a, and also increases as the speed increases. The location of the critical displacement of the car bumper shown in red with its maximum displacement value is 8.59 mm at a speed of 40 kph as shown in Figure 6.a and increase as the speed increases. Davoodi et al. [8] found the energy absorption is 77 J and displacement is approximately 45 mm, larger than displacement results in this study. It means the pineapple fiber reinforced composite has energy absorber higher than 77 J and sufficient for pedestrian impact for low impact collision [8]. The energy absorbing performance is strongly affected by the geometry component [14], in this case, beam way profile or beam corner can serve as stress concentration zone or crack triggering point as shown in Figure 6.

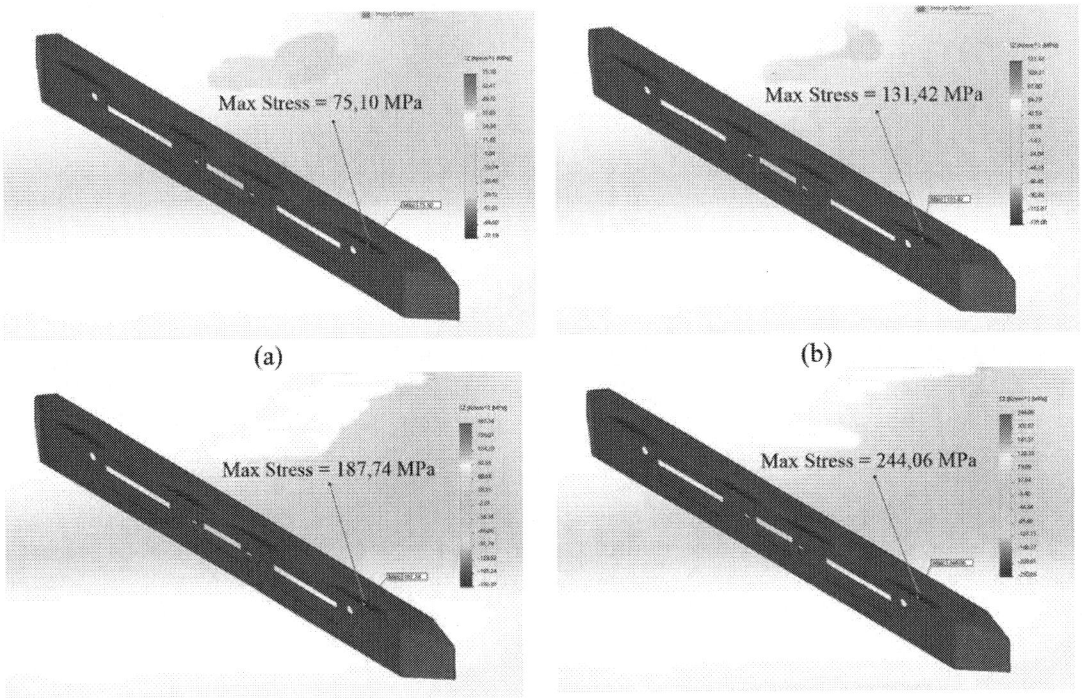

(c) (d)

Figure 4. The location of critical stresses of the bumper at a) 40 kph, b) 70 kph, c) 100 kph, and d) 130 kph.

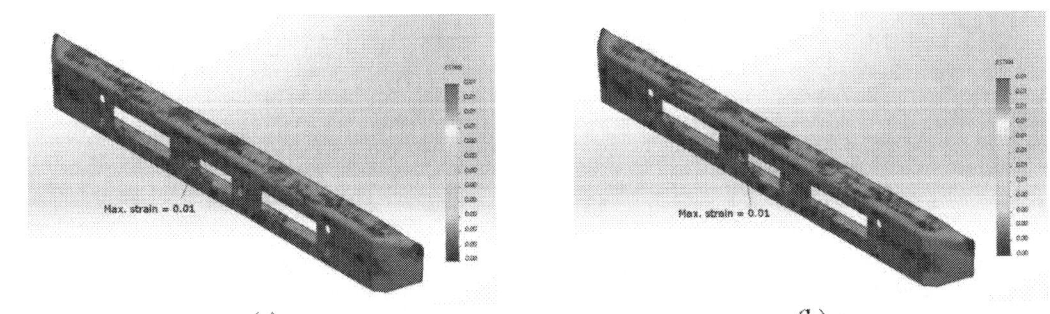

(a) (b)

Figure 5. The location of critical strains of the bumper at a) 40 kph, b) 70 kph, c) 100 kph, and d) 130 kph.

IOP Conf. Series: Materials Science and Engineering **306** (2018) 012038 doi:10.1088/1757-899X/306/1/012038

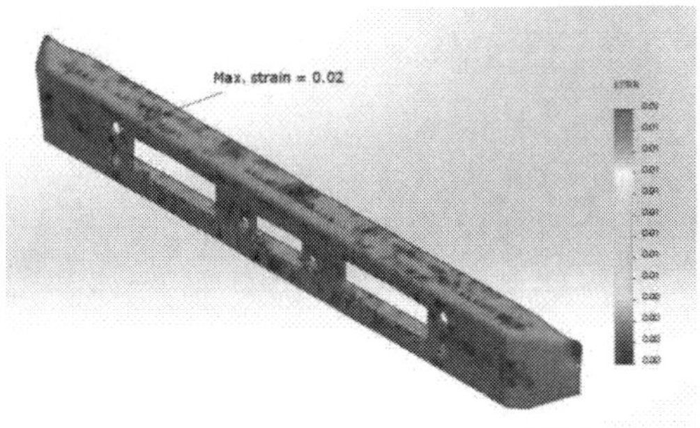

(c)

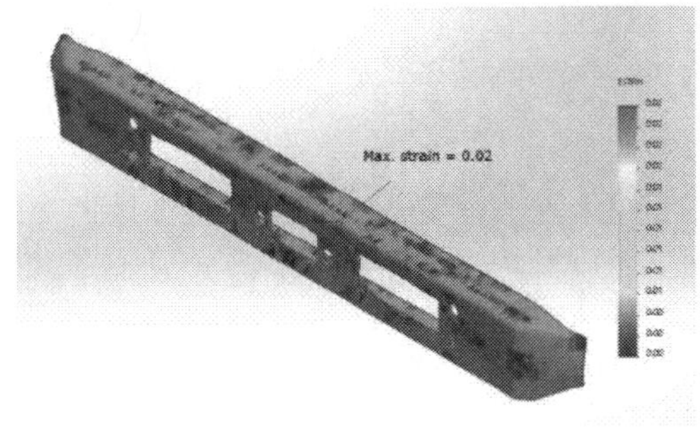

(d)

Figure 5 (Cont.). The location of critical strains of the bumper at a) 40 kph, b) 70 kph, c) 100 kph, and d) 130 kph.

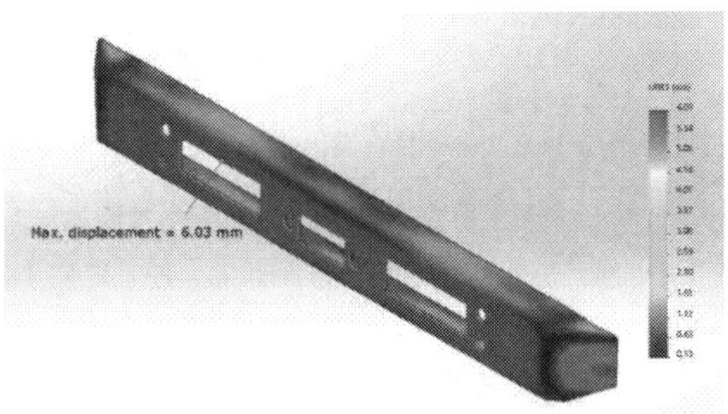

(a)

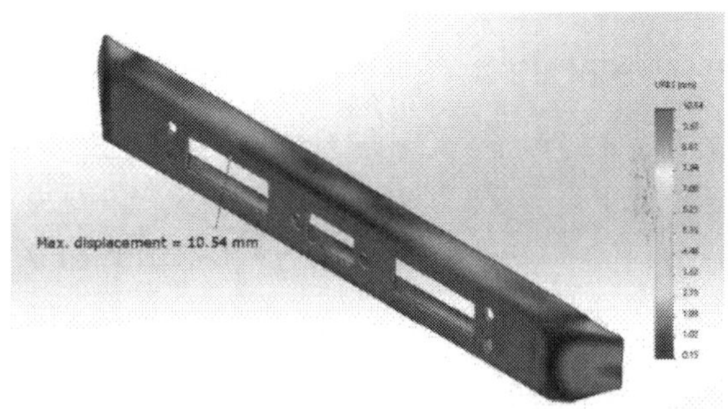

(b)

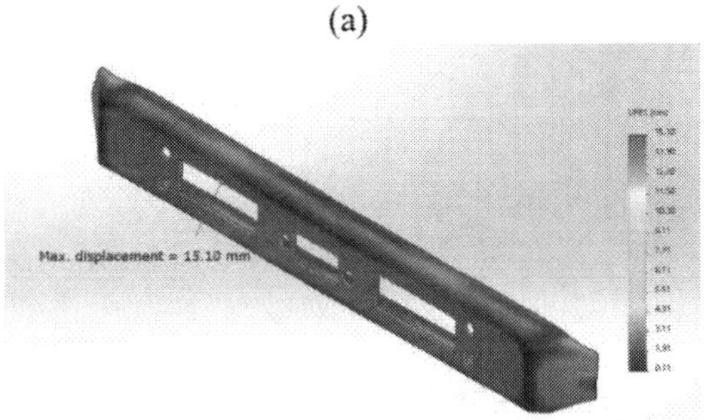

(c)

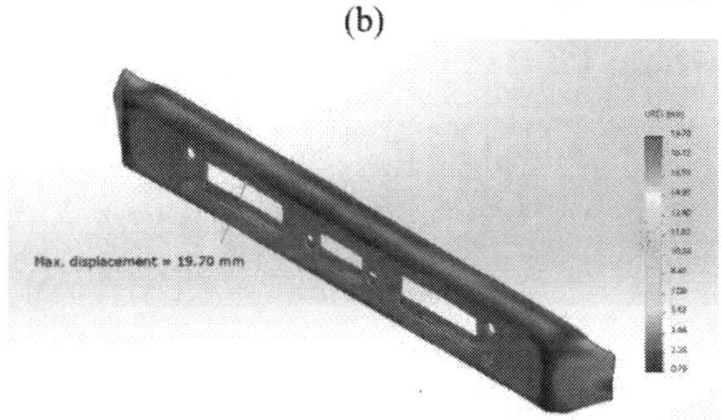

(d)

Figure 6. The location of max. displacements of the bumper at a) 40 kph, b) 70 kph, c) 100 kph, and d) 130 kph.

Figure 7.a. shows the comparison of the maximum stress values generated in the static and drop test simulations. There were no significant differences between the two test results. Figure 7.b. shows the comparison of the maximum strain values generated in the static and drop test simulations. In the static test, the strain tends to increase along with the increase of the speed of the vehicle, while at the drop test, no signification occurs with the increase of collision velocity.

IOP Conf. Series: Materials Science and Engineering **306** (2018) 012038 doi:10.1088/1757-899X/306/1/012038

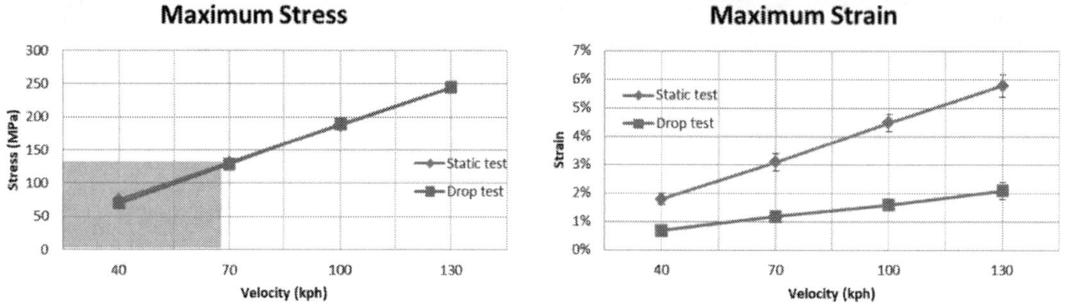

(a) (b)

Figure 7. Graph of maximum stresses and strains versus velocity.

Figure 8 shows the maximum displacement values generated in the static analysis and drop test simulations. The maximum displacement value in the static test is higher than the drop test; the momentum that occurs in the simulation influences this. The momentum of static test and drop test are (e)> 1 and e = 1, respectively. In case of small low-velocity impact, only the bumper beam should be involved and should behave fully elastic (e = 1) [14].

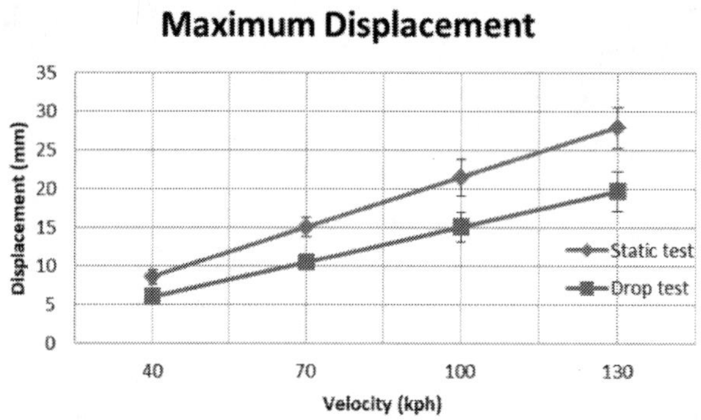

Figure 8. Maximum displacement versus velocity.

Based on data of pineapple leaf fiber-reinforced composite capability of ~ 120 MPa/m [5] as seen in Figure 7.a. as gray block area, and it can be concluded front car bumper using pineapple leaf fiber composite able to withstand impact with speed 40 kph but less than 70 kph. This simulation result needs to be improved by the optimization method as Liu et al. [15] did that using modified particle swarm optimizer. This algorithm was used to accomplish the optimization procedure considering the static strength conditions and dynamic impact the bumper system.

4. Conclusions

Increasing collision speed increases the displacement, strain, and stress. There are significant differences in the displacement and strain results between static analysis and drop test, but not for stress results. Bumpers with pineapple leaf fiber-reinforced composites can withstand collisions at speed less than 70 kph.

References

[1] Kaewpirom S and Worrarat C 2014 Preparation and Properties Leaf Fiber Reinforced Poly(lactic acid) Green Composite *Fibers and Polymers* **15** (7) pp 1469-1477.

[2] Westman M P, Fifield L S, Simmons K L, Laddha S G and Kafenfzis T A 2010 Natural fiber composite: A review *Pacific Northwest National Laboratory Report for U.S. Department of Energy* (Washington: Pacific Northwest National Laboratory) p 7
[3] Joshi S V, Drzal L T, Mohanty A K and Arora S 2004 Are natural fiber composites environmentally superior to glass fiber reinforced composites? *Composites* A **35** 371–76
[4] Rusianto T 2010 Pembuatan papan partikel berbahan baku sabut kelapa dengan bahan pengikat alami (lem kopal) *Jurnal Teknologi* **3** (1) pp 22-32
[5] Arbintarso E S 2015 Tinjauan Komposit Diperkuat Serat Alam Sebagai Bahan Alternatif Untuk Bodi Mobil Urban Concept *Project Report* (Yogyakarta: LPPM Institut Sains & Teknologi AKPRIND) p 33
[6] European Aluminium Association, 2013, *The Aluminium Automotive Manual: Applications – Car body – Crash Management Systems*, European Aluminium Association pp 2-3, 11
[7] Wang T and Li Y 2015 Design and analysis of automotive carbon fiber composite bumper beam based on finite element analysis *Advances in Mechanical Engineering* **7** (6) pp 1–12
[8] Davoodi M M, Sapuan S M, and Yunus R 2008 Conceptual design of a polymer composite automotive bumper energy absorber *Materials and Design* **29** pp 1447-1452
[9] Evolution Ltd 2015 Isuzu Elf NKR 55 Microbus LWB (http://www.the-blueprints.com/) accessed on 12 April 2016
[10] Van Tran A 2006 Chemical analysis and pulping study of pineapple crown leaves *Industrial Crops and Products* **24** (1) pp 66–74
[11] Wahyuni N D 2015 Top 5 Bisnis: Produsen Olahan Nanas Terbesar Dunia Ada di RI (http://bisnis.liputan6.com/read/2261371/top-5-bisnis-produsen-olahan-nanas-terbesar-dunia-ada-di-ri) accessed on 15 June 2017
[12] Asim M Abdan K Jawaid M Nasir M Dashtizadeh Z Ishak M R, and Hoque M E 2015 Review Article: A Review on Pineapple Leaves Fiber and Its Composites *International Journal of Polymer Science* **2015** pp 1-16
[13] Santosh Kumar D S Praveen B A Kiran Aithal S, and Kempaiah U N 2015 Development of pineapple leaf fiber reinforced epoxy resin composites *International Research Journal of Engineering and Technology (IRJET)* **2** (3) pp 2190-2193
[14] Belingardi G, Beyene A T, Koricho E G and Martorana B 2015 Alternative lightweight materials and component manufacturing technologies for vehicle bumper beam *Composite Structures* **120** pp 483-495
[15] Liu Z, Lu J and Zhu P 2016 *Lightweight design of automotive composite bumper system using modified particle swarm optimizer Composite Structures* **140** pp 630-643.

Acknowledgments
The authors would like to thank profusely to Head of LPPM, and Rector of Institut Sains & Teknologi AKPRIND who have given funding part of the research activities.

IOP Conf. Series: Materials Science and Engineering **306** (2018) 012039 doi:10.1088/1757-899X/306/1/012039

Community Participation for Sustainable Tourism Model in Manado Coastal Area

F F Warouw[1,*], F W Langitan[1] and A T Alamsyah[2]
[1]Universitas Negeri Manado, Tondano, 95618, Indonesia
[2]Universitas Indonesia, Depok, Jawa Barat, Indonesia

*ferolwarouw@yahoo.com

Abstract. Manado city with the potential for coastal tourism which is at the center of the world's coral triangle is developing a tourism development policy. Tourism that is being developed should certainly be able to adapt to changing conditions of today's dynamic environment. The extent to which the adaptation process is determined by the communities involved in the development of tourism. Based on data from the population in the city of Manado, there are currently 410 481 thousand inhabitants. This study tried to reveal the extent of community participation in the city of Manado in particular involvement in the development of tourism in coastal Bay of Manado. Level of community participation in the development of tourism in the city of Manado is measured in the form of quantitative research in the form of questionnaires to the community based on the level of community participation developed by Arnstein Sheery by purposive sampling technique. The level of participation was developed by Arnstein Sheery row of low level ketinggi namely: Manipulation, Therapy, informing, Consultation, placation, partnership, Delegaten Power, Citizen Control. Findings community level participation in the development of coastal tourism on Manado bay civilized level of consultation. Consultation rate shows that the government invite the public opinion after the given information to the public and has been a two-way dialogue between government and society. Community has provided input and active discussion by way of the two-way dialogue. Although there has been a two-way dialogue, but the successful rate is low because there is no guarantee that the concerns and ideas of the community will be considered. This requires the development of community-based programs in order to increase the degree of participation so that community participation will increase. To conclude, the program needs to be based on input from the community's needs and it has to involve the public directly to tourism development in accordance with the characteristics and potential of the community itself.

1. Introduction

Manado city with a long coast line of more than 17 km of coastal tourism potential and attractive coral reefs. Not to mention the beautiful sunset views overlooking towards the Pacific ocean in every afternoon to make the coastal city of Manado as a potential area for the development of tourism. This potential is supported by ditelorkannya vision "Manado City Ecotourism Model" by the government of the city of Manado. In addition there is the type of coastal water ecosystems are important in the ecosystem Manado City is famous for it's coral reef dive spots that can be found within close together and easily accessible. Bunaken marine park which is part of the administrative area and right in front of the city of Manado is a marine park that was named the most beautiful marine park in the world. This marine park store with a collection rich biodiversity of coral species is among the most complete in the world and are at the core of the world's coral triangle. This condition is supported by the potential of offshore fisheries are also promising to be developed as part of sport fishing. Besides the local community also has a variety of cultural allure of interest include diverse culinary potential. Referring to the above potential, of the development of tourism in the city of Manado has a very prospective future. This activity directly if managed with the right strategies will improve the welfare of the surrounding community as well will be a part of the preservation of natural resources and the environment which is the strategic issues of global development .

One of the key factors in the development of tourism is community involvement in tourism development process itself. Community participation is voluntary and should take place continuously. Participation in question is the participation of the public view as the subject of any rule of construction rather than as objects of development. The involvement of the community as a whole is done through the development of a mindset that sees society as the subject of regulation by the diversity of behavior. Through the process of engaging public participation from the planning stage, utilization, and control of the utilization of space will appear a system of evaluation of development activities that have been carried out development and become an input for further development process.

The participatory approach is expected to create agreements and rules in society in order to realize social justice in a program that is tailored to their aspirations. It also increases the sense of community (sense of belonging) to development programs in line with their aspirations in the accommodation of the development program, which in turn can manifest the development of efficient and effective [1]. Participation in the development of system development is needed [2] because: (1) at the planning stage, most people know what they need, thereby directing the product plan optimal spatial and proportional to a variety of activities, so as to avoid speculation and excessive distribution for space allocation certain activities, (2) the utilization stage, people will keep the utilization of space in accordance with the time allotment and allocation as well as planned , so as to avoid conflict of space utilization, (3) the control phase, people feel they have and be responsible for maintaining the quality of space comfortable and harmonious as well as useful for future development [3].

For that we need the necessary research to find a model of tourism development which is born of the actual conditions of society itself. Studies to determine the form and level of participation and the factors that influence the development of tourism studies conducted through participation in tourism development in coastal areas Malalayang Manado bay. This study is expected to produce a picture of the community's role in tourism development that can be used as a referral policy development sustainable tourism development .

2. Method

Studies involving community participation in tourism development through the study of participation in tourism development in coastal areas Malalayang Manado Bay is a research focus intensively on one particular object that is studied as a case. Research case study intended to learn intensively about the background of the problem situation and the position of an event taking place at this time, as well as the interaction of the environment of a particular social unit is (given). Research subjects may be individuals, groups, institutions or society. Case study data can be obtained from all parties concerned, in other words, the data in this study were collected from various sources.

Based on the research objectives of this research is quantitative research in order to find the level of participation as well as the function and its implications in the development of sustainable tourism based on local community participation in coastal areas through coastal development studies Malalayang Manado Bay.

2.1. Data collection method

The research method used is through the method of approach to Regulatory Impact Analysis and Survey (Quantitative) with typology community participation [4,5]. Respondents in this study were drawn from the population of the local community involved in the formulation development of tourism in the city of Manado. The criteria selected respondents are local community elders to determine the potential and prospects of the development of tourism in the city of Manado.

Stages of research in order to obtain data regarding the quantification of the development of tourism in the form of a questionnaire totalled 62 responden from local communities in the city of Manado.

Table 1. Research analysis.

Analysis	Method	Description	Result
Level of community participation	RIA analysis with quantitative description through frequency distribution	Analyse the level of community participation in tourism development in Manado city	Magnitude levels of community participation as measured by the typology Arnstein

2.2. Data analyze method

Data processing techniques in question here is the processing of primary data collected directly from respondents through a questionnaire. In processing the data, the respondents' answers from each question will be weighted / predetermined value.

3. Analyse and Interpretation

The coastal area in the Bay of Manado ecoregion Malalayang this region includes three counties - along the Gulf coast city of Manado, but administratively Malalayang most coastal areas are in the administrative area of Manado city.

Manado Bay is located in front of the city of Manado is the capital of Sulawesi, North, Bay of Manado ecoregion is an area which includes the 3 counties and cities within the scope of Bunaken National Marine Park. Administratively Malalayang coastal areas are mostly located in the city of Manado. Manado Bay is the location of the conservation of Bunaken National Marine Park is set in 2010 in the central region segita three coral (Coral Triangle Initiative) [6].

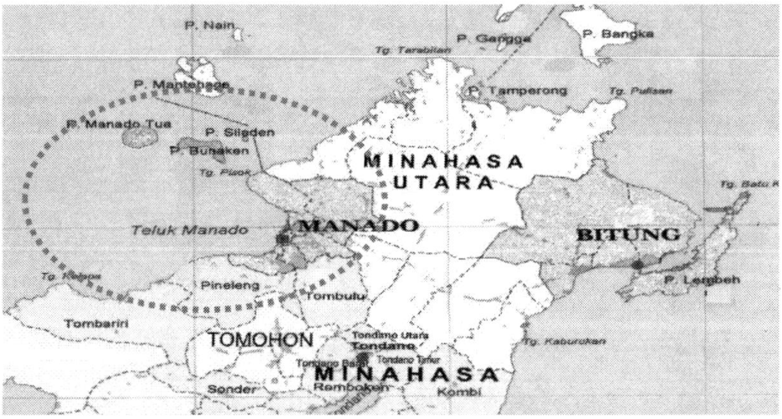

Figure 1. Map of Manado city and surrounding.
Source: Manado in Figures 2016

The use of land in Manado City depicts a mixture of urban and land use functions of the coastal sea and the island in front of the city with views of the mountains including several volcanoes such as Mount Lokon and Mount Klabat. As a city located in the Ring of Fire area of the sea and the gulf coast of Manado It really a rich biodiversity. On the one hand there is an industrial and trading activities while writhing on the economic side, on the other hand there is the sea, as the frontier town itself.

Mixing two contrasting land use characteristics can be potential for the development of the city of Manado is based on ecotourism management [7].

Figure 2. Coral reef one of the potential product for coastal tourism in Manado.
(Source: bunaken.org; wisataindonesia.org; satulayan.net; antaranews.com /clockwise).

The shift from the previous land use into residential estates showed that Manado City continues to grow. Besides the presence of the hotel - an international hotel characterizes today's Manado city. Existing hotels like Gran Kawanua hotel Novotel group, Swiss Bell Hotel Maleosan and development plans Hard Rock Hotel in the coastal region Malalayang. Until now growing tourism in the city of Manado is not concentrated in one place, but in some places according to the tourism potential held each region while developing it. The main centre for eco-tourism development in coastal and marine national park of Bunaken.

Manado bay coast is a haven for coral reef and became a well-known dive sites in various parts of the world [8]. Region of live coral cover are listed where the highest found in the area around Bunaken Island, coastal Malalayang to Awontulap. To determine the potential tourism coastal Bay of Manado, based on the results of research in the field can be known of the potential support for realizing sustainable development of tourism, the natural potential and the potential of coastal marine parks supported by the local socio-cultural potential . Some types of tourism potential of the city of Manado, among others, marine parks, mountain views, a culinary and cultural tour look promising and can be enjoyed from the coastal area of the Bay of Manado. The potential of the unique attractions and famous around the world that the Bunaken Marine Park. Currently foreign tourists (tourists) and domestic travellers (wisdom) who visited the relatively increased. Expenditure per day tourists reached Rp.1.200.000 and the average length of stay 4 days. Domestic tourists each day on average spend 500.000 and the average length of stay 2.5 days. The increase in the number of tourists, the amount of expenditure per day, and length of stay will multiply the benefits of tourism for local development activities. The tourism industry has a high multiplier effect, potentially attracting investment, and prospective development in North Sulawesi [9].

In 2000 Former Governor of North Sulawesi (A.J Sondakh) suggests that the importance of the Bunaken National Park in North Sulawesi, both sustainability and its contribution to the economy through the development of the tourism sector, then together we develop by promoting the principles of conservation through management systems and the management of inter - profitable. Sondakh also cites studies NRM / EPIQ and LPEM - UI on theimpact of the economic crisis in North Sulawesi in 1999, turned out to Bunaken able to contribute greatly to the development of Northern Sulawesi, namely: (1) traditional fishery U.S. $ 3.3 million, (2) commercial fisheries of U.S. $ 1.6 million, (3) seaweed U.S. $ 3.1 million, and (4) dives services U.S. $ 3.2 million. When the U.S. dollar against the rupiah average of USD 8000 alone, the contribution of fisheries and seaweed farming can reap the results of approximately USD 64 billion, while diving services only Rp 25.6 billion [10,11].

Manado city is the centre of the tourism industry on the island of Sulawesi [12]. The tourism industry has been growing for a long time since the potential Bunaken marine park that is a favorite location for divers. Along with the development of the global environment with the main issues of global warming, coral reefs are the main potential Bunaken marine park turned out to be one of the potential that is able to cope with climate change. The potential is then used as the basis for implementing sustainable development by keeping a coral reef in Bunaken marine national park from being damaged and sustainable. In Manado City location potentially ubiquitous tourist area can be developed to prevent environmental degradation can be utilized as a tourist attraction, one of the locations are located in coastal areas and reclamation of the object of study.

4. Results and Discussion

In this section is analyzed on the level of participation in the development of tourism in the city of Manado. The amount of respondents amounted to 62 people where the number of respondents is a community of communities involved in the process of discussion and FGD for development in general especially Manado City Manado City development vision model Ecotourism in Manado Bay coastal region.

Determination of the level of participation categories based on the above table, it can be calculated as follows: There is one variable choice questions with no answers questions 8 choices with each score ranging from 1 to 8. The scores are based on the order of 8 household participation rate of Sherry Arnstein. So the minimum score obtained for each individual (1 x 1) is 1, the maximum scores obtained for each individual (1 x 8) is 8, then when the number of samples 62, it can be seen the minimum score for the level of community participation (62 x 1) is 62 and the maximum score (62 x 8) is 496 .

Table 2. Community participation rate calculation in sustainable tourism development in coastal bay of Manado.

No	Variable	Scale	N	%	B	N X B
1	Level of Participation of Implementation Phase	1) Not participating in the implementation of the program.	8	12,90	1	8
		2) Participate in the implementation of the program but for the government only.	8	12,90	2	16
		3) Participate in the implementation of the program for the benefit of the community only.	13	20,97	3	39
		4) Take an active discussion in the implementation of the program.	14	22,58	4	56
		5) Participate in the implementation of the plan and give some influence to the program implementation.	9	14,52	5	45
		6) Participate in the implementation of the program and share responsibility with the government.	6	9,68	6	36
		7) Participate in program execution and have the authority to make dominant decisions throughout the execution of the plan.	2	3,23	7	14
		8) Participate in program implementation and have the power to plan, implement and oversee the implementation of the program.	2	3,23	8	16
Total			62	100		230

Source: Data processed from the results of research, 2016

By knowing the minimum and maximum scores, it is known the distance of the interval, (496-62) / 8 = 54.25. So when used typology of Arnstein, to determine the level of community participation are:

- Citizen Control, a score of 441.75 to 496.00
- Delegated Power, a score of 387.50 to 441.75
- Partnership, a score of 333.25 to 387.50
- Placation, a score of 279.00 to 333.25
- Consultation, a score of 224.75 to 279.00
- Informing, a score of 170.50 to 224.75
- Therapy, a score of 116.25 to 170.50
- Manipulation, a score of 62.00 to 116.25

From this level will be concluded where participation levels are applicable in the development of tourism in Manado.

To measure community participation in the implementation phase of the program used rating scale that refers to the Public Participation Sherry Arnstein Ladder consisting of 8 ladder as in illustrated in Table 18. Following. Based on the level of participation in the implementation phase, the majority of respondents attended the discussions actively in the implementation of the program, as many as 14 people (22.58 %), followed by respondents participated in the implementation of public interest for just as many as 13 people (20.97 %), respondents participated in the implementation of the plan and give effect to the implementation of the program as many as 9 people (14.52 %), followed by respondents who participate in the implementation of government programs, but for the sake of just as many as 8 people and is involved in the implementation of the program. respectively - each (12.90 %), then Participate in the implementation of the program and share responsibility with the government as much as 6 people (9.68 %) and the rest with the same values Participate in the implementation of the program and has the authority to make decisions on the overall dominant Participate in the planning and implementation of the program and have the power to plan, implement , and oversee the implementation of the program each 2 people (3.23 %).

5. Conclusion

Total score obtained from the analysis is 230, then the level of community participation including category level Consultation. At the level of consultation may mean that the level of attendance at meetings/conferences due to:

- That the government invites public opinion after the given information to the public. Proven with involvement of community representatives in the preparation of the general plan of Manado City Model Ecotourism development.
- There has been a two-way dialogue between the government and the people involved and the community to provide input and active discussion by way of two-way dialogue.
- Although there has been a two-way dialogue, but this way the success rate is low because there is no guarantee that the concerns and ideas of the community will be considered.
- The method used is a public neighborhood meeting and public hearing.
- At the level of this Consultation is included in the degree of tokenism/awards or Degree of Tokenism, a level of participation where people are heard and allowed to argue, but they do not have the ability to get a guarantee that their views will be considered by decision makers.

Tourism development that has been carried out simultaneously that in turn also helped carry implications for various fields. The implication is meant here is the involvement or engagement relationship between at least two variables.

In this context, there are at least three obvious implications of sustainable tourism development based on community participation through the study of the development of ecotourism model of the city of Manado, the implications of economic, social and environmental. Perhaps this could be classed as an implication of the existence of externalities of tourism potential, but - whatever its name - is a decent third implication is recognized as the positive impact of a long history of tourism development area.

References

[1] Kimpraswil 2002 *Pelibatan Masyarakat dalam Penataan Ruang* Makalah dalam Pelatihan Penyusunan Rencana Tata Ruang Propinsi se-Sumbagsel di Palembang 30 September 2002 Available from http: //www.kimpraswil.go.id; INTERNET.

[2] Tosun, C. 2000. Limits to community participation in the tourism development process in developing countries. *Tourism management*, **21** (6), 613-633

[3] Ibrahim, Syahrul 2004. *Paradigma Baru Peran serta Masyarakat dalam Penataan Ruang*. Available from http://www.bktrn.org; INTERNET.

[4] Arnstein, S R 1969 A ladder of citizen participation *Journal of the American Institute of planners*, **35** (4), 216-224

[5] Tritter, J Q, & McCallum, A 2006 The snakes and ladders of user involvement: moving beyond Arnstein *Health policy*, **76** (2), 156-168

[6] Newman, C, & LeDrew, E 2005 Towards community-and scientific-based information integration in marine resource management in Indonesia: Bunaken National Park case study *Environments: a journal of interdisciplinary studies*, **33** (1)

[7] Ross, S, & Wall, G 1999 Evaluating ecotourism: the case of North Sulawesi, Indonesia *Tourism management*, **20** (6), 673-682

[8] KUSEN, J D, & TIOHO, H 2009 The present status of coral reef condition in Bunaken National Park and Manado Bay, North Sulawesi, Indonesia *Galaxea, Journal of Coral Reef Studies*, **11** (2), 219-222

[9] Erdmann, M, Merrill, P, Arsyad, I, & Mongdong, M 2003, September Developing a diversified portfolio of sustainable financing options for Bunaken National Marine Park In *Presentation at the Vth IUCN World Parks Congress, Durban, South Africa*

[10] Sondakh, AJ 2000, *Dewan Pengelolaan Taman Nasional Bunaken Sulawesi Utara Sebagai Model Pengelolaan Para Pihak Kawasan Pelestarian Alam di Indonesia*

[11] Cater, C, & Cater, E 2007 The economic impacts of marine wildlife tourism *Marine wildlife and tourism management: Insights from the natural and social sciences*, 145-162

[12] Alder, J, Sloan, N A, & Uktolseya, H 1994 A comparison of management planning and implementation in three Indonesian marine protected areas *Ocean & coastal management*, **24** (3), 179-198

Acknowledgements

The results of research regarding the level of community participation in tourism development in coastal areas Manado Bay. The authors would like to thank the community and the coastal communities in North Sulawesi young intellectuals who had agreed to be the respondents in this study include FGD involved in order to formulate the results of the research and its implications for the three pillars of sustainable development, namely social, economic and environmental.

A Prospective Method to Increase Oil Recovery in Waxy-Shallow Reservoir

F Hidayat* and M Abdurrahman

Department of Petroleum Engineering, Universitas Islam Riau, Indonesia

*fikihidayat@eng.uir.ac.id

Abstract. Waxy oil has been the main characteristics of The X field. Initial screening criteria studies indicated that cyclic steam stimulation (CSS) would be the optimum option because favorable reservoir condition. Based on this method we would like to know how much oil gain and the effect of steam for the stimulated and surrounding well. The injection of steam was done for 7 days followed by 14 days of soaking period. 39,000 liter of Marine fuel oil was used to generate steam for stimulation with an average produce steam quality about 80%. Average of 255 MMBTU of steam was injected each day with total steam injected was about 1.7 BBTU. The oil production was increased four times from 5 bopd into 21 bopd. Proper well candidate and high permeability are some reason for this method successfully increase oil production. Additional heat from steam reduced the damage near wellbore due to wax deposition. This is verifying by increasing productivity index from 3 bbl/psi to 4 bbl/psi. From results and observation data, this method can be a platform for typical shallow depth reservoir with high paraffinic content especially other reservoir in Sihapas formation.

1. Introduction

Production of oil reservoir with wax problem is challenging due to the characteristic of wax which made the oil not flowing when the temperature below the pour point. To fix this problem, several mitigation and removal techniques have been applied for numerous field with typical wax problems [1]. The common removal techniques are Fused Chemical Reaction, Mechanical Removal, Wax Removing Chemicals which are solvents type and dispersants, Magnetic Fluid Conditioning (MFC) technology, Use of Microbial Products such as marine micro-organisms, and Heat Application [1].

Heat application is one of popular method for fixing the wax problem. Several methods such as hot water injection, electromagnetically near-wellbore heating, and injecting steam, either by cyclic method or continuous, could be utilized to clean the wax in the near-wellbore. Steamflooding is the oldest commercial Enhanced Oil Recovery. It also contribute the most for the world's enhanced oil production [2]. One of its method, cyclic or commonly called as "huff 'n' puff", is widely used in oil recovery. Cyclic Steam Injection gives quick response but its result may not as higher as the other thermal method [3].

Cyclic steam stimulation projects start from decade ago in several different condition of reservoir such as tar sand [4], tight clay-rich reservoir [5], heavy oil [6], and also light oil [7]. In most cases, steam application often use for heavy oil, but some reports suggest that it could be used as an alternative for light oil reservoirs [8–10]. Laboratory and field trial of steam injection also have been reported from same formation, Sihapas formation [7,10,11]. Laboratory test show that 16% of the oil produced by the mechanism of vaporization at temperature of 400 ⁰F [11].

This paper attempted to observe the application of cyclic steam stimulation (CSS) into reservoir with wax problem. Several thermal methods have been applied into reservoir with the same problem, such as Hot Water Injection in Mangala Field [12], In-Situ Heat Generation (Heat Stimulation) [13], and Steam Injection [14,15] but the CSS method has not been applied into this kind of reservoir. Numerous reports suggesting the use of heat for handling the wax problem have been taken into account [13,16,17]. The field reports showed that heat utilization will bring good result regarding both oil recovery and wax deposition. Initial hypotheses based on other fields with other thermal methods suggest the application of cyclic steam stimulation is worth to try.

2. Methodology

A number of screening criteria for the steam application into reservoir is needed to be taken into consideration such as formation thickness, reservoir depth, oil saturations, permeability of rock, and oil viscosity for the success of this project [2,18]. Table 1 shows the screening criteria for steam utilization and its comparison with the field trial project.

Table 1. Comparison between technical screening for steam injection and field properties.

Screening Criteria		**X Field Properties**
	Crude Oil	
API Gravity, ⁰API	8 to 25 ↗	31.9
Viscosity, cp	<100,000	14
Pour Point, ⁰F	-	105-110
	Reservoir	
Net Thickness, ft	>20	30
Depth, ft	<5,000	1100
Permeability, md	>200	500
Oil Saturation, %PV	>40%	77%
Temperature, ⁰F	Not Critical	136 ⁰F

In this trial, 0.05 BBTU/ft pay thickness is used to calculate total steam needed for one cycle. Steam quality range between 70-80%. Recent report suggests that steam quality above 50% is mandatory for the project to be successful [14]. Therefore, 1.5 BBTU of steam was planned to be injected into reservoir for around 7-8 days. Unfortunately, there was lacked of proper equipment to measure rate injection of steam. Hence, we estimate the total steam injected each day by using average injection pressure data. Complete calculation and data during injection is showed by table 2.

Table 2. Steam properties during injection.

Day	Steam Quality, %	Avg. Inj Pressure, psig	Avg. Inj Temp, ⁰F	Water Injected, BCWEPD	Total heat Generated, MMBTU	Cum. Heat Generated, MMBTU	Running time, hours	Est. Steam injected, MMBTU
0	71	171	476	250	71.43	71.43	12	-
1	75	290	539	500	142.86	214.29	13	-
2	77	360	553	814	232.57	446.86	23	25.80
3	77	425	575	530	151.43	598.29	16	56.83
4	77	490	588	427	122.00	720.29	13	77.77
5	77	560	621	848	242.29	962.57	24	206.40
6	77	600	624	848	242.29	1204.86	24	242.29
7	77	600	623	848	242.29	1447.14	24	242.29
8	77	600	624	885	252.86	1700.00	25	252.39
Total				**5950**	**1700**		**7 days 6 hours**	**1103.77**

3. Results and discussion

Out of 1.5 BBTU, around 1.1 BBTU was successfully injected. It was less than the initial volume targeted to be applied into reservoir. Effect of steam can be observed from the incline in oil production. Due to the heat from steam, oil will be less viscous, therefore the mobility of oil will increase. The effect of steam into oil production rate is four times bigger from before steam was injected. It also reduced the production of water. The water cut went down from 99% into 95%. Figure 1 and 2 shows the significant change of oil and water production after the cyclic steam stimulation (CSS).

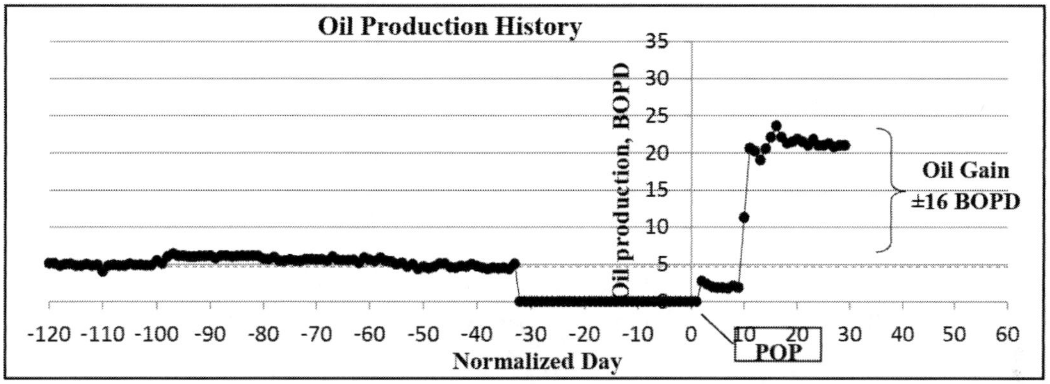

Figure 1. Oil production history.

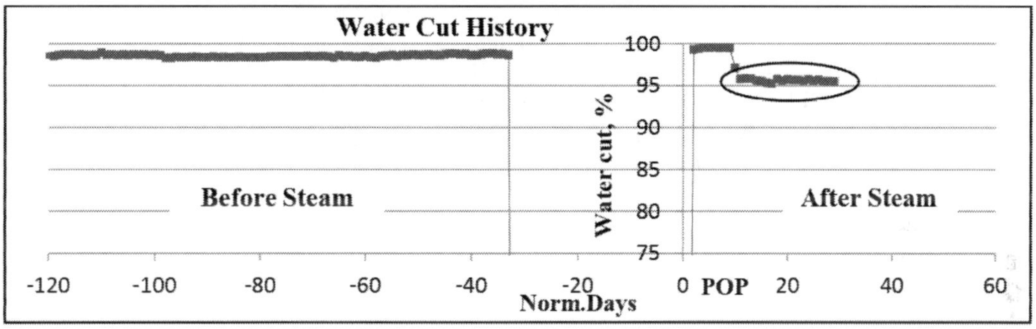

Figure 2. Water cut history.

In the other hand, steam also accountable for the melting of wax deposition near the wellbore. Wax deposited near the wellbore would restrict the flow of oil. With less restriction, it will enhance the ability of reservoir to flow the oil. It could be seen from the increasing of productivity index from 3.09 bbl/psi into 3.97 bbl/psi.

The temperature near wellbore significantly higher during steam injection. However, during soaking period, temperature drop very fast. The dropping temperature due to heat loss could have considerable impact into oil production, as the oil production much likely stay higher when the temperature was maintained high. The heat loss from steam to formation could happen due to time and wellbore heat capacity [19]. Nian et al. pointed that during the early stage of steam utilization, the wellbore heat capacity impact on heat loss is larger [19]. Increasing the injection rate and steam quality can improve the wellbore heat efficiency [20]. Therefore, further test must include the sensitivity of injection rate and steam quality. Optimum condition of those factors will be needed in order to reach the optimum injection criteria for this field. Optimum soaking time will be needed in order to maximize the oil production, thus the wax will not be formed in the near wellbore. The total of injection volume is also play a part in the rapid temperature drop. Only around 70% volume of steam was successfully injected into reservoir. Figure 3 shows the temperature near wellbore after steam injection.

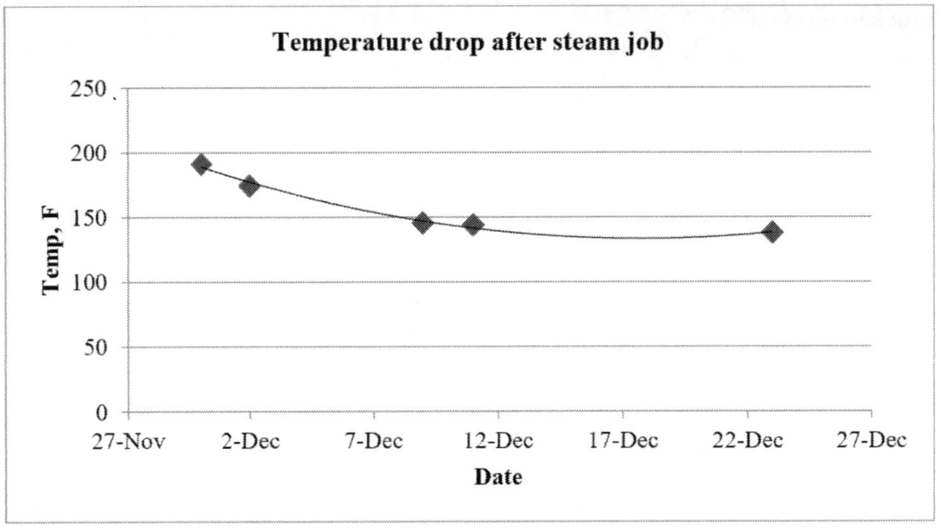

Figure 3. Temperature profile in near wellbore after steam injection.

4. Conclusion

Cyclic steam stimulation proved to be a prospect for rejuvenating oil production from waxy reservoir. The effect of steam not only boasting the oil production, but it also reduced the water production. The wax deposition in near wellbore were flushed out due to the high temperature during steam injection. Positive change from productivity index show that the oil moves freely due to the disappearance of wax. More analysis will be required should the project continue. The soaking period will be the main concern as the optimum time will give considerable effect into the production of oil.

Acknowledgements

Author wish to thank Universitas Islam Riau (UIR) for supporting and encouragement of writing this paper.

References

[1] Al-yaari M, Fahd K 2011 Paraffin Wax Deposition : Mitigation & Removal Techniques *SPE Saudi Arabia Technical Symp. (Dhahran)*

[2] Taber J J, Martin F D, Seright R S 1997 EOR Screening Criteria Revisited - Part 1: Introduction to Screening Criteria and Enhanced Recovery Field Projects *SPE Reserv Eng* **12** 3 189–98

[3] Alvarez J, Han S 2013 Current Overview of Cyclic Steam Injection Process *J Pet Sci Res* **2** 3 116–27

[4] Bott RC 1967 Cyclic Steam Project in a Virgin Tar Reservoir *J Pet Technol* **19** **5** 585–91

[5] Watkins D R, Kalfayan L J, Blaser S M 1987 Cyclic Steam Stimulation in a Tight Clay-Rich Reservoir *SPE California Regional Meeting (Ventura)*

[6] Tewari R D, Abdalla F, Lutfi H G, Keqiang Y, Faroug A, Bakri H 2011 Successful Cyclic Steam Stimulation Pilot in Heavy Oilfield of Sudan *SPE Enhanced Oil Recovery Conf.*

[7] Putra E A P, Rachman Y A, Arsyadanie R, Hafizh G, Firmanto T 2011 Case Study : Cyclic Steam Stimulation in Sihapas Formation *SPE Asia Pacific Oil and Gas Conf. and Exhib. (Jakarta)*

[8] Hanzlik E J 1981 Steamflooding As an Alternative EOR Process for Light Oil Reservoirs *SPE Annual Technical Conf. and Exhib. (San Antonio)*

[9] Gangle F J, Weyland H V, Lassiter J P, Veith E J, Garner T A 1992 Light-Oil Steamdrive Pilot Test at NPR-1, Elk Hills, California *SPE Reserv Eng.* **7** 3 315–20

[10] Dehghani K, Ehrlich R 2001 Evaluation of the Steam-Injection Process in Light-Oil Reservoirs *SPE Reserv Eval Eng.* **4(5)** 395–405

[11] Benard JH, Richardson WC, Sitton GM 1998 Steam Oil Displacement and Vaporization of Minas Oil in Slimtubes *SPE/DOE Improved Oil Recovery Symp. (Tulsa)*

[12] Kumar S, Tandon R, Beliveau D, Kumar P, Vermani S 2008 Hot Water Injection Pilot : A Key to the Waterflood Design for the Waxy Crude of the Mangala Field *International Petroleum Technology Conf. (Kuala Lumpur)*

[13] Tiwari S, Verma SK, Karthik R, Singh AK, Kumar S, Singh MK, et al 2014 In-situ heat generation for near wellbore asphaltene and wax remediation *International Petroleum Technology Conf. (Doha)*

[14] Shuhong W, He L, Wenlong G, Dehuang S, Yu Q, Lihua L 2008 Steam Injection in a Waterflooding , Light Oil Reservoir *International Petroleum Technology Conf. (Kuala Lumpur)*

[15] Shuhong W, Wenlong G, Desheng M, Dehuang S, Jinzhong L, Xiaojin W 2008 Utilizing Steam Injection to Improve Performance of Mature Waterflooding Reservoir *SPE Asia Pacific Oil and Gas Conf. and Exhib. (Perth)*

[16] Chavez MR, Glatz G, Clemens T, Kovscek A 2015 Near Wellbore and Reservoir Processes in In-Situ Combustion *EUROSPEC (Madrid)*

[17] Goenka SK, Singhal J, Kothiyal MD, Parasher A, Tiwari S 2014 Near Wellbore Asphaltene and Wax Remediation Using In-Situ Heat Generation in Both PCP and Non-PCP Wells : Concept , Operational Challenges & Remedial Solutions *SPE Asia Pacific Oil & Gas Conf. and Exhib. (Adelaide)*

[18] Taber JJ, Martin FD, Seright RS 1997 EOR Screening Criteria Revisited—Part 2: Applications and Impact of Oil Prices *SPE Reserv Eng*. **12(3)**199–206

[19] Nian Y, Cheng W, Li T, Wang C 2014 Study on The Effect of Wellbore Heat Capacity on Steam Injection Well Heat Loss *Applied Thermal Engineering* **70(1)** 763-69

[20] Gu H, Cheng L, Huang S, Li B, Shen F, Fang W, Hu C 2015 Steam Injection for Heavy Oil Recovery: Modeling of Wellbore Heat Efficiency and Analysis of Steam Injection Performance *Energy Conversion Management* **97** 166-77

IOP Conf. Series: Materials Science and Engineering **306** (2018) 012041 doi:10.1088/1757-899X/306/1/012041

Analysis of Axial Turbine Pico-Hydro Electrical Power Plant in North Sulawesi Indonesia

F J Sangari* and P T D Rompas
Universitas Negeri Manado, Tondano 95618, Sulawesi Utara, Indonesia

*ferry_sangari@yahoo.com

Abstract. This study presents analysis of pico-hydro electrical power plant in North Sulawesi, Indonesia. The objective of this study is to get a design of axial turbine pico-hydro electrical power plant. The method used the study of literature, survey the construction site of the power plant and the characteristics of the location being a place of study, analysis of hydropower ability and analyzing costs of power plant. The result showed that the design of axial turbine pico-hydro installation is connected to a generator to produce electrical energy maximum can be used for household needs in villages. This analyze will be propose to local government of Minahasa, North Sulawesi, Indonesia.

1. Introduction

North Sulawesi region more specifically in the villages of Minahasa district has a mountainous topography and has many rivers which is a potential source of enormous energy for power plants which, when carefully planned can overcome the problem of electric energy crisis. Problems that have long and at this time every day power outage for about 2-3 hours a day. However, the electricity crisis was not so much solved using the integral energy source potential flow of river water in the area of North Sulawesi. There are still many villages far from urban areas still do not have adequate power supply. In anticipation of that, it is necessary to build small-scale power plants (1 kW - 5 kW). **Figure 1** shows that the electrical energy production by 2013 in the province of North Sulawesi using water power is still very little 9.02% [1]. It shows that the construction of the hydroelectric power plant is still very much needed in the area of North Sulawesi. Shortage of electricity in rural areas is very likely to occur because it is far from the urban and the power grid, but did not rule urban areas are also experiencing the same thing. In fact, many cities and districts that rely on diesel and hard to come by when the oil, resulting in a power outage in rotation may even are expanded. One solution is emerging development in Indonesia at this time is to find a way out through the construction of power plants micro scale with the power source stream flow and more dependable again when rural communities require the construction of power plants as small as possible, namely less than 5 kW and that can be realized through pico hydro power plant (PLTPH).

Based on the analysis of the energy force of the waterfalls and design turbines, all of which were made through experiments in the laboratory and in the field (in the village where the installation location) to the mounting PLTPH and generate electrical energy production is less than 5 kW is an invention/innovation targeted in research this. For the creation of the PLTPH development, it previously had to be done: first, about the theoretical studies as the basis for analysis through literature; second, conduct survey research sites (including survey the village is selected villages that have the potential to obtain characteristics of the villages and communities for the purpose of character education for these communities through socialization to shape the character of the community so that there is a sense of belonging and accept, help build and maintain the power plant in village) in the form of: discharge of river water, river water dam layout, long conductive flow of water to the location of the water falls, the water level fell, and location of turbines; Third, experiments in laboratory experiments in the form of high variation of water falling; Last, economic analysis and supply to the local government Minahasa district for expansion into other villages potentially built PLTPH. The objectives of this research is to get the design of efficient and effective from axial turbine models for installation in PLTPH which generates electrical energy production of less than 5 kW for a household in the village Tinoor Minahasa North Sulawesi, Indonesia.

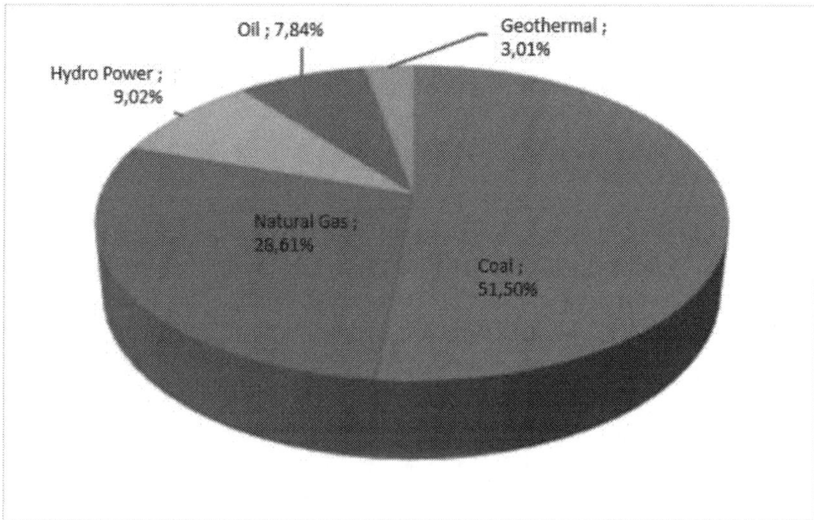

Figure 1. Production of electric energy in North Sulawesi.

Pico hydro is hydro power with a maximum electrical output of 5 kW [2][3][4]. The system is beneficial in terms of cost and simplicity from a different approach in the design, planning and installation from the power applied to the water is greater. The latest innovations in pico hydro technology have made an economic powerhouse even in some of the poorest places in the world and can be accessed. It is also a versatile resource. Alternating current electricity possible can be produced from a standard electrical equipment to be used. Common examples of devices that can be powered by pico hydro are light bulbs, radio and television. Typically, pico-hydro power systems are found in rural areas or hilly [5][4]. This system will operate using a container of water on which a few meters from the ground. From the reservoir, the water flows downhill through the piping system and it allows the water to turn turbines. Thus, the turbine will rotate the alternator to produce electricity. However, this study was conducted to demonstrate the potential of consuming water that is distributed to homes in rural areas as an alternative renewable energy source. The flow of water in the pipe has the potential and kinetic energy will be converted into the potential energy of motion of the turbine which then into electrical energy in generators. Technically, pico hydro has three main components: water (as a source of energy), turbine and generator [6][7]. Pico hydro get energy from the flow of water that has a certain altitude difference. Basically, pico hydro utilized the potential energy of water falling (head). The higher the water falls, the greater the potential energy of water that can be converted into electrical energy. In addition to geographical factors (layout of the river), the height of falling water can also be obtained by stemming the flow of water so that the water level is high. Air flowed through a pipe plant rapidly into the house in general was built on the banks of the river to drive turbines or waterwheels pico hydro. Mechanical energy derived from rotation of the turbine shaft is converted into electrical energy by a generator [8].

Reaction turbines operate under pressure inside the stator (casing). When water passing through the stator in the direction of the turbine shaft helical, causing a whirlpool. The flow was then directed by the blades of the turbine wheel. The angular momentum of the forces in the water rotates the turbine wheel. In contrast to the impulse turbine, the water pressure drops in the stator and the turbine wheel. Examples of a reaction turbine is Propeller (propeller), Kaplan, and Francis, Screw and kinetic turbines water (used to lower head is less than 5 m). Reaction turbines often have houses and geometry turbine blades of a complex which makes it more difficult to process large-scale production of the smallest in the settings in developing countries [9]. However, the reaction turbine can perform well even in low head distance of less than 10 m, thus making it more desirable since the low head of water resources are more accessible and closer location.

Turbine power (Pt) generated:[6][7]

$$Pt = \rho \, g \, H \, Q \eta_t \, (W) \tag{1}$$

Calculation of Electric Power and Energy:

Power turbine shaft

$$Pt = 9.81 \times Q \times H \times \eta turbin \tag{2}$$

Power is transmitted to the generator

$$Ptrans = 9.81 \times Q \times H \times \eta turbin \times \eta belt \tag{3}$$

Power generated generator

$$P' = 9.81 \times Q \times H \times \eta turbin \times \eta belt \times \eta gen \tag{4}$$

2. Methods

The method used literature study, site survey research (including survey the village in order to obtain the characteristics of villages and rural communities Tinoor for the purpose of character education for these communities in receiving, helped build and maintain the power plant), analysis beginning on the ability of electric power, and discussions with local authorities. Conducting preparatory activities before carrying out such research; reflecting the result of socialization, preparation of materials and experimental tools, setting work schedules. Data collection in the village Tinoor form: discharge of river water, water velocity, channel length conductor, high waterfalls, and location of the turbines.

Methods of direct observation in the field through measures such as speed streams and cross-sectional area perpendicular to the water flow of the river to get water discharge flowing river as initial data in the analysis of the ability of river water, and then to analyze the electrical energy taken preliminary data height measurements falling water (planned 1.63 m) including measuring the distance from the dam to the water fall. Technique of direct measurement with the following procedure: first measure the water velocity and the second measuring cross-sectional area perpendicular to the flow of river water in order to get water discharge (cross-sectional area multiplied by the speed of the water, m^3/s), and the last measure the height of falling water to get the length of the aqueduct of dam water to the water fall.

3. Results and discussion

This generator power electric generated will be distributed to users. In planning the required amount of power at the load center should be under the power capacity is raised, so that the power supply voltage is stable and the system becomes more reliable (long). The water flowing with capacity and a certain height distributed to the house installation (casing). At home the turbine, the water plant will pound turbine, turbine ascertained in this case will receive the water energy and convert it into mechanical energy in the form of turbine shaft rotation. The rotating shaft is then connected to the generator by using the tire/belt. Of the generator will produce electricity that will go into the control system before the electrical current supplied to homes or other purposes (load). That briefly the process of pico-hydro, alter energy flow and water level into electrical energy [7].

The results of the analysis of water and high power capabilities of falling water according to calculation on a gross basis before further analysis can be seen in **Table 1** as follows:

Table 1. Results of the analysis of water and high power capability falling water.

Q (m^3/dt)	H_{bruto}(m)	H_{losses}(m)	H_{eff}(m)	P(kW)	P'(kW)
0.236	1.63	0.23	1.4	2.69	1.85

Water discharge (Q) in **Table 1** is a water discharge pipe is planned to enter the rapidly with height of 1.63 m water fall and if we calculate minimum water flow of 20% of the water flow was then obtained 91.6 l / sec. Generates power without taking into account the total efficiency of 2.69 kW and if we take into account the total efficiency of 0.643, the installed power of 1.85 kW.

Table 2. Results of the analysis of electric energy.

$\sum\eta$	H_{eff}(m)	Q_{80} (m^3/dt)	Q_{90} (m^3/dt)	Q_{100} (m^3/dt)	$\sum$E(MWh)
0.643	1.4	0.236	0.207	0.133	19.295

Table 2 shows the results of the analysis of electrical energy which is calculated based on the total efficiency, the force of gravity, high falls and the actual effective water obtained from the difference between the height of fall slop and total loss of height of falling water (0.1 m previously planned 0.23 m). The total energy obtained during a year high real effective water fall of 1.4 m was 19.295 MWh with installed power of 1.85 kW.

Total electrical energy obtained within a year of 19,295 kWh (Table 2). If we calculate the value of selling electricity to PLN by calculating the total cost of expenditure per year of IDR 25 million, the value of the electricity sold at IDR 1295/kWh.

4. Conclusion

The ability of hydroelectric power of 2.69 kW is installed or the electrical power produced by the force of the water. The head of effective of 1.4 m with the generated power of about 1.85 kW with water flow of 0.236 m^3/s. The total energy is obtained within a year was 19.295 MWh.

References

[1] PT PLN (Persero) 2016 Statistik PLN 2013. Sekretariat Perusahaan PT PLN (Persero). ISSN 0852–8179.NO.Publikasi:02601 140722 [Online] *http://www.pln.co.id/dataweb/STAT/STAT2013IND.pdf.* Accessed on July 05, 2017.

[2] Basar M F, Sapiee R, Rahman S, Hamdan Z, Borhan S and Sopian K 2014 Cost Analysis of Pico hydro Turbine for Power Production *Advances in Environmental Biology* **8** 14 147-151

[3] Martin S and Sharma A B 2014 Analysis on Rainwater Harvesting and its Utilization for Pico Hydro Power Generation *International Journal of Advanced Research in Computer Engineering & Technology (IJARCET)* **3** 6 2121-2126

[4] Zainuddin H, Yahaya M S, Lazi J M, Basar M F M and Ibrahim Z 2009 Design and Development of Pico-hydro Generation System for Energy Storage Using

Consuming Water Distributed to Houses *World Academy of Science, Engineering and Technology* **59** 154-159

[5] Maher P and Smith N 2001 *Pico Hidro Potencia para Aldeas: Un Manual Práctico para Instalaciones de hasta 5 kW en Terrenos de Pendientes Fuertes.* Ed. 2.0. [Online] *http://www.riaed.net/IMG/pdf/Pico_Hidro_-_Potencia_para_Aldeas_1.pdf.* Accessed on July 25, 2017

[6] Kapoor R 2013 Pico Power a Boon for Rural Electrification *International Journal of Scientific Research* **2** 9 159-160

[7] Nimje A A and Dhanjode G 2015 Pico-Hydro-Plant for Small Scale Power Generation in Remote Villages *IOSR Journal of Environmental Science, Toxicology and Food Technology (IOSR-JESTFT)* **9** 1 59-67

[8] At-Tasneem M A, Azam W M and Jamaludin U 2014 A Study on the Effect of Flow Rate on the Power Generated by a Pico Hydro Power Turbine *Word Applied Sciences Journal 30 (Innovation Challenges in Multidiciplinary Research & Practice)* 420-423

[9] Sangal S, Garg A and Kumar D 2013 Review of Optimal Selection of Turbines for Hydroelectric Projects. *International Journal of Emerging Technology and Advanced Engineerng* **3** 3 424-430

Project-Based Learning in Programmable Logic Controller

F R Seke[1,*], J M Sumilat[2], D R E Kembuan[3], J C Kewas[4], H Muchtar[5] and N Ibrahim[5]

[1]Electrical Engineering Department of Universitas Negeri Manado. Kampus Tondano 95618. Minahasa.
[2]Primary Education Department of Universitas Negeri Manado. Kampus Tomohon.
[3]Civil Engineering Education Department of Universitas Negeri Manado. Kampus Tondano 95618. Minahasa.
[4]Mechanical Engineering Department of Universitas Negeri Manado. Kampus Tondano 95618. Minahasa.
[5]Education Technology Department of Universitas Negeri. Kampus Rawamangun. Jakarta Timur.

*roykeseke@yahoo.com

Abstract. Project-based learning is a learning method that uses project activities as the core of learning and requires student creativity in completing the project. The aims of this study is to investigate the influence of project-based learning methods on students with a high level of creativity in learning the Programmable Logic Controller (PLC). This study used experimental methods with experimental class and control class consisting of 24 students, with 12 students of high creativity and 12 students of low creativity. The application of project-based learning methods into the PLC courses combined with the level of student creativity enables the students to be directly involved in the work of the PLC project which gives them experience in utilizing PLCs for the benefit of the industry. Therefore, it's concluded that project-based learning method is one of the superior learning methods to apply on highly creative students to PLC courses. This method can be used as an effort to improve student learning outcomes and student creativity as well as to educate prospective teachers to become reliable educators in theory and practice which will be tasked to create qualified human resources candidates in order to meet future industry needs.

1. Introduction

Learning that has more impact on improving learning outcomes are student-centered learning [1]. One of the student-centered learning methods is the project-based learning method. Project-based learning begins with a task to perform one or more tasks that lead to the production of the final product [2, 3]. Student participation in project work encourages creativity, critical thinking, and teamwork [4]. PLC learning discusses the design, assembly, and programming of an electronic-based control system using a fairly complicated logic. But the tendency is the students just memorize the concept alone without a thorough understanding of the subject matter. How to learn like this will not be able to create a candidate electrical engineering teacher who will be tasked to produce quality human resources to answer future needs.

Practical work is needed in the learning of electrical engineering education, because with practical work will make the students directly involved and have knowledge that is more meaningful than learning by memorizing. To overcome this problem some researchers have suggested some suitable learning methods to be applied in the learning of technical education such as problem-based learning [1], cooperative learning [5-8] and project based learning [9]. Although some of the preferred learning methods provide good results the application of such learning methods should be supported with adequate laboratory facilities while existing conditions do not support them.

The aims of this study is to investigate the influence of project-based learning methods on students with a high level of creativity in learning the PLC. The application of project-based learning methods with observe to the level of creativity of students on PLC course is intended to provide experience for students who can become skill as well as knowledge so that after studying PLC courses students can achieve optimal learning outcomes [10] or understand the concept and practice of automatic control process even in the implementation of the learning process has limitations on the availability of laboratory equipment. However, in the application of this method of learning, the problem is found that the availability of practicum tools that have not been adequate. In the application of this methods of learning, to overcome the problem lack of tools in our laboratory make a practical supporting tool in the form of board trainer to facilitate practical work of the students in studying PLC. Unlike other PLC laboratory equipment, this board trainer is so simple that it can help students through the application of project-based learning methods to better understand the completion of PLC design, assembly and programming projects. This board trainer renewal makes the implementation of project-based learning methods suitable for application in developing countries with various limitations in the field of learning facilities and infrastructure.

2. Methods

The method used in this research is an experimental method with care design at level 2 x 2.Dependent variable is Student learning outcomes in the course of PLC, the independent variables are 1) learning methods consisting of project-based learning methods and simulation learning methods, 2) the level of student creativity, differentiated with high and low creativity.

The population in this study is sixth-semester students majoring in electrical engineering education, faculty of Engineering Universitas Negeri Manado academic year 2014/2015 with total 80 people. The sample in this study followed the recommendation from Popham that high and low groups were selected by selecting 30% [11] from 80 students of which 24 students with high creativity divided by 2, so 12 students entered the experimental class with the treatment of project-based learning methods and 12 other students entered the control class with Treatment of simulated learning method, as well as 24 students who have low creativity level. This research was conducted at the Department of Electrical Engineering faculty of Engineering Universitas Negeri Manado in Tondano which lasted for 1 semester which runs on the even semester from January to the beginning of July 2015.

The technique of collecting learning result data is obtained through the test of learning result which consists of 50% written test and 50% practice test. The written test is in form 30 objective questions, with weight 1 for correct answer and 0 for the wrong answer, while for practice test is taken through observation sheet with maximum score 3 for best condition and 0,6 for bad state, as well as techniques of collecting student creativity data obtained through a creativity questionnaire consisting of 30 statements using a scale of 1-5.

Validity test is done in two forms, namely testing the validity of constructs/content and testing of empirical validity. Testing the validity of the constructs/contents is done with expert analysis and empirical validity testing with the correlation of grain scores with the total score of test results test. Team experts for analysis the test of learning outcomes is: Prof. Dr. Ph. Tuerah, M.Si DEA, Prof. Dr. H. Sumual, M.Si, Dr. Ir. V. R. Palilingan, M.Eng, and for the instrument of creativity is Prof. Dr. Yufiarti and Dr. Syuul T. Karamoi, M.Si, Respondents expert review consists of Respondents were asked to provide an assessment of learning outcomes on the following aspects: 1) Question item according to indicator, 2) Problems are clearly defined, 3) The language domain, 4) The formulation of a sentence does not give rise to a double understanding, 5) Material compatibility. After testing the validity of the construct followed by empirical validity testing. The test of the instrument is carried out on the students majoring in electrical engineering education with the number of 30 students in the semester VIII who have passed the PLC courses. Implementation of test questions carried out before the implementation of the experiment and who conducted the trial is a researcher and assisted by a team of lecturers majoring in electrical engineering education. The formula used to test the validity of written test learning result is the correlation of biserial point, and for practice test is product moment correlation, and for creativity instrument is product moment correlation. The conclusion of the test results of validity of the instrument written test learning results are 30 valid questions and 2 invalid questions that are the number 25 and 29, and for practice test is stated all valid items, as well the results of validity test of creativity instrument there are 30 items valid and 6 invalid items are questioned number 6, 7, 8, 11, 30, 34. In this research, internal reliability test is done once and using one instrument the formula used for the written test is Kuder-Richardson 21, as well Reliability test used for product/practice test and creativity instrument is a coefficient of Alpha. The results of the reliability test of the written test show the high reliability because is the value of r11 = 0.87884, for product/practice test show high reliability because is the value of r11 = 0.75369, and creativity test showed the reliability is very high because the value of reliability test is 0.8187656.

Data analysis used in this research include descriptive analysis, prerequisite test analysis, and inferential analysis. Data to be analyzed with inferential statistics are assumed to be normally distributed. Normality test of data distribution is done by using Lilliefors test. Homogeneity measurements were performed to test data of groups A1B1, A2B1, A1B2, A2B2 using Bartlett test. Hypothesis testing was done by variance analysis technique of two variants of ANAVA and continued with the advanced test of Tukey test, with the aim to test the influence of learning method and creativity level to the learning result.

3. Results and discussion

The research data described is the score data of student learning outcomes. The data normality test is done on the data of learning result in eight data groups that are group A1, A2, B1, B2, A1B1, A1B2, A2B1, A2B2. Normality test of this data is done through Lilliefors test with significance level α = 0.05. Result of normality test for group A_1 = 0,1427; A_2 = 0,1269; B_1 = 0,1048; B_2 = 0,1180; A_1B_1 = 0,1468; A_1B_2 = 0,1681; A_2B_1 = 0,1303; A_2B_2 = 0,1683;The normality test results from the eight data sets indicate L count < L table, which means accepting Ho as well as it can be concluded that all data groups are normally distributed or it can be concluded that all groups of data come from normally distributed populations.

The homogeneity test was conducted using Bartlett Test. The data used in this analysis is the value of learning outcomes in eight groups of data that is group A1, A2, B1, B2, A1B1, A1B2, A2B1, A2B2. The result of the homogeneity calculation $\chi^2{}_{count}$ = 1,89952987 < $\chi^2{}_{table}$ = 7,82 so it is concluded that all populations have the same or homogeneous variance, thus the inferential statistical test requirements are met. Thus the requirements of inferential statistical tests are met and can be continued on hypothesis testing. The research hypothesis test used two-factor analysis of variance called Two Way ANOVA, with the aim to test the influence of main effect and interaction effect between the PLC learning result.

The result of calculated by two-lane variance analysis and the results of the calculation of hypothesis testing of the study are presented in the table 1.

Table 1. Two-Line ANAVA calculation result.

Source Varians	Dk	JK	RJK	F_h	F_{Table}	
					α 0,05	α 0,01
Inter Colom (A) (Learning Methods)	1	438,0225	438,0225	33,93**	4,06	7,24
Between Line (B) (Creative)	1	414,1875	414,1875	32,08**	4,06	7,24
Interaction	1	71,53	71,53	5,54*	4,06	7,24
Between Group	3	923,74	307,91	23,85**	2,82	4,26
In Group	44	567,83	12,91		-	-
Total Reduktion	47	1491,57			-	-

Description:*= Significant, ** = Very Significant

Based on the data in table 1, the results of two-way ANOVA calculation shows the results testing the effect of learning method on the learning outcomes of PLC, then obtained the value of F_{count} = 33.93 for the real level α = 0.05 and dk = 1; 44 Obtained F_{table} = 4.06. If both values of F are compared then it is F_{count} = 33.93>F_{table} = 4.06. The interaction hypothesis to be tested is the influence of interaction between the learning method and the level of creativity to the learning outcomes of PLC. Based on the results of interaction hypothesis testing as presented in table 1, which shows the results of testing the interaction of learning methods and creativity to the learning results of obtained F_{count} = 5.54 for the real level α = 0.05 and dk = 1, 44 obtained F_{table} = 4.06. If both values of F are compared then the F_{count} = 5.54>F_{table} = 4.06, which means H_0 is rejected and receives H_1. This shows that there is an influence of learning method interaction with the level of creativity of student to learning result of PLC. These results show that the application of project-based learning methods and simulation learning methods provide different PLC learning outcomes when applied to students who have a high level of creativity and low creativity. Therefore, the hypothesis of interaction tested then further tested further to the simple hypothesis (simple effect).

Further testing of the simple hypothesis (simple effect) done on 2 hypotheses that is a test of simple hypothesis difference of learning result of PLC between all level of factor of a learning method and for each creativity factor. Tukey test calculation results are presented in the table 2.

Table 2. Results of advanced stage analysis with Tukey Test.

Comparison Group	**n**	**K**	**dk**	q_h	q_t	
					α 0,05	**α 0,01**
A_1B_1 and A_2B_1	12	4	3.44	8,187**	3,44	4,28
A_1B_2 and A_2B_2	12	4	3.44	3,472*	3,44	4,28

Based on the values in table 2 that shows the results of testing the effect of learning method based on high level of creativity to PLC learning results obtained q_{count} = 8.187. Using the real level α=0.05 and dk = n - 2=24-2= 22, obtained q_{table} (0.05: 22) = 3.44. If both values of q are compared, then it turns out q_{count} = 8.187>q_{table} = 3.44, which means H_0 is rejected and receives H_1. Thus it can be concluded in the group of students who have high creativity, student learning outcomes PLC taught by project-based learning method is higher than the results of student learning taught by simulation learning method

Based on the values in table 2 which addressed the results of testing the effect of learning method based on the low level of creativity to PLC learning results obtained q_{count} = 3.472 by using the real level α = 0,05 and dk = n - 2 = 24 -2 = 22, obtained q-table (0,05: 22) = 3,44. If both values of q are compared, then it turns out q_{count} = 3.472>q_{table} (0.05: 22) = 3.44 which means H_0 is rejected and receives H_1. Thus it can be concluded that in groups of students who have low creativity, student learning outcomes of PLC taught with project-based learning methods is lower than the learning outcomes of PLC taught by simulation learning method.

The results of the study as described previously basically prove the project-based learning method is one superior learning method to apply on highly creative students to PLC course. The result of this analysis showing that the learning result of PLC group of students taught by project-based learning method is higher than a result of PLC learning group of students taught by simulation learning method. The results of this study prove that project-based learning is a learning that can stimulate learners to think critically [12] and create a flexible learning environment that can trigger the achievement of optimal learning outcomes. In learning of the PLC, students must become familiar with critical thinking because in automatic controllers design and programming suppose a structured logical arrangement to construct a single command in that controller's operation. With project work that is the heart of the application of project-based learning methods, the students has more enthusiastic to completing the project who had gave before, so that students know for themselves how to use the program who they has designed and built [13]. This shows their liveliness in learning has led them to build their skills as well as their knowledge in PLC learning [12], so it can be the engineers who are able to overcome industry problems and be able to face technical challenges in the future [14].

The result on this research is the application of project-based learning methods who treat by students with high creativity better than the application of simulation learning method in PLC learning outcomes. This is indicates that in the application of project-based learning methods are needed on student creativity, to generate brilliant ideas in designing and program in PLC learning. The result of this study are also similar with the research of Frank [15], where in the learning process students are required to complete a mini project and make a report, and the lecturers only provide direction at the initial meeting on how to prepare the mini project report and then to guide them to make a final report. This method can be students more competent to acquire skills through the experiences they discover during the learning process [15].

The challenge as a professional teacher candidate who will be tasked to bring golden generation to fulfil industry demand in the future will be achieved when the learning process that exist in electrical engineering education focuses on effective learning, that is learning who has a core on creativity development and decision making which are all there in project-based learning methods [16] which to focus on the attainment of skills and knowledge through direct student involvement in the learning process not only memorizing theories, but when student directly engage in the knowledge they experience will last longer in their memories and stimulate creativity they are working or completing design and programming PLC project better than just memorize PLC programming instructions.

The successful application of project-based learning methods to students with high creativity also lies in the ability of lecturers [17] to effectively designing the learning process, emphasizing motivation, and supporting their project work, guide them during the learning process that can reduce their doubts [18] to finding solution for finishing their project so that can be meaningful experience on skill or knowledge who will support their learning outcomes [4].

4. Conclusion

The application of project-based learning methods into the PLC courses combined with the level of student creativity enables the students to be directly involved in the work of the PLC project which gives them experience in utilizing PLCs for the benefit of the industry. Therefore, it's concluded that project-based learning method is one of the superior learning methods to apply to PLC courses. This method can be used as an effort to improve student learning outcomes and student creativity as well as to educate prospective teachers to become reliable educators in theory and practice which will be tasked to create qualified human resources candidates in order to meet future industry needs

IOP Conf. Series: Materials Science and Engineering **306** (2018) 012042 doi:10.1088/1757-899X/306/1/012042

References

[1] Prince M 2004 Does active learning work? A review of the research *Journal of engineering education* **93** 3 223-231
[2] Prince M J and Felder R M 2006 Inductive teaching and learning methods: Definitions, comparisons, and research bases *Journal of engineering education* **95** 2 123-138
[3] Helle L, Tynjälä, P, and Olkinuora E 2006 Project-based learning in post-secondary education–theory, practice and rubber sling shots *Higher Education* **51** 2 287-314
[4] Bell S 2010 Project-based learning for the 21st century: Skills for the future *The Clearing House* **83** 2 39-43
[5] Millis B and P Cottell Jr 1998 *Cooperative Learning for Higher Education Faculty* American Council on Education, ORYX Press.
[6] Feden P and R Vogel 2003 *Methods of Teaching: Applying Cognitive Science to Promote Student Learning*, McGraw Hill Higher Education
[7] Johnson D, R Johnson and K Smith 1998 *Active Learning: Cooperation in the College Classroom* 2nd ed. Interaction Book Co. Edina, MN
[8] Johnson D, R Johnson and K Smith 1998 Cooperative Learning Returns to College: What Evidence is There That it Works? *Change* **30** 4 26–35
[9] Mills J E and Treagust D F 2003 Engineering education—Is problem-based or project-based learning the answer *Australasian journal of engineering education* **3** 2 2-16
[10] Rais M 2017 Development of Learning Media Based on Active Learning for Principles Design Subject to Conduct Students Intellectual Skills *International Seminar Reformulating the Paradigm of Technical and Vocational Education* **1** 1 1221-1232 Badan Penerbit Universitas Negeri Makassar
[11] James P W 1981 *Modern Educational Measurement* Englewood Cliff, NJ: Practice Hall, Inc. 296
[12] Doppelt Y 2003 Implementation and assessment of project-based learning in a flexible environment *International Journal of Technology and Design Education* **13** 3 255-272
[13] Horváth I, DUHOVNIK J E, and Xirouchakis P 2003 Learning the methods and the skills of global product realization in an academic virtual enterprise *European Journal of Engineering Education* **28** 1 83-102
[14] Bellmunt O G, Miracle D M, Arellano S G, Sumper A, and Andreu a S 2006 A distance PLC programming course employing a remote laboratory based on a flexible manufacturing cell *IEEE Transactions on Education* **49** 2 278-284
[15] Frank M, Lavy I and Elata D 2003 Implementing the project-based learning approach in an academic engineering course *International Journal of Technology and Design Education* **13** 3 273-288
[16] McLaren S and Dakers J 2011 *A Curriculum for Excellence: Review of research literature* Design and Technology Education
[17] Kokotsaki D, Menzies V and Wiggins A 2016 Project-based learning: A review of the literature *Improving Schools* **19** 3 267-277
[18] Hmelo-Silver C E, Duncan R G, and Chinn C A 2007 Scaffolding and achievement in problem-based and inquiry learning: A response to Kirschner, Sweller and Clark (2006) *Educational Psychologist* **42** 99–107

The Influence of Training Strategy and Physical Condition toward Forehand Drive Ability in Table Tennis

F W Langitan
Universitas Negeri Manado, Tondano, 95618, Sulawesi Utara, Indonesia

fentjewelliamlangitan@gmail.com

Abstract. The objective of this research is to find out the influence of training strategy and physical condition toward forehand drive ability in table tennis of student in faculty of sport in university of Manado, department of health and recreation education. The method used in this research was factorial 2x2 design method. The population was taken from the student of Faculty of Sport at Manado State University, Indonesia, in 2017 of 76 students for sample research. The result of this research shows that: In general, this training strategy of wall bounce gives better influence toward forehand drive ability compare with the strategy of pair training in table tennis. For the students who have strong forehand muscle, the wall bounce training strategy give better influence to their ability of forehand drive in table tennis. For the student who have weak forehand muscle, pair training strategy give better influence than wall bound training toward forehand drive ability in table tennis. There is an interaction between training using hand muscle strength to the training result in table tennis using forehand drive.

1. Introduction

Table tennis is one of the favorite sport in community, especially in school and also in university. This is not only because of this sports is part of school/ university curriculum but table tennis is also an exciting sport that can be played outdoors or indoors (inside a house) with cheapper equipments [1]. It also doesn't required a large space. It can be played by all family member, gives body health for young or old age, entertaining and for the disables can play this kind of sport.

However, Table tennis required complete equipments to achieve a better performance. It needs training strategy, mastering of training technique. A good athlete who wants to achieve a high performance should have a good phisical condition : strength, endurance, muscular power, speed, coordination, flexibility, agility, balance, accuracy, reaction. Moreover, mastering the basic techniques, including forehand strike, needs to be practiced systematically and continously. Harsono stated that "The main objective in training is to maximize the improvement of one's skills" [2]. Thus, with training strategy, physical condition and good basic technique, table tennis could become a dynamic sport and rich of movement.

One of the strikes in table tennis is forehand drive. It's a facing front hand stroke. *Forehand* is commonly used by righthand and hit from right side [3]. The definition of *Forehand drive*according to Larry Hodges in Elementary book of Tabel Tennis, is a stroke using paddle and it's moved to the right elbow direction for the right-handed and to the left for left-handed [4]. Peter Simpson in the book Ping Pong Technique, explain briefly that *forehand drive* is a stroke in right side [5].

The purpose of this research is to test the influence of (wall bounce and pairing) strategy and condition of physical strength of arm muscle towards forehand stroke in table tennis to students of FIK Manado state university. The research of Tjatjo [6], about wall bouncing method and pairing method, is also discussed in my research about the strategy off wall bouncing practice and pairing practice strategy [7].

2. Methods

This research was carried out in sport center of Manado state university in Tondano, Minahasa north Sulawesi. The research duration was 8 weeks with the frequency three times per week, 18 meetings. Each meeting is 120 minutes. The research method is Experimental method with factorial 2x2 design [8]. The variables using in this research were free variable, attribute variable, given variable as described above: Given variable is forehand drive. Free variables were practice strategy [9] e.i. (i) the strategy of wall bounce training and (ii) pairing training. The population of this research is all the students of FIK UNIMA Tondano, in 2016/2017. The main target population of this research is 80 students of sports education department, as the total sampling. The data analyzing technique used in this research is two ways variant analyzing technique (ANAVA), significant $\alpha = 0.05$. The requirements in using this analyzing is normality and homogeneity test. Lillefors test for Normality and Bartlet for homogeneity test. If there is an interaction, the test will continue using Tukey test [10].

3. Results and Discussion

The analyzing using ANAVA technique [11] showed that $F_{observed} = 4.40$, bigger than F_{table}, 4.11 ($F_{observed} = 4.40 > F_{table} = 4.11$), that is why H_0 is rejected and H_1 is accepted. From the calculation above it concluded that overall, there is a significant influence between wall bounce training strategy with paring training strategy in table tennis. In other words it is concluded that the ability of training result of forehand drive in table tennis using wall bounce technique ($X = 29.6$ and $S = 4.25$) is better than the ability using pairing training ($X = 28.45$ and $s = 2.86$). The first research hypothesis states that Glass.

The calculation about different influence of ability using forehand drive in table tennis to a group that has muscular strength using the technique of wall bounce training and pairing training (P1 : P2). The conclusion result using Tukey test as showed in **Table 1**.

Table 1. The conclusion result using Tukey test.

No	Group for Comparison	$Q_{calc.}$	Q_{table}	Notes
1	P1 with P2	4,49	3,79	Significant
2	P4 with P3	0,80	3,79	Significant

The Table 1 shows that $Q_{calc.} = 4.49$ bigger than $Q_{table} = 3.79$ or $Q_{calc.} > Qtable$ in significance level, therefore zero hypotheses (H_0) was denied and alternative hypothesis (H_1) was accepted. It means that the results in forehand in table tennis from the group with strong muscle, trained with the wall bounce strategy ($X = 33.3$ and $S = 1.90$) was higher than pairing strategy ($x = 30.5$ and $s = 1.90$). This proved that the group with strong muscle trained with wall bounce strategy had higher forehand capability compared with paired training strategy.

The next analysis was using Turkey Assessment to compare groups with weak muscle. These two experiments referred to the opinion of Gane V. Glass. It was about calculation on the difference influence to the table tennis forehand for the group with strong muscles, trained pairing and wall-bounce strategy (P4 : P3).

The **Table 1** shows that $Q_{calc.} = 0.80$ bigger than $Q_{table} = 3.79$ or $Q_{calc.} > Q_{table}$ in significance level, therefore zero hypothesis (H_0) was denied and alternative hypothesis (H1) was accepted. It means that the results in forehand in table tennis from the group with strong muscle, trained with the wall bounce strategy ($X = 26.4$ and $S = 2.07$) was higher than pairing strategy ($X = 25.9$ and $S = 2.02$). This proved that the group with strong muscle trained with wall bounce strategy had higher forehand capability compared with paired training strategy.

The conclusion showed that from the variety analysis showed the $F_{addition} = 6.99$ and $F_{table} = 4.11$ or $F_{addition} > F_{table}$. Therefore, zero hypothesis (H_0) found that the interaction between training strategy and arm muscle strength with training results from forehand drive in table tennis was dismissed and the alternative hypothesis was accepted. In other words, level of achievement in forehand drive in table tennis was influenced by the existence of interaction between practice strategy and arm muscle strength.

According to the result of this research, the score rate of forehand drive ability in table tennis from the group of those who had strong arm muscle, trained by using strategy of wall bounce practice is 33.3. The group who had weak arm muscle got 25.9. The average score for the ability of forehand drive from those in group of pairing training got score 30.5 and the group which had weak arm muscle got score 26.4.

The hypothesis based on the fourth research stated that there is an interaction between training strategy and arm muscle strength to the ability of forehand drive in table tennis had been proved.

Forehand drive ability is an important technique in table tennis [12]. The modern table tennis requires every player to be able to posses forehand drive technique to develop various spin drive techniques. Therefore, it needs appropriate practice strategy to develop forehand drive. This research used two practice strategies, wall-bounce practice and pairing, which aimed to see which strategy was better than the other to improve forehand drive in table tennis. It found out that these two strategies had similar goals, but were different in implementation.

Wall-bounce training strategy is a training by bouncing the ball to the wall with forehand drive. Pairing practice in other hand, involves two persons with repeating moves.

Results from the movement analysis of these two trainings were supported by results from variant analysis about the overall different in effectiveness, F, Column observation (FA) = 6.99, greater than F table, 4.11 ($F_0 = 6.99 > F_t = 4.11$). Results from the wall-bounce training strategy was (X = 29.6 and S = 4.25) compares to the pairing practice, (X = 28.45 and S = 2.86). Overall, it could be concluded that the wall-bounced training strategy was better than pairing practice in improving forehand drive in table tennis.

The above results from movement analysis were supported by the results from the group with strong arm muscle in wall bounce training (P1) in comparison to the group with pairing practice strategy (P2): $Q_{calc.} = 4.49 > Q_{table} = 3.79$. In other words, students with strong arm muscle had higher effectiveness level with wall-bounce training strategy, (X = 33.3 and SD = 189) compared to the pairing practice (X = 30.5 and SD = 1.90).

The analysis was also supported by the results from pairing strategy (P4) with the group with weak arm muscle trained in wall bounce practice (P3) with the results was following: $Q_{(0.0)} > Q_{table} = 3.79$. In other words, students with weak arm muscles had better strategy effectiveness (X = 26.4 and SD =2.07).

Results from the 2 x 2 alternative variants stated that interaction between training strategy with arm muscle strength in table tennis which showed $F_{observed} = 6.99 > F_{table(0.05)} = 4.11$. This interaction showed that well-bounce training strategy was more suitable to the students with strong arm muscle compared to the pairing practice strategy, A1 B1 > A2B1. In the contrary, pairing practices was suitable to the students with weak arm muscle compared to the wall-bounce training, A2B2. This finding was furthermore supported from the comparison results between wall-bounce training strategy and pairing practice both with strong arm muscle, A1B1 : A2B1 (P1 : P2, $Q_{calc.} = 4.49 > Q_{table} = 3.79$. In other words, effectiveness of wall-bounce training strategy with strong arm muscle (X = 33.3 and SD = 1.89) was better than the pairing practice strategy (X = 30.5 and SD = 1.90). Pairing practice strategy and wall-bounce training strategy both with weak arm muscle: A2B2: A1B2 (P4 : P3), $Q_{calc.} = 0.80 > Q_{table} = 3.79$. In other words, effectiveness of pairing practice strategy with weak arm muscle (X = 26.4 and sd = 2.07) was better than wall-bounce training strategy (X = 25.9 and sd = 2.02).

4. Conclusion

In summary, wall-bounce training strategy gives more and better influence to the results in practicing forehand drive in table tennis, compares to pairing practice. To the students with strong arm muscle, wall-bounce training strategy gives better influence in the results of forehand drive training in table tennis, compares to pairing practice. In the contrary, for the students with weak arm muscle, pairing practice strategy creates better results in comparison to the wall-bounce training strategy. The research finding proves that there was an interaction between training strategies with arm muscle strength towards the results from forehand drive training in table tennis.

Acknowledgments

The authors wish to thanks for Dirlitabmas, Dirjendikti, Ministry of Education and Culture of the Republic of Indonesia which had finance all of the research activities.

References

[1] Santoso T, Setiono H and Sulaiman 2017 Developing Return Board as an Aid for Forehand Topspin in Table Tennis *The Journal of Educational Development*, JED **5** 2 210 – 223

[2] Kondrič M, Zagatto A M and Sekulić D 2013 The Physiological Demands of Table Tennis *Journal of Sports Science and Medicine* **12** 362-370

[3] Jospin L, Fayt V and Lazzari S 2006 *Influence of Training and Task Difficulty On Efficiency of a Forehand Drive in Table Tennis* IV Congreso Mundial de Ciencia y Deportes de Raqueta 21-23

[4] Raab M, Masters R S W, and Maxwell J P 2005 Improving the 'how' and 'what' decisions of elite table tennis players *Human Movement Science* **24** 3 326-344.

[5] Simpson P 1984 *Teknik Bermain Pimpong*, Pioner Jaya: Bandung

[6] Tjatjo A H 1985 *Metode Latihan Pantulan Tembok dan Berpasangan Melalui Pola Distributes dan Massed terhadap Hasil Latihan Keterampilan Bermain Lawan Tennis* Disertasi Pascasarjana UNJ: Jakarta

[7] Decety J, Grèzes J, Costes N, Perani D, Jeannerod M, Procyk E, Grassi F and Faszio F 1997 Brain activity during observation of actions. Influence of action content and subject's strategy *Brain* **120** 10 1763–1777

[8] Safari I, Suherman A and Ali M 2017 The Effect of Exercise Method and Hand-Eye Coordination Towards the Accuracy of Forehand Topspin in Table Tennis *IOP Conf. Series: Materials Science and Engineering* **180** 012207

[9] Brunnett G, Rusdorf S and Lorenz M 2006 V-Pong: an immersive table tennis simulation *IEEE Computer Graphics and Applications* **26** 4 10-13

[10] Reinoud J, Bootsma R, Houbiers M, Whiting H and Wieringen P 1991 Acquiring an Attacking Forehand Drive: The Effects of Static and Dynamic Environmental Conditions *Research Quarterly for Exercise and Sport* **62** 3

[11] Young K H 2014 Analysis of variance (ANOVA) comparing means of more than two groups *Restorative Dentistry & Endodontics* **39** 1 74–77

[12] Bootsma R and Wieringen P 1990 Timing an attacking forehand drive in table tennis *Journal of Experimental Psychology: Human Perception and Performance* **16** 1 21-29

IOP Conf. Series: Materials Science and Engineering **306** (2018) 012044 doi:10.1088/1757-899X/306/1/012044

Data Model Performance in Data Warehousing

G C Rorimpandey[1*], F I Sangkop[1], V P Rantung[1], J P Zwart[2], O E S Liando[1] and A Mewengkang[1]

[1] Department of Informatics, Universitas Negeri Manado, Jl. Kampus FT-Unima Tondano 95618, North Sulawesi, Indonesia
[2] HAN University of Applied Scieces. Ruitenberglaan 31, 6826 CC Arnhem, Netherlands

*gladlycrorimpandey@unima.ac.id

Abstract. Data Warehouses have increasingly become important in organizations that have large amount of data. It is not a product but a part of a solution for the decision support system in those organizations. Data model is the starting point for designing and developing of data warehouses architectures. Thus, the data model needs stable interfaces and consistent for a longer period of time. The aim of this research is to know which data model in data warehousing has the best performance. The research method is descriptive analysis, which has 3 main tasks, such as data collection and organization, analysis of data and interpretation of data. The result of this research is discussed in a statistic analysis method, represents that there is no statistical difference among data models used in data warehousing. The organization can utilize four data model proposed when designing and developing data warehouse.

1. Introduction

Data model has constant change in storing data therefore consistent and stable interfaces are needed for information that spans for a longer period of time [1]. Because of this, performance in data model became one of the essential criteria of the data warehouse that to be future-proof. It is about response time when the data warehouse is executed to analyze a large amount of data. In the beginning, data warehouses have been designed using 2 approaches. First approach is settled by Ralph Kimbal. He defined the Data Warehouse as a copy of transaction data specifically structured for query and analyzes [2]. Second approach is settled by Inmon. He focuses on building a centralized Enterprise Data Warehouses of which data marts sources their information from [3]. A data warehouse stores all of data in a form that prioritizes query performance, rather than transactional performance or storage volumes [4].

This research is trying to extend a preliminary performance comparison between four data models that has been done in January 2012 by Rael Rutto. Its aim is to measure performances and analyze the results statistically to be able to draw well founded conclusions about differences in performance between the various models [5]. Furthermore, Anchor Model which the one of data model used in data warehouses is a technique recently advocated by Lars Rönnbäck. It uses 6 Normal Form (6NF) databases which are generate expected to perform badly. But, in October 2010 Lars Rönnbäck and friends performed the result of their research that Anchor Model performs substantially better than databases constructed using traditional modeling techniques [6]. Furthermore, they claim however that query optimizers (SQL Server) are so powerful that performance issues are no longer important as for as table designs are concerned. Query performance has an important role to increase the performance of data model in data warehouse [7].

We have decided to test the claim against Star Schema [8], Optimal Normal Form (ONF) [9] and Data Vault [10] by using SQL Server 2008 and with the same facts for the four models. Compared to the previous research where ONF model is populated by 100000 rows and transformed them to the three models by using ETL processes [5], this scenario is populated with 200000 rows for each historized tables and some less number for the small and intermediate size of tables as described. The experiment will be done using processor speed Intel® Core ™ i3 CPU M350 @2.27GHz, 2 GB RAM and 300 GB Space of Hard Drive. The aim of this research is to measure performances and analyze the results statistically to be able to draw well founded conclusions about differences in performance among the various models.

2. Methods

Basically, statistic is the science of data which concerned with processing data, analysing data and collecting, presenting and transforming data to assist decision maker [11]. Therefore, it contains of 3 main tasks, collection and organization; analysis of data; and interpretation of data. In collection and organization task, there are several methods of organizing data, such as graphically and numerically. In analysis of data task, the computation of various quantities associated with data can be done after data is organized. While in interpretation of data task, the information from two tasks before can be used to make assertions about the real world, such as the average of the calculated data. There are 4 kinds of data, which are nominal, ordinal, interval and ratio. Each type requires its own statistical approach or tests. There are two branches of statistics, they are:

a) Descriptive statistics is collecting, summarizing and presenting data in order to describe the situation from which is the data drawn [12, 13].
b) Inferential statistics is drawing conclusions about a population based only on sample data in order to make predictions, generalizations, or other interferences about a larger set of data [14].

In this research, the writer used branches of descriptive statistic which is determined the mean and SD of sample [15]. Also, because of this research is done for more than 2 groups in multiple comparisons, then Analysis of Variances (ANOVA) method will be used as the format of the performance comparison. There are four data models that will be used a performance comparison experiment, which are: Anchor Model, Optimal Normal Form (ONF), Star Schema and Data Vault. The four models are populated with the same facts, and each are queried differently to answer 13 information needs. The information needs are designed in a query for each model as the measurement for those models because each model has different table structures.

3. Results and discussion

There are a number of steps which followed to get the result of this research. First, the four models should be populated with the same facts. Then, the testing queries of each model should be created by having the same result even in the different model. There is one scenario that will be used in this research. This scenario is taken from previous research with extend some part in the elementary IGD of FCO-IM. The domain tables of scenario have been defined. Null value is not allowed in this scenario and the validity in time is included. Therefore, the Historized attributes are defined and this will be shown in FCO-IM model by the fact types with time validity. Based on the GLR-IGD, there are 9 tables will be generated. To implement the business rules in the population, the number of population of these 9 tables will be different. The population of the scenario is grouped based on the explanation in Table 1.

The testing will be done in all the historization tables. Those tables will be tested by using 13 information needs which called suite, which are:

1) Information need 1 (N1): Testing to get all the attributes in the table without any condition for gathering information about the Actor Ethnicity.

2) Information need 2 (N2): Testing to get all the attributes in the table without any condition for gathering information about the Performance Rating Validity.
3) Information need 3 (N3): Testing to get all the attributes in the table without any condition for gathering information about the Actor Performance Earnings.
4) Information need 4 (N4): Testing all the attributes in the table with where condition for gathering information about the Actor Ethnicity.
5) Information need 5 (N5): Testing all the attributes in the table with where condition for gathering information about the Performance Rating Validity.
6) Information need 6 (N6): Testing all the attributes in the table with where condition for gathering information about the Actor Performance Earnings.
7) Information need 7 (N7): Testing some attributes in the table with where condition and having clause for gathering information about the Actor Ethnicity.
8) Information need 8 (N8): Testing some attributes in the table with where condition and having clause for gathering information about the Performance Rating Validity.
9) Information need 9 (N9): Testing some attributes in the table with where condition and having clause for gathering information about the Actor Performance Earnings.
10) Information need 10 (N10): Testing by using more aggregate function in the select part.
11) Information need 11 (N11): Testing by using more aggregate function in select part and more sub query in where clause.
12) Information need 12 (N12): Testing by using operator 'LIKE' in where clause.
13) Information need 13 (N13): Testing by joining tables with nested query.

Table 1. Number of rows in population scenario.

Tables	Group of Tables		
	Small Tables (in rows)	Intermediate Size (in rows)	Historization Tables (in rows)
Gender	2		
Audience	3		
Ethnicity	10		
Rating	10		
Actor		2000	
Performance		10000	
Actor Ethnicity			200000
Performance Rating Validity			200000
Actor Performance Earnings			200000

Moreover, using ETL technique will spend shorter time than using generated queries for each data model. Because all data models used in experiment have totally different structures. So, this project only generated queries for 1 model (ONF model) and then using ETL technique for populating the three models (Star Schema, Anchor Model and Data Vault). The population was checked by using some queries to make sure that all the models have the same facts. Based on the checking result, the generated queries and ETL processes were worked proper to populate the four models. Then, we decided to populate the four models by extended into 200000 rows. It verified that the methods used to populate the models were easy to extend about the number of rows. The queries were executed 12 and 6 times so as to see how the output varies from the first output. The result of the SQL Server 2008 was exported to the GraphPad Prism program to be analyse in the statistic method. All the runs were entered as data and the program calculated the average for the measurements. The runs of the duration were done with several runs (12, 6). Because there is some scatter in the duration of several runs of the same query, every query was run several times and took the averages as a measurement. The scatter was computed, and the SD deviation was typically about 5%. As a check, SPSS yielded exactly the same ANOVA results when only these averages were entered as data. The average for each model for each information needs is shown in table 2.

Table 2. The averages between four models for each information needs.

	N1	N2	N3	N4	N5	N6	N7	N8	N9	N10	N11	N12	N13
AM	3539	3659.67	4890.83	678.5	374.67	669.17	397.5	304.33	3511	666.33	3899.17	6564	32294
ONF	3655..83	3477.67	4820.17	644.83	300.83	668.5	724.17	140.33	5602.33	1209	9668	9359.67	19405.67
STR	3592.5	3614.5	5256.17	642	323.67	725.83	731.33	216.83	15362.5	1076.83	23296.33	24545.17	63414.83
DV	3533	4135.5	4832	630.83	411.33	788.33	1222.33	2038	2219.17	8166.33	7815.17	6154.5	34623.5

For the statistical analysis, figure 1 shows the performance comparison between the four models based on one-way ANOVA. The results of One-way ANOVA analysis has given F(3,36) = 3.32 and p-value = 0.03. It means the null-hypothesis is rejected with a p-value of 0.03, a significant result. Any result of a one-way ANOVA with paired measurements are always reported with their F value, in this experiment the value of F=3.32. The number of (3,36) in F are the degrees of freedom. The number 3 means 4 models – 1 and the number 36 because (13 information needs – 1)*(4 models-1). The rejected null-hypothesis means it was established that some difference between these four models does indeed exist. No information about what difference that is can be obtained in the way however. For that, the post-hoc analysis is needed in this experiment.

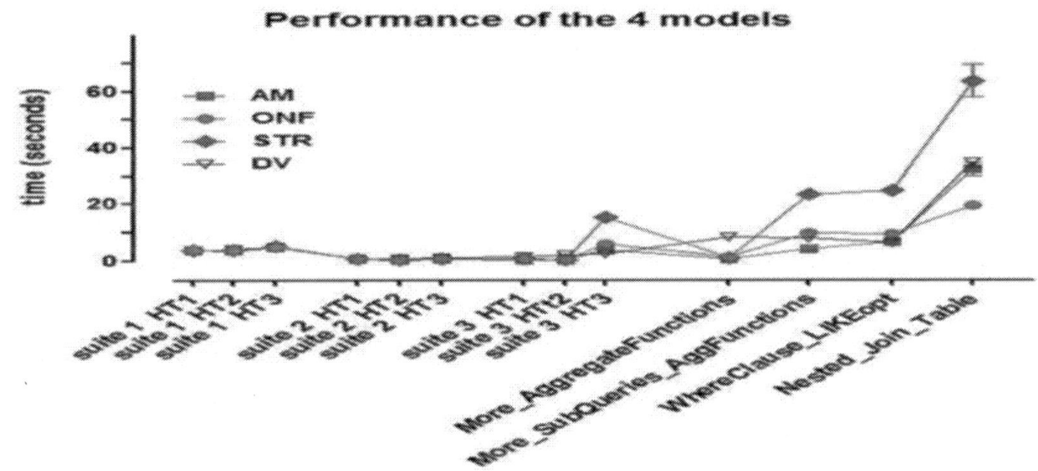

Figure 1. Performance Comparison based on one-way ANOVA.

Based on the one-way ANOVA analysis shown on figure 2, the null hypothesis said there is no difference in performance among the four models. The results of this research experiment presented no statistical significant difference between Anchor Model, ONF, and Data Vault. SQL Server will process the FROM clause first in the SQL statement. When, table the number of join tables are increasing then the rows to be evaluated in the next step (WHERE clause and SELECT clause) become smaller. It means the time execution in the SQL statement with more join tables will be shorter than the SQL statement with less join tables. Furthermore, the experiment has done by comparing all other models with the ONF. The reason is that the ONF is a maximally grouped, redundancy free table structure, which can serve as the point of reference for all other models. This analysis yielded:

1) ONF and Star Schema is significantly different by p-value = 0,03
2) ONF and Anchor Model is not significantly different by p-value > 0.09
3) ONF and Data Vault is not significantly different by p-value > 0.09

Based on the result, the null hypothesis that ONF and Anchor Model do not differ significantly cannot be rejected. So, the claim of the Anchor Model team (Lars Rönnbäck and friends) is not a big issue anymore.

Figure 2. One-way analysis by Graph Prism.

	1way ANOVA			
1	Table Analyzed	models and quiries in 6 fold , models in columns		
2				
3	Repeated Measures ANOVA			
4	P value	0.0304		
5	P value summary	*		
6	Are means signif. different? (P < 0.05)	Yes		
7	Number of groups	4		
8	F	3.323		
9	R squared	0.2169		
10				
11	Was the pairing significantly effective?			
12	R squared	0.7495		
13	F	11.46		
14	P value	< 0.0001		
15	P value summary	***		
16	Is there significant matching? (P < 0.05)	Yes		
17				
18	ANOVA Table	SS	df	MS
19	Treatment (between columns)	350800000	3	116900000
20	Individual (between rows)	4840000000	12	403300000
21	Residual (random)	1267000000	36	35190000
22	Total	6458000000	51	

4. Conclusion

Based on the results of this research the main question can be answered: "Are there any significant performance differences between the four models Anchor Model, Optimal Normal Form (ONF), Star Schema and Data Vault overall, based on all sorts of queries executed on these models?" The results of this research experiment presented no statistical significant difference between Anchor Model, ONF, and Data Vault. However, there is a significant difference between Star Schema and the other three models, with Star Schema performing worse. It can be concluded that lack of redundancy has significant influence to the performance of data model in data warehouse, in terms of accessing it. Generally, the main question and the objectives of this research were met. However, the performance between the four models can be further compared by using more varied scenarios. Because in the world of data warehousing or world of Business Intelligence, each company would face different situation and the data set that were used are not always ideal or stable. For example, a scenario has transactional data are not kept in one database. Other further research for the performance comparison can be experimented in other RDBMS used in data warehousing, such as Oracle.

References

[1] Ribeiro A, Silva A, and Silva A R 2015 *J Soft Eng & App.* **8** 617-634.
[2] Kimbal R, 2002. *The Data Warehouse Toolkit* (Canada: John Willey and Sons, Inc)
[3] Inmon, William H, 2000. *Building the Data Warehouses: Getting Started* (Winston Churchill).
[4] Team Ca, 2008 *Computer Associates* **1** 2
[5] Rutto R 2012 A Performance Comparison between four data models used in data warehouse *Thesis Rev* (*HAN University*)
[6] Rönnbäck L 2010 *Data and Knowledge Engineering* **69** 12 1229-1253
[7] Khan F A, Ahmad A, Muhammad I, Alharbi M, Rehman M, and Jan B 2017 J *SCS* 722
[8] Garani G 2012 *International Journal of Data Warehousing and Mining* **8** 4 22-40

[9] Sunar B and Koc C K 2001 *IEEE Transac on Comp* **50** 1 83-87
[10] Jovanovic V and Bojicic I 2012 *P Southern Ass for Inf Sys Conf.*
[11] Lindstedt D 2011 *Super Change your Data Vault*. Avalaible at: LearnDataVault.com
[12] Hannu 1983 *Statistics and Probability Letters* **1** 6 327-332
[13] Petter J and Erich L 2012 *Springer US* 465-471
[14] Mashall G and Jonker L 2011 *Radiography* **17** e1-e6
[15] Bannister H, Goldys B and Penev S W W 2016 *Automatica* **73** 15-26

Radio Frequency Identification (RFID) Based Employee Attendance Management System

G D P Maramis* and P T D Rompas

Universitas Negeri Manado, Tondano, 95618, Sulawesi Utara, Indonesia

*gmaramis@unima.ac.id

Abstract. Manually recorded attendance of all the employees has produced some problems such as the data accuracy and staff performance efficiency. The objective of this research is to design and develop a software of RFID attendance system which is integrated with database system. This RFID attendance system was developed using several main components such as tags that will be used as a replacement of ID cards and a reader device that will read the information related to the employee attendance. The result of this project is a software of RFID attendance system that is integrated with the database and has a function to store the data or information of every single employee. This system has a maximum reading range of 2 cm with success probability of 1 and requires a minimum interval between readings of 2 seconds in order to achieve an optimal functionality. By using the system, the discipline attitude of the employees and also the performance of the staff will be improved instantly.

1. Introduction

The recording of employee attendance is one of the important factors in human resource management. In-depth and detailed information about an employee's presence can determine work performance, salary, productivity or agency progress in general. However, manually recording the attendance might cause some drawbacks namely time consuming and not being marked due to human error [1][2][3]. RFID (Radio Frequency Identification) technology started to be developed as one of the alternative technology that makes it easy for users to identify for various things [4]. This technology itself consists of a tag in the form of a special chip where the chip has a unique code of information and a reader that serves to read the codes in the tag [4]. This technology has been implemented into several systems such as library management system, parking system [5][6]. The use of RFID technology also has been used widely in such ways for automatic attendance system. In year 2016 and 2017, another student attendance monitoring system using a number of technologies has been presented [7][8]. The process of recording and reporting and employee attendance is a continuous process. Employees come at a certain time to attend the absence then at certain hours also the employee is doing absent return. At a certain times the administrative officer will take the attendance card and do the recapitulation to know the attendance number of an employee. Such a process would be very ineffective because it contains several weaknesses, therefore a system that can replace the process of data collection of employee attendance automatically and computerized with the help of RFID.

To focus to the research, therefore this research has been limited to entry employee profile and attendance of the employee. With reference to the formulation of the above problem, the objective to be achieved in this research is to design an attendance system by using RFID technology that can help the institutions of the Faculty of Engineering, Universitas Negeri Manado to improve the effectiveness in performing employee absenteeism data by minimizing the possibility of errors that will occur and reducing fraud in data collection so that information security can occur.

2. Methods

In the absence system design using RFID technology, the prototype of RFID attendance system is divided into several parts as shown in Figure 1. Although in fact the application of RFID system as a whole is very complex, but in designing RFID prototype this attendance system can generally be divided into three main parts, namely:

RFID part system, which consists of RFID tag and RFID reader module. The next part is the USB port used by the computer to communicate with the RFID reader. The last part is the control unit and display section which is the user interface for the user to use the system. These both units are made by using VB.Net programming language application along with its database.

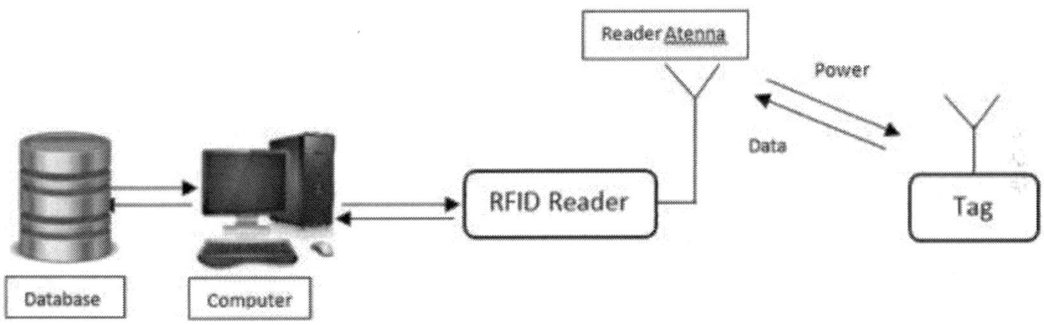

Figure 1. RFID attedance system block diagram.

2.1. Tags

A tag has a microchip with a special unique number, which is stored in the RFID memory, that is useful in identifying objects individually. This memory can be permanent or rewritable, which can be programmed electronically by the programmer to adjust with the needs of the system that is being developed [4]. Tags also have been produced in many variety materials such as plastic, metal and also embedded into coin. In this research, plastic material like a card is occupied. The contactless card attendance system is an option when there are many users that need to be registered. This kind of card also known as RFID tag. Nowadays, this tag has been produced in some variety materials such as metal, plastic. In addition, the length of time users in using the system is also relatively short. Generic attendance machine is an attendance machine that can accept input scan card for attendance. Generally, attendance machines can also be used to control access so that unauthorized parties cannot access a location [5]. Sometimes the identification process is done simultaneously with the process of recording attendance. This attendance system utilizes a non-touch card media in terms of identification. While there is the possibility of lending cards to other parties in the case of identification, this system facilitates the registration of large numbers of participants at short notice [9]. This system also provides convenience in the calculation at the end of the month or the end of the semester because all the data stored in the form of computerization.

2.2. RFID reader

An RFID reader is a special device that take a role to interrogate the tags. This is the most important part of the system. Inside of the reader, there is a small part that takes a very important role in RFID as it generates electromagnetic wave. It reads the number stored in the RFID tag memory and transfer it into the middleware before it is being transferred to the computer. Middleware itself is an interface that takes a role to transfer the number from the tag to the database management system [4].

Frequency that are common in operational used ranges from 125 KHz – 2.4 GHz [4]. Frequency 125 KHz is used in this system as it features low cost, less power consumption, small factor form and easy to use [10].

2.3. Antenna

Antenna is a part of RFID system that is resides inside the reader part. As the name implies, the role of the antenna is to generate electromagnetic field, in which this wave.

2.4. Backend database

The backend database primarily deals with the storage of relevant information recorded by the RFID. It stores the data of the employee and also data of attendance after processed by the computer [4].

3. Results and discussion

As a result of this research, ID Card of every single employee is replaced by an RFID card uses to record their attendance every day. The card has a sequence unique number stored in its memory and uses to identified the member of the faculty. For connecting the RFID reader to the computer, it uses USB port 2.0 as a communication port. This RFID system has a plug n play feature therefore it does not need a driver in order to be recognized by the operating system of the computer. System itself (attendance system) firstly will check the unique number of the card into the database whether the number has been registered into the system or not. If the number of the tag is found in the database, then the system will save the employee id and the name and also the time of attendance into the backend database and display it into the screen of the computer. By the time system found that the number of tag is not recognized in the database, then it will ask the user to contact the administrator to record the number of the tag. System also will check whether the employee has been scanned the tag or not. If the ID has been scanned before than the system will reject the card and give a notification to the user that the card has been scanned on that day. Figure 2 shows the flowchart of the automatic attendance system.

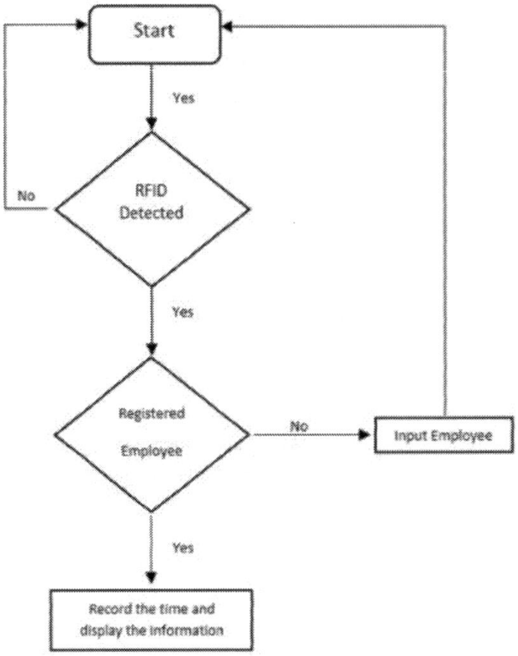

Figure 2. Flowchart of automatic attendance system.

IOP Conf. Series: Materials Science and Engineering **306** (2018) 012045 doi:10.1088/1757-899X/306/1/012045

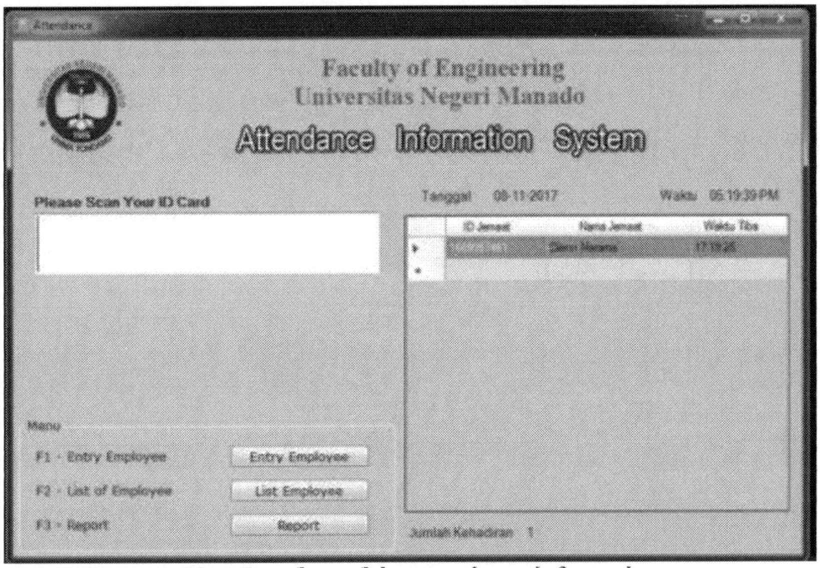

Figure 3. User Interface of the attendance information system.

As shown by the Figure 3, the system will show the id, name and the time of the time in of the employee. The id itself is the special unique number of the tag. There are 3 menus provided by the system, which are entry employee, list of employee and finally the full report of the attendance per month period. When the tag number cannot be found in the database, it will show a dialog box ask the user to contact the administrator to make an entry of new employee as shown by Figure 4.

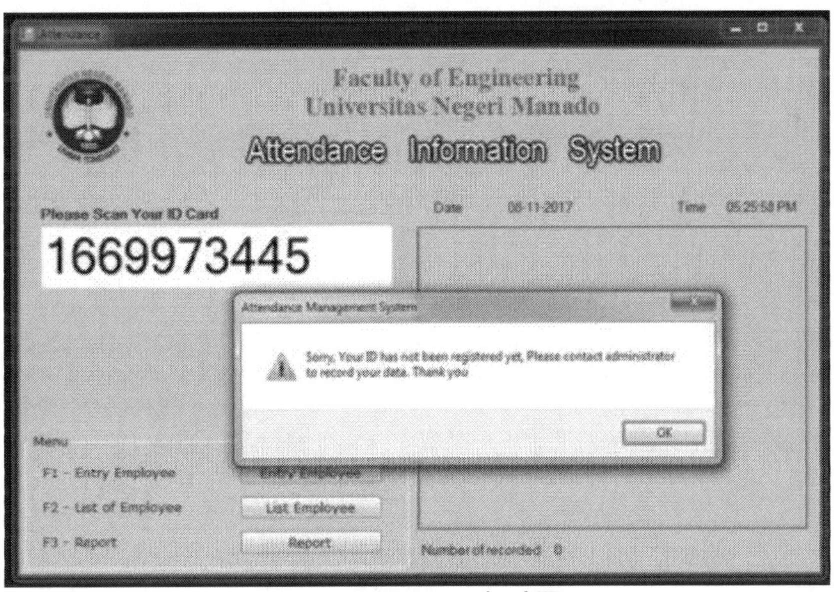

Figure 4. Unrecognized ID.

In addition to this process, Figure 5 shows when the new employee has been added to the system. However, Figure 6 draws the system reject to entry duplicate data when the employee already makes an absence. It is very important to have this function so we can assure that data is valid and readable.

IOP Conf. Series: Materials Science and Engineering **306** (2018) 012045 doi:10.1088/1757-899X/306/1/012045

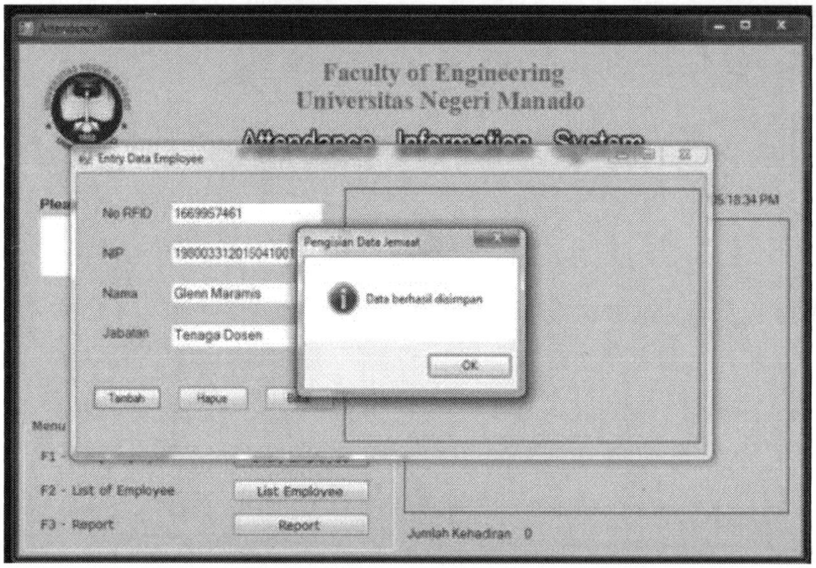

Figure 5. Entry new employee.

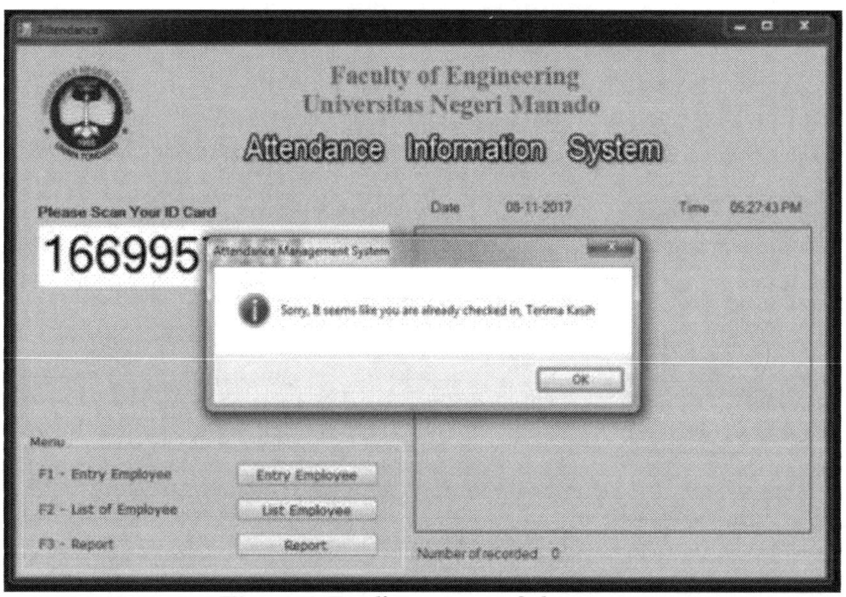

Figure 6. Duplicate entry of absence.

4. Conclusion

Based on the results of analysis, design and implementation has been done, then the conclusions obtained are as follows, which are employee absence process using RFID technology ensures information security from attendance data. Secondly, the existence of databases in data storage facilitate the search data so that it can help the institution or company to improve the effectiveness in performing data absenteeism of employees. Finally, it offers an ease way in recapitulation of employee attendance data for payroll purposes. After that, the system overall increases the efficiency of reporting and evaluating the performance of the employee and the discipline of the employee itself

References

[1] Khan A, Patil A, Viral S and Mansi K 2015 Attendance Monitoring System Using Zig Bee an RFID International Journal for Research in Emerging Science and Technology **2** 2

[2] Shukla A K 2017 Microcontroller Based Attendance System Using RFID and GSM International Journal of Emerging Technologies in Engineering Research (IJETER) **5** 8

[3] Siddique T A M, Shelake P G and Sale H B 2017 Smart Attendance Management and Learning System Imperial Journal of Interdisciplinary Research (IJIR) **3** 3

[4] Dwivedi J K, Tyagi A, Pushkar A,. Tiwari D K, Anand R, Dubey S, Dubey B K and Agnihotri P 2017 RFID Technology Based Attendance Management System International Journal of Engineering Science and Computing (IJESC) **7** 3

[5] Chung C-K and Hsieh Y-K, Wang Y-H, and Chang C-T 2016 Aware and Smart Member Card: RFID and License Plate Recognition System Integrated Applications at Parking Guidance in Shopping Mall 8th International Conference on Advanced Computational Intelligence.

[6] Pawar P, Shegaonkar K and Udane A 2016 Radio Frequency Identification Based Library Management System *International Journal of Innovation In Engineering, Research and Technology (IJIERT)*

[7] Pathak P, Paratnale M M, Khairnar D R, Yadhav P V and Wadgaonkar P R 2016 Student Attendance Monitoring System Via Face Detection and Recognition System *International Journal of Science Technology & Engineering (IJSTE)* **2** 11

[8] Jogiji A and Ghate P 2017 WSN Based Automatic Attendance Monitoring System International Journal of Computer & Mathematical Sciences (IJCMS) **6** 8

[9] Ma R, Woo-Garcia, Lomeli-Dorantes U H, Huerta F L, Herrera-May A L and Castillo J M 2016 Design and Implementation of a System Access Control by RFID Engineering Summit, II Cumbre Internacional de las Ingenierias (IE-Summit), *IEEE International*

[10] Shukla S, Shroff P, Nair V and Kuruvilla R 2016 Access Management and Control using NFC International Journal of Science and Research (IJSR) **5** 3

The Effect of Alkaline Concentration on Coconut Husk Crystallinity and the Yield of Sugars Released

H F Sangian[1] and A Widjaja[2,*]

[1]Jurusan Fisika, Universitas Sam Ratulangi Manado, Sulawesi Utara, Indonesia 95115
[2]Jurusan Teknik Kimia, Institut Teknologi Sepuluh Nopember Surabaya, Jawa Timur, Indonesia 60111

*arief_w@chem-eng.its.ac.id

Abstract: This work was to analyze the effect of alkaline concentration on coconut coir husk crystallinity and sugar liberated enzymatically. The data showed that the employing of alkaline on lignocellulose transformed the crystallinity. The XRD peaks increased highly which indicated that cellulose was more opened and exposed. After pretreatment, the chemical compositions (cellulose, hemicellulose, and lignin) were changed significantly. The employing 1% alkaline, the cellulosic content inclined if compared to that of non-pretreatment. When the alkaline concentration was added to 4%, the cellulose was decreased slightly which indicated that a part of cellulose and hemicellulose was dissolved into solution. It was found the alkaline pretreatment influenced by the biochemical reaction of treated substrates in producing the reducing sugars. The amounts of sugar liberated enzymatically of coconut husk treated by 1% and 4% alkaline increased to 0.26, and 0.24 g sugar/g (cellulose+hemicellulose), respectively, compared to that of native solid recorded at 0.18 g sugar/g (cellulose+hemicellulose).

1. Introduction

The lignocellulose, a renewable- and a recycled carbon material account of 10-90% cellulose which has a linear polysaccharide 1,4-β-D-glucose. The cellulosic structure has a strong connection called a glycosidic linkage β (1→4) and an H bond at the hydroxyl OH serves as a scutum of simple monomers [1]. Prior to the hydrolysis of lignocellulosic materials into sugars, lignin must firstly be degraded and the cellulose structure is changed from crystalline- into amorphous structures so that enzymes can attack easily the cellulose [2, 3].

The initial process to treat the lignocellulosic material prior to hydrolysis is called a pretreatment which is the most crucial step for producing of sugar and ethanol [4]. The objectives of pretreatment (including lignin degradation) were to modify the crystal structure of cellulose for increasing the digestibility of the enzyme work on the hydrolytic process [5]. The weakness of sub- and super-critical methods is difficult to scale up because of cost and safety considerations [6].

However, from the perspectives of sustainable technology and desirable cost, the conventional pretreatment must be improved the waste management. The preparation of sugar derived from lignocellulosic materials requires more environmentally friendly and more efficient method that utilizes a less heat, operates in mild condition, and uses the cheap solvents. The ionic liquid pretreatment is a new method and has many advantages in operation such as low vapor pressure, high thermal stability, easy to dissolve many compounds, and it can be recycled [7]. The lignocellulosic materials have been converted into sugars by ionic liquid pretreatment by delignification [8]. The delignification process is to reduce barriers, mostly caused by lignin, in the enzymatic hydrolysis process. The low lignin lignocellulose, ionic liquid interacts well with cellulose through electrostatic interaction mechanism [9].

The usage of ionic liquids, however, is costly since their prices are very expensive [10]. It is necessary to manufacture the fermentable sugars from lignocellulose treated by the solvent in which it does not harm the environment nor was expensive. This work was to manufacture of reducing sugar hydrolyzed from coconut husk pretreated by alkaline whose concentrations were varied at 1 and 4% and biocatalysts used were the pure cellulase and xylanase.

2. Experimental Setup

2.1. Material preparation

The raw material, coconut husk, was kindly given by the owner of the coconut husk factory in South Minahasa, North Sulawesi Indonesia. Prior to pretreatment, coconut husk was prepared by the procedures as follows: the raw material was heated under sunlight radiation for days and milled and then filtered to the size of 120 mesh using screener apparatus (Retsch GmbH Rheinische strade 362/2781, Haan Germany).

One and four percent alkaline were employed to carry out pretreatment on coconut husk in which method was adapted from previous work [5]. Six grams' lignocellulose was mixed into one percent and four percent alkaline (250mL, 80°C) for 16 h and then the solid was washed by hot water passing through Whatman paper via reduced pressure. An XRD (*Philips X'Pert X-Ray Diffractometer*) was employed to analyze the structure of treated- and untreated substrates of coconut husk as a previously referred study [11]. The chemical composition of coconut husk treated by alkaline was analyzed using the Chesson technique [12].

2.2. Enzymatic hydrolysis

The enzymatic hydrolysis of treated- or untreated- coconut husk and sugar analysis were conducted in the same procedure as previously reported work [13]. Suspension solution was prepared by adding the sodium acetate buffer (pH 3, 30 mL, 0.1M) into one-gram substrates inside Erlenmeyer reactor. A suspension was controlled and incubated in an oil bath at 60°C under slow stirring for 48 h. The enzymatic reaction was begun by adding 0.2 ml of pure cellulase or mixture of cellulase + xylanase from A. Niger. The pH was controlled at 3.0 by adding sodium citrate, or citric acid. Two 0.2mL samples were removed periodically and diluted with 1.8mL pure water and 3 mL DNS solution and then the amount of sugar was measured using spectrophotometer (CECIL 1001). The mixture was isolated in the tube and put in boiling water for ten minutes. The sample tube was moved to an ice-water bath and then was rotated in centrifuge equipment (Hermle Labortechnik GmbH-Z 326 K, Wehingen Germany) to settle the solid and pulp. The measurement of sugars fraction (glucose, xylose and lactose) used HPLC technique (Waters 1515 Isochoric HPLC Pump, 2414 RI and Aminex HPX87P (Bio-Rad, CA) using standard concentrations: as follows glucose: 50.040, 25.020, 12.510; Xylose: 40.070, 20.037, 10.019; and galactose: 20.044, 10.022, and 5.011.

3. Results and Discussion

3.1. The crystallinity analysis

One percent alkaline solution dissolved lignin and hemicellulose so that cellulose I crystal was more exposed. The properties of cellulose I am near to those of pure cellulose, whose crystallinity is high and peaks are vivid at the angles 17.00°, 23.00° and 35.00° in which their Miller indices were (101), (002) and (040), respectively, as previously described works [11]. The XRD curve of 4%NaOH-pretreated substrate, the peak of (002) plane declined slightly whereby a part of cellulose was dissolved as shown in Fig. 1a.

The graph shows the peak symmetry of treated crystal was shifted at 0.5° appertaining to the dilatation of space of the substrate. Some authors reported lignocellulose that has been treated, could increase the digestibility of enzymatic hydrolysis as previously published work [14]. An alkaline solution dissolved lignin and hemicellulose so that cellulose crystal was porous.

3.2. Analysis of chemical compositions

The average chemical compositions of treated- and untreated-coconut husk are shown in Fig. 1b. Chemical compositions of original coconut husk were 26.60% (cellulose), 17.74% (hemicellulose) and 41.18% (lignin), respectively. While cellulosic compositions of 1%NaOH- and 4% NaOH-pretreated substrates were 43.18 percent and 38.30 percent, which is higher lignin removal compared to that of control. The data shows 1%NaOH pretreatment caused the cellulose and hemicellulose improved significantly. When the alkaline concentration was added to 4% for pretreatment the cellulose composition declined because of dissolution of cellulose into solution. The origin, type, and maturity of coconut husk could affect to their compositions [15].

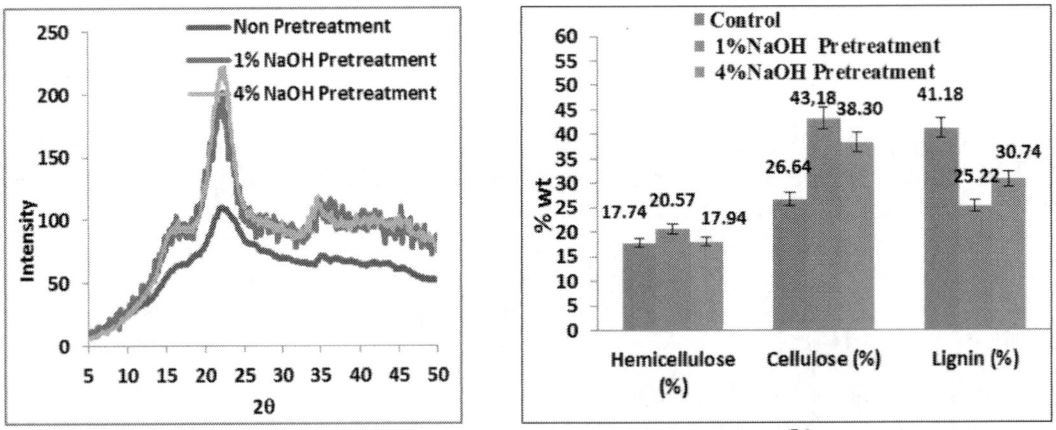

(a) (b)

Figure 1. The XRD patterns (a) and chemical compositions (b) of pretreated coconut husk pretreated by 1 & 4% NaOH (80°C,16h) and compared with non-pretreatment as a control.

3.3. Enzymatic conversion

Preparation of enzyme solution and standard graphs of sugars was adapted from procedures that have been used since 1959 [16]. Fig. 2a shows sugar concentrations of enzymatic hydrolysis of coconut husk pretreated with 1%NaOH, 4%NaOH and original solid as control with using a single pure cellulase.

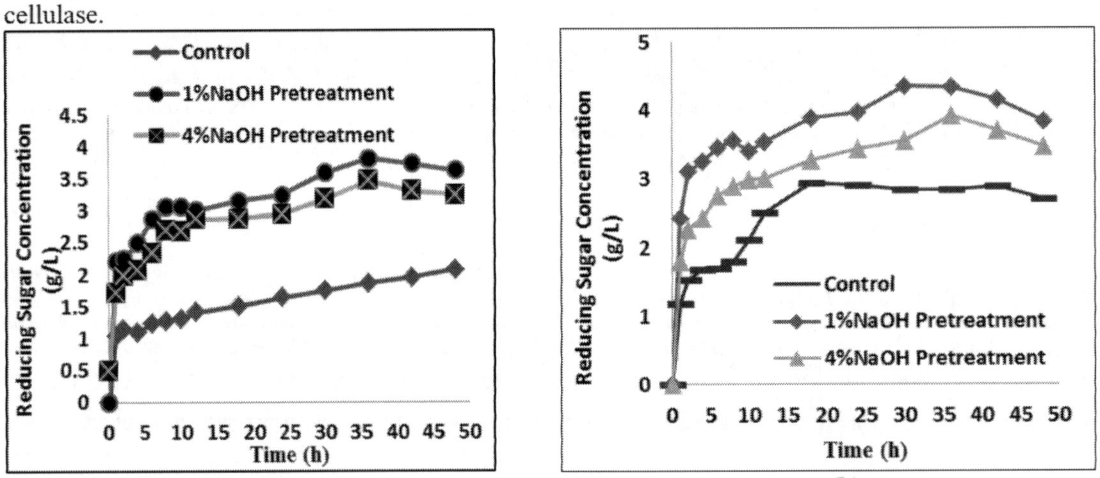

(a) (b)

Figure 2. Time course of enzymatic hydrolysis (60°C for 48 h) using pure cellulase (a) and mixture cellulase+xylanase (b) of coconut husk pretreated with 1%NaOH, 4%NaOH (80°C,16 h) and original coconut husk as a control.

When coconut solid was pretreated by 1%NaOH and 4%NaOH, the maximum sugar concentrations obtained were of 3.814 g/L and 3.464 g/L. respectively. The optimum sugar hydrolyzed from the original substrate as control was 2.074 g/L, which was the least sugar concentration. These results attributed to the chemical compositions and crystalline structure of substrates before and after pretreatment. The higher of lignin content of substrate was the lower of sugar liberated from the substrate. It was found that lignin factor was a major barrier of catalyst or cellulase enzyme to attack cellulose and hemicellulose to liberate sugars as a previously reported study [17, 18]. The second step was the utilization of cellulase+xylanase to enzymatic hydrolysis of pretreated- and unpretreated-substrates. When substrate pretreated by 1%NaOH and 4%NaOH, sugar concentrations were 4.966g/L, 3.928g/L, respectively as shown in Fig. 2b.

Table 1. Comparison of reducing sugars concentration hydrolyzed from pretreated substrates using cellulase and cellulase+xylanase for 48 h.

Pretreatment	Sugar concentration (g/L)		Increase (%)
	Cellulase	Cellulase+xylanase	
1%NaOH	3.629	3.834	9.88
4%NaOH	3.239	3.475	7.30

Table 1 shows the fermentable sugar concentrations hydrolyzed from substrates treated by one and four percent alkaline using cellulase and then compared to those of employing cellulase+xylanase for 48 h. After employing cellulase+xylanase into one percent alkaline-treated substrate, the sugar concentration increased by 9.88% to 3.834g/L compared to 3.629g/L recorded with using single cellulase. While the increase of sugar concentration converted from four percent alkaline-treated substrate, was 7.30 percent to 3.475 g/L with employing cellulase+xylanase from 3.239g/L recorded by applying only cellulase enzyme.

Generally, the amount of sugar liberated from lignocellulose using a mixture of cellulase+xylanase increased only slightly from to those of using single cellulase. Two yield calculations employed as follows: the first yield was to compare an obtained gram of total reducing sugar (TRS) and a gram of dried coconut husk; the second yield was the ratio of a gram of TRS and gram of theoretical total reducing sugar (T TRS). The gram of total reducing sugar is determined by multiplying the TRS concentration by the amount of suspension volume (mL). The theoretical total reducing sugars obtained from the chemical composition (cellulose+hemicellulose) of original coconut husk. For control, T TRS is 0.267 multiplied by the gram of the dried lignocellulose plus 0.177 multiplied by the gram of dried lignocellulose.

Table 2. Yield comparison of reducing sugar hydrolyzed from pretreated substrates using cellulase and mixture of cellulase+xylanase enzymes.

Pretreatment	Yield (%)			
	Cellulase		Cellulase+xylanase	
	Yield (1)	Yield (2)	Yield (1)	Yield (2)
Control	0.06	0.14	0.08	0.18
1%NaOH	0.11	0.25	0.12	0.26
4%NaOH	0.10	0.22	0.10	0.24

When cellulase+xylanase added into one 1% and 4% alkaline-treated substrate, the yield of sugar was 0.26, 0.24 g sugar/g cellulose+hemicellulose which was higher than that of non-pretreatment recorded at 0.18 g sugar/g (cellulose+hemicellulose) as displayed in Table 2. The increase of sugar yield was due to the synergy of cellulase and xylanase to attack cellulose and hemicellulose. This result revealed that the yield of sugar which used the mixture of cellulose and xylanase enzymes was generally larger than that used a single cellulase. The utilization of cellulase+xylanase on enzymatic hydrolysis aimed to improve the sugar yield liberated from substrates since there was a synergy of those enzymes. Cellulase enzyme chopped cellulose chain into glucose and xylanase enzyme cut off hemicellulose (xylan) into [19]. It was expected that yields would improve using cellulase+xylanase compared to those of introducing only cellulase itself.

However, it was found that no significant differences of sugar yields converted from substrates using cellulase and those employing the mixture of cellulase plus xylanase. This result was typical for enzymatic hydrolysis of high-lignin lignocellulose, as coconut husk and due to the low content of hemicellulose in coconut husk as previously described above. The HPLC measurement was performed to verify and to conform the sugar results recorded from DNS method and to analyze the type of reducing sugar which liberated from coconut husk as described in Table 3. It was found the kinds of fermentable sugars measured were dominated by glucose, xylose, and galactose. Galactose concentration obtained was at 2.3142 g/L for four percent alkaline pretreatment.

Table 3. Total reducing sugar concentration measured by HPLC with pretreated substrates using cellulase+xylanase for 48 h hydrolysis.

Pretreatment	HPLC			
	Glucose	Xylose	Galactose	Total
Control	2.1558	0.1141	1.6832	3.9531
1%NaOH	1.6097	0.2723	1.2351	3.1171
4%NaOH	2.1196	0.0863	2.3142	4.5200

HPLC measurement showed and confirmed that amounts of sugar obtained using DNS method are realistic since those figures are close each other.

It was found that the alkaline concentration influenced significantly on the crystallinity and the yield of sugar obtained. One percent (W/V) of alkaline can be decreasing the biomass crystallinity but increasing the yield of sugar released from treated substrate. When the alkaline concentration increased to 4%, the sugars declined slightly from 1% treated solid since a part of cellulose and hemicellulose started dissolved into solution.

4. Conclusion

The alkaline pretreatment was considerable for the preparation of reducing sugars hydrolyzed from lignocellulose-based materials. After pretreatment, the cellulosic structure transformed in which their crystallinity increased attributing to the dissolution of lignin and hemicellulose. By employing an enzymatic reaction, the sugar liberated from 1%NaOH-treated substrate was of 0.26g sugar/g (cellulose+hemicellulose). If alkaline concentration added to 4%, the total sugar obtained was of 0.24g sugar/g (cellulose+hemicellulose) by applying pure cellulase+xylanase). The results showed that alkaline pretreatment influenced on substrate structure and also affected of the yield of sugar obtained via biocatalyst reaction.

Acknowledgments

Authors thank and acknowledge to The Department of Research and Higher Education of Indonesia for financial assistance through the research grants in 2017.

References

[1] Zhao X, Zhang L, and Liu D 2012 Biomass recalcitrance. Part I: the chemical compositions and physical structures affecting the enzymatic hydrolysis of lignocellulose *Biofuels Bioprod Bioref* **6** 465-482

[2] Sangian H F and Widjaja A 2017 Effect of pretreatment method on structural changes of coconut coir dust *BioRes.* **12** 4 8030-8046

[3] Widjaja A, Agnesty S Y, Sangian H F and Gunawan S 2015 Application of Ionic Liquid [DMIM]DMP Pretreatment in the Hydrolysis of Sugarcane Bagasse for Biofuel Production *Bull. Chem. React. Eng. Catal.* **10** 70-77

[4] Cardona C A, Quintero J A and Paz I C 2010 Production of bioethanol from sugarcane bagasse: Status and perspectives *Bioresource Technology* **101** 4754–4766

[5] Asgher M, Ahmad Z and Iqbal H M N 2013 Alkali and enzymatic delignification of sugarcane bagasse to expose cellulose polymers for saccharification and bio-ethanol production *Industrial Crops and Products* **44** *488– 495*

[6] Sangian H F, Ranggina D, Ginting G M, Purba A A, Gunawan S and Widjaja A 2015 Study of the preparation of sugar from high-lignin lignocellulose applying subcritical water and enzymatic hydrolysis: Synthesis and consumable cost evaluation *Scientific Study & Research Chemistry & Chemical Engineering, Biotechnology, Food Industry* **16** 1 13–27

[7] Ha S H, Mai N L, An G and Koo Y M 2013 Microwave-assisted pretreatment of cellulose in ionic liquid for accelerated enzymatic hydrolysis *Bioresource Technology* **102** 1214–1219

[8] Qiu Z and Aita G A 2013 Pretreatment of energy cane bagasse with recycled ionic liquid for enzymatic hydrolysis *Bioresource Technology* **129** 532–537

[9] Feng L and Chen Z 2008 Research progress on dissolution and functional modification of cellulose in ionic liquids *Journal of Molecular Liquids* **142** 1–5

[10] Sangian H F, Kristian J, Rahma S, Dewi H K, Puspasari D A, Agnesty S Y, Gunawan S, and Widjaja A. 2015 Preparation of Reducing Sugar Hydrolyzed from High-Lignin Coconut Coir Dust Pretreated by the Recycled Ionic Liquid [mmim][dmp] and Combination with Alkaline *Bull. Chem. React. Eng. Catal.* **10** 8-22

[11] Park S, Baker J O, Himmel M E, Parilla P A, Johnson D K 2010 Cellulose Crystallinity Index: Measurement Techniques and Their Impact on Interpreting Cellulase Performance *Biotechnology for Biofuels* **3** 10 1-10

[12] Datta and Rathin 1981 Acidogenic Fermentation of Lignocellulose- Acid Yield and Conversion of Components *Biotechnology and Bioengineering* **23** 2167-2170

[13] Zhao D, Li H, Zhang J, Fua L, Liu M, Fua J and Ren P 2012 Dissolution of cellulose in phosphate-based ionic liquids *Carbohydrate Polymer.* **87** 1490-1494

[14] Yuan T Q, Wang W, Xu F and Sun R C 2013 Synergistic benefits of ionic liquid and alkaline pretreatments of poplar wood. part 1: effect of integrated pretreatment on enzymatic hydrolysis *Bio resource Technology* **144** 429–434

[15] Van Dam J E G Martien J A, Van den Oever, Keijsers E R P, Van der Putten J C, Anayron C, Josol F and Peralta A. 2006 Process for production of high density/high performance binderless boards from whole coconut husk Part 2: Coconut Husk Morphology, Composition and Properties *Industrial Crops and Products* **24** 96–104

[16] Miller G L 1959 Use of Dinitrosalicylic Acid Reagent for Determination of Reducing Sugar *Analytical Chemistr.* **31** 3 426-428

[17] Moniruzzaman M and Ono T 2012 Ionic liquid assisted enzymatic delignification of wood biomass: A new 'green' and efficient approach for isolating of cellulose fibers *Biochemical Engineering Journal* **60** 156– 160

[18] Muhammad N, Man Z, Bustam M A, Mutalib M I A, Wilfred C D, Rafiq S 2011 Dissolution and delignification of bamboo biomass using amino acid-based ionic liquid *Application Biochemical Biotechnology* **165** 998–1009

[19] Sangian H F, Kristian J, Rahma S, Agnesty S Y, Gunawan S, Widjaja A. 2015 Comparative study of the preparation of reducing sugars hydrolyzed from high-lignin lignocellulose pretreated with ionic liquid, alkaline solution and their combination *J. Eng. Technol. Sci.* **47** 2 137-148

Industrial Internship and Entrepreneurship Competencies on Vocational High School Students

H F Wendi[*] and I H Kusumah

Postgraduate, Universitas Pendidikan Indonesia, Jl. Dr. Setiabudhi No. 207 Bandung 40154

*hensonfebriwendi@ymail.com

Abstract. The purpose of the research is to explore the influence of internship and vocational skill to student's entrepreneurship competencies. The research used ex post facto approach. Population in this research is all students of Vocational High School in Bandung, Indonesia. The sample of 40 respondents was determined by proportional random sampling technique. The data were collected by instrument questionnaire and test. Data analysis used descriptive statistics and multiple linear regression. The results show that almost half students have a low the competencies of an entrepreneur. The hypothesis testing shows many the influence of factory teaching has a positive and significant effect on the competencies of an entrepreneur. Similarly, vocational skills have positive influence and significant on the competencies of an entrepreneur. Respectively, the influence of factory teaching and vocational skills expertise collectively have the effect on the competencies of an entrepreneur. Therefore, the influence of factory teaching and vocational skills are effective to student's entrepreneurship cap competencies.

1. Introduction

Increasing on the ratio number of Vocational High School, resulted an increase of the number of Vocational High School which is closely significant will resulted on of labor force/employee. Unless the number of Vocational High School graduate workforce is not accompanied by inadequate competences with the needs of the world of the work, the growth of unemployment vocational high school graduates will be greater than high school graduates, junior and college[1] Where in the fact as if a vocational education institution produces some unemployed graduates are unemployed then unit one, ladder or a type of the education can be said irrelevant to the needs for society and national development. With many educated unemployed, vocational the education can also be said to be unable to encourage the growing of the national economy growth may even be a constraint in the growth [2]

According to data of the number of unemployed vocational high school graduates compared with the school and college is the largest of 9.84%. While the level of elementary, junior high, senior high school, diploma and college are respectively getting percentage for elementary school 3.44%, junior 5.76%, 7.22% diploma graduate, 6.22 % [3].

Some Solutions for these problem are how to create a graduates of Vocational High School who have the entrepreneurship competencies that be straight forward readiness for being an entrepreneurship. Some of the encouraging reasons to choose an entrepreneur such as are (1) the solution for graduates oneself and for others aspect because it requires employees on the business to be run well. (2) Reducing unemployment is expected to be increased, and the last one is robe a be to be beneficial (3) can be beneficial in improving welfare [4].

In this case the role of the school is very important to stimulate a competencies and a courage to have entrepreneur [5]. To start a courage is the main begin a prime competency to be had by someone to enter the business world. Not only skill of entrepreneurship but also the courage without cap competencies often forces us into a failure [6].

Vocational High School generally aims to prepare some students to work in a particular field [7]. Therefore, Vocational School is one that plays a role to the students to provide entrepreneurship skills and some skills [8]. Learning outcomes provide by vocational High School, it is wished that students could become an entrepreneurship to minimize the number of unemployed [9]. Model of learning can improve the quality student competence so as to provide the provision of entrepreneurship competencies named teaching factory learning model [10].

So to have the entrepreneurship competencies of vocational high school students after applying teaching factory and vocational skill is essential to have some researchers.

In this research examines the factors that influences the competencies of having an entrepreneur in order that know how far of the influence of teaching factory teaches and vocational skill on the competencies of an entrepreneur for students. By knowing the factors of the competencies of be entrepreneur are expected entrepreneurs in Indonesia can develop in to increasing development and also can reduce the number of unemployed students. Therefore, the authors are interested to raise lift an entrepreneurship theme entitled: The competencies of Entrepreneurship students of Vocational High School at Learning Implementation of Teaching Factory Learning.

2. Research Methods

This research uses ex-post facto research method. Place of Research conducted at Vocational High School in Bandung on student of Motorcycle Engineering skill package. The time of the research was conducted on May 2 to 10, 2017. The population in the study were all students of Motorcycle Engineering skill package. The sample of this research is 40 students using proportional random sampling technique. Data were collected by instrument questionnaire and test. Data analysis used descriptive statistics and multiple linear regression.

3. Results and Discussion

3.1. Data description

The result of calculation through descriptive statistic competencies of having be entrepreneur, role of teaching factory teaching and skill of student of Vocational High School in Bandung competence expertise Technique and Business of Motorcycles presented in Table 1.

Table 1. Distribution of Frequency and Percentage of Entrepreneurship Capacity (Y), Teaching Factory Teaching Role (X1), and Vocational Skill (X2) Student of Vocational High School Competence of Motorcycle Engineering skill package in Bandung.

No	Category	Y		X_1		X_2	
		Frek (student)	%	Frek (student)	%	Frek (student)	%
1	Very low	3	7.50	2	5.00	4	10.0
2	Low	12	30.0	12	30.0	9	22.5
3	Medium	13	32.5	8	20.0	13	32.5
4	High	9	22.5	13	32.5	10	25.0
5	Very high	3	7.50	5	12.5	4	20.0
Jumlah		40	100	40	100	40	100

Based on Table 1, it shows the highest percentage of competencies from entrepreneur (Y) and the role of teaching factory (X1) of 32.5% is in medium category with 13 students. While the highest percentage of vocational skill (X2) is also 32.5% with the number of students 13 students, but in the high category. To strengthen the data there is the influence of the role of teaching factory teach and vocational skills to the competencies of students to become a vocational high school student, then the data is analyzed by hypothesis testing. Hypothetical test was performed by multiple linear regression test. Hypothesis test results are presented in Table 2.

Table 2. Partial Hypothesis Test Results with Linear Regression Test (t test).

Coefficients[a]								
Model		Unstandardized Coefficients		Standardized Coefficients	T	Sig.	Conclusion
		B	Std. Error	Beta			
1	(Constant)t	1.710	3.848		.444	.659	
	Teaching_Factory	.532	.098	.533	5.455	.000	Hypothesis Accepted
	Vocational Skill	.435	.094	.450	4.607	.000	Hypothesis Accepted

a. Dependent Variable: Entrepreneurship

Based on Table 2, the value of learning factory teaching coefficient and vocational skill coefficient value are positive values of 0.532 and 0.435. While the significance value (sig t) for variable of teaching factory teaching variable and practice skills equal to 0.000 is smaller than the significance level of 0.05 ($p < 0.05$). Research hypothesis:

H1: the role of teaching factory learning has a positive and significant impact on competencies having entrepreneur.

H2: the role of vocational skills has a positive and significant impact on competencies having entrepreneur.

H3: the role of teaching factory teaching and vocational skills are jointly influential

Positive and significant to the competencies of student having entrepreneur. Hypothesis decision is to accept hypothesis H1 and H2, which mean of role of teaching factory teaching and vocational skill have positive and significant influence to competencies of having entrepreneur at student of Vocational High School in Bandung Competence of Motorcycle Engineering Skill in Bandung.

The value of teaching factory coefficient in Table 3, positive value of 0.532 has a meaning, if the value of teaching factory teaching roles increases one point then the value of the competencies to having entrepreneur will increase by 0.532 with the assumption that vocational skills remain. The same thing in the value of the coefficient of vocational skills is positive value of 0.435 also means, if the value of the role of vocational skills increased one point then the value of competencies to have entrepreneur will increase by 0.435 assuming that teaching of teaching factory remains.

In order to see the role of teaching factory teaching and the role of vocational skills affect simultaneously on students 'competencies to perform the students' achievement that is based on hypothesis analysis F. The result of hypothesis testing (H3) is presented in Table 3.

Table 3. Simultaneous Hypothesis Test Results with Linear Regression Test (F test).

ANOVA[b]							Conclusion
Model		Sum of Squares	Df	Mean Square	F	Sig.	
1	Regression	10887.619	2	5443.810	96.063	.000[a]	Hypothesis Accepted
	Residual	2096.756	37	56.669			
	Total	12984.375	39				

a. Predictors: (Constant), Vocational Skill, Teaching_Factory

b. Dependent Variable: Entrepreneurship

Based on Table 3, the value of F is 96.063 with significance value (sif.F) of 0.000 ($p < 0.05$). Therefore, it can be concluded that the hypothesis is accepted, meaning the role of teaching factory and the role of vocational skills has an effect simultaneously on the competencies of students' entrepreneurship competency Vocational High School in Bandung Competence of Motorcycle.

Engineering Skill in Bandung. Meanwhile, to see the large percentage of contribution of teaching factory learning and vocational skills to the competencies of be entrepreneur can be seen in Table 4.

Table 4. Summary of Multiple Regression Analysis.

Model Summary

Model	R	R Square	Adjusted R Square	Std. Error of the Estimate
1	.916[a]	.839	.830	7.528

a. Predictors: (Constant), Vocational Skill, Teaching_Factory

Based on table 4 looks big Adjusted R Square value of 0830. This suggests that the variable role of teaching and vocational skills contributes effectively to 83% of the variables of the competencies to have entrepreneur. While the rest (27%) competencies be entrepreneur influenced by factors other than the role of factory teaching and vocational skills.

4. Conclusion

Based on the results of temporary observations, it can be concluded:

- There is a positive influence between the implementation of teaching factory teach on the competencies of the entrepreneur Vocational High School in Bandung Technical Competence Motorcycle Engineering skill package.
- There is a positive influence between the skills of practice to the competencies to have entrepreneur Vocational High School in Bandung Expertise Engineering Competence Motorcycle Engineering skill package.

References

[1] Khan A, Patil A, Viral S and Mansi K 2015 Attendance Monitoring System Using Zig Bee an RFID *International Journal for Research in Emerging Science and Technology* **2** 2

[2] Shukla A K 2017 Microcontroller Based Attendance System Using RFID and GSM *International Journal of Emerging Technologies in Engineering Research (IJETER)* **5** 8

[3] Siddique T A M, Shelake P G and Sale H B 2017 Smart Attendance Management and Learning System *Imperial Journal of Interdisciplinary Research (IJIR)* **3** 3

[4] Dwivedi J K, Tyagi A, Pushkar A,. Tiwari D K, Anand R, Dubey S, Dubey B K and Agnihotri P 2017 RFID Technology Based Attendance Management System *International Journal of Engineering Science and Computing (IJESC)* **7** 3

[5] Chung C-K and Hsieh Y-K, Wang Y-H, and Chang C-T 2016 Aware and Smart Member Card: RFID and License Plate Recognition System Integrated Applications at Parking Guidance in *Shopping Mall 8th International Conference on Advanced Computational Intelligence*

[6] Pawar P, Shegaonkar K and Udane A 2016 Radio Frequency Identification Based Library Management System *International Journal of Innovation In Engineering, Research and Technology (IJIERT)*

[7] Pathak P, Paratnale M M, Khairnar D R, Yadhav P V and Wadgaonkar P R 2016 Student Attendance Monitoring System Via Face Detection and Recognition System *International Journal of Science Technology & Engineering (IJSTE)* **2** 11

[8] Jogiji A and Ghate P 2017 WSN Based Automatic Attendance Monitoring System *International Journal of Computer & Mathematical Sciences (IJCMS)* **6** 8

[9] Ma R, Woo-Garcia, Lomeli-Dorantes U H, Huerta F L, Herrera-May A L and Castillo J M 2016 Design and Implementation of a System Access Control by RFID Engineering Summit, II Cumbre Internacional de las Ingenierias (IE-Summit), *IEEE International.*

[10] Shukla S, Shroff P, Nair V and Kuruvilla R 2016 Access Management and Control using NFC *International Journal of Science and Research (IJSR)* **5** 3

Instructional Model and Thinking Skill in Chemistry Class

H H Langkudi

Department of Natural Science Education Universitas Negeri Manado

*hettylangkudi@yahoo.com

Abstract. Chemistry course are considered a difficult lesson for students as evidenced by low learning outcomes on daily tests, mid-semester tests as well as final semester tests. This research intended to investigate the effect of instructional model, thinking skill and the interaction of these variables on students' achievement in chemistry. Experimental method was applying used 2 x 2 factorial design. The results showed that the use of instructional model with thinking skill influences student's learning outcomes, so that the chemistry teacher is recommended to pay attention to the learning model, and adjusted to the student's skill thinking on the chemistry material being taught. The conclusion of this research is that discovery model is suitable for students who have formal thinking skill and conventional model is fit for the students that have concrete thinking skill.

1. Introduction

The term Chemistry, lately often used in the entertainment world to declare the existence of a harmonious blend among others. The chemistry subject introduced to students in Indonesia during their upper secondary school, and it concerned with the properties and reactions of substances. Chemistry is the study of material and the changes that material undergo. One of the joys of learning chemistry is seeing how chemical principles operate in all aspects of our lives, from everyday activities like lighting a match to more far reaching matters like the development of drugs to cure cancer. Chemistry principles also operate in the far reaches of our galaxy [1]. However, many students say that chemistry is difficult. It has been observed that a lot of students nowadays are losing interest in the science subject such as chemistry [2]. These perceived difficulties are part of the context in which the students develop concepts and problem-solving skills [3].

Students live and operate in the macroscopic world of matter. Unfortunately, they do not perceive chemistry as related to their surroundings. Moreover, they do not easily follow shifts between the macroscopic and microscopic level [4]. Studies in chemistry involve three dimensions, they are dimensional of macroscopic reasoning (relating to what is observed), symbolic dimensions (symbol, formulas, equations), and sub-microscopic dimensions (atom, molecules, ions, molecular structure [5]. Think in three dimensions is a demand chemical disciplines that distinguishes chemistry from other disciplines. Related to this dimensions it is important to the teacher providing models or strategies in order the teaching and learning to be successfully in terms of student learning outcomes.

Here, the purpose of this study was to answer the question is there a difference in the learning outcomes of student that learned with discovery model and direct learning model in chemistry? Moreover, is there a difference in the learning outcomes of students with formal and concrete thinking skills that learned with discovery learning model and the students that learned with direct learning model? This research also examines the interaction between teaching model with the ability to think toward the students learning outcomes.

2. Methods

This study is a quasi experiment with a 2 x 2 factorial design, to compare the effect of discovery learning model and direct learning model or conventional model, directly to the learning outcomes of 80 high school students taken randomly to be the sample based on their thinking ability. The research design is described in Table 1.

Table 1. Research design factorial 2x2.

Thinking Skill (B)	Instructional Model (A)	
	Discovery (A1)	Direct (A2)
Formal (B1)	μA1B1	μA2B1
Concrete (B2)	μA1B1	μA2B2

Data analysis used descriptive statistic and inferential statistic for hypothesis tested. Hypothesis testing is preceded by analysis requirements test, including normality test and homogeneity test.

3. Results and Discussion

Description of result of this study can be seen in the Table 2.

Table 2. Description of research data.

Thinking Ability	Learning Model		Total
	Discovery (A_1)	Direct (A_2)	
Formal (B_1)	X 25,67	X 22,70	X 24,62
	s 2,80	s 1,93	s 3,07
	S^2 7,97	S^2 3,75	S^2 9,36
	N 20	N 20	N 40
Concrete (B_2)	X 19,83	X 20,90	X 20,71
	s 2,08	s 2,47	s 2,49
	S^2 4,35	S^2 6,08	S^2 6,22
	N 20	N 20	N 40
Total	X 23,04	X 22,03	X 22,64
	s 3,86	s 2,29	s 3,29
	S^2 14,87	S^2 5,23	S^2 10,87
	N 40	N 40	N 80

Result of hypothesis testing based on the Analysis of Variance (Anova) 2x2 factorial and the statistical of inferential showed that chemistry learning outcomes of students that learned with the discovery model is higher than that learned by direct learning model.

Teachers' teaching methods are factors that may influence the students' learning outcome. The discovery model applied in teaching colloidal system keeps students active, so the learning outcomes are high. The used of discovery model can encourage motivation, active involvement, and creativity of students, it also promotes autonomy and independence [5].

In this case, the using of student work sheet in teaching of colloidal system gave the student opportunity to expressed their thinking ability. Student with formal thinking ability showed high performance in doing laboratory practical. According to Piaget's theory, during the stage of formal operations probably less well understood than that which occurs during any other stage. At the age of 15-18 years, children become capable of solving problem even when the relationship between the two denominators is not a simple factorial one [6]. In the learning of colloidal system students were expected to find out examples of colloidal in their daily life, how to group colloidal system based on the characteristic of solvents, how to do practical work in the laboratory.

Discovery learning is also strongly tied to problem solving (or learning how to solve problems under a more meta-cognitive perspective): Learning theorists characterize learning to solve problems as discovery learning, in which participants learn to recognize a problem, characterize what a solution would look like, search for relevant information, develop a solution strategy, and execute the chosen strategy [7].

Thinking ability is a matter of attaining capabilities of reasoning including 16 logical operations which make up formal operations [8]. The concrete operational stage is the third in Piaget's theory of cognitive development. This stage lasts around seven to eleven years of age, and is characterised by the development of organized and rationale thinking. Piaget considered the concrete stage a major turning point in the child's cognitive development, because it marks the beginning of logical or operational thought. The child is now mature enough to use logical thought or operations (i.e. rules) but can only apply logic to physical objects (hence concrete operational). Children gain the abilities of conservation (number, area, volume, orientation) and reversibility. However, although children can solve problems in a logical fashion, they are typically not able to think abstractly or hypothetically [9]. Teaching with direct model only can facilitate the way student memorize examples of colloidal system. The students' achievement in chemistry related to their level of interest in the subject. It can be inferred that students who do not have the passion for chemistry may not be able to score high in chemistry exams [10].

4. Conclusion

The results of the study found that instructional such as discovery contributed in chemistry especially in colloidal system teaching. Students' score is higher for them who learned with discovery model compare to direct model. Related to thinking ability, students with concrete thinking skill learned with discovery model get the low scores compare to students learned with direct model. There an interaction between instructional model and thinking skill ability, so teachers have to find out ways to implemented the models.

References

[1] Brown, Theodore, LeMay E, Jr Bursten B E, and Murphy C J 2009 *Chemistry the Central Science* Pearson Prentice Hall, USA

[2] Farhana W Y and Zainun M A 2012 *Urban Students' Attitude towards Learning Chemistry* Elsevier

[3] Tami L N, Avi H, Rahel M N and Ziva B D 2004 Can Final Examination Amplify Students' Misconceptions in Chemistry? *Chemistry Education: Research and Practice* **5** 3 301-325

[4] Johnstone A H 1991 Why is Science difficult to Learn Things are seldom what they seem *Journal of Computer Assisted Learning* **7** 75-83

[5] Christopher P 2014 *Instructional Design Models and Theories: The Discovery Learning Model* https://elearningindustry.com/discovery-learning model

[6] Robbie C 1980 *Intellectual development and Instruction: A Neo Piagetian View* AETS Year Book, The Psychology of Teaching for Thinking and Creativity. Anton Lawson, editor.

[7] Borthick A F and Donald R J 2000 *The Motivation for Collaborative Learning Online and Its Application in an Information System Assurance Course, issues in Accounting Education.* http://aaajournals.org/doi/abs/10.2308/ioce.2000.15.2.181.

[8] Robert M G 1980 Learnable *Aspects of Human Thinking* AETS Year Book, The Psychology of Teaching for Thinking and Creativity. Anton Lawson, editor.

[9] Saul McLeod, Concrete Operational Stage https://wwwsimplypsychology.org/concrete.operational-html

[10] Nordin A and Chin M T 2010 *Pemahaman Konsep Pelajar Tingkatan Empat dalam Tajuk Ikatan Kimia di Skudai Johor 1-10* http://eprints.utm.my/11251/.

How to Improve Interest, IQ, and Motivation of Vocational Students?

H Sumual and D M Ombuh*

Mechanical Engineering Education Departement, Universitas Negeri Manado, Kampus Tondano 95618, Indonesia

*deivy.ombuh@yahoo.com

Abstract. The aim of this research was to study the effect of interest, motivation and IQ of students on the learning result. The survey method with quantitative approach was used in this study. The data were then analysed using path paradigm. Data were collected by questionnaire technique, special tests for IQ and documentation for learning outcomes. The results showed that the interest, IQ and the motivation influence significantly and positively on learning result as well as interest to learning motivation. However, no significant influence of IQ on Learning Motivation was detected in this research.

1. Introduction

Basically, learning result is a change behavior process from unknown to be known, from bad attitude to be better and unskilled to be skilled. The learning result which has not completed yet will become a barrier for the learning process [1]. There are several factors which affect to the learning result, such as interest, motivation and IQ. Interest is the attitude of someone including the three functions of his soul (cognition, conation and emotion), which is fixed on something and related to a strong element of feeling. Ahmadi is more concerned with the condition of interest that arises from the students [2]. The intelligence Questions (IQ) is one of the ways to measure the person ability in understanding something [3]. The motivation and the learning are two things that affect each other. Learning is a relatively permanent change in behavior and potentially occurs as a result of a practice or reinforced practice based on a goal to achieve a particular goal. Nowadays, poor of the learning result become a major problem in educational field. Many previous researchers have tried to investigate the effect of those factors, but still partially [4,5,6]. Therefore, in this research, a combination of some factors (i.e interest, IQ and motivation) was examined to distinguish their influences to the learning result where was applied to some students at Mechanical Engineering Education Department of Manado State University.

2. Research Methods

The method used in this research is survey method with quantitative approach which then analysed by path analysis to determine the direct and indirect effects of learning result. Relationships among all variables can be described in the constellation of the problem can see in figure 1.

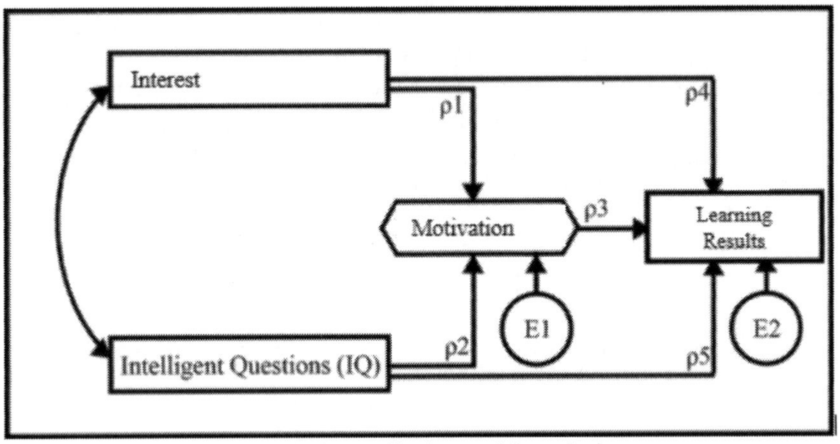

Figure 1. Constellation problem design.

Research was held at Department of Mechanical Engineering Education Faculty of Engineering Manado State University. The data were collected using questionnaire for interest and motivation variable, the special test (physiological test) for IQ and the transcript in each semester for learning results. The samples of this study were 63 respondents. The research variables are Interest (X1), Intelligent Questions (X2), Motivation (X3) and Learning results (X4). To analyse the data in this study used path analysis. The statistics data was calculated by Statistical Product and Service Solutions (SPSS) version 21.

3. Results and Discussion

3.1. The effect of interest on learning results

As shown, the value of t_{count} for learning interest variables to the learning results is greater than t_{table} where $t_{count} = 5{,}508 > t_{table} = 1{,}296$ at $\alpha = 0{,}05$. It means that the interest distributes the positive and significant influence on learning results. This result is agreed to the statement about the interest in learning process will encourages the student to achieve a good learning result [7].

3.2. The effect of intelligent questions on learning results

From the statistical, it can be seen that the value of learning intelligent questions variables to the results of learning is greater than ttable where $t_{count} = 4{,}318 > t_{table} = 1{,}296$ at $\alpha = 0{,}05$. It can be concluded that the influence of intelligent questions on learning result was positive and significant. This emphasizes in the theory of the influence of intelligent questions on learning results is evident that interest in learning encourages the students to have good learning results [8].

3.3. The effect of motivation on learning results

The value of t_{count} for learning motivation variables to the results of learning is greater than t_{table} where $t_{count} = 1{,}763 > t_{table} = 1{,}296$ at $\alpha = 0{,}05$ as displayed. Therefore, the motivation in learning influences positively and significantly on learning results. It corresponds to theory about interest in learning encourages the students to get a good learning results [9].

3.4. The effect of interest to motivation in learning

As shown, the value of t_{count} for learning interest variables to the motivation of learning is greater than t_{table}. Where is $t_{count} = 2{,}547 > t_{table} = 1{,}296$ at $\alpha = 0{,}05$. It means that the interest contributes the positive and significant influence on learning motivation.

This emphasizes the theory of the influence of interest to motivation is evident that interest in learning encourages the students to have good motivation [10].

3.5. The effect of intelligent questions to motivation in learning

In this study Intelligent Questions (IQ) have no direct effect on Learning Motivation. This is proved by the value of t_{count} for Intelligent Questions (IQ) variables on learning outcomes are smaller than t_{table} where $t_{count} = 0{,}843 < t_{table} = 1{,}296$ at $\alpha = 0{,}05$. It can be said that the no significant influence of the Intelligent Questions on Learning Motivation.

Unexpected thing was happened to the Intelligent Questions variable on Learning Motivation. This is possible because not all students who have Superior IQ or IQ have high learning motivation is good, due to over confidence in Intelligent ability which make the students have less motivation to learn. It can be seen in the previous report which describe that the students who have an IQ above average but lacked a strong learning motivation and good enough [11]. It is basically they have good learning result but it is a direct influence of IQ and does not necessarily describe good motivation as well.

Based on the findings, learning results can be influenced by several factors. such as interest, motivation and intelligent questions. From the perspective of data analysis, this paper conducts a research on student learning based on self. It is aimed at putting forward a rule-discovery approach suitable for the student learning result evaluation and applying it into practice so as to improve learning practice skills and finally better serve learning practicing [12].

4. Conclusion

From the above results, it can be concluded that the interest has a direct impact on the student learning result and the motivation of students at Department of Mechanical Engineering Education in Manado State University. Besides, the motivation contributes a direct impact on the learning result. On the other hand, the intelligent question offer a direct impact on the student learning result, but no significant effect on the motivation in learning of students at Department of Mechanical Engineering Education in Manado State University.

Acknowledgments

This research was supported by Technology and Vocational Education Department in Post Graduated Program at Manado State University.

References

[1] Supardi 2015 *Penilaian Autentik Pembelajaran Afektif, Kognitif, dan Psikomotor: Konsep dan Aplikasi* (Jakarta: Rajawali Pers.)

[2] Ahmadi A 2009 *Psikologi Sosial* (Jakarta : Rineka Cipta)

[3] Prasetyono D S 2015 *Yuk, Cari Tahu IQ dan Kepribadianmu* (Yogyakarta: Saufa)

[4] Fatmala 2008 Influence of intellectual intelligence level (IQ) on learning achievement of economic subjects of class X students of SMA Laboratorium Universitas Negeri Malang 2007-2008 academic year *Digital Library State University of Malang* **6** 2

[5] Hidayah 2016 The Influence of Interest and Motivation on Process Skills of Science And Learning Results Of Cognitive Ipa In Students Sd. *Journal Prima Edukasia* **4** 2

[6] Amiluddin R 2016 The Influence of IQ, CQ And EQ On Learning Achievement, And Motivation Student Learning Mathematics Education *Journal of Mathematics Education Research* **3** 1

[7] Krapp A 2009 Interest, Motivation and Learning: An educational-psichologycal perspektive *European Journal of Psichology of Education* **XIV**

[8] Reber A S, Walkenfeld F F, and Hernstadt R 1991 Implicit and explicit learning: Individual differences and IQ *Journal of Experimental Psychology: Learning, Memory, and Cognition*

[9] Carl A B and Edward L D 1984 Quality of Learning With an Active Versus Passive Motivational Set *American Educational Research Journal* **21** 4

[10] Ulrich S 2011 Interest, Learning, and Motivation. Journal Educational Psychologist **26**

[11] Donald L, Terrie M, and Magda S L 1993 Explaining the Relation Between IQ and

Delinquency: Class, Race, Test Motivation, School Failure, or Self-Control? *Journal of Abnormal Psychology* **102**

[12] Hongjie S 2010 Research on Student Learning Result System based on Data Mining *IJCSNS International Journal of Computer Science and Network Security* **10**

IOP Conf. Series: Materials Science and Engineering **306** (2018) 012050 doi:10.1088/1757-899X/306/1/012050

Multimedia Content Development as a Facial Expression Datasets for Recognition of Human Emotions

N E Mamonto*, H Maulana, D Y Liliana, and T Basaruddin

State Polytechnic of Jakarta, Depok, West Java, Indonesia

*Nicolaus.ernest.mamonto@hotmail.com

Abstract. Datasets that have been developed before contain facial expression from foreign people. The development of multimedia content aims to answer the problems experienced by the research team and other researchers who will conduct similar research. The method used in the development of multimedia content as facial expression datasets for human emotion recognition is the Villamil-Molina version of the multimedia development method. Multimedia content developed with 10 subjects or talents with each talent performing 3 shots with each capturing talent having to demonstrate 19 facial expressions. After the process of editing and rendering, tests are carried out with the conclusion that the multimedia content can be used as a facial expression dataset for recognition of human emotions.

1. Introduction

Humans are given the ability to express emotions he was experiencing. Emotions can be expressed through various things such as posture or gesture, voice, and facial expressions. A psychologist named Mehrabian in his research says that facial expressions accounted for 55% in message delivery. While the language and voice of each donate 7% and 38%. Current technological developments had a profound effect on all aspects of human life. Including face detection technology which has been widely used by devices or mobile computers that are used by humans. Face detection or facial recognition is one of biometric technologies that has been widely applied in security systems in addition to the introduction of the retina of the eye, the introduction of fingerprints and iris. In Indonesia until 2016, no researcher has done the research to the recognition of human emotions through facial expressions. Previous research conducted by New York University in 2013 in New York, USA. In the study, the subjects selected were local residents. And the study focuses only on images of human facial expressions and not videos. As a research material, the team needs datasets that contain video and facial images with certain expressions and emotions. In particular, the videos and images required are videos and pictures with Indonesian facial expressions. Talents are needed are 10 people, with each talent would be shot 3 times with emotion predetermined basic emotions are 6 types, 12 types of mixed emotions, and one facial expression without emotion. Shooting results must be edited into videos of each type of emotion and each video will create an image sequence from the first frame to the end. Therefore, authors carry out Multimedia Content Development as A Facial Expression Datasets for Recognition of Human Emotions to support the research team at the Laboratory Machine Learning and Computer Vision (MLCV) Faculty of Computer Science University of Indonesia so that new technologies can be generated that can be used by the public.

2. Literature Study

2.1. Video

According to Kamus Besar Bahasa Indonesia (KBBI) online the word video has meant the recording of live images or television programs to be aired on television. Meanwhile, The video is a sequence of many processing technology of moving pictures produced by the camera [1]. The video consists of a collection of pictures that are displayed in a sequence and in a certain speed so as to give the impression the images are moving. The set of images called frames and a measure of speed is called the frame rate as well as the units called frames per second (fps). Video is an electronic signal processing technology that represents moving images. A common application of video technology is television. Videos can also be used in engineering, scientific, production, and security applications. The term video is also used as an abbreviation videotape, video recorder, and video player. There are currently two categories of video, namely analog video and digital video.

2.2. Image sequence

Image sequences are videos that are rendered into each frame as individual images. Video with 200 frames will be exported as 200 individual pictures, each picture shows the process of the movement [2]. Meanwhile, image sequence is a collection of still images that form a sequence of movements [3]. In Multimedia Content Development as Datasets of Human Emotion Recognition Through Facial Expressions, image sequences generated from the rendering or exporting process of the raw video that has been edited. Image sequences contain the movement of talent expression of each frame.

2.3. Human emotion

Emotions are feelings or affections that can include passion of psychology, conscious experience, and behavioral experience [4]. Emotions are complex. Body, mind, and face play an important role in emotion. Emotions are divided into two categories: basic emotions and mixed emotions. Basic emotions can be defined as six, happy, sad, angry, surprise, disgust, and fear. Mixed emotions are defined as affective experiences that have the characteristic of activating two basic emotions, usually the two opposing valence emotions eg happy and sad.

2.4. Camera settings

The way the camera sees the object based on several things: the aperture (aperture), shutter speed (aperture), ISO or sensitivity [5]. Aperture or diaphragm is a window of light that enters from the lens to the film or sensor camera. Shutter speed is the speed of opening or closing the shutters of light to the sensor. The term ISO in the world of photography is the sensitivity of film or camera sensors. ISO is an acronym of the International Organization of Standarization. Exposure or lighting is the intensity of a light on a photo. Exposure is determined by aperture, shutter speed, and ISO.

2.5. Adobe Premiere Pro CC 2015

Adobe Premiere Pro CC 2015 is a video editing tool favored by video or professional fans and can enhance creativity and creative freedom. Adobe Premiere Pro is the most scalable, efficient and appropriate tool to edit a video. Adobe Premiere Pro CC 2015 can support various video formats including AVCHD, HDV, XDCAM, P2 DVCPRO HD, AVC-Intra, Canon XF, RED, ARRIRAW, QuickTime, and others [6].

2.6. Adobe Media Encoder CC 2015

Adobe Media Encoder is a stand-alone application that can run by itself or can be launched from Adobe Premiere Pro. Can work from Adobe Premiere Pro, Adobe Media Encoder adds the export queue files when the queue button is pressed on the setting of export in export panel settings. Aside from the Adobe Premiere Pro, Adobe Media Encoder can also work from some other source [7].

3. Design and Realization

Multimedia Content Development as A Facial Expression Datasets for Recognition of Human Emotions using the method of multimedia Development-Molina Villamil version. Multimedia content built in the form of video and images.

3.1. Design of multimedia products

Based on Villamil-Molina's multimedia development method, the design is divided into 2 stages, development and preproduction

3.1.1. Development. The idea came from the discussion and interview of the author with a research team at the Machine Learning and Computer Vision (MLCV) Laboratory of the Faculty of Computer Science, University of IndonesiaThe initial goal of this research is to produce multimedia content containing Indonesian facial expressions with 6 basic emotions: happy, sad, disgust, angry, and surprise, 12 mixed emotions are happily surprised, happily disgusted, sadly surprised, sadly fear, sadly disgusted , Sadly angry, fearfully surprised, disgusted surprised, fearfully disgusted, angrily disgusted, and an emotionless facial expression.

3.1.2. Preproduction. Preproduction is the stage of preparation of all kinds of needs required for production processes such as activity schedules, tool preparation, talent preparation, software preparation, and so on.

The required tools are camera and lighting set. The camera used was a Nikon D5100 and Canon EOS M3 obtained by way of rent. While the lighting set obtained by way made by researchers in accordance with the needs of the writer is for video shooting in the room and take done without the movement of the camera. The place required is a studio or location for shooting. Talent required is 10 people with the provisions of 5 men and 5 female gender. The talent must be Indonesian nationals. Talent is determined by the judgment method of the researcher.

3.2. Realization of multimedia products (production)

The production process according to Villamil-Molina involves the process of shooting and editing.

3.2.1. Proses video shooting. Video shooting is done using pre-prepared equipment that is rented camera and lighting set that has been made. The position of the camera, lighting set, and talent position is determined in such a way as to get maximum image results, Placement plan of camera and lighting set can see in figure 1.

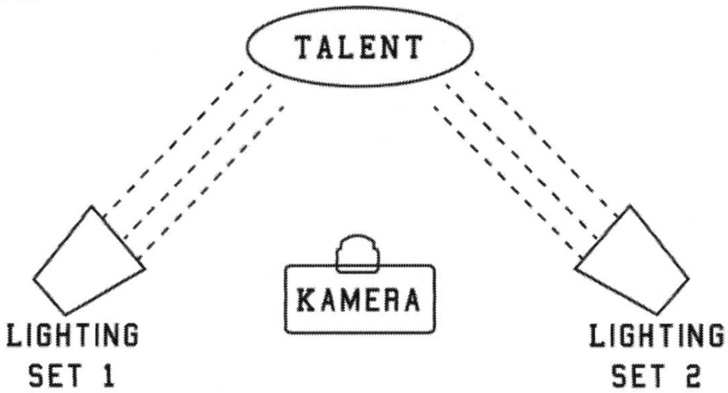

Figure 1. Placement plan of camera and lighting set.

In the first week, the camera used is Nikon D5100. This camera is used to take pictures of 3 talents. Settings performed on this camera are f2.3, ISO 100, and exposure +1.5. While in the second and third weeks, the camera used is Canon EOS M3. This camera is used to take pictures of 7 talents with details of 5 talents in the second week and 2 talents in the third week. Settings performed on this camera are f2.3 and ISO 100 and exposures +1.5. Each talent takes 3 shots. In each shot, talent demonstrates 19 specified emotions. The process of shooting or video shooting is done for 3 days in 3 different weeks. In the first week there are 3 talents or subjects who do the shooting. In the second week there are 5 talents that do take pictures. And in the third week there are 2 talents that do take pictures.

3.2.2. Editing process. The editing process is done to process raw video into a video and image sequences, as well as labeling in accordance with the talent, take, and the expression of the video. In the process of video editing, Adobe Premiere Pro CC 2015 is used. For the queue rendering process, Adobe Media Encoder CC 2017 is used.

a. Raw video editing
At this stage, the video that has been generated from the video shooting process is edited to become a clean video and can be an image sequence. Adobe Premiere Pro CC 2015 is used to perform video editing process. Each sequence is named according to subject and take video shooting. For subject 1 take 1, the sequence is named S1_T1_Master. For Subject 2 take 2, the sequences are named S2_T2_Master, and so on. Then raw video is inserted into the workspace sequence that has been made before. Then the video is trimmed using the Razor Tool according to the tag or marker of the start of the scene seen in the video preview. After the trim process for the video master finished, then made 19 new sequences for each video master. The 19 sequences are named according to the Subject, take, and expression shown. For subject 1, take one, happy expression sequence named S1_T1_Happy, to subject 2, take 2, sadly angry expression sequences named S2_T2_Sad-Birds, and so on. The naming of this sequence matches the label name to be given to the end result of the video and the image sequence. After a new sequence has been created, each previously trimmed video is inserted into a new sequence in accordance with the sequence name and marker on the video. After the video into the sequence has been determined, then do trim technique to remove parts that are not needed are part of the marker at the beginning of the video and transfer section at the end of the video scene. Then after the trim process is complete, the light settings are done by adding Brightness & Contrast effect. The brightness and contrast values for the resulting light are not much different from the light in a stable position in the middle of the video. The process of raw video editing is done to all raw video video shoot results. The total sequence required for the raw video editing process is 600 sequences.

b. Queue rendering process
Once the editing process has finished, the video needs to be rendered to produce video files that can be played on a universal video player. But due to the vast quantity of video, the queue rendering process is done using Adobe Media Encoder CC 2015. Queue rendering is done through Adobe Preimere Pro CC 2015 by setting export presets and then clicking queue button to queue queue to Adobe Media Encoder CC 2015. The format used for video is H.264, this format is selected to generate file with mp4 format that is currently being popular. As for image sequence, the selected format is JPEG.

c. Rendering process through Adobe Media Encoder CC 2015
Once all rendering queues have entered the render queue list, the rendering process is ready. The rendering process is done by clicking the start queue button in Adobe Media Encoder CC 2015 in the case of figure 2. Then automatically, Adobe Media Encoder CC 2015 will perform the rendering process in accordance with the settings that have been made before and in accordance with the queue that has been made.

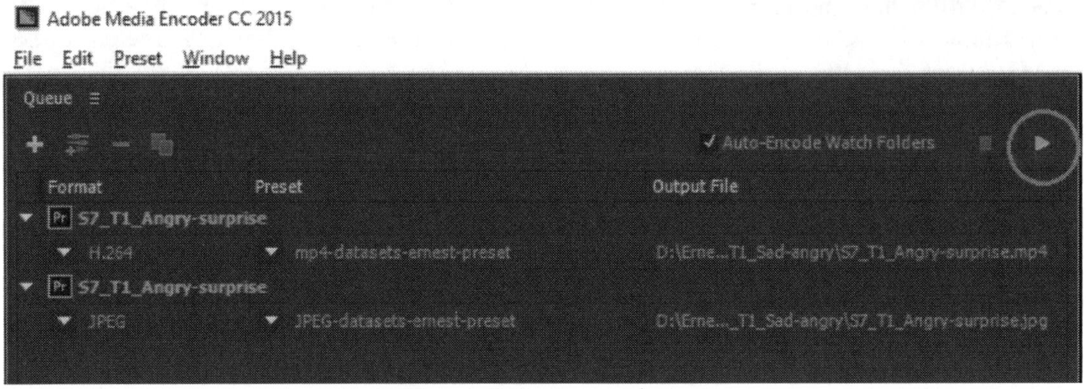

Figure 2. Start queue button.

The total file generated from this rendering process is 67,389 files consisting of 570 videos and 66,819 image sequences.

4. Discussion

After the production process is completed, then the next process is on the method of Villamil-Molina is post production which is testing phase.

4.1. Alpha testing

Alpha testing is done by looking at sample video and image sequences that have been completed in rendering and and will be used in beta testing [8]. The purpose of testing is to make sure no unneeded parts are still present in the rendered video.

4.2. Beta testing

Beta testing is done by testing the video and image sequences that have been produced [8]. Video testing is done by selecting video samples from 2 subjects and selected the same emotional sample. Each emotional sample is selected from 3 take that has been taken. Selected emotional samples are 1 emotionless expression, 1 basic emotion, 1 mixed emotion. While testing the image sequence is done by selecting 2 subject and selected the same emotional sample with the video. The selected expression is 1 take each. Then from the selected expression, the image is selected by selecting 2 images for 1 second of the video with a consistent frame interval or distance of 6 frames. The selected image is the first image and the 7th image in seconds running. In this test, the author chooses subject 8 and subject 9 that will be used as testing material. The selection of this subject is based on the judgment of the researcher with the assumption of using the best and worst subject according to the researcher. While the expression used is the expression without emotion, surprise expression (surprise), and sadly angry expression (sadly angry). As for the image sequence testing, there are 78 images consisting of 41 images for the first subject and 37 images for the second subject. Testing is done by showing the videos to the respondent. Each respondent was asked to fill in the questionnaire enclosed with 2 possible answers, is "Yes" and "No". Total respondents who did beta testing were 25 respondents for video testing and 25 respondents for image sequence testing.

4.3. Results and analysis of alpha testing

From the results of tests that have been done, it can be concluded that the alpha test, video and image sequence sample selected has been produced well and has been in the storage folder should be. The file size of each video is also in line with expectations, which is below 10 MB. All unneeded sections at the beginning of the video as well as at the end of the video are missing on the video's end result.

4.4. Results and analysis of beta testing

Table 1. Results of subject expression.

No	Subject Expression	Answer	
		Yes	No
1	Subject 9; Take 1; Facial expression without emotion (flat / neutral)	88%	12%
2	Subject 9; Take 1; Surprise expression	96%	4%
3	Subject 9; Take 1; Sadly angry expression	80%	20%
4	Subject 8; Take 1 Facial expression without emotion (flat / neutral)	96%	4%
5	Subject 8; Take 1; Surprise expression	36%	64%
6	Subject 8; Take 1; Sadly angry expression	46%	54%
7	Subject 9; Take2; Facial expression without emotion (flat / neutral)	48%	52%
8	Subject 9; Take2; Surprise expression	96%	4%
9	Subject 9; Take2; Sadly angry expression	88%	12%
10	Subject 8; Take2; Facial expression without emotion (flat / neutral)	96%	4%
11	Subject 8; Take2; Surprise expression	64%	36%
12	Subject 8; Take2; Sadly angry expression	32%	68%
13	Subject 9; Take3; Sadly angry expression	92%	8%
14	Subject 9; Take3; Facial expression without emotion (flat / neutral)	28%	72%
15	Subject 9; Take3; Surprise expression	96%	4%
16	Subject 8; Take3; Sadly angry expression	48%	52%
17	Subject 8; Take3; Facial expression without emotion (flat / neutral)	84%	16%
18	Subject 8; Take3; Surprise expression	44%	56%

a. Question number 1, 7, and 14, which is intended to assess the expression without emotion of the subject 9 showed 88%, 48%, and 28% answered yes. Based on these results, the average score is 54.67%. This result shows that the emotionless expression of subject 8 is not perfect, and needs improvement.
b. Questions 2.8 and 15, which were aimed at assessing the surprise expression of subject 9, obtained 96%, 96%, and 96% of the results answered Yes. Based on this result, the average score is 96%. This result indicates that the surprise expression of subject 8 is very good and deserves to be used as the next research data.
c. Questions 3, 9, and 13 aimed at assessing the sadly angry expression of subject 9 resulted in 80%, 88%, and 92% answered Yes. Based on this result, the average score is 86,67%. These results indicate that the sadly angry expression of subject 8 has been very good and feasible to serve as further research data.
d. Questions 4, 10, and 17 aimed at testing the emotionless expression of subject 8 resulted in 96%, 96%, and 84% answered Yes. Based on this result, the average score is 92%. This result shows that the emotionless expression of subject 8 is very good and deserves to be used as the next research data
e. Questions 5, 11, and 18 aimed at testing the surprise expressions of subject 8 resulted in 36%, 64%, and 44% answered Yes. Based on this result, the average score is 48%. This result indicates that the surprise expression of subject 8 is still not perfect and needs improvement.
f. Questions 5, 11, and 18 aimed at testing the sadly angry expressions of subject 8 resulted in 46%, 32%, and 48% answered Yes. Based on this result, the average score is 42%. This result indicates that the sadly angry expression of subject 8 is still not perfect and needs improvement.

These are the analysis of beta image sequence testing.

a. Testing image sequences for emotionless or neutral expressions get high percentage responses above 50% for all images of subject 1 or subject 2. This shows for flat expression can be shown from the first frame of the video to the last frame of the video. It is pronounced emotionless expression is an expression that does not require a stimulus to be expressed so that the appearance of this expression can be seen from the beginning to the end of the video.

b. In the first subject the peak of the surprise expression is in the 30th frame until the 54th frame of the test video is visible from the 90% and above values obtained for the frame. While the transition occurs in the 24th frame. In the second subject, the survey results show that the 18th and 24th frames have reached the peak of the surprise emotions from the second subject from the 12th frame to the 60th frame.
c. In the first subject, the expression of the sadly-angry mixture occurs on the 42nd to 66th frames and the 78th frame up to the 108th frame. The transition occurs on the 72nd frame. In the second subject, the peak of the expression occurs on the 24th to the 36th frame with a value of 92%. Expression continues until the 66th frame with the value decreases from 88% to 68%. This shows that the second subject raises an extreme sadly-angry expression at the beginning of the video and begins to lower the emotional level up to the 66th frame.

Analysis was also conducted to recap back results by dividing respondents into two groups based on education midwives' technology and non-technology [9]. The field of technology is better able to provide a firm assessment of the subject's expression although the image still shows the transitional expression from without emotion to expression with emotion. While the group of respondents with non-technical education field still shows doubt on some expressions. But in general, for other questions, the value obtained is not much different. This shows that the image sequences tested to the respondents can be well received by the respondents in general in accordance with expectations.

4.5. Delivery
Multimedia products are distributed to the research team at the Machine Learning and Computer Vision (MLCV) Laboratory of the Faculty of Computer Science University of Indonesia by the author through in the form of an external hard disk. Furthermore, the video and image sequences will also be distributed through the website to be accessible to the public at large and can be used to conduct similar research.

5. Conclusion
Based on the results of paper creation entitled "Multimedia Content Development as A Facial Expression Datasets for Recognition of Human Emotions" it can be concluded that:
1. Video and image sequences have been successfully created by involving 10 talents or subjects, with 3 shots for each subject, and each subject shoot should display 19 facial expressions.
2. The process of shooting is done for 3 weeks with the prepared equipment and talent that has been determined by the author. After the shooting process is complete, the editing process is done to remove the unnecessary parts of each video, labelling, and rendering process to produce video and image sequences according to the label specified.
3. After going through alpha testing, the entire video and image sequences are in the folder according to the subject, take, and the displayed scenes. Each video has also removed the scene marker at the beginning of the video and the scene transfer section at the end of the video. After going through beta testing, it can be concluded that video with subject 9 is better than video with subject 8, but overall video can be used as research data of human emotion recognition through facial expressions. Testing the image sequence done to 25 different respondents with the respondents on the video test resulted in the conclusion that in the neutral expression, the expression can be found from the first frame to the last frame, but for basic expression and mixed expression can be found on different frames but can be seen Transition expression more clearly on image sequence testing. The analysis was continued by dividing the respondents based on the education field i.e. technology and non technology. Respondents in the field of technology can provide a more assertive assessment of facial expression shown compared to respondents from non-technological areas.

References

[1] Kurniawan E 2014 *Analisis kualitas real time video streaming terhadap bandwidth jaringan yang tersedia* (Medan: Universitas Sumatra Utara)

[2] Smith C 2013 *Video in Photoshop for Photographers and Designers. 1st ed.* (Berkeley, Calif.: Peachpit Press)

[3] Santosa B dan Purnama B 2010 Perancangan Studio Mini Berbasis Multimedia Universitas Surakarta *Journal Speed – Sentra Penelitian Engineering dan Edukasi* **2** 2 17-25

[4] King L A 2011 *The Science of Psychology An Appreciative View (2nd ed.)* (New York: McGraw-Hill)

[5] Kurniawan H 2013 *DSLR Untuk Pemula 1st ed* (Jakarta: Mediakita) 8-15

[6] Jago M 2015 *Adobe Premiere Pro CC Classroom in a Book (2015 Release). 1st ed.* (Indianapolis: Adobe Press)

[7] Adobe Creative Team 2010 *Adobe Premiere Pro CS5 Classroom in a Book* (California: Peachpit)

[8] Windiyani T 2012 Instrumen Untuk Menjaring Data Interval. Nominal, Ordinal Dan Data Tentang Kondisi, Keadaan, Hal Tertentu Dan Data Untuk Menjaring Variabel Kepribadian *Jurnal Pendidikan Dasar* **3** 5 203-208

[9] Darma A A G B 2013 Analisis Sistem Pengadaan Bahan Dan Peralatan Pada Proyek Kontruksi Jembatan Tukad Penet Di Badung Bali *Jurnal Ilmiah Elektronik Infrastruktur Teknik Sipil* **2** 2 XII1-XII7

IOP Conf. Series: Materials Science and Engineering **306** (2018) 012051 doi:10.1088/1757-899X/306/1/012051

Rubric Assessment on Science and Creative Thinking Skills of Students

H Ratnasusanti*, A Ana, P Nurafiati and L Umusyaadah

Universitas Pendidikan Indonesia, Bandung West Java, Indonesia 40154

*henniratna72@gmail.com

Abstract. The result of the monitoring and evaluation of the latest Indonesian curriculum (the 2013 curriculum) implementation at junior high school level year of 2014 showed that one of the difficult things that learners had in implementation 2013 curriculum is doing the result. The characteristic of applying the 2013 curriculum is to emphasize the modern pedagogic dimension of learning, which is using scientific approach, which requires learners to have high-level thinking skills, one of which is creative thinking skills. The aims of this research is to implement performance assessment in measuring the creative thinking of junior high school students on subject Prakarya. The form of the main performance assessment is the task and assessment criteria. The experimental method that been used is the Quasi Experiment with Non-Equivalent Design Group Research. Population in this study is the students of VIII class of junior high school in Bandung, Indonesia which consists of six classes. And two classes are selected for the sample from that six classes and VIII A class were chosen, while VIII F class has been chosen as control class. The result of this research showed that the rubics of performance assessment can be measure or identify the creative thinking skill, its prove by the result of pre-test dan post-test are more dominant. In material of identification student's creative thinking skills are reached an average 85 compare 79 with the control class. while in the presentation the experimental class got an average of 85 bigger than the control class which only reached 79.

1. Introduction

The result of monitoring and evaluation of the implementation of junior high school 2013 curriculum in year of 2014 shows that one of the difficulties of educators in implementing the 2013 curriculum is in carrying the assessment. More less about 60% from the total of the educators who become respondent said that they have not been able to design, implement, processing, report and utilize the result of the assessment, they have not fully understood how to prepare the instrument and rubic skills assessment [1]. Assessments that been measured in the 2013 curriculum are not only in cognitive domain, but also to measuring the sphere of attitude and skill. Aspect skills in the 2013 curriculum are expressed in KI-4. Student's ability can be seen by the teacher through an assessment of the skills aspect. An assessment that can be used to measure the skill aspect is the performance appraisal [2]. Performance assessment is a valuation technique that ask the students to demonstrate and applying the knowledge into real context [2]. The Assessment is used on student performance, behaviour or interaction.

The existence of the 2013 curriculum has change the old paradigm, which is learning that been centered to the teacher (teacher centered) to the new paradigm which is learning that been centered to the student (student centered). The characteristic of applying the 2013 curriculum is to emphasize the modern pedagogic dimension of learning, which is using scientific approach, which requires learners to have high level thinking skills, one of which is creative thinking skills. Creative thinking is a mental activity to develop original, aesthetic, constructive ideas related to conceptual views, and emphasize the rational aspects of thinking [3]. Creative thinking in a learning demands something new created.

Several studies have been conducted with the performance assessment and creative thinking skills such those that have been done by [4] the result of the research showed that the performance assessment instrument or skills can help the teacher to identify student's creative thinking skills. Other study that been done by [5] the result showed that student's opinions about the use of instruments and rubic of skills based on creative thinking in consumer education learning included in the very well category

Prakarya is not a new subject in 2013 curriculum. In 2006 curriculum the name of this is Skills. Broadly speaking the principle between the subjects of skills and workshops is the same, the difference in the workshops has the purposes and basis of "education" to cultivate the sensitivity for the local wisdom products, technological developments and the development of entrepreneurial spirit in accordance with the orientation and 2013 curriculum mission.

The aim of this research is to implemented the performance assessment in measuring the junior high skill creative thinking skill on the skill subject. The main form of this assessment is task and assessment criteria (rubics).

2. Research Methods

The research method that been used is quasi experiment research with the design of research is Non-equivalent group design. The design of this research is been draw like below

Table 1. Non equivalent control group design.

	Pre-test	Treatment	Post-test
Experiment Class	O_1	X	O_2
Control Class	O_1		O_2

Information

O_1 : The beginning test of creative thinking ability and scientific attitude before being given treatment in experimental group and control group

O_2 : The final test of creative thinking ability and scientific attitude before being given treatment in experimental group and control group

X : Applying performance assessment

The location of this research was conducted in SMP Negeri 2 Pameungpeuk kabupaten Bandung. The population in this research is all of the student of clas VIII spread in 6 study group. The sample in this research is 82 people consist of 44 students as the experiment class and 38 students as the control class. The sampling technic used purposive sampling technique, with the criteria of class VIII A and VIII E have the same cognitive abillity. Techniques and instruments of data collection in this research are : (1) giving the test and task; (2) observation; (3) quetionnaire distribution, and (4) documentation.

This instrument is used to measure student's affective and psychomotor aspects with the direct observation. The affetive aspect that been observed were liveliness, critically (asking / answering), coorperation, responbility, discipline, neatness, and cleanliness, while the psychomotr aspect that been observed are skill of using / assembling tools, performing experiments, and observing / recording data. Cognitive teste are performed using a tes method which consist of pre-test and post-test.

3. Results and Discussion

As has been pointed, the principle performance assessment consists of two parts, namely task and criteria (rubics). Performance tasks can be project, exhibits, portfolios and task that requires the students to demonstrate the ability to handle complex things through the application of knowledge and skills about about something in the most real (real-world application). Criteria or rubics are guidelines for scoring, must be clear and agreed by the students and the educators.

The result of the student creative thinking ability in the subject of workshop for cognitive test obtained that the data from the pretest and posttest result. The data on student's pretest and posttest values shown in table 2 below.

Table 2. Distribution of student value acquisition on pretest and posttest.

No	Interval value	category	Experiment class				Control class			
			Pre Test		Post Test		Pre Test		Post Test	
			F	P	F	P	F	P	F	P
1	50 - 74	Low	21	48 %	5	11%	17	45%	16	42%
2	75 - 87	Medium	23	52 %	35	77%	21	55%	20	53%
3	88 -100	High	0	0%	4	11%	0	0%	2	5%
Result			**44**	**100%**	**44**	**100%**	**38**	**100%**	**38**	**100%**

From the table 2 above, it shows that the result for cognitive tests in the application of performance assessment on pretest and posttest show the different result between the experiment class with the control class, and we can see that the student's categories in control class and experimental class are mostly in the 75-78 interval in average category dan its completely done, because the studying object of KKM is 75.

Performance assessment is also conducted on the learner's activity in the implementation based on the criteria that have been set, next are the rubics for performance appraisal to create products on aspects of processing.

Table 3. Assessment rubics creating products on processing aspects.

Rated Aspect		Items Rated	Assessment Guidelines
1. Planning	A. Identification of Needs	Conformity with needs	
	B. Idea	Learners original (own idea)	
II. Implementation / preparation	A. Preparation of material	Complete material	Range of scores per item
	B. Completeness of tool fittings	Complete tools	Rated 0-10.
			Final Value =
	The process of making	The method used	<u>Total score of x 10</u>
III. Presentation	A. Product	level of fineness	Number of grains assessed
	B. Packaging	packing material conformity	

Aspects assessed in the performance appraisal rubic (table 3) include planning, implementation and presentation. Assessment of product that have been done to the product creating by the learners based on the criteria that been set. Next are the average result of the product assessment for the workshop lesson by the processing aspects. The results of material identification, including planning, execution and presentation are outlined in the following table 4 below.

Table 4. Recapitulation of average acquisition of student value on material identification.

No	Description	Experiment class	Control class
1	Planning	85	78
2	Implementation/ Preparation	87	80
3	Presentation	89	78
	Average score	87	79

From the table 4 it can be seen that the average score in creating product in the experimental class obtained 87 and the control class obtained 79, so we can say that in creating product the experimental class has the higher average score than the control class, its mean that the experimental class is more creative than the control class. Performance appraisal is also being done to the activities of learners on the presentation of practicum reporting. Rubics assessment can been seen in the table 5 below

Table 5. Examples of class presentation check list.

No	Student Name	Aspect Assessed										
		1	2	3	4	5	6	7	8	9	10	∑
1	Ahmad	√	√	√	√	√	√	√	√	√	√	10
2	Agil											
3	Budi											
Etc	...											

Score = total result

Description for table 5:

1) The issues discussed clearly formulated
2) There is relevance of the description with the issues discussed.
3) The description is broad and deep
4) The description is clear and not wrong concept
5) The description is delivered smoothly
6) Disclaimer / argumentation is logical and strong
7) Language is good and right
8) Clothes are neat and complete
9) Not Emotional
10) Accepting Opinions of Others

Recapitulation of the average value of the presentation can be seen in table 6 on the next page.

Table 6. Recapitulation of average acquisition of student value in presentation/reporting results of practice/project.

No	Description	Experimental Class	Control Class
1	The issues discussed clearly	80	76
2	There is a relevance description with the issues discussed.	85	76
3	Descriptions are clearly	87	75
4	The description is right without wrong concept	85	79
5	The description is delivered smoothly by the students	90	78
6	Disclaimer / argumentation is logical and strong	80	75
7	Using good and right language	80	80
8	Clothes are neat and complete	87	88
9	Not emotional	85	80
10	Accept others argue	88	80
	Avarege score	85	79

Based on table 6, it can be concluded that the average score for the experimental class presentation is 85 dan control class is 79, so we can have concluded that the experimental class has the higher score than the control class. The experimental class has the higher score at report submission which is 90, meanwhile the control class has the higher score at the clothing level which is 88. The lowest score in experimental class is in the clarity of the formulation problem, the rebuttal and the language usage is 80, while the lowest score for the control class is in the description of the report and the refutation is equal; to 75

4. Conclusion

Based on the results of the study, it can be concluded that the implementation rubics of performance assessments significantly influence the ability of teachers in identifying creative thinking skills and the result of the student that have been learned. This result is then supported by the average value of some activities, which shows the experimental class gets an average value greater than the control class. The results of pre-test and post-test show the result is more dominant. In the creating product, student's creative thinking skills reached an average of 87 to 79 with the control class, in the experimental class presentation having an average of 85 more bigger than the control class which only reached 79.

References

[1] Kementerian Pendidikan dan Kebudayaan Pertama, D J P D dan M D P ekolah M 2016 *Panduan Penilaian Oleh Pendidik dan Satuan Pendidikan Pertama. Kementerian Pendidikan dan Kebudayaan Direktorat Pendidikan Dasar dan Menengah, ed.* (Jakarta: Kementrian Pendididkan dan kebudayaan Direktorat Pendidikan Dasar dan Menengah)

[2] M Mardianah 2014 *Penggunaan Performance Assessment (Penilaian Kinerja Pada Pembelajaran Biologi dengan Kurikulum 2013 (Penelitian Deskriptif di SMAN Kota Tangerang Selatan)* Universitas Islam Negeri Syarif Hidayatullah,

[3] Arnyana I B P 2007 Pengembangan Peta Pikiran untuk Peningkatan Kecakapan Berpikir Kreatif Siswa *Jurnal Pendidikan dan Pengajaran UNDIKSHA* 3 670–683

[4] Marwiyah S 2015 Pengembangan Instrumen Penilaian Keterampilan Berpikir Kreatif pada Mata Pelajaran IPA Terpadu Materi Atom, Ion, dan Molekul SMP Islam Al Falah *Jurnal Edu-Sains* **4** 26–31

[5] Sri W 2014 *The Effectiveness of Using Instrument and Rubric of Creative Thinking Skill-Based Assessment Project in The Learning of Consumer Education* Fakultas Teknik Universitas Negeri Yogjakarta

IOP Conf. Series: Materials Science and Engineering **306** (2018) 012052 doi:10.1088/1757-899X/306/1/012052

Production-Based Education Model for Improving Technical and Vocational Teachers Ability

H Saputro*, Suharno, I Widiastuti and B Harjanto

[1]Department of Mechanical Engineering Education, Universitas Sebelas Maret, Jl. Ahmad Yani No.200, Pabelan, Surakarta, Indonesia.

*hermansaputro@staff.uns.ac.id

Abstract. The vocational high schools (SMK) in Surakarta have applied the Production-Based Education (PBE) in order to improve the student skills. On the other hand, the implementation of PBE model is needed the professional teachers who have the skill qualifications in the production and learning. This study focuses on strengthening of teacher's behavior related to the implementation of PBE model at vocational high schools in Surakarta, especially on teacher's Organizational Citizenship Behavior. This research conducted based on the observational research method that employed 134 teachers of SMK Warga Surakarta as respondents. The data of teachers OCB were analyzed by using the descriptive analysis. The result showed that the PBE model had strengthened the teachers OCB at SMK Warga Surakarta. The measurements of OCB level of SMK teachers had shown the high (H) level and some of the teachers' OCB characters show very high (VH) levels such as altruism and civic virtue. OCB or "good soldier syndrome" could be manifested by the teacher's interaction at school such as loyalty, volunteering and helping others. Therefore, teacher's OCB behavior became important due to the organizational effectiveness at vocational high school (SMK).

1. Introduction

The teacher's reform in Indonesia was started after the enactment of the Law No 14/2005 on Teachers and Lecturers. This law stated that teacher and lecturer in Indonesia should hold an academic qualification and teaching certificate. The teacher reform was facilitated by the government through the teacher professional education or *Pendidikan Profesi Guru* (PPG) in Indonesian. Through this program, the professional teacher gets the teaching certificate. Therefore, the teacher becomes the key point of improving the education quality in Indonesia. In case of Vocational High School (VHS) or *Sekolah Menengah Kejuruan (SMK)*, the professional teacher was required capable in vocational skill and teaching skill [1]. These teacher specifications are related to the main objective of VHS in Indonesia. The SMK is an education for work-oriented at the secondary level of the national education system. The SMK prepared the student's skill for ready to work in the industry.

Production-Based Education (PBE) is one of the teaching models that can be applied at SMK due to the student's skill improvement. PBE model have similarity to the teaching factory model [2]. The teaching factory concept is based on the triangle notion of knowledge i.e. education, research and innovation [3] as shown in Figure 1 on the other hand the PBE concept is integrated education and manufacturing based on industrial demands [2] as shown in Fig. 2. PBE is a new paradigm in manufacturing engineering technology and education. SMK Warga Surakarta is one of the vocational high schools in Surakarta that had applied the Production-Based Education (PBE). As evidenced by the

IOP Conf. Series: Materials Science and Engineering **306** (2018) 012052 doi:10.1088/1757-899X/306/1/012052

implementation of PBE model at SMK Warga Surakarta, the students' skills improved and the graduates acceptable in the job market or industry. This case is interesting and needs to be observed, especially the success story in the implementation of PBE model at SMK.

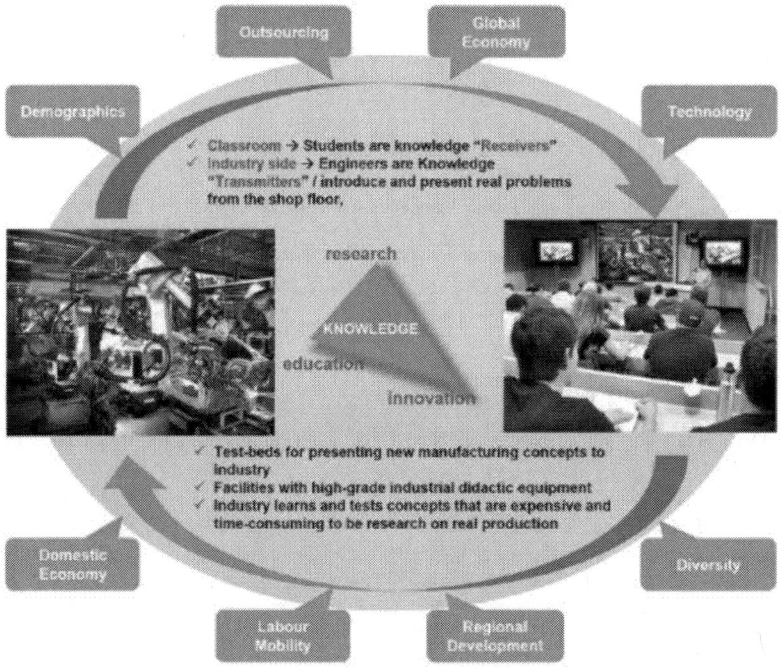

Figure 1. Concepts of Teaching Factory [3].

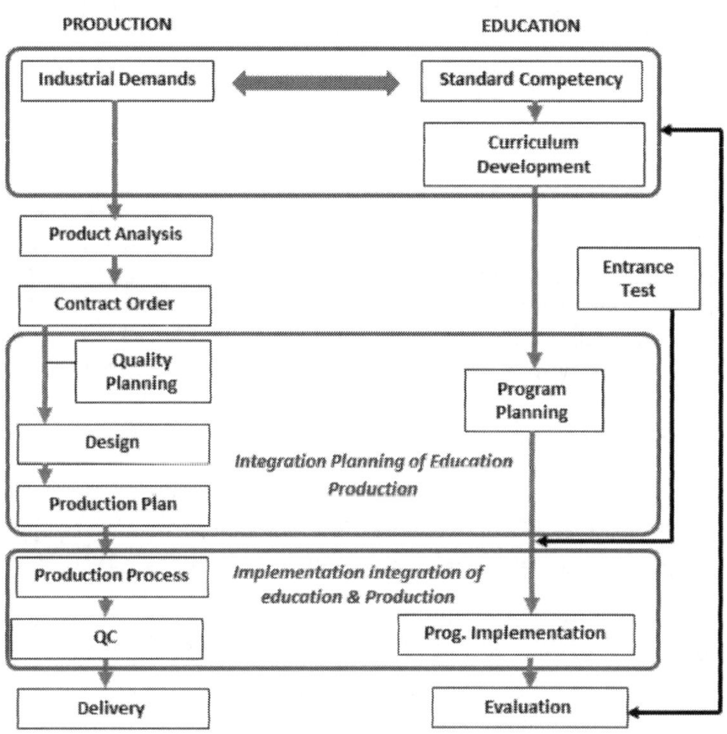

Figure 2. Concepts of PBE [2].

The implementation of PBE model was needed the professional teachers who have the skill qualifications in the production and teaching. Therefore, this study focused on the teacher's behavior related to the implementation of PBE model at SMK Warga Surakarta. This study concentrates on the strengthening the teacher's Organizational Citizenship Behavior (OCB) in the vocational high school through the PBE model. The teacher's OCB behavior became important to study because OCB is a good way to the vocational high school to reach the organizational effectiveness.

Since the Organizational Citizenship Behavior was introduced by the Bateman and Organ, the implementation of OCD Widely used in all discipline, such as social, psychology education, etc. [4]. Smit and Organ was identified the OCB as "good soldier syndrome" [5,6].

Organizational Citizenship Behavior has five categories of character [6] i.e. 1) Altruism is the willingness to provide assistance to co-workers. 2) Conscientiousness means that organizational members have in-role behavior levels above the required minimum standards. 3) Civic virtue is the active participation of members of the organization in thinking of organizational progress, e.g. always looking for the latest info that supports the progress of the organization. 4) Sportsmanship is more emphasis on the positive aspects of the organization than the negative aspects, e.g. disrespectful behavior protests, not complain, and not exaggerate small / trivial problems. 5) Courtesy is doing good and respectful to other members of the organization.

Organizational Citizenship Behavior at school could be manifested by the teachers in school interaction such as loyalty, volunteering and helping others [6]. Many researchers have study the teacher's OCB at school [7], [8,9], and [10]. Investigated the display levels of teacher's OCB at the secondary school in Turkey [7]. Oguz, conducted the study on the relationship between leadership styles and the OCB of teachers in Samsun city, Turkey [8]. Zeinabadi, studied on job satisfaction and organizational commitment as antecedents of teacher's OCB in Tehran, Iran [9]. Runhaar had examined the influence of teachers work context on the relationship between their work engagement and OCB at the secondary school in the Netherlands [10].

This study adopted the teacher's work context that applied production-based education [10]. PBE is a form of learning program by empowering the skills of teachers in Vocational High School. On the other hand, this learning model causes problems or friction between teachers in Vocational High School. It was caused by several factors such as the unequal workload, pressure from industry partners, the pressure of school leaders and teachers' welfare level. This was evident of vocational high schools that became the pioneer of teaching factory program cannot exist until today. Unlike the case of SMK Warga Surakarta that is still consistent and successfully developed production-based learning. This study observed the positive behavior of teachers of SMK Warga Surakarta (OCB's level of teachers). The observation was focused in the strengthening of the teacher's organizational citizenship behavior (OCB) through the PBE.

This study contributes to theories about the impact of organizational and individual factors on organizational citizenship behavior in general and within Vocational High School (VHS) in particular. The practical value of our study is reflected in the guidelines "*How to strengthen the teacher's organizational citizenship behavior (OCB) through the PBE.*"

2. Methods

This research conducted based on the observational research method and employed all teachers in SMK Warga Surakarta (134 teachers) as respondents. The data was collected by the questionnaires. The OCB questionnaires were created based on the 5 categories of OCB character i.e. 1) Altruism, 2) Conscientiousness 3) Sportsmanship 4) Courtesy and 5) Civic Virtue [6]. The indicators were derived from the 5 categories of OCB's character as shown in Table 1. The questionnaire consists of 22 items using 5 Likert-Type Scale (1=very low to 5=very high), 5 item related to altruism, 4 item related to conscientiousness, 4 item related to sportsmanship, 4 item related to courtesy, and 5 item related to civic virtue. The interpretation of arithmetic averages is as follow: quite low (1.00-1.79), low (1.80-2.59), medium (2.60-3.39), high (3.40-4-19) and very high (4.20-5).

IOP Conf. Series: Materials Science and Engineering **306** (2018) 012052 doi:10.1088/1757-899X/306/1/012052

Table 1. The 5 categories of OCB's character and indicators.

No	Categories of character	Indicators
1.	Altruism	1. Teamwork 2. Responsive to the school leader's programs
2.	Conscientiousness	1. Work discipline 2. Effectively work and job responsible
3.	Sportsmanship	1. An organization member who does not complain at work 2. The ability of teachers to solve work problems
4.	Courtesy	1. Keeping the school's image 2. Teacher's contributions to the school
5.	Civic Virtue	1. Looking for the latest info that supports the progress of the PBE 2. Teacher's concern to the PBE

3. Results and Discussions

Descriptive statistics of the data analysis such as frequency, percentage, mean and standard deviation was shown in the Fig. 3. The teachers OCB of SMK Warga Surakarta has highest mean (M=3.98, SD=0.39). It means that the PBE has the positively related to the teachers OCB of SMK Warga Surakarta.

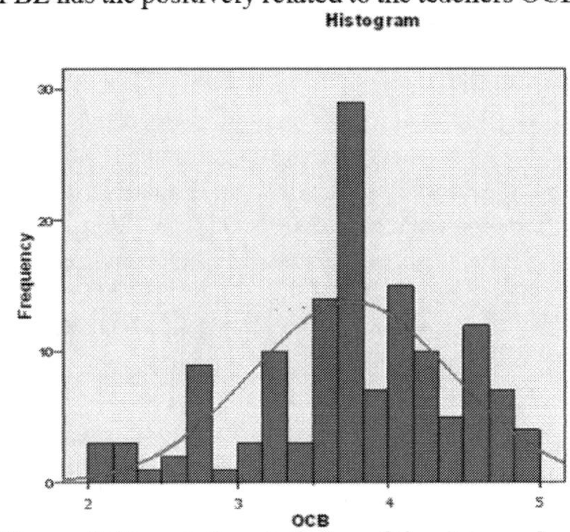

Figure 3. Descriptive statistics of the data analysis.

Table 2 presents the results of data analysis of OCB character and OCB level of SMK Warga Surakarta teachers. The character of altruism and civic virtue of SMK Warga teachers occupies the very high level (VH), while conscientiousness, sportsmanship, and courtesy were high level (H). Thus, the average of OCB level of SMK Warga Surakarta teachers was high level (H). This shows that the existence of PBE learning model at SMK Warga gave the positive impact into the character of teacher's OCB. The following will be discussed each character of teacher's OCB at SMK Warga Surakarta:

Table 2. OCB level of SMK Warga Surakarta teachers.

		Average and Level	
		$\bar{X}$	*Level*
Categories of character	Altruism	4.4	Very High (VH)
	Conscientiousness	3.7	High (H)
	Sportsmanship	3.5	High (H)
	Courtesy	*4.1*	High (H)
	Civic Virtue	4.2	Very High (VH)
	OCB	**3.98**	**High (H)**

3.1. Altruism character of teachers of SMK Warga Surakarta
Altruism character of teachers of SMK Warga Surakarta had shown the very high level (X=4.4). The altruism character of SMK-Warga teachers was shown in the behavior of mutual help among teachers such as in the activities of PBE lesson plans, interpreting the desire of industrial partners and supervising the quality of the student's production. By implementing PBE learning model, the teamwork between teachers at SMK Warga Surakarta becomes solid. The solid teamwork was triggered by the teacher's responsibility for the successful implementation of the PBE learning model. In particular how to maintain the quality of student work was acceptable to partner industry or customer. Evidence from strong teamwork SMK Warga was trusted as CNC machine assembly by PT. Gisma Indonesia. In addition, SMK Warga was trusted by the industries around the Surakarta city to produces the spare parts.

3.2. Conscientiousness character of teachers of SMK Warga Surakarta
Conscientiousness character of teachers of SMK Warga Surakarta had shown the high level (X=3.7). The character of conscientiousness of SMK Warga Surakarta teachers was demonstrated by discipline and also put forward effective working pattern in PBL model. Teachers recognize that PBE-based learning demands fairness and work effectively because the partner industry or customer demands that orders be completed within the scheduled time. This character is supported and evolved with the learning of the PBE model. Thus, the learning model of PBE is able to improve and strengthen the conscientiousness of SMK Surakarta teachers such as discipline and effective work.

3.3. Sportsmanship character of teachers of SMK Warga Surakarta
Sportsmanship character of teachers of SMK Warga Surakarta had shown the high level (X=3.5). The research findings could be seen that school managers (principals and leaders) were able to provide justice for teachers and students in the learning process of the PBE model. This can be seen from the absence of conflict or protests arising from teachers or students related to the implementation of the PBE. The condition of the school becomes conductive and more comfortable for students to learn and for teachers to be optimal in work. Indicators of research that discuss the emergence of conflict between teachers, teachers versus school leaders, and teachers versus students almost pseudo respondents said very rarely. It is evident that the sportsmanship character of teachers of SMK Warga Surakarta increases with the complexity of the work due to the application of the PBE model.

3.4. Courtesy character of teachers of SMK Warga Surakarta
Courtesy character of teachers of SMK Warga Surakarta had shown the high level (X=4.1). This average value was close to the very high level. The results of questionnaire analysis revealed that SMK Warga teachers were very eager to apply the learning model of PBE. This is due to the motivation for self-actualization of their skills and motivation for school image in the community and the workplace. They realize that private schools like SMK Surakarta Warga will be able to survive in the competition if the school has a specificity and positive image in the community and the workplace. This positive image has made SMK Warga easily get good prospective students (input), and the graduates (output) easily got a job. Thus, the PBE learning model had become a positive trigger for increasing courtesy character of teachers in SMK Warga in constructing school positive image.

3.5. Civic virtue character of teachers of SMK Warga Surakarta
Altruism character of teachers of SMK Warga Surakarta had shown the very high level (X=4.2). The result of questionnaire analysis related to the civic virtue character shows that the implementation of PBE learning model affected to the increasing the active participation of teachers at SMK Warga Surakarta. This is due to the innovations that have been conducted by the teachers related to the product development of PBE. Consumer orders are increasingly diverse, such as machining parts, CNC machines to CNC training and heavy equipment from PT. United Tractor and PT. Bisma Indonesia. Therefore the PBE model can improve the civic virtue character of SMK Warga teachers reflected in

the willingness to make innovations for the successful implementation of learning and satisfaction of industry partner or customer.

4. Conclusion

The results showed that the learning model of PBE that was developed by SMK Warga Surakarta suitable with the wishes of the teachers. The PBE learning model has been able to encourage teachers to work hard, even though the teacher's workload becomes more or increases. Overall, the OCB level measurements of Surakarta SMK teachers was high (H) level and some of the teachers' OCB characters show very high levels such as altruism and civic virtue. The real impact of teacher OCB strengthening can be seen in 5 categories of OCB characters as follows:

a. The altruism character of SMK-Warga teachers was shown in the behavior of mutual aid during conducted learning model of PBE.
b. The learning model of PBE can improve and strengthen the conscientiousness of SMK Surakarta teachers such as discipline and effective work.
c. The sportsmanship character of teachers of SMK Warga Surakarta increases with the complexity of the work due to the application of the PBE model.
d. The PBE learning model had become a positive trigger for increasing courtesy character of teachers in SMK Warga in constructing school positive image.
e. The PBE model can improve the civic virtue character of SMK Warga teachers reflected in the willingness to make innovations for the successful implementation of learning and satisfaction of industry partner or customer.

Acknowledgements

This research was partly subsidized by Grant of MRG research PNPB Universitas Sebelas Maret 2017. The researchers gratefully acknowledge financial support from the LPPM Universitas Sebelas Maret, Surakarta, Indonesia.

References

[1] Estriyanto Y, Kersten S, Pardjono, Sofyan H 2017 The Missing Productive Vocational High School Teacher Competency Standard in the Indonesian Education System *J Tech Educ Train* **9** 26-44
[2] Ilyas I P and Semiawan T 2012 Production-based Education (PBE): The Future Perspective of Education on Manufacturing Excellent *Procedia - Soc Behav Sci* **52** 5–14
[3] Chryssolouris G, Mavrikios D, and Rentzos L 2016 The Teaching Factory: A Manufacturing Education Paradigm *Procedia CIRP* **57** 44–8
[4] Bateman T S and Organ D W 1983 Job Satisfaction and the Good Soldier: The Relationship Between Affect and Employee "Citizenship" *Acad Manag J* **26** 587–95
[5] Smith C A, Organ D W, Near J P 1983 Organizational citizenship behaviour - its nature and antecedents *J Appl Psychol* **68** 655–63
[6] Organ D W 1988 *The Good Soldier Syndrome* (Lexington, Mass: Lexington Books)
[7] Polat S 2009 Organizational citizenship behavior (OCB) display levels of the teachers at secondary schools according to the perceptions of the school administrators *Procedia - Soc Behav Sci* **1** 1591–1596
[8] Oğuz E 2010 The relationship between the leadership styles of the school administrators and the organizational citizenship behaviours of teachers *Procedia - Soc Behav Sci* **9** 1188–1193
[9] Zeinabadia H 2010 Job satisfaction and organizational commitment as antecedents of Organizational Citizenship Behavior (OCB) of teachers *Procedia - Soc Behav Sci* **5** 998–1003
[10] Runhaar P, Konermann J, Sanders K 2012 Teachers' organizational citizenship behaviour: Considering the roles of their work engagement, autonomy and leader-member exchange *Teach Teach Educ* **30** 99–108

Entrepreneurship Education through Industrial Internship for Technical and Vocational Students

H Sumual[*] and G J Soputan
Universitas Negeri Manado, Tondano 95618, Sulawesi Utara, Indonesia

[*]herrysumual@unima.ac.id

Abstract. The spirit of entrepreneurship must be formed in an integrated process in learning so that commitment, responsibility and work ethic on the candidate of vocational high school graduates are effectively promoted. Learning teaching factory teaches students how to find problems, build prototypes, learn to make business plan, and learn to present their own solutions. The research objective is to assess the effectiveness of Teaching Factory implementation in improving the competence and skills of students in the field of entrepreneurship. The method used in this research is case study. Key informants were determined purposively in charge of teaching factory, students, and teachers who teach in productive subject. The results achieved in this study are: 1) culinary and hospitality units are the most active groups in producing products for teaching factory, 2) the role of entrepreneurial groups in each unit of production in order to increase productivity 3) students earn revenues of teaching factory and production unit get percentage of product or service sales, 3) the role of each production unit determines the level of teaching factory effectiveness.

1. Introduction

The growth of the number of vocational schools throughout Indonesia is not accompanied by the quality of graduates in accordance with the field of expertise. The quality of Vocational High School (VHS) graduates has not been able to fill the needs in the business world and industry, because the skills are not match in accordance with the standards required by the world of work. Open Unemployment Rate (TPT) of Vocational High School reached 11.11 percent. This number occupies the highest position compared to other workers who graduated from elementary school and university. The second highest position of TPT is from Senior High School (SHS) graduates of 8.73 percent. (BPS, 2016 August). Compared to August 2015 data, 12.65 percent of unemployed were from VHS, 10.32 percent from SHS, 7.54 percent from Diploma and 6.40 percent from University. Looking at the data there was a decrease of 1.54 percent for unemployed VHS graduates.

President Joko Widodo affirmed the government's commitment to encourage Vocational education as an effort to improve the competitiveness of human resources. The Government is paying great attention to vocational education especially with the enactment of the ASEAN Economic Community; Indonesia must be able to provide skilled middle-class workers so that it can compete to fill the demand of the business world and industry [1].

To answer the above problem, the role of VHS really should be improved through collaboration with business world and industry world. This is in line with the learning model implemented in vocational schools that relate between theory and practice as in the industrial world that is Teaching Factory. Learning by Teaching Factory approach, has long been known but in its application not in accordance with the intent and purpose Teaching Factory. Teaching Factory

IOP Conf. Series: Materials Science and Engineering **306** (2018) 012053 doi:10.1088/1757-899X/306/1/012053

Program is a concept of learning in VHS based on production / service that refers to the standards and procedures applicable in the industry and implemented in an atmosphere like that happening in the industry. Implementation of Teaching Factory in VHS can bridge the competency gap between industry requirement and competence produced by school. The implementation of Teaching Factory requires the absolute involvement of the industry as a relevant party assessing the quality of educational outcomes of VHS. Teaching Factory should also involve local government / municipal / municipal / provincial and parents and the community in planning, regulation and implementation. In addition to VHS produces a competent workforce, in the distribution of graduates is expected to also entrepreneurial spirit to not only be able to work but also to create jobs for themselves or for others. This is easy to say but demands seriousness, because entrepreneurial spirit is not easily formed by simply given the science of Entrepreneurship. The soul of entrepreneur must be formed in an integrated process in a learning process so that awakening soul of commitment, responsibility and work ethic on candidate of VHS graduate.

Teaching Factory is a learning method that able to deliver students to achieve industry standard competence through stages of the process of achieving motor, cognitive, and affective standard of mastery and raises the result of learning inspirational - intuitive behavior that is academically described as character learning. Teaching Factory has a strategic value in vocational education and training in improving the competitiveness of vocational institution graduates such as VHS in the local and national as well as regional labor market, as it has a mechanism that always follows the industry's rapid development [2].

Vocational education is expensive; therefore VHS must be managed and executed appropriately by people who are good, creative and responsible. With the proper implementation of learning then the facilities of the expensive facilities can be balanced with the production of many competent personnel are standardized. Professional teachers should be empowered appropriately in order to efficiently and effectively utilize the facilities of facilities, so that "SMK Bisa" can actually produce productive workforce. Vocational High School is one of the educational institutions responsible for creating human resources that have the ability, skills and expertise, so that graduates can develop performance when in the workplace. VHS itself aims to "improve the ability of students to be able to develop themselves in line with the development of science, technology and art, and prepare students to enter the work field and develop a professional attitude. So, vocational education is preparing a person to have certain skills in order to work in the field of work productively in accordance with his field, which is in line with the development of science.

Vocational education is part of an educational system that prepares individuals for a job or group of work [2]. States vocational education is education for a job or some type of work that the individual loves for his social needs. According to the House Committee on Education and Labor (HCEl) vocational education is a form of talent development, basic skills education, and habits that lead to the world of work viewed as skills training [3].

The concept of teaching factory is found for three things: (1) ordinary learning is not enough; (2) the benefit of learners is derived from direct practice experience; And (3) team-based learning experiences involving students, faculty and industry participation enrich the educational process and provide tangible benefits to all. Thus, it can be concluded that teaching factory is a learning activity where students directly perform production activities either in the form of goods or services within the school education environment.

The purpose of teaching factory is to make students realize that teaching should be more than what is in the book [4]. Learners not only practice soft skills in learning, learn to work in teams, practice interpersonal communication skills, but gain hands-on experience and practice work to enter the working world later. Learning factory teaching teaches students how to find problems, build prototypes, learn to make business proposals, and learn to present their own solutions. Teaching-learning process learners learn about skills that are important to master, such as how to meet the level of time and guesses that may arise, build and work in teams and work with diverse people who have the ability and talent are diverse.

Some research results show that teaching-based teaching factory states that the industry is a partner [5]. Learning partnering with industry is a successful experiment [5]. The Learning Factory is a successful experiment which has been pedagogically sound, sustainable, cost effective and transferable [5].

The teaching factory process that has been implemented in VHS RSBI in Yogyakarta area is influenced by supporting and inhibiting factors [4]. Factors supporting the implementation of teaching factory are: (1) Good equipment facilities; (2) Human resources; (3) Products produced; (4) Market influence; (5) Leadership; and (6) Marketing. Factors inhibiting the implementation of teaching factory are: (1) Rules on the legality of production units; (2) Lack of marketing; (3) Parents' perceptions of students; (4) The price of the manufacturer is too high; and (5) The busyness of teachers and students. Supporting factors in the implementation of teaching factory in VHS RSBI DIY is: (1) There is a capital source in the form of a grant, (2) Facilities of production equipment, (3) Teacher ability, (4) Exhibition by local government [6]. While the inhibiting factors are: (1) Operational management, (2) Lack of cooperation with industry, (3) Nature of program in the form of project, (4) Student competence, (5) Marketing.

The purpose of this study is to assess the effectiveness of Teaching Factory implementation in improving the competence and skills of students in the field of entrepreneurship.

2. Methods

This research use case study method, key informant in this research determined by purposive from VHS N 3 (Public Vocational High School) that is manager of unit teaching factory and one person from National Education Office of North Sulawesi Province. While the source of data in this study is the manager of Production Unit, and students and teachers productive. Data collection was conducted with, observation, in-depth interviews, and Focus Group Discussion (FGD). Observation techniques, researchers go directly to school to observe the situation and collect data. In-depth interviews took place directly with informants. The interviews used were using an open system which meant the subject knew they were being interviewed and understood the purpose of the interview. FGD techniques are conducted to collect untapped data from in-depth interviews. Implementation of this technique in the form of discussion groups between researchers and informants who are considered to understand the problem research. Data were analyzed by interactive model through stages: data reduction, data presentation, conclusion and verification [7].

3. Results and Discussion

3.1. Result

Teaching Factory in VHS N 3 Manado consists of several units of production that is the production unit of Culinary, Clothing, Cosmetology, Hospitality and Computer Network Engineering. Products produced in the production unit that includes superior that is included in the product Teaching Factory. Products that have not included good or superior are not included as a result of Teaching Factory. In accordance with the results of interviews with the responsible Teaching Factory that for this unit Culinary already can produce products ordered by the community including offices. These products include catering, lunch box, snack, and hospitality services already requested by several hotels in Manado. Profits from the Teaching Factory result the students earn income and also the production units get a percentage of the sales of products or services. The role of the production unit is expected by the Teaching Factory manager because if the production unit is not productive then no products or services will be sold. This means no income for schools. The results of the sale of products and services help finance the operations in each production unit. Funds provided by the government for Teaching Factory from 2015 is Rp 300.000.000, - to assist the operational of Teaching Factory. The year 2016 is not getting help because there are 3 units of production do not run activities regularly and continuously.

The benefit of Teaching Factory is to support learning. The learning process in VHS demands high skills therefore entrepreneurship groups are effected. In the production unit, there are entrepreneurial groups of students. The group is a way for expanding practicing hours in practice to produce superior products or services. Other benefits gained from the Teaching Factory Teachers' learning implementation can win both regional and national level competitions. In addition, there are 5 students sent by the ministry to study in Japan for 6 months.

Constraints faced in the implementation of Teaching Factory is the cost, because the funds provided to VHS are all the same even when for the production unit Culinary, Clothing, Cosmetology, procurement of materials and equipment to practice more. Lack of funds must be overcome by the revenue earned on the product sales so that the learning activities in the Teaching Factory unit can be sustainable. Another obstacle faced is not all teachers think to improve student productivity through activities in the production unit and in Teaching Factory. Elements of leadership have not fully supported the Teaching Factory activities.

Teaching Factory Learning in VHS N 3 is not fully in accordance with Teaching Factory's own teaching principles because it has not utilized the industry or institution related to Teaching Factory products. Do not have cooperation with the industry either in terms of expert or in terms of marketing Teaching Factory products. Thus, for certain products and services if there is no order then the production is not done.

3.2. Discussion

Teaching factory is a concept that combines learning and a realistic work environment and to bring up relevant learning experiences [4]. This learning is a practice process that integrates application-oriented training with a problem-solving approach. The teaching factory process that has been implemented in VHS RSBI in Yogyakarta area is influenced by supporting and inhibiting factors. Factors supporting the implementation of teaching factory are: (1) Good equipment facilities; (2) Human resources; (3) Products produced; (4) Market influence; (5) Leadership; And (6) Marketing. Factors inhibiting the implementation of teaching factory are: (1) Rules on the legality of production units; (2) Lack of marketing; (3) Parents' perceptions of students; (4) The price of the manufacturer is too high; And (5) The busyness of teachers and students. one of the learning approaches based on production and learning in the world of work is with the factory of learning or known as teaching factory (TEFA).

Based on the Research results of this study, the lack of supporting facilities will hinder the implementation of Teaching Factory properly. Lack of marketing access will hinder continuity of production.

Teaching factory learning model, there are some special parts that distinguish the non-teaching practice model applied in VHS Karsa Mulya Palangka Raya namely: (1) Students do real work according to SOP like in industrial world ; (2) Practice materials that are undertaken i.e. real objects or vehicles owned by consumers; (3) Job based on the problems found in consumer vehicles; (4) Work based on work standards supported by the manual of the type of service the vehicle is working on; (5) Students are required to work on the same time as working in the industry; (6) More enhanced work safety of tools, persons and work pieces; and (7) Planting a greater sense of responsibility, especially on customer satisfaction [4]. The developed teaching factory model is in accordance with the required criteria, namely: (1) The availability of room as a place of practice implementation; (2) There is a production unit as a place of execution of the process; (3) Available supporting facilities such as tools, locks, and machines; (4) Has taken advantage of local environmental conditions in the learning process of practice; (5) Human resources involved in the implementation of teachers / instructors and students; (6) Cooperation between industrial parties and schools; (7) Teachers of practice subjects, instructors have a commitment in implementing and applying the concept of teaching practice model teaching factory; And (8) Students are fully involved in the process. The effectiveness of the teaching factory model development process implemented is shown by the students' ability in completing the work.

The entrepreneurship education given to the students in their implementation in the unit teaching factory is not merely a theory but it is really the level of practice. The students are given the responsibility to carry out the work in the field of expertise with a business oriented. The production unit plays a crucial role in the implementation of entrepreneurship education [8].

4. Conclusions

The research concludes the following points: 1) Production unit in Culinary and Hospitality units are the most active group in producing products for teaching factory, 2) The role of entrepreneurial groups in each unit of production in order to increase productivity 3) Students earn income from teaching factory results and the production unit obtains a percentage of the sales of the product or service, 3) The role of each production unit determines the level of teaching factory effectiveness.

References

[1] Kompas Online, 1 April 2016
[2] Khurniawan A W 2016 *Grand Disign Pengembangan Teaching Factori dan Technopark* Dit. Pembinaan VHS, Kemendikbud
[3] Kurniawan R 2017 Pengaruh Penerapan Model Pembelajaran Teaching Factory 6 Langkah (TF-6M) dan Prestasi Belajar kewirausahaan Terhadap Minat Wirausaha *Innovation of Vocational Technology Education* **10** 1
[4] Siswandi G and Sukoco 2015 Pengembangan Model *Teaching Factory* Di Bengkel Otomotif VHS Karsa Mulya Palangka Raya *Jurnal Pendidikan Teknologi dan Kejuruan* **22** 4
[5] Lamancusa J S, Zayas J L, Soyster A L, Morell L and Jorgensen J 2008 The Learning Factory: Industry- Partnered Active Learning *Journal of engineering education* **97** 1 5-11
[6] Sukardi T, Ibnu S 2008 Faktor Pendukung Dan Penghambat Pelaksanaan Teaching Factory Di VHS RSBI Daerah Istimewa Yogyakarta (online) Available at: http://staff.uny.ac.id
[7] Denzin N K and Yvonna S L 2009 *Qualitative Research Hanbook* (Yogyakarta: Pustaka Pelajar)
[8] Sumual H 2015 *Pengembangan Model Unit Produksi VHS* Laporan Penelitian Hibah Bersaing

Effect of Pertalite-Spiritus Blend Fuel on Performance of Single Cylinder Spark Ignition Engine

H Wibowo*, A A P Susastriawan and D Andrian
Jurusan Teknik Mesin, Fakultas Teknologi Industri, Institut Sains & Teknologi AKPRIND, Yogyakarta 55222, Indonesia

*hary.wibowo@akprind.ac.id

Abstract. This work aims to experimentally investigate an effect of Spiritus and Pertalite blend on engine's performance (brake horsepower and torque, specific fuel consumption rate, and exhaust gas emission). The tests are conducted at 3000-7000 rpm for four different fuels, i.e., pertalite, 95%P-5%S, 90%P-10%S, and 85%P-15%S to obtain an optimum ratio of the Pertalite-Spiritus blend. The blend ratios of Pertalite (P) and Spiritus (S) are measured by volume. The result shows that addition of 15% Spiritus (by volume) on Pertalilte can enhance the combustion process of the blend hence increasing engine's brake horsepower and decreasing specific fuel consumption rate. However, engine's torque is lower when using the blend 85%P-15%S if compared with using Pertalite.

1. Introduction

In recent years, fuel fossil depletion and global warming issues have been a serious problem worldwide. Many renewable fuels have been tried for fossil fuel substitution as well as additional fuel. Alcohol fuels (methanol and ethanol) have a potential for internal combustion engine fuel due to its high antiknock and ability for reducing CO and UHC emission [1]. Bioethanol is also attractive renewable bio-based fuel and its oxygenated characteristics able to reduce particulate emission in CI engine [2]. The use alternate fuels also have been proposed for reducing combustion-related pollution [3].

Due to the advantages of alcohol fuel, i.e., methanol and ethanol, many researchers work intensively in the area of alcohol fuel for internal combustion engine. Vapour pressure of gasoline-alcohol blend increases as alcohol composition in the blend increases, but the pressure decreases after optimum quantity of alcohol added [4]. The ethanol addition increases torque output and fuel consumption, CO and HC dramatically reduced [1]. Emission of CO_2 increases with the use of ethanol-gasoline blends which indicates combustion improvement. Meanwhile, engine performance and exhaust gas emission of ethanol-gasoline blend fueled engine could be affected by ignition timing [5] The same tendency of CO, HC, and CO_2 in exhaust gas was also found for the use of ethanol-unleaded gasoline [6]. Exhaust gas emission as a function of compression ratio from engine fueled by Ethanol-Gasoline blend [7].

Some works were also performed specifically on ethanol-gasoline combustion. Flame propagation of gasoline-ethanol blend in SI engine has been investigated by [8]. Flame speed accelerated when 25% ethanol is added to gasoline. Spray characteristics of the ethanol-gasoline blend were also studied by Gao [9]. In the case of swirl type injector, the spray pattern development is not significantly changing when using ethanol gasoline with a different fraction. However, the use of alcohol as a fuel addition is a required modification on engine carburetor [10]. The effect of ethanol and unleaded

IOP Conf. Series: Materials Science and Engineering **306** (2018) 012054 doi:10.1088/1757-899X/306/1/012054

gasoline blends on engine performance which the ethanol amount was in the range 0-25% [6]. The results showed the engine performance was improved and the engine power was increased significantly.

In this work, the effect of Spiritus addition on Pertalite on engine's performance (brake horsepower and torque, specific fuel consumption rate, and exhaust gas emission) is experimentally investigated. The tests are conducted at 3000-7000 rpm for four different fuels, i.e., pertalite, 95%P-5%S, 90%P-10%S, and 85%P-15%S. The blend ratios of Pertalite (P) and Spirtus (S) are measured by volume. The experiment purposes to obtain an optimum ratio of the Pertalite-Spiritus blend.

2. Experimental Setup

2.1. Fuel characterization

Spiritus is a blend of Ethanol and Methanol and has a high octane number (HON). The Spiritus has a potential to be used as additional fuel of low octane number (LON) fuel, i.e., Pertalite. Properties of Pertalite and Spiritus are shown in Table 1.

Table 1. Properties of Pertalite and Spiritus.

Properties	Pertalite	Spiritus
Lower Heating Value	83.84 MJ/kg	26.90 MJ/kg
Research Octane Number	90 (LON)	123 (HON)
Specific gravity	0.77	NA
Heat of vaporization	343 kJ/kg	NA
Laminar burning velocity (at $\lambda = 1$)	0.5 m/s	NA

2.2. Engine performance setup

Figure 1 shows engine performance setup in this work. Engine test is single cylinder 125 cc gasoline engine. Engine performance is measured in terms of brake horsepower (BHP) and torque, specific fuel consumption rate (SFCR), and exhaust gas emission. The brake horsepower and torque are measured using Dyno test, the specific fuel consumption is determined using the amount of fuel consumed in the buret within particular testing time, and exhaust gas emission is detected with the use of exhaust gas analyzer. The tests are conducted at 3000-7000 rpm for four different fuels, i.e., pertalite, 95%P-5%S, 90%P-10%S, and 85%P-15%S. The blend ratios of Pertalite (P) and Spirtus (S) are measured by volume. For each test, the data taken at interval engine speed of 1000 rpm, i.e., 3000 rpm, 4000 rpm, 5000 rpm, 6000 rpm, and 7000 rpm. The collected data are compared to each blend.

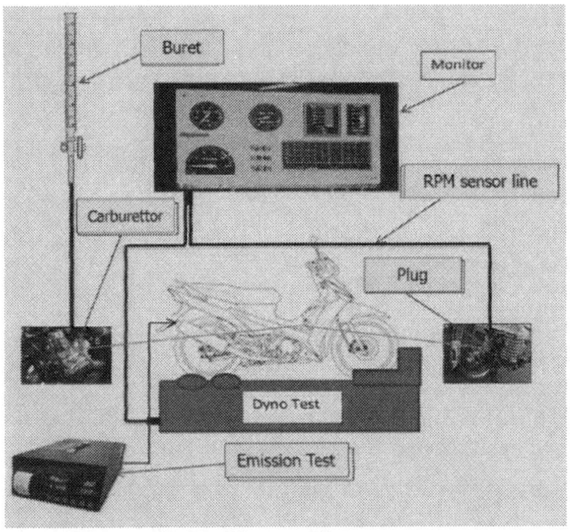

Figure 1. Engine performance setup.

3. Results and Discussion

Figure 2 shows the effect of the fuel blend on Engine's Brake Horse Power (BHP) and Torque base on rpm. When Spiritus content in the blended fuel was increased, the BHP slightly increased for all engine speeds. Engine power was increased gradually as a percentage of ethanol increased in blended fuels (up to 10% [11]) [6,11]. The optimum BHP obtained when 15% Spiritus is added on Pertalite for all engine speed variation. On the other hand, the addition of 15% Spiritus also decreases engine's torque. Increasing methanol content in blended fuel (30-50%) will decrease the engine power for all engine speeds [11].

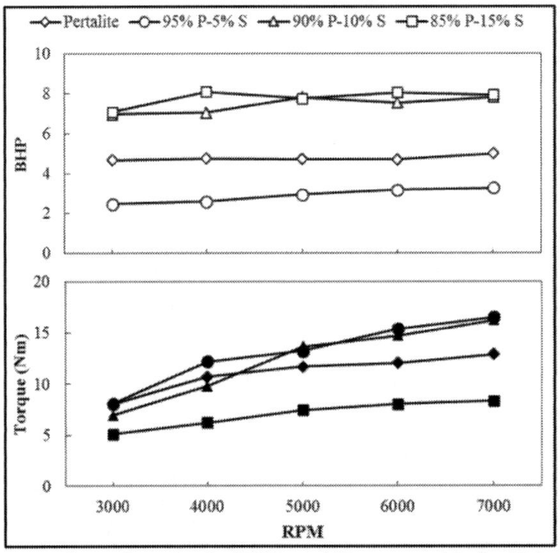

Figure 2. Characterize of engine's BHP and Torque with fuel blend variations.

Specific fuel consumption rate (SFCR) is affected by Spiritus addition on Pertalite as shown in Figure 3. The SFCR reduces when 10% and 15% by volume Spiritus are added on Pertalite but the SFCR increases for a blend of 95% Pertalite-5% Spiritus by volume. This result was not in line with Iliev [11] that SFCR increased as the ethanol (methanol) percentage increased. Increased ethanol up to 20% will reduce the SFCR, similar with the result of this study [6].

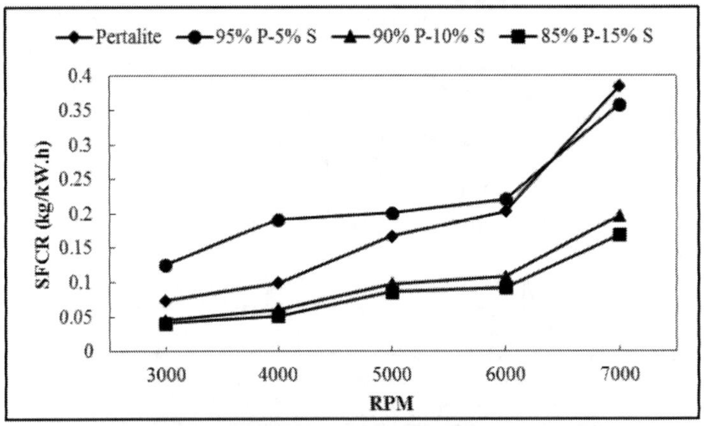

Figure 3. Effect of fuel blend on SFCR.

Meanwhile, Figure 4 presents an average value of CO, CO_2, and HC in exhaust gas emission. The value is engine speed averaging. The graph shows comparison of exhaust gas emission of the blends that a complete combustion occurs when using the blend 85%P-15%S if compared with the use of

95%P-5%S and 90%P-10%S blends. A complete combustion is indicated by higher CO_2 and lower HC in exhaust gas emission as shown by 85%P-15%S blend.

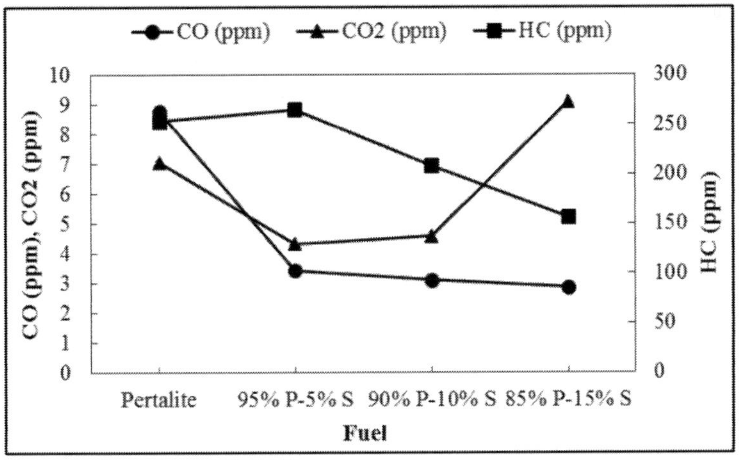

Figure 4. Concentration of CO, CO_2, and HC in exhaust gas emission.

4. Conclusion

Increasing the Spiritus content in the blended fuel will slightly increase the BHP for all engine speeds. The addition of 15% Spiritus (by volume) on Pertalite can enhance the blend combustion hence improve engine's brake horsepower and decrease specific fuel consumption rate. However, engine's torque is lower when using the blend 85%P-15%S if compared with using Pertalite.

Acknowledgments

The author would like to thank to staff of Automotive Laboratory for helping setup preparation and data collection during the experimental work, and to Head of Mechanical Engineering Department Institut Sains & Teknologi AKPRIND who has given funding part of the research activities.

References

[1] Hsieh W D, Chen R H, Wu T L and Lin T H 2002 Engine performance and pollutant emission of an SI engine using ethanol-gasoline blended fuels *Atmos Environ* **36** 3 403–410

[2] Agarwal A K 2007 Biofuels (alcohols and biodiesel) applications as fuels for internal combustion engines *Prog Energy Combust Sci* **33** 3 233–271

[3] Gaffney J S, Marley N A, Martin R S, Dixon R W, Reyes L G and Popp C J 1998 Potential air quality effects of using ethanol-gasoline fuel blends: A field study in Albuquerque, New Mexico *Environ Sci Technol* **31** 11 3053–3061

[4] Pumphrey J A, Brand J I and Scheller W A 2000 Vapour pressure measurements and predictions for alcohol-gasoline blends *Fuel* **79** 11 1405–1411

[5] Topgul T, Yucesu H S, Cinar C and Koca A 2006 The effects of ethanol-unleaded gasoline blends and ignition timing on engine performance and exhaust emissions *Renew Energy* **31** 15 2534–2542

[6] Al-Hasan M 2003 Effect of ethanol – unleaded gasoline blends on engine performance and exhaust emission *Energy Conversion and Management* **44** 9 1547–1561

[7] Yucesu H S, Topgul T, Cinar C and Okur M 2006 Effect of ethanol-gasoline blends on engine performance and exhaust emissions in different compression ratios *Appl Therm Eng.* **26** 17-18 2272–2278

[8] Bayraktar H 2007 Theoretical investigation of flame propagation process in an SI engine running on gasoline-ethanol blends *Renew Energy* **32** 5 758–771

[9] Gao J, Jiang D and Huang Z 2007 Spray properties of alternative fuels: A comparative analysis of ethanol-gasoline blends and gasoline *Fuel* **86** 10-11 1645–1650

[10] Yüksel F and Yüksel B 2004 The use of ethanol-gasoline blend as a fuel in an SI engine *Renew*

Energy **29** 7 1181–1191
[11] Iliev S 2015 A comparison of ethanol and methanol blending with gasoline using a 1-D engine model *Procedia Engineering* **100** 1013-1022

Teacher's Perception about the Use of E-Learning/Edmodo in Educational Activities

H Yanti*, A Setiawan, Nurhabibah and Yannuar

Technology and Vocational Education, Universitas Pendidikan Indonesia (UPI) Jl. Dr. Setiabudhi 207 Bandung 40154 Indonesia

*halidayanti@student.upi.edu

Abstract. This study examined the perception of the teachers about the use of e-learning/Edmodo in their educational activities. The teachers consist of diverse subject. Their perceptions were investigated in terms of three aspects: effects of the use of this technology on their perceived motivation, the perceived usefulness and the perceived ease of use of this technology. Edmodo was set up a Learning Management System (LMS) in an online discussion group of subject. The study was conducted in descriptive method. The data were collected by using a questionnaire, interview, and documentation technique. The findings of the study indicated that the teachers perceived that e-learning/ Edmodo is a useful and also easy to use technology. It was found out that the teachers are satisfied with advantages of the use of this new technology in their LMS.

1. Introduction

Information and communication technology (ICT) has developed nowadays. The popular one is internet. Internet use was not only for communication tool but also a part that include in all side of human life like business, industry, education, social relation, etc. Data of APJII (Asosiasi Penyelenggara Jasa Internet Indonesia) show that the amount of internet user in Indonesia on year 2016 noted as 132.7 million users.

If we research specifically, internet user between 10-24 years old counted as 75.5 %. They usually use internet for uploading, downloading, browsing, v-log making or social media activities. Most of student in Bandung easy to access internet because it was supported by gadget, smartphone, laptop and free wifi anywhere like in the school, campus, public area, etc.

In a little while, internet became the most important guide for distance learning. The use of the internet as an instructional guide forced teacher rethink their ways of instructions offered and administered [1]. Go on the usage of internet in educational activities is virtual learning environment (VLE). A VLE typically provides tools such as those for assessment, communication, uploading of content, return of students' work, administration of student groups, questionnaires, tracking tools, wikis, blogs, chats, forums, etc. over internet. Such e-learning systems are sometimes also called Learning Management System (LMS), course management system (CMS), learning content management system (LCMS), managed learning environment (MLE), learning support system (LSS), learning platform (LP) or e-learning sharing portal (ELSP); it is education via computer-mediated communication (CMC) or online education [2]. E-learning may be defined as instruction delivered electronically via the Internet, Intranets, or multimedia platforms such as CD-ROM or DVD [3]. One of the famous types of LMS that has many features is Edmodo as educational tool. Edmodo was found at 2008 by Nicolas Borg and Jeff O'Hara. Edmodo consist of three accounts can log in, teacher, student and parent.

There are many benefits of using Edmodo as an educational tool. Edmodo help students visualize abstract idea and make it easy to find reliable information. Teachers at any grade level can easily create collaborative activities for students on it. Students' motivation and confidence are increased when technology is integrated into classroom instruction. Computer engagement also improves student's academic achievement. The key component in students' perceived level of intrinsic motivation was directly related to meaningful communication in the course [4].

Many devices are still on the beginning of their using in education, like Edmodo. Research found that Edmodo is the most popular application in the world with 58 million more users (Capterra.com).

Nowadays technology is important for educational process, so the teachers should play the main role in the adoption and integration of technology in teaching-learning process. There are many factors, which are influencing this process. These factors are developing from the previous century till today. Teachers' integration of technology into teaching-learning is also influenced by organizational factors, attitudes towards technology and other factors [5].

Personal characteristics such as educational level, age, gender, educational experience, experience with the computer for educational purpose and attitude towards computers can influence the adoption of a technology [6]. Research found that male teachers have got more positive attitudes toward technology and their using of technology in education process is more frequent than female teachers. Among the factors that influence successful integration of technology into teaching are teachers' attitudes and beliefs towards technology.

If teacher's attitudes are positive towards the use of educational technology then they can easily provide useful insight about integrating of technology into teaching and learning process. Research has shown that teachers' attitudes towards technology influence their acceptance of the usefulness of technology and its integration into teaching and learning [7].

Computer's teacher is shown positive attitudes towards technology because of their computer experience relates positively to their computer attitudes. Computer experience relates positively to their computer attitudes. The more experience teachers have with computers, the more likely that they will show positive attitudes towards computers [8].

A review of the research through observation and interview with teachers who interest and supported e-learning/Edmodo still do not use it. There are schools have carried out a lot of projects in e-learning. However, the teachers had problems about sharing online resources among themselves. There was also a lack of online communication with each other.

2. Methodology

This research investigated the perception of teacher of using e-learning/Edmodo in educational activities. This study uses components of case study by descriptive method with questionnaire, literature and documentation techniques. Therefore, this research focuses on a single case: e-learning/Edmodo application in educational activities that was experienced by teachers. The aim of the use of these methodologies is to gain perceptions of the teachers participated in the study.

The study population was obtained from teachers who know and ever use e-learning/Edmodo from diverse subject in secondary level of school in Pasaman, city of West Sumatera. The samples have chosen by purposive sampling. It consists of 20 respondents. The study was conducted for one week to answer questions in the form of questionnaires.

Table 1. Respondent by subject and gender.

No	Subject	Gender		Number of teachers
		Male	Female	
1	Programming	1	1	2
2	Networking	0	1	1
3	Natural Science and Mathematics	0	3	3
4	Economic Science	1	0	1
5	Computer Science	5	5	10
6	Automotive and Machine	3	0	3

Base on table 1, the mayor samples were from computer teachers. In other word they have base acknowledge about computer/ICT and computer experience.

3. Results and Discussion

Teacher perceptions were investigated in terms of three aspects: Effects of the use of this e-learning/Edmodo in teachers' perceived motivation towards their educational activities, perceived usefulness and perceived ease of use.

Table 2. Abbreviations used for teacher perception questions.

Abbr.	Description
SD	Strongly Disagree
D	Disagree
N	Neutral
A	Agree
SA	Strongly Agree

The scale of the questionnaire was Likert-type and its scales and abbreviations used in this results section was listed in Table 2.

Table 3. Descriptive statistics of teacher perception.

	SD		D		N		A		SA		Mean	Std. Dev
	N	%	N	%	N	%	N	%	N	%		
Perceived Effects on Motivation	0	-	3	15	4	20	2	10	11	55	4.05	1.190
Perceived Usefulness	2	10	2	10	1	5	10	50	5	25	3.70	1.260
Perceived Ease of Use	4	20	3	15	2	10	8	40	3	15	3.150	1.42
Overall	2	10	2.67	13.4	2.34	11.6	6.67	33.3	6.34	31.6	3.634	1.29

As it is seen in the Table 3, 65% of teachers stated positive perception and only 15% of them stated negative perception for Perceived Effects on Motivation. Also, 55% of teacher reported positive perception about Perceived Ease of Use while only 35% of them reporting negative perception. As a result, the mean of the questionnaire data is 3.634 with standard deviation equals to 1.29 and more than a half of teacher reported positive perception about the use of E-Learning/Edmodo in overall.

4. Conclusion

The purpose of this study was to develop a valid and reliable scale for determining teacher's perceptions towards e-learning/Edmodo in teaching-learning process. The developed scale is questionnaire use Likert-type according to the conducted analyses in the study. The scale is thought to be effective in measuring teacher's perceptions towards e-learning/Edmodo in teaching-learning process in terms of effect, usefulness and ease to use [9].

References

[1] Longe F C 2005 Students' Perceptions of Web-Based Learning Tools: A Case Study of the Web Course Tool (WebCT) *An On Line Journal of African Educational Research Network* **21**

[2] Weller M 2007 *Virtual learning environments: Using, choosing and developing your VLE* (London: Routledge)

[3] Hall B 2003 FAQs about e-learning. Retrieved August 15, 2003, (online) available at: www.brandonhall.com/public/faqs2/faqs2.htm

[4] Walker J, Wallace D and Juban R 2004 *Business students online: Profiles of success* Manuscript submitted for publication

[5] Chen C H 2008 Why do teachers not practice what they believe regarding technology integration? *The Journal of Educational Research* **102** 1 65-75

[6] Schiler J 2003 Working with ICT: perceptions of Australian Principals *Journal of Educational Administration* **41** 3 171-185

[7] Huang H M and Liaw S S 2005 Exploring users' attitudes and intentions toward the web as a survey tool *Computers in Human Behavior* **21** 5 729-743

[8] Rozell E J and Gardner W L 1999 Computer-related success and failure: a longitudinal field study of the factors influencing computer-related performance *Computers in Human Behavior* **15** 1 1-10

[9] Albirini A 2006 Teachers' attitudes toward information and communication technologies: the case of Syrian EFL teachers *Computers & Education* **47** 4 373-398

Optimization Placement of Static Var Compensator (Svc) on Electrical Transmission System 150 kV Based on Smart Computation

Hasbullah*, Y Mulyadi, Y Febriana and A G Abdullah

Program Studi Pendidikan Teknik Elektro, FPTK Universitas Pendidikan Indonesia

*hasbullah-elektro@upi.edu

Abstract. To improve voltage profile, we can use FACTS equipment. One of them is SVC (Static Var Compensator). This study aims to determine the location and optimal capacity of SVC and to determine the effect after SVC installation. This research was conducted in 150 kV transmission system, West Java load regulator area of South Bandung and New Ujungberung subsystem. The research method used for power flow simulation is Newton Raphson and to determine the optimal position and capacity of SVC was using genetic algorithm in MATLAB R2014. After the SVC placement is optimized, then the system performance will increase such as the voltage of all buses is at the standard level and the decrease of power losses.

1. Introduction

Nowadays, the need for electrical energy increases along with industrial, commercial and population growth. This is evidenced by the data of PLN and BPPT energy team, that the average electricity demand in Indonesia grew by 6.5% per year [1]. The increasing of load demand will lead to heavy loading of the transmission so that there may be voltage drop [2]. An uncontrolled voltage drop can lead to voltage instability [3].

According to IEEE the allowable voltage standard is 0.95 pu to 1,05 pu [4]. A reliable electric power system must have a voltage that doesn't exceed the tolerance limit and few power losses [5]. The voltage drop affects the power losses. Power losses are unavoidable but can be reduced by improving the voltage profile. The voltage profile setting requires two things: the location and amount of reactive power provided must be precise [6]. Technological developments in the semiconductor industry led to the creation of FACTS (Flexible AC Transmission) equipment. FACTS equipment that commonly used in power grid is SVC (Static Var Compensator) because it is cheaper. SVC is a shunt associated with the generation or absorption of static reactive power to maintain or control special parameters on the electrical system, that is bus voltage. SVC placement should be optimized to have an impact on the power system [7]. Optimization method that used in this research is genetic algorithm. Genetic algorithms are the simplest method of solving optimization problems based on natural selection. Genetic algorithm is simpler, easier in explanation and its ability is fast [8].

1.1. Newton Raphson power flow

Power Flow is a study conducted to obtain information about the power flow or system voltage in steady operating conditions. In each bus, there are four values consist of voltage V |, phase angle (δ), real power (P) and reactive power (Q). On each bus only two kinds of quantities are determined while the other two are the final result of the calculation. Newton Raphson method is an iteration process to achieve the function approach. Newton Rapson uses nonlinear algebraic equations with his own variables. The value of the variable per unit and the phase angle in radians [9]. The equation is shown as follows:

$$P_i = \sum_{j=0}^{n} |V_i||Y_{ij}||V_j| \cos(\theta_{ij} - \delta_i + \delta_j) \quad (1)$$

$$Q_i = -\sum_{j=0}^{n} |V_i||Y_{ij}||V_j| \sin(\theta_{ij} - \delta_i + \delta_j) \quad (2)$$

Where:
V_i = bus i voltage,
V_j = bus j voltage
Y_{ij} = admittance between bus i and bus j
P_i = bus i active power
Q_i = bus i reactive power
θ_{ij} = m admittance polar angle Y_{ij}
δ_j = V_j voltage angle
δi = V_i voltage angle

The power flow from bus i to bus j and the power flow from bus j to bus i will be shown in equation 3-4.

$$S_{ij} = V_i I_{ij}^* \quad (3)$$
$$S_{ji} = V_j I_{ji}^* \quad (4)$$

The i-j bus losses are the algebraic sum of the power flow approaches of equations 3 and 4. Power losses are shown in equation 5.

$$S_{Lij} = S_{ij} + S_{ji} \quad (5)$$

1.2. Static Var Compensator
Static Var Compensator (SVC) is a fast-acting electrical device compensating for reactive power in high voltage transmission lines [10]. In other words, Static Var Compensator is defined as a device which the output can be adjusted and transformed into capacitive or inductive currents to maintain or control special parameters on electrical power systems [11]. This tool is called static because there are no moving components in it, but consists of power electronics components such as a switch. Static switch used is thyristor [12]. SVC consists of TCR (Thyristor Controller Reactor) which is parallel with TCS (Thyristor Switch Capacitor).

Curve of reactive power to bus voltage that are generated from SVC is shown in figure 1 [5].

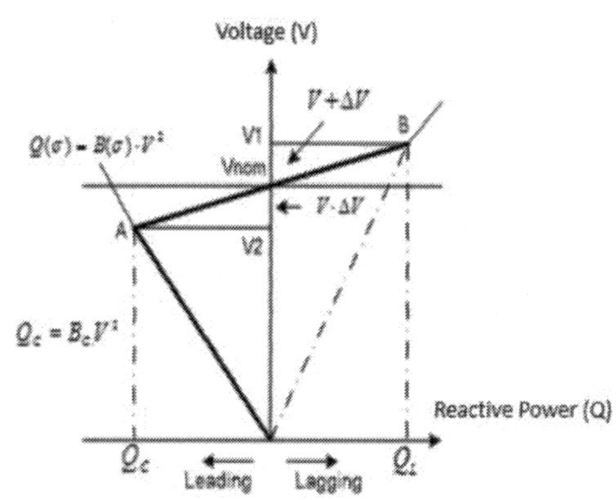

S
Figure 1. Reactive power and voltage curve of SVC.

There are three Static Var Compensator work areas:

1. Between V1 and V2. In this area SVC is capacitive and inductive.
2. When the bus voltage exceeds V1. In this area SVC has inductive characteristics.
3. When the voltage is less than V2. In this area SVC has capacitive characteristics.

There are 2 models of SVC as follows :

1.2.1. Suseptance model. In this model, SVC as reactance is automatically regulated through the suseptance variable to obtain the desired voltage according to the boundary. The suseptance value is obtained from the total of SVC suseptance that is sufficient to maintain the bus voltage at a certain value [13]. The SVC suseptance model is shown in figure 2.

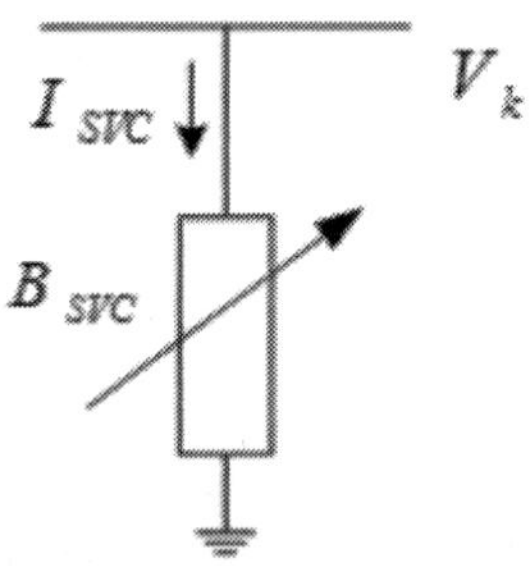

Figure 2. SVC suseptance model.

Based on figure 2, it can be explained that the SVC current can be formulated as follows [14]:

$$I_{svc} = jB_{svc}V_k \qquad (6)$$

jB_{svc} is SVC suseptance value, V_k is voltage

The minimum and maximum restriction of reactive power output QSVC is written as follows [2]:

$$Q_{max} = B_{ind} . V_{ref}^2 \qquad (7)$$

$$Q_{min} = B_{cap} . V_{ref}^2 \qquad (8)$$

Where:
Q_{max} is maximum reactive power
Q_{min} is minimum reactive power
B_{ind} is inductive suseptance. The value is $B_{ind} = 1/X_L$
B_{cap} adalah capacitive suseptance. The value is $B_{cap} = 1/X_C$

SVC Suseptance (BSVC) is expressed as the ignition angle function of thyristor (α) as follows:

$$B_{svc} = B_{cap} - B_{ind}(\alpha) \qquad (9)$$

Reactive power injected by SVC with suseptance model can be formulated as follows [14]:

$$Q_{svc} = Q_k = -V_k^2 B_{SVC} \qquad (10)$$

1.2.2. Firing angle model

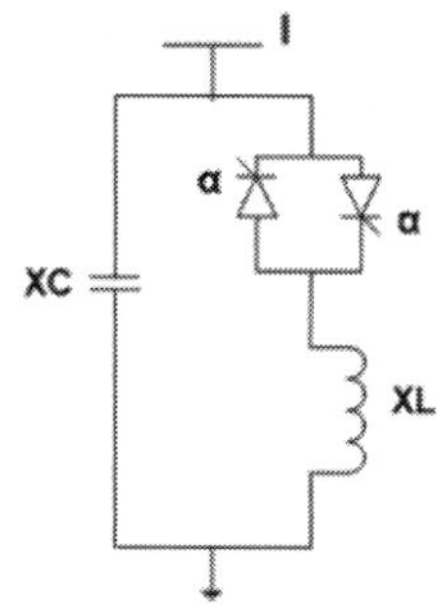

Figure 3. Firing angle model of SVC [3].

Figure 3 shows SVC firing angle model represented by fixed capacitor and thyristor controlled reactor [15]. The thyristor-based switch is used to convert to admittance. The thyristor will be controlled by ignition of the α angle and the change is equivalent to SVC reactance [3]. The reactive power values injected by SVC with the angle of ignition model are as follows [16]:

$$Q_k = -V_k^2 \left\{ \frac{X_c[(\pi-\alpha)+\sin 2\alpha]-\pi X_L}{\pi X_c X_L} \right\} \quad (11)$$

$$X_{svc} = -V_k^2 \left\{ \frac{\pi X_c X_L}{X_c[2(\pi-\alpha)+\sin 2\alpha]-\pi X_L} \right\} \quad (12)$$

Where:
X_c is capacitive reactance, with X_c=1/ωC
X_L is inductive reactance, with X_c=ωL
B_{svc} is SVC suseptance, with
$B_{svc} = -1/X_{svc}$

1.2.3. Genetic algorithm. In artificial intelligence, a genetic algorithm is a search that imitates the natural selection process developed by John Holland and Goldberg in 1970 at the University of Michigan [17]. It is a metaheuristic optimization method that iteratively solves problems by increasing candidate solutions based on certain criteria [18]. Genetic algorithm is a search algorithm based on genetic mechanism and natural selection [19].

The following step will explain how to implement genetic algorithm.

a. Coding Technique
Defines the code representation that used on chromosomes as solution candidates [20]. After encoding, the next step is setting the initial population randomly which is the solution candidate. In general, solution candidates are represented as code numbers according to each variable of the optimization problem, called chromosomes [21].

b. Fitness value evaluation
The fitness function aims to determine the quality of chromosomes associated with objective function [22].

c. Genetic Operator
There are 3 genetic algorithm operators that are selection, crossover and mutation. Before performing the process on the genetic operator, we have to determine the algorithm parameters.

1) Reproduction Selection
Reproduction is the probabilities selection of individu's chromosomes that is chosen to produce offspring based on fitness values [23]. After evaluating an individual with its fitness value then a good individual will have a high probability to do the next iteration [24].

2) Crossover
Crossover process is the process of crossing or exchanging genes from two selected chromosomes of the parent [25]. Crossover's function is to produce a child's chromosome from

a combination of parental chromosome genes. The genetic algorithm will converge faster when using a high probability (0.6 - 0.9).

3) Mutation
Mutation acts as a process of replacing the value of a previous gene with a new value that random-determined with a predetermined range [25]. The selection of mutated genes uses random numbers generated as many as genes in the population. If the result of random number is less than the mutation probability, then the mutation occurs [26].
4) Stop Condition
The optimization process performed using genetic algorithm will stop after a stop condition is met. Some of the stopping conditions used are the number of generations, the computation time limit, the fitness value limit, etc. Towards the end of the solution will converge to a certain value that will be the optimal solution approaching the problem [12].

2. Research Methods

Improved Voltage profiles on transmission systems 150 kV APB West Java subsystems of South Bandung and New Ujungberung is strongly influenced by the location and capacity of the installed static var compensator. It requires optimization to locate the optimal SVC location and capacity. Optimal means that the SVC can reduce power and voltage losses. The research flow diagram is shown in figure 4.

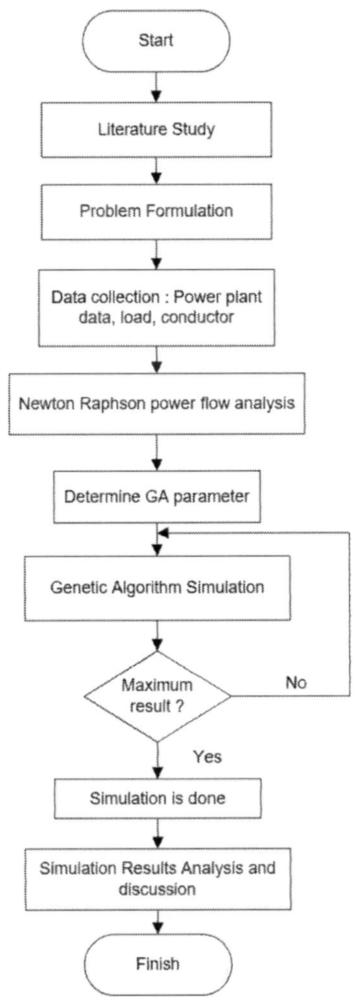

Figure 4. Research flow diagram.

Simulation of power flow and simulation of genetic algorithm using MATLAB R2014A. Simulation of power flow using peak load data, power generated data and conductor data on February 1, 2016. This power flow as input for simulation of genetic algorithm. To determine the location and capacity of the SVC using genetic algorithm with objective function of minimizing the power losses total and considering the voltage at each bus, that is Genetic algorithm parameters is shown in table 1.

Table 1. Genetic algorithm parameter.

Parameter	Value
Population size	10
$P_{crossover}$	0.6
$P_{mutation}$	0.06
Number of genes	16
Max generation	500

3. Results and Discussion

One line diagram of transmission system 150 KV APB West Java subsystem South Bandung and New Ujungberung is shown in Figure 5.

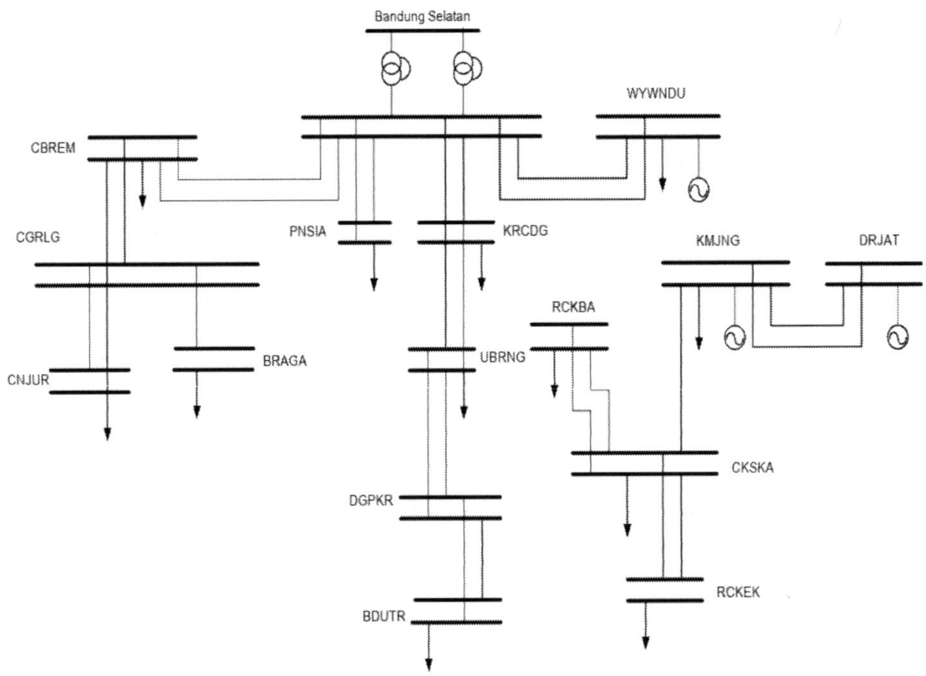

Figure 5. One line diagram of South Bandung and New Ujungberung Subsystems.

Figure 5 shows that the transmission system of 150 KV APB West Java subsystem of South Bandung and New Ujungberung has 16 buses with the following details:

1. Bus slack = South Bandung
2. Bus generator = Kamojang, Wayang Windu and Darajat
3. Load Buses = Cibeureum, Cigereleng, Panasia, Kiaracondong, Ujungberung, Dago Pakar, North Bandung, Rancaekek, Cikasungka, Cianjur, Braga and Rancakasumba

Subsystem channel of South Bandung and New Ujungberung consists of 16 channels. Newton Raphson power flow is done before and after SVC placement to find the voltage profile.

3.1. Power flow before SVC placement

Newton raphson power flow simulation uses MATLAB with 0.0001 accuracy, voltage base 150 kV, MVA base 100 and maximum iteration 500. The power flow results is shown in table 2.

Table 2. Voltage value.

No	Bus	Voltage (KV)	Angle (°)
1	Bandung Selatan	150	0
2	Wayang Windu	150	4.304
3	Kamojang	150	3.469
4	Drajat	150	7.573
5	Cigereleng	144.3	-3.619
6	Kiaracondong	140.85	-4.12
7	Bandung Utara	133.2	-8.065
8	Panasia	149.1	-0.385
9	Cianjur	135.3	-9.529
10	Cibereum	143.7	-4.168
11	Braga	144.15	-3.773
12	Cikasungka	143.55	-1.451
13	Rancaekek	140.7	-3.63
14	Rancakusumba	143.25	-1.903
15	Ujungberung	138.6	-5.191
16	Dago Pakar	135.45	-6.807

Based on IEEE the allowed voltage limits are between 0.95 pu to 1.05 pu [4]. Based on Table 4 it can be seen that the voltage on the South Bandung and New Ujungberung subsystems are outside the limit of 0.95 pu to 1.05 pu, there are 6 buses namely Kiaracondong bus, North Bandung, Cianjur, Rancaekek, Ujungberung and Dago Pakar. The Kiaracondong bus has a voltage of 0.939 pu or 140.85 kV. North Bandung Bus has a voltage of 134.1 kV or 0.888 pu Bus Cianjur has a voltage of 0.902 pu or 135.3 kV. Bus Rancaekek has a voltage of 0.938 or 140.7 kV. Ujungberung bus has a voltage of 0.924 pu or 138.6 kV. Bus Dago Pakar has a voltage of 0.903 pu or 135.45 kV. The power flow results before the placement of SVC are inputted into genetic algorithm simulation.

3.2. SVC location and capacity

In determining the location and capacity of SVC, we used two parameters namely location parameters and capacity parameters. One chromosome consists of 16 genes where 16 genes show the number of buses that exist in the subsystem of South Bandung and New Ujungberung.

Each gene contains two parameters. The first value to determine the SVC location variable encoded in binary numbers (0 and 1) where 0 indicates SVC is not installed and 1 indicates SVC is installed. The second value is to determine the capacity variable of the SVC encoded in a real number between -250 MVAR to 250 MVAR because the upper and lower limit of SVC reactive power in MATLAB genetic algorithm simulation is -250 MVAR and 250 MVAR. Optimal capacity will be obtained through the process of genetic operator with objective function of minimizing power losses total and considering the allowable voltage range, that is $0.95\ pu \leq V_{bus} \leq 1.05\ pu$. The fitness function in MATLAB is the minimized objective function. So it can be said the fitness function as same as the objective function. The location and capacity of SVC using genetic algorithm is shown in Table 3.

Table 3. Location and capacity of SVC using genetic algorithm.

Location	Rating (MVAR)
BDUTR	70.5
CNJUR	62.19
RCKEK	193.53

The iteration process and fitness values comparison is shown in Figure 6.

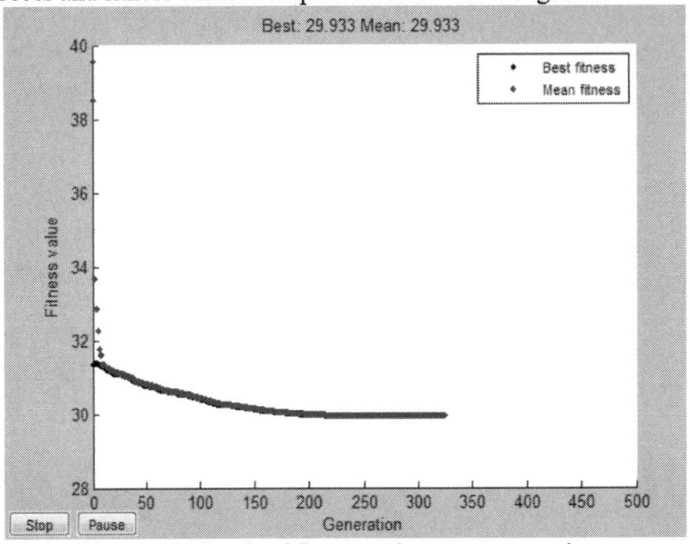

Figure 6. Graph of fitness values per generation.

Figure 6. shows that the algorithm will produce convergent solutions to the best value by increasing the generation or iteration, power losses is smaller than the previous generation.

3.3. Power flow after SVC placement

After SVC added on buses of North Bandung, Cianjur and Rancaekek then voltage profile Subsystem of South Bandung and New Ujungberung is shown in Figure 7. In Figure 7. the voltage profile after placement of SVC is compared with the voltage profile before SVC placement.

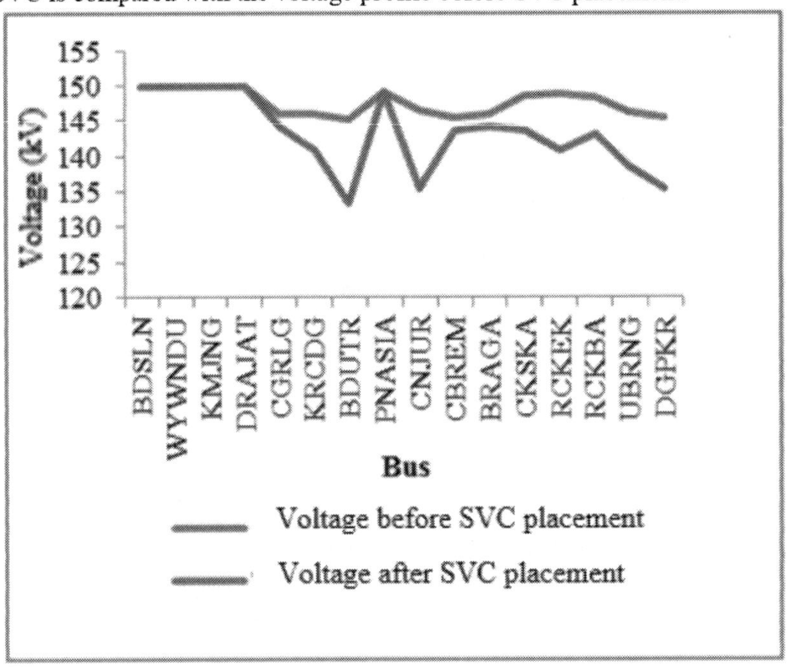

Figure 7. Graph of voltage before and after SVC placement.

In Figure 7, it can be seen that after SVC added on the North Bandung bus, Cianjur and Rancaekek, the voltage profile on the load bus increases while the voltage profile on the generating bus is unchanged. The result of active power loss after SVC placement is shown in Figure 8.

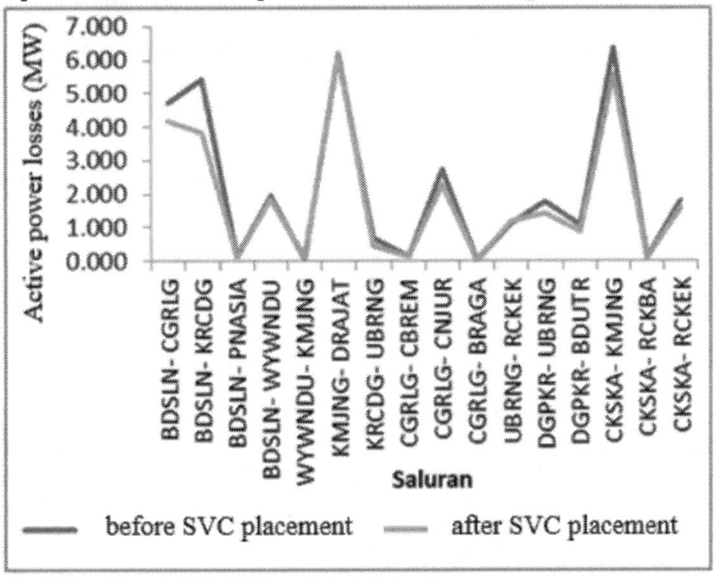

Figure 8. Graph of active power losses before and after SVC placement.

The largest decrease of active power losses occurred in the South Bandung-Kiaracondong channel 1,577 MW or 29.10%. The smallest decrease of active power losses occurred in the Cigereleng-Cibereum channel 0.03 MW or 2.36%. Total active power losses before SVC placement is 34,309 MW and after placement of SVC is 29,933 MW. The comparison of reactive power loss before and after SVC placement is shown in Figure 9.

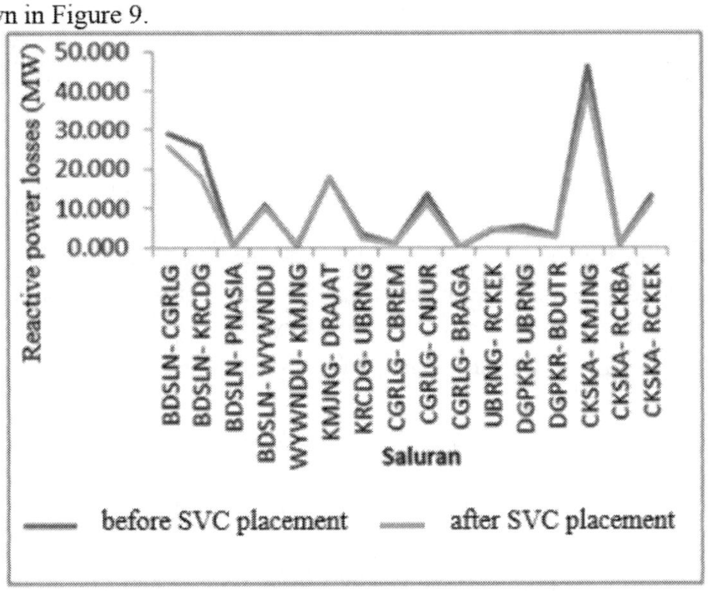

Figure 9. Graph of Reactive power losses before and after SVC placement.

The largest decrease in reactive power losses occurred in the Cikasungka-Kamojang channel 7.46 MVAR or 29.10%. The smallest decrease of reactive power losses occured in Cigereleng-Braga channel

0.02 MVAR or 2%. Total reactive power losses before SVC placement is 174,574 MVAR and after SVC placement is 151,467 MVAR.

4. Conclusion

Optimal Static Var Compensator (SVC) placement on 150 kV transmission system Subsystem of Bandung Selatan and New Ujungberung is placed on North Bandung bus with Rating -70.5 MVAR and 70.5 MVAR, Cianjur bus with Rating -62.19 MVAR and 62.19 MVAR and Rancaekek bus with rating - 193.53 MVAR and 193.53 MVAR. Optimal SVC installation can improve the voltage profile so that the voltage on all buses is within the standard of 0.95 pu to 1.05 pu. In addition, power transmission capacity increases. So, there is a decrease of power losses.

References

[1] Dhi M M dan P darma 2003 PROYEKSI KEBUTUHAN LISTRIK PLN *Pengembangan Sistem Kelistrikan Dalam Menunjang Pembangunan Nasional Jangka Panjang* 19–29
[2] Muslimin, Suyono H, and Hasanah N 2013 Perbaikan Profil Tegangan pada Feeder Harapan Baru Lima (H5) Area Samarinda untuk Pengurangan Susut Energi *EECCIS* **7** 2 131–136
[3] Karami M, Mariun N and Kadir M Z A A 2011 Determining Optimal Location of Static Var Compensator by Means of Genetic Algorithm *IEEE* 172–177
[4] Sullivan D J, Paserba J, Reed G, and Arai J 2006 Design and Application of a Static VAR Compensator for Voltage Support in the Dublin *IEEE*
[5] Liliana I S 2014 Penempatan SVC (Static Var Compensator) Pada Jaringan Distribusi Dengan ETAP 7.5.0. *Jurnal Sains, Teknologi Dan Industri* **12** 1 1–8
[6] Marsudi D 2006 *Operasi Sistem Tenaga Listrik* (Yogyakarta: Graha Ilmu)
[7] Rahman A K M R, Alam S and Hossain Z 2013 Localization of FACTS Devices for Optimal Power Flow Using Genetic Algorithm *IEEE* 0–5
[8] Anwar S, Suyono H, and Soekotjo H 2012 Optimisasi Penempatan SVC untuk Memperbaiki Profil Tegangan dengan Menggunakan Algoritma Genetika *Jurnal Elektro ELTEK 3 1 ISSN 2086-8944* **3** 1 203–208
[9] Cekdin C 2005 *Teori dan Contoh soal Teknik Elektro menggunakan bahasa pemograman Matlab* (Yogyakarta: Penerbit Andi)
[10] Lalit and A K J 2015 Optimization Of Electrical Power Transmission System Using PSO With SVC *International Journal for Technological Research in Engineering* **2** 9 1859–1865
[11] Savic A 2014 Optimal sizing and location of SVC devices for improvement of voltage profile in distribution network with dispersed photovoltaic and wind power plants *Science Direct* **134** 114–124
[12] Skaria N A 2014 Genetic Algorithm Based Optimal Location of SVC in Power System for Voltage Stability Enhancement *IEEE.*
[13] Veramuri S D S 2015 Perfomance Comparison Of SVC and TCSC in power flow control *IEEE*
[14] Pali B S 2012 Newton-Raphson Power Flow Models of Static VAR Compensator *IEEE* 9–13
[15] Kumari M S 2006 A New Decoupled - Quadratic Load Flow Approach for Adjustment of Static *IEEE* 266–270
[16] Hassan M O, Cheng S J, and Zakaria Z A 2009 Steady-State Modeling of SVC and TCSC for Power Flow Analysis *Proceedings of Yhe Internatioanal Multi Conference of Engineers and Computer Scientist* ***II***
[17] Dheebika S K K 2015 Optimal Location Of SVC , TCSC And UPFC Devices For Voltage Stability Improvment And Reduction Of Power Loss Using Genetic Algorithm *IEEE* 1–6

[18] Azim I and Rahman F 2014 Genetic Algorithm Based Reactive Power Management by SVC *International Journal of Electrical and Computer Engineering (IJECE)* **4** 2
[19] Farsangi M M, Nezamabadi-pour H, and Lee K Y 2006 Multi-objective VAr Planning with SVC for a Large Power System Using PSO and GA *IEEE* **1** 274–279
[20] Pisica I, Bulac C, Toma L, and Eremia M 2009 Optimal SVC Placement in Electric Power Systems Using a Genetic Algorithms Based Method *IEEE* **2** 1–6
[21] Marouani I, Guesmi T, Abdallah H, and A O 2011 Optimal Location Of Multi Type FACTS Devices For Multiple Contigencies Using Genetic Algorithms *IEEE*
[22] Bhandari M and Gurav S S 2015 Genetic Algorithm Based Optimal Allocation Of SVC For Reactive Power Minimization In Power Systems *IEEE* 1651–1655
[23] Preethi V A 2011 Application of Genetic Algorithm to Power System Voltage Stability Enhancement Using Facts Devices *IEEEE* 333–338
[24] Dongmei Z, Pei W, and Xu Z 2014 Reactive Power Optimization by Genetic Algorithm Integrated with Reduced Gradient Method *IEEE Workshop on Advance Research Anf Technology in Industry Application* **4** 838–841
[25] Suyanto 2005 *Algoritma genetika dalam matlab* (Yogyakarta: Penerbit Andi)

Gender-Mainstreaming in Technical and Vocational Education and Training

I D A Nurhaeni[1,*] and Y Kurniawan[2]
[1] Public Administration Study Program, Faculty of Social and Political Sciences of Universitas Sebelas Maret, Jl. Ir. Sutami 36A Surakarta, Indonesia 57126
[2] English Study Program, Faculty of Cultural Sciences of Universitas Sebelas Maret, Jl. Ir. Sutami 36A Surakarta, Indonesia 57126

*ismidwiastuti@staff.uns.ac.id

Abstract. Gender differences should be considered in vocational high schools so women and men can develop their potentials without being inhibited by gender bias. Gender mainstreaming in vocational high schools is a strategy to integrate gender differences at all stages in teaching-learning process for achieving gender equality and equity. This research evaluates the implementation of gender mainstreaming in vocational high schools consisting of seven key components of gender mainstreaming. Four vocational high schools in Sragen Regency Indonesia have been purposively selected. The data were obtained through in-depth interviews and documentation studies. The data were analyzed using Kabeer's model of gender analysis. The findings show that not all key components of gender mainstreaming have been implemented in vocational high schools. Most vocational high schools have implemented three of seven key components of gender mainstreaming, namely political will and leadership, policy framework and gender statistics. Meanwhile four of seven key components of gender mainstreaming, namely structure and mechanism, resources, infra structures and civil society have not been well-implemented. In conclusion gender mainstreaming has not been implemented effectively in vocational high schools. Accordingly, the government's education office should continue to encourage and publish guidelines on the implementation of gender-mainstreaming in vocational high schools.

1. Introduction

Achieving gender equality and empowering all women and girls is one of the seventeen goals of Sustainable Development Goals (SDGs). Gender equality is a fundamental human right and necessary foundation for a peaceful, prosperous and sustainable world. Unfortunately, gender inequality practices still occur in nearly all fields of development; one of which is the education development. So, providing women and girls with equal access to education is very important and will fuel the benefits for humanity at large.

Gender differences should be considered in the implementation of education in technical vocational education and training (TVET) especially in vocational high schools so women and men can develop their potential without being inhibited by gender bias. One of the ways to overcome gender bias is implementing gender mainstreaming.

Gender mainstreaming is a strategy to integrate gender differences in all stages of the development and implementation of education policies, plans, programs and projects for achieving gender equality and equity [1]. Gender mainstreaming involves individual and institutional actors, both from within and outside public bureaucracy [2] [3] [4] [5] [6] [7] [8].

Gender mainstreaming in the field of education is regulated through the Regulation of the Minister of the National Education No. 84 of 2008. Gender mainstreaming does not merely supplement the 'female' component or 'gender equality' component into policy, but also encourages the realization of equitable and equal education by taking into account the differences in needs and aspirations of women and men [9] [10].

The importance of applying gender mainstreaming in this field of education due to gender differences often results in gender discrimination in the form of stereotypes, sub-ordination, marginalization, double burden, and violence. This practice is the result of gender ideologies which often reinforces male power and the idea of women's inferiority [5] [11] [12] [13].

This research evaluates the implementation of gender mainstreaming in vocational high schools consisting of seven key components of gender mainstreaming, namely (i) political will and leadership; (ii) policy framework; (iii) structure and mechanism; (iv) resources; (v) gender statistics; (vi) infra structures; and (vii) civil society.

Vocational high school is one of the forms of formal education, which organizes the middle school level that prepares students primarily to work in a particular field [14]. Vocational high school prioritizes the preparation of students to enter employment and develop professional attitudes. In accordance with its kind, vocational secondary schools organize educational programs tailored to the types of employment[15]. Vocational high school students receive education from school on the one hand, and are consciously or unconsciously involved in society on the other hand [16].

Integrating gender mainstreaming into vocational high schools can prepare their graduates to fairly compete in the job markets and have bargaining power when gender discrimination practices take place in their work places [17], [18].

2. Methods

Sragen Regency, Indonesia has been purposively selected because it has the lowest achievement in gender equality index namely Gender-related Development Index (GDI) and Gender Empowerment Measure (GEM) based on the Ministry of Women Empowerment and Children Protection 2013 and 2015 [19] [20]. Four vocational high schools in the regency are purposively selected because they have participated in trainings on Gender mainstreaming in Education, namely gender responsive school program since 2012.

The primary data are obtained through in-depth interviews with principals, students, and parents and the secondary data are gained through documentation studies, such as schools' strategic plans and regulations. The data are then analyzed using Kabeer's model of gender analysis classifying policies into gender-blind policies and gender-awareness policies. Within gender awareness policies, Kabeer further distinguishes into gender-neutral, gender-specific, and gender-redistributive policies. The data validity uses data-source triangulation [12].

What is meant by gender-blind policies is recognizing no distinction between sexes. They make assumptions, which leads to a bias in favor of existing gender relations. Therefore, gender-blind policies tend to exclude women. Then what is meant by gender-awareness policies is recognizing that women are development actors as well as men; that the nature of women's involvement is determined by gender relations which make their involvement different, and often unequal; and that consequently women may have different needs, interests, and priorities which may sometimes conflict with those of men.

Gender neutral is using the knowledge of gender differences in a given society to overcome biases in development interventions, in order to ensure that interventions target and benefit of both sexes effectively meet their practical gender needs. Gender-neutral policies work within the existing gender division of resources and responsibilities.

Gender specific is using the knowledge of gender differences in a given context to respond the practical gender needs of women or men; they work within the existing gender division of resources and responsibilities.

Gender-redistributive policies are to be intended to transform the existing distributions of power and resources to create a more balanced relationship between women and men, touching strategic gender interests. They may target both sexes, or women or men separately.

3. Research Findings

The first key component of gender mainstreaming is political will. The principals make up gender-awareness policies and implement gender-responsive leadership. The empirical evidence in the field indicates that (a) a school owns a female principal through competency-based selection; (b) schools open equal opportunities to both male and female students to occupy positions, such as the heads of intra-school students organization, boy scout, and teenage red-cross; (c) the schools give equal opportunities to male and female teachers and administration staffs to occupy important positions, such as principal or head of administration and to participate in teachers competition; (d) the schools give equal motivation to male and female teachers to achieve higher achievements both in or outside schools; (e) through new-students admission mechanism the schools provide equal opportunity to both male and female will-be students to select their desired department and conduct affirmative action to fill up the gender-gap of study program, for instance vocational schools provide more quota to future female-students to enter engineering department. Conversely, culinary art department provides more quota to future male-students; (f) the schools in cooperation with various construction and engineering companies appeal the companies to recruit their best graduates without considering male or female sexes. Gender-gap cases happening so far are due to the companies' preferences to select male to female graduates in occupying vacant positions although female graduates have higher academic scores than males. Based on the findings the schools' political will and leadership are considered gender-awareness policies, especially classified into the sub-category of gender-redistributive policies.

The second is political framework. It is manifested in the schools' regulations. The research findings indicate that the schools regulations are not gender-bias, for examples: (a) there is no prohibition to male students to participate in fashion and office administration subjects and conversely there is no restriction to female students to occupy the head of school organizations, (b) the types of uniforms for both male and female students are similar, namely trousers; (c) the teaching and learning activities in vocational schools with more male students are more frequently adjusted with the condition of female students that are lesser in number. It can be seen when a male teacher has a class of workshop practice, he will ask the female students first whether they are having period or not. If yes, he will give time dispensation to the female student/s; (d) the schools regulate the teaching and learning activities without making one of the students treated unfairly; (e) the schools give equal awards to both male and female students having good achievements; (f) the schools make equal regulations or prohibitions to male and female students such as wearing school uniforms with the complete attributes in the school areas without triggering sex abuse and sex violence; (g) the schools give permission to male and female students who are sick to rest temporarily at the students health unit; (h) the schools prohibit male and female students to get married during their study period. Based on the findings, the political framework at schools is categorized gender-awareness policies, particularly included into the categories of gender-specific and gender-redistributive policies.

The next is gender statistics. It is defined as statistics that adequately reflect differences and inequalities in the situation of women and men in all areas of life. In term of education, gender statistics could be used to understand about gender differences and inequalities in all level of education then using these data for developing education program or activities for achieving gender equality in education. The research finds that schools already have sex-disaggregated data related to the number of teachers by rank and position, number of students by class and academic achievement, number of education personnel according to rank and position. Thus, schools have owned gender statistics.

The fourth is structure and mechanism. The research finds that schools do not have a gender-responsive organizational structure yet. This can be seen from the strong domination of male in administrative structure of the school, as well as in the organizational structure of school organizations, such as Intra-school students' organization, Boy Scout, and teenage Red Cross. According to the informants, unequal school organizational structures happens because female teachers tend not to take the opportunity to hold important positions in the organizational structure. The reason is because they have a responsibility to take care of the family and they are afraid of not having time for their family. Nevertheless, the principals still encourage female teachers to take on more responsibility, to occupy important positions in school.

The fifth is resources. The research result finds that the schools' human resources, i.e. teachers, administration staffs and students haven't known and understood about gender. Although the principals and teachers have participated in gender training, they still don't understand how to implement gender in school activities. They merely understand about how male and female should respect to each other. Their understanding hasn't come to the divisions of authority, responsibility and just right to men and women.

The sixth is infrastructures. The research result finds that schools' infra-structures have not been gender-responsive. This can be seen from (a) there is no male and female notices at the toilets; (b) there is no separate room for male and female students at the school health unit; (c) the desks for female students have no front cover. This potentially raises the exploitation of female students' body, which can lead to sexual exploitation.

The last is civil society. The research result indicates that the society, particularly the students' parents haven't fully known and understood about the importance of gender-perspective integration into the educational field, both formal and informal education within families. This potentially causes discriminative treatment and tends to harm one the sexes, especially female.

4. Discussion

Implementing gender mainstreaming in education, especially in vocational high school is not a matter of ease although the socialization and training on how to integrate gender perspective has been conducted. Only a part of all key components of gender mainstreaming can be successfully implemented at schools. The difficulty of implementing all the key components of gender mainstreaming at schools is because gender-responsive schools cannot be actualized without well-organized efforts, starting from policies, programs and activities. Therefore, principals' political will and leadership are very necessary, so that schools are able to grow the values of gender responsiveness. Furthermore, gender-responsive schools can be conducted through four aspects, namely internal, institutional, interpersonal and instructional [21]. The four aspects focus on the teachers' understanding about gender and then they implement gender-responsive education to all academics. The difficulties to implement gender mainstreaming also happened in another country because of the misunderstanding of the concept of gender mainstreaming (cases in EU & USA) [8] [22]. The implementation of gender mainstreaming tends to be technocracies and using improper concept of gender mainstreaming (cases in Canada) [23]. At schools, although teachers have been taught about gender they have not fully understood how gender is implemented in schools' life and it is still considered complicated [24]. To support the political will, political framework is required as a foundation in implementing gender-responsive schools.

Political will, leadership and political framework can run optimally if they are supported with gender-statistics. Gender-statistics can become a foundation to formulate policies and used as evaluation towards gender-responsive actions. Gender statistics has important role in wiping out stereotypes, policy formulation and monitoring on the progress of gender-responsive education. Statistics on women and men are needed to raise consciousness, persuade policy makers & promote changes; inspire measures for changes; provide an unbiased basis for policies & measures and monitor and evaluate policies and measures [25]. The low representation of women in the schools' authority structure inhibits the efforts to actualize gender-responsive schools [26].

The availability of separate male and female toilets and the provision of more female toilets than that of male is caused by different biological structures between women and men. Women have period, have pregnancy and usually have to accompany children or old women, so that the time to be spent in toilets is usually longer. Unfortunately, the difference of the needs is not well responded [27] [28]. Whereas toilets for women were set up to enable women and girls to enter public space safely in the system of male domination, in which the female sex caste is subjugated and made vulnerable to sexual assault and harassment on the basis of sex [27] [28]. Eventually, support from the society is important to make the schools capable of providing their students' needs based on their gender identities. Families with children of four years old should help their children choose their sex before entering elementary schools. This becomes a school reference in providing needs fulfillment suitable with the gender identity they have chosen [29].

5. Conclusion

In conclusion gender-mainstreaming has not been effectively implemented in vocational high schools. There are merely three of seven key components of gender mainstreaming which can be implemented well, i.e. political will and leadership, political framework and gender statistics. However, the other four components, i.e. structure and mechanism, resources, infrastructures and civil society, were not well implemented.

Political will and leadership are considered gender-awareness policies, (which is classified as gender-redistributive policies), the political framework at schools is categorized gender-awareness policies, particularly included into the categories of gender-specific and gender redistributive policies, and schools already have owned gender statistics. Whilst the schools do not have a gender-responsive organizational structure yet, the schools' human resources haven't known and understood about gender, schools' infra-structure have not been gender-responsive and the society particularly the students' parents haven't fully known and understood about the importance of gender-perspective integration into the educational field, both formal and informal education within families.

Accordingly, Sragen Regency government, especially the Education Institution should continue to encourage and publish guidelines on the implementation of gender-mainstreaming in vocational high schools.

References

[1] Nurhaeni I D A 2009 *Kebijakan Publik Pro Gender* (Surakarta: UNS Press.)
[2] European Commission 2008 *Programming guide for strategy papers: Gender equality* (online) available at http://ec.europa.eu/development/icenter/repository/F43_genderequality_finen.pdf
[3] Kornegay E 2001 *South Africa's national policy framework for women's empowerment and gender equality*, i-64 (online) available at www.unpan1.un.org//
[4] Radoi C 2012 Theoretical debates on the concept of gender equality *Journal of Community Positive Practices* **12** 5-17
[5] Reeves H and Baden S 2000 *Gender and development: concepts and definitions* (UK: Institute of Development Study)
[6] True J 2003 Mainstreaming gender in global public policy. *International Feminist Journal of Politics* **5** 3 368-396
[7] United Nation Development Program 2001 *Learning and information pack - gender analysis* (New York)
[8] Verloo M 2005 Mainstreaming gender equality in europe: A frame analysis approach *The Greek Review of Social Research* Special issue: *Differences in the framing of gender inequality as a policy problem across Europe* **117b** 11-34
[9] Departemen Pendidikan Nasional 2009 *Pengarusutamaan Gender: Pengalaman dalam Bidang Pendidikan* (Jakarta: Departemen Pendidikan Nasional)

[10] Kementerian Pendidikan Nasional 2010 *Acuan Pelaksanaan Sekolah Menengah Atas Berwawasan Gender* (Jakarta: Kementerian Pendidikan Nasional)

[11] Lee S C 2014 The shaping of gendered school *Journal of Chinese Education and Society* **47** 4 46-52

[12] March C, Smyth I and Mukhopadhyay M 2010 *A guide to gender-analysis frameworks* (Oxford: Oxfam GB)

[13] Kementerian Hukum dan Hak Asasi Manusia Republik Indonesia 2011 *Parameter Kesetaraan Gender dalam Pembentukan Peraturan Perundang-Undangan* (Jakarta: Kementerian Hukum dan Hak Asasi Manusia Republik Indonesia)

[14] *Undang-Undang Nomor 20 Tahun 2003*, Pasal 15

[15] *Peraturan Pemerintah Nomor 29 Tahun 1990*

[16] Chu X, Li Z, Yan B, Han J, and Fan F 2015 Comparative study of reguler and vocational high school students on family socioecoomic status, social support, self-efficacy and well-being *Journal of Social Sciences* **3** 61-68

[17] Oppong A A A and Arthur C 2015 Gender discrimination in the workplace: a study of women's participation in higher education management in Ghana *Afro Asian Journal of Social Sciences* **6** 3

[18] Bilkis A and Habib S B 2010 A review of discrimination in employment and workplace *ASA University Review* **4** 2

[19] Kementerian Pemberdayaan Perempuan dan Perlindungan Anak 2013 *Pembangunan Manusia Berbasis Gender* (Jakarta: Kementerian Pemberdayaan Perempuan dan Perlindungan Anak)

[20] Kementerian Pemberdayaan Perempuan dan Perlindungan Anak 2015 *Pembangunan Manusia Berbasis Gender* (Jakarta: Kementerian Pemberdayaan Perempuan dan Perlindungan Anak)

[21] Gender Spectrum 2017 *Gender in Education* (online) Accessed from www.genderspectrum.org, August 2, 2017

[22] Miske S, Meagher M, and DeJaeghere J 2010 Gender Mainstreaming in Education at the Level of Field Operations: the case of CARE USA's indicator framework. *A Journal of Comparative & International Education* **40** 4 441-458

[23] McNutta K and Bélandb D 2015 Implementing an Integrated Governance Strategy: The Quest for Gender Mainstreaming in Canada *American Review of Canadian Studies* **45** 4 467–485

[24] Gender Spectrum 2017 *Understanding Gender* (online) Accessed from www.genderspectrum.org, August 2, 2017

[25] Hedman B, Perucci F, and Sundström P 1996 *Engendering Statistics: A Tool for Change* (Sweden: Statistic Sweden Publication Service)

[26] European Commission 2010 *Gender Differences in Educational Outcomes: Study on the Measures Taken and the Current Situation in Europe* (Brussel: Education, Audiovisual and Culture Executive Agency)

[27] Jefferys S 2014 The Politics of The Toilet: A Feminist Response to The Campaign to 'Degender' a Women's Place *Women's Studies International Forum* **45** 1 42-51

[28] Kogan T S 2016 *How Did Bathrooms Get to be Separated by Gender in The First Place?* (online) accesed from theconversation.com and theguardian.com, May 26, 2016

[29] Hughes T and Glanfield E 2016 *Now Children as young as four are asked to choose what gender they are before they start schools* (online) Accessed from www.dailymail.co.uk, August 21, 2017

A Multimetric Approach for Handoff Decision in Heterogeneous Wireless Networks

I Kustiawan* **and W Purnama**

Department of Electrical Engineering Education, Universitas Pendidikan Indonesia, Jl. Dr. Setiabudhi No. 207, Bandung 40154, Indonesia

*iwan_kustiawan@upi.edu

Abstract. Seamless mobility and service continuity anywhere at any time are an important issue in the wireless Internet. This research proposes a scheme to make handoff decisions effectively in heterogeneous wireless networks using a fuzzy system. Our design lies in an inference engine which takes RSS (received signal strength), data rate, network latency, and user preference as strategic determinants. The logic of our engine is realized on a UE (user equipment) side in faster reaction to network dynamics while roaming across different radio access technologies. The fuzzy system handles four metrics jointly to deduce a moderate decision about when to initiate handoff. The performance of our design is evaluated by simulating move-out mobility scenarios. Simulation results show that our scheme outperforms other approaches in terms of reducing unnecessary handoff.

1. Introduction

Next generation wireless networks (NGWN) enables a mobile user to roam with a UE (user equipment) with multiple interfaces to access real time or non-real time services. Multi-radio operations in a heterogeneous environment are becoming a norm. The fourth generation (4G) of mobile broadband technology has implemented the integration of multiple networks such as wireless Local Area Networks and mobile telecommunications networks on an all-Internet Protocol basis for broadband mobile capacities. System heterogeneity offers a variety of services and applications with significantly higher data rates anytime and anywhere. However, the Always Best Connected concept allowing seamless mobility and service continuity anywhere at any time is still one of key challenges in such an environment [1]. As NGWN is aimed at ubiquitous computation and communication, the UE should enjoy seamless data transfers via the best access link among all available candidates without perceivable interruption to streaming services [2].

Handoff decision is to select the best prospective access network in the vicinity and decide at any given time whether or not to carry out handoff. Traditional or horizontal handoff involves processing between two adjacent cells of the same radio access technology. Certain link quality parameters such as RSSs (received signal strengths) or signal-to-noise ratios are commonly considered in handoff management. If concerned parameters fall below a predefined threshold, handoff is initiated. Due to different protocol structures and parameters present in the heterogeneous environment, the horizontal handoff mechanism does not cater well for such a networking environment [3].

On the other hand, handoffs among different radio access technologies occur frequently when a mobile user roams about. In particular, handoff involving switching among different types of wireless network is referred to as vertical handoff [4]. Vertical handoff decision warrants closer study. A number of schemes for handoff decision making solutions exist in the literature, that is, solutions based on network conditions using the received signal strength (RSS) or the signal-to-interference-and-noise ratio measured by mobile node, multi attribute decision making algorithms to improve the quality of services, and artificial intelligence using fuzzy logic and neural networks.

The first strategy is based on network conditions using RSSs measured by the UE. Effects of the signal strength threshold combined with an adaptive lifetime metric, available bandwidth, and packet delay has been investigated in [5]. A cross-layer management protocol proposed in [6] uses the speed of the UE and handoff signaling delay information to enhance handoff performance. A layer-3 handoff process is initiated if the RSS of the current base station has dropped below some threshold. However, only RSS alone was evaluated as a metric for considering handoff decision in these schemes.

The second strategy uses a mathematical optimization technique that deals with the problem of choosing the best alternative from a set of options based on their attributes. A performance comparison among vertical handoff algorithms, namely, MEW (multiplicative exponent weighting), SAW (simple additive weighting), TOPSIS (technique for order preference by similarity to ideal solution), and GRA (gray relational analysis) has been proposed in [7]. Comparative results show that MEW, SAW, and TOPSIS provide similar performance irrespective of different traffic classes, while GRA provides a slightly higher bandwidth and a lower delay for interactive and background traffic classes. Network selection in vertical handoff decision, as indicated in [8], can be carried out by using the Multiple Attribute Decision Making strategy. However, this study proposes a handoff mechanism in use by the UE with limited computing powers and resources constraints.

The third strategy refers to artificial intelligence techniques through a fuzzy logic or neural networks combined with several metrics such as network conditions [9], [10]. These approaches consider various input parameters as handoff decision criteria. Provided that RSS is one of the most common metrics accessible to the UE, we still ponder RSS as a determinant to trigger a handoff process. Additionally, in view that a tradeoff among other measures is required for making a vertical handoff decision, we fuse RSSs, data rate, network latency, and user preference that will be taken as multi-criteria inputs to a fuzzy system for reasoning out a single-valued output, indicating whether vertical handoff is imminent.

Our design is indeed complementary to previous studies in that our work can strengthen their operational efficiency. The remainder of this paper is organized as follows. The next section elaborates on related work. Our scheme is provided in Section 2. Section 3 discusses performance results. Lastly, Section 4 draws conclusions.

2. Vertical Handoff Strategy

Under consideration is a two-tier communication paradigm where IEEE 802.11 connectivity is preferred over the cellular network. Accordingly, we are concentrating on WLAN accessibility first. In our architecture, handoff to the cellular network may occur if IEEE 802.11 connectivity cannot sustain.

Handoff decision controller as a decision engine is performed by a fuzzy system. The general architecture of a fuzzy system consists of a fuzzifier, knowledge base IF-THEN rules, fuzzy inference engine, and a defuzzifier [11] as illustrated in Figure 1. In our architecture, each of the input parameters is assigned to one of three fuzzy sets. Values for each type of input consist of linguistic terms: weak, medium, and strong for the RSS parameter; low, medium, and high for the remaining three metrics. These sets are mapped to corresponding membership functions (MF) which refer to characteristic functions defining the fuzzy set as depicted in Figure 2. The general MFs are triangular, trapezoidal, Gaussian, bell, and sigmoidal shapes. The MFs can be chosen arbitrarily according to the user's experience [12]. Here we chose trapezoidal MFs, because this kind of MFs is most frequently used in practical applications [13]. The ranges of inputs for each fuzzy variable RSS, data rate, latency, and user preference are shown in Table 1.

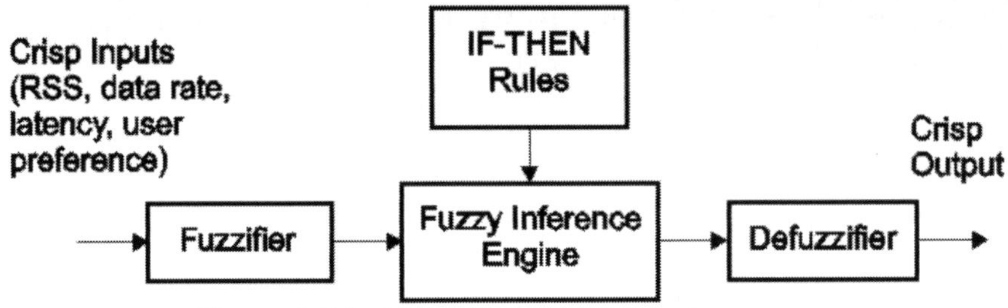

Figure 1. Block diagram of a fuzzy logic system.

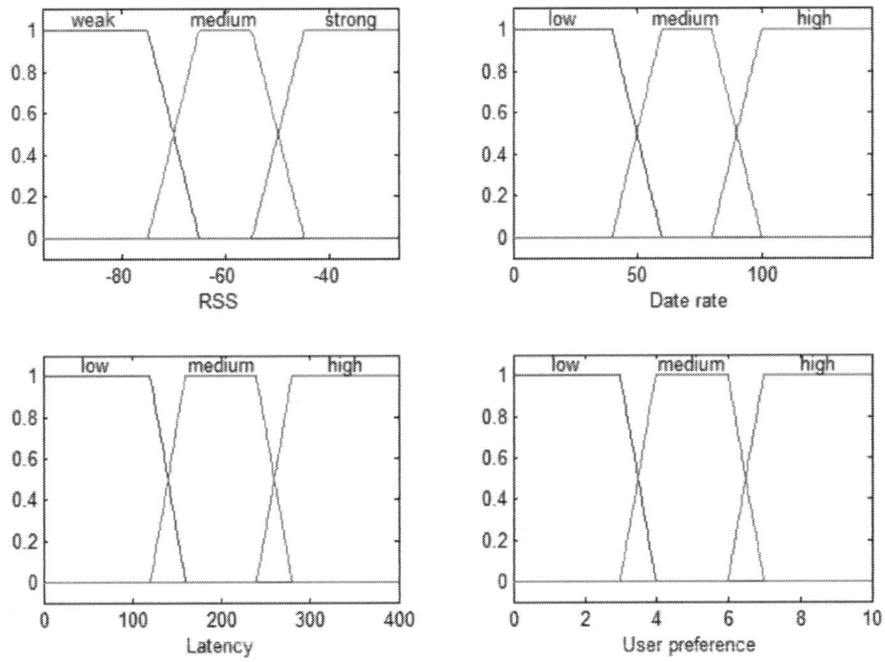

Figure 2. Trapezoidal membership functions of input metrics.

Table 1. Range inputs.

Network	RSS (dBm)	Date rate (Mbps)	Latency (ms)	User Preference	Remarks
WLAN (IEEE 802.11n)	-95 ~ -27	0 ~ 144	0 ~ 400	0 ~ 10	VoIP and video traffic classes
Cellular network (4G LTE)	-115	UL 50/DL 100 Mbps (LTE Cat. 3)	0 ~ 400	0 ~ 10	VoIP and video traffic classes

The rule base contains IF-THEN rules which are required by the Fuzzy Inference System (FIS). Among all possible inference systems, we implement the Mamdani FIS in that it is intuitive, widely accepted and operates in line with human instinct. The database defines membership functions of the fuzzy sets. FIS generates aggregate fuzzified data, based on fuzzy inference methods. The defuzzifier converts the aggregate fuzzified data into a scalar value. The output fuzzy decision sets are aggregated into a single fuzzy set and passed to the defuzzifier to be converted into a precise quantity, the handoff factor, which determines whether a handoff is necessary. We design four fuzzy input variables and three fuzzy sets for each fuzzy variable, so the maximum possible number of rules in our rule base is $3^4 = 81$. The crisp output from the defuzzifier block is used to select the most appropriate network for the UE. The universe of discourse for the handoff factor variable ranges from 0 to 10 as shown in Figure 3.

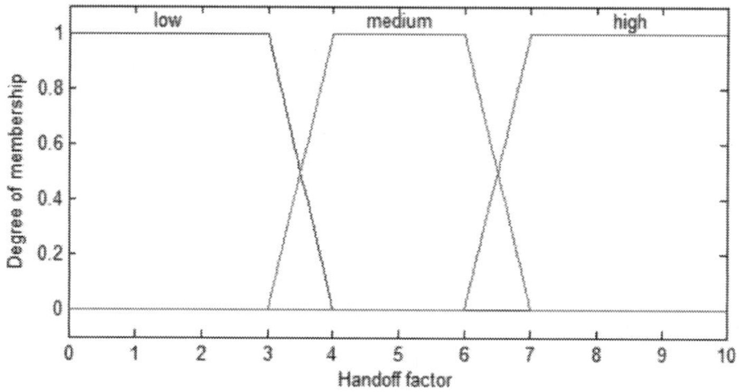

Figure 3. Membership function of the handoff factor.

When the UE is departing from an AP, this is referred to as a move-out scenario, where the UE is expected to perform a vertical handoff to the cellular network soon. On the contrary, if the UE is associating to an AP from the cellular network, a move-in scenario happens, during which another vertical handoff shall occur shortly. In this study, handoff decision is made in a move-out scenario as the following lines:

- If the deduced handoff factor is greater than or equal to 6, then the UE selects the cellular network.
- If the handoff factor is less than 6, then the UE stays connected with its current access network.

The handoff process is diagramed in Figure 4. First, a UE constantly monitors RSSs of a serving point of attachment to the system, either an AP or a base station, and other three parameters to be processed by the FIS. After knowing that the move-out scenario of the UE is determined, the output of the FIS will indicate whether the UE should initiate a vertical handoff or stay connected with the current network.

3. Performance Evaluation

In order to compare our approach with other schemes, we conducted experiments over MATLAB by simulating the network paradigm illustrated in Figure 5. Parameters in use by simulation models are listed in Table 2. β, ε, x_0 are the path loss exponent, a zero-mean Gaussian random variable with standard deviation σ, and a reference distance, respectively for path loss model. Meanwhile Q and R are the covariance of the process and measurement noise for the Kalman filter parameters.

We assume that the UE behaves under the mobile-controlled handoff (MCHO) strategy. In this strategy, a UE periodically monitors the signal of an AP and initiates handoff when certain criteria are satisfied. Suppose that the UE moves away rectilinearly from APs A towards B. The two APs are geographically dispersed, with disjoint coverages overlaid on the cellular network. The trajectory stretches 1000 meters. We shall examine how our schemes perform when the UE experiences move-out scenario.

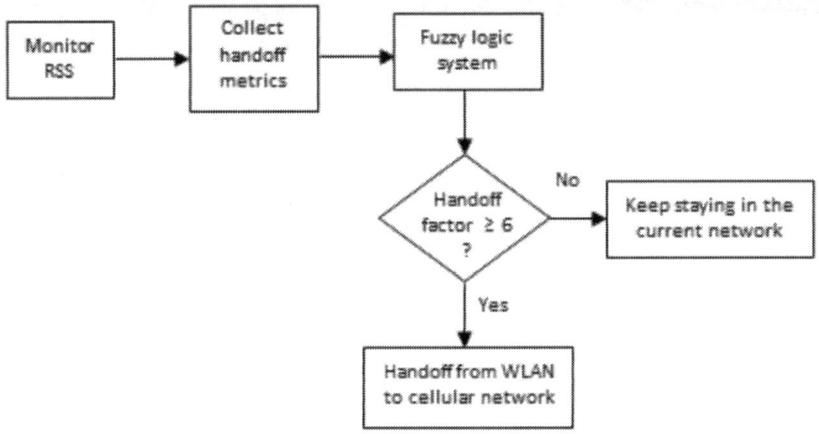

Figure 4. Workflow of our handoff process.

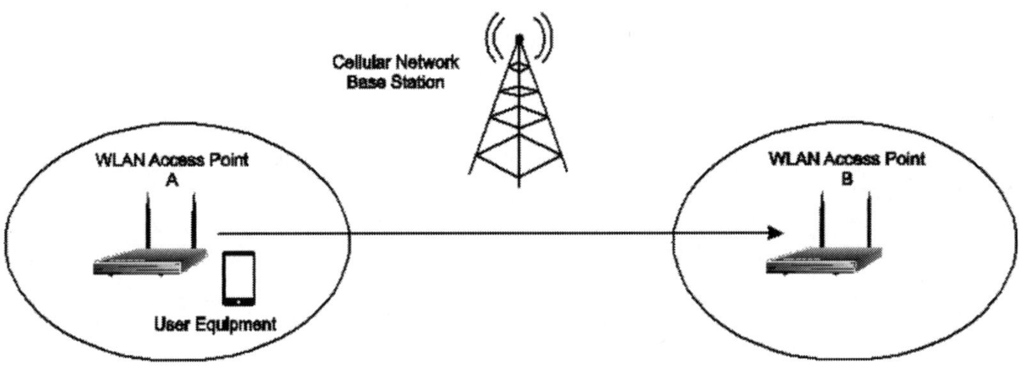

Figure 5. The simulated network model.

Table 2. Simulation parameters.

Parameter	Value	Parameter	Value
β	3.2	Q	0.001
ε	6 dB	R	0.01
x_0	1 m	RSS threshold (move-out)	-75 dBm
Velocity	1 m/s	RSS threshold (move-in)	-76 dBm
AP coverage	120 m	Rx sensitivity level	-90 dBm

Next we focus on how many handoffs are initiated during simulations. In the move-out scenario, the UE performs 75, 55, and 18 vertical handoffs with the traditional fixed-RSS [6], Kalman-filtered RSS [14], and our design, respectively (Figure 6). From Figure 6, it can be seen that the last two approaches reduce handoffs considerably in comparison to the traditional one, with relative reductions reaching 26.67% and 76%, respectively. Relative reductions are assessed below:

$$\frac{(75-55)}{75} * 100 = 26.67\%$$

$$\frac{75-18)}{75} * 100 = 76\%$$

Foregoing experiments show that the Kalman filter [14] contributes to reducing signal fluctuations due to shadowing effects. However, RSS as a single input metric is not enough for considering vertical handoff decision. Therefore, we can still implement a fuzzy logic approach to ponder multimetric without filtering RSSs. This approach brings about fewer handoffs than the others do.

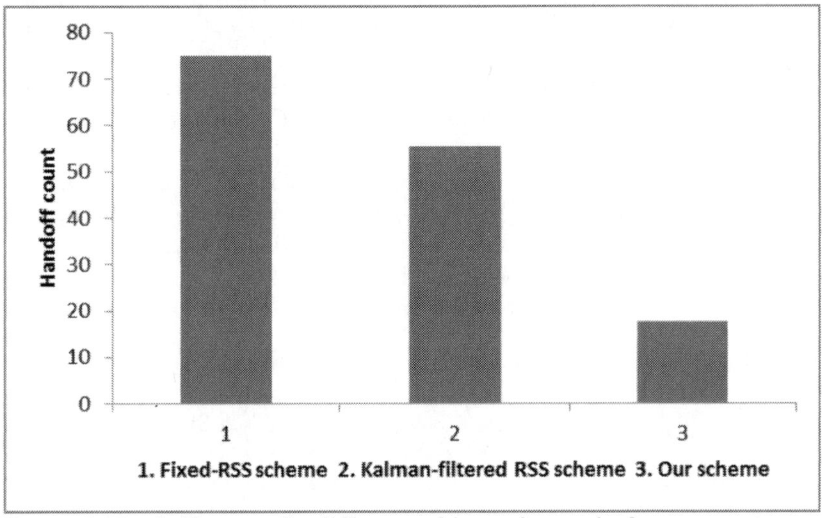

Figure 6. Handoff counts for different subject schemes in the move-out scenario.

4. Conclusion

This paper devised a means to fuse four types of determinants for joint decision making. We developed a Mamdani fuzzy inference system that takes into account multi-attributes of different potency to resolve a moderate decision indicative of handoff initiation. Our approach was compared quantitatively with other schemes, showing the usefulness of our development in all scenarios.

In closing, we conclude that our approach outperforms counterpart schemes in terms of reducing unnecessary handoffs effectively. Our approach enables the UE to avoid more frequent handoffs than necessary, keeping the UE from communication disruption. In another word, the UE is able to maintain seamless connection and enjoy service continuity in heterogeneous network environment.

References

[1] Omheni N, Zarai F, Obaidat M, Hsiao K F, and Kamoun L 2014 A novel media independent handover-based approach for vertical heandover over heterogeneous wireless networks *International Journal of Communication Systems* **27** 811-824

[2] Yan X, Şekercioğlu Y, and Narayanan S 2010 A survey of vertical handover decision algorithms in fourth generation heterogeneous wireless networks *Comput. Networks* **54** 1848–1863

[3] Çalhan A and Çeken C 2013 Case study on handoff strategies for wireless overlay networks *Computer Standard Interfaces* **35** 1 170–178

[4] Kustiawan I, Chi K H 2015 Handoff decision using a Kalman filter and fuzzy logic in heterogeneous wireless networks *IEEE Commun. Lett.* **19** 12 2258–2261

[5] A Zahran, B Liang, and A Saleh 2006 Signal threshold adaptation for vertical handoff in heterogeneous wireless networks *Mob. Networks Appl.* **11** 625–640

[6] S Mohanty and I F Akyildiz 2006 A cross-layer (layer 2 + 3) handoff management protocol for next-generation wireless systems *IEEE Trans. Mob. Comput.* **5** 10 1347–1360

[7] E S Navarro and V W S Wong 2006 Comparison between vertical handoff decision algorithms for heterogeneous wireless networks *Proc. IEEE Veh. Technol. Conf.* 947–951

[8] M Drissi and M Oumsis 2015 Multi-criteria vertical handover comparison between WiMAX and WiFi *Information* **6** 3 399–410

[9] Kunarak S, Suleesathira R 2013 Algorithmic vertical handoff decision and merit network selection across heterogeneous wireless networks *WSEAS Trans. Commun.* **12** 1–13

[10] Q He 2010 A fuzzy logic based vertical handoff decision algorithm between WWAN and WLAN *2010 International Conference on Networking and Digital Society, ICNDS* **2** 561–564

[11] Nkansah G Y and Agbinya J 2008 A vertical handoff decision algorithm for next generation wireless networks *Proc. Third Int. Conf. Broadband Commun. Inf. Technol. Biomed. Appl.*, Gauteng, South Africa 358–364

[12] Mendel J 1995 Fuzzy logic systems for engineering: a tutorial *Proc. of the IEEE* **83** 345–377

[13] Barua A, Mudunuri L, and Kosheleva O 2014 Why trapezoidal and triangular membership functions work so well: Towards a theoretical explanation *J. Uncertain Syst.* **8** 3 64–168

[14] S E Abdelmalek, F Gagnon, C Despins, and H Hu 2011 Vertical handoff algorithm for heterogeneous wireless networks based on scalar Kalman filtering *Proc. IEEE 73rd Veh. Technol. Conf.* 1–5

Evaluation Program on the Implementation of Industrial Apprenticeship (Prakerin) in Electrical Engineering

I Maulana*, Sumarto, P Nurafiati and R H Puspita

Pendidikan Teknologi dan Kejuruan, Universitas Pendidikan Indonesia, Bandung

*indra.md59@gmail.com

Abstract. This research aims to find out the evaluation program of the Industrial apprenticeship (Prakerin) in electrical engineering. This research includes on four variables of CIPP. (1). Context (a). programme planning (b). design. (2). Input (a). readiness of students (b). performance of vocational education teachers (c). Facilities and infrastructure, (3). process (a). performance students (b). performance mentors, (4). Product (a). readiness of student work. This research is a type of program evaluation research with Stake model approach. Data collection methods used are questionnaires with closed questions and frequently asked questions.

1. Introduction

Measuring the effects of apprenticeship training relative to school-based vocational training on labor market outcomes has been a challenge. The problem "arises from the fact that the two vocational routes are rarely available to young people as direct alternatives in the first place. Vocational preparation at sectoral or occupational level typically depends within any one country exclusively on either apprenticeship or full-time schooling [1]

In the apprenticeship programs, participants have a work contract, receive a salary, and are considered as full-fledged employees of a company. They spend half of their time at a company that provides learning in the workplace and the other half in a training center where they are taught the theoretical aspects of the specific trade [2]

In the apprenticeship programs, participants have a work contract, receive a salary, and are considered as full-fledged employees of a company. They spend half of their time at a company that provides learning in the workplace and the other half in a training center where they are taught the theoretical aspects of the specific trade [3]

Program evaluation is a unit or unit of activity that aims to gather information that realizes or implements a policy, takes place in a continuous process, and occurs within an organization that involves a group of people for decision-making. Program evaluation aims to know the achievement of program objectives that have been implemented. Furthermore, the results of the program evaluation are used as the basis for carrying out follow-up activities or for subsequent decision making [4].

The progress of a nation will be determined by the quality of its future generations. To create a formidable generation, it is the core time for development not only oriented to the physical things alone. Rather to human development as a driving factor of development.

Vocational education is growing in line with the development of demands Society, through two social institutions. First, social institutions in the form of job structure with the organization, the division of roles or duties, and behaviors related to the selection, acquisition and stabilization of careers. The second social institution, in the form of education with its double function, is as a media of cultural preservation as well as the media of social change [5].

This policy requires both the school and industry to jointly develop the concept, it is intended that there is a suitability between school and industry. Suitability is meant that the competence gained by students in school is a required competence in the industry. Industry must also play an active role in conveying technological progress to the school so that synchronization between the industrial world and the world of education. Double system education policy is operationalized in the form of Apprenticeship (Prakerin).

Apprenticeship(prakerin) is basically a form of education that involves students directly working in the business world / industrial world so that learners have competencies that match the expectations and demands of the business world / industrial world. In addition, also to obtain work experience as one of the things to improve professional skills. The term Prakerin is not a new thing in the world of education in Indonesia, especially for vocational education or Vocational High School (SMK). This prakerin began popularized in 1994 through the Double Systems Education policy. This is well-founded considering that the industrial world requires a qualified workforce and experts in its field to operate advanced equipment and technology.

The purpose of Prakerin's implementation is to improve the quality of Vocational High School graduates (SMK), both knowledge, skills and work ethics in accordance with the demands of employment, so that SMK graduates are ready to enter the labor market. But in reality the type of expertise and the number of graduates produced by SMK has not been in line with the demand of the labor market

2. Methods

This research categorized as evaluative type, with quantitative descriptive approach. Quantitative descriptive in program evaluation is used to collect, describe and explain the aspects that have been evaluated [6]. The evaluation model used is an evaluation model developed by Stake. Stake Model Evaluation emphasizes the implementation of two main things, namely to do description and judgments. These two main points are obtained through an overview of the evaluation component that includes the input (antecedent), the process (transaction) and the product (outcomes) [7].

Data collection techniques is a major step in the research, because the purpose of the research is to obtain data. Without knowing the techniques of data collection, the researchers will not get the data set Data collection techniques used include:

a. Documentation, this method is used to obtain supporting data about the description of planning and implementation of Prakerin program. The documentation includes Prakerin procedure, Prakerin participants data, list of business world / industry world which become Prakerin partner and Prakerin guidance data available in industry.
b. Observation, this method is used to obtain data on facilities and infrastructure and the state of the school environment.
c. Quizoner / questionnaire, a questionnaire is a set of written questions given to respondents with the intention of revealing the existing situation in the respondent self or outside the respondents who want revealed by the researcher. Conditions to be expressed by researchers is about:
   1) Readiness of learners, teachers' performance of productive subjects, readiness of school facilities and infrastructure.
   2) Performance of learners and supervisor performance during Prakerin implementation.
   3) Readiness of students work after the implementation of Prakerin.

The research instrument is one of the tools used to collect data. Instruments used in this study are documentation, observation and kuisoner. As a guide for collecting data in this study, the lattice is as follows:

a. Documentation, Prakerin implementation guide, list of industries, list of learners and mentors of industry work practices in the industry.

b. Observation is to identify the facilities and infrastructure that exist in the school.
c. Questionnaires / questionnaires, questionnaires used in this study there are two types, namely open and closed. Closed questionnaire consists of questionnaire A for learners and questionnaire B for mentors.

3. Results

Based on preliminary observations then obtained data:

a. Input evaluation: (a) students' readiness in very good category (100%), (b) productive teacher's performance in very good category (68.57%) and good (31.34%), (c) readiness of facilities and infrastructure in very good category (68.58%), good (25.71%) and low (5.71%).
b. process evaluation: (a) the performance of learners in very good category (57.14%) and good (42.86%); (b) performance of the industry's best practices (71.43%) and good (28.57%).
c. product evaluation: the readiness of students to work in very good category (65.71%) and good (34.29%).
d. obstacles: (a) obstacles to learners, lack of direction, insistent orders, subject matter in different schools with the state of the field, (b) supervisory constraints, should guide intensively.

From the results of input evaluation found 7 mentors or 100% states readiness of learners to carry out industrial work practices including into the high or good category. A total of 11 students or 31.34% stated that the productive teachers' performance in teaching learning is included in the high or good category and as many as 24 students or 68.57% stated the performance of productive subject teachers in implementing the learning including into the category very high or very good. A total of 2 students or 5.71% stated that the readiness of facilities and infrastructure included into the low category, 9 students or 25.71% stated the readiness of facilities and infrastructure included into the high or good category and as many as 24 students or 68.58% stated the readiness of the facilities and infrastructure is included into very high or excellent categories.

The result of process evaluation got 3 counselors or 42,86% stated student's performance included into high or good category and as much as 4 counselor or 57,14% stated that student's performance included into very high or very good category. A total of 10 students or 28.57% said the supervising performance included into the high or good category and as many as 25 students or 71.43% said the supervising performance included into the category very high or very good.

Result of product evaluation got 13 students or 34,29% stated readiness of learners work included into high or good category and as much 22 student or 65,71% stated readiness of learners work included into category very high or very good.

4. Conclusion

There are still many obstacles in the implementation of apprenticeship, for the required evaluation to see the results of apprenticeship running well or not. The purpose of apprenticeship to provide students readiness in the world of work must be developed gradually.

From the evaluation data, there are still deficiencies and challenges that give effect of decreasing performance in apprenticeship implementation. Evaluation results exist to see results and data useful for subsequent decision making. So the results of this evaluation will improve the apprenticeship process and form a good output.

References

[1] S Sermsuk, C Chianchana, and P Stirayakorn 2014 5th World Conference on Educational Sciences 2013 A Study of Model of Vocational Curriculum Development Under Vocational Education Commission Using Cross-Impact Analysis *Procedia - Soc. Behav. Sci.* **116** 1896–1901

[2] I Measuring 2016 The effectiveness of apprenticeship training : A within-track comparison of workplace-based and school-based vocational training in Hungary **55**
[3] V Cohen-scali 2014 Career construction in French apprenticeship programs : An exploration of the role of gender in the tutorship *J. Vocat. Behav.* **85** 1 49–56
[4] R Arthur 2015 Learning Approach of Problem Solving for Increase Learning Achievement of the Civil Engineering Evaluation Program **3** 8 964–967
[5] C R and M Yasin 2015 Cultivating learning : a grounded theory of skills acquisition for vocation in modern apprenticeships *Procedia - Soc. Behav. Sci.* **174** 275–282
[6] M Brockmann and I Laurie 2016 Apprenticeship in England – the continued role of the academic – vocational divide in shaping learner identities **6820**
[7] D Irwan and D Mardapi 2014 Evaluasi Program Percepatan Berbicara Bahasa **2** 1 37–49

IOP Conf. Series: Materials Science and Engineering **306** (2018) 012060 doi:10.1088/1757-899X/306/1/012060

Clay Stabilization Using the Ash of Mount Sinabung in Terms of the Value of California Bearing Ratio (CBR)

I P Hastuty*, **R Roesyanto and S M A Napitupulu**

Civil Engineering Department, Universitas Sumatera Utara, Medan, Indonesia

*ika.hastuty@usu.ac.id

Abstract. Most areas in Indonesia consist of clay soils with high plasticity so that to meet technical requirements the soil needs improvement, which is known as soil stabilization.There are three ways of soil stabilization process, i.e. mechanical, physical and chemical. In this study, chemical stabilization was performed, that was by adding stabilizing agents to the soil. The stabilizing agent used was the ash of Mount Sinabung. Since 2010 until now, Sinabung Mountain is still experiencing eruption that produces a lot of volcanic ash and it inconveniences the environment. So, it is expected that this research will be able to optimize the utilization of Sinabung ash. The purpose of this study was to investigate the effect of the addition of Mount Sinabung ash to CBR (California Bearing Ratio) value, to determine the effect of the curing time of one day and fourteen days mixture on the CBR value, and to find the mixed content with effective curing time to produce the largest CBR value. Based on this study, the soil type CL (Clay – Low Plasticity) was obtained, based on the classification of USCS (Unified Soil Classification System) and categorized as A-6 (6) based on the classification of AASHTO (American Association of State Highway and Transportation officials) with the most effective mixed stabilizer material which was the variation of 10% Mount Sinabung ash with fourteen days of curing time. The CBR value resulted from the mixture of 10% Sinabung ash that was cured within fourteen days was 8.95%. By the increase of the content of the Sinabung ash, the CBR value always improved to the level of 10%, Sinabung ash then decreased and became constant at the mixture of higher volcanic ash mixture but remained above the CBR value of the original soil.

1. Introduction

Generally, most of the areas in Indonesia consist of the clay with high plasticity. If the existing base ground is in the form of clay, which has a low shear support capacity, then the construction above it can suffer from damage. Therefore, the base ground must be solid to conform to technical requirements. Clay is the soil that has certain mineral particles that produce plastic properties on the soil when mixed with water. Clay is commonly used as the hoard of the roads. The general properties of the clay soil are extremely hard in dry conditions and plastic in a state of moderate moisture content. However, when the water content is high, clay will be sticky (cohesive) and very soft.Therefore, clay needs to be stabilized. Soil stabilization is the mixing of soil with certain materials, in order to improve the technical properties of the soil, or it may also mean attempts to change or improve the technical properties of certain soils to meet certain technical requirements [1].

Subgrade is the part that will support the pavement thickness. The subgrade is located on the entire width of the road, so it can be in the excavation area, heap, and soil surface. Subgrade materials are taken from local soil. If the materials are in a poor condition (CBR <2%), it is necessary to do soil improvement. There are several ways to increase the value of CBR in subgrade layer, one of them is stabilizing the soil. Stabilization of soil conducted in this research is chemical stabilization by adding chemicals to the soil that is going to be stabilized. Volcanic mountain ash used in this study comes from Mount SinabungFrom the latest data, the volume of the lava of Mount Sinabung that came out has reached 3.3 million cubic meters [2]. The volcanic ash that comes out during the eruption provides long-term benefits like fertilizer. But when the eruption happened, volcanic ash will spread to various regions so that it could damage plants, houses, inhibit water channels, disturb the health of the population, etc. The volcanic ash itself contains aluminum, silica, iron, calcium, and magnesium. Silica (SiO2) contained by the volcanic ash is a major constituent in the manufacture of cement. The pozolanic nature is the binding behavior of other minerals in the clay so that it becomes harder in a certain period of time [3]. For that reason, Mount Sinabung ash can be utilized as the mixture of soil stabilization, in addition to its silica content, it can also reduce untreated waste by residents.

CBR (California Bearing Ratio) is a comparison between test load and standard load which is expressed in percentage. CBR value is a value that states the quality of the base soil compared with standard materials in the form of crushed stone that has a CBR value of 100% in carrying the load. Subgrade on new road construction can be original soil, embankment soil, or dug soil that has been compacted until it reaches 95% of maximum density. Thus, the bearing capacity of the base soil is the value of the soil layer ability to bear the load after the soil is compacted. CBR is called CBR laboratory, because it is prepared in the Laboratory. CBR Laboratories are divided into two types, they are soaked and unsoaked.

In the research of Ronny Hutauruk (2016) analyzed with Unconfined Compression Test conducted on original soil, obtained that compressive strength (qu) of original soil is 1.40 kg / cm^2, meanwhile the compressive strength (qu) of remoulded soil is 0.50 kg / cm^2. The addition of 11% Sinabung ash with 14 days of curing time has the greatest compressive strength (qu) value, which is equal to 2.32 kg / cm^2. There was an increase in the compressive strength of the soil due to the addition of Sinabung ash. [4]

2. Method

The research methodology that was carried out used experimental method at Soil Mechanics Laboratory, Department of Civil Engineering, Universitas Sumatra Utara. The conducted research was the sample of original soil and sample of the soil that has been given additional ash of Mount Sinabung. The material used are: Soil from Patumbak Deli Serdang and Volcanic Ash from Mount Sinabung North Sumatra. Chemical composition of Sinabung Ash SiO_2 is 85. 90%, Fe_2O_3 is 0.007%, Al_2O_3 is 11.93% and CaO is 0.133%

The use of 2% to 20% of Sinabung ash was based on previous research that has been done by Ronny Hutahuruk by analyzing through Unconfined Compression Test (UCT) [4]. This study analyzes the addition of Mount Sinabung ash with the California Bearing Ratio (CBR) laboratory test so that the results of the research can be implemented for a pile of roads. In this study, we use soaked CBR to adjust to the field conditions in case of rain.

2.1. Manufacture of specimens

The test specimen that was utilized in outline consists of the original soil samples and samples using soil that has been mixed with the ash of Mount Sinabung. The percentage of mixture used in this study is 2% to 20%.

2.2. Implementation of testing

Testing is divided into 2 parts namely the test for the original soil and soil testing that has been mixed with stabilizer material. The test was held as follows:

2.3. Data analysis

After all the data were collected, then data analysis was performed. All results obtained will be displayed in the form of tables, graphs and explanations. The flowchart of the research can be seen in **Figure 1** below.

3. Results and Discussion

3.1 Original soil testing and stabilizer testing

The results of the original soil physical properties test are listed in Table 1.

Table 1. Physical properties test data.

No	Testing	Soil	Sinabung Ash
1.	Water Content	14.52%	-
2.	Specific Gravity	2.64	2.62
3.	Liquid Limit	48.64 %	Non Plastic
4.	Plastic Limit	18.81 %	Non Plastic
5.	Plasticity Index	29.82 %	Non Plastic
6.	Sieve Analysis	50.04%	11.40%
7.	Optimum Moisture Content	21.38 %	-
8.	Maximum Dry Density	1.32 gr/cm^3	-

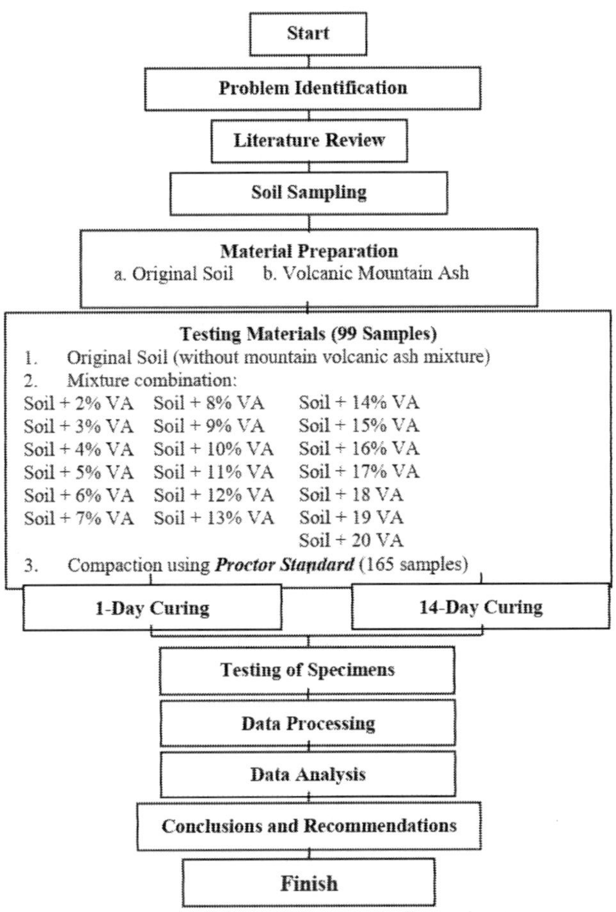

Figure 1. Research flow chart.

3.2 Test of physical properties and soil mechanics with stabilizers

The result showed that the addition of volcanic ash could improve the physical and mechanical properties of the soil. This can be seen in **Figure 2** and **Figure 3** which shows the value of Atterberg boundaries.

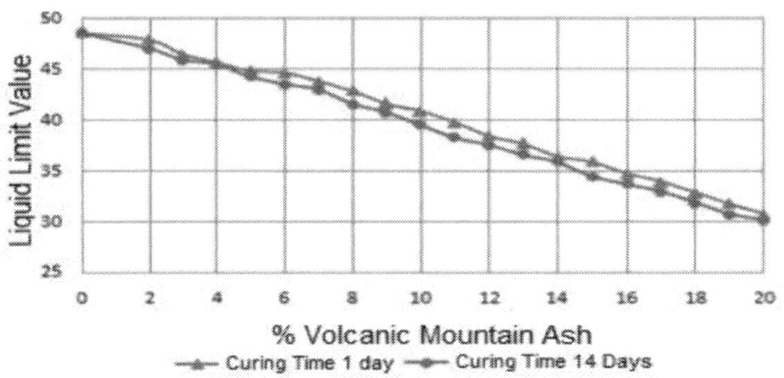

Figure 2. Graph of the relativity between liquid border value and mixed variation of 2% - 20% Mount SinabungAsh for one day and fourteen days curing.

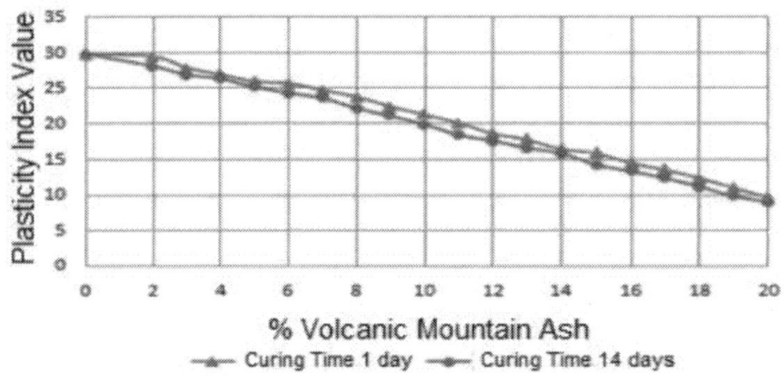

Figure 3. Graph of Plasticity Index value with mixed variation of 2% - 20% Mount Sinabung Ash mixture for one day and fourteen days curing time

In **Figure 2**, it is show that the liquid limit due to the addition of volcanic ash stabilization material has experienced decrement, the greater the percentage of volcanic ash, the smaller the liquid limit. In the original soil liquid limit reached 48.64% while the lowest liquid limit value was in the addition of 20% volcanic ash with 14 days of curing time that was 30.15%.This is due to the soil under goes a cementation process by the ash of Mount Sinabung so that the soil becomes larger grains that make the pull force between particles in the soil decreases **Figure 3** shows that with the addition of stabilizers the plasticity index value will decrease. The decreasing value of the plasticity index may reduce the development potential and depreciation of the soil. This is due to the hydration process of volcanic ash added to the soil. This process strengthens the bonding between the soil particles, resulting in the formation of harder and more stable grain. The filling up of soil pores minimizes the occurrence of seepage on the volcanic ash-soil mixture that affects the decrease of shrinkage potential. Then it is added with stabilizer material in the form of volcanic ash. Silica and alumina from volcanic ash mixed with water form a paste that binds clay particles and covers the pores of the soil. Pore cavities that are surrounded by cementation materials that are more difficult to be penetrated by water will make the soil mixture and volcanic ash become more resistant to water absorption thereby decreasing the plasticity. It can be seen that the decrease of plasticity index from original soil which was initially

29.82%, decreased along with the increasing of volcanic ash mixture until lowest plasticity index value at mixture of 20% volcanic ash at fourteen days curing time, which was 9.04%.

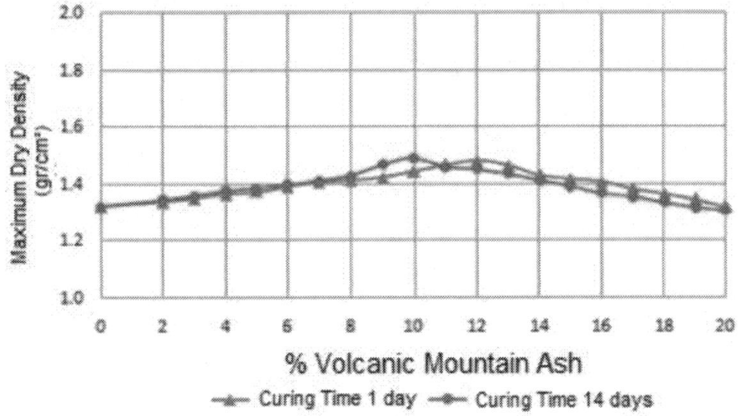

Figure 4. Graph of the relationship between maximum γd of soil with mixed variation 2% - 20% of Sinabung Ash for one day and fourteen days curing

In **Figure 4** it can be seen that the maximum dry content weight of the whole test is in a 10% mixture of volcanic ash with fourteen days curing period was 1.49 g/cm^3, in contrast to the time when the mixture of 10% volcanic ash is only cured in one day with value of maximum dry fill weight of 1.44 g / cm^3. The increasing value of the maximum dry content weight that was affected by the length of the curing caused by the mixture of clay and volcanic ash that was more even. At the optimum moisture content, the 10% volcanic ash mixture in one day curing time that was 19.49% decreased along with the fourteen days curing time to be 19.35%, This was caused by the even mixture of water along with the duration of curing towards the volcanic mountain ash mixture and clay soil as seen in **Figure 5** below. The results of CBR testing performed on each variation of mixed content, the maximum CBR value of all the tests was 8.95% obtained from the mixture of 10% Mount Sinabung ash and clay soil with curing period of fourteen days. The CBR value in the 10% mixture of Mount Sinabung ash increased along with the duration of curing period, in which when the mixture was cured for 1 day, the CBR value was only 8.22%. For the graph of CBR value relationship on the mixing variation of Sinabung ash with the duration of one day and fourteen days curing period can be seen in **Figure 6**

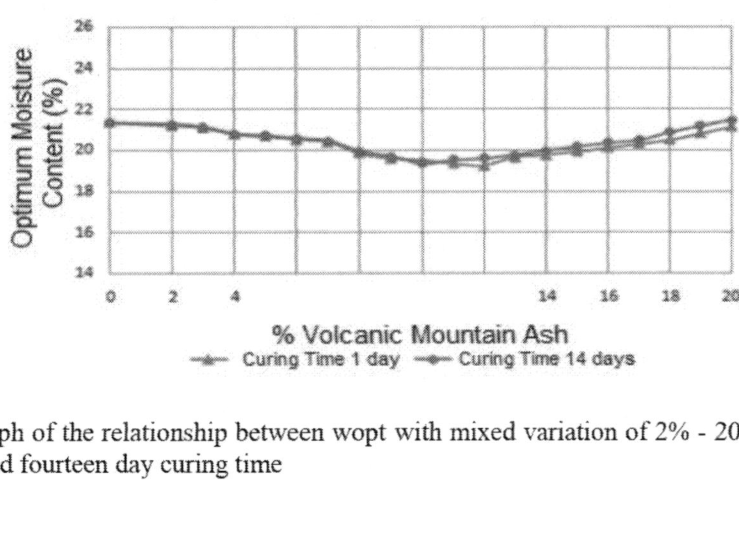

Figure 5. Graph of the relationship between wopt with mixed variation of 2% - 20% of Sinabung ash for one day and fourteen day curing time

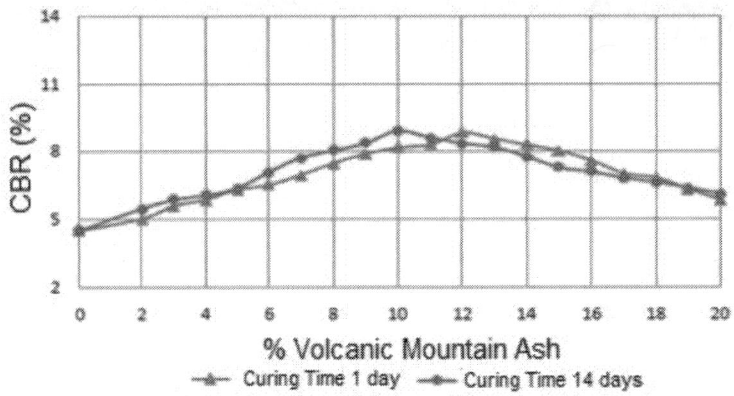

Figure 6. Graph of relationship between CBR value with the mixed variation of 2% - 20% of Sinabung ash for one day and fourteen-day curing time

In the graph it was shown in **Figure 6**, in addition to the increase experienced by the addition of 10% of Mount Sinabung ash along with the duration of curing, it can also be seen that the value of CBR testing for the largest one day curing is in the 12% mixture of Mount Sinabung ash by 8.87%, while for the fourteen days curing, the maximum CBR value is obtained at 10% of Mount Sinabung ash mixture by 8.95%. With the addition of Mount Sinabung ash, the CBR value increases up to 8.95%. The greater the CBR value the greater the soil bearing capacity, so that for the same traffic load it will require a thinner thickness of the pavement.

4. Conclusions

From the research result, it can be concluded that:

1. Based on USCS classification, the soil samples are included in CL (Clay-Low Plasticity) type, which is inorganic clay with low to moderate plasticity and remain on CL type after mixed with Mount Sinabung ash.
2. Based on the AASHTO classification, the original soil samples are included in types A-7-6 (11). At the curing time of fourteen days, the soil mixture and 2% -9% of Sinabung ash are included in type A-7-6, then in the 10% - 18% Sinabung ash mixture included in type A-6, and in the 19% - 20% Sinabung ash mixture included in type A-4. In soil mixture and 10% Sinabung ash included in type A-6 (6).
3. The decline in the plasticity index (IP) value in 20% of Mount Sinabung ash with one day curing time is 9.73% to 9.04% at fourteen days curing caused by the effect of curing period on the chemical reaction of volcanic ash mixing and the soil .
4. From the Proctor Standard results, the optimum moisture content on the original soil is 21.38% and the maximum dry content weight is 1.32 g / cm^3, while the maximum dry content value of all mixture is 10% ash of Mount Sinabung At fourteen days curing at 1.49 g / cm^3, an increase over the previous curing time when only one day of the dry weight of the maximum value of 1.44 g / cm^3, while the optimum water content of 10% mixture of Sinabung ash on the curing one day of 19.49% decreased at the curing of fourteen days to 19.35%.
5. CBR value after the soil was mixed with Sinabung ash, the largest is in 10% of Sinabung ash mixture with fourteen days curing time, that is 8.95%.
6. The curing period resulted in the soil and the Sinabung ash mixture are more even, resulting in higher CBR value.

References

[1] Hardiyatmo, H C 2010 *Mekanika Tanah 1* (Yogyakarta: Gadjah Mada University Press)
[2] Ronny Hutahuruk, 2016 *Stabilisasi Tanah lempung dengan menggunakan abu Vulkanik di tinjau dari Nilai Unconfined Compression Test (UCT)* (Medan: Program Studi Teknik Sipil Universitas Sumatera Utara)
[3] Ika Puji Hastuty, Roesyanto, and Jeriko A 2017 Study of the Effectiveness of the Use of Cement and Bottom Ash Towards The Stability of Clay in Term of UCT Value Science Direct *Procedia Engineering* **171**, 484 – 491
[4] Gunawan, Hendra, and Huda N A 2013 *Pemanfaatan Limbah Karbit Untuk Meningkatkan Nilai Cbr Tanah* (Surakarta: Program Studi Teknik Sipil Universitas Sebelas Maret)

IOP Conf. Series: Materials Science and Engineering **306** (2018) 012061 doi:10.1088/1757-899X/306/1/012061

Three Tier-Level Architecture Data Warehouse Design of Civil Servant Data in Minahasa Regency

I R H T Tangkawarow*, J P A Runtuwene, F I Sangkop and L V F Ngantung

Universitas Negeri Manado, Tondano 95618, North Sulawesi, Indonesia

*irene.tangkawarow@unima.ac.id

Abstract. Minahasa Regency is one of the regencies in North Sulawesi Province. In running the government in Minahasa Regency, a Regent is assisted by more than 6000 people Civil Servants (PNS) scattered in 60 SKPD. *Badan Kepegawaian Diklat Daerah* (BKDD) of Minahasa Regency is SKPD that performs data processing of all civil servants and is responsible for arranging and formatting civil servants. In the process of arranging and determining the formation of civil servants, many obstacles faced by BKDD. One of the obstacles is the unavailability of accurate data about the amount of educational background of civil servants based on rank/class, age, length of service, department, and so forth. The way to overcome the availability of data quickly and accurately is to do Business analytical. This process can be done by designing the data warehouse first. The design of data warehouse will be done by dividing it into three tiers of level.

1. Introduction

Fast and accurate information is highly needed in daily life. Information will be an important element in the development of society for today and the future. However, the needs of high information are sometimes not matched by adequate information presentation. Frequently, the information is still to be rededicated from very large-amounted data.

Civil Servant Agency of Regional Training and Education (*Badan Kepegawaian Diklat Daerah* – BKDD) of Minahasa Regency is a governing body for the management of civil servants (*Pegawai Negeri Sipil* - PNS), especially in Minahasa Regency. The main duties and functions of BKDD of Minahasa Regency refer to the activities that are directly related to the processing and analysis of civil servant data in Minahasa. In its implementation, every semester, BKDD of Minahasa Regency usually gives a circular letter to every Regional Instrument Work Unit (*Satuan Kerja Perangkat Daerah* – SKPD), requesting to enter or report the personnel data related to the SKPD. Additionally, if there is any change of the personnel data, the SKPD concerned should re-update the data of BKDD in Minahasa. The data will later be used to fulfill the processing and management tasks of civil servants in Minahasa Regency, one of which is the task of structuring and formatting civil servants (PNS). To carry out the civil servant arrangement and formation, information of the quantity, quality, composition, and distribution of civil servants in the Minahasa region is required. The data held in the institution of BKDD of Minahasa Regency are in very large quantities. Therefore, it takes time and effort to analyze the data in order to obtain the required information. The number of Civil Servants in Minahasa Regency until 2017 has reached more than 6000 civil servants.

This research was conducted to overcome the problems faced by BKDD of Minahasa Regency in analyzing the data that exist. The design of Three-Tier Level Data Warehouse Architecture for the civil servant data in Minahasa Regency was conducted by integrating the data in the form of spread sheet data into the data warehouse using Pentaho data integration-kettle, doing analysis with MDX query using Pentaho scheme workbench, and displaying dashboard using Pentaho BI server plugin sayku analytic. The process above would be divided into three tiers of level: top tier, middle tier, and bottom tier.

2. Research Methods

2.1. Experimental design

The research design that would be conducted in this research is as follows [1][2]:

a. Bottom-Tier Level, consisting of:
   1) Determining the subject of the data warehouse. One of the characteristics of a data warehouse is subject-oriented. Therefore, the first stage in establishing the data warehouse was defining the subject.
   2) Defining the needs. It was defined all the needs of data sources required by the data warehouse and the information required to be obtained from the data warehouse to support the management decision.
   3) Creating the design of star scheme and designing the process of ETL. Data warehouse generally uses dimensional data model or often called as a star scheme. The star scheme has two parts, namely fact table, and dimension table. Fact table is a table containing the facts of business, generally a table the details of transactions that have occurred [3]. After the scheme was made, the next process was doing the process of ETL (Extraction, Transformation, Loading) from the operational database to the database of the data warehouse. Before the ETL process was conducted, the design for the ETL process was required to be created.
   4) Performing the process of ETL from the operational data/database to the data warehouse database. The process of ETL periodically extracts the data from the source system, transforms it to a general format, and then loads it into a target data source, which is generally a data warehouse or data mart. ETL is very important for data integration and data warehousing. In this research, the ETL was done using Pentaho Data Integration.

b. Middle-Tier Level, in which the researcher created a dimensional data cube for the data warehouse by using MDX-query run on the workbench scheme.

c. Top-Tier Level, displaying the dashboard by using Pentaho BI server plugin sayku analytic.

2.2. Related research

Here are some research related to the design of three-tier level architecture data warehouse:

a. Extract transformation loading from OLTP to OLAP data using Pentaho data integration. This research was conducted at the bottom-tier level, explaining in detail the process of ETL by using Pentaho Data Integration [4].

b. Vertical Information System: A Case Study of Civil Servant Teachers' Data in Manado City [5]. This research used Pentahotool in analyzing and designing the data warehouse on the civil servant teacher data of Manado City.

c. The Team Study Report of Capital Market and Financial Institution Supervisory Agency (BAPEPAMLK) on the Business Intelligence Implementation of the Finance Ministry of the Indonesian Republic of 2007 [6]. This research was aimed to determine the important aspects that must be understood and prepared in implementing the system of Business Intelligence (BI) in the environment of BAPEPAMLK.

d. Business Intelligence Application Design on the Distribution Information System of PT Pertamina Lubricant using Pentaho [7]. This research produced BI application by using Pentaho.There were three phases in conducting this research: The first phase was determining the input. The input was derived from the data that had been recorded in the distribution information system of PT Pertamina Lubricant SR III namely transactional data, master data, and reference data. The second phase was designing the system. In this phase, the input data source would be integrated into one data warehouse through ETL process (Extraction, Transformation, Loading), creating a cube or data mart logically consisting of various data dimensions and fact tables. The third was the process of data analysis, which used an OLAPmethod.

e. Library Warehouse Data Design and Manufacture of STIMIK AMIKOM Yogyakarta [8]. This research was made by using SQL Server Integration Service (SSIS) which is a tool for SQL Server 2005. There were 7 stages implemented in this research.
f. A white paper entitled Microsoft SSIS and Pentaho Kettle: A Comparative Study for Three-Tier Level Data Warehouses [9], explaining the difference between Microsoft SSIS and Pentaho Kettle if viewed from the three-tier data warehouse architecture. This research discusses the difference between the 3 layers namely bottom tier, middle tier, and top tier. However, this research does not emphasize the process and explains the observation of both BI tools in a general way.
g. A Journal of which title was A Service-oriented Architecture for Business Intelligence [10] discussed comparison between service-oriented concepts in BI and the traditional BI concept. The literature study concluded that with the concept of service-oriented in BI, the process of receiving information and technology integration became more effective compared to doing so with the traditional BI concept

3. Results and Discussion

3.1. Three-tier level data warehouse architecture

A data warehouse is not a product but an environment where users can find strategic information [11]. The data warehouse is a collection of logical data separated from the operational database and it is a summary [8]. According to Poniah [11], a data warehouse is:

- Retrieving all data coming from the operating system.
- If applicable, it includes the relevant data derived from outside (external) such as the indicator of industrial benchmarks.
- Integrating all data derived from all diverse data sources
- Eliminating all inconsistent data and transforming the data
- Storing all data in a format that allows the data to be easily accessible for decision making.

In the design of data warehouse architecture, there are three tiers of a level (three-tier level). The three tiers cover the bottom-tier level, middle-tier level, and top-tier level.
It can be seen from the figure below:

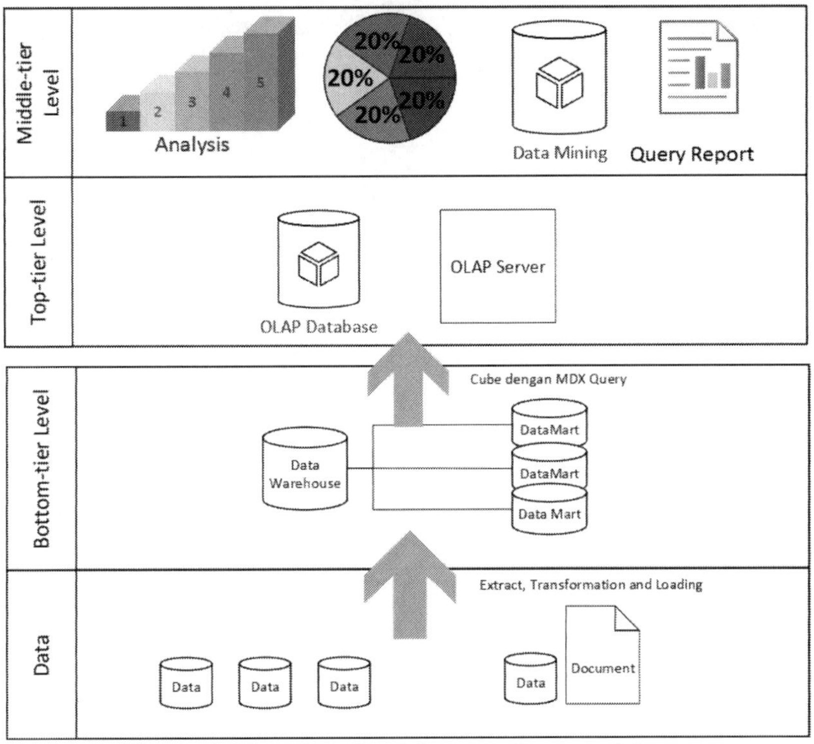

Figure 1. Three-Tier level data warehouse architecture.

3.2. Bottom-tier level

In this research, the design of civil servant data warehouse of Minahasa Regency was created using Open Source Business Intelligence System. To conduct the civil servant structuring and formation, it should be identified the problems occurring in the office of Civil Servant Agency of Regional Training and Education (*Badan Kepegawaian Diklat Daerah* – BKDD) as well as the required and wanted needs. The problems currently faced are the information presentation and data report that are less accurate and qualified, and takes a long time. The data managers find difficulty in sorting and processing the data and obtaining the information needed. This is because the system is inadequate so that, to achieve the needs for the implementation of the civil servant structuring and formation task in Minahasa, it takes a long time. To carry out the structuring and formation of the civil servants (PNS), it needs information for the decision supporting material such as staff distribution (spread) for each SKPD in Minahasa, the number of staffs (civil servants) based on the educational background, the number of positions available in SKPD, the number of staffs distributed [spreaded] based on the position, class, and the echelon level. In its implementation, every semester, BKDD of Minahasa usually gives a circular letter to every Regional Instrument Work Unit (*Satuan Kerja Perangkat Daerah* – SKPD), requesting to enter or report the personnel data related to the SKPD. Additionally, if there is any change of the personnel data, the SKPD concerned should re-update the data of BKDD in Minahasa. The data of the civil servants will later be inputted by each SKPD, then processed in the data processing division to then generate reports and information needed. The data source of civil servants processed by the officers is inputted by each SKPD in Minahasa in the form of excel files with different table format so that the data processing officers experience difficulty in sorting, merging and processing these data. Therefore, the information presentation often meets delays, and it affects the accuracy level of the data.

This stage was made to plan the data warehouse design of the civil servants for the purpose of staff structuring and formation. In this stage, a multidimensional modeling of star scheme was used, as can be seen in the following figure:

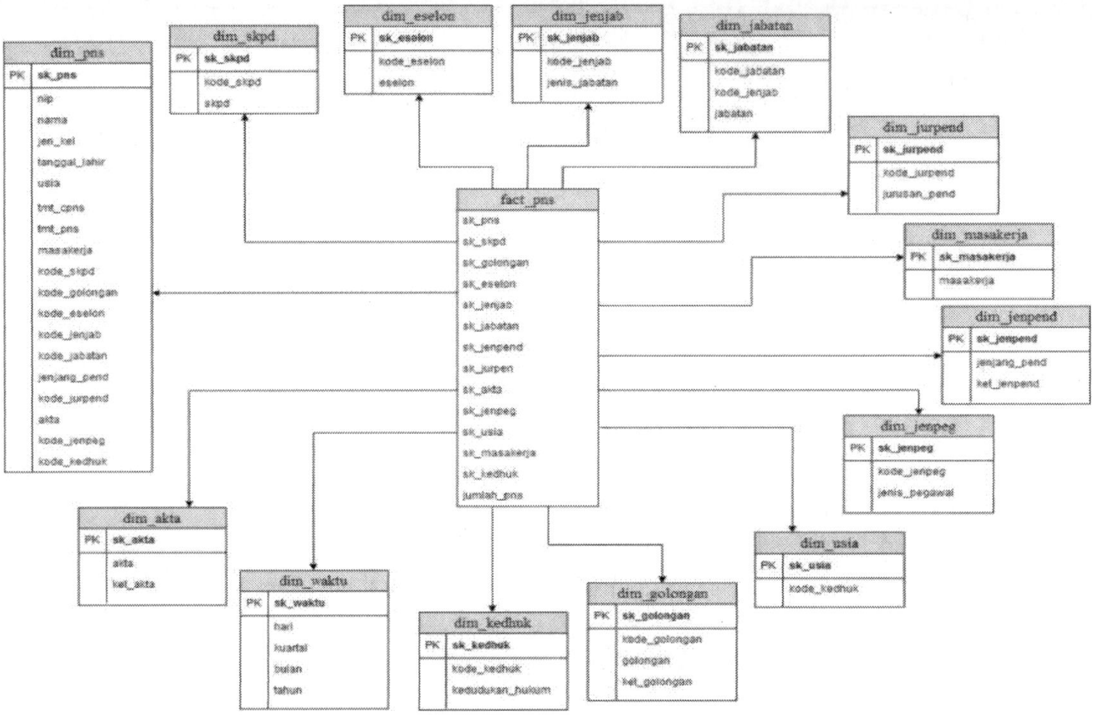

Figure 2. Star schema.

ETL process (Extraction, Transformation, Loading) is a process that must be conducted in establishing data warehouse [2]. In general, the benefits of ETL are as follows:

1. To input data from OLTP to OLAP (data warehouse)
2. To integrate inter-application
3. To clean data
4. To migrate data
5. To export data

As for the process of ETL, it is closely related to data and database sources, including:

1. Extracting data from external sources. The external sources can be in the form of excel, CSV, database, XML, web service, database, etc.
2. Transforming (converting) the data according to the needs (it may include the element of data quality).
3. Loading (inputting) the data into the final target. The final target can be in form of excel, CSV, database, XML, web service, database, etc.

The design of ETL (Extraction, Transformation, Loading) was also created to produce the civil servant data warehouse of *BKKD* of Minahasa Regency. In the process ETL, it was used *Pentaho Data Integration (Kettle)* for the transformation process, which was the real implementation of the star scheme data modelling that had been made previously. The dimension table was one that would establish the civil servant data warehouse of *BKDD* of Minahasa Regency.

3.2.1. Dimension table of civil servants (dim_pns). The data source to perform the transformation process of civil servant dimension table was derived from excel data, resulting **dim_pns.**

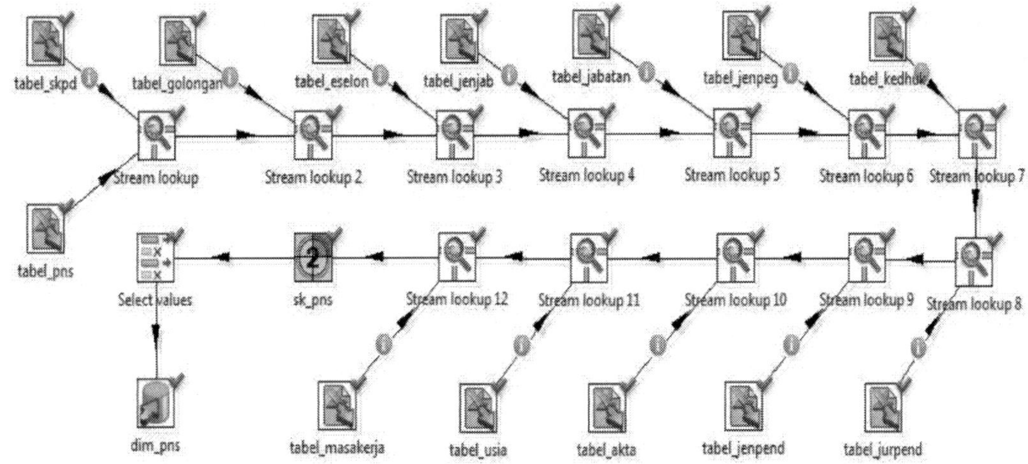

Figure 3. Transformation Process of DIM_PNS.

After the transformation was successfully made, the following table (as can be seen in Figure 4) would be created:

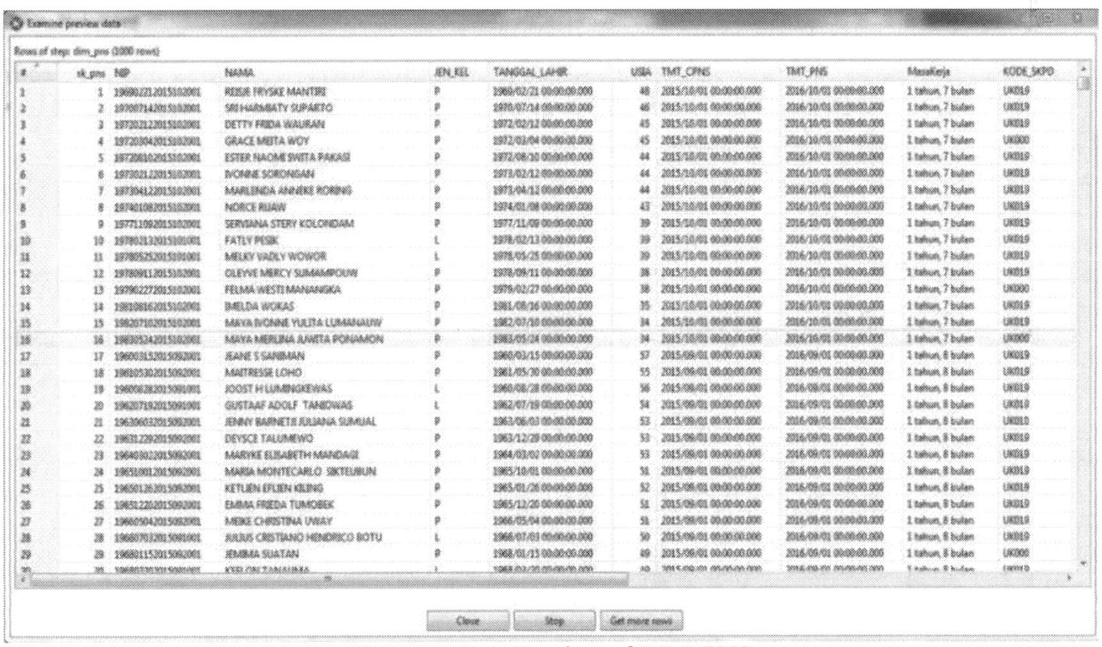

Figure 4. Data Preview of DIM_PNS.

3.2.2. Dimension table of SKPD (dim_skpd). The data source to perform the transformation process of the dimension table of SKPD was derived from excel data, resulting **dim_skpd**.

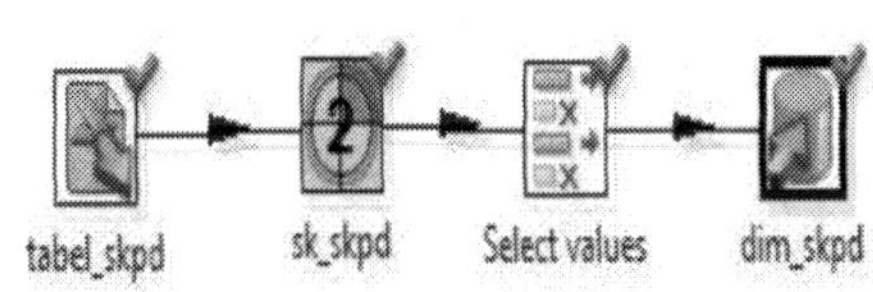

Figure 5. Transformation Process of DIM_SKPD.

IOP Conf. Series: Materials Science and Engineering **306** (2018) 012061 doi:10.1088/1757-899X/306/1/012061

After the transformation process was successfully performed, the following table would be created (as can be seen in Figure 6 below):

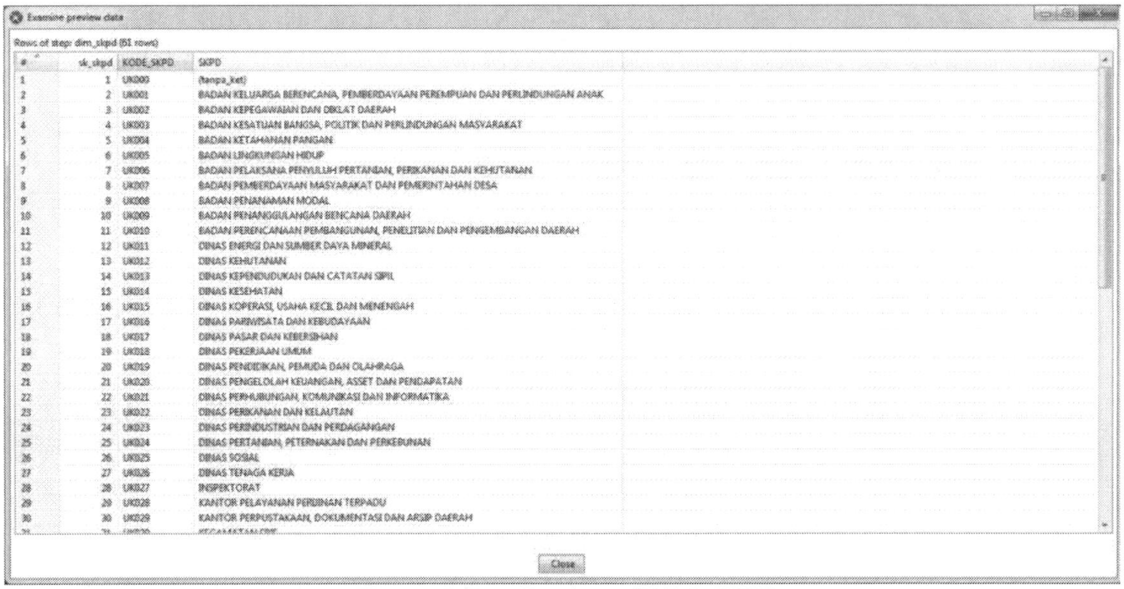

Examine preview data

Rows of step: dim_skpd (51 rows)

#	sk_skpd	KODE_SKPD	SKPD
1	1	UK000	(tanpa_ket)
2	2	UK001	BADAN KELUARGA BERENCANA, PEMBERDAYAAN PEREMPUAN DAN PERLINDUNGAN ANAK
3	3	UK002	BADAN KEPEGAWAIAN DAN DIKLAT DAERAH
4	4	UK003	BADAN KESATUAN BANGSA, POLITIK DAN PERLINDUNGAN MASYARAKAT
5	5	UK004	BADAN KETAHANAN PANGAN
6	6	UK005	BADAN LINGKUNGAN HIDUP
7	7	UK006	BADAN PELAKSANA PENYULUH PERTANIAN, PERIKANAN DAN KEHUTANAN
8	8	UK007	BADAN PEMBERDAYAAN MASYARAKAT DAN PEMERINTAHAN DESA
9	9	UK008	BADAN PENANAMAN MODAL
10	10	UK009	BADAN PENANGGULANGAN BENCANA DAERAH
11	11	UK010	BADAN PERENCANAAN PEMBANGUNAN, PENELITIAN DAN PENGEMBANGAN DAERAH
12	12	UK011	DINAS ENERGI DAN SUMBER DAYA MINERAL
13	13	UK012	DINAS KEHUTANAN
14	14	UK013	DINAS KEPENDUDUKAN DAN CATATAN SIPIL
15	15	UK014	DINAS KESEHATAN
16	16	UK015	DINAS KOPERASI, USAHA KECIL DAN MENENGAH
17	17	UK016	DINAS PARIWISATA DAN KEBUDAYAAN
18	18	UK017	DINAS PASAR DAN KEBERSIHAN
19	19	UK018	DINAS PEKERJAAN UMUM
20	20	UK019	DINAS PENDIDIKAN, PEMUDA DAN OLAHRAGA
21	21	UK020	DINAS PENGELOLAH KEUANGAN, ASSET DAN PENDAPATAN
22	22	UK021	DINAS PERHUBUNGAN, KOMUNIKASI DAN INFORMATIKA
23	23	UK022	DINAS PERIKANAN DAN KELAUTAN
24	24	UK023	DINAS PERINDUSTRIAN DAN PERDAGANGAN
25	25	UK024	DINAS PERTANIAN, PETERNAKAN DAN PERKEBUNAN
26	26	UK025	DINAS SOSIAL
27	27	UK026	DINAS TENAGA KERJA
28	28	UK027	INSPEKTORAT
29	29	UK028	KANTOR PELAYANAN PERIJINAN TERPADU
30	30	UK029	KANTOR PERPUSTAKAAN, DOKUMENTASI DAN ARSIP DAERAH

Close

Figure 6. Data Preview of DIM_SKPD.

3.3. *Middle-tier level*

In the middle tier, we used OLAP Server in which a multidimensional OLAP modelling can be implemented, executing directly the multidimensional data and operations. In this stage, a cube design – a process of creating data modelling logically from one dimension to be multi dimensions – was done. In generating graphs that can be viewed from various dimensions, it needed a process of *MDX Query* by using scheme workbench application.

The process of creating *MDX Query* of the civil servant data scheme on the application of GUI-*Scheme Workbench* containing **dataPNS_cube** and dimensions that have been made such as **dim_pns, dim_skpd, dim_jabatan, dim_jenjab, dim_jenpend, dim_jurpend, dim_kedhuk, dim_jenpeg, dim_usia, dim_masakerja, dim_akta, dim_golongan, dim_eselon, dandim_waktu.** The resulted *Scheme Workbench* can be seen in Figure 7 below:

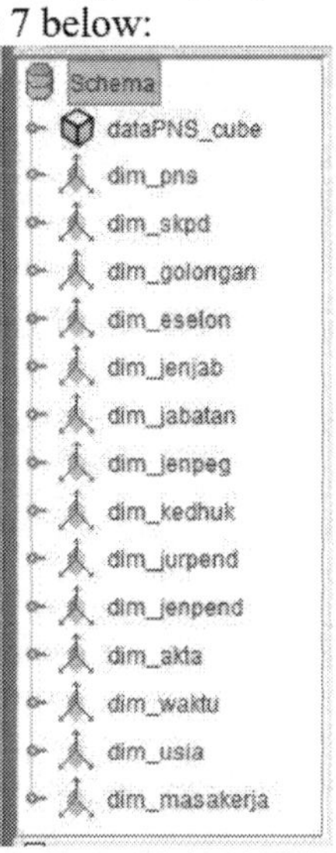

Figure 7. Scheme of Civil Servant Data (DATAPNS) using Workbench Application.

The OLAP data were hierarchically organized and stored in *the cube.* In the dimension of **dim_pns**on the scheme of **dataPNS,** there was a hierarchy of **pns** (civil servants) containing several levels *i.e.* **nama, jenis_kelamin, tanggal_lahir, tmt_cpns, tmt_pns, dantabeldim_pns. Hirarkidalamdim_pns** as can be seen in Figure 8.

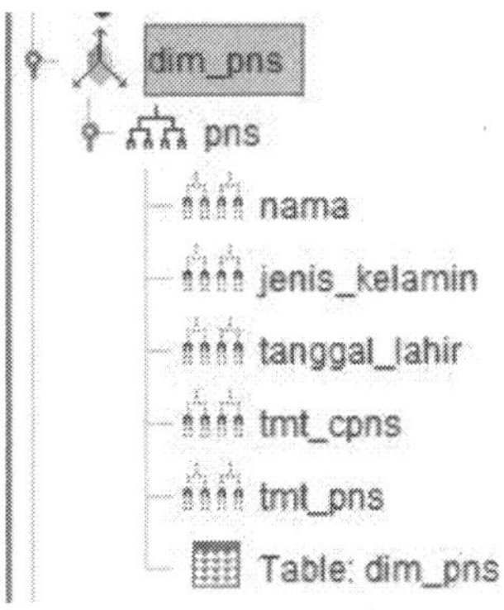

Figure 8. Hierarchy of DIM_PNS.

3.4. *Top-tier level*

These results were formed from the *MDX Query*on *cube* design that would display charts, diagrams, and tables in the *dashboard reporting.* To display the *dashboard* on the panel server *Pentaho BI,* plugins of *Saiku Analytics* was used for the analysis view results. Here are some *reporting* view results using *Saiku Analytics:*

A *dashboard* on the number of civil servants based on the level of education is shown in Figure 9 below.

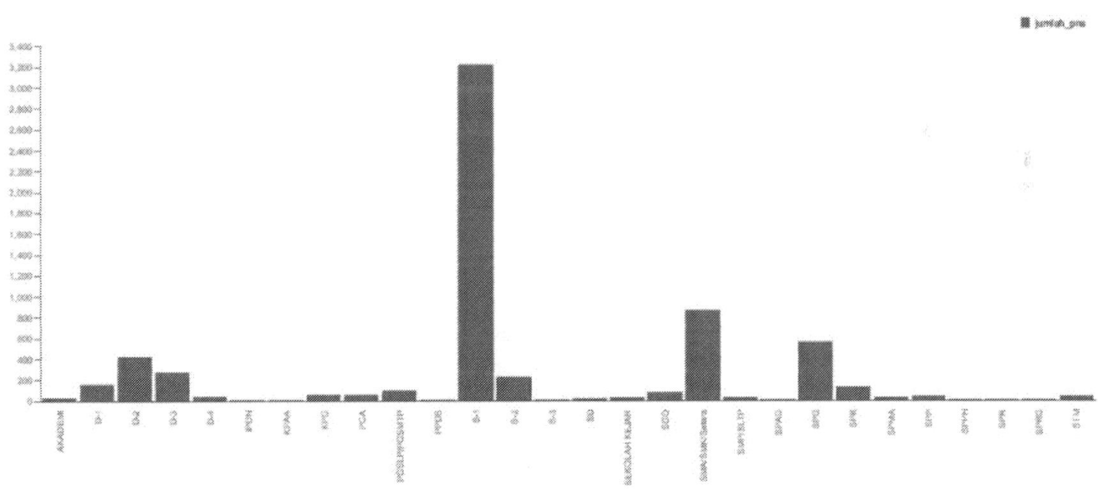

Figure 9. Bar Graph to See the Number of Civil Servants Based on Educational Background Level

Figure 10 below shows the *dashboard* for the number of civil servants based on the major or department (study program) of education in *SKPD* of Tondano Barat District.

Figure 10. Pie Graph on The Number of Civil Servants Based on The Study Program Filtered for SKPD of Tondano Barat.

4. Conclusion

Based on the results of the research conducted in designing a Three Tier Level Architecture in the Civil Servant Data Warehouse of Minahasa Regency using Open Source Business Intelligence System, it can be concluded: data warehouses with the use of Pentaho Open Source application can be created easily and quickly, and free of cost; the design of data warehouses using Open Source Business Intelligence System in Minahasa Regency will helpfully support the government in making a decision for the civil servant structuring and formation held at the office of BKDD of Minahasa. For the report, it can be presented through a dashboard in the form of multidimensional graphs in accordance with the required information; by the existence of civil servant data warehouses using Business Intelligence System, the reporting or presentation of information for the requirement of civil servant structuring and formation task will not experience the problem of delays or will not take a long time to get the required information.

Acknowledments

We would like to thank *Lembaga Penelitian dan Pengabdian Masyarakat (*LPPM) and Engineering Faculty Universitas Negeri Manado for supporting the research.

References

[1] Moss L T and Atre S 2003 *Business intelligence roadmap: the complete project lifecycle for decision-support applications* Addison-Wesley Professional

[2] R Kimball and J Casertan 2015 *The Data Warehouse ETL Toolkit* **1**

[3] J Mulyana 2014 *Pentaho: Solusi Open Source untuk Membangun Data Warehouse* (Yogyakarta: ANDI)

[4] R J Salaki, J Waworuntu and I R H T Tangkawarow 2016 Extract transformation loading from OLTP to OLAP data using pentaho data integration *IOP Conf. Ser. Mater. Sci. Eng.* **128** 12020

[5] J P A Runtuwene and I R H T Tangkawarow 2017 Vertical information system: A case study of civil servant teachers data in Manado city *Indones. J. Electr. Eng. Comput. Sci.* **6** 1 42–49

[6] Tim Studi Implementasi Business Intelligence 2007 *Laporan tim studi tentang implementasi business intelligence (Departemen Keuangan Republik Indonesia Badan Pengawas Pasar Modal Dan Lembaga Keuangan)*

[7] R W Witjaksono, M Wiyogo, and P N Wicaksono 2007 Perancangan Aplikasi Business

Intelligence Pada Sistem Informasi Distribusi Pt Pertamina Lubricant Menggunakan Pentaho *J. Rekayasa Sist. dan Ind.* **2** 2 12–18

[8] A Armadyah 2009 Perancangan Dan Pembuatan Data Warehouse Pada Perpustakaan STMIK AKAKOM Yogyakarta *Tesis. Perpust. MTI* 39–52

[9] Grecol M L 2012 *Microsoft SSIS and Pentaho Kettle: A Comparative Study for Three-Tier Data Warehouses*

[10] Wu L, Barash G, and Bartolini C 2007 A service-oriented architecture for business intelligence *Service-Oriented Computing and Applications, 2007. SOCA'07. IEEE International Conference on* 279-285 IEEE

[11] P Ponniah 2001 *Data Warehousing Fundamentals: A Comprehensive Guide for IT Professionals* **6**

The 3D Digital Story-telling Media on Batik Learning in Vocational High Schools

I Widiaty[1,*], Y Achdiani[1], I Kuntadi[2], S R Mubaroq[2] and D Zakaria[3]

[1] Department of Family Welfare Education, Universitas Pendidikan Indonesia, Jl. Dr. Setiabudhi no 229, Bandung 40154, Jawa Barat, Indonesia
[2] Department of Mechanical Engineering Education, Universitas Pendidikan Indonesia, Jl. Dr. Setiabudhi no 229, Bandung 40154, Jawa Barat, Indonesia
[3] Department of Eletrical Engineering Education, Universitas Pendidikan Indonesia, Jl. Dr. Setiabudhi no 229, Bandung 40154, Jawa Barat, Indonesia

*isma@upi.edu

Abstract. The aim of this research is to make 3D digital Story-telling Media on Batik Learning in Vocational High School. The digital story-telling developed in this research is focused on 3D-based story-telling. In contrast to the digital story-telling that has been developed in existing learning, this research is expected to be able to improve understanding of vocational students about the value of local wisdom batik more meaningful and "live". The process of making 3D digital story-telling media consists of two processes, namely the creation of 3D objects and the creation of 3D object viewer.

1.Introduction

Digital story-telling is a new direction in the study of learning media and a part of the humanities discipline. This media unites computer, science, and visual culture [1]. Digital story-telling is designed in a multimedia form consisting of text, images, audio, and video [2]. Digital story-telling is a new technology in the field of learning media that contains elements of culture pedagogy. The development of this pedagogical aspect is not only a cognitive dimension but also the affective or value dimension that needs to be embraced in life [3].

Introduction aspect and internalization process value can be lifted from local wisdom value. Digital story-telling is believed to be able to uncover the "ideology" or meaning contained within the culture and the unique identity of a community [4]. Identity setting of a community which becomes the focus is shown in the digital story-telling. Digital story telling without the elements of community is less able to express the moral message of a culture powerfully [2].

The digital story-telling elements that are important to display in the learning setting are emotional content, voice and images, economy (brief in presentation), and pacing [5]. Digital story-telling needs to contain elements of information, visual, technology, and media literacy [2]. Digital stories consist of three main elements: personal, informative / instructional, and historical stories. [4] Other components of digital story-telling are digital media software, computers, image capture devices, and audio capture devices [6].

The main approach that underlies the digital story-telling learning media is narrative approach [3] [4]. The narrative approach in the educational context is a means for critical reflection and identity

IOP Conf. Series: Materials Science and Engineering **306** (2018) 012062 doi:10.1088/1757-899X/306/1/012062

constructions. [3] The theories that underlie digital story-telling are social behavior modification [6], spatial humanities [1], and directed to the process of transforming experience in the context of their own voices [2].

Digital story-telling is one of the ways to present intangible culture mapping, relating to people, places, and things / events [7]. Batik Indonesia is one of Indonesia's unique intangible culture that is the world's cultural heritage. Batik is an artwork that contains elements of Indonesian culture and local wisdom displayed in a piece of cloth [8]. Batik as the voice of culture of Indonesian culture is able to become a culture expression [9] and can connect a person with a place, event, and culture of a community [10].

This study examines the development of digital story-telling that can be utilized in batik learning in Vocational High School (SMK). The digital story-telling developed in this research is focused on 3D-based story-telling. In contrast to the digital story-telling that has been developed in existing learning, this research is expected to be able to improve understanding of vocational students about the value of local wisdom batik more meaningful and "live". Batik that has been composed of motifs / patterns in the 2 dimensions (2D) form will be developed in the 3 dimensions (3D) form to make it more interesting and meaningful.

2. Methods

3D digital story-telling media content for batik learning in SMK is Batik Kasumedangan. Batik Kasumedangan is a term for batik from Sumedang, one of the districts in West Java. Batik Kasumedangan is known as batik loaded with elements of history. The two main motifs raised in this paper as the main material for the development of 3D digital story-telling media are a historic site that stands firmly in Sumedang in the form of a statue named "Memorial Cadas Pangeran" and the tomb of one of the national heroes "Tjut Nyak Dien". Both historical relics are adapted in the typical batik motif Kasumedangan which are named the motif "Memorial cadas prince" and Tjut Nyak Dien.

In general, the process of making 3D digital story-telling media consists of two processes, namely the creation of 3D objects and the creation of 3D object viewer. The tool used to display 3D objects is a smartphone (mobile phone or tablet) on an android basis. The use of smartphone is intended to facilitate the learning process by storytelling method. The software that used to create 3D objects is Blender. Blender was chosen because it has many features that support the process of making 3D digital story-telling, including modelling, UV mapping, texturing, rigging, skinning, animation, and particle. Figure 1 shows the flowchart of 3D object creation process using Blender.

3D viewer application creation is done by using Unity3D software. The 3D object will be displayed by the app by moving the marker. The concept of 3D viewer application can be seen in Figure 2.

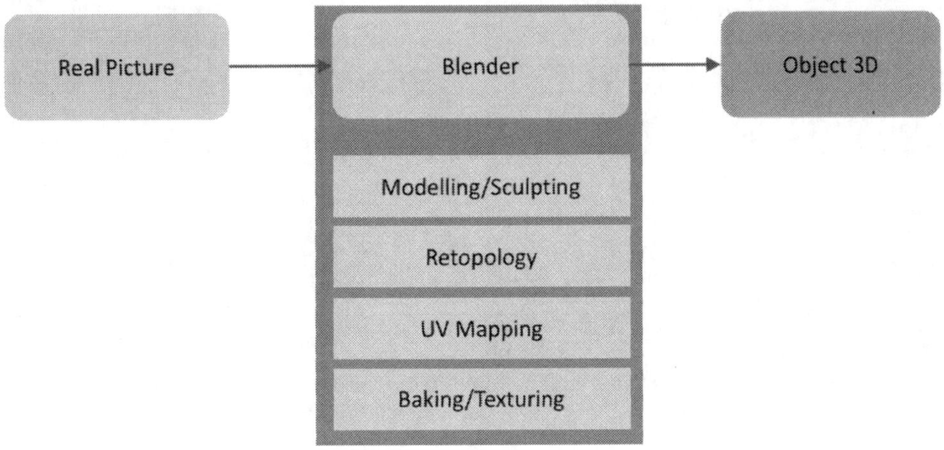

Figure 1. 3D object creation process using Blender.

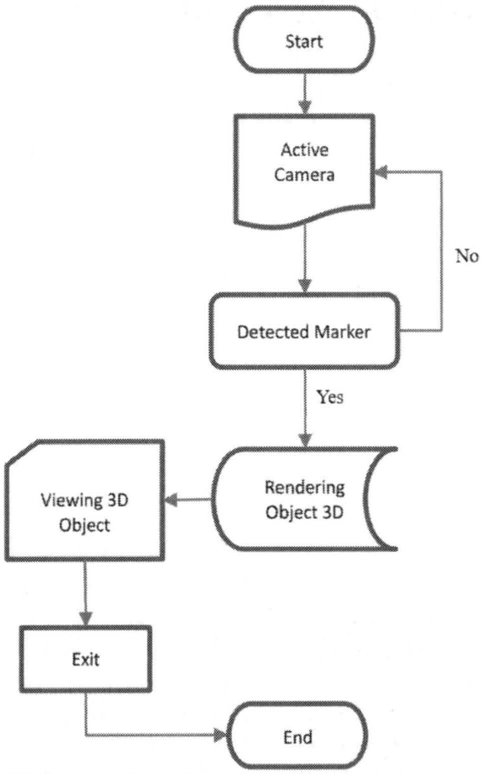

Figure 2. Flowchart of 3D Object viewer app.

3. Results and Discussion

Icon that stands out on Cadas Pangeran batik motif is the silhouette of Cadas Pangeran monument. In Figure 3 (a), (b), the cadas pangeran motif is shown by the silhouette of Cadas Pangeran monument (Figure 3 (c)). Based on this, the 3D digital story-telling media for learning about Cadas Pangeran motif, the 3D object created is a Cadas Pangeran monument, which consists of Pangeran Kornel and Deandless characters who are shaking hands with the left hand.

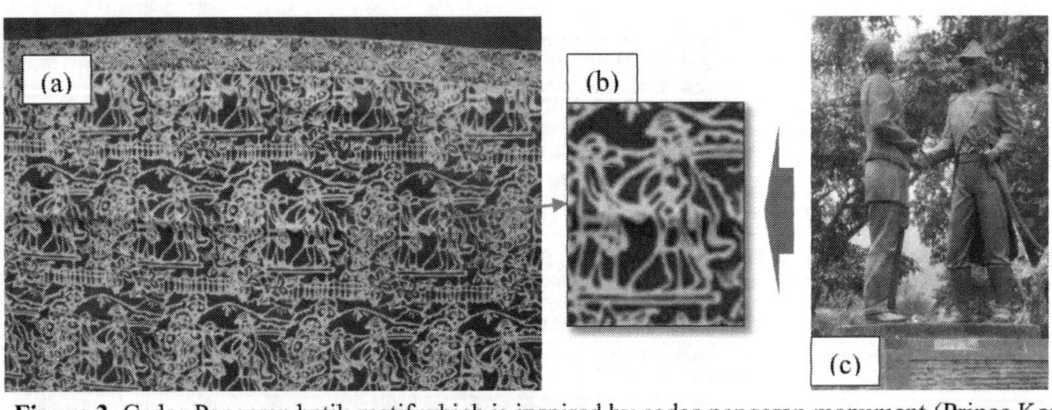

Figure 3. Cadas Pangeran batik motif which is inspired by cadas pengeran monument (Prince Kornel and Deandless characters are shaking hands).

IOP Conf. Series: Materials Science and Engineering **306** (2018) 012062 doi:10.1088/1757-899X/306/1/012062

The result of making Cadas Pangeran Monument 3D object using blender software is shown in figure 4. This 3D object which is displayed in application viewer is used for learning about Cadas Pangeran batik motif with storytelling method. Cadas Pangeran batik motif is inspired by cadas pangeran street located in Sumedang district. Cadas pangeran is a road that has a history of its development. The construction of the Daendels Highway crosses this area. The statue depicts Pangeran Kusumadinata (more popularly known as Pangeran Kornel) as the ruler of the Sumedang region shakes hands with Deandless using his left hand while the right hand holds the dagger, as a form of "stay alert" to the opponent.

Digital Story-telling media is suitable to be used to provide understanding to the younger generation, because it looks more interesting and dynamic. Digital Story-telling is a technology in education based on the theory of social behavior modification. So, it is believed able to provide understanding and change of learners behavior on the response to socio-cultural aspects that occur in the environment [6].

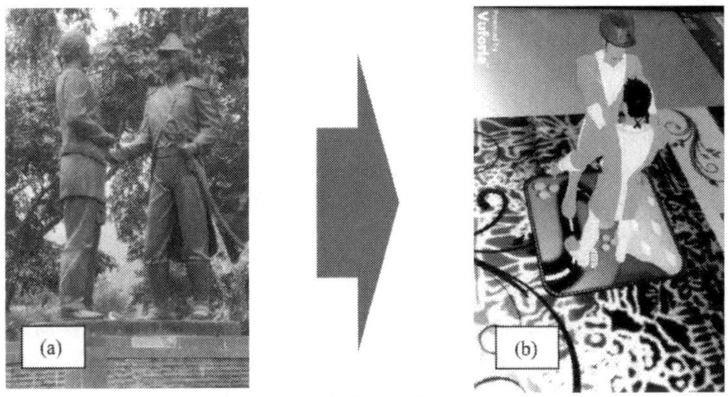

Figure 4. 3D object of Cadas Pangeran monument.

Batik motif "Tjut Nyak Dien" can be seen in figure 5 (a), inspired from the tomb of Islamic women heroes located in Sumedang. In this motif, there is a motif that describes the gate of Cut Nyak Dien's tomb (Figures 5b and 5c), rencong (Figure 5d) as an illustration that Cut Nyak Dien comes from Aceh, as well as a picture of Cut Nyak Dien statue.

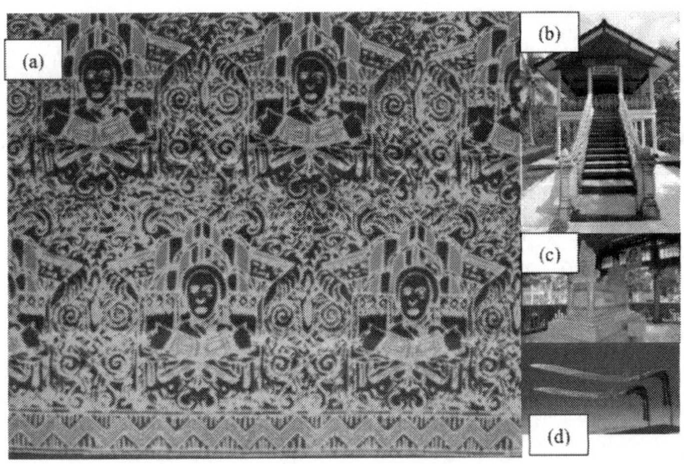

Figure 5. Batik motif "Tjut Nyak Dien".

3D Objects created for learning with 3D digital story-telling of "Tjut Nyak Dien" motifs are 3D tombs (picture 6 (a)) and 3D rencong (figure 6 (b)).

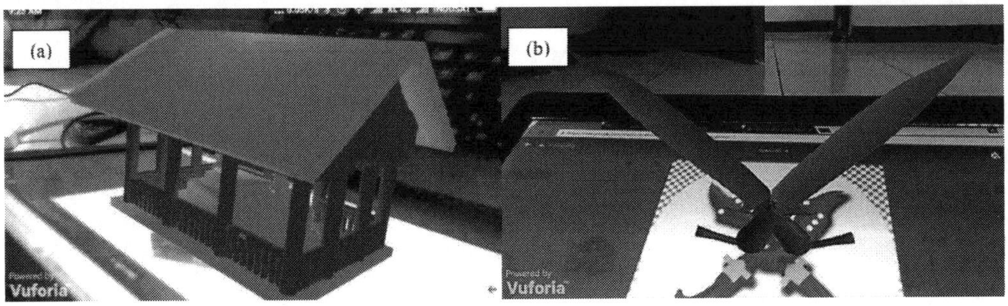

Figure 6. 3D Object of Batik motif "Tjut Nyak Dien".

3D Digital Story-telling provides many benefits for learners in learning local wisdom value of nation's culture. Benefits that learners get are able to explore and communicate their experiences with student's personal voice and as a stimulation that can increase interest in both creator and audience [5]. Digital Story-telling can act as a social mirror [4] and can provide value because the technology tools contain elements of psychological domain that is very useful to improve the sense of belonging and appreciate the elements of local culture [6].

In the context of cultural mapping, Digital story telling will motivate learners to preserve and carry out the process of cultural conservation [11] [12]. Digital Story-telling media will also fully reveal about identity vs knowledge, past vs future, and inside vs outside [10].

4. Conclusions

The process of making 3D digital story-telling media consists of two processes, namely the creation of 3D objects and the creation of 3D object viewer. The tool used to display 3D objects is a smartphone (mobile phone or tablet) on an android basis. The use of smartphone is intended to facilitate the learning process by storytelling method. The software that used to create 3D objects is Blender. Blender was chosen because it has many features that support the process of making 3D digital story-telling, including modelling, UV mapping, texturing, rigging, skinning, animation, and particle.

Digital story-telling in this research is made in 3 dimensional form, so it is expected that the learning process of batik will be more interesting. This 3D story-telling digital display will provide more meaningful and "live" stimulants compared to 2-dimensional media. Visual batik motifs in 2 dimensions is projected into 3 dimensions so it provides more real information and understanding.

Learning batik in Vocational High School will provide insight and extension of competence. Batik learning is no longer just learning the skill of making batik, but it further provides insight into character education and values about the need to love the culture based on local wisdom. This love is expected to evoke a stance to preserve the nation's cultural heritage.

References

[1] T Earley-Spadoni 2017 Spatial History, deep mapping and digital storytelling: archaeology's future imagined through an engagement with the Digital Humanities *J. Archaeol. Sci.* **84** pp. 95–102.

[2] M Kordaki 2014 On the Design of Educational Digital Stories: The Ed-W Model *Procedia - Soc. Behav. Sci.* **116**, pp. 1631–1635.

[3] J A Van Galen 2017 Agency, shame, and identity: Digital stories of teaching *Teach. Teach. Educ.* **61** pp. 84–93.

[4] N E Pappamihiel, D Ousley-Exum and A Ritzhaupt 2017 The impact of digital stories on preservice teacher beliefs about English language learners *Teach. Teach. Educ.* **67** pp. 171–178.
[5] R Bran 2010 Message in a bottle Telling stories in a digital world *Procedia - Soc. Behav. Sci.* **2**(2) pp. 1790–1793.
[6] G Pieterse and R Quilling 2011 The impact of digital story-telling on trait Emotional Intelligence (EI) amongst adolescents in South Africa – a case study *Procedia - Soc. Behav. Sci.* **28** pp. 156–163.
[7] M S Jeannotte 2016 Story-telling about place: Engaging citizens in cultural mapping *City, Cult. Soc.* **7**(1) pp. 35–41.
[8] G Tursyn, K Aktolkyn, S Gulmira, Z Dinara, and Z Z Amanbaevna 2013 National Culture: Tradition and Innovation (On the Basis of Semiotic Analysis of Kazakhstan's Capital) *Procedia - Soc. Behav. Sci.* **81** pp. 32–39.
[9] R Cauchi-Santoro 2016 Mapping community identity: Safeguarding the memories of a city's downtown core *City, Cult. Soc.* **7**(1) pp. 43–54.
[10] K Eräranta, T Leino, T Seppälä, S Viña, and E Timonen 2016 Mapping the Pig Tale Journey: A multidisciplinary design framework for cultural mapping in an old abattoir *City, Cult. Soc.* **7**(1) pp. 25–33.
[11] J P Schägner, L Brander, J Maes, M L Paracchini, and V Hartje 2016 Mapping recreational visits and values of European National Parks by combining statistical modelling and unit value transfer *J. Nat. Conserv.* **31** pp. 71–84.
[12] R Freitas 2016 Cultural mapping as a development tool *City, Cult. Soc.* **7**(1), pp. 9–16.

The Design of Mechatronics Simulator for Improving the Quality of Student Learning Course in Mechatronics

J Kustija*, Hasbullah and Y Somantri

Department of Electrical Engineering Education, FPTK Universitas Pendidikan Indonesia

*jaja.kustija@upi.edu

Abstract. Learning course on mechatronics specifically the Department of Electrical Engineering Education FPTK UPI still using simulation-aided instructional materials and software. It is still not maximizing students' competencies in mechatronics courses required to skilfully manipulate the real will are implemented both in industry and in educational institutions. The purpose of this study is to submit a design of mechatronic simulator to improve student learning outcomes at the course mechatronics viewed aspects of cognitive and psychomotor. Learning innovation products resulting from this study is expected to be a reference and a key pillar for all academic units at UPI in implementing the learning environment. The method used in this research is quantitative method with the approach of Research and Development (R and D). Steps being taken in this study includes a preliminary study, design and testing of the design of mechatronic simulator that will be used in the course of mechatronics in DPTE FPTK UPI. Results of mechatronic design simulator which has been in testing using simulation modules and is expected to motivate students to improve the quality of learning good study results in the course of mechatronic expected to be realized.

1. Introduction

Learning is the process of developing the expertise possessed by each student. Students are required to actively build knowledge and competencies that students have with professors who acts as a facilitator. With this method, students can acquire the knowledge and competence improvement well. The effectiveness of the learning process is influenced by factors of learning methods and media used. Both are intertwined, where the selection of a particular method will affect the type of media that will be used in the learning process, therefore there should be conformity between them to achieve learning objectives. Use of media in learning can generate interest and new talent and stimulate student motivation in the learning process.

In improving student learning and understanding necessary to have a media that can be used as an alternative in implementing the learning process in the classroom. Mechatronic simulator instructional media is an important tool to improve the learning outcomes of students in the subject of mechatronics, because the media is a form of student competency development. In the mechatronics engineering competency standards that are taught in the students of the Department of Electrical Engineering Education FPTK UPI, learning activities carried out by using a medium of learning that is dynamic with simulation methods. Obstacles encountered during the learning engineering is the only student can manipulate visually instead of the original object, so that students learn only through visual analyzes.

The modern advances in information technology and decision making, as well as the synergetic integration of different fundamental engineering domains caused the engineering problems to get harder, broader and deeper [1]. Problems are multidisciplinary and to solve them require a multidisciplinary engineering systems approach, such modern multidisciplinary systems are called mechatronics systems [2-3].

Mechatronics engineer is expected to design products with synergy and integration toward constrains like higher performance, speed, precision, efficiency, lower costs and functionality, also in order to evaluate such concepts and others generated during the design process, without building and testing each one, Mechatronics engineer must be skilled in the modeling, simulation, analysis and control of dynamic systems and understand the key issues in hardware implementation [4].

The design of mechatronic simulator to be tested can be used a variety of engineering systems mektronika, including purely pneumatic systems, systems and pneumatic systems programmed [5]. With the design of mechatronic simulator is expected mastery of the course material can be increased so that the quality of student learning outcomes for the better. The designer of mechatronic has the freedom to modify these values, increase the number of inputs/outputs used and include non-linearities in the subsequent design iterations [6]

2. Literature Review

2.1. Mechatonics

In the narrow sense of mechatronic lead to numerical control technology is a technology using an actuator control mechanism to achieve a certain goal by monitoring the condition of motion of the machine using a sensor and enter that information into the microprocessor. Thus, mechatronics is a combination of basic disciplines such as, mechanics, electrical engineering, and informatics to design, produce, operate, or maintain the system and optimize the product to achieve the desired goal. Figure 1 shows a simple diagram of the science establishment mechatronic.

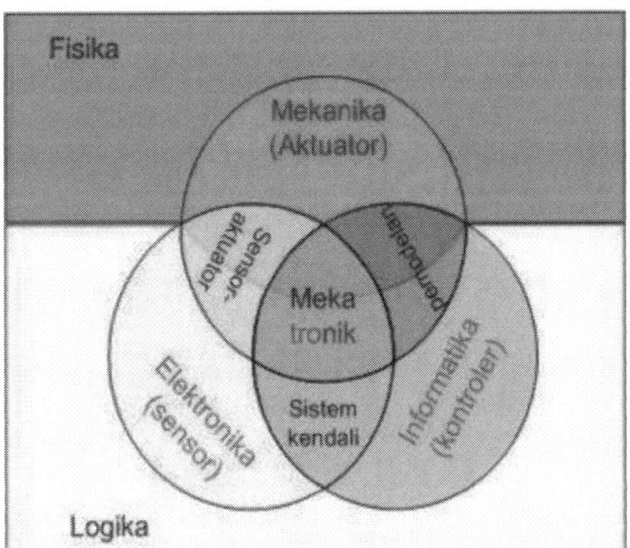

Figure 1. Diagram of formation mechatronic science.

The key element in success of a Mechatronics engineering program, and correspondingly Mechatronics engineering graduates, is directly related to the applied structural design methodology. A guidelines for structural design methodology and tools for the development process of Mechatronic products, that can support educators and help students in solving Mechatronics design integrated tasks with their specific properties and can be applied in educational process is highly required, such guidelines for structural design methodology are proposed in [2-8].

Some of the benefits of applying mechatronics [7], among others:

2.1.1. Increase flexibility. Mechatronics can be applied to improve the flexibility of a piece of equipment, such as industrial robot arm that can perform various kinds of work by changing the software program on a microprocessor as well as a human arm. This is a major factor allowing for the production of diverse type is the relatively small quantities.

2.1.2. Improve reliability.
Mechatronics can improve reliability, such as the use of electronic components to control the movement so that the components of the manual machine motion controllers can be reduced, as well as improving reliability.

2.1.3. Improve the precision and speed. The application of digital control and electronics technology, the level of precision machinery and machine motion speed can be increased even higher to a certain extent. This limit, for example is a machine rigidity that prevents higher speed because of the advent of vibration. The elements of mechatronics include: mechanical machines, sensors, controllers, drive circuit, actuators, and other sources of energy.

Mechatronics system as an overall system, and embedding the sensor, actuators, and microcomputers into the mechanical process, the microcomputers can be integrated with actuators, the process, or sensor or be arranged at several places. Integrated sensors and microcomputers lead to smart sensors, and integrated actuators and microcomputers developed into smart actuators. For large systems bus connections will replace the many cable. Hence, there are several possibilities to build up an integrated overall system by proper integration of the hardware [1].

2.2. Mechatronics system design
Mechatronic system design process addresses these challenges, it is a modern interdisciplinary design procedure, it is the concurrent selection, evaluation, integration, and optimization of the system and all its sub-systems and components as a whole and concurrently, all the design disciplines work in parallel and collaboratively throughout the design and development process to produce an overall optimal design– no after-thought add-ons allowed, this approach offers less constrains and shortened development, also allows the design engineers to provide feedback to each other about how their part of design is effect by others.

Mechatronics system as an overall system, and embedding the sensor, actuators, and microcomputers into the mechanical process, the microcomputers can be integrated with actuators, the process, or sensor or be arranged at several places. Integrated sensors and microcomputers lead to smart sensors, and integrated actuators and microcomputers developed into smart actuators. For large systems bus connections will replace the many cable. Hence, there are several possibilities to build up an integrated overall system by proper integration of the hardware [1-2].

3. Methods
The research method using a combination approach to Research and Development (R & D), where the methods of research conducted in this study required several stages of the preliminary studies, designing, manufacturing tools, product testing is limited to accept the proposed improvements of the user,. The next step is to decide the steps of the improvement of existing products and the improvement of products. After experiencing improvement, product more widely tested and used in the learning process and then analyzed the results of the learning process by using the media created.

Mechatronics design approach challenge conventional sequential design approach, by connecting machine design-test tools and creating a virtual machine prototype before designing the physical machine, to take all advantages that can result from an integrated design, this approach offers less

IOP Conf. Series: Materials Science and Engineering **306** (2018) 012063 doi:10.1088/1757-899X/306/1/012063

constrains and shortened development, also allows the design engineers to provide feedback to each other about how their part of design is effect by others [1,2,7].

4. Results and Discussion

Trainer trials carried out gradually, starting from testing of components, test wiring after component coupled to the stand, and then to test the overall ranging from logic module panel, module temperature control, and stepper motor modules. The following documentation of the results of their trial:

4.1. Logic module panel experiment

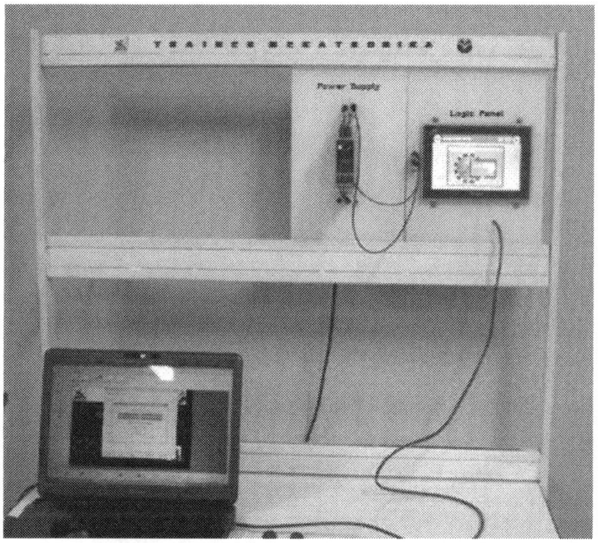

Figure 2. Module logic panel module.

These trials involved testing the wiring, testing the connection between a laptop with logic panel, the testing process of downloading and uploading program. Wiring strung indicated on the module as in figure 3 below:

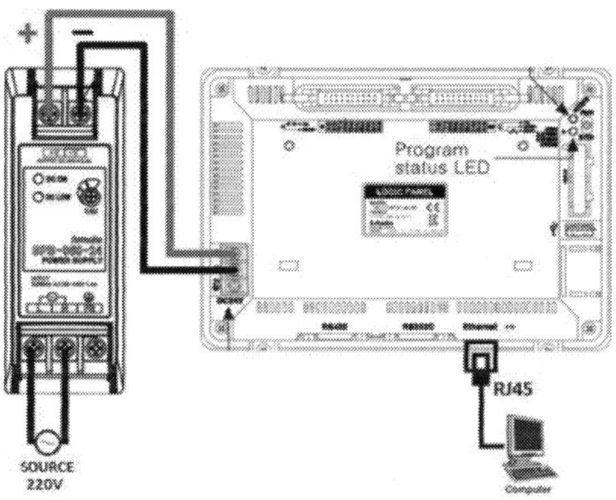

Figure 3. Scheme logic circuit panel.

When the wiring has been completed and no error occurs in the device simulator, further programming HMI / Display using GP Editor software and programming ladder diagram using Smart Studio software.

IOP Conf. Series: Materials Science and Engineering **306** (2018) 012063 doi:10.1088/1757-899X/306/1/012063

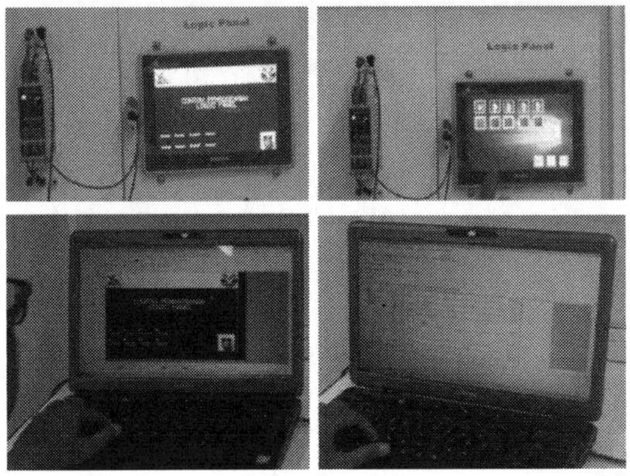

Figure 4. Display panel that has been programmed logic.

In such testing all aspects of the device works fine ranging from components, software, up to the communication device. Not found problems or errors that cause interference with the performance of the device.

4.2. Control module temperature experiment

At the trial of this temperature control module includes the testing of components, wiring test whether it is appropriate or not, the test program, and test sensors. First performed in accordance with the scheme Coupling circuits contained in the module, and then conducted the experiment in accordance jobsheet. Here are some of the documentation of mechatronic simulator experiments.

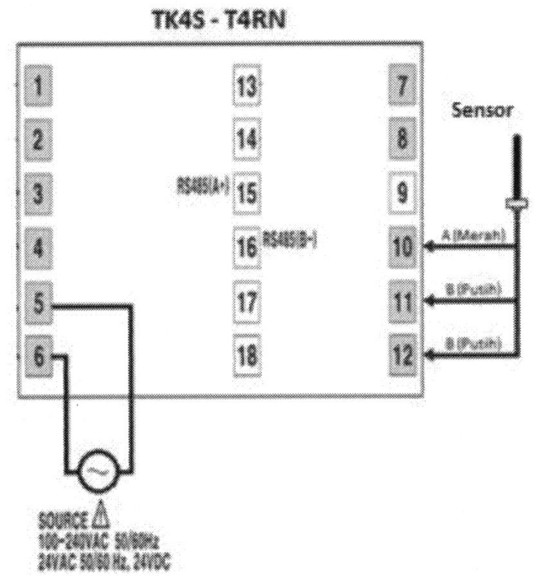

Figure 5. Schematic circuit temperature control module without load.

At trial circuit temperature control all devices are working properly and smoothly and no significant error, this shows that the simulator devices are working properly and in accordance with the given jobsheet [9-11].

5. Conclusion

Based on the results of the design, testing, analysis tools as well as the discussion of the study, the authors came to the conclusion, among others:

a. Design and manufacture of mechatronic simulator that has been equipped with modules and usage guidelines are expected to assist students in learning and supporting subjects mechatronic system so that the quality of learning outcomes for the course is expected to improve both aspects of cognitive and psychomotor aspects.
b. Based on test results that have been carried out, ranging from logic testing panel, temperature control, and stepper motors can be concluded that the trainer designed mechatronic been able to work well without any errors or significant disruption.
c. Trainer mechatronics has been designed has several advantages including easy to use, because it has components that are portable and can be set in accordance with the position that we need and has a plug & play models are more efficient over time with furnished by guidebook or structured modules.

References

[1] Farhan A S 2015 The Role of Modeling, Simulation and Analysis Stage in Mechatronics System Design Education *Journal of Multidisciplinary Engineering Science and Technology* **2**(10).
[2] Farhan A S and Ahmad A M 2014 Mechatronics Design And Implementation Education-Oriented Methodology; A Proposed Approach *Journal of Multidisciplinary Engineering Science and Technology* **1**(3).
[3] Farhan A S and Ahmad A M 2013 A Proposed Approach to Mechatronics Design and Implementation Education-Oriented Methodology *'Innovative Systems Design and Engineering* **4**(10) pp 12-29.
[4] K Craig and F Stolfi 2002 Teaching control system design through Mechatronics: academic and industrial perspectives *Mechatronics* **12**(2) pp. 371-381.
[5] Manfred L 2008 *Design of Mechatronics systems and benefits of open source software tools* 9th International Workshop on Research and Education in Mechatronics September 18th-19th, Bergamo, Italy.
[6] A Saleema, T Tutunji and L Al-Sharif 2011 Mechatronic system design course for undergraduate programmes *European Journal of Engineering Education* **36**(4) pp. 341–356.
[7] Cengange Learning. www.academia.edu/6252278/MATERI_DASARMEKATRONIKA
[8] Sarah B 2008 *Multidisciplinary machine building*, the institute of engineering technology http://eandt.theiet.org/magazine/2008/12/machin e-building.cfm
[9] De Silva and Clarence W 2008 *Mechatronics: An Integrated Approach* (CRC Press).
[10] Jerome L R 1980 *Seri Buku Schaum Dasar* (Jakarta: Erlangga).
[11] Skoog, West, Holler and Crouch 2004 *Fundamentals of analytical chemistry 8ed* (California: Brooks/ Cole Cengange Learning).

Envisioning Science Environment Technology and Society

J Maknun*, T Busono and I Surasetja

Department of Architectural Education, Faculty of Technology and Vocational Education, Universitas Pendidikan Indonesia.

*joharmaknun@upi.edu

Abstract. Science Environment Technology and Society (SETS) approach helps students to connect science concept with the other aspects. This allows them to achieve a clearer depiction of how each concept is linked with the other concepts in SETS. Taking SETS into account will guide students to utilize science as a productive concept in inventing and developing technology, while minimizing its negative impacts on the environment and society. This article discusses the implementation of Sundanese local wisdoms, that can be found in the local stilt house (rumah panggung), in the Building Construction subject in vocational high school on Building Drawing Technique expertise. The stilt house structural system employs ties, pupurus joints, and wedges on its floor, wall, and truss frames, as well as its beams. This local knowledge was incorporated into the Building Construction learning program and applied on the following basic competences: applying wood's specification and characteristics for building construction, managing wood's specification and characteristics for building construction, analyzing building structure's type and function based on their characteristics, reasoning building structure's type and function based on their characteristics, categorizing wood construction works, and reasoning wood construction works. The research result is the Sundanese traditional-local-wisdom-based learning design of the Building Construction subject.

1. Introduction

This research is a multi-year research to 3, the first year of research to produce Learning Devices Building Construction-Based Local Wisdom Traditional Building Sundanese visionary Science Environment Technology and Society (SETS) for Universities that are ready for empirical test. The second year study produced a Learning Device with SETS vision for SMK that is ready for empirical testing.

In this research Local wisdom refers to something that can be a special characteristic of a region either in the form of custom, dance, folksong, or local ceremony. The local potential is always related to the local culture. Culture is not the potential that only deals with art and culture but also everything related to the community's point of view of life that has relation to faith, work, productivity, dietary, value, and norm [1].

The local wisdom of the Sundanese traditional building in West Java has not completely been recognized by the students at the level of vocational senior high school as their treasure, especially in terms of its building structure and construction. The facts are based on several things, for example: the students are not familiar with the Sundanese traditional house. They prefer modern architecture to the local one. Hence, the understanding and awareness of learning traditional building structure and construction should be instilled, for example through the process of learning local things.

The current implementation of the building construction learning has yet to give satisfying result. This results from some reasons. They are building construction is considered as a difficult lesson, lack of practical tools, and teacher still dominates the learning process (teacher-centered). On the other hand, the teacher reveals that the students' ability is considered poor and they are distracted when learning the lesson. Based on those facts, this study explores the development of building construction based on the local wisdom of the Sundanese traditional building with Science Environment Technology and Society (SETS) as the vision.

This study focuses on the Sundanese traditional building architecture, especially the structure and construction as well as the friendly environmental, cheap, and easy created local material. The reason for choosing SETS as the vision is because the approach guides the students to relate the concept of science to the other elements of SETS [2]. The learning model of SETS that has been adjusted to the students' interest and talent is expected to be able to arouse the students' curiosity and support their initiative in the learning activity [3].

2. Methods

This study applies Education Research and Development (R & D) method. That is, the development of the learning equipment of building construction based on the Sundanese traditional local wisdom with the vision of Science Environment Technology and Society (SETS) for the future college students, teachers, and vocational high senior school students focusing on building engineering [4]. In this third year of research, it is the develop stage of the 4D Model procedure.

Table 1. Description of research process year III.

Research Stages	Type of Methods	Research Steps	Output	Information
Develope	Experiment	Vocational teacher training in Building Engineering skills involved in collaborative implementation of KBBKL-BTS Learning Device with vision of SETS	Description of the implementation of SMK teacher training activities in Building Engineering skills involved in collaborative application of Learning Device with vision of SETS	
		Trial of Learning Device with SETS vision in SMK of Building Engineering skill program	Description of implementation, advantages and disadvantages of implementation of Learning Device with vision of SETS	
		Analysis of the effectiveness of the application of Learning Tool SETS vision on students of vocational engineering program of Building	Engineering Description understanding of building structure and building concepts of students SMK building engineering skills program	
	Development	Improvement / Revision of Learning Device with SETS vision for SMK	Learning Device with SETS vision for SMK	
	Descriptive	Dissemination of research results through scientific publications in journals or seminars	Results of research published widely	

3. Results and Discussion

3.1. The local wisdom of the Sundanese traditional building

The most recognizeble Sundanese traditional building architecture is imah panggung. It is a house that has a space under the first floor which is about 40-60 cm in height. Panggung stems from the words pang and agung meaning something that is put on the most upper part [5]. In the Sundanese's point of view, a house symbolizes female because female is the one that does all the actitivites at the house. The form of the Sundanese house is panggung. That is, a house that has a space under the first floor and uses umpak –i.e. a pillar alike- as the foundation. Moreover, panggung is the most important part for the Sundanese people, with the long ceiling and jure. The form of panggung that dominates the building system in the Sundanese culture has technical and symbolic functions. Technically, rumah panggung has three functinons: it does not inhibit the water infiltration, the space under the first floor is used to control air horizontaly –i.e. to make the space either warm or cold, the space is also used to restore firewoods and the other things [6].

Generally, the firmness system in the rumah panggung uses pupurus joinery and paseuk (peg). The blocks in the floor frame, walls, and sawhorse, are set up and connected, either vertically or horizontally, using a tenon called pupurus joinery (tenon and mortise). Meanwhile, the joints are robe fiber, rattan, and pegwood. In the traditional house, the Sundanese custom prohibits the use of nails, nuts, and bolts because the use of those things opposes the ancestors' rule (taboo). Peg has the efficiency level as many as 60% and considered to have a better quality compared to bolts with the efficiency level are 30% or nails with the efficiency level are 50%. Structure and construction have a strong relationship because the joints will not work well if one of them is absent [7].

The division of structure and construction in the Sundanese house –i.e. rumah panggung-is based on the form of stage (panggung). The Sundanese people divide the structure and construction into two types: handap and luhur. Handap refers to the structure that can be found under the house floor. It consists of lelemahan/lemah (ground), and umpak/tatapakan (foundation). Luhur refers to the structure that can be found in the upper part of the house floor such as pangadeg/adeg (wall), lalangit/palapon (ceiling), and rarangka (sawhorse). Structure is the part that cannot be separated from the construction because its function is to support the firmness of a building.

Based on the material, the Sundanese traditional building has already met one of the earthquake resistant building requirements. That is, the building consists of some light materials such as wood and bamboo. The strength of wood as the building material is wood consists of two parts, including the center and outer ones. The center part can resist compression while the outer can resist the tension. The wood that contains less water has a better quality to resist the compression. The wood cell can continue the compression pressure. Wood is a structural material that can support the development of green architecture [8].

The other material that is mostly used for Sundanese traditional building is bamboo. Bamboo has many strengths as the construction material. That is, bamboo is a fast renewable material and has mechanical characteristics like wood. Bamboo fibers are various so that bamboo can be used as either the interior or exterior material [9]. Bamboo is the most flexible source and can be found easily. Bamboo needs to be adopted as the engineering material for house or other buildings [10].

3.2. The competency of building construction for vocational senior high school students focusing on technical drawing

Building construction refers to a lesson that is categorized into the basic programming skills (C2) for technical drawing skill. Building construction is taught in the tenth grade, either in the odd or even semester. Building construction contains some main materials such as the basics for simple building construction science, including natural science, the method of construction implementation, the utility of simple building, and etc. The learning of building construction can be in the form of practice or theory. There are some basic competencies that need to be achieved through the practice because learning without any practice sometimes can be an inhibitor to achieve the required basic competency.

The building learning materials are categorized into stone construction, wood construction, steel construction, road and bridge construction, and hydraulic building construction which can assist the students to analyze, plan, and construct a building. This study analyses a building construction based on the Sundanese local wisdom. The result shows that the material, used in the Sundanes traditional building, is directly made of the natural one such as wood or bamboo. Based on the analysis, the developed building construction competency is limited to the wood construction. The basic competency, main materials, and learning activities can be seen in Table 2.

Table 2. Basic competency, basic material, and learning activities.

Basic competency	Main material	Learning activities
3.1. Implementing the specification and characteristics of wood for building construction 4.1 Processing the specification and characteristics of wood for building construction	The nature and characteristics of wood	• Reading materials about the building materials in accordance with SNI (Standar Nasional Indonesia / Indonesian National Standard) • Observing various building materials • Observing the information about the teachnology and building development • Describing the physical nature of wood • Describing the mechanical nature of wood • Describing the quality of wood • Understanding the checking procedure of wood • Conducting the visual checking of the wood's physics and mechanics • Analysing the visual checking of the wood's physics and mechanics • Presenting the result of the visual checking of the wood's physics and mechanics in the oral and writen forms or through the other media.
	Processed wood	• Identifying the product of processed wood • Identifying the process of creating some processed wood • Presenting the conclusion of how the processed wood is created • Conducting the visual physical and mechanical checking of processed wood • Analysing the result of the visual physical and mechanical checking of processed wood • Presenting the result of the visual physical and mechanical checking of processed wood in the oral and writen forms or through the other media.
3.2. Categorizing various job desks of wood construction 4.9 Understanding job desk of wood construction	Various joints as well as the wood relationship	• Observing various job desks of wood construction • Collecting data about job desks of wood construction • Identifying types of wood joints • Identifying wood joinery tool • Observing various joints and wood relationship • Creating various types of wood joints • Presenting the result of the creation process of various types of wood joints
	Various wood frame	• Identifying various wood frame • Identifying parts of wood frame • Observing the procedure of settip up door and wood frames • Presenting the result of the procedure of settip up door and wood frames
	Wooden trust construction	• Identifying parts of wooden trust • Observing the procedure of wooden trust setting up • Presnting the result of the procedure of wooden trust setting up
	Ceiling construction	• Identification of ceiling function • Identifying the shape of ceiling construction • Observing the procedure of ceiling construction setting up • Presenting the result of the procedure of ceiling construction setting up
	Wood attic construction	• Indetifying the material used for wood attic production • Observing the procedure of wood attic setting up • Presenting the result of the procedure of wood attic setting up
	Wood wall construction	• Identifying the material used for wood wall production • Observing the procedure of wood wall setting up • Presenting the result of the procedure of wood wall setting up

3.3. The learning design of building construction for vocational senior high school with SETS as the vision

The learning with SETS as the vision refers to the combination of science, environment, technology, and society. In some literature reviews, the SETS learning is known as Science, Technology, and Society approach (STS). Basically, STS approach contributes to the understanding of the relationship among science, technology, and society and the sensitivity training of student assessment toward the environmental effect as the result of science and technology development [11]. The use of technology is usually needed by society when they are going to put their plan into action. In fact, society and science use technology as the medium to restore information. Technology functions as the medium to implement and observe Science, Technology, and Society (STS) approach. The data also imply the nature of knowledge as a field that all people have.

The SETS learning stages are:

a. **Introduction** explores the issues in society that the students reveal. Yet, the teachers can reveal the issues if their students do not give any response. This stage can be called as the initiation –i.e. starting- or also can be called as invitation –i.e. to invite the students to make them focus on the learning.
b. **The process of the concept making** can be conducted through the use of various approaches and methods. For example, the process skills approach, historical approach, life skills approach, demonstration approach, experiments in laboratory, group discussion, role play, and etc.
c. The students will be able to analyze the issues or to resolve the problems that are known as the **concept application** if they understand the concept.
d. During concept-making, problem solving, and problem analysis, teachers need to guide students, so there is no misunderstanding. This activity is referred to as the consolidation concept.
e. **Assessment**

4. Conclusion

The firmness system in a Sundanese traditional building called rumah panggung is supported by the use of rope, pupurus joinery, and paseuk (peg). The blocks, that are used for the frame of floor, wall, and sawhorse, are set up and connected either vertically or horizontally using pupurus joinery (tenon and mortise). Meanwhile, the blocks are strung up using rope fibers, rattans, and pegwoods. None of the nails, nuts, and bolts are used because the use of nails, nuts, and bolts is not in accordance with their ancestor (taboo).

The learning material development is based on the syllabus of building construction lesson. This study inserts the concepts of structure and construction based on the local wisdom into the material that the teachers develop. The concepts of the Sundanese traditional building structure and construction can be implemented in some materials / concepts such as the wood beam joinery construction, the wood board joinery construction, wood and door frame construction, wood stairway and floor construction, and wood framed ceiling construction. The learning media developed for supporting the learning is the learning media that is based on Information Technology and Communication (ITC). The learning development is realized through the use of SETS approach with some stages including introduction, the process of the concept making, concept application, concept consolidation, and assessment.

References

[1] Arowolo 2010 The Effects of Western Civilisation And Culture on Africa *Afro Asian Journal of Social Sciences* **1** 1 IV

[2] Binadja A 2005 *Pedoman praktis pengembangan bahan pembelajaran bervisi SETS* (Semarang: Laboratorium SETS UNNES)

[3] Rusilowati A, Supriyadi, Binadja A dan Mulyani S 2012 Mitigasi Bencana Alam Berbasis Pembelajaran Bervisi Science Environment Technology and Society *Jurnal Pendidikan Fisika Indonesia* **8** 1 51-60

[4] Thiagarajan S, Semmel D S, Semmel M L 1974 *Instructional Development for Training Teacher of Exceptional Children* (Minnesota: Indiana University)

[5] Nuryanto 2006 *Kontinuitas dan Perubahan Pola Kampung dan Rumah Tinggal dari Kasepuhan Ciptarasa ke Ciptagelar, di Kab. Sukabumi (selatan), Jawa Barat* Tesis Riset Magister Arsitektur, Program Studi Arsitektur SAPPK-ITB, Bandung

[6] Adimihardja K 2008 *Dinamika Budaya Lokal* (Bandung: Pusat Kajian LBPB)

[7] Felix 1999 *Konstruksi Kayu* (Bandung: CV Trimitra Mandiri)

[8] Brostow W, Datashvili T, and Miller H 2010 Wood and Wood Derived Materials *Journal of Materials Education* **32** 3-4 125 – 138

[9] Sharma B, Gatoo A, Bock M, dan Remage M 2015 Engineered bamboo for structural applications *Construction and Building Materials* **81** 66–73

[10] Thomas and Ganiron 2014 Investigation on the Physical Properties and Use of Lumampao Bamboo Species as Wood Construction Material *International Journal of Advanced Science and Technology* **72** 49-62

[11] Poedjiadi A 2005 *Sains Teknologi Masyarakat Model Pembelajaran Kontekstual Bermuatan Nilai* (Bandung: PT Remaja Rosdakarya)

How to Improve Engineering Competencies for Students with Special Needs?

J Maknun*, M S Barliana and D Cahyani

Universitas Pendidikan Indonesia, Jl. Dr. Setiabudhi No. 207 Bandung 40154, Indonesia

*joharmaknun@upi.edu

Abstract. The problem of vocational education for Children with Special Needs (CSN) is not only about the service profile, spectrum relevance and competency level, but also the carrying capacity of educational facilities. In this regard, two important things are highlighted. First, the different characteristics of the design of educational facilities between regular and exceptional schools. Second, the distinctive characteristics of the design of the school facilities are extraordinary for general education (academic) and vocational education. The purpose of this study is to describe the level of suitability of the architectural design of educational facilities with the needs of vocational learning behavior for children with special needs in West Java, Indonesia. The entire research used the Education Research and Development (R & D) method of Developing the Architectural Facility Design Guide to Support the Vocational Competence of Crew Competence in accordance with the stages developed by Thiagarajan (1974) known as Four-D Model. To achieve the above objectives, then the stages of the R & D method that is done is the define, design and develop stage. Evaluation results show the infrastructure of education of Special School (SLB) Cicendo, Indonesia has met the standards set by the government, especially on aspects of land and building areas have met the standards. Most aspects of accessibility such as the basic size of space, pedestrian pathways, and doors have been met. But other aspects such as guiding lines, ram, ladders, toilets, showers, sinks, furniture and signs do not meet accessibility requirements. The conclusion is the educational infrastructure of the school in general has met the standards set by the government.

1. Introduction

Children with physical handicap and mental retardation are classified into Children with Special Needs (CSN). They go to special school called as special school. Special school aims at encouraging Children with Special Needs (CSN) to be independent so that they can continue their life. The limitations that the children have will influence their academic performance. They seem to have difficulties in adjusting to their surroundings. Thus, they need school that can support their academic activities [1].

Special education serving the needs and problems while they are studying. Students with special need are categorized into their specific conditions including (1) learner who vary in their interactions; (2) learner who vary in accessing the enviroment; (3) learner who vary in their learning styles and rates. Those specific conditions are re-categorized by the problems the children have. Children with behavior problem, children with visual impairment, and children with mental retardation [2]. In the field of education, the existence of individual with special needs leads to the discussion of how they are still able to obtain education as normal individuals do. Based on the educational concept, individual with special needs refers to someone who needs special educational service and the other things that support the service so that they can develop their skills that are suitable for their potential [3].

In accordance with the educational aims for CSN, education in special school does not only deal with the aspect of knowledge but also pre-vocational and vocational programs. Vocational education can help CSN prepare themselves to face their lives; they can be more independent. The problems of vocational education for (CS) not only associate with service profile, spectrum relevance, and competency level but also educational facilities. In relation to this case, there are two things that need to be focused on. First, the difference of design characteristics between regular school and special school. Second, the difference of design characteristics of educational facilities of special school in terms of academic and vocational education.

In some developed countries, people with special needs have already been trained to have skill for their life. In America, the government provides carriers for students with special needs since they are at school age by guiding and helping the students with the carrier exploration, of course the participant of their parents are necessary here. The government also will give skill training, skill assessment, and help them get alternative jobs when they graduate. German government has prepared the regulation and provided job vacation for younger people with special needs by mapping and providing vocational education for them. Vocational education for students with special needs in Finland has changed the concept of inclusive education into vocation based education so that students with special needs can easily be accepted by the community.

Indonesia, as the developing country, has not had the same policy as the developed countries do, especially in considering and providing vocational education for people with special needs. In Indonesia, students with special needs start to receive skill training when they go to special school at the level of senior high school, reflected in the special educational curriculum.

The results of the study show there are exsisting guidance of the building accessibility for CSN, namely the Minister of Public Works Regulation No. 30 of 2006. However, this guideline is not comprehensive enough to regulate what is needed by CSN. This rule is not sufficiently organized to be implemented in building design. For accessibility of educational facility building, there is Permendiknas Number 33 year 2008, which also does not adequately regulate comprehensively the accessibility of education facilities and infrastructures for CSN, except for the standard of completeness. Moreover, when it comes to specialized vocational education facilities, it seems to have not been touched at all.

Research on the development of educational facility design guidelines has not been implemented by other researchers. Selection of special education as the focus of the study, because it is related to increased access to education for all. Then the problem of this study wichh is going to solved is how is the existing condition of vocational education facilities for CSN at SMALB in West Java? Does the architectural design of educational facilities meet the needs of vocational learning behavior for children with special needs in West Java?

2. Methods

This paper is part of research uses the method of Education Research and Development (R & D) namely Development of Architectural Facility Design Guide to Support the Achievement of Vocational Competence of CSN. Therefore, this development research is oriented towards product development where the development process is described as thoroughly as possible and the final product is evaluated. Meanwhile to describe the level of suitability of the architectural design of educational facilities with the needs of vocational learning behavior for CSN in West Java objectives, then the stages of the R & D method that is done is the define and design stage which will be presented with descriptive method.

Define stage is by analyzing vocational learning problems and educational infrastructure facilities for ABK at SMALB. The output of this stage is the description of vocational learning and facilities problem for CSN in SMALB. Meanwhile design stage is by surveying facilities vocational education for CSN in SMALB. This section produces description of Vocational education facilities condition for ABK.

This study is conducted in Bandung, with the analysis unit is a special school at the level of senior high school. The research object selection is based on the purposive consideration. That is, a special school at the level of senior high school providing vocational education for students with special needs. The subjects are children with physical and sensory disability. The data are collected using observation, content analysis, interview and documentation study..

3. Result and Discussion

3.1. Vocational educational for CSN

Education for Children with Special Needs (CSN) refers to the learning that is designed to respond the unique characteristics for Children with Special Needs. This learning cannot be accommodated by standard curriculum applied in regular school [4]. Hence, education for students with special needs is the education that serves various characteristics of students with special needs. Of course, this special education uses various approaches, strategies, and learning methods, including the educational facilities that can meet the students' needs.

Education for CSN is explained in Permendikbud No. 157 Year 2014 regarding the curriculum for special education. In general, the aim of special education is to help students with physical handicap or mental problem develop and improve behavior, knowledge, and skill as an individual or members of the community so that they can access the social, cultural, and natural interplays in the community as well as have a job or continue their school. Based on this reason, the achievements of special education include independence, social integrity, life choice, and ability to meet economic need based on age, potential, and capacity of CSN [5].

A basic thing to get the achievements is having life skills needed by individual to survive. Life skills are attributed to the process of learning. The concept of life skills in the school system is categorized into two types. Those are (1) general life skills/GLS including personal skills (self awareness and thinking skills), social skills, and (2) spesific life skills (SLS) including academic and vocational skills [6].The management of vocational learning for Children with Special needs is not easy if it does not relate to

The various and individual potential of CSN. The skills learning for CSN is divided into children with low mental retardation, normal, and severe mental retardation as well as the inhibitor complexity that is attributed to their specialty. The most fundamental functional skill learning CSN is the school's courage and strict policy to implement skill learning based on interest, talent, and needs after school life, for example job. That is, the learning that is not limited by school hour in the attempt to achieve the result of learning skill at the level of vocational independence/economic activities (advanced level). Skill learning is conducted in a real situation by activating stakeholder's role. Parent's role is also necessary to follow up the practice of skill learning in daily life, especially functional skills (self-protective ability) for children with mental retardation. The empowerment of CSN independence through skill learning -i.e. dealing with Children with Special Needs' competency and performance-needs the community's recognition [7].

The result and analysis of vocational educational facilities in a special school in Bandung is described in the following points. The types of vocation in the special school are:

1. *Food and beverage program*

   Food and beverage refers to the lesson related to household things that is given in special school at the level of junior and senior high schools .The aim of food and beverage major is to develop students' potential so that they can have life skills and independence, especially a business in the field of food and beverage. The proposed materials include:

- How to make browine, cookies, and coffee served in an attractive package
- How to serve drink and food with a good standard service and presentation
- How to make innovative cake and coffee

2. *Computer (IT)*
   Computer skill is an extracurricular lesson, since 2009 – 2010 academic years; that the students need to take especially in special school at the levels of junior and senior high schools. The aim of computer skill is to create life skills and independence. The materials that will be taught are:
   - Introduction to computer theories
   - Introduction to computer parts
   - How to display programs (hardware and software)
   - The process of data print
3. *Automotive*
   Automotive skill is a new skill provided by special school. The skill will still be developed. The materials that will be taught:
   - Introduction to the basic theory
   - Introduction to motorcycle/car parts
   - How to wash motorcycle/car
   - The process of repairing motorcycle/car
   - The process of motorcycle/car assembly

   The vocational education facilities in the special school can be seen in Figure 1.

Food and beverage skill

Computer skill (IT)

Automotive skill

Figure 1. The vocational activity in Cicendo private special school.

Basically, every school has educational facilities in which the needs are adjusted to the available activities or curriculum. It is not different from special school that has special educational facilities for their students. Beside its different curriculum and learning, the design of class room, toilet, and the other rooms have specialty. The class room is not only about how to create a more conducive atmosphere for studying but also -considering the needs of the students- a communicative, comfortable, and educative atmosphere. Therefore, the implementation of special accessibility to educational facilities design is needed, especially for special school.

3.2. The facility of vocational education for CSN

The domination of normal population over the population with special needs contributes to the lack of building design for people with special needs. In an attempt to create architectural design which is responsive and accessible for people with special needs, a comprehensive study of the architectural design aspect that can respond the perceptive, cognitive, affective, and spatial behavior aspects is needed. Environmental spatial behavior is influenced by the environmental experience stimulus. First, the information of this behavior is received by the perceptive process. Second, the information of this behavior is recognized in the cognitive process. Third, it is responded using the affection process. Finally, it is followed up through an action. Individual with special needs has different psycho-spatial process from normal individual.

In relation to the statement above, an architectural design especially the one that can be used by people with special needs has to pay attention to the environment that can be responded by human sensing. There are six environments that can be caught by human sensing. They are visual, audial, tactile, olfactual, thermal, and kinestetic environment [8].

The condition of education infrastructure for CSN is described in the following sections:

Table 1. Condition of Cicendo.

Aspects	Provision	Result
Baisc size of room	Buildings that can be used by many people at once and using the maximum base size of room	It already facilitates the need for wheelchair and "crutches" users
Pedestrian lane	• maximum slope of 7° at every 9 m • Light intensity: range of 50-150 lux • the minimum width of the pedestrian path is 136 cm for one-way and 180 cm for both directions	The slope is generally less than 7°, but it does not seem to be specially designed, especially from the parking lot. The width of the road is adequate
Guiding route	• provided for the blind by using warning and steering bloktexture of the steering wheel to explain the situation • around the user	Guiding route is not found
Parking lot	• Disabled parking spaces are on the closest route to the building • The parking area is close to the pedestrian lane and the main door • Parking area must have free space, so that people with disabilities can use it. • Parking areas for disable completed with special signs	Lack of appropriate parking and pedestrian line.
Door	• The gate should be easily opened and closed by persons with disabilities • The main exit / entrance door has an opening width of at least 80cm • The area around the entrance is avoid from ramp or floor height	Most of SLBN Cicendo Building are the Dutch relics, the size of the door is quite adequate
Ramp	• The maximum slope should not exceed 7° • • Horizontal length should not be more than 900 cm • • Ramps should be free and flat with the size of a drink 160cm • The minimum width of the ramp is 95 cm without a safety edge and 136 cm with safety edge. • Must be textured • The ramp safety edge (low curb) is 10cm • Must be equipped with handrail handle	Ramp is not found
Stairs	• Must have same sized *antrede* and *optrede* dimensions • Must have a slope of less than 600 • There is no hollow incline that can endanger the ladder. • Must be equipped with handrails on both sides • Handrails should be easy to hold with altitude 70-90 cm from the floor and free from disturbing construction elements and the edges should be rounded or deflected properly towards the floor, wall or pole • Handrails should be added in length at the end of the tip (top and bottom) with 10-15 cm. • For ladders located outside the building, the ladder should be designed so that no stagnant rainwater.	Only single storey building, no stairway

Table 1 Cont.

Lavatory	• in general, toilet is equipped with "disable" • Have enough space • Must be equipped with handrail • Put other furniture in an affordable place • The sink should use a leverage valve • The floor should not be slippery	Do not have signs for the disabled, do not have enough space, no handrail
Shower	• Should have adjusted wide and altitude seat • Equipped with handrail and alarm botton • Using a type that can be opened from the outside under emergency Conditions • Using out door openings • Free of sharp and slippery elements • leverage system faucet	No seating area, no handrail
Sink	• easy to reach • enough free space • have free space under the sink • Ruang gerak bebas yang cukup, than wheelchair user is easy to use it • Adjusted height of mirror for user • leverage system faucet	Not easy to reach, not enough free space
Furniture	• Can be use by CSN • Accessible • Safe for children • Disguise sharp corners with curved shapes • Not using hazardous chemicals	There is no spesifik furniture and equipment for CSN
Sign and markers	• Embossed or braille signs • International sign and symbol • easy to see and easy interpreted • Made from not dazzled materials	Signage is available but not embossed or braille

The evaluation shows that the educational facilities of Cicendo public special school, in general met the government standard, esspecially in area and the building size. The building quality such as class room, skill room, library, and the other rooms have a good quality. Most of the accessibility aspects such as the basic size of a room, pedestrian route, and doors are met. However, the other aspects such as guiding route, ramp, stairway, lavatory, shower, sink, furniture and sign have yet to meet the accessibility requirement.

4. Conclusion

The evaluation result shows that the educational facilities of Cicendo public special school, Bandung generally have met the standard that the government upholds. The building quality such as class room, skill room, library, and the other rooms have a good quality. The school provides vocational skills including food and beverage, computer (IT), and automotive program. Most of the accessibility aspects such as the basic size of the room, pedestrian route, and doors have met the requirement. Yet, the other aspects such as guide route, *ram*, stairways, toilet, washbasin, fountain, equipment and sign have yet to meet the requirement of accessibility.

Vocational skills for Children with Special Needs (CSN) have been given in special school –i.e. at the junior high school and senior high school levels. The implementation of vocational skill learning still needs improvement. Kinds of vocational skill need to be more various so that Children with Special Needs (CSN) can have a bigger chance to get various skills. In Indonesia, there has yet to be found vocational special school at the level of senior high school that specifically provides full vocation.

References

[1] Damayanti P A 2015 Sekolah Dasar Luar Biasa (SDLB) di Kota Semarang dengan Pendekatan Desain Universal *Canopy: Jurnal of Architecture* **4**(2)

[2] Shea, Thomas M, Bauer, and Anne Marries 1977 *Special Education: A Social Systems Perspective* (USA. Brown & Benchmark. A Times Mirror Company)

[3] Hallahan D P, Kauffman J M and Pullen P C 2009 *Exceptional Children: Introduction to Special Education* 12th Edition (New Prentice Hall International. Inc.)

[4] Mahabbati A 2014 Kebijakan, Implementasi dan Isu Strategis Pendidikan Bagi Individu Berkebutuhan Khusus *Jurnal Pendidikan Islam* Volume **4**, Nomor 1, Juni 2014/1435.

[5] Peraturan Menteri Pendidikan dan Kebudayaan Republik Indonesia Nomor 157 Tahun 2014 *Tentang Pendidikan Khusus*

[6] Anwar 2004 *Pendidikan Kecakapan Hidup: Life Skills Education* (Bandung: Alfabeta)

[7] Isharwati 2010 *Pembelajaran Keterampilan Untuk Ppemberdayaan Kemandirian Anak Berkebutuhan Khusus* (Dinamika Pendidikan: Majalan Imu Pendidikan No. 02/ Th. XVII/ Oktober 2010)

[8] Lang and Jon 1987 *Creating Architectural Theory* (New York: Van Nostrand Reinhold Company)

A Comparative Analysis of Extract, Transformation and Loading (ETL) Process

J P A Runtuwene, I R H T Tangkawarow*, C T M Manoppo and R J Salaki
Universitas Negeri Manado, Tondano 95618, Sulawesi Utara, Indonesia

*irene.tangkawarow@unima.ac.id

Abstract: The current growth of data and information occurs rapidly in varying amount and media. These types of development will eventually produce large number of data better known as the Big Data. Business Intelligence (BI) utilizes large number of data and information for analysis so that one can obtain important information. This type of information can be used to support decision-making process. In practice a process integrating existing data and information into data warehouse is needed. This data integration process is known as Extract, Transformation and Loading (ETL). In practice, many applications have been developed to carry out the ETL process, but selection which applications are more time, cost and power effective and efficient may become a challenge. Therefore, the objective of the study was to provide comparative analysis through comparison between the ETL process using Microsoft SQL Server Integration Service (SSIS) and one using Pentaho Data Integration (PDI).

1. Introduction

Business Intelligence (BI) is concept of utilizing large number of corporate data (Big Data) processed in such a way to produce useful information. Companies used such information to improve decision-making quality based on data-based existing system. In order to run BI, one should have data warehouse that can store the entire data from various data source and different types of data well-known as Big Data [1]. To develop data warehouse, data source should go through a process called Extract Transformation and Loading (ETL).

In the current development of BI, many vendors have developed tools that are able to perform ETL easily. This study will use 2 tools, one is developed by Microsoft called Sql Server Integration Service (SSIS) and the other is developed by Pentaho called Pentaho Data Integration (PDI) or Kettle. In practice, users find it difficult to select which tolls that can run the ETL to get data warehouse effectively and efficiently. In fact, a lot of developers are wasting too much labor, time and cost because they select unsuitable ETL tools.

The objective of the study is to compare between the use of SSIS and PDI. Using Northwind database as the data source, the researchers will observe ETL process in both BI tools mentioned previously. Comparison scenario is determined to facilitate the comparative analysis between the two tools. This research was conducted to give recommendation for BI developer based on the result of comparative study between Microsoft SSIS and PDI. The goal of this research is to compare Microsoft SQL Server Integration Service and Pentaho Data Integration in running ETL process and integrating Northwind database by using Microsoft SQL Server Integration Service and Pentaho Data Integration.

2. Research Methods

The following was the scenario that compares between the ETL process from Microsoft SQL Server Integration Service and Pentaho Data Integration (Kettle). Before the comparative study was conducted, the researchers prepared the specimen/ object for comparison. The specimen was architecture design using Microsoft SQL Server Integration Service and one using Pentaho Data Integration (Kettle). Furthermore, data warehouse that would be used to store the result of integration should be prepared. Figure 1 described the design of both BI tools architecture.

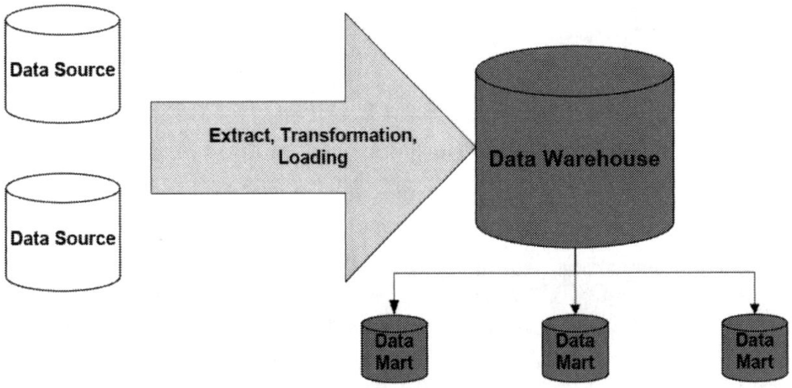

Figure 1. BI Architecture.

Comparative scenario conducted using Microsoft SQL Server Integration Service were as follow:

a. Data Collection on Northwind Database (Extract)
System data input was in the form of Northwind Inc data sales. This data were stored in SQL Server 2008 database.

b. Data Transformation Process
The data from this source was extracted into the same data format, SQL Server 2008. Before data were stored into the data warehouse, some adjustment was conducted in order that the data meet the requirement of the data warehouse designed previously.

c. Loading Data Processes into Data Warehouse
The following procedure was to store/ import (loading) the data into data warehouse using data import feature in SQL Server 2008.

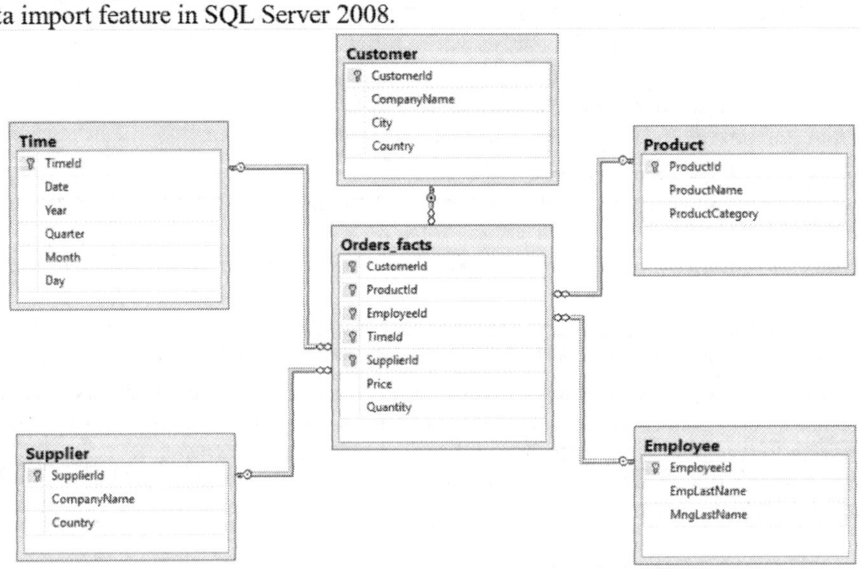

Figure 2. Star Schema Northwind DB.

Comparative scenario conducted using Pentaho Data Integration were as follow:

a. Data Collection on Northwind Database (Extract)
System data input was in the form of Northwind Inc data sales. This data were stored in SQL Server 2008 database.
b. Data Transformation Process
The data from this source was extracted into My SQL data format. Before data were stored into the data warehouse, some adjustment was conducted in order that the data meet the requirement of the data warehouse designed previously.
c. Loading Data Processes Into Data Warehouse
The following procedure was to store/ import (loading) the data into data warehouse using data import feature in Pentaho Data Integration.

The following was a star schema for Extract, Transform and Loading process using Microsoft SQL Server Integration Service and Pentaho Data Integration.

2.1. Comparative criteria

Somya [2] mentioned 5 criteria for comparative analysis, namely:

2.1.1. Comparison from developer perspective. Comparison from developer perspective would compare implementation of both architectures. The perspective would compare what process researchers should take to apply each of the architectures. The finding was the more user-friendly type of architecture.

2.1.2. Comparison from user perspective. Comparison from user perspective would compare how users add and integrate data into the data warehouse. The finding was which type of architecture of which integration process was more user-friendly.

2.1.3. Comparison from system development perspective. The type of comparison would contrast how each type of architecture be developed in the future. The result was the type of architecture of which system was easier to develop.

2.1.4. Comparison from system scalability. It would compare support for various data sizes that were integrated into the data warehouse.

2.1.5. Comparison from system stability. It would compare how stable system was in supporting data integration process, more particularly data integration of various different data.

This study focused on the comparison between the developer, user and system development perspectives.

2.2. Previous related studies

The following was list of previous studies related to data warehouse and business intelligence.

1. A journal entitled Microsoft SSIS and Pentaho Kettle: A Comparative Study for Three-Tier Data Warehouses [3], described the differences between Microsoft SSIS and Pentaho Kettle based on the three-tier data warehouse architectures. It discussed differences taking place at the 3 layers, bottom tier, middle tier and top tier. The study did not emphasize on the process or explain general observations of both existing BI tools.
2. A study entitled *Perbandingan Traditional Business Intelligence dan Service-oriented Business Intelligence (SoBI) untuk Integrasi Data Akademik dan Keuangan (Studi Kasus: Universitas Kristen Satya Wacana)* [2] or Comparison between Traditional Business Intelligence and Service-oriented Business Intelligence (SoBI) for Integration of Academic and Financial Data (Case Study: Satya Wacana Christian University) compared the use of

SoBI and traditional BI for academic and financial data of SWCU . The purpose of the study was to produce a graphical dashboard.

3. A Journal of which title was A Service-oriented Architecture for Business Intelligence [4] discussed comparison between service-oriented concepts in BI and the traditional BI concept. The literature study concluded that with the concept of service-oriented in BI, the process of receiving information and technology integration became more effective compared to doing so with the traditional BI concept.
4. A previous study on data warehouse and data mining was a study entitled *Penggunaan Data Warehouse dan Data Mining untuk Data Akademik Sebuah Studi Kasus Pada Universitas Nasional*, the use of Data Warehouse and Data Mining for Academic Data: A Case Study in National University [5]. It examined the extraction of operational data into data warehouse, and then continued with data analysis using data mining techniques. The finding was data warehouse and web-based information reporting application. In addition, the results of the data mining were pattern of student characteristics taking a particular interest.
5. Extract transformation from OLTP to OLAP data using pentaho data integration was a research on bottom tier level which explained in detail ETL process using Pentaho Data Integration [6].
6. Vertical Information System: A Case Study of Civil Servant Teachers' Data in Manado [7] was a study using Pentaho tool in analyzing and designing data warehouse on data related to teachers in Manado
7. A Report on Implementation of Business Intelligence in the Ministry of Finance, Capital Market and Financial Institution Supervisory Agency in 2007 determined important aspects to understand prior to preparing for and applying Business Intelligence (BI) system within the BPPM and Financial Institutions [8].
8. Developing Business Intelligence Application Design for Distribution Information Systems in Pertamina Lubricant LLC using Pentaho (6) resulted in BI applications using Pentaho [9]. The first phase was to determine input. The input was derived from data that had been recorded in the distribution information system of Pertamina Lubricant SR III LLC, namely transactional data, master data, and reference data. The second phase was system design where input data source was integrated in one data warehous through ETL (Extract, Transformation, Load) process, making cube or data mart logically that consisted of various dimensions of data and fact tables. OLAP was used as the method for data analysis.
9. A study entitled Data Warehouse Design and Processing in the Library at STIMIK AMIKOM Yogyakarta [10] used SQL Server Integration Service (SSIS), SQL Server 2005 tool. The study consisted of 7 stages. A study entitled Big Data, Data Analysis, and Development of Librarian Competence was a library research [1].
10. White paper entitled Microsoft SSIS and Pentaho Kettle: A Comparative Study for Three-Tier Data Warehouses [3] described the differences between Microsoft SSIS and Pentaho Kettle based on three-tier data warehouse architecture. It discussed the differences in 3 layers, bottom tier, middle tier and top tier. It did not emphasize on the process or explain general observations towards both existing BI tools.

3. Results and Discussion

3.1. Comparative scenario using Microsoft SQL Server Integration Service

In the section, the Northwind database was transferred into data warehouse using Microsoft SQL Server Integration Service. The steps were as follow:

a. Selecting data from the Northwind Database (Extract)
b. Data Transformation
c. Loading Data into the Data Warehouse

IOP Conf. Series: Materials Science and Engineering **306** (2018) 012066 doi:10.1088/1757-899X/306/1/012066

Figure 3 described the transformation design developed in the study.

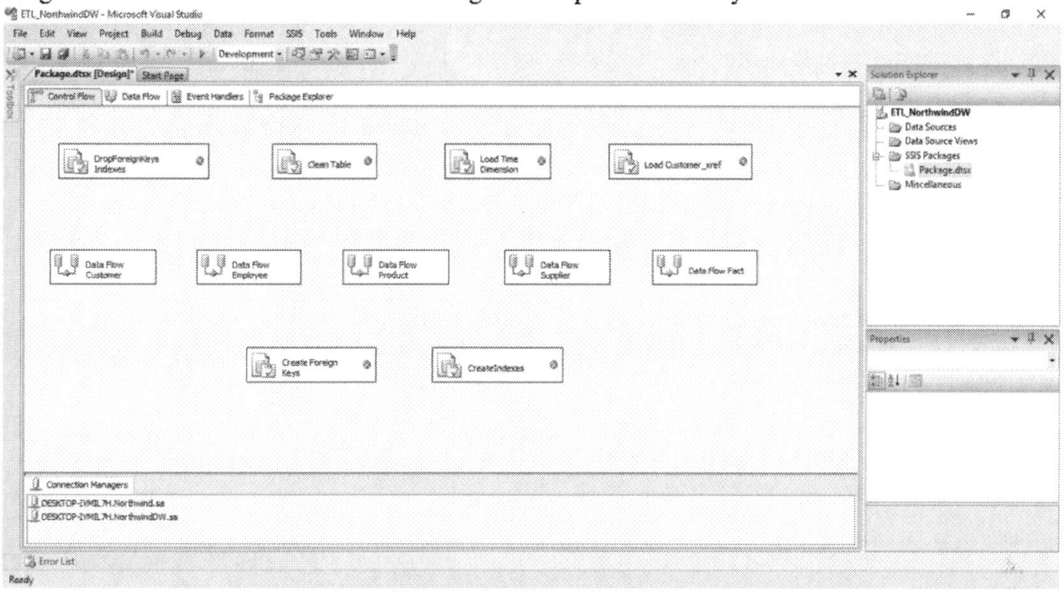

Figure 3. ETL with SSIS.

Error! Reference source not found. had 6 Execute SQL Task, namely DropForeignKeys Indexes, Clean Table, Load Time Dimension, Load Customer_xref, Create Foreign Keys, and CreateIndexes. In addition, there were 5 Data Flow Task, Data Flow Customer, Data Flow Employee, Data Flow Product, Data Flow Supplier, and Data Flow Fact.

3.2. Comparative scenario using Pentaho Data Integration

In the section, the Northwind database was transferred into data warehouse using Pentaho Data Integration. The steps were as follow:

a. Selecting data from the Northwind Database (Extract)
b. Data Transformation
c. Loading Data into the Data Warehouse

Figure 4 described how new file transformation using list of content was developed.

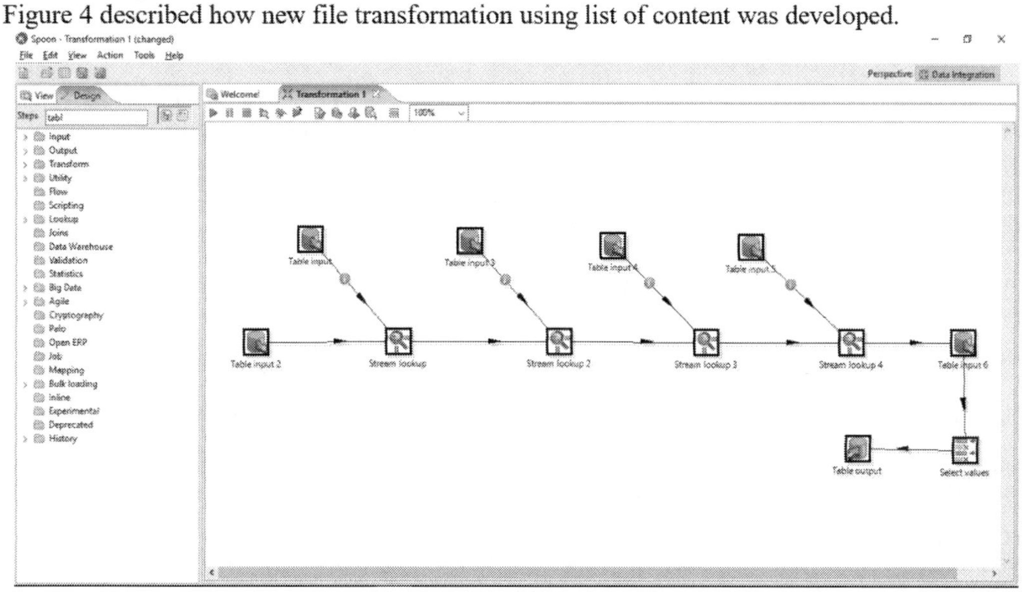

Figure 4. ETL with PDI-KETTLE.

Based on the experiments related to the implementation of ETL using SQL Server Integration (SSIS) 2008 and Pentaho Data Integration (PDI) Kettle, the following section discussed several comparative criteria used to distinguish the difference between ETL process using SSIS and one using PDI.

3.3. Comparison using developer perspective

Implementation of both tools with 1 similar architectural database (Northwind star) requires developers to carry out several steps. Table 1 described the implementation of ETI using SSIS and PDI.

Table 1. ETL Implementation using SSIS and PDI.

SQL Server Integration Service	Pentaho Data Integration
1. Identifying data from data source,	1. Identifying data from data source,
2. Extracting data manually,	2. Developing database as data warehouse,
3. Transforming data manually,	3. Developing dimension table (extract and transform),
4. Loading data manually.	4. Loading data using application.

Table 2. Loading Data with SSIS and PDI.

SQL Server Integration Service	Pentaho Data Integration
1. Selecting "Import Data" menu,	1. Selecting which type of data to add,
2. Selecting data source, a Microsoft excel file,	2. Loading data
3. Selecting destination for data storage,	
4. Loading data.	

3.4. Comparison using user perspective

Comparative analysis using the user perspective may be seen from loading data to data warehouse. Table 2 described process of loading data with SSIS and PDI.

3.5. Comparative analysis based on system development system

Having finished the experiment, one may see the difference both architecture had. PDI simplified the data integration process and was able to integrate data from multiple sources. Data sources with different databases may be integrated using PDI easily. In addition, the nature of the reusable web service made it possible to reuse the pre-made web service for future reuse [2].

SSIS may also integrate data but the process was more complicated as the system development should understand the structure of database as data source and operate or work with SQL Server. SSIS requires complex integration process for data integration from data source from different database.

4. Conclusion

Having compared ETL process using both SSIS and PDI, the conclusions: based on the experiment, data integration using SSIS was more complicated and took longer time compared to doing so using PDI. Using SSIS, data integration process began with pretty complicated ETL (Extract Transform Loading) process, while data integration process using PDI was simpler with web service; SSIS supports data source from different formats, such as excel (.xls/.xlsx), Dbase *File* (.dbf), *Text File* (.txt), MS Access *Database* (.mdf) and others, even though the ETL should be conducted manually. PDI has also supported data source that directly came from database of or manual one wih various data source file for instance Traditional Business Intelligence; PDI is more effective in solving technology integration system developer may encounter in the future.

Acknowledgments

We would like to thank Lembaga Penelitian dan Pengabdian Masyarakat (LPPM), Universitas Negeri Manado for supporting the research.

References

[1] A P Narendra 2016 Big Data, Data Analyst, and Improving the Competence of Librarian *Rec. Libr. J.* **1** 2 83–93

[2] R Somya 2012 *Perbandingan Traditional Business Intelligence dan Service-oriented Business Intelligence (SoBI)* Salatiga

[3] Grecol M L 2012 *Microsoft SSIS and Pentaho Kettle: A Comparative Study for Three-Tier Data Warehouses*

[4] Wu L, Barash G, and Bartolini C 2007 A service-oriented architecture for business intelligence Service-Oriented Computing and Applications, 2007. SOCA'07. IEEE International Conference on 279-285 IEEE

[5] A Azimah and Y G Sucahyo 2007 Penggunaan Data Warehouse dan Data Mining untuk Data Akademik *J. Sist. Inf. MTI UI* **3** 2 1–7

[6] R J Salaki, J Waworuntu, and I R H T Tangkawarow 2016 Extract transformation loading from OLTP to OLAP data using pentaho data integration *IOP Conf. Ser. Mater. Sci. Eng.* **128** 12020

[7] J P A Runtuwene and I R H T Tangkawarow 2017 Vertical information system: A case study of civil servant teachers data in Manado city *Indones. J. Electr. Eng. Comput. Sci.* **6** 1 42–49

[8] Tim Studi Implementasi Business Intelligence 2007 *Laporan tim studi tentang implementasi business intelligence (Departemen Keuangan Republik Indonesia Badan Pengawas Pasar Modal Dan Lembaga Keuangan)*

[9] R W Witjaksono, M Wiyogo, and P N Wicaksono 2007 Perancangan Aplikasi Business Intelligence Pada Sistem Informasi Distribusi Pt Pertamina Lubricant Menggunakan Pentaho *J. Rekayasa Sist. dan Ind.* **2** 2 12–18

[10] A Armadyah 2009 *Perancangan Dan Pembuatan Data Warehouse Pada Perpustakaan STMIK AKAKOM Yogyakarta* Tesis. Perpust. MTI 39–52

Higher Education Students' Behaviour to Adopt Mobile Learning

J R Batmetan[1*] and V R Palilingan[2]

[1]Information Engineering Department, Universitas Sari Putra Indonesia Tomohon, Tomohon, 95416
[2]Information Technology and Communication Education Department, Universitas Negeri Manado, Tondano,95618

*john.reimon@gmail.com

Abstract. Mobile phone is an electronic device most often used by Y generation in Indonesia. This ages have become an important part in the growth of higher education in this country. The problem raised in this study is that very few students in higher education are adopting and accessing digital learning content using mobile phones. The objective of this study is to investigate the higher education students' behaviour in using mobile learning. The research method used is Structural equation models (SEM) method to analyse the factors that influence higher education students' behaviour in using mobile learning. The results of this study indicate tends of this student 85% to keep internet access in privacy. The majority of respondent is 78% having behaviour to keep adopting mobile learning and still use it in the future. Why? because this study shows that on the level of usability, easy to use, easy to learn, in various devices have a significant effect on the level of adoption of mobile learning. Implication of this study is higher education students' behaviour of especially Y generation tends to prioritize the usability towards mobile learning and will continue to adopt mobile learning in the future.

1. Introduction

Indonesia is a large country with an increasing rate of technological development. Nielsen survey in 11 major cities in Indonesia in 2017 with a total population of 54.8 million inhabitants showed that the internet penetration rate of Indonesia reached 44% with the number of internet users as much as 24.2 million users in 11 major cities of Indonesia [1]. The survey also showed that 77% of Internet users are Y and Z generations and they still in higher education student. Other data revealed is where 91% of internet accesses from home and only 33% access from schools, libraries, and other educational establishments [1]. This shows that Y generation less access the internet while in school. Other comparable data can be seen from survey published by Jakpat Indonesia, in 2016 the population of Indonesia as much as 253 million people and 50% among the Y generation and Z [2]. Jakpat also reported that internet users in Indonesia of 83.6 also users with 69.3% are Y and Z generation. While active users of mobile devices are actually 282 million users or 13% more than the population of Indonesia [2]. This means 13% mobile phone users have more than one device. Jakpat also reported that in 2016, Indonesia users 89% using mobile phones to access the internet. As many as 84% use the internet to access social media, 81% access Facebook, 48% Instagram, the rest access to other social media [2]. This data indicates that internet users Indonesia is still not productive and just looking for entertainment only [3]. Mobile learning is still very little accessed by mobile phone users in Indonesia although 60% of Y generation is in high education [4][5][6]. Mobile learning (m-learning) becomes an important learning media in higher education student. M-learning applications had an effect on improving higher education student learning outcomes in college [7][8]. M-learning has not been optimally adopted by higher education students because it is not all compatible with the smartphone used by higher education students [9]. Content m-learning also still not in accordance with the needs of higher education students [10].

Limitations of learning resources and access to learning resources using mobile devices have always been a major problem in teaching and learning. This limitation has a direct impact on higher education students to gain good learning achievement and gain knowledge [11] [12]. Indeed, the learning process will become suboptimal when the smartphone device is very useful, easy to use, and easy to learn. For this reason, in some universities in developing countries, the process of teaching and learning has been done by directly taking learning materials from the internet. However, the use of learning materials from the internet using m-learning is even more difficult to manifest in developing countries because smartphone devices are less compatible, limited internet access, and m-learning application that is still difficult to use [13]. Indeed, one of the strategies developed in most developing countries is accessing learning resources from the internet using a computer or laptop device. However, some problems appear as difficult to use because of their large size and weight making it difficult to use mobile. these devices can only be accessed in certain places that have limited internet access. The adoption of m-learning cannot explain in detail about its application to describe students about realistic conditions. Furthermore, m-learning is complex and incompatible with some smartphones, creating conflicts with limited adoption of m-learning. While this strategy is quite effective, problems with more mobile usage and faster Internet access, easy to use, easy to learn applications are directly correlated with the adoption of m-learning.

To solve the problem, some researchers have suggested the following strategy: the adoption of m-learning should look at aspects of the compatibility that can be accessed by all mobile devices, internet speed on mobile networks and simpler, user-friendly and lightweight content so that mobile devices can display content well and perfectly [14]. The suggested strategy is effective and can be implemented and easy to adopt by higher education students [15] [16]. However, some problems still persist. m-learning cannot explain in detail about how to use the application so that higher education students have difficulty [17] [18]. Furthermore, m-learning becomes complicated and sophisticated [19], thus creating conflicts with limited use to adoption by higher education students.

Here, the purpose of this study was to investigate the higher education student's behavior to adoption mobile learning. These novelty makes the higher education students behavior for developing countries that have many limitations using m-learning. The higher education student's behavior to adoption m-learning, including usability, easy to use and easy to learning m-learning.

2. Methods

This research was conducted in North Sulawesi with participants 18-35 years old and was studying in higher education. The distribution of respondent is in several universities, polytechnics, institutes and college located in North Sulawesi Indonesia. This study was conducted in the even semester of academic year 2016/2017. Participants involved in this research were 150 Participants. Participant distribution data based on age and sex can be shown in table 1, out of 150 participants, 89 were male and 61 were female. Participants' ages ranged from 18 years to 35 years. Some students were simultaneously enrolled in more than one class in which they used the mobile devices.

Table 1. Range of participants' ages.

Age	18 – 20	21 – 23	24 – 26	27 – 29	30 – 32	33 – 35	Total
Male	35	30	15	6	2	1	89
Female	26	25	8	2	-	-	61
Total	61	55	23	8	2	1	150

Participants varied in their levels of experience as a teacher and came from North Sulawesi Indonesia. Response at level 2 to level 8 actively follow the lecture, as shown in table 2.

Table 2. Range of participants' level.

Level	2	4	6	8	Total
Male	26	24	23	16	89
Female	20	21	16	4	61
Total	46	45	39	20	150

Participants actively use m-learning in each lecture activity either in face-to-face or in project projects. Participants use smartphones to access m-learning and use various learning content. All respondents access m-learning through cellular network ie 3G and 4G. Participants were given an opportunity to assess m-learning for 3 months, then measured. Data were collected using questionnaires with questionnaire techniques. The item related to Intent to use mobile learning and factors Likert scale. 1 represented Strongly Disagree, 2 as Disagree, 3 as Not Sure, 4 as Agree, and 5 as Strongly Agree. Sampling is done by purpose sampling with age and level criteria.

This research using a quantitate method. This study using above theoretical model and hypotheses by using the questionnaire and Structural Equation Model (SEM). The variable in the model are adopted from existing research literature home and abroad, the participant was requested to grade item on five-point scale. Intention to use mobile learning is the dependent variable in this study and other independent variables used in this study are perceived usefulness, perceived ease of use, perceived easy to learning, and adoption m-learning.

Based on SEM and the researches from learning literature, the present study develops a research model that examines factors that determines undergraduate students' intention to adopt m-learning. Figure 1 depicts the research model.

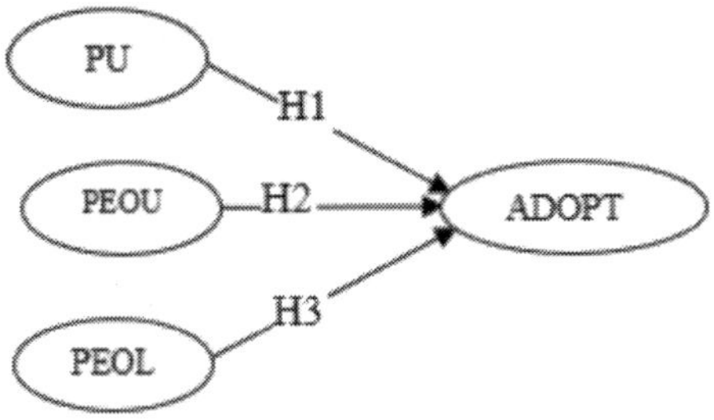

Figure 1. Model Research

The independent variables of this study are perceived usefulness (PU), perceived ease of use (PEOU), and perceived easy of learning (PEOL). These independent variables may be the determinants that influence dependent variable, i.e. intention to adopt (ADOP) M-learning among Higher Education Students in North Sulawesi. A self-administered questionnaire was used in this research, which consists of two sections.

2.1. Perceived usefulnes

The results of previous research indicate that usability has a strong relationship in the adoption of m-learning [5]. Batmetan findings show that adoption will occur against mobile applications also have strong usability [3]. Other research also shows that usability has a strong influence on the adoption of mobile applications [4] [20]. Perceived is usefulness as the degree to which a person believes that using a particular system would enhance his or her job performance [3] [13]. The importance of perceived usefulness has been widely recognized in the field of mobile learning. Perceived Usefulness of a system may be defined as the extent to which individuals believe that using the new technology will enhance their performance. Batmetan explain that perceived usefulness can influence the intention to accept and adopt mobile learning directly or indirectly. Perceived usefulness is a prominent factor which is widely

used in explaining consumer behavior towards a new technology. Perceived Usefulness of a system may be defined as the extent to which individuals believe that using the new technology will enhance their performance [3]. Therefore, perceived usefulness can influence the intention to accept and adopt mobile learning directly or indirectly.
According to value theory, hypotheses H1 in this study:

H1: usability has a relationship of Higher Education Students' behavior and determines in adopting M-learning.

2.2. Perceived easy to use

The results of previous research indicate that easy of use has a relationship in the adoption of m-learning [5]. Batmetan also findings show that adoption will occur against mobile applications have easy to use [3]. Similarly, other research also founded that PEOU has a positive effect on behavioral intention to use the m-learning [21] [22] [23]24]. The other study founded that PEOU had a significant positive effect on behavioral intention to use m-learning in East Africa[25]. Perceived Ease of Use (PEOU) refers to the degree to which a person believes that using a particular system would be free of effort. Chung[5] found that PEOU has a determinant of customer acceptance on adoption mobile application. perceived Ease of Use (PEOU) defined as the degree to which the prospective user expects a new technology to be free of effort. It is further described as the internal belief of mental effort involved in using a system. This study defines it as the degree to which the user expects the system to be user friendly.
According to value theory, hypotheses H2 in this study:

H2: Ease of use has a relationship of Higher Education Students' behavior and determines in adopting M-learning.

2.3. Perceived easy of learning

Previous research has shown that the adoption of m-learning is also influenced by easy of learning [3]. The same thing is pointed out by Chung that the adoption rate of mobile apps should be easy to learn so that they are easy to use [5]. It is also evident in previous studies that show that m-learning used by higher education students should be easy to learn and takes a short time to understand it [25] [26]. Easy of learning can be defined as the ability to understand and use an application with a relatively short time [3]. Easy of learning shows that a mobile learning application can be learned quickly and the higher education students does not require a long time to be able to use it at the time of first use. M-learning is easy to learn and quick to be used can provide a higher level for a return visit and used continuously by the higher education students. From the research that has been done previously showed that the level of learning that is both easy can increase the rate of adoption of m-learning.
Thus, the current study added price value as a predictor of intention to use mobile learning. Then, we hypothesis that:

H3: easy of learning has a relationship of Higher Education Students' behavior and determines in adopting M-learning.

2.4. Higher education students behavior

Higher education students' behavior is the concept is all about how people make their decisions on personal or household products, with the use of their available resources such as time, satisfaction and effort[27][28]. Higher education students were high in adoption and diffusion due to better quality mobile devices, user-friendly m-learning interfaces, more useful applications and services, lower prices, better security, better coverage and higher speed[29].

3. Results and Discussion

Analysis of measurement model, was used to build conceptual model of the measurement model in this study, the fit indexed between in data and proposed measurement model was generated by SPSS 20. So only fully completed 155 questionnaires were processed using SPSS version 20. Response rate was 99.98 percent linear regression was used for statistical testing of data. Variables were tested for reliability using SPSS and all values were above 0.7 and more. This is each variable Cronbach's alpha values are given below at Table 3.

Table 3. Result of reliability test.

Variable name	Table Column Head	
	No. items	Cronbach's Alpha
Perceived usefulness	9	0.793
Perceived easy to use	12	0.872
Perceived easy to learning	5	0.787
Adoptions	8	0.789

Structural Equation Modeling (SEM) using SPSS AMOS 20 revealed sufficient evidence of goodness-of-fit between the measuring models and the sample data (factor validity test). Structural model created by latent variables to check the hypotheses of research was tested by utilizing the maximum-likelihood calculation technique Subsequent structural test produced good fit statistics $\chi 2$/df =2868.769, CMIN/df = 1.768 RMSEA=0.042, CFI=0.874 GFI=0.852 NFI= 0.865.

This present study focused on how the Higher education students' behavior in North Sulawesi to the adoption of M-learning. The main motivation for this research was that there was no study conducted that involved Higher education students' behavior intention to adopt M-learning in North Sulawesi.

Result of hypotheses show this study, the H1: usability has a relationship of Higher education students' behavior and determines in adopting M-learning generated the highest regression weight of 0.776 a significant regression, generated the highest regression and amongst the various determinants. This result is compatibility affects behavioral intention to adopt M-learning. The H2: Ease of use has a relationship of Higher education students' behavior and determines in adopting M-learning generated a significant regression weight of 0.759. Respondents indicated that it would be easy for them to learn (mean=7.543). These findings reflected the techno savvy characteristics of the Higher education students'. The H3: easy of learning has a relationship of Higher education students' behavior and determines in adopting M-learning (regression weight=0.157), possibly because M-learning is currently in its infancy in North Sulawesi. This study showed a variable usability has a highest significant support to Higher education students' behavior to adopt m-learning in North Sulawesi. The second to support is variable ease of use to adopt m-learning in North Sulawesi and the third variable is easy of learning. This study showed Higher education students' behavior to adopt m-learning in North Sulawesi.

Based on analysis and previous research from learning literature, the current study examined factors that determine Higher education students' behavior intention to adopt m-learning. The results of structural equation modeling analysis highlight several interesting findings.

First, the perceived of usability on intention to use m-learning is greater than that of performance expectancy. This is an interesting finding. In fact, among the all significant paths, perceived usefulness is the most significant factor, as indicated by its path loadings and significance levels. The findings highlight the importance of perceived usefulness in determining Higher education students' behavior intention to use m-learning which is often not considered by adoption studies in the university or higher education [3] [5]. The results indicated that the perceived usefulness is the most significant factor in the Higher education students' behavior to adopt mobile learning.

Second, we also found that perceived ease of use has a relationship on Higher education students' behavior intention to adopt m-learning [3] [17]. This finding suggests that influences from mobile application and important interface and content connections are a critical determinant and has a positive effect for Higher education students' behavior intention to adopt m-learning. The implication is straight, m-learning educators should pay attention to the importance of content of m-learning. We also found the newly added simple content, simple interface and short text a positive impact on Higher education students' behavior intention to adopt m-learning. The current study highlights the importance of perceived of easy to use in Higher education students' decision making regarding m-learning adoption

[16] [18]. This finding indicates that the simple application and short structure of use m-learning have a significant influence on Higher education students' behavior intention to accept m-learning.

Finally, we found that perceived of easy to learning has positive effects on Higher education students' behavior intention to adopt m-learning. This is consistent to results of some prior studies [3] [29]. One possible reason is that the inherent features of mobile devices, such as small screen and complex keypad, will complicate user input. We also found that the influence of habit of using mobile phone on students' intention to use m-learning was not significant. This indicates that students with a stronger automaticity level of using mobile phone did not mean that they will more likely to use m-learning. The reason may be that higher education students' use their mobile phone primary for connecting with people, and the fitness between mobile devices and learning activities is relative low than between mobile devices and communications. Indeed, there are at least three reasons for students with a higher education students' will not likely to use mobile devices for learning. First, many studies argue that some inherent technical limitations of mobile devices, such as the user interface are difficult to use, slow network speeds, small screens and keyboard, and inadequate memory, will hinder higher education students' to use m-learning; Second, the pedagogical limitations of m-learning also impede students' to use m-learning, content is not simple, hard to understand, for instance, using mobile phone in class may obstruct student concentration and interrupt normal learning process; Third, students usually use mobile devices for hedonic purpose such as gaming or texting with friends and communication with other people.

4. Conclusion

This study showed that perceived higher education students' behavior with perceived usefulness, easy to use, easy to learning, are positive relationship of behavioral intent to adopt M-learning. This study showed to us a perceived usefulness, easy to use, easy to learning, has important determinants of behavioral intent to adopt M-learning. The implications for teacher, lecturer, higher educational organization, marketers and mobile service providers are discussed. The limitations of this study are that it has not yet explore other potential factors (technology enabler, network aggregator, content provider and wireless operators) that may determine behavioral intention towards the adoption of M-learning.

Acknowledgments
We would like to thank Universitas Sari Putra Indonesia Tomohon and Universitas Negeri Manado is supported by athe research fund from the Information Engineering Department, Faculty of Engineering, Universitas Sari Putra Indonesia Tomohon and Information Technology and Communication Education Department, Universitas Negeri Manado

References

[1] T N Company 2017 *The New Trend among Indonesia's Netizens, How and where digital consumers are watching content online*

[2] Jakpat D S id 2016 *Indonesia's Digital Consumer Behavior Report 2016*

[3] S J R Batmetan and J D C L Suares 2016 An Empirical Investigation on Customer Behavior to Adopt Mobile Commerce among the Y Generation in Indonesia *Sriwijaya International Conference on Engineering, Science and Technology 2016 (SICEST 2016)* 333–337

[4] R Mohamad, A Building, and N A Ismail 2010 An Empirical Study of Factors Influencing Adoption of Internet Banking among Students of Higher Education: Evidence from Pakistan *J. Internet Bank. Commer.* **15** 1 1–11

[5] K C Chung 2013 Gender, culture and determinants of behavioural intents to adopt mobile commerce among the Y Generation in transition economies: evidence from Kazakhstan *Behav. Inf. Technol.* **33** 7 743–756

[6] L Roback, K Steinbernreiter, D Halvarsson, and B Degree 2016 *It ' s the future !, A qualitative study exploring Generation Y's attitude towards m-commerce* The Linnaeus University

[7] H Z Mehdipour and Yousef 2013 Mobile Learning for Education : Benefits and Challenges *Int. J. Comput. Eng. Res.* **3** 6 93–101

[8] S M Jacob and B Issac 2007 Mobile Learning Culture and Effects in Higher Education *IEEE Multidiscip. Eng. Educ. Mag.* **2** 2 19–21

[9] Linawati 2012 Performance Of Mobile Learning on GPRS Network *Teknol. Elektro* **11** 1 1–6

[10] Sunismi 2015 Developing Guided Discovery Learning Materials Using Mathematics Mobile Learning Application As An Alternative Media For The Students Calculus II *Cakrawala Pendidik.* **XXXIV** 3 334–346

[11] W A Syafei 2013 Design Of Streaming Education Content On Android Based Mobile Device For Long Life E-Learning *J. Inf. Syst.* **9** 1 26–31

[12] S A Ariffin 2011 Mobile Learning in the Institution of Higher Learning for Malaysia students : Culture Perspectives *Proceeding of the International Conference on Advanced Science, Engineering and Information Technology 2011* 283–288

[13] D Kim, D Rueckert, D Kim, and D Seo 2013 Students ' Perceptions And Experiences Of Mobile Learning *Lang. Learn. Technol.* **17** 3 52–73

[14] C Cavanaugh 2013 Substitution to augmentation : faculty adoption of iPad mobile learning in higher education *Interact. Technol. Smart Educ.* **10** 4 270–284

[15] L Byrne-davis 2015 Just-in-time research : a call to arms for research into mobile technologies in higher education *Res. Learn. Technol.* **23** 1–10

[16] M Pegrum, G Oakley, and R Faulkner 2013 ascilite Schools going mobile : A study of the adoption of mobile handheld technologies in Western Australian independent schools *Australas. J. Educ. Technol.* **29** 1 66–81

[17] S Yang 2013 Understanding Undergraduate Students ' Adoption of Mobile Learning Model : A Perspective of the Extended UTAUT2 *J. Converg. Inf. Technol.* **8** 10 969–979

[18] M Jambulingam 2013 Behavioural Intention to Adopt Mobile Technology among Tertiary Students *World Appl. Sci. J.* **22** 9 1262–1271

[19] S M Jacob and B Issac 2007 Mobile Learning Culture and Effects in Higher Education *IEEE Multidiscip. Eng. Educ. Mag.* **2** 2 19–21

[20] O Viberg and Å Grönlund 2013 Cross-cultural analysis of users ' attitudes toward the use of mobile devices in second and foreign language learning in higher education : A case from Sweden and China *Comput. Educ.* **69** 169–180

[21] Á Fernández-lópez, M J Rodríguez-fórtiz, and M L Rodríguez-almendros 2013 Mobile learning technology based on iOS devices to support students with special education needs *Comput. Educ.* **61** 77–90

[22] S Onlinecourses, A Abu-al-aish, and S Love 2013 Factors Influencing Students' Acceptance of M-Learning : An Investigation in Higher Education *Int. Rev. Res. Open Distance Learn.* **14** 5 82–107

[23] K F Hashim, F B Tan, and A Rashid 2014 Adult learners' intention to adopt mobile learning: A motivational perspective *Br. J. Educ. Technol.* 1–10

[24] S O M y F J G-P José Carlos Sánchez Prieto 2013 Understanding mobile learning: implications and research lines devices *pedagogical, Teor. la Educ. Soc. la Inf.* **15** 1

[25] R R J S Mtebe 2014 Investigating students ' behavioural intention to adopt and use mobile learning in higher education in East Africa *Int. J. Educ. Dev. using Inf. Commun. Technol.* **10** 3 4–20

[26] S M Jacob and B Issac 2008 The Mobile Devices and its Mobile Learning Usage Analysis *Proceedings of the International MultiConference of Engineers and Computer Scientists* **I** 19–21

[27] J P B M Ally 2014 What is the future of mobile learning in education ? *RUSC - Rev. Univ. y Soc. del Conoc. Univ. Knowl. Soc. J.* **11** 1 142–151

[28] L S and K G T D Thomas 2013 The utility of the UTAUT model in explaining mobile learning adoption in higher education in Guyana *Int. J. Educ. Dev. using Inf. Commun. Technol.* **9** 3

[29] K M Callum and L Jeffrey 2013 The influence of students â€TM ICT skills and their adoption of mobile learning *Australas. J. Educ. Technol.* **29** 3 303–314

IOP Conf. Series: Materials Science and Engineering **306** (2018) 012068 doi:10.1088/1757-899X/306/1/012068

E-Learning Development Process for Operating System Course in Vocational School

J R Tuna[1], C T M Manoppo[2], D R Kaparang[2] and A Mewengkang[2,*]

[1]Department of Language and Art, Universitas Negeri Manado, Tondano 95618, North Sulawesi, Indonesia

[2]Department of Information Technology and Communication Education, Universitas Negeri Manado, Tondano 95618, North Sulawesi, Indonesia

*mewengkangalfrina@unima.ac.id

Abstract. This development research aims to produce learning media in the form of E-Learning media using Edmodo which is interesting, efficient and effective on the subjects of operating system for students of class X TKJ in SMKN 3 Manado. The development model used was developed by S. Thiagarajan et al., Often known as the Four-D model, but this research only uses (define, design, and develop). Trial of the product is done twice (limited and wide). The experimental design used was the before-after experimental design. Data collection techniques used are interview techniques, questionnaires, and tests. The analytical technique used in this development research is descriptive qualitative. These include analysis of attractiveness test, efficiency and effectiveness of E-Learning media using Edmodo. The media attractiveness test was measured using a student response questionnaire. Media efficiency test was obtained through interviews of researchers with operating system subjects teachers and students of class X TKJ 1 at SMKN 3 Manado. While the media effectiveness test obtained from student learning outcomes before and after applying E-Learning media using Edomodo. Then tested by paired sample t test formula. After the media was piloted on the subject of trials (limited and broad), and the results show that E-Learning media using Edmodo is interesting, efficient and effective. It is shown on average student response score of 88.15% with very interesting interpretation. While the average value of student learning outcomes increased from 76.33 to 82.93. The results of differential test (paired sample t-test) the value of $t = 11\ 217 \geq t_{table} = 2{,}045$ with significant value $= 0.000 < \alpha = 0.050$ showing the media E-Learning using Edmodo is effective.

1. Introduction

The rapid development of information and communication technology has a direct impact/influence on education in general. Along with these developments, learning methods are also more advanced, which include personal learning methods, learning media and learning process. One form of implementation of information technology in education is the model of E-Learning.

E-Learning media that is implemented in this research is a web-based application named Edmodo. Edmodo is a private web-learning media that prioritizes the privacy of students and teachers (separately). Edmodo can be directly accessed from the web (computer-based), also available in mobile versions for smartphone / tablet users.

There are some problems in learning in SMKN 3 Manado, among others; Lack of student interest in previous media, ie powerpoint media. Direct learning process that exceeds allotted time. And learning outcomes are less satisfactory, so that learning objectives have not been achieved optimally.

One of the constraints of conventional learning model is the necessity of direct meeting (in class) between teacher and student. E-Learning model using Edmodo can overcome these obstacles, ie learning process can be continuously done without direct face-to-face between teacher-student. this will make students more active and improve interaction among fellow students to exchange information. Students can also practice on tasks that have been prepared by the teacher. This media trains students' independence, while maintaining the interaction between teachers and students as well as between students.

According to Rusman, learning media is used by teachers to deliver lesson material not only as a means, but also to provide stimulus for students to learn, as well as develop the intellectual and emotional aspects of students. A learning medium requires good planning before being used in the learning process. The media used must be completely mastered by teachers, so that knowledge can be transferred to the students clearly and effectively.

Based on the background of the issues stated above, we are interested to conduct a study of E-learning development process for operating system subjects. Specifically, developed e-learning is tailored to the learning needs of students of class x, with reference to previous learning design. Methods.

2. Methods

2.1. Model

The development process is related to activities at each stage of development. The final product of the development result will be evaluated based on the aspect of product quality specified. The development stages of the Four-D model are: (1) Define, (2) Design, (3) Develop, and (4) Disseminate or adapted into 4-D model, ie definition, design, development, and deployment. The following stages of development using the Four-D model [1].

2.1.1. Define. The defining stage is the stage for defining and defining development requirements, consisting of; analysis of development needs, product development requirements in accordance with student characteristics, teacher needs analysis, as well as appropriate research and development models to develop the product.

2.1.2. Design. The design stage consists of 4 main activities, namely:

- Criterion-test construction, as the first action to determine the learner's initial ability and as an evaluation tool after implementation.
- Media selection, choosing a learning model that matches the material and characteristics of the students.
- Format selection, the choice of presentation model in accordance with the learning media
- Initial design, complete the prototype of the application before the trial run.

2.1.3. Develop. Development phase is done through two steps, namely expert appraisal followed by revision and developmental testing. The purpose of this development stage is to finalize learning devices after revisions based on validator evaluations and test results data.

2.1.4. Disseminate. The dissemination stage is carried out to promote the development product in order to be accepted by users, whether individuals, groups or systems. Distribution stage is divided into three activities, namely: validation testing, packaging, diffusion and adoption.

In this development study, researchers did not undertake the entire stages of development. Researchers only do 3 stages of define, design, and develop without implementing dissemination.

Reason Phase 4 (dissemination) is not implemented because the use of this learning media only in SMKN 3 Manado. There is no important value for dissemination to a wide audience or even a market [2-6].

2.2. Trial test
This research applies before-after (pretest-postest) experimental design, comparing the values and perceptions of students before and after using Edmodo E-Learning media. The following pretest-postest experimental design is shown in the following figure 1:

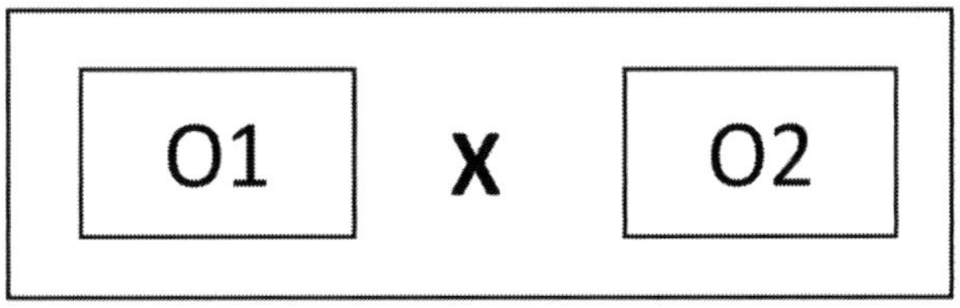

Figure 1. Experiment design using Edmodo [2].

O1 = Pretest score
X = Treatment (Edmodo E-Learning)
O2 = Postest score

Edmodo E-Learning media testing is done by involving several categories of respondents in SMKN 3 Manado, namely students of class X and teachers TKJ.

- Target Segment
  Limited Trial : 10 randomly selected student of class TKJ at SMKN 3 Manado
  Open Trial : All students of class TKJ 1 at SMKN 3 Manado.
- Material validator, ie teacher of operating system subjects
- Validator media display, the teacher of TKJ subjects in SMKN 3 Manado.
- Prior to conducting the experimental activities, the researchers socialized the E-Learning media using Edmodo to the test subjects. Limited Trial Test is done by using smartphone.

2.3. Data organizing

2.3.1. Attractiveness of Edmodo E-learning media. The level of attractiveness of Edmodo E-Learning media is obtained from the questionnaire results of the student responses obtained from two stages of testing (limited and open).

2.3.2. The efficiency level of Edmodo E-learning media. Testing efficiency of Edmodo E-Learning, conducted through interviews with teachers and students operating system subjects in class X TKJ 1 SMKN 1 Manado. Interviews are structured, and completed with checklists.

2.3.3. The effectivity level of Edmodo E-learning media. The level of effectiveness of Edmodo E-Learning is seen from the presence or absence of differences in student learning outcomes (operating system structure material), before and after the implementation of Edmodo E-Learning. The paired sample t test is used to assess whether there is a significance difference of the two paired samples.

3. Results and Discussion
This development research succeeded in producing Edmodo E-Learning media on the subjects of Operating System, for students of class X TKJ 1. This media can be accessed using computer and handphone through www.edmodo.com, As for Edmodo application available on PlayStore and AppStore on mobile device . The resulting product consists of some content, including:

IOP Conf. Series: Materials Science and Engineering **306** (2018) 012068 doi:10.1088/1757-899X/306/1/012068

- Instruction guide.
- Learning materials, about the structure of the operating system. This material is adopted from Abdul Munif's book.
- Video learning, which contains the lesson of the operating system structure.
- Powerpoint Presentation
- Exercises and test, as a means of training students' abilities after learning with Edmodo E-Learning.

3.1. Trial test results

The percentage of scores obtained from content validation by validator is 81.82%. The score shows the media is feasible, but there are some revised criteria according to the suggestion of the validator. while the percentage of scores obtained from design validation is 71.43%. The score shows the media is feasible, but there are some revised criteria according to the suggestion of the validator.

3.1.1. Trial test results; Attractiveness of Edmodo E-learning media. The level of attractiveness of Edmodo E-Learning media was measured based on the results of questionnaires among the ten selected students. The mean percentage of response score obtained from this stage is 91%.

3.1.2. Trial test results; The efficiency level of Edmodo E-learning media. Trial test results show nine out of ten students can quickly understand the material with Edmodo E-Learning media. Learning time used in the learning process of the operating system that is for 3 x 45 minutes, without any reduction in study time. Based on the results of the narrative of the students, it can be interpreted that the Edmodo E-Learning media is efficiently used in the learning process, because it can help students to quickly understand the material.

3.1.3. Trial test results; The effectivity level of Edmodo E-learning media. The hypothesis proposed in the effectiveness test is:

H0 : Average learning outcomes before using edmodo media $\geq$ Average learning outcomes after using edmodo e-learning media

H1 : Average learning outcomes before using edmodo media $\leq$ Average learning outcomes after using edmodo e-learning media

The results of paired sample test analysis showed significant differences in pretest and posttest values, so it can be said that edmodo e-learning media is effective in improving learning outcomes. This can be seen from thitung = 6.708 $\geq$ ttabel = 2.262 or significance value = 0,000 $<\alpha = 0.05$. The mean postest score was higher (81.5) than the pretest (74), then the hypothesis (H1) was accepted. It can be concluded that there is an increase in student learning outcomes after using edmodo e-learning media (see Figure 2).

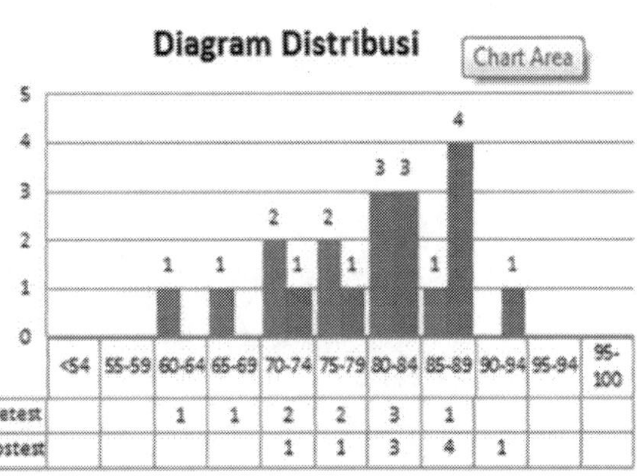

	<54	55-59	60-64	65-69	70-74	75-79	80-84	85-89	90-94	95-94	95-100
∑Pretest			1	1	2	2	3	1			
∑Postest					1	1	3	4	1		

Figure 2. Distribution chart of trial pretest and postest value.

3.2. Test results

3.2.1. Full test results; Attractiveness of Edmodo E-learning media. Overall classroom testing performed after revision of the trial test results. Edmodo E-Learning is further implemented in all 30 students of class X TKJ 1.

Measuring the level of attractiveness of Edmodo E-Learning media in the class test, still using closed questionnaires. The average percentage of students' scores on media responses was 88.15%. The value can be interpreted as a category: very interesting, with the terms of the acquisition value range of 80% - 100% (see Table 1).

Table 1. Scores per indicator.

No.	Indicator	Score
1	Learning content completeness	80,00%
2	Comprehensive language	80,00%
3	Media content completeness	80,00%
4	Clarity of the question	90,00%
5	Easy access	93,33%
6	Attractive media design	96,67%
7	Positive affects	93,33%
8	Encouraging to learn	90,00%
9	Interest in using media for other subjects	90,00%
	Mean	88,15%

3.2.2. Full test results; The efficiency level of Edmodo E-learning media. The percentage of students who more quickly understand the material using Edmodo E-Learning media has increased, ie from 40% up by 80% compared to previous media. The amount of time spent in the learning process on this full test which initially 3 x 45 minutes reduced to 2 x 45 minutes, so that the remaining 45 minutes can be utilized by teachers for question and answer activities and practice questions.

3.2.3. Full test results; The effectivity level of Edmodo E-learning media. The average pretest value of 76.33 with standard deviation of 6.970, while the average posttest value of 82.93 with standard deviation of 6.987. It shows an increase in student learning outcomes. While standard deviations indicate the distance between the top and the bottom of the student's average score, the smaller the value may well represent the real data of the observed value. The result of paired sample test shows the significance value p = 0,000 <α = 0,050, which means there is significant difference between before and after using Edmodo E-Learning media. Thus, the use of Edmodo E-Learning media is effective in improving student learning outcomes.

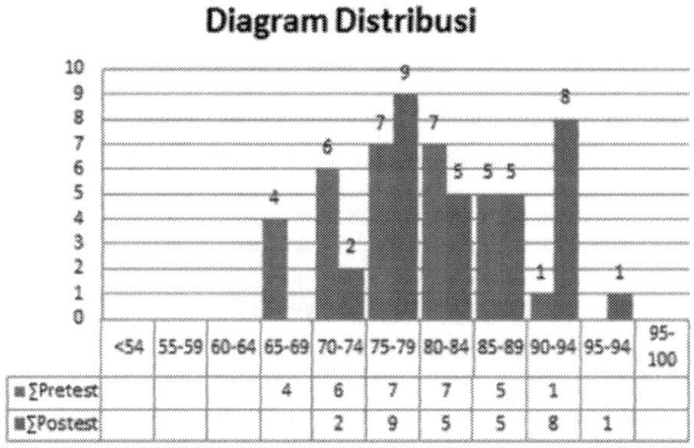

Figure 3. Distribution chart of full pretest and postest value.

Based on Figure 3, The graph above shows an increase in the value of learning results obtained by students. the results show there are no students with a value of 65 (low) at postest. Improvements occur in most students, such as those exposed in the graph.

3.3. Discussion

The results of Edmodo E-Learning media development on the subjects of the operating system for students of class X TKJ, successfully meet the criteria of attractiveness, efficiency, and effectiveness of the product. Judging from the attractiveness level, the average acquisition score of student response questionnaire on the limited trial is 91.11% and the full test is 88.15%, indicating that the developed media is very interesting if used and it is very helpful for the students in the learning process. The use of interesting learning media can reflect the smooth interaction between teachers and students during the learning process and help students learn optimally [7, 8].

Edmodo E-Learning Media Operations does not need to go through complicated steps, so students are not confused in using E-Learning media using Edmodo. The completeness of the content available makes it easier for students to use the media.

Considering the efficiency level, Edmodo E-Learning Media can be used efficiently in the learning process in SMK N 3 Manado, which is shown by the teacher to be faster delivering the learning content and objectives. The results of limited trials show that by using Edmodo E-Learning media the learning time used is still in accordance with the allocation of time allocation in the syllabus, with the students' responses expressing fast understanding as much as 90%. While the test of media efficiency level at full test shows that the utilization of media developed more efficient. Even the use of time in the learning process can be reduced. And the positive response of students to the utilization of Edmodo E-Learning media, because it makes students more quickly understand the learning content. The results show that students can understand the material optimally, with less time than the allocation specified in the syllabus. The use of IT-based learning media in the learning process can help teachers in saving time, and students can also understand the material more easily [9, 10].

The development of Edmodo E-Learning is effectively applied in the learning process at SMK N 3 Manado, which is shown by the increasing of learning result obtained, compared to learning result before using Edmodo E-Learning. This is supported by the results of different test calculations (paired sample t-test). The improvement of students' learning outcomes is triggered by the ease of operating the media, so that students feel happy and enthusiastic in using it, making students motivated to learn the material of the operating system. The high motivation of the students makes the students able to master every indicator to be achieved. It proves that E-Learning media use Edmodo effectively, because it is able to achieve the learning objectives shown by the improvement of student learning outcomes from before.

4. Conclusions

In general it can be stated that the use of Edmodo E-learning media successfully improve student achievement in learning, especially in the material operating system. More specifically, the results of this study can be summarized as follows;

- The development of integrated learning media in Edmodo E-learning, managed to generate interest in students to learn operating system material. Students consider Edmodo E-learning is interesting and practical.
- Media developed more efficiently, successfully helping students more quickly understand the learning materials, or in other words make students able to understand the material in a shorter time. The results of this research show that students can understand the material optimally, with less time than the specified allocation.
- The development of integrated learning media in Edmodo E-learning, succeeded in improving student learning outcomes. Compared to before using new learning media. These results were marked by a higher postest score than the pretest ($p < 0.05$).

References

[1] Thiagarajan S, Semmel D S, and Semmel M I 1974 *Instructional Development for Training Teachers of Expectional Children* (Minneapolis, Minnesota: Leadership Training Institute/Special Education, University of Minnesota).

[2] Sugiyono 2016 *Metode Penelitian Kuantitatif, Kualitatif dan R&D* (Bandung: Alfabeta).

[3] Arikunto S 2006 *Prosedur Penelitian Suatu Pendekatan* (Jakarta: Rineka Cipta).

[4] Azhar A 2015 *Media Pembelajaran* (Jakarta: PT. Rajagrafindo Persada).

[5] Rusman D, dkk. 2011 *Pembelajaran Berbasis Teknologi Informasi dan Komunikasi (Mengembangkan Profesionalitas Guru)* (Jakarta: PT.Rajagrafindo Persada).

[6] Widyawati P T 2012 *Pengembangan Media Pembelajaran Matematika Berbasis Web dengan MOODLE Pada pokok Bahasan Himpunan Kelas VII SMP Al Muttaqin Jember* (Tidak Diterbitkan. Skripsi. Jember: Universitas Jember).

[7] Wahyuningsih A 2015 *Pengembangan Media Pembelajaran E-Learning Menggunakan Edmodo Pada Materi Sistem Moneter Untuk Siswa Kelas X IPS Di MAN Jember* (Tidak Diterbitkan. Skripsi. Jember: Universitas Jember).

[8] Sudjana N, and A Rivai 2005 *Media Pengajaran:Penggunaan dan Pembuatannya* (Bandung: Sinar Baru).

[9] Supardi 2013 *Sekolah Efektif (Konsep Dasar dan Praktiknya)* (Jakarta: PT. Rajagrafindo Persada).

[10] Suwarna, dkk. 2006 *Pengajaran Mikro, Pendekatan Praktis Menyiapkan Pendidikan Profesional* (Yogyakarta; Tiara Wacana).

IOP Conf. Series: Materials Science and Engineering **306** (2018) 012069 doi:10.1088/1757-899X/306/1/012069

Spatial Modeling of Tsunami Impact in Manado City using Geographic Information System

J C Kumaat[1,*], S T B Kandoli[2] and F Laeloma[3]

[1] Department of Geography, Universitas Negeri Manado, Jalan Tondano, Koya, Tondano Selatan., Kabupaten Minahasa, Sulawesi Utara 95618, Indonesia

[2] Department of Geography Education, Universitas Negeri Manado, Jalan Tondano, Koya, Tondano Selatan., Kabupaten Minahasa, Sulawesi Utara 95618, Indonesia

[3] Student at Department of Geography, Universitas Negeri Manado, Jalan Tondano, Koya, Tondano Selatan., Kabupaten Minahasa, Sulawesi Utara 95618, Indonesia

*joykekumaat@unima.ac.id

Abstract. Manado City is a coastal area in the shape of a bay. Manado Bay is a water body that protrudes in the area of Manado City where the condition of this region is likely to have a tsunami threat. Manado Bay is home to several rivers such as Tondano River has a geological history of both land and sea. There are several active faults, such as in the sea, subduction of subplate in the north of the island, Mayu mountain plate, and Sangihe plate east of North Sulawesi. The purpose of this study is divided into two parts: General purpose is to describe GIS-based disaster mitigation that can be done to minimize disaster risk if Tsunami disaster occurs in coastal area of Manado Bay, while special purpose consists of 3 parts, namely: 1. mapping of zone- Tsunami vulnerability zone of Manado Bay; 2. mapping the distance and time of the scenario of the Manado Bay Tsunami evacuation route; 3. mapping of the number of buildings and roads exposed to the Manado Bay Tsunami. Data collection techniques use secondary data collection techniques. Secondary data comes from related institutions or institutions, libraries, or individual archives. The data collection is also continued by direct observation. Direct observation is meant by direct observation by using a checklist for secondary data adjustment and then the determination of coordinate point with Global Position System (GPS) at some tsunami location.

1. Introduction

The island of Sulawesi has 24 out of 136 districts/municipalities that enter the priority of disaster management in the Rencana Pembangunan Jangka Menengah Nasional (RPJMN) 2015-2019 by the National Disaster Management Agency (National Board for Disaster Management / BNPB). A disaster is an event or series of events that threaten and disrupt people's lives and livelihoods caused by both natural factors and /or non-internal factors and human factors resulting in the occurrence of human casualties, environmental damage, property loss, and psychological impact (UU No. 24 Tahun 2007 according to Disaster Mitigation). This shows the island of Sulawesi has a high risk disaster in the central region of the economy. One of them on the northern part of Sulawesi Island is Manado City which is the capital of North Sulawesi Province. This is evidenced by the events of recent years Manado city hit by natural disasters ranging from a flash flood [1], landslides [2], earthquakes [3], to the potential tsunami [4-6]. Manado city is shaped coastal bay area. Manado Bay is a water body that protrudes in the area of Manado City which is classified as tsunami hazard. Tsunami is a series of waves that are capable of spreading at speeds up to 900 km / h, mainly caused by the earthquake which occurred on the seabed. This threat exists because of its geological history from the mainland there are several active faults and in the sea, there are sub plates on the north of the island, Mayu mountain plates, and Sangihe plate in the east of North Sulawesi. Then the volcano tertiary and quarter. The precipitate quarter is decomposed, loose, and yet compact so that the geological conditions can amplify the effect of earthquake shocks [7], required the geographical location and demographic circumstances Manado Bay as coastal areas which became the center of government, economy, education and maritime tourist attraction. Centralized community activities increase the population density inhabiting low land contour areas that are increasingly adding to the needs of development in coastal areas. This condition makes the Gulf of Manado experiencing rapid development changes one of the only beach reclamation. The region has the potential tsunami should have the awareness of disaster preparedness. This alert is necessary because the tsunami is a natural disaster that can devastate a large area [8, 9] and result in exposure effects of people [10], buildings, and roads [11], in the coastal area of Manado Bay. Therefore, disaster preparedness can be applied with disaster mitigation. Disaster mitigation can be managed by using the Geographic Information System (GIS). The purpose of this study is divided into two parts: The general objective is to describe GIS-based disaster mitigation that can be done to minimize disaster risk if Tsunami disaster occurs on the coast of Manado Bay, whereas the special purpose consists of 3 parts, namely: 1. mapping of zones tsunami vulnerability Manado bay; 2. mapping the distance and time scenario of the Manado Bay Tsunami evacuation route; 3. mapping the number of buildings and roads exposed to the Manado Bay Tsunami.

2. Methods

The research location is along the coastal areas of the Bay of Manado North Sulawesi province with astronomical layout is 1°33'25.79 " North Latitude and 124°48'30.31" East Longitude to 1°27'24.81" North Latitude and 124°47'47.26" Longitude East. Tools and materials used are Global Positioning System (GPS), ironing board, computer, software Quantum GIS, OpenStreetMap software [11], software InaSAFE, map city of Manado scale of 1: 20,000, Landsat 8 OLI, satellite image Aster and Bing Map around Manado Bay. Retrieving data using secondary data collection techniques that are quantitative. Secondary data comes from related institutions or institutions, libraries, or individual archives. The data collection is also continued by direct observation. A direct observation which meant that direct observation by using a list of records for the adjustment of secondary data and then determining the coordinates [12] with a Global Position System (GPS) on several locations of areas to be modeled the Tuminting District. Application of Secondary Data Analysis (ADS), which is managed by Quantum GIS software (QGIS) which provides InaSAFE plugin toolbar that has the function of analyzing spatial data of tsunami disaster [11]. So the Bay of Manado tsunami vulnerability can be classified based on the classification of the tsunami by Bandung Institute of Technology (ITB) namely (see Table 1):

Table 1. Tsunami classification.

Zone Name	Color	Value (meters)
dry zone		$>0.00\leq0.10$
low hazard		$>0.10\leq1.00$
medium hazard)		$>1.00\leq3.00$
high hazard		$>3.00\leq8.00$
very high hazard		$>8.00\leq9999.00$

IOP Conf. Series: Materials Science and Engineering **306** (2018) 012069 doi:10.1088/1757-899X/306/1/012069

3. Results and Discussion

3.1. Tsunami threat zone in the Bay of Manado

As shown in Figure 1, the Hazard Zone of the Manado Bay Tsunami is divided into 4 zones: dry, low, medium, high, and very high.

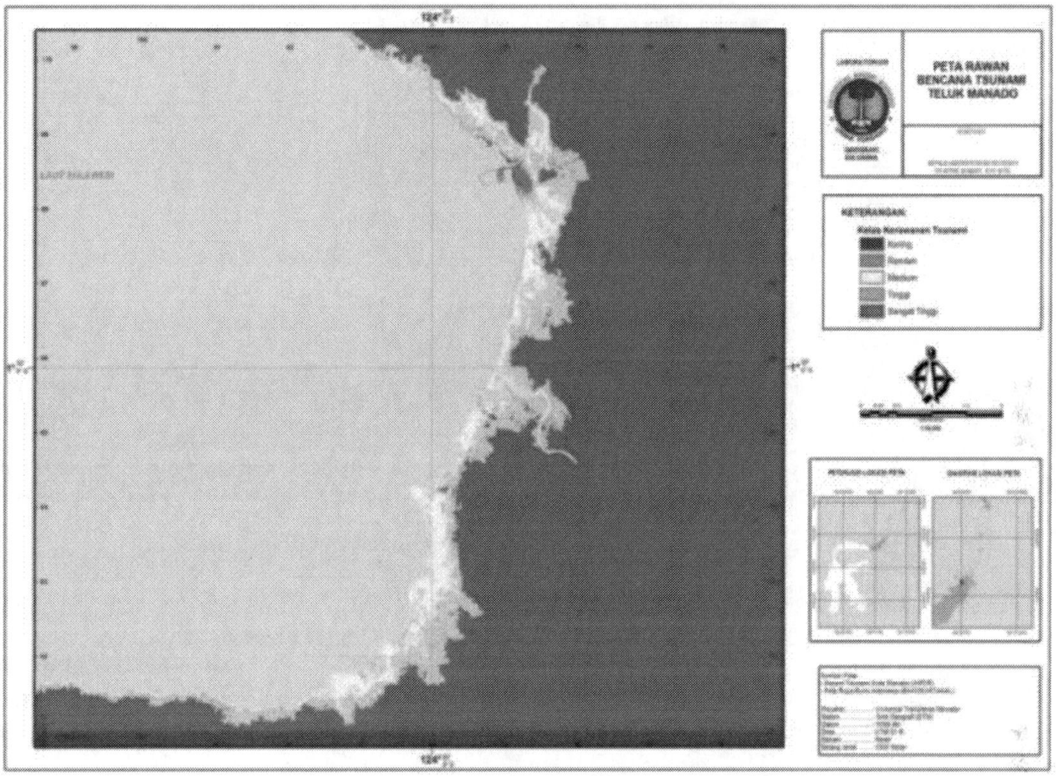

Figure 1. Manado Bay Tsunami Hazard Map.

3.2. Run up

Tsunami-prone zoning then overlapped with a 1meter IFSAR contour resulted in a contour dry zone of 14 - 19 meters, contour low zone 8 s / d 13 meters, medium contour 5 - 7 meter, contour zone height 2 - 4 meters, and very high zone contour <1 meter. So that can be identified wave height (run up) when the tsunami happened that reaches 7 meters. Contour identification can be assumed that run-up at the tsunami can reach 7 meters, this is because very high zones, high zones, and middle zones are classified as impacted areas whereas low and dry zones are classified as unaffected (Figure 2).

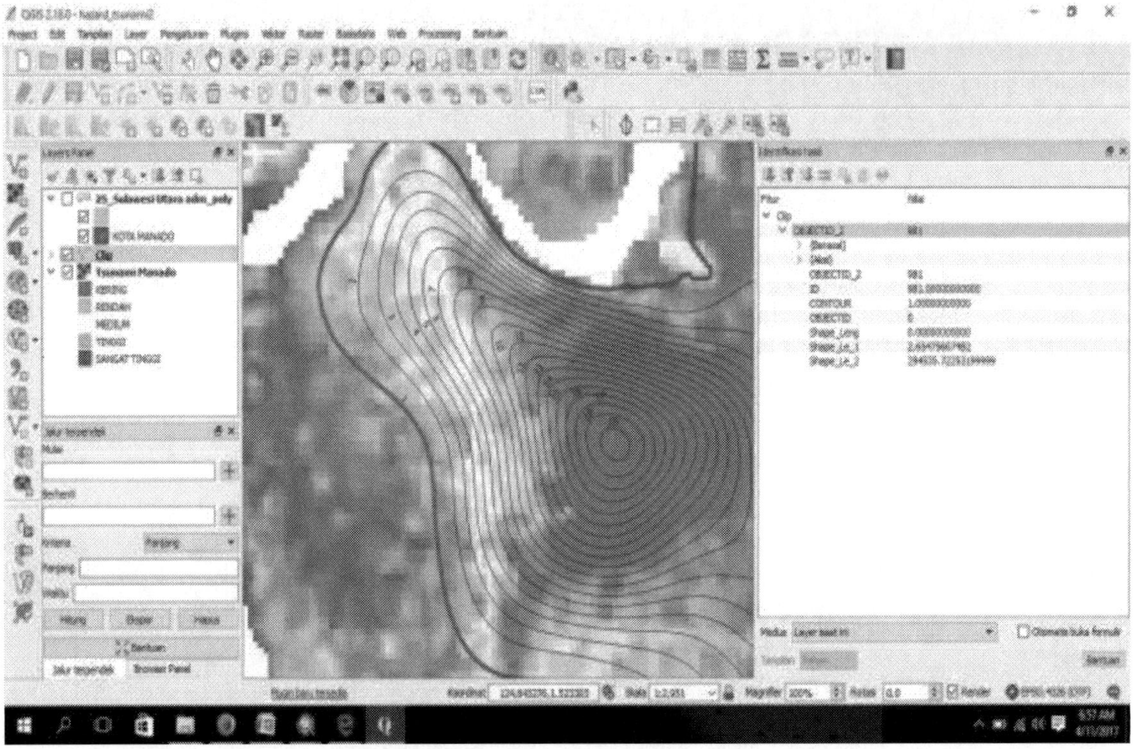

Figure 2. Results of image panel identification quantum GIS software.

3.3. Building exposure analysis

Building exposure (Figure 3) or tsunami hazard analysis in QGIS software must first complete the InaSAFE keyword then focused Tuminting District which has been analyzed to produce 6,900 buildings classified impacted 3,800 buildings, not affected 530 buildings, and not exposed 2,600 buildings, for roads totaling 90,600 meters classified impacted 25,700 meters, not affected 5,600 meters, and not exposed 59,400 meters.

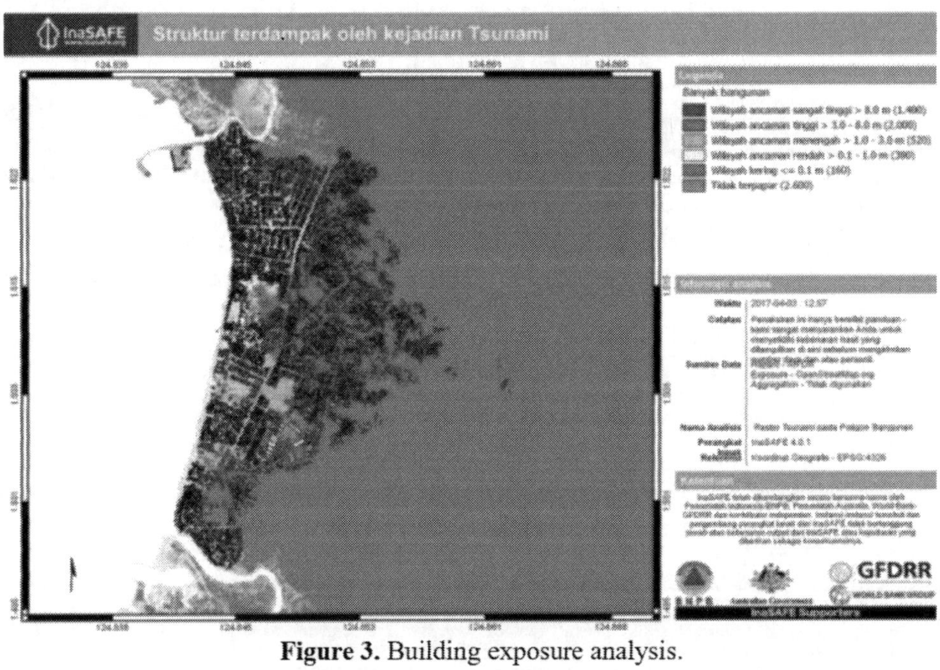

Figure 3. Building exposure analysis.

3.4. Road exposure analysis

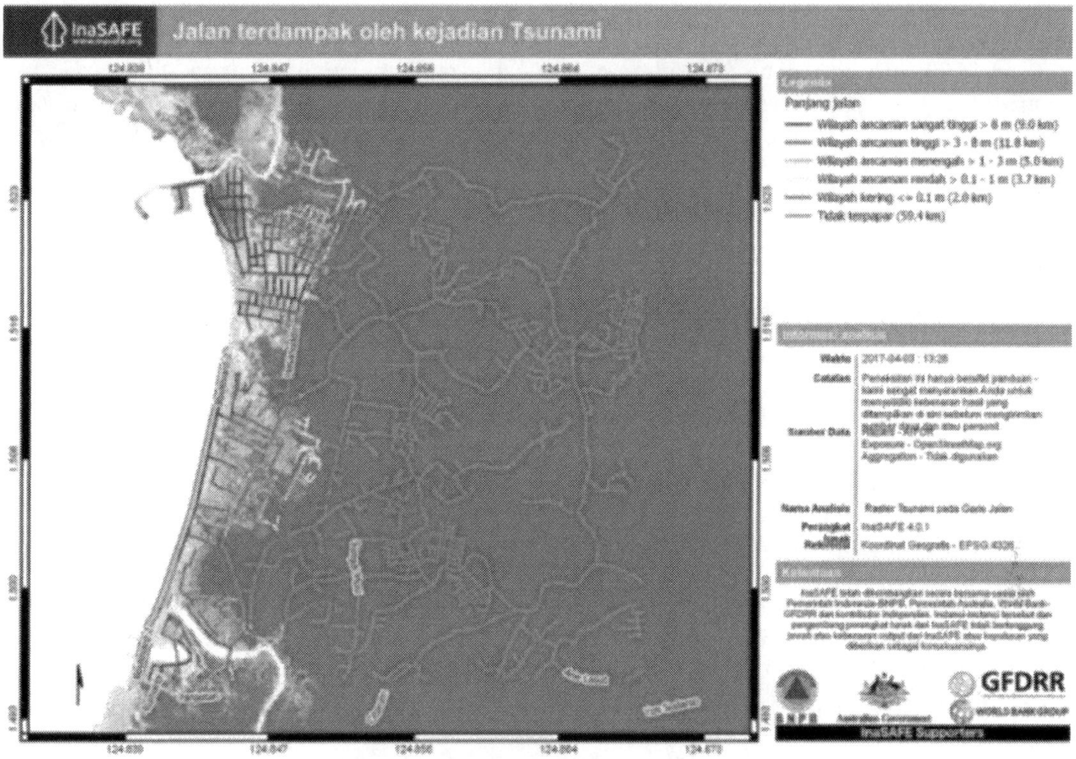

Figure 4. Road exposure analysis.

In figure 4, the analysis of road exposure using QGIS software basically has the same stages as the tsunami hazard analysis and the previous building exposure analysis. It's just that there are still some differences in keywords for information on tsunami threats and road exposure. In InaSAFE analysis obtained for roads have a length of 90,600 meters classified impacted 25,700 meters, not affected 5,600 meters, and not exposed 59,400 meters.

3.5. Tsunami evacuation route

The determination of the evacuation path is the selection of the fastest path (time) and the shortest path (distance). In QGIS software has been equipped with Road Graph plugin (road graph). The Short Route with the starting point of Tarus Road 1 and the stop point of Nike Road is then adjusted to the local tsunami evacuation time 30 minutes to obtain the formula:

V: 45 km / h; T: 30 minutes = 0.5 hours;

Then: s = v.t

= 45 km x 0.5 hours

= 22.5 km / h or 23 km/h

Speed 45 km/h then obtained the road length of 1.44 km and time of 0.032 hours. Based on 45 km / h, within 30 minutes (0.5 hours), the mapping of the road evacuation path is 1.43 km and the time is 0.062 hours.

4. Conclusions

Based on the results of research and discussion that has been described previously, then got some conclusions as follows: Tsunami-prone zone of Manado Bay is classified into 5 zones of dry zone, low zone, medium zone, high zone, and very high zone; Overlapping results of hazard data of Manado Gulf tsunami source AIFRD with 1 meter IFSAR contour vector data show that potential tsunami threats could devastate buildings, inundate many roads and casualties along the coastal areas of Manado Bay. This is because the sea level (run up) of Manado Bay tsunami can reach 7 meters which is the highest contour in the affected area; Based on the analysis of OpenStreetMap vector data of buildings and roads in InaSAFE in Manado Bay area Tuminting Sub-district resulted in building (structure) exposed 530 of 3,600 and 25,700 meters of exposed road from 90,600 meters; Mapping the scenario of the evacuation route from the very high threat zone of Tarus 1 Road to the low threat zone of Nike Road with the Road Graph plugin calculation(graph of the road) within 30 minutes after the spice to escape. Human running speed at a maximum of 45 km / h so that within 30 minutes (0.5 hours) obtained the evacuation route 1.43 km long road in 0.062 hours which is the fastest and shortest path as the initial act of self-salvation to go to the existing shelter.

References

[1] Sleeter B M, Wood N J, Soulard C E, and Wilson T S 2017 Projecting community changes in hazard exposure to support long-term risk reduction: A case study of tsunami hazards in the US Pacific Northwest *International Journal of Disaster Risk Reduction* **22** pp 10-22.

[2] Xu J, Li L, and Zhou Q 2017 Spatial-Temporal Analysis of Open street map Data After Natural Disasters: A Case Study Of Haiti Under Hurricane Matthew *International Archives of the Photogrammetry, Remote Sensing & Spatial Information Sciences* **42**.

[3] van den Homberg M 2017 Toward a Balkans' Data for Disaster Management Collaborative? *In Implications of Climate Change and Disasters on Military Activities* pp 11-18 (Springer, Dordrecht).

[4] Dewi C A, dan Romi F 2014 *Analisis Pembuatan Peta Zona Rawan Bencana Tsunami Pada Daerah Pesisir (Studi Lokasi: Pesisir Kota Bandar Lampung)* (Fakultas Teknik, Jurusan Teknik Sipil, Prodi D3 Survei dan Pemetaan, Universitas Lampung).

[5] Tomaszewski B 2014 *Geographic Information System (GIS) for Disaster Management* (New York, USA, Rochester Institute of Technology, September).

[6] Mohammadi H, Delavar M R, Sharifi M A, and Pirooz M D 2017 Spatiotemporal visualization of tsunami waves using kml on google earth *International Archives of the Photogrammetry, Remote Sensing & Spatial Information Sciences* **42**.

[7] Lynett P J, Borrero J C, Weiss R, Son S, Greer D, and Renteria W 2012 Observations and modeling of tsunami-induced currents in ports and harbors *Earth and Planetary Science Letters* **327** pp 68-74.

[8] Hoppe M W 2010 *Pengetahuan Tentang Risiko* (Jakarta GITWES Capacity Building in Local Communities).

[9] Latif S K M Rakibul I, Monjurul I K dan Syed I 2011 *An OpenStreetMap for the Disaster Management in Bangladesh* (Dhaka, Bangladesh: Department of Computer Science and Engineering Bangladesh University of Engineering and Technology).

[10] Rogers B D, and Dalrymple R A 2008 SPH modeling of tsunami waves *Advanced numerical models for simulating tsunami waves and runup* **10** pp 75-100.

[11] Ulaganathan M N 2016 *Building a volunteered geographic information system (VGIS): A mobile application for disaster management* (California State University, Long Beach).

[12] Fleischhauer S, Behr F J, and Rawiel P 2017 Concept And Implementation Of An Architecture For The Immediate Provision Of Geodata In Disaster Management *International Archives of the Photogrammetry, Remote Sensing & Spatial Information Sciences* **42**.

Designing Low-Income Housing Using Local Architectural Concepts

K Trumansyahjaya* and L S Tatura
Faculty of Engineering, Gorontalo State University

*trumansyahjaya@gmail.com

Abstract. The provision of houses for low-income people who do not have a home worthy of being one of the major problems in the city of Gorontalo, because the community in establishing the house only pay attention to their wants and needs in creating a healthy environment, the beauty of the city and the planning of the home environment in accordance with the culture of the people of Gorontalo. In relation to the condition, the focus of this research is the design of housing based on local architecture as residential house so that it can be reached by a group of low income people with house and environment form determined based on family development, social and economic development of society and environment which take into account the local culture. Stages of this research includes five (5) stages, including the identification phase characteristics Gorontalo people of low income, the characteristics of the identification phase house inhabited by low-income people, the stage of identification preference low-income households, the phase formation house prototype and the environment, as well as the stage of formation model home for low-income people. Analysis of the model homes for low-income people using descriptive analysis, Hierarchical Cluster Analysis, and discrimination analysis to produce a prototype of the house and its surroundings. The prototype is then reanalyzed to obtain the model home for low-income people in the city of Gorontalo. The shape of a model home can be used as a reference for developers of housing intended for low-income people so that housing is provided to achieve the goals and the desired target group.

1. Introduction

The community is part of the development of the city of Gorontalo as a whole, with a population growth rate of about 2-3% per year. The people of Gorontalo City are based on the socio-economic condition of the community, grouped into three levels of community groups, i.e. high-income communities, middle income and low-income communities. These low-income groups still do not have permanent houses and some still live in undue environments that can disrupt the beauty and spatial layout of Gorontalo City.

Government policies on the development of "Very Simple Homes" and "Simple Homes" (RSS / RS) intended for low-income communities have not fully met the needs of low-income people. This is because part of the community cannot provide the down payment and monthly instalment as a requirement to get the House ownership credit. In addition, various government efforts in cooperation with developers have been implemented in order to meet housing needs for low-income people at affordable prices, but in the construction of housing is not noticed the arrangement and model of the form of houses and infrastructure facilities such as sewerage, septic tank, a garden that formed an irregular and unhealthy housing area and seemed to be slums.

The need for a home for low-income peoples is the target to create a decent and healthy home design based on local architecture so as to create a well-organized and organized residence.

2. Research Methodology

To be able to see the dynamics that occur in the housing needs of low-income communities in the city of Gorontalo with the object of the target is the community group who do not have a house and want to have a decent house, this research is done qualitatively to get the design house with local architecture concept based on the desire of low-income peoples. Data collection methods used in this study with random sampling technique, that is taking sampled data taken randomly from a population of 160 peoples in low-income groups in 9 (nine) regions in the city of Gorontalo.

The stages of this study include five (5) stages, including identification of the characteristic stages of low-income Gorontalo peoples, the identification phase characteristics of houses occupied by low-income households, the identification phase of low-income household preference, the prototype, and environmental phases for low-income peoples. Design analysis of home models for low-income peoples using descriptive analysis, Hierarchical Cluster Analysis, and analytical discrimination to generate a prototype from the house and environment. The prototype was then re-analyzed to get a house model design for low-income peoples in the city of Gorontalo.

3. Theoretical Review

Low-income peoples are groups of peoples who experience long-standing economic, social, cultural and political pressures resulting in a culture called poor culture [1]. The low-income peoples are caught in its poor culture so that the community can no longer see its potential to be developed. Determining the low-income people's group is not easy because of the uncertainty of income [2].

The behavior of low-income people is strongly influenced by external stimuli that are the physical environment and social environment. The physical environment in the form of an environment that is less and minimal while the social environment occurs a very tight competition to fight and even maintain something, so there is a strong internal stimulus in order to fight something and maintain something. This external stimulus takes place within the framework of strong economic pressures. This economic pressure limits the accessibility of low-income communities to environmental facilities and infrastructure [1].

Home as a place of residence is a place used to live and will develop in line with the psychological development, social development of the inhabitants, and economic development so that the house will develop in accordance with the developments in the community where the fulfillment of community needs as the core of human relationships against the desire of a home [3]. The desire for homes is influenced by the level of society that is divided into low-income, middle-income and high-income communities [4]. The condition of low-income people is more concerned with the priority scale of the location of housing close to the workplace than the status of land ownership and the quality of housing [4].

Low-income peoples are more concerned about opportunities than identity and security, increasing the level of life for the community then the community's preference for homes will undergo changes that can lead to changes to the fulfillment of space needs. The fulfillment of home preferences is influenced by housing characteristics consisting of environmental and occupancy characteristics [5] and accessibility factors, achievements (transportation), and distance to urban facilities [6]. Changes that occur in the home are closely related to the stages of family development, especially those associated with changes in the number of family members that will affect the environmental changes, the way family members live in conducting activities in a residential environment [3]. In fact, the culture of low-income communities in meeting the needs of homes will shape their own architecture so that people living with national culture will be able to produce an architecture that has an identity that is the local architecture [7].

Based on the literature review related to housing development for low-income people, it can be concluded that the need for continuous study in the form of housing model for low-income people based on the analysis of community characteristics and the characteristics of the people's desire for the shape, composition of space and building materials as appropriate with community culture.

4. Results and Discussion

A study to obtain an overview of the housing needs of low-income people based on the wishes and family development in Gorontalo City by firstly analyzing the characteristics of the community and identifying the condition of houses inhabited by low-income communities located in the city of Gorontalo.

The Government of Gorontalo City through the Office of Housing and Settlement Housing has classified the community in accordance with income in several levels, ie income below Rp. 2.400.000 belonging to the poor, income between Rp. 2,400,000 up to Rp. 4,200,000 belong to a low-income community (MBR), and income above Rp. 4,200,000 belong to non-MBR. However, based on the Central Bureau of Statistics of Gorontalo City 2014, the category of poor people on the poverty line with monthly income is between Rp. 400,000 up to Rp. 500.000 so from the data obtained then the priority of providing homes for low-income people who have monthly income less than Rp. 1,000,000.

The survey results show the percentage of average monthly income of low-income peoples (Figure.1) affected by the job (figure.2) and the level of education (figure.3) due to the increasing skill gained from education can affect the income received by the community low-income.

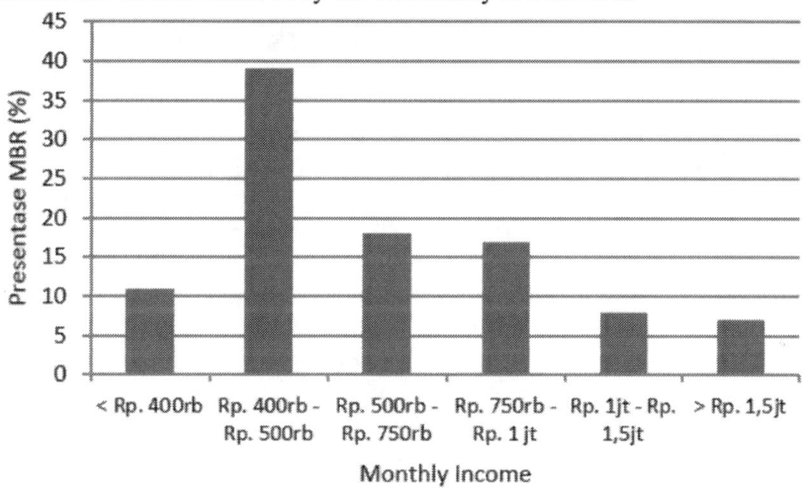

Figure 1. Percentage of Monthly Income for Low-Income Communities in Gorontalo City
Source: Field survey results

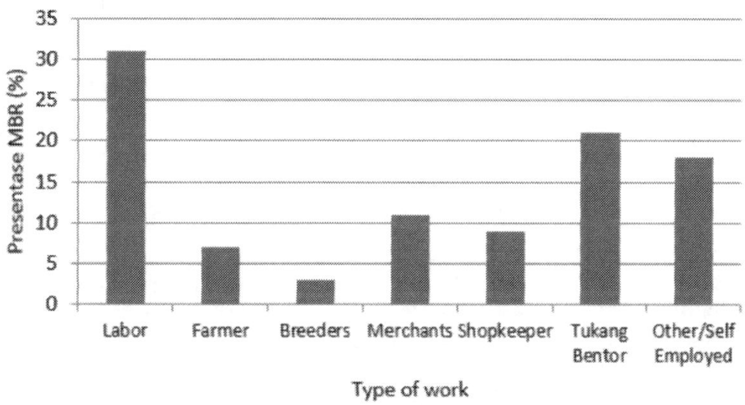

Figure 2. Percentage of Type of Work for Low-Income Communities in Gorontalo City
Source: Field survey results

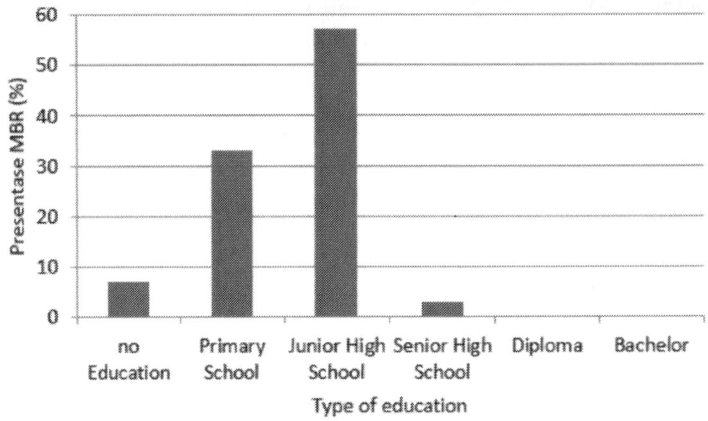

Figure 3. Percentage of Type of Education for Low-Income Communities in Gorontalo City
Source: Field survey results

The house needs for low-income communities in Gorontalo City are influenced by socio-economic level and differences in community characteristics to the needs of the community at the level of needs of space functions, the extent of space and building materials. The level of requirement for the function of the space is adjusted to the activities performed routinely from day to day by low-income communities, so that will form a difference with the desire of spatial function in every family that exist in low-income communities. The results of the survey show that families in low income communities are required to meet the needs of space functions in the bedroom and family room due to the location of sleeping space according to Gorontalo culture can determine the smoothness of sustenance for residents while the family room is used as a place of family deliberation before an event. For the function of the dining room and living room should not have its own space because the existing activities in the dining room and living room can be done in another room that is in the family room. It can create a configuration of space functions in maximizing space utilization for some activities.

The character of society towards the extent of space occurs because to accommodate all sorts of activities undertaken by low-income communities. Space needs are influenced by the activity and the number of space users, so that with the number of activities and the number of users ie family members then the needs of the extent of space in the house more spacious. In addition, the space requirement is influenced by grouping of space functions so that the number of spatial functions present in one space causes the overall area of space to increase. The formation of the extent of space is adjusted by the socio-economic conditions of low income communities in Gorontalo City.

The need for building materials is the character of low-income communities in meeting the overall quality of the house. Interior and exterior quality at home can be formed well due to the quality of building materials and the availability of building materials in the area so that the needs of building materials is closely related to the use of building materials on walls, floors, ceilings, roofs and frames. The availability of building materials in the area provides similarities in the use of building materials such as coconut wood for roof trusses, batako and wooden boards / triplex for walls, zinc for roof coverings. The similarity of use of building materials to low-income houses in the city of Gorontalo shows the same level of importance to the home as a residence to protect against all kinds of disturbance. The use of building materials especially local building materials not only as a complementary material but will make as an aesthetic element that can reflect as Gorontalo architecture.

Based on the economic level on the community and the character of society on the function of space, the extent of space and building materials then created a home configuration that will be used as a benchmark in determining the design of the house. Configuration of the house as a result of the desired analysis by low-income communities in accordance with economic level, space function, the extent of space and building materials (table 1).

Table 1. The configuration of the house for Low-Income Communities in Gorontalo City.

No	Socio-Economic Condition	Space Needs	Building Material Needs	Home Configuration
1	▪ Income Rp. 400rb-Rp. 750ribu ▪ Type Education No education and Primary School ▪ Type Job Labor, Farmers, Breeders, *Tukang Bentor*	▪ Living Room + Business Room Uk. 6 – 9 m² ▪ Family Room + The dining Room Uk. 6 – 12 m² ▪ Bedroom Uk. 6 – 9 m² ▪ Kitchen Uk. 4 – 6 m² ▪ Bathroom Uk. 3 – 4 m²	▪ Floor Ceramics 30x30 ▪ Wall Batako ▪ *Kusen Kayu kelas* 3 ▪ Plafond Triplex ▪ Roof Seng	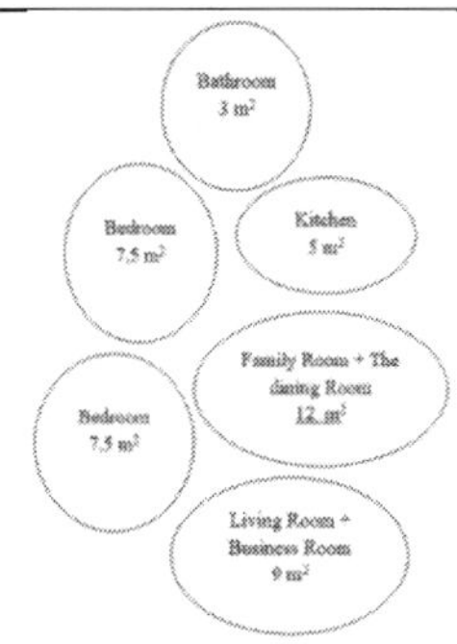
2	▪ Income Rp. 750rb-Rp. 1jt ▪ Type Education Junior High School and Senior High School ▪ Type Job Shopkeeper, Entrepreneur	▪ Business Room Uk. 6 – 9 m² ▪ Living Room Uk. 6 – 9 m² ▪ Family Room Uk. 9 – 12 m² ▪ The dining Room + Kitchen 6 – 9 m² ▪ Bedroom Uk. 9 – 12 m² ▪ Bathroom Uk. 3 – 6 m²	▪ Floor Ceramics 30x30 ▪ Wall Batako ▪ *Kusen Kayu kelas* 2 ▪ Plafond Triplex ▪ Roof Seng	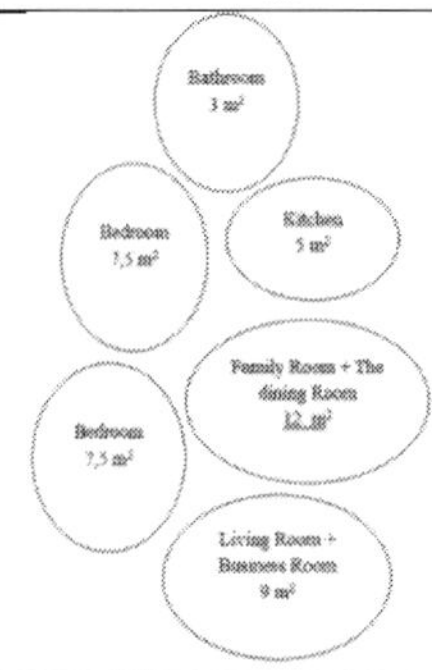
3	▪ Income Rp. 1jt – Rp. 1,5jt ▪ Type Education Junior High School and Senior High School ▪ Type Job Shopkeeper, Merchants, Entrepreneur	▪ Business Room Uk. 6 – 9 m² ▪ Living Room Uk. 6 – 9 m² ▪ Family Room Uk. 9 – 12 m² ▪ The dining Room 6 – 9 m² ▪ Bedroom Uk. 9 – 12 m² ▪ Kitchen Uk. 6 – 9 m² ▪ Bathroom Uk. 3 – 6 m²	▪ Floor Ceramics 40x40 ▪ Wall Batako ▪ *Kusen Kayu kelas* 2 ▪ Plafond Triplex ▪ Roof Seng	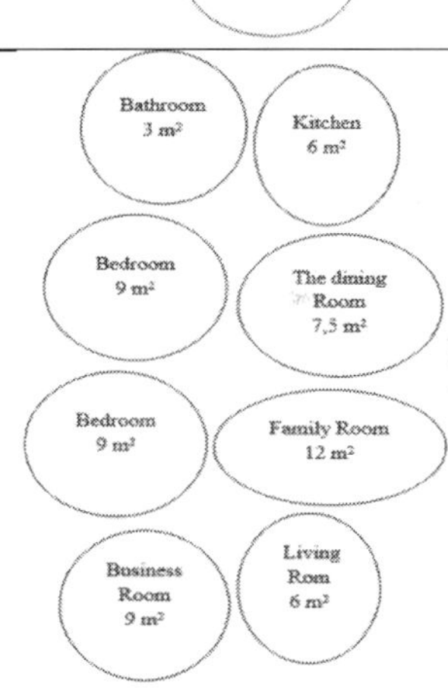
4	▪ Income > Rp. 1,5jt ▪ Type Education Senior High School ▪ Type Job Merchants, Entrepreneur	▪ Business Room Uk. 6 – 9 m² ▪ Living Room Uk. 6 – 9 m² ▪ Family Room Uk. 9 – 12 m² ▪ The dining Room 6 – 9 m² ▪ Bedroom Uk. 9 – 12 m² ▪ Kitchen Uk. 6 – 9 m² ▪ Bathroom Uk. 3 – 6 m²	▪ Floor Ceramics 40x40 ▪ Wall Batako ▪ *Kusen Kayu kelas* 2 ▪ Plafond Triplex ▪ Roof Seng	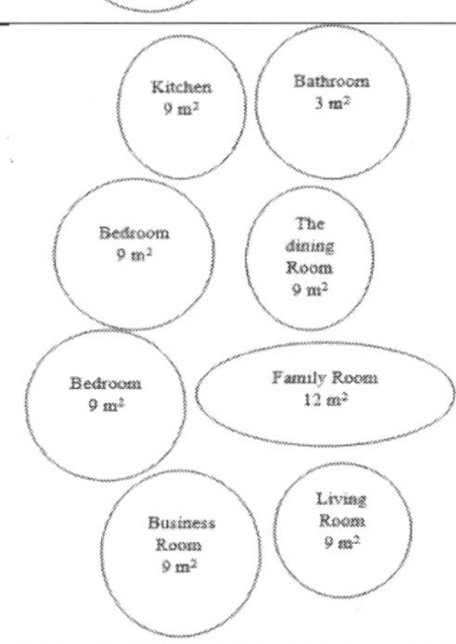

Fulfilling the needs of homes for low-income people is a target to do a decent home design and healthy design based on local architecture so as to create a residential area arranged neat and organized and cultured. Configuration of the house is the result of the desire of low-income people so that the composition of the space provided a reflection of the thinking needs of the community. Thus the process of creation in architecture, especially homes for low-income people must see the needs and conformity with the environment of the community so that the needs of society into consideration in realizing the physical form of culture.

5. Conclusion

The desire of low-income communities in Gorontalo City to the needs of decent and healthy homes greatly affect the development of life perspective both in terms of socio-economic and cultural life. The development of housing needs for low-income people is adjusted to varying socio-economic conditions based on the level of income earned from employment and the educational background of the community. Low-income communities in Gorontalo City are more dominated by working as laborers, traders, *Tukang Bentor*, and entrepreneurs with education level only to elementary and junior high school so that the income earned per month is very minimal that is between Rp. 400,000 up to Rp. 750,000.

Socioeconomic conditions and character of the society towards the function of space, the extent of space and building materials that exist in low-income communities will affect the needs of space in designing a home model that will create a variety of home form configuration needed by low-income communities. The character of the society on the function of space has a clear relationship with the space extent, so less the provision of space function in the house to accommodate family activities then the need for the extent of space in each space will increase. The desire for space for a low-income house can be grouped by main activities and additional activities. The main activity is based on the routine that the low-income peoples always do, such as family rooms, bedroom, kitchens and bathroom so that the rooms can reflect the daily culture of the society in Gorontalo in general. Additional activities as a sparse activity so that existing space will be able to accommodate some activities, such as receiving guests and eating activities can be done in the family room or can be used as a place of business in increasing revenue.

The form of house model that is allocated for the low-income peoples in Gorontalo city is classified based on socio-economic condition, namely (1) house form using the same space function by combining living room function with business room, kitchen function with dining room and combine dining room function with family room, (2) house form which has been done separation of spatial function based on activity type done by low income peoples. The forms of house models provided for the low-income communities must be adapted to the local physical form so that they reflect local architecture as a pride for the people of Gorontalo.

References

[1] Lewis O 1984 *Kebudayaan Kemiskinan dalam buku Kemiskinan Perkotaan, DR. Pasurdi Suparlan* (Jakarta: Sinar Harapan)

[2] Budihardjo E 1991 *Arsitektur dan Kota di Indonesia* Alumni. Bandung.

[3] Habraken N J 1976 *Variations: The Systematic Design of Support* the Laboratory of Architecture and Planning at MIT. Cambridge Mass

[4] Turner J F C 1972 *Housing by People* (London: Marion Boyars Publisher Ltd.)

[5] Lodhi A and A P Hafiz 1991 Housing Demand in Developing Countries: A Case Study of Karachi in Pakistan *Urban Studies* **28** 4

[6] Rappoport A 1970 *Human Aspect of Urban Form.* (California: Pergamon Press.)

[7] Ettehad S, Azeri K, Amir R, and Kari G 2014 The Role of Culture in Promoting Architectural Identity *European Online Journal of Natuiral and Social Sciences* **3** 4

Osteoarthritis Severity Determination using Self Organizing Map Based Gabor Kernel

L Anifah[1*], M H Purnomo[2], T L R Mengko[3] and I K E Purnama[2]

[1]Informatics Department, Faculty of Engineering, Universitas Negeri Surabaya, Kampus Unesa Ketintang, Jl. Ketintang, Surabaya East Java 60231, Indonesia
[2]Electrical Engineering Department, Institut Teknologi Sepuluh Nopember, Surabaya, Kampus ITS Keputih Sukolilo Surabaya East Java 60111, Indonesia
[3]Electrical Engineering Department, Institut Teknologi Bandung, Jl. Ganesha 10/12 - Bandung, West Java, Indonesia

*lilikanifah@unesa.ac.id

Abstract. The number of osteoarthritis patients in Indonesia is enormous, so early action is needed in order for this disease to be handled. The aim of this paper to determine osteoarthritis severity based on x-ray image template based on gabor kernel. This research is divided into 3 stages, the first step is image processing that is using gabor kernel. The second stage is the learning stage, and the third stage is the testing phase. The image processing stage is by normalizing the image dimension to be template to 50□200 image. Learning stage is done with parameters initial learning rate of 0.5 and the total number of iterations of 1000. The testing stage is performed using the weights generated at the learning stage. The testing phase has been done and the results were obtained. The result shows KL-Grade 0 has an accuracy of 36.21%, accuracy for KL-Grade 2 is 40,52%, while accuracy for KL-Grade 2 and KL-Grade 3 are 15,52%, and 25,86%. The implication of this research is expected that this research as decision support system for medical practitioners in determining KL-Grade on X-ray images of knee osteoarthritis.

1. Introduction

Approximately 13% of women and 10% of men at the age 60 years and older have symptomatic knee OA [1]. Based on data at the Ministry of Health of Indonesia, the number of patients with Osteoarthritis disease (OA) is quite high. From 1995 to 2007 the prevalence is 30 percent. So one of thirty Indonesians suffer osteoarthritis disease [2]. This is because the disease cannot be cured and the meet of tibia and femur bones causes pain for the sufferer. Knowing the status of the patient so it can be treated hopefully that the pain can be reduced [3-6]. Classification by Kellgren-Lawrence, based on the severity of osteoarthritis (normal, doubtful, minimal and moderate) [7].

X-rays, magnetic resonance imaging (MRI), osteo CT and arthroscopy are the usual methods used to determine the status of osteoarthritis [8, 9]. But in Indonesia has commonly used X-ray and Magnetic Resonance Imaging (MRI) to know the status of OA. Osteoarthritis has been studied by researchers all over the world. Research on the localization of the junction space area using delineation of cortical bone plates (active shape model) [10].

The classification using 20 pre-selected images of 150 × 150 is used as gold standard, to search for ROIdone by scanning used eucledian distance to predict junction area, while to find the status of osteoarthritis used fisher score from feature data [11], another research finding the distance between the femur and the tibia [12]. But the research that has been done is still difficult to classify KL-Grade 4. Another study is a novel hybrid of S2DPCA and SVM for knee osteoarthritis classification performed by Rima [13].

Some other studies are Osteoarthritis X-Ray Image Analysis Using Gray Level Co-occurrence Matrices Based Wavelet performed by Haryanto [14]. Automatic Segmentation of Impaired Joint Space Area for Osteoarthritis Knee on X-ray Image Using Gabor Filter Based Morphology Process [15]. Osteoarthritis Classification Using Self-Organizing Map Based on Gabor Kernel and Contrast-Limited Adaptive Histogram Equalization [16]. Research about performance evaluation of osteoarthritis X-ray image is investigated by Anifah [17].

The research that has been done above discussed ROI segmentation with various methods. Research that discussed the classification of osteoarthritis severity and classified image into KL-Grade 0, KL-Grade 1, KL-Grade 2, KL-Grade 3 and KL-Grade 4 but however the research is mostly texture feature-based. This research is done with feature-based template, hopefully this research can improve the result of previous research and become one of method to determine KL-Grade on x-ray image of knee osteoarthritis. The aim of this paper to determine osteoarthritis severity based on x-ray image template based on gabor kernel. This research is expected to contribute as decision support system for medical practitioners in determining KL-Grade on X-ray images of knee osteoarthritis. The novelty of this research is to determine KL-Grade based on template feature.

2. Methodology

2.1. Data

Please follow these instructions as carefully as possible so all articles within a conference have the same style to the title page. This paragraph follows a section title so it should not be indented. The data used in this research are 303 x-ray images, which consists of 25 data for training process and 258 data for testing. All data obtained from Osteoarthritis Initiative (OAI). The processed data is x-ray image of the knee fixed- flexion with 10□ image capture angle [18].

2.2. Methods

The diagram block of this research is illustrated in Figure 1. This research is divided into 3 stages, the first step is image processing that is using gabor kernel. The second stage is the learning stage, and the third stage is the testing phase. The image processing stage is by normalizing the image dimension to be template to 50□200 image. Learning stage is done with parameters initial learning rate of 0.5 and the total number of iterations of 1000. The testing stage is performed using the weights generated at the learning stage.

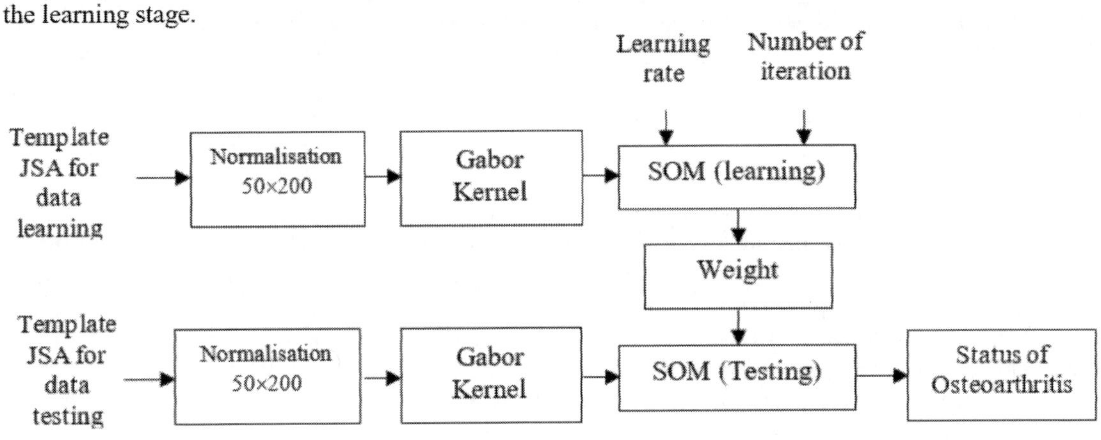

Figure 1. The diagram block of this research.

IOP Conf. Series: Materials Science and Engineering **306** (2018) 012071 doi:10.1088/1757-899X/306/1/012071

SOM algorithm used is illustrated in Figure 2. The learning step uses SOM with the following stages: initialization, determining the number of classes, setting up the learning rate parameters, determining the number of iterations and calculating the distance between the random data and the predetermined weights and obtaining the new weight (update weight). The purpose of using the gabor kernel here is to bring up the special features of x-ray images that have been convoluted to the gabor kernel. With the selection of the right parameters then get the typical image features.

The gabor complex formula is:

$$g(x,y) = s(x,y)w_r(x,y) \tag{1}$$

Where s (x, y) is a complex sinusoid (carrier), and w_r (x, y) a two-dimensional Gaussian-shaped function (envelope).

The sinusoid complex is defined as:

$$s(x,y) = \exp(j(2\pi(u_0x + v_0y) + P)) \tag{2}$$

Where (u_0, v_0) is defined as the spatial frequency and P as the phase of the sinusoid.

The real and imaginary part of the sinusoid is:

$$\begin{aligned} \Re(s(x,y)) &= \cos(2\pi(u_0x + v_0y) + P) \\ \Im(s(x,y)) &= \sin(2\pi(u_0x + v_0y) + P) \end{aligned} \tag{3}$$

Parameters u_0 and v_0 are defined in Cartesian coordinates. While spatial frequency is expressed in polar coordinates with peak F_0 and direction ω_0 as follows:

$$\begin{aligned} Fo &= \sqrt{u_0^2 + v_0^2} \\ \omega_0 &= \arctan\left(\frac{v_0}{u_0}\right) \\ u_0 &= Fo\cos(\omega_0) \\ v_0 &= Fo\sin(\omega_0) \end{aligned} \tag{4}$$

Complex sinusoid becomes:

$$s(x,y) = \exp(j(2\pi Fo(x\cos\omega_0 + y\sin\omega_0) + P)) \tag{5}$$

And gaussian envelope expressed as,

$$w_r(x,y) = K\exp(j(-\Pi(a^2(x-x_0)_r^2 + b^2(y-y_0)_r^2)) \tag{6}$$

Where K is the magnitude of the gaussian envelope, (x0, y0) is the peak of the function, a and b are the scale parameters of the envelope gaussian, whereas r is denoted as the rotation operator [17]. So:

$$\begin{aligned} (x-x_0)_r &= (x-x_0)\cos(\theta) + (y-y_0)\sin(\theta) \\ (y-y_0)_r &= (x-x_0)\sin(\theta) + (y-y_0)\cos(\theta) \end{aligned} \tag{7}$$

IOP Conf. Series: Materials Science and Engineering **306** (2018) 012071 doi:10.1088/1757-899X/306/1/012071

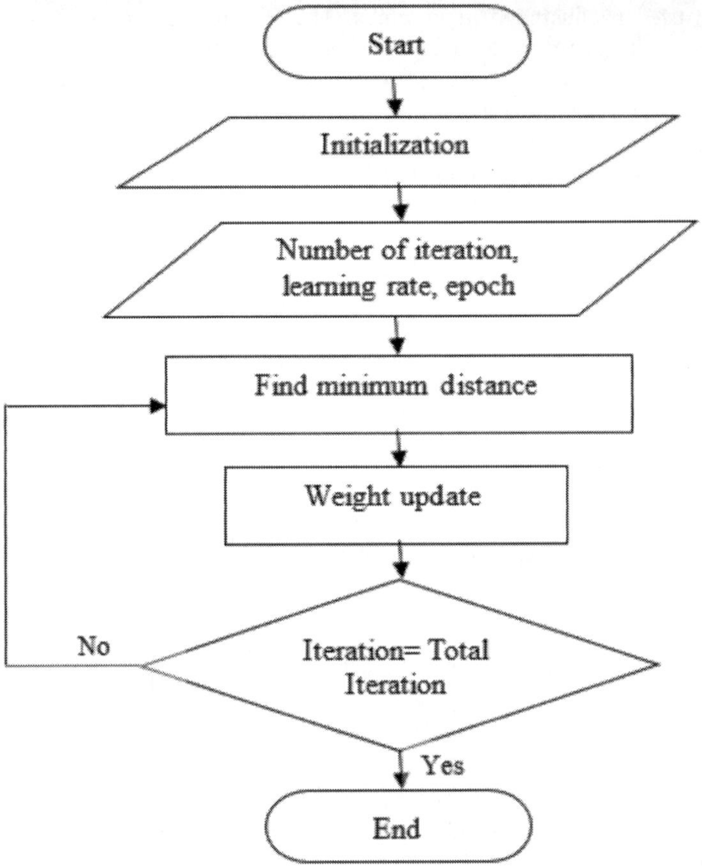

Figure 2. SOM algorithm used in this research.

3. Results and Discussion

Figure 3 is a template used in training process. Experimental results are strongly influenced by the intensity of the processed image, so that required normalization is not just normalization dimensions (size only). The first experiment using the image normalization of 10×40 is too small and possible the amount of missing image information, so image information has not been represented if using this normalization. The third experiment used templates based on gabor as a standardized gold standard to 50×200 and SOM parameters used initial learning rate 0,5 and total iteration 1000.

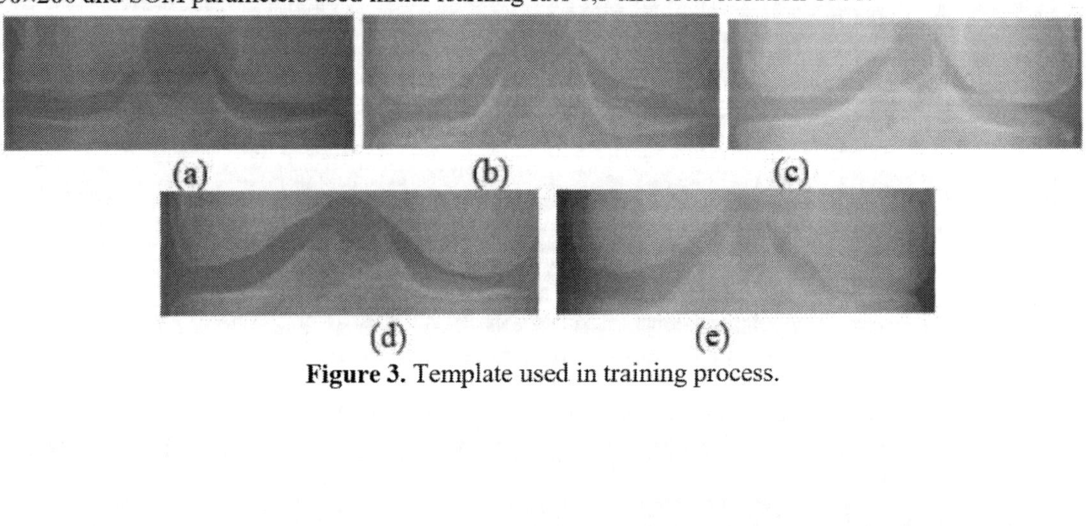

Figure 3. Template used in training process.

The experimental results still have not produced good results. Figure 4 is the result of the normalization of 50□200 previously edge detection using the Gabor Kernel, the five KL-Grades appear almost the same, allowing some information lost after the normalization process.

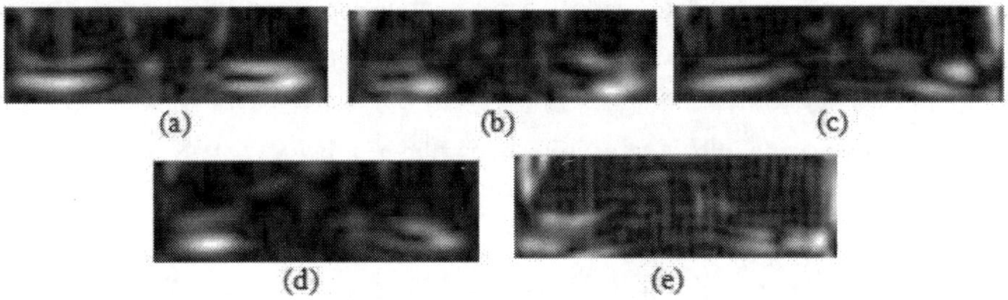

Figure 4. Result of 50×200 image using Gabor Kernel.

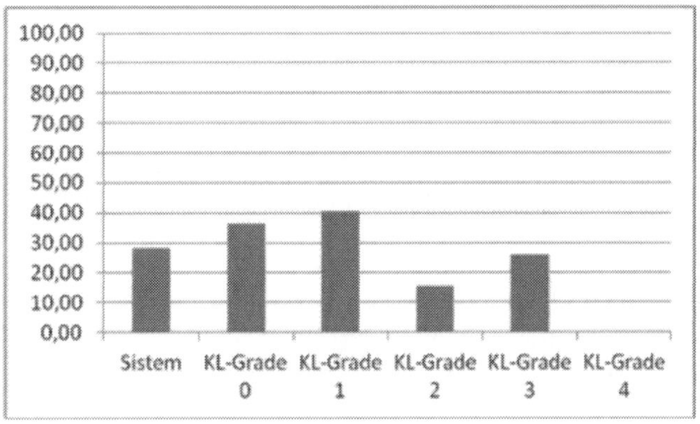

Figure 5. Accuracy of the system.

Figure 5 is illustrated of the system accuracy. KL-Grade 0 has an accuracy of 36.21% because 31.9% is classified as KL-Grade 1, 12.93% are classified as KL-Grade 2 and 18.97% as KL-Grade 3. Classified KL-Grade 1 as KL-Grade 0 is 14.66% while those read as KL-Grade 2 and KL-Grade 3 are 12.07% and 32.76% respectively. KL-Grade 2 which is read as KL-Grade 0 is 26.72%, and which is known as KL-Grade 1 is 44.83% and the remaining 12.93% is known as KL-Grade 3. KL-Grade 3 is recognized as KL-Grade 0, KL-Grade 1, and KL-Grade 2 were 14.66%, 44.83%, and 14.66%. While using this method has a weakness, it still cannot recognize KL-Grade 4, but KL-Grade 4 is known as KL-Grade 0 20%, known as KL-Grade 1 20% and is recognized as KL-Grade 3 32%. Some of the causes that cannot be detected correctly are some images of objects or medical instruments contained in the x-ray image, such as knee bond plate. This will affect the outcome of the kernel gabor process. This can be solved by re-checking the x-ray image before it is inserted into the system.

4. Conslusions

Osteoarthritis severity determination using Self Organizing Map based Gabor Kernel does not produce excellent results. KL-Grade 0 has an accuracy of 36.21%, accuracy for KL-Grade 2 is 40,52%, while accuracy for KL-Grade 2 and KL-Grade 3 are 15,52%, and 25,86%. The weakness of this system is still not able to classify KL-Grade 4. Based on this result, further research is needed to obtain higher accuracy.

Acknowledgments

This research was supported by Universitas Negeri Surabaya and DIKTI Indonesia. We would also like to show our gratitude to Osteoarthritis Initiative for sharing their data and biomarker for this research.

References

[1] B Heidari 2011 Knee osteoarthritis prevalence, risk factors, pathogenesis and features: Part I *Caspian Journal of Internal Medicine* 2(2) p 205–212.

[2] Http://health.liputan6.com/read/2356525/1-dari-10-orang-indonesia-terkena-penyakit-sendi-osteoarthritis.

[3] Woolf A D, and Pfleger B 2003 Burden of major musculoskeletal conditions *Bull World Health Organ* **81**p 646-56.

[4] Elders M J The increasing impact of arthritis on public health *J Rheumatol* **60** p 6-8.

[5] Martel-Pelletier J, Lajeunesse D, Fahmi H, Tardif G, and Pelletier J P 2006 New thoughts on the pathophysiology of osteoarthritis: one more step toward new therapeutic targets *Curr Rheumatol Rep* **8** p 30-6.

[6] Brandt K D, Radin E L, Dieppe P A, and van de Putte L 2006 Yet more evidence that osteoarthritis is not a cartilage disease *Ann Rheum Dis* **65** p 1261-1264.

[7] Buckland-Wright 1997 Current status of imaging procedures in the diagnosis, prognosis and monitoring of osteoarthritis *Bailliere's Clinical Rheumatology* 11(4).

[8] Sing-Tze Bow 2002 *Pattern Recognition and Image Preprocessing* (Marcel Dekker).

[9] Gour Karmakar, Laurence Dooley, M. Manzur Murshed Dengsheng Zhang, Guojun Lu *Image Segmentation and Searching, Gippsland School of Computing & Info Tech Monash University*.

[10] P Podsiadlo, M Wolski, and G W Stachowiak 2008 Automated selection of trabecular bone regions in knee radiographs **35** p 1870–1882.

[11] Lior Shamir, Shari M. Ling, William W. Scott, Angelo Bos, Nikita Orlov 2008 Knee X-ray image analysis method for automated detection of *Osteoarthritis* p 1–10.

[12] Tati L. Mengko, Rachmat G. Wachjudi, Andriyan B. Suksmono, and Qonny Danudirdjo 2005 Automated Detection of Unimpaired Joint Space for Knee Osteoarthritis Assessment p 400–403.

[13] R T Wahyuningrum, Purnomo, L. Anifah, and Purnama 2016 A novel hybrid of S2DPCA and SVM for knee osteoarthritis classification, Computational Intelligence and Virtual Environments for Measurement Systems and Applications (CIVEMSA) *2016 IEEE International Conference on, 27-28 June 2016 Hungaria*.

[14] Haryanto, E M Rohman, Anifah 2016 Osteoarthritis X-Ray Image Analysis Using Grey Level Co-occurrence Matrices Based Wavelet *4th UPI International Conference on Technical and Vocational Education and Training* (Available at: http://tvet.conference.upi.edu/2016 15-16 November 2016).

[15] Anifah L, Purnomo, Hariadi, Purnama 2013 Osteoarthritis Classification Using Self Organizing Map Based on Gabor Kernel and Contrast-limited Adaptive Histogram Equalization *The Open Biomedical Engineering Journal* **7** p 18-28.

[16] Anifah L, Purnomo, Hariadi, Purnama 2011 *Automatic Segmentation of Impaired Joint Space Area for Osteoatritis Knee on X-ray Image using Gabor Filter Based Morphology Process IPTEK* **22**(3) p 117-176.

[17] Anifah L, Purnomo, Hariadi, Purnama 2012 Performance Evaluation of Osteoarthritis x-Ray Image Enhancement Using Clahe and Gabor Kernel *Proceeding of International Conference on Biomedical Engineering and Madical Applications (ICBEMA) BME-Days 2012*.

[18] Osteoarthritis Initiative 2009 *Osteoarthritis* Available at: www.OAI.com.

IOP Conf. Series: Materials Science and Engineering **306** (2018) 012072 doi:10.1088/1757-899X/306/1/012072

Vocational Students' Motivation for Professional Skills

L Sojow[*], A Wajong and N Sangi
Universitas Negeri Manado, Tondano, 95618, Sulawesi Utara, Indonesia

[*]lucks_59@yahoo.com

Abstract. The purpose of this research is to know whether there is a positive and significant correlation between readiness work and achievement motivation with performance of vocational education graduates. Expost facto research method by taking data with instrument was employed in this study. To obtain data on work attitude variables, achievement motivation is taken from primary data through employee respondents in Bitung City, Indonesia industry with a sample of 37 people. Results of the research show that: (i) good work readiness is the dominant element in determining the performance level of vocational graduate worker. (ii) There is correlation between achievement motivation with graduate performance, (iii) there is correlation between work readiness and achievement motivation together with graduate performance. The contribution of work readiness and achievement motivation together with graduate performance was 73%.

1. Introduction

Vocational education has a very important role in the era of development today to meet the demands of job competence. Vocational high school which is a formal education is expected to fill the needs of business and industry. Vocational schools should sharpen their function and role in preparing their students in order to have certain skills that can be directly beneficial to enter the world work [1]. So to be ready to work need to link and match premises work. That in fact, schools that apply Honda Motor Engineering Curriculum (KTSM) Honda already have standardization qualified for graduates ready to plunge into the world of work [2]. But in fact, there are still a number of vocational graduates who have not got a job. This is one of the causes of the lack of synchronization of science taught in schools with the skills needed in the workplace.

The Vocational Device in preparing the wasted personnel in certain areas of expertise as a provision for sustainable self-development into a national asset that can play a role to face the competition of the MEA region and at the same time utilizes job opportunities. Because, it needs prepared a reliable human resources and able to master the knowledge, technology, and skills in certain areas of expertise. The establishment of a professional workforce should be established through an educational program appropriate to the needs of employment. Vocational secondary education is essentially a labor market to meet employment seekers who are the place of projection of their graduates, meaning to be able to carry out their duties properly and achieve their goals effectively so vocational secondary education cannot be separated from employment. Vocational education is designed and implemented and evaluated in relation (link) to the demands of employment needs so that the results are really suitable or match with the demands of employment needs. In order to develop human resources implemented through vocational secondary education can be highly efficient and able to increase productivity it is necessary efforts towards improving the quality of education so that learners can prepare to enter the world of work. Providing career guidance services in schools is expected to learners have the skills in making decisions about his career in the future, and can provide a variety of information related to the world of work in accordance with the field of expertise. This will be very useful for learners to add insight as capital entered the business world. The form of successful learning in schools can be seen from the achievement of student learning, because the achievement of learning shows the results of efforts achieved by students during their learning activities in schools that are generally shown in the form of value. Similarly, the readiness of the vocational graduate worker in the Industrial World. The unpreparedness of the workforce that we have apart from the limited employment opportunities that are not comparable with the labor force graduate performance [3].

Vocational High School in Bitung City continues to strive to prepare students in order to become graduates who are ready to work, which becomes the main priority of vocational graduates that answer the needs of workers and open employment, or to continue to higher education, of course that suits the field of expertise of graduates while in vocational school. Industrial Work Practices and the provision of Career Guidance services is a mandatory activity of Vocational High School which is useful for preparing graduates. The Career Guidance Services at the school also assist students in preparing graduation as per the qualifications implied by the World Work World of Industries. The graduate of Vocational High School in Manado City, North Sulawesi Province (Sulut) is prepared to enter the world of work, especially retail industry [4]. We started this preparation by giving lessons about the retail industry to students' children in vocational high school Manado, through Alfamart class. So, also in the city of Bitung, the program Alfa mart Class to meet the needs of ready-to-use labor in the retail industry. With this educational program, it can assist local governments in providing employment for vocational high school graduates.

Readiness is an important key before someone enters the workforce. Someone who already has a job readiness will be more successful in pursuing his career. Readiness of work is all the individual conditions that include physical readiness, mental, broad insight and experience and the willingness to perform a job or activity. In addition to these three aspects, one's success towards his work is also supported by his love of work or often referred to as passion. Someone who loves his work will work diligently, passionately, creatively, and not depressed. In preparing the student as a workforce also need to provide motivation to learn that is contained in self and outside students to make changes in behavior. How important to cultivate student learning motivation, because its existence is very meaningful for the act of learning to the goals to be achieved. A student, who has been motivated to learn something, will try to learn it well and diligently, in the hope of getting good results. Good results will affect the readiness of the student's work. This should be nurtured early, which is to optimize the readiness of work in the field of Mechanical Engineering expertise.

Achievement Motivation can affect student learning activities so that in the end is an attempt to achieve the learning achievement field of mechanical engineering expertise optimally. Achievement Motivation can encourage workers / employees to further improve and maintain Achievement work in the field of mechanical engineering skills. When workers/employees have high Achievement Motivation will show higher value in their performance compared with vocational high school graduate workers who have low Achievement Motivation. But all that is not easy to achieve if from within the employees, in order to, have the impetus to carry out quality work. Empirically, the lack of motivation achievement of graduate worker of vocational high school competence of mechanical engineering skill can be known by the lack of readiness to work when given the task by the company in accordance with the competence of its expertise, in addition to the seriousness of the worker at the beginning of the job. At the time of the sub-manager direction of the field of mechanical engineering workers less responsive / attention to the explanation of the sub-manager, so when the assignment to the worker, the concerned does not maximize the task or the work is not complete.

2. Literature Review

2.1. The performance of vocational high school graduates

Performance is a manifestation of work done by employees who are usually used as a basis for assessment of employees or organizations [5]. Good performance is a step toward achieving organizational goals. Therefore, efforts should be made to improve performance. But this is not easy because many factors affect the high and low performance of a person. Performance is a result achieved by a person according to the size applicable to the work in question [6]. Something done or product / service produced or given a person or group of people [7]. Performance is a record of acquisition resulting from the function of a particular job or activity during a given period of work [8]. Performance refers to the level of achievement of tasks that make up a job employees. Performance reflects how well employees meet the requirements of a job [9]. Performance is a real behavior that displayed every person as work performance generated by employees in accordance with its role in the company [10]. Employee performance is a very important thing in the company's efforts to achieve its goals. From some of these descriptions, it can be argued that performance is the real work achieved by a person in carrying out the tasks assigned to him in accordance with the criteria and objectives set by the organization. Almost all means of performance appraisal consider the following: (1) Quantity is the amount to be completed; (2) Quality is the quality produced; (3) The timeliness in accordance with whether or not the time has been planned [7]. Furthermore, Performance appraisals should not be understood narrowly, but can produce a wide range of performance types that are measured in various ways. The key is to frequently measure performance and use that information for mid-period correction [9]. The performance of employees is actually assessed on five dimensions such as: (1) Quality, (2) Quantity, (3) Project completion, (4) Cooperation, (5) Leadership. Elements of performance assessed as follows: (1) Loyalty, (2) Job performance, (3) Responsibility, (4) Obedience, (5) Honesty, (6) Initiatives, (7) Leadership. From some of these descriptions, it can be argued that performance is the real work achieved by a person in carrying out the tasks assigned to him in accordance with the criteria and objectives set by the organization [11].

2.2. Achievement motivation

Motivation comes from the Latin *Movere* which means to push or move. The word motivation and motive are used alternately in explaining the impulse that drives a person to behave. Motivation is the desire that there is an individual who stimulates him to perform the actions or something that becomes the basis or reason someone behaves. Motivation work can be interpreted as a desire or need that lies behind a person so he was motivated to work. Therefore, various motives appear in someone, so motivated to do something. As it is understood that motives are internal conditions that move, control and direct individuals toward certain goals. Motivation is the whole process of encouraging / improving one's motives in such a way that he is motivated to do something (behave) with eager to achieve the desired goal in an effort to satisfy his needs [12]. A person's motivation is determined by the intensity of his motive. An important question of job for a managerial leader is: "How to motivate the work of his group members". Motivation is a psychic process that encourages a person to do something. Motivation is a tool superior for subordinates willing to work hard and work smart as expected. Knowledge of motivation patterns helps managers understand the work attitude of their employees. Motivation can come from within and outside of a person. Motivating yourself let alone motivating others or subordinates which is not an easy task. Job routines often lead to deep saturation that can degrade the motivation of achievement, which is exacerbated by unsupportive working conditions. Willingness can be overcome to conduct training. The basic factor is the need that must be met. In line with the development of motivation developed theories of motivation. Driving factors are things that drive the achievement that is intrinsic, which means sourced from within a person for example: cleanliness, recognition, creativity, responsibility. And maintainer factor is extrinsic factor which means sourced from outside oneself e.g.: Work environment, work incentive, working relationship, work safety. Motivation can be interpreted as a process of psychology that reflects the interaction between attitudes, needs, perceptions, and decisions that occur in a person [13].

The process of psychology arises due to the factor within a person itself is called intrinsic and extrinsic. Factors within a person can be personality, attitude, experience and education, or expectations, aspirations that reach into the future while factors from outside the self can be generated by sharing other factors are very complex. But stimulation, subordinate behavior in organizational life is essentially task-oriented. That is, that the behavior of subordinates are usually driven by the desire to achieve goals must always be observed, supervised, and directed in the framework of execution of tasks in achieving organizational goals. A person's motivation is influenced by the intrinsic power stimuli present in the individual / individual, the external stimuli may also affect motivation, but the motivation itself reflects the individual's reaction to the stimuli. Furthermore, Motivation is closely related to learning theory [14]. Before argues that many needs are derived from culture. Three of the needs are: 1) the need for achievement (need of achievement) abbreviated, 2) the need for affiliation (need of affiliation) abbreviated as Aff and 3) the need for power (abbreviated as Power1q). Achievement motivation is an inner drive to overcome all challenges and obstacles in an effort to achieve goals. The motivation of affiliation is the urge to connect with others or the urge to have as many friends as possible. Motivation in power is the urge to influence others to submit to their will. Work motivation is something that raises the spirit or encouragement and work. Therefore, Motivation work in psychology as a morale booster. Motivation is giving power that creates the excitement of one's work, so they will cooperate, effectively and integrated with all its efforts to achieve satisfaction [5]. Also, the motivation is the desire that is in the individual an individual who stimulate him to perform the actions.

2.3. Readiness work

Readiness according to the dictionary of psychology is the level of development of maturity or maturity that is profitable to practice something [15]. It is also suggested that readiness includes the ability to position itself if it will initiate a series of movements related to mental and physical readiness. Defines readiness as follows: Readiness is the whole condition that makes it ready to respond in some way to a tendency to respond. Conditions include at least three aspects: (1) physical, mental and emotional conditions, (2) needs, motives and goals, (3) skills, knowledge and other understandings learned [16]. Readiness is a fairly good ability physically and mentally [17]. Physical readiness means adequate energy and good health, while mental readiness, has enough interest and motivation to perform an activity, Readiness is the level or circumstance that must be achieved in the process of individual development at the level of growth mental, physical, social and emotional [18]. Job Readiness is the ability of a person to carry out work well within and outside the employment relationship to produce goods or services. Readiness Work is the ability of a person to complete a particular job, Job Readiness is the ability, skill, and work attitude that is in accordance with the demands of society and in accordance with the potentialities of students in various types of specific jobs that are directly applicable [19]. Readiness A person's work is not just what he has done, but a job or position that is truly appropriate and compatible with the potentials of those who hold him, so that everyone who holds the job he holds will be happy to shake him and then they will do their best to improve their performance, develop their own potential, their environment, and the necessary infrastructure to support their work. Based on the above opinion, it can be concluded that Readiness Work is the overall condition of individuals who cover physical, mental and experience and the willingness and ability to perform a job or activity. Job Readiness includes the desire and ability to perform a job and to seek a certain activity, in this case depending on the level of maturity, past experience, mental state and emotions of a person. Before passing maturity and behavior, Readiness Work cannot be owned though through intensive training and quality.

The main purpose of vocational learning is to prepare learners to become successful workers in the workforce [20]. Therefore, graduate of Vocational High School is expected to be a successful worker in the world of work, both as a worker and entrepreneur. The Work Readiness Program is a program-based competency that utilizes the learning experience to enable students to work well while being supervised by their work components [21]. This program should be done by all vocational education, especially vocational education so that the main purpose of vocational high school can be realized. Based on the above opinion, it can be concluded that Readiness Work is the main goal of vocational high school. So this research is done in vocational high school Bitung City which also has a goal which creates graduated from vocatinal education ready for work.

3. Methods

This research was conducted on vocational graduate workers in Bitung City industry from January to November 2016. Purposive samples were 37 people and distribution of sample members (vocational graduate workers) in Bitung City with a population of 5 industries of mechanical engineering as many as 47 people. PT Unggul Setia Abadi as many as 15 people, PT Dok Kelapa Dua as many as 7 people, PT Pathemaang Dock Yard as many as 3 people, PT Industri Kapal Indonesia as many as 10 people, PT Samudra Puranable as many as 12 people.

4. Results and Discussion

The results of the three research hypotheses proved to reject Ho. Readiness work variables (X1), and achievement motivation (X2), are positively and significantly correlated with graduate performance (Y) can be explained as follows: First, through simple regression and correlation analysis on work attitude variable with performance of vocational graduate (worker performance) we get the equation of regression line $\hat{Y} = 26.406 + (0.764)\ X1$ and correlation coefficient equal to 0.764 which each significant $\alpha = 0.01$. Both of these variables are positively and significantly correlated, where 53.8% variance in SMK graduate performance can be explained by Readiness Work of regression equation $\hat{Y} = 26.406 + (0.764)\ X1$. This proves that Readiness Work is one of the predictors of graduate performance.

The result of regression analysis and simple correlation on Readiness Work variable correlated positively and significantly with the performance of graduates obtained by equation of regression line $\hat{Y} = 26.406 + (0.764)$ X1 and correlation coefficient 0.734 each significant at $\alpha = 0.01$. This proves that Readiness Work is one of the predictors of graduate performance. Both of these variables are positively and significantly correlated, where 28% variance that occurs in the performance of SMK graduate workers can be explained by Readiness Work through regression line equation $\hat{Y} = 26.406 +$ (0.764) X1. This shows that a good working attitude is a dominant element in determining the level of performance of vocational graduate workers. The results research showed that graduate program completers who had an internship and an assistantship were significantly different from those who did not in terms of their perceived work readiness [22]. The results of the vocational graduates job readiness is influenced by the experience, insight of the world of work and vocational competence through employability skill [23]. Second, through simple regression and correlation analysis of achievement motivation variable with the performance of graduates obtained equation of regression line $\hat{Y} = 16.665 + (0.123)$ X2 and correlation coefficient equal to 0,811 which each significant $\alpha =$ 0.05. Both of these variables are positively and highly significant, where 65.8% of the variance that occurs in the performance of graduates can be explained by the level of motivation preceding the equation of the regression line $\hat{Y} = 16.665 + (0.123)$ X2. This proves that achievement motivation is one of the predictors of graduate performance.

Third, from result of regression analysis and double correlation on work attitude and achievement motivation together with graduate performance obtained by regression line equation = 14.45 + 0.28X1 + 0.21X2 and correlation coefficient 0.854 each significant at $\alpha = 0.05$. The independent variables and dependent variables are positively and significantly correlated, where 73% of the variance in the performance of graduates can be explained together through work attitude and achievement motivation through the regression equation.

5. Conclusions

The results of research can be concluded as follows: (i) there is a significant relationship between work attitude with the performance of vocational graduates, so a good work attitude can determine the performance level of vocational graduates; (ii) there is a significant relationship between achievement motivation and the performance of vocational graduates; and (iii) there is a significant relationship between job readiness and achievement motivation with the performance of vocational graduates in Bitung City.

References

[1] Sasongko A 2017 *Lulusan SMK Harus Berkualitas dan Siap Kerja* http://www.republika.co.id/berita/pendidikan
[2] Apriliananda D 2017 *Penyebab Lulusan SMK Binaan Honda Siap Kerja* http://ekonomi.kompas.com/read/2017/08/02/penyebab-lulusan-smk-binaan-honda-siap-kerja
[3] Siregar D P 2015 *Pengangguran dan Kesiapan Kerja* https://www.kompasiana.com
[4] Tigauw N L 2015 *Lulusan SMK Manado Siap Kerja Industri* https://manado.antaranews.com/berita/23958/lulusan-smk-manado-siap-kerja-indu
[5] Hasibuan, Malayu S P 2011 Manajemen Sumber Daya Manusia (Jakarta: Bumi Aksara, Jakarta).
[6] As'ad M 2012 *Psikologi Industri* (Yogyakarta: Libery).
[7] Dharma A 2001 Manajemen Prestasi Kerja (Jakarta: Rajawali).
[8] Bernardin H J, Russel J E A 2007 *Human Resources Management: An Experimental Approach* (New York: Mc Graw Hill).

[9] Simamora H 2004 *Manajemen Sumber Daya Manusia* (Yogyakarta: STIE YKPN).
[10] Veithzal R 2010 *Manajemen Sumber Daya Manusia untuk Perusahaan: dari Teori ke Praktik* (Jakarta: Raja Grafindo Persada).
[11] Tohardi A 2002 *Pemahaman Praktis Manajemen Sumber Daya Manusia* (Jakarta: CV. Mandar Maju).
[12] Kambey D C 2006 *Landasan Teori Administrasi/Manajemen (Sebuah Intisari)* (Manado: Yayasan Tri Ganesha Nusantara).
[13] Wahjosumidjo 2010 *Kepemimpinan Kepala Sekolah* (Jakarta: PT Rajagrafindo Persada).
[14] Husaini U 2008 *Manajemen* (Jakarta: Bumi Aksara).
[15] Chaplin J P 2006 *Kamus Lengkap Psikologi* (Jakarta: PT. RajaGrafindo Persada).
[16] Slameto 2010 *Belajar dan Faktor-faktor yang Mempengaruhinya* (Jakarta: Rineka Cipta).
[17] Dalyono 2015 *Psikologi Pendidikan* (Jakarta: Rineka Cipta).
[18] Hamalik O 2014 *Psikologi Belajar Mengajar* (Bandung: Sinar Baru Algesindo).
[19] Sofyan H 2006 *Implementasi Pembelajaran Berbasis Proyek pada Bidang Kejuruan* (Cakrawala Pendidikan Yogyakarta: LPM UNY).
[20] Finch, Curtis R, Crunkilton J R 1999 *Curriculum Development in Vocational and Technical Education Planning, Content, and Implementation* (5thed) (Bostom: Allyn and Bacom).
[21] Danielson 2008 *Pendidikan Kejuruan* (Jakarta: Gramedia).
[22] Doe R 2015 *Work Readiness among Graduate Students Louisiana State University and Agricultural and Mechanical College*, (rdoe1@lsu.edu, Doctoral Dissertations). https://digitalcommons.lsu.edu/
[23] Putriatama E, Patmanthara S, Sugandi R M 2016 Work readiness by vocational school graduates viewed from industrial work practice's experience and vocational skills *AIP Conference Proceedings* **1778(**1).

Performance of Savonius Blade Waterwheel with Variation of Blade Number

L Sule[1*] and P T D Rompas[2]
[1]Universitas Hasanuddin, Makassar, Sulawesi Selatan, Indonesia
[2]Universitas Negeri Manado, Tondano 95618, Sulawesi Utara, Indonesia

*luther.sule@yahoo.co.id

Abstract. The utilization of water energy source is mainly used as a provider of electrical energy through hydroelectric power. The potential utilization of water flow energy is relatively small. The objective of this study is to know the best blade of Savonius waterwheel with various variables such as water discharge, blade number, and loading. The data used the efficiency of waterwheel, variation of blade number, variable water discharge, and loading in the shaft. The test results have shown that the performances of a top-water mill with the semicircular curve where the variation in the number of blades are 4, 6, and 8 at discharge and loading of 0.01587 m^3/s and 1000 grams respectively were 9.945%, 13.929%, and 17.056% respectively. The blades number of 8 obtained the greatest performance. The more number of blades the greater the efficiency of the waterwheel Savonius.

1. Introduction

The condition of Indonesian topography which has many mountains and hills as well as the stretch of river is almost found in every region [1]. Water energy is the energy that is suitable and the most potential to be developed in Indonesia. The rate of growth of hydropower in Indonesia is very slow, whereas the potential of Indonesia's hydro power is quite large reach 75.000 MW. Utilization through the national electricity supply only reached 10.1% or 7,572 MW. Based on data from the Ministry of Energy and Mineral Resources, the potential of hydropower energy is spread by 15,600 MW in Sumatra, 4,200 MW in Java, 21,600 MW in Kalimantan, 10,200 MW in Sulawesi, 620 MW in Bali-NTT-NTB, 430 MW in Maluku and 22,350 MW in Papua. While in the world, the potential for water energy is estimated to reach 657 million HP or 489,924.8156 MW, but the utilization to 15%. The potential of water energy in each continent is different [2] and we can be seen in the Table 1.

The greatest potential energy of water is in the African continent [3]. It can be seen with the river Congo in Africa which became the largest potential energy of water in the world as well as several other rivers that are used as a producer of electricity. Electric power is obtained from the conversion of hydropower that turns the waterwheel or water turbine that utilizes the waterfall or stream in the river [4].

Fine the optimum performance of each specific amount of blade, i.e. 4, 6 and 8 blades on the water (discharge) and the constant end equal head for Savonius model blade of capacity used for hydropower. Determine the maximum power coefficient of the change in the number of Savonius model drops used for hydropower [5]. As reference in the design of water wheel with the Savonius model savings benefits research. The research results can be applied in the field of renewable energy conversion, especially in the design of small-scale hydro power plant to micro [6].

Table 1. Potential and developed hydropower in the world

Continent	Potential (in million horse power)	Percentage of total	Developed (in million horse power)	Percentage of total
Africa	212	41.4	0.6	0.6
Asia	151	23.0	13.7	13.6
North America	87	13.2	41.1	40.8
Europe	69	10.5	40.8	40.5
South America	55	8.4	3.1	3.1
Oceania	23	3.5	1.4	1.4

2. Methods

This study used experiment in fluid machineries laboratory of Mechanical Engineering Department, Universitas Hasanuddin with the installation as in Figure 1. The data retried by repeating the test/retrieval procedure at least 5 repetitions for analysis. The data collection procedure/testing such as we conducted by: the initial, check the state of the tool to be used on the waterwheel test and check the valve is in good condition; connecting the pump with a power source; set the valve opening to determine the discharge; allow the water to flow for 2 minutes to obtain stable flow conditions in the channel; calculate the amount of first discharge by using bucket and stopwatch; install one of the waterwheels with the number of 4 pieces of blade in position; record the water level and water temperature in the channel; after that calculate the number of turns of the waterwheel using a tachometer without loading until loading until the wheel cannot spin or stop; after that record the amount of rounds generated in each loading; after obtaining the data from the waterwheel test the 4 pumps are turned off by disconnecting the electric current source; and the finally, repeat procedure 2-10 for waterwheels 6 and 8 (Figure 2).

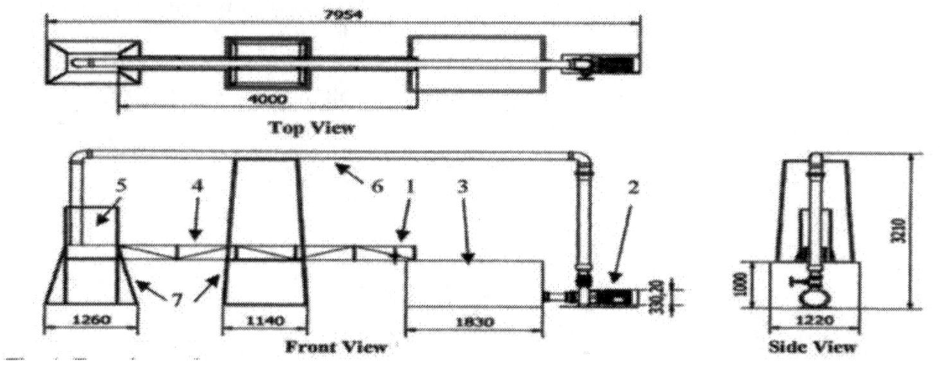

Figure 1. Testing installation

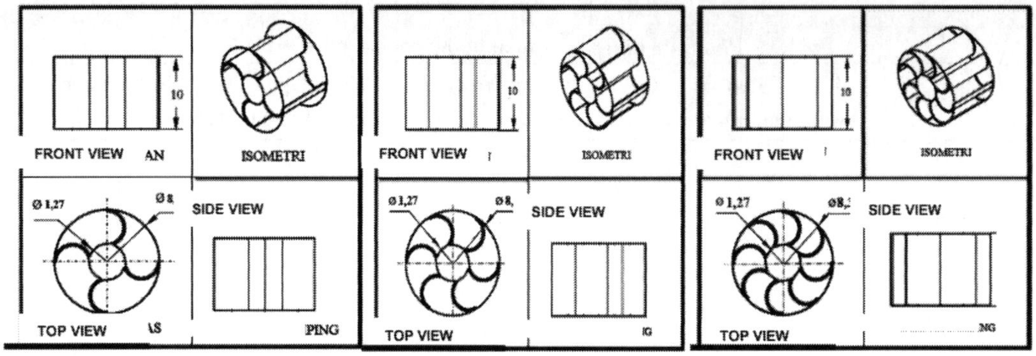

Figure 2. Parameters of Blades (4, 6, and 8)

3 Results and Discussion

3.1. Relation of efficiency to load for 4 blades

The Figure 3 and Table 2 showed the relation of efficiency to the waterwheel load with variation of discharge i.e. 0.01124 m^3/s (Q1), 0.01358 m^3/s (Q2), and 0.01587 m^3/s (Q3) with waterwheels of 4 blades. The maximum efficiencies occurred at 600 gram loading to discharge of 0.01124 m^3/s and 0.01358 m^3/s are 6.13% and 5.95% respectively. While, at discharge of 0.01587 m^3/s and loading of 1000 gram, the maximum efficiency is 9.66%.

In Figure 5, for 0.01124 m^3/s of discharge, it is seen that the efficiency increases from no load to 600 gram loading and decreases the efficiency of the waterwheel as the addition of loading, from load 600 to 1200 gram loading. The waterwheel will produce power because the wheel can offset the given torque [7][8][9]. When a given torque is equal to zero, the waterwheel will not produce power because the windmill will spin very quickly as a result of the absence of loading given so that no incubation occurs [10][11]. Conversely, after reaching the critical point along with the increase in loading then braking occurs in the pulley resulting in angular velocity and the power of the waterwheel decreases [12][13][14]. The phenomenon that occurs for discharges of 0.01124 and 0.01358 m^3/s is caused by the same thing that occurs at a discharge of 0.01587 m^3/s for the same blade.

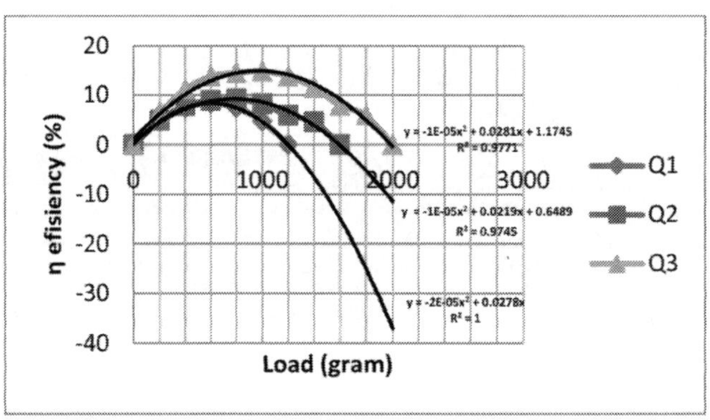

Figure 3. Relation of efficiency to load with variation of discharge on waterwheel with 4 blades

In Figure 4 and Table 2, it is found that the discharge of 0.01587 m^3/s is the largest discharge between the variations of the discharge used where the efficiency of 9.95% to loading of 100 gram. This is because the amount of water flow that flows directly proportional to the water speed in the channel makes the mass of the waterwheel lighter so that the resulting round is also large and requires a large load to brake [15].

IOP Conf. Series: Materials Science and Engineering **306** (2018) 012073 doi:10.1088/1757-899X/306/1/012073

Table 2. η efficiency to load and discharge

Load	Variation for 4 blades			Variation for 6 blades			Variation for 8 blades		
(gram)	η (Q1)	η (Q2)	η (Q3)	η (Q1)	η (Q2)	η (Q3)	η (Q1)	η (Q2)	η (Q3)
	0	0	0	0	0	0	0	0	
0	3,213	3,176	3,552	3,797	3,970	4,691	4,674	4,975	
200	5,258	5,161	6,251	6,427	6,749	8,244	8,179	9,097	
400	6,134	5,955	8,099	7,887	8,338	11,087	9,640	12,366	
600	4,674	6,352	9,661	8,179	8,735	12,508	10,516	14,782	
800	2,921	5,955	9,946	7,303	8,933	13,503	10,224	16,346	
1000	0	4,764	9,377	5,258	8,338	13,645	8,763	17,056	
1200		4,169	7,956	4,089	6,948	13,929	6,134	16,914	
1400		0	6,820	0	7,764	13,645	4,674	15,919	
1600			5,115		3,573	12,792	0	15,351	
1800			2,842		0	9,950		2,792	
2000			0			7,814		9,381	
2200						5,117		6,823	
2400						0		3,696	
2600								0	
2800									

Note: variation of discharge i.e. 0.01124 m3/s (Q1), 0.01358 m3/s (Q2), and 0.01587 m3/s (Q3

3.2. Relation of efficiency to load for 6 blades
The relation of efficiency to the waterwheel load with variation of debit i.e. 0,01124 m3 / s, 0,01358 m3 / s and 0,01587 m3 / s with waterwheels 6 blades is shown in Figure 5 and Table 3.

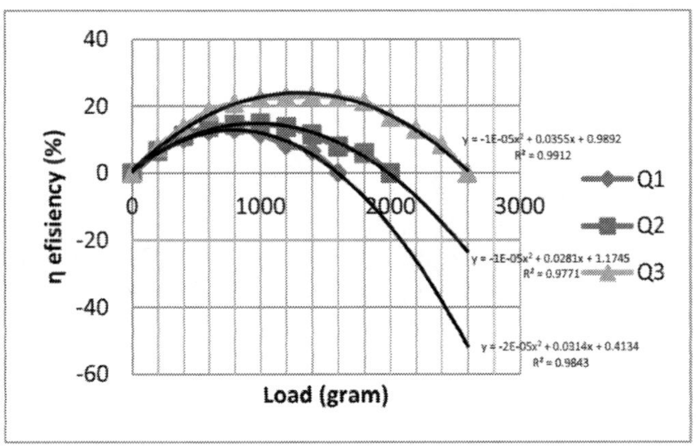

Figure 4. Relation of efficiency to load with variation of discharge on waterwheel with 4 blades

Based on Figure 4 and Table 3 were found that for discharge of 0.01124 m^3/s, the maximum efficiency is 8.17920093% which occurs at 800 gram loading. While at discharge of 0.01358 m3 / s maximum efficiency is obtained 8.93% with loading 1000 gram. While at 0.01587 m^3/s of discharge, the maximum efficiency at loading of 1400 gram is 13.93%.

In Figure 5, for 0.01124 m^3/s of discharge, it was seen that the efficiency increased from no-load to 800 gram loading and decreased the efficiency of the waterwheel as the addition of loading, from a load of 800 to 1600 gram loading. The waterwheel will produce power because the wheel can offset the given torque [7][8][9]. When a given torque is equal to zero, the waterwheel will not produce power because the windmill will spin very quickly as a result of the absence of loading given so that no incubation occurs [16]. Conversely, after reaching the critical point along with the increase in loading then braking occurs in the pulley resulting in angular velocity and the power of the waterwheel decreases [17]. The phenomenon that occurs for discharges of 0.01124 and 0.01358 m^3/s is caused by the same thing that occurs at a discharge of 0.01587 m^3/s for the same blade.

3.3. Relation of efficiency to load for 8 blades

Based on Figure 5 and Table 4 it is found that for debit 0.01124 m^3/s maximum efficiency is 10.52% which occurs at 800 gram loading. While at debit 0.01358 m^3/s maximum efficiency is obtained by 13.90% with loading 1000 gram. While at 0.01587 m^3/s of discharge, the maximum efficiency at loading of 1200 gram is 17.06%.

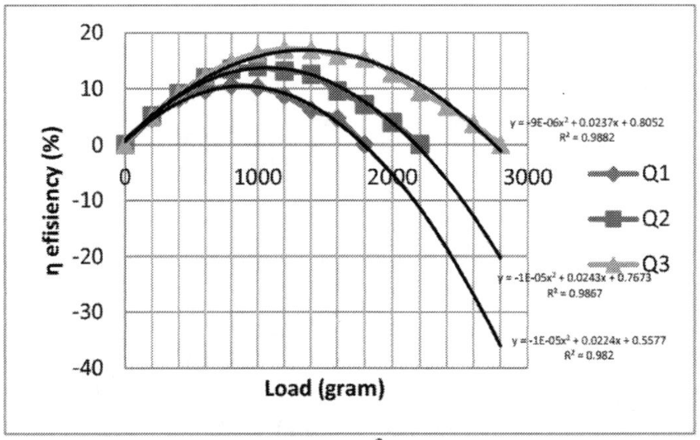

Figure 5. Distributions of the available power per m^2 at seawater column of 20 m when low tide currents at flow rate of 0.3 Sv in the Bangka strait.

In Figure 6 for a 0.01124 m^3/s of discharge, it is seen that the efficiency increases from no-load to 800 gram loading and decreases the efficiency of the waterwheel as the addition of loading from a load of 800 to 1600 gram. The waterwheel will produce power because the wheel can offset the given torque [7][8][9]. When a given torque is equal to zero, the waterwheel will not produce power because the windmill will spin very quickly as a result of the absence of loading given so that no incubation occurs [16]. Conversely, after reaching the critical point along with the increase in loading then braking occurs in the pulley resulting in angular velocity and the power of the waterwheel decreases [17][18][19]. The phenomenon that occurs for discharges of 0.01124 and 0.01358 m^3/s caused by the same thing that occurs at discharge (0.01587 m^3/s) for the same blade [20].

4. Conclusions

The efficiency produced by the waterwheel Savonius blade is affected by the discharge and the amount of the blade. In the waterwheel of 4 maximum efficiency blade obtained 9.95% with debit 0.01587 m^3/s at loading 1000 gram. While on the waterwheels 6 blades, the greatest efficiency of 13.93% is obtained at discharge of 0.01587 m^3/s at 1400 gram loading. While on the waterwheels 8 blades, the greatest efficiency of 17.06% is obtained at a discharge of 0.01587 m^3/s for with a loading of 1200 grams. The maximum power of the wheel produced by the water wheel on the flat plate blade occurs at the opening of the discharge III (Q3 = 0.01587 m^3/s) with the number of 8 blades of 2.403984 W. The greatest efficiency was obtained at 0.01587 m^3/s discharge of 17.06% with 1200 gram loading. So the best blade is a waterwheel with 8 blades.

IOP Conf. Series: Materials Science and Engineering **306** (2018) 012073 doi:10.1088/1757-899X/306/1/012073

References

[1] Patty O F 1995 *Tenaga Air* (Erlangga: Jakarta)
[2] http://www.yourarticlelibray.com/powers/distribution-of-hydro-power-in-the-word/25579/
[3] Robert H D 2002 *Water Wheel,* E-Maill : *www.waterhistori.org*
[4] Szytula A and Leciejewicz J 1989 *Handbook on the Physics and Chemistry of Rare Earths* **12** ed K A Gschneidner Jr and L Erwin (Amsterdam: Elsevier) 133
[5] Nakajima M, Iio S and Ikeda T 2008 Performance of Savonius rotor for environmentally friendly hydraulic turbine *Journal of Fluid Science and Technology* **3** 3 420-429
[6] Kyozuka Y 2008 An experimental study on the Darrieus Savonius turbine for the tidal current power generation *Journal of Fluid Science and Technology* **3** 3 439-449
[7] Esty J 2005 *Water wheels* (Peabody museum of natural history: Yale University)
[8] Samman M 2005 *Rediscovering the waterweel "Noria Al-Muhammadiyah"* (Asme Internasional Histori Lanmark; Asme Northwest Houston Subsection Meeting)
[9] Kusnaidi 1999 *Karakteristik Kincir Air* (online) available at http://anonime.co.id
[10] Gerholdt M 2009 *Modeling the chaotic waterwheel* Department of Mathematics Graduate Student Presentation, Spring
[11] Gerald M, Cristian W 2004 *The breastshot waterwheel: design and model tests* (Berlin, Germany: Researcher, Leibniz-Institute of Freshwater Ecology and Inland Fisheries)
[12] Hansen R D 2002 *Water wheel* (online) available at www.waterhistory.org
[13] Pelz P F and Bearrf T 2006 *Studies on the maximum harvestable hydropower in open channel-aplication to a water wheel and a turbine* Technische Universitat Germany
[14] Radheika 2008 *Kincir pusar untuk irigasi* (online) available at http//radheika.com
[15] Tevata A and Inprasit C *2011* The effect of paddle number and immersed radius on water wheel performance *Energy Procedia* **9** 359-365
[16] Pujol T, Sola J, Montoro L and Pelegri M 2010 Hydraulic performance of an ancient Spanish watermill *Renewable Energy* **35** 387-396
[17] Ibrahim A G, Haron C C H and Ashari H C 2002 *Traditional water wheels as a renewable rural energy.* Department of Mechanichanical and Material Engineering, Unversiti Kebangsaan Malaysia, Bangi Darul Ehsan Selangtor 436000: Malaysia
[18] Pujol T and Montoro L 2010 High hydraulic performance in horizontal waterwheels *Renewable Energy* **35** 2543-2551
[19] Denny M 2004 The Efficiency of overshot and undershot waterwheels *Eur.J.Phys.* **25** 193-202
[20] Choi Y D, Lim J I and Lee Y H 2008 Performance and internal flow characteristics of a cross-flow hydro turbine by the shapes of nozzle and runner blade *Journal of Fluid Science and Technology* **3** 3 398-409

IOP Conf. Series: Materials Science and Engineering **306** (2018) 012074 doi:10.1088/1757-899X/306/1/012074

Designing an Elderly Assistance Program Based-on Home Care

L Umusya'adah*, A Juwaedah, Y Jubaedah, H Ratnasusanti and R H Puspita

Technical and Vocational Education, Universitas Pendidikan Indonesia (UPI)

*liskaumusyaadah@student.upi.edu

Abstract. PKH (Program Keluarga Harapan) is a program of Indonesia's Government through the ministry of social directorate to accelerate the poverty reduction and the achievement of Millennium Development Goals (MDGs) target as well as the policies development in social protection and social welfare domain or commonly referred to as Indonesian Conditional Cash Transfer (CCT) Program. This research is motivated that existing participants of the family expectation program (PKH) that already exist in Sumedang, Indoensia, especially in the South Sumedang on the social welfare components is only limited to the health checking, while for assisting the elderly based Home Care program there has been no structured and systematic, where as the elderly still need assistance, especially from the family and community environment. This study uses a method of Research and Development with Model Addie which include analysis, design, development, implementation and evaluation. Participants in this study using purposive sampling, where selected families of PKH who provide active assistance to the elderly with 82 participants. The program is designed consists of program components: objectives, goals, forms of assistance, organizing institutions and implementing the program, besides, program modules include assisting the elderly. Form of assistance the elderly cover physical, social, mental and spiritual. Recommended for families and companions PKH, the program can be implemented to meet the various needs of the elderly. For the elderly should introspect, especially in the health and follow the advice recommended by related parties

1. Introduction

The success of health development indicated by the increase of several aspects, for example human resources quality, life quality, family and community welfare, and life expectancy. One of the concerns is the increase of community life expectancy that will influence the increase number of elderly. The increase of life expectancy has an impact on the increase of elderly in the age of 60 and above. Population ageing has become a notable and common demographic phenomenon in most countries. The proportion of the population aged 60 years or over is growing substantially, with the number of elderly in the world projected to grow by 56%, from 0.9 billion to 1.4 billion, between 2015 and 2030. Research has also shown that there will be 2.0 billion elderly adults by 2050, constituting an even larger share of society, with about 80% of them living in developing countries [1].

The increase of elderly population leads to the elderly problems includes problems physically, socially, psychologically, and economically. Actually, family plays a very important role to return elderly's trust to feel that they still needed and capable of something, both in family or society, so they will reach prosperity both physically and mentally for the rest of their life, in other words they can perform their social function well and improve their life quality [2]. In fact, many elderly families who do not understand the elderly needs, considering that elderly needs not only on eating, drinking, and maintaining physical health, more than that, family awareness is really needed in completing other needs [3]. Therefore, to increase elderly's life quality and social function, elderly needs to be accompanied by someone who has certain characteristic and approach that can be well received by the elderly. To overcome the elderly problems above, the government through the Ministry of Social Directorate of Elderly Social Service (Dit. PSLU) in Republic Indonesia has conducted a pilot program of elderly assistance and care in family environment (home care), adopted a program of home care help age in Korea. The trial of this program (home care) supported by the Decree of the Minister of Social Affairs No. 67/HUK/2006 on Guidelines for elderly assistance and social care in family environment (Home care).

Home care is a program of elderly assistance and social care outside orphanage. Based on Ministry of Social Affairs in Republic Indonesia, home care is 'a form of elderly care and social service in family environment/home as a form of concern to elderly by prioritizing the role of family-based community'. Home care as a form of elderly care and social service in family environment or home can meet the elderly needs because family environment actually has a significant role in returning elderly's trust, so they feel still needed and capable of something, both in family and society, so the elderly can perform their social functions well and improve their life quality [3].

PKH (Program Keluarga Harapan) is a program of Indonesia's Government to accelerate the poverty reduction and the achievement of Millennium Development Goals (MDGs) target as well as the policies development in social protection domain. PKH is a cash grant program to RTSM (Very poor Household) as long as they meet the requirements related to effort of the improvement human resources quality, include education and health. PKH is an indirectly program that provides home care services has responsibility to manage elderly assistance activities conducted by family as a primary companion in conducting home care based elderly assistance [4].

The holder of main role for home care program is the family members of the elderly [5]. If there are no family members of the elderly, it may involve the community members living in the same neighborhood as the elderly who need assistance or care in the family environment. PKH can be said as an empowerment program that has a companion agency, responsible to assist the community [3].

Companions are needed not only as program owners but also facilitators and educators for the community. They are also responsible to help community in giving information so the community can make the right decision. Companions as agents of renewal are not only responsible to share new ideas to community. They also can be called as professionals or extension officers from agencies or institutions that wish to empower the community [4].

Companions in PKH are the executors of the program and accompany the community in activities of every program. A community facilitator must have strategy competence and mastery in helping community. This is important for the sustainability of PKH empowerment program. In order to the community to participate in the empowerment program, PKH companions have to be able to motivate PKH participants to achieve independence and self-empowerment. Particularly to motivate participants to fulfill the obligations in education, health, and social welfare components which is improving the participation of compulsory primary education for nine years, health checking for children and elderly regularly and assisting the elderly life quality and social function.

The participation of PKH participants that already exist in Sumedang Region, especially in the South Sumedang District on the social welfare components is only limited to the health checking, to the program of home care based elderly assistance, there is no structured and systematic program yet, whereas elderly needs assistance, especially assistance from the elderly family and surrounding community. Home care based program are designed to be more specialized for PKH participants who have a non-potential elderly target, because the non-potential elderly according are elderly that find difficulties to earn living, so that their life is depend on the help of others, therefore, the family role is very important, because family is the greatest source of support which is useful to meet the basic human needs of the elderly [6].

Home care based elderly assistance is one of the solutions to improve the elderly functions in the family environment [5]. However, Home care is not fully known and understood by the elderly family and community, so the design of home care based assistance program is needed which is suitable with the characteristics and needs of the elderly. The design of home care based elderly assistance program starts from identifying the needs that suitable with the elderly characteristics, and then it will be go on designing programs that can meet the needs of the elderly through various forms of Home care services by involving the role of the elderly family, PKH companions, and the surrounding community. Home care based assistance for the elderly is a research subject, related to the great benefits provided to the elderly needs through assistance with the involvement of the elderly family, PKH companions, and community environment. The selection of this problem is in line with the researcher's knowledge and understanding as an educator of social worker, especially when the researchers learn about an elderly concept and its services which are obtained from the family welfare education major, Ministry of Family Welfare Education, FPTK, UPI Bandung.

Based on the study of the above literature can be divided the concept of the brief that the concept in research as follows:

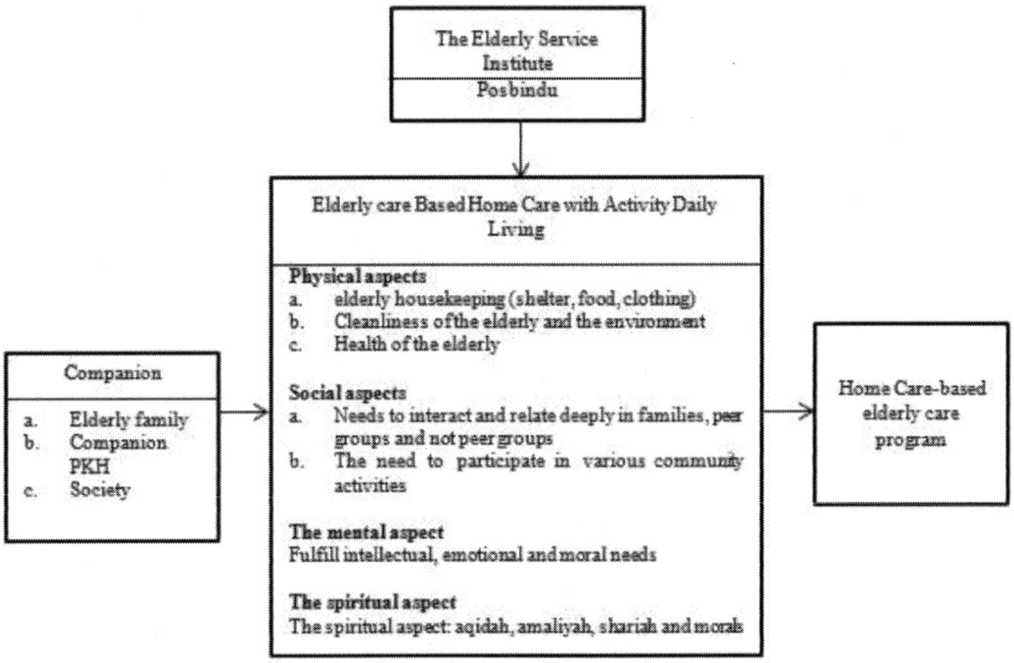

Figure 1. Research Framework

2. Methods

The method that is used in this research is Research and Development method in Addie Model. The stages of the research that will be conducted using Addie model are as follow:

2.1. Analysis

The analysis stage in this research is the activity of identifying the needs of home care based assistance program that is appropriate to the elderly characteristics of PKH participants in South Sumedang District.

2.2. Design

The design stage in this research is organizing the design of home care based assistance program based on the result of the program needs identification.

IOP Conf. Series: Materials Science and Engineering **306** (2018) 012074 doi:10.1088/1757-899X/306/1/012074

2.3. Development
The development stage in this research is developing home care based assistance program regarding to the elderly needs of PKH participants in South Sumedang District. After the program is designed, it will be validated using expert judgments that experienced to assess the designed program, so we will find both the advantages and disadvantages of the program. Consequently in this stage, if the program has been validated so the disadvantages can be found, so the improvement design will be made according to the direction of experts.

2.4. Implementation
The implementation stage is conducted to apply the home care based assistance program for the elderly.

2.5. Evaluation
The evaluation stage in this research is used for seeing the responses and the user ratings after the implementation of home care based assistance program for the elderly of PKH participants in South Sumedang District.

Designing a home care based elderly assistance program can be described as follows :

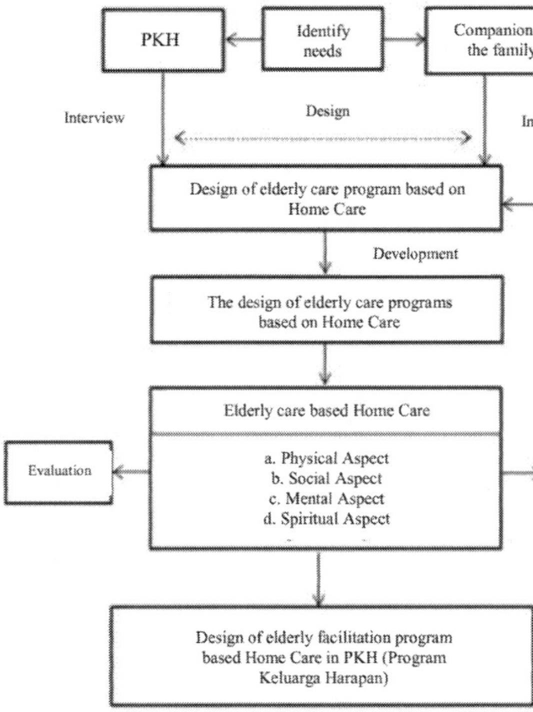

Figure 2. Research design

The participants of this research used purposive sampling, which is devoted to the family of PKH participants who provide active assistance to the elderly. Interview will be conducted to 82 families of elderly scattered in South Sumedang District.

Home care based elderly assistance program has been designed, then the program design will be validated using the expert judgment by elderly experts, include academics; three lectures and three PKH assistants who have understanding about elderly and have experiences as well as active in social service activities, so the advantages and the disadvantages will be gathered.

The research will be located in South Sumedang District, Sumedang regency, so the trial of home care based elderly assistance program will be conducted by the elderly family and PKH assistants in providing assistance to elderly.

The research instruments used are interview guidelines, expert judgment, and program observation guidelines. Interviews were conducted to the elderly family and PKH assistants as the identification subjects of the program design needs.

Validation format of expert judgment is used to assess the program conducted by expert academics and practitioners of elderly service institutions. Observation guidelines are used when piloting the program by the elderly family and PKH assistants.

Data analysis used are reduction and data display to summarize the interview results as identification subjects of the program design needs, then validate data using expert judgment. The program test results are then explained and interpreted.

3. Conclusion

The form of home care based of elderly assistance program is formulated by considering the elderly needs as someone who has experienced a change that will affect the economic and social of elderly, so generally it will influence the activity of daily living. The form of home care based of elderly assistance program that is designed include physical, social, mental, and spiritual aspects where the family roles in elderly care includes: keeping or caring the elderly, maintaining and improving mental status, anticipating socio-economic status changes and giving motivation and facilitating spiritual need for the elderly.

The main executor of home care based of elderly assistance program is the elderly family. Family has a significant role in providing assistance for the elderly because family is the closest person who will spontaneously take a part as a care giver, when the family needs it. Family helps the elderly to adapt themselves to the development tasks that must be done, such as preparing self for a declining condition, fostering good relationships with the other elderly and society in comfort, preparing a new life as an elderly and preparing for the death of the spouse as well as the death of oneself.

References

[1] Healy J 2004 *The benefits of an ageing population* Canberra: Australia Institute

[2] Maryam S 2008 *Mengenal Usia Lanjut dan Perawatannya* Jakarta: Salemba Medika

[3] Nindika, Nova N 2011 *Hubungan Kompetensi Pendamping Program Keluarga Harapan (PKH) dengan Keberlanjutan Pendidikan Anak Peserta: Kasus di Kelurahan Balumbang Jaya, Kecamatan Bogor Barat, Kota Bogor* [Online]. Retrieved from ejournal.ipb.ac.id

[4] Kementerian Sosial R. I. 2014 *Modul Pendampingan Pelayanan Sosial Lanjut Usia* Jakarta: Direktrat Pelayanan Sosial Lanjut Usia

[5] United Nation 2012 *Department of Economic and Social Affairs Population Division* World Population Prospects Available at: http://esa.un.org/wpp/index.htm.

[6] Widyakusuma N 2013 Peran Pendamping Dalam Program Pendampingan Dan Perawatan Sosial Lanjut Usia Di Lingkungan Keluarga (Home Care): Studi Tentang Pendamping Di Yayasan Pitrah Sejahtera, Kelurahan Cilincing, Kecamatan Cilincing Jakarta Utara *Sosio Informa* **18** 2 211-2

IOP Conf. Series: Materials Science and Engineering **306** (2018) 012075 doi:10.1088/1757-899X/306/1/012075

The Attitude of Construction Workers toward the Implementation of Occupational Health and Safety (OHS)

L Widaningsih*, I Susanti, and T Chandra

The Department of Architecture Engineering Education, The Faculty of Technical and Vocational Education, Universitas Pendidikan Indonesia, Bandung, Indonesia

*Liswida@upi.edu

Abstract. Construction industry refers to one of the industries dealing with high accident rate. Besides its outdoor workplace involving many workers who usually work manually, the workers' work culture and less awareness of occupational health and safety (OHS) are attributed to the high accident rate. This study explores some construction workers who are involved in some construction projects in big cities such as Bandung and Jakarta. The questionnaire-given to the construction workers focusing on stone construction, wood construction, and finishing session-reveals that the construction workers knowledge and understanding of nine Occupational Health and Safety (OHS) aspects reach above 50%. However, does not appear to reflect their knowledge and understanding of Occupational Health and Safety (OHS). The results of Focus Group Discussion (FGD) and an in-depth interview show that the fallacious implementation of Occupational Health and Safety (OHS) is attributed to their traditional "work culture".

1. Introduction

The implementation of Occupational Health and Safety (OHS) in an industry is the main requirement that can influence the investment value, the product quality, the construction industry's survival, and competitiveness in a country. Globally, in a global industrial competition, industrial workers are required to pay attention to the Occupational Health and Safety (OHS). This is an attempt to maintain quality, productivity, and occupational health [1].

In fact, the implementation of Occupational Health and Safety (OHS) in Indonesia still obtains less attention. The high occupational accident rate reflects this condition. Many occupational accidents occur in construction service. Besides its outdoor workplace involving many workers who usually work manually, the workers' work culture and less awareness of occupational health and safety (OHS) are attributed to the high accident rate.

This study explores some occupational workers, the members of *Paguyuban Kuli Bangunan* (PAKUBA) community in Sudalarang village, Sukawening, Garut as the participant. They are often involved in some construction projects in big cities such as Bandung and Jakarta [2]. They are categorized by the field that they are focusing on. They are divided into a group focusing on stone construction, wood construction, and finishing session. The participant reveals that their knowledge and understanding of nine Occupational Health and Safety (OHS) aspects in terms of construction occupation mostly reaches above 50%. However, the implementation of Occupational Health and Safety (OHS) in practice does not appear to reflect their knowledge and understanding of

IOP Conf. Series: Materials Science and Engineering **306** (2018) 012075 doi:10.1088/1757-899X/306/1/012075

Occupational Health and Safety (OHS) [3]. The results of Focus Group Discussion (FGD) and an in-depth interview show that the fallacious implementation of Occupational Health and Safety (OHS) is attributed to their traditional "work culture".

2. Method

2.1. Participant

A survey, in terms of the field of work and educational background, involves 79 active PAKUBA members. An interview is used to measure the knowledge level of 29 participants. Also, another in-depth interview involves 3 participants who's over 30-years work experiences.

2.2. Data collection

The data are collected through a questionnaire and in-depth interview [4]. The questionnaire is used to measure the construction workers' knowledge about the condition of Occupational Health and Safety (OHS) [4]. Meanwhile, the in-depth interview is used to explore the construction workers' understanding of and attitude to the implementation of Occupational Health and Safety (OHS) in practice [5].

2.3. Data analysis

In Indonesia, most of the construction workers do not have the vocational educational background that is suitable for their skill and interest. The reason for them to be the construction workers is due to the lack of job vacancy related to their skill. Statistics show that the construction workers –i.e. the members of PAKUBA community – are senior high school graduates (47%) (see Figure 1). This statistic is higher than the percentage of national workers, who mostly are elementary school graduates and even do not have the formal educational background (45.66%) [6].

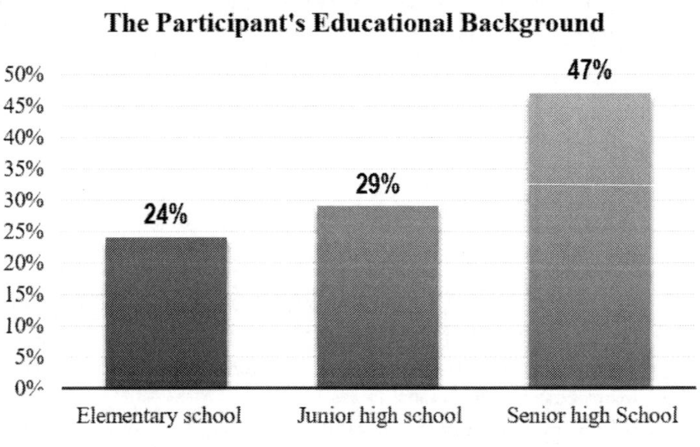

Figure 1. The Participant's Educational Background
Source: Research Document, 2016

There are 79 construction workers who are reported as the active members of PAKUBA. They are categorized by the field of work that they are focusing on, including bricklayer, carpenter, painter, blacksmith, electrician, executor, and labor. Most of them deal with construction work, focusing on building (see Figure 2).

IOP Conf. Series: Materials Science and Engineering **306** (2018) 012075 doi:10.1088/1757-899X/306/1/012075

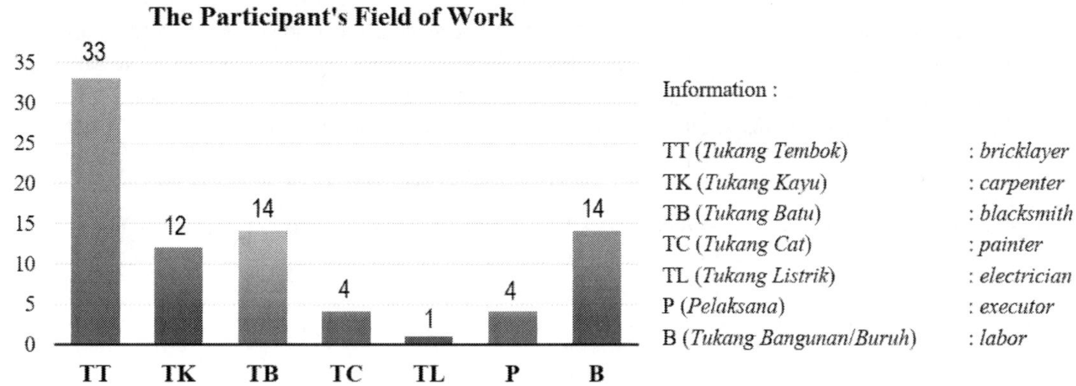

Figure 2. The Participant's Field of Work
Source: Research document, 2016

The conditions of the construction workers' Occupational Health and Safety (OHS) include Construction Safety Equipment (CSE) and Construction Personal Protective Equipment (PPE). There are nine aspects of knowledge that are measured (see Figure 3). They are 1) the use of safety boot, 2) the use of safety helmet, 3) the use of safety glasses, 4) the use of dust mask, 5) the use of long sleeve T- shirt, 6) the use of safety gloves, 7) the use of safety belt, 8) rechecking of Construction Safety Equipment (CSE) and Construction Personal Protective Equipment (PPE), 9) the understanding of the Occupational Health and Safety (OHS) regulation.

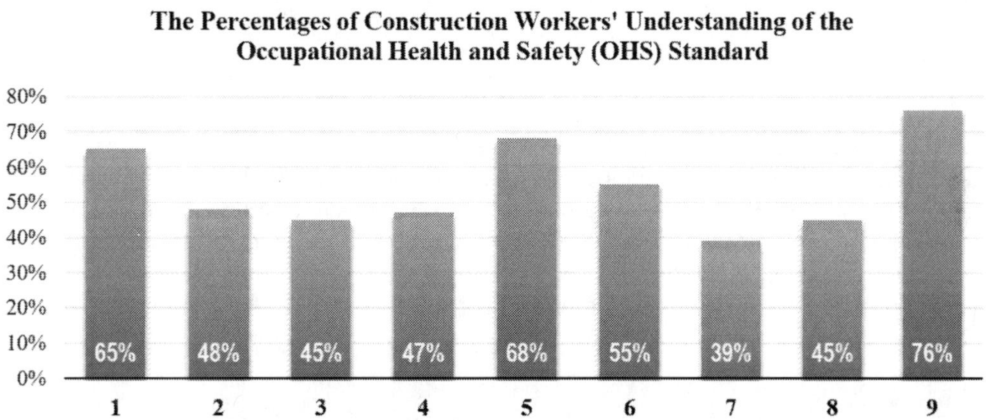

Figure 3. The Percentages of Construction Workers' Understanding of the Occupational Health and Safety (OHS) Standard
Source: Research document, 2016.

3. Analysis and Discussion

Construction industry refers to the heavy industry that needs many workers who can deal with their job manually where the job can put them into dangerous conditions [7]. Lacks of education and training of the Occupational Health and Safety (OHS) result in construction workers' less awareness of their occupational health and safety (OHS) [8]. Moreover, the construction industry, hiring the workers, has the responsibility to take care of the Occupational Health and Safety (OHS). The reason for the construction industry to take care of the Occupational Health and Safety (OHS) is because OHS deals with many problems related to the aspects of human right, law, and cost which become the responsibility of the industry itself [9].

The data analysis of the construction workers as informants depicts that their knowledge regarding Occupational Health and Safety (OHS) in nine aspects analyzed in this study is actually sufficient. It means that the workers know about the Occupational Health and Safety (OHS) regulation which have to be implemented during their work hours. Problems that arise is the knowledge is not implemented when they work due to various reasons. Most of the Indonesian construction workers' attitude and behavior still disregard the implementation of Occupational Health and Safety (OHS), they do not care about their own safety and health. For example, smoking. Most of the construction workers are heavy smokers and smoking has become a habit even when they are working. In the aspect of works safety, the use Construction Safety Equipment (CSE) and Construction Personal Protective Equipment (PPE) still becomes a burden which, in their perspective, will interrupt their work. The result of the analysis is supported an informant's statement:

"Personally, every Occupational Health and Safety (OHS) requirement and rule is very important and has to be implemented because it will protect the workers from accidents. It is just that when working in the field, it is still hard to implement it because we are not used to it and do not feel comfortable if we have to wear the required uniform" (Kholidin).

The proportion of work accidents in Indonesia in the construction sector becomes the biggest contributor along with manufacture industries which are 32%. Several types of work accidents experienced by the construction workers include falling from a high place (26%), concussion (12%), and crushed (9%) [10]. Based on the information gathered from informants, work accidents in construction projects are often experienced by their team whether it is a minor or major accident.

"When undergoing a skyscraper project in Mangga 2, there was one an accident that a worker (subordinate) fell from the 28th floor, he does not use the safety belt, but thank God he is saved because the safety net was installed" (Rahmat Sihabudin).

Another informant expresses that work accidents happen several times in construction projects and it has taken its toll because one of their team members from Desa Citangtu was dead. Among the workers, work accidents are only considered as a risk and even fate that they have to accept. The awareness that the Occupational Health and Safety (OHS) implementation is an important part which is not supposed to be separated from their work activities has not been well planted. As a matter of fact, currently, one of the characteristics and work demands is Occupational Health and Safety (OHS) related issues which can change the culture of work safety as well as the power of regulation which controls it both at national and local levels as an important agenda for every country [11].

4. Conclusion

From the aforementioned explanation, it can be concluded that :

1. Occupational Health and Safety (OHS) for construction workers in Indonesia still becomes a fundamental problem found in some workspace both in simple building work and construction projects with a high level of complexity.
2. The construction workers' level of knowledge of Occupational Health and Safety (OHS) is quantitively enough with an above average percentage of 50%, but on the implementation in the field, the knowledge is not well implemented when they are working.
3. The lack of the Occupational Health and Safety (OHS) training, as well as work culture, developed among the workers, is the cause of the lack of awareness to apply Occupational Health and Safety (OHS) while working.
4. Attitudes and behaviors in the workspace are traditional habits which are considered as prevalent, even work accidents are considered as a risk and fate that they have to accept.

IOP Conf. Series: Materials Science and Engineering **306** (2018) 012075 doi:10.1088/1757-899X/306/1/012075

Reference

[1] Keputusan Menteri Tenaga Kerja Dan Transmigrasi 2014 *Republik Indonesia Nomor 31 Tahun 2014* Tentang Penetapan Standar Kompetensi Kerja Nasional Indonesia Kategori Konstruksi Golongan Pokok Konstruksi Gedung Pada Jabatan Kerja Tukang Bangunan Gedung.

[2] Pangkey F, Malingkas G Y, and Walangitan D O R 2012 Penerapan Sistem Manajemen Keselamatan Dan Kesehatan Kerja (Smk3) Pada Proyek Konstruksi Di Indonesia (Studi Kasus: Pembangunan Jembatan Dr. Ir. Soekarno-Manado) *Jurnal Ilmiah MEDIA ENGINEERING* **2** 2 100–113.

[3] Prastowo A 2011 *Metode Penelitian Kualitatif dalam Perpektif Rancangan Penelitian* (Yogyakarta: Ar Ruzz Media)

[4] Creswell J W 2014 *Research Design: Qualitative, Quantitative and Mixed Methods Approaches* SAGE Publications, Inc.

[5] Denzin N K and Lincoln Y S 2000 *Handbook of Qualitative Research* (New Delhi: SAGE Publications)

[6] Undang-Undang Nomor 13 Tahun 2003 Tentang Ketenaga Kerjaan. Badan Pusat Statistik Nasional, 2015. Badan Sertifikasi Keterampilan

[7] Eaves S, Gyi D E, and Gibb a G F 2016 Building healthy construction workers: Their views on health, wellbeing and better workplace design *Applied Ergonomics* **54** 10–18

[8] Piri S, Sompie B F, and Timboeleng J a 2012 Pengaruh Kesehatan, Pelatihan Dan Penggunaan Alat Pelindung Diri Terhadap Kecelakaan Kerja Pada Pekerja Konstruksi Di Kota Tomohon *Jurnal Ilmiah MEDIA ENGINEERING* **2** 4 219–231

[9] Wicaksono I K and Singgih M l 2011 Manajemen Risiko K3 (Keselamatan dan Kesehatan Kerja) Pada Proyek Pembangunan Apartemen Puncak Permai Surabaya) *Prosiding Seminar Nasional Manajemen Teknologi XIII 3*

[10] Public Communication Bureau of the Ministry of Public Works and Housing 2015

[11] Kim Y, Park J, and Park M 2016 Creating a culture of prevention in occupational safety and health practice *Safety and Health at Work* *7* 2 89–96

Motivation, Compensation, and Performance for Science and Technological Teachers

R M Abast*, N M Sangi, M S S S Tumanduk and R Roring
Universitas Negeri Manado, Tondano 95618, North Sulawesi, Indonesia

*rio_abast@unima.ac.id.

Abstract. This research is operationally aimed to obtain the result of analysis and interpretation about: relationship of achievement motive, compensation with performance at a junior high school in Manado, Indonesia. This research applies a quantitative approach with correlation analysis method. The research was conducted at one junior high school in Manado, Indonesia. The results showed achievement motive at the school teachers is quite high. This result means that, generally, the teachers of the school have a desire to improve achievement; the performance at the school is good enough. This result means that in general, the performance of teachers at the school is increasing, there is a linkage degree and determinative power between the achievement motive with the performance of teachers at the school amounted 0.773% or 77.3%, compensation for the school teachers in Manado is good enough. This result means that the compensation received is satisfactory, there is a linkage degree and determinative power between compensation and performance of the school teachers in Manado amounted to 0.582 or 58.2%.

1. Introduction

Education is an essential process of empowerment, which is a process to reveal the potential that exists in humans as individuals, which in return may contribute to the society empowerment within the surrounding environment. In fact, the level of educational development activities keeps on changing follows the dynamics of development achieved by human beings both in the form of mindset advancement and technological progress that accompany it, therefore it is not surprising that with those changes, it has also produced more methods and advance techniques in the implementation of educational activities itself. Indonesian government itself has given enough attention for the development of national education in producing output quality. This aspect is very important because lately the problem of quality and competitiveness of the national education system is quite a polemic. Which is understand able because in the high competition of globalization era, the level of productivity of Indonesian human resources is still low.

Indonesian human resource quality standards on average are at a low number of four [1]. This figure has achieved by the quality of Thai human resources in 1970. This low figure illustrates the lowness productivity of Indonesian human resources. Furthermore, it is argued that this low number is related to the low quality of national education therefore it is not surprising that many undergraduate graduates are unemployed because of lack of creativity and low of competitiveness. This issue should be seen from laying of foundation of national education in the existing educational level, if the basic education is given well at a higher level, logically it will produce a qualified output. Therefore, one aspect that must be studied in this weakness is by seeing how well and how qualified the education activity at the basic level given by educator in national education environment.

The quality of human resources is basically determines the performance of an organization [2]. However, the quality hierarchically on the one hand has not been able to guarantee the performance of individuals for all employees or existing employees. It is said that because no matter how good the quality and capabilities possessed if it is not supported by motivation, especially that comes within itself, it is very difficult for these employees to be able to improve their job performance. In fact, this phenomenon is often encountered in an organization, which is true that the motivation of work is the basic framework that builds one's work performance. This is understandable because in reality, somebody, in doing something must be built from the desire to act in order to achieve or meet a need.

Theoretically a teacher will have a good performance if there are motives of achievement. This achievement motivation will build the spirit of work to continue advancing and developing so that in the educational activities will be implemented through teaching and learning process which will able to grow direct student learning behavior with the improvement achieved as a lesson. Many examples are often encountered that there are less teachers do not even have a plan, and goals during learning activities. In this case the teacher implements teaching principle which is limited to the activities based on the teaching mindset alone without considering the forward aspect to be achieved. In such a concept, the teacher gives the image that the teacher no longer has a motive for achievement and associated with the teaching performance and learning process it is said to have a low performance that will affect student learning behavior which this teaching patterns tend to be monotonous and not varied will then affects student's interest in learning. In this condition, students' learning behavior will not be improved and will even lead the students to the passive and non-creative learning behavior characterized by the lack of student response with the lessons received.

Furthermore, if people are satisfied with the compensation received in both material and non-material form, it will be able to improve the performance because basically the activities carried out by humans through the organization aimed to the fulfillment of needs, which in other words, in order to live worthily on the basis of humanity, man has a need to fulfill [3].

2. Research Methods

This research uses a quantitative approach with correlation analysis method. The research was conducted at SMP Negeri in Manado. There are 14 public junior high schools in Manado, those are: SMP Negeri 1, SMP Negeri 2, SMP Negeri 3, SMP Negeri 4, SMP Negeri 5, SMP Negeri 6, SMP Negeri 7, SMP Negeri 8, SMP Negeri 9, SMP Negeri 10, SMP Negeri 11, SMP Negeri 12, SMP Negeri 13, SMP Negeri 14. The research process took three months.

The population in this study is all the characteristics related to professional competence variables and teacher performance. While the members of the population are all teachers amounted to 600 people. Determination of sample is taken by Proportional random sampling technique, with Krejcie Table and nomogram of Harry King, if the sample of population is 95% or error rate of 5%, then the number of samples taken from each member of the population is 234 teachers.

To measure the research variables that include the variables of achievement motif (X1) and Compensation (X2) and Performance (Y) as well as to describe, to determine and to analyze the relationship both associative relationships and the influence of the variable between the variables requires research data collected by certain techniques. In this study, the technique used to collect data is a questionnaire technique (questionnaire) which is a list of questions in the form of Likert Scale.

3. Results and Discussion

3.1. SMP Negeri Teachers Achievement Motive in Manado

Based on the results of Z test obtained it can be said that the achievement motive of SMP Negeri teachers in Manado is quite high. This means that in performing their duties as educators, the Public SMP Negeri teachers in Manado have a motive, desire or drive for achievement. Teachers are the main drivers of educational activities implemented in a school institution. Therefore, the declining output of education produced by school institutions is strongly influenced by the ability of teachers in managing all activities of school education.

These concepts apply as well in the role of the teacher as the implementer who regulates educational activities. Those teachers are considered to have managerial skills to able to carry out various activities as stated at the top. Teacher managerial ability is also seen from the ability to formulate strategy. In the concept of strategy formulation, a teacher must have a strategy stratification that can examine what are the strengths, weaknesses, opportunities and threats of the scope of school education activities so that it can formulate the formulations designed for the school's own progress [4].

All the above concepts could be implemented if the teacher has a motive for achievement. If there is an achievement motive, then it will grow a desire to improve performance itself and have the ability in terms of:

a. To Define goals and objectives
b. To Formulate a strategy
c. To Plan
d. To Determine the work program
e. To Organize
f. Mobilizing human resources
g. To Monitor operational activities
h. To Carry out supervision
i. Carry out an assessment

It is fully realized that achievement motive is the most determining factor of teacher performance because it is a human problem related to psychological aspects personally. That can be said because no matter how well and how qualified human resources and existing supporting facilities, if the implementing element itself does not have the desire or motivation to carry out all the work responsibilities given it will not produce a good performance and optimal performance.

3.2. Relationship achievement motive with the performance at SMP Negeri teachers in Manado.

Based on the results of correlation analysis obtained that the level of relationship between achievement motive variables (X1) and performance (Y) is 0.727. The value means that there is a functional relation between achievement motive with teacher performance at SMP Negeri teacher in Manado

States that the achievement of a person's performance is determined with how much the thrust or motive that drives the individual for further improve performance [5]. This aims to encourage people or employees to be enthusiastic and reach the desired outcome of those people. Achievement motives can be generated within the power of an individual which is known as internal or intrinsic and can also be sourced from outside a person or the power that exists outside the individual which stimulates or affects him to voluntarily willing carrying out activities or work, known as external or Extrinsic. This understanding is also encountered to every SMP Negeri teacher in Manado where psychologically have a desire to always improve performance in accordance with their ability and expertise, in hope it will be able to improve their job status and positions to achieve their desire carrying out the duties and responsibilities with full discipline and thoroughness.

States that the desire to progress and develop, the fulfillment of economic needs and the desire to gain promotion at a higher level are the main motivators for individuals to improve performance [6]. Fulfillment of all needs is important to grow the achievement motive with the assumption that in each individual there is a reservoir of capabilities that need to be developed to change its nature from potential to something tangible so that with a clear final goal, the employees are encouraged to develop potential for then later can be proven through the implementation of work more effectively and productively together with positive behavior to be able to achieve various goals and objectives and on the other hand will achieve a higher level of satisfaction. In addition, the desire to improve performance is also a driving force to further develop not only in terms of mental intellectual, but also in a professional sense. This benefit is very important because an individual will be able to achieve progress if the individual is trying to grow and develop in all aspects of life and livelihood which then leads to the motive of a person to become the best worker in his work even though the conditions are not supported by appropriate facilities and infrastructure which in other words, through the existing potential it will be able to optimize all the existing supporting elements.

3.3. Compensation relation with the performance SMP Negeri teachers in Manado

The functional contributive relationship between compensation and teacher performance is 0.905. This value means that there is a constitutive functional relationship between compensation and performance. Compensation affects a person's performance at work. The value of compensation in both material and non-material forms has a relationship with the performance level at which the high compensation value is likely to have a high productivity effect whereas in the medium and low compensation values it will produce the effect of moderate and low performance outcome. This is understandable because basically every human being has a need in life that must be met every day whether it is primary needs or secondary needs, which in order to obtain these needs, people works in the hope that the wages or compensation given as a payment is in accordance with the amount of responsibility in work undertaken, in other words economic needs and self-actualization is one thing that is valuable to encourage someone to give his effort in realizing what is the purpose of the organization, in other words can be said that someone doing activities or work between Driven by economic needs. When the stimulus is given in accordance with the rules set it will affect the productivity of work generated.

3.4. Relationship of achievement motive and compensation together with teacher performance SMP Negeri in Manado.

The relationship between achievement motive and compensation is together to the performance of the teacher that is equal to 0.912 or 91.2%. This result means that there is a contextual relationship between the motivation of achievement and compensation (collectively) with the performance SMP Negeri teachers in Manado.

All activities carried out by humans are based on the principle of meeting the needs of both primary and secondary needs. In this case every human individual needs money as a means of fulfilling all the necessities in supporting diverse daily life needs according to the circumstances that exist in their lives. This necessity then causes why every human individual will always look for a job to earn wages. The problem of fulfillment is a crucial and very complex that greatly affects the implementation of the work. The fulfillment of this need was very influential and most fundamental in motivating the employees. This condition motivates teachers strongly to always carry out the job responsibilities in the hope that if they perform the task well then by itself they will get compensation in return, especially the material compensation from the organization.

3.5. Degree of associative linkage and determinative power between achievement motives with SMP Negeri teachers in Manado.

The results of simple regression analysis obtained the coefficient of determination (R2) of 0.773 in other words that the 77.3% of teacher performance is determined by achievement motive. This value means that the degree of associative associates and determination of achievement motive to teacher performance is 77.3%.

States that encouragement given aims to encourage individual workers to be enthusiastic and be able to achieve the desired outcomes where the motivation given is essential in the effort to mobilize and to direct all the potential power of the existing workforce towards optimal utilization, within the limits of human capability with the help of tools and other facilities [7]. Achievement motives within the scope of the organization include the teachers in educational institutions which greatly determine in how high performance can be generated. Without the achievement motive the performance is unlikely to increase.

3.6. Degree of associative associate and determinative power between compensation and performance SMP Negeri teacher in Manado.

The result of simple regression analysis resulted in the coefficient of determination (R2) of 0.582 in other words there is 58.2% of teacher performance is determined by compensation. This value means that the degree of associative associate and determination of compensation with teacher performance is 58.2%.

In fact, humans formed organization to work in order to meet their needs, while on the contrary that needs also is the reason for human to formed organization. In order to work to meet his needs and vice versa based on his ability to meet the needs in quality and quantity respectively, humans obtain socioeconomic status in the society, because basically humans are not only different by the ability to do something but also different by the terms of desire or motivation. In this case, one's motivation according to [8] Depends on the strong or weakness of the one's motive for the needs, desires, impulse where the motive itself is directed at a particular goal. This is the main reason why humans have the motivation to increase work productivity.

4. Conclusions

Based on results of the research, the conclusions are: the achievement motive SMP Negeri teacher in Manado is quite high. This result means that in general teachers of SMP Negeri in Manado have a desire to improve achievement; the performance SMP Negeri teacher in Manado is good enough. This result meant that in general the performance SMP Negeri teacher in Manado is increasing; there is a positive relationship between achievement motives, compensation with the performance SMP Negeri teachers in Manado; there is a degree of linkage and determinative power between the achievement motive with the performance SMP Negeri teachers in Manado of 0.773 or 77.3%; compensation for SMP Negeri teachers in the city of Manado is good enough. This result means that the compensation received is satisfactory; there is a degree of linkage and determinative power between compensation and performance SMP Negeri teachers in Kota Manado amounted to 0.582 or 58.2%.

References

[1] Tilaar H A R 2004 Permasalahan Pendidikan Indonesia Perlu Dipetakan Kembali.Kompas. 12 Oktober 2004.
[2] Siagian S P 2003 *Teori dan Praktek Kepemimpinan* (Jakarta: Rineka Cipta)
[3] Dubin R 1998 *Industrial Society and Human relation* (New York: Prentice-Hall)
[4] Dephilipps F 2003 *Human Resources and Organization Homewood* Richard D Irwin Taronto
[5] Sedarmayanti 2001 *Sumber Daya manusia Dan Produktivitas Kerja* (Bandung: Mandar Maju)
[6] Siagian S P 2004 *Teori Motivasi & Aplikasinya* (Jakarta: Rineka Cipta)
[7] Manullang M 1983 Organisasi pendidikan Dalam Era Reformasi *Jurnal Ilmu Pendidikan*

[8] Blanchard K 2007 *Leading a Higher Level Upper sadle River* (New Jersey: Prentice Hall)

A Comparative Study of the Traditional Houses Kaili and Bugis-Makassar in Indonesia

M F Suharto*, R S S I Kawet and M S S S Tumanduk

Universitas Negeri Manado, Tondano, 95618, North Sulawesi, Indonesia

*fachruddinsuharto@unima.ac.id

Abstract. In this study, I compared the physical elements of two Indonesian traditional houses between a Kaili tribe (Central Sulawesi) and a Bugis-Makassar tribe (South Sulawesi). If we viewed of the name, meaning and function from both traditional houses have similarities, namely the Souraja/Saoraja house (House of the King), however, observed more detail the physical elements of architecture also show the differences. The spatial, physical and stylistic systems (N. John Habraken's theory) were applied to analyze their differences and the similarities of the physical elements of architecture on those two traditional houses. The results of the analysis identified that the physical elements of architecture such as the orientation, the function and distribution of rooms (the spatial system), the constructions and materials of floor, wall and roof (the physical system) and the opening types of the door and window as well as ornaments used showed similarities. Meanwhile the physical elements of architecture such as the arrangement of columns, form and spatial pattern as well as the placement of the stairs (the spatial system), the constructions and materials of foundation, column and beam (the physical system) as well as the form of the roof and façade found differences of both traditional houses.

1. Introduction

The diversity of Indonesian traditional architectures which spread out through all areas in Indonesia becomes unlimited sources for knowledge. The traditional architecture of Indonesian is closely related to the local customs because a house is not only a place of residence, but also a center of social relations, traditional laws, religions, taboos and myths. J.Crawfurd [1] divided the human settlements in the era of the Dutch East Indies (Indonesia) into two categories. Firstly, the houses of the maritime tribes (Sumatera, Kalimantan, and Sulawesi) which stand on columns located on the riverside or seaside and generally called as a stilt house..Secondly, the houses of the agricultural tribes which are found in Java, Bali and other islands in Indonesia.

Generally, the Indonesian traditional houses have many similarities, each other, however, in principle, they have any significance differences, especially in the cultural background, geography, concept and architectural form in detail. The Kaili and Bugis-Makassar tribes are two tribes in Indonesia located on the Sulawesi Island. These two tribes have the traditional houses which are made from wood with a similar name called Souraja/Saoraja*, means a big house or house of the king and his family (table 1). Although they are similar in names, philosophy and etc. there are also many differences in their characteristics. Therefore, the author interested to explore the characteristics of these two traditional houses in order to obtain their similarities and differences. Besides that, there were only a few reports studied about the Souraja/Saoraja house mainly in comparing two objects with the similar name and meaning. This paper is expected to become a reference for other researchers whose same object or any things that want to investigate from the Saoraja/Souraja house..

2. Methods

2.1. Research objects

Traditional House of Kaili Tribe (Souraja) and Traditional House of Bugis-Makassar Tribes (Saoraja). Kaili tribe is one of the ethnic majorities in the territory of the province of Central Sulawesi. It has several types of the traditional houses and one of them is a Souraja house (Figure 1). According to Walter Kaudern in Masiming Mariani (1999) [2], the typology of traditional architecture of Central Sulawesi is classified into three types. Firstly,type A, the house with a rectangular form that is divided into two parts; the bottom part as a barrier of the house body and the top part as the house body which only consists of the elements of roof. Secondly, type B, the house with a rectangular form which is divided into two parts; the bottom part as a barrier that is supported by columns, and the top part as the house body which consists of the elements of wall and roof. Third, type C,the house with a rectangular form which is composed of a number of columns to support the house body and roof and Kaudern made an analogy as a stilt. From the classification above, the Souraja house can be put in *type C*.

The Souraja house in Kaili is divided horizontally into three rooms [3]: the front room *(Lonta Karavana)* which has functions for receiving guests and as a bedroom for male guest who want to stay overnight, the middle room *(Lonta Tatangana)* which is used to receive close relatives with the bedrooms and dining room, and the back room *(Lonta Rorana)* which is used for dining room, *Avu (kitchen)* and *Pakuntu* (an open space as a place for relaxing.). Similarly, the Souraja house is classified vertically into three parts: a. The top part *(Landue/attic)* as a place to store heirlooms, b. The middle part as a dwelling place, and, c. The bottom part *(Kolong/empty room underneath the house)* as a place for storing the agricultural products.

The Bugis and Makassar tribes are located in the province of South Sulawesi. The traditional houses of the Bugis and Makassar are often called in one name as the traditional house of Bugis-Makassar because they have similar morphology, philosophy and culture. The traditional house of Bugis-Makasar is classified into three categories regarding to social statuses (strata) of its residents; firstly, *Sao-raja (sallasa)*, the big house for the royalties *(Anakarung)* (Figure 2), secondly, *Sao-piti*, the small house for ordinary people *(Maradeka* or *Tosama)* and thirdly, *Bolla*, the a house for the civilians *(Ata or Suro)* [4]. According to Abu Humid [5], the concept of the traditional architecture of Bugis-Makassar came from a philosophy called as *"Sulapa Appa"* where all aspects of human life will become perfect if they are in a rectangular form. This philosophy is reflected in the form of houses, the walls and the layout plan which are oriented to the four natural elements such as earth, fire, water and wind. In their activities, the Bugis-Makassar peoples are always guided by their customs and beliefs that are implemented in the building process of the house. The house should be built under a supervisor who is known as *PanritaBolla* [6].

By viewing vertically, the Saoraja house (Bugis-Makassar) consists of three parts [7]. They are *Rakeang/Pammakkang (attic)* is a place to store food supplies, *Allo-bolla/kale-balla* is a place used for a living room, bedroom,dining room, and *Awaso/siring (Kolong)* is a place to store agricultural tools, carpentry and also a place for livestock. Meanwhile, by seeing horizontally, the souraja house is divided into three parts, such as *Lontang ri saliweng/padaserang dallekang* (the front room) is a place to receive guests and guest bedrooms, *Lontang retengngah/padaserang tangnga* (the middle room) is a place that has function as as a bedroom for family and children, and *Lontang rilaleng/padaserang riboko* (the back room) used as a bedroom for girls and the elderly parents [8]. Besides that, there are two additional rooms, namely *dego-dego*, the open space in front of the house which has function as a waiting place for guests or used for relaxing place and *dapureng* (kitchen) that is located at backside of the house.

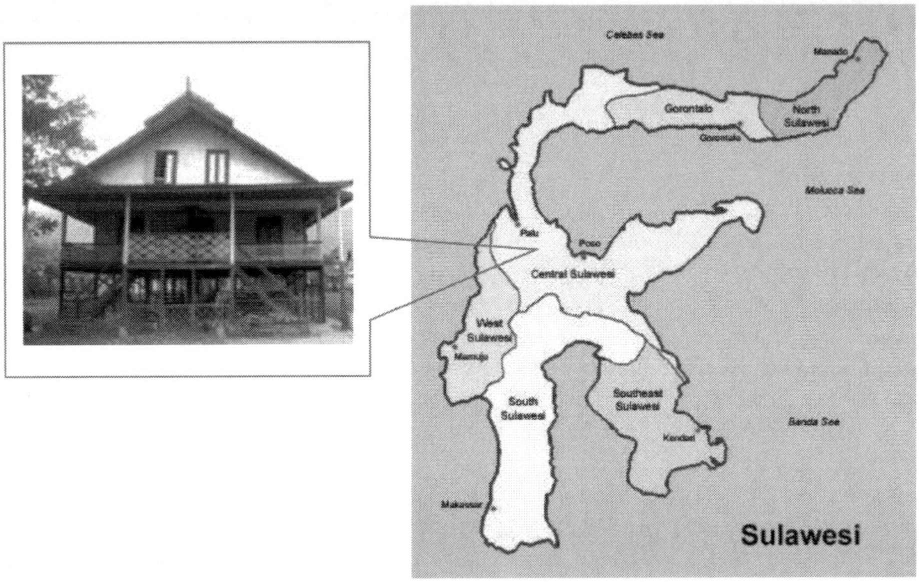

Figure 1. The traditional house of Kaili, Central Sulawesi

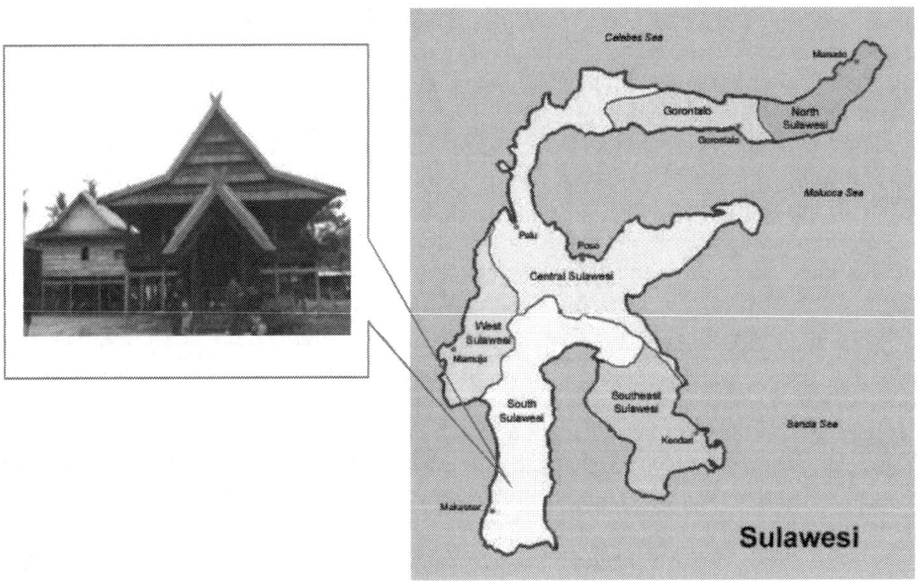

Figure 2. The traditional house of Bugis-Makassar, South Sulawesi

2.2. Research methods
This research used the purposive sampling method and literature study, which focused on the physical elements of the architecture as the sp atial, physical, and stylistic systems that classified by N. John Habraken. This study was started by introducing the traditional architecture of the Kaili and Bugis-Makassar, then a comparative study of both the Souraja/Saoraja houses in detail, and ended with the conclusion.

3. Results and discussion
(Comparative study of the traditional houses of Kaili and Bugis-Makassar)

Any different ways can be used to see the type of the house and N. John Habraken [9] classified these ways into three systems, namely *the spatial, the physical, and the stylistic systems*. In this study, I used those three systems in comparing the traditional houses of Kaili (Souraja) and Bugis-Makassar (Saoraja) houses in order to observe their differences and similarities. These three systems will be explained detail in the following part (table 2).

3.1. The spatial system

In this system, we identify the spaces characteristics of the house and the relationship between one space to another. The arrangement of columns, orientation, function and distribution of rooms, form and spatial pattern as well as the placement of the stairs were the physical elements of architecture observed as the spatial system in this study. As results, the grid pattern was applied in the arrangement of column for two traditional houses (the Souraja house and the Saoraja house) and the distance between the columns was similar in size with ranges from 1.5 m to 3 m. In contrast, the differences were detected on the total number of columns and the arrangement of columns where the Souraja house has approximately 41 columns by arrangement of 4 columns for each row started from the front to the back of the house. However, the Saoraja house was built with approximately 63 columns with 6 columns for each row in its arrangement (Figure 3 and 4). In their orientation, both traditional houses were shown a clear similarity where the direction of the house faced to the street. Although there was a rule believed by the Bugis-Makassar peoples about the orientation of the house that should face to the north or the east, but it has been rarely applied and more dominantly in facing to the street (Figure 5 and 6).

Both traditional houses have the similar function and distribution of rooms viewed from horizontal and vertical side. By horizontally, the rooms were mainly distributed into three areas, such as *the public area* which was located at the front of the house, *the private area* that consists of bedrooms and living room in the middle side of the house, and *the service area* consisting of the open space and a kitchen at the back of the house. Meanwhile, by vertically, the distributions of rooms were classified into *the attic* (the top part of the house) which was analogous to the human head and used for storing of heirlooms. The middle part of the house has functioned as the rooms for residence and analogous to the human body, and the bottom part of the house or *the Kolong* (the empty room underneath the house) which is analogous to the human foot and used as a place to store agricultural products and livestock (Figure 7 ~10).

In the form and spatial patterns, although the basic form of room in the both traditional houses was rectangular but it was extended from the front to the back of the house on the Souraja house, however, it more varied such as the form of the letter "L" on the Saoraja house. The differences also were found in the arrangement of rooms where the Souraja house was classified in *The terrace ~ The rooms of dwelling ~ Open space ~ The kitchen with the* open space (the relaxing room) located in the back of the house. On the other hand, the rooms of Saoraja house was arranged from *The terrace~The open space~The rooms of dwelling~The kitchen and back terrace with the* open space or the relaxing room was located at the front of the house (Figure 11 and 12). During investigating the stairs, I found that these two traditional houses have similarities in the usage of the stair materials (wood) and the number of the stairs located at the back of the house (only one stair). Meanwhile, the difference was obtained in the amount of the stairs located at the front of the house where there were two stairs faced each other with the number of steps of 7 or 9 every stairs on the Souraja house and only one stair that facing towards the house with 11~15 steps on the Saoraja house (Figure13 and 14).

3.1.1. The arrangement of columns

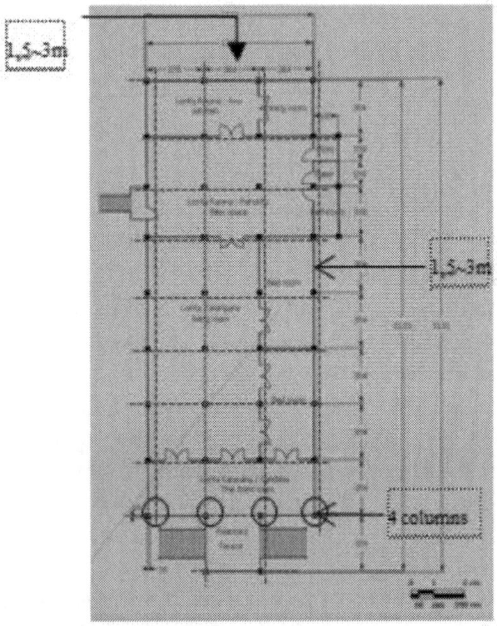

Figure 3. The floor plan of the Souraja

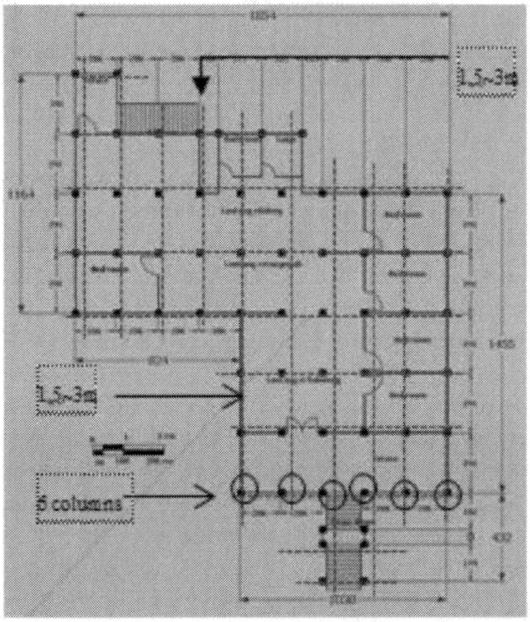

Figure 4. The floor plan of the Saoraja

3.1.2. The orientation

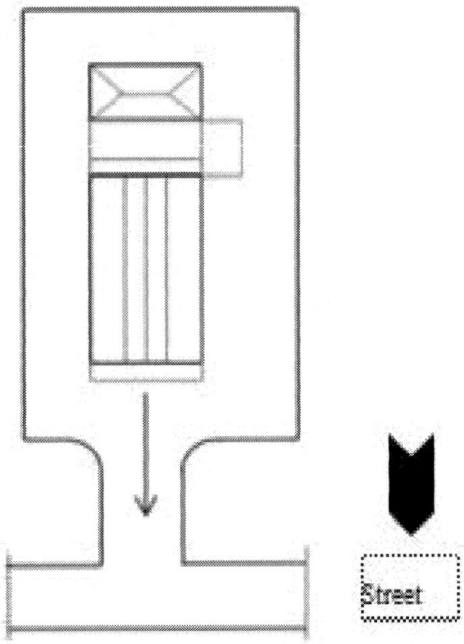

Figure 5. The orientation of the Souraja

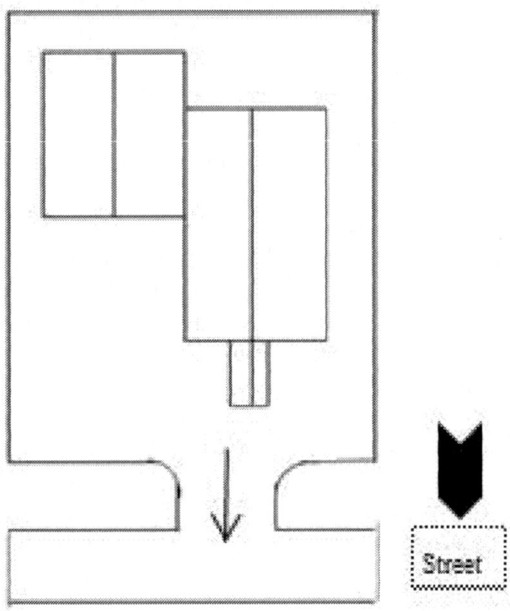

Figure 6. The orientation of the Saoraja

3.1.3. The function and distribution of rooms

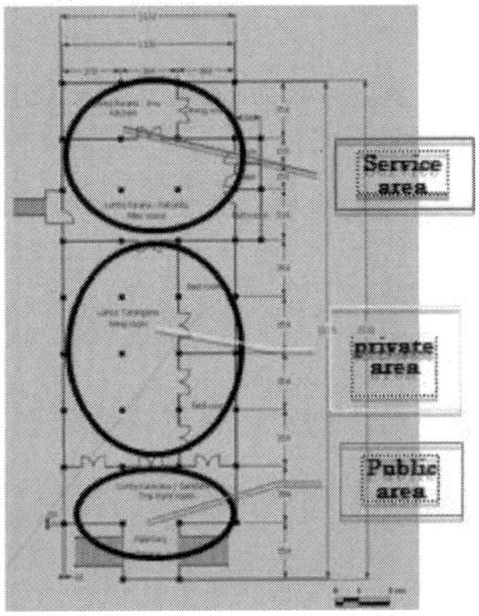

Figure 7. Distribution of rooms horizontally

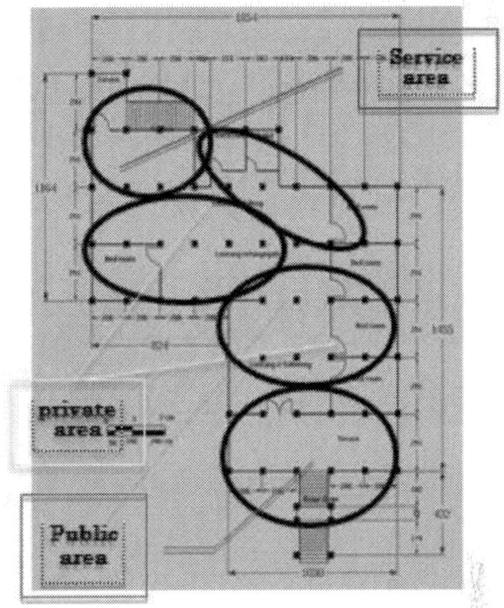

Figure 8. Distribution of rooms horizontally

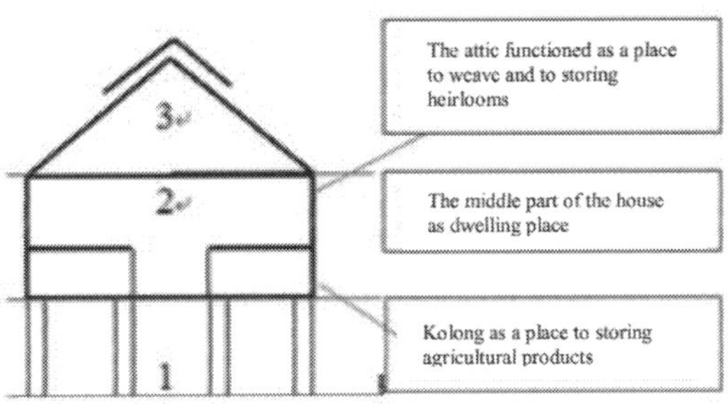

Figure 9. Distribution of rooms horizontally

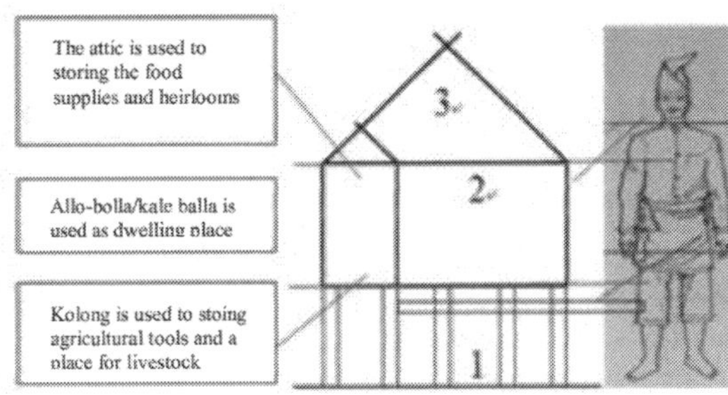

Figure 10. Distribution of rooms horizontally

3.1.4. The form and spatial pattern

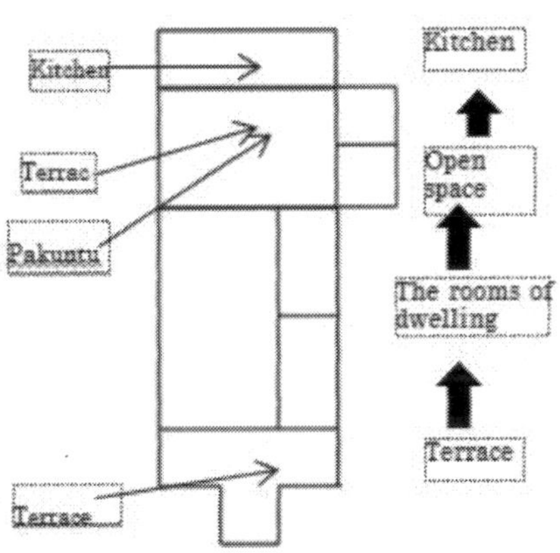

Figure 11. The spatial pattern of the Souraja

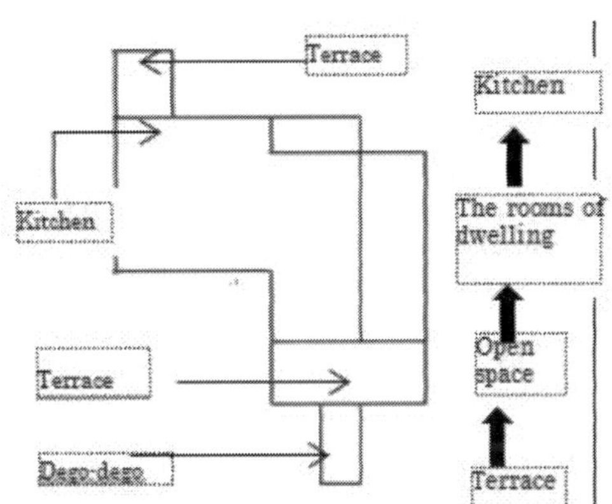

Figure 12. The spatial pattern of the Saoraja

IOP Conf. Series: Materials Science and Engineering **306** (2018) 012077 doi:10.1088/1757-899X/306/1/012077

3.1.5. The placement of the stairs

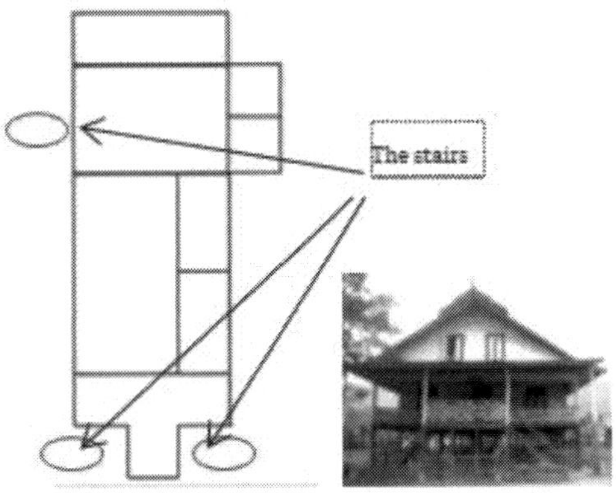

Figure 13. The placement of the stairs of the Souraja

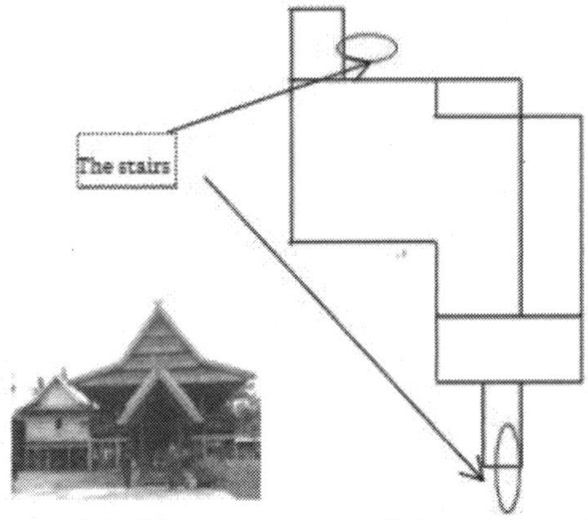

Figure 14. The placement of the stairs of the Saoraja

3.2. The physical system

In the physical system, we observed any aspects that related to the constructions and materials used for build a house, such as roof, floor, wall, foundation, and beam. In this study, it was detected that the middle part of these two traditional houses which were consisted of the floor and the wall has the similar construction (using pegs and pens in their joint system) and similar material (wood type). Besides, these two traditional houses also have the similar construction of roof, called as the gable roof that were used the similar woods (ulin wood, bayan or merbabu wood) and covered by rumbia leaves (Figure 15 ~ 20). In contrast, the bottom part of the house (the Kolong) which consists of foundations, columns and beams has difference constructions and materials used on both the traditional houses. The Souraja house used the foundation system with the columns were planted directly into the ground around +/- 1m, and then covered by cast concrete. Meanwhile, the foundation of the Saoraja house used the natural stones or the cast concrete in a cubic form (Figure 21 and 22).

The Souraja house used the beams with dimensions of 0,07m x 0,11m and 0,06m x 0,15m and the columns with a rectangular cross section with dimensions of 0,16m x 0,16 m and 2 ~ 2,25 m in high. The columns and beams were made from the iron wood or the merbabu wood. The joint system of columns and beams was made by passing the beams from two different directions through a hole in the column until they come out about 20 cm. Assuming this system is applied to the modern houses with the new materials usage, the detail construction will be performed more artistically. On the other hand, the cross section of columns on the Saoraja house was a rectangular form with dimension of 0,15m x 0,15m and has a high dimension between of 2,1 ~ 2,25m. Meanwhile, the dimension of the beam was 0,04m x 0,125m. The construction of the Saoraja house has two stacked beams made from the jackfruit wood or the durian wood. The construction of the Saoraja house used two types of the beams, such as pattodo/pattolo and palangga/arateng. The pattodo is the beams located at the width side of the house and have a function to connect one column to another column. Meanwhile, the pallangga is the flat beams with their lengths are longer than the length of the house and it has the functions as a place of the floor beams and also as a buffer to the columns of the house (Figure 23 and 24).

3.2.1. The construction and materials of the roof

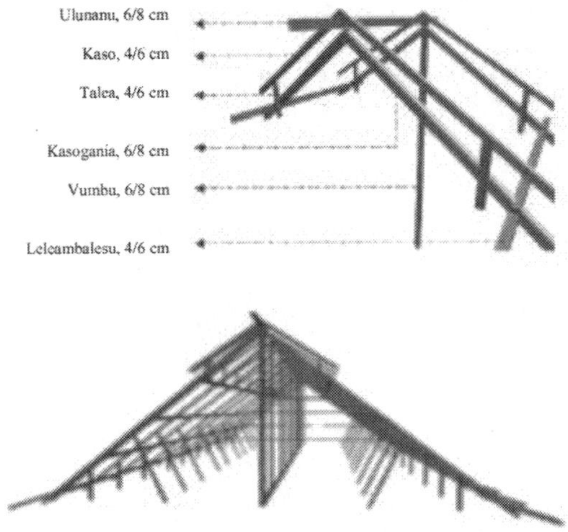

Figure 15. The construction of the roof of the Souraja

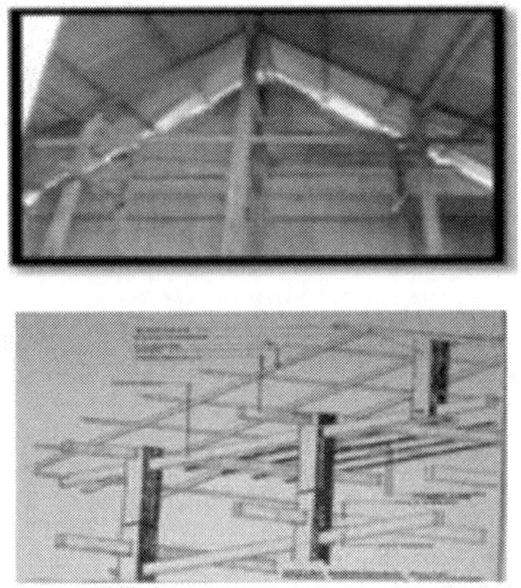

Figure 16. The construction of the roof of the Sauraja

3.2.2. The construction and materials of the floor

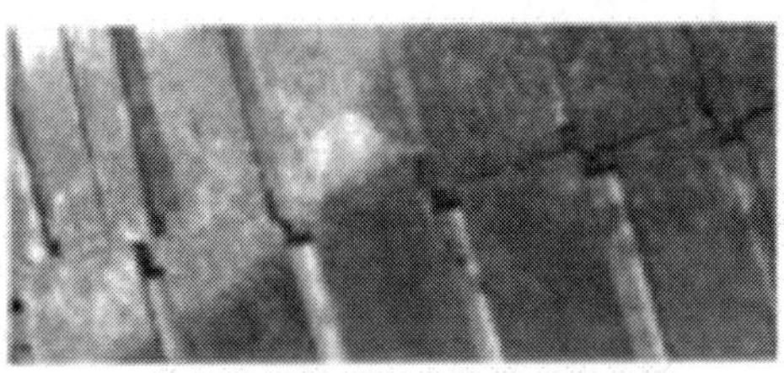

Be apaced +/- 1-1.5 cm

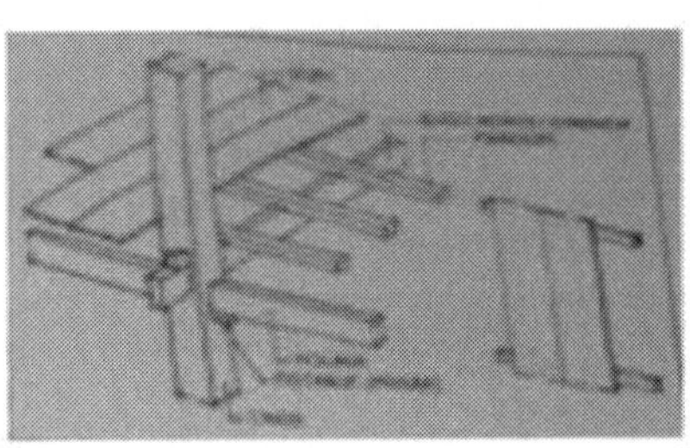

Figure 17. The construction of the floor of the Souraja

Be apaced +/- 1-1.5 cm

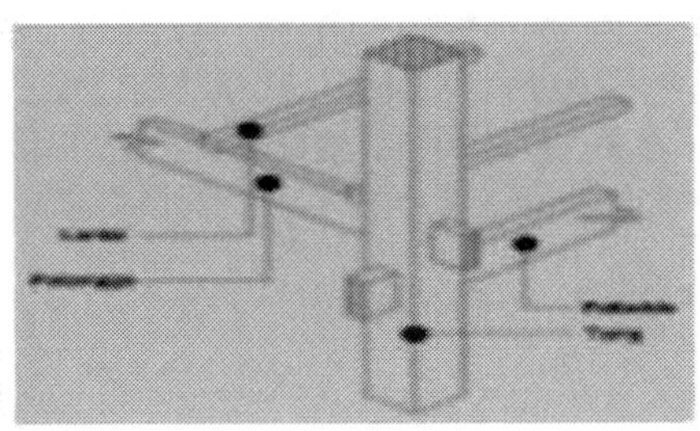

Figure 18. The construction of the floor of the Sauraja

IOP Conf. Series: Materials Science and Engineering **306** (2018) 012077 doi:10.1088/1757-899X/306/1/012077

3.2.3. The construction and materials of the walls

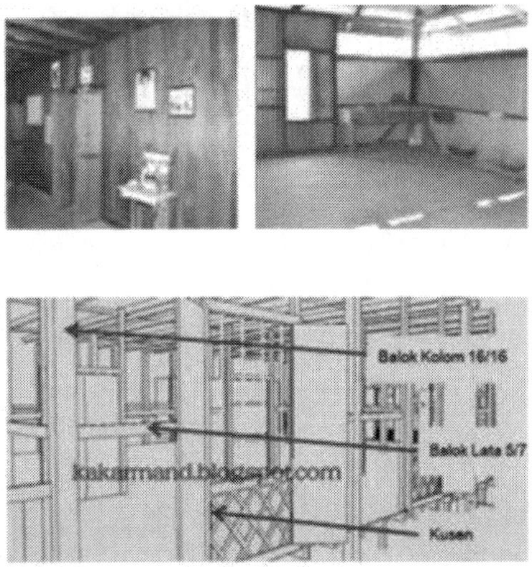

Figure 19. The construction of the wall of the Souraja

Figure 20. The construction of the wall of the Sauraja

3.2.4. The construction and materials of foundation

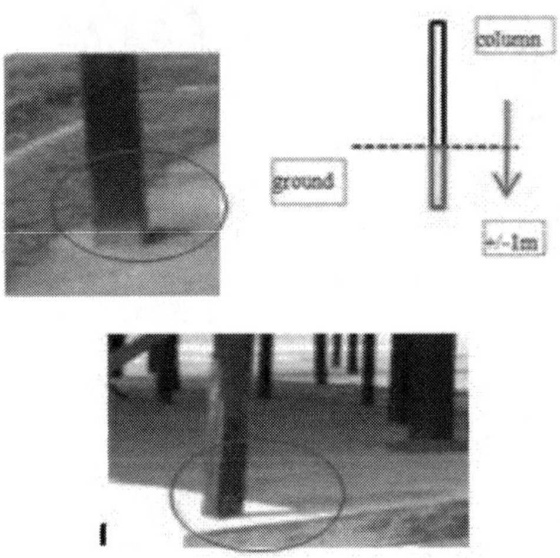

Figure 21. The system of foundation of the Souraja

Figure 22. The system of foundation of the Sauraja

IOP Conf. Series: Materials Science and Engineering **306** (2018) 012077 doi:10.1088/1757-899X/306/1/012077

3.2.5. The construction and materials of column and beam

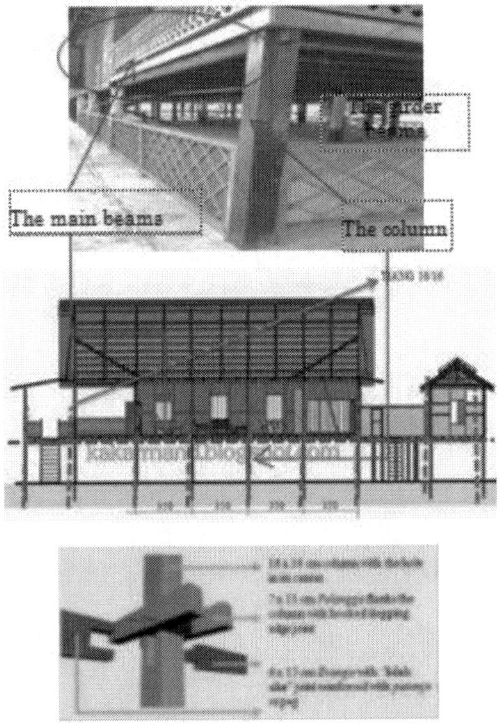

Figure 23. The construction of the column-beam of the Souraja

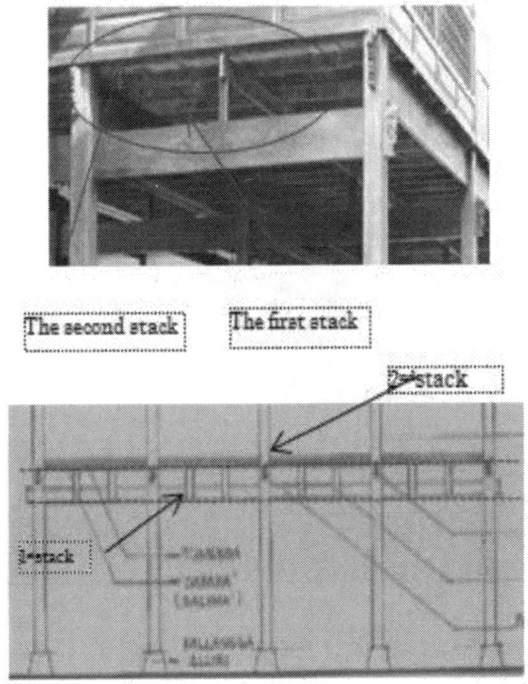

Figure 24. The construction of the column-beam of the Sauraja

3.3. The stylistic system

As the stylistic system, the elements of architecture correlated with models and styles of a house are observed. In this study, I investigated the form of roof, facade, opening types of the door and window and ornaments used as the stylistic system. As results, there were two main similarities between these two traditional houses. Firstly, these two traditional houses have the similar opening type of the door (the swinging french door) and window (casement window), where two door or window leaves can be opened or closed until 180 degrees. In addition, there was trellis on the window of both traditional houses made from wood bar with the amount of odd number and put vertically side by side. Secondly, the ornaments of these two traditional houses used the similar motifs, such as animals and plant motifs and also Islamic motif or Arabic calligraphy on the door or window, or carving pompeninie on walls, attics, and eaves outskirts (Figure 29 ~ 32).

In contrast, the differences can be found on the placement of the doors and the windows. There was a rule in placing of the doors and the windows in the Saoraja house where they should be put on a measured location with even numbers. In contrast, no regulation for the Souraja house. In addition, the difference was also clearly visible on the form of the roof. The roof of the Souraja house has a slope of +/- 30° and the basic form of the roof was a triangle with the gable roof type used and made in two storeys. The triangle part was covered by sheets of wood occupied by two small windows. On the other hand, the Saoraja house has the roof with the isosceles triangle as the basic form. Although this house also used the gable roof type, there were the frames, which crossed each other at the end of the ridge. The triangle part of the roof has a slope of 30°~40° which was covered by the sheets of wood made in 3~5 stacks and represented the status of the owner. Another difference was shown at the front view of the house or the facade. The Souraja house has a symmetric of view between the left side and right side of the house, despite that the Saoraja house looks asymmetric (Figure 25 ~ 28).

IOP Conf. Series: Materials Science and Engineering **306** (2018) 012077 doi:10.1088/1757-899X/306/1/012077

3.3.1. The form of the roof

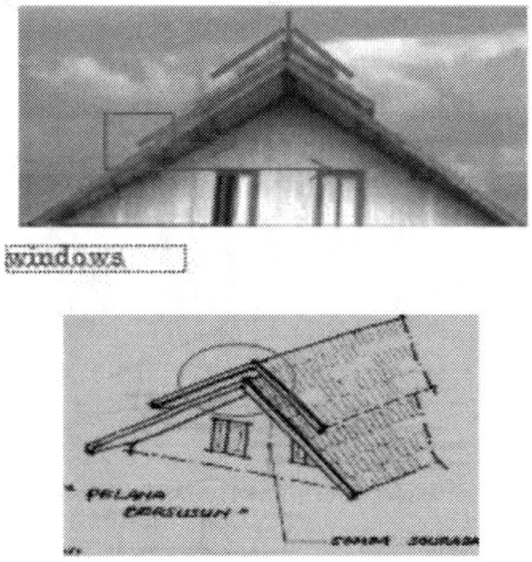

Figure 25. The construction of the column-beam of the Souraja

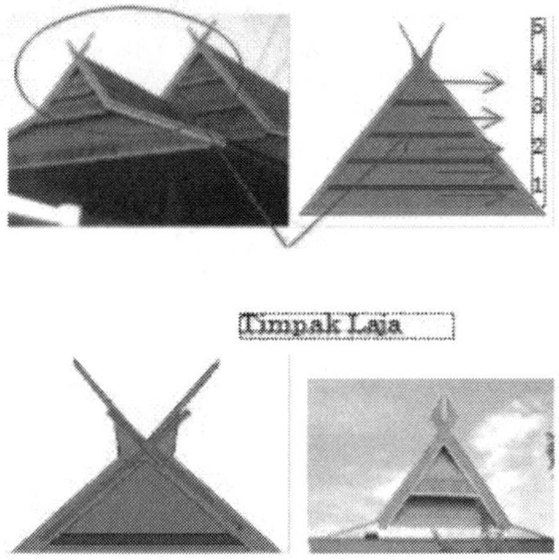

Figure 26. The construction of the column-beam of the Sauraja

3.3.2. The façade

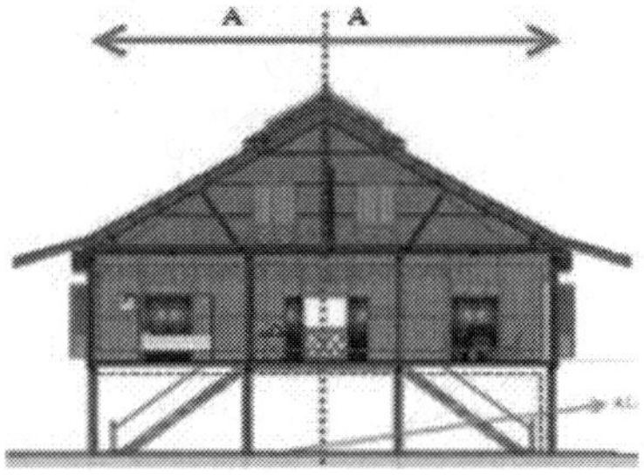

Figure 27. The facade of the Souraja

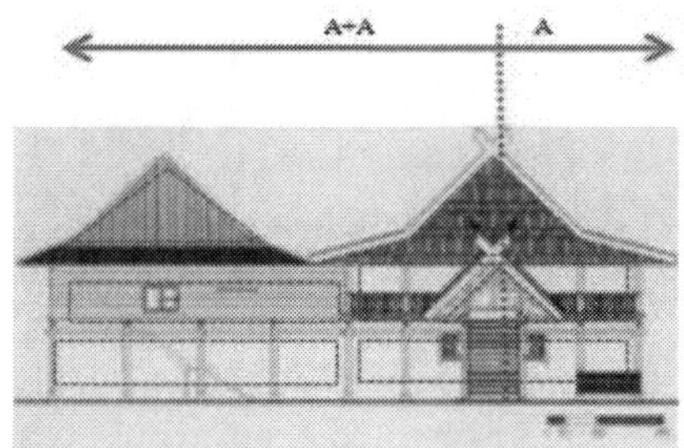

Figure 28. The facade of the Saoraja

3.3.3. The form of doors and windows

Figure 29. The form of door and window of the Souraja

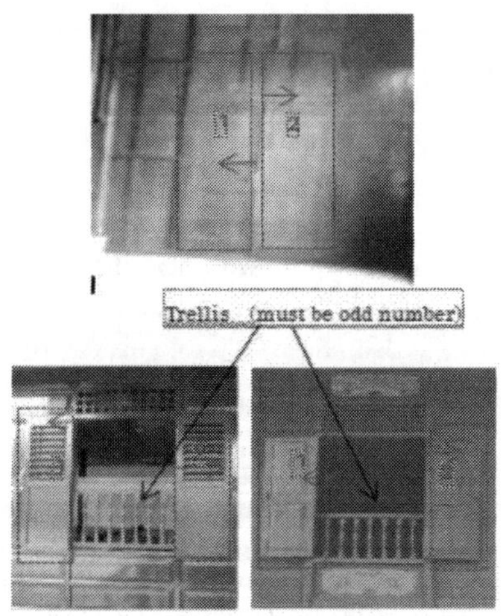

Figure 30. The form of door and window of the Sauraja

3.3.4. The form of doors and windows

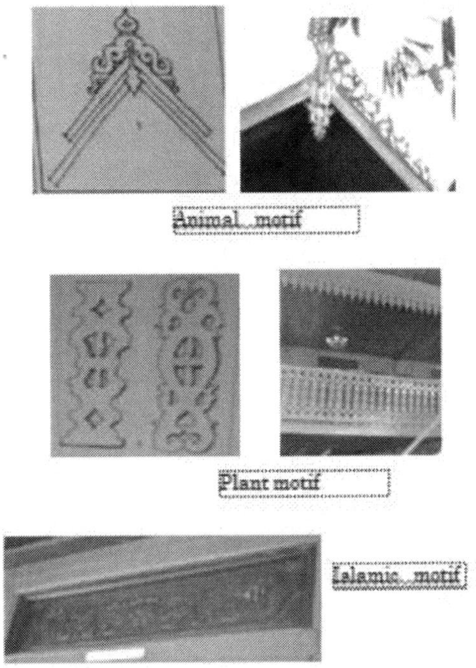

Figure 31. The ornaments of the Souraja

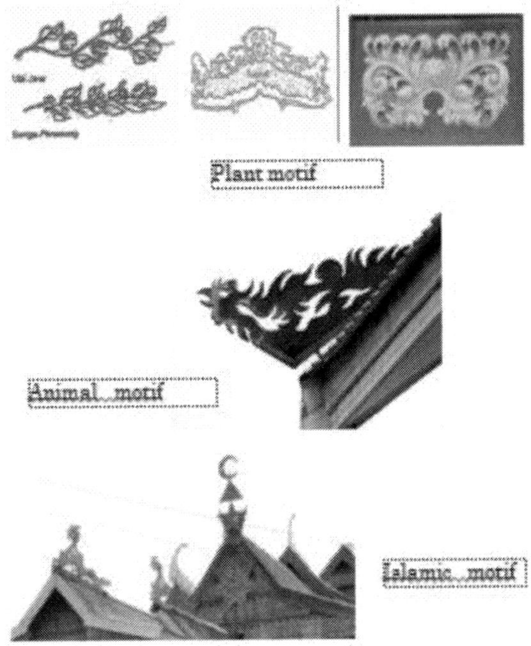

Figure 32. The ornaments of the Sauraja

4. Conclusion

Comparison of the Souraja house (Kaili) and the Saoraja house (Bugis-Makassar) based on observations of the physical elements of the architecture of the three systems are classified by N. John Habraken, can be summarized as follows; On the spatial system, I found the similarities of the physical elements of architecture on both traditional houses such as the orientation and the function as well as distribution of rooms. Meanwhile, on the physical elements such as; the arrangement of columns, form and spatial pattern as well as the placement of the stairs found differences of both traditional houses. Construction and material on the middle part of the house (floor and wall) and top part of the house (roof) shown similarities, whereas, construction and material on the bottom part of the house (foundation, column and beam) found differences on the physical system elements. The similarity shown from the physical elements of architecture on the stylistic system contained in the opening types of the door and window as well as ornaments used, while the differences was found in the physical elements of the form of the roof and facade.

References

[1] Crawfurd J 1820 *History of the Indian Archipelago* **II** (Edinburgh: Archibald Constable)

[2] Masiming and Mariani Z 1999 *Tipologi Arsitektur Tradisional Sulawesi Tengah, Penerapannya pada Arsitektur Masa Kini* Laporan Penelitian, Fakultas Teknik Universitas Tadulako, Palu-Sulawesi Tengah

[3] Rumagit and Ricky F S 2011 *Arsitektur Tradisional Orang Kaili* Departemen Kebudayaan dan Pariwisata Balai Pelestarian Sejarah dan Tradisional, Manado

[4] Wunas S, Amin S, and Oman W W 2005 *Morfologi Rumah Tradisional Makassar di Pesisir Pantai Galesong Kabupaten Takalar* Hibah Penelitian, Jurusan Arsitektur Fakultas Teknik Universitas Hasanudin, Makassar-Sulawesi Selatan 12-16

[5] H Syahriartato *Arsitektur Tradisional Sulawesi Selatan Pusaka Warisan Budaya Lokal Indonesia*, Persidangan Buku Antara Bangsa, Malaysia,
[6] Ibid
[7] Samingoen S 1984 *Album of Traditional Architecture* Ministry of Education and Culture Directorate general of Culture Project of Culture Media 63-64 1983/1984
[8] Ibid
[9] Habraken N J 1988 Type As A Social Agreement *A paper presented at The Asian Congress of Architects, Seoul, Department of Architecture Massachusets Institute of Technology* 5
[10] Dawson B and Gillow J 1994 The traditional architecture of Indonesia *Thames and Hudson*
[11] Schefold R, Nas P, and Domenig G 2004 *Indonesian Houses: Tradition and transformation in vernacular architecture* **207** NUS Press

IOP Conf. Series: Materials Science and Engineering **306** (2018) 012078 doi:10.1088/1757-899X/306/1/012078

Improved Information Retrieval Performance on SQL Database Using Data Adapter

M Husni*, S Djanali, H T Ciptaningtyas, and I G N A Wicaksana

Department of Informatics, Faculty of Information Technology,
Institut Teknologi Sepuluh Nopember, Indonesia

*mhusni43@gmail.com

Abstract. The NoSQL databases, short for Not Only SQL, are increasingly being used as the number of big data applications increases. Most systems still use relational databases (RDBs), but as the number of data increases each year, the system handles big data with NoSQL databases to analyze and access data more quickly. NoSQL emerged as a result of the exponential growth of the internet and the development of web applications. The query syntax in the NoSQL database differs from the SQL database, therefore requiring code changes in the application. Data adapter allow applications to not change their SQL query syntax. Data adapters provide methods that can synchronize SQL databases with NotSQL databases. In addition, the data adapter provides an interface which is application can access to run SQL queries. Hence, this research applied data adapter system to synchronize data between MySQL database and Apache HBase using direct access query approach, where system allows application to accept query while synchronization process in progress. From the test performed using data adapter, the results obtained that the data adapter can synchronize between SQL databases, MySQL, and NoSQL database, Apache HBase. This system spends the percentage of memory resources in the range of 40% to 60%, and the percentage of processor moving from 10% to 90%. In addition, from this system also obtained the performance of database NoSQL better than SQL database.

1. Introduction

Currently the flow of information through the internet has been very heavy and for the next few years, it is predicted to continue. As a result, the stored data is also getting bigger. This triggers a new field of knowledge known as big data. Along with this, there is also database to manage big data, known as Not Only SQL (NoSQL). Big data became more famous, after the development of cloud computing services. The United Nations Economic Commission for Europe (UNECE), predicts if the increase in the amount of data will reach 350% by 2019 when compared to 2015 [1].

The technique for managing the most widely used database is the relational database (RDB) with the Structured Query Language (SQL) method, where MySQL is one of the well-known Database Manager Systems for web-based application. This system is reliable and easy to use, so MySQL has become the main choice database used by high profile companies like Facebook, Twitter, and YouTube [2].

But with increasing amount of data, big data management for data retrieval operation, using SQL method is felt less economical and less efficient, because the manager must provide server with high I/O bus specification. NoSQL database is emerging as a result of growing data transactions over the internet exponentially using web applications. NoSQL itself has a schema-less specification and does not need to define statically how data should be modeled [3].

System which use NoSQL database, do not use the same syntax as retrieving data in SQL databases because the query syntax is different. Instead of changing the source code or changing the SQL database to NoSQL, this paper shows how to integrate two databases using data adapter.

NoSQL is designed to address the performance and scalability requirements of web applications that can't be handled by relational database. Many open source and commercial NoSQL are developed by various vendors. There are currently more than 150 NoSQL databases[4]. Ying-Ti Liao compares data adapter systems to support hybrid database architectures, including relational databases and NoSQL databases, using three different query approaches[5].

Data consistency issues are often experienced for data synchronization in an application that uses more than one database. Husni uses model-view-controller method to synchronize distributed application [6]. Jue Wang and Dong-song Zhang present the use of middleware to create a synchronization system for distributed databases [7]. Slightly different from middleware, Ramesh Dharavath and Chiranjeev Kumar researched three-tier architecture with distributed middleware to support atomic transactions in heterogeneous NoSQL databases [8]. Yubiao Wang presents an incremental web-based incremental synchronization mechanism on a heterogeneous database [9].

In this paper, a research conducted to synchronize SQL databases with NoSQL databases using data adapters for the case of database usage expansion in applications. The application is connected with SQL to handle small and medium-sized data, the NoSQL database server as a back-end system for analyzing data and performing a number of read or write operation, or periodically performing data backups from SQL.

2. Methodology

2.1. General system design

The system built is to synchronize two SQL and NoSQL databases. MySQL is used for Database Management System (DBMS) in SQL database. Apache HBase is used for NoSQL database. The total number of servers used in this study is three servers, one server for MySQL, one server for Apache HBase and one server for Data Adapter. Figure 1 shows the General Data Adapter System Design.

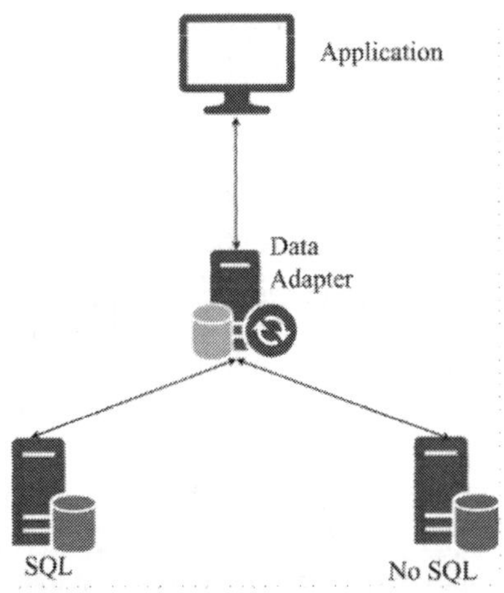

Figure 1. System Architecture

Data adapter is the link between the application and the database. Data adapters play a role in receiving requests from applications and transforming data from MySQL to Apache Hbase. The main component of this data adapter system consists of four parts: the relational database, NoSQL database, DB Adapter and DB Converter. DB Adapter serves to receive requests from applications such as insert data, update data, delete data and retrieve data. The data adapter provides an interfaces that can be accessed by the application to be able make changes and access the data in database. The interface provided by the data adapter return value in Application Programming Interface (API). This interface value result is JSON data type which made using Flask. Flask is a micro framework that runs in Python programming language. Application can accessed through port 5000. With this interface, applications can make changes and access data in the database without change the SQL query syntax.

Meanwhile, DB Converter is responsible for transforming data and reporting the transformation results recorded in the database. Data transformation is done from the RDB database to the NoSQL database. To receive requests from DB Adapter, DB Converter runs on port 5001. Apache Phoenix is ude to help the transformation process. In DB Converter is also installed a database using MySQL to store the synchronization log. The main components of this data adapter are built using Python programming language.

2.2. Data adapter design

Data adapter has the most important role in the system. Any request made by the application will be processed first in the data adapter, which will then be forwarded to the database. There are several examples of interfaces used by the application, in this research the interface used is API (JSON) [10].

Data adapter consists of two main components: DB Adapter and DB Converter. DB Adapter serves to receive all requests from applications such as query submission, data retrieval, data updates, and data deletion. In order to do this, the data adapter provides an interface in API form. API generates output data in JSON data type. All data communication performed by the application is through this interface. Flask is a micro framework used to create interface features [11].

DB Converter transform data from the SQL database to the NoSQL database. In this case, the transformation is from MySQL to Apache HBase. By default, requests for changes to data from the application will be redirected to MySQL, while for data access, the requests from applications will be directed to HBase. The data transformation process is using Apache Phoenix. Apache Phoenix allows system to translate SQL queries, compiles them into a series of HBase commands, and executes them to HBase. Python [12] is used to implement all existing processes in the data adapter.

2.3. SQL database design

The SQL database is used to store all relational application data. In the design and development of an application demanded the ability of the application to be accessible by many users. Many developers separate the application server with database server on the basis of this need. In this paper, a separate server from the application server is used as the SQL database. We use MySQL for SQL database. In order to use the application, it is necessary to create user accounts that have the authority to make data changes on the database server. All requests from applications related to data changes such as insert, update, delete, and retrieval of data can be executed in MySQL.

2.4. NoSQL database design

The NoSQL Database stores data in a different format with the SQL database. This database is an additional database because of the growing amount of data. We use Apache HBase for NoSQL database. To be able to use Apache HBase, Apache Hadoop, a file system that specifically handles big data problems, must install first on the server because Apache Hbase runs on top of Apache Hadoop layer [13] [14].

The installation of the NoSQL database is performed on one separate server from the SQL application and database. HBase will accept the transformed data from MySQL that is controlled through the data adapter. This transformation process needs help from Apache Phoenix as a SQL query translator.

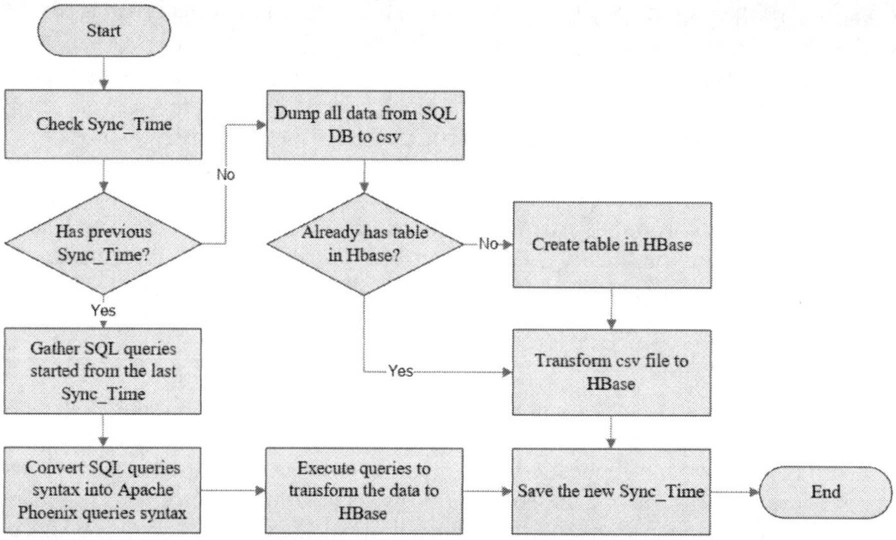

Figure 2. Sync Process

2.5. Transformation and synchronization process

The transformation process is done by running the application through the terminal manually. The first thing to do when synchronizing is check the synchronization log database, whether system ever do synchronize before. If it has not been done before, then system will do the initialization process. Flow diagram of the general synchronization process can be seen in Figure 2.

The initialization process is the process of copying all data in the SQL database to the NoSQL database. This process begins by dumping all data in a MySQL table into a .csv file format. Then the system will create a new table similar to the existing table in MySQL. Once completed, the next process is to transform all data to HBase. This data transformation process is handled using Apache Phoenix.

If it has been synchronized before, then the next process is the system retrieves all queries from the logs in the MySQL database. Remote MySQL database system, then read the log list in */var/log/mysql/log_query.log* file. Logs that are read are based on the last synchronization time. From the log, system checks all queries which changes the data such as insert, update and delete query syntax. The query is taken and then converted to Apache Phoenix query syntax. All new syntax then stored in a single file with the .sql format. With Apache Phoenix, all queries in the .sql file will be transformed to HBase, so all the data changes that occur in the SQL database will occur also in the NoSQL database. If the synchronization process succeeds, the system will record the log and sync status to the synchronization log database. The database used for logging uses MySQL and is separate from SQL database server, NoSQL database server or application server.

2.6. Interfaces design

The application accesses the database by running SQL queries through the interface or routes provided by the data adapter. With query direct access approach, the data adapter will directly forward the query syntax to the intended database and return the value of the database output. From the route accessed by the application, the system will first check whether the query is changing data in the database or retrieving data. If the query is insert, update or delete, then the query command will be executed to the MySQL database. If not, or in this case the query is select, then the system will make sure whether the synchronization process is in progress or not. By default, the query select will be directed to the HBase database, but if the synchronization process is in progress then the query will be redirected to the MySQL database. The system then prints the output obtained from the database in the form of JSON.

3. Implementation

In this research, we uses data set from *openflight.org* which contains data about aviation around the world [15]. This data consists of three tables, there are routes, airlines and airports. The number of rows and tables is not fully used but changed according to the scenario to be performed.

3.1. Database initialization

Initialization of the database will create a new table in the NoSQL database. If the table in HBase is already created, then the system will export data in the SQL database into .csv file format. This file then executed to the NoSQL database. This test will ensure that this process runs when the synchronization process in the SQL and NoSQL databases has never been done before, or in other words the system run the synchronization process for the first time. Figure 3 shows the print screen when the initialization process run.

3.2. Database transformation

The transformation process takes place when there are data changes in the SQL database. Changes occur from the query of applications such as insert, update and delete. If the previous system has already done the transformation, then the initialization process will not take place. The system will take a list of queries that are executed in the SQL database, then run into the NoSQL database so that the data in both databases is same. Figure 4 shows the print screen when the transformation process is in progress.

```
sinkron terakhir:  0
waktu sekarang:  2017-05-22 15:11:15.602675
time stamp:  2017-05-22 15:11:15
sql :  INSERT INTO log_sinkronisasi (waktu, ip_src, ip_dst, status) VALU
17-05-22 15:11:15', '10.151.36.22', '10.151.36.29', '0')
Belum pernah sinkronisasi
sql:  SELECT * FROM routes2
file_output:  file/csv/fetch_all_routes2.csv
Berhasil, tabel  routes2  berhasil di dump ke fetch_all_ routes2 .csv
sql:  SELECT * FROM airline2
file_output:  file/csv/fetch_all_airline2.csv
Berhasil, tabel  airline2  berhasil di dump ke fetch_all_ airline2 .csv
sql:  SELECT * FROM airport2
file_output:  file/csv/fetch_all_airport2.csv
Berhasil, tabel  airport2  berhasil di dump ke fetch_all_ airport2 .csv
Data berhasil di export!
Inisialisasi data awal..
Mentransformasi data ke HBase...
Tabel berhasil dibuat di HBase
Tabel routes2 berhasil diimport ke HBase...
Tabel airline2 berhasil diimport ke HBase...
Tabel airport2 berhasil diimport ke HBase...
Proses inisialisasi selesai..
Durasi sinkronisai : 33.0158581734 seconds
Done
EXIT!
```

Figure 3. Initialization process.

```
sinkron terakhir:  2017-05-23 10:13:25
waktu sekarang:  2017-05-23 10:15:06.701635
ime stamp:  2017-05-23 10:15:06
ql :  INSERT INTO log_sinkronisasi (waktu, ip_src, ip_dst, status) VALUES ('2017-05-
3 10:15:06', '10.151.36.22', '10.151.36.29', '0')
inkronisasi terakhir:  2017-05-23 10:13:25
Mengambil data Query dari log ...
Cek aksi INSERT, UPDATE, DELETE pada log...
[sync] INSERT  airport2
[sync] INSERT  airport2
[sync] INSERT  airport2
[sync] UPDATE  airport2
nama tabel:  airport2
PK:  airport_id
[sync] DELETE  FROM
[sync] DELETE  FROM
Sinkronisasi, DELETE, INSERT dan UPDATE file on progress sync to HBase...
Update status sinkronisasi berhasil
Sync berhasil dilakukan.
Durasi sinkronisai : 4.45966291428 seconds
Done
EXIT!
```

Figure 4. Transformation process.

3.3. Data adapter interfaces

The data adapter interface provides a route that can be accessed by the application to perform a process to the database. This feature also assigns to select from database where application request will be processed. This test is expected each request from the application in the form of data changes such as insert query, update and delete, will be directed to the SQL database and for data retrieval requests will be directed to the database NoSQL.

In addition to find out whether or not the interface is running, it is necessary to record the response time required by the system until the process ends in the table.

4. Experimental Result

The response time is recorded as many as five experiments, then the average value is taken. Interface details can be seen in Table 1 (column 2) and the experimental result is shown in Table 1 (column 3).

Table 1. Functionality test result.

Interface	Explanation	Result
/sinkron	Synchronize database	Success
/insert_routes	Add new line to routes table	Success
/delete_routes_by_id	Delete a row data in routes table	Success
/update_routes_by_id	Edit data in routes table	Success
/select_all_routes	Retrieve all data rows in routes table	Success
/select_route_by_id/<id>	Retrieve one row data in routes table	Success
/insert_airline	Add new line to airlines table	Success
/delete_airline_by_id	Delete one row data in airline table	Success
/update_airline_by_id	Edit data in airline table	Success
/select_all_airline	Retrieve all data rows in airline table	Success
/select_airline_by_id/<id>	Retrieve one row data in airline table	Success
/insert_airport	Add new line into airport table	Success
/delete_airport_by_id	Delete one row data in airport table	Success
/update_airport_by_id	Edit data in airport table	Success
/select_all_airport	Retrieve all data rows in airport table	Success
/select_airport_by_id/<id>	Retrieve one row data in airport table	Success
/select_routes_country/<id_route>	Retrieve one row data from route table with its country detail	Success
/select_all_number_routes_by_airline	Retrieve all airline's data and routes number	Success
/select_all_number_routes_more_than/<number>	Retrieve all flight number as per the option	Success
/select_all_airline_in_routes	Retrieve all available airline's data on routes table	Success
/select_all_airline_source	Retrieve all airlines data and its source flight	Success

4.1. Comparison number of rows in table

This test is performed to find out how many resources are used and the response time from the MySQL database to HBase during the initialization process takes place by comparing the number of lines. Transformation times can be seen on figure 5, figure 6 and figure 7. The ratio of the number of rows used in the table for each experiment is as follows:

- Airplane: 6.000, 12.000, 18.000, 24.000, 30.000
- Airport: 7.000, 14.000, 21.000, 28.000, 35.000
- Route: 60.000, 120.000, 180.000, 240.000, 300.000

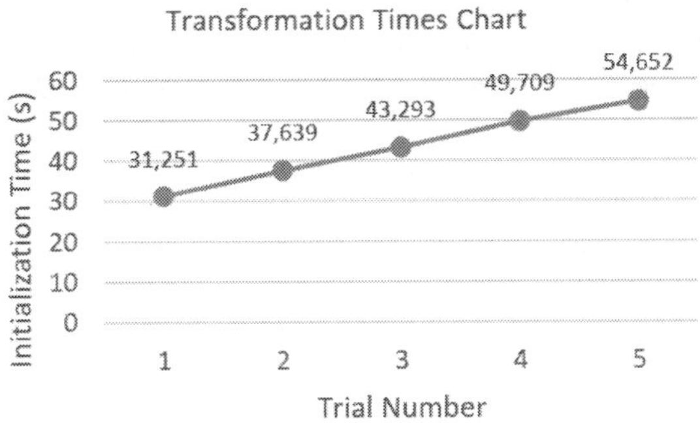

Information:
- *AL : Airline's table*
- *AP : Airport's table*
- *RT : Route's table*

No	1	2	3	4	5
Number of row	AL : 6.000 AP : 7.000 RT : 60.000	AL : 12.000 AP : 14.000 RT : 120.000	AL : 18.000 AP : 21.000 RT : 180.000	AL : 24.000 AP : 28.000 RT : 240.000	AL : 30.000 AP : 35.000 RT : 300.000

Figure 5. Transformation time chart based on initialization time.

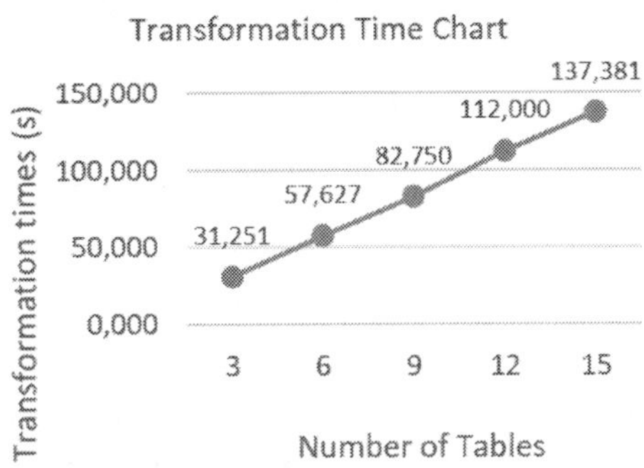

Information:
- *AL : Airline's table*
- *AP : Airport's table*
- *RT : Route's table*

Tables	3	6	9	12	15
Detail	AL : 1 AP : 1 RT : 1	AL : 2 AP : 2 RT : 2	AL : 3 AP : 3 RT : 3	AL : 4 AP : 4 RT : 4	AL : 5 AP : 5 RT : 5

Figure 6. Transformation times chart based on number of tables.

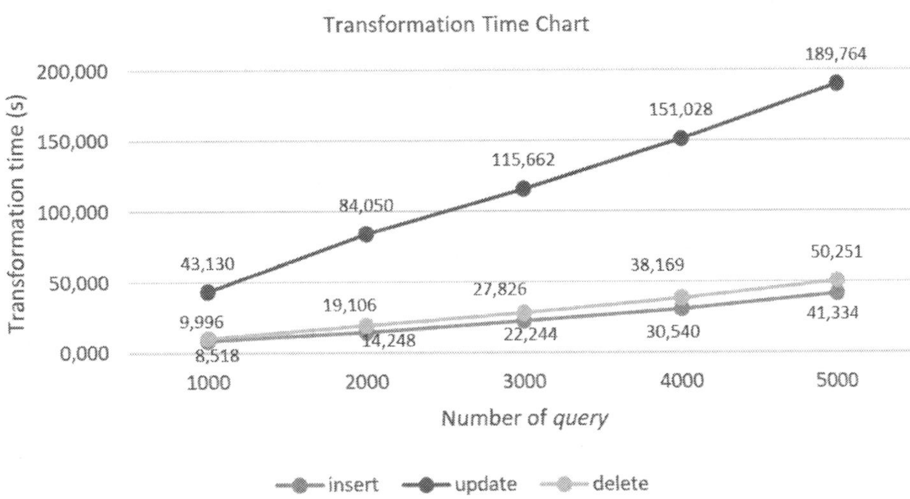

Figure 7. Transformation time chart based on number of query.

4.2. Comparison number of table in databases
This test's aim is to find out how much resources used and the response time of the data adapter system during the initialization of data from the MySQL database to Hbase. The comparison number of tables used for each experiment is 3, 6, 9, 12, and 15 tables. Transformation times can be seen on Fig. 4.

4.3. Comparison query number insert, update, and delete
This test's aim is to find out how much resources and time spent by the system to execute three different query operations such as insert, update and delete from MySQL to HBase on the data transformation process. In each query operation, the comparison of query numbers used is starting at 1.000, 2.000, 3.000, 4.000 and 5.000 queries. Transformation times can be seen on Fig. 5.

4.4. Evaluation
From functionality testing, all functions can run as expected. However, for the data retrieval process on MySQL, there is one interface with a response time that exceeds the limit. MySQL is not good enough to handle queries that require retrieval data to more than one table in tables that have large amounts of data. On the other hand, HBase is able to handle all query processing requests from any tested interface.

In terms of capacity and performance, for all test scenarios, the percentage of RAM memory resources used showed a linear increase, no changes soaring or decreasing significantly, ranging from 1.736,6 MB to 2.431,8 MB. While the processor resources used show a decrease in percentage for each increase in the number of transformed queries. This decrease is also shown in the test results for the ratio of the number of rows and the ratio of the number of tables at the time of the initialization process. However, the difference occurs at fourth attempt. This is because Apache Phoenix only requires a high percentage of processors when the process begins or when Apache Phoenix starts executing queries. For the rest, during the synchronization process takes place, the percentage value of use tends to be lower and moves at a consistent rate. Apache Phoenix requires more RAM memory resources to perform the transformation process than the percentage of processor usage.

The average percentage of the processor used for the initialization process is 37.389%. Query transformation process spent less processor resource average than initialization process, that is 26,546% for insert, 16,208% for update and 24,943% for delete. The update query transformation process has the smallest percentage of processor usage, but in contrast to its longest transfomation time between other operations. The average transformation time required for each operation is 23.4 seconds for insert, 116.7 seconds for update and 29.0 seconds for delete.

5. Conclusion
From the experimental results, we can conclude:
1. Data adapters can be used as a system to synchronize between SQL databases and NoSQL databases. By using direct access query approach, the application can still access the database while the synchronization process is in progress.
2. Based on the test results, for each increase in the amount of data or query, the time required to initialize is linear, both in the initialization process and the transformation.
3. NoSQL database, i.e. Apache HBase, has a better capability when compared with relational database, which is MySQL, in terms of query analysis for data retrieval of multiple tables with a large number of rows.
4. Data synchronization between SQL databases with NoSQL databases using data adapter improves information retrieval performance on SQL database.

Suggestions given for further development of this system are as follows:
1. Data adapter mechanism needs to be developed by comparing other NoSQL database type.
2. Further research is needed to synchronize the database with more number of nodes, and how the system is capable of handling failures which occur on the database server or data adapter server.

Acknowledgments
This research was supported by Lab Research Grant (Non PNBP Fund) 2017, Institut Teknologi Sepuluh Nopember (ITS), Surabaya, Research ID 869/PKS/ITS/2017.

References

[1] G Audin 2017 *No Jitter* (Online) available at http://www.nojitter.com/post/240170228/the-network-impact-of-big-data

[2] Oracle 2017 *Oracle* (Online) available at https://www.oracle.com/mysql/index.html

[3] Couchbase 2017 *Couchbase* (Online) available at https://www.couchbase.com/nosql-resources/why-nosql

[4] D G Chandra 2015 BASE analysis of NoSQL database *Future Generation Computer Systems* **52** 13-21

[5] C Liao and C Ying-Ti 2016 Data adapter for querying and transformation between SQL and NoSQL database *Elsevier* **65** 111-121

[6] Muhammad H, Henning T C, and I G N A Wicaksana 2016 Model View Controller for Distributed Application *The International Conference on Engineering, Technology, and Industrial Application (ICETIA 2016)* Solo

[7] D Jue and S Z Wang 2015 Research and Design of Distributed Database Synchronization System *8th International Conference on Intelligent Computation Technology and Automation*, China

[8] C Ramesh and K Davarath 2015 A Scalable Generic Transaction Model Scenario for Distributed NoSQL Databases *The Journal of Systems and Software* **101** 43-58

[9] X Junhao, R Wen, W Yubiao, and F Wang 2015 Research on Incremental Heterogeneous database Synchronization Update Based on Web Service *International Conference on Computational Intelligence and Communication Networks*, India

[10] Oracle 2016 *MySQL* (Online) available at https://www.mysql.com/about/

[11] A Ronacher 2015 *Flask* (Online) available at http://flask.pocoo.org/

[12] Python Software Foundation 2017 *Python* (Online) available at https://www.python.org/

[13] Apache HBase Team 2017 *The Apache Software Foundation* (Online) available at http://hbase.apache.org/book.html#_read_hbase_shell_commands_from_a_command_file

[14] Apache Software Foundation 2017 *Apache Software Foundation* (Online) available at https://phoenix.apache.org/who_is_using.html

[15] OpenFlights 2017 *OpenFlights* (Online) available at https://openflights.org/data.html

IOP Conf. Series: Materials Science and Engineering **306** (2018) 012079 doi:10.1088/1757-899X/306/1/012079

The Evaluation of Industry Practical of Mechanical Engineering in Vocational Education: A CIPP Model Approach

M Kamaludin, W Munawar, D Mahdan, M V Simanjuntak, and H F Wendi
Technical and Vocational Education, Universitas Pendidikan Indonesia
Jl. Dr. Setiabudhi 207 Bandung, 40154 Indonesia

*mkamaludin@live.com

Abstract. The learning system is not only studied on campus but also practicing in the world of work. Industry Practical aims to enable students to develop their skills in accordance with the real world of work. To know the success of the implementation of industry practical program then held evaluation. The evaluation of the program in this study used the CIPP evaluation approach (Context, Input, Process, Product). The purpose of this research is to know the extent of achievement and success of industry practical program at vocational school in Bandung with descriptive research method using mix method approach. The sample in this research is students majoring in mechanical engineering in the city of Bandung who have done industry practical.

1. Introduction

Implementation of vocational education is currently entering an important stage, the stage where vocational education graduates will be at stake in the workforce in the Asian region. Efforts must be made is to make the arrangements and improvements as much as possible in the vocational education sector, both arrangement in the pattern of recruitment, development of education and training programs, innovation of education and training process, evaluation and certification development.

One of the efforts undertaken among others is by cooperating with the business world / industry intensively so that together develop the program and curriculum of education and vocational training and its supporting tools are education and training standard, competency standard, test standard and certification.

In the 2013 curriculum there are some things that need to be focused on *Prakerin* / Industry practical, such as areas of expertise must be in accordance with global needs, the addition of life and career skills, the need to involve users (in this case related industries) in preparing curriculum, project-based learning and schools open for students for longer periods of lessons, balance of hard skills and soft skills and also the need to establish a conducive school culture [1,2,3,4].

The purpose of Industry practical implementation is to prepare students to become productive people, able to work independently, to fill the existing job vacancy in Business World and Industrial World (DUDI) as middle-level worker, in accordance with the competence in the program of expertise of choice, to equip the students to be able choose a career, tenacious and persistent in competing, adapt in the work environment and develop a professional attitude in the field of expertise of interest and equip learners with science, technology, and art in order to develop themselves in the future either independently or through higher education.

If the Industry Practical goes according to the stated goal, then it will affect job creation and reduce unemployment. The dual system has proved its success, as it is shown by its ability to react quickly and effectively to the many changes in the economy and in the Society, and the fact that youth unemployment is low [5]. The number of unemployment is influenced by the quality of human resources while the quality of human resources is influenced by the education system. In Industry Practical Learners get enough knowledge in school and gain sufficient skills from industry.

In an advanced company Industrial practice is implemented by applying the Practice participants together with the employee themselves. The positive aspect of education undertaken at the factory is that education is close to the reality of Practice, education can be quickly adjusted to new requirements arising in the labor market, and lower tuition costs compared to institutional costs others because learners come to work [6].

It is a classic problem for SMK education in Indonesia in general that the link and match between the output of vocational education with the business world / industry (DU / DI) as the user of SMK education output has not been achieved. One of the problems lies in the quality of SMK graduates who are not in accordance with the standard of skills required by the labor market. Various efforts have been made but until now the implementation of Industry Practical is still not optimal benefits because of many obstacles encountered in the field.

There are five main issues that tend to hinder the success of training programs are: (1) Training is not intended to refer to a specific purpose, (2) The purpose of the training program is not clearly identified, (3) The instructor is not prepared or the material does not fit the training objectives, (4) Training participants are not interested in joining the program, (5) There is no provision for evaluating the results of training results [7].

One of the competencies of the vocational education expertise in Bandung that carries out *Prakerin's* activities is on mechanical engineering skills. Mechanical engineering specializes in discussion or learning about mechanical engineering matters. The industrial practice work program that students have done needs to be evaluated to see the suitability between the planned program and its implementation. This is intended as a basis for preparing a follow-up program that must be done on the achievement of student competence as well as on Industry Practical program.

The effectiveness of the implementation of the program is not only seen from the students only factors but other factors must also be considered. For example, teachers, curriculum, facilities and infrastructure, teaching and learning activities in schools, industry practical, industrial relations or partner institutions and other factors. Based on the existing problems, it is necessary to evaluate the Industry Practical program of mechanical engineering competence. Evaluation to be conducted is evaluation of student, productive teacher, Industrial Practical guidance teacher, curriculum, infrastructure, teaching and learning activity at school, practice activity in industry.

2. Objective

This research aims to see how far the achievement and achievement of Industry Practical program in terms of context, input, process and product.

3. Research Methodology

This research was conducted to evaluate the implementation of industry practical program vocational school students in Bandung. Therefore, this research is included in the type of evaluation research with descriptive method with mix method approach. This research data is obtained through questionnaire given to the student and teacher mentor after *prakerin* done. In addition to using questionnaires, research data also obtained through observation, interviews and documentation. This research is population research, where there are 91 respondents and to support the respondent's answer the researcher also conducted interview with 5 speakers consisting of vice principal of public relations, head of study program and productive teacher.

The conceptual framework refers to a set of related concepts to each other logically with respect to the phenomena [8]. Conceptual frameworks are used as a guide for conducting research [9]. For example, to identify a research question. Research questions are built within a conceptual framework [10].

Prakerin programs that have been done by students need to be evaluated to see the suitability between the program and its implementation. This is intended as a basis for the preparation of follow-up programs that must be done both to the achievement of student competence and to *prakerin* program.

To carry out this research, the researcher has developed a conceptual framework based on the Stufflebeam model evaluation study [9], while the evaluation used is CIPP as shown in the following figure:

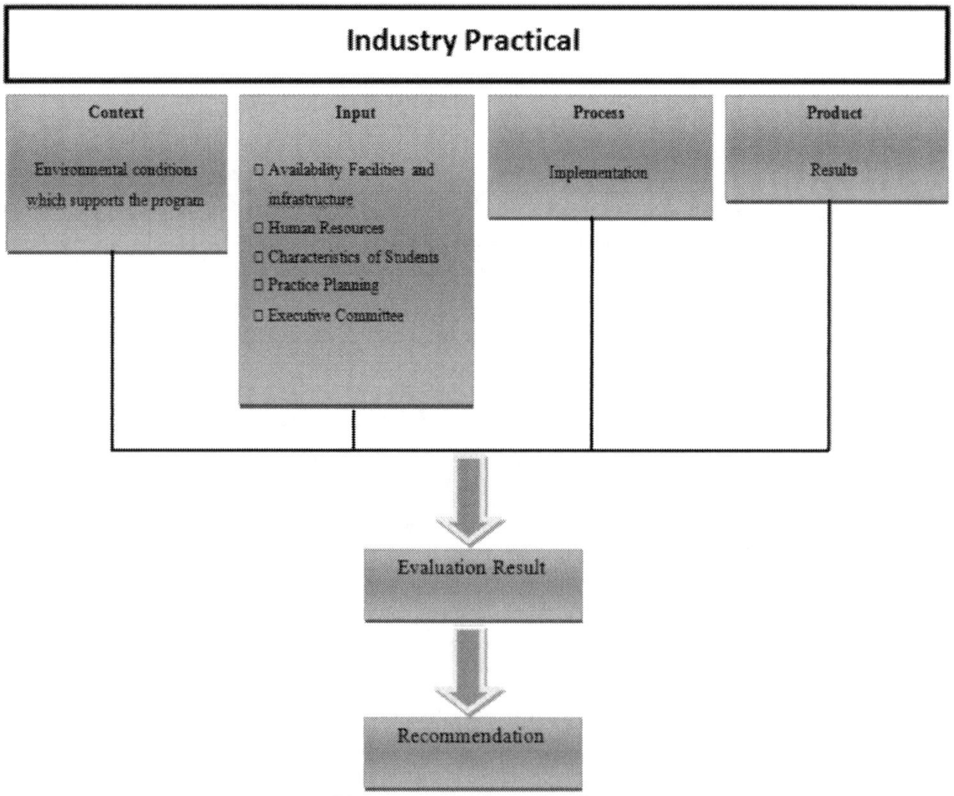

Figure 1. CIPP evaluation

According to the figure above, there are four dimensions that must be considered in the implementation of *prakerin* program that focuses on the dimensions of context, input, process and product. The Stufflebeam Evaluation Model [11] involves a process evaluation of context, input, process and product. The assessment context refers to needs, problems and opportunities as a basis for determining goals and priorities in determining whether a product is significant or not. Input evaluation involves various approaches and program planning inputs needed to implement the program. The assessment process focuses on the implementation of activities that can help achieve targets. Product evaluation is to identify the results that have been achieved or not to know how far the effectiveness of the program achieved.

This study was conducted to find out how successful Prakerin program focuses on educational goals and learning outcomes for vocational school students in Bandung Indonesia in terms of context, input, process and product dimensions [12,13].

IOP Conf. Series: Materials Science and Engineering **306** (2018) 012079 doi:10.1088/1757-899X/306/1/012079

4. Results and Discussion

The results showed that the average aspect of the context in program is 32.54% which is included in very good category. The average that occurred on the input aspect of 48.07% included in either category. Aspects of the process in program showed an average of 33.65% included in either category. While for average aspect of product in program equal to 24,79% which is included in high category. There is a positive relationship between the results of the evaluation Context, Feedback, Process together to the evaluation of the Product, contribution relationship = 82.1% the remaining 17.9% contribution of other factors; a positive relationship between context evaluation and product evaluation = 0.656 occurs strong relationship; the relationship between input evaluation with product evaluation = 0.449 occurs a strong enough relationship; the relationship between process evaluation and product evaluation = 0.895 occurs a very strong relationship.

5. Conclusions

Prakerin programs that have been done learners need to be evaluated to see the suitability between the program and its implementation. Components of the organization Industry practical include students, productive subject teachers, mentors, and mentors from DUDI. When compared to other scoring models, CIPP contains context, Input, process and product dimensions can be researched using four dimensions or using any combination of CIPP dimensions. This depends on the program requirements. The CIPP appraisal model is more emphasized to collect information in order to facilitate decision-making. This is intended as a basis for the preparation of follow-up programs that must be done both to the achievement of competence of students and to *Prakerin* program. This information will also be used as a guide to further improve the program both on planning, structuring, implementation and outcome. Consequently, future decisions will be made whether to proceed, modify, add or discontinue the program.

References

[1] Depdikbud 2013 *Pemendikbud RI Nomor 54 tahun 2013* tentang Standar Kompetensi Lulusan

[2] Depdikbud 2013 *Pemendikbud RI Nomor 65 tahun 2013* tentang Standar Proses

[3] Depdikbud 2013 *Pemendikbud RI Nomor 66 tahun 2013* tentang Standar Penilaian

[4] Depdikbud 2013 *Pemendikbud RI Nomor 70 tahun 2013* tentang Kurikulum SMK

[5] Rainer 1992 *Vocational Training in The Dual System* Hamburg: Hamburg Institute

[6] Helmut N 1983 *Pendidikan Kejuruan* Jakarta: Gramedia.

[7] Donaldson L and Scanell, Edward E 1993 *Pengembangan Sumber Daya Manusia Panduan Bagi Pelatih Pemula* Jakarta: Gaya Media Pratama

[8] Merican S 2006 *Penyelidikan Sains Social. Pendekatan Paragmatik* Kedah: Hexa

[9] Stufflebeam D L 2000a *The CIPP model for evaluation. In Stufflebeam, D.L., Madaus, G.F. Kellaghan, T. (Eds). Evaluation models. Viewpoints on Educational and Human Service Evaluation* 279-317 2nd edition Boston: Kluwer Academic

[10] Marshall C and Rossman G B 2006 *Designing Qualitative Research, Fourth Edition* London: SAGE Publications

[11] Stufflebeam D L 2002 *CIPP Evaluation Check List. A Tool for Applying the Fifth Installment of The CIPP Model to Assess Long-Term Enterprises*. Retrieved 11 Agustus 2017.http://www.wmich.edu/evalctr/checklists/cippchecklist.htm

[12] Stufflebeam D L, McKee H, McKee B 2003 The CIPP Model for Evaluation *the Annual Conference of the Oregon Program Evaluators Network 2003*

[13] Stufflebem D L 2003 The CIPP Model for Evaluation *Journal of International Hanbook of Education, International Hanbook of Education* Kluwer International Hanbook of Educational **9** 31-62

Job and Workload Analysis System for Civil Servants in North Sulawesi Province, Indonesia

M Krisnanda*, **A Mewengkang, P T D Rompas and P V Togas**

Department of Information Technology and Communication Education, Universitas Negeri Manado, Jalan Kampus Tondano 95618, North Sulawesi, Indonesia

*madekrisnanda@unima.ac.id

Abstract. This study aims to create automation system to assist sub division of job analysis and workload of North Sulawesi, Indonesia in completing its annual tasks to record all civil servants. The method used in this research is prototyping where a system will be developed according to the revision supplied by the user. The result of this research is a system used to generate reports and calculations based on user filled forms. The result gathered from the interviews, users can make reports faster and easier. Users can also optimise reports on time using the built system.

1. Introduction

In the effort to build clean and good government, the improvement of usability, effectiveness, transparency and accountability of government administration becomes very important and urgent [1]. But it would be a utopian joke if the governance work is still abstract, unspecified, and overlapping [2]. As a result, work aspects such as workload, work, implementation mechanisms and workflows become irregular and unpredictable, which ultimately raises major doubts about the accuracy, precision and reliability of government performance [3].

The provincial government of North Sulawesi is one of the provinces experiencing difficulties in conducting job analysis. The problems faced by North Sulawesi government today are the job analysis model and the workload is still done manually, which is very time and energy consuming. As a result, the level of achievement of North Sulawesi Government position analysis does not reach 100%, for example achievement analysis and work load in 2015 which is still at 90% level even though it has been done for full year. This makes the North Sulawesi government difficult to obtain complete information to make adjustments to the amount of remuneration and performance allowances for civil servants.

To solve the problems, several researchers have suggested the following strategis: using metadata analysys [4] assessment by simulation [5], managed acceleration for In-Memory database analytic workloads [6], and classifies the current cognitive task analysis methods for job or task design and analysis [7]. However, several problems still persisted. So building a modern system, effective and easy to use in conducting job analysis becomes an urgent need [8] for the Government of South Sulawesi. Information technology strategy is proposed as a solution to the problems that occur. An information technology-based system will be built to link human components and performance, hardware, and software. The mechanism of job analysis is stipulated in a standardized, consistent and reliable program, which is expected to be a useful application of information technology model to support the performance of local government, especially in North Sulawesi .

Specifically this research aims to build a system that helps the implementation process of job analysis and analysis of civil servant's workload in North Sulawesi Provincial Government and its report of recapitulation.

2. Methods

Tools (software) designed in this study, is expected to overcome the problems obtained from problem analysis by providing the following functions; (1) Provide easy-to-fill job analysis forms and easily accessible guides and dictionaries. (2) Establish inspection methods performed during charging, in an effort to build a consistent database. (3) Produce reporting results right after completion of the filling, so that the inspection and reporting time can be shortened.

System development is done using prototyping method which is used stages to build it [9]. These stages including: (1)Interviewing users in Job and Workload Analysis Department for new system requirements, (2) A preliminary design is created for the new system, (3) A first prototype named Sistem Informasi Analisis Jabatan dan Beban Kerja (SINJAB) built based on the design, (4) The users thoroughly evaluate the first prototype, noting its strengths and weaknesses, what needs to be added, and what should to be removed. Revision made mostly at interface design and form, (5) The second prototype is evaluated in the same manner as was the first prototype, (6) The preceding steps are iterated three times, until the users are satisfied that the prototype represents the final product desired, (7)The final system is thoroughly evaluated and tested. Routine maintenance is carried out on a continuing basis to prevent large-scale failures and to minimize downtime.

There are two types of positions that need to be analyzed in accordance with the structure of North Sulawesi provincial government. Both types of positions are structural positions and general functional positions. Process Analysis on Sub Division of Position Analysis is done to know the flow of each process of implementation analysis of civil service positions in the Government of North Sulawesi province.

The design of information technology-based systems using Windows platform with Visual Studio media [10]. Visual Studio.NET is an integrated environment to build and perform Testing and Debugging of various applications. The Visual Studio .NET platform, will be easier to create applications because in Visual Studio .NET there are new support facilities added, including Integrated Development Environment (IDE), Microsoft Intellisense, better debugging and capabilities in XML Web Services.

Table 1 shows analysis of needs to be built in accordance with the needs of Sub Division Job Analysis.

Table 1. Needs analysis.

Data source	Process	Outcome
Job description and specification in each regional level work units, Dinas, and Bureau	Conducting suitability checks and similarity of input data according to position analysis guidance	Job analysis document and workload analysis
Workload in each regional lever work units, Dinas, dan Bureau		Database of job analysis and workload

Then it can be concluded that the main data needed is the job description information, job specifications and workload of regional level work units, service, and bureau. As per the form, the filled data consists of 34 points. Each point must be a process of examination both in terms of conformity and similarity of data input. This is necessary because the result of filling in the job analysis form will be attached along with the recapitulation of job analysis report and work load. This report will then determine the performance allowances and remuneration of civil servants.

3. Results and Discussion

The result of the research is a system called Sistem Analisis Jabatan dan Beban Kerja (SINJAB). This system needs to be installed on the user's computer with Windows operating system with version 7.0 and above. Figure 1 shows the menus of SINJAB. SINJAB can be used after the user enters the username and password. The SINJAB system consists of 4 menus, but the analysis uses only 1 menu.

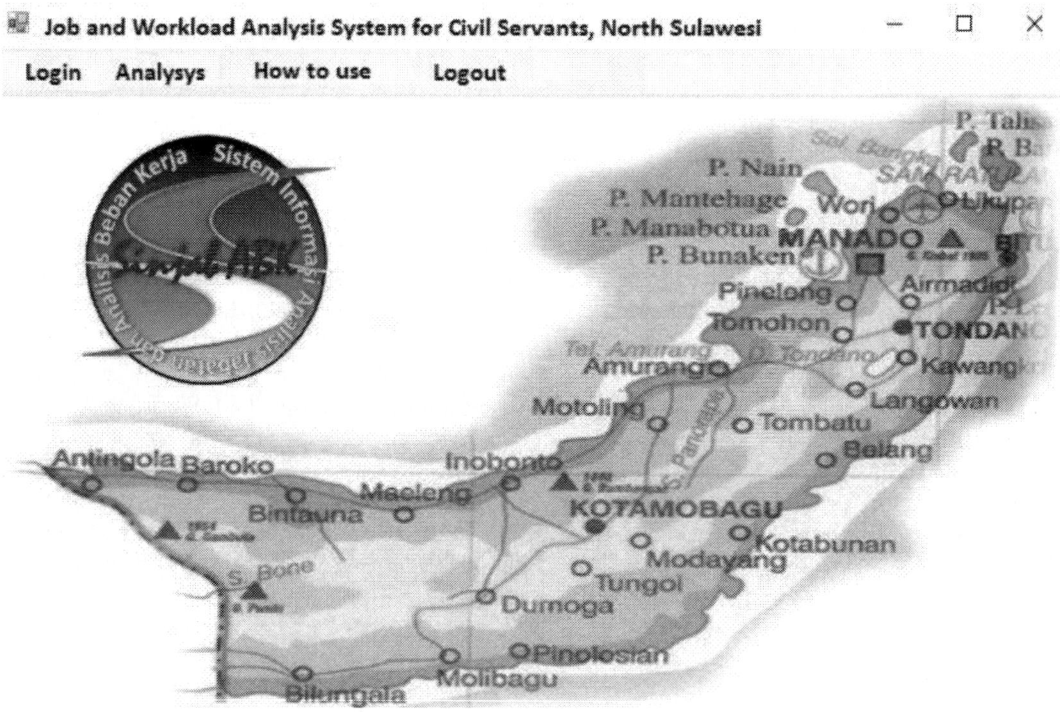

Figure 1. SINJAB's home screen.

SINJAB consists of the following functions : (1) Provide easy-to-fill job analysis forms and easily accessible guides and dictionaries, (2) Make a method of inspection conducted during charging takes place so that the data input entries can be guaranteed, and (3) Make reporting results right after the completion of the filling, so that the examination time and preparation of reports can be shortened.

Table 2 shows how the system works can be described as follows:

Table 2. SINJAB's works.

Input	Process	Output
a. Position data of Work Unit Regional, Department, and Bureau b. Data on workload of Work Unit Regional, Department, and Bureau	- Conduct fit and suitability inspection of input data according to dictionary and position analysis guide	- Job analysis document and workload analysis. - Recapitulation Report of job analysis result and work load

Based on Table 2, in Job analysis document and workload analysis, the test is divided into two phases, where the first phase will be tested by the sub-division head and staff of job analysis and workload of the North Sulawesi governor's office. The second phase will be tested by staff of job analysis and workload of several work units in North Sulawesi.

In the first phase the tester acknowledges that the system shortens the processing time because the form is populated digitally and the report automatically fills out based on form fields. however, some user interfaces improvements are made to facilitate form filling.

The second phase was conducted directly on the staff of various work units in North Sulawesi Province, inviting 50 analytical staff from various municipalities and districts. Before using SINJAB, the participants were trained and assisted with the installation on their respective laptops.

Most of the test participants in the second phase (48 out of 50) felt helped by the new system being used. The average time required to fill both forms (occupation and workload) is 30 minutes, compared to the previous 5 hours. The participants were also satisfied that the work could also be taken home. Tab order makes the data charging process more organized and not confusing. Report templates in the form of word and excel automatically generated after the data is stored in the application.

4. Conclusions

The results of this study indicate that information technology can speed up the analysis and documents required in government. Job analysis and workload are part of the North Sulawesi provincial government offices that benefit from this. While still under consideration, it is likely that this system will be diverted through the web so that it is easily accessible to all parts of government in the province. Also implementation in other provinces needed to improve national performance

Acknowledgments

The authors wish to express their appreciation to DRPM, KemenRisTekdikTi of the Republic of Indonesia which had given funding all of the research activities, and to Rector of Universitas Negeri Manado, Indonesia who has agreed our research.

References

[1] Department of Economic and Social Affair 2016 *United Nations E-Government Survey 2016* https://publicadministration.un.org/egovkb/Portals/egovkb/Documents/un/2016-Survey/Executive%20Summary.pdf

[2] World Bank 2014 *Corporate Governance of State Owned Enterprises: A Toolkit* (Washington, DC: World Bank. License: Creative Commons Attribution CC BY 3.0 IGO).

[3] Russell N, van der Aalst W M P, ter Hofstede A H M, Edmond D 2005 Workflow Resource Patterns: Identification, Representation and Tool Support. In: Pastor O, Falcão e Cunha J. 2005 (Eds) *Advanced Information Systems Engineering* Lecture Notes in Computer Science vol 3520 (Springer, Berlin, Heidelberg).

[4] Marcus J F, Justin P, Rodger W G 2017 Job demands, control and support: Meta-analyzing moderator effects of gender, nationality, and occupation *Human Resource Management Review* **27**(1) pp. 39-60.

[5] Matthias T, Mark S, Cristovao S, Ting Q 2017 Drum-buffer-rope and workload control in High-variety flow and job shops with bottlenecks: An assessment by simulation *International Journal of Production Economics* **188** pp. 116-127.

[6] Matthias T and Mark S 2016 Workload control in job shops with re-entrant flows: an assessment by simulation *International Journal of Production Research* (DOI: 10.1080/00207543.2016.1156182).

[7] June W and Gavriel S 2004 The cognitive task analysis methods for job and task design: review and reappraisal *Journal Behaviour & Information Technology* **23**(4).

[8] Sparrow P 2000 New Employee Behaviours, Work Designs and Forms of Work Organization: What Is in Store for Future of Work? *Journal of Managerial Psychology* **15**(3) pp 202–218.

[9] M Carr and J Verner 2004 Prototyping and Software Development Approaches *Prototyp. Softw. Dev. Approaches* **3** pp 1–16.

[10] Chatterjee A 2017 *Building Apps for the Universal Windows Platform: Explore Windows 10 Native, IoT, HoloLens, and Xamarin* (Apress).

IOP Conf. Series: Materials Science and Engineering **306** (2018) 012081 doi:10.1088/1757-899X/306/1/012081

Performance Analysis of a Static Synchronous Compensator (STATCOM)

M M Kambey and J D Ticoh*

Electrical Engineering Education Department, Universitas Negeri Manado

*jdticoh@unima.ac.id

Abstract. Reactive power and voltage are some of the problems in electric power supply and A Gate Turn Off (GTO) Static Synchronous Compensator (STATCOM) is one of the type of FACTS with shunt which can supply variable reactive power and regulate the voltage of the bus where it is connected. This study only discuss about the performance characteristic of the three phase six-pulse STATCOM by analysing the current wave flowing through DC Capacitor which depend on switching current and capacitor voltage wave. Simulation methods used in this research is started with a mathematical analysis of the ac current, dc voltage and current equations that pass STATCOM from a literature. The result shows the presence of the capacitor voltage ripple also alters the ac current waveform, even though the errors to be not very significant and the constraint of the symmetry circuit is valid if the source voltages have no zero sequence components and the impedances in all the three phases are identical. There for to improve STATCOM performance it is necessary to use multi-pulse 12, 24, 36, 48 or more, and/or with a multilevel converter.

1. Introduction

Variables of complex electrical power system always changes. The indicator can be seen in the change of voltage, current, active power, reactive power or frequency in the power system. The amount of reactive power reserves in the power system is an indicator of testability of the voltage [1]. Recent development of power electronics introduces the use of FACTS devices in power systems. FACTS devices are capable of controlling the network condition in a very fast manner and this unique feature of FACTS devices can be exploited to improve the transient stability of a system [2-4]. Reactive power compensation is an important issue in electrical power systems and shunt FACTS devices play an important role in controlling the reactive power flow to the power network and hence the system voltage fluctuations and transient stability [5, 6]. Depending on the power rating of the STATCOM, different topologies are used for the power converter. High power STATCOM (several hundreds of Mvars) normally use GTO-based, square-wave voltage-sourced converters (VSC), while lower power STATCOM (tens of Mvars) use IGBT based (or IGCT-based) pulse-width modulation (PWM) VSC.

There were some technique are used to enhance the dynamic performance of the STATCOM [7, 10-16]. However, the analysis of the basic form of STATCOM with three-phase six-pulses is still very useful for understanding other models. In this paper the model of a GTO-based STATCOM on an electrical power transmission system [17-19]. There were some problems according with enhancing STATCOM performance, but because the limitation of the resources so the purpose of this study only

IOP Conf. Series: Materials Science and Engineering **306** (2018) 012081 doi:10.1088/1757-899X/306/1/012081

to analyzing the AC current waveform, current wave flowing through DC Capacitor (i_{dc}) which depend on switching current and voltage wave capacitor (v_{dc}).

2. Methods

The methods used in this research are literature study, started with a mathematical analysis of the voltage and current equations that pass through the three phase six-pulse STATCOM. It's begun first by construct GTO STATCOM mathematical equation with three phase six pulse to express the characteristic of AC current waveform, current wave flowing through DC Capacitor (i_{dc}) which depend on switching current and voltage wave capacitor (v_{dc}). There was three assumes is taken in paper study, first the capacitor size is infinite and therefore the DC side voltage is constant, the second the losses in the circuit are neglected and the third each switch is turned on only once in a cycle of supply voltage and conducts for 180^0 each.

The STATCOM consists of a voltage-sourced converter (VSC), a magnetic circuit (MC), a shunt coupling transformer, a shunt breaker, and a control and protection unit. In a VSC, a number of square wave voltages are generated at fundamental frequency by operating the controllable semiconductor switches once per cycle of the fundamental frequency. A solid-state voltage source inverter with several Gate Turn Off thyristor switch-based valves, a DC link capacitor, magnetic circuit, and a controller [13-16], the number of thyristors and the various configurations of the magnetic circuit depend on the desired quality of AC waveforms generated by the controller. The DC link capacitor is switched through the inverter circuit and the required reactive power is injected to the transmission circuit [15]. The basic building block circuit of a high power GTO based STATCOM with six pulse used in this paper shown in Fig. 1. The circuit consists of six switches, made up of six GTO thyristors with antiparallel diodes connected as a six pulse bridge.

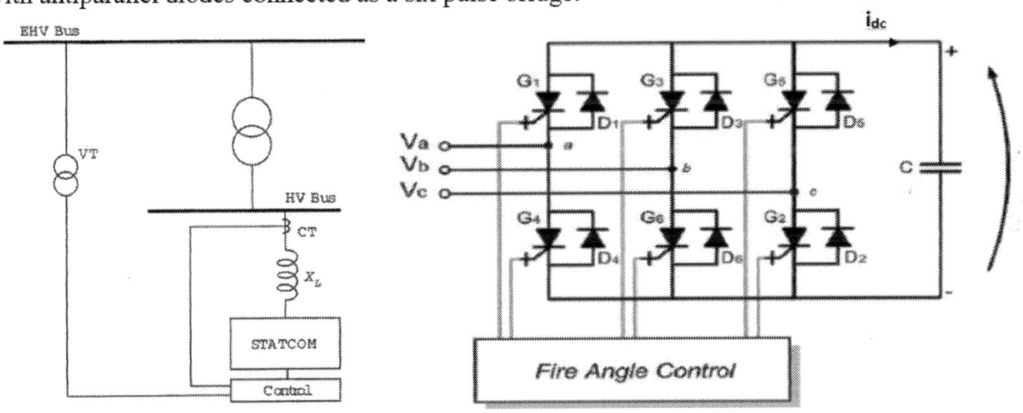

(a) **(b)**

Figure 1. (a) STATCOM connection diagram. And (b) a three-phase six-pulse VSC circuit.

3. Results and Discussion

The waveform of the voltage (E_{aN}) as shown in Fig.2. The waveforms of E_{bN} and E_{cN} are also similar except that they are displaced from one another by 120^0. (E_{bN} lags E_{aN} by 120^0 dan (E_{cN} lags E_{bN} by 120^0) [19].

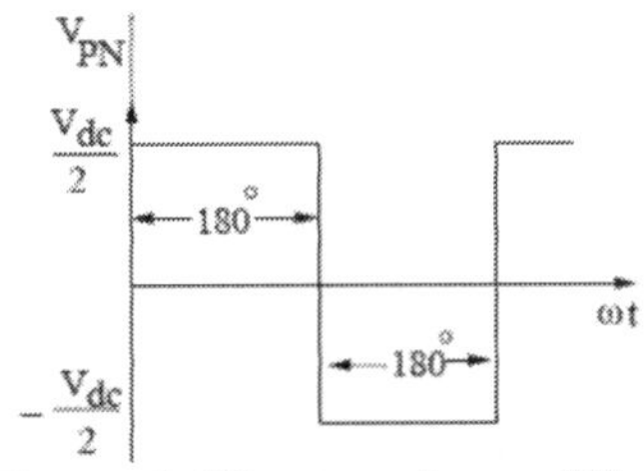

Figure 2. The waveform of V_{PN}.

The voltages E_{an}, E_{bn} dan E_{cn} respect to the source neutral can be obtained from the following equations

$$E_{an} = E_{aN} + V_{Nn} \quad (1)$$

$$E_{bn} = E_{bN} + V_{Nn} \quad (2)$$

$$E_{cn} = E_{cN} + V_{Nn} \quad (3)$$

Where the sum of the symmetry circuit E_{an}, E_{bn} *and* E_{cn} is zero, and can be expressed as below

$$E_{an} + E_{bn} + E_{cn} = 0 \quad (4)$$

Substituting Eq. (4) in (1) to (3), we get

$$E_{Nn} = -\frac{E_{aN}+E_{bN}+E_{cN}}{3} \quad (5)$$

And

$$E_{aN} = \frac{2E_{aN}}{3} - \frac{E_{bN}}{3} - \frac{E_{cN}}{3};\ E_{bN} = \frac{2E_{bN}}{3} - \frac{E_{cN}}{3} - \frac{E_{aN}}{3};\ \text{and } E_{cN} = \frac{2E_{cN}}{3} - \frac{E_{aN}}{3} - \frac{E_{bN}}{3} \quad (6)$$

The waveform of E_{an} is shown in Fig. 3 (which also shows the supply Voltagev_a). The fundamental frequency component (rms value) of E_{an} is obtained as

$$E_{a1} = \frac{4}{\pi\sqrt{2}} \int_0^{\pi/2} E_{aN}\, sin\theta d\theta$$

$$= \frac{4}{\pi\sqrt{2}} V_{DC} = 0.45\, V_{dc} \quad (7)$$

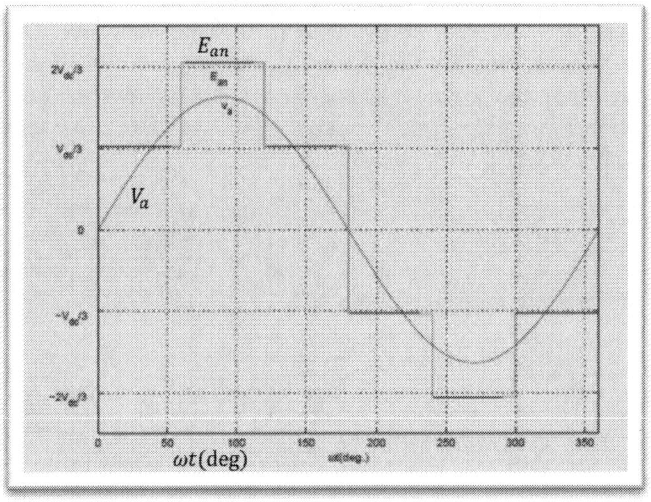

Figure 3. Waveforms E_{an} and v_a.

The harmonic component (E_{ah}) is obtained as

$$E_{ah} = \frac{E_{a1}}{h} = \frac{0.45V_{dc}}{h}, \qquad h = 6k \pm 1, \quad k = 1,2,3 \quad (8)$$

The rms value of the fundamental component of (reactive) current, I_r is calculated from

$$I_r = \frac{0.45V_{dc}}{\omega L} \quad (9)$$

The harmonic current (rms) is obtained as $I_h = \frac{0.45V_{dc}}{h^2\omega L}$ (10)

3.1 Analysis of AC current waveform

The instantaneous current in phase a (ia) is obtained from

$$L\frac{dia}{dt} = v_L(t) = v_a(t) - E_{an}(t) \quad (11)$$

Since $E_{aN}(t)$ varies depending on the interval in whichωt, lies, we get deferent expressions for $i_a(t)$ depending on the interval. However, in all intervals, $v_a(t) = \sqrt{2}\, V \sin\omega t$.

Interval 1, $0 \leq \omega t \leq 60^0$,

$$i_a(t) = \frac{V}{\omega L}\left[-\sqrt{2}\cos\omega t + \frac{V_{dc}}{V}\left(\frac{2\pi}{9} - \frac{\omega t}{3}\right)\right] \quad (12)$$

Interval 2, $60^0 \le \omega t \le 120^0$,

$$i_a(t) = \frac{V}{\omega L}\left[-\sqrt{2}\cos\omega t + \frac{V_{dc}}{V}\left(\frac{\pi}{3} - \frac{2\omega t}{3}\right)\right] \tag{13}$$

Interval 3, $120^0 \le \omega t \le 180^0$,

$$i_a(t) = \frac{V}{\omega L}\left[-\sqrt{2}\cos\omega t + \frac{V_{dc}}{V}\left(\frac{\pi}{9} - \frac{\omega t}{3}\right)\right] \tag{14}$$

The current in the interval, $180^0 \le \omega t \le 360^0$ is given by, $i_a(t) = i_a\left(t - \frac{\pi}{\omega}\right)$ (15)

Figure 4 shows the AC current waveforms (as calculated from Eqs (12) to (15) for V = 1:0, ωL=0,2, i_r=± 1,0. Both leading and lagging current waveforms are shown. As discussed earlier, the lagging current flows through the GTO thyristor switch 1 till $\omega t = 90^0$ and the reverse current flows through the antiparallel diode. Thus, the current °owing through the switch 1 is zero when it is turned off. On the other hand, switch 1 current starts from zero at $\omega t = 90^0$ when it is leading and is at a peak value when it is turned off. The peak GTO thyristor turn-off current (in the capacitive mode) is obtained from Eq. (17) (substituting $\omega t = 0^0$) expressed by

$$I_{Tpeak} = \frac{1}{\omega L}\left[\frac{2\pi V_{dc}}{9} - \sqrt{2V}\right] \tag{16}$$

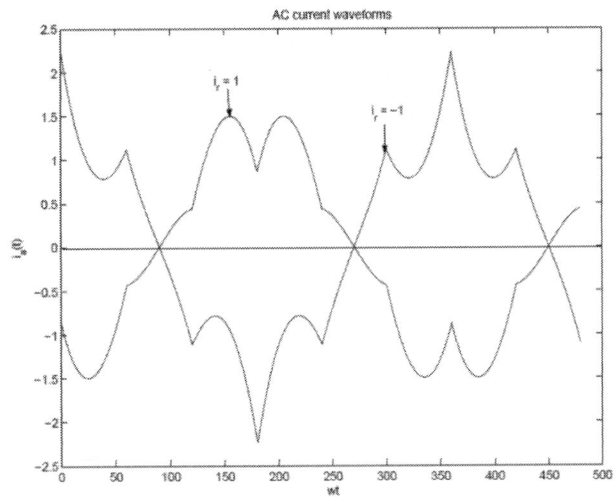

Figure 4. AC current Waveform.

3.2 Analysis of DC capacitor current and voltage

The current flowing through the DC capacitor (i_{dc}) is dependent on the switch currents. If we consider the interval, $0^0 < \omega t < 60^0$, the switches 5, 6 and 1 are conducting. In general, for any combination of switches, the following equation applies (neglecting losses in the switches).

$$V_{dc}i_{dc} = E_{an}i_a + E_{bn}i_b + E_{cn}i_c \tag{17}$$

The above is applicable at all times. Since E_{an}, E_{bn}, and $E_{cn}$are linearly dependent on V_{dc} (see Fig. 3), we can express them as

$$E_{an}(t) = S_a(t)V_{dc} + S_b(t)V_{dc} + S_c(t)V_{dc} \tag{18}$$

Where $S_a(t)$, $S_b(t)$ and $S_c(t)$ are termed as switching functions with constant values dependent on the interval considered. The switching functions are shown in Fig. 5. For the interval,$0 \le \omega t < 60^0$, values of the switching functions are

$$S_a = \frac{1}{3}, S_b = -\frac{2}{3}, S_c = \frac{1}{3} \tag{19}$$

If we substitute Eq. (18) in (17), we get

$$i_{dc}(t) = S_a(t)i_a(t) + S_b(t)i_b(t) + S_c(t)i_c(t) \tag{20}$$

For the interval 1 ,$0 \le \omega t < 60^0$, we get

$$i_{dc}(t) = \frac{1}{3}i_a(t) + -\frac{2}{3}i_b(t) + \frac{1}{3}i_c(t) \ = \frac{1}{3}[i_a(t) + i_b + \ i_c(t)]-i_b \quad (21)$$

As the sum of phase currents,i_a, i_b and i_c s zero by Kirchcho's Current Law (KCL), we can simplify Eq. (21) as

$$i_{dc}(t) = \ -i_b(t)\ , 0 \le \omega t < 60^0 \quad (22)$$

It can be shown that (from symmetry)

$$-i_b(\omega t) = i_{a2}(\omega t + 60^0) \ = \frac{V}{\omega L}\left[\sqrt{2}sin(\omega t - 30^0) + \frac{V_{dc}}{V}\left(\frac{\pi}{2} - \frac{2\omega t}{3}\right)\right] \quad (23)$$

Where $i_{a2}(\omega t)$ is the expression for $i_a(\omega t)$ in the interval, $60^0 \le \omega t < 120^0$, given in Eq. (15). The capacitor current waveform in a cycle is obtained as the repetition of the waveform (from $0 \le \omega t < 60^0$) over the remaining 5 intervals. It can be shown that the average value of the capacitor current is zero in steady state. (Otherwise the capacitor voltage cannot remain a constant). The fundamental component of this periodic waveform has the frequency of 6ω.

The capacitor voltage (v_c) was assumed as constant in the above analysis (due to infinite capacitance). However, with infinite capacitance, the capacitor voltage will have ripples. Assuming that the capacitor current $i_{dc}(t)$ is not affected by the voltage ripple, v_c can be obtained from

$$v_c(t) = V_0 + \frac{1}{C}\int i_{dc}(t)dt \quad (24)$$

Where V_0 is the value of$V_{dc}(t)$ at t=0. Substituting Eq. (23) in (24), we get (for $0 \le \omega t < 60^0$)

$$v_c(t) = V_0 - \frac{\sqrt{2}cos(\omega t-30^0)}{\omega^2 LC} + \frac{V_{dc}\pi t}{9\omega LC} \ - \frac{2V_{dc}t^2}{6LC} + \frac{\sqrt{6}V}{2\omega^2 LC} \quad (25)$$

The average value of $v_c(t)$ is $V_{dc}(t)$. This enables the calculation of V_0 as

$$V_0 = V_{dc}(t) - 0{,}061\frac{V_{dc}}{\omega^2 LC} + \frac{0{,}126V}{\omega^2 LC} \quad (26)$$

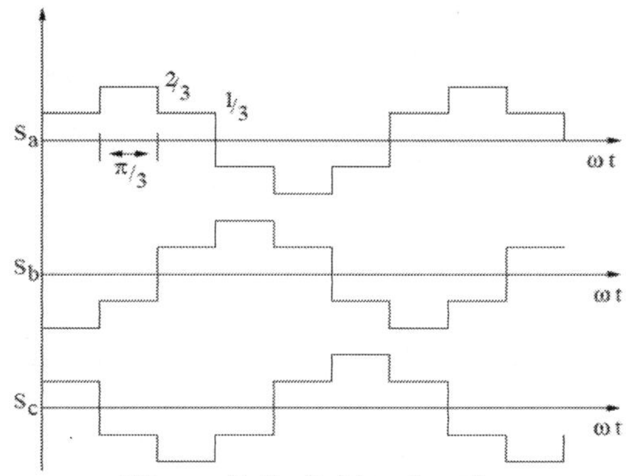

Figure 5. Switching functions.

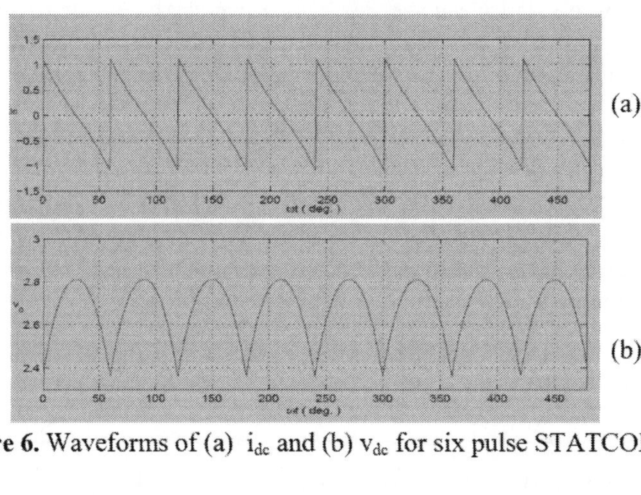

Figure 6. Waveforms of (a) i_{dc} and (b) V_{dc} for six pulse STATCOM.

The waveform of $v_c(t)$ for the interval, $0 \leq \omega t \leq 60^0$ is obtained after substituting Eq. (26) in (25). The waveform of $v_c(t)$ over the entire cycle is a repetition of the waveform for the interval $0 \leq \omega t \leq 60^0$. The waveforms of $i_{dc}(t)$ dan $v_c(t)$ over a complete cycle are shown in Fig.6 above. For V=1,0 pu, i_r= -1,0, ωL=0,2 and ωC=0,6.

And to determine the GTO thyristors and diodes rating voltages it is necessary to calculate the peak value of the capacitor voltage (V_{pk}), by substituting $\omega t = 30^0$ in the expression given by Eq. (25). We get,

$$V_{pk} = V_{dc} + \frac{0{,}03V_{dc}}{\omega^2 LC} + \frac{0{,}06V}{\omega^2 LC} \quad (27)$$

4. Conclusions

The presence of the capacitor voltage ripple also alters the AC current waveform. However, the errors in the expressions of the instantaneous ac current waveform in phase a, i_a, on the interval 1, 2 and 3 can be assumed to be not very significant. As the capacitor voltage (v_c) was assumed as constant in the above analysis (due to infinite capacitance). However, with infinite capacitance, the capacitor voltage will have ripples. Assuming that the capacitor current idc (t) is not affected by the voltage ripple. The objective in deriving these analytical expressions is to gain insight. Such approximate analysis is quite common in power electronics as the first step towards a detailed analysis based on more exact models. There for to improve STATCOM performance it is necessary to use multi-pulse 12, 24, 36, 48 or more, and/or with a multilevel converter.

References

[1] Weedy B M, Cory B J, Jenkins N, Ekanayake J B and Strbac G 2012 *Electric power systems* (John Wiley & Sons).
[2] Flourentzou N, Agelidis V G and Demetriades G D 2009 VSC-based HVDC power transmission systems: An overview *IEEE Transactions on power electronics* **24**(3) pp. 592-602.
[3] Rashid M H 2009 *Power electronics: circuits, devices, and applications* (Pearson Education India).
[4] Murali D, Rajaram M and Reka N 2010 Comparison of FACTS devices for power system stability enhancement *International Journal of Computer Applications* **8**(4) pp. 30-5.
[5] Dixon J, Moran L, Rodriguez J and Domke R 2005 Reactive power compensation technologies: State-of-the-art review *Proceedings of the IEEE* **93**(12) pp. 2144-64.
[6] Panda S and Patel R N 2006 Improving power system transient stability with an off-centre location of shunt FACTS devices *Journal Of Electrical Engineering-Bratislava* **57**(6) pp. 365.
[7] Norouzi A H and Sharaf A M 2005 Two control schemes to enhance the dynamic performance of the STATCOM and SSSC *IEEE Transactions on Power delivery* **20**(1) pp. 435-42.
[8] Mienski R, Pawelek R and Wasiak I 2004 Shunt compensation for power quality improvement using a STATCOM controller: modelling and simulation *IEE Proceedings-Generation, Transmission and Distribution* **151**(2) pp. 274-80.
[9] El-Moursi M S and Sharaf A M 2005 Novel controllers for the 48-pulse VSC STATCOM and SSSC for voltage regulation and reactive power compensation *IEEE Transactions on Power systems* **20**(4) pp. 1985-97.
[10] Townsend C D, Summers T J, Vodden J, Watson A J, Betz R E and Clare J C 2013 Optimization of switching losses and capacitor voltage ripple using model predictive control of a cascaded H-bridge multilevel StatCom *IEEE Transactions on Power Electronics* **28**(7) pp. 3077-87.
[11] W Pan, Z Xu and J Zhang 2007 Novel configuration of 60-pulse voltage source converter for STATCOM application *International Journal of Emerging Electric Power Systems* **8**(5).
[12] M S El-Moursi and A M Sharaf 2005 Novel Controllers for the 48-Pulse VSC STATCOM and SSSC for Voltage Regulation and Reactive Power Compensation *IEEE Transactions on Power Systems* **20**(4) pp 1985-1997.

[13] Sen K 1999 STATCOM-STATic synchronous COMpensator: theory, modeling, and applications. In Power Engineering Society 1999 *Winter Meeting* IEEE 1999 Feb (Vol. 2, pp. 1177-1183). IEEE.
[10] Rashid M H 2010 *Power electronics handbook: devices, circuits and applications* (Academic press).
[14] Arvind P and Nitin S 2013 Transient Stability Improvement by Using Shunt FACTS Devices (STATCOM) with Reference Voltage Compensation (RVC) Control Scheme *International Journal of Electrical, Electronics and Computer Engineering, IJEECE* **2**(1) pp 7-12.
[15] Peng F Z, Qian W and Cao D 2010 *Recent advances in multilevel converter/inverter topologies and applications*. In Power Electronics Conference (IPEC) 2010 International 2010 Jun 21 (pp. 492-501). IEEE.
[16] Dávalos R, Ramirez J M and Tapia R 2005 Three-phase multi-pulse converter StatCom analysis *International Journal of Electrical Power & Energy Systems* **27**(1) pp. 39-51.
[17] Chakrabartis R, Chakraborty S, Mitra S, Choudhury R R, Chowdhury, Mondal S S and Halder S SVC Operation in Electric Transmission Line *International Journal of Emerging Technology and Advanced Engineering* **5**(8).
[18] Kowsalya M, Ray K K and Kothari D P 2009 Positioning of SVC and STATCOM in a Long Transmission Line *International Journal of Recent Trends in Engineering* **2**(5) pp. 150-154.
[19] Padiyar K R *FACTS controllers in power transmission and distribution*.

Utilization of Multimedia Laboratory: An Acceptance Analysis using TAM

M Modeong* and V R Palilingan
Universitas Negeri Manado, Tondano 95618, Sulawesi Utara, Indonesia

*merriamisty.modeong@gmail.com

Abstract. Multimedia is often utilized by teachers to present a learning materials. Learning that delivered by multimedia enables people to understand the information of up to 60% of the learning in general. To applying the creative learning to the classroom, multimedia presentation needs a laboratory as a space that provides multimedia needs. This study aims to reveal the level of student acceptance on the multimedia laboratories, by explaining the direct and indirect effect of internal support and technology infrastructure. Technology Acceptance Model (TAM) is used as the basis of measurement on this research, through the perception of usefulness, ease of use, and the intention, it's recognized capable of predicting user acceptance about technology. This study used the quantitative method. The data analysis using path analysis that focuses on trimming models, it's performed to improve the model of path analysis structure by removing exogenous variables that have insignificant path coefficients. The result stated that Internal Support and Technology Infrastructure are well mediated by TAM variables to measure the level of technology acceptance. The implications suggest that TAM can measure the success of multimedia laboratory utilization in Faculty of Engineering UNIMA.

1. Introduction

Rapid development of current technologies are increasingly encouraging younger generation to follow in its footsteps. Even as educators, teachers also continues to renew their knowledge about technology in order to be able to present the newest learning method and material. Current education problems is that educators are faced with challenges and problems on how to find the best way to convey the concepts to students in order to be useful as an important information. The same thing happened in some instances constrained education like presentation as an interesting learning concept. Like a case in the Faculty of engineering in *Universitas Negeri Manado* (UNIMA), educators are always trying to present a theory that packed with catchy design media.

In the process of teaching and learning, technology connects all elements of learning –teachers, students, and the teaching materials– so that, can make the output of the education process [1]. Multimedia is a type of technology that is commonly used in the learning process. A combination between text, images, graphics, sounds, animations, and videos with artistic design, multimedia change the learning process become more interesting, allow the user to receive information 60% easier [2]. The process of learning using multimedia doesn't intend to replace the traditional learning process that has long been applied, but this method is used as a learning solution that attract the attention of students by combining some elements of technology in learning.

The laboratory is a place for student experiment to increase their active and vital participation in the learning process [3]. Herrington & Nakhleh [4] also mentioned that it is important to assess the students' perceptions toward learning instructions in the laboratory to find out the effectiveness about the learning process that has been done. This assessment is conducted to determine the user's response to the technology that being run. So, to answer the response, it needs a theory to measure the level of user's acceptance about technology. Assessment on laboratory utilization is done using the Technology Acceptance Model (TAM) that developed by Davis [5].

TAM is the intention-based models that are specifically intended to explain and/or predict user acceptance of technology [6]. TAM has core variables such as perceived usefulness, perceived ease of use, intention, and actual use. But TAM does not eliminate the possibility of situational variables that indicated can measure the level of technology acceptance. According to the conditions of Faculty of Engineering UNIMA, indicated that internal support and technology infrastructure become situational variables which can be applied as the determinant variable that connected to student acceptance of technology.

Internal support includes support from university's higher ranked officials and lecturers to direct and maximize practice-oriented learning, as well as provide the human resources that functioning as a laboratory assistant. The problems in the Faculty of engineering are some lecture materials that should be balanced with practice but only filled with theory or oral illustration. This is due to the unregulated laboratory utilization schedule and the lack of support from university officials towards the fulfillment of the laboratory needs and encouraging the practice-oriented learning.

Furthermore, the technology infrastructure related with the availability of technology tools, as well as the availability of laboratory support facilities. According to the condition of the multimedia laboratory in the faculty of engineering UNIMA, multimedia equipment is not well balanced with the number of students who often practice in the laboratory. In addition, the other support equipment such as digital camera, video recorder, LCD projector, etc. still need to be held in sufficient quantities for its users.

2. Research Model

In TAM theory, perceived about usefulness and ease of use are mediate the external variables' influence on behavioral intentions and then the actual use of the system can be known [7]. This opinion is evidenced by several studies that use TAM as a reference, among others: an open source e-learning system such as moodle [7, 8], valid measurement scales for predicting user acceptance of information technology [5], Malaysian student and teachers' integrating their technology in teaching and learning [9], interactive learning technologies using clickers in the macroeconomics classes [10], and so on. The proposed research model is presented in **Figure 1**.

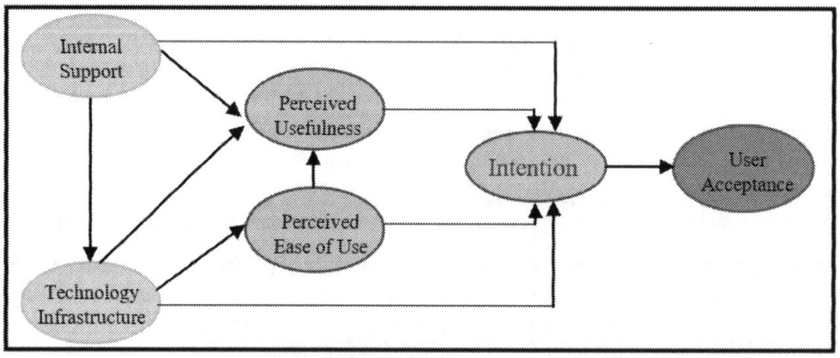

Figure 1. Proposed research model.

2.1 TAM variables

A system can be said to be successful if the system could provide a positive impact for its users. Positive impacts such as usability and ease of use that exist on the system, can support a particular work for individuals who use it. PU and PEOU are the key factors that affect the behavioral intention to use technology [11]. Thus the hypotheses related with PU, PEOU, and intention are as follows:

Hypothesis 1: Perceived Usefulness have a positive direct effect on Intention
Hypothesis 2: Perceived Ease of Use have a positive direct effect on Perceived Usefulness
Hypothesis 3: Perceived Ease of Use have a positive direct effect on Intention
Hypothesis 4: Intention have a positive direct effect on User Acceptance

2.2 Internal support (IS)

The physical and social environment variables are selected as situational variables that are believed to be capable affecting the variable of perceived usefulness, ease of use, and user acceptance [12]. The physical environment and the social environment describe the support of university officials and all lecturers to applying strategy on systematic learning. The hypotheses related with IS are provided below:

Hypothesis 5: Internal support have a positive direct effect on Perceived Usefulness
Hypothesis 6: Internal support have a positive direct effect on Intention
Hypothesis 7: Internal support have a positive direct effect on Technology Infrastructure
Hypothesis 8: Internal support have a positive indirect effect on Intention
Hypothesis 9: Internal support have a positive indirect effect on User Acceptance

2.3 Technology infrastructure (TI)

Akande [13] revealed that learning can occur through a person's interaction with his/her environment. The environment here refers to the facilities that are available to facilitate the student when learning. Because the facilities are believed to be able to help students to develop a problem solving ability [14]. Thus the hypotheses related with TI are as follows:

Hypothesis 10: Technology Infrastructure have a positive direct effect on Perceived Usefulness
Hypothesis 11: Technology Infrastructure have a positive direct effect on Perceived Ease of Use
Hypothesis 12: Technology Infrastructure have a positive direct effect on Intention
Hypothesis 13: Technology Infrastructure have a positive indirect effect on Intention
Hypothesis 14: Technology Infrastructure have a positive indirect effect on User Acceptance

3. Methods

This study used the quantitative method with an ex-post facto approach. The study was conducted in April until June 2017 to find students' perception about the factors that influence the students' acceptance about usefulness of multimedia laboratory. The subject of this study was students who take Multimedia expertise in the Faculty of Engineering UNIMA with 185 total populations. Using Proportional Random Sampling, so the samples are: (i) 4th Semester: 23 respondents; (ii) 6th Semester: 30 respondents; and (iii) 8th Semester: 27 respondents.

The instrument of data collection in this study is questionnaire using Likert scoring, with four choice answers as follows: Strongly Agree (SA), Agree (A), Disagree (DS), and Strongly Disagree (SDS). Path analysis used as data analysis to measuring the relationship between variable and testing the hypothesis. Before testing the hypothesis, first, we take the classical assumptions test to ensure that the data been collected is worth enough to be processed using Path analysis.

3.1 Classical assumption test

Before performing multiple linear regressions testing, it needs to be done a classical assumption test first (**Table 1**), consisting of: (a) normality test, (b) linearity test, and (c) multicollinearity test. The model that formulated in this study was broken down into 5 substructures and then conducted a classical assumption test. The five substructures models are designed to facilitate the researcher in the process of data analysis.

The data is said to be normally distributed and has a linear relationship when the significance value is greater than the alpha value (Sig. > α), the alpha value is set at 5% or 0.05. Furthermore, the multicollinearity test is done by looking at the magnitude of the VIF value, if VIF <10, then there is no multicollinearity between the independent and the dependent variable.

Table 1. Classical assumption test result.

Variables		Normality	Linearity	Multicolinearity
Dependent	Independent			
Perceived Usefulness	Internal Support		0.109	1.261
	Technology Infrastructure	0.093	0.153	1.212
	Perceived Ease of Use		0.241	1.045
Technology Infrastructure	Internal Support	0.373	0.554	1.000
Perceived Ease of Use	Technology Infrastructure	0.292	0.078	1.000
Intention	Internal Support		0.369	1.108
	Technology Infrastructure	0.982	0.378	1.130
	Perceived Usefulness		0.257	1.061
	Perceived Ease of Use		0.278	1.103
User Acceptance	Intention	0.637	0.104	1.000

3.2 Multiple regression analysis

After the classical assumption test has been met, then conducted regression analysis to answer the research hypothesis. The decision criteria used if the significance value higher than 0.05 then H_0 is accepted, otherwise if the significance lower than 0.05 then H_0 is rejected.

Table 2. Summary of the influence between dependent and independent variable.

Variables		*A*	R^2	Sig.
Dependent	Independent			
Perceived Usefulness	Internal Support			0.012
	Technology Infrastructure	1.242	0.313	0.569
	Perceived Ease of Use			0.000
Technology Infrastructure	Internal Support	8.223	0.172	0.000
Perceived Ease of Use	Technology Infrastructure	19.602	0.055	0.036
Intention	Internal Support			0.039
	Technology Infrastructure	6.118	0.252	0.231
	Perceived Usefulness			0.024
	Perceived Ease of Use			0.039
User Acceptance	Intention	10.855	0.079	0.012

The multiple regression analysis in **Table 2** and **Table 3** stated that there are two hypotheses were rejected, the 10th hypothesis and the 12th hypothesis (**Table 4**). Both hypotheses were rejected because they have a significance value greater than 0.05. While the other 12 hypotheses get a significance value lower than 0.05 (see **Table 4**).

3.3 Path analysis

Path analysis that used in this research is a trimming model. The trimming model is a model that's been used to improve a Path analysis structure by removing insignificant exogenous variables, then retesting.

Based on the trimming model analysis, the insignificant path coefficient are TI to PU (Sig = 0.569) and TI to IN (Sig = 0.231) which means greater than 0.05. Therefore, this variable path is excluded from the research model that has been formulated. After elimination, we conducted the regression test again. The improvement of the acceptance model by using trimming model illustrated in **Figure 2**.

Then, the answer of the hypothesis in this study are described in **Table 4**, where the value obtained the result of the second regression analysis after eliminating the insignificant exogenous variables. **Table 4** also describes the acquisition of path coefficient values that have an indirectly effect i.e. H8, H9, H13, and H14.

Table 3. Partial influence of dependent and independent variable.

Variables		Correlation coefficient				Sig.
		Before elimination		After elimination		
Dependent	Independent	β	t-value	β	t-value	
Perceived Usefulness	Internal Support	0.275	2.576	0.299	3.094	0.003
	Technology Infrastructure	0.056	0.533	-	-	-
	Perceived Ease of Use	0.418	4.304	0.415	4.299	0.000
Technology Infrastructure	Internal Support	0.415	4.024	0.415	4.024	0.000
Perceived Ease of Use	Technology Infrastructure	0.235	2.136	0.235	2.136	0.036
Intention	Internal Support	0.221	2.099	0.245	2.362	0.021
	Technology Infrastructure	0.128	1.209	-	-	-
	Perceived Usefulness	0.237	2.300	0.236	2.289	0.025
	Perceived Ease of Use	0.221	2.102	0.254	2.499	0.015
User Acceptance	Intention	0.280	2.579	0.280	2.579	0.012

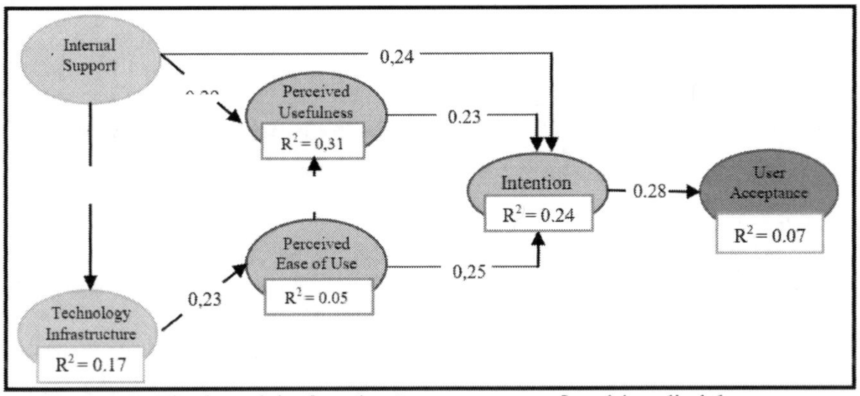

Figure 2. Final model of student's acceptance of multimedia laboratory.

Table 4. Summary of hypotheses tests.

Hi	Relationship	Effects	Sig. Value	B	Decision
H1	PU → IN	Direct	0.025	0.236	Accepted
H2	PEOU → PU	Direct	0.000	0.415	Accepted
H3	PEOU → IN	Direct	0.015	0.254	Accepted
H4	IN →UA	Direct	0.012	0.280	Accepted
H5	IS → PU	Direct	0.003	0.299	Accepted
H6	IS → IN	Direct	0.021	0.245	Accepted
H7	IS → TI	Direct	0.000	0.415	Accepted
H8	IS → IN	Indirect	-	0.100	Accepted
H9	IS → UA	Indirect	-	0.100	Accepted
H10	TI → PU	Direct	0.596	0.056	Rejected
H11	TI → PEOU	Direct	0.036	0.235	Accepted
H12	TI → IN	Direct	0.231	0.128	Rejected
H13	TI → IN	Indirect	-	0.080	Accepted
H14	TI → UA	Indirect	-	0.022	Accepted

4. Results and Discussion

4.1 TAM variables

This research aimed to make TAM as measurement about perceived usefulness, perceived ease of use, intention and situational variables to increasing users' number on multimedia laboratory utilization. The results of the analysis stated that Perceived Usefulness, Perceived Ease of Use, and the Internal Support are the variables that directly affect to the student's intention to use multimedia laboratory. PU has a positive influence on IN with path coefficient of 0.236 (Hypothesis 1) and PEOU has a positive influence on IN with path coefficient of 0.254 (Hypothesis 3). In addition PEOU also has a positive correlation value on PU (Hypothesis 2) of 0,415. Futhermore, IN gives a direct positive effect on UA with coefficient value of 0.280 (Hypothesis 4).

Besides having a direct influence, these four TAM variables have a determinant coefficient (R^2) or an effective contribution based on it substructures. In the first substructure, 31% of PU are influenced by IS and PEOU. This means that 69% of PU is influenced by other factors outside the model formulated in this study. In the third substructure, 5% of PEOU is positively influenced by TI, which means that TI has very little effect on PEOU. There are three variables that together gives a positive influence to the intention in fourth substructure, i.e. IS, PU, and PEOU. These three variables provide a coefficient of determination of 24%. This means 76% IN influenced by other variables that are not formulated in this model. Furthermore UA has a coefficient of determination of 7% from IN, which means that 93% of UA is determined by other factors outside this study.

4.2 Internal support

In the research model design, IS has a direct and indirect effect on user acceptance decisions about the technology that has been used. IS has a direct positive effect on the other 3 variables, that is PU, IN, and TI. The path coefficient of IS to PU is 0.299 (Hypothesis 5), IS to IN of 0.245 (Hypothesis 6), and IS to TI of 0.415 (Hypothesis 7). IS also has an indirect effect on IN and UA, with the acquisition of each path coefficient is 0.100 (Hypotheses 8 and 9).

4.3 Technology infrastructure

In the design model on this study, TI has three direct and two indirect effects. However, based on the analysis that has been done, there are two paths from TI that are insignificant, that is direct influence of TI to PU and TI to IN (Hypothesis 10 and 12). Therefore, both path are removed from model. On a significant paths, TI has a direct positive effect on PEOU with a path coefficient of 0.235 (Hypothesis 11), the indirect effect of TI to IN has a path coefficient of 0.080 (Hypothesis 13), and the indirect effect of TI to UA with a path coefficients of 0.022 (Hypothesis 14).

Besides having a direct and indirect effect, TI also has a coefficient of determination (R^2) or an effective contribution to the second substructure. There is 17% TI is influenced by IS, which means there are still 83% other factors outside the IS that are expected to greatly affect the percentage of TI.

5. Conclusions

This research aimed to make TAM as measurement about perceived usefulness, perceived ease of use, intention and situational variables to increasing users' number on multimedia laboratory utilization. Internal support and Technology infrastructure variables which act as a situational variable in this research. IS affected the acceptance of multimedia laboratory utilization indirectly through four lines of calculation, namely: (1) through PU and IN variables; (2) through the IN variable; (3) through TI, PEOU, and IN variables; and (4) through TI, PEOU, PU, and IN variables.

Furthermore, TI variable affected the acceptance of multimedia laboratory utilization indirectly through two lines of calculation, namely: (1) through PEOU and IN variables; and (2) through PEOU, PU, and IN. Thus these two variables have no direct influence on the acceptance of multimedia laboratory utilization, so this study is in line with the explanation of the actual TAM theory which is stated that the intention as a single variable that directly affected the acceptance of multimedia laboratory utilization [15-17].

References

[1] Al Musawi A, Ambusaidi A, Al-Balushi S and Al-Balushi K 2015 *Effectiveness of e-lab use in science teaching at the omani schools* (Turkish: TOJET) 14 p 45-52.
[2] Vaughan T 2011 *Multimedia: Making it work* (New York: McGraw Hills) p 2.
[3] ICSU 2011 *Report of the ICSU, Ad-hoc revieaw panel on science education* (Paris: International Council for Science) Accessed: 30.5.2014, At: http://www.icsu.org/publications/report.pdf.
[4] Herrington D G and Nakhleh M B 2003 What Defines Effective Chemistry Laboratory Instruction? Teaching Assistant and Student Perspectives *Journal of Chemical Education* **8** p 1197.
[5] Davis F D 1989 Perceived usefulness, perceived ease of use, and user acceptance of information technology *MIS Quarterly* **13** p 319-340.
[6] Hu P J, Chau, P Y K, Sheng, O R L and Tam, K Y 1999 Examining the technology acceptance model using physical acceptance of telemedicine technology *Journal of Management Information Systems* 16 p 91-112.
[7] Hsu H and Chang Y 2013 *Extended TAM model: impacts of convenience on acceptance and use of moodle* (US: China Education Review) 3 p 211-218.
[8] Al Adwan A, Al-Adwan, A and Smedley 2013 Exploring students acceptance of e-learning using technology acceptance model in Jordanian universities *IJEDICT* **9** p 4-18.
[9] Wong K T, Osman R, Goh, P S C and Rahmat M K 2013 Understanding student teachers' behavioural intention to use technology: technology acceptance model (TAM) validation and testing *International Journal of Instruction* **6** p 89-104.
[10] Wu X and Gao Y 2011 Applying the extended technology acceptance model to the use of clickers in student learning: some evidence from macroeconomics classes *American Journal of Business Education* **4** p 43-50.
[11] Coşkunçay D F and Őzkan S 2013 *A model for instructors' adoption of learning management systems: empirical validation in higher education context* (Turkish: TOJET) 12 p 13-25.
[12] Venkatesh V, Morris M G, Davis G B, and Davis F D 2003 User acceptance of information technology: toward a unified view *MIS Quarterly* 27 p 427.
[13] Akande O M 1985 *Hints on teaching practice and general principles of education* (Lagos: OSKO Associates).
[14] Owoeye J S and Yara P O 2011 *School facilities and academic achievement of secondary school agricultural science in ekiti state* (Nigeria: Asian Social Science) 7 p 64-74.
[15] Davis F D, Bagozzi R P and Warshaw P R 1989 User acceptance of computer technology: A comparison of two theoretical models *Management Science* **35**, p 982-1003.
[16] Venkatesh V and Davis F D 2000 A theoretical extension of the technology acceptance model: four longitudinal field studies *Journal of Management Science* **46** p 186-204.
[17] Tan P J B, Potamites P R and Wens-Chi L 2012 Applying the TAM to understand the factors affecting use of online banking in the Pescadores *Journal of Science and Technology* **2** p 1022-1028.

Web-Based Virtual Laboratory for Food Analysis Course

M N Handayani*, I Khoerunnisa and Y Sugiarti

Study Program of Agro-industrial Technology Education, Fakultas Pendidikan dan Teknologi Kejuruan, Universitas Pendidikan Indonesia, Jalan Dr. Setiabudhi No 207 Bandung 40154

*mustika@upi.edu

Abstract. Implementation of learning on food analysis course in Program Study of Agro-industrial Technology Education faced problems. These problems include the availability of space and tools in the laboratory that is not comparable with the number of students also lack of interactive learning tools. On the other hand, the information technology literacy of students is quite high as well the internet network is quite easily accessible on campus. This is a challenge as well as opportunities in the development of learning media that can help optimize learning in the laboratory. This study aims to develop web-based virtual laboratory as one of the alternative learning media in food analysis course. This research is R & D (research and development) which refers to Borg & Gall model. The results showed that assessment's expert of web-based virtual labs developed, in terms of software engineering aspects; visual communication; material relevance; usefulness and language used, is feasible as learning media. The results of the scaled test and wide-scale test show that students strongly agree with the development of web based virtual laboratory. The response of student to this virtual laboratory was positive. Suggestions from students provided further opportunities for improvement web based virtual laboratory and should be considered for further research.

1. Introduction

Laboratory is one of the most important learning infrastructure facilities, in the academic process in agro-industrial technology education study program, especially in the subjects of expertise that require learning of lab work in it. Currently, the laboratory conditions are insufficient to accommodate the large number of students so that the lesson is less than optimal. In addition, the equipment in the laboratory is still incomplete to undertake a wide variety of practicum activities in subjects such as food analysis.

Food analysis is a course that studies the principles of chemical analysis of food, both qualitative and quantitative analysis. After following the course of food analysis, students are expected to be able to explain the working principle and carry out foodstuff analysis, including proximate analysis. Implementation of learning courses in food analysis, one of which applies practicum method in the laboratory as an effort to achieve student's competence, both cognitive, affective and psychomotor. Many studies have shown that learning in the laboratory can improve student's ability and learning motivation in science learning [1,2]. In addition laboratory activities serve to link theory or concepts and practices, increase attractiveness or interest, can correct misconceptions, and develop analytical and critical attitudes to learners [3].

Implementation of practical study of food analysis courses, in the laboratory of agro-industrial technology education study program, still faces the problems. The limited resources of practicum of food analysis course become one of the causes of less than optimal learning process, thus affecting the achievement of students' cognitive and psychomotor competence. In addition, the absence of self-learning means to be one of the causes of less optimal learning of food analysis courses such as the water content analysis and ash content analysis, because only supported by literature study, as an effort to achieve student competence.

Information and Communication Technology integration in teaching and learning process is no longer a luxury but has become a vital requirement to develop educational infrastructure/ structures. Technology offers qualitative leap in reformulating all educational process components [4]. Information and Communication Technology has had a positive impact on the innovation of learning style and teaching. Some other studies reported that computer simulation experiments are more affective [5-7]. Virtual lab provides some advantages in education. Virtual laboratories exist as a complement and substitute for real laboratories in overcoming the limitations of equipment and lab testing in real laboratories [8]. Virtual laboratory is a computing system that allows to share the physical resources available in a laboratory with remorw users connected on the internet [9]. Virtual laboratory helped student to understand concepts and theory in online course. It is particulary useful wheb some expriments involve equipment that may cause harm to human beings [10].

Therefore, virtual laboratory is expected to be an alternative solution of the limitations of agro-industrial technology education study program. Moreover, the potential of information communication technology owned by the students and the ease of internet access in the campus. This study aims to develop a virtual laboratory for food analysis courses on water content analysis and ash content analysis, as the development of instructional media for food analysis courses.

2. Methods

This research applied research and development approach which will use Borg & Gall model. This model is used to develop a product and test its effectiveness [11]. The research stages consist of needs analysis, product development, expert validation-revision, product trial-revision. Product trials conducted on students who follow lecture of food analysis as user of this product (web-based laboratory).

2.1 Data collections

The instrument used in this research is non test instrument in the form of questionnaire. Questionnaires used in this study is a closed questionnaire, which is the choice of answers already provides by researchers. Questionnaires for experts use Guttman scale, and questionnaires for respondents use Likert scale.

2.2 Data analysis

Data analysis used to determine the feasibility of virtual labs in this research is descriptive statistics. The result of data analysis interpreted based on predetermined interval.

3. Results and Discussion

3.1 Needs analysis of virtual lab development

Implementation of learning for water content analysis and ash content in food analysis courses were not only in classroom, but also in laboratory. Availability of practicum facilities was not proportional to the number of students so that laboratory could not be implemented in one lesson. In addition, students did not have self-learning media for food analysis courses. On the other hand, Agro-industrial Technology Education Study Program had potency to support the development of web-based virtual laboratory, namely availability of wireless hotspot areas or wireless computer network services around campus. Moreover, student literacy in information communication and technology was very good. It became an opportunity for the development of learning media that can improve the quality of learning in the laboratory.

Web-based virtual laboratory was developed in accordance to syllabus of the food analysis course, in which there was an indicator of the achievement of water content analysis material and ash content analysis. The result of interview with laboratory assistant in Study Program of Agro-industrial Technology Education, it was known that it needs a self-learning media for students to improve understanding in food analysis course. Based on the indicator of material achievement in the syllabus of food analysis course, also interview result with laboratory assistants and lecturer, obtained information that component of virtual laboratory that need to be developed, consist of safety and job safety in laboratory, introduction of laboratory equipment and water content analysis method, as well as development of a simulated how to analysis watercontent of oven method and analysis of ash content of dried ash method.

3.2 Development of web-based virtual laboratory

Development of web-based virtual laboratory consists of three stages, namely design, development procedure and product testing (evaluation). This virtual lab development research refers to the development model of Borg and Gall [11], with modifications tailored to the field research conditions. Development of an effective virtual laboratory is used to present a practicum simulation with a scientific method. Simulation in a virtual lab can transform the abstract concept into concrete experience so that representatives in conveying the message [12]. Development of web-based virtual laboratory for food analysis courses aims to overcome the problem of limited laboratory facilities in real laboratory and lack of self-learning media for students. Designing was the first stage in the development of this web-based virtual lab. This stage consists of making flowcharts, storyboards, simulation and website layout designs, and content compilation. Making a flowchart aims to describe the flow of virtual lab processes that would be made as presented in Figure 1.

Making a story board aims to provide load information on simulation of water content analysis and ash content analysis. Preparation of learning materials aims to prepare the material to be loaded on the website. The composition materials consists of work safety in laboratory (Personal protective equipment, laboratory rules, Hazard Symbols, First Aid), introduction of lab tools for Analysis of water content and ash, lecture materials of food analysis. Making applications asset aims to prepare the required assets in the simulation component. This asset application was in the form of drawing tool design, materials and sample of practicum needed on analysis of water content and ash content. Design tools and materials tailored to the tools and materials contained in the laboratory of Agro-industrial Technology Education Study Program. In addition, created a virtual lab logo that serves as an identity as presented in Figure 2.

The second stage in the development of virtual web-based lab was a development procedure where there was a process of integrating laboratory components that have been designed into a complete media in accordance with the story board that has been made. Web-based virtual lab were created using wordpress.com with upi.edu domains so it has the URL *http://labvirtual.agroindustri.upi.edu*. Virtual laboratory outlook is presented in Figure 3.

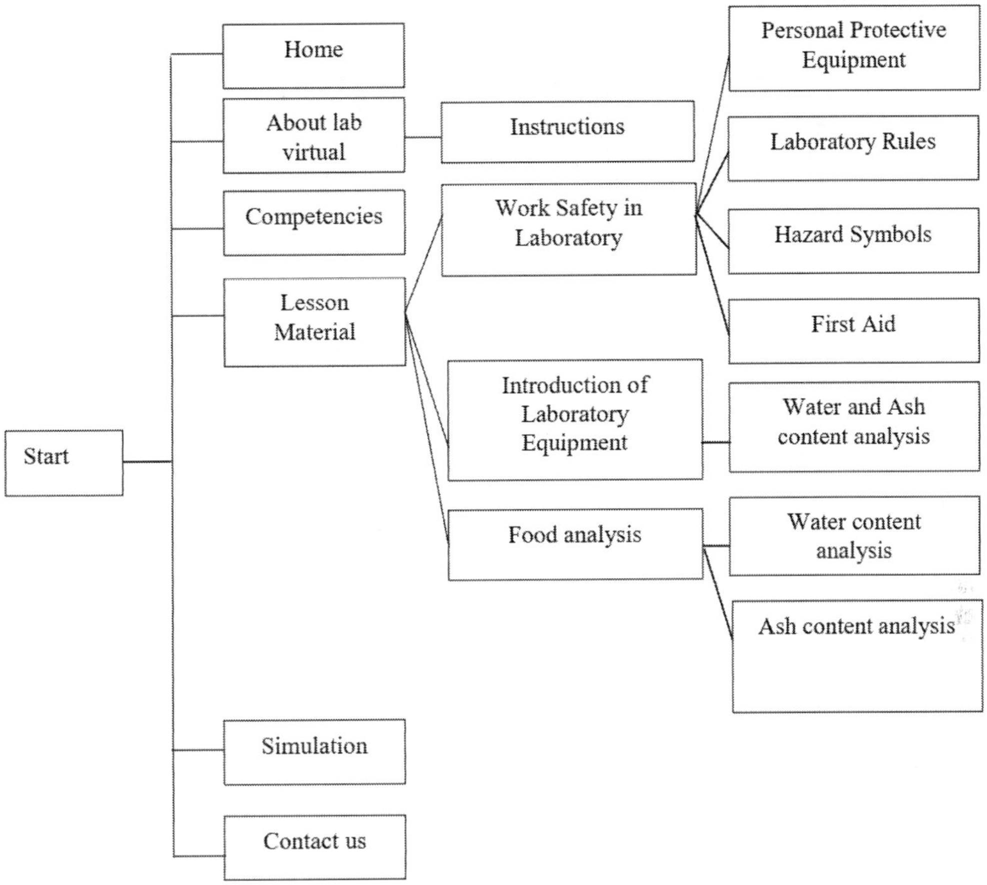

Figure 1. Flowchart of web-based virtual laboratory.

Figure 2. Virtual laboratory logo.

While simulation of water content analysis and ash content analysis in virtual lab using unity application. One of the characteristics of multimedia learning must be interactive that has the ability to accommodate user response [12]. Virtual lab as computer-based learning media can be considered as a complete messaging system such as drills and exercises, reference tools, tutorials, games, simulations and complex calculations [13]. Virtual laboratory facilitates students like studying in real laboratory. Providing feedback on a virtual laboratory can create two-way communication resulting in student learning motivation [14]. Evaluation was the final step undertaken in the development of a virtual laboratory. This stage consists of previewing and editing prior to validation by expert validator.

Figure 3. Outlook of web-based virtual laboratory *http://labvirtual.agroindustri.upi.edu*

3.3 *Expert validation-revision*

Expert validation on development of virtual laboratory, conducted to find out expert opinion and measure feasibility of developed virtual laboratory, before being tested to respondents. Assessment of media experts on the development of virtual laboratory, includes aspects of software and visual communication, referring to the media feasibility assessment guidelines. Media outlook is the result of visualization of investigations, simulations and materials so that integration of media and material view is a mutually supportive unit [15]. The expert's assessment of the feasibility of virtual laboratory is conducted using a questionnaire.

The results showed that media expert judges appropriately on web-based virtual laboratory, on aspects of software and visual communication. In addition, media experts provide suggestions for improvements such as adding program menu hints, adding a left-right scroll bar to the tool display, increasing the size of the logo on virtual laboratory home page display, moving the widget about the program to the main menu parallel on Home page and calendar menu deletion. Revisions have been made to create an interactive virtual laboratory as learning media. In addition, interactive learning media must have ability to accommodate user responses [12]. Moreover, one of criteria of software assessment on learning media is readability aspect, ease of use, display quality and program management quality.

Expert assessment for materials of web-based virtual laboratory is conducted to test the feasibility language of virtual laboratory materials, in terms of material relevance, material adequacy and language quality used. The results showed that the material experts stated feasible for web-based virtual laboratory, from material relevance aspect, adequacy of benefits and quality of language. Nevertheless, the material expert provides suggestions for material improvements in virtual laboratory, such as addition of an animated video tutorial on analysis of water content and ash content; addition of learning evaluation tools, addition of simulations on various methods of analysis of water content and ash content; addition of Good Laboratory Practice (GLP) materials; and addition of modules in the virtual lab. Revisions have been made to create an effective virtual laboratory as learning media. The principle that must be fulfilled in development of virtual laboratory as learning media is relevance. Developed media must have benefit as a tool in learning process, able to attract more students' attention so that it can generate learning motivation. In addition, the use of language in learning media such as virtual laboratory can affect effectiveness of information.

3.4 Product trial-revision

Product trials were conducted in two stages: on a limited scale and on a wider scale, which was performed after revision-expert validation. Web-based virtual laboratory was tested in limited scale to students who had been following food analysis course. This limited-scale trial was conducted to obtain information on feasibility of developed virtual laboratory that were measurable from student responses after trying to use virtual laboratory. The results showed that in terms of software, more than 75% of respondents stated strongly agree that virtual laboratory run smoothly when used (or not easily error), keeps running well even if used for a long time, easy to use. Obviously, the navigation keys used were consistent and work properly. This virtual laboratory is an innovation product that has never been developed before in agro-industrial technology education study program. It affected the responses of respondents. The majority of students responded positively after trying to use virtual lab. Web-based virtual laboratory, able to visualize experiments of water content analysis and ash content analysis conducted in real laboratories. In addition, respondents expressed their feasibility of the virtual lab software aspect and stated strongly in agreement with development of virtual lab, also very interested in learning to use Virtual lab.

The results of the study also showed that majority of respondents (70-80%) agreed to use of language and sentence in virtual laboratory because it was easy to understand, the text presented can be read clearly, the quality of color brightness was good, the animation was displayed well, the audio used nice to hear, and virtual lab display was pretty good. Virtual laboratory developed were equipped with simulation analysis that require input from user so that user can run the simulation own. Effectivity and efficiency of virtual laboratory was reflected from its ability to overcome the limitations of objects so as to provide ease, usefulness and security at an affordable cost [16].

The results also showed that students strongly agree if virtual lab can increase spirit and desire of students to learn. Respondents also stated that application of language and flow of virtual laboratory presentation was easy to understand and the material presented systematically, so it can help the students in understanding the content of water content analysis and analysis of ash content in food analysis courses. Increased motivation of attention aspect can cause interest in student learning that is the tendency to feel interested in learning or certain subject and feel happy to learn the material. In this trial, obtained suggestions from the respondents include software improvements, visual communications and learning materials.

Large-scale trial for virtual laboratory were performed after a limited-scale trial revision. This trial was conducted to determine level of virtual laboratory feasibility developed. The results of large-scale virtual laboratory trials seen from the software aspects were at 75% -82% intervals. It showed that students strongly agree that virtual laboratories (not easy to error when used), easy to use, user manual used is clear, navigation key used consistently and can function well. The result of research also showed that student very agree with development of virtual laboratory and very interested to learn using Virtual laboratory content analysis of water content and analysis of ash content of food analysis courses.

4. Conclusions

Web-based virtual laboratory for food analysis courses was developed through several stages: needs analysis; development of web-based virtual laboratory; expert validation; product trial- revisions. Expert validation results, both media experts and material experts, in terms of software engineering, visual communication, material relevance, adequacy of benefits and quality of language, indicated that web-based virtual laboratory, categorized as proper to use. As well as students as product trial respondents, expressed satisfaction with web-based virtual laboratory since it was easy to use, improved their learning motivation, also made it easier to understand food analysis course.

Acknowledgments

The work is funded by UPI (Universitas Pendidikan Indonesia) Research Grant, Budget Year 2017

References

[1] Bagci N and Simsek S 1999 The Influence of Different Teaching Methods in Teaching Physics Subjects on Student's Success *The Journal of Gazi Education Faculty* **19**(3) p 79-88.

[2] Bryant R J and Marek E A 1987 They Like Lab-Centered Science *Science Teacher* **54**(8) p 42-45

[3] Maknun D, Surtikanti R H K, Munandar A and Subahar S 2012 Keterampilan Esensial dan Kompetensi Motorik Laboratorium Mahasiswa Calon Guru Biologi dalam Kegiatan Praktikum Ekologi *Jurnal Pendidikan IPA Indonesia* **1**(2) p 141-148.

[4] Al Musawi A, Ambusaidi A, Al-Balushi S and Al-Balushi K 2015 Effectiveness of E-Lab Use in Science Teaching at the Omani Schools *TOJET: The Turkish Online Journal of Educational Technology* **14**(1).

[5] Russell J W, Kozma R B, Jones T, Wykoff J, Marx N and Davis J 1997 Use of Simultaneous-Synchronized Macroscopic, Microscopic, and Symbolic Representations to Enhance the Teaching and Learning of Chemical Concepts *J. Chem. Educ* **74**(3) p 330.

[6] Svec M T 1995 Effect of Microcomputer-Based Laboratory on Students' Graphing Interpretation Skills and Conceptual Understanding Of Motion.

[7] Redish E F, Saul J M and Steinberg R N 1997 On the Effectiveness of Active-Engagement Microcomputer-Based Laboratories *American journal of physics* **65**(1) p 45-54.

[8] Domingues L, Rocha I, Dourado F, Alves M and Ferreira E C 2010 Virtual Laboratories in (bio) Chemical Engineering Education *Education for chemical engineers* **5**(2) p e22-e27.

[9] Carnevali G and Buttazz Q G 2003 A Virtual Laboratory Environment for Real-Time Experiments *IFAC Proceedings* **36**(12) p 31-36.

[10] Yang B 1999 Virtual Lab: Bring the Hands-on Activity to Online Courses. In *American Society for Engineering Education Annual Conference, USA*

[11] Gall M D, Borg W R and Gall J P 1996 *Educational research: An introduction* (Longman Publishing).

[12] Daryanto 2013 *Media Pembelajaran: Peranannya Dalam Mencapai Tujuan Pembelajaran.* (Yogyakarta: Penerbit Gava Media)

[13] Carl L E and Anderson T 1994 *The Virtual Reality Casebook* (New York: Van Nostrand Reinhold).

[14] Yuniarti, et al. 2012 Pengembangan Virtual Laboratory Sebagai Media Pembelajaran Berbasis Komputer Pada Materi Pembiakan Virus *Journal of Biology Education* **1**(1) p 86-94.

[15] Dobrzański L A and Honysz R 2010 The Idea of Material Science Virtual Laboratory *Journal of Achievements in Materials and Manufacturing Engineering* **27**(2) p 207-210.

[16] Saleh K F, Mohamed A M and Madkour H 2009 Developing Virtual Laboratory Environment for Engine Education *International Journal of Arts and Sciences* **3**(1) p 9-17.

IOP Conf. Series: Materials Science and Engineering **306** (2018) 012084 doi:10.1088/1757-899X/306/1/012084

The Opinions about Relationship between Students and Teachers in the Class of Hands-on

M Pigultong*

Department of Educational Technology and Communications, Faculty of Technical Education, Rajamangala University of Technology Thanyaburi, Thailand

*mtpphd@rmutt.ac.th

Abstract. This research has the purpose to study on 1) Relationship between Students and Teachers in the Class of Hands - on and 2) Class Management at Rajamangala University of Technology Thanyaburi. The research consists of collecting information from 400 students who have valid student status in 2016 at Rajamangala University of Technology Thanyaburi. This research uses content analysis technique, Average ($\bar{X}$) and Standard Deviation to interpret the information. The results of the research focus on 2 topics 1) The Human relationship between Students and Teachers. The samples group had high expectations of human relationship (x=3.87). 2) Class Management. The samples group had high expectations of Class Management (x=3.88).

1. Introduction

Higher education has many missions to make graduates who can go out in the work force, work with knowledge and skills for each organization, and propel the national economy forward. The important mission that all of technological universities in the world must hold on the first priority is hands on instruction with high quality. The success factor for make the important mission complete is the human relationship between students and teachers. In this way, the good human relationship with warmly and reliability can make happiness atmosphere in their class. However, if the human relationship is in opposite way or full of stress and distrustful about professionally status of the teacher. Both the students and the teachers will feel discouraged and this will not be beneficial to learning [1]. In addition, Class management is strategy that important to helps students and teachers to be keep right way for right activities that good for the brain to function properly for learning [2]. Management of the good environment around classrooms is directly affected by the university's ability to setup and control [3, 4].

The research Questions are focus in two topics 1) Which the opinions of the students are to show that how they feel about their teacher's relationship with their students? 2) Which the opinions of the students that they want the teachers to do about class management for make them learning successful? Then, this research has the purpose are 1) to study about student's opinion of Human relationship between Students and Teachers in the Class of Hands – on and 2) to study about student's opinion of Class management.

2. Methods

In this research, the researcher divides the research process into 2 steps as follows.

2.1. The study of information from student leaders
Collect information from the primary source. The researcher conducts the following processes.

2.1.1. Create a tool for collecting data. The researcher studies related documents both in Thai and foreign language to determine the scope and then create an open-ended questionnaire for the student leaders concerning the management of relationship between students and Teachers and at Rajamangala University of Technology Thanyaburi. The researcher brought the open-ended questionnaire to 3 specialists for content validation process. The Item Objective Congruence (IOC) calculation was used for content validation, the result of content validation = 1.00. After finished, the researcher brought the open-ended questionnaire to Population for collecting data.

2.1.2. Population. The researcher specifies the population to be used for collecting data by Purposive Sampling. All student club presidents of 11 faculties who are leaders to provide information.

2.1.3. Data collecting. The researcher uses the open-ended questionnaire in step 2.1.1. launch to the student club presidents of 11 faculties. In this opportunity, the researcher interviews each student club president for more information about relationship between Students and Teachers in their class.

2.1.4. Data processing. The researcher uses the answers received from step 2.1.3. and processes it with the Content Analysis method for creating the next research tool.

2.2. Collecting data from the students
Collecting data from secondary data source includes the following processes.

2.2.1. Determine the population and samples. The population used in this research is students who are in their second semester of 2016, from a total of 23741 students, at Rajamangala University of Technology Thanyaburi.

2.2.2. Determine the size of the sample. The researcher calculates the size of the sample with the Taro Yamane formula with a margin of error of 5% is 399.98 people. Therefore, the researcher must collect 400 data sets.

2.3. Create research tool for questioning the students in step 2.2
- The researcher uses the data from the analysis in step 2.1.4 to create a close-ended questionnaire consisting of 2 Groups of topics, within 24 questions.
- Before collecting the data from the sample group. The researcher brought the close-ended questionnaire to 3 specialists for content validation process. The Item Objective Congruence (IOC) calculation was used for content validation, the result of content validation = 0.96.

2.4. Collecting data from the sample group
After the researcher creates the close-ended questionnaire from the data obtained from the student leaders in step 2.1, the researcher provides the close-ended questionnaires to the 400 Students. The sampling is carried out by method of Accidental Sampling for the convenience of data collecting by researcher and consists of the following steps.

2.4.1. Find volunteers. Who are third year students of the Educational Technology and Communications Department, to collect the data with the purpose of giving these students educational research experience.

2.4.2. The researcher explains to the volunteer. The purpose of the research, the method of creating the questionnaire and answers to be careful about when administering the questionnaire to the sample group.

2.4.3. Send the volunteers to collect the data from sample group. The researcher will oversee the process and advise the volunteers while collecting the data.

2.4.4. When the sample group completes the questionnaire. The data collector must check the completeness of the answers in the questionnaire. If it is incomplete, the data collector will follow up the details until the questionnaire is complete.

2.5. Data Analysis
After receiving the questionnaire back, the researcher brings all the questionnaires to analyze data by using a few standard statistics such as Average ($\bar{x}$) and Standard Deviation. The data analyze was used by Microsoft Excel.

3. Results and Discussion
The result of study about student's opinion of Human relationship between Students and Teachers in the Class of Hands – on as table 1.

Table 1. Shows the opinions about relationship between students and teachers in the class of hands-on: Human relationship.

		$\bar{x}$	SD
1.1	Ask the teachers must teach knowledge from sources that can be clearly proven accurate.	4.03	0.813
1.2	Ask the teachers teach by using real examples.	3.94	0.789
1.3	Ask the teachers choose the right learning media.	3.89	0.813
1.4	Ask the teachers must let students practice on their own without being disturbed.	3.87	0.782
1.5	Ask the teachers to teach students with proven knowledge that is reasonable and practical.	3.86	0.831
1.6	Ask the teachers must teach students with a psychology to realize that "We can create innovation".	3.86	0.755
1.7	Ask the teachers must trust the students to create their own innovation.	3.84	0.737
1.8	Ask the teachers to listen to the opinions of students in the class seriously.	3.84	0.821
1.9	Ask the teachers must create empirical innovation to gain the faith of students.	3.81	0.712
1.10	Ask the teachers to display empirical professionalism to be examples for students	3.79	0.752
		3.87	0.780

From table 1, The Opinions about Human Relationship between Students and Teachers in the Class of Hands - on, the results indicate that the students expect the teachers must teach knowledge from sources that can be clearly proven accurate, the teachers to Teach by using the real examples, The teachers must let students practice on their own without being disturbed, the teachers to teach students with proven knowledge that is reasonable and practical, the teachers to listen to the opinions of students in the class seriously, improve the facilities in the classroom according to teaching activities. In all the points mentioned, the students of the samples group had high expectations (x=3.87).

Table 2. Shows the opinions about relationship between students and teachers in the class of hands-on: Class management.

		$\bar{x}$	SD
2.1	Provide information in basic form and electronic.	4.18	0.814
2.2	Provide more SMART classrooms	4.06	0.827
2.3	Provide start-up programs for the students	3.96	0.776
2.4	Provide training for filtering information to the students.	3.96	0.806
2.5	Provide the opportunity for students to meet businesses for the exchange of professional ideas.	3.93	0.767
2.6	Set up student clubs that provide activities for developing careers.	3.86	0.757
2.7	Promote all subjects to teach details that can be used in real-life work.	3.85	0.819
2.8	Provide a more relaxed atmosphere in classes.	3.84	0.765
2.9	Develop practical tools in classrooms that match real-life conditions in the work place.	3.81	0.800
2.10	Provide innovation contests.	3.81	0.784
2.11	Improve the facilities in the classroom according to teaching activities.	3.80	0.768
2.12	The required tools which support creating limitless innovation.	3.80	0.792
2.13	Create activities to exchange and learn between student and society.	3.79	0.776
2.14	Provide measures for teachers to use educational innovation for teaching in classrooms.	3.70	0.763
		3.88	0.787

From table 2, The Opinions about Human Relationship between Students and Teachers in the Class of Hands – on: Class Management, the results indicate that Provide information in basic form and electronic, Provide more SMART classrooms, Provide start-up programs for the students, Provide training for filtering information to the students, Provide the opportunity for students to meet businesses for the exchange of professional ideas, Set up student clubs that provide activities for developing careers, Promote all subjects to teach details that can be used in real-life work, Provide a more relaxed atmosphere in classes, Develop practical tools in classrooms that match real-life conditions in the work place, Provide innovation contests, Improve the facilities in the classroom according to teaching activities, The required tools which support creating limitless innovation, Create activities to exchange and learn between student and society, Provide measures for teachers to use educational innovation for teaching in classrooms, In all the points mentioned, the students of the samples group had high expectations (x=3.88).

4. Conclusions

The Opinions about Relationship between Students and Teachers in the Class of Hands – on finds that the students of the sample group have high expectations that 1) the teachers must teach knowledge from sources that can be clearly proven accurate, 2) the teachers to Teach by using the real examples, 3) the teachers must let students practice on their own without being disturbed, 4) the teachers to teach students with proven knowledge that is reasonable and practical, 5) the teachers to listen to the opinions of students in the class seriously, 6) improve the facilities in the classroom according to teaching activities, 7) provide information in basic and electronic form, 8) provide more SMART classrooms, 9) provide start-up programs for the students, 10) provide training for filtering information to the students, 11) provide the opportunity for students to meet businesses for the exchange of professional ideas, 12) Set up student clubs that provide activities for developing careers, 13) promote all subjects to teach details that can be used in real-life work, 14) provide a more relaxed atmosphere in classes, 15) develop practical tools in classrooms that match real-life conditions in the work place, 16) provide innovation contests, 17) improve the facilities in the classroom according to teaching activities, 18) the required tools which support creating limitless innovation, 19) create activities to exchange and learn between student and society, 20) provide measures for teachers to use educational innovation for teaching in classrooms.

References

[1] Vichit T 2009 *Learning Environment* [Available on] https://www.l3nr.org/posts/314570%20%5B4
[2] Jeraphan P and Naruemon N 2016 Environment for Learning [Available on] http://www.nareumon.com/index.php?option=com_content&task=view&id=24&Itemid=50&limit=1&limitstart=0
[3] Suchin W 2013 Learning Environment for Science Education 2013 [Available on] http://edu.stou.ac.th/EDU/UploadedFile/22758-11.pdf
[4] Yamane T 1967 *Statistics, An Introductory Analysis, 2nd Ed* (New York: Harper and Row).

IOP Conf. Series: Materials Science and Engineering **306** (2018) 012085 doi:10.1088/1757-899X/306/1/012085

Priority Determination of Underwater Tourism Site Development in Gorontalo Province using Analytical Hierarchy Process (AHP)

M Rohandi[1*], M Y Tuloli[2] and R T Jassin[3]

[1,3]Department of Informatics Engineering, Faculty of Engineering, Universitas Negeri Gorontalo, Indonesia
[2]Department of Civil Engineering, Faculty of Engineering, Universitas Negeri Gorontalo, Indonesia

*manda.rohandi@ung.ac.id

Abstract. This research aims to determine the development of priority of underwater tourism in Gorontalo province using the Analytical Hierarchy Process (AHP) method which is one of DSS methods applying Multi-Attribute Decision Making (MADM). This method used 5 criteria and 28 alternatives to determine the best priority of underwater tourism site development in Gorontalo province. Based on the AHP calculation it appeared that the best priority development of underwater tourism site is Pulau Cinta whose total AHP score is 0.489 or 48.9%. This DSS produced a reliable result, faster solution, time-saving, and low cost for the decision makers to obtain the best underwater tourism site to be developed.

1. Introduction

Geographically, Gorontalo province lies between the sea of Celebes and the Gulf of Tomini. This geographical position gives the province highly potential marine tourism spots. Given this potentiality, both the coastal and underwater, efforts should be made to tap these tourism resources for economic gains. To this end, efforts have been done by Gorontalo Province office of Tourism to develop and manage the potential marine tourisms spots in Gorontalo province. Regrettably, even though there are many potential underwater tourism sites in Gorontalo province, the development and management of marine tourism are mostly focused on the coastal tourism only. The underwater tourism resources except for the Olele marine park, draw little attention from government due to the limited of budget allocation. There are five essential elements a tourism spot can offer for the tourists to be satisfied when enjoying a tourism spot namely Attractions/charm, facilities, infrastructure, transportation/accessibility, and hospitality/security [1]. These five elements can be used as criteria in determining the potential of underwater tourism to be developed. Therefore it is necessary to define the best priority of underwater tourism sites to be developed among many sites in Gorontalo province.

One approach that can be used to assist the local government in determining the priority of potential underwater tourisms development is a *Decision Support System* (DSS) that implements *Multi-Attribute Decision Making* (MADM). MADM is a decision-making method that suitable to establish the best alternative from a number of underwater tourisms alternatives to be developed based on certain criteria. There is plenty of DSS definition, including [2], DSS is a system that integrates database and model base in a dialogue display and is used to interact by the users to assist decision-making. DSS is a set of model-based procedures for data processing and assessment to assist the managers in decision-making [3].There are several DSS methods that implement the MADM in decision-making, one of them is AHP method. The AHP is originally proposed by Thomas Saaty, 1980 [4-6] which is a structured technique for dealing with complex decision. In this research, AHP method is used to determine the priority of underwater tourism sites development in Gorontalo province. AHP is used because it provides a comprehensive and rational framework for structuring a decision-making problem, for representing and quantifying its elements, for relating those elements to overall goals, and for evaluating alternative solutions.

2. Methods

2.1. Data collection

In this research, the data is obtained by interviewing the officials from the department of tourism, staff in the diving centers, and tourists that usually dive in Gorontalo province. In addition, the research data is collected using questionnaires and literature review from books, tourist brochures, and articles on the internet. Based on data collected there are 28 potential underwater tourism sites listed along with 5 criteria and 5 sub-criteria. These data is then modeled as shown in table 1.

Table 1. Model/Alternatives of underwater tourism sites.

Location	Attractions/ Charm	Facilities	Infrastructure	Transportation/ accessibility	Hospitality/ security
Tri Corner	Very Good	Good	Less	Less	Good
Tanjung Barat	Very Good	Good	Enough	Less	Good
Sponge Wall	Good	Good	Less	Less	Good
Tanjung Putih	Very Good	Good	Less	Less	Very Good
Mystic Point	Very Good	Good	Less	Less	Good
Tjendrawasih Barge Wreck	Good	Good	Very Good	Very Good	Good
Japanese Cargo Wreck	Very Good	Good	Good	Very Good	Very Good
Mirabela	Good	Very Good	Very Good	Very Good	Good
Swirling Step	Good	Enough	Enough	Good	Good
Sand Bowl	Very Good	Enough	Good	Good	Very Good
Kurinae Beach	Less	Enough	Good	Good	Good
Sunken Island	Good	Enough	Very Good	Very Good	Good
Sand Castle	Less	Good	Good	Good	Good
Shadow Land	Good	Less	Good	Good	Good
Honey Comb	Good	Good	Good	Good	Very Good
Sentinel	Good	Good	Good	Good	Very Good
Traffic Circle	Very Good	Good	Good	Good	Very Good
Traffic Jam	Very Good	Good	Good	Good	Very Good
Jinn Cave	Very Good	Good	Good	Good	Very Good
Tanjung Kerbau	Very Good	Good	Good	Good	Very Good
Helicopter Bay	Very Good	Good	Good	Good	Very Good
Mas Island	Very Good	Good	Good	Good	Very Good
Bitila Island	Good	Enough	Good	Good	Very Good
Lahe Island	Good	Very Good	Very Good	Very Good	Good
Pulau Cinta	Very Good	Very Good	Good	Very Good	Very Good
Sudimampir	Good	Very Good	Good	Good	Very Good
Abu Nawas	Good	Enough	Less	Enough	Good
Oluhuta	Good	Good	Good	Good	Very Good

2.2. The AHP steps
The AHP can be implemented in five simple consecutive steps referring to [4-8], as follows:
- Defining Objective
- Structuring elements in criteria, sub-criteria, alternatives, etc
- Making pair-wise comparison of elements in each group by determining the coefficient of importance (weight) with respect to the objective
- Calculating weighting and consistency index
- Evaluating alternatives according to weighting.

3. Results and Discussion
The implementation of AHP is describes as follows:

3.1. Defining the goal/objective
The goal of this DSS is to obtain the best underwater tourism sites to be developed in Gorontalo province.

3.2. Structuring the elements in criteria, sub-criteria, and alternatives
The hierarchy consists of 3 level, from the complex issues (objective from the decision-maker's view point) to the minor elements (criteria, sub-criteria, alternatives) that are linked together to form a hierarchy. Level 0 is the goal of the analysis, while level 1 is multi-criteria that consist of several criterions, to which we can also add several other levels of sub-criteria. Finally, level 2 is the alternative choices. There are 5 criteria and 5 sub-criteria that are used to obtain the potential of underwater tourism to be developed. These 5 criteria are Attractions/charm, facilities, infrastructure, transportation/accessibility, and hospitality/security. While the sub-criteria used are very good, good, enough, less, and bad. With respect to alternatives, there are 28 alternatives location of underwater tourism sites as shown in table 1.

3.3. Making pair-wise comparison of elements in each group by determining the coefficient of importance (weight) with respect to the objective
Construct a set of pair-wise comparison matrices (*m x m)* as shown in table 2, where *m* is the number of evaluation criteria considered. Each entry a_{jk} of the matrix represents the importance of the *j*th criterion relative to the *k*th criterion. ifa_{jk}>1, then the *jth* criterion is more important than the *k*th criterion. On the contrary, if a_{jk}<1, then the *jth* criterion is less important than the *k*th criterion. If two criteria have the same importance, then the entry a_{jk} is 1. The pair-wise comparison are done in terms of which elements dominates the other. There are *n (n-1)* per-judgement required to develop the set of matrices, and the number of comparison depends on the number of criteria *n,* with $(n^2-n)/2$ comparison. The relative scale measurement is taken from Saaty [7]. After pair-wise comparison matrix of criterion is created, the next step is to create a standardized matrix of criterions as shown in table 3. The standardized matrix is obtained by dividing the input value for each criterion C_{th} with the total sum of all criteria columns. As a result, we have the higher to the lower criteria that are used to obtain the priority of underwater tourism sites.

Table 2. Pair-wise comparison matrix of criterions.

Item Description	Attractions/Charms	Facilities	Infrastructure	Transportation/Access	Hospitality/Security
Attractions/Charms	**1.00**	9.0	9.0	7.0	7.0
Facilities	0.11	**1.00**	5.0	0.7	0.7
Infrastructure	0.11	0.20	**1.00**	0.7	0.7
Transportation/Access	0.14	1.43	1.43	**1.00**	0.7
Hospitality/Security	0.14	1.43	1.43	1.43	**1.00**
Sum	**1.51**	**13.06**	**17.86**	**10.83**	**10.10**

Table 3. Standardized matrix of criterions.

	Attractions/Charms	Facilities	Infrastructure	Transportation/Access	Hospitality/Security	Weight
Attractions/Charms	0.66	0.69	0.50	0.65	0.69	**63.9%**
Facilities	0.07	0.08	0.28	0.06	0.07	**11.3%**
Infrastructure	0.07	0.02	0.06	0.06	0.07	**5.6%**
Transportation/access	0.09	0.11	0.08	0.09	0.07	**8.9%**
Hospitality/Security	0.09	0.11	0.08	0.13	0.10	**10.3%**

3.4. Structuring elements in criteria, sub-criteria, alternatives

The consistency is determined by using the Eigen-Value, λ_{max}, to calculate the Consistency Index (CI) as follows: CI = (λ_{max} –n) / (n-1), where n is the matrix size. Judgement consistency can be checked by taking the consistency ratio (CR) of CI with the appropriate value according to [7]. The CR is acceptable, if it does not exceed 0.10. If it is bigger than that, the judgement matrices are inconsistent. To obtain a consistent matrix, judgements should be reviewed and improved [5]. The inputs of this matrix are obtained by multiplying the scale of each criterion in table 2 with the weight of each criterion in table 3. Whereas, sum/weight of column is obtained from the sum of column in table 4, then it is divided by the weight column in table 3.

Table 4. CI and CR matrix of criterions.

	Attractions/ Charms	Facilities	Infrastructure	Transportation/access	Hospitality/ Security	SUM	SUM/ Weight
Attractions/Charms	0.64	1.02	0.50	0.62	0.72	3.50	5.48
Facilities	0.07	0.11	0.28	0.06	0.07	0.60	5.29
Infrastructure	0.07	0.02	0.06	0.06	0.07	0.28	5.09
Transportation/access	0.09	0.16	0.08	0.09	0.07	0.49	5.53
Hospitality/Security	0.09	0.16	0.08	0.13	0.10	0.56	5.46
						Total	**26.85**

In table 4, Number of criteria or n = 5, λ_{max} = sum of weight / n = 26.85 / 5 = 5.372, CI = (λ_{max}–n)/(n-1) = (5.372-5)/(5-1) = 0.093, CR = CI/1.12 = 0.08 (1.12 is taken from average random consistency proposed by Saaty [7]). Because the consistency ratio (CR) is ≤ 0.10, then the CR is acceptable or the matrix is consistent. By repeating the steps of 3.3 to 3.4 to calculate the sub-criteria and we have:

Table 5. Pair-wise comparison matrix of sub-criterions.

Item Number	1	2	3	4	5
Item Description	Very Good	Good	Enough	Less	Bad
Very Good	**1.00**	3.0	5.0	7.0	9.0
Good	0.33	**1.00**	3.0	5.0	7.0
Enough	0.20	0.33	**1.00**	3.0	5.0
Less	0.14	0.20	0.33	**1.00**	3.0
Bad	0.11	0.14	0.20	0.33	**1.00**
Sum	**1.79**	**4.68**	**9.53**	**16.33**	**25.00**

Table 6. Standardized matrix of sub-criterions.

	Very Good	Good	Enough	Less	Bad	**Weight**
Very Good	0.56	0.64	0.52	0.43	0.36	**50.3%**
Good	0.19	0.21	0.31	0.31	0.28	**26.0%**
Enough	0.11	0.07	0.10	0.18	0.20	**13.4%**
Less	0.08	0.04	0.03	0.06	0.12	**6.8%**
Bad	0.06	0.03	0.02	0.02	0.04	**3.5%**

Table 7. CI and CR matrix of sub-Criterions.

	Very Good	Good	Enough	Less	Bad	SUM	SUM/Weight
Very Good	0.50	0.78	0.67	0.47	0.31	2.74	5.46
Good	0.17	0.26	0.40	0.34	0.24	1.41	5.43
Enough	0.10	0.09	0.13	0.20	0.17	0.70	5.20
Less	0.07	0.05	0.04	0.07	0.10	0.34	5.03
Bad	0.06	0.04	0.03	0.02	0.03	0.18	5.09
						Total	**26.21**

$n = 5$, $\lambda_{max} = 26.21/5 = 5.243$, CI = (5.243–5) / (5-1) = 0.061, CR = 0.061/1.12 = 0.05. Because the consistency ratio (CR) is ≤ 0.10, then the CR is acceptable or consistent.

3.5. The final step is to evaluate the alternatives by weighing the sub-criteria according to the weight of main-criteria

The formula to evaluate the alternatives is as follow:

$$A^{i}_{AHP} = \sum_{j=1}^{N} a_{ij}\, w_{ij}, \quad \text{for } i = 1,2,3,\dots..N. \quad (1).$$

Where: A^{i}_{AHP} is the result of evaluation, a_{ij} are ratings for the decision alternatives for each criterion, w_{ij} are criteria of weight. The final result of an alternatives evaluation shows that Pulau Cinta was the best underwater tourism site to be developed with total score of 48.9% or 0.489. It is the highest score among other locations as shown in table 8.

Table 8. Result of AHP calculation.

No.	Location	Attractions/Charm	Facilities	Infrastructure	Transportation/ accessibility	Hospitality/ security	Total Score
1	Pulau Cinta	0.321	0.057	0.015	0.045	0.052	0.489
2	Japanese Cargo Wreck	0.321	0.029	0.015	0.045	0.052	0.462
3	Traffic Circle	0.321	0.029	0.015	0.023	0.052	0.440
4	Traffic Jam	0.321	0.029	0.015	0.023	0.052	0.440
5	Jinn Cave	0.321	0.029	0.015	0.023	0.052	0.440
6	Tanjung Kerbau	0.321	0.029	0.015	0.023	0.052	0.440
7	Helicopter Bay	0.321	0.029	0.015	0.023	0.052	0.440
8	Mas Island	0.321	0.029	0.015	0.023	0.052	0.440
9	Sand Bowl	0.321	0.015	0.015	0.023	0.052	0.426
10	Tanjung Putih	0.321	0.029	0.004	0.006	0.052	0.412
11	Tanjung Barat	0.321	0.029	0.008	0.006	0.027	0.391
12	Tri Corner	0.321	0.029	0.004	0.006	0.027	0.387
13	Mystic Point	0.321	0.029	0.004	0.006	0.027	0.387
14	Mirabela	0.166	0.057	0.028	0.045	0.027	0.323
15	Lahe Island	0.166	0.057	0.028	0.045	0.027	0.323
16	Sudi mampir	0.166	0.057	0.015	0.023	0.052	0.312
17	Tjendrawasih Barge Wreck	0.166	0.029	0.028	0.045	0.027	0.295
18	Honey Comb	0.166	0.029	0.015	0.023	0.052	0.285
19	Sentinel	0.166	0.029	0.015	0.023	0.052	0.285
20	Sunken Island	0.166	0.015	0.028	0.045	0.027	0.281
21	Bitila Island	0.166	0.015	0.015	0.023	0.052	0.271
22	Swirling Step	0.166	0.015	0.008	0.023	0.027	0.239
23	Shadow Land	0.166	0.008	0.015	0.023	0.027	0.238
24	Sponge Wall	0.166	0.029	0.004	0.006	0.027	0.232
25	Abu Nawas	0.166	0.015	0.004	0.012	0.027	0.224
26	Oluhuta	0.022	0.029	0.015	0.023	0.052	0.141
27	Sand Castle	0.043	0.029	0.015	0.023	0.027	0.137
28	Kurinae Beach	0.043	0.015	0.015	0.023	0.027	0.123

4. Conclusions

The underwater tourisms in Gorontalo are highly potential to be developed, due to the geographic position of the province which lies between the sea of Celebes and Tomini bay. Due to these potentials, the development of underwater tourism should be managed well. Given the limited funding, prioritizing scheme on developing underwater tourism spots is needed. DSS is a helpful tool to help the government determine the priority of underwater tourism site development in Gorontalo province. AHP is one of many DSS methods that apply MADM bases judgments on knowledge and experience of the decision-makers. AHP method reduces complex decision-making process using pair-wise comparison, and then synthesizes the result. In order to determine the best underwater tourism sites to be developed, 5 criterions and 28 alternatives were used. The result of AHP calculation shows that Pulaucinta was the best underwater tourism site to be developed. Despite of all the strengths of AHP, this method is not reliable to evaluate a large number of judgements [9]. Further research is needed to combine AHP with other DSS method, such as SAW that uses cost benefit attribute to find the weighted sum of performance ratings on each alternative on all criteria [10].

References

[1] J J Spillane 1994 *Tourism Indonesia: Economic Strategy and Engineering culture* (Yogyakarta: Canisius).
[2] Trisnawarman D and Nico S W 2007 *DSS Applications for Food Menu Selection for People with Diabetes* (Bandung: Informatika).
[3] Turban E, Aronson J E, and Liang T P 2005 *Decision Support Systems and Intelligent Systems-7th Ed Volume 1*. (Yogyakarta: CV. ANDI Offset).
[4] Ameri A 2013 Application of the Analytic Hierarchy Process (AHP) for Prioritize of Concrete Pavement *Global Journal of human social science interdisciplinary* **13**(3).
[5] Al Harbi K M A 2001 Application of the AHP in project *Management International Journal of Project management* p 19-27.
[6] Geoff C 2017 *The Analytic Hierarchy Process (AHP) Introduction. Practical Strategy* Open access Material. Pearson Education Limited. 2004. Access February 5 2017 http://www.booksites.net/download/coyle/student_files/AHP_Technique.pdf
[7] Saaty T L 1980 *The Analytic Hierarchy Process* (New York: Mcgraw-Hill International).
[8] Saaty T L 1990 How to make a decision: The Analytical Hierarchy Process *European Journal of Operational Research* (North-Holland).
[9] Belton V and Gear T 1983 Short coming of saaty mehtod of analytic hierarchies *Omega* **1** p 143-144.
[10] Stajanovic M 2013 Multi-Criteria Decision-Making for Selecting of Renewable Energy System *Safety Engineering Journal* **3**(3) p 115-120.

IOP Conf. Series: Materials Science and Engineering **306** (2018) 012086 doi:10.1088/1757-899X/306/1/012086

Numerical Simulation by using Soldiers Pile of the Embankment on Semarang-Solo Highway

M S S S Tumanduk*, T S Maki, T U Y Pangkey and Y C Pandeiroth
Universitas Negeri Manado, Tondano 95618, Sulawesi Utara, Indonesia.

*morristumanduk@gmail.com

Abstract. Semarang-Solo highway works section II Gedawang-Penggaron constitutes a labile area. It is thought to be effect of the existence of coat clay shale which have moulded. For the purpose of anticipating the embankment mass movement it is placed line bored pile and stringed up (soldiers pile). The objective of this research is to know the efficient use of soldier's pile of the embankment on Semarang-Solo highway section II Gedawang-Penggaron pursuant based upon numerical simulation. The result of analysis depicts that original slope in a stabil state with horizontal displacement which equal to 0.06 m and safety factor (SF) which equal to 1.31. The strengthened embankment with bored pile is not effective to give am SF improvement at slope so that, at this phase, the slope cannot be slid to be safe enough from landslide namely with horizontal displacement which equal to 0.20 m and SF which equal to 1.09. The effect of traffic load horizontal displacement is which equal to 0.21 m with SF which equal to 1.00. The earthquake simulation results horizontal displacement which equal to 0.75 m with SF which equal to 1.00. Long variation of bored pile of phase II by neglecting bored pile phase III at the depth 35 m yields horizontal displacement which equal to 0.03 m and SF optimum which equal to 2.17. The variation of pile location by placing bored pile under embankment slope foot with distance from the location of bored pile of phase II which equal to 20 m without changing the profile of the existing bored pile creates the horizontal displacement which equals to 0.02 m with SF which equal to 2.29. The result of the horizontal displacement and SF of the two alternative is safer compared to the existing condition (SF>1.5).

1. Introduction

Semarang-Solo highway is a main regional road connecting East-West area and North-South area and contributing to economic growth in the Central Java and surrounding areas. Semarang-Solo highway with Section II Semarang-Bawen Space of Gedawang-Penggaron was constructed [1], starting from Sta.3+525-Sta.8+475 (4950 m). In general, this work of construction consisted of excavation, heap, overpass, box culvert and two bridges, Susukan and Penggaron [2].

Performance of highway construction in area Sta.5+600 – Sta.6+200 is labile area and frequently experiencing creep landslide. It is supposed as result soft clay and weathered clay shale layers. The area is location of embankment construction as high as 20-25 m with embankment soil characteristics such as clay silt put on sub grade layer in 9° – 10° slope. This embankment is worried to experience mass landslide if there are changes in soil characteristics in rainy season [3]. The mass heap landslide is coped with soldiers pile. Dimensions of soldiers pile used are 1 m pile diameter, minimally 10 m and maximally 18 m length of pile; 2 m pile distance and 500 m length of soldiers pile (Sta.5+600-Sta.6+100).

The objective of this research was to find efficiency of soldiers pile application to embankment of Semarang-Solo highway of Section II Gedawang-Penggaron with Numerical simulation. To analyze slope stability of embankment and effectiveness of placement of soldiers pile, software of version 8.2 Plaxis would be used [4].

2. Methods

2.1 Location
Location which was used for this research object was Sta.5+739 of Semarang – Solo High way of phase I of Semarang – Bawen space of Section II Gedawang – Penggaron in Semarang, Central Java. Map of research object location of Sta.5+739 can be seen in **Figure 1**.

Figure 1. Map of research object location of Sta.5+739 (Cipta Strada, 2009).

2.2 Materials
The object in Sta.5+739 can represent from Sta.5+600 to Sta.5+800 and the location is vulnerable to mass movement. Mass movement landslide can be coped with placement of soldiers pile. To analyze efficiency of soldiers pile placement, secondary data were used.

2.3 Tools and collection of data
Tools used in this research were a set of hardware unit, namely, Pentium IV based-computer with 1 Gb memory and software of version 8.2 [4]. Plaxis data which were needed in this research were primary and secondary data. Primary data of this research consisted of data of interview and field observation, and secondary data used in this research were:

- Map of topography, bestek picture, and research sketch;
- In situ data (SPT, CPT, Bore Log, Geoelectric);
- Laboratory data.

2.4 Traffic load and earthquake load
In simulation of traffic load, 16 kPa was used based on design parameters in of Semarang - Solo Wighway [1]. Based on earthquake zoning map issued by Public Work Ministry published in 2010, research location existing on Semarang City, Central Java Province, is included in Zone 5 with rock peak speed (PGA) 0.2g – 0.20g. Earthquake data used were earthquake spectrum files of SMC format which were downloaded from USGS, with peak speed (PGA) 0.2399g or 239.9 Club of Multimillionaires/s^2 on February 28 of 1990 in Upland, California [5].

3. Results and Discussion

3.1 Numeric simulation using Plaxis software
Analysis of slope in Sta.5+739 was simulated in some conditions according to field condition with various water surface depths, -4m depth and -8m depth based on machine bore test data and assumed representing land condition in rainy and summer seasons. The analysis was conducted to find

displacement tendency behavior [6], examine bored pile placement efficiency, potential fall mechanism, and safe rate in Sta.5+739.

Numerical analysis of stability was classified into three models:

1) Model 1, analysis of existing slope condition before there was highway construction;
2) Model II, analysis of existing condition in site. Model II has three analytic reviews, namely:
   a. Slope condition with only embankment and it had been conducted by sub grade replacement under embankment;
   b. Embankment slope condition with reinforcement system application of *bore pile* installed parallel to embankment foot; and
   c. Condition with loading simulation.
3) Model III, analysis of bored pile reinforcement system use efficiency. This model based on variation of length and location of bored pile to improve or increase rate of slope safety factor to be > 1.5.

3.2 Validation based on measurement data

Considering the results of field test and instrumentation data, changes in parameter data inputs were made, also geometry of model according to inclinometer 3 data as shown in **Table 1**. Based on consistence of land parameter data, modeling of validation geometry is classified into five land layer clusters: first, embankment layer (clay silt); second, silt clay layer which is outset layer; third, weathered clay shale; fourth, clay shale; and fifth, fresh clay shale. Geometry of Sta.5+739 can be seen in **Figure 2** [4].

Based on the results of simulation in validation phase, value of horizontal displacement was 0.14 m and data of horizontal displacement in field in inclinometer 3 data is 0.14 m, so the simulation could be continued to modeling phase. Graphic of horizontal displacement of validation results in point inclinometer 3 is shown in **Figure 3**.

Table 1. Land parameter data after making trial and error.

No	Parameter	Type of Land				
		Fresh Clayshale	**Clayshale**	**Weathered Clayshale**	**Silt clay**	**Heap**
1	Model	MC	MC	MC	MC	MC
2	Type	*Undrained*	*Undrained*	*Undrained*	*UnDrained*	*UnDrained*
3	γ_{unsat}(kN/m^3)	19.0	17.0	15.5	16.0	17.5
4	γ_{sat} (kN/m^3)	20.0	19.0	17.5	18.0	18.0
5	k_x (m/day)	0.00315	0.00315	0.0315	0.00315	0.00315
6	k_y (m/day)	0.00315	0.00315	0.00315	0.00315	0.00315
7	E_{ref} (kN/m^2)	370000	9000	560	8500	20000
8	V	0.35	0.35	0.35	0.35	0.35
9	C_{ref} (kN/m^2)	320.00	80.00	15.00	70.00	65.0
10	Φ	26	10	9	10	23
11	Ψ	0	0	0	0	0
12	*R inter*	0.7	0.7	0.7	0.7	0.7

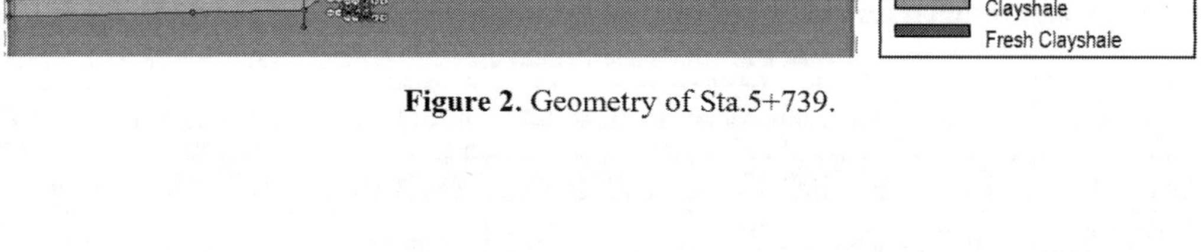

Figure 2. Geometry of Sta.5+739.

IOP Conf. Series: Materials Science and Engineering **306** (2018) 012086 doi:10.1088/1757-899X/306/1/012086

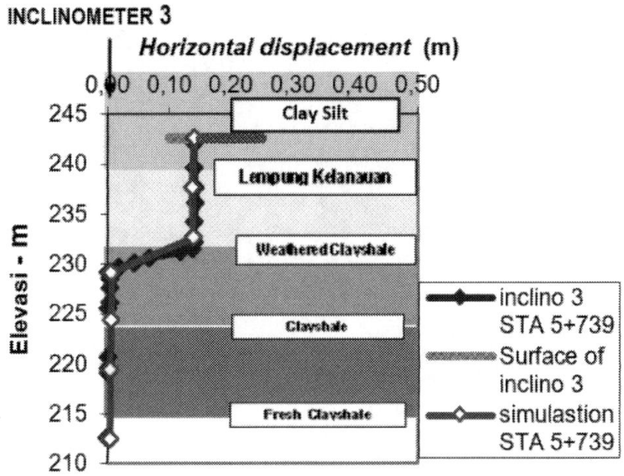

Figure 3. Graphic of horizontal displacement of validation results in point inclinometer 3.

Table 2. Data of recapitulations of horizontal displacement (cross section in location of inclinometer 3) and safety factor in Plaxis software.

No.	Simulation of Plaxis	Horizontal Displacement (m)	SF
1	Existing condition of slope	0.06	1.31
2	Condition of Stripping/Digging, Heap of Phase I, and Installation of Phase II *Bore Pile* (Validation Phase)	0.14	1.09
3	Condition of Phase II Heap	0.20	1.08
4	Condition of Phase III *Bore Pile* Installation	0.20	1.09
5	Condition of Traffic Load	0.21	1.00
6	Condition of Earthquake Load	0.75	1.00
7	Alternative 1 (Length variation of *Bore Pile*)	0.03	2.17
8	Alternative 2 (Position/Distance Variation of *Bore Pile*)	0.02	2.29

3.3 Analysis of horizontal displacement and safety factor (SF)

From the results of simulation for various conditions, data of horizontal displacement and safety factor deformations were obtained as shown in **Table 2**. Recapitulation of horizontal displacement can be seen in **Figure 4** and recapitulation safety factor can be seen in **Figure 5** [7].

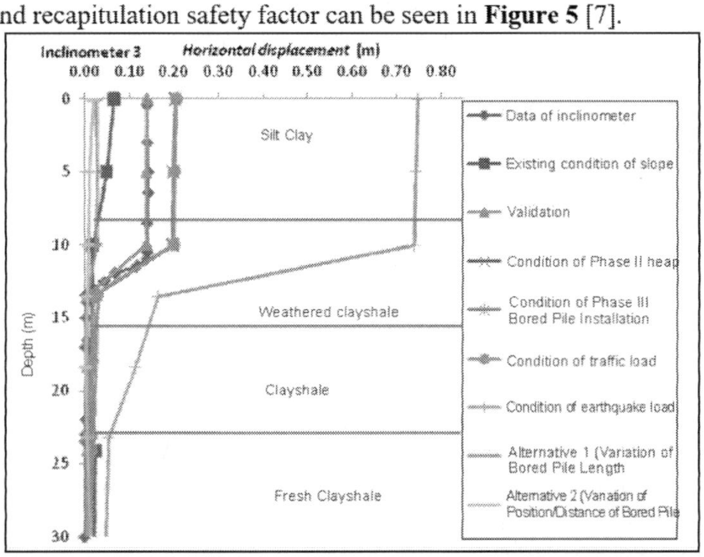

Figure 4. Recapitulation of horizontal displacement.

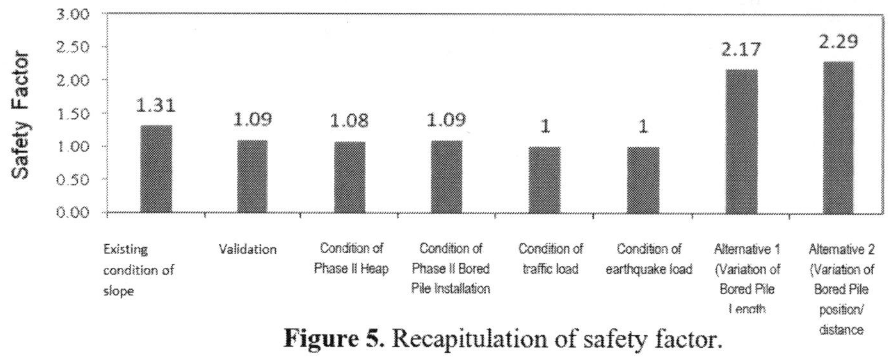

Figure 5. Recapitulation of safety factor.

4. Conclusions

Based on results of analysis conducted in this research, it can be concluded that: analysis of Model I (simulation of sub grade slope condition) generated horizontal displacement = 0.06 m with safety factor = 1.31. the safety factor indicates that the embankment slope condition of Sta.5+739 was stable (SF > 1.25); analysis of Model II (simulation of existing condition) indicates that road heap which had been reinforced by bore pile was ineffective to increase safety factor in slope, so that, in this phase, slope had not been found sufficiently safe for slide, namely, horizontal displacement = 0.20 m and safety factor = 1.09, as result of traffic load, the horizontal displacement = 0.21 with safety factor = 1.00. Simulation of earthquake generated horizontal displacement = 0.75 m with safety factor = 1.00. Based on the results of depth analysis, bore pile was not flanked perfectly by layer of fresh clay shale; mass movement in embankment was affected by land layer of weathered clay shale, so that it reduced landslide strength parameters resulting in cracking on road body. It caused less stability of embankment slope resulting in mass movement embankment; pattern of mass movement in slope was slide where slip plane tended to curve with rotational motion; analysis of Model III (analysis of embankment reinforcement by bored pile showing variations of Phase II bored pile length in depth 35 m by ignoring Phase III bored pile generated horizontal displacement = 0.03 m and optimum safety factor = 2.17. Variations of Phase III bored pile location in condition under body embankment slope foot with distance from Phase II bored pile location = 20 m without changing profiles of bored pile and Phase II bored pile remained as in existing condition generating horizontal displacement = 0.02 m, with safety factor = 2.29. The results of horizontal displacement and safety factor of both alternatives were safer than existing condition; proposal of embankment reinforcement alternatives to improve or increase safety factor was to increase Phase II bored pile length to be 35 m without changing profiles by ignoring Phase III bored pile. This alternative was more efficient and economic because it ignored installation of Phase III bored pile construction; and effect of pile location and distance between piles, gave great effect to safety factor increase.

Acknowledgments

The authors wishing to acknowledge assistance or advisor from Dr. Ir. Ahmad Rifa'i, M.T. as a lecturer in Department of Civil and Environmental Engineering, Faculty of Engineering, Gadjah Mada University and Rector of Manado State University.

References

[1] PT Cipta Strada 2009 *Analisa dan Perencanaan Area Clay Shale pada Sta.5+600 s.d Sta.6+200* Project of Semarang-Solo Tolroad Construction, Semarang, Indonesia.

[2] Nurdiana A, Wibowo M A and Hatmoko D 2015 Sensitivity Analysis of Risk from Stakeholders' Perception Case Study: Semarang-solo Highway Project Section I (Tembalang-Gedawang) *Procedia Engineering* **125** 12-17

[3] Murakami K, Tokida K-I, Tachibana A 2012 *Simulation Of Sliding Failure Occurred At Highway Embankment In 2011 Great East Japan Earthquake* Proceedings of the

International Symposium on Engineering Lessons Learned from the 2011 Great East Japan Earthquake, March 1-4, 2012, Tokyo, Japan.

[4] Plaxis 2D – *Material Models Manual* 2015 [Online] https://www.plaxis.com.

[5] S´aez E and Ledezma C 2012 Earthquake-induced pressures on discontinuous piling support on Santiago gravel *Soil Dynamics and Earthquake Engineering* **41** 72–83

[6] Mujah D, Ahmad F, Hazarika H and Watanabe N 2013 Numerical analysis of the multirow arrangement of small diameter steel piles for landslide prevention *International Journal of Current Engineering and Technology* **3**(1) 1-20

[7] Laxminarayana A and Naresh M 2017 Analysis of Embankments with Different Fill Materials using Plaxis-2D *International Journal of Engineering Trends and Technology* **45**(6) 280-284

IOP Conf. Series: Materials Science and Engineering **306** (2018) 012087 doi:10.1088/1757-899X/306/1/012087

The Development of Indonesian Labour Market Information System (LMIS) for Vocational Schools and Industries

M T Parinsi[1*], V R Palilingan[2], Sukardi[3] and H D Surjono[3]

[1]Department of Technical and Vocational Education, Doctorate Program Student, Universitas Negeri Yogyakarta, Indonesia
[2]Department of Information Technology and Communication Education, Faculty of Engineering, Universitas Negeri Manado, Indonesia
[3]Department of Technical and Vocational Education, Doctorate Program Lecturer, Universitas Negeri Yogyakarta, Indonesia

*marioparinsi@yahoo.com

Abstract. This paper is aim to developing Information system for the labor market which specifically linking vocational schools (SMK) graduates and industries. The methods of this application using Research and Development (R&D) from Borg and Gall conducting in North Sulawesi Province in Indonesia. The result are reliable and acceptable for the graduate students. The Labor Market Information system (LMIS) can help the industries to find a labor/graduates that matched with the company requirement at a real time. SMK may have a benefit by extracting the Information from the application, they can prepare their students for the specific work in the industries. The next development of the application will designed to be available not only for the SMK graduate students.

1. Introduction

Vocational School (SMK) is a High School formal education institutions. This SMK organizes vocational education at secondary level as a continuation of the first secondary School or equivalent. Vocational education is education that prepares students to work in certain occupations, such as engineering, catering and fashion, hospitality, crafts, administration offices, and others [1]. According to what its states on Indonesia Law Number 2 in the Year of 2003, the law it is about Indonesia education law that concerning the national education system mentioning about how the education intended to preparing student to work based on the what he/she mastered for. It also had the similarity with the regulation that created by the government of Indonesia that support the meaning of the vocational education.

The government, in this case, realize the importance of development of SMK, with some research conducted in 2006 on the benefits of SMK than SMA (general senior high school) provides conclusions for the government to take a decision long-term plan to make the ratio of Vocational higher compared with the SMA to a ratio of 70% SMK and 30% SMA [2].

The results of the education and training that are specific for macro human capital is to create a student who has a skilled needed by the world of work/Industries. DPSMK (Indonesian Directorate of TVE) conducted a research, by using quantitative methods to correlate the proportion of students of

SMK:SMA with RGDP (Real GDP) and economic growth rate. The primary data from BPS and PSP (center for education statistics) from 2006 to 2010, the result showed that there is a positive correlation between the ratio of SMK students and the RGDP. The result showed that almost in every province, conclude that the higher the SMK student ratio the higher the RGDP ratio.

Indonesia is an archipelago, SMK spread across all provinces in Indonesia as well as with industries. Geography condition is one of the factor that vocational graduates cannot get jobs in different areas, lack of information makes the Industries with different areas cannot be mutually beneficial with SMK (see table 1).

Table 1. Industries precentage distribution.

No	Island	Percentage
1	Java	75
2	Sumatera	17
3	Kalimantan	3.41
4	Sulawesi	2.6
5	Bali, NTT,NTB	2.6
6	Papua, Maluku	0.27

(Sources: Ministry of trade and industry Indonesia)

The government has made an initiative to establish a LMIS for job Seeker to find a job. But Indonesia's LMIS currently lacks a labor market signalling framework, which can provide detailed analysis of fluctuations in labor market supply and demand. In the LMIS it is very important that there is a signal or flexibility of labor Market Information with the attribute of information is maintained the originality [3]. Signals of this information are very important to provide guidance to prospective workers who will take the decision to apply for a job in accordance with the capabilities. In Indonesia LMIS problems sometimes do not provide the information to the Job seeker, the data held is not volatile and flexible. The ability of LMIS in Indonesia is still very questionable because many major features can not be used, and some information can not be analyzed completely.

2. Literature Review

The literature review will based on the perspective of the expert and researcher on how to define the LMIS concept with the mutual advantages for vocational education and industries, this review is very important to develop a framework of the concept. The literature is divided into vocational education and Labour Market Information System review.

2.1 Vocational education

Vocational education is one part of the national education system in Indonesia that purposes to help the student to develop professional attitude and able to compete in mastering skilled to entering the industries (world of work). Today economy has little room for those who cannot solve problems using technology, manage resource, work in team and continue to learn on the job. SMK also part of education system that preparing individual to work on specific job. Vocational education has the main purpose is to helping student to prepare what needed to be in the world of work, the job will be inside or outside from home, the other meaning is to work in every particular area [4].

2.2 Labour market information system

The Concept and definition of the LMIS may be very various in every country, the concepts have the same meaning but have a different translation. The perspectives of the concepts can be developed in some overall sense toward the definition of the LMIS. The International Labour Organization (ILO) defines a Labour Market Information as information relating to the functions, problems, opportunities contained in the Labour Market, comprising the relationship between the job seekers and the industries who are looking for employees. The perspective from the ILO shows that once the Information related with the employee who is needed for a job or in the industries that needed for employee must be define as the Labour Market Information.

The concept has a different perspective raised in Northern Ireland, the use LMIS be describe as Labour Market Intelligence with the same functions, Labour Market Intelligence has the distinction of purpose and leads or focus toward how to provide solutions in the decision, to determine what work the most desirable, most suitable and in line with their expertise. Labour Market Intelligence reports on these decision making and also make the report that shows how the effectiveness of the Labour Market Intelligence [5]. LMIS in Northern Ireland has transformed into multi functioned as a decision support system. LMIS have the capability to collecting the data that concern about job vacancy in single window, it is also evaluate and disseminate all about labour market information to all the stakeholders that needed the for making decision [6].

Since the development of LMIS, it occurs to have a step forward by the changes views on the Labour Market, LMIS theory leads to the relationship between the Labour Market Planning and a good decision making. LMIS also supports career development for people who need jobs, Information Systems that assist in planning the career of a person should be integrated into the Labour Market [7]. Different terms also on LMIS are in the United Stated of America (USA), the government uses the term Workforce Information to provide guidance to the Labour Market and the information itself. The system is then defined by the government of the USA (Workforce Information Council) as a provider of information are the times to keep up to date on all matters that are closely related to each rate salary of jobs available, the specification needs to get a job, and to provide wide assistance to decision makers in choosing a job for the future of the customer.

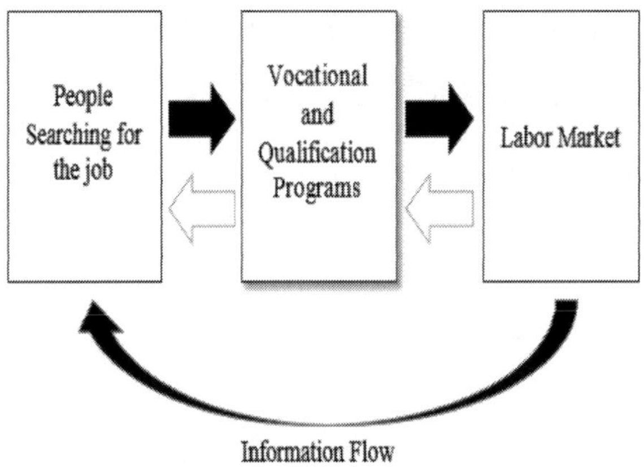

Figure 1. Information flow on LMIS.

The View of LMIS is an active Labour Market that have a policy of collecting data, evaluating and providing everything in the form of Information Systems to both parties to the supplier and the demand side [8], the purpose of the optimal LMIS as an active labour shown in Figure 1, the meaning of the Figure is to make the cycle in form of systems that connect the supply and demand of the market and especially the SMK as actively. The information flow must be align with the people who was searching for the job qualifications program. The Labour Market will run with the labour profile, and job placement based on information flow in Figure 2.

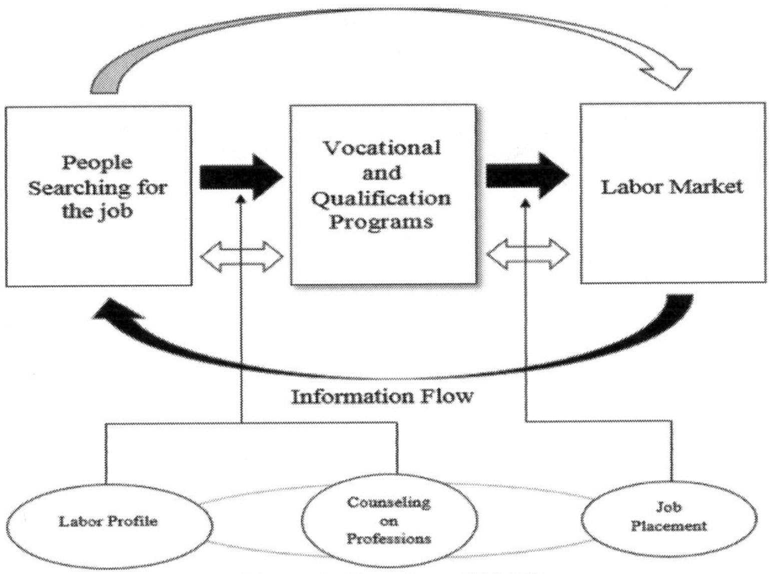

Figure 2. Function of LMIS.

There has been a various studies regarding the LMIS in many countries, the outline of a number of common constraints in the development of these systems. The strengthening of labour market Information Systems will allow for employment outcomes to be monitored more closely in Indonesia. In this regard, it may be useful to assess of the utility of producing a "job vacancy rate" or "employment projections" as part of employment monitoring systems. The distinction about Labour Market becomes clearer, The concept is more than a connection between employers and job seeker but also make a model that run for short term and long term with the interlink of the Institutions, from the larger labour market transaction, Information System could be the most important one for creating the opportunity for each job seeker to recognized the flow of employers requirements for short or long period of time [9]. The LMIS is created to make link between SMK and industries in Indonesia with the information as the bridge. The research framework is based on three stakeholders that interact with system directly, which are; SMK, Government/Labour Department, and Relation between SMK and industries in Indonesia. The feature in the LMIS should including the carrier decision support for the job seeker.

This LMIS not a new idea for several country (from the explanation above), but for linking directly to job seeker and the industries from mobile is kind a breakthrough of the information system for the labor market, the remodeled of the general LMIS by using mobile based application in several unique ways.

3. Methods

The development Method using Research and Development (R&D) Method, R&D is the research that purposes to produce a product to solve problem that occur in workplace. Development Research is a process to develop and validate product that used in education and learning [10].

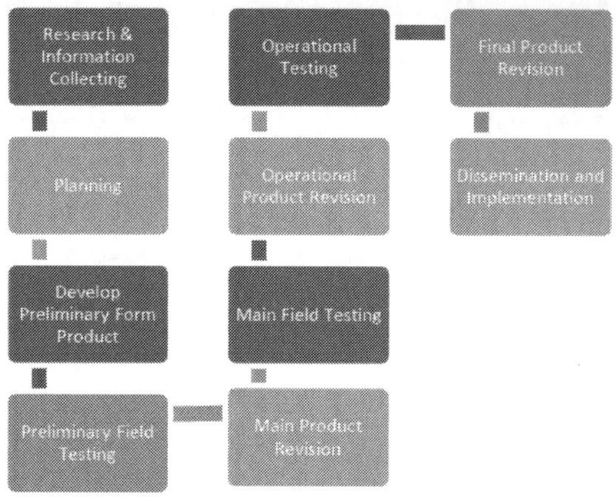

Figure 3. Research and development design.

Research and Development is a process used to develop and validate a product of education [10]. The steps of this process is usually referred to as the cycle of R & D, which consists of studying the research findings related to the product and developing the products based on these findings. The finding of this research are based on literature study and field observation based on Indonesia SMK and labour market conditions. The 10 phase on the figure 3 will be simplified in 3 phase by the researcher, the phase are:

3.1 Information collecting
Beside the literature study, the researcher collecting information to develop the LMIS by the government, industries and SMK point of view in North Sulawesi Province.

3.2 Planning and design

At this phase researcher converted the information into the Application framework, this include develop the alfa product (pre product), this preparation needed a form of concepts framework, the rich picture, and the application modeled (all include the application user interface completely).

3.3 final product main testing and revision

Product tested in advance to obtain input from SMK students response by spread questionnaires. The data were obtained and analyzed use as reference to look at the correctness, efficiency and usability of the application has been made. The research sample comprised 20 SMK students from 4 different schools. we informed to the participants that we were testing the application and they have to pay attention and follow the instruction. After operated the application, they requested to completed questionnaires related to correctness, efficiency and usability of the application.

The research measurements consisted of three sets of categories which are (a) correctness, (b) efficiency, (c) usability. We employed pen-end-paper questionnaires to conduct the test. All were ranked using 4-point scale (1 = Very Less to 4 = Very Good) and indicates for all items were reliable. For validity criterion, item instrument must meet r Product Moment Correlation coefficient table for N = 20 of 0.444 in 5% significance. In this test used benchmark r Product Moment of 0.361 with a significance level of 5%. So the item price has r count > 0,361 declared valid and item has price r count <0.361 disqualified. To measure the validity of instrument used interpretation of the correlation coefficient table (see Table 2). Then test the reliability of research instrument in the form of a questionnaire using Cronbach Alpha formula (see Table 3). SPSS 21.0 software used to analyze both validity and reliability of instruments. Further, Evaluation System by Students for a third aspect of the assessment using the data conversion table of the quantitative data into quantitative data scaled 5 (see Table 4) [11]. The detail of data analysis shows in result section.

Table 2. R coefficient correlation (validity).

Raw Score	Interpretation
0.800-1.00	Very High
0.600-0.799	High
0.400-0.599	Moderate
0.200-0.399	Low
0.000-0.199	Very Low

Table 3. R coefficient correlation (reliability).

Raw Score	Interpretation
0.800-1.00	Very High
0.600-0.799	High
0.400-0.599	Moderate
0.200-0.399	Low
< 0.200	Very Low

Table 4. Qualitative to quantitative data.

Interval	Grade	Interpretation
$x > 3.40$	A	Very Good
$2.80 < x \geq 3.40$	B	Good
$2.20 < x \leq 2.00$	C	Moderate
$1.60 < x \leq m_1 -2.20$	D	Bad
$x \leq 1.60$	E	Very Bad

4. Results and Discussion

The main purpose of the LMIS is to make information about job description that company needed, the reason is to have a better information show what the qualification needed, the qualification of a job has a similar with the SMK output, the SMK eager to make the graduate have a specific skill that can help the person to have a job. Optimal LMIS is simple when we know how to designing an information flow with the job description specification, the optimal system will be automatically sort by whom is qualified to enter the industries job interviewed (see figure 4).

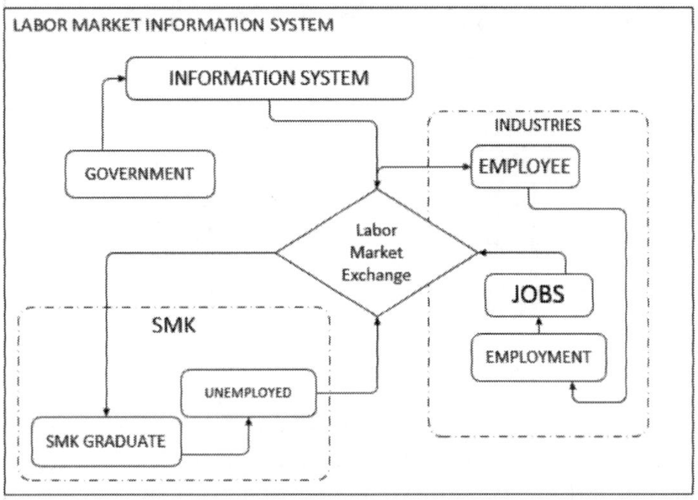

Figure 4. Indonesian LMIS concept for SMK.

From Figure 4 Show the result of the concept based on literature into the framework and can be integrated into mobile Application LMIS that has the capability of the key modules of the framework. The purposes to create Mobile based Application to make the system can be used more flexible and agile. The design of the LMIS should know what the most needed in this technology based era. User Interface (UI) should be more realistic and make people interested to access the Information System. The ideal application is the design should be like hierarchy structured from top to bottom, the monitoring of the system should be step by step in order to response with the changing of the LMIS model from time to time. Figure 5 show the information system can be flow to the SMK graduates and the Schools so the SMK teacher can be prepared their students skilled based on the Industries needed.

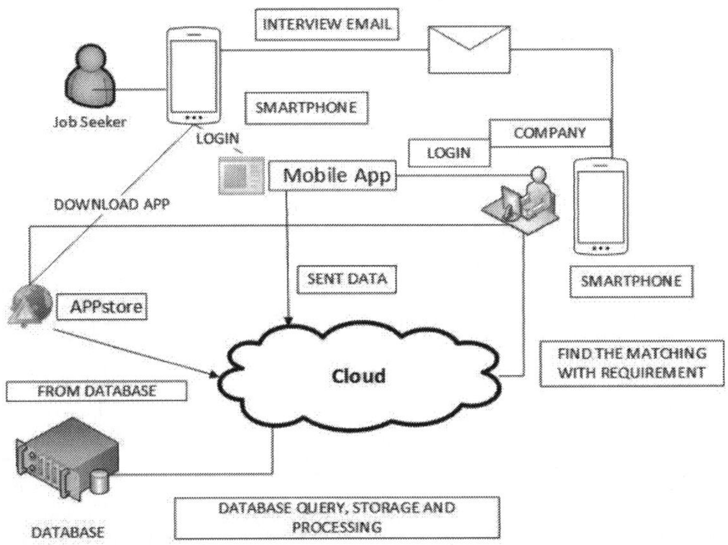

Figure 5. Rich pictures of proposed LMIS.

The final product (LMIS application) screen can be seen from this picture below.

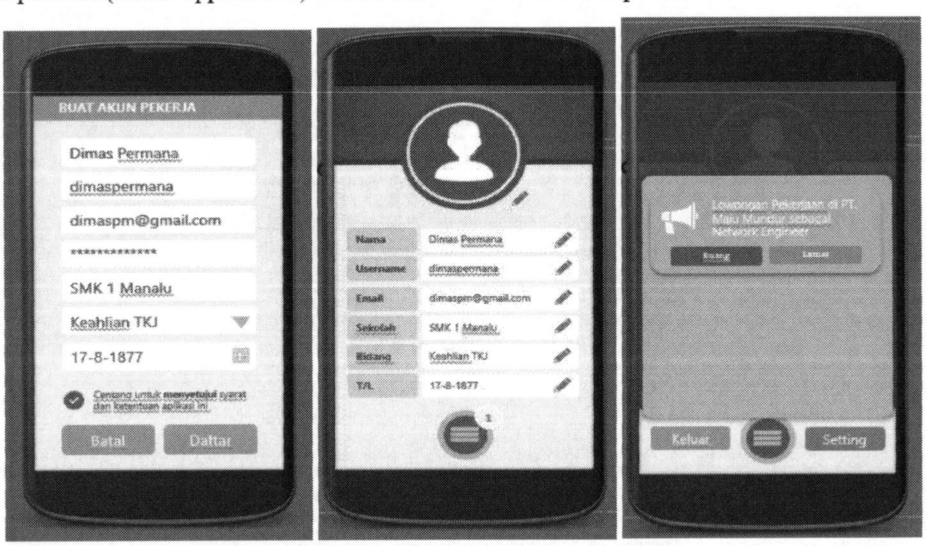

Figure 6. User Interface LMIS for user.

On the figure 6 explain the register screen from (the left picture), the register consist the data (curriculum vitae) that need to be filled by the job seeker (skilled need to be filled), on the middle picture are the next screen shows after the student has registered, the system automatically search for the job that match with the industries that looking for the employee with the specific skills. When the system found the suitable job for the job seeker, the message will show as a popup in the student smartphone (right picture). The student can view the company profile before choose to apply for job or to ignored the company offers, once the student apply for the job, the system will directly sent the student data to the company private email.

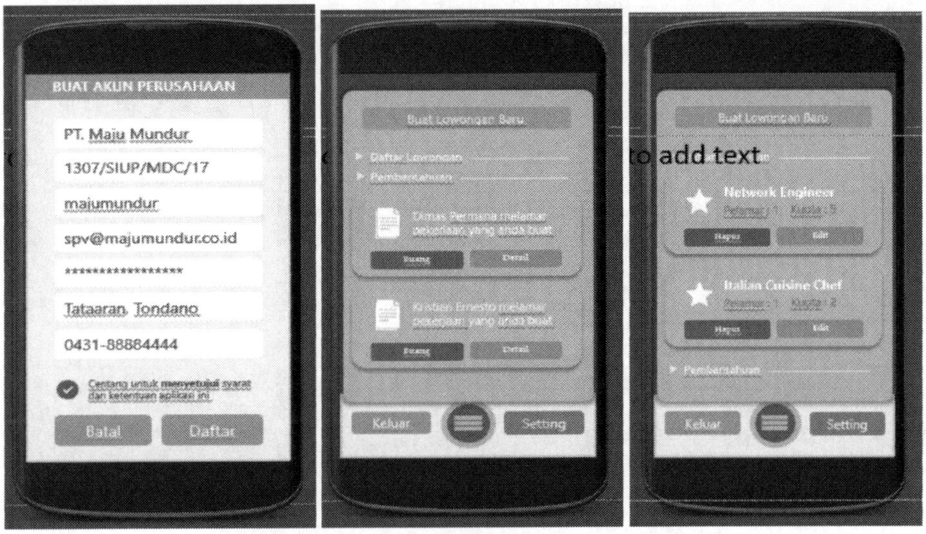

Figure 7. User Interface LMIS for Company

On the figure 6 show the user interface from company point of view. the register screen (left picture) same on the process that explain in the figure 7, the different is the company must filled the company profile as the main company data to be saved in database on server, the middle picture show on the company that filled the employee available from with the job description that company needed. The response of the job seeker will be in the private email of the company. Based on figure 8 show the user interface LMIS for Government on PC.

Figure 8. User Interface LMIS for Government on PC

Table 5 describes the validity and reliability score of 3 sets of items. The analysis result shows that all instruments were reliable and valid based on measurement reference.

Table 5. Validity and reliability score of 3 sets of items.

Category	Item	Validity	Interpretation	Reliability	Interpretation
Correctness	1	.628	High	.996	Very High
	2	.609	High		
	3	.804	Very High		
	4	.865	Very High		
	5	.812	Very High		
	6	.835	Very High		
	7	.595	Moderate		
	8	.852	Very High		
Efficiency	1	.639	High	.977	Very High
	2	.545	Moderate		
	3	.492	Moderate		
	4	.467	Moderate		
	5	.622	High		
	6	.556	Moderate		
Usability	1	.659	High	.976	Very High
	2	.488	Moderate		
	3	.572	Moderate		
	4	.508	Moderate		
	5	.541	Moderate		
	6	.598	Moderate		

Table 6 shows the result of student's assessment toward this application based on spread questionnaires divided into 3 aspects which are correctness, efficiency and usability. The number of question for each aspect is 8, 6, 6 respectively.

Table 6. Validity and reliability score of 3 sets of items.

M-Learning in Social Media Platform	Average	Std. dev
Correctness of	3.45	0.53
Efficiency of	3.46	0.35
Usability of	3.45	0.35

To sum up, the evaluation study was successful because it gave us some good feedback on what the students think about our LMIS. It should be pointed out that the small sample size of the study limits the generalization (or external validity) of the results. Nevertheless, this gives the first picture of the usefulness of the applications that have been created.

5. Conclusions

The Concept of LMIS will be more flexible and more accurate for SMK Graduates to gain Information that specific with the skill of the Graduates The needed of the system for link and match, to create a connection from SMK and Industries is possible nowadays, it is possible by review on every aspect that concern with the stakeholders of the system, which is requires a features that can provide information on the vocational needs of the industries. The schools are able to prepare graduates according to the needs of the working world, this system should be online and can be updated, and there will be a direct relationship between SMK and the world industries through the system from the internet using application from mobile phone. LMIS specific for SMK and industries will have many benefit for the stakeholders in specific ways:

- For the government: the information system will help government to mapping what is the specification of the human resources that needed for industries, displaying which SMK competency that match with the Industries needed, and can be the next step to developing the partnership between industries and SMK employee.

- For SMK; the Information system will help the school to :prepared their student to work on Industries based on what that Industries needed, gain a lot of information about the world of work, have a better partnership with Industries.
- For the industries; the information system will help the industries to have a human resources with the skilled matched with the needed and create a partnership with government and SMK to prepared for the skilled labor.

References

[1] Soenarto 2005 *Metode Penelitian Pengembangan Untuk Peningkatan Kualitas Pembelajaran* (Bali: Departemen Pendidikan Nasional).
[2] Departemen Pendidikan dan Kebudayaan 1995 *Keterampilan Menjelang 2020 untuk Era Global* (Jakarta: Departemen Pendidikan dan Kebudayaan).
[3] Pardjono 2011 *Peran Industri dalam Pengembangan SMK Makalah disajikan dalam Workshop peran Industri dalam Pengembangan SMK, di SMKN 2 Kasihan Bantul* (SMKN 2 Kasihan Bantul).
[4] Hoaclander G E, Kaufman P, and Levesque K 1992 *Vocational Education in United States* (Washington: US Department of Education).
[5] Kirpal S R 2011 *Labour Market Flexibility and Individual Careers* (London: Springer).
[6] Mahesh Maurya N S 2014 Labor Market Information System *International Journal of Application or Innovation in Engineering & Management* (IJAIEM).
[7] Woods J F 2006 *Conceptual Framework for an Optimal Labour Market Information System* (Canada: Upjohn Institute Technical).
[8] Woltermann S 2012 *The Labor Market Information System as an Instrument of Active Labor Market Policies 2012* (Germany: University of Göttingen).
[9] Yavitz B M 1973 *The Labour Market: An Information System* (New York: Praeger Publishers).
[10] Borg, W. R., Gall, M. D., & Gall, J. P. 2003 *Educational Research: An Introduction.* (USA: Allyn and Bacon).
[11] Suharsimi A 2005 *Manajemen penelitian* (Jakarta: Rineka Cipta).

IOP Conf. Series: Materials Science and Engineering **306** (2018) 012088 doi:10.1088/1757-899X/306/1/012088

Industrial Student Apprenticeship: Understanding Health and Safety

M V Simanjuntak*, A G Abdullah, R H Puspita, D Mahdan and M Kamaludin

Pendidikan Teknologi dan Kejuruan, Sekolah Pascasarjana Universitas Pendidikan Indonesia, Bandung

*mariavsimanjuntak@yahoo.com

Abstract. The level of accident in industry is very high caused by lack of knowledge and awareness of workers toward the health and safety. Health and Safety are efforts to create a comfortable and productive atmosphere to accomplish a purpose or goal as maximum risk in the workplace. Vocational Education students must conduct training on business and industry, prior to that they should have a clear understanding on occupational health and safety. The purpose of this research is to analyze the understanding, preparation, and implementation of work health and safety of the students. Method used is descriptive method and data are collected using instrument, observation and interview. The result of study is conclusion of understanding occupational health and safety of vocational education students.

1. Introduction

Vocational education is education that prepares learners to work in DUDI (Dunia Usaha & Dunia Industri) or education training that provides practical experience in a particular occupational field, as algiculture, home economics, or industry [1]. Four purposes that are most central to its project are educational provisions focused on a) the preparation for working life including informing individuals about their selection of an occupation; b) the initial preparation of individuals for working life, including developing the capacities to practice their selected occupations; c) the ongoing development of individuals throughout their working life as the requirements for performance of work transform over time; and d) provisions of educational experiences supporting transitions from one occupation to another as individuals either elect or are forced to change occupations across their working lives [2].

Vocational education especially electrical engineering should understand health and safety. Electricity is a familiar and necessary part of everyday life, but electricity can kill or severely injure people and cause damage to property. Electrical injuries can be caused by a wide range of voltages but the risk of injury is generally greater with higher voltages and is dependent upon individual circumstances. Torch batteries can ignite flammable substances. Electrical injuries: electric shock, electrical burns, loss of muscle control and thermal burn [3]. Students who do practical work must understand the importance of health and safety

The level of accident in industry is very high caused by lack of knowledge and awareness of workers toward the health and safety [4]. In the workplace unseparated from risk of work. Accident is event that happens unexpectedly, especially causing injury or damage. The number of accidents at work in Indonesia is still high. Citing data of Badan Penyelenggara Jaminan Sosial (BPJS) Employment, until the end of 2015 there have been work accidents as many as 105,182 cases. While, for the case of serious accidents that resulted in deaths were recorded as many as 2,375 cases of the total number of work accidents. The main cause of ill health and accidents at work is the low awareness of the importance of Health and safety application in industry and society. Implementation of Health and safety is often regarded as cost, not as an investment to prevent accidents [5].

Vocational Education students must conduct training on business and industry, prior to that they should have a clear understanding on occupational health and safety [6]. In the improvement of students' skills, students undertake internship (training) for three months with the aim to improve the ability and recognition of industry. Vocational schools are becoming more aware of the safety and health challenge [7]. The first should we know is how to managed health and safety effectively is: a) know about the risks in work, b) control the risks that need it, c) make sure the risks stay controlled [3], but still many students ignore the existence of the importance of health and safety in the industrial world. This makes researchers want to analyze what causes students to ignore the importance of health and safety. This study analyzes from understanding, preparation and application of occupational health and safety in doing technical internship. There can be a number of ways to control the risks you identify as a priority for action. Make sure the risks stay controlled. Based on the problem, then conducted how to fine out understanding student knowledge about health and safety?

2. Methods

This study was conducted to find out the understanding of young workers. Using a descriptive survey model, this study takes the population of children who will undertake technical internships, method and data collected using instrument, observation and interview. This research was conducted for vocational education students. The questionnaire contains questions of knowledge related to occupational health and safety, how they apply health and safety during their practice in the industry, how to manage health and safety and how to control the risks.

3. Results and Discussion

The result of study is conclusion of understanding occupational health and safety of electrical engineering students. Based on reliable sources it is found that the cause of occupational accidents is a lack of awareness of the importance of occupational health and safety. Everyone in the workplace, from the employer to the newest worker, has different but important duties to keep the workplace safe. Based on International Labour Organization Jakarta (ILO) data 40 % of young workers are more at risk of accidents than adult workers. Workplace differences have different work risks [8]. So the handler must be appropriate.

Research instrument the important thing that must be understood by the electronic learners in the application of occupational health and safety:

- Recognize the hazard: Students must recognize the dangers of the workplace.
- Asses the hazard: Students understand the risks that will cause injury
- Control the hazard: Students can control the hazard by looking for ways to get rid.
- Evaluate: Evaluate how well the hazard controls are working.

4. Conclusions

Accidents occur due to lack of understanding so that the level of awareness of health and safety becomes low. Environmental cultures that do not require learners to apply health and safety in accordance with the procedure make the lack of awareness of the application of health and safety. Four steps to inculcate the importance of occupational health and safety: Recognize, Asses, Control and Evaluate.

References

[1] Press O U 2003 *Dictionary* (New Edition).
[2] Billet S 2011 *Vocational Education: Purposes, Traditions and Prospects* (New York).
[3] Hse S E 2006 *Essentials of health and safety at work* p 1–105.

[4] Reese C D 2009 *Handbook of Safety and Health for the service industry : Industrial Safety and Health for Goods and Materials Services (First)* (Boca Raton, London, NewYork: Taylor & Francis Group).
[5] Ketenagakerjaan B 2016 *Jumlah kecelakaan kerja di Indonesia masih tinggi* Retrieved from http://www.bpjsketenagakerjaan.go.id/berita/5769/Jumlah-kecelakaan-kerja-di-Indonesiamasih-tinggi.html
[6] Voc U N E 2003 *Occupational Health and Safety.*
[7] Brauer R 2006 *Second edition safety and health* Second Edition (A Jhon Wiley & Sons, Inc).
[8] ILO 2015 *International Labour Organization* (Occupational safety and health country profile: Indonesia).

IOP Conf. Series: Materials Science and Engineering **306** (2018) 012089 doi:10.1088/1757-899X/306/1/012089

Rambutan Seed (*Nephelium Lappaceum L.*) Optimization as Raw Material of High Nutrition Value Processed Food

M Wahini*, M G Miranti, F Lukitasari and L Novela

Universitas Negeri Surabaya

*medawahini@unesa.ac.id

Abstract. Rambutan (*Nephelium Lappaceum L.*) is a plant that identical with Southeast Asian countries, in some areas of Indonesia no exception, but rambutan seed is considered as a waste. Therefore, it needs to be optimized into raw materials of food and processed with high nutritional value and has economic value. The purpose of this research were: 1) to find the best rambutan seed immersion formula; 2) to know the nutritional value of the best immersed rambutan seed; 3) to produce raw material and various processed of rambutan seed product. The research method was quasi experiment with 6 treatments and 2 factorial design, materials for immersion was NaCl and $Ca(OH)_2$. The results showed that: 1) the best rambutan seed immersion formula was using $Ca(OH)_2$; 2) the best rambutan seed contains 1,6 ash, 31,2 protein, 26,9 fat; 3) the best rambutan seed produce flour and processed of seasoned nuts. This research indicates that rambutan seed is very potential to be an alternative high-value raw materials.

1. Introduction

Rambutan (*Nephelium lappaceum L.*), family of Sapindaceae is important tropical fruit in Southeast Asian region, no exception in Indonesia, Thailand and Malaysia. Rambutan is a tropical plant which grown in warm, humid and low evaporation rates with high rainfall. Rambutan fruit is potential to be used as functional food due to its capability to provide beneficial health effects. Underutilized part of rambutan (rind and seed) contain some active components which are reported to prevent some diseases (see Figure 1) [1].

Figure 1. Rambutan seed [2].

Indonesia has 22 varieties of rambutan derived from pure variety and grafting from two different varieties [3]. Rambutan fruit production has increased significantly in this past 7 years. It can be seen from the extensive rambutan season in 2015, which reached 107119 ha, production amounted to 906438 tons, and the level of productivity of 18.04 tons/ha [4]. East Java province is the second largest rambutan producer after West Java province [3]. The largest rambutan fruit production is in Subang with a total production of 87 thousand tons, the second is in Bogor as many as 24 thousand tons, the third is in Semarang as many as 16 thousand tons, the fourth is in Jember with total production 15 thousand tons, and the fifth is in Blitar with totally 13 thousand tons [5]. But, rambutan's rind and seed is considered as a waste. Rambutan are consumed, fresh, canned or processed and its consumption results in the production of vast amount of rambutan waste from their seeds and rind, disposal of these rambutan wastes can have a serious environmental impact which is becoming harder to solve. Much effort will therefore be needed to develop the nutritional and industrial potential of by-products waste and these underutilized agricultural products [6].

The seeds and rind which are often the waste part of the fruits have not generally received much attention with a view to being used or recycled rather than discarded. Interestingly, the seed and rind of some fruits have higher vitamins, fibres, minerals and other essential nutrients activity than the pulp fractions [7]. Rambutan seeds are safe for consumption because it is not toxic and contains carbohydrate, fat, and protein which consider as the body's needs [8]. Rambutan seed contains quite high polyphenols. Polyphenols are very astringent compounds, consisting of anthocyanins, leukoantosianin 3%, katekhin 3% and complex polyphenols [9]. Main polyphenol compounds are tannins and flavonoids which were identified as geraniin, coraligin, and ellagic acid [10, 11]. That various polyphenols such as antioxidant, antiinflammatory, anticarcinogenic, and other bioactivities demonstrated suggest that they may have beneficial effects on human health and provide protection against such chronic diseases as cardiovascular diseases, neurodegenerative disorders, and cancers [12]. Having this desirable attribute, there is a need to find an appropriate way to remove bitterness from the seed in order to develop flour and seasoned nuts product that can be consumed as valuable nut. Therefore, immersion is one of the preliminary techniques to reduce the amount of tannin, saponin and total phenol content as a cause of bitterness in the seed in this study. Immersion ingredients in this study used 2 type ingredient and 3 different concentration formula which intended to determine the best quality of rambutan seed flour and seasoned nuts from rambutan seed.

2. Methods

The main material used research is Rapiah rambutan seed. Rambutan seeds are obtained from the sellers of rambutan and from collectors of seeds. Seeds that have been obtained by the collectors, wash and then dried under the sun, in order to avoid the growth of fungi that can cause decay in the seeds. Preparation takes ± 3 weeks. Preparing rambutan seed as raw material to produce flour and seasoned nuts is started with peeling rambutan's rind. Then separate the seeds with the flesh of the fruit, rambutan good seed weight is 6-7 g. In 1 kg of rambutan 100-135 gram rambutan seed can be collected. Immerse the seeds in whiting water for 1 hours, then wash with water and drain it. Drain the seed up, then peel the seed if want to immediately use, and store in an airtight container if saving the seed is desired. Save the seeds in the refrigerator. Rambutan seeds contain fat and protein so it will be easy to quickly rancid, which means damaged if stored at room temperature for a long time. Rambutan seed is a source of protein and carbohydrate can be considered, especially in food application [13].

This study used a quasi-experimental method and implemented in February-July 2017. This study consisted of 2 treatments and each treatment consisted of 3 factors. The materials used are rapiah seeds, $Ca(OH)_2$ and NaCl. The experiment was conducted in the laboratory of science and food management laboratory of PKK FT Unesa and Balai Penelitian dan Konsultasi Industri (BPKI) Surabaya. The experimental design are shown in Table 1.

Table 1. Experimental design.

R \ P	P1	P2	P3
R1	R1P1	R1P2	R1P3
R2	R2P1	R2P2	R2P3

Note:
P : Percentage of $Ca(OH)_2$ and NaCl
R : Type of immersion
R1 : $Ca(OH)_2$
R2 : NaCl
P1 : 10%
P2 : 20%
P3 : 30%

3. Results and Discussion

Rapiah rambutan seeds have solid criteria, rather round, the outside of the seeds coated with gray-brown leather with a color in a young cream. Seed immersion was done in IPA laboratory of PKK FT Unesa, while laboratory test was done at BPKI on July 26th, 2017. The following result of lab test of nutritional content of rambutan seed of rapiah type done at BPKI is shown at Table 2.

Table 2. Composition of Rambutan nutrition substances.

No.	Analysis Various	Levels of 100% BK
1	Ash content	1,70
2	Water	14,20
3	Carbohidrate	64,19
4	Lipid	6,01
5	Protein	11,38
6	Fiber	2,51
7	Vit. B	0,33
8	Mineral (Ca-Fe-P)	62,50

Table 2 gives the proximate analysis of the rambutan seeds. The seeds contained 11,38% protein. These nutrient can be used as flour (starch) which its value isn't lower than other starches. The protein content of defatted rambutan flour (10.07%) was in agreement with that of commercial available all-purpose wheat flour, being normally between 9 and 12% [14]. Another study has shown that amylose content value for defatted rambutan flour (32.16%) differed and was higher than that from all-purpose wheat flour with a value of 17.47% [15]. Therefore, rambutan seeds can be used as flour that can be processed into industrial product such as seasoned fried nuts or chips.

Directly fried rambutan seeds have a bitter taste and if too much to consume it will cause dizziness effects. Rambutan seeds flour without immersion treatment, produce a bitter taste in the product it produces. This is because the seeds have the content of flavonoid. Many flavonoids and isoflavonoids have an undesirable bitter taste, which hampers their use as food bioactives [16]. Therefore, to eliminate the bitter taste on the rambutan seeds, immersion treatment is necessary by using $Ca(OH)_2$ (whiting) and NaCl (salt). To get the best quality result of rambutan seed, this study used some 3 different formula percentage of $Ca(OH)_2$ and NaCl, which is 10%; 20%; and 30% for each formula. Another study use Ethanol to dissolve flavonoid [17], Ethanol is a solvent that belongs to polar group. Polar solvent is a solvent that can dissolving soil flavonoids and tannins contained in the skin of the fruit and rambutan seed [18].

Percentage of immersion treatment other than used to remove bitterness, also used to find the best nutritional content of rambutan seeds. After immersion, rambutan seeds analyzed its nutritional content to find the best nutrition. The results of nutrient analysis of rambutan seeds conducted in the Research and Consulting Industry:

Tabel 3. Nutrition of rambutan seeds type rapiah after immersion.

Various Analysis	$Ca(OH)_2$ Immersion			NaCl Immersion		
	10%	20%	30%	10%	20%	30%
Ash content	1,82	1,90	1,98	2,01	2,24	2,26
Water	14,82	15,11	15,30	15,05	15,90	16,10
Carbohidrate	63,01	62,36	61,78	62,99	62,08	61,85
Lipid	6,12	6,34	6,41	6,04	6,02	5,96
Protein	11,82	11,90	12,08	11,52	11,48	11,41
Fiber	2,41	2,38	2,35	2,36	2,33	2,31
Vit. B	0,31	0,38	0,35	0,36	0,35	0,34
Mineral (Ca-Fe-P)	61,85	62,11	62,86	60,05	60,15	60,05

From the results of nutrition analysis in table 3 can be seen that the best nutritional content of rambutan seeds is on treatment 3; the immersion by using $Ca(OH)_2$ as much as 30%. This can be seen from the content of protein and lipid respectively 12.08 and 6.41. As the result of the product that is the reduction of bitterness because Alkaline water (such as aqueous Na_2CO_3, NaOH and $Ca(OH)_2$ solution) can be used as flavonoid extraction [19]. High lipid (fat) content in rambutan seed has a possible industrial use. The rambutan seed kernels provide a considerable yield of fat and the high arachidic acid content makes that fat highly stable to oxidation. Extracted fat from rambutan seed has a potential to be a source of natural edible fat with possible industrial use [20].

4. Conclusions

From the research result, it is found that that rambutan seed is very potential to be an alternative high-value raw materials and have quite high economic value. This is because rambutan seeds have nutritional value required by the body, including; 11.38% protein, 6.01% crude fat, 64.19% carbohydrate, 2.51% fiber, 0.33% vitamin B, and 62.50% minerals. The most appropriate way to get rid of bitter taste in the seeds is by using immersion formula of 30% $Ca(OH)_2$.

Acknowledgments

The authors wish to express their gratitude to the Institute for Research and Community Service (LPPM) Unesa who have supported this research; and Directorate of Research and Community Service, Minister of Research, Technology and Higher Education Republic of Indonesia as providers of research funding scheme MP3EI.

References

[1] Rohman A 2017 Physico-chemical properties and biological activities of rambutan (*Nephelium lappaceum L.) fruit. Res. J. Phytochem.* **11** pp 66-73.
[2] Rambutan Seed Figure 2016 *Healthy Benefits Times* Available at: www.healthbenefitstimes.com
[3] Rahayu S L, Sasmito M D, Rudijanto A, and Fatchiyah 2013 Production and Potency of Local Rambutan at East Java as a Candidate Phytopharmaca *Agrivita* **35**(3) pp 270-276.
[4] Badan Pusat Statistik dan Direktorat Jendral Holtikultura 2015 *Data Produksi,Produktivitas, dan Panen Nasional Buah Rambutan.*
[5] Badan Pusat Statistik 2011 *Data Produksi Rambutan.*
[6] Fila W O, Johnson J T, Edem P N, Odey M O, Ekam V S, Ujong U P, and Eteng O E 2012 Comparative anti-nutrients assessment of pulp, seed and rind of rambutan (Nephelium Lappaceum) *Annals of Biological Research* **3**(11) pp 5151-5156.
[7] Jayaprakasha G K, Singh R P, and Sakariah K K 2001 *Foodchemistry* **73** pp 285-90.

[8] Polanditya P 2007 Biji rambutan sebagai alternatif makanan baru *J. Ilmu kimia FPMIPA Universitas Islam Indonesia* pp 1 – 4.

[9] Susanto B 1994 *Teknologi Pengolahan Hasil Pertanian* (Surabaya: PT. Bina Ilmu).

[10] Ling L T, A K Radhakrishnan, T Subramaniam, H M Cheng and U D Palanisamy 2010 Assessment of Antioxidant Capacity and Cytotoxicity of Selected Malaysian Plants Molecules *Basel, Switzerland* **15**(4) pp 2139–2151.

[11] Thitilertdecha N, A Teerawutgulrag, J D Kilburn and N Rakariyatham 2010 *Identification of Major Phenolic Compounds from Nephelium lappaceum L. and Their Antioxidant Activities. Molecules* **15** pp 1453–1465.

[12] Ovaskainen M, Riitta T, Jani M K, Harri S, Jarkko H, Heli R, and Pirjo M 2008 Dietary Intake and Major Food Sources of Polyphenols in Finnish Adults *The Journal of Nutrition* **10** pp 563-566.

[13] El-adawy T A, K M Taha 2001 Characteristic and Composition of Different Seed Oils and Flours *J. Food Chem* **74** pp 47-54.

[14] Jirawat E, Sorada W, Tuanta S, Sareeya R, Wimolsri P, Nantaprecha H, and Sukit K 2015 Rambutan (Nephelium lappaceum) Seed Flour Prepared by Fat Extraction of Rambutan Seeds with SC-CO_2 *The International Conference on Herbal and Traditional Medicine* **49** pp 138-146.

[15] Ng K F, Abbas F M A, Tan T C, and Azhar M E 2014 Physi cochemical, pasting and gel textural properties of wheat-ripe Cavendish banana composite flours *IFRJ* **21**(2) pp 655-662.

[16] Wibke S U, Leo B, Harry G, Marianne D, Robin J G, Gerrit S, Jean-Paul V 2013 Bitter Taste Receptor Activation by Flavonoids and Isoflavonoids: Modeled Structural Requirements for Activation of hTAS2R14 and hTAS2R39 *J. Agric. Food Chem.* **61**(44) pp 10454–10466.

[17] Ibrahim A, dkk 2013 *Potensi Ekstrak Kulit Buah dan Biji Rambutan (Nephelium lappaceum) sebagai* Senyawa *Anti Bakteri Patogen pada Ikan* (Laporan Penelitian, Lampung: Universitas Lampung).

[18] Khasanah A N 2011 *Uji Aktivitas Penangkap Radikal Ekstrak Etanol, Fraksi-Fraksi dari Kulit Buah dan Biji Rambutan (Nephelium lappaceum L.) Serta Penetapan Kadar Fenolik dan Flavonoid Totalnya* (Skripsi. Universitas Muhammadiyah Surakarta).

[19] Xu R, Yang Y, and Weimin Z 2011 *Introduction to Natural Products Chemistry* (CRC Press).

[20] Wanrada S, Wannee J, and Utai K 2011 Response surface optimization and characteristics of rambutan (Nephelium lappaceum L.) kernel fat by hexane extraction *LWT Food Science and Technology* **44** pp 1946-1951.

A Design of Innovative Engineering Drawing Teaching Materials

Mujiarto[1,2*], A Djohar[2], M Komaro[2]

[1]Department of Mechanical Engineering, Universitas Muhammadiyah Tasikmalaya
[2] Department of Mechanical Engineering, Faculty of Vocational Education and Technology, Universitas Pendidikan Indonesia

*mujiarto@umtas.ac.id

Abstract. Good teaching is influenced by several things such as effective school leaders and skilled teachers who are able to use information communication technology as a medium of learning. The purpose of this research in general is to develop innovative teaching materials in the form of multimedia animation for engineering drawing in the field of technology and engineering at vocational high school. Research method used research and development (research and development / R & D). The results showed that the E-book Multimedia Animation Engineering Drawing (E-MMAED) is easy to possess and contains complete material. Students stated that the use of E-MMAED adds to learning motivation and improves learning outcomes (student competencies). We recommend that teachers apply E-MMAED as a learning medium and create other innovations to improve student competences.

1. Introduction

Education in Indonesia is divided into primary, secondary, and higher education. One of the types of secondary education in Indonesia is vocational high school. Vocational high school is a form of formal education unit that provides vocational education at the level of secondary education as a continuation of education in junior high school/madrasah tsanawiyah (Islamic equivalent) or other equivalent schools. Vocational high school offers many expertise programs. One of the expertise programs offered is mechanical engineering drawing.

Problem encountered in the teaching and learning of the competence standard of drawing in the department of mechanical engineering lies in the lack of teaching materials so that learning objectives are not optimally attained. Due to a lack of teaching materials, learning goals have not been achieved according to expectation [1]. Meanwhile, the application of inappropriate learning models and the coverage of materials that is not in accordance with the syllabus can result in low student achievement [2]. In addition, the materials for the topic of Basic Networking have not been compiled comprehensively. The learning model implemented especially by vocational teachers is not innovative and more teacher-centered. The facilities in the forms of computers/laptops owned by students have not been fully optimized to support the learning process.

Teachers play an important role in the attainment of learning objectives, in addition to preparing teaching materials. The role includes being a tutor, mentor, instructor, and motivator who is able to encourage students to increase their knowledge through experience. To be a good teacher, learning facilities are necessary for the ease of learning process. Teachers should develop students' interest and invoke their desire to explore all the knowledge they want to gain in the best possible way [3]. One of the best ways is to use the development of information and communication technology (ICT).

The rapid development of ICT is used by various sectors, ultimately in this case education, to facilitate learning. Currently, many multimedia animations are developed to improve student success in learning. The application of Computer-Assisted Instruction (CAI) based on multimedia proved to be appropriate to be a learning medium [4]. The multimedia application developed in her study is in the form of names of parts of the human body in 2 languages (bilingual), namely Indonesian and English. The application is intended for young children of 5-6 years to introduce the name of the body parts with the concept of playing while learning. The development of interactive and innovative multimedia-based chemistry materials for senior high school or its equivalent Islamic school helps students in attaining competence and improving learning outcomes [5]. The bilingual website "Close to Radioactivity" as a learning medium is found to be appropriate to be used as a self-learning resource for students of senior high school or its equivalent Islamic school and can improve student learning outcomes [6].

Furthermore, multimedia animation teaching materials could improve learning achievement, retention, and satisfaction [7]. On a different note, the results of research conducted to employees of an electric company show that training methods based on multimedia animation technology and equipped with computer software and database technology could improve the learning efficiency of the employees and greatly reduce the cost of training [8]. The results of previous research thus suggest that the use of multimedia animation has many advantages compared to conventional learning. These advantages include improving students' achievement, learning satisfaction, and learning autonomy.

The purpose of this research in general is to develop innovative teaching materials in the form of multimedia animation for engineering drawing in the field of technology and engineering at vocational high school.

2. Methods

The research employed the research and development (R & D) method. R & D in education as a process used to develop and validate educational product [9]. The product developed in this research is in the form of innovative teaching materials for engineering drawing for vocational high school students of the technology and engineering expertise, mechanical engineering expertise program. The use of R & D method in this research is to create, develop, and implement innovative teaching materials to vocational high school students of the mechanical engineering expertise program (see in figure 1)

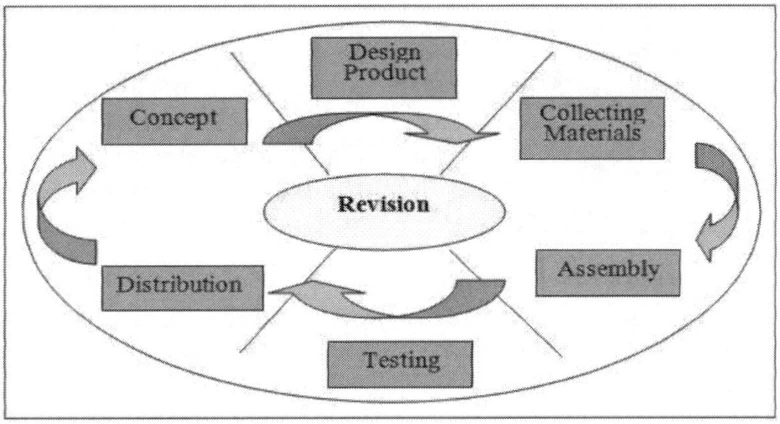

Figure 1. Model development of E-book based multimedia animation engineering drawing.

3. Results and Discussion

This study begins with a preliminary study, e-book design creation, limited testing, expert validation, and extensive trials. the results of extensive trials can be seen in table 1.

Table 1. Results of the test data calculations more widely.

Data	score	
Pre test	The highest	60
	The lowest	10
	Average	33,00
Post test	The highest	100
	The lowest	51
	Average	80,39
N-gain	The highest	100
	The lowest	34,67
	Average	72,38

From the broader trial it can be seen that the average N-Gain is 72.38 in 100 or at 0.7238. This is evidence that this medium can improve students' ability in high category. The reliability of this media increased from the previous that the category is becoming high-categorized after experiencing the process of repair or development. The process of improvement is done by pay attention student response about the legibility of the media as a result of media testing to the user. In addition, other improvements are also made after considering the test results of media experts to obtain the perfect media according to Media Experts, and Material Experts.

At the beginning of the opening E-MMAED will appear as shown in Figure 2, which consists of: Syllabus and RPP, Instructions for Use, Play button.

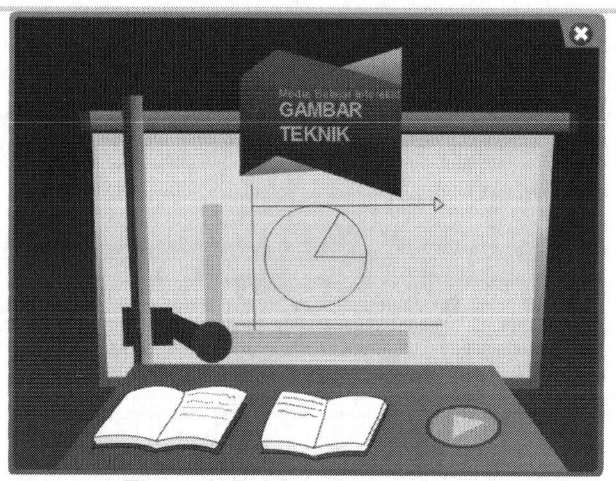

Figure 2. E- MMAED preview.

Students will start using E-MMAED by clicking on the instructions, as in Figure 3. When students finish reading and understanding the instructions for use, click the "Back" button, to return to the start screen. To start using E-MMAED click the "Play" button.

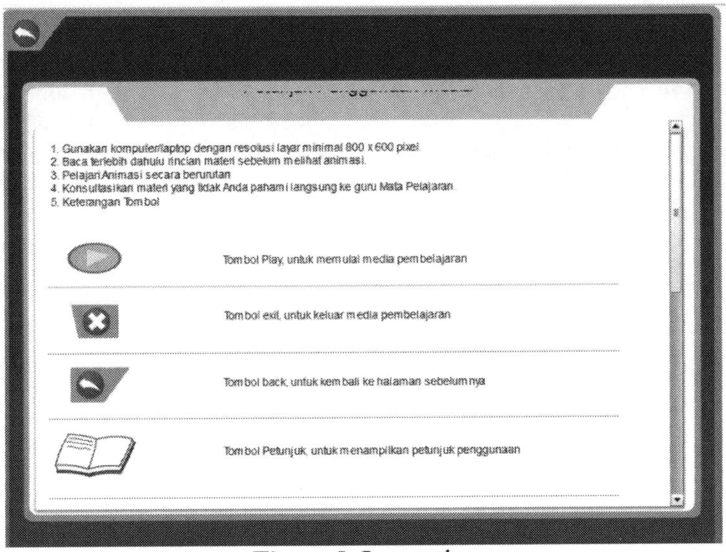

Figure 3. Instructions.

Education aims to optimize and develop the ability of learners, especially the intellectual ability. To realize this aim, the assistance in the form of appropriate media is needed. Media are an integral part of the teaching and learning process in order to achieve the goals of education in general, and the objectives of teaching and learning in schools in particular [10].

The development of science and technology further encourages renewal efforts in its utilization in the learning process, which in this case is in the form of media utilization. The media used can be the ones that are already available or developed in accordance with the needs of learning. Teachers are required to be able to use the tools available in the school to meet the expected teaching objectives. In addition, they are required to develop learning media when necessary, so that teachers are required to have sufficient knowledge and understanding of learning media.

The rapid development of information technology today has given a new breath toward usage of computer in education. One of the increasingly popular nowadays is a multimedia technology that merges a variety of media such as text, graphics, animation, video and audio controlled by a computer. With this technology, a wide range of multimedia element can be developed to improve the quality of education. Interactive multimedia learning system with augmented reality technology can improve the effectiveness and efficiency of teaching and learning behavior [11]. Multimedia element for animation video was use significant have capable to increase imagination and visualization of student [12]. The implications of this study provide information of use of multimedia element will student effect imagination and visualization. In general, these findings contribute to the formation of multimedia element of materials appropriate to enhance the quality of learning material for engineering drawing. Students are very interested in learning using multimedia because students feel satisfied and learning outcome is achieved [13]. Students who use multimedia animation make students feel helpful in learning the practice and innovation in singing [14].

Based on the description of learning resources above, especially about learning media, it is important to link learning media to development of learning strategies. It means that learning media are one of the means to make the delivery of teaching materials more effective and efficient. Meanwhile, based on the preliminary study data, the type of media used by the teachers to deliver the materials on engineering drawing in general was textbooks, and the teaching method used in delivering the learning materials was mostly lecturing.

The existing condition discovered from the preliminary study shows a gap between the media required that should be in accordance with the characteristics of the teaching materials and the media that have been used. The finding about inappropriate media used in the schools was evidenced by the students' responses to the clarity of materials delivered by teachers. Some students responded that the materials were not clear. In addition, there is evidence of low learning outcomes in the topic. The results of this preliminary study became the basis for the development of the alternative learning media as well as learning source called Bilingual Multimedia Animation for Engineering Drawing. The characteristics of the e-book based on multimedia animation (Indonesian abbr., E-MMAED) implemented in the field of expertise of technology and engineering in SMK Teknologi YAF Banjar City were formulated based on literature study and research. The main point of the use of E-MMAED on the topic of engineering drawing is that the medium serves as a learning resource and learning medium to improve vocational students' learning outcomes in the forms of conceptual mastery and problem-solving skills.

4. Conclusions

On the basis of the main point and objectives, then E-MMAED has seven groups of characteristics, as follows: 1) E-book characteristics; 2) characteristics of learning resources required based on the preliminary study; 3) characteristics of animation multimedia; 4) characteristics of learning media: 5) characteristics of the required learning media based on the preliminary study; 6) characteristics of the engineering drawing subject; and 7) characteristics of the learning media developed on the basis of indicators. The E-book based on multimedia animation (E-MMAED) for engineering drawing is created in a Flash format, a format that can integrate images, texts, animations, videos, and sounds. The stages of the making of E-MMAED include material identification and material creation based on several goals or indicators to be achieved.

References

[1] Soma I M, I M Candiasa, and I M Tegeh 2014 Pengembangan Bahan Ajar Menggambar Dengan Perangkat Lunak Berbantuan Program Aplikasi Autocad *E-Journal Program Pascasarjana Universitas Pendidikan Ganesha Prodi Teknologi Pembelajaran* **4**.
[2] Tirta N N, Santyasa I W, and Warpala, I W S 2014 Pengembangan Multimedia Interaktif Berbasis Proyek untuk Pelajaran Kejuruan Jaringan Dasar di SMK Negeri 3 Singaraja *Jurnal Teknologi Pembelajaran* **4**(1).
[3] Ahmad A, Y Kamin, and H D Minghat 2013 A Conceptual Model for Vocational Teaching Method as AnApproach to Enhance Students Learning *2nd International Seminar on Quality and Affordable Education (ISQAE 2013)*.
[4] Firdania M S 2016 Aplikasi CAI Berbasis Multimedia Untuk Pengenalan Bagian Tubuh Manusia Pada Anak Usia Dini *untan*.
[5] Situmorang M, M Sitorus, and Z Situmorang 2015 Pengembangan Bahan Ajar Kimia SMA/ MA Inovatif Dan Interaktif Berbasi Multimedia *Prosiding Semirata 2015*.
[6] Agusta I 2015 Pengembangan Media Pembelajaran Website Bilingual 'Close to Radioactivity' Sebagai Sumber Belajar Kimia Secara Mandiri Siswa SMA/ MA.
[7] Chiou C.-C., L.-C. Tien, and L.-T Lee 2015 *Effects on Learning of Multimedia Animation Combined with Multidimensional Concept Maps Computers and Education*.
[8] Liao S Z, J H Liang, and Y D Li 2012 Application of Multimedia Animation Technology in Learning Power Updating Knowledge *Applied Mechanics and Materials* 239–240.
[9] Borg W R, and Gall M D 1971 *Educational research: An introduction* (McKay).
[10] Munir 2010 *Kurikulum Berbasis Teknologi Informasi Dan Komunikasi* (Bandung: Alfabeta).
[11] Lin, Edgar Chia-Han, Yang-Chia Shih, and Rong-Chi Chang 2017 A Research on Integrating AR and Multimedia Technology for Teaching and Learning System Design *International Conference on Frontier Computing* pp 725-731.
[12] M E, Ismail et al 2017 The Use of Animation Video in Teaching to Enhance the Imagination and Visualization of Student in Engineering Drawing The Use of Animation Video in Teaching to Enhance the Imagination and Visualization of Student in Engineering Drawing *IOP Conference Series: Materials Science and Engineering*.

[13] Daly C J, et al. 2016 A Comparison of Animated versus Static Images in an Instructional Multimedia Presentation *Advances in Physiology Education* **40**(2).

[14] Lin-li Liu, Yi Pang, and Zhuang-li Hu 2016 Application of Spectrogram Analysis in Traditional Vocal Music Teaching and Multimedia Animation Vocal Music Teaching *International Journal of Emerging Technologies in Learning* **11**(11) p 64-67.

How does Socio-Economic Factors Influence Interest to Go to Vocational High Schools?

N F Utomo* and D Wonggo
Universitas Negeri Manado, Tondano 95618, Sulawesi Utara, Indonesia

*nurfadliutomo@outlook.com

Abstract. This study is aimed to reveal the interest of the students of junior high schools in Sangihe Islands, Indonesia, to go to vocational high schools and the affecting factors. This study used the quantitative method with the ex-post facto approach. The population consisted of 332 students, and the sample of 178 students was established using the proportional random sampling technique applying Isaac table's 5% error standard. The results show that family's socio-economic condition positively contributes 26% to interest to go to vocational high schools thus proving that family's socio-economic condition is influential and contribute to junior high school students' interest to go to vocational high schools.

1. Introduction

Vocational education is needed to support human resources (HR) to become skilled worker and entrepreneur. The background of human resource development through vocational education is also driven by *UU No. 20 Tahun 2003 Tentang Sistem Pendidikan Nasional, Bab II Pasal 3* [1] which is contains National Education's foundation to develop the ability and form the character and civilization of dignified nation in order to educate the nation, aims to the development of the potential of learners to become a faithful and pious to God Almighty, noble, healthy, knowledgeable, capable, creative, independent, become a democratic and responsible citizen. Based on Badan Pusat Statistik (BPS) Sangihe [2], the main obstacles of education today, especially in Sangihe, namely the ease of access to education regardless of economy class where only 63% people who participate in education.

According to McMillan & Western [3], socio-economic status consists of three dimensions: education, employment, and income. Berns [4] stated that a family's socioeconomic status is its rank or position within society based on social and economic factors such as income, occupation, and education of the parents. Socio-economic status is a position in society based on economic factors such as income and parents' education level. The people's socio-economic status can be viewed as the family's economic background. The family's socio-economic condition may affect the child's interest about something.

The higher the socio-economic status of their families, the greater their interest in achieving the highest education level. Like on Adelaide, where adolescents with low socio-economic status are usually located in the suburbs and remote areas, as stated by Rothman in Curtis's report at Flinders University Research Report [5] which shows high reports about the difficulties with low socio-economic status students on completing education in public schools and universities. Socio-economic conditions compel and provide few life choices to pursue education and obtain decent work with minimal time and cost because universities are optional, vocational education is an alternative way with middle and lower socio-economic societies, these factors will form a good interest to arise either perforce or voluntarily within the students to continue their education to vocational school.

Stated by Al-Dajeh [6], concerning interest in vocational education is highly dependent on the teacher's knowledge and ability on what vocational education is, sometimes student ignorance can also be caused by a lack of understanding of teachers in junior high schools on vocational education. Supported by the study of Bunker and Shadbolt [7] in Australian Family Physician Journal stating that there is a significant potential effect on the delivery about career information and its options during primary and junior secondary education in enhancing career interest through vocational education. Of course, it also should be supported by teachers' knowledge and ability on vocational education and in shaping students' perceptions about vocational schools. In Sangihe Islands Regency itself, the interest of junior high school students to go to vocational schools can be seen from BPS's data that is 54.48% for Public High School and 45.58% for vocational high schools, which is not meet the national target, 30% for public high school and 70% for vocational school.

Research's purpose is to know the description of the interest of junior high school students to continue to vocational high school, correlation, and influence that occurs between the family's socio-economic conditions on the junior high school students' interest to go to vocational high school in Sangihe Islands, Indonesia.

2. Methods

This research is a quantitative research with an ex-post facto approach that is the research in which independent variable sets have taken place, when the researcher began to observe the dependent variable. This study conducted in March till April, 2016. The total population are consisted of 332 9th grade students. Using Proportional Random Sampling, the number of samples based on the Isaac table are: (1) SMPN 1 Tahuna: 97 respondents, (2) SMPN 2 Tahuna Timur: 47 respondents; and (3) SMPN 5 Tahuna: 34 respondents. Using Likert for scoring, has four answer choice as follows: Strongly Agree (SA), Agree (A), Disagree (DS), and Strongly Disagree (SDS). The data is analyzed using correlation's contingency table, and regression test. Before data analysis been conducted, we testing the data validity using normality, heterocedasticity, and linearity test (see in Figure 1).

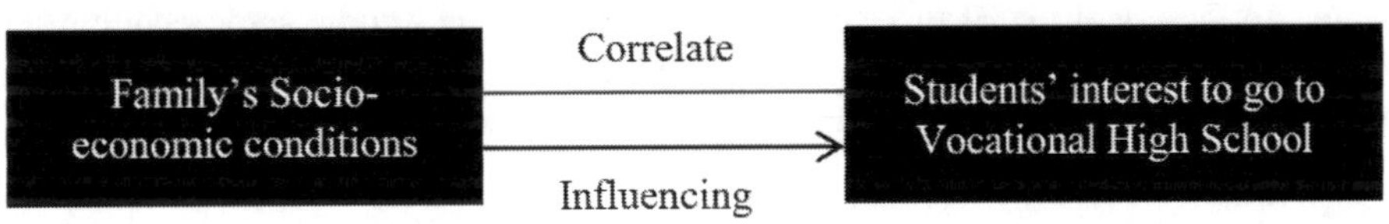

Figure 1. The framework.

3. Results and Discussion

3.1 Data description

Description of data of this study based on the variables of family's socio-economic condition (X or independent) and unior high school students' interest to go to vocational high school (Y or dependent), can be seen in Table 1 and Table 2.

Table 1. Data description.

Variable	Min	Max	Mean	Median	Mode
X	8	50	29,18	29	29
Y	22	86	59,27	61	44; 46

Table 2. Frequency distribution table of family's socio-economic condition.

No	Interval Class	Interval Class Limits	Frequency	Percentage (%)
1	5-7	4.5-7.5	14	7.8
2	8-10	7.5-10.5	22	12.4
3	11-13	10.5-13.5	59	39.9
4	14-16	13.5-16.5	36	20.2
5	17-19	16.5-19.5	27	15.2
6	20-22	19.5-22.5	8	4.5
	Total		178	100

The objective assessment criteria of the parents' socio-economic condition according to histogram in Figure 2 are as follows: (1) data range 3-10 has criteria of low class socio-economic condition, (2) data range 11 and 16 has a criteria of middle class socio-economic conditions, and (3) data range 17-23 has a criteria of high or top class socio-economic condition.

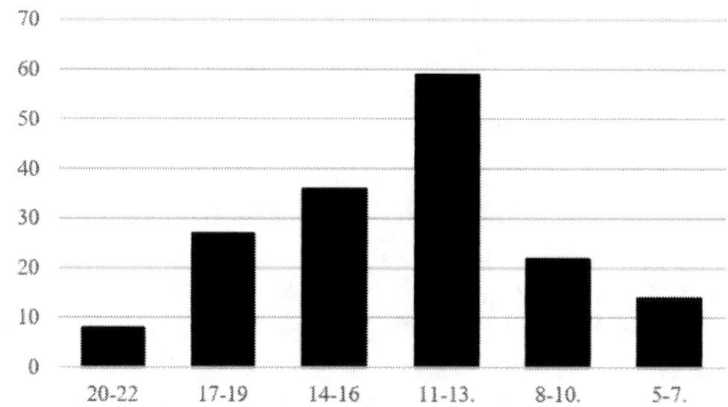

Figure 2. Family's socio-economic condition's histogram.

The socio-economic condition of the students in Sangihe is the middle socio-economic level, seen from data-range gathered between 11 and 16. The data descriptions in Table 1 then processed into the frequency distribution for the junior high school students' interest to go to vocational high school variable, which can be seen in Table 3.

Table 3. Frequency distribution table of junior high school students' interest to go to VET.

No	Interval Class	Interval Class Limits	Frequency	Percentage (%)
1	22-30	21.5-30.5	7	3.9
2	31-39	30.5-39.5	13	7.3
3	40-48	39.5-48.5	45	25.3
4	49-57	48.5-57.5	16	9
5	58-66	57.5-66.5	26	14.6
6	67-75	66.5-75.5	32	18
7	76-84	75.5-84.5	34	19.1
8	85-93	84.5-93.5	5	2.8
	Total		178	100

The objective criteria for junior high school students' interest in continuing education to vocational high school according to histogram in Figure 3 are as follows: (1) data range 22-57 has low interest criteria for junior high school students to continue to vocational high school (2) Data range 58-93 has high interest criteria for junior high school students to continue to vocational high school.

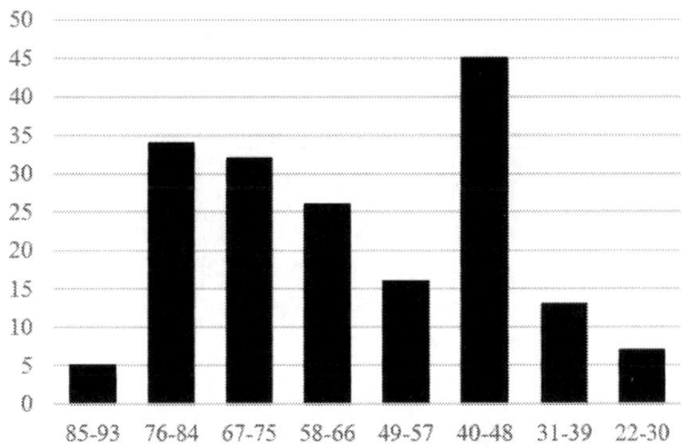

Figure 3. Students' interest to go to vocational high school's histogram.

The Figure 3 shows that the junior high school students' interest to go to vocational high school has a high interest, seen from most data gathered between 58 and 93.

3.2 Normality test
Done by comparing Z statistic table 5% from total N on this research is 1,645 with skewness value (curve) or kurtosis (tilt curve). The variable is normally distributed if Z α < Z table (see in Table 4).

Table 4. Skewness and kurtosis.

Variable	Skewness	Kurtosis
Family's Socio-Economic Condition	0.075	-0.551
Students' interest on VET	-0.177	-1.047

3.3 Heteroscedasticity test
If the scatterplot has a neat pattern there has been heteroscedasticity, and if it happens otherwise, it will be homocedasticity (see in Figure 4).

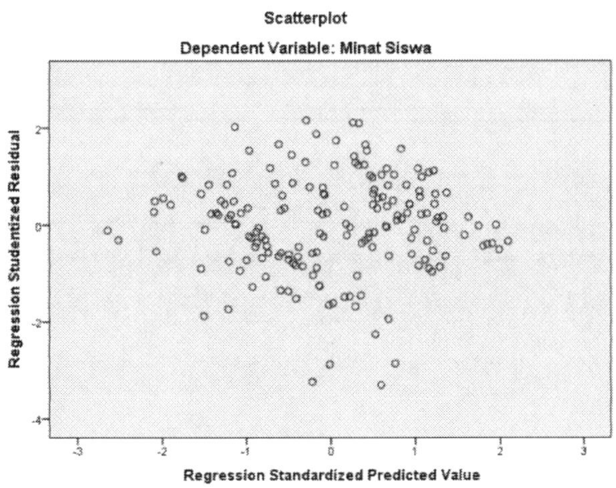

Figure 4. Heteroscedasticity test's scatterplot.

3.4 Linearity test
If sig value greater than the value of α is 0.05 then linearity occurs between the independent and dependent variables, conversely, if the sig < α (where α = 0.05), then there is no linearity between independent variables, family's socio-economic conditions (X) and Students' interest to go to vocational high school (Y).

Table 5. Linearity test.

Variable	*Sig. Value Deviation from Linearity*	Conclusion
X	0.516	Linear

Based on Table 5, the family's socio-economic condition (X) meets the linearity criterion where the sig value > α (0.05) so it can be concluded that the independent variables have linearity to the dependent variable, the junior high school students' interest to go to vocational high school (Y).

3.5 Correlation analysis

The correlation test results meet the *Pearson* criteria where the value is said to have perfect correlation if the variable correlation has value by 1, and the two-tailed criterion, all the variable correlation value where 0.000 < α (0.05) so that the family's socio-economic condition variable (X) can be related to the dependent variable, junior high school students' interest to go to vocational high school (Y).

The correlation between the family's socio-economic condition and the junior high school students' interest to go to vocational high school is also represented in the contingency table based on the parents' job and the level of student's interest.

Table 6. Contigency table about student's interest to go to high school based on father's job.

Father's Job	Student's Interest to go to Vocational High School		
	Low	Medium	High
Labor	2	9	28
Service personnel	0	0	3
Entrepreneur	1	6	7
Officer, employee	21	57	30
Experts, professional staff	0	1	0
Senate's staff, government's higher ranked officer	1	0	0
Others/Unemployment/ Freelancer	2	4	6

Table 7. Contigency table about student's interest to go to high school based on mother's job.

Mother's Job	Student's Interest to go to Vocational High School		
	Low	Medium	High
Labor	0	0	4
Service personnel	0	3	3
Entrepreneur	3	10	13
Officer, employee	17	45	24
Experts, professional staff	2	0	0
Senate's staff, government's higher ranked officer	0	0	0
Others/Unemployment/ Freelancer	5	19	30

Based on Table 6, if father's job i.e. laborers and employees, then has a high tendency level of students' interest to go to vocational high school. Then, in Table 7, if mother's job such as temporary employment (housewives, freelances or unemployed) has a high tendency level of student's interest to go to vocational high school. If the mother's job is an employee or officer, then has a medium tendency level of student interest to a vocational high school.

3.6 Simple regression analysis

Hypothesis's acceptance criteria, if sig > α (0.05) then H_0 accepted and H_a rejected, otherwise if sig value < α (0.05) then H_0 rejected and H_a accepted.

Table 8. Regression analysis results

Var	R	R^2	Constanta	Coefficient	t	Sig.
X → Y	0.514	0.264	89.050	-2.27	-7.95	0.00

Based on Table 8, means that it meets H_a acceptance criteria and the rejection of H_0, where sig <α (0.05). The final decision is H_a hypothesis accepted that 'there is influence between the family's socio-economic condition on the junior high school students' interest to go to vocational high school in Sangihe Islands'. This is inlined with the results of a study from Windarto (2013) which suggests that the family's socio-economic conditions have a significant effect on student interest variables to go to vocational high school in Bantul.

Table 9. Variable regression's equation

Variable	Equation
X	Y = 89.050 + (-2.274X)

If X variable add up one unit, it will increase the junior high school students' interest to go to vocational high school (Y) by -2.274. Based on the regression equation from Table 9, the socio-economic condition variable contributes 26% to the interest of junior high school students to go to vocational high school, and the remaining percentage of contribution is influenced by other variables.

4. Conclusions

Initial description of the interest of junior high school students in Sangihe Islands to go to vocational high school, 97 respondents from 178 respondents stated will be interested to continue to vocational high school or 54.5% of the total respondents. Proved by Pearson's analysis scale is 1 which shows a high correlation rate. So, there is an influence between the family's socio-economic condition (X) and the junior high school students' interest to go to vocational high school (Y) with 26% influential contribution percentage.

The middle and lower class socio-economic students in Sangihe Islands tend to take Vocational High School as a safe option, as if, in case they're not going to continue their education to university. The result same as study been conducted in Adelaide [5], they tend to take vocational high school to acquire working skill and some extra skill like entrepreneurship, so when they graduated, they can get a job as soon as possible. But, the culture difference between Indonesia and Australia, make the factors can get different either, i.e. family decision, when commonly, in Indonesian family, either the children willing to take education at University or not, it's up to the parents to make the decision, not the children in most case.

References

[1] Undang-Undang Republik Indonesia No. 20 Tahun 2003 National Education Act.
[2] Badan Pusat Statistik Kabupaten Kepulauan Sangihe, available at: sangihekab.bps.go.id
[3] McMillan J and Western J 2000 Measurement of The Socio-Economic Status Of Australian Higher Education Students *Journal Department of Anthropology and Sociology* p 223.
[4] Berns R M 2004 *Child, Family, School, Community Socialization and Support 6th Edition* (Melbourne: Thomson Learning. Inc.) **6** p 119.
[5] Curtis D Drummond, A Halsey J and Lawson M J 2012 *Peer-Mentoring Of Students In Rural And Low Socio-economic Status Schools: Increasing Aspirations For Higher Education* (Adelaide: Flinders University Research Report/National Vocational Education And

Training Research Program) p 9.
[6] Al-Dajeh 2012 Assessing Environmental Literacy of Pre-Vocational Education Teachers In Jordan *College Student Journal University of Jordan/Proquest* p 3.
[7] Bunker J and Shadbolt N 2009 Choosing General Practice As A Career: The Influences Of Education And Training *Australian Family Physician Journal* **38** pp 341-344.

What are the Perspectives of Indonesian Students to Japanese Ritual during Solar Eclipse?

N Haristiani*, A Rusli, A S Wiryani, A B D Nandiyanto, A Purnamasari, T N Sucahya and N Permatasari

Universitas Pendidikan Indonesia, Bandung, Indonesia

*nuriaharist@yahoo.com

Abstract. In this globalization era, many people still believe the myths about solar eclipse. The myths about solar eclipse are different between one country or are to another. In this context, the aim of this study was to investigate the perspective of Indonesian students in viewing how the Japanese people face their believing myths in solar eclipse. This research also investigated the student belief on several mythical stories in Indonesia, their understanding of the Islamic view, and their knowledge based on science concept relating to the solar eclipse phenomenon. To understand the Indonesian students' perspective about the solar eclipse myths in Japanese, we took a survey to Indonesian students which are studying Japanese culture and language. Based on the results, the Indonesian student think that there is no significant difference between Indonesian and Japanese people in facing the solar eclipse.

1. Introduction

Total solar eclipse is a phenomenon which occurs when the position of the moon, the earth and the sun are aligned. Therefore, the moon shadow through the earth and covered the sunlight into the earth [1-4]. Because of this phenomenon turns the day becomes dark, many people relate this phenomenon with mystical event [5-7]. Believes in the myths are not only exist in under develop countries but also developed countries such as Japan.

Japan is a well-known country that is one of the pioneers for developing technology and modernizing social culture [8]. Japan is one of the countries that maintain its culture, including its myths facing the natural event (such as solar eclipse). That makes Japanese culture to be learned and studied in all over the world [9].

Here, the purpose of this study was to investigate the perspective of Indonesian students (who are focusing study in Japanese culture and language) in viewing the Japanese people in facing solar eclipse phenomenon. The study was also compared with Indonesian myths. This research also investigated the student belief on several mythical stories, their understanding of the Islamic view, and their knowledge based on science concept relating to the solar eclipse phenomenon.

2. The Cultural View in Japan and Indonesia, Comparing to Islamic View and Science Concept

2.1. Japanese culture facing the solar eclipse
During the solar eclipse, ancient Japanese believe in Shinto religion (worship to the God of Sun) that poison is spread out. Therefore, ancient Japanese close their wells for avoiding the poison to come to the water. Further, they also hang up amulet (made by a stone) above the tree [10].

2.2. Pre-Islamic Indonesian culture facing the solar eclipse
In pre-Islamic Indonesian in Java, they believe that the solar eclipse occurs because Batara Kala demon devouring the sun. Therefore, they must do several rituals, such as pregnant women and children should not go outside. Further they must hide to avoid the anger of Batara Kala. Further, they must not do several activities, such as eating, cooking, looking at the sun, constructing a house, and celebrating marriage [5, 11, 12].

2.3. Islamic view about eclipse
In Islam, a solar eclipse phenomenon has been explained in these following hadiths (translated from Arabic) [13, 14]:

Narrated Al-Mughira bin Shu'ba (In Bukhari, Book: 16; Hadith: 1043):
"The sun eclipsed in the life-time of Allah's Messenger on the day when (his son) Ibrahim died. So the people said that the sun had eclipsed because of the death of Ibrahim. Allah's Messenger said, "The sun and the moon do not eclipse because of the death or life (i.e. birth) of someone. When you see the eclipse offer Salat (prayer) and invoke Allah."

Narrated Abu Bakr (RA) (In Bukhari, Book: 16; Hadith: 1040):
"We were with Allah's Messenger when the sun eclipsed. Allah's Messenger stood up dragging his cloak till he entered the Mosque. He led us in a two-Rak'a prayer till the sun (eclipse) had cleared. Then the Prophet (p.b.u.h) said, "The sun and the moon do not eclipse because of someone's death. So whenever you see these eclipses offer Salat and invoke (Allah) till the eclipse has cleared."

Aishah narrated that (In Sunan Abu Dawud, Book of The Prayer for Rain; Hadith: 1191):
"The Prophet (p.b.u.h) said: "The sun and the moon do not eclipse due to the life (birth) or death of anyone. So when you witness it (an eclipse), supplicate to Allah, the Mighty & Sublime, say the Takbir and give charity."

Asma narrated (In Sunan Abu Dawud, Book of The Prayer for Rain; Hadith: 1192):
"The Prophet (p.b.u.h) would command that slaves be freed during the eclipse prayer."

Based on the above hadiths, we can conclude that in Islamic views, the solar eclipse phenomenon has no relation to the birth or the death of someone. This phenomenon is a reminder from Allah to His devotees about the end of times (Armageddon or *Qayamah*). Therefore, during the solar eclipse, Muslims are recommended to offering *Ibadah*, such as: (i) praying (*Salat*), (ii) giving a charity (*Sadaqa*), (iii) freeing slaves, and (iv) asking the God for forgiveness from the torments of the grave (iv).

2.4. Scientific views about solar eclipse
Solar eclipse is a natural phenomenon that the sun is covered by the moon for several minutes [1-3]. During the solar eclipse, several incidents can be observed, such as the changing color of the sky, the moving shadow, the shadows bands, the corona and its prominences, the sun's chromospheres, the diamond ring formation. Further, planets and stars suddenly become visible against the darkening sky [2]. The solar eclipse happens at least twice a year. Solar eclipse has almost no impact to the human.

The effect of the solar eclipse is found only when human is looking sun during by naked eyes [11, 12, 15-20].

3. Research Methodology

To investigate Indonesian students' perspective to the Japanese myth, we took a survey to 51 students who are focusing study in Japanese Language in Department of Japanese Language Education in Universitas Pendidikan Indonesia. We took subjects with different ages and grades, which are between 18 and 22 years old. The survey was taken from February 27 to March 10, 2016.

To identify the survey for the student opinion about the total solar eclipse myth, the student completed a short questionnaire about the solar eclipse fact and myth. The questionnaire consists of 8 items of statements on three-point likert scale. The questionnaire subjects were allowed to fill out the questionnaire for 15 minutes. The result of the survey was then analyzed using a SPSS software to get a precise trend about the students' opinion about total solar eclipse myth in Indonesia and Japan.

4. Results and Discussion

Table 1 shows result of students responses to the questionnaire about their opinion regarding solar eclipse myth. Based on this table, 94% of students know that total solar eclipse is one of natural phenomena, whereas 65% of respondents do not believe the myth about total solar eclipse. We also found that 50% of respondents do not believe the myth that eating, constructing a house and celebrating marriage are not allowed during total solar eclipse. However, 55% of respondents believed that during the solar eclipse, people should do religious activity such as going to worship, giving charity, and praying.

Based on the above results, research subjects believed that the total solar eclipse is a natural phenomenon. Students have understood regarding correlation between solar eclipse and mystical activities (such as prohibition to eating, constructing and celebrating marriage during the solar eclipse event). The paradigm of the student in the university was more realistic and logic. As a result, they do not observe a phenomenon only based the cultural belief without rationalizing [21]. This is confirmed by the first statement in Table 1 that 94% of respondents believe about the disconnection between solar eclipse and myth.

Table 1. Student responses to the questionnaire about Indonesian total solar eclipse myth.

Statement	Response (%)		
	Believe	Not believe	Not know
Total solar eclipse is one of natural phenomenon	94	2	4
Total solar eclipse is related with mistique things	0	65	35
Total solar eclipse myth : should not eating, constructing, and marriage	6	50	44
Should doing religious activity (e.g: going to worship, giving charity, and praying)	55	22	23

In addition to myth, we found the correlation be-tween the solar eclipse and religious activity. Most of research subjects give positive recommendation to do worshipping during the total solar eclipse. This is because majority of Indonesian people is a Moslem, in which this has impacts to the people daily activity [22]. Thus, 55% of respondents believe that people should do religious activity during the solar eclipse.

Table 2 shows students responses to the questionnaire about their opinion how the Japanese people face the solar eclipse myth. According to this table, 78% of respondents understand that Japanese people face the solar eclipse as a natural process. 37% of students consider the disbelief of Japanese people regarding Japanese myth during the solar eclipse. And, 40 % of respondents consider that Japanese people do not hang up the amulet above the tree during the solar eclipse. In addition, 55% of respondents believed that Japanese people do worshipping during this event.

From the data in Table 2, Indonesian students assume that Japanese people do not believe myths during the solar eclipse, such as (i) closing the wells, and (ii) hanging up the amulet above the tree. This is the respondent assume that since Japanese people are able to develop world technologies, the Japanese will have realistic mindset and always think based on the fact [23]. Regarding the religious activity, Indonesian students consider that Japanese people do religious activity during the solar eclipse. This consideration is because of the mindset of Indonesian students that come from the Islamic culture [24].

Table 2. Student responses to the questionnaire about Indonesian total solar eclipse myth.

Statement	Response (%)		
	Believe	Not believe	Not know
Total solar eclipse is one of natural phenomenon	78	12	10
Total solar eclipse myth : should closing wells during solar eclipse	31	37	32
Total solar eclipse myth : should hanging amulet above the three	32	40	28
Should doing religious activity (e.g: going to worship, giving charity, and praying)	55	13	32

5. Conclusions

Based on the literature review, there are several myths regarding the solar eclipse. However, since there is a boost in science, knowledge, and technology for facing the solar eclipse phenomena, the Indonesian student think that there is no significant difference between Indonesian and Japanese people in facing the solar eclipse. However, the present experimental results are still incomplete. Further study that focused directly to the Japanese people must be done. Therefore, Indonesian students' perspective can be compared correctly.

Acknowledgments

We acknowledged RISTEK DIKTI (Grant: Penelitian Unggulan Perguruan Tinggi) and SPS UPI (Grant: Penelitian Lintas Disiplin Ilmu).

References

[1] Nishanth T, N Ojha, M S Kumar and M Naja 2011 Influence of solar eclipse of 15 January 2010 on surface ozone *Atmospheric environment* **45**(9) pp 1752-1758.

[2] Fabian P, M Winterhalter, B Rappenglück, H Reitmayer, A Stohl, P Koepke, H Schlager, H Berresheim, T Foken and B Wichura 2001 The BAYSOFI Campaign–Measurements carried out during the total solar eclipse of August 11, 1999 *Meteorologische Zeitschrift* **10**(3) pp 165-170.

[3] Foken T, B Wichura, O Klemm, J Gerchau, M Winterhalter and T Weidinger 2001

Micrometeorological measurements during the total solar eclipse of August 11, 1999 *Meteorologische Zeitschrift* **10**(3) pp 171-178.

[4] Nandiyanto A B D, D Sofiani, N Permatasari, T N Sucahya, A S Wiryani, A Purnamasari, A Rusli, E C Prima 2016 Photodecomposition Profile of Organic Material during the Partial Solar Eclipse of 9 March 2016 and Its Corelation with Organic Material Concentration and Photocatalyst Amount *Indonesian Journal of Science & Technology* **1** (2) pp 132-155.

[5] Mohapatra J 1991 The interaction of cultural rituals and the concepts of science in student learning: a case study on solar eclipse *International Journal of Science Education* **13**(4) pp 431-437.

[6] Evans C A 1993 Life-of-Jesus research and the eclipse of mythology *Theological Studies* **54**(1) pp 3-36.

[7] Gerasopoulos E, C Zerefos, I Tsagouri, D Founda, V Amiridis, A Bais, A Belehaki, N Christou, G Economou and M Kanakidou 2008 The total solar eclipse of March 2006: overview *Atmospheric Chemistry and Physics* **8**(17) pp 5205-5220.

[8] Cousins S D 1989 Culture and self-perception in Japan and the United States *Journal of Personality and Social Psychology* **56** pp 124-131.

[9] Kubota R 1999 Japanese culture constructed by discourses: Implications for applied linguistics research and ELT *TESOL quarterly* **33**(1) pp 9-35.

[10] Sugimoto M, Swain D L 2016 *Science and culture in traditional Japan* (Tokyo: Tuttle Publishing).

[11] Hope-Ross M, S Travers and D Mooney 1988 Solar retinopathy following religious rituals *The British journal of ophthalmology* **72**(12) pp 931.

[12] Keeler W 1988 Sharp rays: Javanese responses to a solar eclipse *Indonesia* (46) pp 91-101.

[13] Dawud S A 1984 English translation with explanatory notes by Prof. *Ahmad Hasan* 1: lll.

[14] Bukhâri M i I and M M Khan 1997 The translation of the meanings of Sahih al-Bukhari: Arabic-English, *Dar-us-Salam.*

[15] Dhir S, A Gupta and I Jain 1981 Eclipse retinopathy *British Journal of Ophthalmology* **65**(1) pp 42-45.

[16] Michaelides M, R Rajendram, J Marshall and S Keightley 2001 Eclipse retinopathy *Eye-London-Ophthalmological Society of the United Kingdom Royal College of Ophthalmological* **15**(2) pp 148-151.

[17] Codenotti M, F Patelli and R Brancato 2003 OCT findings in patients with retinopathy after watching a solar eclipse. *Ophthalmologica* **216**(6) pp 463-466.

[18] Schatz H and F Mendelblatt 1973 Solar retinopathy from sun-gazing under the influence of LSD *The British journal of ophthalmology* **57**(4) pp 270.

[19] MacFaul P 1969 Visual prognosis after solar retinopathy *The British journal of ophthalmology* **53**(8) pp 534

[20] Ben H.-A. W, W Laabidi, A Maalej, R Rannen and S Gabsi 2008 Solar retinopathy: optical coherence tomography findings *Bulletin de la Societe belge d'ophtalmologie* (311) pp 17-22.

[21] Piaget J, Inhelder B, & Piaget J 2013 *The growth of logical thinking from childhood to adolescence: An essay on the construction of formal operational structures* (British: Routledge).

[22] Vickers A 2013 *A history of modern Indonesia* (New York: Cambridge University Press).

[23] Davidson B W 1995 Critical thinking education faces the challenge of Japan *Inquiry: Critical Thinking Across the Disciplines* **14**(3) pp 41-53.

[24] Calverley E E 2004 *Worship in Islam: Being a Translation, with Commentary and Introduction of Al-Ghazzáli's Book of the Iḥyá' on the Worship* (New Delhi: Gorgias Press LLC).

Solar Eclipse: Concept of "Science" and "Language" Literacy

N Haristiani*, R Zaen, A B D Nandiyanto, A N Rusmana, F Azis, A A Danuwijaya and A G Abdullah

Universitas Pendidikan Indonesia, Bandung, Indonesia

*nuriaharist@yahoo.com

Abstract. The purpose of this study was to evaluate the concept of science and language literacy of solar eclipse. The study was conducted through a survey to 250 students with different ages (from 17 to 23 years old), grades, and majors in Universitas Pendidikan Indonesia. The survey was completed with a questionnaire consisting of 41 questions. In the case of the language literacy, experimental results showed that various expressions in facing the solar eclipse phenomenon are found. Relating to the science literacy, most students have good science understanding to the solar eclipse phenomenon. In conclusion, the understanding about the solar eclipse is affected by formal science education and religion understanding that they have been accepted since their childhood. These factors have also influenced the belief of Indonesian people to the solar eclipse myth and the way of expressions a language literacy.

1. Introduction

Solar eclipse is a spectacular phenomenon [1, 2], which occurs when the position of the moon is between the sun and the earth (virtual solar system project: building understanding through model building) [3]. Solar eclipse causes the day becomes dark; therefore, many people have an enthusiasm to face this event [4]. There are many interesting expressions that found around the people related with their understanding about solar eclipse [5]. These expressions are communication language that well known in around society [6].

Many language literacy research such as in education [7, 8] and sociocultural [9]. Combining science and language literacy also have been developed in these fields [10]. This method empowering research. [11]. However, there was no research using this method in facing solar eclipse.

Here, the purpose of this study was to evaluate the concept of Indonesian students (who are studying in Universitas Pendidikan Indonesia) about science and language literacy of solar eclipse on the necessity for science education. Beside science and language literacy, this research also investigated the education that influences student understanding about solar eclipse phenomenon and student belief on knowing solar eclipse myths in Indonesia.

2. Research Methodology

To evaluate the concept of Indonesian students about science and language literacy of solar eclipse, we took a survey to students who are studying in Universitas Pendidikan Indonesia that located in Bandung, where Bandung experienced a partial solar eclipse of 9 March 2016. The number of subject in this research was 250 students. We also took subjects with different ages, levels and majors, which were between 17 and 23 years old.

The survey was completed with a questionnaire consisting of 41 questions. The questions were divided into two sections. 26 questions relates to language literacy, whereas the other is science literacy and understanding. The questionnaire was divided into two types. First, questions with four-point multiple choice (right-wrong questions to assess student concept) and the other one is questions with three-point multiple choice were described their understanding of solar eclipse. The result of the survey was then analyzed using a SPSS software to get a precise trend about the students' responses about solar eclipse. The survey was taken from 25 to 28 March 2016.

3. Results and Discussion

Figure 1 shows student responses to the questionnaire about language literacy related the solar eclipse. Figures 1a and b are the result for vocabulary used when facing solar eclipsed in English and Indonesian local language, respectively. We selected several Indonesian local language, such as Javanese (for "gerhono"), Sundanese (for "samagaha"), and Batak (for "holom").

Based on Figure 1a, 46.4% of students used vocabulary "see" when solar eclipse occurred. This vocabulary was well-known among people [12].

Regarding Indonesian local language (see Figure 1b), 40.8% of students knew vocabulary "gerhono". The respondents selected this word because the structure of Javanese language is not too different with Indonesian [13]. Usually, it only needs to change "a" to "o", for example "jawa" "jowo" [14, 15]. The term of "samagaha" is not popular among respondent [16]. Term "samagaha" was heared strange even for sundanese people [17].

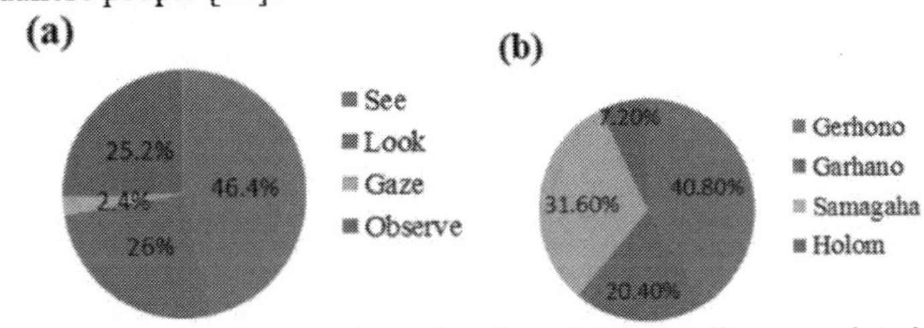

Figure 1. Student responses to the questionnaire about language literacy related solar eclipse. (a) vocabulary which is used when solar eclipsed and (b) vocabulary of solar eclipse in local language.

Figure 2 shows student responses to the questionnaire about expression relating solar eclipse myths. Based on this figure, 69,6% of students selected frowned expression when they heard myths about the solar eclipse. This gesture expression is face expression that show a confusion and doubtful about information that they got [18, 19]. We also found that they have less believe regarding the solar eclipses myth that spread out in around them.

In addition to the disbelied on myth, the students have already get knowledge about correlation solar eclipse and myth by formal science education and religion understanding since their childhood [20, 21]. This results was also supported by the fact that majority of Indonesian people is a Moslem [22]. Thus, they would rather believe to do religious activity during this event [23]. Based on this experimental result, we can conclude that nowadays myths of solar eclipse have lost among Indonesian people [24, 25].

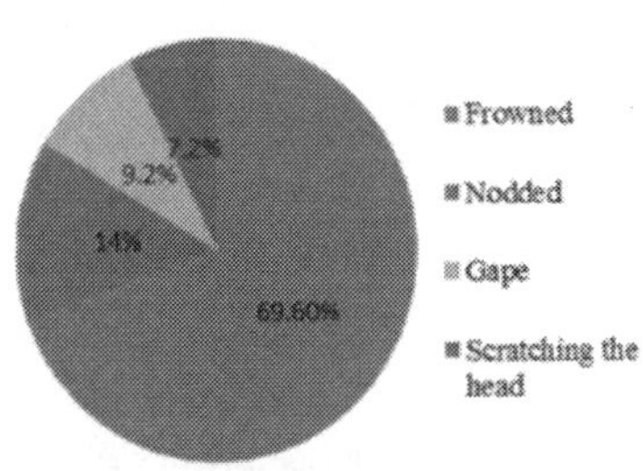

Figure 2. Student responses to the questionnaire about expressions related solar eclipse myths.

Table 1 shows student responses to the questionnaire about Indonesian total solar eclipse myth. 98% of respondents nodded to show their gesture in facing solar eclipse. 92.8% of students have already understood the vocabulary of "see" in showing expression to face solar eclipse. 92% of students used the vocabulary of "observe" in facing solar eclipse. 26.8% of respondents knew the lexical vocabulary of "gerhono".

From the data in Table 1, students tended to use the common word to show their expression in facing solar eclipse phenomenon. As discussed above, the term of "Gerhono" was well known lexical term than the others [26].

Table 1. Student's responses to the questionnaire about understanding total solar eclipse's vocabularies.

Question	Responds (%)		
	Agree	Disagree	Not Know
In childhood, a solar eclipse has been told as a sign that a syaithon came.	19.6	22	61.6
In childhood, a solar eclipse has been told as a sign that an angel came.	8	25.6	67.6
Did you understand the vocabulary of "melihat"?	95.2	5.6	1.2
Did you understand the vocabulary of "memandang"?	90.8	8.8	0.8
Did you understand the vocabulary of "menonton"?	93.2	6.8	0.4
Did you understand the vocabulary of "mengamati"?	96.8	3.6	0
Did you understand vocabulary of "see"?	92.8	7.2	0.4
Did you understand the vocabulary of "look"?	90	9.6	0.8
Did you understand the vocabulary of "gaze"?	46.8	3.6	14
Did you understand the vocabulary of "observe"?	92	8.4	1.6
Did you understand the vocabulary of "gerhono"?	26.8	47.6	30
Did you understand the vocabulary of "garhano"?	17.6	48.8	36.8
Did you understand the vocabulary of "samagaha"?	26	39.2	39.2
Did you understand the vocabulary of "holom"?	10.4	45.2	46.8
Did you understand the gesture of "mengernyitkan dahi?"	94.8	4.8	2
Did you understand the gesture of "menganggukkan kepala?"	98	1.6	0.4
Did you understand the gesture of "menggaruk-garuk kepala?"	96.4	3.2	0.4
Did you understand the gesture of "melongo-longo tidak karuan?"	90.4	7.6	2
Did you understand the different of solar eclipse and total solar eclipse?	90.4	8.8	0.8

Table 2 shows students responses to the questionnaire about the total solar eclipse in science perspective. Based on Table 2, 53% of respondents answer the answer correctly. This experimental data showed that student's knowledge were high. It was supported with the understanding of students about the question relating science literacy [9, 27].

Table 2. Student's responses to the questionnaire about total solar eclipse in science perspective.

Question	Responds (%)	
	True	False
How the position of the moon, the earth, and the sun in the event of a solar eclipse?	87.2	12.8
What have caused a total solar eclipse?	86.8	13.2
When did the occurrence of a solar eclipse?	32	68
When did the occurrence of a total eclipse of the sun?	6.8	93.2
When a total solar eclipse could happen the same place?	83.6	16.4
Why do we not allowed to see a total solar eclipse directly?	74.8	25.2
Where the total solar eclipse on march 9 2016 happened?	75.2	24.8
How long duration the total solar eclipse on the march 9 2016 in Indonesia?	4	96
Which the most dangerous rays in during the total solar eclipse?	66.8	33.2
How the earth condition when total solar eclipse happen?	96	4
How the intensity of light sun on earth when the total solar eclipse happen?	91.6	8.4

Table 3 shows students responses to the questionnaire about understanding of the total solar eclipse. According to this table, 65% of students have good science understanding in facing solar eclipse phenomenon.

Table 3. Student's responses to the questionnaire about understanding of the total solar eclipse.

Statement	Responds (%)		
	Believe	Not Believe	Not Know
Did you know that a solar eclipse has occurred in every 6 months?	33.2	22	47.2
Did you know that a total solar eclipse can happen in the same place within 300 years?	73.2	15.2	14.8

4. Conclusions

Based on the experimental result, we conclude that there were various expressions in facing the solar eclipse phenomenon. Showing effect of their language literacy. Relating to the science literacy, most students have science understanding to the solar eclipse phenomenon (more than 65%). Indeed, the student understanding about the solar eclipse is affected by formal science education and religion understanding that they have been accepted since their childhood. These have also influenced the belief of Indonesian people to solar eclipse myths. Therefore, when the solar eclipse experienced, they have high enthusiasm in facing and observing this event.

Acknowledgements

We acknowledged RISTEK DIKTI (Grant: Penelitian Unggulan Peguruan Tinggi) and SPS UPI (Grant: Penelitian Lintas Disiplin).

References

[1] Emde C, and Mayer B 2007 Simulation of solar radiation during a total eclipse: a challenge for radiative transfer *Atmospheric Chemistry and Physics* **7**(9) pp 2259-2270.

[2] Nandiyanto A B D, Sofiani D, Permatasari N, Sucahya T N, Wiryani A S, Purnamasari A, Rusli A, and Prima E C 2016 Photodecomposition profile of organic material during the partial solar eclipse of 9 March 2016 and Its correlation with organic material concentration and photocatalyst amount *Indonesian Journal of Science and Technology* **1**(2) pp 132-155.

[3] Barnett M, and Morran J 2002 Addressing children's alternative frameworks of the moon's phases and eclipses *International Journal of Science Education* **24**(8) pp 859-879.

[4] Bakich M E 2016 *Solar Eclipses Throughout History* In Your Guide to the 2017 Total Solar Eclipse pp 43-72 (Springer International Publishing).
[5] Kavanagh C, Agan L, and Sneider C 2005 Learning about phases of the moon and eclipses: A guide for teachers and curriculum developers *Astronomy Education Review* **4**(1) pp 19-52.
[6] Carey J W 2008 *Communication as culture, revised edition: Essays on media and society* (Routledge).
[7] Smith M W, and Dickinson D K 2002 Early Language & Literacy Classroom Observation (ELLCO) Toolkit, Research Edition [with] User's Guide.
[8] Hyland K 2007 Genre pedagogy: Language, literacy and L2 writing instruction *Journal of second language writing* **16**(3) pp 148-164.
[9] Bybee R, McCrae B, and Laurie R 2009 PISA 2006: An assessment of scientific literacy *Journal of research in Science Teaching* **46**(8) pp 865-883.
[10] Wellington J, and Osborne J 2001 *Language and literacy in science education* (UK: McGraw-Hill Education).
[11] Yore L D, and Treagust D F 2006 Current realities and future possibilities: Language and science literacy empowering research and informing instruction *International Journal of Science Education* **28**(2-3) pp 291-314.
[12] Dale E 1965 Vocabulary measurement: Techniques and major findings *Elementary English:* **42**(8) pp 895-948.
[13] Anderson B 1966 The languages of Indonesian politics *Indonesia* (1) pp 89-116.
[14] Geertz C 1957 Ritual and social change: a Javanese example *American anthropologist* **59**(1) pp 32-54.
[15] Geertz C 1960 The Javanese Kijaji: The changing role of a cultural broker *Comparative Studies in society and history* **2**(02) pp 228-249.
[16] Jackson K D, and Pye L W 1978 *Political power and communications in Indonesia* (California: University of California Press).
[17] Horikoshi H 1975 The Dar ul-Islam movement in West Java (1948-62): an experience in the historical process *Indonesia* (20) pp 59-86.
[18] Campos J J, Sorce J F, Emde R N, and Svejda M 2013 Emotions as Behavior Regulators: Social Referencing in Infancy *Emotions in early development* **2** pp 57.
[19] Darwin C, Ekman P, and Prodger P 1998 *The expression of the emotions in man and animals* (USA: Oxford University Press).
[20] Buaraphan K 2012 Embedding nature of science in teaching about astronomy and space *Journal of Science Education and Technology* **21**(3) pp 353-369.
[21] Mohapatra J K 1991 The interaction of cultural rituals and the concepts of science in student learning: a case study on solar eclipse *International Journal of Science Education* **13**(4) pp 431-437.
[22] Hefner R W 2011 *Civil Islam: Muslims and Democratization in Indonesia* (New Jersey: Princeton University Press).
[23] Küçüközer H 2007 Prospective Science Teachers' Conceptions about Astronomical Subjects *Science Education International* **18**(2) pp 113-130.
[24] Dorson R M 1955 The eclipse of solar mythology *The Journal of American Folklore* **68**(270) pp 393-416.
[25] Reddy F 2012 Eclipses of the Sun and Moon *In Celestial Delights* pp 95-153 (Springer New York).
[26] Krishnan S K 2008 Code-switching in the teaching of science to lower secondary students.
[27] Stacey K 2014 The PISA view of mathematical literacy Indonesia *Journal on Mathematics Education* **2**(02) pp 95-126.

IOP Conf. Series: Materials Science and Engineering **306** (2018) 012094 doi:10.1088/1757-899X/306/1/012094

Vocational High School Students' Profile and their English Achievement

N V F Liando*, D M Ratu and V Sahentombage
Universitas Negeri Manado, Tondano 95618, Sulawesi Utara, Indonesia

*nihtaliando@unima.ac.id

Abstract. Vocational education has been given more attention in Indonesian education in the recent years. There have been many projects for vocational education since Jokowi Widodo took his presidential office in October 2014. In supporting government actions, vocational high school students need to improve their profile. Living in the global worlds requires the ability to interact with people from all over the world. The ability to communicate using English as the lingua franca is important. The purpose of the research reported here is, to prove whether the direct method is effective in improving vocational high school students' English pronunciation or not. This research design is a true experimental using post-test only. The population is students from one vocational high school in North Sulawesi. Sample of this research was year 11 students consisting of two classes class A (27 students) and class B (27 students). The instrument used in collecting data is tests. The results showed that the mean of the experimental group (36.99) statistically describes the students' improvement in pronouncing English words in which have been compared by result of the $t_{observed}$ (2.897) exceed $t_{critical}$ (1.943) at the level of significance 0.05. It means that there is a significant difference between the mean score of experimental group and control group regarding students' English pronunciation. This then supports the claim that 'rejects' *Ho* and 'accept' *Ha*. Based on the result, it could be concluded that the direct method is considered effective in improving students' English pronunciation.

1. Introduction

English teaching can be viewed as a system. It comprises some elements among others: teachers, students, curriculum, materials, methods, media and testing. These components are called as the system of English teaching. The material of teaching as an important component in the system of English teaching is indispensable. It is important because it is the core to be learned in the classroom. In the learning activities, there are many things that influence students in mastering foreign language, such as language skills, motivation and time spent for learning. Those are presented to support the students' ability in understanding the language. To maintain interaction of integrated-skills approach to language teaching, the emphasis should be on the interrelationship of skills. For example; reading ability will best be developed in association with writing, listening, and speaking activity. Even in those courses that may be labeled "reading", the goals will be best achieved by capitalizing on the interrelationship of skills.

In the implementation of teaching and learning the students are expected to know and master the materials, so that they can use English either in oral or in writing. By the interrelation of English skills, the pronunciation item is one of the most significant factors to influence students' progress in reading a kind of text. It is considered difficult because the English sound system differs with other language, and it can be presented explicitly as far as it is related to the theme or sub-theme being discussed. Reading in the target language should be taught from the beginning of language instructions; however, the reading skill will be developed through practice with speaking. In this context, appropriate method and techniques are needed to help students develop their pronunciation. Several research studies have looked at English pronunciation [1-4] and suggested ways of overcoming problems.

Direct Method is one of the methods which is effective to improve students' English pronunciation. The teacher and students play their role as partners in the process of teaching and learning using direct methods [5]. It means the initiation of the interaction goes both ways, from teacher to students and from the students to teacher. Although it works on the four skills from the start, oral communication is considered as a basis. Thus, the reading and writing exercises are based upon what the students practice orally first. Pronunciation also receives attention right from the beginning of a course. This is clear that the goals of teacher who use the direct method intend that the students learn how to communicate, associate meaning and the target language directly. When the teacher introduces a new target language, word or phrase is demonstrated using *realia*, pictures, pantomime or gestures. The teacher never translates the words into students' native language. Students speak in the target language a great deal and communicating as if they are in real situations. Further, Larsen asserts that "In fact, the syllabus used in the direct method is based upon situations.

The oral communication skills will be built up in a carefully graded progression organized around question-and-answer exchanges between teachers and students in small or intensive class [5]. In practice, therefore, classes based upon the direct method tradition are small, use only the second language as both a means of communication and instruction. Theoretically, the approach is also underpinned by a belief that language teaching should be based around phonetics and accurate pronunciation [6]. This then supports that how important students' English pronunciation is, including their articulation. Almost any 'method' can succeed when clients are willing to pay high prices for small class, individual attention, and intensive study. The direct method places teachers at the center of classroom activity, with critics arguing that learners, and success of the method itself, are too dependent on a teacher's individual skill and ability [6]. It means that use of the direct method depends on the teacher's performance how to apply it.

In applying the direct method, there are some techniques that are adapted when process of teaching and learning is done. They are reading aloud, question and answer exercise, getting students to self-correct, conversation practice, the use of minimal pairs, dictation, and the others [5]. This gives a signal that role of the techniques is indispensable in teaching and learning process. One of the techniques which most frequently cited barriers to reading aloud are lack of time [7]. As an English teacher, this certainly constitutes a challenge to spend more time to apply. The lack of time for using the technique, the teacher does not sometimes care this case; and it can influence student's improvement particularly on English pronunciation. Indeed, English teacher should have enough and or special time to make up this case. Because how other people can speak and understanding the target language if user of the language is not based on the pronunciation. This is a key to gaining full communicative competence and it is specifically to build a learner's articulatory competence [8].

In reading activity, when the students are asked to read a kind of reading text, the most emergent problem gained is difficult for them to read the reading passages, either sentence by sentence or paragraph, even when they are asked to pronounce the certain English words, they are not able to pronounce the words, although the teacher frequently employ the techniques but the students' improvement concerning English pronunciation has not still been proven. After the teacher observes deeper the students' problem, it was found that they are hard to pronounce English words. These are caused by differences between the English and Indonesian sound systems. The problem has existed for a long time since the first time taught at the Junior High School for a long time up till now the problem is still categorized as a crucial problem. Therefore, the teacher is motivated to investigate further including how to solve the problems. For instance: problems arise when students produce sounds of the English consonants such as the phoneme /ʃ / in *she* with /s/ in *see,* and /z/ as in *zoo;* differentiation between consonant sounds */f/, /p/, /v/* as in *fish, enough, play, photo, vast;* consonant sounds */θ/* and */ð/*, verbs that end with consonant sounds */t/ and /d/* as in *expect* becomes *expected, need* becomes *needed,* vowel /ə/ and /e/ as in *above, went*; and the others. Those are also caused by lack of understanding basic concept of

English particularly in teaching and learning pronunciation. It is proven that almost all indicators of school-based curriculum do not handle the theories of English pronunciation, the students are only asked to imitate or following the teacher's model without giving basic concept of the English. Thus, the purpose of conducting this research is, to prove whether the direct method is effective in improving students' English pronunciation or not. The hypothesis is "the students who are taught pronunciation using direct method, will get mean score higher than that of those who are not taught using direct method", with the level of significance 0.05 (***H***a: $\mu 1 \geq \mu 2$).

2. Methods

This research design is a true experimental using post-test only. In this design, initial differences between the groups are controlled by the random selection and random. The population of this research is students from one vocational high school in North Sulawesi. Sample of this research was at year 11 students consisting of two classes class A (27 students) and class B (27 students). The instrument used in collecting data is tests. Post-test was given to the experimental group after treatment, and to the control group without treatment. In *reading aloud practice* the students were given a kind of reading text for reading practice without underlining certain words with choosing 30 words of the text that became target of this study. In minimal pairs, they were given 30 words, and both were tested orally; next, the words were chosen to be items for dictation namely 30 words including verbs ended in consonant /t/, /d/, and suffix '-es' and '-ed'. In this study, the obtained data were analyzed by using *t*-test.

3. Results and Discussion

Six meetings were conducted to implement the three techniques in teaching and learning activity. In the first meeting, the experimental group was given a kind of reading text by underlining certain words which are assigned as target of the research. In meeting 2, reviewing the materials in meeting 1 including the emphasis on sounds of English words which are targeted in the study. Then, the use of minimal-pairs was given to know the difference of English sounds in detail either consonant [s, ʃ, z], [θ, ð], [f, p, v] or vowel sound [e] and [ə]; including the verbs ended in consonant *t* and *d* in meeting 3. In meeting 4, reviewing the materials in meeting 2 including emphasis on different sounds of the English words. In the fifth meeting, dictation was given by taking some words that had been taught on reading-aloud and the use of minimal-pairs above including verbs ended in consonant /t/, /d/, and verbs that added by suffix '-es' and '-ed'. Above to be as dictation drill. In meeting 6, reviewing all words that become target of this study from the beginning up to the end. Finally, the post-test was administered.

In this study, a one-tailed test at ∝: .05 (1.943) was planned before the study was conducted. From the result of calculation that $t_{observed}$ is 2.897 at the .05. It falls between 2.132 and 2.352 of the observations. Since the absolute value of $t_{observed}$ (2.897) exceeded the critical value of t (1.943), the null hypothesis is rejected, and the alternative hypothesis is accepted. It could be said that while the difference between means was in the expected direction, there is sufficient evidence to lead us to believe that this difference represents anything but a chance occurrence. This supports the claim that the direct method using three techniques is effective in improving students' English pronunciation.

4. Conclusions

The discussion leads to strongly conclude that the mean of the experimental group (36.99) statistically describes the students' improvement in pronouncing English words in which have been compared by result of the $t_{observed}$ (2.897) exceed $t_{critical}$ (1.943) at level of significance 0.05. This is then support the claim that 'rejects' *Ho* and 'accept' *Ha* which means that there is a significant difference between the mean score of experimental group and control group regarding students' English pronunciation. The students' difficulties in pronouncing English words can be overcome using the three techniques in the direct method namely reading aloud, minimal-pairs, and dictation. The students' errors are minimalized by treatment of the various techniques used. Thus, it could be concluded that direct method is effective in improving vocational high school students' English pronunciation.

Acknowledgments
The authors wish to express their appreciation to the Research Institute of Universitas Negeri Manado which had financed the research activities.

References

[1] Han L N 2011 *A Study on Common Mistakes Made by Vietnamese Beginners of English in Pronouncing and Possible Solutions* (Unpublished thesis. VNU University of Languages & International Studies Post-Graduate Faculty).
[2] Tergujeff E 2012 English Pronunciation Teaching: Four Case Studies from Finland *Journal of Language Teaching and Research* **3** 599-607.
[3] Zhu D 2012 Using Games to Improve Students' Communicative Ability *Journal of Language Teaching and Research* **3** 801-805.
[4] Hasan E M I 2014 Pronunciation Problems: A Case Study of English Language Students at Sudan University of Science and Technology *Journal of English Language and Literature Studies* **4** 31-44.
[5] Larsen D 1986 *Techniques and Principles in Language Teaching* (Oxford: Oxford University Press).
[6] Brown H D 2001 *Principles of Language Learning and Teaching* (San Francisco State University: Prentice Hall Regents Englewood cliffs, NJ 07632).
[7] Smith J P 1989 Children among the Poor *Demography* **26** 235–248.
[8] Brown H D 1994 *Teaching by Principles: An Interactive Approach to Language Pedagogy* (USA: Tina Carver).

IOP Conf. Series: Materials Science and Engineering **306** (2018) 012095 doi:10.1088/1757-899X/306/1/012095

Machine Maintenance Scheduling with Reliability Engineering Method and Maintenance Value Stream Mapping

N Sembiring* and A H Nasution
Departemen Teknik Industri, Fakultas Teknik, Universitas Sumatera Utara, Jalan Almamater Kampus, Medan 20155

*nurhayati4@usu.ac.id

Abstract. Corrective maintenance i.e replacing or repairing the machine component after machine break down always done in a manufacturing company. It causes the production process must be stopped. Production time will decrease due to the maintenance team must replace or repair the damage machine component. This paper proposes a preventive maintenance's schedule for a critical component of a critical machine of an crude palm oil and kernel company due to increase maintenance efficiency. The Reliability Engineering & Maintenance Value Stream Mapping is used as a method and a tool to analize the reliability of the component and reduce the wastage in any process by segregating value added and non value added activities.

1. Introduction

Due to the increasing satisfying customer's need in manufacturing, the company must make continuous improvement to keep the company's product quality requirement. Good quality products and can meet the needs and the desires of consumers is a goal of all corporate. Achieving that, manufacturing companies are focused to optimize their production [1, 2]. The life time of using machinery can be extended through scheduled maintenance activities. The maintenance is used for several actions taken to keep the machine's condition as before or fix it until a condition that is qualified. Pressing the delay of completion of work, increase production efficiency, and prevent the decrease in production volume is one of the benefit of maintenance. It is important for the company to further increase the efficiency of its operations. Through good planning and maintenance, the machine damage can be minimized. By a good maintenance scheduling, machine will always be in ready condition. If machine maintenance is done regularly, it is easier to predict the possibility of future damage. This can be reduce losses due to non-functioning of production equipments.

The high downtime is due to the industry using corrective maintenance [3]. If this method is applied as the maintenance strategy, it will result in the high impact of unscheduled maintenance activities of replacement components. Of course it will cause the machine to stop operating at the time of production process takes place. This leads to a loss of production process time when there is a repair process of machine components [4, 5].

The observed factory maintenance system is still corrective maintenance. Due to that situation, it is important to apply preventive maintenance. Scheduled maintenance planning to perform component replacement so that the machine can be used without any damage during production process [6, 7].

The reliability of the machines and the prevention of machinery from breakdown will assist the smoothness of production. One of definition of system reliability is the capability of the system performing its intended function by utilizing probabilistic expressions, like as the probability that the system will fail in a certain time, the expected period of time to break down and the expected number of failures for a certain period [8].

It must be known that reliability analysis is conducted and then the results are presented and also reported to the leader. System Reliability analysis has purposes that all the aproximation process of the system's probability will be able to achieve its function. Reliability of system's part affected by the main of system reliability [9].

Value Stream Mapping (VSM) as a tool has been widely used in various research and applications in the real world. Some development from VSM is energy/environmental VSM, sustainable VSM, quality VSM and VSM 4.0 [10-13]. Widely application of VSM i.e.: automobile, camshaft, water heater manufacturing, process industry, and coffee industry. VSM also applicated in various support function as well sales process, truck transportation, supply chain of cottonseed industry, electrical manufacturing service, healthcare, project management and new product development [14]. It also used for residential construction [15].

It is applied for global system of observations. In Indonesia, previous research about maintenance VSM had been done. The result of both researches is conduct increasing the maintenance efficiency [16, 17].

In this paper, specifics VSM applications in the field of maintenance is used and there for it called Maintenance Value Stream Mapping. It's focus on repair and maintenance of the critical component of critical machine. The goal of this research is to design preventive maintenance machine system with Reliability Engineering and Maintenance Value Stream Mapping (MVSM). Benefits of this research are the company can eliminated non value activities in maintenance, has a scheduled calendar for critical component that observed and increasing it's maintenance efficiency.

2. Methods

A descriptive research is applied, which aim to represent facts and properties of an object. The result is to give the proposal of replacement schedule of the machine component. The method begins with data collection, i.e. data collected from company documentation (machine downtime and breakdown, the frequency of damage, interval damage).

Data collected from direct observation and interview with the head of maintenance. All of that data is important to applicative the method. Further testing of distribution, parameters, and Mean Time To Failure (MTTF) values, and calculation of maintenance efficiency at MVSM.

VSM contains all non value added and value added activities. The objective of VSM is to eliminate wastes [18]. Value Stream Mapping (VSM) method is very effective to redesign of value streams. As a tool VSM help us to understand product's flow activity. By analizing the difference between the current and the future state , the performance of the system could be achieved [19].

The idea of Mean Maintenance Lead Time (MMLT) is being used for maintenance measurement. One of MMLT defenition is the time between analizing the need for maintenance on a particular component to the actual performance of such maintenance and the repair of the component [20]. MMLT has the maintenance activities into account from an operational level. It is a useful tool to analize the maintenance system and activities. The schema of MMLT is shown in Figure 1.

MMLT is given by the following equation:

$$MMLT = MTTO + MTTR + MTTY \quad (1)$$

Where,
MTTO = Mean time to organize (Time required to distribute tasks to initiate the maintenance repairs)
MTTR = Mean time to repair (Time required to repair the component)
MTTY = Mean time to yield (Time required to yield a good component after repair)

IOP Conf. Series: Materials Science and Engineering **306** (2018) 012095 doi:10.1088/1757-899X/306/1/012095

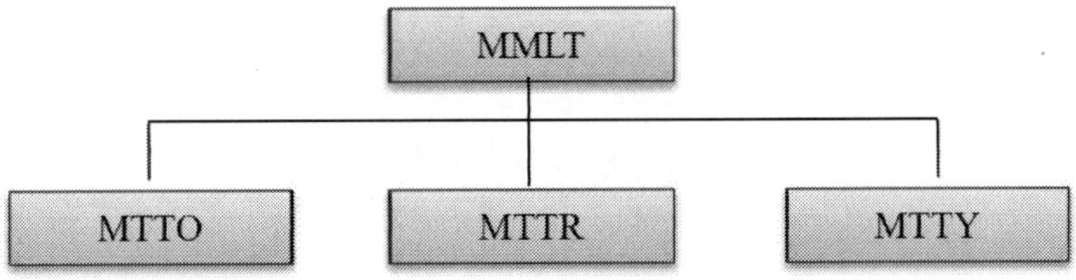

Figure 1. Schema of MMLT.

Within the MMLT schema, the only period component that adds value to the maintenance operations is MTTR, since this is the only period component that contains the actual performance of the maintenance repair task. The other period components MTTO and MTTY are non-value added time. Hence value added time and non value added time are given by,

$$\text{Value added time} = \text{MTTR} \quad (2)$$
$$\text{Non-value added time} = \text{MTTO} + \text{MTTY} \quad (3)$$

Maintenance Efficiency is calculated as the percentage of MMLT that is happen on repairing the component.

$$\% \, Maintenance \, Efficiency = \frac{V \quad A \quad Ti}{M \quad (M \quad M \quad L \quad Ti\ )} \, x \, 100\% \quad (4)$$

3. Results and Discussion

Based on historical data for July 2016 - June 2017 period, in Figure 2 we can see machine's damage frequency. *Screw Press* machine have the highest damage frequency. In Figure 3 we can see Screw Press's component damage frequency. The highest frequency of component damage is *Left Handed Shaft*. So the observed object is *Left Handed Shaft* as a critical component of *Screw Press* machine.

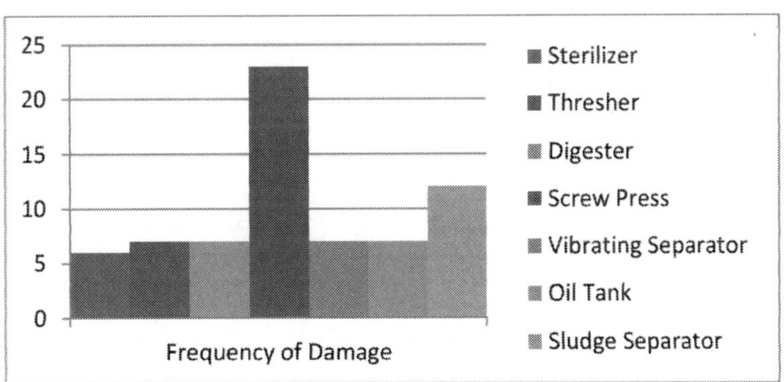

Figure 2. Machine's damage frequency.

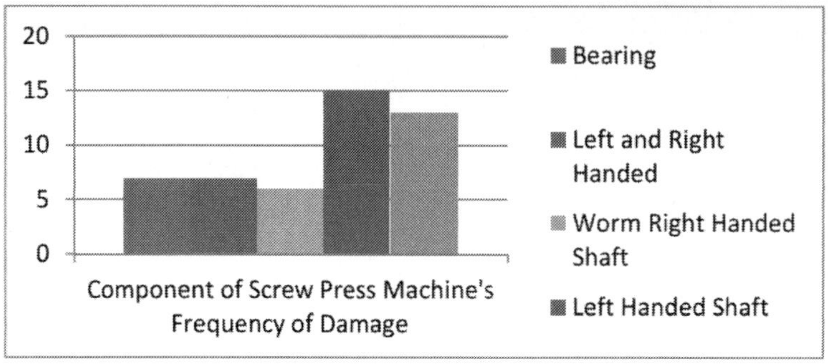

Figure 3. Component of screw press's damage frequency.

3.1. Testing distribution

The first step in the Reliability Engineering Method is testing the distribution of critical machine component. The results of the selected distribution pattern test for *Left Handed Shaft* is Log normal. The preferred distribution pattern is the distribution that gets the largest Index of Fit (Correlation Coefficient) value.

3.2. Calculation of parameters and MTTF machine components

Calculating MTTF (Mean Time To Failure) is used as a parameter for determining the replacement of machine components. The result of MTTF calculation, hence got the interval of change of *Left Handed Shaft* is 64 days. It means that after using that part for 64 days, *Left Handed Shaft* must be replaced [21].

3.3. Reliability calculation on component replacement interval schedule

The calculation of the reliability value of the critical machine component on the proposed replacement schedule is used to determine the reliability value of the machine components when the proposed component replacement schedule is performed [8, 9]. Reliability calculation for *Left Handed Shaft* is 0,2622. It means that reability of *Left Handed Shaft* too small and if not replaced, the *Screw Press* machine will breakdown because it's component broke. Component *Left Handed Shaft* should to be replaced after 64 days of using. Schedule of replacing *Left Handed Shaft* component for August 2017 - July 2018, can be seen in Table 1.

Table 1. Schedule of replacing component *Left Handed Shaft*

No	Schedule of Replacing
1	24th August 2017
2	27th October 2017
3	30th December 2017
4	5th March 2018
5	8th May 2018
6	11st July 2018

3.4. Comparison of activities of Left Handed Shaft for current state map with future state map.

The comparison can be seen in Table 2.

IOP Conf. Series: Materials Science and Engineering **306** (2018) 012095 doi:10.1088/1757-899X/306/1/012095

Table 2. Comparison of activities current state map dan future state map for *Left Handed Shaft*

Current State Map	*Future State Map*
Equipment breakdown	*Equipment breakdown*
Communicated the problem	*Communicated the problem*
Delay, Because no maintenance team	*Indentification of problem*
Indentification of problem	*Indentification of the manpower*
Indentification of the manpower	*Allocated of the manpower*
Delay of no component available	*Preparing work to be done*
Allocated of the manpower	*Maintenance*
Preparing work to be done	*Running to machine after repair*
Maintenance	
Running to machine after repair	

From Table 2 for future state map, non value activities such as delay because no maintenance team and delay of no component available that shown in current state map should be eliminated by allocated of the manpower and preparing stock of components [14, 17, 20].

3.5. Calculation of maintenance efficiency on MVSM

Calculation of maintenance efficiency on MVSM could be done by calculating the value of MMLT based on MTTO, MTTR and MTTY [14, 20]. The comparison of MTTO, MTTR and MTTY value for Current and Future State Map can be seen in Figure 4.

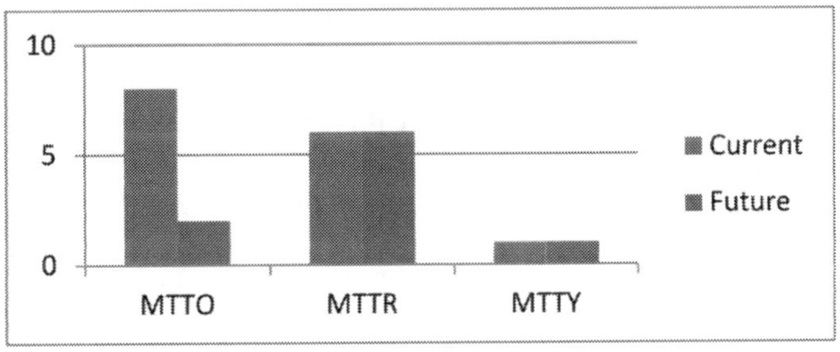

Figure 4. The comparison of MTTO, MTTR and MTTY value for current and future state map.

With this equation:

$$\%\ Maintenance\ Efficiency = \frac{V \quad A \quad Ti}{M \quad (M \quad M \quad L \quad Ti \;)} x\ 100\%$$

The value of % Maintenance Efficiency can be calculated. The comparison Maintenance Efficiency of Current State Map with Future State Map for *Left Handed Shaft* component can be seen in Figure 5.

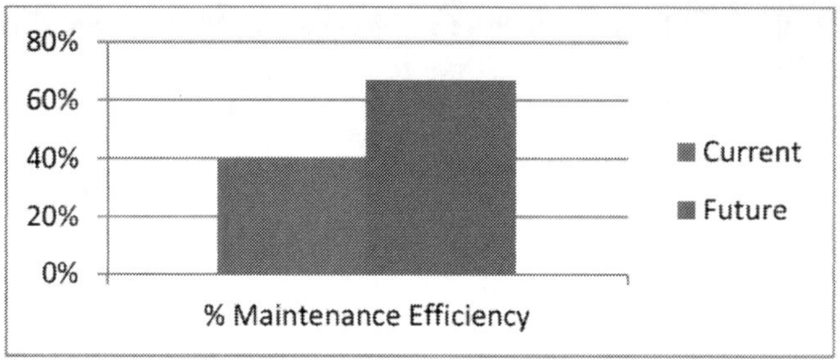

Figure 5. The comparison maintenance efficiency of current state map with future state map for *Left Handed Shaft* component.

4. Conclusions

Mapping activity based on the maintenance activities in terms of the components of MMLT, namely, MTTO, MTTR, and MTTY, Non-Value added time, and Value added time. Maintenance Value Stream Mapping (MVSM) resulted in the percentage of maintenance efficiency for critical component analyzed is increased. This increase is derived from the comparison of current state map application value to development with future state map. The percentage of maintenance efficiency in future state map for *Left Handed Shaft* is 66,66% increased from maintenance efficiency in current state map (40 %). Based on calculation of Mean Time To Failure is 64 days with Reliability value 0.2622 , the schedule of replacing component *Left Handed Shaft* for period August 2017 – July 2018 are 24th August 2017, 27th October 2017, 30th December 2017, 5th March 2018, 8th May 2018 and 11st July 2018. From this schedule, preventive maintenance will be applied in this manufacturing company. The whole step for this critical component *Left Handed Shaft* should be done for all components and machines in this manufacturing company due to apllied preventive maintenance.

Acknowledgments

The authors wish to express sincere gratitude to the Crude Palm Oil (CPO) and Kernel company that gives permission to make this research.

References

[1] S V Sriraam, Sridhar J, and Srivanth V S 2017 Process Time Reduction in an Electrode Manufacturing Plant using Value Stream Mapping Technique *IJERT* **6** p 190–203.

[2] J P Rishi, Ramachandara C G, and T R Srinivas 2016 Keys to Succeed in Implementing Total Preventive Maintenance (TPM) and Lean Strategies *IJMTST* **2** p 25-30.

[3] Kumar S, Raj B, and Shubham S 2017 Study of Total Productive Maintenance & it's Implementation Approach in Steel Manufacturing Industry: A Case Study of Equipment Wise Breakdown Analysis *IRJET* **4** p 608-613.

[4] Haefner B, Alexandra K, Torsten S, and Gisela L 2014 Quality Value Stream Mapping *Procedia CIRP* **17** p 254-259.

[5] Yang Li, Xiaobing Ma, Rui Peng, Qingqing Zhai, and Yu Zhao 2017 A Preventive Maintenance Policy Based On Dependent Two-Stage Deterioration And External Shocks *Reliability Engineering And System Safety* **160** p 201-211.

[6] Wang Zhenyou, Cuntao Xiao, Xianwei Lin, And Yuan-yuan Lu 2017 Single Machine Total Absolute Differences Penalties Minimization Scheduling with a Deteriorating and Resource-Dependent Maintenance Activity *British Computer Society* p 1-11.

[7] Maleki Hamed, and Yang Yiangjie 2017 An Uncertain Programming Model For Preventive Maintenance Scheduling *Emerald insight* **7**(1) p 111-122.

[8] Aven T 2017 Improving the Foundation and Practice of Reliability Engineering *Journal of Risk and Reliability* p 1-11.

[9] Compare, Michele, Michele Bellora, and Enrico Zio 2017 Aggregation of Importance Measures For Decision Making In Reliability Engineering *Journal of Risk and Reliability* **1** p 1-13.

[10] Verma, Neha, and Vinay Sharma 2016 Energy Value Stream Mapping a Tool to develop Green Manufacturing *Procedia Engineering* **149**, 526-534.
[11] Faulkner, William, and Fazleena Badurdeen 2014 Suistainable Value Stream Mapping (Sus-VSM): Methodology to Visualize and Assess Manufacturing Suistainability Performance *Journal of Cleaner Production* p 1-11.
[12] Edtmayr Thomas, Alexander Sunk, and Wilfried Sihn 2016 An Approach to Integrate Parameters and Indicators of Sustainability Management Into Value Stream Mapping *Procedia CIRP* **41** p 289-294.s
[13] Meudt, Tobias, Joachim Metternich and Eberhard Abele 2016 Value Stream Mapping 4.0: Holistic Examination Of Value Stream And Information Logistic In Production *CIRP Annals-Manufacturing Technology* p 1560-1563.
[14] Chaple Anup P, and Narkhede Balkrishma E 2017 Value Stream Mapping in a Discrete Manufacturing: A Case Study *IJSCM* **6**(1) p 55-68.
[15] R Barathwaj, R V Singh and Gunarani G I 2017 Lean Construction: Value Stream Mapping For Residentials Construction *IJCIET* **8** p 1072-1086.
[16] Christian T 2016 *Evaluasi Penjadwalan Perawatan Untuk Meningkatkan Realibility Mixer Seanima* (Sidoarjo: Universitas Muhammadiyah Sidoarjo).
[17] Sania Angelita 2017 *Study Analisa Preventive Maintennce Dengan Metode Realibility Engineering dan MVSM di PT. XYZ* (Medan: Universitas Sumatera Utara).
[18] Rohani, Jafri Mohd, and Seyed Mojib Zahraee 2015 Production Line Analysis Via Value Stream Mapping: A Lean Manufacturing Process Of Color Industry *Procedia Manufacturing* **2** p 6-10.
[19] De Steur, Hans, Joshua Wesana, Mnoj K. Dora, Darian Pearce and Xavier Gellynck 2016 Applying Value Stream Mapping to Reduce Food Losses and Wastes In Supply Chains: A Systematic Review *Waste Management* p 1-10.
[20] Soundararajan Kannan, Yanzhen Li, Naveed Ahmed, and Zeid El-Akkad *Developing A Maintenance Value Stream Map* (Department of Industrial and Information Engineering The University of Tennessee).
[21] Mayatra Mehul, Mr. N.D. Chauhan, Mr. Parthiv Trivedi, Mr. M.N. Qureshi 2016 Implementation of Value Stream Mapping Methodology in Bearing Industry *IJARIIT* **2** p 1-19.

Technical and Sociological Approaches for Curriculum Innovation on Clothing Education Department

N Tristantie

Universitas Negeri Medan, Indonesia.

n1n4tristan@gmail.com

Abstract. Education in the context of social and technical development is defined as the main factor in the learning process which is implied into curriculum. It needs to be anticipated responsively how the goals of Clothing Education Department should be achieved. The sociological and technological through curriculum innovation at Clothing Education Department aims to gain good profile of the professional graduates in the future. By using the literature study, it is found out that sociological development and technological approach are the main foundation for sustainability of Clothing Education Department.

1. Introduction

Clothing Education Department can be observed on how the graduates are employed in the labor market. Relating to sociological studies, Clothing Education Department contributes a large impact to the social and its environment. In fact, the profile of graduates could be learned by part of society as a career being a vocational teacher. Contrast with career opportunity in Indonesian fashion industry, the graduates are still considered as teachers. Indonesia plays an important role in world's fashion industry in order to improve the global garment sector for the future [1]. 2015 Index Export of Batik was at the level of US$ 156 million. It increased by 10 % from the previous year. Ministry of industry, Saleh Husein, said that the main market of Batik Export are Japan, US and Europe [2]. This is a fact that the development needs of society on clothes are predicted to be increased with the condition by considering the creativity and innovation in creating a product.

Globalization needs accelerations in several things, they are information system, and cultural, social and technology. It needs to be anticipated by Clothing Education Department as well. As the impact of globalization, there will be a lot of changing in the life of society and it also happens in the fashion learning process. The development of technology and science are regarded as triggers and has privilege to determine the human future. In fact, technology is formulated by social resources and is given new meaning by the society. The enthusiasm of product by users is produced and constructs product orientation that contents variety of quality and quantity [3]. Fashion product is expected to cover the technology needs to adapt the social developments related to life style such as clothes that could make the users slim, fire resistant for fire brigade. The questions are how a technology product can be adapted and anticipated by Clothing Education Department and how to prepare the learners who can anticipate the technology and social changing related to the market. As it was done by China and India which has the largest textile industry surviving the economic system. The life of society proceeds becoming industrial society as the technology is being used. Clothing Education Department should improve recent curriculum to create the more qualified fresh graduates in adapting knowledge implied in the global economy.

2. Methods

2.1. Social development and clothing industry needs

Globalization is indicated by how people can access fashion and adopt it entire the world. The technological modernism brings a spectacular impact on clothing. As the natural fabrics are substituted by synthetic fiber, it causes a new understanding what 'clothes' is called. In the social context, clothe could group people based on different status symbol and could manipulate fashion as general consumption. It is the capitalism propaganda attributed by mass production which Jean Baudrillard calls as Consume Culture. The rapid cycle of fashion is one of pillar of the consumer culture. The society is driven to keep on consuming for being up date. The rapid changing of trends symbolizes consumer society. Fashion is becoming ideology which has a great power to colonize group of people to consume fashion and things for showing up as a fashionable identity.

The needs of clothes will keep on increasing adhere the rising consumption. One of which is indicated by the needs of clothing in Indonesia 7.5 kg per year, including batik [4]. The indication of high consume can be seen through the movement of the fashion industry with expensive branded produced by big famous company for mid-level market orientation with low price. Fashion cycle dynamic causes well known fashion branded can be bought by society even mid - low level which is called Fast Fashion. Big industries grow well employ Indonesian cheap laborers. Sociologically, it causes social problems. Although unemployment is under controlled, but the wages are paid very unfit to meet the standard of welfare. Besides, these conditions bring impacts to the local creative industry [2]. The question is how to prepare young professional through vocational education to address these social issues.

2.2. Clothing education department in curriculum context

Clothing education Department is under management of Economic Welfare Department as a part of Faculty of Engineering. Historically, Clothing Education Department discussed about home economy subject. This term is generally reversed to how optimize the family skill to cover the economic problems. Applying the managing concept which stressing on the skill based on economic family economic source. Those skills are including fashion design, culinary, cosmetology and tourism. The concept of home economy is resolved by emphasizing on Clothing Education Department program which goes along with the administrative order. The graduates have area of occupation which will be employed as: 1) Vocational High School Teacher, 2). Instructors at Informal Institutions, 3). Fashion Business Management

The description of occupation area of the graduates is still obviously struggling in the house hold industry. Meanwhile, the concept of link and match between the needs of education and industry to solve the problem of job opportunity needs to be reviewed. The presence of various of Clothing industry makes Fashion Education Department review kinds of proper industry character made as a reference.

Sociological approach is applied to meet these problems supposing by examining the social needs and community response about industry and technology. The questions addressed are linked to the quality of graduates, competency of educators, and government role to excel them. Numbers of factor will appear to be the fundamental of basic thinking to proceed an action plan for the next step. However, the sociological approach does not seem to be fully implemented to overcome the problems that exists in the role and task of education. The traditional cultural approach still inherent in Clothing education which is tied to the emergence history of this major. Conservative thinking causes this institution left behind to meet the needs of global. Clothing Education Department should have a quick response by examining the curriculum.

Social changes occurs in a response of diverse of changing in the social and non-social environment. Education could initiate social changes by bringing them in attitudes and outlook. Curriculum of Clothing Education Department should refer to the social development and the acceleration of technology in the field of fashion industry [5]. If social turmoil follows the accompanying impacts must be seen based on facts and reviewed into curriculum, will provide some solutions and continue to the actual learning implementation. This effort makes Clothing Education Department will graduate the professional generation.

2.3. Sociological education approach

It is time for education to be seen from various disciplines, especially from Sociology. By using this approach, a fundamental answer will be driven because sociological approach supplies significant contribution to education. The contributions include philosophical, theoretical and methodological ground. Theoretically, social science has contributed theoretical critics known as critical pedagogy. Methodogically, science education adopts methods of qualitative and quantitative social science.

Education is an absolute rule which is essentially able to deliver the thinking development based on in depth study of research, especially viewed by human as users. Sociological studies in Indonesia [6], is an essential approach for Indonesian future education development. This is an understanding how sociological of developmentalism perspective sees technology as a tool that can briefly address education needs yet it is different when it is seen from critical perspective that provides solutions to technical problems. In perspective critics context, technology is analyzed from various functional aspects and consequences which contributed useful values for social life.

Sociology for Clothing Education Department is expected to reach essential aspect of technology loaded with critical science. Fashion Education Department employs a traditional curriculum. The theoretical perspective critics is hoped to imply themselves understand the ethical system and create it as a new culture for institution and community. Considering these conditions, perspective developmentalism will not be enough as it must be urgently accompanied by the study of critical perspective in order to improve the quality of education and formation of new culture by abandoning the traditional cultural.

3. Results and Discussion

3.1. Clothing education department, social change and technology anticipation

Practically Sociology constructs a new culture. Learning activities are conducted primarily to preserve what already exists in society and culture, which is owned as a cultural heritage [6]. Education science interprets concept of dynamism, it involves many factors economic, social, cultural and technological. Technology can not be separated from the sociologically concept as an approach that is used to improve the education quality.

The application and using of technology is wisely required by Clothing Education Department, actually students play the main role and are required to respond the development of both local communities and the world in the context of global economy [3]. If education is able to make assumptions about the future in which there are economic vision and encompasses global knowledge, it will be easy for institutions to prepare students for a professional future. In this case, Clothing Education Department can act as a resource for students to build an alternative future that offers real hope and a sustainable way of life with clear career.

However, it does not happen naturally, it requires maximum effort to achieve that goal. By using the sociological approach, the character of the community with the premise of culture, will be known quickly by identifying the symptoms that occur in the loop problems. Culture that has been patterned or not been shaped yet will be found, though it is still a symptom, but it will be immediately anticipated by providing alternative approaches. For example, how the needs of fashion and textiles industry society deals with professional work, how the desire of society as users on clothes, how is effective marketing sales and other forms of predictions will have implications on dressmaking skills learning that must be given to the students. Educational institutions are under the control of different cultural groups which reflect the values of the groups that support and control education [5]. Clothing Education Department can take over the role.

3.2. The curriculum for future

Social changes occurs along with human needs, including economic issue accompanied by the central world issues such as energy scarcity, and climate and global warming. Meanwhile, other issues such as international financial and competition culture that has power to change. It is impossible those conditions will move back. That is why Clothing Education Department needs to think about its future.

Those factors have influence on the fashion industry and have implication on the learning process. It means these conditions must be highly responsed and made as a basic concept to design new curriculum. The future of education arises from educational instrument concept to response the economic problem likely what kind of business competition will be faced in the future?, what skills are needed in the workplace? How the fashion program can guarantee the competitive advantage for the nation technically in the socio context. Education must involves society to solve these problems. Clothing Education Department should act and play role as sources for students to build their future, real expectation and sustainable way of life by organizing campus as a public space for students, lectures and society [3]. The Components participate in the future conversation and sustainable development. The activities are all carried out in the context of formal and informal. The components of the community involve various groups such as industry, economy, parents who share the commitment for future development.

Campus is positioned as sustainable economic development machine that ultimately has impact on local area, by doing work practice and make a mini project research to promote economic activity. Students can intervene and shape the future of their own work especially in making career decision based on strong labor market information. Unlikely when they are adult, they use work experience to get professional goals. In this case people begin to consider about job opportunity and will make this opportunity as the basis for a possible career when they are getting older. For an apprentice Clothing Design Students, the work place is generally used to learn the works are skilled, choose what area of works can satisfy them in the long term. They learn to find the good and bad things of them, realization that often leads to failure or disappointment [7].

3.3. Clothing education to answer the future

Clothing Education is one of vocational educations which in its application has multifunction role that if well implemented will contribute greatly to the nation achievement development goal. These functions include (1) socialization, namely, transmission and concretization of economic values, solidarity, religion, values, solidarity, religion, art and services; (2) social control that is behavioral control with the norms they are cooperation, orderliness, hygiene, discipline, honestly and openness; (3) selection and allocation, preparing, selecting and placing prospective workers accordance with labor market demand; (4) assimilation and cultural conservation that is, the intercultural absorption society and the maintenance of local culture; (5) promoting change for the sake of improvement. Department Vocational education does not just educate and train existing skills, but it also should serve as a driving change. Vocational education serves as a process of acculturation or adaptation to the changes and enculturation or carrier change for the community. Therefore vocational education is not only adaptive but also must be anticipatory [8]

Achievement of Technology in the discussion of curriculum of Clothing Education Department includes an understanding the needs of industrial society. Incorporating current issues concerns important technology to do as an awareness attempt to the base technology in fashion learning. Departing from the issue, it is then used as the concept served as the vision and mission and as a 'vehicle' to get to the future goal for Pendidikan Tata Busana. Hornby defines sociology as a scientific study of the nature, social development and social behavior. Sociological overview of vocational technical education is the public opinion of the vocational technical education. The importance of vocational education in sociological point of view can not be undermined. Unemployment, juvenile delinquency, adult crimes, unstable and unsatisfactory home have been associated with lack of vocational education and training [7].

4. Conclusions

Sociological context sees the aspects of technology and knowledge should be utilized for the best interests of the development-oriented economic growth Clothing Education Department should not ideally focus only on traditional emphasis on learning, it must open its scientific insights to global challenges. The role of institutions in improving their quality can be nurtured through innovative curriculum which is based on the social and technological change. Preparation of Clothing Education Department through innovated curriculum based on the real goal will give students the opportunity to engage a challenging experience. Students can see how the relationships between subjects by viewing the world through such an important dimension of globalization, technology and sustainability that starts from creating a sense of learning.

References

[1] Noor F 2004 *Pemberdayaan Pendidikan Teknik Busana di Perguruan Tinggi Untuk Pengembangan Industri Garmen di Pasar Global* Available at: staff.uny.ac.id/.../pemberdayaan%20pendidik... Accessed on 20 April 2015.
[2] Apriliani Tri 2016 Hubungan Kerja dalam Industri Fast Fashion: Analisis isi terhadap fenomena eksploitasi (studi kasus film the true cost dan nike sweatshops) *Informasi Kajian Ilmu Komunikasi* **46**(1).
[3] Facer K 2011 *Learning Futures.Education, Technology and Social Change* First edition (Routledge).
[4] Faerm S 2012 A Future Pedagogy: The Evolution of Fashion Design Education *International Journal of Humanities and Social Science* **2**(23).
[5] Patil M, Namita P 2012 Role Of Education In Social Change *International Educational E-Journal* **2**(1).
[6] Wahyono, S Bayu 2007 *Mengkaji Fenomena Pendidikan Persekolahan dari Perspektif Sosiologi Kritis* Available at: core.ac.uk/download/pdf/11059180.pdf. Accessed: on 3 juni 2015.
[7] Kennedy O O 2011 Philosophical and Sociological Overview of Vocational Technical Education in Nigeria *International Journal of Academic Research in Business and Social Sciences* **1**.
[8] Kisfaludy M 2008 Fashion and Innovation *Acta Polytechnica Hungarica* **5**(3) pp 55 – 59.

Analysis of ICT Literacy Competence among Vocational High School Teachers

Nurhabibah*, A Setiawan, H Yanti, Y Z Miraj and Yannuar

Technology and Vocational Education, Universitas Pendidikan Indonesia, Jl. Dr. Setiabudhi 207 Bandung 40154 Indonesia

*habibahn09@gmail.com

Abstract. To realize learning quality in Vocational Secondary School, and to achieve educational goal, teacher competence is needed. The application of ICT Literacy in vocational secondary school is intended to upgrade teachers' knowledge, skill and competence in ICT. This paper is aimed to describe the process of teachers' competence adaptation to ICT integrity in learning in Vocational Secondary School. This study use descriptive method with literature study and documentation technique. The source in this study is research journal and research report book. The study result showed that teachers lack of self-confident in using ICT, and gender factor influence ICT integration in which the level of ICT literacy in male is higher than female. The group of young teachers aged 21-40 have higher level of ICT literacy compared with the older group. Demographic factor in ICT literacy competence are gender, education level and age. This study suggest that teachers enhance the ability in ICT literacy competence, increase their knowledge and knowledge creation in each aspect of ICT literacy competence.

1. Introduction

Quality learning reflect teachers and school preparation to create learning environment in enhancing students achievement [1]. It require teachers' ability and competence. Until now, learning still very formal and teacher-centered. Learning system now is demanded to use technology application and using ICT to overcome the difficulty in learning material [2]. Instructional technology such as this has many advantage and benefit, but the use of facility is still less optimal and need implementation strategy to enhance effectiveness and efficiency [3]. There are few literatures which attentive to instructional technology particularly on how teachers can apply technology in learning process [4].

This study will discuss how teacher obstacles in applying ICT literacy in learning process which are divided into two categories, namely external and internal obstacle. External obstacle consist of) lack of access toward resource, b) lack of time, c) lack of effective training and d) technical problem. In the other side, internal obstacle consist of: a) Lack of self-confident, b) Resistance toward change and negative attitude, c) There is no perception about advantage. These two kind of obstacles play big role in forming teacher competence and finally will influence students' learning achievement [5, 6].

Teachers competence influence students' learning achievement particularly in vocational secondary school. To achieve education goal, it is regulated in Standard of Educational National that educators should possess academic qualification and competence as learning agent, physically and mentally health in ability to realize national education goal. The implication of teachers competences which need to be enhanced namely pedagogical, personality, social and professional competence [7]. By using efficient ICT facility, government through Directorate of Vocational Development had developed Center of Vocational Education Service for vocational secondary school teachers. The information center give instrument for teachers to compare their competence in ICT learning. Until 2010, there are 431 vocational schools which had been appointed as center of ICT service distributed in all city regencies in Indonesia. Because each school can establish Center of ICT Vocational Service, it is expected that teachers can develop ICT based-learning and empower students. However, there are issues of teacher competence in achieving technology mastery such as their experience and knowledge of ICT facility, websites, and teaching task. Five factors which influence teacher competence in mastering technology are (a) teacher skill in using ICT actually in learning, (b) ICT resource, (c) teachers' attitude toward ICT usage in learning, (d) technical support and necessity, and (e) staff development in using ICT. Lack of teacher competence in using ICT integration in classroom can result in students' distrust to teachers [8, 9]. Integrity of ICT facility depend on teacher ability to arrange learning environment from traditional pedagogic to non traditional [10, 11]. Teacher ability has big influence to computer usage, competence in class management, and finally to pedagogic skill.

Information Communication and Technology (ICT) is technology to process data, including to process, obtain, arrange, store, and manipulate data with various ways to produce quality information. ICT had been grown rapidly in various fields particularly education [12-14]. ICT also has broad experience comprising everything related to process, tool, manipulation, and management of information. ICT also media or tool to transfer data, to acquire data and information, and can be used for communication tool both one-way and two-way communication [15]. ICT has various aspects involving technology, technique and management technique used in controlling and processing information and its usage, computer and human and human relation, and everything related to social, economic and culture [16]. The kind of ICT which commonly used are Computer, Laptop, printer, LCD projector, Internet. ICT consist of Hardware and Software to operate hardware in the form of Operating System (OS), application or content.

ICT literacy is basic competence for teachers in utilizing ICT to prepare students in order to be able to master new technology as equipment for them in developing themselves as long life learner [17]. This stage focus on technology literacy development in teachers to integrate ICT tool.

Teachers' competence and understanding need skill, behavior and professional value [18]. It means that teachers' competence can comprise fact and concept repetition until advance motor skill toward learning behavior and professional value [19, 20], that competence is one's unique characteristic and the way of behave and thinking in every situation which is effected on their ability to use tool and facility in workplace. From this opinion, competence refer to one's performance in a job which can be seen from thinking, attitude and behavior.

The enhancement of teachers' competence is a necessity and comprehensive training is needed for each teacher. To develop skill and technology, learning is needed [15]. Teacher can use learning facility to enhance competence test, online teacher education and online performance assessment. It means that teachers should learn in using ICT [8]. Besides, there are three important things which should be thought again concerning education modernization: (a) how we learn (how people learn); (b) what we learn (what people learn); and (c) when and how we learn (how and when people learn). By seeing the answer to three questions, and the potential of ICT which can be exploited as had been explained before, the role of ICT in nation education modernization can be formulated.

This study use literature study. The study result concluded that teachers understanding toward ICT literacy competence form knowledge and skill in teaching [11]. The mastery of ICT literacy competence started from planning preparation, learning presentation, evaluation and analysis of evaluation result and means to search and download learning source [15, 21]. Therefore, teachers in all levels should be ready to continuously learn in ICT to fulfill the need. The earlier study had proposed various enhancement in teachers' ICT literacy competence which should be suited with adequate facilities. However, learning quality enhancement through ICT sometimes is not synergic and need integrated vision to develop competence. In vocational, education and training (VET), ICT integrity is not only an option but also a need to make the educational process more interesting. ICT research findings are not only a tool for creating lesson materials, but also as a means of conveying, cultivating, and discussing even though they are limited [22]. ICTs play an important role to assist teachers in teaching. Change is essential to VET programs in support of labor development [23]. ICT in education has been identified as the top trends and issues in VET [24]. There are significant differences in ICT and computer usage scores by gender and age [25].

2. Methods

This study use descriptive survey research methodology. Thirty teachers who are selected randomly from Public and Private Vocational Secondary School had been taken as sample. Research instrument consist of test and non-test mode. Test instrument is ICT basic competence test whereas non test instrument is questionnaire including ICT literacy questionnaire, observation, and interview with Vocational Secondary School teachers. Based on table 1 about sample of study.

Table 1. Sample of study.

No	Name of SMK	Amount
1	SMKN 1 Pekanbaru	3
2	SMKN 2 Pekanbaru	3
3	SMKN 3 Pekanbaru	3
4	SMKN 4 Pekanbaru	3
5	SMKN 5 Pekanbaru	3
6	SMKN 6 Pekanbaru	3
7	SMKN 7 Pekanbaru	3
8	SMKS Muhammadiyah	3
9	SMKS Taruna	3
10	SMKS Ibnutamiyah	3
11	SMKS Hasanah	3

2.1. *Characteristic of respondent studied in research*

Data is collected from Vocational Secondary School teachers. They fill questionnaire which is distributed and explain their demography. Next, they collect data to obtain description of respondent characteristic such as gender, age and education level. Respondents are 30 persons who work as teachers in Vocational Secondary School in Pekanbaru as summarized in table 2.

Table 2. Respondents characteristic.

Respondents	**Total**	
	Person	%
Respondents by gender		
• Men	17	57%
• Women	13	43%
Respondents by age		
• 25-30 years old	5	17%
• 30-35 years old	12	40%
• 35-40 years old	6	20%
• 40-45 years old	5	17%
• 45-50 years old	2	6%
Respondents based on education level		
• D3	4	13%
• S1	17	57%
• S2	6	20%
• S3	3	10%

Source: interview results processed author.

3. Results and Discussion

In this study, it is found level of ICT literacy competence among teachers based on ICT literacy questionnaire.

3.1. Gender and ICT implementation in Vocational Secondary School in Pekanbaru

Analysis of questionnaire answer of respondents showed that gender influence teacher ability in ICT integration. It is predicted that male teachers have higher percentage in ICT integration in class learning by 59% whereas the percentage of female teachers is 41%, therefore male and female is different biologically. This difference certainly has perspective and treatment difference toward male and female.

Gender also influence emotional development and intellectual capacity in which female generally is better in memory and male is better in logical thinking [26, 27]. Female is more interested in daily life problem which is practical, whereas male is interested in abstract aspect.

Regarding the issue of gender and ICT integration in class, gender is factor which very influence the use of ICT integration in class, that gender difference will be effected on teacher competence and finally on teacher performance [28, 29]. Male teachers will be more easy to use ICT than female teachers in ICT integration in learning process [30]. In this case, female is more interested in another issue so they don't have opportunity to use ICT facility [31]. The finding of study showed that there is gender gap in which female teacher is differ in frequency of using ICT because historically, computation is generally thought as masculine activity. The tendency of male in dominating mathematics and computation than female in computer industry shows gender gap in technology. This argument is the same with our analysis that male teachers use more computer than their counterpart.

3.2. The influence of age in ICT implementation in Vocational Secondary School in Pekanbaru

Analysis of questionnaire answer shows that teacher competence also influenced by age. It means that ICT application is effected on respondents aged 25-30 years old. Our observation shows that the percentage of teachers who use ICT only reach 35%. Only 43% of teachers aged 30-35 who use ICT compared with teachers aged 40-4 years old.

It is supported by age influence on ICT integration in class by reporting that young teacher utilize ICT facility more than older teacher [32]. Based on study, young teachers in the range of 21-40 years old more often use ICT facility than another age group. It is confirmed by study result in Philipina which emphasize that young teachers use ICT more than older teachers [33, 34].

3.3. *Implementation of education and ICT in Vocational Secondary School in Pekanbaru*

Analysis of questionnaire answer viewed from teachers competence is influenced by education background particularly in ICT integration in learning process. The result of observation showed that ICT application among respondents in education level of D3 has percentage of 17% in ICT usage. This figure is increase if education level of respondents is higher in which S1 is 21%, whereas S2 is 32% which is higher compared with S3.

Education level has strongest influence to ICT integration because most teachers in school use ICT to work [35, 36]. It means that teachers with higher education level tend to use ICT more often because they probably possess more skill. Formal education also form new mindset and work style among teachers, because they has higher burden and pressure to finish their job, and enhance skill, knowledge and mastery in ICT, with academic discipline as demographic factor to determine teachers' adoption and use of ICT integration in learning [37, 38]. The result is there is significant difference among teachers' science disciplines particularly technology science, social and humanity science toward ICT integration.

4. Conclusions

Accessing various ICT resource, that access describe to what extent level of ICT literacy competence among vocational school teachers. Besides, the distance of location will determine whether a teacher will integrate ICT in learning process. Teachers' lack of self-confidence in using ICT facility in learning process is because of teacher demographic. It is evidenced from analysis result which showed that the percentage of ICT literacy among teachers in learning process is higher for male teachers than their counterpart. Besides, age also influence teachers in using ICT particularly for them in the age range of 21-40 years old who have ICT competence higher than another age group. Age and education level should become main attention of teachers and school administrator in order that ICT implementation can be more optimal in learning process. Based on discussion result about analysis of ICT literacy competence among teachers, it can be concluded that ICT literacy competence among teachers had reached good category, but they need to increase their knowledge and knowledge creation in ICT integration in learning process. Instrument in ICT literacy competence need to be developed in order to develop instrument which is empirically can measure ICT literacy competence among teachers.

References

[1] Barr R B, and Tagg J 1995 From teaching to learning—A new paradigm for undergraduate education *Change: The magazine of higher learning* **27**(6) pp 12-26.

[2] Jung I 2005 TIK-pedagogy integration in teacher training: Application cases worldwide *Educational Technology & Society* **8**(2) pp 94-101.

[3] Koschmann T D, Myers A C, Feltovich P J, and Barrows H S 1994 Using technology to assist in realizing effective learning and instruction: A principled approach to the use of computers in collaborative learning *The journal of the learning sciences* **3**(3) pp 227-264.

[4] Zhao Y, Pugh K, Sheldon S, and Byers J L 2002 Conditions for classroom technology innovations *Teachers college record* **104**(3) pp 482-515.

[5] Khan M, Hossain S, Hasan M, and Clement C K 2012 Barriers to the introduction of TIK into education in developing countries: The example of Bangladesh *Online Submission* **5**(2) pp 61-80.

[6] Davis H A 2003 Conceptualizing the role and influence of student-teacher relationships on children's social and cognitive development *Educational psychologist* **38**(4) pp 207-234.

[7] Sercu L 2002 Autonomous learning and the acquisition of intercultural communicative competence: Some implications for course development *Language culture and curriculum* **15**(1) pp 61-74.

[8] Levin T, and Wadmany R 2006 Teachers' beliefs and practices in technology-based classrooms: A developmental view *Journal of Research on Technology in Education* **39**(2) pp 157-181.

[9] Townsend T, and Bates R 2007 Teacher education in a new millennium: Pressures and possibilities *Handbook of teacher education* pp 3-22.

[10] Maker C J, Jo S, and Muammar O M 2008 Development of creativity: The influence of varying levels of implementation of the discover curriculum model, a non-traditional pedagogical approach *Learning and Individual Differences* **18**(4) pp 402-417.

[11] Bingimlas K A 2009 Barriers to the successful integration of ICT in teaching and learning environments: A review of the literature *Eurasia Journal of Mathematics, Science & Technology Education* **5**(3).

[12] Hennessy S, Ruthven K, and Brindley S 2005 Teacher perspectives on integrating ICT into subject teaching: commitment, constraints, caution, and change *Journal of curriculum studies* **37**(2) pp 155-192.

[13] Woodrow J E 1992 The influence of programming training on the computer literacy and attitudes of preservice teachers *Journal of research on Computing in Education* **25**(2) pp 200-219.

[14] Heeks R 2010 Do information and communication technologies (ICTs) contribute to development? *Journal of International Development* **22**(5) pp 625-640.

[15] Weippert A, Kajewski S L, and Tilley P A 2002 Internet-based information and communication systems on remote construction projects: a case study analysis *Construction Innovation* **2**(2)

pp 103-116.
[16] Kalay Y E 2006 The impact of information technology on design methods, products and practices *Design studies* **27**(3) pp 357-380.
[17] Fouad N A, Grus C L, Hatcher R L, Kaslow N J, Hutchings P S, Madson M B, ... and Crossman R E 2009 Competency benchmarks: A model for understanding and measuring competence in professional psychology across training levels *Training and Education in Professional Psychology* **3**(4S) pp S5.
[18] Winterton J, Delamare-Le Deist F, and Stringfellow E 2006 Typology of knowledge, skills and competences: clarification of the concept and prototype (Luxembourg: Office for Official Publications of the European Communities).
[19] Lawton M P 1982 Environmental press, and the adaptation of older people. In M P Lawton, P G Windley, and T O Byerts (Eds.) Aging and the environment: Theoretical approaches *Competence* **7** pp. 33-59.
[20] UNESCO Office in Bangkok: ICT in ducation. http://www.unescobkk.org/education/ict/
[21] Wang Q, and Woo H L 2007 Systematic planning for ICT integration in topic learning *Educational technology & society* **10**(1) pp 148-156.
[22] Paryono and Quito B G 2010 *Meta-Analisysof ICT Integration in Vocational and Technical Educatio in Southeast Asia*.
[23] Buntat Y, Saud M S, Dahar A, et al. 2010 *Computer Technologi Application and Vocational Education: A review of Literatur and Risearch European Journal of social sciences*.
[24] Omar M S H and Paryono 2008 *Current trends and issues in VTET : Seameo Vectech's response*.
[25] Summak M S and Samancioglu M 2011 Assessment of Technology Integration Vocational Education and Training School *International Journal of Education and Develoment using Information and Comuniaction Technology*.
[26] Miller N B, Silvermany L K, and Falk R F 1995 Emotional development, intellectual ability, and gender *Talents and Gifts* **18**(1) pp 20-38.
[27] Gallagher A M, and De Lisi R 1994 Gender differences in Scholastic Aptitude Test: Mathematics problem solving among high-ability students *Journal of Educational Psychology* **86**(2) pp 204.
[28] Fu J S 2013 ICT in education: A critical literature review and its implications *International Journal of Education and Development using Information and Communication Technology* **9**(1) pp 112.
[29] Nicaise V, Cogérino G, Bois J, and Amorose A J 2006 Students' perceptions of teacher feedback and physical competence in physical education classes: Gender effects *Journal of Teaching in Physical Education* **25**(1) pp 36-57.
[30] Hammond M, Reynolds L, and Ingram J 2011 How and why do student teachers use ICT? *Journal of Computer Assisted Learning* **27**(3) pp 191-203.
[31] Van Dijk J, and Hacker K 2003 The digital divide as a complex and dynamic phenomenon *The information society* **19**(4) pp 315-326.
[32] Guo R X, Dobson T, and Petrina S 2008 Digital natives, digital immigrants: An analysis of age and ICT competency in teacher education *Journal of educational computing research* **38**(3) pp 235-254.
[33] Prensky M 2005 What can you learn from a cell phone? Almost anything *Innovate* **1**(5)
[34] Passey D, Rogers C, Machell J, McHugh G, and Allaway D 2004 *The motivational effect of ICT on pupils* (Department of Educational Research).
[35] Afshari M, Bakar K A, Luan W S, Samah B A, and Fooi F S 2009 Factors Affecting Teachers' Use of Information and Communication Technology *Online Submission* **2**(1) pp 77-104.
[36] Agyei D D, and Voogt J 2011 ICT use in the teaching of mathematics: Implications for professional development of pre-service teachers in Ghana *Education and Information Technologies* **16**(4) pp 423-439.

[37] Kurga S J 2014 The influence of teachers" age, gender and level of training on attitudes towards the use of integrated e-learning approach to the teaching and learning of Business Studies in Kenyan secondary Schools *Journal of Emerging Trends in Educational research and policy studies (JETERAPS)* **5**(2) pp 190-198.
[38] Aramide K A, Ladipo S O, and Adebayo I 2015 Demographic Variables And Ict Access As Predictors Of Information Communication Technologies' Usage Among Science Teachers In Federal Unity Schools In Nigeria Library *Philosophy and Practice* **1**.

Mobile-Based Dictionary of Information and Communication Technology

O E S Liando, A Mewengkang*, D Kaseger, F I Sangkop, V P Rantung and G C Rorimpandey
Universitas Negeri Manado, Tondano 95618, Sulawesi Utara, Indonesia

*mewengkangalfrina@unima.ac.id

Abstract. This study aims to design and build mobile-based dictionary of information and communication technology applications to provide access to information in the form of glossary of terms in the context of information and communication technologies. Applications built in this study using the Android platform, with SQLite database model. This research uses prototype model development method which covers the stages of communication, Quick Plan, Quick Design Modeling, Construction of Prototype, Deployment Delivery & Feedback, and Full System Transformation. The design of this application is designed in such a way as to facilitate the user in the process of learning and understanding the new terms or vocabularies encountered in the world of information and communication technology. Mobile-based dictionary of Information And Communication Technology applications that have been built can be an alternative to learning literature. In its simplest form, this application is able to meet the need for a comprehensive and accurate dictionary of Information And Communication Technology function.

1. Introduction

For many students, there is often a difficulty in learning when dealing with new vocabulary or terms they have recently encountered in Information and Communication Technology (ICT) subjects [1]. One attempt to overcome these difficulties, is to build an application contained the insights and explanations of various terms in the world of information technology [2]. Which can be run and accessed online and offline[3]. It is expected to address the need for more flexible and easy-to-use knowledge access, meaning that information must be available at all times and at any time in the absence of internet connectivity [4].

Based on research conducted by Nielsen, in 2011, 91% of the population owns mobile phones, 38% of them are smartphones, and Nielsen is expected to grow 50% from the previous year. Furthermore, in the results of research Nielsen On Device Meter (ODM) in February 2014, noted that smartphone users in Indonesia spend an average of 140 minutes per day to use their smartphones. While the results of research from StatCounter in Indonesia during 2013 to 2014, found that 60.79% of them use smartphones with android system [5]. The youngsters became one of the most popular segments of android-based smartphones, in addition to having the flexibility of the use of applications, also became the basis of various smartphone operating systems with specifications, as well as varying prices. Currently in the smartphone market, we can easily find android-based smartphones from the price of one million dollars to a dozen million rupiah. Although users at the youngsters level are very high, not so with the level of understanding of vocabulary or terms in the world of information technology.

To solve the problems, several researchers have suggested the following strategies: dictionary learning based reconstruction algorithm for DCVS [6], mobile learning [7] and user-experience design considerations [8].

The Dictionary Of Information And Communication Technology database targeted by this research, defined as a collection of integrated information, is organized and stored in a way that facilitates retrieval. As information providers, database design needs to be engineered in such a way as to avoid data repetition [9]. So that the role as a source of information and knowledge can be achieved well. Furthermore this study aims to design and build mobile-based Dictionary Of Information And Communication Technology applications to provide access to information in the form of glossary of terms in the context of information and communication technologies.

2. Methods

This research is designed based on prototyping software development methods [10]. According to Howard, prototyping is one approach in software engineering that directly demonstrates how a software or software component will work in its environment before the actual construction stage is done [11]. This study uses a reuseable prototype model that makes it easy to transform prototype into a fully running application without removing the prototype used for testing / evaluation. Furthermore the device used for the development of this application include: (1) Windows 7 Ultimate 32-bit (operating system), (2) ECLIPSE Juno Release 4.2.0, (3) Android SQLite (relational database management system), (4) JAVA (programming language), (5) Android Virtual Device Google Nexsus S (AVD as a medium for running applications that have been created).

3. Results and Discussion

The results of this study will be discussed based on the stages development of dictionary of Information And Communication Technology applicaton. The prototyping stages used in this study are described as follows: (1) Communication, (2) Quick Plan, (3) Modeling Quick Design, (4) Construction of Prototype, (5) Deployment Delivery & Feedback (6) Full System Transformation

At communication stage, the needs of the system to be built are categorized, by conducting direct data retrieval, ie listening to complaints and requests from customers or prospective users, this process can also be referred to as identification of needs, or initial requirements.

Quick Plan stage is done by examining the results of initial requirements, so that it can define what input variables will be used, along with the output that will be provided system, (3) Modeling Quick Design . As shown as Figure 1, this stage includes the creation of mockup or wireframe which is the initial design or user interface such as menu, dialog, input and output, major transaction files, and simple processing functions. To identify the parts of the software tested then this stage is accompanied by modeling the system flow, by using Unified Modeling Language (UML).

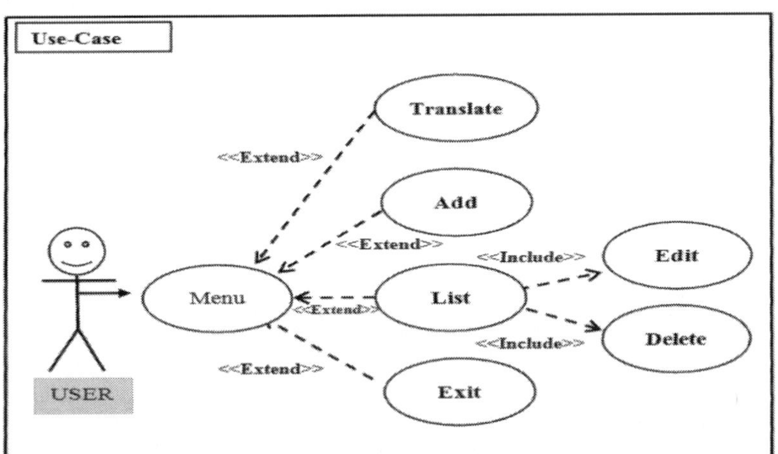

Figure 1. Use case diagram.

Figure 2 illustrates the flow sequence of user activity Dictionary Of Information And Communication Technology, this activity is divided into several categories according to the capabilities the system provides for user access.

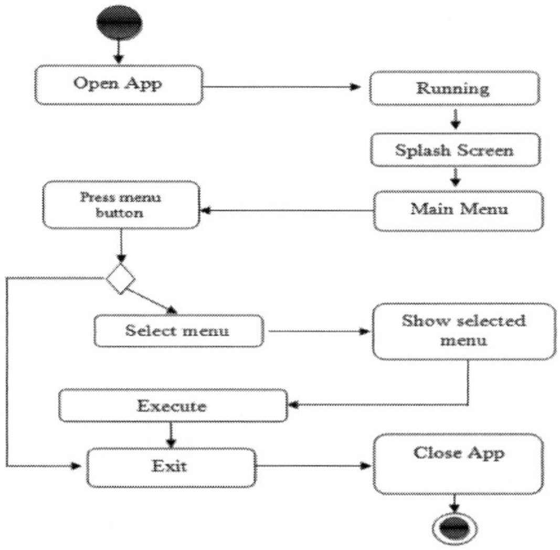

Figure 2. Activity diagram.

In Construction of Prototype stage, with the help of Integrated Development Enviroment (IDE) computer program, to perform coding and realize user interface design in previous modeling stage. Figure 3 present database design created with SQLite using ECLIPSE applications can be viewed using SQLite Manager addons in Mozilla Firefox applications such as:

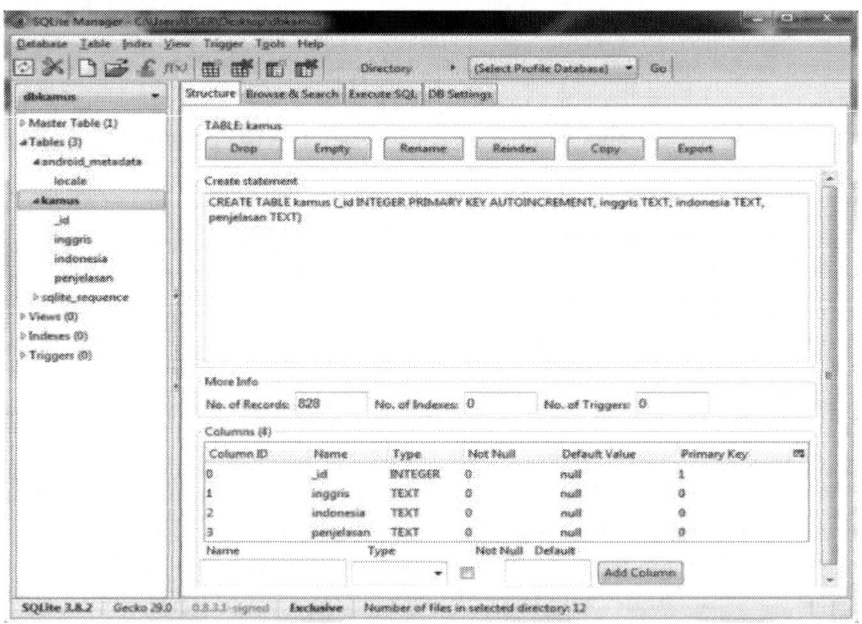

Figure 3. Dictionary database.

After the analysis phase and system design is completed, then the next step is making the application begins by making a application logo splash screen, the results after the application run can be seen in the figure 4 below:

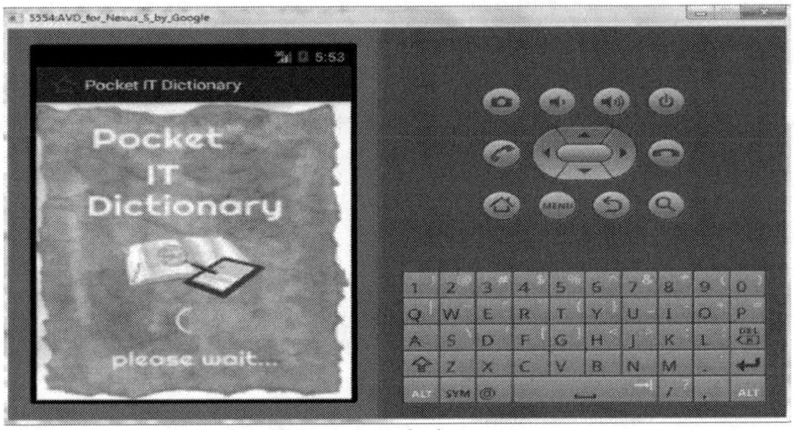

Figure 4. Splash screen.

In Deployment Delivery & Feedback stage, prepare prototype to user in order to test and evaluate where user do test to prototype. Some changes are made based on feedback given by the user about the system, this test is done several times until the user can really effectively use the application.

Last stage is Full System Transformation, transform prototypes into fully operational software, by eliminating unneeded codes, adding necessary codes and programs, repairs and software testing. After seeing the test results, it can be concluded that the application can run quite well and able to adjust to user expectations, where Graphical User Interface (GUI) or user interface can be understood and used easily by the user.

Dictionary of information and communication technology application built in this research, able to provide convenience for students in understanding various terms, especially on english terms. as well as prototype tests, students who experiment do not experience difficulties in applying the program, and can find the various terms they need.

Simplicity in this application can certainly reap a complaint, such as graphics that are too simple or limited options. It is actually fully realized when building modeling, which only provides access to information terms Information and Communication Technology and its translation and explanation. This application is case sensitive, does not contain suggestion feature in term search and does not load index for term list.

On the other hand, the simplicity in this application is based on the consideration that the basic goal to be achieved is the functioning of the application as an easy to use and comprehensive dictionary of information and communication technology. Abundant features and features may be the next option when doing further development.

Furthermore, future application development frameworks may focus on (1) Dictionary Of Information And Communication Technology application development by adding features such as auto suggestion to simplify and speed up the term translation process. (2) Conversion of letters not case sensitive can be added so that users can easily search terms without having to worry about writing the letters. (3) Adding features to give warning to the user when they want to add data and data are already in the database need to be done in order to avoid redundancy of data and consuming storage memory. (4) Create an index on the word list to make it easy to navigate and search for the term in question without having to scroll down the list.

4. Conclusions

Mobile-based Dictionary Of Information And Communication Technology applications that are built can be an alternative to learning literature. In a simple form, this application is able to meet the need for the function and role of dictionary of information and communication technology is comprehensive and accurate. However, as an android-based application, the development framework is certainly needed to improve the quality (both graphical and feature visuals), as well as the flexibility of use.

References

[1] Livingstone Sonia 2004 Media literacy and the challenge of new information and communication technologies *The Communication Review* **7**(1) p 3-14.
[2] Barish, Greg, et al 2000 Theaterloc: Using information integration technology to rapidly build virtual applications *Data Engineering, Proceedings. 16tssh International Conference on. IEEE.*
[3] Brockway, Brian, et al. 2013 Legal compliance, electronic discovery and electronic document handling of online and offline copies of data.
[4] Legris Paul, John Ingham, and Pierre Collerette 2003 Why do people use information technology? A critical review of the technology acceptance model *Information & management* **40**(3) pp 191-204.
[5] Gifary S and Kurnia I 2015 Intensitas Penggunaan Smartphone Terhadap Perilaku Komunikasi *Jurnal Sosioteknologi* **14**(2) pp 1-9.
[6] Haixiao Liu, Bin Song, Hao Qin and Zhiliang Qiu 2013 Dictionary learning based reconstruction for distributed compressed video sensing. 2013 *Journal of Visual Communication and Image Representation* **24**(8) pp 1232-1242.
[7] Zahra Taleb and Amir Sohrabi 2012 Learning on the Move: The use of Mobile Technology to Support Learning for University Students *Procedia - Social and Behavioral Sciences* **69**(24) pp 1102-1109.
[8] Fisher M and Derek E B 2006 Making mLearning work: Utilizing mobile technology for active exploration, collaboration, assessment, and reflection in higher education *Journal of Educational Technology Systems* **35**(1) pp 3-30.
[9] Basili Victor R and David M W 1984 A methodology for collecting valid software engineering data *IEEE Transactions on software engineering* **6** pp728-738.
[10] Bowman D A and Hodges L F 1997 An Evaluation of Techniques for Grabbing and Manipulating Remote Objects in Immersive Virtual Environments *Proc. Symposium on Interactive 3D Graphics (I3D '97), Providence, RI* pp 35–38.
[11] Antonelli M 2017 Design, Fabrication and Demonstration of a Computer Numerically Controlled Rapid Prototyping Machine for Additive Manufacturing, CNC Machining and Laser Cutting

Factors Affecting Optimal Surface Roughness of AISI 4140 Steel in Turning Operation Using Taguchi Experiment

O Novareza*, **D H Sulistiyarini and R Wiradmoko**
Industrial Engineering Department, Faculty of Engineering, Universitas Brawijaya, Malang, Indonesia

*novareza15@ub.ac.id

Abstract. This paper presents the result of using Taguchi method in turning process of medium carbon steel of AISI 4140. The primary concern is to find the optimal surface roughness after turning process. The taguchi method is used to get a combination of factors and factor levels in order to get the optimum surface roughness level. Four important factors with three levels were used in experiment based on Taguchi method. A number of 27 experiments were carried out during the research and analysed using analysis of variance (ANOVA) method. The result of surface finish was determined in Ra type surface roughness. The depth of cut was found to be the most important factors for reducing the surface roughness of AISI 4140 steel. On the contrary, the other important factors i.e. spindle speed and rake side angle of the tool were proven to be less factors that affecting the surface finish. It is interesting to see the effect of coolant composition that gained the second important factors to reduce the roughness. It may need further research to explain this result.

1. Introduction

Various machine tools i.e. lathes, millings, grindings, and drilling are used in conventional machining. Hence, the turning operation by using a lathe machine is one process that includes in traditional or conventional machining [1]. It removes material form a work piece by cutting and produce finished part and chips as the waste.

The friction during the turning process is created amount of heat that can shorten to the tool life [1-3]. The wear of tool raise the vibrations that increase the surface roughness of the finished product. The coolant or lubrication lately reported to be used in conventional machining to increase the tool life [1, 4]. Mineral based oil can be used as coolant or lubrication during turning process. However, this type of coolant may increase the pollution to the environment [1, 4, 5]. Recently, the development of the water-based lubricant becomes favourable to the researcher worldwide. Aqueous copolymer solutions can be used as lubricants that not harm the environmental [6-8].

The AISI 4140 steel is subject commonly in turning process [4, 9]. Aslan et al. and Asiltürk and Akkußs conducted study using AISI 4140 steel in turning operation in dry condition. Taguchi method was chosen to do various experiments with the surface roughness as their final parameter of the turning performance. Both of authors firstly increased the hardness of the AISI to approximately above 55 HRC. The counterpart hardness during turning was approximately above 63 HRC [10, 11]

Determine the untreated AISI 4140 steel quality after turning process is found less attracted to the researcher. However, the composition of AISI 4140 that contains Cr and Mo permitted this steel to have good toughness and ability to create surface harder [9, 12, 13]. Hence, it is interesting to see the lowest possible roughness by combining kinds of cutting parameter and comparing between dry and wet turning condition.

2. Taguchi Method

The Taguchi method for engineering has the goal for improving the quality of the product and the process, simultaneously [11, 14-16]. It also can be used for optimizing cost and resources. It makes the product and the process have robust properties against the noise factor. The noise factor identified as the cause of a large number of output variation of a process. Therefore, this method is also referred to as robust design.

In a process using turning machine, surface roughness and tolerance, are playing important role to determine the quality of the part [14-16]. The smaller number of roughness as initial roughness also confirmed as one factor to reduce wear when the two parts is used as tribo-pairs. Therefore, smaller is the better is used in this study. The further research in lubricated surface of copolymer on bare surface of metal also mention that the surface finish created different thickness and behavior of the layer [6-9].

3. Methods

3.1. Experiments

This study used work piece made from AISI 4140 steel. The condition of the material was proceed without any further processes e.g. heat treatment. It has carbon content of 0.380% - 0.430%, the hardness of the material is 197 BHN and has tensile strength approximately of 655 Mpa [17]. It is a solid cylinder work piece with the dimensions of each work piece was 30.0 mm long and 32.5 mm in diameter.

Table 1 shows the important factor levels that consists of three level of depth of cut, three levels of spindle speed, three levels of coolant and three levels of side rake angle. Common coolant namely Shell Dromus B is used in this research. The combination of the coolant that mixed with demineralized water shown in Table 1 was get from Shell recommendation. The $L_9(3)^4$ was assigned as orthogonal array for Taguchi experiment.

Due to the experiments were carried out in lubricated condition, the work piece was cleaned prior the roughness test. The metal was rinsed using demineralized water followed by cleaning using mild soap to remove the oil dirt. The work piece was dried in room temperature condition. Prior the test, it swabbed with acetone in order to make sure the surface free form any dust.

Table 1. Important factor levels.

Items	Factor Levels		
	1	2	3
Depth of cut [A]	0.125 mm	0.25 mm	0.375 mm
Spindle speed [B]	420 rpm	620 rpm	1000 rpm
Coolant combi. [C]	20:1	30:1	40:1
Side rake angle [D]	5°	10°	15°

3.2. Turning

This research conducted at Industrial Engineering System Manufacturing Laboratory of Faculty of Engineering, Universitas Brawijaya. The experiment used equipment, a manual lathe, Krisbow KW15-604 type. The maximum swing over bed of the lathe is 210 mm and the maximum swing over gap is 320 mm. It can be used to create metric thread approximately of 0.4 until 3 mm. The spindle bore is 20 mm with the range of its speed approximately of 125 to 2000 rpm, longitudinal feed per spindle revolution is 0.1 to 0.2 mm, tool post cutting area is 14 mm square, a motor of 550 W and 220 V-380 V used as a prime mover to the lathe.

A 7.5 mm turning process is performed to each of work piece with the size of the end product was 30 mm long and 25 mm in diameter. The tool was made from High Speed Steel with three different side rake angle (see Table 1); 5°, 10° and 15°.

3.3. Surface roughness

Surface roughness is a surface that has irregular shortwave patterns occurring due to the vibration of the cutting tool or the improper proportion of the cutter feed in the process [10, 11, 14-16]. Surface roughness measurement is aided by a device called surface roughness tester. The data is obtained from a diamond-shaped stylus motion signal to move along a straight line on the surface. The working principle of this tool is to use a transducer and processed with a microprocessor. The value is expressed in micrometer (μm).

In addition, data collection for roughness conducted at the Industrial Metrology Laboratory of Mechanical Engineering of Engineering Faculty of Universitas Brawijaya. It using SFT-210 Series portable surface roughness tester Mitutoyo brand with the following specifications: measuring range (17.5 mm in X axis and 360 μm in Y axis). The measuring speed is varied from 0.25 mm/s, 0.5 mm/s and 0.75mm/s. The stylus tip force is 0.75 mN/2 μm with radius of 60°. It can be used with the skid force less than 400 mN. The manufacturer's mention that it applicable to the various standards e.a. JIS '82, JIS '94, JIS '01, ISO ' 97, ANSI and VDA. The sampling length is 0.08, 0.25, 0.8, and 2.5mm.

Three types of surface finish were used in this research. The first one is arithmetic average deviation surface roughness. It is an arithmetic average of the absolute value between the measured profile value and the middle profile. The value of the average arithmetic roughness (Ra) has a tolerance price. Each roughness price has a roughness class from N1 to N12.

4. Results and Discussion

There were 27 specimens produced that refers to the number of experiments and replication given by orthogonal array matrices. The surface roughness, Ra, results from experiment is presented in Table 2. The equation to measure the Ra is shown in equation below [16, 18]:

$$R_a = \frac{1}{L}\int_0^L |y(x)|\, dx \qquad (1)$$

Arithmetic average deviation surface roughness is denoted with Ra, L is the sampling length and y is the coordinate of the profile. The Microsoft Excel was used to analysis the data to get Analysis of Variance (ANOVA) table as shown in Table 3.

The Sum of Square denoted as SS in Table 3, while the Degree of Freedom is DF and MS is for mean square. Experiments can be assumed to be good when the percentage of contribution value (Rho %) generated is lesser than 50%. It means that there was no significant factor missing from the experiment. On the other hand, if the value is more than 50% means that it can be assumed if there were several factors that have a significant effect on the experiment being lost [19]. In Table 3 it can be seen that the error value generated of 37,45%, which means there is no significant factor missing from the experiment.

After processing data with ANOVA from the data, which applies either to mean value or value of Signal to Noise Ratio, got a result that there are two factors from four factors that have an influence and big contribution to the level of surface roughness. Meanwhile the other two factors have an impact but in small quantities.

Table 2. Results of the Taguchi experimental surface roughness measurements.

Depth of Cut (A)	Spindle Speed (B)	Coolant Composition (C)	Side Rake Angle (D)	R1	R2	R3
0,125 mm	420 rpm	1:20	5°	2,879	2,139	2,627
0,125 mm	620 rpm	1:30	10°	2,815	2,894	2,797
0,125 mm	1000 rpm	1:40	15°	1,785	1,913	2,214
0,25 mm	420 rpm	1:30	15°	2,301	1,835	2,339
0,25 mm	620 rpm	1:40	5°	1,801	1,792	1,758
0,25 mm	1000 rpm	1:20	10°	1,089	1,118	0,86
0,375 mm	420 rpm	1:40	10°	2,288	1,215	2,706
0,375 mm	620 rpm	1:20	15°	1,202	1,646	1,977
0,375 mm	1000 rpm	1:30	5°	2	2,874	2,366

Table 3. Analysis Of Variance (ANOVA) from experiments.

Source	*SS*	DF	MS	*F-Ratio*	*SS'*	*Ratio%*	F tabel 0.05 (2;18)
A	2,86	2	1,43	10,48	2,59	27,32	3,55
B	0,95	2	0,48	3,49	0,68	7,16	3,55
C	2,63	2	1,31	9,64	2,36	24,89	3,55
D	0,57	2	0,29	2,10	0,30	3,18	3,55
Error	2,45	18	0,14	1	3,55	37,45	
SST	9,47	26	4,53		9,47	100	
Mean	112,98	1					
SStotal	122,44	27					

There is only one factor that has value F-Ratio ≥ F-Table (F 0.05 (2:18) = 3.55) (see Table 3). So it can be concluded that depth of cut factor is statistically has an effect on the level of surface roughness of the shaft. The SNR value calculation is done on each experiment with three replication, so that nine SNR data is then processed in ANOVA calculation of Signal to Noise Ratio value.

The depth of cut in this research was proven to be the most factors to reduce the Ra. Similar conclusion was reported from Thamizhmanii et al. [16] that conducted similar experiment. However, they reported that the experiment was carried out using dry condition and the metal was much harder than this research reported. Almost ten times higher of depth of cut value was used for Thamizhmanii et al. [16] compared to this setup. Higher depth of cut may increase heat and wear but proven to be useful to produce better quality of finished part. The chosen parameters and the ability of the lathe used in this research may affect to the results. Other conditions i.e. the hardness of material and the coolant needs to investigate further.

Figure 1 shows one of evaluation profile of Ra, the sixth trial and the third replication with value Ra 0,86 μm. It shows the result of surface roughness testing with the lowest value from the 27 experiments for Ra. It was one and only result below 1. On the contrary, Figure 2 shows a clearly different profile from the first trial and the first replication. It comes from the result of test with the highest roughness value with value Ra 2,879 μm.

A significant variation is detected from the results shown in Figure 1 and Figure 2. It may come from the depth of cut that proves as the most affected factors to surface roughness. However, Asiltürk and Akkußs [11] that was doing similar experiment and similar method with this study have different opinion. The feed rate was mentioned as the most factor affected to the best of roughness when the AISI 4140 is processed with turning machine.

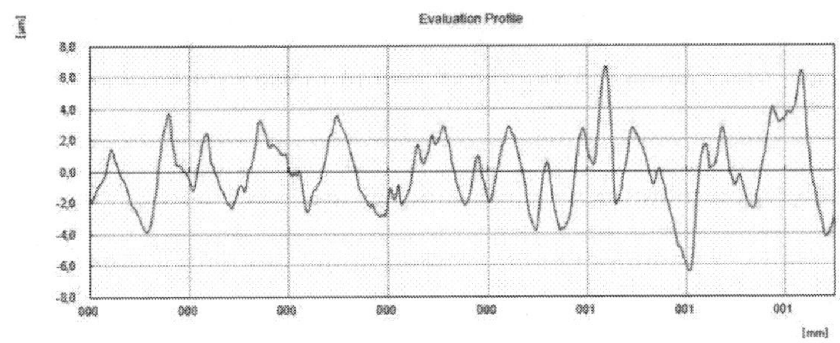

Figure 1. Evaluation profile of Ra 0,86 µm.

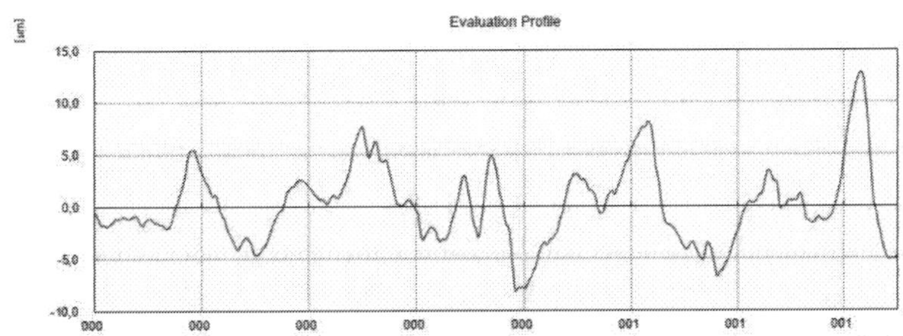

Figure 2. Evaluation profile of Ra 2,879 µm.

It is interesting to see the data of Ra from other researchers. Much lower Ra value was gained during their experiment in dry condition [11, 14-16]. The result shows in their experiment was carried out in the different parameters especially it done without cooling or dry condition. The cooling may not the most affected factor in this paper, however, it is interesting to do future experiment to see the effect of cooling process and the type of coolant to get the smallest value of surface roughness.

5. Conclusions

Based on the results of research conducted, the most factors that have an effect on the roughness during wet condition turning process on AISI 4140 untreated steel is depth of cut factor with a contribution value of 27.32%. Coolant composition factor with a contribution value of 24.89%, spindle speed factor with value contribution of 7.16%, and the last factor depth of cut with contribution of 3.18%.

Acknowledgments

Authors wishing to acknowledge assistance or encouragement from the staff of Industrial Engineering System Manufacturing Laboratory and Industrial Metrology Laboratory of Engineering Faculty of Universitas Brawijaya.

References

[1] Liew P J, Shaaroni A, Sidik N A C and Yan J 2017 An overview of current status of cutting fluids and cooling techniques of turning hard steel *International Journal of Heat and Mass Transfer* **114** p 380–394.

[2] Sharma V S, Dogra M and Suri N 2009 Cooling techniques for improved productivity in turning *Int. J. Mach. Tools Manuf.* **49** p 435–453.

[3] Bhuiyan M, Choudhury I and Dahari M 2014 Monitoring the tool wear, surface roughness and chip formation occurrences using multiple sensors in turning *J. Manuf. Syst.* **33** p 476–487.

[4] Hadad M and Sadeghi B 2013 Minimum quantity lubrication-MQL turning of AISI 4140 steel alloy *Journal of Cleaner Production* **54** p 332-343.

[5] Cambiellaa A, Benitoa J M, Pazosa C, Cocaa J, Ratoi M, and Spikes H A 2006 The effect of emulsifier concentration on the lubricating properties of oil-in-water emulsions *Tribology Letters* **22** p 53.

[6] Kosasih B, Novareza O, Tieu K and Zhu H 2014 Friction and anti-wear property of aqueous tri-block copolymer solutions in metal forming *International Journal of Surface Science and Engineering* **8**(2/3) p 109-123.

[7] Novareza O, Kosasih B, Tieu K and Zhu H 2014 Tribological Performance of Aqueous Copolymers Solutions in Metallic Contact *Materials Science Forum* **773** p 678-686.

[8] Kosasih B, Novareza O, Zhu H, Taheri R, Lin B and Tieu A K 2016 Thickness and scratch resistance of adsorbed film formed by triblock symmetrical copolymer solutions *Lubrication Science* **28**(5) p 299-315.

[9] Ulutan M, Celik O N, Gasan H and Er U 2010 Effect of Different Surface Treatment Methods on the Friction and Wear Behavior of AISI 4140 Steel *J. Mater. Sci. Technol.* **26**(3) p 251-257.

[10] Aslan E, Camußscu N and Bingören B 2007 Design optimization of cutting parameters when turning hardened AISI 4140(63 HRC) with Al2O3 + TiCN mixed ceramic tool *Mater. Des.* **28** p 1618–1622.

[11] Asiltürk I and Akkußs H 2011 Determining the effect of cutting parameters on surface roughness in hard turning using the Taguchi method *Measurement* **44** p 1697–1704.

[12] Kessler O, Herding T, Hoffmann F and Mayr P 2004 Microstructure and wear resistance of CVD TiN-coated and induction surface hardened steels *Surf. Coat. Technol.* **182**(2-3) p 184.

[13] Sohi M H, Ebrahimi M, Raouf A H and Mahboubi F 2010 Effect of plasma nitrocarburizing temperature on the wear behavior of AISI 4140 steel *Surface & Coatings Technology* **205** (2010) p S84–S89.

[14] Bilga P S, Singh S and Kumar R 2016 Optimization of energy consumption response parameters for turning operation using Taguchi method *Journal of Cleaner Production* **137** p 1406-1417.

[15] Kumar R, Bilga P Sand Sehijpal Singh S 2017 Multi objective optimization using different methods of assigning weights to energy consumption responses, surface roughness and material removal rate during rough turning operation *Journal of Cleaner Production* **164** p 45-57.

[16] Thamizhmanii S, Saparudin S and Hasan S 2007 Analyses of surface roughness by turning process using Taguchi method *Journal of Achievements in Materials and Manufacturing Engineering* **20**(1-2).

[17] AZO Materials. *AISI 4140 Alloy Steel* (UNS G41400). http://www.azom.com/article.aspx?ArticleID=6769. Accessed: 2017-05-19

[18] Arbizu I P and Prez C J L 2003 Surface roughness prediction by factorial design of experiments in turning processes *Journal of Materials Processing Technology* 143- 144 390-396.

[19] Belavendram N 1995 *Quality by Design Taguchi Techniques for Industrial Experimentation* (London: Prentice Hall International (UK) Limited).

Voltage Analysis Improvement of 150 kV Transmission Subsystem Using Static Synchronous Compensator (STATCOM)

P A Akbar*, D L Hakim, and T Sucita

Universitas Pendidikan Indonesias

*pamungkas.achmadi@student.upi.edu

Abstract. In this research, testing improvements to the distribution voltage electricity at 150 kV transmission subsystem Bandung Selatan and New Ujungberung using Flexible AC Transmission System (FACTS) technology. One of them is by doing the control of active and reactive power through the power electronics equipment Static Synchronous Compensator (STATCOM). The subsystem is tested because it has a voltage profile are relatively less well when based on the IEEE / ANSI C.84.1 (142.5 - 157.5 kV). This study was conducted by analyzing the Newton-Raphson power flow on the simulator DigSilent Power Factory 15 to determine the profile of the voltage (V) on the system. Bus which has the lowest voltage to be a reference in the installation of STATCOM. From this research is known that the voltage on the conditions of the existing bus 28, as many as 21-23 still below standard buses (142.5 kV), after the installation is done using STATCOM, voltage on the buses improved by increasing the number of tracks that follow the standard / is in the range 142.5 kV -157.5 kV as many as 23-27 buses or 78.6% - 96%, with the optimum mounting on a bus Rancaekek STATCOM II with a capacity of 300 MVA.

1. Introduction

An electric power transmission system has a very important role for electrical energy distribution. With the extremely high requirements will require good quality electricity for distribution. As is the case in this study is based on data in the field in 2016 are sourced from PT. PLN (Persero) *Area Pengontrol Beban* (APB) *Jawa Bagian Tengah*, the sub-system 150 kV transmission area Bandung Selatan on February 3 at 19:00 and February 15 at 19:00 if used PLN Standard [1], the value of the voltage buses has the appropriate value, because the value is in the range 135 kV - 157.5 kV (-10%, + 5%). However, in the present study used comparisons with the IEEE / ANSI C84.1 with rated voltage ± 5% of nominal voltage from 28 the existing buses, 21-23 are below the limit of 142.5 kV (0.95 p.u).

Solutions that can be used to improve the quality of the installation of electric power systems by using control technology are qualified. FACTS technologies allow to improve the operation of the transmission system to the minimum of infrastructure investment costs, environmental impact and execution time compared with the construction of new transmission lines [2]. One of the FACTS which are connected in parallel, has an efficiency of reactive power and voltage is STATCOM. Installation STATCOM needed to reduce losses and the minimum amount of reactive to generated [3].

The ability to overcome the problems STATCOM reactive power to make such equipment as an option for use. Moreover, with the development of power electronics today, STATCOM is also experiencing growth in reliability in operation. So based on reliability in the operation of the discussion of the STATCOM be an interesting thing to be examined

2. Flexible Alternating Current Transmission System (FACTS)

Flexible AC Transmission System (FACTS) is a power electronics-based equipment systems that provide control of one or more AC transmission lines to improve the control and improvement of power distribution. Where there are 3 types of FACTS ie, parallel, series and series-parallel [4].

The using of reactive power compensators in the electric transmission system has been proven to improve the quality and reliability of the electrical system. This is because in addition to be able to regulate the amount of reactive power in the system, reactive power compensators also has a function for setting and balancing of voltage levels. [5]

2.1. Static Synchronous Compensator (STATCOM)

Static Synchronous Compensator (STATCOM) is a plant Volt Amp Reactive (VAR) static, which output is varied to maintain or set the type of reactive impedance, using thyristor arrangement, reactors and capacitor switch, or the type of source synchronous voltage using pen-switch power converter[1]. In Figure 1 shows the topology of the STATCOM, which consists of a capacitor DC power converters are alternated by itself and step up transformer or transformer circuit.

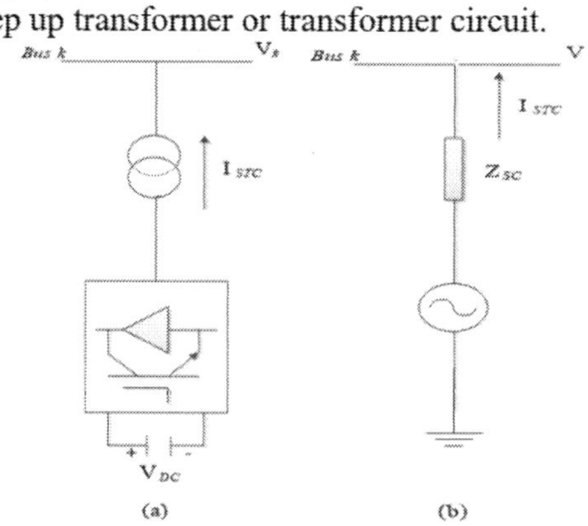

Figure 1. STATCOM equivalent circuit diagram of the circuit: (a) Schematic diagram of STATCOM; (b) equivalent circuit STATCOM. [6]

2.2. Working principle and control of STATCOM

STATCOM is a Voltage-Source Converter (VSC), which converts the DC input voltage into AC output voltage to compensate for active and reactive power needed by the system [7]. By using VSC controller and coupling transformer, STATCOM performance depicted in the figure below[8]

Basic operating principle of generator reactive power by using VSI is similar to a conventional synchronous machines. Reactive currents drawn by the synchronous compensator voltage depending on the magnitude of system voltage (V), the converter Vo and the overall circuit reactance (reactance leaking transformer with transformer coupling reactance) X : [7]

$$I = \frac{V - Vo}{X} \qquad (1)$$

And reactive power exchange found in expression in:

$$Q = \frac{1 - \frac{Vo}{V}}{X} V^2 \qquad (2)$$

STATCOM is FACTS controller that based on voltage source converter (VSC). Where VSC generates synchronous voltage at the fundamental frequency, large control voltage and phase angle. If a VSC is connected in parallel to the system by combining transformers, STATCOM can generate and absorb reactive power at bus paired and adjust the voltage [9].

3. Research Methods

In completing the necessary research framework / research stages of processing from start to start to finish in order to facilitate the writer and the reader in understanding the stages of execution of this study. Here is a flow diagram of the stages in performing research.

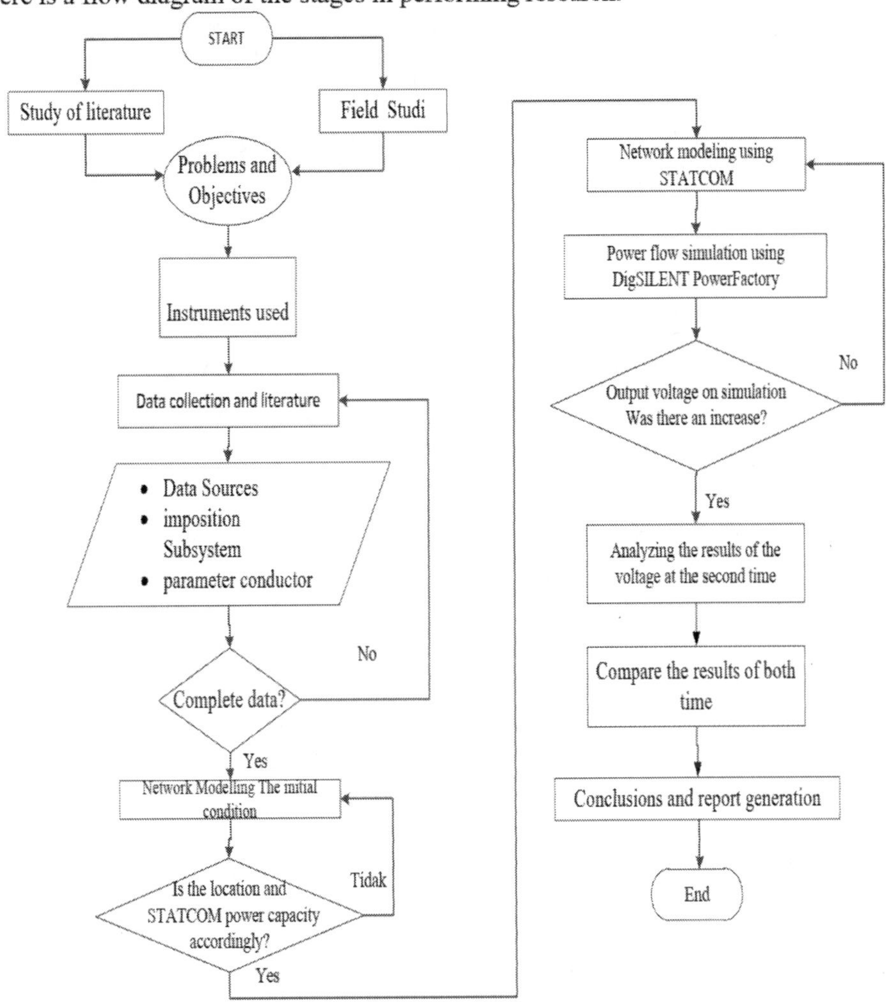

Figure 2. Research flowchart diagram

Based on the flow chart, technical data which includes such as data source of electricity, charging sub-systems and conductor parameters which can be described as follows:

Data loading is on subsystem consists of a static load of the sub-system of 150 / 70kV and dynamic load on the transformer, but the system was changed to a static load (lump load). The data consists of the capacity of the power, active power and reactive power. The data retrieved is the data loading that occurs in South Bandung subsystem 150 kV, contained on a single line diagram susbsitem South Bandung 150 kV and New Ujung Berung The loading of data to be tested was taken on February 3, February 15th at 19:00 and 19:00.

4. Results and discussion

4.1. Profile voltage 150 kV transmission subsystem Bandung Selatan and new Ujungberung after installation STATCOM

Results obtained from the process of calculation and experimental approaches in determining the capacity and the injection of the STATCOM reactive power to maintain voltage value most prevalent in the overall bus placement STATCOM obtained by using STATCOM capacity of 300 MVA (+250 MVAr; -250 MVAr) on February 3, 2016, with the results shown in table 1 and table 2 as follows:

Table 1. Voltage profile simulation results after the installation of STATCOM in Rancaekek II bus dated February 3, 2016 at 19:00

No	Substation	V operating(kV) simulation without STATCOM	V with STATCOM
1	Bandung Selatan I	144,93	146,11
2	Bandung Selatan II	144,93	146,11
3	Bandung Utara II	**137,23**	144,09
4	Braga I	**141,47**	142,69
5	Braga II	**141,47**	142,69
6	Cibereum I	**141,32**	142,54
7	Cibereum II	**141,32**	142,54
8	Cigereleng I	**141,61**	142,83
9	Cikasungka I	**140,11**	148,51
10	Cikasungka II	**140,11**	148,51
11	Cianjur I	**134,30**	**135,59**
12	Dago Pakar I	**138,04**	144,85
13	Dago Pakar II	**138,04**	144,85
14	Darajat I	143,61	150,79
15	Darajat II	143,61	150,79
16	Kamojang I	**141,91**	149,17
17	Kamojang I	**141,91**	149,17
18	Kiaracondong I	**140,96**	145,53
19	Kiaracondong II	**140,96**	145,53
20	Panasia	144,36	145,54
21	Panasia II	144,36	145,54
22	Rancakasumba I	**139,75**	148,18
23	Rancakasumba II	**139,75**	148,18
24	Rancaekek II	**139,53**	148,63
25	Ujung Berung I	**139,13**	145,89
26	Ujung Berung II	**139,13**	145,89
27	Wayang Windu I	143,83	147,89
28	Wayang Windu II	143,83	147,89

According to the table 1 above, there are 27 standardized buses and one bus is not appropriate based on IEEE / ANSI C84.1 standard. In addition to the improved results by using STATCOM 3 =

300 MVA (+284.27 MVAr; -284.27 MVAr) on the buss on February 15, 2016 at 19.00, shown in the table 2 below:

Table 2. Voltage profile simulation results after the installation of STATCOM In bus Rancaekek II dated February 15, 2016 at 19:00

No	Substation	V operating(kV) simulation without STATCOM	V with STATCOM
1	Bandung Selatan I	143,40	144,75
2	Bandung Selatan II	143,40	144,75
3	Bandung Utara II	**135,39**	143,21
4	Braga I	**140,87**	**142,24**
5	Braga II	**140,87**	**142,24**
6	Cibereum I	**140,66**	**142,04**
7	Cibereum II	**140,66**	**142,04**
8	Cigereleng I	**141,00**	**142,37**
9	Cikasungka I	**138,61**	148,16
10	Cikasungka II	**138,61**	148,16
11	Cianjur I	**132,82**	**134,29**
12	Dago Pakar I	**136,15**	143,94
13	Dago Pakar II	**136,15**	143,94
14	Darajat I	**142,25**	150,37
15	Darajat II	**142,25**	150,37
16	Kamojang I	**140,66**	148,87
17	Kamojang I	**140,66**	148,87
18	Kiaracondong I	**138,94**	144,17
19	Kiaracondong II	**138,94**	144,17
20	Panasia	143,00	144,35
21	Panasia II	143,00	144,35
22	Rancakasumba I	**138,26**	147,83
23	Rancakasumba II	**138,26**	147,83
24	Rancaekek II	**137,85**	148,23
25	Ujung Berung I	**137,23**	144,94
26	Ujung Berung II	**137,23**	144,94
27	Wayang Windu I	142,90	147,49
28	Wayang Windu II	142,90	147,49

Results obtained voltage improvement as much as 22 pieces and buses are still under the standard IEEE / ANSI C84.1 as much as 6 pieces. So with the comparison between the two time ranges the number of bus improvements bus voltage range between 23-25 and the mean number of improvements of 82.14% - 96.4% of the total number of 28 buses.

4.2. Analysis effect of STATCOM most optimized placement

After the trial the effect of the installation of STATCOM, it is necessary to analyze the most optimal placement STATCOM, in order to produce the best voltage value improvement, which can be determined based on the table.

Based on the table 1, the result of improved value of the voltage buses on February 3, 2016 at 19:00 based on the value of the average voltage bus generated the overall order of placement on the bus STATCOM consist of Rancakasumba, Cikasungka, Rancaekek, Bandung Utara, Dago Pakar and Ujungberung. Whereas based on the result in table 2 of improved buses voltage value at the date of February 15, 2016 at 19.00, has a placement optimization results are quite different, which has four locations namely placement STATCOM there is in Rancaekek, Bandung Utara, Dago Pakar and Ujung Berung.

With the consideration of a STATCOM called optimal position when the voltage profile improvement occurs evenly and have an average value of a relatively large after a comparison between two different times, then the installation position STATCOM on buses that are considered the most optimal located on the Rancaekek II bus.

Based on the simulation results are show that STATCOM provide an effective additional voltage on the bus is connected directly. STATCOM placed as close to the bus closest to the loads.

5. Conclusion

Voltage profile in Bandung Selatan subsystems and New Ujungberung, which is dated February 3, 2016 and February 15, 2016 at 19:00 before using STATCOM overall when using SPLN have a value according to standards but if based on standard IEEE / ANSI C84.1 have conditions voltage relatively poorly characterized by the bus below the standard range between 21-23 buses below 0.95 p.u or 142.5 kV.

After installation of Static Synchronous Compensator (STATCOM) voltage profile in the subsystem Bandung Selatan and New Ujungberung, which is dated February 3, 2016 and February 15, 2016 19:00 have improved voltage of 28 buses that have so increasing the number of tracks that follow the standard IEEE / ANSI C84.1 as many as 27 buses on 3 February and 22 buss on February 15, with the installation of the bus Rancaekek II with a capacity of 300 MVA. Positioning and capacity STATCOM most optimal in maintaining the voltage profile of the most effective currently on Rancaekek II bus with a capacity of 300 MVA STATCOM.

References

[1] Z Aini 2012 Analisis Kestabilan Tegangan Dengan Menggunakan Static Synchronous Compensator (STATCOM) *J. Sains, Teknol. dan Ind.* **9** 2 9–14

[2] J J Paserba 2003 How FACTS Controllers Benefit AC Transmission Systems *IEEE* **3** 1–8

[3] S U Mageswaran and N O G Sekhar 2013 Reactive Power Contribution of Multiple STATCOM using Particle Swarm Optimization *Int. J. Electr. Technol.* **5** 1122–126

[4] M A B Ardi 2013 Power Factor Improvement in Power System with The Integration of Renewable Energy *University Tun Hussein Onn Malaysia*

[5] A S Hidayat 2010 *Analisa Dan Perancangan Sistem Pengendalian Multilevel STATCOM (Static Syncronous Compensator) Dalam Mengurangi Gangguan Tegangan Kedip* Fakultas Teknik,

[6] T Ogbomoso 2011 Analysis and Modelling of Static Synchronous Compensator (STATCOM): A comparison of Power Injection and Current Injection Models in Power Flow Study *Int. J. Electr. Technol.* **36** 65–76

[7] S S Wibowo, H Suyono and R N Hasanah 2013 Analisis Implementasi Fixed Capacitor, SVC, dan STATCOM untuk Perbaikan Performansi Stabilitas Tegangan pada Sistem Petrochina *EECCIS* **7** 2 147–152

[8] S B Aruna 2013 Power Flow Control by Using Optimal Location of STATCOM *IJPaper* **04** 305–310

[9] N G Hingorani and L Gyugyi 2000 *Understandig FACTS*. Institute of Electrical and Electronics Engineers, Inc.

IOP Conf. Series: Materials Science and Engineering **306** (2018) 012101 doi:10.1088/1757-899X/306/1/012101

4D Model on Assessing Psychomotor Aspect in Continental Food Processing Practice

P Nurafiati*, A Ana, H Ratnasusanti and I Maulana

Pendidikan Teknologi dan Kejuruan, Sekolah Pascasarjana Universitas Pendidikan Indonesia, Bandung, Indonesia.

*pursitanurafiati02@gmail.com

Abstract. This research aims to develop and find out the response of observers for the assessment instrument of student's psychomotor aspect on continental food processing practice. This research belongs to development research with 4P model that confined till the definition, design, and development stages. The data that gained during the research is analyzed descriptively. Research's product is assessment instrument rubric form that consists of performance's aspect which should be assessed and performance's quality which stated in gradation score with 0-4 level and performance description that completed with picture illustration in every single score. Product was validate and responded based on material, construction, language, objectively, systematic, and practicability aspects. The result show that assessment instrument of student's psychomotor aspect on continental food processing practice which developed gain very good response with percentage of 84,47%.

1. Introduction

Education is a learning activity that aims to develop and actualize students' self-potential. Education as a standard establishes national standards as a minimum quality of citizens broken down into standard contents, process standards, graduate competency standards, educator standards and education personnel, and educational assessment standards [1]. In the implementation, the goal to be achieved in the learning activities is a form of student ability that occurs because of the interaction between teachers, students, and learning environment. Achievement of learning objectives can be known through assessment activities that are able to provide information about the ability of students who are divided into aspects of cognitive, affective, and psychomotor [2]. The assessment process is carried out by steps; planning assessment, gathering information through a number of evidence showing achievement of learning outcomes, reporting, and use of information about learning outcomes. Prior to carrying out an assessment of the learning process and outcomes, the teacher must first make the tools so that the assessment is done in accordance with the competence to be tested [3].

Permendikbud No. 66 of 2013 on Education Assessment Standards provides that "assessment of learners' learning outcomes includes competence of attitudes, knowledge, and skills performed on an equal basis ..." [4]. But in fact, the success of education has been measured more than the assessment of cognitive aspects related to knowledge competence. This is evidenced by the dominance of the implementation of formative tests, summative tests, and even the National Exam in the form of a written test that is only able to provide information about students' concept of understanding only.

Implementation of a judgment focused solely on one competency cannot describe the student's ability objectively, accurately, and comprehensively.

The assessment of skills competencies in the Curriculum 2013 is obtained through a performance assessment defined in Permendikbud Number 66 Year 2013 on Education Assessment Standards as "... an assessment that requires learners to demonstrate a particular competency by using practical tests, projects, and assessments portfolio ". Performance appraisal has the advantage of revealing the potential of students in solving problems, reasoning, and communication in written and oral form [5]. Performance appraisal is done by observing directly the performance shown by students during practicum activities. Quality assessment is one component in the organization of quality education system [6].

Several studies on the implementation of the assessment on practice activities gave results that the assessment on practice activities was mostly done by looking at the final product in the form of a practice report, while performance evaluation through observation was rare [7]. One way to assess skills competence is through direct observation of student performance during learning activities [8]. Psychomotor ability in practicum activity is shown in the form of student performance in the form of real action that can be observed during follow practice activity. Implementation of an assessment that only assesses the final product and does not observe the performance of students directly is not able to provide a description of the students' psychomotor aspects capabilities are valid and objective.

The previously described description forms the basis for the need to fulfill the needs of assessment instruments that can be reached by all education activists, especially teachers as the main subject in conducting the assessment in the learning activities. Thus, it is necessary to develop a psychomotor aspect assessment instrument on a continental food processing practice as an effort to achieve the objective of the assessment activity that is to provide students' ability information validly, objectively and comprehensively in every aspect of ability. The main problem arises from several discussion of the research is how to develop assessment instruments on the psychomotor aspects of the practice of continental food processing?

Based on the problems found, then conducted research on how the development of psychomotor aspects assessment instruments on the practice of continental food processing, so that the formulation of the problem can be formulated into:

- How is the development of psychomotor aspects assessment instruments on continental food processing practice?

2. Methods

The research method used is research and development method with 4D model, "This model consists of four development stages, namely define, design, develop, and disseminate, or adapted into 4P model, that is defining, designing, developing, and spreading". Based on the predetermined research objectives, the 4P model applied in this study is limited to the development stage because the activities undertaken to the development stage have been able to meet the objectives of the study.

The defining, designing, and developing phases are then described each by examining the activities, characteristics, changes, and relationships, both in terms of similarities and differences and the interlinking of the activity steps. In its implementation, each stage of the development research is always related to revision activities aimed at evaluating and improving the designs made based on the inputs and assessments obtained from the learning device validation activities. Validation activities are conducted by requesting an assessment and feedback on the accuracy of learning tools or in this study in the form of assessment instruments developed to experts. Assessment and input is then a reference for revision.

3. Results and Discussion

This research is oriented towards the development of products whose processes are described and the results are tested in a limited way to obtain user responses to the use of the product. Product development is done through three stages, namely definition, design, and development. Based on the research that has been done, the data obtained in the form of interview results of teachers and students, suggestions validator during the development process and user response. The results of research on three stages of product development are as follows:

3.1. Definition stage

The defining stage is performed to establish the basic problem in learning so that it is necessary to develop the assessment instrument and the criteria as the reference in developing the assessment instrument. The defining stage is divided into 5 steps of activity that is the analysis of the front end through an interview with the teacher. The result of the interview shows that the psychomotor aspect assessment in the practicum activity is done in group with the assessment instrument that assess the psychomotor ability in general, not the detail, and does not differentiate the point of assessment between one practicum material with other material. Based on the results of interviews also note that the assessment instrument used is not equipped with the gradation of quality is expressed by the score and description of performance in each score. The next stage is the analysis of students, the results of student analysis through literature review indicates that the cognitive development of high school students included into formal - operational stage according to Piaget learning theory. The results of the interviews of the three XI students gave information about their experiences during practicum activities, which in their practicum activities were divided into preparation, implementation and closing. The next stage is the task analysis, the results of the literature review students' psychomotor skills acquired during practical activities that include manipulative skills in using practicum tools and procedural skills in doing the job in a particular order. Phase of concept analysis is divided into two namely the analysis of material and analysis of core competence and basic competence. From result of learning indicator, conducted instructional goal specification. The format of writing instructional objectives is composed of the concept and objectives of instructional products and processes.

3.2. Desain stage

Some the design stage is done to prepare the initial prototype of the assessment instrument. The design phase is divided into 4 steps of activities, namely Standard Reference Test Preparation, Media Selection, Selection Formats and Preliminary Draft. In the step of preparing the benchmark reference test, the grid of the assessment instrument is prepared based on the instructional objective formula. After obtaining a grid of assessment instruments, then carried out the stage of media selection. Selected media are worksheets and tools and materials used. The next step is the selection of the format, the instrument format of assessment is divided into three stages of activity that is opening, execution and closing. In the last stage of the preliminary draft, was performed in preparation of worksheets practicum and assessment instruments text.

3.3. Development stage

At the development stage a revaluation instrument is generated and a revised user response questionnaire based on the suggestion of the validator for later use in the limited trial stage. The development stage is divided into Validation and Limited Trial Stages. Overall, the user's response to the developed assessment instrument got a percentage of 84.47% with very good category. This suggests that the developed assessment instrument is worthy of being used as an instrument of psychomotor aspects assessment in continental food practice.

4. Conclusions

This study aims to develop the students' psychomotor aspect assessment instrument on a continental food practice and to know the user's response to the developed assessment instrument. Based on these objectives, it is summarized as follows:

- Generated student psychomotor aspect assessment instrument on continental food practice through 4P development model covering definition, design and development stage. The assessment instrument is an assessment rubric consisting of criteria and quality. The criteria show the quality of performance expressed in the form of gradation scores with a range of 0-4 and a performance description that has illustrated drawings in each score.

- Student psychomotor aspect assessment instrument on continental food practice gets user percentage of 84,47% with very good category.

References

[1] Yulianti N 2014 Pengembangan Instrument Penilaian Psikomotorik pada Pokok Bahasan Suhu dan Kalor SMP *Jurnal Inovasi Dan Pembelajaran Fisika* Available at: ejournal.unsri.ac.id.

[2] Pengembangan Instrumen Penilaian Unjuk Kerja (Performance Assesment) Laboratorium pada Mata Pelajaran Fisika Sesuai Kurikulum Tingkat Satuan Pendidikan / IK Susila - Jurnal Penelitian dan Evaluasi pendidikan, 2012 - 119.252.161.254

[3] Pengembangan Instrumen Performance Assesment sebagai bentuk penilaian berkarakter kimia / RLP Sari - Makalah Seminar MIPA, 2010 - staff.uny.ac.id

[4] Salinan Lampiran Peraturan Menteri Pendidikan dan Kebudayaan Republik Indonesia Nomor 66 tahun 2013 tentang Standar Penilaian. 2013. Jakarta: Kementrian Pendidikan dan Kebudayaan Nasional Republik Indonesia

[5] Iryanti P 2004 *Penilaian Unjuk Kerja* (http;//p4tkmatematika.org/downloads/ppp/PPP04_UnjukKerja.pdf).

[6] Sudria I B N, Siregar M 2009 Pengembangan Rubrik Penilaian Keterampilan Dasar Praktikum dan Mengajar Kimia pada Jurusan Pendidikan Kimia. *Jurnal Pendidikan dan Pengajaran* Available at: ejournal.undiksha.ac.id.

[7] Sapriati A 2006 Pengembangan Instrumen Penilaian Praktikum Fotosintesis *Jurnal Pendidikan* **7**(1) pp 1-11.

[8] Sofyan A, Feronika T, and Milama B 2006 *Evaluasi pembelajaran IPA berbasis kompetensi* (Jakarta: UIN Jakarta Press).

IOP Conf. Series: Materials Science and Engineering **306** (2018) 012102 doi:10.1088/1757-899X/306/1/012102

Validation of a Numerical Program for Analyzing Kinetic Energy Potential in the Bangka Strait, North Sulawesi, Indonesia

P T D Rompas[*], **H Taunaumang and F J Sangari**
Universitas Negeri Manado, Tondano, 95618, Sulawesi Utara, Indonesia

*parabelemrompas@unima.ac.id

Abstract. The paper presents validation of the numerical program that computes the distribution of marine current velocities in the Bangka strait and the kinetic energy potential in the form the distributions of available power per area in the Bangka strait. The numerical program used the RANS model where the pressure distribution in the vertical assumed to be hydrostatic. The 2D and 3D numerical program results compared with the measurement results that are observation results to the moment conditions of low and high tide currents. It found no different significant between the numerical results and the measurement results. There are 0.97-2.2 kW/m^2 the kinetic energy potential in the form the distributions of available power per area in the Bangka strait when low tide currents, whereas when high tide currents of 1.02-2.1 kW/m^2. The results show that to be enabling the installation of marine current turbines for construction of power plant in the Bangka strait, North Sulawesi, Indonesia.

1. Introduction

The potential of kinetic energy in Bangka Strait, North Sulawesi, Indonesia can analyzed by approach of a numerical program that analyzes the velocity of the ocean currents necessary to calculate the kinetic energy of ocean currents.

The study of ocean currents through numerical modeling has performed [1]. Who implicated the numerical model in San Francisco Bay California and the Lagoon of Venice, Italy by using semi-implicit finite difference method for 3D shallow water flow, Rompas and Manongko [2] simulated the marine currents in the Bunaken strait, North Sulawesi, Indonesia with the numerical model, Rompas et al [3-5] studied the marine currents in the Bangka strait, North Sulawesi, Indonesia with the modelling and numerical simulation which predicted the velocities and the kinetic energies but the numerical program has not been validated, Rodrigues-Cuevas et al [6] researched a numerical model with difference turbulence models, O'Donncha et al [7] were applied the model-3D of hydro-environmental code to a designated tidal energy test site on the East Coast of the United States, Gonzales-Gorbena et al [8] optimized the hydrokinetic turbine array layouts by surrogate modelling, Zangiabadi et al [9, 10] developed the a CFD model by using the bathymetry of a potential tidal stream turbines deployment site and presented the tidal stream turbines for tidal power production. Martins et al [11] used 3D modelling in the Sado estuary by using a new generic vertical discretization approach. Luquet et al [12] tested design and model of an optimized ducted marine current turbine. The 3D hydrodynamic model used [13] for knowing influence of the Aral Sea negative water balance on its seasonal circulation patterns, whereas [14] studied climate with ocean modelling for eliminated short time scales in long-term and high resolution of ocean circulation. The simulation of electrical power potential in the Alderney Race has been successfully which conducted by Myers and Bahaj [15] for marine current turbine arrays. On the other hand, Thiebot et al [16] developed the effects of large arrays of tidal turbines with dept-average Actuator Disks by modelling. Pinon et al [17] predicted wake of marine current turbines with a particle method by using the numerical simulation. Whereas, Elhanafi et al [18] used the numerical simulation with CFD to analyzing the offshore stationary-floating oscillating water column-wave energy converter. Also, Ho and Riddette [19] applied CFD to evaluate hydraulic performance of spillways in Australia.

The purpose of numerical program validation is to obtain a numerical program that can generate data in accordance with the measurement results [20]. Validation done by comparing the numerical programs with the data of measurement results and there are some researcher have conducted it. Maters et al [21] have validated a numerical program to analyzing tidal stream turbine. The study of numerical and experimental has conducted by Mycek et al [22] for analyzing interaction between two marine current turbines.

This paper studied on the validation of the numerical program computation results to the measurement results of the velocities of marine current when the low tide and high tide in the Bangka Strait. The objectives of the study are getting validation results of the numerical program computation result to measurement result and analyzing the available power potential of marine current kinetic energy in the Bangka strait, North Sulawesi, Indonesia.

2. A Numerical Program

2.1 A three-dimensional semi-implicit numerical method

The three-dimensional semi-implicit numerical method used by Casulli & Cheng [1] and Chen [23] in the computation of shallow water flows. The method the finite differences is used for simplicity of its implementation. The basic idea consists with [24]:

- To reduce the field of resolution of the differential equation to a limited field.
- To define a grid (or grid) finished points of this field.
- To approach the derivative this appears in the equation using a development of Taylor around the points of the grid. For the points located at the edges of the field of calculation, we will write the boundary conditions in an exact or possibly approximate way.

Concerning the first and second phase, there are no general methods [25]. The determination of the field of calculation depends on the problem to approach. For the third stage, it is enough to recall that if f is a sufficiently regular function of the variable reality x in a vicinity of the point x_0, then we have the development of following Taylor [25]:

$$f(x)=\sum_{k=0}^{n}(x-x_0)^k\frac{f^{(k)}(x_0)}{k!}+(x-x_0)^{n+1}\frac{f^{(n+1)}[\theta x+(1-\theta)x_0]}{(n+1)!} \tag{1}$$

Where $f^{(k)}$ indicates the derivative of f of order k and $\theta \in \{0,1\}$. The order to which the development is truncated gives the space order of approximation of the quantity f.

2.1.1. Fractional steps method. The basis algorithm consists of three fractional steps. We write [25]:

$$\frac{\partial f}{\partial t}=\frac{f^{n+1}-f_D+f_D-f_C+f_C-f^n}{\Delta t} \tag{2}$$

This leads to a successive resolution of three steps:

- An advection step:

$$\frac{f_C-f^n}{\Delta t}+\text{advection terms}=0\ \ (f=U,V) \tag{3}$$

- A diffusion step:

$$\frac{f_D - f_C}{\Delta t} + \text{diffusion terms} = \text{source terms} \; (f = U, V) \tag{5}$$

- A pressure-continuity step:

$$\frac{f^{n+1} - f_D}{\Delta t} + \text{pressure terms} = 0 \tag{6}$$

This last step includes also the resolution of the continuity equation $div(\vec{U}) = 0$ that helps to deduce the water depth h and the vertical component of velocity W [25]. Each of these steps will be resolved in detail one after the other. Prior to this, however, discretization in space and building up the mesh to specified.

2.1.2. Spatial discretization. The spatial discretization selected by a discretization with the differences finished in parallelepiped elements [1]. The parallelepiped makes it possible to build a three-dimensional grid starting from a two-dimensional grid. It is enough to net in rectangles the two-dimensional field, then to duplicate this grid on the vertical. It is possible to make with the same grid a calculation in 2D and 3D [26]. The grid makes it possible to discretize the physical field in a whole of material points to which we will apply the finite difference method. For the calculation of the variables, those are in a fixed mesh with different positions. Into particulate, the scalar sizes cantered.

An Arakawa C-grid [26] is used. For the calculation of the variables, those are in a fixed mesh in different positions (staggered concealment), so much on the horizontal levels as on the vertical levels. The velocities defined on the edge of the mesh; we guessed virtual meshes to write the limiting conditions with the walls; we decorate the free surface with the grid.

2.1.3. Advection step. In advection step we are using Eulerian-Lagrangian discretization of convective and viscous terms. This discretization is one of the major difficulties in the numerical treatment of the shallow water equations. Consider then the following convection-diffusion equation in three space dimensions [1]:

$$\frac{\partial C}{\partial t} + \bar{u}\frac{\partial C}{\partial x} + \bar{v}\frac{\partial C}{\partial y} + \bar{w}\frac{\partial C}{\partial z} = \mu\left(\frac{\partial^2 C}{\partial x^2} + \frac{\partial^2 C}{\partial y^2}\right) + \frac{\partial}{\partial z}\left(v_{eff}\frac{\partial C}{\partial z}\right) \tag{7}$$

Where μ and v_{eff} are non-negative diffusion coefficients and for the time being, the convective velocities $\bar{u}$, $\bar{v}$ and $\bar{w}$ are assumed to be constants.

The equation (7) can solve numerically in a variety of ways. A convenient semi-implicit finite difference method, whose stability does not depend upon the vertical eddy diffusivity, is obtained by discretizing the convective terms by explicit upwind finite differences, the horizontal eddy diffusivity by explicit central differences and the vertical eddy diffusivity term by an implicit finite difference.

For non-negative $\bar{u}$, $\bar{v}$ and $\bar{w}$ the resulting finite difference equation is:

$$\begin{aligned}
&\frac{C^{n+1}_{i,j,k} - C^{n}_{i,j,k}}{\Delta t} + \bar{u}\frac{C^{n}_{i,j,k} - C^{n}_{i-1,j,k}}{\Delta x} + \bar{v}\frac{C^{n}_{i,j,k}C^{n}_{i,j-1,k}}{\Delta y} + \bar{w}\frac{C^{n}_{i,j,k-1}}{\Delta z} \\
&= \mu\left(\frac{C^{n}_{i+1,j,k} - 2C^{n}_{i,j,k} + C^{n}_{i-1,j,k}}{\Delta x^2} + \frac{C^{n}_{i,j+1,k} - 2C^{n}_{i,j,k} + C^{n}_{i,j-1,k}}{\Delta y^2}\right) \\
&+ \frac{v_{eff_{k+1/2}}\dfrac{C^{n+1}_{i,j,k+1} - C^{n+1}_{i,j,k}}{\Delta z_{i,j,k+1/2}} - v_{dif_{i-1/2}}\dfrac{C^{n+1}_{i,j,k} - C^{n+1}_{i,j,k-1}}{\Delta z_{i,j,k-1/2}}}{\Delta z_{i,j,k}}
\end{aligned} \tag{8}$$

For every i and j this method requires the solution of a symmetric, positive definite, tri-diagonal system. The necessary and sufficient stability condition of scheme equation (8) is

$$\Delta t \le \left[\frac{|\bar{u}|}{\Delta x} + \frac{|\bar{v}|}{\Delta y} + \frac{|\bar{w}|}{\Delta z} + 2\mu\left(\frac{1}{\Delta x^2} + \frac{1}{\Delta y^2} \right) \right]^{-1} \tag{9}$$

In convection-dominated problems, the stability condition equation (9) is not very restrictive. This method, however, is only first-order-accurate in space and the truncation error is in the form of a diffusion term. This artificial viscosity is directionally dependent. Hence, in convection-dominated problems, not only the artificial viscosity will prevail over the physical viscosity, but also drastically different numerical predictions can obtained simply because of different spatial orientations of the computational grid.

In order to improve the stability and accuracy of an explicit finite difference method, consider again equation (7) in the Lagrangian form

$$\frac{\partial C}{\partial t} = \mu\left(\frac{\partial^2 C}{\partial x^2} + \frac{\partial^2 C}{\partial y^2} \right) + \frac{\partial}{\partial z}\left(v_{eff} \frac{\partial C}{\partial z} \right) \tag{10}$$

Where the substantial derivative d/dt indicates that the time rate of change is calculated along the streak line defined by

$$dx/dt = \bar{u}, \quad dy/dt = \bar{v}, \quad dz/dt = \bar{w} \tag{11}$$

A natural semi-implicit discretization of equation (10) is simply given by

$$\frac{C^{n+1}_{i,j,k} - C^{n+1}_{i-a,j-b,k-d}}{\Delta t} = \frac{v_{eff\,k+1/2} \dfrac{C^{n+1}_{i,j,k+1} - C^{n+1}_{i,j,k}}{\Delta z_{i,j,k+1/2}} - v_{eff\,k-1/2} \dfrac{C^{n+1}_{i,j,k} - C^{n+1}_{i,j,k-1}}{\Delta z_{i,j,k-1/2}}}{\Delta z_{i,j,k}}$$
$$+ \mu\left(\frac{C^n_{i-a+1,j-b,k-d} - 2C^n_{i-a,j-b,k-d} + C^n_{i-a-1,j-b,k-d}}{\Delta x^2} + \frac{C^n_{i-a,j-b+1,k-d} - 2C^n_{i-a,j-b,k-d} + C^n_{i-b,j-b-1,k-d}}{\Delta y^2} \right) \tag{12}$$

Where $a=\bar{u}\,\Delta t/\Delta x$, $b=\bar{v}\,\Delta t/\Delta y$ and $d=\bar{w}\,\Delta t/\Delta z$ are the grid Courant Numbers.

Then $C^n_{i-a,j-b,k-d}$ in equation (12) approximated by Casulli and Cheng [1] become:

$$\begin{aligned} C^n_{i-a,j-b,k-d} = {} & (1-r)\{(1-p)[(1-q)C^n_{i-l,j-m,k-n} + qC^n_{i-l,j-m,k-n}] \\ & + p[(1-q)C^n_{i-l-1,j-m,k-n} + qC^n_{i-l-1,j-m-1,k-n}]\} \\ & + r\{(1-p)[(1-q)C^n_{i-l,j-m,k-n-1} + qC^n_{i-l,j-m-1,k-n-1}] \\ & + p[(1-q)C^n_{i-l-1,j-m,k-n-1} + qC^n_{i-l-1,j-m-1,k-n-1}]\} \end{aligned} \tag{13}$$

The stability condition for the scheme equation (12) as follow [1]

$$\Delta t \le \left[2\mu\left(\frac{1}{\Delta x^2}+\frac{1}{\Delta y^2}\right)\right]^{-1} \tag{14}$$

Which is much less restrictive than equation (9). Clearly, when μ=0, this scheme becomes unconditionally stable.

2.1.4. Diffusion step. A general semi-implicit discretization of the momentum equations can write into form as [1]:

$$\overline{u}^{n+1}_{i+1/2,j,k} = (F\overline{u})^{n}_{i+1/2,j,k} - g\frac{\Delta t}{\Delta x}\left(\eta^{n+1}_{i+1,j}-\eta^{n+1}_{i,j}\right) + \Delta t\frac{\nu_{k+1/2}\dfrac{\overline{u}^{n+1}_{i+1/2,j,k+1}-\overline{u}^{n+1}_{i+1/2,j,k}}{\Delta z_{i+1/2,j,k+1/2}} - \nu_{k-1/2}\dfrac{\overline{u}^{n+1}_{i+1/2,j,k}-\overline{u}^{n+1}_{i+1/2,j,k-1}}{\Delta z_{i+1/2,j,k-1/2}}}{\Delta z_{i+1/2,j,k}}, \tag{15}$$

$$\overline{v}^{n+1}_{i,j+1/2,k} = (F\overline{v})^{n}_{i,j+1/2,k} - g\frac{\Delta t}{\Delta y}\left(\eta^{n+1}_{i,j+1}-\eta^{n+1}_{i,j}\right) + \Delta t\frac{\nu_{k+1/2}\dfrac{\overline{v}^{n+1}_{i,j+1/2,k+1}-\overline{v}^{n+1}_{i,j+1/2,k}}{\Delta z_{i,j+1/2,k+1/2}} - \nu_{k-1/2}\dfrac{\overline{v}^{n+1}_{i,j+1/2,k}-\overline{v}^{n+1}_{i,j+1/2,k-1}}{\Delta z_{i,j+1/2,k-1/2}}}{\Delta z_{i,j+1/2,k}} \tag{16}$$

Where $\Delta z_{i+1/2,j,k}$ and $\Delta z_{i,j+1/2,k}$ are in general the thickness of the kth water layer more simply denoted by Δz_k.

The finite difference operator F in equations (15) and (16) can define as [1]:

$$\begin{aligned}(F\overline{u})^{n+1}_{i+1/2,j,k} &= \overline{u}^{n}_{i+1/2-a+1,j-b,k-d} \\ &+ \mu\Delta t\left(\frac{\overline{u}^{n}_{i+1/2-a+1,j-b,k-d}-2\overline{u}^{n}_{i+1/2-a,j-b,k-d}+\overline{u}^{n}_{i+1/2-a-1,j-b,k-d}}{\Delta x^2}\right. \\ &+ \left.\frac{\overline{u}^{n}_{i+1/2-a,j-b+1,k-d}-2\overline{u}^{n}_{i+1/2-a,j-b,k-d}+\overline{u}^{n}_{i+1/2-a,j-b-1,k-d}}{\Delta y^2}\right) \\ &+ f_{cor}\Delta t\overline{v}^{n}_{i+1/2-a,j-b,k-d},\end{aligned} \tag{17}$$

$$\begin{aligned}(F\overline{v})^{n+1}_{i,j+1/2,k} &= \overline{v}^{n}_{i-a,j+1/2-b,k-d} \\ &+ \mu\Delta t\left(\frac{\overline{v}^{n}_{i-a+1,j+1/2-b,k-d}-2\overline{v}^{n}_{i-a,j+1/2-b,k-d}+\overline{v}^{n}_{i-a-1,j+1/2-b,k-d}}{\Delta x^2}\right. \\ &+ \left.\frac{\overline{v}^{n}_{i-a,j+1/2-b+1,k-d}-2\overline{v}^{n}_{i-a,j+1/2-b,k-d}+\overline{v}^{n}_{i-a,j+1/2-b-1,k-d}}{\Delta y^2}\right)\end{aligned}$$

$$- f_{cor}\Delta t\overline{u}^{n}_{i-a,j+1/2-b,k-d} \tag{18}$$

The boundary conditions at the free surface and at the sediment-water interface are following [1]:

$$\nu_{M+1/2}\frac{\overline{u}^{n+1}_{i+1/2,j,M+1}-\overline{u}^{n+1}_{i+1/2,j,M}}{\Delta z_{i+1/2,j,M+1/2}}=\tau^{w}_{x},\quad \nu_{M+1/2}\frac{\overline{v}^{n+1}_{i,j+1/2,M+1}-\overline{v}^{n+1}_{i,j+1/2,M}}{\Delta z_{i,j+1/2,M+1/2}}=\tau^{w}_{y} \tag{19}$$

$$\nu_{m-1/2}\frac{\overline{u}^{n+1}_{i+1/2,j,m}-\overline{u}^{n+1}_{i+1/2,j,m-1}}{\Delta z_{i+1/2,j,m-1/2}}=-\frac{g\sqrt{\left\{\left(\overline{u}^{n}_{i+1/2,j,m}\right)^{2}+\left(\overline{v}^{n}_{i+1/2,j,m}\right)^{2}\right\}}}{C_z^2}\overline{u}^{n+1}_{i+1/2,j,m},$$

$$\nu_{m-1/2}\frac{\overline{v}^{n+1}_{i,j+1/2,m}-\overline{v}^{n+1}_{i,j+1/2,m-1}}{\Delta z_{i,j+1/2,m-1/2}}=-\frac{g\sqrt{\left\{\left(\overline{u}^{n}_{i,j+1/2,m}\right)^{2}+\left(\overline{v}^{n}_{i,j+1/2,m}\right)^{2}\right\}}}{C_z^2}\overline{v}^{n+1}_{i,j+1/2,m} \tag{20}$$

The equations (15) and (16) with the respective boundary conditions equations (19) and (20), we can write in the compact matrix form as [1, 5]:

$$\mathbf{A}^{n}_{i+1/2,j}\mathbf{U}^{n+1}_{i+1/2,j}=\mathbf{G}^{n}_{i+1/2,j}-g\frac{\Delta t}{\Delta x}\left(\eta^{n+1}_{i+1,j}-\eta^{n+1}_{i,j}\right)\mathbf{\Delta Z}^{n}_{i+1/2,j} \tag{21}$$

$$\mathbf{A}^{n}_{i,j+1/2}\mathbf{V}^{n+1}_{i,j+1/2}=\mathbf{G}^{n}_{i,j+1/2}-g\frac{\Delta t}{\Delta y}\left(\eta^{n+1}_{i,j+1}-\eta^{n+1}_{i,j}\right)\mathbf{\Delta Z}^{n}_{i,j+1/2} \tag{22}$$

Where **U**, **V**, **ΔZ**, **G** and **A** are defined as:

$$\mathbf{U}^{n+1}_{i+1/2,j}=\begin{bmatrix}\overline{u}^{n+1}_{i+1/2,j,M}\\ \overline{u}^{n+1}_{i+1/2,j,M-1}\\ \overline{u}^{n+1}_{i+1/2,j,M-2}\\ \vdots\\ \overline{u}^{n+1}_{i+1/2,j,m}\end{bmatrix},\quad \mathbf{V}^{n+1}_{i,j+1/2}=\begin{bmatrix}\overline{v}^{n+1}_{i,j+1/2,M}\\ \overline{v}^{n+1}_{i,j+1/2,M-1}\\ \overline{v}^{n+1}_{i,j+1/2,M-2}\\ \vdots\\ \overline{v}^{n+1}_{i,j+1/2,m}\end{bmatrix},\quad \mathbf{\Delta Z}=\begin{bmatrix}\Delta z_{M}\\ \Delta z_{M-1}\\ \Delta z_{M-2}\\ \vdots\\ \Delta z_{m}\end{bmatrix},$$

$$\mathbf{G}^{n}_{i+1/2,j}=\begin{bmatrix}\Delta z_{M}(F\overline{u})^{n}_{i+1/2,j,M}+\Delta t\tau^{w}_{x}\\ \Delta z_{M-1}(F\overline{u})^{n}_{i+1/2,j,M-1}\\ \Delta z_{M-2}(F\overline{u})^{n}_{i+1/2,j,M-2}\\ \vdots\\ \Delta z_{m}(F\overline{u})^{n}_{i+1/2,j,m}\end{bmatrix},\quad \mathbf{G}^{n}_{i,j+1/2}=\begin{bmatrix}\Delta z_{M}(F\overline{v})^{n}_{i,j+1/2,M}+\Delta t\tau^{w}_{y}\\ \Delta z_{M-1}(F\overline{v})^{n}_{i,j+1/2,M-1}\\ \Delta z_{M-2}(F\overline{v})^{n}_{i,j+1/2,M-2}\\ \vdots\\ \Delta z_{m}(F\overline{v})^{n}_{i,j+1/2,m}\end{bmatrix},$$

$$\mathbf{A}=\begin{bmatrix} \Delta z_M+\frac{(\nu_{eff})_{M-1/2}\Delta t}{\Delta z_{M-1/2}} & \frac{-(\nu_{eff})_{M-1/2}\Delta t}{\Delta z_{M-1/2}} & & 0 \\ \frac{-(\nu_{eff})_{M-1/2}\Delta t}{\Delta z_{M-1/2}} & \Delta z_M+\frac{(\nu_{eff})_{M-1/2}\Delta t}{\Delta z_{M-1/2}}+\frac{(\nu_{eff})_{M-3/2}\Delta t}{\Delta z_{M-3/2}} & \frac{-(\nu_{eff})_{M-3/2}\Delta t}{\Delta z_{M-3/2}} & \\ \vdots & \vdots & \vdots & \vdots \\ 0 & & \frac{-(\nu_{eff})_{m-1/2}\Delta t}{\Delta z_{m-1/2}} & \Delta z_m+\frac{(\nu_{eff})_{m+1/2}\Delta t}{\Delta z_{m+1/2}}+\frac{g\Delta t\sqrt{(\overline{u}^2+\overline{v}^2)}}{C_z^2} \end{bmatrix}$$

Equations (21) and (22) are linear tri-diagonal systems which are coupled to the seawater surface elevation (η^{n+1}) at time (t_{n+1}).

2.1.5. Pressure-continuity step. For determine $\eta_{i,j}^{n+1}$ and for numerical stability, the new velocity field has to satisfy for each i,j the finite difference analogue of the seawater surface elevation equation:

$$\eta_{i,j}^{n+1}=\eta_{i,j}^{n}-\frac{\Delta t}{\Delta x}\left[\sum_{k=m}^{M}\Delta z_{i+1/2,j,k}\overline{u}_{i+1/2,j,k}^{n+1}-\Delta z_{i-1/2,j,k}\overline{u}_{i-1/2,j,k}^{n+1}\right] -\frac{\Delta t}{\Delta y}\left[\sum_{k=m}^{M}\Delta z_{i,j+1/2,k}\overline{v}_{i,j+1/2,k}^{n+1}-\Delta z_{i,j-1/2,k}\overline{v}_{i,j-1/2,k}^{n+1}\right] \quad (23)$$

Or we can be written in the compact matrix form:

$$\eta_{i,j}^{n+1}=\eta_{i,j}^{n}-\frac{\Delta t}{\Delta x}\left[\left(\mathbf{\Delta Z}_{i+1/2,j}\right)^{\mathbf{T}}\mathbf{U}_{i+1/2,j}^{n+1}-\left(\mathbf{\Delta Z}_{i-1/2,j}\right)^{\mathbf{T}}\mathbf{U}_{i-1/2,j}^{n+1}\right] -\frac{\Delta t}{\Delta y}\left[\left(\mathbf{\Delta Z}_{i,j+1/2}\right)^{\mathbf{T}}\mathbf{V}_{i,j+1/2}^{n+1}-\left(\mathbf{\Delta Z}_{i,j-1/2}\right)^{\mathbf{T}}\mathbf{V}_{i,j-1/2}^{n+1}\right] \quad (24)$$

Since $\mathbf{A}$ is positive definite, $\mathbf{A}^{-1}$ is also positive definite and therefore $(\mathbf{\Delta Z})^{\mathrm{T}}\mathbf{A}^{-1}\mathbf{\Delta Z}$ is a non-negative number. Hence equation (25) constitutes a linear five-diagonal system of equations for $\eta_{i,j}^{n+1}$ which is symmetric and strictly diagonally dominant with positive elements on the main diagonal and negative ones elsewhere. Thus the system is positive definite and has a unique solution. In practice, this five-diagonal system can be solved very efficiently by a conjugate gradient method. Once the new free surface location has been determined, equations (21) and (22) are readily applicable to yield the new velocities $\overline{u}$, $\overline{v}$ at time t_{n+1}.

$$\begin{aligned}\eta_{i,j}^{n+1}&-g\frac{\Delta t^2}{\Delta x^2}\left\{\left[(\mathbf{\Delta Z})^{\mathbf{T}}\mathbf{A}^{-1}\mathbf{\Delta Z}\right]_{i+1/2,j}^{n}\left(\eta_{i+1,j}^{n+1}-\eta_{i,j}^{n+1}\right)-\left[(\mathbf{\Delta Z})^{\mathbf{T}}\mathbf{A}^{-1}\mathbf{\Delta Z}\right]_{i-1/2,j}^{n}\left(\eta_{i,j}^{n+1}-\eta_{i-1,j}^{n+1}\right)\right\}\\ &-g\frac{\Delta t^2}{\Delta y^2}\left\{\left[(\mathbf{\Delta Z})^{\mathbf{T}}\mathbf{A}^{-1}\mathbf{\Delta Z}\right]_{i,j+1/2}^{n}\left(\eta_{i,j+1}^{n+1}-\eta_{i,j}^{n+1}\right)-\left[(\mathbf{\Delta Z})^{\mathbf{T}}\mathbf{A}^{-1}\mathbf{\Delta Z}\right]_{i,j-1/2}^{n}\left(\eta_{i,j}^{n+1}-\eta_{i,j-1}^{n+1}\right)\right\}\\ =\eta_{i,j}^{n}&-\frac{\Delta t}{\Delta x}\left\{\left[(\mathbf{\Delta Z})^{\mathbf{T}}\mathbf{A}^{-1}\mathbf{G}\right]_{i+1/2,j}^{n}-\left[(\mathbf{\Delta Z})^{\mathbf{T}}\mathbf{A}^{-1}\mathbf{G}\right]_{i-1/2,j}^{n}\right\}\\ &-\frac{\Delta t}{\Delta y}\left\{\left[(\mathbf{\Delta Z})^{\mathbf{T}}\mathbf{A}^{-1}\mathbf{G}\right]_{i,j+1/2}^{n}-\left[(\mathbf{\Delta Z})^{\mathbf{T}}\mathbf{A}^{-1}\mathbf{G}\right]_{i,j-1/2}^{n}\right\}\end{aligned} \quad (25)$$

Finally, the vertical component of the velocity $\overline{w}$ at the new time level can discretized from the continuity equation becomes:

$$\overline{w}_{i,j,k+1/2}^{n+1} = \overline{w}_{i,j,k-1/2}^{n+1} - \frac{\Delta z_{i+1/2,j,k}^{n}\overline{u}_{i+1/2,j,k}^{n+1} - \Delta z_{i-1/2,j,k}^{n}\overline{u}_{i-1/2,j,k}^{n+1}}{\Delta x} - \frac{\Delta z_{i,j+1/2,k}^{n}\overline{v}_{i,j+1/2,k}^{n+1} - \Delta z_{i,j-1/2,k}^{n}\overline{v}_{i,j-1/2,k}^{n+1}}{\Delta y} \quad (26)$$

Where $k=m,m+1,...,M$, and the no-flux condition across the bottom boundary is assumed by taking $\overline{w}_{i,j,m-1/2}^{n+1} = 0$.

2.2 The available power of marine current and the boundary conditions

The available kinetic energy in this study is the available power per m^2 (kW/m^2). We used the available power that is equation of the marine current power in the Bangka strait from [2, 5]:

$$P = \frac{E_k/dt}{A} = \frac{1}{2}\rho\,(\mathrm{v}_{i,j,k}^{n+1})^3 10^{-3} \quad (27)$$

Where P is the marine current power in the Bangka strait in kW/m^2, E_k is kinetic energy, dt is time, and $\mathrm{v}_{i,j,k}^{n+1} = \sqrt{\overline{\mathrm{u}}^2 + \overline{\mathrm{v}}^2 + \overline{\mathrm{w}}^2}$ is velocity resultant with $\overline{\mathrm{u}} = \frac{1}{2}(\overline{u}_{i,j,k}^{n+1} + \overline{u}_{i+1,j,k}^{n+1})$, $\overline{\mathrm{v}} = \frac{1}{2}(\overline{v}_{i,j,k}^{n+1} + \overline{v}_{i,j+1,k}^{n+1})$ and $\overline{\mathrm{w}} = \frac{1}{2}(\overline{w}_{i,j,k}^{n+1} + \overline{w}_{i,j,k+1}^{n+1})$ are scalars, respectively.

We used the boundary conditions for the simulations of model 2D and 3D such as:

- On the free surface the effect of the wind, supposed negligible which is not taken into account, which translated by $\nu\frac{\partial \vec{U}}{\partial n} = 0$.
- At the bottom, the coefficient of friction (law of Chezy) is $C_z = 48$.
- At open boundary condition, we can radiation equation from Treguier et al [27].

The values of parameter in the computational of numerical program, we used:

- The domain of computation is discretized of 55332 elements (for 2D) where 174 elements in direction x and 318 elements in direction y.
- Each element of horizontal is 60 m x 60 m. In 3D, there are 221328 elements where 174 elements in x-direction, 318 elements in y-direction and 4 elements in z-direction.
- Each element is parallelepiped dimensions of 60 m x 60 m x 20 m.
- The flow rate is 0.3 Sv (300000 m^3/s where 1 Sv = 1 x 10^6 m^3/s [28]) with two conditions of marine currents e.g. when low and high tide.

3. Results and Discussion

The comparisons of numerical model results and measurement results have been obtained by specify the velocity measurement results at area inlet and outlet as boundary conditions. Then, the numerical model results (2D and 3D) are studied and compared mainly focused at points location of P1 to P9 (see Figure 1). Finally, the numerical model results for the study on marine currents in the Bangka strait has been obtained both in 2D and 3D with four flow rates variations which are as the inlet boundary conditions.

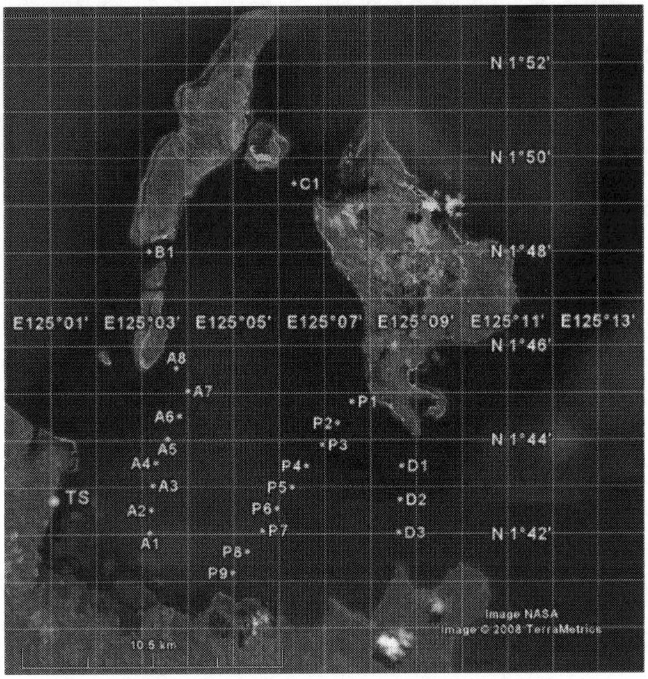

Figure 1. The locations of measurement.

3.1. The comparisons of the numerical program results and the measurement results

Figure 2 shows the comparisons of the minimum velocities from the measurement (observation) results and the numerical program (2D and 3D model) results when low tide currents. The velocities in locations of P3 and P8 are come near with results of observation but the other locations are rather different. The difference values of velocities for both 2D and 3D to observed results in P3 are 0.1 m/s and its directions are 11^0 (Figure 3), whereas in P8 of 0.02 m/s for 3D and there is not for 2D and only 2^0 for both 2D and 3D respectively. On the contrary, in P1, there is 0.34 m/s for 2D and 0.38 m/s for 3D, whereas its directions for both 2D and 3D of 18^0 respectively.

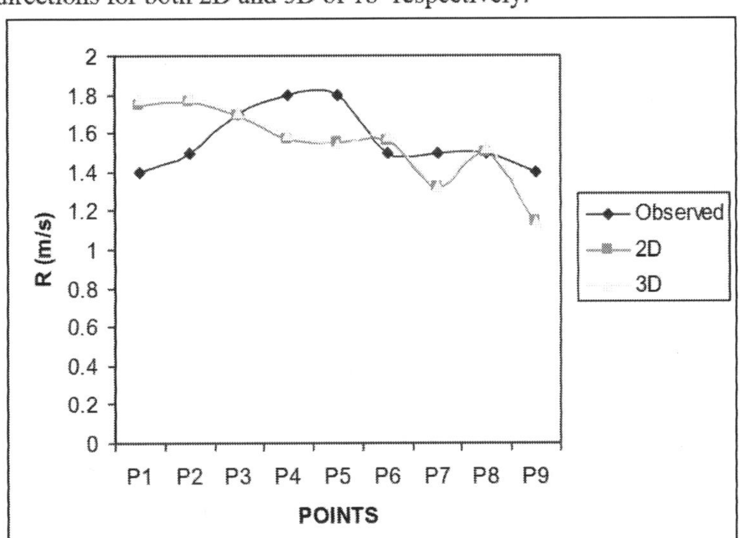

Figure 2. Current velocities at the minimum velocities boundary condition when low tide currents.

The current directions from the numerical model results are not so differ with the observation results mainly P4 to P8 (Figure 3) and also points of P5 to P7 in Figure 5 with boundary conditions of the maximum velocities whereas the other locations are rather different. The biggest differences are in locations of P1, P2, P5, and P9. If we see in Figure 4, the velocities at P5 to P7 are not so differ between observation results and numerical model with the boundary conditions. Whereas velocities from the calculation results of 2D and 3D models are not so differ.

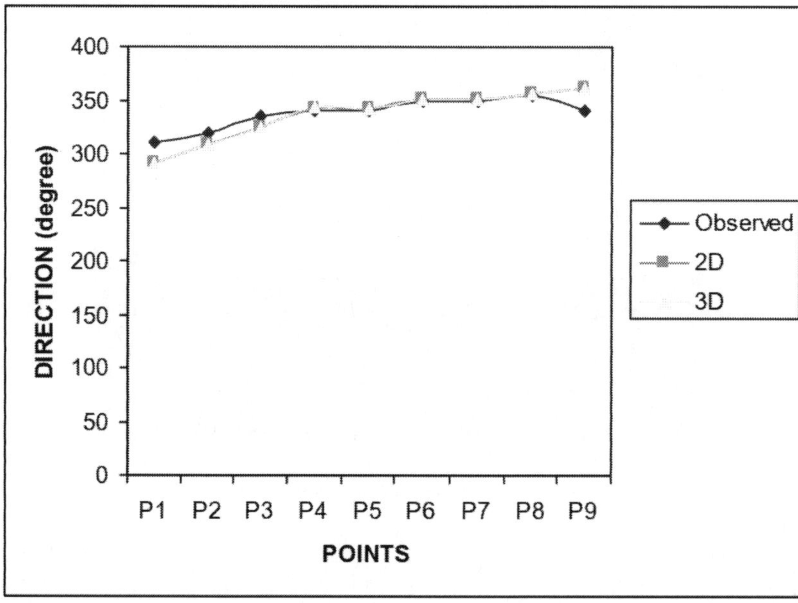

Figure 3. Current directions at the minimum velocities boundary condition when low tide currents.

The difference values of velocities for both 2D and 3D to observed results in P5 are 0.04 m/s (Figure 4) and its directions are 3^0 in Figure 5, then in P7 of 0.04 m/s for 2D and 0.02 m/s for 2D and there are not differ with its directions for both 2D and 3D respectively. On the contrary, in P1, there is 0.55 m/s for 2D and 0.58 m/s for 3D, whereas its directions for both 2D and 3D of 20^0 respectively.

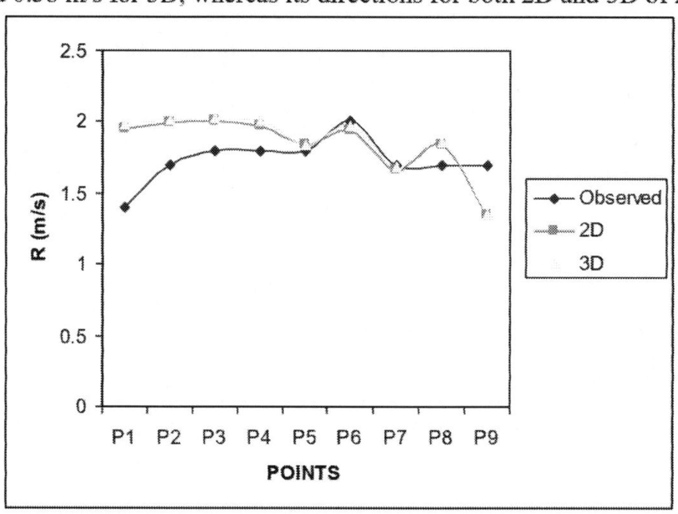

Figure 4. Current velocities at the maximum velocities boundary condition when low tide currents.

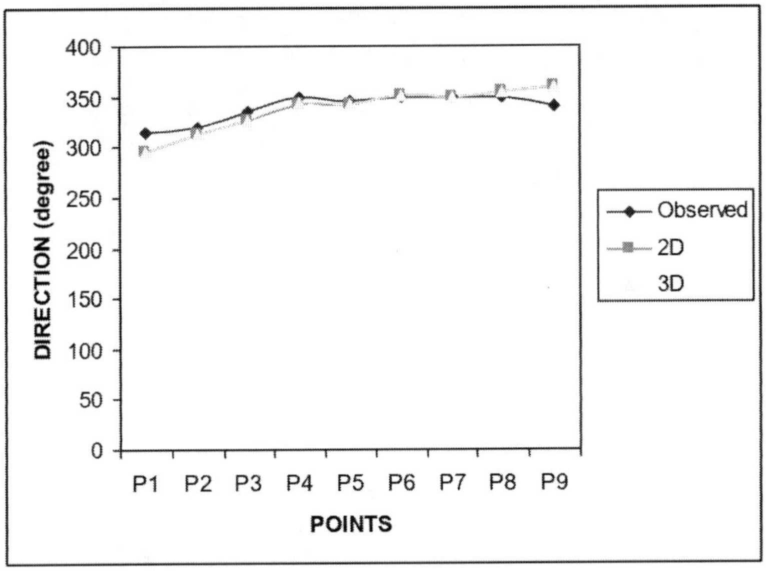

Figure 5. Current directions at the maximum velocities boundary condition when low tide currents.

Figure 6 shows comparisons of the minimum velocities from observation results with the numerical model results when high tide currents. At point P4 where the velocity of 2D and 3D models result compared to the observation result is almost same and also at points of P4 to P6 in Figure 7. Whereas current directions rather near between models (2D and 3D) and observation in Figure 8 except at points P1 to P3.

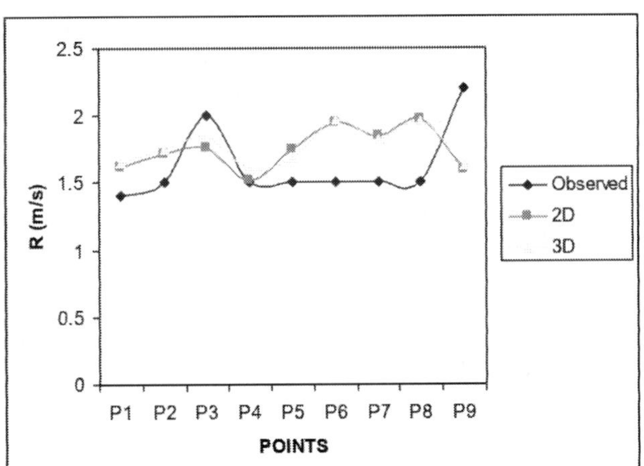

Figure 6. Current velocities at the minimum velocities boundary condition when high tide currents.

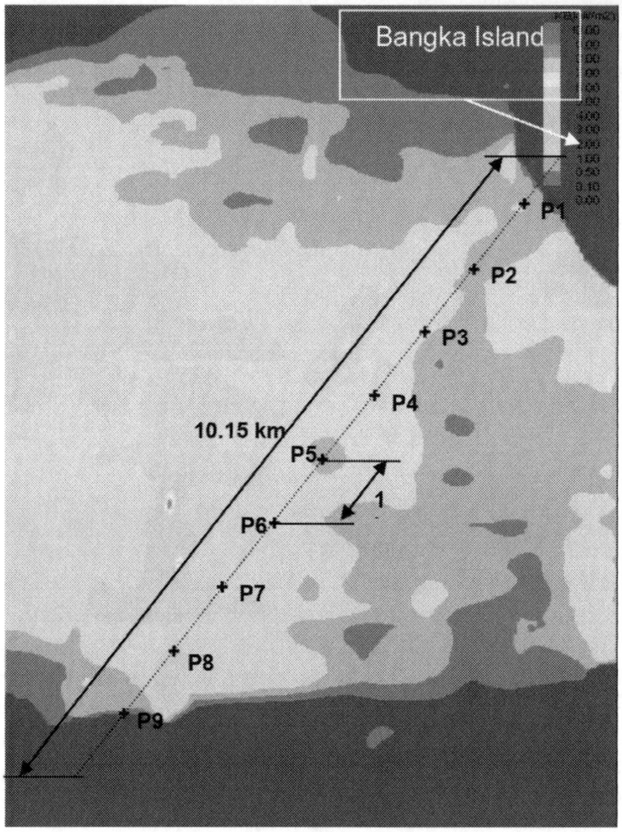

Figure 7. Distributions of the available power per m^2 at seawater column of 20 m when low tide currents at flow rate of 0.3 Sv in the Bangka strait.

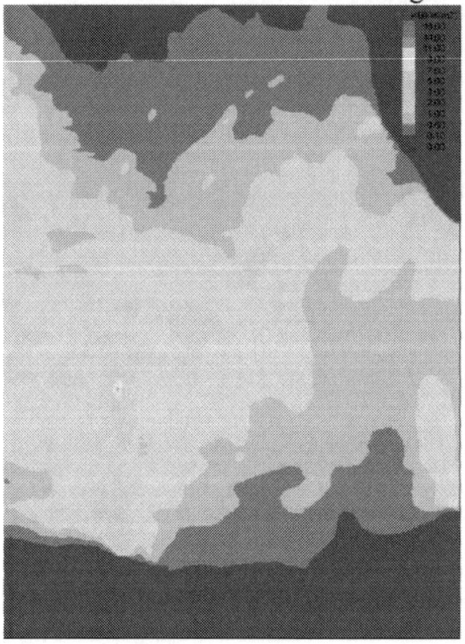

Figure 8. Distributions of the available power per m^2 at seawater column of 20 m when high tide currents at flow rate of 0.3 Sv in the Bangka strait.

3.2. The kinetic energy potential in the Bangka strait

The kinetic energy potential in this study that generated in form the available power of marine current in the Bangka strait have been analyzed. We can see the distributions of the available power per unit of area (kW/m^2) at seawater column of 20 m in Figures 7 and 8 which described the potential in the Bangka strait when low and high tide currents at flow rate of 0.3 Sv (1 Sv = 1 x 10^6 m^3/s [27]).

Figure 7 shows distributions of available power per m^2 at seawater column of 20 m when low tide currents at of 0.3 Sv in the Bangka strait. There is 2.22 kW/m^2 of the available power at P4 and its value is biggest compared at the other points. It caused by nearby the point is existed a manger with deepness of ~5 m. On the contrary, the value of available power per m^2 is smallest at P1 (0.97 kW/m^2). We can also see that available power per m^2 in South area of Bangka island (in the enter channel) where around 3-5 kW/m^2 bigger than the other area in around that of 1.5 kW/m^2. Also, in West area, especially at center area where power availabilities around 2-7 kW/m^2. Whereas in North and South area where available power per m^2 still less unless near point of P9 about 2-3 kW /m^2. If we see in Westside of P6 and P7 where there are power availabilities biggest around 9-10 kW/m^2. That thing caused by existence of manger and average depth in the place of ~5 m [3, 5].

The distributions of the available power per m^2 at seawater column of 20 m when high tide currents at flow rate of 0.3 Sv in the Bangka strait showed in Figure 8. The available power per m^2 counted of 1.02 kW/m^2 at point of P1 and that value is smallest if compared with the other points. On the contrary, there is 2.1 kW/m^2 available of the biggest available power per m^2 at P8 which has the biggest velocity.

4. Conclusions

The validations of the numerical program through the comparisons of the numerical computation results and the measurement results have been analyzed and the results are not so far different. The kinetic energy potential in the form the distributions of available power per area in the Bangka strait when low tide currents of 0.97-2.2 kW/m^2, whereas when high tide currents of 1.02-2.1 kW/m^2. The results show that to be enabling the installation of marine current turbines for construction of power plant in the Bangka strait, North Sulawesi, Indonesia.

Acknowledgments

The authors express gratitude to DRPM, Ministry of Research, Technology, and Higher Education of the Republic of Indonesia that had finance all of the research activities, and to Rector of Universitas Negeri Manado, Indonesia through the head of research institution who has approved this research.

References

[1] Casulli V and Cheng R T 1992 Semi-implicit finite difference methods for three-dimensional shallow water flow *International Journal for Numerical Methods in Fluids* **15** 629-648

[2] Rompas P T D and Manongko J D I 2016 Numerical simulation of marine currents in the Bunaken Strait, North Sulawesi, Indonesia *IOP Conf. Series: Materials Science and Engineering* **128**(012003) 1-7

[3] Rompas P T D, Taunaumang H, and Sangari F J 2017 A numerical design of marine current for predicting velocity and kinetic energy *Indonesian Journal of Electrical Engineering and Computer Science* **5**(2) 401-409.

[4] Rompas P T D, Taunaumang H, and Sangari F J 2017 A numerical model of seawater volume and velocity dynamic for marine currents power plant in the Bangka Strait, North Sulawesi, Indonesia *IOP Conference Series: Materials Science and Engineering* **180**(012100) 1-7.

[5] Rompas P T D, Sangari F J, and Tanunaumang H 2017 Study on marine current with approach of a numerical model for marine current power plant (PLTAL) in the Bangka Strait North Sulawesi *Proceedings-2016 International Seminar on Application of Technology for Information and Communication, ISEMANTIC 2016. IEEE* **1** 104-110.

[6] Rodriguez-Cuevas C, Couder-Castaneda C, Flores-Mendez E, Herrera-Diaz I E, and Cisneros-Almazan R 2014 Modelling shallow water wakes using a hybrid turbulence *Journal of Applied Mathematics* **1** 1-10.
[7] O'Donncha F, James S C, and Ragnoli E 2017 Modelling study of the effects of suspended aquaculture installations on tidal stream generation in Cobscook Bay *Renewable Energy* **102** 65-76
[8] Gonzalez-Gorbena E, Qassim R Y, and Rosman P C 2016 Optimisation of hydrokinetic turbine array layouts via surrogate modelling *Renewable Energy* **93** 45-57
[9] Zangiabadi E, Masters I, Williams A J, Croft T N, Malki R, Edmunds M, Mason-Jones A, and Horsfall I 2017 Computational prediction of pressure change in the vicinity of tidal stream turbines and the consequences for fish survival rate *Renawable Energy* **101** 1141-1156
[10] Zangiabadi E, Edmunds M, Fairley I A, Togneri M, Williams A J, Masters I, and Croft N 2015 Computational fluid dynamics and visualisation of coastal flows in tidal channels supporting ocean energy development *Energies* **8** 5997-6012
[11] Martins F, Leitao P, Silva A, and Neves R 2000 3D modelling in the Sado estuary using a new generic vertical discretization approach *Oceanologica Acta* **24** (1) 1-12
[12] Luquet R, Bellevre D, Fréchou D, Perdon P, and Guinard P 2013 Design and model testing of an optimized ducted marine current turbine *International Journal of Marine Energy* **2** 61-80
[13] Sirjacobs D, Gregoire M, Delhez E, and Nihoul J C J 2004 Influence of the Aral sea negative water balance on its seasonal circulation patterns: use of a 3D hydrodynamic model *Journal of Marine System* **47** 51-66
[14] Balasubramanya T N, Taylor M, and Lorenz J 2006 Ocean modelling for climate studies: Eliminasi short time scales in long-term, high resolution studies of ocean circulation *Mathematical and Computer Modelling* **44** 870-886
[15] Myers L and Bahaj A S 2005 Simulated electrical power potential harnessed by marine current turbine arrays in the Alderney Race *Renewable Energy* **30** 1713-1731
[16] Thiebot J, Guillou S, and Nguyen V T 2016 Modelling the effects of large arrays of tidal turbines with dept-averaged Actuator Disks *Ocean Engineering* **126** 265-275
[17] Pinon G, Mycek P, Germain G, and Rivoalen E 2012 Numerical simulation of wake of marine current turbines with a particle method *Renewable Energy* **46** 111-126
[18] Elhanafi A, Fleming A, Macfarlane G, and Leong Z 2017 Numerical hydrodynamic analysis of an offshore stationary-floating oscillating water column-wave energy converter using CFD *International Journal of Naval Architecture and Ocean Engineering* **9** (1) 77-99
[19] Ho D K H and Riddette K M 2010 Application of computational fluid dynamics to evaluate hydraulic performance of spillways in Australia *Australian Journal of Civil Engineering* **6** (1) 81-104
[20] Georgia Tech Research Corporation 2013 *Assessment of energy production potential from ocean currents along the United State Coastline* (Final project report September 15, 2013, Award number: DE-EE0002661, [Online]. Available: https://energy.gov/sites/prod/files/2013/12/f5/energy_production_ocean_currents_us_0.pdf [Accessed: 25 March 2017])
[21] Masters I, Williams A, Croft T N, Togneri M, Edmunds M, Zangiabadi E, Fairley I, and Karunarathna H A comparison of numerical modelling techniques for tidal stream turbine analysis *Energies* **8** 7833-7853
[22] Mycek P, Pinon G, Germain G, and Rivoalen E 2013 Numerical and experimental study of the interaction between two marine current turbines *International Journal of Marine Energy* **1** 70-83
[23] Chen X 2003 A free-surface correction method for simulating shallow water flows *J. Comput. Phys.* **189** 557-578
[24] Ferziger J H and Peric M 2002 *Computational methods for fluid dynamics* (Springer 3rd rev.ed., ISBN 978-3-540-42074-3, XIV-423. 128 illus)

[25] Hervouet J M 2007 *Hydrodynamics of free surface flows: Modelling with the finite element method.* (England: John Willey & Sons, Ltd.) xiv-341
[26] Arakawa A and Lamb V R 1977 Computational design of the basic dynamical processes of the UCLA general circulation model *Methods of Computational Physics* **17** 173-265
[27] Treguier A M, Barnier B, and De Miranda A P 2001 An eddy-permitting model of the Atlantic circulation: Evaluating open boundary condition *J. Geophy. Res. Oceans* **106** (C10) (22115-22129) 1-23
[28] Siedler G, Church J, Gould J 2001 *Ocean Circulation and Climate, Observing and Modelling the Global Ocean* (Academic Press, San Diego, xx-715)

IOP Conf. Series: Materials Science and Engineering **306** (2018) 012103 doi:10.1088/1757-899X/306/1/012103

Students Perception on the Use of Computer Based Test

R A Nugroho[1*], N S Kusumawati[2] and O C Ambarwati[2]

[1]Public Administration Department, Faculty of Social and Political Science, Universitas Sebelas Maret, Ir. Sutami street 36 A, Kentingan, Jebres, Surakarta, Indonesia
[2]Graduate School of Public Administration, Universitas Sebelas Maret, Ir. Sutami street 36 A, Kentingan, Jebres, Surakarta, Indonesia

*rino.nugroho@staff.uns.ac.id

Abstract. Teaching nowadays might use technology in order to disseminate science and knowledge. As part of teaching, the way evaluating study progress and result has also benefited from this IT rapid progress. The computer-based test (CBT) has been introduced to replace the more conventional Paper and Pencil Test (PPT). CBT are considered more advantageous than PPT. It is considered as more efficient, transparent, and has the ability of minimising fraud in cognitive evaluation. Current studies have indicated the debate of CBT vs PPT usage. Most of the current research compares the two methods without exploring the students' perception about the test. This study will fill the gap in the literature by providing students' perception on the two tests method. Survey approach is conducted to obtain the data. The sample is collected in two identical classes with similar subject in a public university in Indonesia. Mann-Whitney U test used to analyse the data. The result indicates that there is a significant difference between two groups of students regarding CBT usage. Student with different test method prefers to have test other than what they were having. Further discussion and research implication is discussed in the paper.

1. Introduction

Information technology (IT) advancement has changed the way education delivered from lecturers to students. Electronic learning (e-Learning) was introduced as new way of delivering lecture notes, giving assignment and doing test. By using e-Learning student can have plenty benefits such as timely lecture notes, easiness in accessibility and flexibility in time of study [1, 2]. Assessment is critical in measuring student's learning process achievement [3]. With the increasing number of students in recent years, education needs new ways of doing assessment for more efficient [4]. In doing so, Computer Based Test (CBT) is introduced. As part of learning conducted electronically (e-Learning), CBT has some advantages such as easiness in accessibility, flexibility, efficiency, and more consistent result than PPT [1, 2, 4].

Even though plenty benefits have been found by prior studies, some researcher doubt the superiority of CBT compare to more traditional Paper and Pencil Test (PPT). Prior studies have indicated that CBT provide a faster way of collecting test result than its counterparts but it did not have significantly different result [5, 6]. In addition to that recent studies strengthened the argument by stating both test CBT and PPT have similar performance with notes on that computer administered test need more attention on the technical considerations [7] such as its acceptance by the intended users.

CBT adoption has been studied by various studies to seek the predictors of its success. The acceptance and success factor has been studied with various independent variable to predict e-Learning acceptance [8-10]. Meanwhile other studies had probing the instructor's perception and the complexity on the e-Learning system where CBT is included [11].

Having described some of the prior research, it is known that the debates of CBT were focused on the acceptance, performance and the complexity of the test itself. Prior studies rarely discuss about student perception toward both test types [12]. Therefore this preliminary study will fill the gap by providing insight on students' perception on both tests. Such understanding will help academics in providing a better test for students. Having that aim this study will seek the difference between two groups of students on their perception on CBT and PPT. In order to obtain the result, this study examined a hypothesis as follows:

H0: There is no difference between students in perceiving CBT as tools of assessment

2. Methods

This paper is quantitative in nature. It used quantitative technique to describe student's perception towards CBT. Data collection in this study was using online-survey with close-ended questionnaire by asking students in a higher degree vocational education that is pursuing diploma certificate in administration science. Sample was taken from two classes: A and B. The classes were grouped as follows: A-class with PPT as the mean of final exam and B-class with CBT as the mean of final exam). Both classes have exactly the same subject that is computer application for administration science. All of the class members have had experience in using CBT at the mid exam of the same subject. All 30 members of students in each class were invited to fill the questionnaire. The questionnaire was developed based on prior studies regarding e-Learning use [13] with adjustment to the CBT context. The question that used in this preliminary study was "I prefer computer based test to paper-pencil test" using five point likert scale ranging from strongly disagree to strongly agree. After the data was collected statistical procedure was conducted to obtain the correct test. First Shapiro-Wilk tests of normality to test the nature of the data, followed with Levene's test for checking the homogeneity and ends with Mann-Whitney U to examine the difference between the two groups. Finally to enrich the meaning descriptive statistics was used to obtain the complete understanding about the two group difference.

3. Results and Discussion

The survey was conducted a month after the final exam was taken and the final score has been released, so it was not intervene the students' final marks in the subject. Average final scores of these two classes were relatively equal, i.e. 3.7 in 1 – 4 scale. This survey was voluntary so students may or may not fill–in the survey on their convenience. Among 30 of students who were invited 22 students from A-class and 29 students from B-class were responded (see Table 1).

Table 1. Tests of Normality.

		Shapiro-Wilk		
	Class	Statistic	df	Sig.
CBT	A	.896	22	.025
Preferences	B	.914	29	.021

The normality test using Shapiro-Wilk technique indicates non normal distribution for both A and B class with p-Value $< .05$. The result indicated that Mann-Whitney U test is the correct procedure for testing the difference between the two groups. The next step is testing the Homogeneity of the data using Levene's test as part of the Mann-Whitney U test procedure (see Table 2).

Table 2. Test of homogeneity of variance.

		Levene Statistic	df1	df2	Sig.
Attitude	Based on Mean	.161	1	49	.690
	Based on Median	.251	1	49	.619
	Based on Median and with adjusted df	.251	1	48.999	.619
	Based on trimmed mean	.227	1	49	.636

The Levene's test indicate that the data variance is homogenous (p-Value=0.690 > 0.05). This result justified and strengthened the decision to use of Mann-Whitney U test.

Table 3. Test statistics.

	CBT Preferences
Mann-Whitney U	209.000
Wilcoxon W	644.000
Z	-2.157
Asymp. Sig. (2-tailed)	.031

The Mann-Whitney U Test presented in table 3 shown that the p-Value is less than the critical value (.03 < .05). This result indicates that the H0 should be rejected, meaning that there is significant difference between the two groups on the student's perception of the preferences of CBT. This finding is interesting to probe further, thus descriptive statistics was used for deeper understanding on how the agreements were differ. As mentioned before question about test preferences: "I prefer computer based test to paper-pencil test" was asked in the survey. The results are presented in figure 1 and 2.

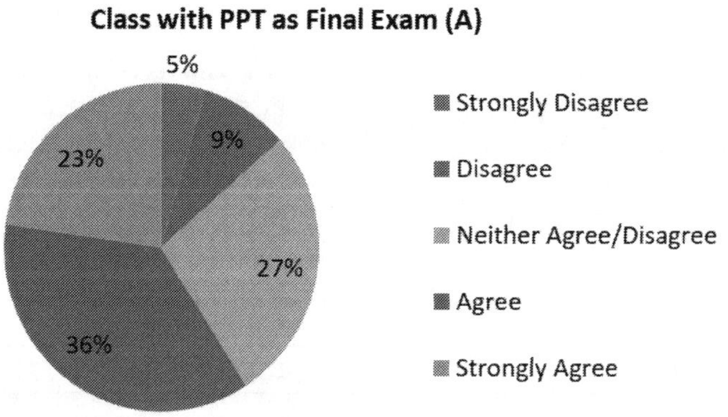

Figure 1. Diagram of A-class students' preferences on CBT.

Figure 1 shown that 59% of students that using PBT as their final exam prefers to use CBT as the means to conduct a test. Meanwhile only 14% of them were disagree of CBT usage for conducting a test. The other students have neutral opinion towards the CBT test (27%). It means that most of the students that used PPT prefers CBT as the tools for final assessment.

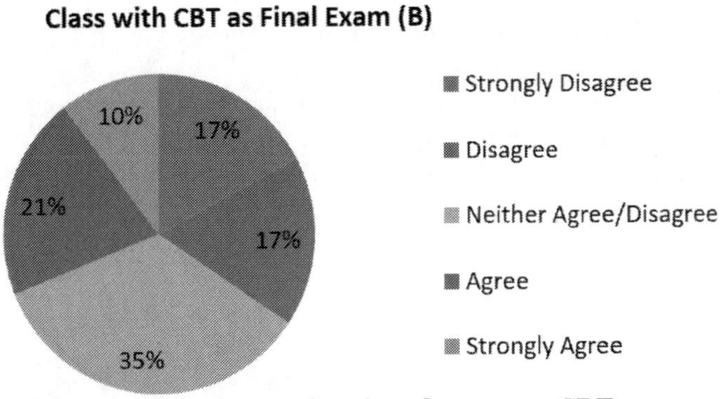

Figure 2. Diagram of B-class students' preferences on CBT.

Figure 2 shown that 31% of students that using CBT as their final exam prefers to use CBT as the means to conduct a test. Meanwhile only 34% of them were disagree of CBT usage for conducting a test. The other students have neutral opinion towards the CBT test (35%). This result contradicts A-class opinion where students with PPT prefers CBT while those students who use CBT were mostly prefers PPT.

The hypothesis test shows that there was significant difference between two groups in this preliminary research. By having this result it can be stated that the student's perception on two modes of test: CBT and PPT is different. Such finding was then probed further by using descriptive statistics as the second step of analysis. The data presented in figure 1 and 2 indicated an interesting result, where most of students that used PPT as their final exam were more favouring CBT while on the contrary most of the students with CBT were not favouring CBT as the tools of assessment. This inconsistent finding is interesting. In the condition where both classes have experience in having CBT in the prior test and relatively have similar performance in their final exam, they have different perception toward the use of CBT.

Prior studies found that computer unfamiliarity as the reason of students' negative perception toward CBT [14, 15]. However this finding might not relevant with the case in this study, where samples were taken from class with computer related subject. Therefore they must be familiar with computer usage. Moreover they have taken CBT on their mid-exam with most students have good result. Another studies found that social anxiety disorder (SAD) has hindered students in performing their best during an examination using CBT method [15]. It might be the reason why the students in B-class did not favour CBT as the tool for final assessment. On the contrary students that having PPT as the way of doing final assessment might also having similar anxiety so they imagining of having CBT as their tool preference. This finding argues that most of those students having SAD, so they tend to imagine other forms other than what they were having.

4. Conclusions

This preliminary study result opposed prior studies that mostly conclude there was no difference between the paper pencil test (PPT) and computer based test (CBT) [16-20]. This might be based on the condition where most prior studies were primarily focused on the test performance whereas this study focused on the students' perception. Interestingly those students who have different perception were reversed, those who have PPT preferring CBT on the contrary those who were having CBT prefers other forms of test. This condition might be based on the presence of social anxiety disorder (SAD) as has been found by Tilfors [15]. The result strengthened prior studies [15] as well as completing the result on how other forms of test (PPT) also trigger SAD for the students. Practically, this result suggests that preparing student to control their anxiety before test is essential before examination, neither by using PPT or CBT. For further studies this research suggests the use of anxiety concept as the variable that influence students' perception towards CBT and PPT use.

References

[1] Cantoni V, Cellario M and Porta M 2004 *Journal of Visual Languages and Computing & Computin* **15** pp 333–45
[2] Kelly T andBauer D 2004 *Handbook on knowledge management 2: Knowledge directions* (Berlin, Germany : Springer) pp 511–32.
[3] Chaczko Z, Klempous R,Nikodem J and Rozenblit J 2016 *IEEE 14th International Symposium on Applied Machine Intelligence and Informatics* pp 263 -67.
[4] Karay Y, Schauber SK, Stosch C and Schuttpelz-Brauns K 2015 *Teaching and learning in medicine* **27**(1) pp 57-62.
[5] Bodmann, Shawn M., and Daniel H. Robinson 2004 *Journal of Educational Computing Research* **31**(1) pp 51-60.
[6] Wang S, Jiao H, Young MJ, Brooks T and Olson J 2008 *Educational and Psychological Measurement* **68**(1) pp 5-24.
[7] Bugbee Jr AC 1996 *Journal of research on computing in education* **28**(3) pp 282-99.
[8] Liaw S S 2008 *Computer and Education* **51**(2) pp 864-73.
[9] Liu S H, Liao H L and Pratt J A 2009 *Computer and Education* **52** (3) pp 599-607.
[10] Pituch K and Lee Y 2006 *Computers and Education* **47**(2) pp 222–44.
[11] Sánchez-Franco M J, Martínez-López F J and Martín-Velicia F A 2009 *Computers and Education* **52**(3) pp 588–98.
[12] Lee B C , Yoon J O and Lee I 2009 Learners' acceptance of e-learning in South Korea: Theories and results *Computers & Education* **53**(4) pp 1320-29.
[13] Bugbee Jr A C 1996 The equivalence of paper-and-pencil and computer-based testing *Journal of research on computing in education* **28**(3) pp 282-99.
[14] Lee, J., Moreno K E and Sympson J B 1986 *Educational and Psychological Measurement* **46** pp 727–33.
[15] Tillfors M, Furmark T, Carlbring P and Andersson G 2015 *Journal of anxiety disorders* **33** pp 103-9.
[16] Mills C N (Ed.) 2002 *Computer-based testing: Building the foundation for future assessment* (NJ: Lawrence Erlbaum).
[17] Smith B and Caputi P 2004 *Journal of Educational Computing Research* **31**(4) pp 407–22.
[18] Wise S L and Plake B S 1990 *Measurement and Evaluation in Counseling and Development* **23** pp 3–10.
[19] Wise S and Plake B 1989*Educational Measurement: Issues and Practice* **8**(3) pp5–10.
[20] Akdemir O and Oguz A 2008 *Computers & Education* **51** pp 1198-204.

Design Learning of Teaching Factory in Mechanical Engineering

R C Putra*, I H Kusumah, M Komaro, Y Rahayu, E P Asfiyanur

Technology and Vocational Education, Sekolah Pascasarjana Universitas Pendidikan Indonesia, Jl. Dr. Setiabudhi 207 Bandung 40154 Indonesia.

*rickycsp@student.upi.edu

Abstract. The industrial world that is the target of the process and learning outcomes of vocational high school (SMK) has its own character and nuance. Therefore, vocational education institutions in the learning process should be able to make the appropriate learning approach and in accordance with the industrial world. One approach to learning that is based on production and learning in the world of work is by industry-based learning or known as Teaching Factory, where in this model apply learning that involves direct students in goods or service activities are expected to have the quality so it is worth selling and accepted by consumers. The method used is descriptive approach. The purpose of this research is to get the design of the teaching factory based on the competency requirements of the graduates of the spouse industry, especially in the engineering department. The results of this study is expected to be one of the choice of model factory teaching in the field of machinery engineering in accordance with the products and competencies of the graduates that the industry needs.

1. Introduction

The world of industry makes use of high technology to improve its production in this emerging market era. To take advantage of these high technologies, the industry requires quality human resources [1]. Recent research indicates the need for a teaching curriculum that can meet the growing and growing industry needs [2]. Lessons learned need to use facilities and infrastructure along with real situations in accordance with the industry [3].

Teaching Factory is a learning based on modern practices and facilities for the realization of industrial products aimed at providing a better educational experience that emphasizes the business world and the industrial world especially the manufacturing industry [4]. Teaching Factory is expected to be an alignment between learning with the modern industry world and for two-way communication between academia and industry [5]. A harmonized learning becomes one of the efforts to develop skills so as to produce professional graduates in the field of engineering, especially manufacturing [4].

Teaching Factory is an effort to complement the knowledge of emerging modern industry, to produce graduates who can compete in the industry today, update the existing curriculum, to be a solution to the dynamics of challenging technology as a whole, transferring technology and information from the couple's industry [6]. Therefore, it is necessary to design teaching factory alignment of the school with the industry so that the competence of graduates in accordance with industry needs, especially majoring in machining engineering.

1.1. Teaching factory

Teaching Factory or *Pabrik Pengajaran* is a concept of learning in the real atmosphere, so as to bridge the gap of competence between industry needs and knowledge of the school. Innovative learning technology and productive practice is the concept of educational methods oriented to the management of student management in learning to be in tune with the needs of the industrial world [7].

In a simple concept teaching factory is the development of the double unit production and education system that has been implemented in SMK - SMK. The concept of teaching factory is one form of development from vocational school to model of production school. SMK is still difficult to implement production-based education (production based education and training) as implemented in ATMI (Akademi Teknik Mesin Indonesia) [8]. Therefore, the term teaching factory requires SMK to implement a business unit or unit of production as a place for student learning. In the business or production unit, students directly practice by producing goods or services that can be sold to consumers. The implementation of teaching factory for learning by establishing a business unit or production in school is in contrast to the learning process taking place in Germany. The practical activities of vocational school students in Germany are conducted within a factory or company, while the government teaches theoretical materials at school for one to two days per week [9].

In another sense, production-based learning is a process of skill or skill learning that is designed and implemented based on real-time work procedures and standards to produce goods or services appropriate to market or consumer demands. In other words, the manufactured goods can be a production that can be sold or that can be used by the community, the school or the consumer. Thus, teaching factory is a learning activity where students directly perform production activities either in the form of goods or services within the school education environment. Goods or services produced have the quality so it is worth selling and accepted by the public or consumers. Teaching factory as one learning strategy has several goals. In a paper published by the American Society for Engineering Education Annual Conference and Exposition, the objectives of the teaching factory are: to produce professional graduates in their fields, to develop curricula that focus on modern concepts, demonstrate appropriate solutions for the challenges facing the industrial world, as well as the transfer of technology from industries that are partners with students and educational institutions [10]. While the development of the teaching factory at Penn State University, The University of Puerto Rico-Mayagues, The University of Washington and Sandia National Labs aims to provide real experience in the design, manufacture and realization of designed products and develop a curriculum that has a balance between knowledge theory and analysis with professional manufacturing, design, business, and skill [11].

2. Methods

This research uses descriptive research method with the flow of research activities as follows (see figure 1):

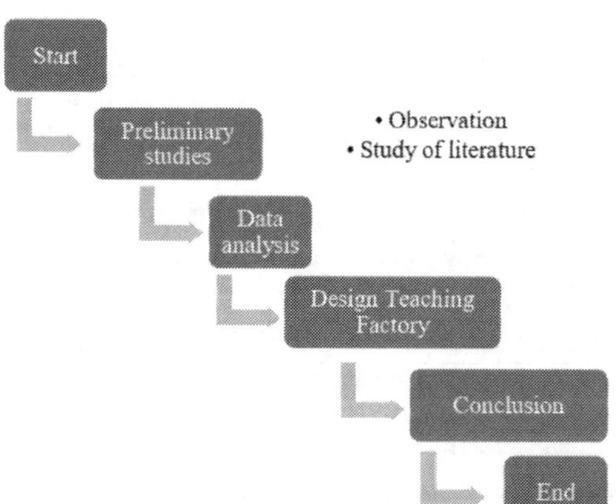

Figure 1. Flowchart of research activities.

3. Results and Discussion

3.1. Concept teaching factory

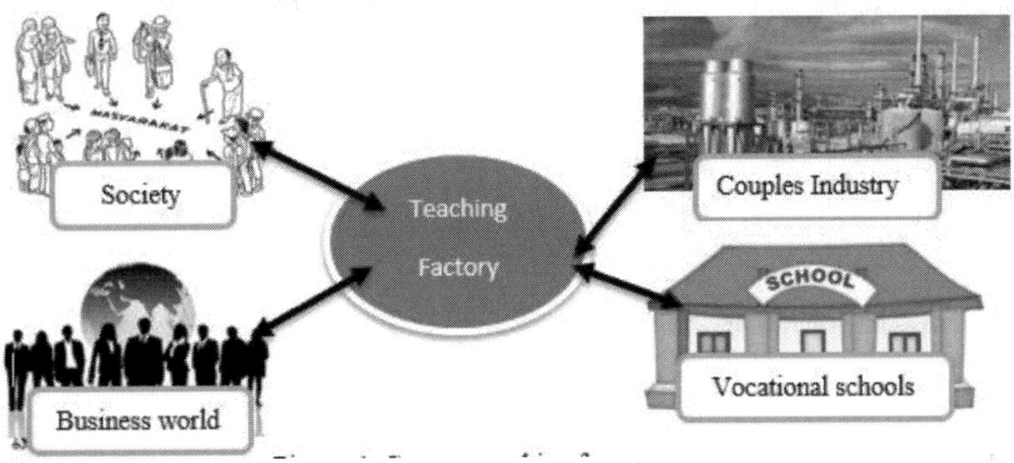

Figure 2. Concept teaching factory.

Based on figure 2, there are 4 main factors that become the basis of competence that must be achieved in teaching factory teaching is a society that represents aspects of soft skill and attitude that must be owned by graduates who will join the community itself, Business World representing entrepreneurial competence expected graduates become self-sufficient and entrepreneurial, Industry partner is the main target that represents the hard skill and soft skill competencies to work, and the vocational school itself has a major role in teaching facilities and facilities itself.

3.2. Frame work

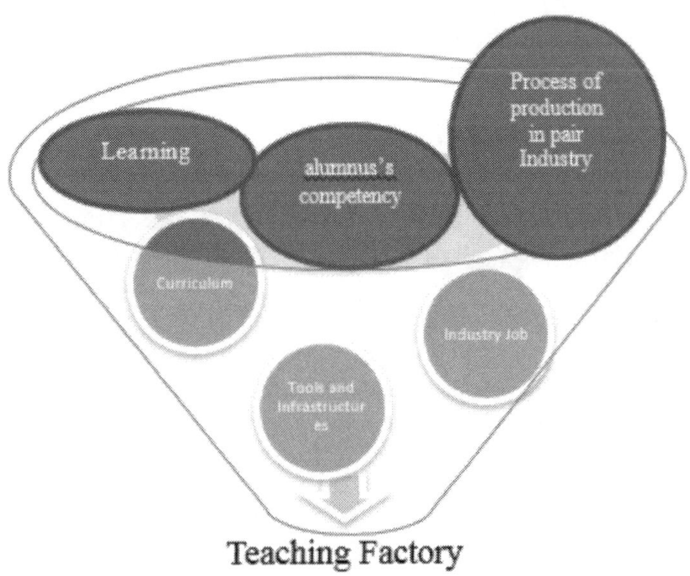

Figure 3. Frame work of teaching factory.

Based on figure 3, there are six factors according to researcher that has to be attention to get the teaching factory concept right, they are: learning process, alumnus's competency, process of production in pair industry, curriculum, job in Industry, and Tools & Infrastructures.

3.3. Design teaching factory
Based on figure 4 in appendix, there are explains about design teaching factory machining technique.

4. Conclusions
The design of this factory enables the pair industry's products especially the machining engineering department which can be produced and done by the students in the school through the learning process. In the planning of the educational institutions and the industry to negotiate to determine the products that can be done students with existing school facilities, but the product has a sale value to be accepted by consumers and the industry itself. In the process of learning that is implemented also allows third parties can order special order products that can also be done or produced students with existing facilities at school through analysis first. This design is expected to be one of the bridge gap trimmer between the competence of graduates with the competence of industry needs, especially industry couple.

References
[1] Nuryake F 2012 Evaluasi pelaksanaan teaching factory smk di Surakarta *Jurnal Pendidikan Vokasi* **2**(3).
[2] Rentzos L A, et al. 2014 Integrating Manufacturing Education with Industrial Practice usingTeaching Factory Paradigm: A Construction Equipment Application *Procedia CIRP 17* pp 189 – 194.
[3] Dimitris M, et al. 2017 A Web-based Application for Classifying Teaching and Learning Factories *7th Conference on Learning Factories, CLF 2017: Procedia Manufacturing* **9** pp. 222 – 228.
[4] John S L, et al. 1997 The Learning Factory—A New Approach to Integrating Design and Manufacturing into the Engineering Curriculum *Journal of Engineering Education.*
[5] G Chryssolouris, et al 2016 The Teaching Factory: A Manufacturing Education Paradigm *49th CIRP Conference on Manufacturing Systems (CIRP-CMS 2016): Procedia CIRP 57 (2016)* pp 44 – 48.
[6] Sema E A 2001 Teaching Factory *Proceedings of the 2001 American Society for Engineering Education Annual Conference & Exposition.*
[7] Orlich D C, et al. 2007 *Teaching Strategies: A guide to Effective Instruction* (New York:Houghton Mifflin Company).
[8] Triatmoko S J 2009 *The ATMI story, rainbow of excellence* (Surakarta: Atmipress).
[9] Moerwishmadhi 2009 *Teaching factory suatu pendekatan dalam pendidikan vokasi yang memberikan pengalaman kearah pengembangan technopreneurship* (Makalah disajikan dalam Seminar Nasional Technopreneurship Learning for *Teaching factory* di Universitas Negeri Malang).
[10] Alptekin S E, et al 2001 Teaching factory *Proceedings of the 2001 American Society for Engineering Education Annual Conference and Exposition, Cal Poly, San Luis Obispo.* Available at http://digitalcommons.calpoly.edu.
[11] Jorgensen J E, et al. 1995 The learning factory *Proceedings of the Fourth World Conference on Engineering Education, St. Paul, Minneapolis, USA.*

Appendix

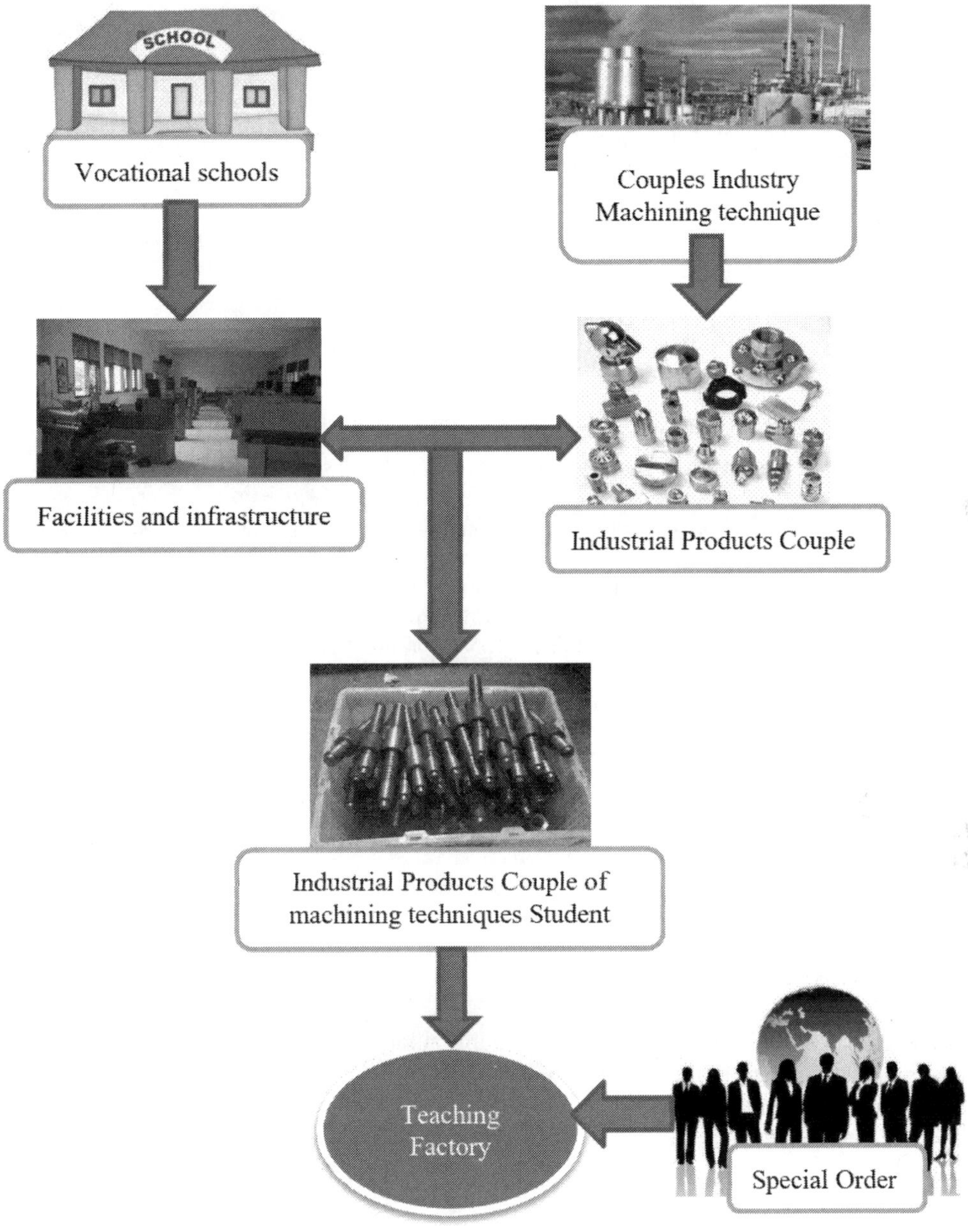

Figure 4. Design teaching factory machining technique.

IOP Conf. Series: Materials Science and Engineering **306** (2018) 012105 doi:10.1088/1757-899X/306/1/012105

People with Disability in Vocational High Schools: between School and Work

R H Haryanti

Department of Public Administration, Universitas Sebelas Maret, Surakarta, Indonesia

rinaherlinaharyanti@staff.uns.ac.id

Abstract. Vocational education is positioned within the framework of Vocational Education for All. Therefore, the alignment between the world of education and the world of work is an issue that is always actual within the framework of vocational education, including being an actual issue for people with disabilities. This article aims to map how the state frames disability and vocational education issues within the framework of public policy. The research was conducted using qualitative research method in which the data obtained from the study of documentation. Analysis of the data using content analysis. The results of the study show that the State Policy has not fully framed the issue of vocational education for the disabled into special policies. The vocational education policy for the disabled is still integrated in the major policies in certain institutions. No policy innovations have yet significantly provided a special place for the disabled.

1. Introduction

Vocational education as part of the national education system that prepares its graduates to enter the world of work is required to produce learning outcomes that are in harmony with the needs of the world of work. [1]. In addition, vocational education has a strategic role in spurring economic growth and change agents in increasing the resilience, competitiveness, and glory of the nation. To achieve this goal vocational education needs to be positioned within the framework of Vocational Education for All (VoEFA), one of which is decent and decent vocational education for disable people.

In some developed countries, people with disabilities are prepared to have life skill skills to support their independence and livelihood later as adults. In their research in the United States show that the government prepares the careers of special needs students of school age by conducting career guidance and exploration with parental participation, skills training, skills assessment and helping to find alternative employment for them after graduation [2]. Spichtinger [3] in his research mentions that the German government has prepared regulation and provision of the labor market for disables by mapping and preparing vocational education for young people with disabilities. Hirvonen [4] in his research also mentioned that vocational education for disabled students in Finland has been able to change the concept of inclusive education into vocational-based education so that students with special needs are more easily accepted in the community.

Indonesia as a developing country certainly does not yet have a policy like in developed countries in thinking and implementing vocational education programs for the disabled to prepare for their future.

How in Indonesia? In Indonesia work is a crucial issue for people with disabilities to survive and gain social recognition.

The Ministry of Manpower and Transmigration states that the number of disability workers or persons with disabilities who were at productive age in Indonesia in 2010 reached 7,126,409 people [5], but 63% of persons with disabilities are not working. In 2016 the Central Bureau of Statistics (BPS) issued a national employment survey (Sakernas). This allows for a deeper analysis of the conditions of persons with disabilities in the Indonesian labor market. Head of Research Team of LPEM FEB University of Indonesia, Halimatussadiah explains the estimated number of persons with disability in Indonesia by 12.15 percent. The average category is 10.29 percent and the weight category is 1.87 percent. Meanwhile, the prevalence of provincial disability in Indonesia is between 6.41 percent to 18.75 percent. The three highest prevalence provinces are West Sumatra, East Nusa Tenggara and South Sulawesi. From 12.15 percent of people with disability 45,74 percent education level of disabilities never or not pass elementary school, far compared to non- disabled with 87,31 percent educated elementary to above. And it turns out that number of disability is more woman that is 53, 37 percent. While the remaining 46.63 percent were male.

2. Research Methods

This article aims to analyze how the state through its policies regulates the vocational education for the disabled. This article attempts to analyze policy content on INPRES No9 of 2016 on Vocational Education Revitalization Policy. The study was conducted by literature study. The analytical techniques used in the literature study are (1) synthesize, comparing, (2) summarize, summarizing the results of expert research on vocational education policy for the disabled and (3) criticize, How vocational education is in state policy.

3. Vocational Education for Difabel

Traditionally, according to Pavlova [6] vocational education is education with the main purpose of preparing to work by using competency-based education approach. Furthermore, according to Pavlova education work is an educational program with three interrelated components: learning for work, learning about work. Components of learning to work include work-related knowledge and practice, the learning component of work includes setting and condition, and components of understanding of the nature of work relating to socio-cultural, economic and political pressures that affect employment.

Referring to the formulation to develop the quality of vocational school education (vocation) can be formulated as a whole (characteristic) ability graduates who can meet the needs of the world of work. Quality control of education, concerning the control of educational components that support the fulfillment of the quality of education required by the world of work. These components consist of the quality policy of education, curriculum, learning, educational facilities, learners, and educators. The result of the educational process is the ability of graduates, while the graduate quality criteria is a description of the capability (performance) demanded by the world of work. Quality control is a technique and operational used to meet quality requirements. Vocational education is an amalgamation of theory and practice in a balanced manner with an orientation to the readiness of its graduates.

The curriculum in vocational education, concentrated on the apprenticeship of learning system on specific trades. The advantages of this vocational education, among others, learners can directly develop their skills tailored to the needs of the field or field of tasks that will be faced. According to Iswari [7] the purpose of vocational education is: 1. To improve the skills of students with special needs / disabled persons in doing a job that is in accordance with the interests, abilities and needs of the community, and areas of arable according to disability. 2. To improve the skills of learners in achieving and creating the type of work that is appropriate ability and not hindered by disability. 3. At the same time instilling a high entrepreneurial attitude and spirit to continue entering the world of work both as a worker and as an entrepreneur. 4. To increase the confidence of the business world and industry in order to be able to employ them as a provision to face a decent life as any other normal person. Vocational education is a learning process that prepares students to enter the work field after completing their studies. This means that vocational education is a real condition that is established to realize knowledge in accordance with the values expected in the work. Thus, in the organization of vocational education, the vocational education curriculum (in the sense of metaphysics) should be structured according to the reality required for work, the method of teaching and learning (in the sense of epistemology) is also adapted to conditions such as work, and has a result value (in the sense of axiology) Which are expected to match the demands of the labor market.

According to The Vocational Education in Democracy [8] there are 16 things to consider in vocational education: 1. The training environment is the working environment or a replica of the working environment. 3. The trainee is trained specifically in the manipulative habits and thinking habits required in the occupation itself 4. The training is the trainee to capitalize his interests and abilities to the highest possible Degree 5. The training is given to those who need it, want it, and are able to profit by it 6. Adequate repetitive training in experiences from the occupation fixes right habits of doing and thinking necessary for employment. The instructor is himself master of the skills and knowledge he teaches 8. Training is carried to the point where it can secure employment or hold employment. 9. Training meets the market for any given occupation. 10. Training is given on actual jobs and not in exercises or pseudo jobs. 11. The content of the training which is taught is obtained from the masters of the occupation, not theorists. 12. This teaching contents applies so much and specifically to the occupation that it has the functioning value for this occupation only. 13. The training needs 14. The particular characteristics of those it serves are considered --both in methods of instruction and in personal relations with learners. 15. The administration is elastic and fluid. 16. The funds expended on training are at least sufficient to permit good training to be done.

4. Map of Vocational Policy for Difabel in Indonesia: Analysis of Presidential Regulation Number 9 Year 2016 on Vocational High School Revitalization

Jokowi-JK government commitment in education, especially vocational education is interesting to be observed and deserves to be appreciated. By mid-2016, President Jokowi has even made a clear affirmation of the importance of revitalizing vocational education. Next end of 2016, precisely on September 9, 2016 President Jokowi publish Policy Revitalization of Vocational Education. This means that President Jokowi's policy and commitment to vocational education is really affirmative. As stipulated on the institution on September 20, 2016, mentioned in the framework of strengthening synergy among stakeholders and to improve the quality and competitiveness of Indonesian human resources, President Joko Widodo on September 9, 2016 has signed the Presidential Instruction Number 9 Year 2016 on the revitalization of Vocational High School in order to Improve the Quality and Human Resources of Indonesia.

In general, the Presidential Instruction is addressed to: 1. The Ministers of the Cabinet of Work: 2. The Head of National Agency for Certification of Profession (BNSP); And 3. The Governors, in the Presidential Instruction President Jokowi also gave special assignment to 11 Ministries / Institutions. In the Instruction, in detail, the tasks and instructions given to 11 ministries and institutions are: Ministry of State-Owned Enterprises, Ministry of Industry, KKP, Ministry of Labor, Ministry of Transportation, Ministry of Energy and Mineral Resources, Ministry of Health, Ministry of Finance and National Agency Professional Certification. The 11 ministries and institutions have been firmly granted duties and powers as their respective domains. Of course, in it includes arranging roadmap for revitalization of vocational education.

Very interesting and need to be appreciated from this Presidential Instruction is the president involves so many ministries and institutions even explicitly the governors of Indonesia as part of this policy implementer. Unfortunately, this policy did not discuss the revitalization of vocational education for the disabled. Unfortunately, the authors do not see the social ministries involved and are mentioned in the policy. Whereas when speaking disable, the Ministry of Social Affairs is usually an inseparable part of the disability. The great hope is that the revitalization instruction of vocational education can run smoothly. However, this policy roadmap requires governments and stakeholders to put first what we call the paradigm of vocational education.

The key to success lies in how the central and regional governments (provinces) have collaborative action frameworks and synergies in realizing the revitalization of vocational education. Policy on revitalization of vocational education involving 11 (eleven) ministries is actually used as a momentum to put the fundamentals of vocational education. The key variable is integrating economic development and vocational education in one national policy breath. The government no longer sees education as a supporting system of national development. Education, especially vocational education is the backbone of economic development and essential for the sustainability of national development. However, the vocational education paradigm in Indonesia is actually built on the framework of Vocational Education for All (VoEFA), one of which is decent and decent vocational education for disable people.

5. Conclusions

Basically career and employment are the rights of every citizen including a population with special needs (disabled). Failure is not an obstacle for a person to choose and run a career according to its potential. There have been many examples and evidence showing that disable is able to succeed in accordance with their career choices. Therefore, vocational education is required to create a design education and training that is friendly and feasible for the disable. The design should be covered in state policies included in Presidential Instruction No. 89 of 2016

Acknowledgments

The authors wish to acknowledge the University of Sebelas Maret, the Ministry of Education and Culture, the Ministry of Technology and Higher Education and All the Parties who assist the completion of this article.

References

[1] Wagiran 2008 *Butir-butir Pemikiran Pengembangan Pendidikan Vokasi Secara Holistik.* Universitas Negeri Padang,

[2] Mynatt, dkk. 2011. *Preparing Students with Disabilities for Their Future Careers. American Counseling Association Conference*, March 25-27, 2011, New Orleans, LA. (online), in (http://counselingoutfitters.com/vistas/vistas11 /Article_08.pdf)

[3] Spichtinger, dkk. 2013. Prevocational Training Situation of Young People with Special Educational Needs (SEN) In Germany. Social Welfare Interdisciplinary Approach 3 (1). (online), at http://www.su.lt/bylos/mokslo_leidiniai/Social_welfare/3013_3_1/spichtinger_valaike.pdf

[4] Hirvonen, Maija 2010 From Vocational Training to Open Learning Environments: Vocational Special Needs Education During Change. Journal of Research in Special Educational Needs. 11 (2). (online), dalam (http://onlinelibrary.wiley. com/doi/10.1111/j.1471-3802.2010.01159.x/abstract)

[5] Sulistyawati, L 2015 *Difabel Desak RUU Disabilitas Disahkan Tahun ini.* Jakarta: republika.co.id.

[6] Pavlova, M 2009 *Technology and Vocational Education for Sustainable Development Empowering Individuals for The Future.* Queensland: Springer.

[7] Iswari, Mega 2007 *Kecakapan Hidup Bagi Anak Berkebutuhan Khusus*. Jakarta: Departemen Pendidikan Nasional Direktorat Jendral Pendidikan Tinggi Direktorat Ketenagaan

[8] Prosser. C.A & Quigley. T.H 1950 *Vocational Education in a Democracy.* Great Britain: American Technical Society

IOP Conf. Series: Materials Science and Engineering **306** (2018) 012106 doi:10.1088/1757-899X/306/1/012106

The Impact of Internet Use for Students

R H Puspita* and D Rohedi

Universitas Pendidikan Indonesia, Bandung.

*hayu.rona@yahoo.co.id

Abstract. Development of Internet technology increasingly modern and sophisticated not only benefit users but also have an effect that is not good for users, especially among students, from a study of 120 students sampled, Internet usage will become addicted and more likely to ha-case negative and less support in learning activities. This is evident from the results of research using the Internet for social media as much as 82 respondents or 68.33% answered always and 50 respondents or 41.67% answered frequently, use for online gaming activity, response of 120 respondents 65 respondents or 54, 17% answered always utilize the internet for online games, 50 respondents or 41.67% answered frequently use the internet for online games, while utilizing the Internet for watch youtube, 56 respondents or 46.67% answered frequently use the Internet to watch youtube and 62 respondents or 51.67% always use the Internet to watch youtube, answers of 120 respondents 41 respondents or 34.17% answered rarely use the Internet to learn and seek information

1. Introduction

Internet is a collection of millions of computers around the world that are connected between each other, the Internet is a medium that does not have the limitation of information on each user, many people are highly dependent on the Internet, because the Internet has many advantages that can simplify a lot of work, the internet is helpful for effectiveness and efficiency. The shape of the effectiveness and efficiency of a given Internet such as Internet has many services for its users so much to give ease and pamper users. With the presence of the internet which is the development of information and communication technologies.

Online activity on social media such as Facebook in Indonesia has reached 42.5 million, the fourth highest in the world, about 18% of the population of Indonesia has had the most popular social media accounts in this world. The number of Twitter users in Indonesia amounted to 5.7 million, the third largest in the world. The number of mobile phone users in Indonesia has reached 266 million numbers, exceeding the population of Indonesia itself Indonesian population as a whole reached 245 million people, including one of the countries with the largest population in the world. The number of Internet users reached 55 million people. Number of Internet users in Indonesia dominate Asia amounted to 22.4%, after Japan Technological developments are increasingly modern and sophisticated this is not the only benefit users but also have an impact that is not good for users, especially among students [1].

Is perceived very well that the development of Internet technology has changed the pattern of community interaction, both business interaction, the interaction of economic, social interaction, and cultural interaction. Internet has contributed so much to society, company, industry, education and government, communication, publications, as well as a means to obtain the information needed by everyone, especially among students who are essentially still in the process of self, is still full of flavor curious, and unstable. With this fact it is possible to abuse the use of the internet [2].

2. Literature Review

Internet or Internetworking is generally defined as the largest computer network in the world to connect all existing computer network (Intranet, Wide Area Network, Metropolitan Area Network, Personal Area Network, etc.) along with all computers, connected devices (Smartphone, Tablet , Switches, routers, hubs, and other connecting devices), as well as the computer itself, into a single container world's computer networks. The time spent using e-mail and surfing online has increased significantly between years 1996 and 2001, various types of Internet use has been associated with positive and negative effects, symptoms of dysphoric has been associated with several types of internet use like shopping, playing games, or doing research, conclusions drawn from this study is that the use of the Internet is one of the main factors which affects academic performance and social life of students [3, 4].

Social media sites have become common, giving people new ways to interact with each other and communicate with the world. Social networking became popular between 2004 and 2006, after Facebook and MySpace were created. Facebook, for example, has more than 500 million members and is still growing and about 85% of students who use facebook, figures are expected to grow because Facebook users will continue to grow [5].

Ghana declared 60% agree with the scenario of 2020, education high will be very different from the way of learning now. There will be mass adoption of teleconferencing and distance learning to leverage technology resources, a large number of learning activities will move to individualized, just-in-time, there will be a transition to a "hybrid" classes that combine components of online learning with the lack of lectures on campus, Requirements for graduation will be significantly shifted to the adjusted results [6].

At this time the increasingly widespread use of the Internet, not only in developed countries, but also in many developing countries, including Indonesia, India and China. Although the time spent on people in general to perform online access can be very productive, compulsive Internet use can be a bad influence on the life, work and relationships within the family. If your child or your child feel more comfortable with friends online than with friends in the real world, or you can not refrain from playing online games or unlock your smart phone or other gadget, it is possible that you or your children already using the Internet too much (the term: Internet addiction) [7].

3. Methods

This research uses descriptive quantitative research, data obtained from a population sample is analyzed according to the statistical methods used then interpreted Qualitative research was also described as a model that goes that occur in natural settings that allow researchers to develop the level of detail from high involvement in the actual experience. One identifier qualitative research is a social phenomenon studied from the perspective of the participants [8]. This type is chosen because the researcher intends to describe the intensity of the impact of Internet use on student vocational high school in the city of Pekanbaru both positive and negative impacts arising without testing the hypothesis, descriptive study can provide a picture or description on a state as clearly as possible without any treatment of the object researched.

3.1. Population, sample and sampling techniques

Random sampling or sampling probability sampling method is something that provides the same opportunity or opportunities to be taken to every element population. Is a sampling technique selected randomly, how this can be taken when analyzing the study tended to be descriptive or generic nature. Each element of the population shall have equal opportunity to be selected into the sample [9]. In this study, the authors define the study population are students of Vocational High School located at SMKN 4 Pekanbaru with the number of students in 1120, will be taken a number of samples by random sampling, number of samples recommended in using random sampling techniques, with the estimation

IOP Conf. Series: Materials Science and Engineering **306** (2018) 012106 doi:10.1088/1757-899X/306/1/012106

procedure the sample is 100-150 samples. But others argue that the sample size is ideal for random sampling that is five times the number of indicators / questions. In this study used 120 samples that meet both of these assumptions.

4. Results and Discussion

4.1. Internet usage time

The survey conducted by the author of the 120 respondents said they have all been familiar with the internet and always use. They start waking up to sleep again did not seem to exist apart from the internet on their phones, they use the Internet in learning activities, recess school, after school, before bed even getting out of bed. Frequency at 4 table above shows more than 55% of them always use the internet from morning till night, the frequency is decreased visible in Table 5 that when I wake up, although there are some respondents who answered always use the internet when I wake up.

4.2. Impact of internet use

4.2.1. Internet for social media. Deviations social behavior is felt most noticeably at this time include less or do not want to hang out with peers. Students are more likely to confine or cool to enjoy interacting via social media than hanging out with friends around, they are more focused in the virtual world, regardless of what was happening around him, This is evidenced by the questioner that has been done with the data as follows:

Table 1. Question: Do you use the Internet for social media?

		frequency	Percent	valid Percent	Cumulative Percent
valid	Rarely	1	, 8	, 8	, 8
	Often	37	30.8	30.8	31.7
	Always	82	68.3	68.3	100.0
	Total	120	100.0	100.0	

From the table 1, the answers of 120 respondents 82 respondents or 68.33% answered always use the Internet for social media, 37 respondents or 30.83% answered frequently use the Internet for social media and only 1 person or 0.83% of respondents who answered rarely utilize for social media.

4.2.2. Internet for online games. Lying or cheating is done to cover up his mistake. Some students lied to get into internet cafes (cafe) opens an online gaming site. For hours he take the time to play this game. In addition to the time that should be used to help parents for students who need help parents even used ostensibly to the cafe to look for material assignment. In terms of looking for material for the task just a little while, but the opening hours of gaming sites, maybe parents do not know if their children skip school because of farewell from home as usual wearing a school uniform and wearing school bag. But it was in the bag already there is a change of clothes so as not to get caught out later if to cafe Students of nowhere. Parents are only realized after a call from the school if their children skipped school often and when questioned admitted, online games search.

Table 2. Question: Do you use the Internet for online games?

		frequency	Percent	valid Percent	Cumulative Percent
valid	Rarely	5	4.2	4.2	4.2
	Often	50	41.7	41.7	45.8
	Always	65	54.2	54.2	100.0
	Total	120	100.0	100.0	

From the table 2, the answers of 120 respondents 65 respondents or 54.17% answered always utilize the internet for online games, 50 respondents or 41.67% answered frequently use the Internet for online games and only 5 people or 4.17% of respondents who answered rarely utilize internet for online games.

4.2.3. Internet for online shopping. International Data Corporation (IDC) Indonesia said Internet users in Indonesia is still much to do online shopping transactions. In a recent survey of IDC themed e-commerce, found the fact that the number of new online shopping activity by 13.3 percent compared to the number of Indonesian Internet users as a whole.

Table 3. Question: Do you use the Internet for online shopping?

		frequency	Percent	valid Percent	Cumulative Percent
valid	No	42	35.0	35.0	35.0
	Rarely	42	35.0	35.0	70.0
	Often	36	30.0	30.0	100.0
	Total	120	100.0	100.0	

From the table 3, the answers of 120 respondents 42 respondents or 35% answered not use the Internet for online shopping, 42 respondents or 35% said rarely use the Internet for online shopping and 36 respondents or 30% answered frequently use the Internet for online shopping.

4.2.4. Internet to learn and seek information. The internet is the easiest to meet the needs in finding the information we want, news and a variety of necessary information can be found on the internet, the way in which it was very easy, simply by typing the name of a website address or search for it with the search engines, in seconds the information that we find we get. This happens because basically their individual needs are very diverse, so that their needs are what caused the motive to find information on a medium that most considered accurate and fast, in this case the Internet is widely used because it offers a variety of literature and references of knowledge in accordance with user needs.

Table 4. Question: Do you use the Internet to learn and seek information?

		frequency	Percent	valid Percent	Cumulative Percent
valid	Rarely	41	34.2	34.2	34.2
	Often	41	34.2	34.2	68.3
	Always	38	31.7	31.7	100.0
	Total	120	100.0	100.0	

IOP Conf. Series: Materials Science and Engineering **306** (2018) 012106 doi:10.1088/1757-899X/306/1/012106

From the table 4, the answers of 120 respondents 41 respondents or 34.17% answered rarely use the Internet to learn and seek information, 41 respondents or 34.17% answered frequently use the Internet for online shopping and 38 people or 31.67% of respondents who answered always use the internet to learn and seek information

4.2.5. Internet to watch youtube. Youtube is a website that provides a wide range of videos from the video clip to the movie, as well as videos made by users themselves youtube, youtube is one of the largest video service provider at this time, the negative effects can be used to search youtube porn videos, video- violent video, the video can broken person's name.

Table 5. Question: Do you use the Internet to watching youtube?

		frequency	Percent	valid Percent	Cumulative Percent
valid	No	1	, 8	, 8	, 8
	Rarely	1	, 8	, 8	1.7
	Often	56	46.7	46.7	48.3
	Always	62	51.7	51.7	100.0
	Total	120	100.0	100.0	

From the table 5, Answers from 120 respondents 1 or 0.83% of the respondents answered not use the Internet to watch youtube, 1 person or 0.83% of respondents answered rarely use the Internet to watch youtube, 56 respondents or 46.67% answered frequently use the Internet to watch youtube and 62 respondents or 51.67% always use the internet to watch youtube.

4.2.6. Internet to access pornography. Pornography on Internet media in Indonesia thrives caused by the ease of use of the Internet itself, weak supervision and security, as well as the rise of online communities that participate in the spread of` pornography on the Internet media in Indonesia, Action porno already widely reported in the print and electronic media about sexual crimes caused by the internet, especially those that occur through social networking site facebook, the victims are usually women of school age, easily offender abducted, molested and even raped the victim.

Table 6. Question: Do you use the Internet to access pornography?

		frequency	Percent	valid Percent	Cumulative Percent
valid	No	56	46.7	46.7	46.7
	Rarely	60	50.0	50.0	96.7
	Often	4	3.3	3.3	100.0
	Total	120	100.0	100.0	

From the table 6, as many as 56 respondents or 46.67% answered not use the Internet to access Pornography 60 people or 50.0% of respondents answered rarely use the Internet to pornography That means they've access to pornography sites and4 people or 3.3% of respondents answered frequently utilize the internet to pornography, Of the data is very alarming because they as students who should not know considered taboo but once and there is even frequently use the Internet to access pornography.

4.2.7. Internet to keep up with technology. The trend among teenagers was once spread through television or radio, is now living in an era where all information is available on the Internet makes the Internet a forum for distributing trend, many trend created by young people for the world's attention. Later, several trends in social media widely heard in the ears of society, keeping abreast of increasingly sophisticated technology is one of pleasure, especially for adolescents who happy to explore all the latest things they think are very attractive, whether it was spotted and just part of the trend or until lifestyle, moral decadence among students, who always seeks the fulfillment of material desires, has led some students to be rich in material but poor in spirit.

Table 7. Question: Do you use the Internet to keep up with technology and trends?

		frequency	Percent	valid Percent	Cumulative Percent
valid	No	41	34.2	34.2	34.2
	Rarely	35	29.2	29.2	63.3
	Often	44	36.7	36.7	100.0
	Total	120	100.0	100.0	

From the table 7, the answers of 120 respondents 41 respondents or 34.17% answered not use the Internet to follow technological developments and trends, 35 respondents or 29.17% answered rarely use the Internet to follow technological developments and trends and 44 respondents or 36.67% who answered frequently use the Internet to follow technological developments and trends.

5. Conclusions

The conclusion that can be drawn from Study Use of Internet Among Students In Vocational School District 4 Pekanbaru internet is a negative impact because it has become an addiction, use of the Internet are more likely to ha-negative and less support in learning activities. This is evident from the results of a study of 120 respondents use the Internet for social media as much as 82 respondents or 68.33% answered always and 50 respondents or 41.67% answered frequently, use for online gaming activity, response of 120 respondents 65 people or 54.17% of respondents answered always utilize the internet for online games, 50 respondents or 41.67% answered frequently use the internet for online games, while utilizing the Internet for watch youtube, 56 respondents or 46.67% answered frequently use the Internet to watch youtube and 62 respondents or 51.67% always use the Internet to watch youtube.

The third indicator is the indicator most use of the Internet by students at SMK 4 Pekanbaru three indicators above even greater than the indicator use the Internet to learn and seek information.

References

[1] Agus T 2015 Implementasi Teknologi Informasi Dalam E-Commerce Menuju Mea (Masyarakat Ekonomi Asean) *the proceedings of the National Seminar on Communication and Information Technology Industry (SNTIKI)* ISSN: 2085-9902.

[2] Rusno 2010 Pengaruh Penggunaan Internet Sebagai Salah Satu Sumber Belajar Terhadap Prestasi Mahasiswa Pendidikan Ekonomi Universitas Kanjuruhan Malang *Modernization Journal* **6**(2).

[3] Agus T 2016 Rancang Bangun Jaringan Wirelless di Politeknik Negeri Bengkalis Menggunakan MAC Filtering *National Seminar on Innovation and Technology Applications in Industry (SENIATI)* ISSN: 2085-4218.

[4] Muhammad M A, et al. 2010 Effect Of Internet On The Academic Performance And Social Life Of University Students In Pakistan *Journal of Education and Sociology* ISSN: 2078-032X.

[5] M O Acheaw, and A G Larson 2015 Use of Social Media and its Impact on Academic Performance of Tertiary Institution Students: A Study of Students of Koforidua Polytechnic,

Ghana *Journal of Education and Practice* **6**(6).

[6] Quitney J A, et al. 2003 *The future impact of the Internet on higher education: Experts expect more efficient collaborative environments and new grading schemes; they worry about massive online courses, the shift away from on-campus life, the Pew Research Center's Internet & American Life Project, an initiative of the Pew Research Center in 1615 L St.* (NW - Suite 700 Washington, DC 20036 202-419-4500).

[7] Helpguide.org 2017 *Internet and Computer Addiction: Signs, Symptoms and Treatment* [article on-line]; http: //www.helpguide.org/mental/internet_cybersex_addiction.htm; Internet; accessed June 6, 2017.

[8] Carrie W 2007 *Research Methods, Journal of Business & Economic Research* **5**(3).

[9] Charles T and Fen Y 2007 Sampling A Typology of Mixed Methods With Examples *Journal of Mixed Methods Research* **1**(1).

How does Entrepreneurship Education Develop Soft Skills?

R Humsona* and S Yuliani

Universitas Sebelas Maret, Surakarta, Indonesia

*rahesli64@gmail.com

Abstract. The objective of research was to identify the comprehensive information on the entrepreneurship education program in Olifant School, Yogyakarta, Indonesia. The research method employed was descriptive qualitative one. The sampling technique used was purposive; the data collection was carried out using observation, in-depth interview and document. The data validation was carried out using source triangulations, and data analysis using an interactive model. The result of research showed that the headmaster implemented entrepreneurship education program through holding some training for the teachers, admitting the students with special needs, modifying curriculum and providing school infrastructure in accordance with the needs of the students. The constraints with entrepreneurship education programs included the limited number of special assistant teachers and the response of parents. The result of learning process can be seen from the students' creativity, independency, and optimism as mentioned in entrepreneurship skill.

1. Introduction

About 2% of populations should be entrepreneurs for a country to be prosperous [1]. It means that out of 240 Indonesian populations, 4.8 million should be entrepreneur. But, currently there have only been 400,000 (0.16%) of total Indonesian population categorized as entrepreneur [2]. It is believed as related to education system in Indonesia. For a long time, education system in Indonesia tends to create education only constituting the process of transferring knowledge. The students were not given adequate opportunity of developing their creativity and independency [3].

Creativity and independency are very important to an entrepreneur. There are two characters of entrepreneur: as creator and as innovator [4]. As a creator, entrepreneur creates an actually new business. As an innovator, entrepreneur initiates reformation in the terms of production, marketing, and management for the existing business. Through creativity, bravery, independency and adequate skill, the failure in business can be minimized. Despite failure, it will be able to analyze the factors causing it, persistent and get up immediately, more creatively and innovatively. The increased number of educated unemployed populations over years may result in social problems such as poverty, thuggery, drug abuse, radicalism, trafficking, and other social problems.

It begins to make the government and a few of Indonesian people aware of the importance of entrepreneurship, that is, to grow and to develop the prospect entrepreneurs [4]. Entrepreneurship education begins to be taught by education institution from elementary school to college. Many researches on entrepreneurship education have been found widely but generally at

High school and college levels only. The study on education entrepreneurship at elementary school in Japan, for example, showed the need for cooperation between education and economic ministries to develop entrepreneurship in Japan [5]. Another research at senior high school in East Java showed the students' enthusiasm not only in the classroom but also in apprenticeship activity [3, 6]. A research conducted in Nigeria suggested that entrepreneurship education could be used as a strategic approach to empower the youth [7]. This article contributed finding about entrepreneurship education at elementary level.

One of schools explicitly mentions its program to develop its student's soft skill through entrepreneurship education is Olifant School in Yogyakarta. This research studies the entrepreneurship education program in Olifant School, the constraints encountered and the strategy taken.

2. Methods

The subject of research was Olifant School in Yogyakarta. This study employed a qualitative method with case study strategy. The informants of the research were headmaster, class teacher supervisor, special teacher assistant, parents and students. The data was collected directly from its source using observation, in-depth interview and document techniques. To validate the data, source triangulation was used [8]. The data analysis was conducted using an interactive model with three components: data reduction, data display, conclusion drawing [9]. Analysis was also conducted along with the informants involved in data collection.

3. Results and Discussion

3.1. *Entrepreneurship education to develop soft skills*

Entrepreneurship education is related to teaching practices aiming to create and to enhance the students' abilities of acting responsibly, being active, creative, and seizing opportunities, assessing and taking controlled risks, and planning and managing the suitable-size project. Entrepreneurship education, according to the Finnish National Core Curriculum, is also to help pupils to observe society from the different players' point of view, and is linked to enhancing the students' entrepreneurial attitudes as characterized by pro-activity, independence, innovativeness, and initiative [10].

Entrepreneurship education can be defined as an education for prospect entrepreneurs in order to have bravery, independency and skill thereby minimizing the failure in business. Entrepreneurship education is not the one on marketing or selling educating an individual to be a trader (seller). Entrepreneurship is far broader than trader [4]. There is a wide variety of needs and need levels for entrepreneurship education, starting from international level needs and going down to national, regional, or individual level. The variety of needs shows the differences between policy-makers' strategic level views and individual teachers' or students' needs, not to mention needs presented for different education levels [10].

Social entrepreneurship is different from traditional one focusing only on material benefit and customer satisfaction, and its significance to the society life [11]. Social entrepreneur sees the problem as the opportunity of creating a new business model beneficial to surrounding community empowerment [5]. Social entrepreneurship is the utilization of entrepreneurial behavior oriented more to the achievement of social objective rather than to profit gaining or the profit gained is used for social interest [12]. However, social entrepreneur- ship education should be based on and start with entrepreneurship education.

Entrepreneurship education can develop the students' soft skill. The students should be trained and equipped with soft skills in order to undertake their tasks responsibly. Soft skills include all aspects of generic skill involving cognitive elements related to non-academic skill. The elements of soft-skills to be mastered are: communicative skills; thinking skills and problem solving skills; team work force; life-long learning and information management; entrepreneur skill; ethics, moral and professionalism; and leadership skills [13].

Those elements of skill, by Jain, are categorized into 2: must-have and good-to-have skills. Meanwhile the good-to-have skill in entrepreneurship is the ability to identify job opportunity. Meanwhile, the good-to-have skill is the ability to propose business opportunity, ability to build, explore and seek business opportunities and job, and the ability to be self-employed. The elements of entrepreneur skill can be developed through entrepreneurship education adjusted with students' need levels.

3.2. *Entrepreneurship education at Olifant School*

Olifant School lives in Cendrawasih Street 4 A-B, Demangan Baru, Yogyakarta with the concept of construction architecture different from that of common school. From the front, Olifant seems to be like an elephant that is drinking water from a cup [14]. It has a philosophy as suggested by Dewi Setyorini, SE, Head of Preschool Olifant that: "from the front it is like a family of elephants consisting of 2 prime elephants (purple) and 3 elephant kids (yellow, blue, and red). Personally, the elephant is strong, has mind so that it can be trained, and socially, it is a protective animal, that will cooperate with its band when there is a danger threatening them, while the cup is as if a container of our knowledge attempting to create foundation for every child so that it can absorb and digest entire knowledge given. Different colors of elephants represent the multi-cultural ethnic, unity in diversity".

Olifant School is one of Yogyakarta leading innovative national educational systems offering a wide range of programs for children from as young as 1 year of age (preschool), all the way to elementary, daycare, courses and high school. The School was established 9 years ago by families and professionals who are deeply concerned with education and who are themselves practitioners in the field of education in Yogyakarta.

When one hears the word "entrepreneur", and reads a formal definition, one will mostly find references to an economic activity, such as: an entrepreneur is one who organizes and manages any entrepreneur, especially a business, usually with considerable initiative and risk. They assume all the risks and rewards of an idea, or goods or services, and always look for opportunities.

As a school, Olifant prefers to look at the entrepreneur spirit and the qualities of a person's character allowing them to be creative, to take risks, to be resourceful, to take responsibility, to follow through with an initiative, to promote cooperation and teamwork, to respect the contributions of others and to look at both social and economic impact of any initiative. It is this educative definition that Olifant applies within its school's curriculum and program.

Olifant's integrated curriculums for preschool and elementary are designed to balance all aspects of child development: academic, creative, physical, social, cultural, and psychological.

These development skills are delivered using the national education system but applied through an international thematic teaching method Olifant enhances the curriculum with unique intra-curricular programs such as "Umbrella (integrated learning process) activity", field trips, and experiential activities, as well as extra-curricular activities such as arts, sports, applied science programs, cultural programs and etc to encourage and enhance creativity and innovation.

Olifant's one-of-a-kind Character building program is designed to instill morals, healthy daily habits, and manners at an early age the program teaches children the importance of healthy routines, respect, independence, perseverance, responsibility, and spirituality in a simple and effective way. Every aspect of the Olifant building is specially designed to enhance the learning experience of the children through the "Building as the second teacher" theory. The Olifant building was designed with three primary things in mind: creative and cognitive stimulation, discovery through fun and educational spaces, and most importantly safety. The facilities accommodate all aspects of child development with educational signage (edusign) on the walls and floors, specialized interactive play areas and class rooms, stunning labs and studios, libraries and IT rooms, canteens, multifunction sports and performance halls, and medical/dental facilities.

Teaching and learning methods at Olifant school refer to 3E (Exploration, Enrichment and Entrepreneurship). Exploration: Enables children to discover new things, new knowledge and understanding through performing, practical daily activities. Enrichment: Integrated with relevant subject matter, this extracurricular program offers a variety of applied activities to enrich student's learning experience. Entrepreneurship: allow children to build curiosity, perseverance, independence, compassion, and awareness of their surrounding environment.

To meet the teacher's need, Olifant School utilizes social and electronic network. By posting online recruitment advertisement, about 500 applications are received for the teacher position yearly. Olifant School recruits new teacher annually through a tight selection to get the teachers consistent with the school's mission and vision. To prepare the teachers, routine training is held both inside and outside school. The need for obtaining appropriate teacher makes some of teacher's work contract is not extended due to either the schools or the teacher's will.

In practical activity, the students are involved in designing activity and evaluation. The activity program is usually conducted in group adjusted with the students' grade. Mrs. Iin, the parent of a student explained: "Parents are also involved in school activity through parents committee. Parents contribute to supporting any school programs such as deepening the faith of children, religious holiday celebration existing in Indonesia, umbrella activity and graduation"

Deepening the faith of children is an optional Enrichment program for children from kindergarten a through elementary grade 6. It introduces them to the wonders of creation and the history behind their perspective faith while instilling religious morals and gratitude. Children learn to pray according to their religion and the traditions that follow. PIA is held every Saturday from 9:00-10:00 at Olifant. Teachers for PIA are provided from pondok pesantren (Islam), church community (Christian), and conventual (Catholic). Once every 2 months all of the religions will join in the one day to do some common activity likes games, making crafts, storytelling about sharing, tolerance, friends and family, etc. This activity is to encourage children to be tolerant and respectful.

To introduce entrepreneur world, Olifant School held an event called Kidpreneur Bazaar. In this event, the students sold food and product they made. This activity is an umbrella activity intended to the fifth graders. The activity is initiated with the establishment of committee consisting of chief, secretary, treasurer, and divisions. The students held meeting and discussion to prepare technical matters such as product selection to be sold, how to obtain product, how to package and to serve, and how to serve the consumer. The students also discussed the name of product to be sold. The capital for the event was collected from the students gradually. The students prepared the material of presentation to get sponsorship from parents, teachers, and community around the school. They also prepared and designed invitation themselves for parents and community. The stands opened in Kidpreneur Bazaar are named Heln Fish & Chip, Nachos & Shake, Eat & Watch selling chicken Katzu and Fruit Salad, Sweet Stand selling mini martabak (thick folded crepe filled with spices and pieces of meat) and juice, Pop Shop selling snack, DIY craft and the last, I Love Stationary. The products were sold with price ranging between IDR 2500 to IDR 10,000. This event was welcomed by other students, parents and community surrounding. In the end of event, the students prepared the report of event organization from each division, intended to make the students responsible for the event held.

As a relatively new school, Olifant School encounters some constraints. Deasy Andriani, one of the founders of Olifant School suggested them: "Firstly, no similar school model can be a complete reference, so that the school should develop itself. Secondly, perception and attitude of stakeholders (headmaster, teachers, parents and students) should be equated in some school activities. The difference usually arises in understanding learning portion, form and method appropriate to the students. Thirdly, it is not easy to get teacher corresponding to the school's vision and mission.

The first generation of Olifant Elementary was graduated in 2016. The first graduation book organized by the students along with teacher team was entitled Junior Entrepreneur [15]. All of the 6th graders wrote their understanding on junior entrepreneur. While the learning process used English and Indonesian, and Javanese languages as local content, for this task, they wrote in English. Here are some opinions of students indicated by a teacher:

- A junior entrepreneur is about creativity, learning and confidence we can learn and be creative to make an art project or in our work we need confidence to be on a stage. He is also brave. We need this bravely to defend ourselves and our work.
- A junior entrepreneur is a person who loves to explore the new things and to make things creative. It is a person who is brave and enjoys the things that she does.

- Junior means younger and "entrepreneur" means to be with activity… Activities like exploring, having fun, enjoying learning, being creative, and being brave. When we use this approach, we can make our life dreams come true.

In 2016, Olifant High School (OHS) was opened, that in its process developing social entrepreneurship education named socio entrepreneurs. Olifant School prepares their students to embrace the future and face the global challenges between nations in the fields of technology, social issues, and commerce. As Agents of change, it is vital to empower their students with mature characters and well developed life skills.

4. Conclusions

The result of learning process one of them can be seen from the students' ability of completing their tests and works. Reading comprehension the students wrote on graduation book reveal their creativity, independency, and optimism as mentioned in entrepreneurship skill [13]. If on subsequent level they get social entrepreneurship education, they would have adequate soft skill to be social entrepreneur. The constraint the school encounters related to difference perception among stakeholders should be treated as a learning process for all. If this process can be passed through, Olifant School would be a school model that can develop both hard and soft skills of students. As such, the opportunity of increasing the number of social entrepreneurs in Indonesia is greater.

References

[1] McClelland D C and Winter D G 1969 *Motivating Economic Achievement* (NY: Free Press).
[2] Ciputra 2009 *Ciputra Quantum Leap Entrepreneurship: Mengubah Masa Depan Anda dan Masa Depan Bangsa* (Jakarta: Elexmedia).
[3] Hadi C, Ismail S W and Andi C 2015 Entrepreneurship and education: creating business awareness for students in east java indonesia *Procedia - Social and Behavioral Sciences* **177** pp. 459 – 463.
[4] Nurseto 2011 *Pendidikan Berbasis Entrepreneur* Available at http://staff.uny.ac.id/sites/default/files/penelitian/tejonurseto-mpd/pendidikan-berbasis-entrepreneur-copetejonurse
[5] Utomo H 2013 *Menumbuhkan Minat Kewirausahaan Sosial* Available at Stieama.ac.id/index.php/ama/article/99
[6] Fitrianto A and A'la A 2014 Developing entrepreneurial curriculum in indonesian islamic higher education: a transformation Process *Social Sciences &Humanities Journal.*
[7] Maigida J F and T M Saba 2013_Entrepreneurial skills in technical vocational education and training as a strategic approach for achieving youth empowerment in Nigeria *International Journal of Humanities and Social Science* **3**(5).
[8] Moleong L J 1995 *Metode Penelitian Kualitatif* (Bandung: Remaja Rosdakarya).
[9] Miles Matthew and A Michael Huberman 1992 *Analisis Data Kualitatif* (UI Press: Jakarta).
[10] Ruskovaara E 2014 *Entrepreneurship education in basic and secondary education-measurement and empirical evidence* (Finland: Lappeenranta University of Technology).
[11] Wulleman M and Mark Hudon 2015 Models of social entrepreneurship: empirical evidende from mexico *Journal of Social Entrepreneurship* **7**(2).
[12] Hibbert S A et al. 2005 Social entrepreneurship: understanding consumer motives for buying the big issue *Journal of Consumer Behavior* **4**(3).
[13] Jain V 2009 *Importance of Soft Skills Development in Education* Available at http://schoolofeducators. com/ 2009/02/ importance-of-soft-skills-development-in-education/
[14] Ganang 2013 *Bermain dan Belajar Berkarakter di Preschool Jendela* 1(12).
[15] Arnold T et al. 2016 *Junior Entrepreneur 2015-2016.*

IOP Conf. Series: Materials Science and Engineering **306** (2018) 012108 doi:10.1088/1757-899X/306/1/012108

Community Governance and Vocational Education

R Martasari*, R H Haryanti and P Susiloadi

Department of Public Administration, Universitas Sebelas Maret, Surakarta, Indonesia.

*riezkymartha@gmail.com

Abstract. Vocational education is required to create a design of education and training that is friendly and feasible for disabled people. The state has a responsibility for it, but with all the limitations, the state can not always be present. This article aims to analyze the capacity of community governance in passing vocational education for people with disabilities in Ponorogo, Indonesia. Articles are the results of research for approximately two years by using data collection techniques through interviews, documentation and observation. Source Triangulation with analysis technique using interactive model is used for data validation. The results show that there are two large capacities owned by Organisasi Sosial Rumah Kasih Sayang as community governance in conducting vocational education for the disabled person namely community credibility and community vigilance.

1. Introduction

Currently, 15% of the world's population are people with disabilities who live in poverty [1]. The phenomenon of poverty and disabilities is also happening in Ponorogo, Indonesia. The number of poor people in Ponorogo in the last 5 years tends to fluctuate, while in 2015 the number of the poor people in Ponorogo has increased to 12,19% [2]. It is ranked 19th among all cities in East Java Province (going up a level from the previous year). It is undeniable that poverty occurs because of some disadvantages areas, where there are many citizens with disabilities. In this case, 3 areas are highlighted: Karangpatihan, Sidoharjo, and Krebet villages, look at table 1.

Table 1. The most villages with persons with disabilities.

Villages	Number of disabilities
Karangpatihan Village	326
Sidoharjo Village	210
Krebet Village	132
Total	**668**

Initially, Sidoharjo and Krebet village were on one area, but in 2007 the government made the expansion of this region, which aimed to maximize the handling and service for person with disabilities. The phenomenon of disability in both villages has occurred since many years ago, many citizens are living in the poverty line and it has not been shown any real changes through the years. Many of them are low-educated, with farmers as their main job while the land of both villages is barren. Consequently, this causes them to have low income, with only IDR 300.000 per month.

This problem should be given more attention and immediate handling by the government, especially handling for improving economic capability for example, by giving vocational education to improve people's skill; and giving knowledge or skill training as one of the efforts to improve their life quality. If this situation is not immediately addressed, then the situation will be protracted. This is government's duty as a form of their responsibility to provide the public service for citizens, including people with disability. Moreover, people with disabilities have equal rights and duties in all aspects of life and livelihood as well as other person.

This research becomes interesting because of the vocational education that should be done by government such as: the educational and training department, social department, and other departments. However, the limitations of the government have made it difficult to do so. Where many programs are provided by the government, but have not shown any change. Even many programs that do not work effectively, one of them is the posyandu for person with disabilities from public health department [3]. In order to cover the government, the vocational education program has been done by Organisasi Sosial Rumah Kasih Sayang (OSRKS). Organisasi Sosial Rumah Kasih Sayang (OSRKS) is a community of civil society that participates in public services by providing skill training for person with disability. In the study of public administration, this is known as community governance. It corresponds with the governance paradigm in the study of public administration, the institutions of governance include 3 elements (domain) which their respectives roles, that is state or government, private, and society [4]. So it means the communities/society can participate in providing public service to resolve public issues that can not be solved by other sector [5].

In the process of implementing vocational education as a form of educational service that is performed by community governance as a part of governance, they also need things that support them. However, not all these communities can do community governance. According to Innes and Booher, only a community that has the capacities can do the community governance [5]. Therefore in this article we analyze what capacity which is owned by OSRKS as a community governance, so they can provide a vocational education to person with disabilities.

2. Literature Review

Nowadays, study of public administration has experienced a paradigm shift from government to governance, where within this governance paradigm does not only involve one sector. Institutions of the governance paradigm include 3 elements, which are state, private and society sector [4]. The 3 sectors of the governance can participate in the administration of public administration according to their respective roles. It means that the community as a society sector can participate in the settlement of public issues that can not be solved by individuals or private and state sector [6]. Therefore, the society or the community can manage its community to participate in public problem solving or better known as community governance.

Community governance has been defined as an arena of participation which extends beyond the parish community council to embrace all activities that involve either the provision of public service within the community, or the representation of community interest to external agencies [7]. This means that as the domain of governance, the community can participate in order to fulfill the interests of the group or to solve the public issue as a form of public service.

Instead of the problems that happened in the society, only a community with capacity can do the community governance. It is an interaction that occurs in human capital, organizational resources, and social capital. Here, social capital is an effort to be able to perform collective action, there are 3 forms of social capital which consist of: bonding social capital, bridging social capital and linking social capital [4]. If the community has these capacities, it is expected to support the realization of a good community governance. Additionally, to be able to conduct community governance, it also needs to be supported with the trust of the public so the community is considered as a capable community and also must be supported by a good arrangement as a form of vigilance of the community [8]. Both of these are the dimensions of community governance.

Nowadays, what citizens need the most from the public service is education, an education that can help to increase and master a skill. Furthermore, vocational education is the best choice to fulfill the requirement. It is like what UNESCO has noted that vocational education is designed to add or master a skill for someone, and help them in getting better in working field [9]. The vocational education is aimed to improve people's knowledge and competence, including people with disabilities. People with disabilities, indeed, need special attention in term of competence. By giving vocational education to them, it is expected to improve their ability to be more productive.

Furthermore, the vocational education given by OSRKS as community governance to people with disabilities is aimed to improve their knowledge and skills to be more productive.

3. Methods

This paper examines the capacities owned by Organisasi Sosial Rumah Kasih Sayang (OSRKS) as community governance, as they provide a vocational education for people with disabilities. This research is conducted in OSRKS Krebet Village, Jambon, Ponorogo. This research belongs to qualitative descriptive research and uses case study as its approach. Moreover, the sampling technique used in this research is purposive sampling with data collection done by in-depth interview, observation, and documentation. While data validity is checked by using source triangulation with interactive technique analysis.

4. Results and Discussion

Rumah Kasih Sayang is a building that was established by government in 2011 in Krebet Village. This is a form of government response to the needs of people with disabilities, because many of residents around the villages have disabilities. The management is submitted to the local residents so they formed a social community or social organization, named Organisasi Sosial Rumah Kasih Sayang (OSRKS). The aim of OSRKS is to improve the service and empowerment of people with social welfare issues, including people with disabilities. Since its establishment, there have been many programs and effort undertaken OSRKS, by doing public service as a form of their participation. So what they do is complement to improve the public service by 2 other governance sectors.

There are some programs that have been run by OSRKS as a form of public service community, especially for people with disabilities. The programs are training for cadres to serve and assist people with disabilities; additional nutrition food service; providing guidance and skill training (vocational education); improving religious quality; providing health therapy; counseling about the importance of education and health, the danger of drugs, counseling of entrepreneurship and work motivation; conducting special posyandu for disabled people, teaching disabled people to be independent; and cadres self-reliance through various skill training and insights.

4.1. OSRKS as community governance

OSRKS is an independent social organization, which tries to participate in addressing existing public issues by providing public service for people with disabilities. As a community governance the decision-making process is done by deliberation to get mutual agreement, and it has the division of labor system. In addition, OSRKS also acts as a mean of exchange of information and knowledge that can be useful in achieving its goals. However, not all communities are able to carry out community governance, only communities with the capacity can do that. Then how can OSRKS provide public service like the state? This can be seen from the following points.

4.2. Capacity of OSRKS

There are 3 aspects studied in analyzing the capacity of community governance, these are human capital, organizational resources and social capital, and the capacity of the OSRKS:

4.2.1. Human capital OSRKS. Human capital OSRKS is managed by 26 social workers, with Mr. Jemiran as its head and also as the Chief of Krebet Village. He is also assisted by Mr. Meswan, S.Pd.I as the secretary of the organization and Drs. Djaenuri as treasurer who is also active in OSRKS development. SRKS is assisted by 23 social workers, which consist of 9 women and 14 men. On average, they are over 30 years old and mostly work as farmers and housewives. There are only 23% of all OSRKS cadres that have bachelor degree background, while the rest are senior high school graduates. Although not all cadres are highly educated and rich, they have high motivation and passion to guide and train people with disabilities. They also have been equipped with skills, such as processing hand-drawn batik, crops, making liquid fertilizer, farming, making catfish cultivation, doormat, sulak, plastic bags, key chains, accessories, home industries, etc. In addition, they are also equipped with knowledge for people with disabilities assistance. So that with the education of skills and abilities, they can teach the people with disabilities by doing guidance and skills training. There are 5 villages which are scouted by OSRKS, look at table 2:

Table 2. Data on disabled people which built OSRKS in 5 villages.

Villages	Number of disabilities		Total
	Male	Female	
Dayakan Village	17	17	34
Krebet Village	56	76	132
Sidoharjo Village	50	54	104
Pandak Village	7	17	24
Karangpatihan Village			
Total	**130**	**164**	**294**

Until today, they are still doing skills training and mentoring activities of disabled people in Rumah Kasih Sayang building every Sunday although the number of social workers and also people with disabilities who built it decreased from the beginning of OSRKS establishment.

4.2.2. Organizational resources. In addition to human resources OSRKS also requires other resources, especially money to support all operational activities such as the fulfillment of the needs of the person with disabilities, the provision of tools and training materials, etc. At first, OSRKS received regular funding support from the government, but due to limited government budget, in 2012 the support was stopped. Therefore, OSRKS is seeking other funding support from both the private sector and from government. From the private sector, it is usually in the form of goods and it is directly given to people with disabilities, while for the government sector OSRKS must submit proposals. From the proposal, the government then also provides assistance, although rarely, in the form of funds or other assistance such as skills training assistance. The government involved is from city government of Ponorogo, the provincial government and Central Government. So that, with the availability of resources of the Organization in OSRKS, it can support the operational activities of the OSRKS so it can achieve its goal.

4.2.3. Social capital. Social capital from these two points before it can be said, to carry out community governance, a community can not do it alone. They need to form networks with other communities to cooperate with the collective action as the achievement of the goal. In OSRKS, there are 2 forms of social capital that are very strongly perceived which consist of: first, Bonding Social capital, OSRKS tries to keep building the cohesiveness and trust of its members. This can be seen from the attention of the OSRKS management to the social worker by having a meeting each week to provide skills training, the cadres are also given motivations so that their spirit will not recede, recreation, skills training, study production center in the village of appeal so that it can be utilized with the ability to generate extra income. And the second linking social capital, there are collaboration between the OSRKS and governance. The Government provides assistance in the form of funds or in other forms such as training assistance or relief goods. To maintain good relationship with Government, OSRKS always gives reports in the form of a written report or report in person with photo and video form when the activity takes place. So with both of these, relationships are maintained and the exchange of information between the OSRKS and government can still take place.

4.3. Dimension of community governance in OSRKS

In running the organization, OSRKS also gets support from the public, the support is in the form of public confidence to OSRKS, both the general public and people with disabilities. The trust was formed on the basis of two things, that is the first, because basically they have the same identity as locals so that they know each other and the second, because of the program of OSRKS that helps the community to improve the quality of life people with disabilities. Consequently, it can be said that the OSRKS is credible. This trust is very important, because trust can build a good relationship and support the achievement of organizational goals.

Besides that, OSRKS is also supported by the management of members by doing the division of labor system. The OSRKS is divided into several divisions which consist of education and training, counselling and spiritual, social, cultural, economic and welfare, information and communication field, the field of research and development, health and the environment, the field of logistics and infrastructure, line of business and marketing, and public relations fields in which each of these areas is managed by two or more cadres. This division of labor system is one form of organizational awareness, so that with the system of division of labor it can make easier for the chairman to control and supervise its members. Thus, the performance of the organization still runs according to its original purpose. Both of these are the most important dimensions of community governance, because without them a community will not be able to carry out community governance properly.

4.4. Discussion

Nevertheless, it turns out that in the implementation of OSRKS as a community governance there are also problems such as the inability of OSRKS' to oversee the 5 villages because of the location which is quite far and not easy to be reached. Consequently the program becomes inefficient and this inefficiency makes some members and the public believes they are not credible anymore. People assume that this is because they are not social workers who have the skills and special background to handle the person so that their spirit easily fell and for now only 50% of the initial number and while the people who scouted currently only 10% of the initial number, resulting in many unsustainable programs. And for now, OSRKS just focuses on the handling in the village of Krebet. In addition, human resource issues owned by OSRKS are also not balanced with the number of people with disabilities and are not sufficient enough to guide incidents of disability, therefore to handle the person with disabilities requires the human resources who have special abilities and skills to provide guidance for people with disabilities characterized by appropriate educational background. Other than that, it needs to be an enhancement in the skill of social workers so that with the skills possessed, there will be more variety of training and they can gain the public trust again. Another problem with budget is that the private sector has not involved yet, so the managers should encourage the private sector to better cooperate or other CSRs to get involved. The findings in this study are also a weakness which is also an input in this study. So with this input, it is expected that OSRKS to improve the performance to regain the trust from the community, so that services can effectively run.

5. Conclusions

From the previous discussion, it can be concluded that the community can also participate in solving public issues by conducting public services (community governance) to complement the shortcomings of both other sectors and public services, to be more effective. But the community that will do this community governance requires some capability or capacity that is owned by human resources in the organization, generated organizational resources to support the creation of the organization, and also, the need for collective action from various parties that make up the network. In this case, it can be seen that it is necessary to maintain the trust of the public and management of the organization so they remain credible to do their work, if both things are bad it will affect the sustainability of the organization or community concerned.

Acknowledgments
A special acknowledgement in this research should go to the board OSRKS, chairman, and the member of OSRKS for their excellent support, and also for the head dusun of Sidowayah, the head village of Krebet and the head village of Sidowayah, citizen of Krebet and Sidowayah who have supported this whole research.

References

[1] Groce N E, London J and Stein M A 2014 *Inheritance, Proverty, and Disability Disability and Society* **29**(10) p 1554-68.
[2] Badan Pusat Statistik Kabupaten Ponorogo dalam Angka 2016 Available at: Ponorogokab.bps.go.id [update 2016 August 29; cited 2017 marc 31].
[3] Putri M 2014 Peran Dinas Kesehatan Kabupaten Ponorogo dalam Pelayanan Kesehatan Prima Bagi Penderita Intellectual Disability (Studi di Kecamatan Jambon Kabupaten Ponorogo) *Jurnal Mahasiswa Ilmu Pemerintahan* **1**(1)
[4] Syafri W 2012 *Studi tentang Administrasi Publik* (Jakarta : Erlangga) p 177.
[5] Sudarmo 2015 *Menuju Model Resolusi Konflik Berbasis Governance* (Surakarta : UNS Press) p 170-82.
[6] Bowles S and Gintis H 2001 *Social Capital and Community Govenance Economic Journal*
[7] O'toole K and Burdess N 2004 New Community Governance in Small Rural Town: The Australian Experience *Journal of Rural Studies* **20** p 433-43.
[8] Kayhan V O 2015 The nature dimesionality and effect of perceptions of community governance *Informasi and Manajemen Journal* **52** p 18-29.
[9] Razzak A 2015 Designing a Model of Vocational Programs for Disables in Pakistan *Academi Reseach International* **6**(2) p 308-16.

IOP Conf. Series: Materials Science and Engineering **306** (2018) 012109 doi:10.1088/1757-899X/306/1/012109

Promoting Creative Thinking Ability Using Contextual Learning Model in Technical Drawing Achievement

R Mursid*

Mechanical Engineering Education, Faculty of Engineering, State University of Medan.

*mursid.tp@gmail.com

Abstract. The purpose of this study is to determine whether there is influence; the differences in the results between students that learn drawing techniques taught by the Contextual Innovative Model (CIM) and taught by Direct Instructional Model (DIM), the differences in achievement among students of technical drawing that have High Creative Thinking Ability (HCTA) with Low Creative Thinking Ability (LCTA), and the interaction between the learning model with the ability to think creatively to the achievement technical drawing. Quasi-experimental research method. Results of research appoint that: the achievement of students that learned technical drawing by using CIM is higher than the students that learned technical drawing by using DIM, the achievement of students of technical drawings HCTA is higher than the achievement of students who have technical drawing LCTA, and there are interactions between the use of learning models and creative thinking abilities in influencing student achievement technical drawing.

1. Introduction

Technical drawing means to express an idea or ideas engineers. Therefore, a technical drawing is often also referred to as technical language or languages for the technical experts. Read technical drawing is one of the vocational competence in SMK (Vocational High School) courses mechanical engineering skills that must be mastered by the student. Students of vocational graduates should have the competence expected by the industry or the business world. But in fact the student competence in understanding these subjects is still lack. Students do not understand the subject matter, misread images, wrong to make the image size and so on.

Drawing is the "engineering language". Therefore, it should be able to depict and describe the explanations precisely and objectively. Particulars images accurately and objectively. The information in the form of visual images, symbols, emblems and image standards. Such as information is of ideas, then abstract concepts are realized by the image [1].

SMK, particularly in the Department of Mechanical engineering, the learning process technical drawing is not optimal, because the dominant learning using conventional models. The use of technology and constructivist theories together is more effective and integrated learning devices and usage models designed learning that focuses on student [2]. The main task of the teacher is not only to acquaint the students with information but also to use the information in a proper way in the real world [3]. The innovation of education should have a high impact technical and complexity through technology that includes education by considering future goals [4].

Learning model has five characteristics, namely: the syntax, the social system, the principle of the reaction, and the impact of instructional support system and escort [5]. The learning model is a frame of mind to direct the teacher to design, implement, and guide so that the interaction of teaching and learning that is more focused. Learning is a process of thought, means that knowledge does not come from outside but is formed by individuals themselves in cognitive structure. Learning process be categorized into three types, namely teaching of thinking, teaching for thinking and teaching about thinking [6]. Teaching of thinking is more emphasis on the process of thinking while teaching for thinking emphasis on how the thought process that occurs when learning while teaching about thinking emphasis on how teaching methods can motivate what to think. Constructivist learning in the school setting is strongly supported by the technology, talent, facility, democratic environments, interactive learning, and experience the learning needs of the community of interest [7].

An innovative learning model of learning is by organizing the learning content and learning activities resulting in innovative activity in learning technical drawing. CTL innovative learning is learning, cooperative learning, mastery learning, problem based learning and project based learning. Contributing to integrate creative thinking with critical thinking are very supportive thinking and learning by engaging students thinking skills in learning strategies [8].

Creative thinking is one kind of thinking that directs gained insight are new, new approaches, new perspectives, or new ways of understanding things [9]. Creative thinking occurs triggered by tasks or challenging problems. Creative thinking as the process of obtaining an idea that emphasizes the aspects fluently, flexibility, originality and elaboration in thinking [10].

Creativity can also be seen from the process. Generating a creative product of any kind certainly preceded by the construction of creative ideas [11]. These creative ideas generated through thinking that involves cognitive activity. This process is referred to as creative thinking. A person's ability to think creatively can be improved by understanding the creative thinking process and the various factors that influence and through proper exercise. The factors influencing creative thinking is: contextual, cultural, school, social, economic, acculturation of parents have a significant pathway that leads to creative thinking indexes [12].

This study was formulated as follows: (1) whether there are differences in the results between students that learn drawing techniques taught by contextual Innovative Model (CIM) with Direct Instructional Model (DIM)?; (2) whether there are differences in achievement among students of technical drawings that have High Creative Thinking Skills (HCTA) with creative thinking ability is low (LCTA)?; and (3) whether there are interactions between the learning model with the ability to think creatively to learn the results of technical drawing?

2. Methods

The population in this study were taken from the two schools namely SMK Negeri 5 Medan and SMK Negeri Binaan Provinsi Sumatera Utara. The sampling technique study with cluster random sampling technique. Areas of study technology and technical expertise, courses mechanical technical expertise, competence machining technical expertise, the technical drawing subject, class XI, 3rd semester, and competency standards explain how to read the technical drawing, and the basic competence work piece image complete with size. This study uses a quasi-experimental research by conducting experiments in the classroom that is already available as they are, without making changes to classroom situations and learning schedule.

The technique of analysing the data is descriptive statistic and inferential techniques. In hypothesis testing, the statistic inferential technique which is two-way analysis of variance (2x2 factorial design) with significance level of 0,05 is used in the study. Before two-way analysis of variance (ANOVA) being carried out, an analysis requirement such as Lilifors test for Normality is determined. Fisher (F) and Barlette tests are also employed to try out the Homogeneity requirements in the study. Once the test on all analysis requirements has been done, the two-way ANOVA is possible to be applied. Furthermore, post hoc test is done if the results indicate that the two-way ANOVA is significant. If ANOVA two significant lines, then conducted a further test (post hoc test). Further trials will be

IOP Conf. Series: Materials Science and Engineering **306** (2018) 012109 doi:10.1088/1757-899X/306/1/012109

conducted with the Scheffe test 'because they sample of each cell is not the same (n not equal). Based on figure 1, explain about Contextual Innovative Learning Model (CILM) with constructivist approach.

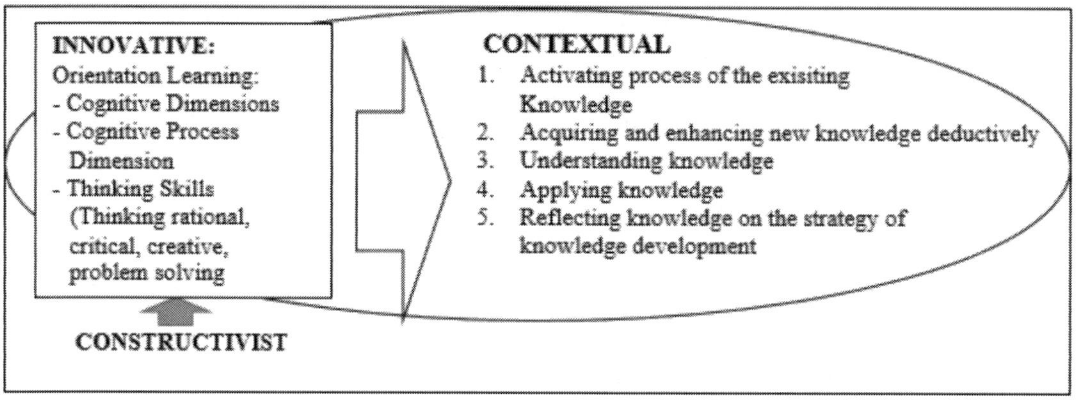

Figure 1. Contextual Innovative Learning Model (CILM) with constructivist approach.

3. Results and Discussion

3.1. Results

Hypothesis testing is done by using analysis of variance (ANOVA). For purposes of analysis of variance, the complete data are provided in table 1.

Table 1. The data of the students' technical drawing achievement.

Creative Thinking Ability (B)	Learning Model (A)		Total
	Contextual Innovative (A_1)	Direct Instructional (A_2)	
High BK (B_1)	n_{A1B1} = 19 $\sum X$ = 565 $\sum X^2$ = 16875 $\bar{X}$ = 29.73	n_{A2B1} = 15 $\sum X$ = 422 $\sum X^2$ = 11950 $\bar{X}$ = 28.13	n_t = 34 $\sum X$ = 987 $\sum X^2$ = 28825 $\bar{X}_t$ = 28.93
Low BK (B_2)	n_{A1B2} = 14 $\sum X$ = 385 $\sum X^2$ = 10651 $\bar{X}$ = 27.5	n_{A2B2} = 17 $\sum X$ = 477 $\sum X^2$ = 13499 $\bar{X}$ = 28.05	n_t = 31 $\sum X$ = 862 $\sum X^2$ = 24150 $\bar{X}_t$ = 27.77
Total	n_t = 33 $\sum X$ = 950 $\sum X^2$ = 27526 $\bar{X}_t$ =28.61	n_t = 32 $\sum X$ = 899 $\sum X^2$ = 25449 $\bar{X}_t$ = 28.09	n_t = 65 $\sum X$ = 1849 $\sum X^2$ = 52975 $\bar{X}_t$ = 28.35

The calculation results of ANOVA described in table 2 is the summary of 2x2 factorial analysis.

Table 2. The summary of 2x2 factorial analysis.

Variance Source	JK	dk	RJK	$F_{calculation}$	F_{table}	Explanation
Learning Model	90.95	1	90.95	4.48	3.96	Significant
Creative Thinking Ability	24.25	1	24.25	16.82	3.96	Significant
Interaction	262.86	1	262.86	48.61	3.96	Significant
Inter-group	48.20	3	16.06			
Intra-group	329.86	61	5.41			
Total	378.06	64				

The results of analysis of variance for both learning models indicate that $f_{cal.} = 4.485$ is greater than $f_{table(0,05)} = 3.96$ at a significant level $\alpha = 0.05$. Thus, the hypothesis which says that the students' technical drawing achievement using CIM is higher than DIM is true and verified.

The result of analysis of variance for both creative thinking abilities show that $f_{cal.} = 16{,}820$ is greater than $f_{table(0,05)} = 3.96$ at a significant level $\alpha = 0.05$, so that h_0 is rejected at significance level of $\alpha = 0.05$. Therefore, the hypothesis which says that the technical drawing achievement of HCTA students are higher than LCTA students is true and verified.

In the study, a mean difference test of the two proportions is necessary to be performed since the result of the third hypothesis testing indicates that there is an interaction between the learning models and the creative thinking abilities. Assessed from the effect and the interaction of the learning models and creative thinking abilities, it is noted that the average outcomes of the students' technical drawing learning using CIM are higher than DIM. The study also corroborates that creative thinking ability is one of the students' characteristics which needs to be optimised further, considering the fact that it has significant impact on technical drawing achievement.

3.2. Discussion

Results learn drawing techniques emphasis on cognitive and psychomotor abilities requires students to actively read pictures, and drawing techniques correctly according to ISO standards. Therefore CIM provides convenience for students' active and motivated towards learning can construct itself to evolve, so get learning results increased compared with DIM only gives examples in working on technical drawing without constructing their students to be more active. CIM learning on achievement can be improved because students are taught in the classroom providing a climate conducive to the development of reasoning power, the power of inquiry, and creativity of students. The factors that influence in drawing techniques are: inadequate facilities draw, the factors teacher, and the student background [13].

Contextual innovative learning makes the students have the capacity to think creatively and skilled in solving problems. Students are able to use this kind of reasoning is clear in the process of understanding something, and skilled in making choices, and making decisions, especially in understanding technical drawing, and able to read pictures, and drawing techniques correctly. In addition, learning CIM is also reflected in the results shown students communicative, and collaborative in articulating thoughts, and ideas clearly, and effectively through speech / oral, written, and pictures [14]. Learning technical drawing using the media power point better than with conventional learning, more effective and have higher benefits for students [15].

Innovative learning characterized encourage students to discover new ideas and encourages learner to make new things. Learning technical drawing given appropriate basic competence in the appointment of the material and the size of the projection image is in desperate need of work accuracy and precision of placement and use line drawings properly, so that required students were able to draw correctly. Practice drawing were repeatedly given to the students to provide a good skill ability, so that the drawing be good. Owners drawing tools are not enough to guarantee good performance of students in drawing technique, this should be dated into a fun learning to achieve student success [16].

The learning model should be able to construct the students to active learning. CIM requires students to engage mutually exchange ideas, collaborate, and communicate to achieve the desired learning, so expect pupils to develop communication skills in their technical drawing. Constructivist approach, teachers can engage students in learning activities, they can arrange instruction to meet the different learning levels, and styles, and they can extend the range of resources available for learning [2].

From observation, and discussion of researchers together with teachers in the assessment of learning strategies in the classroom innovative contextual, and DIM, for the treatment classes CIM teacher just give the problem, or problems are different to look for completion by students while in class treated DIM more teachers completing the problems presented in the example. This is evidenced by the length of time the presentation of the subject matter in the classroom DIM compared with a presentation of the material in the CIM class. Reviewed by CIM classes through the guidance of teachers, students find their own knowledge of the problems set by the teacher so that such knowledge will be more meaningful. This lead into the cause of high achievement technical drawing on CIM classes. It is quite reasonable for students who have HCTA will have the curiosity of the ways to find, or solve problems relating to technical drawings until they find the correct answer, the spirit of great learning in order to achievement, it means that the student is not easy influenced by things that are beyond the reach of logic. The ability of the technique to draw the students to the achievement obtained can be increased with the process of creative thinking of students at the stage expected, so with a mastery of the material that has been taught, students will be able to do well.

The study's findings indicate that there was an interaction between the learning strategies and creative thinking abilities on achievement technical drawing. Students who have HCTA has a higher achievement than students who have LCTA taught using CIM as well as with students who have LCTA has a higher achievement than students who have HCTA in class by DIM. This indicates the interaction between the uses of learning strategies with creative thinking skills to the achievement of students of technical drawing.

Students who have HCTA will be motivated to study harder, and be able to control themselves because they are always optimistic to be able to find out information on learning objectives technical drawing in their daily life, and improve achievement. While students who have LCTA may be worried that they are not able to follow the learning drawing techniques considered difficult by most students, considered a bogey in a series of subjects that occupied so will result in the achievement are achieved. DIM is a learning-oriented teacher who is linear, so it is more suitable for students who have LCTA, this is due to the students who have LCTA tend to passively wait for information from the teacher. Thus teachers have many opportunities to provide motivation, encouragement, and guidance so that students are more motivated, and driven in knowing, and understanding the application of technical drawing information in everyday life, and improve achievement, but if learning DIM is given to students who have HCTA will cause boredom and burnout. This is due to the learning process seem monotonous and lacking in variation in the presentation as well as learning about providing opportunities for students to actively and creatively.

4. Conclusions

It was concluded that: (1) The students learn drawing techniques that learned by using CIM get the higher achievement than the students that learned technical drawing by using DIM; (2) The student's achievement of technical drawing HCTA is higher in the study of technical drawing than the students who have the ability LCTA; and (3) There are interactions between the use of models of learning and creative thinking abilities in influencing student achievement technical drawings. The real gaps in teachers is the performance of students in learning, partly due to the shortage of qualified teachers and inadequate resources that will ultimately lead to poor achievement. Teachers continually develop the capacity, motivation, and effective coordination, and control the principals to teachers, teachers' morale, and commitment, is a major determinant of the quality of student achievement [17]. A strong partnership with the industry, making the increased quality of students in technical drawing, and useful for the students, because it can identify the effectiveness of learning drawing techniques. Teachers will be able to evaluate the students, monitor, and improve their competence in learning technical drawing [18].

References

[1] Sato T, and Sugiarto N 2008 *Drawing Machine According to ISO Standards* (Jakarta: Pradnya Paramita).

[2] Pourhosein A and Mei L 2013 Teachers' Use of Technology and Constructivism. *I.J.Modern Education and Computer Science* **4** pp 49-63.

[3] Khalid A and Azeem M 2012 Constructivist Vs Traditional: Effective Instructional Approach in Teacher Education *International Journal of Humanities and Social Science* **2**(5).

[4] Adams R and Evangelou D et al. 2011 Multiple Perspectives on Engaging Future Engineers. *Journal of Engineering Education* **100**(1) pp 48–88.

[5] Joice B, dkk. 2009 *Models of teaching* (Cetakan ke-8: Yogyakarta: Pustaka Pelajar).

[6] Costa L A 1988 *Developing minds* (Virginia: ASC).

[7] Ng'ang'a S I and Otii L O 2013 Constructivism and the Likert Scale on the Perception of Teaching/Learning Creativity at the University Level *Journal of Sociological Research* **4**(1).

[8] Salim A and Nizam H 2014 The Effects of Integrating Creative and Critical Thinking on Schools Students' Thinking *International Journal of Social Science and Humanity* **4**(6).

[9] McGregor D 2007 *Developing Thinking Developing Learning* (Poland: Open University Press).

[10] Grieshober W E 2004 *Dictionary of Creativity* (New York: International Center for Studies in Creativity. State University of New York College at Buffalo).

[11] Dickhut JE 2007 *A Brief Review of Creativity* http://deseretnews.com/dn/view/0,1249,5100 54502,00.

[12] Meintjes H and Grosser M 2010 Creative thinking in prospective teachers: the status quo and the impact of contextual factors *South African Journal of Education* **30** pp 361-386.

[13] Dieaso D K, Manabete S S, and Amalo K, et al. 2013 Evaluation of students' performance in technical and engineering drawing towards an effective career choice in engineering and technical and vocational education *International Journal of Educational Research and Development* **2**(4) pp 089-097.

[14] Hamied F A 2009 Model Pembelajaran Inovatif di Era Global (Suatu Kajian Perbandingan di Negara Maju) *Jurnal Ilmiah Pendidikan* **1**(2) pp 95-110.

[15] Isiaka A and Taiwo H 2015 Effectiveness of Powerpoint Presentation On Students' Cognitive Achievement in Technical Drawing *Malaysian Online Journal of Educational Technology* **3**(4).

[16] Medupin R O, Abubakre O K and Adebayo S A 2015 Students' Academic Performance in Engineering Drawing in Nigerian Polytechnics: A Case Study of the Federal Polytechnic Bida, Nigeria *Journal of Multidisciplinary Engineering Science and Technology (JMEST)* **2**(2).

[17] Joshua A and Ayandoja C 2014 Assessing Principals" Coordinating and Controlling Strategies for Effective Teaching and Quality Learning Outcome. International *Journal of Learning, Teaching and Educational Research* **7**(1) pp 180-200.

[18] Elisha N E 2014 Effective Teaching and Learning in Technical Colleges: Challenges of Technical Drawing *Journal of Educational Policy and Entrepreneurial Research (JEPER)* **1**(1) pp 76-86.

Incident Management in Academic Information System using ITIL Framework

V R Palilingan[1] **and J R Batmetan**[2*]

[1]Universitas Negeri Manado, Tondano 95618, Sulawesi Utara, Indonesia
[2]Universitas Sari Putra Indonesia, Tomohon, 95416, Sulawesi Utara, Indonesia

*john.reimon@gmail.com

Abstract. Incident management is very important in order to ensure the continuity of a system. Information systems require incident management to ensure information systems can provide maximum service according to the service provided. Many of the problems that arise in academic information systems come from incidents that are not properly handled. The objective of this study aims to find the appropriate way of incident management. The incident can be managed so it will not be a big problem. This research uses the ITIL framework to solve incident problems. The technique used in this study is a technique adopted and developed from the service operations section of the ITIL framework. The results of this research found that 84.5% of incidents appearing in academic information systems can be handled quickly and appropriately. 15.5% incidents can be escalated so as to not cause any new problems. The model of incident management applied to make academic information system can run quickly in providing academic service in a good and efficient. The incident management model implemented in this research is able to manage resources appropriately so as to quickly and easily manage incidents.

1. Introduction

The academic system is an important part of the university. Academic systems in various universities have been managed with information-based systems to make it easier, faster and more effective in winning large academic data. The academic information system has become a backbone in 60% of large and advanced universities in Indonesia. More than 80% of the services provided by the University can be served using an academic information system. These services include Study Plan Card (KRS), Study Result Card (KHS), lecture schedule, teaching materials, assignments and other academic services. Academic information systems have been used to improve productivity [1] academic services at universities [2]. In conducting good academic services, universities need to provide adequate technology infrastructure and meet the needs of users [3] who use the service. The hope is to have a good and continuous service guarantee [4] with good quality.

Academic services require quality and accurate service management including service to incidents that arise. The incident process needs to be set so that no service interruption is provided [5]. Many of the incident issues that arose have not been adequately addressed, 70% of incidents are still handled without clear procedures and still rely on individuals who work only on the basis of previous experience. This is precisely 60% of them have not been able to resolve the incident and leave only the problems that make the academic system becomes disrupted and not running properly. The absence of operational standards of incident management and incident management procedures makes 80% of incidents only handled manually and very slow to complete while 20% of incidents are left unresolved. This then creates a new problem in the academic system so that the system becomes disturbed.

To solve the problem, some researchers have suggested the following strategy: information system incident should look at aspects of the management by ITIL framework or some framework to solvem the problem [6]. The suggested strategy is effective and can be implemented and easy to adopt by framework of e-governanca [7, 8]. However, some problems still persist. Incident management in information system cannot explain in detail about how to use the application so that some incident have difficulty to solved [9, 10]. Furthermore, incident management becomes complicated and sophisticated [11], thus creating conflicts with limited use to adoption by ITIL framework.

Here, the purpose of this study was to find the right way of incident management in academic information system. These novelty makes the academic information system incident for developing countries that have many limitations using framework to solved the problem. The incident management in academic information system, including indentification, Incident logging, Incident categorization, Incident Priorization, Initial diagnosis, incident escalation, Investigation and diagnosis, resolution and recovery, Incident closure, Incident management report, and Incident Management evaluation. Every Incident faced should be handled so it does not matter much. The method of treatment to be used in this study is to implement incident management in accordance with the Information Technology infrastructure library (ITIL) framework.

2. Methods

There are many model frameworks [6] IT governance [7, 8] one of which is Information Technology Infrastructure Information Service (ITIL) [8]. The ITIL framework contains the best IT governance framework in use today [9]. This framework has stages that are systematically [8] explained in its entirety. Stages are as follows Service Strategy: the stage of strategic transformation of IT management services into a strategic strategy of the organization. Service Design: this stage contains the direction and direction of the IT service management guide in accordance with the strategy that has been made before. This stage is a continuation of the previous service strategy. This stage is also built into the satisfaction of customer satisfaction. Transition Services [10]: This stage contains the transition process from the old governance model to the new governance model created in the Service Design stage. Service Operations [10]: This stage contains best practice steps in IT service management. Continual Service Improvement [11]: This section contains the management of feedbacks obtained from customers [12]. The inputs obtained in the analysis can then be implemented in a pre-made stage so that the system becomes better and the results are improved [13]. Thus the system can improve the results of the Strategy, Service, Transition Services, and Service Operations (see Figure 1).

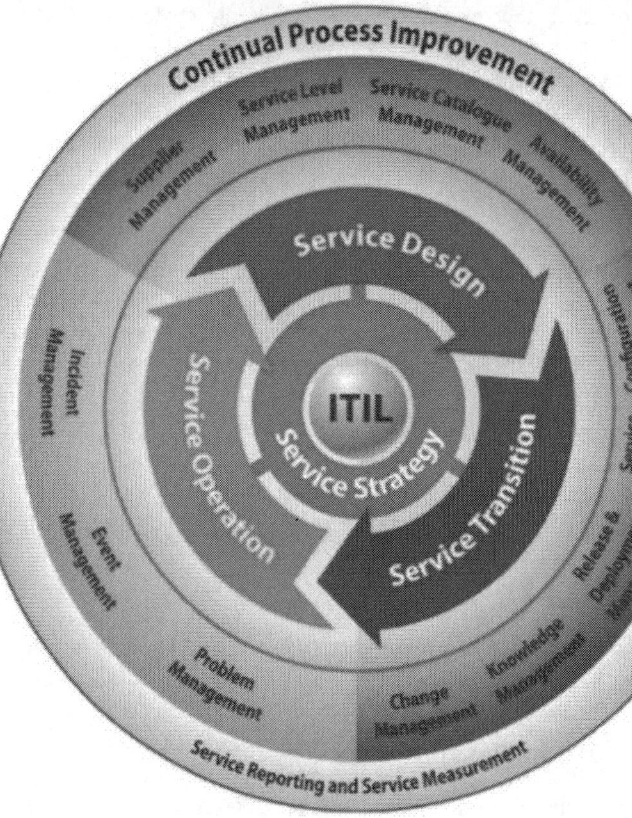

Figure 1. ITIL framework.

Incidents can be defined as an interruption or quality reduction of IT services [9]. Even the slightest incident has not caused significant problems in the system such as a system configuration error can be called an incident. Problem solving an incident may be called incident management. Incident processes can be detected more quickly through automatic detection of an event management tool, and can also be through technician reports, and service desks [14]. The ITIL v3 framework has incident management and is placed on the Service Operation cycle [15]. After an incident occurs, the system should be able to restore the IT service conditions to normal as soon as possible without leaving new problems that have a greater impact on the system. Thus an incident management is required that minimizes the negative impacts of the organization's main business activities. The normal state of IT services can be called a predefined state in an SLA (Service Level Agreement).

According to the ITIL v3 framework, the activities in incident management are 1). Incident identification the incident management process begins with identification. 2). Incident logging this step is required for each type of incident, both large and small. 3). Incident categorization in making incident categories requires a special process between the IT manager and the organization management [16, 17]. It aims to generate incident categories and priority handling in line with the organization's business processes. 4). Incident priority the incident priority step is based on a pre-made categorization. 5). Initial diagnosis The initial diagnosis of an incident must be carried out by any person initially connected with the incident, whether it is a service desk, a technical staff, or an automated device such as event management. 6). Incident escalation Incident escalation is an act of raising the level of incident handling. 7). Investigation (investigation and diagnosis) Investigation measures are conducted to find the source of the problem from the incident. 8). Resolution (resolution and recovery) this step is an action taken to resolve an incident. 9). Closing (incident closure) the closing step is the steps undertaken by the service desk and associated technician staff to ascertain whether the incident has been properly addressed (see Figure 2).

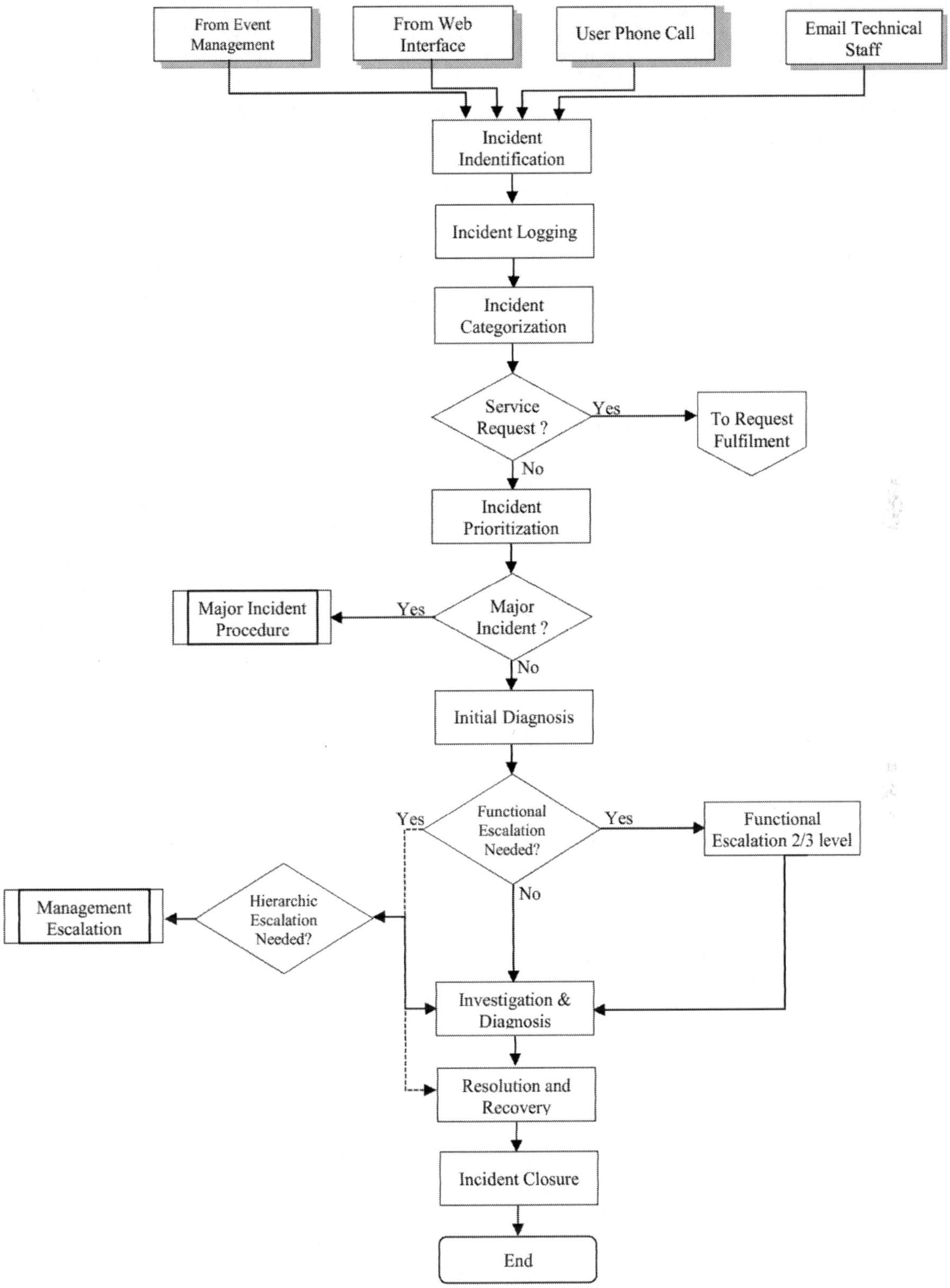

Figure 2. Incident management with ITIL.

This study uses the incident management method of service operation on ITIL v3 framework. The steps taken are as follows: 1). Preliminary; In this stage, the definition of background, problem formulation, problem definition, research objectives and benefits and the methodology used to solve the problem, namely the making of document management of incident management process. 2). Information Collection and Analysis; at this stage, the review activity of the University Information Technology governance document is carried out. Selin also conducted a study of ITIL framework literature. 3). Preparation of Governance Documents; in this step, the preparation of governance documents is based on the results of the analysis in the previous step. The documented procedures will consist of details of incident management activities, and incident category, incident priority, metric and CSF (Critical Success Factor), SLA (Service Level Agreement) and RACI charts. In addition to the attachments above, will also be created a flow chart that describes each activity undertaken. 4). Document Verification; in this stage will be verified each part of the document management to determine whether each activity in the document is in accordance with its purpose and can be implemented. 5). Document Validation; in this phase, validation of the governance document will be conducted to determine whether the main objective of the incident management process has been fulfilled with this document. 6). Conclusion; in this stage will be the formulation of conclusions of the overall steps taken and the results obtained.

3. Results and Discussion

The results of IT governance document analysis show that not all programs in the IT Strategic Plan document have supporting document management. For programs that already have document management, it is known that the document has not been standardized and the development is done solely by each sub-section. This resulted in the implementation of the program is often not maximal and its performance can not be measured. Table 1 below summarizes the help desk management program documents and IT support performed by the IT Support sub-section.

Completion of incidents occurring, the settlement using best practices on the ITIL framework according to table 1 shown below.

Table 1. Document details incident management.

No	Activity	Product
1	Organize service centers and technical assistance for all IT users	IT services and technical assistance
2	Coordinate the implementation of IT operational services and technical assistance with IT Helpdesk solver	Implementation of services and technical assistance
3	Evaluate the operational issues of system, application, network and communication system received	Operational evaluation reports on systems, applications, networks and communications systems
4	Deliver IT service information to those who need it for the issues that have been submitted to the IT Helpdesk	Information IT services
5	Compile and manage database handling/solutions to IT operations problems	Database of troubleshooting
6	Provide, administer access rights and authority associated with the domain (intra / xnet), internet, email and desktop (administrator)	Implementing IT System Administrators associated with domains, email and desktops
7	Manage group archives and documents	Management of archives and documents

After the analysis to refine the document management by rebuilding the details of each activity in the program. These details define each activity objective of the program, the performance indicators for the purpose, the forms, and documents required to carry out the activity, the details of the activity execution steps, the RACI diagram of activity and the activity flow diagram. Details of the management document display the following activity details (see Table 2).

Table 2. Activity details incident management.

Activity	Objective	Performance Indicator
1.Incident identification	1.1 Ensure that each incident can be identified before it has negative implications for the ongoing business process.	Incidents that have negative implications before they can be reported.
	1.2.Ensure incident cards with "Not completed" status are opened and distributed	Incident cards reopened each day.
2.Incident logging	2.1 Ensure that recording of information from incident reports is entered as a basis for the implementation of the incident handling process.	a) The time required to record incident information. b) Complain from U because HO is not in place at time of reporting c) reports entered by phone d) incoming reports via email e) reports that come in the helpdesk table
	2.2 Ensure the information is fully recorded and the source verified.	a) the incident card is fully recorded b) incident cards with verified sources
	2.3 Ensure a summary of incidents and incident card search keywords is made	a) an incident card that has the right keywords
3.Incident categorization	3.1 Ensure the categorization of incident reports is appropriate and done in a short period of time.	a) Incident reports falling into the correct category. b) the time required to categorize incident reports
4. Incident priorization	4.1Ensuring incoming incident reports is given priority of appropriate handling and is done in a short period of time.	a) Incident reports given the appropriate priority. b) The time required to prioritize incident reports.
	4.2 Ensure the delegation of incident handling has the right staff to handle it.	a) Events where staff at HO level are not available. b) escalation of HO to HS due to staff's inability
5. Initial diagnosis	5.1 Ensure early diagnostic action is performed at HO level in a short time.	a) Time required in the initial diagnostic action. b) HO error in initial diagnosis.
	5.2 Ensure that leaders / officials receive priority on-site handling.	a) Availability of HO staff for on-site handling.
	5.3 Ensuring early diagnostic measures may provide input for incident handling Overall and if possible it could provide a solution to the incident	a) Incident reports found the solution in the initial diagnosis.
6. incident escalation	6.1 Ensure the escalation process is done in a short time to meet the target SLA Time of incident handling.	a) The time required to establish an incident escalation. b) Incidental reports that are late in escalation.
	6.2 Ensure that escalation is carried out with due consideration to the action taken.	a) Escalation through consideration of IM.
	6.3 Ensure the selection of persons responsible for post-incident handling escalation	a) Availability of staff to handle combined incidents.
7.Investigation and diagnosis	7.1 Ensure thorough and in-depth investigations to find the source of incident problems.	a) The time required for the investigation of incident reports. b) A complete incident description prior to the investigation. c) Investigative action carried out to meet U again to complete the description of the incident.
	7.2 Ensure that investigative and diagnostic activities are carried out according to standards and meet targeted SLAs of handling time.	a) The time it takes to conduct investigative and diagnostic activities.

Table 2. Cont.

	7.3 Ensuring that solutions are found is appropriate for the intended incident.	a) A solution found after investigation and diagnostic activities by HS. b) Solutions found after investigation and diagnostic activities by SM. c) Solutions found after investigation and diagnostic activities by NM. d) Solutions found after investigation and diagnostic activities by MM. e) handling is given to suppliers
8. resolution and recovery	8.1 Ensure solutions to incidents are tested and can be implemented	a) The solution being implemented is the true solution to the incident. b) Solutions that can not be implemented.
9. Incident closure	9.1 Ensure closure activity is carried out.	a) Incident card with "Not completed" status but actually already found and already implemented the solution. b) Appropriate incident handling of target SLA handling time.
	9.2 Ensure complaints of U are accepted.	a) A complaint from the Incident Reporter (U) on the given solution.
10. Incident management report	10.1 Ensure daily recapitulation is performed. 10.2 Ensure monthly recapitulation is performed.	a) completeness of recapitulation
	10.3 Ensure that an incident response report is prepared as an evaluation of future action measures.	a) Completeness of the report. b) timeliness of report submission
11. Incident Management evaluation	11.1 Ensure that evaluations are undertaken on a monthly basis to improve the quality of incident handling.	a) Evaluation meetings conducted during the working year. b) Surveys conducted during the working year.
	11.2 Ensure evaluation results are followed by each party at the level of incident handling.	a) Evaluation results are not followed up.

This result shows the entire incident management process, displays the incident management process of helpdesk and IT support along with the input to implement it, and the resulting output is:

- Input
  1. Incident report from IT employee / user leader, 2. Interview reports from inspection staff (reporting via telephone, email or directly to the helpdesk desk)
- Process
  1. Incident identification, 2. Incident logging, 3. Incident categorization, 4. Incident priority, 5. Initial diagnosis, 6. Incident escalation, 7. Investigation and diagnosis, 8. Resolution and recovery, 9. Incident closure, 10. Incident management report, 11. Incident management evaluation
- Output
  1. Solution, 2. Problem Management, 3. Configuration Management, 4. SLA Management, 5. Capacity Management, 6. Availability Management, 7. Monthly reports of incident handling, 8.9. Performance evaluation report.

This study aimed to answer the research question: what types of requirements should be taken account in building an incident management system in academic information system. Basic functional requirements of the incident management system are: Submit request, Check the request status, create request, Assign request, Update request, Relate request, Configure settings, Create reports, Maintain knowledge base, Maintain workflow, and Maintain registers. Besides incidents, the system should handle service requests, problems and requests for change. Regarding data requirements, 23 incident attributes were identified. The main contribution of this study lies in showing how IT service management (ITIL) concepts affect the tool specification and requirements.

Both case study and action research methods were used in this study. The case organization was an IS academic department of a university. However, there are some limitations to this study. A case study as a research method has been criticized for a number of reasons. It has been claimed that a case study lacks the academic rigor (the control of research), research results based on case studies cannot be generalized, and that a case study requires a lot of resources and experienced case study researchers. To add the amount of control to the study, we used a project diary where notes of each meeting were stored. Regarding the generalization of results, it is true that we cannot draw statistical conclusions from case study results. Data for this study was collected from one case organization. However, we obtained a good overview of the case organization by using multiple data collection methods, such as discussions with persons who had different perspectives to customer support and maintenance work: an IS manager, process managers, designers responsible for the incident management tool development, and service desk workers. Incident management in academic information system.within IT service management is a new and interesting research field that requires more academic research. Further research (with more case organiza- tions) is needed to investigate the design, deployment and the introduction of incident management systems in academic information system.

4. Conclusions

The conclusions made based on the research that has been done are as follows: The governance document developed for some of the objectives of the help desk management program and IT support, i.e. the incident management process. This governance document contains 11 (eleven) activities consisting of 9 (nine) activities based on the ITIL framework and 2 (two) additional activities as required by the organization i.e. reporting and evaluation. Implementing the program is divided into 7 (seven) parties namely Incident Reporter (U), Helpdesk Operator (HO), Helpdesk Specialist (HS), Incident Manager (IM), Software Manager (SM), Network Manager (NM) And Maintenance Manager (MM). The RACI diagram shows the duties and responsibilities of each of the above-mentioned parties in each step of the activity. Activities developed from the ITIL framework, the implementation is continuous and continuous. While the reporting and evaluation activities are carried out at the beginning and end of the month only. The governance matrix is built to be the conclusion of the entire program process. The matrix contains each activity in the following programs, objectives, performance indicators, forms and documents required for the execution of activities, and the RACI diagram.

Acknowledgments

We would like to thank Universitas Sari Putra Indonesia Tomohon and Universitas Negeri Manado is supported by a the research fund from the Information Engineering Department, Faculty of Engineering, Universitas Sari Putra Indonesia Tomohon and Information Technology and Communication Education Department, Universitas Negeri Manado.

References

[1] B Phillips 2013 Information Technology Management Practice: Impacts upon Effectiveness *J. Organ. End User Comput.* **25**(4) pp 50–74.
[2] R Ajami and I Technology 2013 IT Governance in Higher Education Institutions in UAE *Int. J. It/bus. Alignment Gov.* **4**(2) pp 1–18.
[3] R. Pereira 2012 Designing a New Integrated IT Governance and IT Management Framework Based on Both Scientific and Practitioner Viewpoint *Int. J. Enterp. Inf. Syst.* **8**(4) pp 1–43.
[4] S H Parisa Aasi, and Lazar Rusu 2014 Culture Influence on IT Governance *Int. J. It/bus. Alignment Gov.* **5**(1) pp 34–49.
[5] A Nabiollahi, R A Alias, and S Sahibuddin 2010 A Service Based Framework for Integration of ITIL V3 and Enterprise Architecture *IEEE* pp 1–5.
[6] N Mohamed 2012 A conceptual framework for information technology governance effectiveness in private organizations *Int. Manag. Comput. Secur.* **20**(2) pp 88–106.
[7] M Vicente and N Gama 2014 A Business Motivation Model for IT Service Management *Int. J. Inf. Syst. Model. Des.* **5**(1) pp 83–107.
[8] M Mora and R V O Connor 2014 An Extensive Review of IT Service Design in Seven

International ITSM Processes Frameworks *Int. J. Inf. Technol. Syst. Approach* **7**(2) pp 83–107.
[9] B C Potgieter 2002 *Evidence that use of the ITIL framework is effective* pp 160–167.
[10] S Sahibudin and M Ayat 2008 Combining ITIL , COBIT and ISO / IEC 27002 in Order to Design a Comprehensive IT Framework in Organizations Mohammad Sharifi Centre for Advanced Software Engineering (CASE), University Teknologi Malaysia *Second Asia International Conference on Modelling & Simulation* pp 749–753.
[11] J Järveläinen 2013 International Journal of Information Management IT incidents and business impacts : Validating a framework for continuity management in information systems *Int. J. Inf. Manage.* **33**(3) pp 583–590.
[12] M Marrone, F Gacenga, and A Cater-steel 2014 IT Service Management : A Cross-national Study of ITIL Adoption IT Service Management : A Cross-national Study of ITIL Adoption *Commun. Assoc. Inf. Syst.* **34** pp 865–892.
[13] A Cater-steel and M Toleman 2007 *The Role of Universities in IT Service Management Education* pp 369–382.
[14] M Jäntti 2009 Defining Requirements for an Incident Management System : A Case Study *Fourth International Conference on Systems* pp 184–189.
[15] A A Norita Ahmad, Noha Tarek Amer, and Faten Qutaifan 2013 Technology adoption model and a road map to successful implementation of ITIL *J. Enterp. Inf. Manag.* **26**(5) pp 553–576.
[16] I A Tøndel, M B Line, and M G Jaatun 2014 Information security incident management: Current practice as reported in the literature *Comput. Secur.* pp. 1–27.
[17] M Marrone and L M Kolbe 2011 Uncovering ITIL claims : IT executives ' perception on benefits and Business-IT alignment *Inf Syst E-Bus Manag.* **9** pp 363–380.

IOP Conf. Series: Materials Science and Engineering **306** (2018) 012111 doi:10.1088/1757-899X/306/1/012111

What are the Dominant Factors of Students' Productive Skills in Construction Services?

R R Oroh*, **Haris A S, R M Sugandi and Isnandar**

Mahasiswa Pascasarjana Universitas Negeri Malang, Malang.

*rolly.oroh@gmail.com

Abstract. The purpose of this study to determine the dominant factors of students' productive skills in doing the work of concrete structures that fit the needs of construction services. Sample of the respondents is vocational high school students from several districts and cities in North Sulawesi, Indonesia. Data are obtained through the performance test instruments of student. Whereas, data analysis is performed using factor analysis. The result of this research show the dominant factors of the students' productive skills in doing the work of concrete structures that is according to the need of construction services, namely: (a) factor the working of concrete casting consists of making scaffolding from good materials and conducting concrete casting according to working method; and (b) factor the working of concrete reinforcing consists of read the working drawings for concrete reinforcement and make the concrete formwork from good material. Some of the respondent's students in doing some concrete structure work have done well, but not yet according to working drawings, working methods and technical specifications of the work. The learning is done in accordance with the competency-oriented school curriculum but the teaching materials given have not been maximized in accordance with the needs of productive skills required construction services industry. The results have an impact on the low absorption of graduates in the implementation of the construction services industry.

1. Introduction

The need for skilled students as graduates of vocational high school (SMK) of competency in stone and concrete engineering skills (TKBB) will be greater in line with government policy to improve sustainable infrastructure development. Graduates of SMK TKBB are expected to fill positions in jobs as skilled workers in the implementation of construction services as operators or implementers on building work. This means that students of SMK must have a standard of work competence with adequate skills in doing the job according to field of expertise. Thus, students must equip themselves with job skills in the implementation of building work with various elements of building work. Construction of buildings in the construction process has several elements of work that are integrated in the overall construction of buildings [1, 2].

The competence of skilled workers is reflected in the Indonesian national working competence standard (SKKNI) on the job of implementer of building work [3]. The competence is described in several units of competence, namely: (a). Carry out preparatory work, (b). Carry out foundation work, (c). Carry out structural work, (d). Carrying out architectural work, and (e). Create a job implementation report. For each unit of competence, it consists of several elements of competence and performance criteria. Assessment for each of these competencies should be consistently disclosed in 3 (three) aspects, namely aspects of knowledge, skills and attitudes. Furthermore, it is mentioned in the Indonesian National Qualification Framework (KKNI) [4] that SMK graduates are equivalent to level 2 (two) levels, with job description: (a). Be able to perform a specific task, using commonly tools, use information, work procedures, and demonstrate performance with measurable quality, under the direct supervision of their superiors; (B). Have basic operational knowledge and factual knowledge of specific work areas, so as to be able to choose available solutions to common problems; And (c). Responsible for the work itself and can be given the responsibility of guiding others.

Skills are defined as the students's ability to think and act effectively and creatively in the abstract and concrete area as a development from those studied in schools independently. Students' skills are directed at the ability to observe carefully, imitate activities, adapt (imitate by making adjustments), organize or shape, work according to procedures, work meticulously, work naturally [5]. Skills to be one aspect of the formation of competencies directed at the development or mastery of certain skills. Skills are an important aspect for skilled workers or construction personnel, so that students who are prepared to become skilled workers or executing personnel must equip themselves with skills in the field of expertise. The importance of skills for a workforce becomes the basis of the workforce's success in carrying out its work tasks, because a construction project will be successful if it is done by skilled and trained workers [6]. Skill aspect is an aspect that is gained or obtained by the worker in this case the students such as working atmosphere at work place. This is in line with Willert, etc [7] that students will succeed with real skills like at real work. The skill aspect refers to the ability to perform work under various conditions and possibilities [8]. It is further stated by Maclean and Wilson [9] that the skill aspect is one of the standards required to demonstrate competence. Thus, it can be stated that any skilled workforce will always succeed in such work in carrying out a construction project.

The work skills provided to students in schools are still considered not appropriate or relevant to the needs of the construction industry. Some of the skills competencies given through the subjects in school, it has not fully accommodated all the needs of productive competence of construction services. There are several skills required in the field of construction work, but have not yet been presented in the skill competency documents issued by schools [10]. The relevance of the skills of vocational students in Indonesia to the needs of the business world and the industrial world tends to be low [11].

Students of SMK TKBB who have skill according to field of expertise at job position of building worker, can be seen on work competence unit that is dimension of foundation work and concrete structure work. According to the curriculum structure of SMK, the subjects are divided into three groups, namely normative subjects, adaptive subjects, and productive subjects. Productive subjects equip students with the mastery of productive skills that refers to the minimum standard of work skills applicable in the world of work. The purpose of this study to determine the dominant factors of students' productive skills in doing the work of concrete structures that fit the needs of construction services.

2. Methods

This research uses quantitative research approach. The sample of respondents are students of SMK TKBB taken from several vocational schools in North Sulawesi. The total population of students of SMK TKBB in North Sulawesi in 2016 only amounted to 107. The distribution of students of SMK TKBB was in a wide area, then the sample was taken randomly using Yamane formula [6, 12] with 10% error rate obtained 52 samples of students. This research instrument uses the performance test given to the students. The performance value of the student is focus on 5 types of work: making scaffold, assemble of concrete reinforcement, making of concrete formwork, installing of concrete reinforcement, and concrete casting. The research instrument was developed from theoretical reference and adapted to SKKNI. According to the purpose of this research, the data analysis use factor analysis with the principal component analysis method and varimax rotation method to get the value of communality and eigenvalues. The use of factor analysis is intended to obtain a minimal factor with a simple principle that can generate correlation between indicators measured [13]. The value of communality and eigenvalue obtained by testing the data feasibility requirements first through the value of KMO (Kaiser-Meyer-Olkin Measure of Sampling) and MSA (Measure of Sampling Adequacy).

3. Results and Discussion

Based on the data obtained shows that the range of scores obtained by students through the performance test instruments of students' productive skills on the work of concrete structures is relatively vary from minimal score to maximal score. Some respondents showed performance results that had relatively high levels of skill in performing concrete structure work. When viewed from the score of student achievement of respondents through the scale of the value of performance, it can be seen that most students of respondents showed excellent score on performance assessment. But the findings of the study showed that there were 1,92 percent of the respondents gave unfavorable scores on some focus on performance assessment. This indicates that there are some of the respondent's students in performing some concrete structure work such as making scaffold, assemble of concrete reinforcement, making of concrete formwork, installing of concrete reinforcement, and concrete casting have been done well, but not yet fit working drawings, work methods and job technique specifications. This is supported by preliminary survey results, limited observations and interviews with some vocational teachers and construction service providers that learning in school has been in accordance with the competency-based curriculum but found some work competence has not maximally accommodate the needs of construction services. This is in line with Sutrisno, Dardiri and Sugandi [10] stating that there are some productive competencies required in the field of construction work, but not yet presented in the school curriculum documents. The learning is done in accordance with the competency-oriented school curriculum but the teaching materials given have not been maximized in accordance with the needs of productive skills required construction services industry. The results have an impact on the low absorption of graduates in the implementation of the construction services industry. Therefore, the implementation of vocational education in all building engineering programs must always be dynamic and in accordance with the standards of job skills required by the construction services industry.

According to the purpose of this study, then based on the data obtained conducted factor analysis. After testing the feasibility of data obtained KMO value greater than 0.5. Furthermore, the MSA value is calculated for each instrument component, and the value of all instrument components is greater than 0.5, it can be stated that the available data is feasible to be analyzed through factor analysis. After the process of factor extraction with the main component analysis method, the value of communalities of students' skill in performing the concrete structure work (see Table 1) is sorted from the largest value to the smallest values.

Table 1. Communalities value.

Instrument components	Communalities value
Read the working drawings for concrete reinforcement	0,722
Make the concrete formwork from good materials	0,521
Make scaffolding from good material	0,436
Implementing of concrete casting in accordance with the work methods	0,387

Based on the value of communalities in Table 1, then the factor rotation. Factor rotation stage is done with the aim to find the factor that is able to maximize the relationship between indicator component of student productive skill variable. Furthermore, the result of factor rotation gives eigenvalue of student skill in doing concrete structure work, as given in Table 2.

Table 2. Eigenvalue.

Instrument components	Eigenvalue
Make scaffolding from good material	2,110
Implementing of concrete casting in accordance with the work methods	2,110
Read the working drawings for concrete reinforcement	1,138
Make the concrete formwork from good materials	1,138

Based on the eigenvalue in Table 2. obtained after the factor rotation, then from the instrument component can be formed two factors, namely: (a). Factor 1 or the factor of concrete casting consists of: making scaffolding from good materials and conducting concrete casting stages according to the working method; and (b). Factor 2 or the factor of concrete reinforcing consists of: read the working drawings for concrete reinforcement and make the concrete formwork from good material.

The results of this study reveal the dominant factors of students' productive skills in doing the work of concrete structures according to the needs of construction services. Through the results of this study also revealed that students as candidates for skilled workers or implementers must have minimum standards of skills in the work of concrete structures include the standard skills of concrete reinforcing consists of components of the work of installing concrete reinforcement in accordance with the working drawings, the assembling of concrete reinforcement in accordance with the working drawings, and make the concrete formwork from good material. This is in line with the results of study by Almira et al [14] that the students' productive skills of SMK TKBB that the construction industry needs are skill in drawings, structural work and installation of scaffolding. Therefore, the aspects of productive skills in the work of concrete structures are important for students' success in their work projects. As mentioned by Pavlova [15] that skills are key to the student's performance that industry expects as in the implementation of construction services.

The students' productive skills in carrying out the concrete structure work is important to be mastered by the students as the basic capital to do the job task according to the competence of the skill. Because with good skills the students become competent and expert in their field, as mentioned by Winterton, et al [16] that skill can lead to performance with accuracy and speed in work. Also mentioned by Winch [17] that productive skills learned make the worker can act in a certain way in relation to tasks. Therefore, the skills become important so that the students as a candidate skilled personnel to be competent and expert in the field of expertise.

The importance of skills for a workforce is the basis of the success of the workforce in carrying out the work tasks, as Elfaki and Alatawi [6] notes that a construction project will success if it is done by skilled and trained workers. Because the skill aspect is an aspect gained by the workforce in this case the student is like a workplace atmosphere on real work, as Willert et al [7] mentioned that students will succeed with real skills such as workplace. Thus, it can be stated that any skilled workforce will always succeed in such work in carrying out a construction project.

4. Conclusions

The result of this research show the dominant factors of the students' productive skills in doing the work of concrete structures that is according to the need of construction services, namely: (a) factor the working of concrete casting consists of making scaffolding from good materials and conducting concrete casting according to working method; and (b) factor the working of concrete reinforcing consists of read the working drawings for concrete reinforcement and make the concrete formwork from good material. Some of the respondent's students in doing some concrete structure work have done well, but not yet according to working drawings, working methods and technical specifications of the work. The learning is done in accordance with the competency-oriented school curriculum but the teaching materials given have not been maximized in accordance with the needs of productive skills required construction services industry. The results have an impact on the low absorption of graduates in the implementation of the construction services industry.

The results of this study reveal the students' productive skills in doing the work of concrete structures according to the needs of construction services is the work of concrete casting consisting of making scaffolding from good materials and conducting concrete casting stages according to the working method. Through the results of this study also revealed that students as candidates for skilled workers must have minimum standards of skill in the work of concrete structures that is the work of concrete reinforcing consists of reading working drawings for concrete reinforcement and make the concrete formwork from good material.

References

[1] Allen E and Iano J 2009 *Fundamentals of Building Construction: Materials and Methods* (John Wiley & Sons Inc., New Jersey) **5** p 16-23.

[2] Ching F D K 2014 *Building Construction Illustrated* (John Wiley & Sons Inc., Hoboken New Jersey) **5** p 2.03-2.05.

[3] Decree of the Minister of Manpower of the Republic of Indonesia Number 205 of 2015 on SKKNI (Indonesian National Work Competency Standards) on job title of building construction worker.

[4] Israel G D 2013 *Determining Sample Size* (IFAS Univ. of Florida).

[5] Regulation of the Minister of education and culture of the Republic of Indonesia No. 54/*2013 about tentang Competency Standards of Primary and Secondary Education Graduates*.

[6] Elfaki A O and Alatawi 2015 *J. of Building Construction and Planning Research* **3** p 189-195.

[7] Willert S, Keller H D and Stegeager N 2011 Academic Vocational Training: Bridging the gap between Educational Space and Work Space *Elec. J. of Knowledge Management* **9**(2).

[8] Trinder J C 2008 *Competency Standards - a Measure of the Quality of a Workforce* School of Surveying and SIS (UNSW SYDNEY NSW 2052, Australia).

[9] Maclean R and Wilson D 2009 *International Handbook of Education for the Changing World of Work Bridging Academic and Vocational Learning* (Springer Science+Business Media B.V).

[10] Sutrisno, Dardiri A dan Sugandi R M 2015 *J. Teknologi dan Kejuruan* **38**(1) p 51-64.

[11] Sudira P 2012 *Strategic Issues of Decentralization of Vocational Education in Indonesia,* Proceedings of the ISPI National Seminar (Yogyakarta, Indonesia).

[12] Singh A S and Masuku M B 2014 *Int. J. of Economics, Commerce and Management* **11** p 1-22.

[13] Widarjono A 2015 *Analisis Multivariat Terapan, dengan Program SPSS, AMOS, dan SMARTPLUS Edisi kedua* (Yogyakarta: UPP STIM YKPN).

[14] Almira D, Dardiri A, and Isnandar 2016 *J. Pendidikan: Teori, Penelitian, dan Pengembangan* **1** (4) p 673-680.

[15] Pavlova M 2009 *Technology and Vocational Education for Sustainable Development, Empowering Individuals for the Future* (Amsterdam: Springer).

[16] Winterton J, Delamare F L D, and Stringfellow E 2006 *Typology of Knowledge, Skills and Competences: Clarification of the Concept and Prototype* (Luxembourg: Cedefop Reference series).

[17] Winch Ch 2010 *Dimensions of Expertise, a Conceptual Exploration of Vocational Knowledge* (London: Continuum International Publishing Group).

IOP Conf. Series: Materials Science and Engineering **306** (2018) 012112 doi:10.1088/1757-899X/306/1/012112

Design Control Systems of Human Machine Interface in the NTVS-2894 Seat Grinder Machine to Increase the Productivity

S Ardi* and D Ardyansyah

Production Engineering & Manufacturing Processes, Polytechnic Manufacture Astra
Jl. Gaya Motor Raya No.8, Sunter II, Jakarta 14330, Jakarta

*syahril.ardi@polman.astra.ac.id

Abstract. In the Manufacturing of automotive spare parts, increased sales of vehicles is resulted in increased demand for production of engine valve of the customer. To meet customer demand, we carry out improvement and overhaul of the NTVS-2894 seat grinder machine on a machining line. NTVS-2894 seat grinder machine has been decreased machine productivity, the amount of trouble, and the amount of downtime. To overcome these problems on overhaul the NTVS-2984 seat grinder machine include mechanical and programs, is to do the design and manufacture of HMI (Human Machine Interface) GP-4501T program. Because of the time prior to the overhaul, NTVS-2894 seat grinder machine does not have a backup HMI (Human Machine Interface) program. The goal of the design and manufacture in this program is to improve the achievement of production, and allows an operator to operate beside it easier to troubleshoot the NTVS- 2894 seat grinder machine thereby reducing downtime on the NTVS-2894 seat grinder machine. The results after the design are HMI program successfully made it back, machine productivity increased by 34.8%, the amount of trouble, and downtime decreased 40% decrease from 3,160 minutes to 1,700 minutes. The implication of our design, it could facilitate the operator in operating machine and the technician easer to maintain and do the troubleshooting the machine problems.

1. Introduction

This research was conducted in a company engaged in the manufacture of Automotive Spare Parts valve engine for the purposes of motor vehicle engines. Valve engine manufacturing process are generally consists of forging, satellite welding, heat treatment, and machining. Increased sales of vehicles resulted in increased demand for production of engine valve of the customer. Referring to this, the company plans to conduct the addition of 1 new line bringing the total Machining Line that there be 18 lines. In addition, The Company is also does improvement or overhaul on some machines. Under these conditions, we do the overhaul and improvement. One machine overhaul is performed the NTVS-2894 seat grinder for the seat grinder on machining line 7. Overhaul is based on the growth of developing technology and also because of the addition of lines in the area of machining lines. Improvement made on human machine interface in the absence of back-up program HMI NTVS-2894 seat grinder also to facilitate the operator to operate the seat grinder machine.

In our previous researches [1-3], we have designed and analysis the various sensor applications on the machine automation PLC-based in manufacturing industry, and also using the HMI (Human Machine Interface). We have designed automatic control system according to the working specification to improve the working efficiency of the machine, time reduce of the machine trouble, the addition of pokayoke system sensor (early warning error detection), and modification of control system in manufacturing machine [4-7]. Relatively our new research is control systems modification in height lever checker machine using the laser sensor. G. Valencia-Palomo et al. have design an auto-tuned predictive control based on minimal plant information using PLCs [8]. Putman et al. have design virtual fusion by integration virtual components into a physical manufacturing system [9]. While Buhrer et al. have discussed about changeability of manufacturing automation systems using an Orchastration Engine for PLCs [10]. Milik have discussed about PLCs control program hardware about mapping and scheduling [11].

In this paper, we design and manufacture the human machine interface program. The advantage is that the company does not have to pay for the services HMI programming and also the advantages of using HMI that allows an operator in operating machines as well as facilitates the troubleshooting to minimize lost production time due to the operation and troubleshooting the machine by operator and maintenance. We design how do communicate Human Machine Interface GP-4501T with PLC and how to design a program of the Human Machine Interface GP-4501T of the seat grinder machine. In addition, we also design how to facilitate the operator in operate machine and maintenance in do troubleshooting; accordingly trouble and downtime on the NTVS-2894 seat grinder machine is decline and impact on enhancement productivity machine.

2. Methods

2.1 Seat grinder machine

Seat grinder machine is a machine that is in the area of machining process. Seat grinder machine is a machine that uses a grinding stone as media process works. Millstone that used is NORTAKE GRIND 510 x 12 x 20 2 CX100R. Seat machine grinder is in machining section. In the seat grinder machine, part processed valve is part of the valve seat. The results of the processing valve which have a standard, owned by the valve. The following are the standard that must be considered after going through the process on the Seat Grinder machine: the dimensions of the distance from the seat to the tip end; seat angle; seat run out; seat roundness; and seat roughness.

Dimensions and tolerances of the seat grinder machine are micron; consequently it very high precision. If one of the dimensions is not in accordance with the standard of tolerance, the valve will be declared not good (NG). Figure 1 shows the NTVS-2894 seat grinder machine.

Figure 1. The NTVS-2894 seat grinder machine.

3. Results and Discussion

3.1. Control system design

Figure 2 shows control system design that includes: input, output, and the process of PLC.

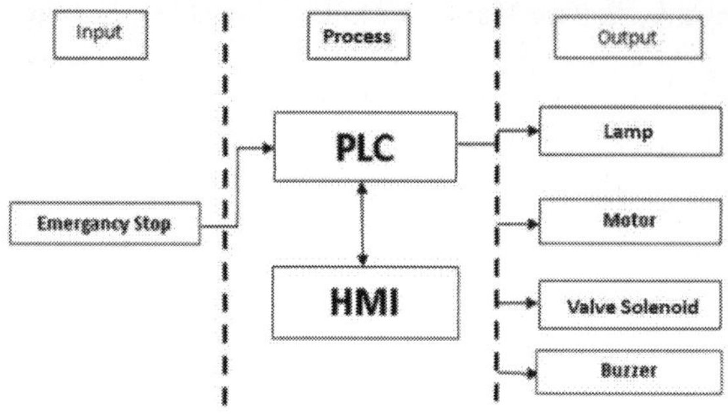

Figure 2. Control system design.

3.2. Process control system

3.2.1. PLC. In this design, PLC serves as the control center that controls output based on input received either from the operator or sensor. In the design of PLC as a control center not only process the data in the form of digital data but also PLC process data in the form of serial data. Binary serial data in the form of numbers, Hexadecimal, Decimal, or Text are sent via the serial communication in the form of a serial cable that is communicated to the HMI Preface.

3.2.2. HMI. HMI is used as an intermediary between the users (in this case the operator) with the machine. So users can interact well with machine, it means the HMI makes it easy for users to operate the machine and know the system is in progress. HMI communicating is using two-way serial communication that can transmit data in the form of binary, hexadecimal, decimal, and text.

3.3 Part input

3.3.1. Emergency stop button. Emergency stop button is used to stop the machine or cycle off when the trouble or unwanted circumstances.

3.4 Part output

3.4.1. Lamps. Lamps are used for indicators related to the engine. To be used as an indicator lamp and lamp power master on.

3.4.2. Motor. The motor is used as the driving of the spindle, and oscillations in the engine as well as a driver for playback millstone.

3.4.3. Solenoid Valve. Solenoid valve is used as the driving piston. Both for forward and backward.

3.4.4. Buzzer. Buzzer is used as a warning sign in the form of sound when there is trouble on the engine Seat Grinder NTVS-2894.

3.5 Design of the programming

Before design a program in advance, we create a flow program of the seat grinder machine. Here will be explained how the diagram flow or process flow from the NTVS-2894 seat grinder machine. Flow process on the NTVS-2894 seat grinder machine consists of a flow process preparation, flow process manually, and the process flow Auto. Here is an explanation of each process flow that exists on the NVS-2894 seat grinder machine.

3.5.1. Preparation of process flow. Process flow preparation in which the operator presses the master key on, engine life was marked with monitors HMI lights up, after the operator pressed the buttons on HMI start of lubrication on to oscillation on and selects the auto or manual. Figure 3 shows the process flow preparation.

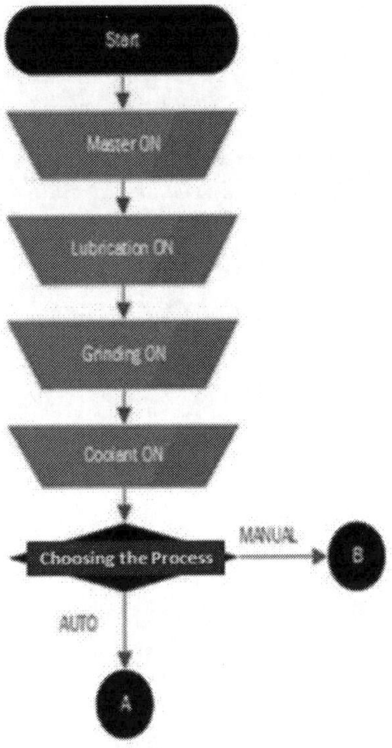

Figure 3. The process flow preparation.

3.5.2. Auto process flow. When the operator presses the Auto button on the HMI, then the NTVS-2894 seat grinder machine works automatically. But the requirement to begin the process of auto first pressing cycle on that machine will work automatically. After pressing, cycle on feeding machine will proceed automatically, by starting the gate off the chuck open later ejector forward and ejector backward then loader down the clamp open after the pusher forward andfinger down and then chuck close later pusher backward and spindle on and finger up the loader up and then the gate open and close clamp. When the process is finished, the spindle off and re-started from the gate off.
3.5.3. Manual process flow. When select a manual process, then process the NTVS-2894 seat grinder machine will work manually. Conditions typically used to perform the manual setting process of the machine. When the maintenance or cleaning and while there is a problem. On this manual process each sequence is controlled by a button on the HMI. For example cylinder that serves as a loader, loader will be open or closed will have its own button on the HMI and so did the other sequences.

3.6 Design of HMI programming
The design of the HMI control program of the NTVS-2894 seat grinder machine is based on the design of component specifications and designs that have been made. The design of this HMI programming is in accordance with the work process of the NTVS-2894 seat grinder machine. The HMI program that we use is GP-Pro software version 4.0. Figure 4 shows the screen preparation of HMI programming.

Figure 4. Screen preparation.

3.7 Results analysis

3.7.1. Machine Trouble. One of the causes of the overhaul on the NTVS-2894 seat grinder machine is due to trouble on the NTVS-2894 seat grinder machine to reach 176 times troubles. With 18 months with 176 times of troubles, average the NTVS-2894 seat grinder machine has trouble of ± 10 troubles per month. After the overhaul of the NTVS-2894 seat grinder machine, the amount of trouble on the NTVS-2894 seat grinder machine is decreased. With the decrease of the amount of trouble on the machine it also affects the downtime of the machine. The downtime of the NTVS-2894 seat grinder machine has decreased. The decreased of downtime machine is 1,700 minutes, from the previous is 3,160 minutes.

Figure 5 shows the production achievement of the NTVS-2894 Seat Grinder machine every day for a month. The red line is production target and the blue bar is production actual.

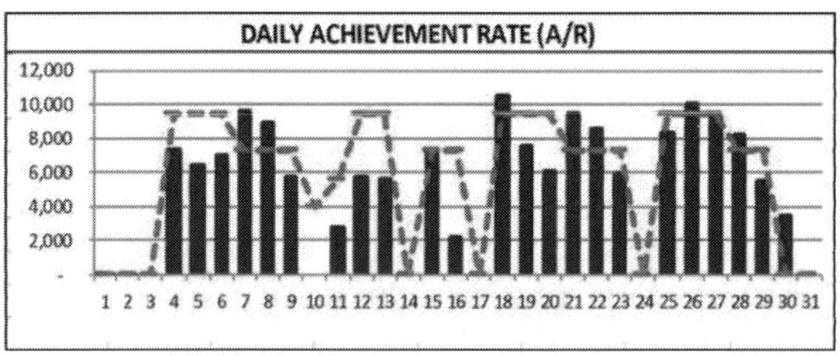

Figure 5. Achievement of the production.

The total of valve production on the previous NTVS-2894 Seat Grinder machine is 160,724 pcs per month, with a production target is 143,838 pcs per month. Thus, the actual machine production capacity exceeds a predetermined target. Figure 6 shows the total Achievement of Production.

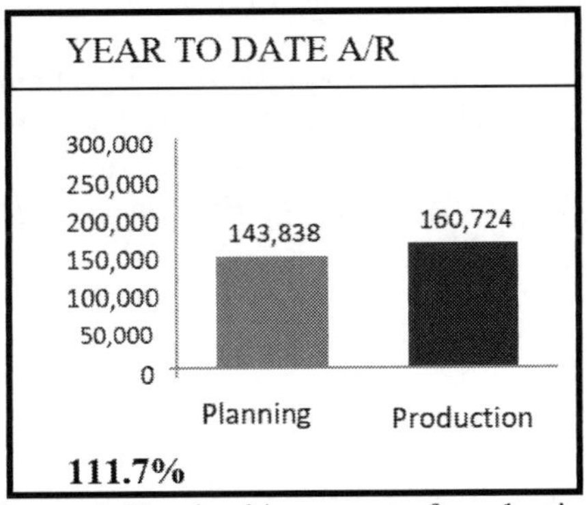

Figure 6. Total achievement of production.

3.8 Discussion

One benefit after overhaul is to optimize the performance of the NTVS-2894 seat grinder machine. Production data before the overhaul, NTVS-2894 seat grinder machine has production achievement of averaged only 77% per month of total production targets. After the overhaul, machine production reached 111.7% calculated in the first month of the production. So the achievement of this machine production increased by 34.8%. Further program development can be done in accordance with work function, to meet production needs, for maintenance, and for machine repair.

4. Conclusions

In this paper, we have discussed about the design control systems of Human Machine Interface in the NTVS-2894 seat grinder machine to increase productivity, to decrease trouble and downtime of the machine. For Pro-Face HMI communicates GP-4501 T (Analog) with PLC is using RS-232 cable D-sub 9-pin and design Human Machine Interface programming covers the main screen and alarm screen with use GP-Pro EX software ver. 4.0. After do overhaul and modification system control, the HMI occurs productivity machine increased 34.8% from 77% to 111.7%, declining trouble from an average of 10 times per month. The downtime on the NTVS-2894 seat grinder machine decline of 3,160 minutes is 1,700 minutes. It could facilitate the operator in operating machine and the technician easer to maintain and do the troubleshooting the machine problems. In addition, for the future research, further program development can be tailored to the operator's work function and to meet changes in production targets, machine maintenance, and machine troubleshooting.

References

[1] Ardi S, Prasetyani L, and Budianto R G 2013 Pokayoke Control System Design using Programmable Logic Controller (PLC) on Station Final Check Propeller Shaft *Proceeding Annual Engineering Seminar 2013* pp C-74 – C-80.

[2] Ardi S, Ponco A, and Kurnia A F 2013 Design Control Systems of the Out Diameter Finish Machine Based on Programmable Logic Controller *Proceeding International Conference On Instrumentation, Communication, Information Technology and Biomedical Engineering 2013* pp 307 – 312.

[3] Ardi S, Djoko Subagio, and Sidik M 2014 Automatic Detection Machine on the OLP (Outer Link Plate) Cam Chain Using Camera Sensor and Programmable Logic Controller *Proceeding of the International Conference MICEEI 2014*.

[4] Ardi S, Jimmy M, and Agustono R 2015 Design of Pokayoke Sensor Systems in Engraving Machine to Overcome Upside Defect Production using Programmable Logic Controller *Proceeding International Conference on QiR 2015 University Indonesia*.

[5] Ardi S, and Al-Rasyid A 2016 Design of Pokayoke Sensor Systems in Drill Oil Hole Machine to Detect the Presence of Drill using Programmable Logic Controller *Advanced Science*

Letters **22**(7) pp 1813-1816.

[6] Alphonsus E R, and Abdullah M O 2016 A review on the applications of programmable logic controllers (PLCs) *Renewable and Sustainable Energy Reviews* **60** pp 1185-1205.

[7] Ardi S, Rusuk Ginarto R, and Putro AP 2017 Control Systems Modification in Height Lever Checker Machine using The Laser Sensor *Journal of Engineering and Applied Sciences* **12**(8) pp 2148-2153.

[8] G Valencia-Palomo, and J A Rossiter 2011 Programmable logic controller implementation of an auto-tuned predictive control based on minimal plant information *ISA Transactions* **50** pp 92-100.

[9] Putman N M, Maturana F, Barton K, and Tilbury D M 2015 Virtual Fusion: Integration Virtual Components into a Physical Manufacturing System *IFAC-PapersOnLine* **48**(3) pp 904-909.

[10] Buhrer U T, Legat C, and Vogel-Heuser B 2015 Changeability of Manufacturing Automation Systems using an Orchastration Engine for Programmable Logic Controllers *IFAC-PapersOnLine* **48**(3) pp 1573-1579.

[11] Milik A 2015 On PLCs Control Program Hardware Implementation Selected Problems of Mapping and Scheduling *IFAC-PapersOnLine* **48**(4) pp 354-361.

IOP Conf. Series: Materials Science and Engineering **306** (2018) 012113 doi:10.1088/1757-899X/306/1/012113

Development of Learning Management in Moral Ethics and Code of Ethics of the Teaching Profession Course

S Boonsong*, S Siharak and V Srikanok

Faculty of Technical Education, Rajamangala, University of Technology Thanyaburi, Pathumthani, Thailand.

*sutboon@hotmail.com

Abstract. The purposes of this research were to develop the learning management, which was prepared for the enhancement of students' Moral Ethics and Code of Ethics in Rajamangala University of Technology Thanyaburi (RMUTT). The contextual study and the ideas for learning management development was conducted by the document study, focus group method and content analysis from the document about moral ethics and code of ethics of the teaching profession concerning Graduate Diploma for Teaching Profession Program. The main tools of this research were the summarize papers and analyse papers. The results of development showed the learning management for the development of moral ethics and code of ethics of the teaching profession for Graduate Diploma for Teaching Profession students could promote desired moral ethics and code of ethics of the teaching profession character by the integrated learning techniques which consisted of Service Learning, Contract System, Value Clarification, Role Playing, and Concept Mapping. The learning management was presented in 3 steps.

1. Introduction

Moral Ethics (ME) is the good issues for personal characters and Code of Ethics (CE) can refer to the personal characters in various careers that people should follow and involves moral principles and ethics. ME and CE also should be taught to teacher student which are needed in both government and non-government schools teachers. ME and CE are the appropriate behaviours of teacher professionals that all teachers should be aware of. The necessity and importance of ME and CE for teachers are as follows: 1) ME and CE can control the standard and guarantee the quality of teaching profession, 2) ME and CE can control the ethics of teaching professions, 3) ME and CE can decrease the problems of teachers which may going on teaching professions. Thai governments have been aware of these problems in Thailand for many years. The information has shown some social problems came from a lack of thinking and development of ethical behavior [1]. These problems are the result of inadequately addressing these issues in the educational system. Previous method training in ethics are not very useful today. They involve a 'lecturing' approach which is not acceptable or appropriate in modern times. It is necessary for educators to improve the teaching of moral ethics and code of ethics of the teaching profession, especially for the teacher students [2].

It is not expected that the results of this study will be able to solve all of the problems of society. But this study will attempt to show the necessity of improving ME and CE teaching in the Thai education system and the probability of its success given the use of particular methodologies. This

IOP Conf. Series: Materials Science and Engineering **306** (2018) 012113 doi:10.1088/1757-899X/306/1/012113

study looks at the teaching of ME and CE for the teaching profession, in particular the in designing the learning management in moral ethics and code of ethics of the teaching profession for teacher students. Rajamangala University of Technology Thanyaburi (RMUTT) has provided Graduate Diploma for Teaching Profession Program since 2007. The students of this program have been expected to the teacher who have good intelligence, character, and competency to support the development of young people in school on the new standards of Teachers' Council of Thailand [3].

The ideas of learning were used in this research will show that the utilization of techniques were Service Learning, Contract System, Value Clarification, Role Playing, and Concept Mapping.

2. Methods

2.1. The purposes

The purposes of this study is to present the development of the learning management, which will prepared for the enhancement of moral ethics and code of ethics of the teaching profession for Rajamangala University of Technology Thanyaburi (RMUTT).

2.2. Methodology

The methodology for contextual study and learning management was conducted by document study, the focus group method and content analysis from the document about moral ethics and code of ethics of the teaching profession concerning Graduate Diploma for Teaching Profession Program. The action research methodology was used in the studies also [4].

2.3. Sample

The samples consisted of several documents about moral ethics and code of ethics of the teaching profession concerning Graduate Diploma for Teaching Profession Program.

2.4. Steps of the study

The steps of this study were; 1) Studied the several policies need and some information for moral ethics and code of ethics of the teaching profession, and also the theories and ideas about learning management for expected characteristics. 2) Developed the learning management for the enhancement of students' moral ethics and code of ethics of the teaching profession (see figure 1).

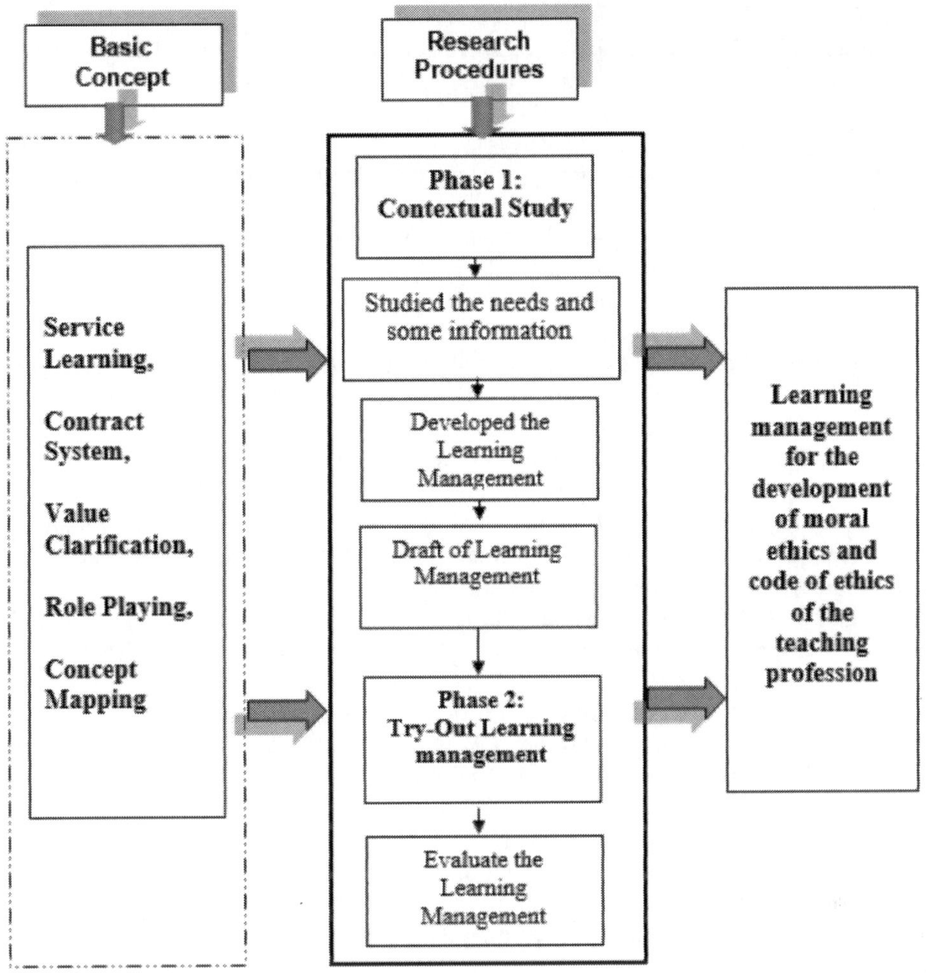

Figure 1. Steps of the study.

2.5. Instrument
The main tools of this study were the summarized papers and analysed papers.

2.6. Data Collecting
The steps of data collecting were;
- The information and data for design the learning management were collected from a lot of documents, and some information from experts, teachers and some teacher students of Graduate Diploma for Teaching Profession Program in the academic year 2016-2017.
- The information concern the learning management improved was collected for the consideration by researcher and recorded them.

2.7. Data analysis
- Based information for using to develop the learning management were analysed and concluded for the descriptive report.
- Information of the learning improving that had been evaluated, were analysed, interpreted, and grouping.

2.8. *Place and time*
Faculty of Technical Education, Rajamangala University of Technology Thanyaburi (RMUTT) Pathumthani, Thailand between August 2016 to June 2017.

3. Results and Discussion
The result from contextual study and learning management development was conducted by document study and the focus group method found the current problem situations concerning the teaching of ethics in Thai education. Many problems can be found from several Thai educational researches. Most of these researches found the university education need to be increased in every field of study, because there is not enough appropriate ethical teaching, especially in education profession. Teachers also need to be trained to know more about the content and teaching of ethics. The course guidelines in ethics and professional ethics should be improved at the same time. Many teachers have agreed to integrate content about ethics and professional ethics into previous ethical or moral subjects from other courses. They have also agreed that the important issues that should be emphasized are honesty and responsibility. In Thailand now there have been s situation similar the situation of teaching ethics in Australia. The goals in Australian teaching of ethics are that they need students to learn the rights and responsibilities and to have a correct attitude toward work and employment. Teachers are aware that they need to teach students with a good sense of ethics, which includes justification, individual feeling, and respect for others and for social institutions [5].

Faculty of Technical Education, Rajamangala University of Technology Thanyaburi (RMUTTT) provide learning management for the teacher students in Graduate Diploma for Teaching Profession Program. The policy of Faculty of Technical Education prepares teacher students of RMUTT to meet the expectations of them at work [6]. They must be able to adapt themselves to technological changes, discipline, workmanship, the particular ME and CE associated with their work and their responsibility to society. RMUTT students of Graduate Diploma for Teaching Profession program have been taught morals and ethics in "Moral Ethics and Code of Ethics of the Teaching Profession Course" which there is moral ethics and code of ethics content is included in this course [7]. But there is also the problem with teaching methodology. As a result of the lack of knowledge of the appropriate way to teach moral ethics and code of ethics, the teaching of these subjects needs to be improved [8, 9].

The results of development showed the learning management for the development of moral ethics and code of ethics of the teaching profession for RMUTT Graduate Diploma for Teaching Profession program students could promote desired moral ethics and code of ethics of the teaching profession character by way of the integrated of several learning methods which consisted of Service Learning, Contract System, Value Clarification, Role Playing, and Concept Mapping. The integrated learning management was presented 3 steps which can be showed as following (see figure 2, 3, 4, and 5).

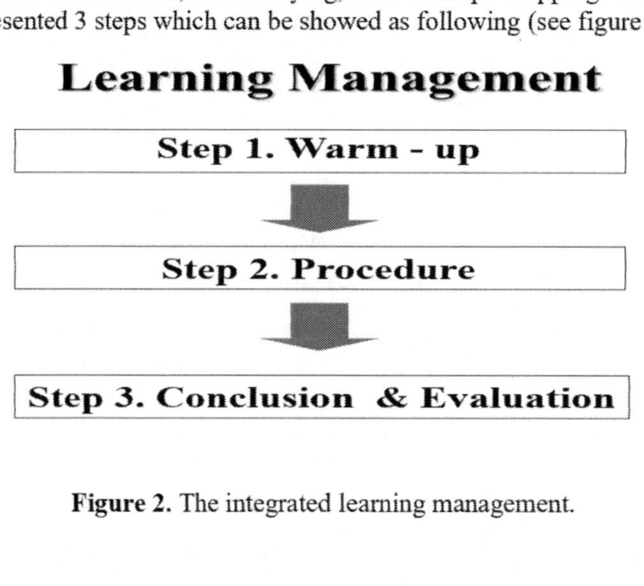

Figure 2. The integrated learning management.

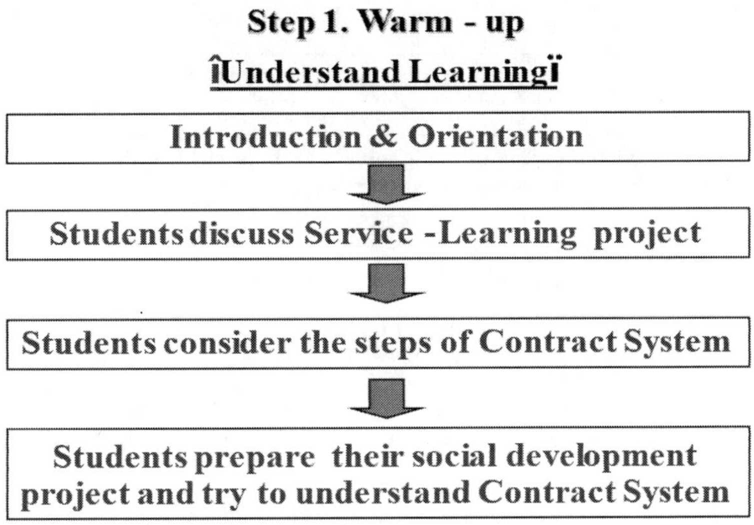

Figure 3. Step 1: Warm-up (understand learning).

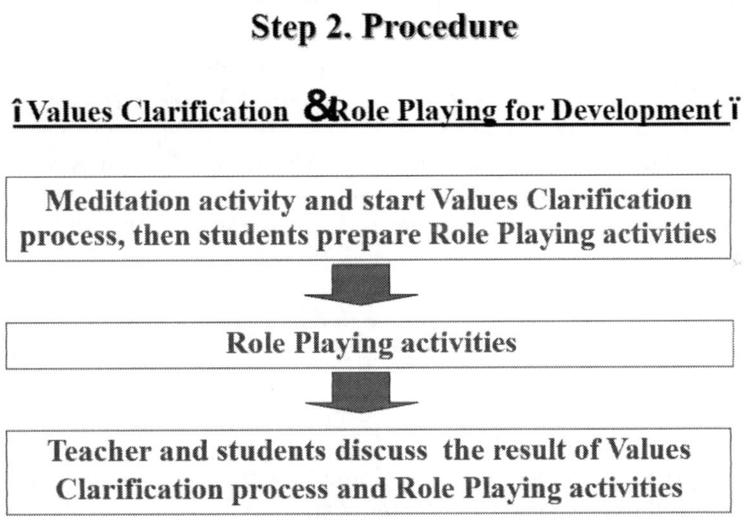

Figure 4. Step 2: Procedure (values clarification & role playing for development).

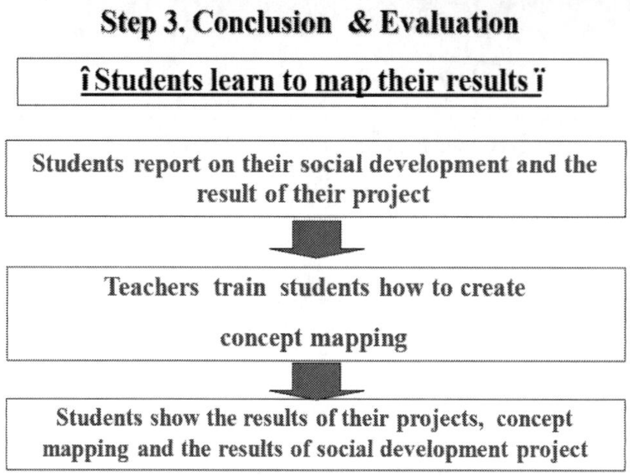

Figure 5. Step 3: Conclusion & evaluation (students learn to map their result).

4. Conclusions

The results of development showed the learning management for the development of moral ethics and code of ethics of the teaching profession for Rajamangala University of Technology Thanyaburi (RMUTT) could promote desired learning management in learning moral ethics and code of ethics of the teaching profession by way of the integrated of several learning techniques which consisted of Service Learning, Contract System, Value Clarification, Role Playing, and Concept Mapping. The integrated instructional model was presented in 3 steps. The next studies should study the result of this learning management in RMUTT and others institute to see any results, and should have any research which study using the others learning management to develop characteristics of teacher students next.

References

[1] Boonsong S 2015 Instructional model for the development of moral ethics and code of ethics of the teaching profession *1st International Symposium on Innovative Education and Technology, 24th to 25th of July 2015 Rajamangala University of Technology Thanyaburi (RMUTT), Pathum Thani,Thailand.*

[2] Steven R L and Allen M G 2012 *Teaching Public Values: The Instructional Approaches* (Social Education 1: Reader. Faculty of Education, Geelong: Deakin University).

[3] Teachers' Council of Thailand 2015 *Qualification Requirements for Education Professional License Renewal* http://www.ksp.or.th/ksp2013/service.php?t=2, 18 Feb 2015.

[4] Howard K 1995 *100 ways To Enhance Values and Morality inschools and Youth Settings* (MA: Allyn & Bacon Needham Heights).

[5] Boonsong S 2006 *The Development of Instructional Model for the Developing of Professional Ethics for Rajamangala Institute of Technology Students* (Thailand Educational Counsil Seminar , September 2006. ,Bangkok).

[6] RMUTT 2015 *Rajamangala University of Technology Thanyaburi Strategy of Policy and Planning 2557-2561* (Department of Policy and Planning, Rajamangala University of Technology Thanyaburi).

[7] Fraenlrel J R 2007 *How to Teach about Value* (New Jersey: Treatise – Hall).

[8] Grassian V 1981 *Moral Reasoning: Ethical Theory and Some Contemporary Moral Problem* (New Jersey: Prentice – Hall).

[9] Rath L E, Harmin M, and Simon S B 1966 *Values and Teaching* (Ohio:Charles E. Merrill Publishing. United Stratigies).

IOP Conf. Series: Materials Science and Engineering **306** (2018) 012114 doi:10.1088/1757-899X/306/1/012114

Does Vocational Education Model fit to Fulfil Prisoners' Needs Based on Gender?

S H Hayzaki* and I D A Nurhaeni

Program Studi Ilmu Administrasi Negara Fakultas Ilmu Sosial dan Politik
Universitas Sebelas Maret

*safirahyzki13@student.uns.ac.id

Abstract. Men and women have different needs, based on their gender or the socio-cultural construction. The government has issued a policy about accelerating the equivalence of gender since 2012 through responsive planning and budgeting. With the policy, every institution (including the institutions under the ministry of law and human rights) must integrate its gender perspective on planning and budgeting, then it can fulfill the different needs between men and women. One of the programs developed in prisons for prisoners is vocational education and technology for preparing the prisoners' life after being released from the prison cells. This article was made for evaluating the vocational education and training given to the prisoners. Gender perspective is employed as the analyzing tool. The result was then used as the basis of formulating vocational education model integrating gender perspective. The research was conducted at the Prison of Demak Regency, Indonesia. The method used in the research is qualitative descriptive with data collection techniques using by in-depth interviews, observation and documentation. The data analysis uses statistic description of Harvard's checklist category model and combined with Moser category model. The result shows that vocational education and training given have not considered the differences between men and women. As a result, the prisoners were still not able to understand their different needs which can cause gender injustice when they come into job market. It is suggested that gender perspective must be included as a teaching material in the vocational education and training.

1. Introduction

Men and women have different needs due to their different gender and socio-cultural constructions. Since 2012, the government has issued a policy on accelerating gender mainstreaming through gender responsive planning and budgeting. With this policy, every institution (including those under the Ministry of Justice and Human Rights) should integrate gender perspective in their planning and budgeting so as to meet the different needs of men and women. One of the programs developed by prisons is vocational education and technology in order to prepare the prisoners to go back to the community.

There are many detained women experiencing criminal acts, having different needs fulfilment during their prison, something that would require policies targeted to meet the needs of gender-responsive of prisoners [1][2]. It is necessary to adopt gender needs in all aspects including the fulfilment of the rights of the prisoners in vocational education and techniques which must also focus on the problems of men and women.

There are complex and various problems occurring in prison. Overcapacity is one of them as proven by the data from a recorded humanitarian database system totalling 211 prisoners which exceed the capacity available in the second-class prison B like the case in Demak. Besides, the prisoners are not separated based on their sex even though they have different needs. Inmates and prisoners with specific crime rates have to be mixed. This can be seen from the data obtained from the data below.

Below is the latest data of the total number of UPT residents in Second Class at Regional Office of Demak, Central Java.

Table 1. Number of UPT residents in Second Class at Regional Office of Demak, Central Java

Category	Adults		Children		Total
	Men	Woman	Men	Woman	
Inmates	103	6	4	0	113
Prisoners	94	4	0	0	98
Total	197	10	4	0	211
Capacity					100
Over Capacity					111

Source: Directorate General of Corrections, 2017 [3].

With the overcapacity condition and the obligation that a prison should integrate gender perspective, the government and the prison caretakers must work together in fulfilling gender needs. This article focuses on providing Vocational Education and Technical assistance (VET) to prisoners and inmates in gender-needs fulfilment.

The former research discovered that despite the inmates and prisoners' lack in freedom, they still have the right to get the equal treatment in education and training to reduce recidivism [4][5]. The contribution of VET programs was considered valuable by most respondents, on the grounds of acquiring additional knowledge and skills and subsequently enhancing their employability [6][7]. Vocational education also reduces the risks of youth unemployment in Russian regions [8]. VET program is used for preparing the capability of students to step into working life in companies by giving an opportunity to learn directly by understanding gender differences in school and at work [9] [10].

This paper update is meant to conducting vocational education programs and techniques that integrate gender perspectives for prisoners, so that they can understand the difference in gender needs, in the hope that they would not experience gender inequality once they leave prisons and enter a new business life. Therefore, this article examines the fulfilment of vocational and technical education needs with a gender perspective for all the prisoners at Demak Class II Prison by focusing only on gender needs.

2. Literature Review

Gender is a social category that indicates the increase of women status so that it is equal to that of men. Gender is not only limited to the issue of role equality but also related to awareness and commitment. Therefore, gender does not focus on the interest of women but also focuses in the balance between women and men [11][12][13][14].

An important concept that needs to be understood in order to address the issue of women is to distinguish between the concept of sex and gender. The interest will appear to raise the issues of social injustice that especially affect women. The appearance of social injustice is caused by the lack of gender roles in development and the injustice that has been rooted in the ideological system of society [15].

According to Fakih [16] the manifestations of gender injustice are ranging from stereotypes (sex labeling is detrimental to other sexes), marginalization (attempting to limit and / or marginalize processes caused by different sexes), subordination (assuming that the roles by one sex, perceived below than the other sex), violence (physical, psychological, economic, or sexual acts that cause feelings of discomfort to one sex), and many burdens (division of tasks and responsibilities that incriminate one of the sexes). While the results of Nurhaeni's research finds that people often encounter confusion when they create the classifications of gender injustice. For that reason Nurhaeni suggests the use of four classifications of gender injustice: stereotypes, marginalization, double burdens, and violence [17].

Vocational education is an education for the mastery of knowledge and skill that have economic value, according to market needs with high education labor coefficient, so it creates a quality workforce [18][19][20]. In addition, with aspects developed through innovation model and entrepreneurship, it is able to encourage individuals to create their own business [21]. Vocational education is the total process of education aimed at developing the competencies needed to function effectively in an occupation or group of occupations. The meanings implied in this definition are: (1) the development of competence, (2) the competencies required, (3) the competencies developed to function effectively, and (4) the competencies developed in relation to a job-or group of work. Vocational education is a special education (specialized) and including all types and levels of work. The education is not only concerned in the manual skills, but also in mental, manual skills, values, and attitudes [22].

Prisoners are people who are deprived of their freedom to move as a consequence of the offense which has been committed. While the other rights as human beings are still given by the state and regulated in the applicable Laws and Regulations, One of the rights regulated in UU RI No 12 Year 1995, about Pemasyarakatan (penitentiary) Chapter I Article 3 is obtaining education and knowledge [23]. Prison is a place for persons detained, both women and men for investigation, prosecution, and probation in court [24][25].

By assessing the fulfilment of gender needs, this paper uses the Moser's model of gender analysis tools by proposing practical and strategic needs aimed at improving the access and control for tangible resources through community organizations. The practical need is an immediate need that has specific and short-term contexts. Where as strategic need is the need enabling women to transform power imbalance between women and men. The strategic need also responds to long-term need in an effort to change a more equal gender division of labor, power, and control, including issues related to legal rights, domestic violence, wage equality, and women's control over themselves [26].

3. Methods

This article evaluates the vocational education and training provided to prisoners. The analysis tool used is a gender perspective. The results of this research are used as the basis for formulating a vocational education model that integrates a gender perspective. The research is conducted at the Demak District Prison, Indonesia. The method used in this research is descriptive qualitative by employing data collecting techniques, i.e. in-depth interviews, observation, and documentation.

Harvard checklist model is used in data analysis, particularly on women's dimension in project identification (women are counted for in the identification phase through the assessment of women's needs, the definition of project objectives and the identification of negative-effect opportunities), project design (women are considered from the project impact on women's access, activity, and control), project implementation (women are considered from individuals, operations and logistics, finance, and flexibility), Project evaluation (considering women from data requirement, analysis, and data collection on specific projects) for data analysis. Harvard is considered the most appropriate model because this study is aimed at designing more efficient programs and improve overall work productivity with an optimal level of gender justice as well as to map vocational training programs for men and women in the prisons [27].

Harvard model is combined with Moser's gender analysis to see the fulfilment of practical needs, specific needs, and gender strategic needs. This model was appropriately used because it is able to see the remoteness of the guidance of women and men. With this model it can be known that the gender needs of men and women are different [26].

4. Results and discussion

This research identifies four aspects of Harvard checklist category model at women's stage dimension in project identification, project design, project implementation and project evaluation and combined with Moser's category model at practical needs, strategic and gender specific.

Table 2. Harvard and Moser Analyses

Analyzed Unit	Practical Needs of Gend	Strategic Needs of Gender	Specific Needs
Women's dimension in project identificati	Based on aspiration to identify their needs	X	Specific needs of men and women prisoners already identified
Women's dimension in project design	Training plan already designed by pay attention on activity access and control for women prisoners	The training is designed to improve the prisoners' skill when re entering the labor market	a specific training plan for both male and female prisoners'
Women's dimension in project implementatio	X	X	X
Women's dimension in project	X	X	X

4.1. Women's dimension in project identification

The data shows that women's needs in vocational education only meet on specific needs. It can be seen from the identification of various training which will be addressed to male prisoners the same statement is explained by two respondents in the interview. While a training to make accessories is given to female prisoners, but it has not been implemented. Gender practical needs can be seen from identifying its needs; prisoners are free to express their opinions directly to the prison officers or using suggestion boxes which is still active. Gender strategic needs cannot be fulfilled because there is no involvement from female prisoners in making decision to join the training [28] [29].

4.2. Women's dimension in project design

In this stage, strategic, practical and gender specific needs are fulfilled. The data show the statement of the security officer as one of the respondents in the interview. He explained that the training is planned to meet the needs of vocational education to improve skills and expertise as well as activities reduce recidivism. To minimize the occurrence of planning failure noticing the failure of the previous training to make crackers is done [30].

4.3. Women's dimension in project implementation

Based on the statement of the two prisoner's respondents and the security officer, it was known that in the implementation phase female prisoners have not been empowered. Since the training intended to them is still in the planning stage. Here, the budget planning there hasn't been separation of budget for male and female prisoners. Moreover, male prisoners are still more dominant in their participation in the training [28] [29].

4.4. Women's dimension in project evaluation

The data show that the evaluation and monitoring have not fulfilled the needs of the targeted people, in the strategic, practical and gender specific needs. According to the security officer evaluations are only performed by authorized prison officers, prisoners are only as program targets. Improvements continue to be made because there are some obstacles found in the result of previous training program evaluations that still found various obstacles in the implementation. Prisons do not restrict if there are criticisms and suggestions from the prisoners. However, the decisions are on the authority of men having higher position and power in employment structure [29].

From the training activities in the fulfillment of vocational education the prisoners of the prison imply that the needs of women have not been fulfilled because gender integration in the dimensions as Harvard intended has not all fall into the category of practical, strategic or gender specific needs of the Moser model. New women are integrated into the dimensions of identification and planning design dimensions. Later in the implementation and evaluation dimensions do not meet the needs of the three because there is no participation and there is still a gap in the fulfillment of the needs for the female prisoners. In fact, the prisoners are still not aware of the importance of gender needs that must be fulfilled in this training program. So when there is a gap between the fulfillment of the needs of the male and female prisoners, they are not aware of gender injustice.

5. Conclusion

Prison Vocational Education Training (VET) for prisoners and inmates have not yet considered the need differences between men and women at the stage of project identification, project design, project implementation, and project evaluation. Women have not become an integral part of the entire stage of the program's dimensions proposed by Harvard as the research analytical tool. There is women dominance in the stage of women's dimension in project identification and project design, while on other stage women dominance is not found. So, the implication of this training program has not been able to meet gender needs. As a result, the prisoners and inmates have not understood their specific, practical and strategic gender needs as described by Moser as also a research analytical tool and thus potentially have the experience of gender inequality as they renter the labor market. It is recommended that gender perspectives should be included as teaching materials in vocational education and training.

References

[1] Zaplin R 2008 Female Offenders: Critical perspectives Interventions *Jones and Barlett Publisher, Inc, Sudbury*

[2] Joiner M N 2011 What is the Impact of Gender-Responsive Treatment on Women Offenders? *College of Professional Studies Professional Projects* **31**.

[3] Direktorat Jenderal Pemasyarakatan 2017 *Data Bulanan Jumlah Rutan Kelas II A Jakarta Timur Kanwil DKI Jakarta* (online) available at http://smslap.ditjenpas.go.id/public/grl/detail/monthly/upt/db5ce9b0-6bd1-1bd19dae313134333039

[4] Natália F G and María J H S 2014 Are inmates different higher education students? *Procedia-Social and Behavioral Sciences* **112** 114 – 123

[5] Hadi M, Wan A W M 2015 Reducing Recidivism Rates through Vocational Education and Training *Procedia - Social and Behavioral Sciences* **204** 272-276

[6] Eugenia A P 2012 Vocational education and training of unemployed women in Greece: An initial approach *Procedia - Social and Behavioral Sciences* **69** 1729 – 1736

[7] Hilal R 2012 Vocational Education and Training for women and youth in Palestine: Poverty reduction and gender equality under occupation *International Journal of Educational Development* **32** 686-695

[8] Tatiana B, Svetlana B, and Victor R 2015 Vocational Education in the System of Determinants of Reducing Youth Unemployment: Interregional Comparisons *Procedia - Social and Behavioral Sciences* **214** 526-534

[9] Hellen C 2007 Learning as Becoming in Vocational Education and Training: class, gender and the role of vocational habitus *Journal of Vocational Education and Training* **55** 471-498

[10] Ricarda S and Ursula K 2017 Good at school = successful on the job? Explaining gender differences in scholastic and vocational success *Personality and Individual Differences* **105** 107-115
[11] Hubeis, Aida and Vitayala S 2010 *Pemberdayaan Perempuan dari Masa ke Masa* (Bogor: IPB Press)
[12] Dewi K H 2007 Demokrasi dan Dekonstruksi Ideologi Gender Orde Baru *Jurnal Penelitian Politik* **4** 59-67
[13] Puspitawati H 2012 *Gender dan Keluarga: Konsep dan Realita di Indonesia* (Bogor: PT IPB Press)
[14] Handayani dan Sugiharti 2008 *Konsep dan Teknik Penelitian Gender* (Malang: UMM Press)
[15] Fakih M 2010 *Analisis Gender & Transformasi Sosial* (Yogyakrata: Pustaka Pelajar)
[16] Fakih M 1999 *Analisis Gender dan Transformasi Sosial* (Yogyakarta: Pustaka Pelajar) 12- 23
[17] Nurhaeni, Ismi, Monika S Y dan Rino A N 2016 Measuring Gender Differences in Information System Use: Case study of e-Government *AIP - Conference Proceedings* **1746**
[18] Muljani A N *2008 Strategi Efisiensi Pembiayaan Pendidikan* Materi kuliah EkonomiS3, Pendidikan Teknologi dan Kejuruan, Universitas Negeri Yogyakarta (UNY)
[19] Marit B, George L O, Jacques Z 2013 Non-formal vocational education in Uganda: Practical empowerment through a workable alternative *International Journal of Educational Development* **33** 88-97
[20] Koudahl, Peter D 2010 Vocational education and training: dual education and economic crises *Procedia - Social and Behavioral Sciences* **9** 1900-1905
[21] Munastiwi E 2015 The Management Model of Vocational Education Quality Assurance Using 'Holistic Skills Education (Holsked)' *Procedia - Social and Behavioral Sciences* **204** 218-230
[22] Wenrich R C dan Wenrich J W 1974 *Leadership in Administration of Vocational and Technical Education* (Columbus: Charles E MerrillPublishing Company, A Bell & Howell Company)
[23] Republik Indonesia 1995 *Undang-Undang No. 12 Tahun 1995 tentang Pemasyaraktan* (Jakarta: Sekertaris Negara)
[24] Prasetyo T 2012 *Hukum Pidana Revisi* (Jakarta: PT Raja Grafindo Persada)
[25] Yayan M G 2003 *Perawatan Tahanan Rumah Tahanan Negara* Departemen Kehakiman dan Hak Asasi Manusia RI, Pusat Pendidikan dan Pelatihan Pegawai, Akademi Ilmu Pemasyarakatan, Jakarta.
[26] Nurhaeni *Pedoman teknis penyusunan Gender Analysis Pathaway (GAP) dan gender Budget Statement (GBS)* (Jakarta: AIDP)
[27] March C, Ines S and Maitrayee M 2003 *A Guide to gender Analysis Framework* Oxfam
[28] Respondent 1, personal interview. Dated 31 Mei 2017
[29] Respondent 2, personal interview. Dated 31 Mei 2017
[30] Security Officer, personal interview. Dated 31 Mei 2017

IOP Conf. Series: Materials Science and Engineering **306** (2018) 012115 doi:10.1088/1757-899X/306/1/012115

Intercultural Communication Training in Vocational and Industrial Education Training

S Hastjarjo[1,*] and A Nuryana[2]

[1] Dept. of Communication Science, Sebelas Maret University, Surakarta, Indonesia.
[2] ISP Cruiseship and Hotel School Surakarta, Indonesia.

*sri.hastjarjo@staff.uns.ac.id

Abstract. The globalization and free trade between countries and nations has created demands for the knowledge and skills in the area of intercultural interaction and transaction. Intercultural Communication Competences (ICC) is one of the capabilities that need to be possessed by workers and professionals who want to have a bigger role in the business and industries in international level. Vocational education institutions are demanded to provide their students with a certain degree of competences in multicultural interaction and communication. This paper aims to address the effectiveness of trainings in a vocational education institution in equipping its students with the intercultural communication skills. Using a sample of students from the ISP Cruiseship and Hotel School Surakarta, Central Java, this study will analyses the differences of ICC between groups of students who have undergone various forms of training in intercultural communication, in order to determine the effectiveness of the training in equipping the students with the necessary intercultural communication skills. The study incorporates a quantitative approach, using survey method. The data will be analyzed using descriptive statistics and analysis of variations between groups. The result shows that the intercultural communication training increase the level of ICC especially in the intercultural confidence dimension.

1. Introduction

The globalization and free trade between countries and nations has created demands for the knowledge and skills in the area of intercultural interaction and transaction. Intercultural Communication Competences (ICC) is one of the capabilities that need to be possessed by workers and professional who want to have a bigger role in the business and industries in international level. Vocational education institutions, especially in the hospitality industries, are demanded to provide their students with a certain degree of competences in multicultural interaction and communication.

The implementation of ASEAN Economic Community (AEC) in 2015 raises many questions about challenges and impacts faced by enterprises in this geographic territory, which in turn questions the readiness of education/training institutions in preparing and equipping their students with necessary competences to meet the demands of the new-multinational trades and competitions. Education institutions need prepare their students to recognize their own culture and respect other cultures different from their own; in other word the students must be equipped with some intercultural communication skills [1].

1.1. Intercultural communication competence

Intercultural communication, as the study of communication from the cultural perspective, focusing on the value of effective interaction between individuals of different cultures [2]. Seven categories for effective cross-cultural communication, including display of respect, orientation to knowledge, empathy, role behavior, interaction management, tolerance for ambiguity, and interaction posture [3]. Other scholar proposed instruments to measure the intercultural communication competence (ICC). Four dimensions in intercultural communication competence: personal attributes, communication skills, psychological adaptation, and cultural awareness [4]. To measure those dimensions, an instrument they include five traits, which are: intercultural confidence, intercultural attention, intercultural empathy, ethnocentrism, and intercultural tolerance [5].

1.2. Intercultural communication training

Intercultural communication training aims to equip participants with awareness, knowledge and skills to engage intercultural situations Pusch [6]. The literature has discussed the characteristic of effective intercultural communication training Paige [7] Porter & Samovar [8]. Although no study has been able to determine the most effective model for intercultural communication training, the literature agrees that the experiential and interactive approach is more effective than the didactic approach.

The ISP Cruiseship and Hotel School Surakarta, Indonesia, is one of vocational training institution focused on the hospitality business, which offer a 1-year program in hotel and cruiseship training. Established in 2009, the school has become one of the best vocational training centre in Indonesia. The school was awarded with Indonesia Innovative Quality Award 2015 as The Best Institution Training and Education of Cruises Ship & Hotel and in 2016 the school was awarded The Best Vocational Training Center Management by the Indonesian Ministry of Education and Culture.

The ISP Cruiseship and Hotel School uses two approaches in the intercultural communication training: (1) didactic approach in the form of intercultural communication subject in the classroom; and (2) interactive and practical approach in the form of internship program with international partners (hotels and cruiseship operators). Both training methods are conducted in second semester of the training program.

2. Methods

In order to address the main issues of this paper, a quantitative approach was used in the data collection and analysis. The sample (N=210) was drawn from the students of The ISP Cruiseship and Hotel School Surakarta, Indonesia from classes of 2016 and 2017. Data collection was conducted using self-administered questionnaires. The research instrument comprised three main variables: demographic, attendance to intercultural communication training, level of intercultural exposure, and intercultural communication competence (ICC) (using Likert's 1-5 scale). The data then was analyzed using descriptive, simple linear regression, and Chi-square statistics.

3. Results and Discussion

The result of the study will be divided into three main parts: the description of the variables, the influence of demographic factors to the level of ICC, and the influence of intercultural communication training to the level of ICC.

3.1. Description of the variables

Table 1 presents the descriptive statistic of each variables measured in this study, including the breakdown of the ICC dimensions. The media use variable is comprised of 7 items. The ICC scale is comprised the total of 24 items, divided into five dimensions: IC attention (3 items), IC confidence (4 items), IC empathy (3 items), ethnocentrism (3 items), and IC tolerance (4 items).

Table 1. Descriptive of the research's variables (n=210).

Demographics	Male = 108 (51.4%)		Female = 102 (48.6%)	
Enrolled to Intercultural Communication Training	Yes = 105 (50.0%)		No =105 (50.0%)	
Variables	Min.	Max.	Mean	S.D.
Intercultural Exposure	8	25	12.23	3.040
Intercultural Communication Competence (ICC)	51	80	63.13	5.967
Intercultural Attention	7	15	11,06	1,534
Intercultural Confidence	10	20	14,09	1,845
Intercultural Empathy	6	14	10,35	1.453
Ethnocentrism	6	15	11,09	1,881
Intercultural Tolerance	12	20	16.53	2.004

From Table 1, we can see that the composition of the sample is almost equal between male and female students, as well as between those who have enrolled to the internship program and those who have not. The average score of Intercultural Communication Competence (ICC) among the students is in the medium level (mean=63.13). When the ICC is broken-down into its five dimensions, this study finds that intercultural attention (mean=11.06), intercultural empathy (mean=10.35), and ethnocentrism (mean=11.09) are in the medium level; while intercultural confidence (mean=14.09) and intercultural tolerance (mean=16.53) are in the high level. We can conclude that the students have a high level of confidence to conduct intercultural communication as well as a high-level tolerance when encountering people from other culture background.

3.2. Variation of ICC based on demographic variable

Table 2 presents the results of Chi-square statistic for determining the differences of ICC scores based on the demographic variable and attending the intercultural communication training.

Table 2. Results of Chi-Square (X^2).

	Sex	Intercultural Communication Training
Intercultural Communication Competence (ICC)	1,103 sig. 0,576	0,332 sig. 0,847
Intercultural Attention	1,495 sig. 0,474	0,953 sig. 0,621
Intercultural Confidence	4,980 sig. 0,083	3,909 sig. 0,142
Intercultural Empathy	0,352 sig. 0,839	1,009 sig. 0,604
Ethnocentrism	2,486 sig. 0,289	0,421 sig. 0,810
Intercultural Tolerance	0,861 sig. 0,650	2,771 sig. 0,250

As we can see from Table 2, the results of Chi-square tests do not demonstrate any significant differences or variations of both total ICC scores and ICC dimensions among the students. We can conclude that based on the sex and enrollment in intercultural communication training, there is no difference of ICC level. All the students, regardless of their demographic backgrounds and their attendance in the international communication training, possess a medium level of Intercultural Communication Competence.

3.3. Influence of intercultural exposure on the ICC level

Table 3 presents the results of linear regression statistic to determine the influence of intercultural exposure among the students on their ICC level.

Table 3. Linear regression with intercultural exposure as the predictor.

	R	R Square	F	B	t	sig.
Intercultural Communication Competence (ICC)	0,251	0,063	19,059	0,520	4,366	0,000
Intercultural Attention	0,190	0,036	10,603	0,101	3,256	0,001
Intercultural Confidence	0,372	0,139	45,542	0,238	6,748	0,000
Intercultural Empathy	0,195	0,038	11,220	0,098	3,350	0,001
Ethnocentrism	0,001	0,000	0,003	-0,002	-0,540	0,957
Intercultural Tolerance	0,121	0,015	4,223	0,084	2,055	0,041

From Table 3 we can observe that intercultural exposure can be used as predictor of the level of intercultural communication competence (t=4,366), although the influence is quite small (6,3%) in increasing the ICC. There are still other variables that influence the level of ICC, those variables are not included in this particular study. Although new media use is low, but we can conclude that the model we used in this study is sufficient to propose that intercultural communication training and exposure has certain impacts on increasing the level of intercultural communication competence.

Comparing the results of each five dimensions of ICC, this study finds that the intercultural exposure does not have influence on the ethnocentrism among the students. The intercultural exposure has various impacts on the other four ICC dimensions, with the highest on the IC confidence (13,9%), followed by IC empathy (3,8%), IC attention (3,6%), and IC tolerance (1,5%). The intercultural exposure significant impacts in boosting the student's confidence in communicating with other people from different cultural background.

4. Conclusions

This paper aims to describe the level of Intercultural Communication Competence (ICC) among the hospitality students and to examine the influence of demographics and intercultural communication training and intercultural exposure on the students' ICC. This study concludes that:

- The level of ICC among the students in this study is in the medium level, which means that they still have to be more prepared and equipped in the intercultural communication skills in order to face the demands and challenges of the new-multinational industries and enterprises.
- The intercultural communication training and exposure has proven impact for increasing the level of ICC, especially in boosting the confidence of the students to engage communication/ interaction with other people from different cultural background.
- From the conclusion point b, this study suggest that the students need to be encouraged and trained to use expand their intercultural exposure to initiate, engage, and maintain communication and interaction with people from other culture, in order to increase the students' intercultural communication competence.

References

[1] Tantiniranat S 2015. Some Intercultural Implications of ASEAN and Thai Educational Policies for Thai Higher Education. *CORERJ: Cambridge Open-Review Educational Research Journal* **1** (2) pp 154-165.

[2] Hall E T 1956. Orientation and Training in Government for Work Overseas. *Human Organization* **15**(1) pp 4-10.

[3] Ruben B D 1976. Human communication and cross-cultural effectiveness. *International and Intercultural Communication Journal* (4) pp 95-105.

[4] Chen G 1989. Relationships of the dimensions of intercultural communication competence. *Communication Quarterly* **37** (2) pp 118-133.

[5] Chen & Starosta W J 2000. The development and validation of the intercultural sensitivity scale. *Human Communication* **3**(1) pp 1-15.

[6] Pusch M 1981 *Learning across cultures* (Yarmouth: Intercultural Press).

[7] Paige R M (ed.) 1993 *Education for the intercultural experience* (Yarmouth: Intercultural Press).

[8] Porter R & Samovar L 1991 *Intercultural communication: A reader, 6th edn* (Belmont, CA: Wadsworth Publishing).

IOP Conf. Series: Materials Science and Engineering **306** (2018) 012116 doi:10.1088/1757-899X/306/1/012116

Developing Traditional Food Service: A Portrait of Women in Culinary Industry

S M D Maukar*, F W Langitan, T F S Tangkere and A Dondokambey
Universitas Negeri Manado, Tondano 95618, Sulawesi Utara, Indonesia

*maukar.sylvana@yahoo.co.id

Abstract. The purpose of this research is to obtain data about development of traditional food service for small woman business in Minahasa Toulour, Indonesia. The type of research used is descriptive qualitative method. The result of this research is to show that the profile data of the development of the service quality of the catering service business and the traditional home industry of the small business women at the grassroots around Lake Tondano, Minahasa, is in desperate need of rocks and guidance, because although it has the strength and opportunities such as traditional food products, the taste is quite good and popular consumer, the main raw material is the main agricultural products Minahasa so the price is relatively affordable, the role of print media and electronics to support the socialization of traditional foods Regional, National, International so it can be exported abroad, but on the other hand is also faced with weaknesses both internally and externally such as the lack of traditional entrepreneurial knowledge of Minahasa's traditional cuisine, suffering from a lack of capital, and the impact of lack of knowledge and lack of capital result in the following items being prepared ,as limited and monotonous as well lack of innovation, inadequate food business equipment, lack of clear health insurance, information dissemination and limited development, although facilities exist but cannot be implemented due to lack of funds, poor sanitation is noticed.

1. Introduction

Small business women who are grassroots who are engaged in catering business or household food industry around Lake Tondano, are quite a lot, with superior traditional food products that taste good enough, and make people like to consume them, but unfortunately the dishes produced are monotonous, and have not yet explored all the traditional foods of Minahasa's Toulour, and show little new innovations in terms of appearance, composition, recipe, cooking methods, decoration, ethics of traditional Minor's Toulour food, and situations like this very influential to the smoothness of business, and this problem is as a result of several factors, including: lack of capital, low knowledge of women's network of small businesses associated with traditional food entrepreneurs typical Toulour Minahasa The Anh Phan and Thi Huong Thanh Nguyen in research on An Analysis Of Factors Impact On Customer Satisfaction In Vietnam Restaurants: Case Of Fast Food Restaurants, In this study, the scale is measured on four main factors as Service quality, Product quality, Price, and Environment affecting to all common fast-food restaurants in Vietnam market [1].

On this basis the development of a model of empowering the quality of traditional food business services typical Toulour Minahasa network of women both in the form of catering services and food industry class household is badly needed. The development model includes the provision of education and training on the quality of the excellent service of the entrepreneurs (catering business, household food industry) related to the traditional foods of Minahasa cuisine, both the staple food, the side dishes up to the delicious, a network of small women at the grassroots. This effort is in addition to raising and improving the food business of women entrepreneurs is also preserving local culture Minahasa Toullu, especially traditional foods Toulour Minahasa with raw materials seedling Toulour farmers are rice, freshwater fish. Sara Ghezzi et al., in a study on Food Safety in the US catering industry: empirical findings, the results of his research, explained that the results suggested that employees in the catering industry who work part-time need more training and development. Management was seen as more knowledgeable than non-management personnel and was seen to have more training. Overall the study found that there is a need to focus on training for employees, with even greater emphasis for new employees [2].

Based on the problems mentioned above, it motivates the writer to conduct research with topic "development of empowerment model of quality of service business of traditional food network of women small business in order to preserve local cultural tourism object in Minahasa Regency, with the first stage is descriptive qualitative research, and the research targets are related to traditional food business development profile Toulour Minahasa network of women small businesses, challenges, obstacles, advocates the development of traditional food business Toulour Minahasa network of women small business in Kabupaten Minahasa. The second stage is the research of development of traditional food empowerment model of Toulour Minahasa through the writing of entrepreneurship instruction book of traditional cuisine of Minahasa and Minor cuisine, and third step is experimental research by giving counseling and training of entrepreneurship of various traditional foods Minahasa Toulour with main ingredient of rice and freshwater fish to women's network of small business catering and home industry.

2. Methods

The type of research used to obtain data on the Quality of Service Traditional food business typical Toulour Minahasa in order to preserve local cultural attractions in Minahasa Regency is using descriptive qualitative research methods.

The informants in this study were women of small businesses around Lake Tondano Minahasa District. In determining the informant is using snowball sampling technique. The procedure of data analysis in this study using interactive analysis with the steps as follows: data collection, data reduction, data display, conclusions: drawing/verifying. The validity of data is tested using triangulation technique of data source.

3. Results and Discussion

3.1. Profile data development of quality of traditional food business services typical Toulour Minahasa

3.1.1. Type of food business and menu prepared. Existing food business types include catering and home industry, with a prepared menu consisting of modern food and traditional food. Types of traditional foods typical of Minahasa grove that exist in catering and home industry business is very limited, monotonous, and has not yet explored all traditional foods Minahasa Toulour, and show less new innovation, in the sense of in need of development, digging further so that there is a change that is: making traditional foods Minahasa Toulour coloring the excellent food in every food business both catering and home industry; making traditional Minahasa Toulour food is superior to modern food, which is starting to shift traditional food in various food business; growing, promoting traditional types of recipes and food recipes that had previously disappeared, to be promoted in every culinary business, so that the traditional food resources of Toulour Minahasa can be socialized not only at the local level but also to the National and even

International level, in order to preserve the cultural attractions local in Kabupaten Minahasa. Conducting new innovative activities will be traditional Minahasa Toulour cuisine.

In connection with the preparation of the menu, then, Bahattin Ozdemir, Osman Calisk in his research on A review of literature on restaurant menus: Specifying the managerial issues explained that based on the conceptual and empirical findings of menu literature, the major menu management issues are menu planning, menu pricing, designing, operating menu and development menu [3]. Furthermore Kimberly Mathe-Soulek et al in his research with the title The Impact of Price-Based and New Product Promotions on Fast Food Restaurant Sales and Stock Prices research results are suggested Results suggest that new product promotions can have a significant and positive effect on same store sales, whereas price-based promotions tend to be lower than the same-store sales changes and changes in stock price [4].

3.1.2. Food business device. Food business device that is cooking utensil, equipment & supporting facilities, location, and kitchen, is entered standard category, simple, and not yet complete, in the sense of need attention from entrepreneurs themselves to be held improving or completion adjusted to each requirement, food business. Large-scale enterprises require a larger business device than small-scale enterprises.

In connection with the food business apparatus, Alice J Chow, Abel D Alonso, Alecia C Douglas, Martin A O'Neil in his research entitled Exploring open kitchens 'impact on restaurateurs' cleanliness perceptions explain Overall, respondents whose kitchens are both open and closed viewed open kitchen restaurants as the most hygienic, with open kitchen restaurant operators being more emphatic about the cleanliness differential. In addition, open kitchens were thought to positively influence employee behavior, particularly in regards to being more exposed to customers' scrutiny. Finally, the provision of an entertaining and engaging atmosphere, as well as enhanced cleanliness perception factors that why open kitchen restaurants were constructed in that manner [5].

3.1.3. Marketing. Marketing of catering business products, household food industry in general can be sold out, only the consumers are mostly around the location near the business, and the maximum is the District level, although for some traditional food products have penetrated overseas through the family line but not routine and continue, and only temporary, and for example red-colored products have arrived in the Netherlands, through the family line, and the nature of marketing is only temporary, in the sense that a new order is produced. The main cause of the traumatic food products typical Toulour Minahasa only circulated the regional level, although in terms of taste quite good is due to the way socialization only using traditional techniques of mouth to mouth, and not using electronic equipment or other mass media. Also the other main cause of home industry products in general does not have a business license and expiration. Most consumers, especially tourists, are more likely to buy / choose foods that have been listed on labels about the expiry date, the Health Department permit.

As the solution is to provide understanding, awareness to small business women network of how important every food business, product have business license, and sign of expiration period, so that consumer have health insurance if want to buy, consume traditional food product Toulour Minahasa. Martin Hingley, Julie Boone, Simon Haley in his research on Local Food Marketing as a Development Opportunity for Small UK Agri-Food Businesses explained that the study explores local food as a marketing opportunity for small food producers and identifies barriers to development [6].

3.1.4. Packaging. The taste of traditional food products is good enough, just the unfortunate specialty of the home industry's manual of packaging techniques that are less hygienic, and the lacks model of consumer appeal and which primarily lacks the labels of expiration and does not include permission from the health department.

Findings from S.T. Wang Edward et al., in his research on the visual packaging design on perceived food product quality, value, and brand preference also suggested that. Based on the study findings, food firms should emphasize the visual packaging factors such as color, typeface, logo, graphics, and size to form consumers' positive perceptions and brand preference, originality, value [7]. Sarah Mari, John Safari, James Lwelamira, in the research on Consumers' Perceptions on Processed Food Products in Dodoma Municipality, Tanzania, In conclusion, packaging affects preferences of food products and contributed influences purchase decisions of the products. The findings of this study underscore the importance of the packaging and packaging of designs for the fulfillment of many functions related to the logistics and marketing of food products. The main implication drawn from the study is that entrepreneurs need to be cognizant of consumers' specific packaging preferences for which they can use as a strategic marketing tool [8].

Now it has been sold freely in the market for tools, packaging machines. This tool is very helpful for home industry entrepreneurs to produce packaging products that are varied and hygienic, so as to be a capital to make consumers more interested to get, buy existing products.

3.1.5. Cleanliness. Food entrepreneurs sometimes neglect, forget the hygiene factor, and these unfavorable habits should be eliminated, especially those associated with food handlers, facilities and processing, and the food itself. For example, many food handlers do not wear work clothes when processing food. As food handlers have to wear work clothes. Clothes that must be worn by the workforce are: a) Clothes worn by labor (apron), b) hair cover, c) cempal, napkin. Work clothes should be white.

Likewise with the cleanliness of facilities and food processing and the environment, for example, in general the disposal of wet and dry wastes is united, and or has no waste bins in the treatment room, and these habits should be prevented and avoided. Dry and wet bin should be separated. Further related to the foodstuff, it is less, not good use of food that is not worth using. Such hygiene efforts are able to minimize the contamination of food by impurities through food handlers, facilities and processing, and the food itself, so that the health of consumers who enjoy the results of the business, the product is guaranteed

Nelson Barber, Joseph M Scarcely, in research on clean restrooms: how important are they to restaurant consumers? Research has shown that consumers are concerned with restroom cleanliness, with a functioning restroom having a positive influence on customers' perception of the restaurant; and has also been shown to the impact of where to eat or to return to a restaurant. This study tested these assumptions by using a data set from the southwestern US, finding that consumers are concerned with the cleanliness of an eating establishment [9].

3.1.6. Labor. Manpower is still limited, which means only the cost of financing additional personnel. Basically a small-scale enterprise is definitely less labor-intensive, but if there is an increase in business, it would require additional labor. Also, the workforce generally lacks the traditional entrepreneurial knowledge of the Minahasa's traditional foods.

Zorana Antic, Srđan Bogetic in the Food Industry Workers Attitudes on the Importance of Factors Affecting Food Stuff Quality Management, argued that an organization's success depends on the knowledge, skills, creativity and motivation of the company's workers and partners. Focus on its employees enables a company's development and improvement, whereas business ethics ensures public health and safety protection, environmental protection and life quality improvement. The company management's responsibility lies foremost in education, worker training and development, thus enabling a direct and indirect influence on the foodstuff quality characteristics [10].

3.1.7. Knowledge of traditional foods Minor's Toulour. Based on the data indicates that the level of knowledge of traditional food entrepreneurs network of women small business has sufficient experience in relationship with food entrepreneurs, but because the level of entrepreneurial knowledge of traditional foods typical Minahasa Toulour into the category less, is a hindrance to the progress of a food business.

Applied research and educational needs in food service management, explained that food safety management, food quality management and product development; equipment and facility layout, design; operational planning and modeling; as well as market and consumer related aspects. Underpinning scientific disciplines and operational, strategic benefits of the described studies as well as the three tiers in the educational pathways in food production are described [11].

3.1.8 Business capital. In general, food entrepreneurs, both catering business and home industry, network of women of small business, are lack of capital and this element can hinder the progress of the business. An examination of capital structure in the restaurant industry, the results show that using ratios from publicly-traded restaurant firms in the USA and ordinary least squares regression models, the results tend to support the notion that both the pecking-order and the financial growth cycle influence financing decisions [12].

3.2. Supporters
Proponents of food business development with Traditional Toulour Minahasa food menu are as follows:

- Traditional food that is currently good in catering business or in home industry the taste is quite good and favored by consumers, Minahasa people and outside Minahasa, and therefore the opportunity for the development of a variety of traditional foods seeds more open.
- The main ingredients of traditional food are the main agricultural products of the Minahasan people, and these circumstances are very helpful to entrepreneurs to obtain raw materials, and at the same time the price is relatively affordable.
- Increasing the role of the media in both print and electronic media in the development of recipes and promotions of traditional foods Toulour Minahasa can be socialized both local, national, international, and this business is very helpful to the success of culinary entrepreneurs, home industry women small business Toulour Minahasa.

3.3. Inhibitors
Inhibition of food business with Traditional Toulour Minahasa food menu is as follows:

- Lack of management knowledge of traditional food entrepreneurship services typical of Minahasa Toulour women's network of small business.
- Women's network of small businesses that are trying both catering services and food industry typical of households have lack of capital.
- The impact of the lack of knowledge and lack of capital resulted in the following as a menu that was prepared limited and monotonous and lack of innovation, food business device is not adequate, health insurance is less clear, information dissemination and development is limited, although there are facilities but cannot be implemented due to lack of funds, sanitation is underestimated.
- Marketing of business results is sometimes constrained as a result of product competition among food entrepreneurs themselves. One way out is that the business owner must have one advantage, unique to the resulting product so as to make the business become a bestseller. Because of its uniqueness resulted in competitors difficult to follow. If the product is the same as a competitor it will run along with a competitor or will be crushed by a competitor.

Thus, the elements of the food business services among which the type of food business and prepared menus, food business devices, marketing, packaging, cleanliness, labor, knowledge and capital is in need of serious attention and development, because it greatly affects the smooth operation of traditional food business padpa catering business or home industry small business women. The better these elements the better the quality of traditional food business services network of women small businesses. The development of traditional food business typical of Minahasa groovy both for catering business and food industry of household class of small business woman business in grass roots around Lake Tondano is not yet enter the category of entrepreneurs who are able to manage food business well and means in the process of growth less show rapid progress

Then Education including the knowledge of food service business management is very helpful for small business women in managing food business for the better. The better the level of education including the knowledge of food service business management the better the quality of food service business.

Sunghyup Sean Hyun in his research on Predictors of Relationship Quality and Loyalty in the Chain Restaurant Industry explained that based on the literature review, five dimensions influence restaurant patrons' behavior: food quality, service quality, price, location, and environment. Theoretical derived from the literature review. Data analysis indicated that these five attributes influence loyalty formation, with impact mediated by relationship quality. Then Annie Gay Barlan-Espino in her research on Operational Efficiency and Customer Satisfaction of Restaurants: Base for Business Operation Enhancement, explains that it is concluded that majority of the employees are sufficient seating capacity that accommodate large volumes of customers [13]. Restaurants are efficient on the aspect of kitchen and early operations and sometimes encountered problems. Customers are satisfied in terms of 5 P's. Product, Policies, People, Processes and Proactivity [14]. Hui Key Lee, et al., in research on the Assessment of Food Safety Knowledge, Attitude, Self-Reported Practices, and Microbiological Hand Hygiene of Food Handlers, the results of the study are in the conclusion, the suggestion of this study was that the food handlers had adequate food safety knowledge [15].

4. Conclusions

The conclusions of the study are: (i) the profile data of quality service development of catering service business and traditional household food industry of small business women at the grassroots around Lake Tondano Minahasa, is in desperate need of rocks and guidance that the elements of quality of foodservice services are traditional food, food, marketing, packaging, hygiene, labor, knowledge and capital are in need of serious attention and development. The better these elements the better the quality of traditional food business services network of women small business. Furthermore, education including the level of knowledge of food service business is affecting the quality of food service business, therefore the better the level of education and knowledge of food service business then the better quality of service for small business women; (ii) small business woman Toulour Tondano faced also with several obstacles between them: (a) lack of traditional entrepreneurial food knowledge typical of Toulour Minahasa network of small business women; (b) small business women's networks, both in catering and in the household food industry, are lack of capital; (c) marketing of business results is sometimes constrained as a result of product competition among food entrepreneurs themselves; (d) the impact of the lack of knowledge and lack of capital results in the following matters being prepared as limited and monotonous and lacking innovation, inadequate and incomplete food business devices, lack of clear health insurance, limited promotion of information dissemination and development, facilities but cannot be implemented due to lack of funds, sanitation less attention by business owners; (iii) supporters such as: (a) types of traditional food produced today both in catering business or in home industry tastes quiet and favored by consumers, Minahasa community and outside Minahasa, therefore, the opportunity for the development of a variety of traditional foods seeds more open; (b) the main ingredients of traditional food are the main agricultural products of the Minahasan people, and these circumstances greatly assist entrepreneurs to obtain raw materials, and at the same time their prices are relatively affordable; (c) increasing the role of the media in both print and electronic media, in the development of recipes and promotion of nutritious traditional foods, so that traditional cuisine products of Toulour Minahasa can be socialized at Local, National and

International levels so that the home industry product network of small-business women (SME) Toulour Minahasa can be exported abroad. The level of development needs of the traditional food development model of Minor Food Supply Minahasa network of small business women is quite high and development activities are planned through Writing, publishing textbook entitled: Guidance of traditional food business service of Toulour Minahasa as reference book in giving education and training of traditional cuisine toulour minahasa to the network of women small businesses around Lake Tondano in Minahasa District. And no less important is the cooperation with the government in providing funding assistance as a catering business venture capital, household industry small business women in Tonano Minahasa.

References

[1] Phan T A and Nguyen T H T 2016 An Analysis Of Factors Impact On Customer Satisfaction In Vietnam Restaurants: Case Of Fast Food Restaurants *International Journal of Business and Management Review* **4** (6) p 1-17.

[2] Ghezzi S and Baker Ayoun 2013 Food safety in the US catering industry: empirical findings *International Journal of Contemporary Hospitality Management* **25** (3) p 365-382.

[3] Ozdemir B and Caliskan O 2013 *A review of literature on restaurant menus: Specifying the managerial issues.*

[4] Mathe-Soulek K, Krawczyk M, Harrington R J and Ottenbacher M 2015 *The Impact of Price-Based and New Product Promotions on Fast Food Restaurant Sales and Stock Prices* p 110-117.

[5] Chow A J, Alonso A D, Douglas A C and O'Neill M A 2010 *Exploring open kitchens' impact on restaurateurs' cleanliness perceptions.*

[6] Hingley M, Boone J and Haley S 2010 *Local Food Marketing as a Development Opportunity for Small UK Agri-Food Businesses* **1** (3).

[7] Edward S T W 2013 The influence of visual packaging design on perceived food product quality, value, and brand preference *International Journal of Retail & Distribution Management* **41** (10) p 805-816.

[8] Mmari S, Safari T J and Lwelamira J 2015 Consumers' Perceptions on Packaging of Processed Food Products in Dodoma Municipality, Tanzania *Social Sciences* **4** (4) p77-81.

[9] Barber N and Scarcely J M 2009 *Clean restrooms: how important are they to restaurant consumers?* DOI: 10.1111/j.1748-0159.2009.00155

[10] Antić Z and Bogetić S 2015 *Food Industry Workers' Attitudes on The Importance of Factors Affecting Food Stuff Quality Management.*

[11] Rodgers S 2005 Applied research and educational needs in food service management *International Journal of Contemporary Hospitality Management* **17** (4) p 302-314.

[12] Upneja A and Dalbor M C 2001 An examination of capital structure in the restaurant industry *International Journal of Contemporary Hospitality Management* **13** (2) p 54-59.

[13] Hyun S S 2010 Predictors of Relationship Quality and Loyalty in the Chain Restaurant Industry *Research Article* First Published March 11, 2010.

[14] Barlan-Espino A G 2017 Operational Efficiency and Customer Satisfaction of Restaurants: Basis for Business Operation Enhancement *Asia Pacific Journal of Multidisciplinary Research* 5 (1) p 122-132

[15] Lee H K, Halim H A, Thong K L and Chai L C 2017 *Assessment of Food Safety Knowledge, Attitude, Self-Reported Practices, and Microbiological Hand Hygiene of Food Handlers*, Published online 2017 Jan 10. Doi: 10.3390/ijerph14010055 PMCID: PMC5295306.

A Review of Soft-skill Needs in in Terms of Industry

S Prihatiningsih*

Technical and Vocational Education Studies Program, Graduate School of Universitas Pendidikan Indonesia, Jl. Dr. Setiabudhi 229 Bandung 40154, Jawa Barat, Indonesia.

*ucisricimahi@gmail.com

Abstract. World of work or industry requires a number of characters that are owned by employees working in work activities, it is intended that the industry can easily achieve business success through achievement indicators through the contributions of employees in work activities. For employees who are graduates of vocational education level, should this character have with him since they completed vocational education, but in fact the character is not fully owned by the students in learning in vocational, only 56.4% of characters / soft skills given in vocational learning, so that it becomes soft skill gaps vocational graduates with soft skills required by industry.

1. Introduction

Asean Economic Community (AEC) or the ASEAN Free Trade which has started since 2015 making goods, services and labor is getting easier for people passing in the ASEAN member countries. For Indonesian, this deal could be a double-edged sword. If processed properly, product and labor Indonesia has the potential to dominate the Southeast Asian market, otherwise if not ready to compete then Indonesia will be a market for other ASEAN member countries. This will certainly affect the labor conditions in Indonesia.

Working Conditions is a set of conditions or circumstances in the work of a company that is a place of work of the employees who work in that environment. What is meant here is the good working conditions that is comfortable and supportive workers to be able to carry out its activities properly [1]. Covering everything in the environment can affect the employee's performance, as well as safety and job security, temperature, ventilation, lighting, cleanliness and others.

Learning in vocational demanding process / activities that can bring the ability of learners to work competence appropriate to the world of work (DU / DI), this is in line with the purpose of establishing the SMK is to prepare students to enter the world of work in accordance with the field of expertise chosen by the participants learners and develop a professional attitude in certain professions [2].

Data by BPS of 2014 recorded the number of unemployed vocational school graduates more than 1.6 million people (17.26%) of 9.39 million. That condition is a misnomer because in the current government to continue to promote vocational education and building a new department for vocational but it turned out to be the largest contributor of educated unemployment in Indonesia. This phenomenon shows that secondary vocational school which is expected to be a bridge link and match was also not met expectations. High unemployment vocational graduates also showed that the level of relevance of vocational education with real life is still low. The other side does not absorb graduates, most graduates of vocational schools in Indonesia are not only less able to adapt to the development of science and technology but also less able to develop himself and his career in the workplace [3].

Their weakness resulting vocational graduates are not absorbed in the world of work needs to be change in the quality of graduates of vocational orientation, during which only now oriented hard skill inclusion of elements indispensable development soft skill DU / DI. Needs for industrial in soft skills and soft skills to the role a person's success in life can be seen in the figure 1 below.

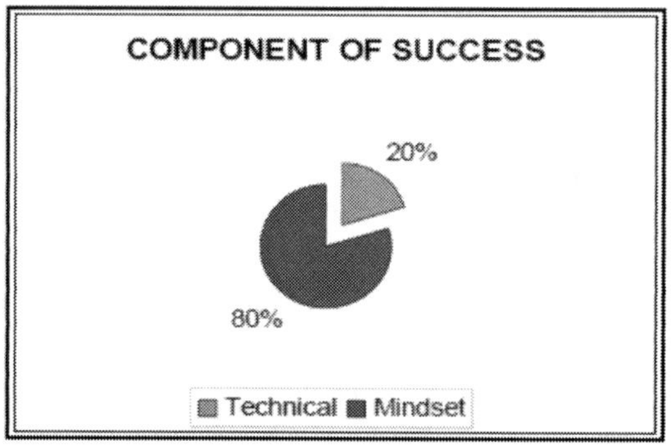

Figure 1. Ratio Requirement soft skill and hard skills in the world of work / industry [4].

Research and the actual facts on the ground show that soft skills have an important role in determining a person's success in the work. Hard skill is a minimum requirement for a person to enter certain occupations, while soft skill will determine development in the job. It is therefore education, including vocational challenge to integrate both kinds of components in an integrated manner and not biased to be able to prepare a complete human resources that have the ability to work and grow in the future [5].

The demands of the relevance of education to the world of work in a broad sense hinted need mastered a number of competencies that can be demonstrated while working. Vocational education is education that prepares graduates to master knowledge and competence for the field / department. Vocational graduates are not enough to master the hard skills alone, but also have to master the soft skills as support hard skill to be better able to work productively, and quality [6].

2. Methods

Soft skill the needs of industry in the conduct of business activities is obtained as a result of the research study using a qualitative approach in the form of descriptive research using soft skill analysis of a part of the world of work and the phenomenon results from empirical observation on labor vocational graduates. The implications of this research are expected to be a contribution to learning in SMK [7]. The research process begins with a preliminary study to obtain a correlation between soft skills needs in the workforce with vocational learning.

Research continued to conduct a survey in the form of observations in the field of electronics industry in the production and technical section which refers to the key success factors in a career in the field of electronics, especially the working positions are grouped in soft skills [8]. This component will be grouped to obtain fulfillment of priority, so that will be a proposal for strengthening the components soft skill on learning in SMK. These results become in learning in vocational orientation which is used as the location of research examining the learning process.

The next stage in the research is to conduct a survey to teachers organize learning related to vocational education, in the form of data collection to get the phenomenon soft skill in learning in vocational preparation to enter the job / industry.

IOP Conf. Series: Materials Science and Engineering **306** (2018) 012117 doi:10.1088/1757-899X/306/1/012117

3. Results and Discussion

Research the number of personnel and the technical part of the production of industrial electronics field indicates a number of data including (see table 1):

Table 1. Softskill some personnel and the technical in company.

No	Industrial Soft skill	Dominance Results
1	The need of English Communication Skill in a company	SS 59.2%
2	Honesty needs n Attitudes Behaviour	SS 89.1%
3	Needs Sense of Responsibility Employees	SS 82.2%
4	Discipline needs time	SS 64.6%
5	Needs Teamwork	SS 75.9%
6	Needs Work Safely	SS 54.1%
7	Creative needs in Problems Solving	SS 66.1%
8	High Commitment needs in Keeping Promises	SS 65.3%
9	Needs Capabilities Managing Information	S 53.4%
10	Needs Ethics and Morals in Words and Deeds as Norma Employees	SS 64.9%
11	Organize yourself needs well in work	SS 55.5%
12	Spirit of needs of work	S 49.7%
13	In deference to the needs of older people as a culture of employee	S 47.4%
14	The need toughness in work	SS 54.9%
15	Needs to Overcome Stress	SS 46.8%
16	Needs of Self Work	S 36.8%
17	Concern for the needs of others	SS 51.1%
18	Needs Easily receive input	S 55.2%
19	Care needs to clean environment	SS 71.1%
20	Troubleshooting Needs at Work	SS 54.9%

Research into the effect of soft skills in teaching vocational students generate the following data (see table 2):

Table 2. Soft skill toward the teaching and learning activity in Vocational School.

No	Students Soft Skill Conditions	Dominance Results
1	Honesty state and Attitude Behavior in Learning	S 57.1%
2	The state of moral ethics in word and deed	S 60.3%
3	High state in keeping promises	S 58,.4 %
, 4	Discipline state time in learning	S 53.2 %
5	Sense of Responsibility circumstances	53.8 %
6	Safe Working state	S 57.1 %
7	Circumstances ability Communicate in Learning	S 62.3 %
8	Circumstances Have respect	S 48.7 %
9	Creative attitude and state of mind Many	S 64.5 %
10	The state of Information Management	S 68.8%
11	State responsibility / persistent in learning	S 26.9 %
12	Concern for the state of others	S 57.1%
13	Set yourself up with good state	S 60.3 %
14	Needs Easily receive input	S 64.9 %
15	Morale in the state of learning	S 61.5 %
16	Circumstances Able to cope with stress	S 58.4 %
17	Concerned about the state of the environment	S 51.9 %
18	Able state of Troubleshooting	S 56. 4 %

Source: [9].

4. Conclusions

By comparing the condition of soft skills in vocational students in learning with the needs soft skill the world of work / industry on the research data, obtained the data that the number of targets soft skill the orientation of industry in conducting its activities had not fully met by learning at SMK. The gap soft skill must be loaded in learning in vocational include personal ability to communicate, honest, responsible, disciplined, working, job security, the ability of problem solving, commitment, manage information, ethics, self-management, passion, respect, toughness in work, overcoming stress, open and concerned [10]. In the vocational learning is very important to be submitted regarding the fulfillment indicator soft skill world of work / industry, it is intended that graduates of vocational competence can have a good working either on fulfillment of the components of Knowledge, Skills and Attitudes work.

References

[1] Kuppermann M, Lubeck D P, Mazonson P D, Patrick D L, Stewart A L, Buesching D P, and Filer S K 1995 Sleep problems and their correlates in a working population *Journal of General Internal Medicine* **10**(1) pp 25-32.

[2] Gillard S 2009 Soft skills and technical expertise of effective project managers *Issues in informing science & information technology* **6**.

[3] Badan Pusat Statistik 2015 Dalam http://bps. go. id/ (diakses 14 Januari 2015).
[4] Thomas J N, James M C, and Paul B B 1999 *Lessons from the top: The search for America's best business leaders*. (Broadway Business).
[5] Sharma M 2009 How Important Are Soft Skills from the Recruiter's Perspective *ICFAI Journal of Soft Skills* **3**(2).
[6] McCale C 2008 It's Hard Work Learning Soft Skills: Can Client Based Projects Teach the Soft Skills Students Need and Employers Want? *Journal of Effective Teaching* **8**(2) pp 50-60.
[7] Id. wikipedia, Qualitative Research, 2017
[8] Creswell J W 2013 *Research design: Qualitative, quantitative, and mixed methods approaches* (Sage publications).
[9] Suryanto D, Kamdi W, and Sutrisno S 2014 Relevansi Soft Skill yang Dibutuhkan Dunia Usaha/Industri dengan yang Dibelajarkan di Sekolah Menengah Kejuruan *Teknologi dan Kejuruan* **36**(2).
[10] Tulgan B 2015 *Bridging the Soft Skills Gap: How to Teach the Missing Basics to Todays Young Talent*. (John Wiley & Sons).

Measurement of Employability Skills on Teaching Factory Learning

S Subekti[1,*] and A Ana[2]

[1]Culinary Education Study Program, Fakultas Pendidikan Teknologi dan Kejuruan, Universitas Pendidikan Indonesia, Jalan Dr. Setiabudhi No 207 Bandung 40154.
[2]Home Economic, Technology and Vocational Education, Universitas Pendidikan Indonesia, Jalan Dr. Setiabudhi No 207 Bandung 40154.

*sri_subekti@upi.edu

Abstract. Vocational High Schools as one of the educational institutions that has the responsibility in preparing skilled labors has a challenge to improve the quality of human resources as a candidate for skilled labors, to compete and survive in a changing climate of work. BPS noted an increase in the number of non-worker population (BAK) in 2015-2017 on vocational graduates as many as 564,272 people. The ability to adapt and maintain jobs in a variety of conditions is called employability skills. This study purpose to measure the development of employability skills of communication skills, problem-solving skills and teamwork skills on the implementation of teaching factory learning in SMK Negeri 1 Cibadak, THPH Skills Program on bakery competency. This research uses mixed method, with concurrent triangulation mix methods research design. Data collection techniques used interviews and questionnaires. The result shows that there are increasing students' employability skills in communication skills, problem solving skills, and teamwork skills in teaching factory learning. Principles of learning that apply learning by doing student centering and learning arrangements such as situations and conditions in the workplace have an impact on improving student employability skills.

1. Introduction

One indicator of the era of globalization is the emergence of free trade, starting from ASEAN free trade (AFTA) in 2003, followed by Asia Pacific free trade in 2010 for developed countries and 2020 for developing countries, and the ASEAN Economic Community (AEC) in end of 2015. This raises some concerns about ASEAN professional workers who will enter the Indonesian labor market, while the competitiveness of Indonesian workers is still low compared to other ASEAN countries. In 2017, Global Innovation Index (GII) recorded Indonesia ranked 87th out of 143 countries, lower than Singapore, Malaysia, Thailand and Vietnam.

In addition, the condition of Indonesian citizens aged 15 years and over who are not labor force (BAK) on SMK graduates in 2015-2017, has increased from 3,294,510 people to 4,158,782 people. While the number of BAK in the population that has never gone to school has decreased from 3,987,167 people to 3,158,028 people [1]. SMK as one of the educational institutions that has the responsibility in preparing skilled labors has a challenge to improve the quality of human resources as a candidate for skilled labors, to compete and survive in a changing climate of work. Work climate change is caused by the rapid development of technology, so that the prepared labor must have the ability to adjust to the conditions of the labor market and the ability to maintain its work.

The ability to adapt and maintain jobs in a variety of conditions is called employability skills. United Nations Educational, Scientific and Cultural Organization (UNESCO) in 2012, establishes employability skills as a skill that a person possesses to retain his work. International Labor Organization (ILO) by 2016, defines employability skills as a skill, knowledge and competence that can improve the ability of workers to secure and retain jobs, demonstrate performance skills, overcome changes, and secure other jobs when stopped or dismissed and easily enter the labor market at another period of its life cycle.

The work skills required by vocational graduates in Indonesia should be adjusted to the needs of the job market on the job skills of prospective workers. Some studies have shown that an important aspect of measuring employability skills required by students to fill the needs of the industrial job market and the ability to persevere for achievement in the workplace is the ability of communication skills [2-7], problem solving skills [2, 5-9] and teamwork skills [2-4, 6, 7]

Demand for qualified labors who have employability skills in accordance with the needs of the job market should be responded quickly by SMK, because it has implications for the preparation of the quality of qualified graduates of SMK. The effort to anticipate the missmatch between labor market demand and the preparation of labor from SMK graduates is by developing SMK curriculum and implementing learning approach using work oriented learning model [10]. Work-Based Learning (WBL) is a learning model used to gain experience through work. This allows students to adapt to the work environment and workplace reality, so that students can make the best career choices for the future [11]. One of WBL implementation is teaching factory learning.

This study purpose to measure the development of employability skills of communication skills, problem-solving skills and teamwork skills on the implementation of teaching factory learning in SMK Negeri 1 Cibadak, Agricultural Processing Technology (TPHP) skills program on bakery Competency.

2. Methods

2.1. Research design

The research design used is a mixed triangulation concurrent method, which is qualitative and quantitative data collected in one time, then the data obtained compared to know the existence of convergence, difference, and / or combination. The process of mixing the data is done at the stage of interpretation and discussion. Mixing of data is done by integrating or comparing the results side by side in the discussion.

2.2. Participants, population, and sample

Participants in this study were productive teachers on teaching factory learning, management of teaching factory, and students who participated in teaching factory learning activities. The population of this research is Skills Program in SMK Negeri 1 Cibadak which has implemented teaching factory learning model. The sample of this research is TPHP skills program on bakery competency. The sample is determined by purposive sampling technique, which is sample determination technique specially based on research objectives. Respondents in this study are students who are implementing teaching factory learning.

2.3. Data collection

Qualitative data obtained by interview with manager of teaching factory, while the quantitative data obtained with a questionnaire filled by productive teachers in teaching factory. The approach used for quantitative data collection is quasi experimental with single-group experimental design of interrupted time-series design.

2.4. Research instruments

The research instrument used is a questionnaire. The employability skills aspects measured are described in the indicators. The determination of the indicators was adapted from the employability skills assessment [6] and employability skills assessment [10]. The employability skills indicators are described [10] and then elaborated in the form of analytic rubrics used as a guide to employability skills measurement of students and adjusted to research needs. The rubric developed in the study consisted of 75 items covering all indicators measured of communication skills, problem solving skills and teamwork skills with criteria for measuring score 4 for highest score, and score 1 for lowest score. Preparation of items in accordance with Indonesian National Work Competency Standards (SKKNI) no. KEP. 45 / MEN II / 2009 on Food and Beverage Industry Sector for selected competence, curriculum structure of specialization group (C2) basic of skill program for agricultural and fishery processing, and skill package (C3) production of vegetable product and stage of teaching factory implementation.

2.5. Data analysis research

Analysis of research data used is descriptive statistical analysis. Qualitative and quantitative data were analyzed by descriptive statistical analysis technique.

3. Results and Discussion

3.1. Description teaching factory at SMK Negeri 1 Cibadak

SMK Negeri 1 Cibadak is one of the formal educational institutions of vocational secondary education in agriculture in West Java. TPHP is one of the skills program at SMK Negeri 1 Cibadak which has implemented teaching factory in learning process. Teaching factory learning is designed and implemented in accordance with real work standards, so that the learning process setting is like the situation and conditions at work. Teaching factory learning's orientation is industry, so that the resulting product is an industry-standard products. It is implemented in habituation in applying the principle of occupational Health and Safety (K3) during learning, and works in accordance with standard operational procedure (SOP) that has been set at each stage of the process.

Implementation of teaching factory learning is student-centered, in student work groups and product groups. Student involvement in teaching factory learning starts from grade 10 to grade 12, with different learning loads, and different roles. Grade 10 students served as helper, grade 11 served as producer, and grade 12 served as manager. Some of the products in teaching factory are bread, yogurt, fresh drinks, chips, and fruit juice, with the main product being produced is bread.

The minimum production target of bread in one day is 15 kg of raw material, with 10 kg of raw material for production of sweet bread, and 5 kg of raw material for white bread production. In the implementation of the production process, students work in groups with a maximum of six students. If failure is found in the production process, the student as the producer is responsible, because it can be ascertained that the production error occurs due to lack of accuracy in implementing the SOP, both the personnel SOP, the material quality manual, the certainty of the formula, and the mode of operation of the equipment. In this case, as a preventive effort, the optimization of the manager in ensuring the readiness of students before the production process.

In addition of the production process, students are involved in the product marketing process. Products marketed still use the traditional practice of door to door with the target market is the students SMK Negeri 1 Cibadak and the surrounding community. In addition, there are several partners SMK Negeri 1 Cibadak in marketing are Berkah department store and alumni association SMK Negeri 1 Cibadak (Agrika) which has its own market. The number of products marketed by agrika in one day as many as 1500 pcs of bread.

Implementation of evaluation in teaching factory teaching is done by conducting competency test which is associated with Profession Certification Institution (LSP). In this case the assessment is done directly to each student to measure the ability to produce bread independently. The implementation of teaching factory in SMK Negeri 1 Cibadak is not only implemented to improve the cognitive, psychomotoric and affective ability but employability skills include the ability of teamwork, problem solving and communication to become dominant during the learning process because students are required to do learning with the principle of learning by doing.

3.2. Measurement of employability skills of communication skills

Measurement of communication skills consists of six indicators, are reading skill, writing skill, counting skill, listening skill, speaking skill, and serving customer skill. Result of employability skills' measurement of communication skills in SMK Negeri 1 Cibadak TPHP skills Program Bakery Competency, is shown in figure 1.

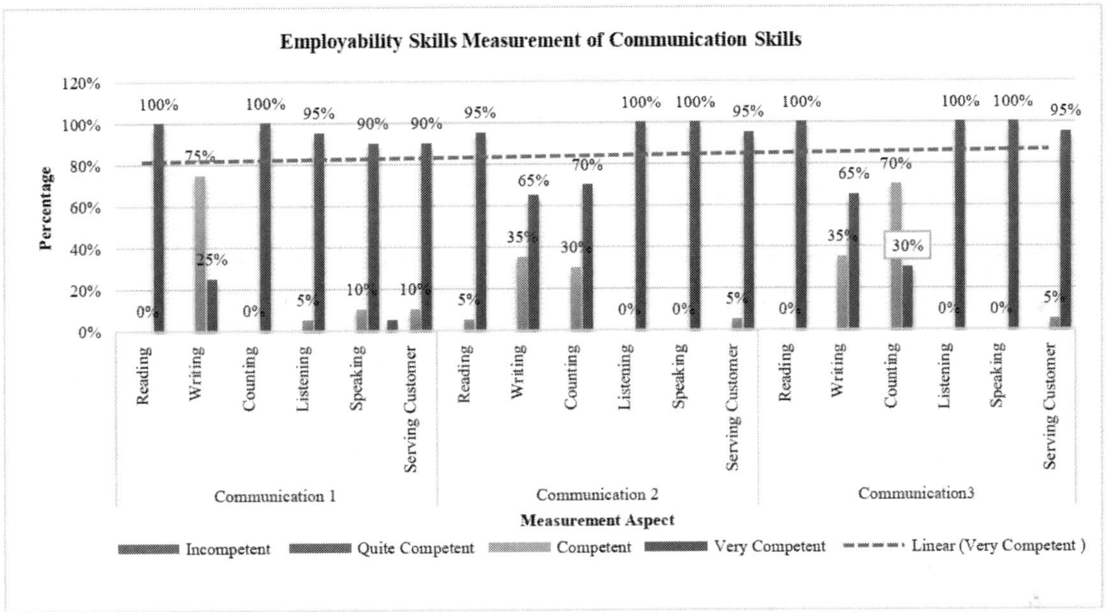

Figure 1. Employability skills measurement of communication skills.

Reading skill is measured by the ability of students in placing, understanding, and interpreting written information to perform daily tasks. Writing skill is measured from the ability of students in conveying opinions, ideas, information, and news in written form. Counting skill is measured by the ability of students to demonstrate a good calculation base by using a practical approach to solving technical problems at work. Listening skill is measured by the ability of students to receive, follow, interpret and reply information verbally or with appropriate body language. Speaking skill is measured by the ability of students to convey verbal messages appropriate to the work situation, participate actively in discussions or convey ideas and ideas within the workgroup. Serving customers is measured by the ability of students in doing service and communicate to consumers with empathy to give satisfaction in accordance with customer expectations.

3.3. Measurement of employability skills of problem solving skills

Measurement of problem solving skill consists of five indicators, are implementation of safety principle, facility management skill, decision making skill, knowing how to learn skill, and problem solving skill. Result of employability skills' measurement of problem solving skills in SMK Negeri 1 Cibadak TPHP Skills Program Bakery Competency, is shown in figure 2.

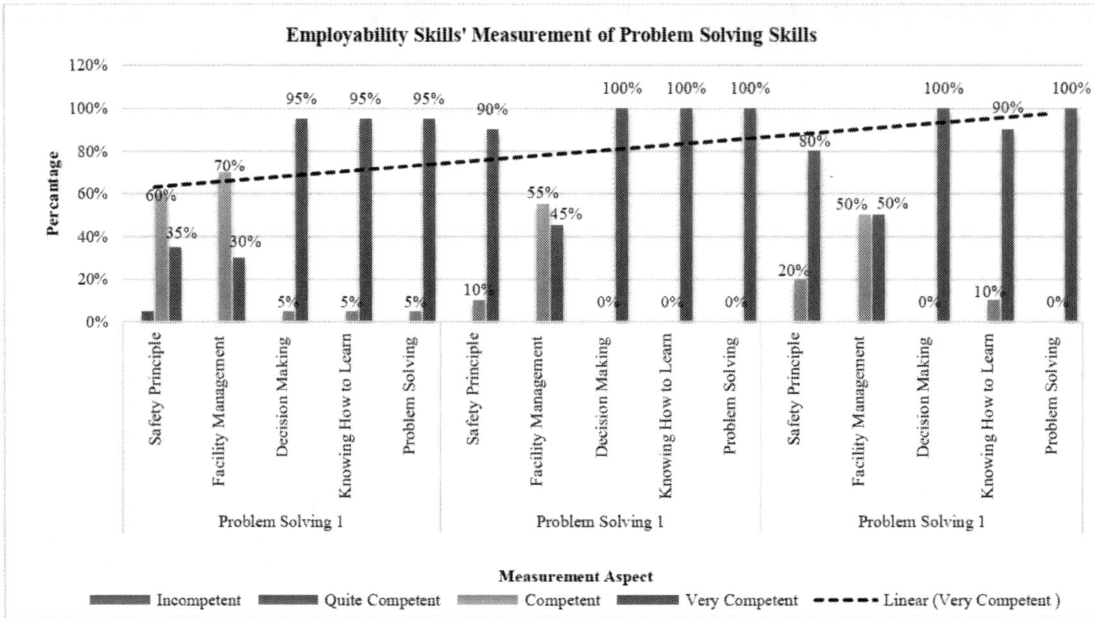

Figure 2. Employability skills measurement of problem solving skills.

Implementation of the safety principle is measured from the students 'awareness of the procedures and practices of occupational safety and personal or group health, and can perform well the rules of work. Facility management skill is avoided from the students' ability to store, organize, and distribute materials, components, equipment, or final product results in order to be used properly. Making decisions skills is measured by the students' ability to set goals, make choices, calculate risks, assess, and make best choices. Knowing how to learn is measured by the students' ability to use appropriate analytical techniques and implement knowledge, and skills in ordinary situations or unfamiliar new situations. Problem solving skill is measured by the students' ability to understand the problem, identify possible different reasons, complete, and execute action plans to address, assess, and monitor progress, and improve work plans based on field data.

3.4. Measurement of employability skills of teamwork skills

The measurement of teamwork skills consists of four indicators, are participation as members, teaching friends, leading group friends, and working with different cultures. Result of employability skills' measurement of teamwork skills in SMK Negeri 1 Cibadak TPHP skills Program Bakery Competency, is shown in figure 3.

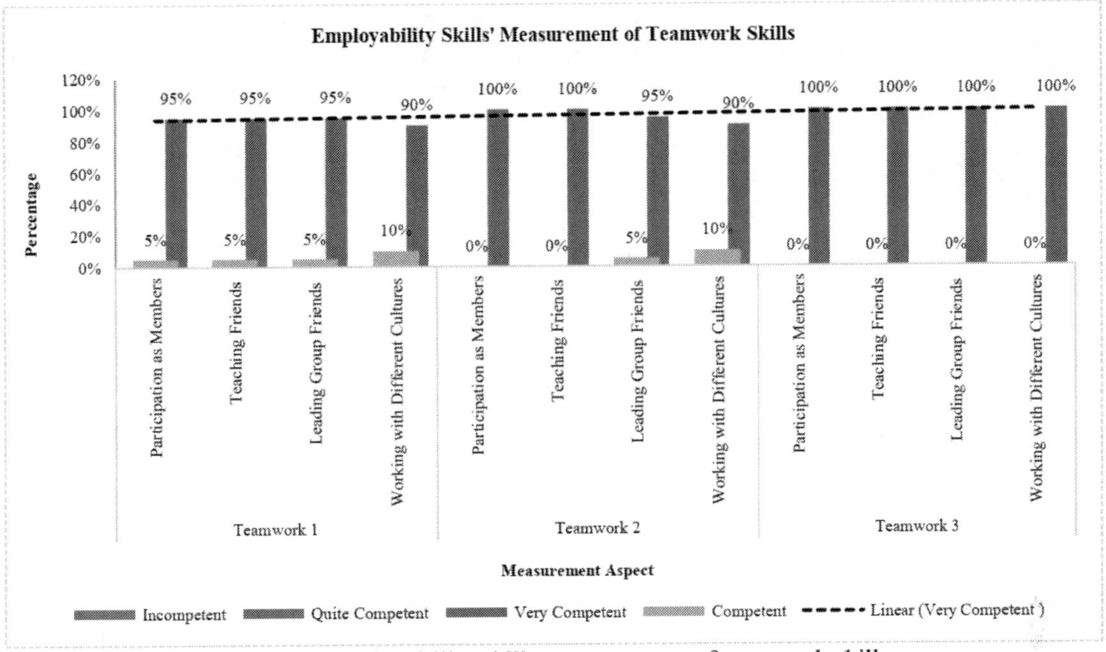

Figure 3. Employability skills measurement of teamwork skills.

Measurement of participation as member's indicator is measured by the ability of student's cooperation with friends in groups and to contribute in the form of ideas, proposals or business. The ability to teach a friend is measured by the ability of students to help others learn the necessary knowledge and skills, identify training needs and provide job information to help friends or groups. The ability to lead is measured by the ability of students to convey thoughts, feelings, and ideas to give support, confidence, trust or motivation to individuals or groups, responsible for job regulations based on guidelines and authority given. The ability to work with different cultures is measured by the collaborative ability of students with fellow men or women or with various tribes, social, and educational backgrounds.

4. Conclusions

Implementation of teaching factory learning in SMK Negeri 1 Cibadak can improve students' employability skills on aspects of communication skills, problem solving skills, and teamwork skills. Learning process that is dominated by learning by doing principle, centered on students or students active learning, with conditioned learning setting like situation and condition at work location have implication on existence of employability skills development of students.

References

[1] BPS 2017 *Survei Angkatan Kerja Nasional (Sakernas)* [Online] accessed on the page https://www.bps.go.id/linkTabelStatis/view/id/1909 on 19 Juni 2017.

[2] Yusof H M, Mustapha R, Mohamad S A M S, and Bunian M S 2012 Measurement Model of Employability Skills Using Confirmatory Factor Analysis *Procedia-Social and Behavioral Sciences* **56** pp 348-356.

[3] Buntat Y, Jabor M K, Saud M S, Mansor S M S S, and Mustaffa N H 2013 Employability Skills Element's: Difference Perspective Between Teaching Staff and Employers Industrial in Malaysia *Procedia-Social and Behavioral Sciences* **93** pp 1531-1535.

[4] Suleman F 2016 Employability Skills of Higher Education Graduates: Little Consensus on a Much-Discussed Subject *Procedia-Social and Behavioral Sciences* **228** pp 169-174.

[5] Lapiņa I, and Ščeulovs D 2014 Employability and Skills Anticipation: Competences and Market

Demands *Procedia-Social and Behavioral Sciences* **156** pp 404-408.

[6] Rasul M S, Rauf R A A, Mansor A N, Yasin R M, and Mahamod Z 2013 Graduate Employability for Manufacturing Industry *Procedia-Social and Behavioral Sciences* **102** pp 242-250.

[7] Arfandi A 2013 Relevansi Kompetensi Lulusan Diploma Tiga Teknik Sipil di Dunia Kerja *Jurnal Pendidikan Vokasi* **3**(3).

[8] Ismail S, and Mohammed D S 2015 Employability Skills in TVET Curriculum in Nigeria Federal Universities of Technology *Procedia-Social and Behavioral Sciences* **204** pp 73-80.

[9] Suarta I M, Hardika N S, Sanjaya I G N, and Arjana I W B 2015 Model *Authentic Self-Assessment* dalam Pengembangan *Employability Skills* Mahapeserta didik Pendidikan Tinggi Vokasi *Jurnal Penelitian dan Evaluasi Pendidikan* **19**(1) pp 46-57.

[10] Hanafi I, dkk. 2014 *Buku Panduan Model Integrasi dan Penilaian Employability Skills pada Pendidikan Kejuruan dan Vokasi* (Jakarta: Fakultas Teknik UNJ).

[11] Ismail S, Mohamad M M, Omar N, Heong Y M, and Kiong T T 2015 A Comparison of the Work-based Learning Models and Implementation in Training Institutions *Procedia-Social and Behavioral Sciences* **204** pp 282-289.

IOP Conf. Series: Materials Science and Engineering 306 (2018) 012119 doi:10.1088/1757-899X/306/1/012119

Indonesian Teacher Engagement Index (ITEI): An Emerging Concept of Teacher Engagement in Indonesia

Sasmoko[1,*], F Doringin[2], Y Indrianti[3], A M Goni[4] and P Ruliana[5]

[1,2]Primary Teacher Education Department, Faculty of Humanities, Bina Nusantara University, Indonesia
[1,3]Research Interest Group in Education Technology, Bina Nusantara University, Indonesia
[4] Universitas Negeri Manado, Indonesia
[5] Communication, STIKOM Interstudi, Indonesia

*sasmoko@binus.edu

Abstract. This paper presents a new concept of teacher engagement in Indonesia. The various studies in this paper examine various perspectives and even criticize the initial research on teacher engagement, so as to build the concept of different teacher engagement and in accordance with the Indonesian context so that it can be implemented and has direct impact as a guideline on improving the quality of teachers and education personnel in Indonesia. The method used in this paper is the Neuroresearch research method focused on exploratory research. The conclusion of this research is the development of Indonesian Teacher Engagement Index concept (ITEI) as a concept that describes the condition of teachers who experienced various psychological conditions positively, actively participate in building positive education, able to show good performance, have supportive competence, have national character as Characteristic of Indonesia and able to show the nationalism leadership engagement.

1. Introduction

Research on teacher engagement is mostly done because teacher engagement proved to give positive impact in the process of organizing education. Teachers who engage are able to make students become more engage so as to create a more effective school atmosphere. Engaged teachers also show the degree to working harder to make classroom activities more meaningful. Engaged teachers are willing to work with students in extracurricular activities so that students will be tied to school and improve their achievements. Furthermore, four types of engagement based on studies from previous studies, 1) Engagement with the school as a social unit, where the teacher is able to create community and concern for the school so that it integrates between his personal life and his working life; 2) engagement with school academic goals, where teachers are integrated with high demands that are influenced by the quality factor of their work life thereby increasing teacher commitment to provide meaningful feedback; 3) engagement with students as individuals so that teachers show attention to the personal lives of the students, more sensitive and aware of student development to become more mature; And 4) engagement with discipline that affects the formation of effective learning [1].

In 2014, a study attempted to expose a new construct in measuring teacher engagement. This research states that teacher engagement is formed by six indicators, namely 1) designing and organizing which is how teachers manage various responsibilities so that teachers can meet the needs of students; 2) facilitating discourse where the teacher is a facilitator for students and parents; 3) instructing is how teachers provide instruction for students with attention to the diversity of students so as to assist their cognitive development; 4) nurturing that the teacher is a very important socialization agent that is able to maintain and provide support both social and emotional; 5) motivating is how teachers have motivation as well as pass the motivation so that it can increase student involvement; And 6) monitoring, where teachers are able to manage their time and progress so as to effectively monitor student behavior and learning to achieve learning objectives [2].

Various research on teacher engagement more to explore the aspects of engagement related to learning, while looking at the role of teacher engagement that can improve student engagement and student achievement [3] then the teacher engagement of course not only centered on the learning process but on the whole Aspects of the teacher in carrying out his profession. Teacher engagement is closely related to the teacher's internal factors in the entire educational process whereby a teacher has high energy levels and strong identification of his work [4]. This concept is close to the concept of engagement in general. Engagement is conceptualized as a whole person's self-investment into a role so as to explain a comprehensive range of relationships with its performance [5]. Engagement is closely related to the source of work and performance of a person at work because when employees have control over their decisions and work and have good social support then they will provide feedback through optimal performance [6]. Engagement is also able to have a positive impact by surpassing the negative impact of work fatigue so as to increase the satisfaction of one's life and well-being [7, 8]. A person who engages is personally able to express himself and his work so that while working is able to present themselves fully and play an active role both physically, cognitively and emotionally [9].

By examining the overall concept of engagement, then the appropriate engagement teacher for the context in Indonesia is an engagement that is not only related to the learning process. Teachers in Indonesia are professional educators with the main task of educating, teaching, guiding, directing, training, assessing, and evaluating learners in early childhood education formal education, basic education and secondary education [10].Existence of teachers in Indonesia is crucial in improving the nation's life by involving all aspects of spiritual, emotional, intellectual, physical and other aspects [11]. Even the government of Indonesia is currently undergoing the transformation of education so that it can improve the feasibility of teachers as learning agents, improve the dignity of teachers and professionalism of teachers so that ultimately can improve the process and quality of education [12]. Therefore this research attempts to build a new construct on teacher engagement that is appropriate to the Indonesian context.

2. Methods

This research uses Neuroresearch method, which is a research method that synthesizes quantitative and qualitative research. Procedures in the Neuroresearch method are carried out through three processes: exploratory research, explanation research, and confirmatory research. This research only applies exploratory stage. An exploratory stage is the phase of a literature study as a reference in constructing theoretical construct. This stage involves the process of thinking qualitatively [13].

Exploratory research is a qualitative step in which neuroresearch research is done through literature review and study of various theories that will produce theoretical construct. This stage will result in a deeply variable review so as to find the construct. Construct is the final conclusion of a variable inspired by research journals, textbooks, theories, and other references that are contextualized to population conditions whose conclusions are ultimately conceptual definitions, dimensions and indicators. The construct position in neuroresearch is the academic forecast of the researcher. The theoretical construct of variables along with dimensions and indicator is then become the main reference of research instrument preparation [13-15].

3. Results and Discussion
The literature study on teacher engagement with the basic concept of engagement and study in the Indonesian context finds several findings as follows:

3.1. Engagement is a concept in positive psychology
A person engaged with his work is someone who performs his organizational role optimally both physically, cognitively, emotionally and psychologically characterized by high energy at work, dedication and absorption [7, 9, 16, 17]. Various research issues about positive psychology focus on work engagement is a positive, satisfying and affective motivational work situation [18]. The positive goal of psychology is to catalyze the focus shift in the field of psychology towards building individual quality in a positive way [19]. Interventions in the positive psychology should target individuals with less engage as this means that there is a lot of undeveloped potentials [20]. Therefore, teacher engagement is also included in the realm of positive psychology.

3.2. Engagement is a concept in positive education
In the world of education, teachers who engage will seek to apply the best methods and focus on student progress [4, 21, 22]. While education that seeks to improve skills and happiness is part of the concept of positive education that synergies between learning and positive emotions [23]. So that teacher engagement in Indonesia becomes more comprehensive when included in the concept of positive education.

3.3. Engagement relates to performance
A study of Dutch teachers attempts to test teacher engagement models in which job resources are related to the level of work and engagement and are able to predict their performance on a weekly basis [24]. This suggests that engagement can not be detached from the resulting performance and has been proven through various studies that find that there is an influence of determinant engagement on one's performance [25]. Therefore, performance becomes one of the important factors in the formation of teacher engagement in Indonesia.

3.4. Analysis of teachers in Indonesia
The law states that the national development in the field of education is the effort to improve the life of the nation and improve the quality of Indonesian people who believe, pious, and noble and master the science, technology, and art in realizing a society that is advanced, fair, prosperous and civilized based Pancasila and the Constitution of the State of the Republic of Indonesia Year 1945. This means that every teacher in Indonesia must practice the nation's philosophy as a characteristic character of the Indonesian nation.

In addition, the law also states that teachers in Indonesia must have competence, namely a set of knowledge, skills, and behaviors that must be owned, experienced, and mastered by teachers in performing professional duties. Teacher competence includes pedagogic competence, personality competence, social competence, and professional competence.

The existence of teachers in Indonesia who helped determine the intellectual life of the nation shows the existence of three fundamental things that must be imbued by teachers in Indonesia, the character of nationality which is reflected through Pancasila, four basic competencies and nationalism attitudes related to the love of the fatherland so that Always upholds social justice for all students in Indonesia [11, 26, 27].

4. Conclusions
This research is a preliminary research in constructing teacher engagement constructs in accordance with teacher context in Indonesia. The concept of teacher engagement is considered appropriate to describe Indonesian teachers with complex tasks and responsibilities. To improve the quality of education in Indonesia it requires teachers who engage in accordance with the Indonesian context and

absorb all positive aspects in the concept of engagement as a whole. The existence of teachers in Indonesia is very decisive efforts to educate the life of the nation became the main axis in this study which was captured through the Indonesian Teacher Engagement Index (ITEI). The concept of teacher engagement is defined as integrating the teacher's capacity as a "professional educator-mentor" characteristic of Indonesia by considering the positive aspects of psychology, positive education, teacher performance, teacher competence, national character and nationalism spirit. This research is a preliminary research so it still needs to continue the study to find the construct building on teacher engagement, namely the unification of various capacity of teachers.

References

[1] Louis K S, and Smith B A 1991 Restructuring teacher engagement and school culture: Perspective on school reform and the improvement of teacher's work *Educational Resources Information Center*.

[2] Borup J, Graham C R, and Drysdale J S 2014 The nature of teacher engagement at an online high school *British Journal of Educational Psychology* **45**(5) p 793–806.

[3] Basikin 2007 Vigor, dedication and absorption: Work engagement among secondary school English teachers in Indonesia *AARE International Conference, Perth, Australia, 27-28 November 2007* p 25–29.

[4] Cardwell M E 2011 *Patterns of Relationships Between Teacher Engagement and Student Engagement* (Fisher Digital Publication).

[5] Rich B L, Lepine J A, and Crawford E R 2010 Job Engagement: antecents and effects on job performance *Academy of Management Journal* **53**(3) p 617–635.

[6] Chung N G, and Angeline T 2010 Does work engagement mediate the relationship between job resources and job performance of employees ? *African Journal of Business Management* **4**(9) p 1837–1843.

[7] Hakanen J J, and Schaufeli W B 2012 Do burnout and work engagement predict depressive symptoms and life satisfaction? A three-wave seven-year prospective study *Journal of Affective Disorders* **141**(2–3) p 415–424.

[8] Schaufeli W B 2012 Work engagement: what do we know and where do we go? *Romanian Journal of Applied Psychology* **14**(1) p 3.10.

[9] Kahn W A 1990 Psychological conditions of personal engagement and disengagement at work *The Academy of Management Journal* **33**.

[10] Undang Undang Republik Indonesia No 14 Tahun 2005 tentang Guru dan Dosen.

[11] Suparlan D, and Ed M 2005 *Menjadi Guru Efektif* (Yogyakarta: Hikayat).

[12] Sujanto B 2009 *Cara Efektif Menuju Sertifikasi Guru* (Jakarta: Raih Asa Sukse).

[13] Fios F, Sasmoko, and Gea A A 2016 Neuro-Research Method: A Synthesis Between Hermeneutics and Positivism *Advanced Science Letters* **22**(9) p 2202–2206.

[14] Sasmoko, and Ying Y 2015 Construct Validity in NeuroResearch *Advanced Science Letters* **21**(7) p 2438–2441.

[15] Sasmoko, and Anggriyani D 2016 Neuroresearch (A Model of Research Method). In A Khan, M N A Ghafar, A R Hamdan, and R Talib (Eds.) *Research on Educational Studies* p 33–45 (New Delhi: Serial Publications PVT. LTD).

[16] Hu Q, Schaufeli W, Taris T, Hessen D, Hakanen J J, Salanova M, and Shimazu A 2014 East is east and west is west and never the twain shall meet: Work engagement and workaholism across Eastern and Western cultures *Journal of Behavioral and Social Sciences* **1**(1) p 6–24.

[17] Shimazu A, Schaufeli W B, Kamiyama K, and Kawakami N 2014 Workaholism vs. Work Engagement: the Two Different Predictors of Future Well-being and Performance *International Journal of Behavioral Medicine* p 18–23.

[18] Bakker A B, Schaufeli W B, Leiter M P, and Taris T W 2008 Work engagement : An emerging concept in occupational health psychology **22**(3) p 187–200.

[19] Seligman M E P, and Csikszentmihalyi M 2000 Positive psychology: An introduction *American*

Political Science Review **55**(1) p 5–14.

[20] Ouweneel E, Blanc P M Le, and Schaufeli W B 2013 Do-it-yourself An online positive psychology intervention to engagement at work *Career Development International* **18**(2) p 173–195.

[21] McLaughlin M W, and Pfeifer R S 1986 Teacher evaluation: Learning for improvement and accountability *ERIC*.

[22] Rutter R A, and Jacobson J D 1986 Facilitating Teacher Engagement *Educational Research* (3).

[23] Seligman M E P, Ernst R M, Gillham J, Reivich K, and Linkins M 2009 Positive education: positive psychology and classroom interventions *Oxford Review of Education* **35**(3) p 293–311.

[24] Bakker A B, and Bal P M 2010 Weekly work engagement and performance: a study among starting teachers *Journal of Occupational and Organizational Psychology* **83** p 189–206.

[25] Campbell J P, and Wiernik B M 2015 *The Modeling and Assessment of Work Performance Annual Review of Organizational Psychology and Organizational Behavior* **2**.

[26] Kusumawardani A, and Psikologi B 2004 *Nasionalisme* **61**(2) p 61–72.

[27] Koopmans L, Bernaards C M, Hildebrandt V H, Schaufeli W B, de Vet Henrica C W, and van der Beek A J 2011. Conceptual Frameworks of Individual Work Performance *Journal of Occupational and Environmental Medicine* **53**(8) p 856–866.

IOP Conf. Series: Materials Science and Engineering **306** (2018) 012120 doi:10.1088/1757-899X/306/1/012120

Preserving *Calung Banyumasan* through Vocational Education and its Community

Suharto* **and Indriyanto**

Faculty Languages and Arts, Universitas Negeri Semarang
Kampus Sekaran Gunungpati, Semarang 50229, Indonesia

*suharto@mail.unnaes.ac.id

Abstract. Traditional music in Banyumas especially *Calung Banyumasan* is very famous for its very distinctive style. But now there is a tendency of quantity and the quality of the music began to decline. Vocational high school of Banyumas is one potential that can be used in this art conservation effort. The purpose of this research is to find the form of preservation of *Calung* art through education. The result of the research shows that the effort of defence of the art of *Calung* tradition is part of the effort to preserve the art of tradition and culture of Banyumas in general. The *Calung* material spread in several subjects, both theoretical and practical manifested in the form of Basic Competence and Basic Material of each subject in the syllabus. All *Calung* learning materials are classic (Banyumas original style) without influence from other musical styles. Meanwhile, to popularize this type of music, this school performs an off-campus musical performance with a mixed-style, classic and contemporary type. To develop the competence of students and to contribute to the community of artists of this music, the school performs activities *Prakerin* (Practice industrial activities).

1. Introduction

Traditional music is part of a cultural product. As a result of cultural products, art will disappear if it is no longer part of the culture itself as art as part of ritual events, social customs, and social communication tools [1]. *Calung*, for example, is still preferred by the people of Banyumas although the number is limited. Unfortunately, the music used to accompany *Lengger* and *Ebeg* is not the type of music used as in the golden age in the 1980s. *Garap* that are played and the way they play has changed a lot. Many elements affect this change. What a great influence *Campursari*, so sometimes have to sacrifice and get rid of other art that is considered too "missed". The use of electronic instruments and the entry of Western musical instrument on the device without considering aesthetic *Calung* often makes the rumors to use the term "modern", so that eventually the term *Calung Campursari* or *Lengger Campursari* [2]. Actually this is a very common thing, because the art cannot be separated from the social and cultural life of the community. People always follow the dynamics of art that takes place in the community. Sometimes they have to follow in order not to be considered outdated [3].

Changes in play styles and musical instruments used can be caused by players who understand *gending Calung* is no longer bequeathed to his successor. So, the play tends to *balungan* style, just as a complement [3]. Technological developments are one of the reasons for this change or shift. Pop songs or *dangdut* became one of the reasons for the shifting of some of this *Calung Banyumasan* style.

In a various *Calung* performance in Banyumas society, *Calung* music became accompaniment of these types of songs.

One aspect of *Calung Banyumasan* is songs which are sung in this music. *Banyumasan* songs are always sung from the beginning of *Calung* played in 1970'an until now where *Calung* music used as accompaniment *Lengger*, Wayang and Ebeg. With its own peculiarities the songs seemed to not change in character either in terms of musical lingual (language in the lyrics of the song). With its own peculiarities, the songs seemed to not change in character either in terms of musical lingual (language in the lyrics of the song). The songs played by the groups of *Calung* always follow the circumstances of Banyumas people themselves are seen from the structure of the song, lyric characters (in terms of style of language), and performance style.

Banyumasan songs concludes that many songs in *Calung* performance depict the character of the Banyumas community [4]. Banyumas songs in the form of *wangsalan, parikan, senggakan* is an eternal literary work. It is proved that the songs are always sung in every *Banyumasan* music performances which actually contain excellent local content for character education as well as preservation, conservation and development so that the music as an accompaniment of some Banyumas arts will remain sustainable and liked by the Banyumas people. Unfortunately, art life such as *Lengger, Ebeg, Jemblung* puppeteer is still limited in certain communities on the outskirts of Banyumas area.

Universitas Negeri Semarang, one of which has the mission of cultural conservation has made a strategic plan of research (Renstra), research topics, and roadmap. One of them is the education of conservation, policy and development of conservation. Traditional art, one of which is *Calung* music in Banyumas region is a very important national cultural heritage. The artistry of *Karawitan*, for example, the style *Banyumasan* called *gagrak Banyumasan* including in *Calung*, has colored *Karawitan* and puppet play. Likewise, the songs that have *Banyumasan* uniqueness has become quite serious attention in the academic world. Local governments are still searching for the identity of art that became peculiarities that describe the Banyumas people [5].

Banyumas unique character model that describes the Banyumas society [6]. This is very interesting and considered important to be examined in order to identify carefully about the typical art Banyumas especially *Calung* which has long been a part of Banyumas society culture that is viewed from various aspects both musically and culturally. In terms of language for example, is an important aspect of the Banyumas society that continues to be preserved. The local wisdom that exists in the art of songs Banyumas describe the character of the people of Banyumas. The equation of degrees known as *blakasuta* is the local wisdom highly respected by the Banyumas [7] [6]. However, this music has experienced degradation that causes some of the peculiarities of music *Calung* has been influenced by the music of other areas such as Sunda [3]. That is why the efforts of defence, preservation and development of this *Calung* music need to be done.

It needs to be made for efforts to preserve the culture of Banyumas especially *Banyumasan Calung* art this. The uniqueness, which is contained in the lyrics of the song as well as the music style and the local wisdom contained in the support society, forced the local government to make this conservation effort. One of the most strategic undertakings is on the education path. The path of formal education is one way in an effort to crystallize the culture as a result of the influence of cultural globalization is so great [8]. This effort according to Sutton can be an effort to increase confidence Banyumas community from the current globalization. This is so that people do not run away from their distinctive culture to popular culture so that it eliminates its identity. Society should not be inferior to other cultures considered higher but should instead be proud [8].

Secondary School of Indonesian *Karawitan* of Banyumas (SMKI) is one of the vocational schools that have a different curriculum with art vocational schools in Indonesia. This school is now called as Vocational High School 3 (SMK 3) Banyumas. This school is also known as an institution that organizes traditional art education, especially the art of *Calung Banyumasan* until now. Many alumni who become artists and create educated *Calung* artists who are also expected to be the heirs and preservers of this traditional art. The purpose of this study is to describe the vocational school efforts of SMK 3 Banyumas in preserving the art of *Calung Banyumasan*.

2. Methods

This research uses descriptive qualitative approach. Data were collected by interview, observation, and document study. The collected data were analyzed using interactive analysis flow from Miles and Huberman, using measures including data reduction, data display, and conclusion drawing/verification [9].

3. Results and Discussion

3.1. Calung in Banyumas between needs, idealistic, pragmatic

Researchers identify that there are still many *Calung* groups in the region of Banyumas Regency. There are at least 80 *Calung* groups registered in the regency of Banyumas [10]. These groups are scattered almost across the region. Many groups are domiciled in areas that tend to belong to the city people. Nevertheless, the performances are more prevalent in suburban districts. This can be because as traditional art such as *lengger Calung* is still part of the social life of the community [1]. People in the suburbs still consider *lengger* music that uses *Calung* has become part of the social life of the community. The show held in one of the residents named Sukarto (65 years old) in *Pangadegan Banyumas* Village, for example, is not solely aesthetic needs as said by *RM Soedarsono* [11] but because of the social status that a perceived person of the show perceives. Someone will feel proud and praised by the people around because this art is still a part of social status in the villagers of the scene. Though the cost to hold the show was quite expensive about 30 million. On the other hand, Pak Sukarto is not a wealthy citizen who is seen from the physical condition of the house.

There is an indication of the formation of this group because of the business interests that are called pragmatics rather than music as an idealistic aesthetic needs, usually held by true artists. Artists who include pragmatists are those who use art as part of an instrument to meet economic needs, benefiting from entertainment arts performances. Some owners of the *Calung* group can serve as manager and coordinator of some *Calung* dancers and *l*

Lengger dancers to do *Calung* show business. This coordinator performs recruitment of both *Lengger* dancers, *Calung* players and rental equipment *Calung*. The quality and quantity of the show are tailored to the order that will perceive the show. Prabowo Laras (55 years old) is one example of the best-selling *Lenger Calung* group owners in the western part of Banyumas. He is a pragmatic or realistic artist because he himself is not a person who can play *Calung* or sing. This art group is the one that tackles business like players that are widely available in the region of Banyumas, and the high artistic needs for society, technology, and art that is currently favored by contemporary society such *as campursari, organ solo* and so on.

Other artists are idealists type. Sujiman Bawor (60 years old) is one of the one. He is not solely material but there is a social side in his attitude of art. The music games tend to be classic like the use of musical instruments and the original *Calung* performance. Meanwhile, Prabowo is free to adjust the circumstances so that the use of modern tools such as keyboards and other types of music are common.

Figure 1. Sujiman Bawor (left) and Prabowo Laras (Right)

Both types of artists live side by side in the Banyumas community. Supporting societies do not care about this. Both are needed by the community to meet their needs of art.

3.2. Preservation of Calung music through education by SMK 3 Banyumas

In the area of Banyumas, traditional art is now becoming better known and appreciated by the inhabitants of Banyumas and Indonesians in general. In cultural centers such as Solo and Yogyakarta previously known as negaragung recognize the existence of this typical art Banyumas [12]. Many of the art traditions such as *Karawitan*, *Wayang*, and dance that have uniqueness are recognized and learned in art education in Indonesia. Many also the alumni of SMK 3 Banyumas who continue in higher education and become academics, artists who develop the art of this *Banyumasan Calung*. They ultimately become prominent figures supporting this *Calung Banyumasan*. However, there are efforts made in developing and maintaining this art. One of them is conservation efforts.

Conservation covers all preservation activities based on the local situation and condition [10]. Further Rachman state that preservation activities can also cover the conservation scope, restoration, reconstruction, adaptation and revitalization. In terms of cultural conservation, especially the music of *Calung*, schools are a very appropriate means of conservation and development. The most strategic effort is the curriculum development. According to the Head of the vocational school, Nurainy Haryati (54), curriculum development is a strategic way for this conservation effort. Since recent years, especially after the 2013 curriculum implemented, SMK 3 Banyumas has developed an existing curriculum from the center into a solid curriculum. Local content, especially those related to Banyumas culture, is very dominant. Things related to Banyumas culture such as *Calung*, *Lengger*, *Banyumasan* style, are included and explicitly in Basic Competence of the syllabus. The implementation and reinforcement of traditional *Calung* music in the curriculum at SMK 3 Banyumas can be seen in Table 1.

Table 1. The form of *Calung* implementation in all subjects on the curriculum of SMK 3 Banyumas

No.	Subject Name	Number of hours in 3 years	Strengthening forms related *Calung*
1.	*Simulasi dan Komunikasi Digital*	6	The use of digital notation for the recitation of *Calung* songs
2.	*Tinjauan Seni*	4	Knowledge Culture *Banyumasan* with music *Calung*
3.	*Dasar-Dasar Kreativitas*	4	Knowledge of *Calung* creativity
4.	*Pengetahuan Karawitan*	6	Knowledge of Javanese karawitan and *Calung*
5.	*Teknik Instrumen Pokok*	10	Techniques basic music *Calung*
6.	*Teknik Karawitas Dasar*	10	Karawitan and its implementation in

			Calung Banyumasan,
7.	*Titi Laras dan Vokal dasar*	4	Understanding and practicing *Banyumasan* song notation
8.	*Praktik Vokal*	24	Singing styles, types of *Banyumasan* songs
9.	*Praktik Individu Instrumen Pokok*	26	Practicing individual *Calung* performance
10.	*Praktik Karawitan Bersama*	28	Practicing *Calung* ansamel together
11.	*Praktik Karawitan Iringan*	20	Using *Calung* as song accompaniment
12.	*Produk Kreatif dan Kewirausahaan*	20	Creating *Calung* work

(Source: Interview with Kuntowibowo, 2017)

According to one Karawitan teacher in the school, Kuntowibowo (35), the *Calung* material is integrated into all subjects, both theoretical and practical. This is done so that students understand and skillfully in playing *Calung* music integrally, from the cultural point of view to the social. This will give students the awareness to feel their own traditions and awareness to maintain the existence of traditional art in order to remain sustainable.

One of conservation effort is to pay attention to a number of prospective students by originating from areas in the district. The effort has been made by obligating every sub-district to send one prospective student to study in SMK 3 Banyumas. Now, without that obligation, the applicants who enroll in the school are quite high (Interview with Kuntowibowo, 2017).

The curriculum that needs to be considered is the addition of subjects or the addition of learning hours for the music learner. According to one of the alumni of SMK 3 Banyumas, Dimas Nurseto (19 years), lessons of *Calung Banyumasan* given by the school is quite a lot. Besides that, it is very pleasantly for the students.

Special lessons, *Calung*, for 3 hours of lessons each week for three years of lessons are enough to get the material *Calung* Banyumas in depth. Students are also added for the *Prakerin* in areas that have the artists the *Calung Banyumasan* (interview with Kuntowibowo 2017). However, the extracurricular lessons that were originally in the 10th grade should still be done because many students who want to deepen this Banyumas art. Dimas Nurseto goes on to say:

> "Actually, we are happy when there are extracurricular lessons of *Calung* music, including *lengger Calung*, let alone the teacher from art practitioners *lengger Calung*. Many of them want to become *Lengger* dancers and professional *Calung* players who want to deepen *Calung* after graduating from the high school "

This expression indicates the great desire of the students to develop themselves. Even he who is now a student in a non-arts institution has developed art skills, especially *Calung* music in the new place of learning. The extracurricular lessons are generally followed by students who really interest so the lessons are followed seriously [13]. Generally, they will excel because of a high motivation for the activity. Some students who are active in extra-curricular activities will feel confident both on a small stage and stage performances place. Figure 2 shows some student activities of SMK 3 Banyumas.

IOP Conf. Series: Materials Science and Engineering **306** (2018) 012120 doi:10.1088/1757-899X/306/1/012120

Figure 2. Activities and achievements of SMK 3 Banyumas students

The things that Dimas discloses and the school achievements indicate that the great potential of the performing arts and its local wisdom can be developed through education.

3.3. Preservation to strengthen identity

Banyumas is a unique culture like any other Javanese culture. Even though Banyumas including Central Java is sometimes referred to as "liyan" because of its uniqueness [11]. This uniqueness should not be eroded by the rapid flow of globalization. The intensive art encroachment of Banyumas *Calung* tradition will be a very effective fortress to stem the flow (interview with Sugito, 2017).

Calung Banyumasan is a cultural identity of Banyumas. Identity, such as Barker [15] says can be seen from people's feelings, beliefs, attitudes and lifestyles. Personal and social identity cannot be separated from one another [16]. A lot of research related to Banyumas culture, including Banyumas art which gives the peculiarity of Banyumas society which is actually a Banyumas identity itself. In *Calung Banyumasan* for example, many of which contain local content which needs to be preserved like a song lyric, playing style, *cablaka* attitude, the values system in the show or in the process of any activities and interactions during the *Calung* performances Banyumas unique and not found in other art performances [12] [8]

Cablaka and *blakasuta* attitude are often pinned to the people of Banyumas on the one hand to make the community into the inferior in comparison, but also can be want to improve and even want to rebel in order to overcome the attitude that is considered superior. That is why Banyumas people are also called "rebellious" and hardworking people [17], [18]. As long as Banyumas society, including *Calung* is considered non-standard art, it is only because it only imitates Gamelan and rough like the character of folk art in general. This is another attitude as Broh [19] say, as opposition to the inferior views shown to the Banyumas community. Another attitude is what causes the people of Banyumas still maintain the art of *Calung Banyumasan* until sustainable until now.

The interesting thing about the effort made by Sujiman Bawor as an artist who belongs to an idealist artist is the coaching to the younger generation around his home, as well as his art studio. For several years he coached children who are interested in *Lengger Calung* for *lengger* dance practice accompanied by *Calung* music. What he did was that children should not forget the traditional art.

Figure 3. Sujiman Bawor trains the children to dance *lengger Calung*

Children who live around this house also have the ideals to continue education in art high school that is SMK 3 Banyumas is already known to produce professional *Lengger* dancers. Sujiman Bawor activity in fostering children in practicing *lengger* dance as well as community support in empowering society around in terms of education and economic aspect. Generally, those who follow the training are middle and lower economic who want to improve their economics by taking formal and non-formal education in the arts.

3.4. Empowering the community and preserving the culture through Calung performances

Some points in the interview to some informants and stakeholders are, first, the empowerment of *Calung* music to improve the welfare of the community. This effort can be done by cooperation between local government (educational unit like SMK 3 Banyumas), tourism office, art owner community, formal education institute, and non-formal), improvement of quality and quantity through educational channel, and second, guidance to groups *Calung* art Banyumas that already exist or not.

Education and the Community are two components that cannot be separated from each other. Society needs education and vice versa. Without education society will not run well because in education there are elements of society such as teachers, learners and others. And vice versa without education society will be ignorant and have no knowledge. In addition, the community is also seen as a "laboratory where children learn, investigate and participate in community efforts that contain elements of society [12]

Empowerment of local government can be done for example to provide opportunities to *Calung* groups to fill regular events such as in precisely-tourist attractions. Of course, there must be a guarantee that they will earn money from the show. For those who already exist and quality should be given the opportunity to be published either in the form of recording, appearing on television and documentary. While those that do not exist must be given root assistance can compete with the given assistance in the form of training and coaching fund assistance.

The policy of regional tourism development should be based on the paradigm that developed in the region [13]. Therefore, it is very logical if the areas that have a *Calung* group that has the potential to become the sources of production as the basis of strategy. Empowerment strategy is a development strategy based on people's ability that is oriented to excavation and development of potential that exists in society. The purpose of this strategy can be to free the community from its backwardness [14]. Traditional arts such as *lengger Calung* is a potential both economically and culturally to empower the community of potential owners. So, gradually increasing the welfare of society can be realized. One empowerment effort is to send gifted children to follow formal education in the field of art that is available in Banyumas in SMK 3. Of course after the government provides special funds in organizing can be maximized.

4. Conclusion

Calung Banyumasan music is a unique music that is born from a unique culture. Location of Banyumas which is between West Java and Central Java to make music traditions are unique, not the same as the music in Central Java in general. In *Banyumasan* songs, for example, contains many elements that reflect the character of Banyumas society like the style of the game, as well as the style of language in the lyrics.

Conservation efforts that are considered strategic are in the field of education. One of the educational units that still holds a special education music *Calung Banyumasan* is Vocational High School 3 Banyumas. This school hosts three hours of lessons each week for three years. However, this activity only occurs in the *Karawitan* program. As a special art school, of course, this is considered still lacking.

Vocational high schools that has special music tradition is considered very strategic to organize conservation efforts, especially music *Calung Banyumasan*. Especially in a very strong global era that can erode the values of tradition, especially Banyumas which has a local wisdom that is very useful for civilization of Banyumas society itself. *Calung* seen from the musical elements, lingual through the lyrics of his songs, performances and its supporter's society contains a positive element in supporting the Banyumas society in appreciating the community's self-esteem and cultural sustainability is deemed.

The circumstances of community actors and supporters of *Calung* that tend to come from the middle to lower society require the intervention of local and central government to empower this community through *Calung* that has become a source of products in the empowerment. If empowerment can take place, then *Calung* that has become part of this socio-cultural life will remain sustainable.

References

[1] S Ambarwangi and Suharto 2014 Reog As Means of Students' Appreciation and Creation in Ats and Culture Based on the Local Wisdom *Harmon. J. Arts Res. Educ.* **14** 1 37–45

[2] H B S Suraji 2013 Imbal: sebuah Ide Penciptaan Komposisi Musik dalam Perangkat *Calung Keteg* **13** 1 71–87

[3] Darno 2006 Revitalisasi Gending-gending *Banyumasan* dalam gamelan *Calung*, sebuah tawaran karya seni *Keteg* **6** 2 121–153

[4] S Suharto 2016 *Banyumasan* Songs As Banyumas People's Character Reflection *Harmonia: Journal of Arts Research and Education* **16** 1 49–56

[5] Yusmanto 2006 *Calung (Kajian Identitas Kebudayaan Banyumas)* (Surakarta:Institut Seni Indonesia Surakarta)

[6] S Priyadi 2007 Cablaka sebagai Inti Model Karakter Manusia Banyumas *Diksi* **14** 1 11–18

[7] S Priyadi 2003 Beberapa karakter orang banyumas *Bhs. DAN SENI* **31** 1 14–36

[8] R Sutton 1986 The Crystallization of a Marginal Tradition: Music in Banyumas, West Central Java *Yearb. Tradit. Music* **18** 115–132

[9] M B and H Milles 1984 *Qualitative Data Analysis* (London: Sage Publication)

[10] M Rachman 2012 Konservasi nilai dan warisan budaya *Indones. J. Conserv.* **1** 1 30–39

[11] I Meliono *Lengger Banyumas and Padhepokan Banyu Biru As Model Community Empowerment : A Case Study In The Village Of Plana , Soma Gede District , Banyumas*

[12] A Ahmadi 2004 *Sosiologi Pendidikan* (Jakarta.: Rineka Cipta)

[13] C Fandeli 2002 *Perencanaan Kepariwisataan Alam* (Yogyakarta: Fakultas Kehutanan, Universitas Gadjah Mada)

[14] A Khoirul 2003 *Agama Tradisional* (Yogyakarta: LKIS)

IOP Conf. Series: Materials Science and Engineering **306** (2018) 012121 doi:10.1088/1757-899X/306/1/012121

Optimizing the Information Presentation on Mining Potential by using Web Services Technology with Restful Protocol

T Abdillah*, R Dai and E Setiawan
Department of Informatics Engineering, Faculty of Engineering, Universitas Negeri Gorontalo, Indonesia

*tajuddin@ung.ac.id

Abstract. This study aims to develop the application of Web Services technology with RestFul Protocol to optimize the information presentation on mining potential. This study used User Interface Design approach for the information accuracy and relevance as well as the Web Service for the reliability in presenting the information. The results show that: the information accuracy and relevance regarding mining potential can be seen from the achievement of User Interface implementation in the application that is based on the following rules: The consideration of the appropriate colours and objects, the easiness of using the navigation, and users' interaction with the applications that employs symbols and languages understood by the users; the information accuracy and relevance related to mining potential can be observed by the information presented by using charts and Tool Tip Text to help the users understand the provided chart/figure; the reliability of the information presentation is evident by the results of Web Services testing in Figure 4.5.6. This study finds out that User Interface Design and Web Services approaches (for the access of different Platform apps) are able to optimize the presentation. The results of this study can be used as a reference for software developers and Provincial Government of Gorontalo.

1. Introduction

Gorontalo Province has several gold mining locations, among others: (1) Pohuwato Regency (Pani and Bulontio Mountains); (2) Boalemo Regency (Bilato); (3) Bone Bolango Regency (Tulabolo); (4) North Gorontalo Regency (Hulawa and Ilangata Villages). In general, the mining location is traditionally managed by the community, so that the mercury level in the river in the mining area is as high as 0.5 µg / g [1] or above the safe level declared by the World Health Organization (WHO).

The information related to this mining location has not been well-presented so that the investment in Gorontalo mining sector remains very small. Truthfully, the potential of metal mineral (gold), Non-metallic mineral (Granite), *Batu Gamping*, Toseki, Felspar, and an Andesite with their reserve amounts are about 528.29 tons, 470.000.000 m^3, 12,414,500 m3, 6,000,000 m^3, 1,000,000 m^3, and 1,050,000 m^3 respectively [2].

Currently, the method of presenting the mining area information is still applied manually, so that the investors from other regions cannot access it directly. In consequence, the mining potential is not well-managed [3]. Therefore, this study will present the information by using Chart for the accuracy and relevance of the information [4].

Web Services Technology with the RESTFul protocol is employed to present the mining information to be accessible by using mobile devices and other wireless devices. It is because RESTFul protocol is lighter and suitable to be accessed by the resource-limited devices [5].

This study is intended to optimize the information presentation of mining potential in Gorontalo Province by using Web Services technology with Restful Protocol as well as two following approaches: (1) the information accuracy and relevance; (2) the reliability in presenting the information.

2. Methods

2.1. User interface design model

In some cases, the application users reject the use of the application since the User Interfaces design is not really practical, and therefore the suitability of the User Interface design can give a good impact on the application itself. An application with a good User Interface design must take into account the response time, command and action structure, error handling, and *help* facility in that application [6]. Several things need to be considered in designing the user interface in web-based applications are [7]:

2.1.1. Interface design. An excellent interface design considers the appropriate colors and objects in which each coloring should offer a meaning to the things. The users can easily use the in-app navigation objects. The users' interaction with the application is utilizing symbols and languages understood by the users.

2.1.2. Information design. The data are presented in simple forms such as figure display and textual form in minimizing the information. The results of the design can predict what the user is doing in the application, and then the application delivers messages related to what he is doing. The availability of information that matches the character of the users is very crucial because the users who are already experienced with computers usage will be unsatisfied with long messages. Conversely, inexperienced users require long and detailed messages.

2.1.3. Interaction design. The application users' habit, age, and computer knowledge should be considered in interaction design, including the computers used to run the application. Other considerations are age, gender, physical ability, educational background, culture or ethnicity, motivation, target and personality of the addressed system users [8]. There are three best rules for designing a user interface, which are: (1) Placing the controlled user; (2) Reducing the user's memory load; (3) Creating a consistent interface [9].

2.2. Web services model

The Internet has changed the way people live today, such as how to communicate, share knowledge, trade, and entertainment [10]. It is the part of the Information and Communication Technology (ICT) that has a positive impact on various sectors of life, including human development. However, ICT cannot improve people's lives [11].

The current technology of presenting information has entered the age of web services, which is according to [12], web service is defined as "*a software system identified by a URI, whose public interfaces and bindings are defined and described using XML. Its definition can be discovered by other software systems. These systems may then interact with the Web service in a manner prescribed by its definition, using XML based messages conveyed by Internet protocols*."

This research utilizes the RESTFul protocol on Web Services so that the information presentation on mining potential can be accessed by various devices (mobile and computer) in which the results [5, 13] reveal that RESTFUL protocol is more suitable than SOAP to be accessed by using Wireless devices (measured by small message size and fast response time). On the contrary, SOAP is better than the RESTFul protocol for its security.

Another reason for using RESTFul in this research [14] is that RESTFul has a faster response time than SOAP because of the few throughput of RESTFul, which reduces the whole Web Services performance.

3. Results and Discussion

3.1. The accuracy and relevance of the information
This study employs approaches [4, 7] to optimize the information presentation of mining potential.

3.1.1. The implementation of user interface in the applications
3.1.1.1. The appropriate colors and objects consideration

The coloring in the application uses a combination of colors that can be distinguished (e.g., black is combined with white). The consideration of this colors combination is due to male dominance as the application users (7% of men have *Deuteranomaly*). Figure 1 displays the main menu of the application that uses the appropriate colors combination.

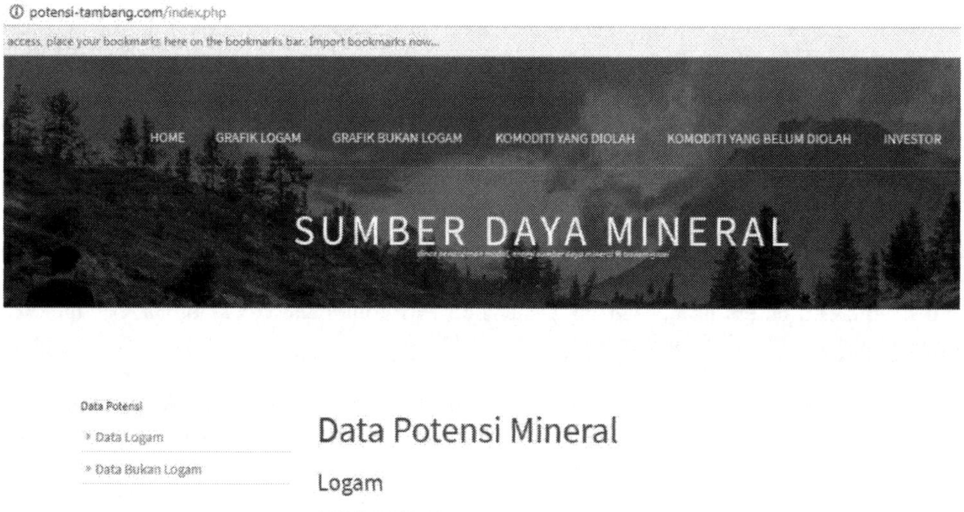

Figure 1. The Main Menu by using distinguished colors combination.

3.1.1.2. The easiness of using the navigation
The navigation created in the application makes it easier for the users to use it, such as to determine the closest distance from the current position to the mining location. Figure 2 shows the map with the path to be passed through. The application users can choose the display mode: map or satellite.

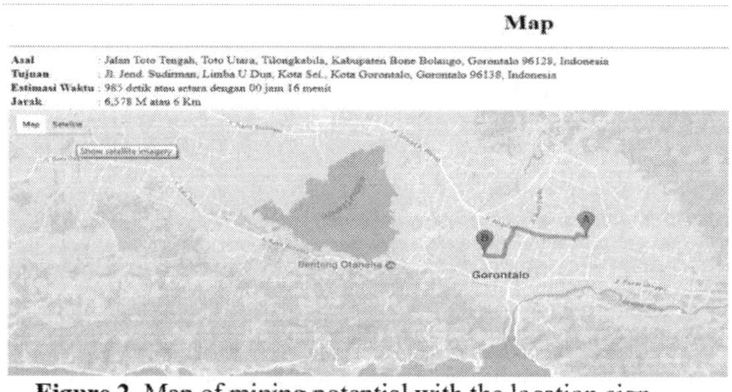

Figure 2. Map of mining potential with the location sign.

IOP Conf. Series: Materials Science and Engineering **306** (2018) 012121 doi:10.1088/1757-899X/306/1/012121

3.1.1.3. The users' interaction with the application is using symbols and languages understood by the users

The symbols in the application are commonly used and known in the computers area, while the language used is the Indonesian language.

3.1.2. The presentation of information. In presenting the information, the charts are used to facilitate the users in understanding the information. Tool Tip Text is provided in the application to help users understand the charts/pictures by placing the mouse pointer over the object. Figure 3 displays the information for the investors about the presentation of metal and non-metal contents in Gorontalo Province.

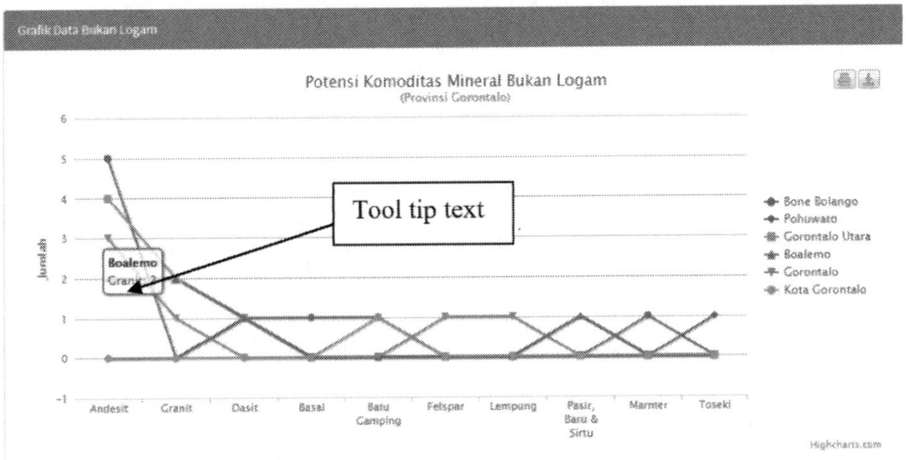

Figure 3. The presentation of graphical information with Tool Tip Text.

3.1.3. The users' interaction with the application. This application is created for users who have the ability to operate a computer and mobile device. It is because the result of the interview finds out that the application users are commonly proficient with a computer.

3.2. The reliability in presenting the information

The reliability in presenting the information is tested by employing Response Web Service with two test scenarios, including 'Testing' that is conducted by accessing the Web Service Client by utilizing the Add-Ons from Mozilla Firefox which is SOA Client. This is to find out if the WSDL created can work properly when the client accesses it. The first step to test the web service in this library is by inputting the URL of web service that will be tested in the column of WSDL address. In this study, the URL of the web service to be tested is http://sippt-glto.com/config/service/index.php?wsdl.

After that, the URL WSDL is inputted to the RestFul Client, and then click the *Parse WSDL* button, so that it will show a display as presented in Figure 4.

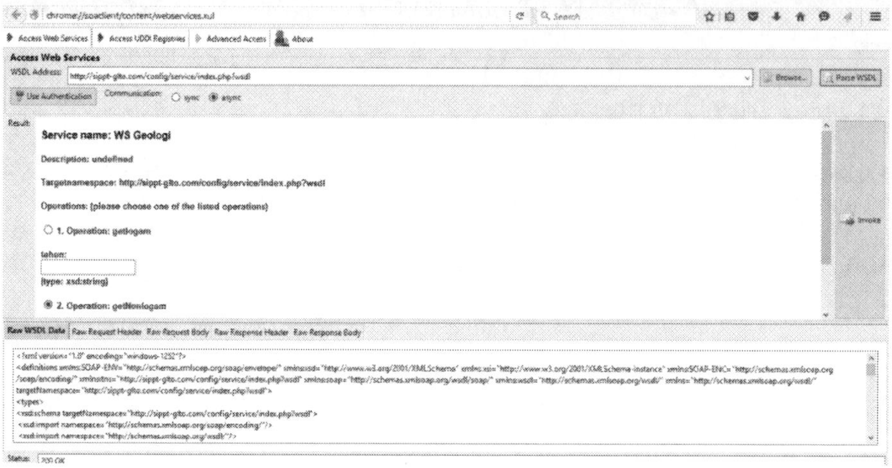

Figure 4. The result of parse WSDL.

Figure 4 is the display result after clicking the Parse WSDL. It includes the operations that we will choose. After selecting the available operation, input the parameters or code to be requested. There are two operations that have been provided, namely getlogam and getNonLogam. In that operation, the year parameters have been provided. Therefore, input the year parameter to request the existing data. For example, put 2016 in the getlogam operation. After clicking the *invoke* button, the response from the getlogam operation with the parameter of 2016 will be displayed in the *Result* column as shown in figure 5.

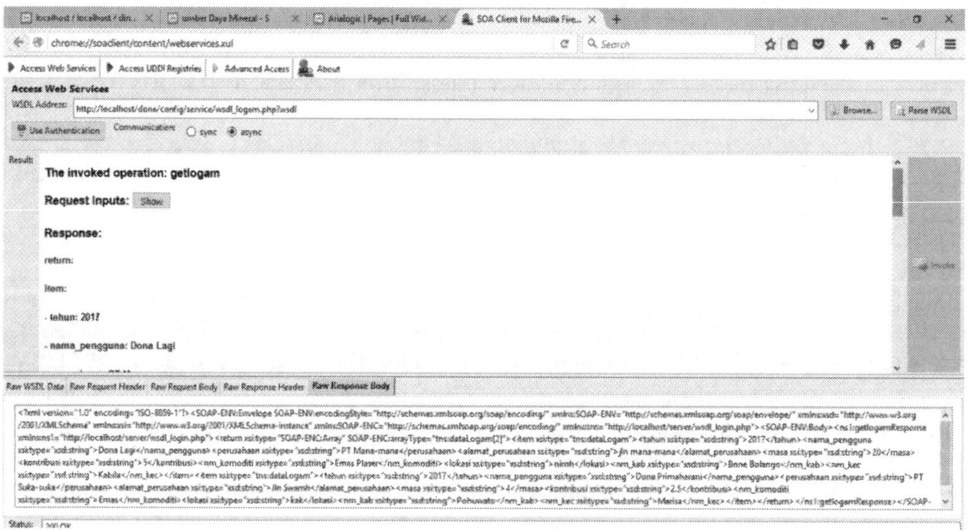

Figure 5. The testing result of getlogam operation with parameter of 2016/

Based on figure 5, it can be seen that getlogam operation is successfully executed and gives a response as created. The response of the getlogam operation consists of username, company, company address, job title, uploaded approval file, commodity name, regency name, sub-district name, and location. Figure 6 shows the list of investors to be used by the other information system.

IOP Conf. Series: Materials Science and Engineering **306** (2018) 012121 doi:10.1088/1757-899X/306/1/012121

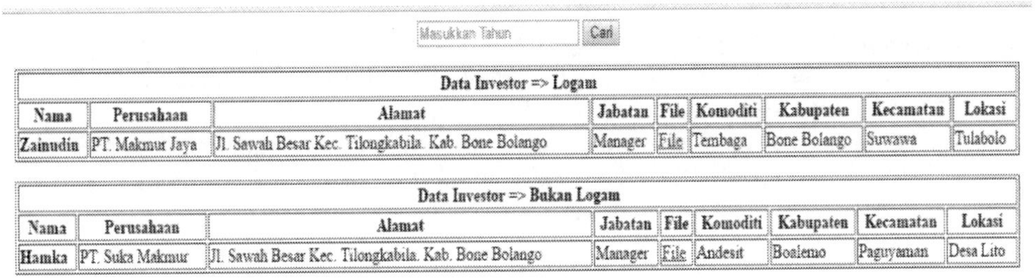
Masukkan Tahun Cari

Data Investor => Logam								
Nama	**Perusahaan**	**Alamat**	**Jabatan**	**File**	**Komoditi**	**Kabupaten**	**Kecamatan**	**Lokasi**
Zainudin	PT. Makmur Jaya	Jl. Sawah Besar Kec. Tilongkabila. Kab. Bone Bolango	Manager	File	Tembaga	Bone Bolango	Suwawa	Tulabolo

Data Investor => Bukan Logam								
Nama	**Perusahaan**	**Alamat**	**Jabatan**	**File**	**Komoditi**	**Kabupaten**	**Kecamatan**	**Lokasi**
Hamka	PT. Suka Makmur	Jl. Sawah Besar Kec. Tilongkabila. Kab. Bone Bolango	Manager	File	Andesit	Boalemo	Paguyaman	Desa Lito

Figure 6. The list of investors.

4. Conclusions
Optimizing the information presentation of mining potential by using Web Services Technology with the Restful Protocol has been explained. The application is created by following the rules of web-based application design.

Acknowledgments
The authors express their gratitude and appreciation to the Directorate General of Higher Education, Ministry of Research, Technology and Higher Education of the Republic of Indonesia that has funded this research as well as the Rector of Universitas Negeri Gorontalo and the Head of Research and Community Service Institution (LPPM) Universitas Negeri Gorontalo who have proposed the research funding.

References
[1] Arifin Y I, Sakakibara M, and Sera K 2015 Impacts of Artisanal and Small-Scale Gold Mining (ASGM) on Environment and Human Health of North Gorontalo Regency, Gorontalo Province, Indonesia *Geosciences Journal* **5** p 160-176.
[2] BPMPTSP 2015 Potensi Pertambangan dan Energi Provinsi Gorontalo [Mining and Energy Potentials in Gorontalo Province] (Online) Available at: *http://bpmptsp.gorontaloprov.go.id/potensi-pertambangan-dan-energi.* Accessed on May 6, 2017.
[3] D Roviana, A Tajuddin and S Edi 2016 Geographical Information System Model for Mining Potential Data Management Presentation in Kabupaten Gorontalo *Proceedings IOP Conference Series* **180**(1).
[4] A R Montazemi and S Wang 2015 The Effects of Modes of Information Presentation on Decision-Making: A Review and Meta-Analysis *Journal of Management Information Systems.*
[5] Kishor Wagh and Ravindra Thool 2012 A Comparative Study of SOAP Vs. REST Web Services Provisioning Techniques for Mobile Host *Journal of Information Engineering and Applications* **2**(5).
[6] Sridevi S 2014 User Interface Design *International Journal of Computer Science and Information Technology Research* **2**(2) pp 415-426.
[7] Sandhya I and Suchithra 2017 User Interface Design in E-commerce Website *Imperial Journal of Interdisciplinary Research (IJIR)* **3**(5).
[8] Shneiderman B 1997 *Designing the User Interface (3rd ed)* (Addison-Wesley).
[9] Mehdi B and Bahrami M 2012 *A Review of Software Architecture for Collaborative Software's* (Advanced Materials Research).
[10] Adeoye O S 2012 The Internet And The Needs Of The Society *International Journal of Computer Science and Information Technology & Security (IJCSITS)* **2**(3).
[11] Usman-Hamza F E 2012 Information And Communication Technology In Relation To Human Development *International Journal of Computer Science and Information Technology & Security (IJCSITS)* **2**(3).

[12] W3C 2002 *Web Services Architecture Working Group Charter* (Online) Available at: http://www.w3.org/2002/01/ws -arch-charter
[13] Hamad H, Saad M, and Abed R. 2010. Performance Evaluation of RESTful Web Services for Mobile Devices *International Arab Journal of e-Technology* **1**(3).
[14] Bora A and Bezboruah T 2015 A Comparative Investigation on Implementation of RESTful versus SOAP-based Web Services *International Journal of Database Theory and Application* **8**(3).

IOP Conf. Series: Materials Science and Engineering **306** (2018) 012122 doi:10.1088/1757-899X/306/1/012122

Non Destructive Test Dye Penetrant and Ultrasonic on Welding SMAW Butt Joint with Acceptance Criteria ASME Standard

T Endramawan* and A Sifa

Mechanical Engineering Department, Politeknik Negeri Indramayu, Lohbener-Indramayu.

*tito@polindra.ac.id

Abstract. The purpose of this research is to know the type of discontinuity of SMAW welding result and to determine acceptance criteria based on American Society of Mechanical Engineer (ASME) standard. Material used is mild steel 98,71% Fe and 0,212% C with hardness 230 VHN with specimen diameter 20 cm and thickness 1.2 cm which is welded use SMAW butt joint with electrode for rooting LB 52U diameter 2.6 mm, current 70 Ampere and voltage 380 volt, filler used LB 5218 electrode diameter 3.2 mm with current 80 Ampere and 380 volt. The method used to analyze the welded with non destructive test dye penetrant (PT) method to see indication on the surface of the object and Ultrasonic (UT) to see indication on the sub and inner the surface of the object, the result is discontinuity recorded and analyzed and then the discontinuity is determine acceptance criteria based on the American Society of Mechanical Engineer (ASME) standards. The result show the discontinuity of porosity on the surface of the welded and inclusion on sub material used ultrasonic test, all indication on dye penetrant or ultrasonic test if there were rejected of result of welded that there must be gouging on part which rejected and then re-welding.

1. Introduction

The rapid industrial market in various fields in the country is characterized by the increasing infrastructure development in various region. In the infrastructure development is not spared from the role of the metal welding process. The welding process in this metal is of many kind include Shielded Metal Arc welding (SMAW), Gas Metal Arc Welding (GMAW), Gas Tungsten Arc Welding (GTAW), Flux-cored Arc Welding (FCAW), and Submerged Arc Welding (SAW).

Welding technology at this time is experience a very advanced development with a variety of new methods that have been found both using conventional and automatic. Behind the benefits of the welding process so much, but in the process of metal welding the results are not always good, this happen is influenced by many factor both intentionally and unintentionally. To handle it, it is necessary to inspect the welding results to be in accordance with the wishes and safe to use. Methods to inspect the welded results are one with non-destructive test (NDT) method of dye penetrant and ultrasonic.

The method of Liquid Penetrant Test (dye penetrant) is the simplest method of NDT but the advantage of speed and accuracy in detecting defects on the surface. This method is used to find defects in the open surface of solid components, both metal and non-metal, such as ceramics and fiber plastic. Through this method, defects in the material will be seen more clearly by looking at the indication on

IOP Conf. Series: Materials Science and Engineering **306** (2018) 012122 doi:10.1088/1757-899X/306/1/012122

the surface of the test object after sprayed developers who then in skeet to be used as reporting work that will be translated to acceptance criteria.

This test can also be used in the aircraft industry for routine maintenance in checking of aircraft components, in which the test procedure, the surface preparation of the components, the component protection during the test and all pertaining to the penetrant test [1].

Ultrasonic Testing (UT) is one of the non-destructive test on the material. Ultrasonic test is that uses the sound frequency to detect defects in solid objects. This is based on the fact that solid objects represent good sound waves. This Testing Principle is performed by reflecting ultrasonic sound waves an object then the reflection of the wave from the object is captured by the probe. The difference in surface depth is an indication of a defect in an object [2].

Ultrasonic waves can be generated by changes in electrical energy to the mechanical energy of a transducer called a probe, via a piezoelectric effect. The piezoelectric effect is a reversible effect means that if there is a change of electrical energy to mechanics, the mechanical energy changes to electrical energy occurs. To check the thickness of the material and the presence of defects in a material with ultrasonic waves can be done in three ways, namely resonance techniques, transmission techniques, and echo techniques. Of the three techniques, echo techniques are most commonly used primarily on field checks [2].

The basic principles of the three techniques are:

1.1. Resonance technique
The thickness of the material can be measured by measuring the ultrasonic frequency or wavelength that can cause maximum resonance on the material. The presence of defects can be detected by the occurrence of resonance change as the distance of the resonating material changes.

1.2. Transmission technique
The existence of defects in the material can be known from the decrease of wave intensity. Ultrasonic received by the receiver probe while the thickness of the material can not be measured by the transmission technique.

1.3. Echo technique
The thickness of the material, location and magnitude of the defect can be determined from the time of propagation and wave amplitude received by the probe.

The velocity (v) of ultrasonic wave in a material depends on the type of material through which the wave modes pass as defined in the following equation [2].

Longitudinal Mode:

$$V_L = \sqrt{\frac{E\,(1-\sigma)}{\rho(1+\sigma)(1-2\sigma)}} \qquad (1)$$

Transversal Mode:

$$V_T = \sqrt{\frac{E(1-\sigma)}{2\,\rho(1+\sigma)}} \qquad (2)$$

Couplan serves to facilitate the propagation of wave from the probe into the specimen. Because if between the probe and the test object is air then almost 100% of the wave will be reflected back into the probe and the reflected wave will spread more than focus on the defects in the material. In order for the thick coupling between the probe and the fixed assay, the pressure applied to the probe must be constant so as not to affect the amplitude of the indication on the screen [2].

Ultrasonic aircraft have in common with oscilloscopes where measurements are made based on time and voltage measurements. Time measurements are presented on a horizontal scale as ultrasonic wave distance measurements. Voltage measurements are presented on a vertical scale as an amplitude measurement to determine the attenuation coefficient of the wave through the medium. This horizontal and vertical scale must be linear in order and produce an accurate output value [2].

There are 3 types of probes:

- Single Normal Probe
- Single Angle Probe (450, 600, 700)
- Normal Probe and Twin Corner

The purpose of this research is to determine the specimen can be accepted based on acceptance criteria of ASME standard. The expected result of this research is to provide input to take holder the welding test about non destructive test ultrasonic methods and provide standard description for acceptance criteria of defect indication on weld result based on ASME standard.

2. Literature Review

The metal welding process has been known for a long time, various techniques have been developed till now, and one of the most applied welding techniques in the field is Shield Metal Arc Welding (SMAW). The SMAW welding process is included in the arc welding process group, which includes Tungten Arc Welding Gas and Metal Arc Melding Gas. In some previous studies stated that the welding arc welding technique can be examined the results weld using ultrasonic testing [3]. The assessment process was developed using an ultrasonic method of a transducer conducted on the welded joint area by direct contact technique.

The methodology for qualifying and demonstrating ultrasonic test technique to determine the results of welded joints on pressure vessels. The pressure vessel is made of welded steel plate welded joint with customized shape and size. The inspection process is carried out from only one part i.e. the outer part of the welded joint [4] .

Ultrasonic tests by comparing SMAW and GTAW welding results in AISI 316L steel. The austenitic character is divided into grain orientation distribution and anisotropy shows that on GTAW specimen is more isotropic than SMAW for grain orientation. B-scan results show SMAW inspection easier than GTAW process [5] .

Ultrasonic applications for conventional processes include casting and welding. They reported ultrasonic wave on austenitic stainless steel (AISI 316) with an investigation system of 20 kHz ultrasonic wavelength effect on the material microstructure during the welding process [6]. The complex vibration system in which ultrasonic welding is effectively applied to different types of specimens include the same metal specimen as well as to specimen with different metals.

The inspection of the thickness of the welded joint using an ultrasonic laser SAFT. Detecting from the defects on the butt connections on the welding is required to reduce the cost and rework time. The specimen were used from steel with thickness 25 mm and 50 mm [7].

2.1. ASME standards

Standard American Society of Mechanical Engineers (ASME) Section IX is the most commonly used standard for welder qualification. Committees and subcommittees of volunteer workers are interested in advancing the quality and efficiency of the welding industry by developing this code. ASME is specialized for welder qualification and welding procedures. A "construction standard" such as ASME part VIII Division I must be used in conjunction with Part IX for fabrication [8].

3. Methods

In this study using material from mild steel which then made SMAW welding process of 3G butt joint position on plate material with size 30 cm x 20 cm x 1.2 cm which bevel 300 on each specimen. The welding process uses a maxstar 200 miller with LB 52U diameter electrode diameter of 2.6 mm using 70 Ampere current and 380 volt voltage, while for filler using LB 5218 diameter 3.2 mm electrodes with 80 Ampere current and 380 volt voltage [9].

Hardness testing process using innovates engine type verzus 700AS vickers method with the use of 20 Kgf load and indentation time 10 seconds. The hardness test method according to ASTM E-384 (see figure 1), sets the diamond indentor to the angle between opposing surfaces is 136° [10].

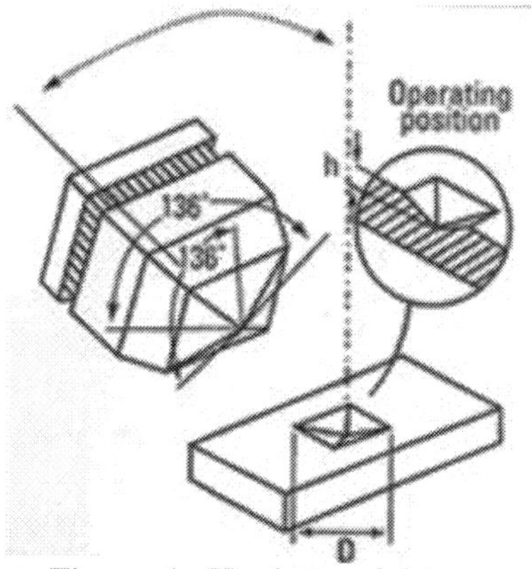

Figure 1. Hardness vickers.

After the force is removed then it is measured diagonally, so the hardness of vickers can be formulated by the equation [10]:

$$H = \frac{1.8544\,P}{d^2} \quad (3)$$

NDT of dye penetrant test spotcheck method using the visible Magnaflux type visual marker with SKC-S No. code cleaner. 170404, Penetrant SKL-SP2 No. 160408 and developer SKD-S2. 170401. Tools used include wire brushes, coats, and duster.

3.1. Dye penetrant test

Non Destructive Test Dye Penetrant process is done with the following stages:

- Check the light intensity of at least 1000 lux
- Check penetrant sensitivity using standard blocks
- Clean the surface using a wire brush to remove the dirt that covers the defect
- Clean the surface by using solvent to remove dirt in the form of dust, oil and others
- Spray the penetrant to surface evenly
- Allow the penetrant (dwell time) for 10 minutes
- Remove excess penetrant using solvent on fabric in one direction
- Dry the surface after use of solvent
- Spray the developer to surface evenly
- Wait for the developer to react (dwell time) about 30 minutes
- Evaluation of discontinuity indication (see figure 2)
- Record indication of discontinuity (indication of length, diameter, discontinuity of reference point
- Post cleaning using solvent until clean.

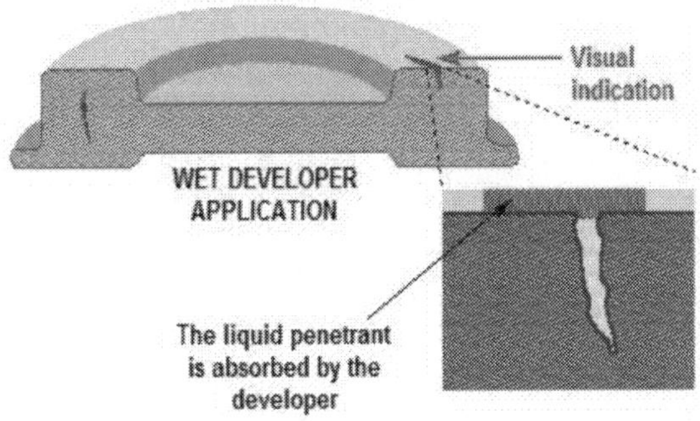

Figure 2. Indication of discontinuity [11].

3.2. Ultrasonic test

This Ultrasonic NDT Test uses the Karl Deutch 1090 series flaw ultrasonic detector with 700 probes, 4 Mhz frequency and coupl grease. The Tools needed such as pencils and long gauges (Ruler).

The method used in the Determination of Disabled Dimensions is done by the 6 dB drop method is when the probe position on the edge of the defect can be determined if 50% of the wave is continued while the other 50% is reflected back to the probe. Therefore, the probe can be said to be in a defective edge position. By shifting the probe across the surface of the specimen, the edge of the defect will be determined to obtain its dimensions. In this method also performed the addition of 6 dB from the previous calibration gain. 6dB increment occur because to detect defects should look for pulses that are in the position of 50% full height screen. When the initial pulse position 100%FSH changes to 50%FSH there is a dB reduction of 6dB. Therefore, the maximum amplitudes are obtained to determine the edge defect. To detect the end limit of the long edge of the defect then the probe must be moved back to find the pulse in the same position that is 50%.

Test steps are as follows:

1) Measure base material and configuration drawing
2) Check the visual root
3) Normal probe calibration
4) Inspection of laminate defects in base material
5) Inspection on weld result
    - Check the angle index point of the probe
    - Determine calibration using blocks V1 and V2
    - Perform mileage calibration
    - Place the DAC curve at 700 angles
    - Setting gain sensitivity
    - Analyze defects on the ultrasonic screen detector tool
    - After you see any defect indication, find the maximum pulse then measure from the point of probe indices to the middle of the weld (X0) and read M1 patch on the tool screen (S0)
    - Find the defect length by shifting the probe up down to down to 20% DAC and marked on the center of the probe face.
    - Plotting determines the position and location of defects in scan B
    - Measure the long bar of defects (L)
    - Measure position (Y) from the nearest zero point
    - Specify acceptance criteria see ASME sec. VIII app.

IOP Conf. Series: Materials Science and Engineering **306** (2018) 012122 doi:10.1088/1757-899X/306/1/012122

4. Results and Discussion

4.1. Composition test
This material composition test uses optical emission spectrometer with brand ARL 3460 with the following results (see table 1):

Table 1. Composition base material.

No	Element	Value (%)
1	Carbon	0,212
2	Silicon	0,189
3	Sulfur	0,014
4	Phosporus	0,021
5	Manganese	0,716
6	Nickel	0,02
7	Chromium	0,06
8	Molybdenum	0,005
9	Ferro/iron	98,713

The composition test results can be seen on the table 1 that the main elements contained in the material is Iron (Fe) that is equal to 98.713 percent and with carbon content of 0.212 percent. With this steel having carbon content of less than 0.25 percent it is classified into low carbon steel [12, 13].

4.2. Hardness test

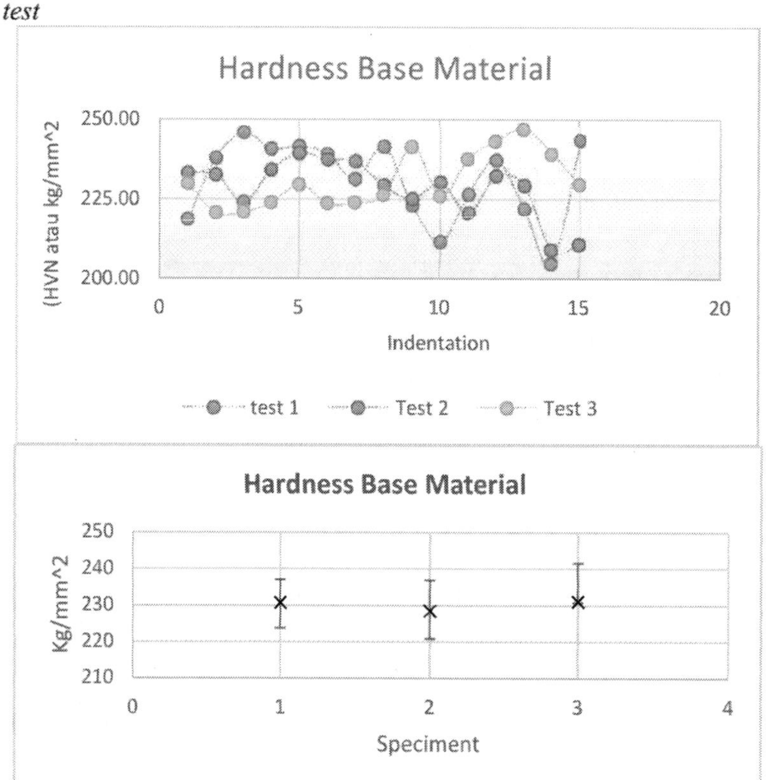

Figure 3. Hardness of base materials.

Based on figure 3, the material hardness that contains three materials tested the lowest hardness value is 209.05 VHN and the highest hardness value is 247.00 VHN. As for the average value of the three specimens is 230.16 VHN.

4.3. Dye penetrant test

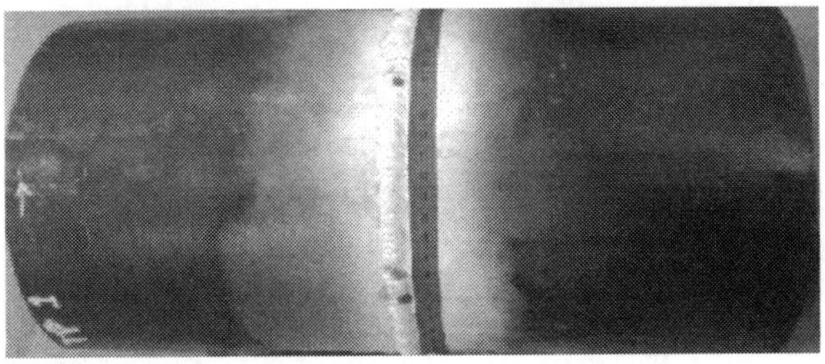

Figure 4. Specimen 1.

Table 2. Indication discontinuity specimen 1.

No	Size of discontinuity (mm)	Distance (mm)	Indication	Remark
1	8	57	porosity	Rejected
2	6	73	porosity	Rejected
3	4	77	porosity	Accepted
4	6	171	porosity	Rejected
5	3	182	porosity	Accepted
6	2	185	porosity	Accepted
7	3	188	porosity	Accepted
8	4	233	porosity	Accepted
9	1	246	porosity	Accepted
10	5	250	porosity	Accepted

Based on figure 4 and table 2 above, according to the standard ASME Section VIII Division 4 mandatory appendix 8 on the liquid penetrant testing method states that the standard criteria for discontinuity contained in the welding results shall be independent of linear indication, rounded indication greater than 5 mm and there are 4 or more Rounded indication with a distance of 1.5 mm. So based on the existing data then on this specimen 1 indication that there are some that still fall within the limits of the existing standard so that is stated Accepted. But there are some that are rejected on indications 1, 2, and 4 so it must be repaired welding results by way of gouging then re-welded.

Table 3. Indication discontinuity specimen 1.

No	Size of discontinuity (mm)	Distance (mm)	Indication	Remark
1	2	23	porosity	Accepted
2	7	102	porosity	Rejected
3	2	111	porosity	Accepted
4	3	195	porosity	Accepted
5	1	201	porosity	Accepted
6	3	234	porosity	Accepted

IOP Conf. Series: Materials Science and Engineering **306** (2018) 012122 doi:10.1088/1757-899X/306/1/012122

Based on table 3, according to the standard ASME Section VIII Division 4 mandatory appendix 8, then based on the available data then in specimen 3 this indication is still within the permitted tolerance limits so that the specimen 3 is declared accepted.

4.4. Ultrasonic test

Table 4. Specimen 1 ultrasonik test.

Indication nomer	Probe/Angle	From Surface/side	Reference Level (dB)	% DAC of Discontinuity	Position, Location and size (mm)								Discontinuity			
						Sweep Reading S0	SOD stand of Distance	L Length (mm)	Y Position (mm)	d Dept(mm)	Pos (x)(mm)	M1 pd(mm)	Type	Accepted	Reject	Remark
					Skl	S_0 (mm)	X_0 (mm)									
1	70	I	46,5	32	2,6	26,9	21	10	52	9,2	5,9	25,1	inc		V	gouging
2	70	I	46,5	34	2,8	27,8	22	6	93	9,5	5,8	25,2	inc	V		
3	70	I	46,5	34	2,8	26,7	24	5	102	9,1	2,7	25,1	inc	V		
4	70	I	46,5	43	2,8	28,2	24	11	123	9,6	4,2	25,1	inc		V	gouging
5	70	I	46,5	40	2,8	26,9	23	14	150	9,2	3,9	25,3	inc		V	gouging
6	70	I	46,5	38	2,6	25,4	20	10	205	8,7	5,4	23,3	inc		V	gouging

The test on specimen 1 table 4, six discontinuities were found, where the defect location was spread along the weld. In general, the location of the defect is located on the root of the weld is around the depth of 8.7 mm to 9.6 mm. The shortest defect length is 5 mm and the longest defect length is 14 mm. Referring to the Mandatory Appendix 12 Ultrasonic Examination Of Weld (UT) then the accepted defect (two) is located at a distance, 93 mm, 102 mm. While the rejected there are 4 that is located at a distance of 52 mm, 123 mm, 150 mm and 205 mm.

Table 5. Specimen 2 ultrasonik test.

Indication nomer	Probe/Angle	From Surface/side	% DAC of Discontinuity	Position, Location and size (mm)								Discontinuity			
					Sweep Reading S0	SOD stand of Distance	L Length (mm)	Y Position (mm)	d Dept (mm)	Pos (x)(mm)	M1 pd(mm)	Type	Accepted	Reject	Remark
				Skl	S0mm	X0(mm)									
1	70	2	40	2.8	26.1	26	14	288	9.1	0.1	25	Inclusi		✓	gouging
2	70	2	40	3.3	27.9	21	6	260	9.5	6.9	27	Inclusi	✓		
3	70	2	25	3	28.5	23	4	240	9.7	5.5	26.8	Inclusi	✓		
4	70	2	30	3.4	31.9	30	4	170	9.1	1.9	30	Inclusi	✓		
5	70	2	35	3.2	32.5	29	5	69	8.9	3.5	30.5	Inclusi	✓		
6	70	2	35	3.2	29.8	26	5	40	9.8	3.8	28	Inclusi	✓		

In the test on the specimens 2 table 5, six discontinuities were found, where the defects were scattered throughout the weld. In general, the location of the defect lies in the root of the weld is ranging from a depth of between 8.9 mm to 9.8 mm. The shortest defect length is 5 mm and the longest defect length is 14 mm. Referring to the Mandatory Appendix 12 Ultrasonic Examination Of Weld (UT) then the defects received (accepted) there are 4 pieces located at a distance of 40 mm, 69 mm, 170 mm and 240 mm. While the rejected there is 1 that is located at a distance of 288 mm.

5. Conclusions

Based on the results of the above discussion, it can be drawn some conclusions include: Base material contains composition Iron (Fe) 98.713 percent and carbon 0.212 percent. The base material has hardness was 230.16 VHN. All indication on dye penetrant or ultrasonic test if there were rejected of result of welded that there must be gouging on part which rejected and then re-welding.

References

[1] Callister W 2001 *Fundamental of Materials Science and Engineering* (John Wiley & Sons Inc).
[2] Anonim 2014 *Modul Pelatihan NDT Level I dan II* (Balai Besar bahan dan Barang Teknik (B4T)).
[3] Tiago Vieira da Cunha and Carlos Enrique 2014 Ultrasound in arc welding: A review *Ultrasonic.*
[4] Rafael Martínez et al. 2013 Ultrasonic inspections in the fabrication of the ITER vacuum vessel sectors *Fusion Engineering and Design* **88** pp 2155– 2159.
[5] Tabatabaeipour, and F Honarvar 2010 A comparative evaluation of ultrasonic testing of AISI 316L welds made by shielded metal arc welding and gas tungsten arc welding processes *Journal of Materials Processing Technology* **210** pp 1043–1050.
[6] Habib Alavi, Sandip P Harimkar 2015 Melt expulsion during ultrasonic vibration-assisted laser surface processing of austenitic stainless steel *Ultrasonics* **59** pp 21–30.
[7] Levesque et al. 2016 Inspection of thick welded joints using laser-ultrasonic SAFT *Ultrasonics* **69** pp 236–242.
[8] ASME Boiler and Pressure Vessel Committee 2013 *ASME Boiler and Pressure Vessel Code Section VIII Division I* (New York).
[9] T Endramawan, Emin H, Felix D, and Yuliana P 2017 Aplikasi Non Destructive Test Penetrant Testing (Ndt-Pt) Untuk Analisis Hasil Pengelasan Smaw 3g Butt Joint *Jurnal Teknologi Terapan* **3**(2).
[10] Mechanical Testing and Evaluation Volume 8 of the ASM Handbook Committee 2000.
[11] Xu Guirong, Guan Xuesong, Qiao Yuliang , GaoYan 2015 Analysis and Innovation for Penetrant Testing for Airplane Parts *Procedia Engineering* **99** pp 1438 – 1442.
[12] Carlisle, R "Carl" Smith 2008 *ASME and AWS welding codes—Similarities and differences* (A Publication of the Fabricators and Manufacturers Associations, Intl).
[13] Sifa A, and Endramawan T 2017 Identification of the Quality Spot Welding used Non Destructive Test-Ultrasonic Testing:(Effect of Welding Time) *In IOP Conference Series: Materials Science and Engineering* **180**(1) pp 012101.

IOP Conf. Series: Materials Science and Engineering **306** (2018) 012123 doi:10.1088/1757-899X/306/1/012123

Professionalism of Lecturers at Faculty of Education

T F S Tangkere*, F W Langitan, S M D Maukar and R F Roring
Universitas Negeri Manado, Tondano 95618, North Sulawesi, Indonesia

* tellytangkere16@gmail.com

Abstract. The main objective of this research was to get the picture pertaining to the professionalization of Lecturers at Faculty of Education in Manado State University, Indonesia. The research method was naturalistic inquiry with qualitative approach. The research techniques were: deep interview, participative observation and document study. The data were analyzed by: data reduction, data display and conclusions, while the validation of data was done by four criteria, namely: credibility, transferability, dependability and conformability. The collecting procedure and data recording were done through observation and interviews. From the findings and conclusions, it can be identified that professionalization of Lecturers at Faculty of Education in Manado State University has been well processed. This can be proved by fulfillment of the minimum academic standard Ninety-one out of the total 112 lecturers has been certified. Based on conclusions, the researcher recommends that the teacher always develop their capability through increasing their academic qualification, self-development through attending educational trainings, conducting more research and publishing those researches through accredited journals. Dean of every Faculty and also execute supporting activities which will support self-development of the lectures and increase the budget for research of the lecturers.

1. Introduction

At the college environment, lecturer is one of the needs. It is just like a driving machine to all things related to scientific and academic activities. Lecturers are determining the position of education and graduates which are produced by such college. Having high quality of lecturers gives impact on the high quality of the college or otherwise, however high the designed programs without being supported by qualified lecturers, the results will be unsatisfactory [1]. By having good and high-quality lecturers, the college is able to formulate programs with modern curriculum to assure high quality graduates with very satisfactory achievement [2]. Based on this, the professionalism of the lecturers becomes important effort to increase the quality of the college.

In Indonesia, the program of lecturer's quality development has been well-known since 1970s. Colleges have carried out activities including lecturer coaching activity such as special seminars for all new lecturers. And even certain universities established center of training faculty and staff and organized some kinds of lecturer coaching in regional and national levels. It is interesting to the lecturers to do collaboration with the industry in the form of technology transfer as well [3].

Professionalization of lecturers has been running for almost four decades; however there have not been any exhilarating results. The data of Research and Development Institution of National Education

Department (2014) shows that 50.65% of 120,000 permanent lecturers of both Private and State-Owned Colleges in Indonesia were undergraduate or Bachelor. 50% of 240,000 lecturers of State Owned Colleges have not had master degree [4]. Among such number, 15% of them were holding doctorate degree. If compared to colleges in Malaysia, Singapore, and the Philippines whose doctorate degree could reach more than 60%, then the lecturers in Indonesia colleges seem to be left behind. Having 50% of lecturers with undergraduate degree, it is hard to attain the ideal. Furthermore, it seems there is no significant effort to increase their professionalism. It is very important to build an awareness that the profession of lecturers relate so much to the knowledge, skills, and ethics which are necessary to be developed [5][6]. Fortunately, these dimensions are not much becoming the intention of the lecturers, so no wonder the critics are consistently addressed to them.

This condition shows that there is a wide gap between the ideal goals and real condition of the college lecturers in Indonesia generally and in North Sulawesi specifically. This is influenced by some factors such as educational management, economy, social reality, and other factors. Toward the profession, the characters encourage in exhibiting commitment and responsibility to do self-improvement as the real contribution for students learning [7]. However, establishing an appropriate system of standards can help educators work towards both high standards and the same professional status. Therefore, in order to fix it, it is imperative to have comprehensive lecturer professionalism which includes different parties or stakeholders such as the college itself, government, and community. In association with the problems the researcher found in her initial study, she is interested in further study about professionalization of college lecturers at Faculty of Education in Manado State University.

1.1. Focus of the Study
Based on the background of the problem, this study focuses more on the professionalization of lecturers at Faculty of Education in Manado State University.

1.2. Problem Formulation
Based on the study focus, the problem can be formulated as follows:

a) How is the professionalization of lecturers at Faculty of Education in Manado State University?
b) What are the factors hindering the professionalization of lecturers at Faculty of Education in Manado State University?
c) What efforts have been made to solve those hinders against the professionalization of lecturers at Faculty of Education in Manado State University?

1.3. Purposes of the Study
The purposes of the study are to discover the results of analysis and description about:

a) Professionalization of lecturers at Faculty of Education in Manado State University.
b) Factors which hinder professionalization of lecturers at Faculty of education in Manado State University.
c) Efforts which have been made to solve those hinders of professionalization of lecturers at Faculty of Education in Manado State University.

2. Methods
This study uses qualitative approach and descriptive method of naturalistic inquiry. The study is taken place at Faculty of Education in Manado State University.

2.1. Data Resources

a) Data used in this qualitative study are subjective words both spoken and written including behavior which was under the observation with written data in the forms of documents and others.
b) The resource of the study was from the informant, and the rest taken from the documents and others,

2.2. Techniques of Data Collection

Data collection technique is a way used by researchers to obtain data in their field study. In this study, we choose the qualitative method, which data have to be deep, clear and specific. Further the techniques of collecting data as follows:

a) Observation
Observation activities were done through direct observation the condition of lecturer's professionalization.
b) Interviews
The interviews were done through structured and unstructured ones. The structured interview was used to get general understanding about a topic. Steps of the structured interview was intended to focus on certain topics according to the problems
c) Documents
The documents assessed in this study include written documents in addition to other notes.
d) Recording Document Data in the Site Record
The researcher made site record containing what we heard and saw.

2.3. Data Analysis Techniques

Qualitative data analysis is an effort done by working with data, organizing data, sorting them into manageable units, synthesizing, searching and discovering patterns, discovering the important things which have learned, and presenting to the others [8]. In analyzing data, we used four steps: (1) reading, analyzing, and scrutinizing data, (2) data reduction (3) data presentation (4) validity assessment

2.4. Checking of Data Validity

There are four criteria used checking of data validity in this study as follows:

a) Level of Credibility
There are ways which can be equaled with validity test so that the truth of the results can be trusted:
    - Prolong of the observation duration
    - Perseverance of observation
    - Triangulation
    - Counterpart crosschecking
    - Sufficiency of references
b) Transferability
The researcher endeavored to build value of study transfer by reporting the results of the study accurately and precisely.
c) Dependability
This is done by valuing whether or not the process of the study has quality.
d) Confirmability
In research, it is a scientific effort done by the researcher to value the quality of the research findings by assessing whether the results are stable.

3. Results and discussion

3.1. Professionalization of lecturers at Faculty of Education in Manado State University

Viewing from the theoretic perspective, professionalization refers to the process of improving qualifications and competence of the professionals in meeting the standardized criteria in their performance as professionals [9]. Basically, professionalization constitutes a series of process of professional development, in which it is done through either education or training of both "pre-service" and in-service training.

Concept of professionalization includes two dimensions. They are the improvement of status and the improvement of practice [10]. This idea suggests that the first dimension encompasses the organized efforts to meet the ideal professional criteria. The second dimension, according to Holey that is to perfect the improvement (improvement of practice). It can be said that this matter has been in accordance with understanding of the first dimension.

3.2. Hindering Factors on Professionalization of Lecturers at Faculty of Education in Manado State University

Factors which hinder the professionalization of lecturers regarding the improvement of practice such as the lack of training enforcement which is in accord with expertise of study program. This is caused by the lack of the allocated funds for different kinds of training. By looking at the numbers of students with the ratio of lecturer service to students, it is found that there is a lack of permanent lecturers whose expertise is in line (linear) with the study program. The other finding shows that there is lack of research quantity carried out by lecturers. Most of the lecturers do the study mostly twice a year. It is felt lacking in terms of continuous implementation on skill as well as knowledge improvement.

Relating the development of human resources, the following are some obstacles found: a) management of DIKTI scholarship might be said complicated and insufficient, 2) lacking of continued education interest for doctorate due to flurries of the lecturers, c) study program will be lacking the lecturers due to doctorate upgrading, and d) lacking of the will or willingness to arrange the administrative requirements in terms of proposing professor status. The lecturers who have accomplished their master and doctorate degrees are found not in line with the needed expertise. Additional facilities to meet minimum standard of managing the study program is lacking such as laboratory room and other supporting facilities which are not functioning or damaged, and even the study program does not have supporting tools.

3.3. Efforts made to solve those hinders of professionalization of lecturers at Faculty of Education in Manado State University

The following are the efforts made to solve the hinders: a) design periodic doctorate program so the lecturers in the study program will not drastically decreased, b) Faculty will be more selective in determining the emphasis to take in line with the expertise. The expertise which is inappropriate or unsuitable with home base study program can be recommended to change the study program as home base. Lecturers who have not met the teaching/lecturing certification will be adjusted their recommendations to the requirement for them. The effort to deal with indiscipline is to apply punishment in accordance with level of mistakes done by lecturers and yet the instructions are given with the spirit of deliberation to reach a consensus. In order to deal with the lack of permanent lecturers, there must be recruitment of permanent lecturers based on the necessary expertise of study program. Regarding the lack of facilities, the procurement should be planned and implemented. The same thing regarding research activities of the lecturers, it is important to maximize the forms of cooperation as well as motivate the lecturers to develop research through independent research.

IOP Conf. Series: Materials Science and Engineering **306** (2018) 012123 doi:10.1088/1757-899X/306/1/012123

4. Conclusion and suggestion

4.1. Conclusion

Based on the research findings and discussion, the conclusion can be drawn as follows:

1. Professionalization of lecturers has run well. Minimum limitation shows that there are only two lecturers with undergraduate degrees who are pursuing their graduate level (master degree). There have been 82.25% of 112 the lecturers certified.
2. Hindering factors against the professionalization of lecturers: lacking of permanent lecturers whose expertise is unsuitable with their study, problem of in disciplinary, lacking of facilities, evaluation results are still regarded as formality activities.
3. Efforts done to solve the hinders such as developing the academic qualifications on doctorate program particularly, enforcing lecturers to do research and learning in accord with expertise of the last education, recommending the recruitment the expertise suitable with study program, motivating lectures to attend training related to their competence development.

4.2. Suggestion

Based on the conclusion, the researcher suggests as the following:

1. that the lecturers develop competence through improving academic qualification, attending the training, publishing the researches on the accredited or indexed media.
2. that the administrators of the Faculty and rector pay attention to, plan, and implement providing facilities which are supporting the learning process, as well as supporting activities for self-development and increasing allocated funds to conduct studies.

References

[1] Y Hill, L Lomas and MacGregor 2003 Students' perceptions of quality in higher education, *Quality Assurance in Education* **11** 15-20

[2] M A Barakat 1998 *Al-Tathwir al-Mahnity li A'dla'I Hay'at al-Tadris* (Tunis: Al-Munazhzhamah al-'Arabiyah li al-Tarbiyah)

[3] J Lee and H N Win 2004 Technology transfer between university research centers and industry in Singapore *J Technovation* **24** 433-422

[4] N Syam 2016 *Kualifikasi Pendidikan Dosen* Quoted on June 14, 2017 via *http://nursyam.sunan-ampel.ac.*

[5] G S Akerlind 2003 Growing and Developing as a University Teacher--Variation in Meaning *Studies in Higher Education* **28** 375-390

[6] M Parsell, T Ambler, C J Trawonger 2012 Ethics in higher education research *J Studies in Higher Education* **39** 166-179

[7] H Tanang and B Abu, 2014, Teacher Professionalsm development Practices in South Sulawesi, Indonesia, *Jornal of Curricculum and Teaching* **3** 33-42

[8] R C Bogdan and K S Biklen 1982 *Qualitative Research for Education: An Introduction to Theory and Methods* (Boston London: Allyn and Bacon, Inc.)

[9] D Satori, S Kartadinata, Y Syamsu and S Makmun 2006 *Profesi Keguruan* (Jakarta: Universitas Terbuka)

[10] E Hoyle 1980 *The Rule of The Teachers* (London: Owe and Brydone)

IOP Conf. Series: Materials Science and Engineering **306** (2018) 012124 doi:10.1088/1757-899X/306/1/012124

Design and Simulation of Microstrip Hairpin Bandpass Filter with Open Stub and Defected Ground Structure (DGS) at X-Band Frequency

T Hariyadi*, S Mulyasari and Mukhidin

Universitas Pendidikan Indonesia

*tommi.hariyadi@upi.edu

Abstract. In this paper we have designed and simulated a Band Pass Filter (BPF) at X-band frequency. This filter is designed for X-band weather radar application with 9500 MHz center frequency and bandwidth -3 dB is 120 MHz. The filter design was performed using a hairpin microstrip combined with an open stub and defected ground structure (DGS). The substrate used is Rogers RT5880 with a dielectric constant of 2.2 and a thickness of 1.575 mm. Based on the simulation results, it is found that the filter works on frequency 9,44 - 9,56 GHz with insertion loss value at pass band is -1,57 dB.

1. Introduction

The radar development has been increasingly rapid ever since the World War II. Radar is used as a supplement to war vessels and fighter planes for detecting enemy's presence. Currently, radar system usage is broader, not only for military but also for commercial purposes such as for Air Traffic Control and weather or road monitoring.

One type of radar works at X-band frequency which is frequency range from 8 GHz to 12 GHz. X-band frequency is highly suitable for naval radar, airborne weather avoidance radar, airborne doppler navigation radar and police speed radar. Moreover, X-band frequency is also applied in marine radar, coastal surveillance radar, and weather radar.

Radar is a system consisting of three main components, namely antenna, transmitter and receiver. Receiver in a radar system has the capability of filtering the signal to make it suitable for detection as well as the capability of amplifying weak object signal and forwarding it to data and signal processor and then displaying it on the monitor screen [1]. Radar frequency is expected to work at 9-10 GHz, while X-band itself has frequency range of 8-12 GHz. With intervention in the frequency band due to wide spectrum of X-band frequency, filter is required to reject undesirable frequency.

Filter is an electronic circuit having a function of passing a desirable signal and/or not passing an undesirable signal [2]. Filter can be designed using many methods that have their respective advantages and disadvantages. These methods include hairpin, edge-coupled, parallel coupled, interdigital, and so forth. Hairpin has a relatively smaller size in comparison with other methods [3]. The selection of element order in hairpin method affects the dimensions and frequency response. The lower the number of element order (less than 5), the more sloping the frequency response and worse the selectivity. By contrast, the higher the number of filter element order (greater than 5), the steeper the frequency response and the bigger the filter dimensions [3].

In the research of hairpin microstrip bandpass filter, the addition of open stub in the first and last resonators may narrow the bandwidth to meet the desired specification [3]. Besides, to suppress the harmonics to increase the stop-band rejection without affecting the working frequency and insertion loss in bandpass filter, DGS (Defected Ground Structure) is added [4]. The selection of DGS may also increase the value of return loss [5].

The advancing development has increased the need for small, integrated passive component (filter). A component in large size physically will take larger space, thus attention should be put on minimizing filter dimensions to enable easy application of the filter.

In this research, filter is designed by using six order hairpin method and using open stub that can narrow the filter bandwidth as well as DGS to increase the stop-band rejection. The center frequency of this bandpass filter is 9.5 GHz. The simulation is carried out using 3D electromagnetic simulator software to figure out the performance of the filter by testing the necessary parameters.

2. Filter design

There are several key parameters used as a reference in BPF design. The specification of the BPF designed is expected to meet the parameters as shown in Table 1. The type of substrate used is rogers RT5880. Table 2 shows the specification of the substrate rogers RT5880.

Table 1. Specification of BPF designed

No.	Specification	Values
1.	Working frequency	9.45 – 9.55 GHz
2.	Bandwidth -3 dB	100 MHz
3.	Centre frequency	9.5 GHz
4.	FBW (Fractional BW)	0.01
5.	Insertion loss	> -2 dB
6.	Return loss	< -10 dB
7.	Matching Impedance	50 ohms
8.	VSWR	1<VSWR<1.5
9.	Filter poles	6
10.	Frequency response	Chebyshev

The initial stage of filter design is creating BPF using hairpin method with six resonators as shown in Figure 1. In the design of this bandpass filter, the dimensions of the filter should be obtained first. The dimensions of the filter calculated such as the width of resonator by using equations (1) and (2), the length of resonator by using equations (3) – (5), and the distance between resonators by using equations (6) and (7).

Table 2. Specification of the Substrate Rogers RT5880

Specification	Value
Type of Substrate	Rogers RT5880
Dielectric Constant (ε_r)	2.2
Dielectric Loss Tangent (δ)	0.0009
Thickness of Substrate (h)	1.575 mm

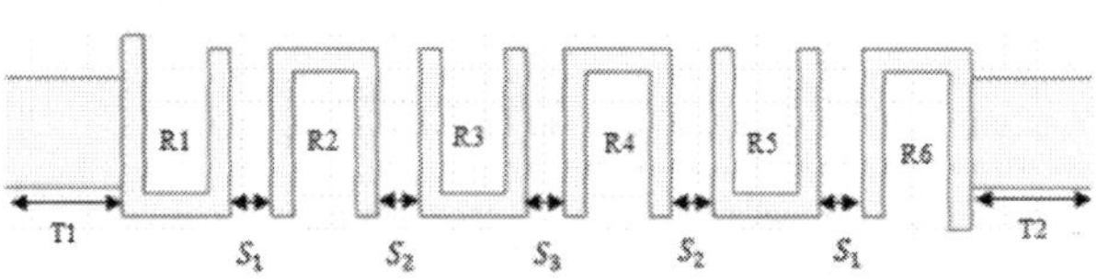

Figure 1. Initial design of bandpass filter

To find the width resonator line, the following equation is used:

$$\frac{W}{h} = \frac{2}{\pi}\left\{(B-1) - In(2B-1) + \frac{\varepsilon_r - 1}{2\varepsilon_r}\left[In(B-1) + 0.39 - \frac{0.61}{\varepsilon_r}\right]\right\} \quad (1)$$

with

$$B = \frac{60\pi^2}{Z_c\sqrt{\varepsilon_r}} \quad (2)$$

thus,

$$B = \frac{60\pi^2}{Z_c\sqrt{\varepsilon_r}} = \frac{60\pi^2}{50\sqrt{2.2}} = 7.9825$$

$$\frac{W}{h} = \frac{2}{\pi}\left\{(7.9825-1) - \mathrm{In}(15.965-1) + \frac{2.2-1}{2(2.2)}\left[\mathrm{In}(7.9825-1) + 0.39 - \frac{0.61}{2.2}\right]\right\} = 3.08$$

Since W/h > 2, the width of resonator line is obtained as follows:

$$W = \frac{W}{h} \times h = 3.08 \text{ x } 1.6 \text{ mm} = 4.9 \text{ mm}$$

To obtain the length of resonator, equation (3) is used.

$$L = \frac{\pi/180^0}{\sqrt{\varepsilon_e K_0}}\, \theta^0 \ (in\ meter) \quad (3)$$

As hairpin bandpass filter uses a line of half of the wave length, $\theta = 180^0$, with

$$\varepsilon_e = \frac{\varepsilon_{r+1}}{2} + \frac{\varepsilon_{r-1}}{2}\sqrt{\frac{1}{1+12\left(\frac{h}{W}\right)}} \quad (4)$$

$$K_0 = \frac{2\pi f}{c} \quad (5)$$

f is the center frequency of bandpass filter, while c is the speed of light in the air, with $c = 3 \cdot 10^8$ m/s.
Therefore,

$$\varepsilon_e = \frac{2.2+1}{2} + \frac{2.2-1}{2}\sqrt{\frac{1}{1+12\left(\frac{1.6}{4.9}\right)}} = 1.87$$

$$K_0 = \frac{2\pi\ x\ (9.5)x10^9}{3\ x\ 10^8} = 198.96\ m^{-1}$$

Thus, the length of hairpin resonator line (L) can be found by using equation (3).

$$L = \frac{\pi/180^0}{\sqrt{1.87\ x\ 198.96}}\, 180^0\ m = 0.0115\ m = 11.5\ mm$$

The design of the filter is conducted by using low pass filter parameters based on Chebyshev table with ripple of 0.01 dB as the design parameter to get coupling coefficient [6] – [9]. It is used to determine the distance between hairpin resonator. Based on Chebyshev table, the six-order filter have the following parameters.

g1 = 0.7814	g2 = 1.3600
g3 = 1.6897	g4 = 1.5350
g5 = 1.4970	g6 = 0.7098
g7 = 0.9085	

By using the parameter values above, the coupling coefficient between resonators can be calculated using the following equations [7]:

$$K_n = \frac{\mathrm{FBW}}{\sqrt{g_n x g_{n+1}}} \quad (6)$$

IOP Conf. Series: Materials Science and Engineering **306** (2018) 012124 doi:10.1088/1757-899X/306/1/012124

$$K = \frac{f_H^2 - f_L^2}{f_H^2 + f_L^2} \quad (7)$$

As FBW is the fractional bandwidth of BPF, g_n is the low pass filter parameter based on the six order Chebyshev table.

$$K_n = \frac{FBW}{\sqrt{g_n x g_{n+1}}}$$

$$FBW = \frac{112.8}{9500} = 0.01$$

$$K_{12} = \frac{0.01}{\sqrt{0.7814 \, x \, 1.3600}} = 0.0097$$

$$K_{23} = \frac{0.01}{\sqrt{1.3600 \, x \, 1.6897}} = 0.0065$$

$$K_{34} = \frac{0.01}{\sqrt{1.6897 \, x \, 1.5350}} = 0.0062$$

$$K_{45} = \frac{0.01}{\sqrt{1.5350 \text{ x } 1.4970}} = 0.0065$$

$$K_{56} = \frac{0.01}{\sqrt{1.4970 \text{ x } 0.7098}} = 0.0097$$

$$K_{67} = \frac{0.01}{\sqrt{0.7098 \text{ x } 0.9085}} = 0.0124$$

$$K_{12} = K_{56}$$

$$K_{23} = K_{45}$$

After the coupling coefficient has been obtained, the next step is finding out the distance between resonators that matched with the coupling coefficient as calculated above. To find it, a simulation is carried out by observing S21 chart from the simulation results of cut-off frequencies, namely lower and upper cut-off frequencies.

The calculation conducted yields results: resonator width = 3.5 mm, resonator length = 4.7 mm and the distance between resonators = 1.2 mm. The next step is conducting simulation and optimization of the dimensions of the filter until the parameters of BPF match or near with the desired specification is obtained.

3. Initial Simulation

In this stage, the initial simulation of BPF is conducted, with results of dimensions obtained from literature study. The parameters of the filter without open stub and DGS can be seen in Table 3.

Table 3. Filter Initial Parameters

Parameters	Dimensions
L	12 mm
Lr	4.7 mm
Lt	4 mm
W	35 mm
Wr	3.5 mm
Wt	4.9 mm
Wsr	0.7 mm

The simulation results do not show a good Band Pass Filter response shape, in which a considerable number of ripples are found, making the value of bandwidth unknown. Figure 2 and 3 show the initial design of the filter and results of the initial simulation based on calculation, respectively.

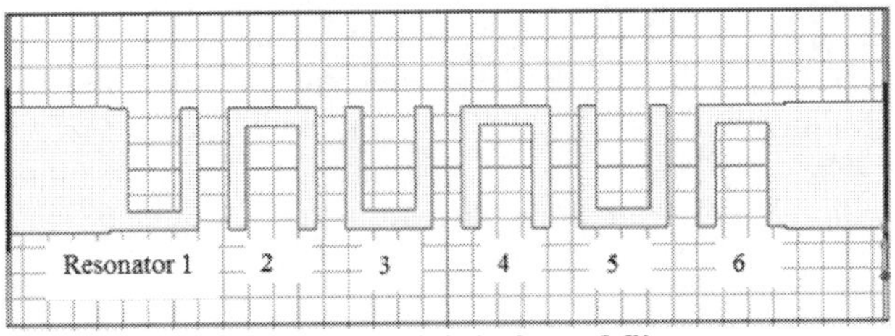

Figure 2. Initial design of filter

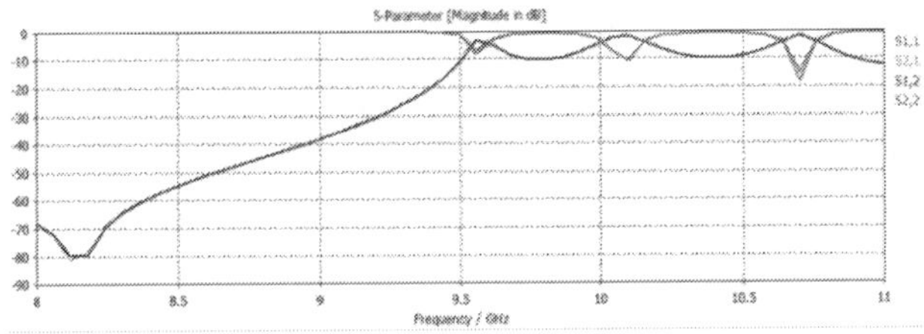

Figure 3. Results of initial simulation

As the results of the initial simulation do not meet the expected specification, optimization is conducted by changing related parameters such as the distance between resonators (S), length of resonator (Lr), width of tap (Wt), DGS dimensions and length of open stub.

The results of the simulation based on the calculation do not show any suitable bandpass filter, so another simulation is conducted to obtain results meeting the specification. After conducting an analysis, this can be overcome by changing the distance between resonators 3 and 4 into zero mm, or in other words, the distance between resonators 3 and 4 is eliminated, and the sizes of the substrate and ground are also changed according to the reduction of distance between resonators 3 and 4. Figure 4 shows the design of change of distance between resonators 3 and 4, while Figure 5 shows the results of the simulation of parameter S from the change of distance between resonators 3 and 4.

Optimization of distance between resonators can also shift the working frequency and change the bandwidth. The larger the distance between resonators, the higher the working frequency and the narrower the bandwidth, or vice versa.

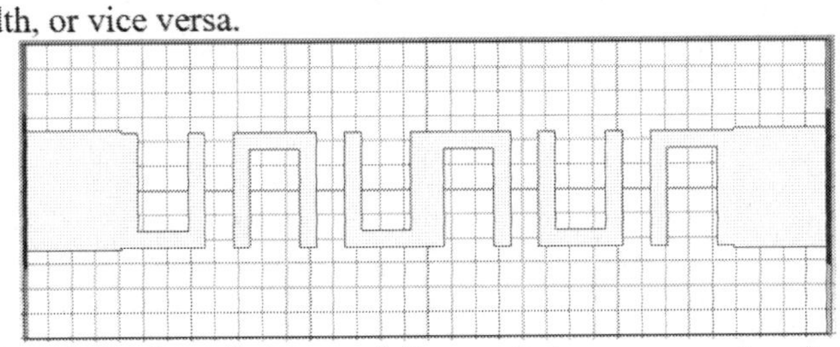

Figure 4. Design of change of distance between resonators 3 and 4

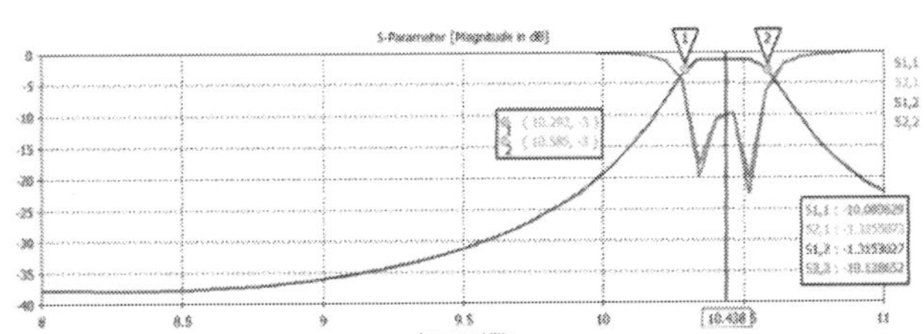

Figure 5. Results of simulation of parameter S, optimized distance between resonators 3 and 4

Figure 6 shows that the shorter the length of the resonator, the higher the working frequency of the filter. On the other hand, the longer the length of the resonator, the lower the working frequency of the filter.

Table 4 shows the resulted difference after optimization is carried out on the length of tap. It is different in that the lower the width of tap, the higher the value of impedance.

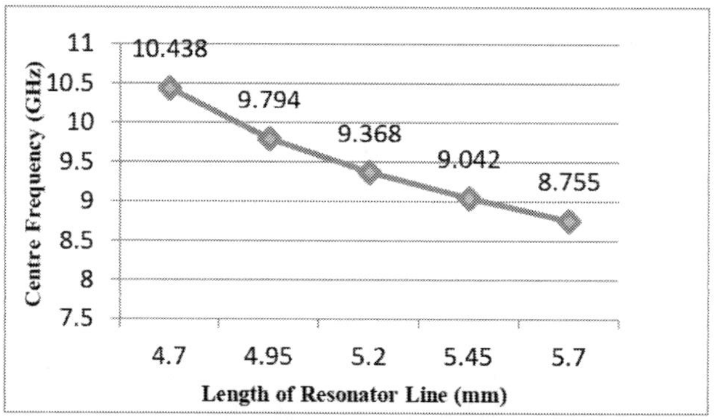

Figure 6. The effect of length of resonator line on working frequency

Table 4. Results of Optimization of Tap Width

W_t (mm)	Impedance (ohm)
4.9	38.14
4.4	43.05
3.9	48.19
3.4	53.86
2.9	60.40

The results of the optimization of DGS are shown in Table 5, namely the return loss value of BPF with DGS is lower than the return loss of BPF without DGS. Figure 7 shows that the longer the open stub, the narrower the bandwidth, and vice versa.

Table 5. Comparison of S11 in the center frequency

BPF	S11 (Return Loss) at Center Frequency
Without DGS	-11.70
With DGS	-20.58

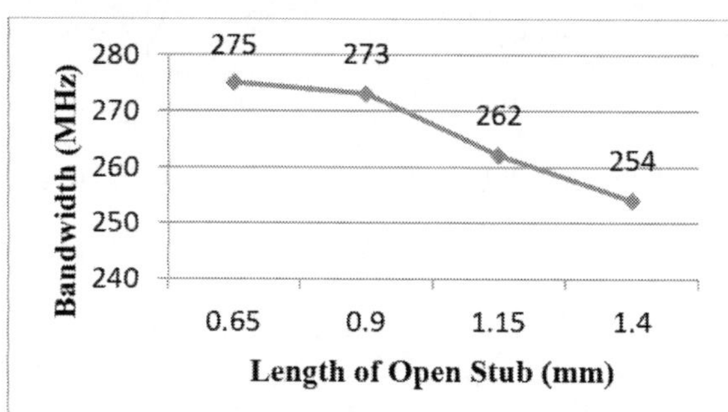

Figure 7. The effect of open stub length on bandwidth

4. Results and discussion

After calculating the dimensions of the filter, and then some optimization is conducted on the dimensions of the bandpass filter, the dimensions of BPF meeting the desired specification are achieved. After the dimensions are optimized several times, the final dimensions of bandpass filter are obtained. Table 6

IOP Conf. Series: Materials Science and Engineering **306** (2018) 012124 doi:10.1088/1757-899X/306/1/012124

shows the change in dimensions after and before optimization. The final design of BPF for front and back views are shown in Figures 8 and 9, respectively.

Table 6. Change in Dimensions Before & After Optimization

Parameters	Dimensions Before Optimization	Dimensions After Optimization	Description
L	12 mm	12 mm	Length of Substrate and Ground
L_r	4.7 mm	5.23 mm	Length of Resonator Line
L_t	4 mm	4 mm	Length of Tap
W	35 mm	35.76 mm	Width of Substrate and Ground
W_r	3.5 mm	3.5 mm	Width of Resonator
W_t	4.9 mm	3.47 mm	Width of Tap
W_{sr}	0.7 mm	0.7 mm	Width of Resonator Line
h	1.6 mm	1.6 mm	Thichness of Substrate
S_f	2.1 mm	2.1 mm	Slide factor
s	1.2 mm	1.69 mm	Distance between Resonators
L_1	-	7 mm	Length of DGS
L_2	-	0.5 mm	Lengtho of DGS Line
W_1	-	0.7 mm	Width of DGS
W_2	-	0.5 mm	Width of DGS Line
S_1	-	0.5 mm	DGS Distance from SMA connector
L_{os}	-	0.45 mm	Length of Open Stub

After optimization is conducted on the distance between resonators (s), the length of resonator line (Lr), width of tap, DGS size and length of open stub, the results of final simulation of bandpass filter are obtained as shown in Figure 10.

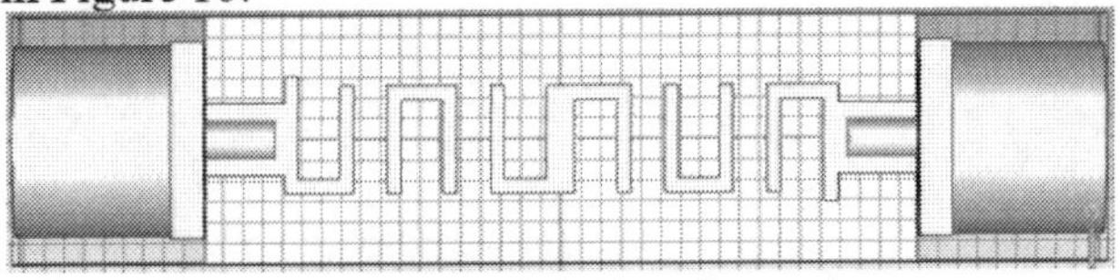

Figure 8. BPF final design, front view

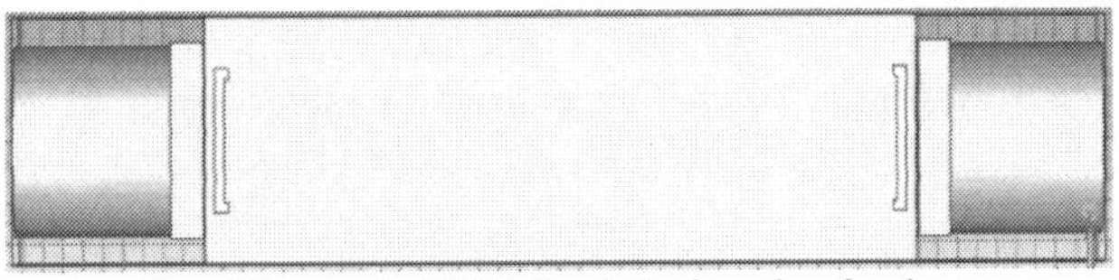

Figure 9. BPF final design, back view

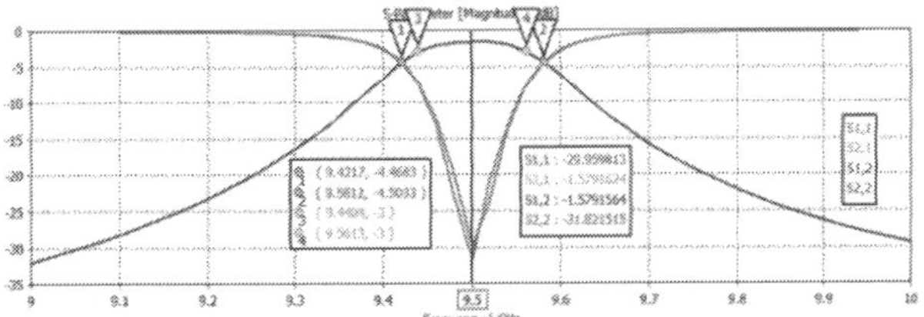

Figure 10. Final Bandpass Filter Simulation Results

Figure 10 shows the final results of bandpass filter simulation. From the simulation obtained that the return loss (S11) at the 9.5 GHz is -29.95 dB and the insertion loss (S21) is -1.57 dB. The BPF working frequency is 9.44 – 9.56 GHz, with bandwidth of 120 MHz. In addition to S11 and S21, the VSWR at a frequency of 9.5 GHz is also found that is 1.06. The results obtained are close to the specification determined in Table 1.

5. Conclusion

To obtain simulation results of parameter S in bandpass form, the distance between resonators 3 and 4 is changed into zero mm, thus eliminating the distance between resonators 3 and 4, or both resonators merged. To shift the working frequency of a bandpass filter, the length of the resonator line is changed. The longer the resonator line, the working frequency shift to the lower frequency. The use of open stub in the first and last resonators of BPF can narrow the bandwidth of the BPF designed. However, in addition to using open stub, narrowing the bandwidth can also be done by increasing the distance between resonators, except for the distance between resonators 3 and 4. The use of DGS may improve the value of return loss due to the difference in the value of return loss before and after the installation of DGS. The final results of bandpass filter simulation shown that filter parameters are close to the specification so that it can be used for x-band radar.

References

[1] D M Pozar 2012 *Microwave and RF Design of Wireless Systems* New York: J. Wiley & Sons, Inc.
[2] D M Pozar 2012 *Microwave Engineering 4th Edition* New York: J. Wiley & Sons, Inc.
[3] Y Fauzi 2012 *Rancang Bangun BPF untuk Aplikasi Radar X-band Menggunakan Resonator Mikrostrip Hairpin dengan Open Stub dan Square Groove* (Thesis) Universitas Indonesia, Jakarta
[4] K Vidhya and T Jayanthy 2011 *Design of Microstrip Hairpin Bandpass Filter Using Defected Ground Structure and Open Stub* IEEE (International Conference on Information and Electronics Engineering). 2011.
[5] Chaudhary G, Y Jeong and J Lim 2013 *Dual-band bandpass filter with independently tunable center frequencies and bandwidths* IEEE Transactions on Microwave Theory and Techniques **61** 1 107-116
[6] S Seghier, N Benabdallah, N Benahmed, F T Bendimerad and K Aliane 2012 Design and Optimization of Parallel Coupled Microstrip Bandpass Filter for FM Wireless Applications *The Computing Science and Technology International Journal* **2** 1 207-211
[7] N Toledo 2001 *Practical Techniques for designing Microstrip tapped hairpin resonator filters on FR4 laminate* in 2nd National ECE Conference, Manila, Philippines
[8] S Y Lee and C M Tsai 2000 New cross-coupled filter design using improved hairpin resonators *IEEE Transactions on Microwave Theory and Techniques* **48** 12 2482 – 2490
[9] G L Matthaei 2003 Narrow-Band, Fixed-Tuned, and Tunable Bandpass Filters with Zig–Zag Hairpin Comb Resonators *IEEE Transactions on Microwave Theory and Techniques* **51** 4

The Effect of Learning Based on Technology Model and Assessment Technique toward Thermodynamic Learning Achievement

T Makahinda*
Universitas Negeri Manado, Tondano 95618, Sulawesi Utara, Indonesia

* tinekemakahinda6@gmail.com

Abstract: The purpose of this research is to find out the effect of learning model based on technology and assessment technique toward thermodynamic achievement by controlling students intelligence. This research is an experimental research. The sample is taken through cluster random sampling with the total respondent of 80 students. The result of the research shows that the result of learning of thermodynamics of students who taught the learning model of environmental utilization is higher than the learning result of student thermodynamics taught by simulation animation, after controlling student intelligence. There is influence of student interaction, and the subject between models of technology-based learning with assessment technique to student learning result of Thermodynamics, after controlling student intelligence. Based on the finding in the lecture then should be used a thermodynamic model of the learning environment with the use of project assessment technique.

1. Introduction

Creating learning process in order to optimize learners' learning outcomes is the task of an educator / lecturer. Teaching learning strategies is one important aspect of the consistently claimed promotion of self-regulated learning in classrooms [1]. One of the determinants of the achievement of learning outcomes is the ability of lecturers in designing (designing), applying teaching materials and organizing lectures. Learning is how to build and maintain productive professional relationships with the people in one's care is no simple matter, yet many assume that this is a natural rather than learned capacity [2]. Teacher is always dissatisfied with the learning outcomes achieved by the learners, so the result of the assessment is always reviewed to seek new businesses and ways of teaching to obtain better learning outcomes of learners. Effective instructors commonly pride themselves on having positive student interactions in and out of the classroom, provide prompt feedback, and encourage teamwork amongst students [3].

Problem of learning in thermodynamics at the Department of Physics FMIPA UNIMA is that there is no laboratory for physical phenomena of the laws of thermodynamics. The learning process of thermodynamics is only emphasis on theory, not practicum. The model of environmental utilization learning (technology product) and simulation animation is a learning model / method that involves the students taking an active role. Environmental empowerment (technology product) is an approach / model of learning to increase the involvement of learners through the utilization of the environment (technology products) as a source of learning. This learning model assumes that learning activities will attract the attention of learners if what is learned is removed from the environment [4].

The model of thermodynamic with the utilization of the environment is intended: to provide an opportunity for students to observe the existing symptoms in the surrounding environment related to the concept of the laws of Thermodynamics. The model with the student technology-based learning model involved two stages of the learning process are: learning in the classroom and learning outside the classroom (service garage, in PLTP, etc) supported by the lecture activities.

The quality of the assessment system in learning process cannot be separated from the use of assessment techniques, because the use of inappropriate assessment techniques will greatly affect learners, both in terms of learning motivation, or interest in learning that will ultimately affect the learning outcomes for the coming students. Assessment can help learning is not new, but what is new is a growing body of evidence that suggests that attention to what is sometimes called formative assessment, or assessment for learning, is one of the most powerful ways of improving student achievement [5][6]. Clearly stated learning intentions are an essential component of formative assessment strategy. They help teachers to be mindful of what their goals are to effectively plan and deliver lessons and they facilitate student learning by communicating expectations about the desired outcomes for each lesson [7]. Environmental Impact Assessment (EIA) is a process to assess the environmental consequences of any project and design proper mitigation plans to minimize the possible adverse impacts [8][9].

A portfolio is a collection or option file that can provide information for an assessment. Characteristics of student portfolio reflects important changes in a student's intellectual process. The portfolio is a systematic collection of one's work. In portfolio education refers to a systematic collection of student work [10]. Portfolio is a collection of student work that can show them (also for others) for their efforts, progress, and achievements in a particular field of study [11].

Factors of learning success in addition to learning models and assessment techniques. There are other factors that contribute to determine the success of learning from learners that is Intelligence. Intelligence is a mental ability that involves rational thought processes of learners or in other words intelligence is the capacity or skills of learners to be able to adjust his mind to the situation / situation faced by learners.

This study aims to determine the effect of technology-based learning model and assessment techniques on thermodynamic learning outcomes after controlling student intelligence, and the influence of their interaction on thermodynamic learning outcomes of third semester students Department of Physics FMIPA UNIMA.

2. Methods

Method used in this research is the experimental method with factorial design and dependent variable is the learning result of student thermodynamics.

Table 1. Research design

Assessment Technique	Learning Model (A)	
	Environmental Utilization (Technology Product) (A_1)	Animation Simulation (A_2)
Project Assessment (B_1)	$(X,Y)_{11k}$ $k=1,2 .. n_{11}$ (A_1B_1)	$(X,Y)_{21k}$ $k=1,2 .. n_{21}$ (A_2B_1)

Population in this research is Physics student of FMIPA UNIMA academic year 2011/2012 with the target of population is the third semester students of Physics education FMIPA UNIMA number of 4 classes. Sample of this research is 107 students. Sampling is done by cluster random sampling. Instrument for this research is validated through a panelist consisting of 17 items of the description made recommended 15 items based on the calculation of Aiken's validity index and obtained a reliability coefficient (Hyot) of 0.92. It given to 20 students who are not treated by reliability coefficient (alpha Cronbach) of 0.95.

Measurement of student intelligence was conducted by Department of Psychology UNIMA with CIFT (Culture Fair Intelligence Test) scale 3A and B with permission no. 144 / BKIN / PP / 2006. Data analysis techniques consists of: descriptive analysis, prerequisite test analysis (including: normality test, homogeneity test, linearity test, significance test, line alignment test), inferential analysis for hypothesis test is covariance analysis (ANKOVA).

3. Result and discussion

3.1. Result

Score of learning outcomes of thermodynamics of students taught by model of environmental utilization learning as follows with Number of respondents 40 as follows: minimum score 85 and maximum score 146 so the data range 146 - 85 = 61, average 122.1, standard deviation 14.55. Further data are presented in the form of frequency distribution tables with many classes using Sturgess in can 7, interval class length 10. These results are presented in the form of a histogram graph (Figure 1).

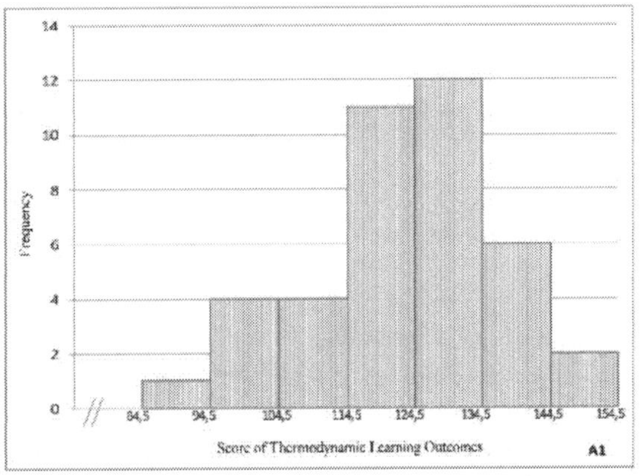

Figure 1. Histogram Score of Thermodynamics Learning Outcomes Students Taught Learning Model Environmental Utilization

Score of learning result of thermodynamic student taught by simulation animation learning model with number of respondent 40 as follows: minimum score 80 and maximum score 135 so the data range 135 - 80 = 55, average score 114,55, standard deviation 12,35. Further data is presented in the form of frequency distribution table with many classes 7, interval class length 9. Results are presented in Histogram (Figure 2).

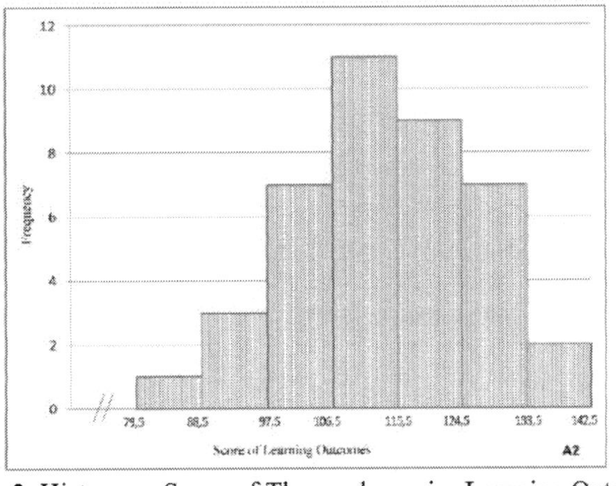

Figure 2. Histogram Score of Thermodynamics Learning Outcomes Students Taught by Simulated Learning Animation Model

Score of learning outcomes of thermodynamics of students who were given project assessment with the number of respondents 40 as follows: minimum score 80 and maximum score 146 so the data range 146 - 80 = 66, average score 121.15, standard deviation 14.93. Furthermore, data is presented in the form of frequency distribution table with many classes 7, interval class length 11 (Figure 3).

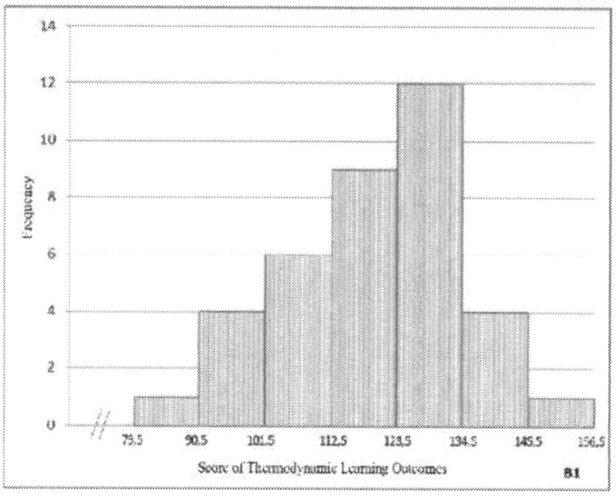

Figure 3. Histogram Score of Thermodynamics Learning Outcomes Student Assessed Technical Assessment Project

Scores of student learning outcomes Thermodynamics are taught by model of environmental utilization learning with project assessment (A1B1). with the number of respondents 20 as follows: minimum score 96 and maximum score 146 so the data range 146 - 96 = 50, average score 128.4, standard deviation 12.68. The data are then presented in the form of frequency distribution tables with many classes of 6, the interval class length 10 (Figure 4).

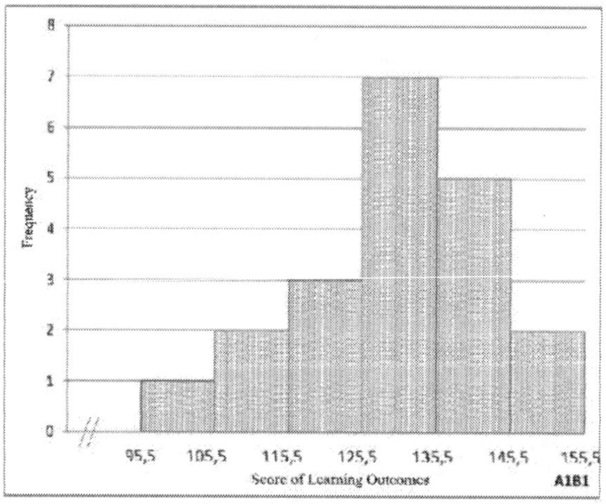

Figure 4. Histogram Score of Thermodynamics Learning Outcomes Students Taught by Environmental Utilization Learning Model

3.2 Discussion

The results showed that the learning outcomes of thermodynamics between students taught by the model of learning environment utilization is higher than the learning result of students' thermodynamics taught by simulation animation learning model, after controlling student intelligence.

In the thermodynamic lecture model of environmental utilization model is very helpful for students in understanding the concept and physical principles. Lecture activities with the utilization of the environment can serve as a medium of learning but also as an object of study. Some objects in our environment can be utilized as learning resources, whether used directly or designed first.

Indispensable creativity and innovation of lecturers so as to utilize the environment as a source of learning, because learning is essentially an interaction between individuals and the environment. The environment provides stimulus (individual stimulus) and the individual responds to the environment. In the interaction process it can happen changes in the individual in the form of behavior change, where learners get information based on direct experience, therefore teaching will become more meaningful and interesting and become more concrete.

Model of learning with the utilization of the environment, the application of science in daily life becomes easier and in accordance with the problems faced by learners, because the environment around us are able to demonstrate the concept and principles of Physics and can involve students develop motivation and the principle of "learning how to learn "Based on the scientific method and the development of process skills that require an understanding born from direct observation of the phenomena or thermodynamic phenomena present in the environment.

The result of the research shows that students' thermodynamic learning outcomes assessed by project assessment have high scores, after controlling student intelligence. Assessment of the project as one of the class assessment, which has a scientific procedure that essentially is observation, collection of information or data from what is observed, analyzing data, and concluding. These stages are very relevant to the procedure in understanding the concepts of thermodynamics, so it will greatly help the process of discovery or construction of knowledge and understanding of the concepts. The results of the research analysis showed that the students' learning achievement of the thermodynamic class of students who taught the model of environmental utilization learning with the project assessment technique has a high result. Learning Thermodynamics has material characteristics that contain phenomena or phenomena that are closely related to events related to the natural environment, therefore as a good lecturer should choose the model of learning and assessment techniques in accordance with the characteristics of the material and characteristics of learners. The combination of learning model of environmental utilization with project assessment technique was very influential on student learning outcomes in the learning of thermodynamics. Learning of the use of the environment enriches the insight and knowledge of learners because the environment is a learning resource that can be used as a natural laboratory that can demonstrate the concept and principles of thermodynamic physics, while the project assessment techniques are very appropriate where students are trained to think scientifically in carrying out tasks in the form of projects, and the formation of skills in students about knowledge, attitude and skills.

4. Conclusion

The result of learning of thermodynamics of students who taught the learning model of environmental utilization is higher than the learning result of student thermodynamics taught by simulation animation, after controlling student intelligence. There is influence of student interaction, and the subject between models of technology-based learning with assessment technique to student learning result of Thermodynamics, after controlling student intelligence. This shows that between learning model and assessment technique is very influential to each other and cannot be separated in the learning process. This means that in choosing the learning model should be tailored to the appropriate assessment techniques by looking at the characteristics to be taught. Students 'thermodynamic learning outcomes given the project assessment techniques have high results, as well as in the group of students who were given the project assessment techniques, the students' thermodynamic learning outcomes that were taught the environmental utilization learning model (A1B1).

References

[1] S Kistner 2015 Teaching learning strategies: The role of instructional context and teacher beliefs *Journal for Educational Research* **7** 1 176–197

[2] D Lowenberg 2010 The Work of Teaching and the Challenge for Teacher Education *Journal of Teacher Education* **60** 497
[3] A Paolini 2015 Enhancing Teaching Effectiveness and Student Learning Outcomes *The Journal of Effective Teaching* **15** 1 20-33
[4] A Poedjiadi 2002 *Community Technology Science* Bandung: Remaja Rosdakarya
[5] D Wiliam 2013 *Assessment: The Bridge between Teaching and Learning* National Council of Teacher English. Voices from the Middle **21** 1
[6] C Stringher 2016 Assessment of Learning to Learn in Early Childhood: An Italian Framework *Italian Journal of Sociology of Education* **8** 1 102-128
[7] Hanover 2014 *The Impact of Formative Assessment and Learning Intentions on Student Achievement* Hanover Research, District Administration Practice
[8] A Pratibha 2015 Environmental Impact Assessment (EIA)-Decision Making Tool
[9] V K Tiwari 2016 A review on Environmental Impact Assessment of Construction Projects *IOSR Journal of Environmental Science, Toxicology and Food Technology (IOSR-JESTFT)* **10** 1
[10] W J Popham 1995 *Classroom Asesment What Teachers Need to Know.* Boston: Allyn and Bacon
[11] Genesee, Fred and John A. Upshur 1997 *Classroom-Based Evaluation in Second Language Education* Cambridge: Cambridge University Press

IOP Conf. Series: Materials Science and Engineering **306** (2018) 012126 doi:10.1088/1757-899X/306/1/012126

Instructional Package of Development of Skill in Using Fine Motor of Children for Children with Intellectual Disabilities

T Sangsawang

Rajamangala University of Technology Thanyaburi

sthosporn@rmutt.ac.th

Abstract. This research has the following purposes: 1) to find the efficiency of the self-learning activity set on development of skill in using fine motor of children with intellectual disabilities., 2) to compare the abilities to use the small muscles after the study more than before the study of children with intellectual disabilities, who made study with the self-learning activity on development of small muscles use., 3) to study the satisfaction of the children with intellectual disabilities using the self-learning activity on development of small muscles use. The sample groups on the research are the children with intellectual disabilities of the special education Maha Chakri Sirindhorn Provincial Nakhon Nayok Center in the school year 2016, for 7 children. The tools used on the research consist of the self-learning activity on development of small muscles use for the children with intellectual disabilities of the special, the observation form of abilities of small muscles before and after using the activity set and the observation form of satisfaction of the children with intellectual disabilities of the special towards the self-learning activity set on development of small muscles for the children with intellectual disabilities of the special. The statistics used on the research include the percentage, mean value, standard deviation and the t-test for dependent sample. From the research, it was found that the self-learning activity set on development of small muscles use for children with intellectual disabilities of the special is efficient based on the criteria in average equal to 77.78/76.51, the educational coefficient of the student after the study higher than before the study with average points before the study equal to 55.14 and S.D. value equal to 3.72. The average points after the study equal to 68.86, S.D. value equal to 2.73, t-test value before and after the study equal to 7.94, which are different significantly on statistics at the level 0.05 and the satisfaction observation form of the student towards the self-learning activity on small muscles use for he down syndrome children with average value equal to 4.58 in the considerable level.

1. Introduction

The instructional package can be change behaviours and solving problem, activities embedded in contextualized, or real-life, scenarios; use of manipulative techniques to reinforce concepts; and a self-monitoring component, a critical behaviour management component to instructional interventions for students. Its focus, to the effectiveness of each approach to enhance the learning performances of students. [1] The development of skill in using fine motor of children for Down Syndrome Children has been for educational use in the form of Down syndrome use of Down syndrome as an educational tool has provided learners and educators with a wider range of new and interesting learning experiences and teaching environments that are not possible in traditional in-class education. Down syndrome has been developed mainly by instructional designers using traditional instructional design models such as instructional systems design, cognitive flexibility theory and constructivist learning environments [2]. However, many of these approaches still lack two important considerations needed for implementing learning based on the instructional package: 1) integration of the learner participant with instructional design and social context, 2) development of a process and evaluation framework to improve the overall quality of Down syndrome. This study addresses these two weaknesses from the perspective of the overall quality of Down syndrome, education environment. This research aims to synthesize an instructional design framework from mental process, learning by doing approach, and social context, and to create an instructional design framework for special education. According to Thailand's National Education Act 1999, 'In organizing the learning process, educational institutions and agencies concerned shall provide training in thinking process, management, how to face various situation and application of knowledge for obviating and solving problems'. This is to make Down syndrome, Thai learners capable of developing their own self and able to compete with the world's knowledge-based economy. Moreover, section 66 also reads 'learners shall have the right to develop their capabilities for utilization of technologies for education as soon as feasible so that they shall have sufficient knowledge and skills in using these technologies for acquiring knowledge themselves on a continual lifelong basis'. The aim is to make quality of Down syndrome, Thai learners capable of using technology to seek knowledge. [3] Although quality of Down syndrome in the past relied heavily on teachers in the classroom to guide their study, learning demands greater self-regulation and discipline from learners. As such, teachers need to come into center stage to guide their learners, and such action might be different from the ideals of supervision and facilitation of the past [4]. To reflect on the current issue about the learning which requires self-regulation from learners, researchers need to think of ways which can comply with the learner-based pedagogy in the Thai education environment with the hope of developing an instructional model for teachers and educators in dealing with such issue. It is a fact that any instructional design model is intended to prepare learners for the world of work and 'link[s] competence and competency to performance' Both competence and competency are derived from professional standards which are essential to what learners are required to do. Courses in education provide Down syndrome with opportunities and experience in hands-on learning [5]. When instructional package is introduced to quality of Down syndrome education, practical or hands on elements are transformed into conceptual understanding [6]. Students can neither see nor touch the elements of their subject. When teachers introduce abstract concepts into vocationally based learning [7] there can be erroneous notions about the concepts taught in class when only traditional materials are available [8].

Therefore, instructional package enhanced and student-centered learning environments can facilitate the learning and understanding of abstract concepts due to the fact that students can notice graphically displayed changes of concrete experience [9]. Effective instructional packages environments allow students to work socially with each other. To achieve it, the core elements are the presence of the teacher, the availability of resources and the purposive interaction with media by students [10]. Although general education courses are widely used in all educational levels, it should be noticed that quality of Children with intellectual disabilities education needs both academic and practical approaches. Each approach requires different instructional design frameworks. There are three frameworks of instructional packages which will be discussed in this paper: mental processes, learning by doing approach, and social context. With these three models as a guide, a questionnaire was developed to submit to educational experts. Based on the Delphi technique, these three methods were synthesized in order to create a new method for vocational training, which will be applied to the teaching of vocational [11]. The researcher synthesized the three theories mentioned above by using Delphi technique so that experts were able to express their opinion on what they agreed in order to create an instructional framework that complies with psychological principles according to instructional design framework for autonomy. This framework will help learners achieve their learning objectives effectively and efficiently, as well as help learners to understand in a faster and more stable way. [12] Theoretical frameworks provide a myriad of ways in which Instructional Package-based educational practicums may be utilized, based upon the instructional and pedagogical needs of the vocational course. The development of instructionally effective learning environments that meet these pedagogical needs requires the application of appropriate instructional design principles. Upon designing an instruction, the underlying pedagogical philosophy and application of learning theories, including constructivism and constructionism, will influence decisions regarding what instructional strategies may be adopted [4]. Furthermore, the relationships between the design process framework and the effectiveness of the media-based environment will be discussed.

2. Review of the Literature

This study develops a framework based on three theories which include learning theories in mental processes, learning by doing approach, and social context. Each of the theories involves the psychology learning as following.

2.1. Mental Processes theories

Many experts focus on mental processes and consider the ways in which thoughts occur to the ways in which behaviorism arose from mental processes. This system examines the outcomes or behaviors of learning. Instructional design needs to be applicable to cognitive, behavioral, and attitudinal learning in that the strategies work together to create understanding [14], are applicable [9], for creating cognition and following elements of situated cognition, for building on cognitive or mental phenomena [16], cognitive activity [17], and cognitive organizers [18], and operating on of processes of memory. The mental processes of this instructional design framework are applicable to cognitive, behavioral, and attitudinal learning. Therefore, as an example of a micro theory, it provides a set of procedures to follow for each instructional event to enhance learning. The learning processes identify nine events of instruction, which are tied to cognitive strategies learning theory. Note that the first event in the theory of instructional design framework is 'attention', which parallels that of cognitive strategies learning theory. The next event, 'informing of the objectives' further activates a process of getting the trainees' 'attention' focused. The stimulation of the recall of requisite learning ties in with the activation of memory. The nine events of instruction will be examined in more detail with reference to the relevant sections of social learning theory for additional information [19]. Gagné developed a nine-step process called the events of construction. These details how an instructional event corresponds with an internal mental process in the learner. The interactive content in an instructional package course can keep the learners' attention as described in Gagné's nine events of instruction [20].

Table 1. Gagné's nine events of instruction

Instructional Event	**Internal Mental Process**
1. Gain attention	Stimuli activates receptors
2. Inform learners of objectives	Creates level of expectation for learning
3. Stimulate recall of prior learning	Retrieval and activation of short-term memory
4. Present the content	Selective perception of content
5. Provide "learning guidance"	Semantic encoding long term memory
6. Elicit performance (practice)	Responds to questions to enhance encoding and verification
7. Provide feedback	Reinforcement and assessment of correct performance
8. Assess performance	Retrieval and reinforcement of content as final evaluation
9. Enhance retention and transfer to the job	Retrieval and generalization of learned skill to new situation

2.2. Learning by doing theories

The most popular and effective process of learning ever adopted by psychology and the teaching-learning process is that of learning by doing. In these learning processes the activities are based on practical activities to create skills for occupations or vocations. There is reference to experience in group participant of techniques or technology. There is a focus on constructivist learning theory, the ways of doing and thinking, [26], and humanist activities in which thoughts occur [15]. The literature agrees that learning by doing, which focuses on knowledge construction based on the learner's previous experience, is a good fit for instructional packages [27][28]. Entrenched in the learning processes theories advanced by Dewey (1916), Piaget (1972), and Bruner (1990), constructivism learning theory is defined as the active construction of new knowledge based on a learner's prior experience [28]. Woolfolk states, 'the key idea is that students actively construct their own knowledge: the mind of the student mediates input from the outside world to determine what the student will learn. Learning is active mental work, not passive reception of teaching'. [30] Honebeins advanced a set of goals to aid the design of constructivism in learning settings. There are seven goals. These are: to provide experience with the knowledge construction process; to provide experience in and appreciation of multiple perspectives; to embed learning in realistic and relevant contexts; to encourage ownership and voice in the learning process; to embed learning in social experience; to encourage the use of multiple modes of representation and to encourage self-awareness in the knowledge construction process [29].

2.3. Social context theories

There is focus on constructionism [26]; knowledge-building, situating constructionism, society and situation [31], situated learning, social negotiation, social constructionism and constructivism, and social contexts. These theories draw upon the premise that there must be a social context in order for the learner to build and develop his ideas [33]. An idea must be created. It cannot be transmitted by a teacher, peer, or book. An individual makes an idea. Constructionism then adds to this fundamental notion, that knowledge is constructed more effectively when the individual is engaged in the construction of 'personally meaningful products'. In constructionism, the effect of is as important as gaining knowledge. When placed in the context of vocational Internet-based training, the curriculum must be carefully designed to incorporate the learner and excite in him the desire for the outcome. The training that occurs online should foster a desire to make meaningful online tools. Otherwise, it is argued, the learner will not engage. Socially, an instructional design model is intended to prepare learners for the world of work and 'link[s] competence and competency to performance'. Both competence and competency are derived from professional standards which are essential to what learners are required to do. Courses in quality of Children with intellectual disabilities education provide learners with opportunities and experience in hands-on learning [6]; [7]). When instructional packages introduced to quality of Children with intellectual disabilities education, practical or hands on elements are transformed into conceptual understanding. This might be a significant drawback due to the fact that students can neither see nor touch the elements of their subject. There are reports which discuss how teachers introduce abstract concepts into vocationally based learning [8] yet there can be erroneous notions about the concepts taught in class when only traditional materials are available.

3. Methods

This study aims 1) to synthesize an instructional design framework from mental process, learning by doing approach, and social context, and create an Instructional package instructional design framework for quality of Children with intellectual disabilities education. The following data were used to conduct analyses to develop valid and reliable perceived attribute measures. 2) to find the efficiency of the instructional package of development of skill in using fine motor of Down Syndrome Children., 3) to compare the abilities to use the small muscles after the study more than before the study., and 4) to study the satisfaction of Down syndrome children using the instructional package.

Seventeen experts were chosen through the purposive sampling method. Seven experts were qualified in educational psychology and ten in educational technology. They all had a doctoral degree and had worked for over five years in at least the position of Assistant Professor. The sample groups on the research are the Down syndrome of the Wat Bang Choad School, for 7 children.

The operation was done by using three types of tools: 1) Brainstorming, 2) Evaluation and 3) Re-evaluation.

3.1. Brainstorming
Researchers conducted semi-structured interviews with 17 experts for the first round: brainstorming of experts' opinions would be related to the framework developed from mental processes, learning by doing approach and social context. Researchers' analyzed the interviews of the experts' opinions about each idea. The details interview form was in four parts as follows: 'key ideas and principles, teaching - learning activities strategies, teaching-learning environments, stages of the instructional sequence. Researchers synthesized the first round of opinions of the experts using a Liker's five-point rating scale. After that an instructional design framework was prepared covering mental processes, learning by doing, and social context.

3.2. Evaluation
Researchers evaluated the ideas using a Likert five-point rating scale questionnaire for the second round as shown in the evaluation of the 17 experts' ideas on mental processes, learning by doing, and social context concerning an instructional design framework for quality of Children with intellectual disabilities education. Researchers selected the items from the results of questionnaire I. This meant that all key ideas principles, teaching-learning activities strategies, teaching-learning environments, and stages of instructional sequence comprising mental processes, learning by doing, and social context were pooled together as similarities or differences. The similarities meant that most of the 17 experts agreed while the differences meant the reverse. The results of the synthesis of similarities and differences led the researchers to develop a diagram chart Then, the 17 experts were required to respond "Yes" or "No" to questionnaire II.

3.3. Re-evaluation
Researchers selected the items from the results of questionnaire II. These included all key ideas and principles, teaching-learning activities strategies, teaching-learning environments, and stages of instructional sequence comprised from mental processes, learning by doing, and social context. The findings were pooled together as similarities or differences. The similarities meant that most of the 17 experts agreed while the differences meant the reverse. The results of the synthesis were used to develop questionnaire III. The researchers created an instructional design framework for quality of Children with intellectual disabilities education. The research involved a framework with the sample for the study consisting of 100 instructors who developed instructional packages in the quality of Children with intellectual disabilities education fields of electrical engineering, electronics, civil engineering, and mechanics from several faculties of technical education in universities in Thailand. After the researchers concluded questionnaire III the framework for Instructional Package Instructional Design Framework for Quality of Children with intellectual disabilities education was developed.

4. Data Analysis
An initial study was conducted with the experts and instructors. The survey was on a 5-point Likert-type scale. Data collection was done by questionnaires which were analyzed to determine the results. Data analysis was done using SPSS/FW (Statistical Package for Social Science/for Windows) software. The part with selection items was analyzed using frequency and percentage. The part with five scales was analyzed using mean (X), standard deviation (S.D.) and correlation. The levels of agreement from respondents were as follows: Average Score of 1.00 -1.49 means strongly disagree whereas average score of 4.50-5.00 means definitely agree. Data analysed by using Mind Manager Application. Statistical procedures for data analysis included, the E1/E2, mean, standard deviation, and t-test for dependence samples.

IOP Conf. Series: Materials Science and Engineering **306** (2018) 012126 doi:10.1088/1757-899X/306/1/012126

5. Data Collection

The research conducted in the following steps:

- Review the literature regarding Comparison of Selected Psychology Theories as in Gagne's, Constructivism, and Constructionism.
- Developed a semi-questionnaire regarding Comparison of Selected Psychology Theories as in Gagne's, Constructivism, and Constructionism.
- Collect data and interview using the questionnaire developed.
- Analyze the data and conclude the results (a) the instruments were Instructional package according to GCC model in instructional package according to GCC model. (b) an achievement test and a questionnaire to assess students' opinions toward the Instructional package according to GCC model.
- From the results of this GCC model perspective on instructional package of development of skill in using fine motor of children for Down syndrome children, developers as well as instructors can apply the GCC model to the development of Instructional Package.
- The Down syndrome children using the instructional package of development of skill in using fine motor for improve vocational training.

6. Results and discussion

The results of this research focus on three clusters that create the framework for the theories, and also include aspects of the learning process. The study allowed for the expression of experts' opinions, and similarities and differences could be described in the framework. This is presented in the three parts that follow [33].

- The GCC model orders to increase learner's thinking attributes, Encouraging interests, informing expected to learn results of students, Building conditions to arouse interests to give rise to learning procedure, giving useful advice for learning, finding answers and altering ideas, finding own wit thinking and abilities to transfer knowledge.,[7] These four steps as following: Principle: Creating situations, organization knowledge from experience.(Instructors), Encouraging learners to originate learning. (Instructors), Creating atmosphere of opinions exchanging. (Instructors), Linking students' ideas. (Instructors), the Creating experience of learning new things. (Instructors).
- Stages of teaching-learning activities, as following: a) Activate pre-knowledge on one's own b) Exchange ideas and improving mistakes. c) Building organization knowledge by ways of understanding, memorizing, analyzing and transfer d) Creating ideas, reflexive thinking and initiate thinking.
- Components of teaching-learning management: Learners, Instructors,
- Teaching-learning models; a) Independent Learning. b) Drills and practices. c) Instruction. d) Tests instruction. e) Games education. f) Dependent Learning. g) Collaborative learning. h) Project-based Learning, Group Process Learning. [8]

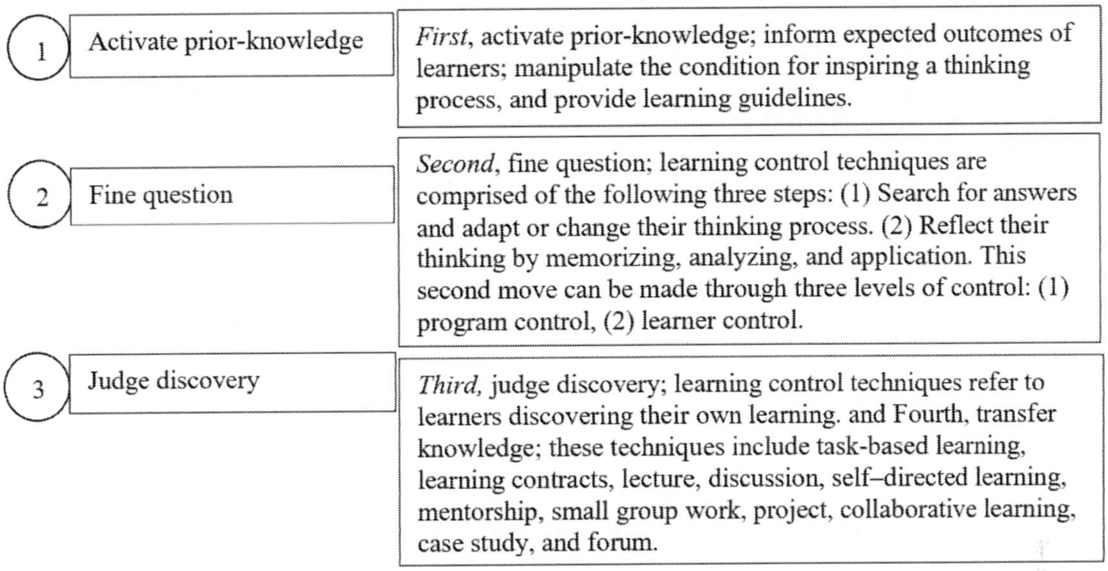

Figure 1. The GCC model. [35]

The process of model synthesis, as well as operant conditioning and learning control techniques which are keys of the new paradigm, will be explained and discussed. The development of skill in using fine motor for improve vocational training is the systematic development of instructional specifications, using learning and instructional theory to ensure the quality of instruction. It is the entire process of analysis of learning needs and goals and the development of a delivery system to meet those needs. It includes the development of instructional materials and activities and tries out an evaluation of all instruction and learner activities. [32], [33]. The development of skill in using fine motor for improve vocational training according to GCC model in Down syndrome children should facilitate Self–Discovery Learning (SDL), because of these reasons, based on experts' opinions of selected psychology theories as in three theories and as shown in figure 3. The GCC model. [14]

In this article, the researchers have offered a framework and design process for the Instructional Package environment. The implementation of the GCC model involved three steps including a consideration of various aspects of information, conceptual development, psychology theories and evaluation of the overall quality of the system environment. In particular, the research aims to improve the design process and usability of the Internet-based environment. The study also confirms that for Instructional Package for Quality of Children with intellectual disabilities education to be successful, various aspects of the online environment should be considered such as the application of domain knowledge, conceptual theory, psychology theories, and evaluation of the overall quality of the design process.

References

[1] C Herron and H York 2006 A Comparison Study of the Effects of a Story-based Video Instructional Package Versus a Text-based Instructional Package in the Intermediate-level Foreign Language Classroom *CALICO journal* **23** 2 281-307

[2] W Dick, L Carey and J O Carey 2005 *The Systematic Design of Instruction (6th ed)*. Boston: Pearson, Allyn and Bacon

[3] The National Education Act B.E. 1999 *Office of the National Education Commission, Constructivism and Constructionism, Bangkok, National Education Act B.E.* 2542 7-179

[4] T Sangsawang, K Jitgarun and P Kaittikomol 2006 *A learning strategy suitable for learners' self-discovery* EDU-COM 2006 Engagement and Empowerment: New Opportunities for Growth in Higher Education, November 22-24, 2006, NongKhai Campus of KhonKaen University, NongKhai, Thailand 619
[5] P Cobb and L P Steffe 2011 The Constructivist Researcher as Teacher and Model Builder. A Journey in Mathematics Education Research *Mathematics Education Library* **48** 1 19-30
[6] C A Conroy and N J Walker 2000 An Examination of Integration of Academic and Vocational Subject Matter in the Aquaculture Classroom *Journal of Agricultural Education* **41** 2 54-64
[7] C Velde and T Cooper 2000 Students' perspectives of workplace learning and training in quality of Children with intellectual disabilities education *Education - Training* **42** 2 83-92
[8] G Attwell 1997 Pressures for change in the education of quality of Children with intellectual disabilities education and training personnel. In A. Brown (ed.) *Promoting Quality of Children with Intellectual disabilities education and Training: European Perspectives* 107-122
[9] M J Marr, E W Thomas, M R Benne, A Thomas and R M Hume 1999 Development of instructional systems for teaching an electricity and magnetism course for engineers *American Journal of Physics* **67** 9 789-802
[10] M J Hannafin and S M Land 1997 The foundations and assumptions of technology-enhanced student centered learning environments *Instructional Science* **25** 167-202
[11] C Powell The Delphi technique: myths and realities *Journal of Advanced Nursing* **41** 4 376–382
[12] B Alexander 2001 Youth organizing comes of age *Youth Today* **10** 3 16–18
[13] R Phillips 2005 Pedagogical, institutional and human factors influencing the widespread adoption of educational technology in higher education *the ascilite 2005 conference Proceedings* 541-549
[14] J Taylor 1996 Piagetian perspectives on understanding children's understanding *Childhood Education* **72** p 259
[15] S Merriam and Caffarella 1991 *Learning in Adulthood. A comprehensive guide* San Francisco: Jossey-Bass
[16] M Lynch and D Bogen 2005 *My memory has been shredded." A non-cognitivist investigation of "mental" phenomena*. In H. te Molder & J. Potter (Eds.), Coversation and cognition: Discourse, mind and social interaction. Cambridge University Press pp 226
[17] B Jetin 2006 The cognitive international division of labour hypothesis: what possible effects for Thai labour and education? *Journal of education* **33** 1-2 1-45.
[18] K Tan, V Dawson and G Venville 2008 Use of cognitive organizers as a self-regulated learning strategy *Educational Research* **18** 2 194
[19] P N Blanchard and J W Thacker 2007 *Effective Training: Systems, Strategies, and practices* (3 rd ed.) Upper Saddle River, NJ: Pearson Education, Inc.
[20] J A Muzio, T Heins and R Mundell 2002 Experiences with reusable Instructional Packages objects: From theory to practice *The Internet and Higher Education* **5** 1 21-34
[21] K Kruse 2002 *Beginner Basics: Gagne's Nine Events of Instruction: An Introduction"*. http://www.e-learningguru.com/articles/art3#.htm.
[22] T Sangsawang 2006 Comparison of Selected Psychology Theories as in Gagne's, Constructivism, and Constructionism *The 4th international conference on developing real-life learning experiences: education reform through Performance-Based Learning*, July 13-14 2006 KMITL, Bangkok, Thailand, p 327-328
[23] R M Gagné 1985 *The Condition of learning* New York: Holt, Rinehart and Winston
[24] Harman & Koohang, 2005; Hung, 2001; Hung & Nichani, 2001; Koohang & Harman, 2005
[25] J Dewey 2005 *How we think*. Boston, MA: Houghton Mifflin Company. Dick, Carey & Carey
[26] J Piaget 1972 *To Understand Is to Invent* New York: The Viking Press Inc. p. 20
[27] J Bruner 1990 *Acts of meaning* Cambridge, MA: Harvard University Press. Case, Harris & Graham

[28] A Koohang, L Riley, T Smith 2008 Instructional packages Constructivism: From Theory to Application *Interdisciplinary Journal of Instructional packages and Learning Objects* **5** 1-109

[29] P Honebein 1996 *Seven goals for the design of constructivist learning environments. In B. Wilson, Constructivist learning environments* New Jersey: Educational Technology Publications 17-24

[30] K Harman and A Koohang 2005 Discussion board: A learning object *Interdisciplinary Journal of Knowledge & Learning Objects* **1** 67-77

[31] A Bandura 2000 *Social foundation of thought and action: A social cognitive theory* Englewood Cliffs, N J: Prentice Hall

[32] Baptista and Fowell 1996 Effective training systems, strategies and practices 3rd *Pearson education international* 198-199

[33] Y Kafai and M Resnick 1996 *Constructionism in practice: Designing, thinking, and learning in a digital world* Mahwah, NJ: Lawrence Erlbaum 71-96

[34] R Likert 1932 *A technique for the measurement of attitudes* Archives of Psychology **140** 1-55

[35] T Sangsawang 2017 Constructivism perspective on multimedia games for hearing impaired children *New Trends and Issues Proceedings on Humanities and Social Sciences* **3** 7 100-108

IOP Conf. Series: Materials Science and Engineering **306** (2018) 012127 doi:10.1088/1757-899X/306/1/012127

Maintenance Policy in Public-Transport Involving Government Subsidy

U S Pasaribu[1], Y Bayuzetra[1], L E Gunawan[1] and H Husniah[2*]

[1] Mathematic, Bandung Institute of Technology
Jalan Ganesa 10, Bandung 40132, Indonesia
[2] Industrial Engineering, Langlangbuana University
Jalan Karapitan 116, Bandung 40261, Indonesia

*hennie.husniah@gmail.com

Abstract. A public transport with government subsidy is considered to encourage the sustainability of the transportation. The transportations revenue is determined by the maximum of the uptimes of the vehicle. In this paper, we study a one-dimensional maintenance policy for new vehicle which is characterized by age parameter. We consider that the degradation of the vehicle is affected by the age of the vehicle, and modelled by using a one-dimensional approach. The owner performs both preventive and corrective maintenance actions, and the preventive maintenance action will reduce the vehicle failure rate and hence it will decrease the corrective maintenance cost during the life time of the vehicle. The decision problem for the owner is to find the optimal preventive maintenance time of the vehicle of each subsidy option offered by maximizing the expected profit for each subsidy.

1. Introduction

As a capital city, Jakarta faces a problem with the increase of air pollution. Transportation contributes 70% pollutant in Jakarta [1]. In order to cope with this condition government tends to promote the use of public transport. Nowadays, to increase the passenger of public transport government introduces subsidy scheme. Since 2016, Indonesia's transportations ministry give subsidy for Damri to supply buses in 11 big cities, that is Medan, Padang, Palembang, Bandar Lampung, Bandung, Yogyakarta, Surakarta, Surabaya, Denpasar, Mataram, and Makassar. With this subsidy people can have other alternative for they transport way.

This paper deals with government subsidy model for Damri. There are two subsidy models that will be studied, that is subsidy for buying buses and for reducing ticket price. We consider a non-cooperative game theory solutions.to maximize the expected profit. In non-cooperative solution, manufacturer act as a leader and take policy to maximize profit first. Damri, as a follower, maximize profit based on manufacturer's policy.

This paper is organized as follows. Section 2 gives model formulation and section 3 gives solution for the model. Section 4 gives case study to see which model is better and finally conclude in section 5.

2. Model Formulation

The following notations will be used in model formulation

- q = Number of passengers per year
- Q = Total number of passengers
- n = Bus demand
- PM = *Preventive Maintenance*
- CM = *Corrective Maintenance*
- K = Bus operating year
- N = Number of PM
- δ = Degree of repair
- $\lambda_0(t)$ = Failure rate without PM
- $\lambda(t)$ = Failure rate after PM
- p = Ticket price
- C_m = Expectation total cost CM
- C_f = Cost per CM
- C_p = Expectation total cost PM
- C_r = Production cost of manufacture
- τ = PM's period
- u = Subsidy amount per year
- w = Bus price
- Ψ_d = Damri's profit
- Ψ_m = Manufacture's profit
- Ψ = Total profit

In this model formulation, assumed that wholesale price of bus is determined by manufacturer and ticket price is determined by Damri. Otherwise, assumed that every failure item of the bus only need minimal repair so that time between failures is negligible [2].

2.1. *Damri's Revenue*

In *ceteris paribus* condition the demand's law said that if the product price increase then demand will decrease and if the product price decrease then demand will increase. In this case, if the number of passengers per year is q and ticket price is p have a linear relation, then the demand function is

$$q(p)=\gamma_0+\gamma_1 p \text{ with } \gamma_1<0 \text{ and } \gamma_0>0 \tag{1}$$

The number of buses per year n can be obtained by dividing q with bus capacity m so that $n=q(p)/m$; $n,m>0$. If Damri operate in K year and the number of passengers per year constant, then the total number of passengers is

$$Q(p,K)=q(p).K \quad ;K>0 \tag{2}$$

Damri's revenue, R_d, is obtained from total passengers multiplied by ticket price

$$R_d(p,K)=Q(p,K).p \tag{3}$$

In this paper, we will use two government subsidy models. First, subsidy for buying bus from manufacturer and the second one is subsidy for reducing ticket price. For simplicity, we use index 1 for first model and 2 for second model function.

For the first model, subsidy doesn't influence demand function so $q(p_1)=\gamma_0+\gamma_1 p_1$. For the second model, subsidy amount u influence demand function so $q(p_2,u)=\gamma_0+\gamma_1 p_2+\gamma_2 u$ where $u,\gamma_2>0$. Thus, based on (3) we have two revenue functions R_d

$$R_d=\begin{cases}(\gamma_0+\gamma_1 p_1)p_1 K & \text{; for first subsidy model}\\ (\gamma_0+\gamma_1 p_2+\gamma_2 u)p_2 K & \text{; for second subsidy model}\end{cases} \tag{4}$$

And we have two number of bus functions

$$n=\begin{cases}\left[\left(\gamma_0+\gamma_1 p_1\right)/m\right]\\ \left[\left(\gamma_0+\gamma_1 p_2+\gamma_2 u\right)/m\right]\end{cases}$$

2.2. Damri's Expense

The main Damri's expense is preventive maintenance (PM), corrective maintenance (CM), and buying buses. Maintenance is purposed to reduce failure intensity.

2.2.1. Preventive Maintenance. Let Damri do N times PM in K-year period. Then, the interval time between two PM τ is formulated by $\tau=\left(K/N+1\right)$ year. According to Kijima-Type 1 model, PM turns age of bus t into virtual age $v(t)<t$. Assumed every PM has the same degree of repair $0\le\delta\le 1$ where $\delta=1$ means minimal repair and $\delta=0$ means perfect repair [3].

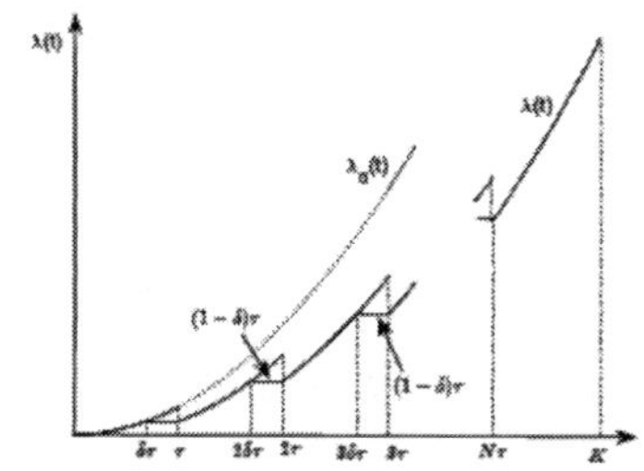

Figure 1. Failure intensity curve after PM

As we see in figure 1, PM reducing failure intensity function become $\lambda(t)<\lambda_0(t)$ and normal age by $(1-\delta)\tau$. As a result (Hamidi et al, 2016), the bus virtual age for $i\tau\le t<(i+1)\tau$ is $v(t)=i\delta\tau+t-i\tau$. So, the failure intensity function of bus will become

$$\lambda(t)=\lambda_0\left(v(t)\right)=\theta\left(i\delta\tau+t-i\tau\right)$$

If every cost PM is $C_p^*=a+(1-\delta)\tau b$ then total cost for N times PM is

$$C_p=Na+N(1-\delta)\tau b=Na+\left[NbK(1-\delta)/(N+1)\right] \quad (5)$$

2.2.2. Corrective Maintenance. While bus may have failure at a random time. When failures occur, bus need to be repaired. In this paper, every failure is assumed minimally repaired so that the failure intensity just the same as that just before the failure. Without any PM, failure is Non-Homogenous Poisson Process (NHPP) with failure intensity $\lambda_0(t)$ [4][5]. After PM, failure process in interval $\left[i\tau,(i+1)\tau\right)$ for $i=0,1,2,\ldots,N$ is still NHPP with intensity function $\lambda(t)=\lambda_0\left(v(t)\right)$ (Kim, et al.,2004). The expected total number of failure is

$$\sum_{i=0}^{N}\int_{i\tau}^{(1+i)\tau}\lambda_0\left(v(t)\right)dt=\sum_{i=0}^{N}\int_{i\tau}^{(1+i)\tau}\theta\left(i\delta\tau+t-i\tau\right)dt=\frac{K^2\theta(N\delta+1)}{2(N+1)}$$

If cost for every PM is C_f, then the expected total amount of PM C_m for K-year operating is

$$C_m=\left[C_f K^2\theta(N\delta+1)/2(N+1)\right] \quad (6)$$

2.2.3. Bus Price. Another A expense for Damri is bus price w. For first subsidy model Damri has government subsidy for buying bus, so that Damri must pay $w_1 < w$. For the second model, Damri must pay $w_2 = w$ for buying bus from manufacturer.

2.2.4. Damri's Profit Function. Profit function is the difference between revenue (4) and expense for PM (5), CM (6), and bus price. Damri's profit function for first subsidy model is given by

$$\begin{aligned}\Psi_{d_1}(p_1, w_1, K, N, \delta) &= R_{d_1}(p_1, K) - (w_1 + C_p + C_m) n_1 \\ &= (\gamma_0 + \gamma_1 p_1) p_1 K - (w_1 + C_p + C_m)\frac{(\gamma_0 + \gamma_1 p_1)}{m}\end{aligned} \tag{7}$$

And for the second model is

$$\begin{aligned}\Psi_{d_2}(p_2, w_2, K, u, N, \delta) &= R_{d_2}(p_2, K) - (w_2 + C_p + C_m) n_2 \\ &= (\gamma_0 + \gamma_1 p_2 + \gamma_2 u) K p_2 - (w_2 + C_p + C_m)\frac{(\gamma_0 + \gamma_1 p_2 + \gamma_2 u)}{m}\end{aligned} \tag{8}$$

2.2.5. Manufacture's Profit Function. If the production cost for every is c_m then manufacture's profit function for the first model is

$$\Psi_{m_1}(p_1, u, w_1, K) = uK + (w_1 - C_r) n = uK + (w_1 - C_r)(\gamma_0 + \gamma_1 p_1 / m) \tag{9}$$

And for second model will be

$$\Psi_{m_2}(p_2, w_2, u) = (w_2 - C_r) n = (w_2 - C_r)(\gamma_0 + \gamma_1 p_2 + \gamma_2 u / m) \tag{10}$$

3. Non-cooperative Model Solution

In the non-cooperative solution, manufacturer will act as a leader and make profit policy first. Damri will act as a follower and make profit policy based on manufacturer's policy [6][7]. In first subsidy model, we determine ticket price p_1 that maximize profit function (7) by differentiating $(\partial \Psi_{d_1} / \partial p_1) = 0$ and $\partial^2 \Psi_{d_1} / \partial p_1^2 < 0$, yields

$$p_1 = \frac{1}{2mK}(w_1 + C_p + C_m) - \frac{\gamma_0}{2\gamma_1} \tag{11}$$

Substitute (11) into (8), we have

$$\Psi_{m_1}(p_1, u, w_1, K) = uK + \frac{\gamma_1}{2m^2 K}(w_1 - C_r)\left(w_1 + C_p + C_m + \frac{\gamma_0 m K}{\gamma_1}\right) \tag{12}$$

To get manufacturer's maximum profit, determine w_1 so that $\partial \Psi_{w_1} / \partial p_1 = 0$ and $\partial^2 \Psi_{w_1} / \partial p_1^2 < 0$, yields

$$w_1 = \frac{1}{2}\left(C_r - C_p - C_m - \frac{\gamma_0 m K}{\gamma_1}\right) \tag{13}$$

Substitute (13) to (12), then we have manufacturer maximum profit is $\Psi_{m_1}(\max) = uK - (\gamma_1 / 8m^2 K) A^2$; $\gamma_1 < 0$ with $A = \left[C_r + C_p + C_m + (\gamma_0 m K / \gamma_1)\right]$. For Damri's profit, substitute (13) to (11) we have

$$p_1 = \frac{1}{4mK}\left(C_r + C_p + C_m - \frac{3\gamma_0 mK}{\gamma_1}\right) \tag{14}$$

Substitute (14) to (7), we have $\Psi_{d_1}(\max) = -\left(\gamma_1/16m^2K\right)A^2$.

With the same way, we can find maximum profit for Damri and manufacturer for the second model. We have manufacturer's maximum profit is $\Psi_{m_2}(\max) = -\left(\gamma_1/8m^2K\right)B^2$ and Damri's maximum profit is $\Psi_{d_2}(\max) = -\left(\gamma_1/16m^2K\right)B^2$ with $B = \left[C_r + C_p + C_m + \left(\gamma_0 mK/\gamma_1\right) + \left(\gamma_2 umK/\gamma_1\right)\right]$.

3.1. Proposition 1: For the second model, Damri's profit is better than the first model $\left(\Psi_{d_1} < \Psi_{d_2}\right)$.

3.2. Proposition 2: Degree of repair and the number of PM that make optimum profit are $\delta = 0$ and $N = \left(K\left(C_f K\theta - 2b\right)/2a\right)^{-1/2} - 1$ with N is integer. The value $\delta = 0$ means Damri do perfect repair PM.

4. Case Study

We use bus first failure data which is collected from Damri's maintenance workshop (see. Fig 2) and fit data with Weibull(α, β, γ) distribution. We apply MLE to find the estimation parameter values. By using Minitab 18, the estimation values be as follow $\hat{\alpha} = 91.49388$, $\hat{\beta} = 1.31733$, and $\hat{\gamma} = 84.87987$.

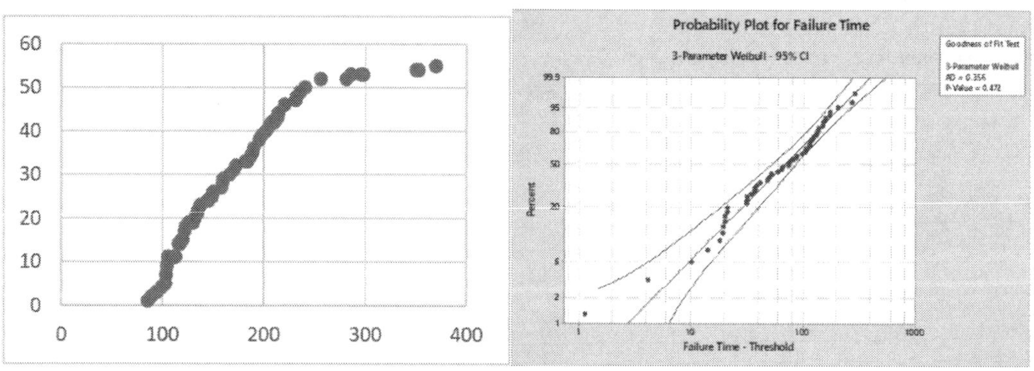

Note: Levenne test, KruskallWalliis test, MIL HDBK189 test are conducted before using Goodness of fit test [6]

Figure 2. First Failure Data with 3-parameters Weibull Distribution.

The failure intensity function of Weibullis $\lambda(t) = (\beta/\alpha)(t - \gamma/\alpha)^{\beta-1}$. After PM, the failure intensity function become $\lambda(v(t))$ and the total expected failure is $C_m = \alpha^{-\beta} C_f \left[\left(\frac{N\delta K}{2} + \frac{K}{N+1} - \gamma\right)^{\beta} - \left(\frac{N\delta K}{2} - \gamma\right)^{\beta}\right]$.

We have a linear model for the number of passengers per year $q(p,u) = \gamma_0 + \gamma_1 p + \gamma_2 u$, where $\gamma_0 = 1975000$, $\gamma_1 = -208$ and $\gamma_2 = 0{,}00006$. If bus capacity for one year is $m = 54000$ passengers then we have number of buses is $n = q(p,u)/54000$. Let failure intensity function of bus is Weibull with $\alpha = \sqrt{0{,}5}$ and $\beta = 2$ so the failure intensity function become $\lambda_0(t) = 4t$. Cost for every CM is

$C_f = Rp700.000,-$ and for every PM is $C_p^* = a + (1-\delta)\tau b$ with $a = Rp300.000,-$ and $b = Rp200.000,-$.

According to proposition 2, we have $\delta = 0$ and $N = \left[K\left(C_f K\theta - 2b\right)/2a\right]^{-1/2} - 1$ to maximize the profit. Here is results profit of Damri and manufacture for $K = 3$ years.

Table 1. Expected profit for Model 1 and Model 2

Subsidy		Non-Cooperative Solution							
		Model 1				Model 2			
		Ψ_d1		Ψ_m1		Ψ_d2		Ψ_m2	
Rp	150.000.000,00	Rp	1.384.930.704,20	Rp	3.219.861.408,39	Rp	1.614.838.227,29	Rp	3.229.676.454,57
Rp	200.000.000,00	Rp	1.384.930.704,20	Rp	3.369.861.408,39	Rp	1.622.085.442,83	Rp	3.244.170.885,66
Rp	250.000.000,00	Rp	1.384.930.704,20	Rp	3.519.861.408,39	Rp	1.629.348.884,34	Rp	3.258.697.768,68
Rp	300.000.000,00	Rp	1.384.930.704,20	Rp	3.669.861.408,39	Rp	1.636.628.551,81	Rp	3.273.257.103.61
Rp	350.000.000,00	Rp	1.384.930.704,20	Rp	3.819.861.408,39	Rp	1.643.924.445,24	Rp	3.287.848.890,47
Rp	400.000.000,00	Rp	1.384.930.704,20	Rp	3.969.861.408,39	Rp	1.651.236.564,63	Rp	3.302.473.129,25
Rp	450.000.000,00	Rp	1.384.930.704,20	Rp	4.119.861.408,39	Rp	1.658.564.909,98	Rp	3.317.129.819,96
Rp	500.000.000,00	Rp	1.384.930.704,20	Rp	4.269.861.408,39	Rp	1.665.909.481,29	Rp	3.331.818.962,59
Rp	550.000.000,00	Rp	1.384.930.704,20	Rp	4.419.861.408,39	Rp	1.673.270.278,57	Rp	3.346.540.557,14
Rp	600.000.000,00	Rp	1.384.930.704,20	Rp	4.569.861.408,39	Rp	1.680.647.301,81	Rp	3.361.294.603,61

From the examples above, clearly, we can see that profit for Damri from second model is higher than first model.

5. Conclusion

We have studied two model government subsidies for one of the public transportations, that is bus. This subsidy is purposed to increase people's interest in using bus. By this way, government is hoped it can decrease the traffic jam and reduce air pollutions in big city. After analyze both model, we have conclusion as follow.

a) PM can reduce failure rate and make bus operating time longer
b) For manufacturer, government subsidy model for buying bus is better
c) For Damri, government subsidy model for reducing ticket price is better

References

[1] Xu, Ying, Zhang, Xuemei and Y Hong 2016 The game model between government subsidies act and green supply chain *International Journal of u- and e- Service, Science and Technology* **9** 11 121-130

[2] B P Iskandar, H Husniah and U S Pasaribu 2014 Maintenance service contract for equipment sold with two-dimensional warranties *Quality Technology and Quantitative Management* **11** 3

[3] M Hamidi, H Liao and F Szidarovszky 2016 Non-cooperative and cooperative game theoretic models for usage-based lease contracts *European Journal of Operational Research* **4** 64

[4] J Jaturonnatee, D Murthy and R Boondiskulchok 2006 Optional preventive maintenance of leased equipment with corrective minimal repairs *European Journal of Operational Research* **174** 1 201-215

[5] D N P Murthy and N Jack 2016 Game theoretic modelling of service agent warranty fraud *Journal of the Operational Research Society*

IOP Conf. Series: Materials Science and Engineering **306** (2018) 012127 doi:10.1088/1757-899X/306/1/012127

[6] S E Marshall and S Chukova 2010 On Analysing Warranty Data from Repairable Items *Quality and Reliability Engineering International* **26** 43-52

[7] M S Ross 1996 *Stochastic Processes second edition* Canada: John Wiley and Sons

IOP Conf. Series: Materials Science and Engineering **306** (2018) 012128 doi:10.1088/1757-899X/306/1/012128

Auto Drain Valve Water Separator inside the Unit of Komatsu HD 465-7R

V A T Manurung*, Y T Joko W and R I Poetra

Politeknik Manufaktur Astra Jakarta

*vuko.manurung@polman.astra.ac.id

Abstract. Water separator is a component that separate water from fuel, so the circulating fuel in the fuel system is not contaminated by water. If there is water inside the water separator, it will be carried by into the fuel system and then impacting to the engine performance. It's such as lowering engine power because the fuel filter is clogged due to the fuel mix with water. Then the real danger is in case of the fuel mixes with the water. It will damage the fuel system components such as blockage of injectors due to corrosion and wear of fuel supply pump. As informed from daily maintenance record data, we have found that the low power engine trouble was caused by the fuel filter that was clogged high enough. Using the fishbone analysis, we got the main problem is there was water in the fuel separator at maximum level and did not discharge. In this condition, it is need optional device to automatically discharge the water from the water separator while maximum level reached, so the operator does not need to drain the water manually. The operator will be warned by buzzing active alarm and flashing caution lamp inside the cabin. By this method, the potential risk of mix up water with fuel would be avoided and the loss of others component failure would be mostly avoided. By using this tool, we can save net quality income around IDR (Indonesia Rupiah) 11,673,519,800.

1. Introduction

HD 465-7R is a huge dump truck material carrier, from medium to long distance, where material brought can be filled by excavator, wheel loader, or shovel. Seeing its function then Dump truck is very suitable to be used in mining [1].

At Komatsu HD 465-7R after observed then there is a weakness in the fuel system that is on the system of separation of water with fuel. In this unit fuel separator system use manual drain plug so that required attention from the operator when daily checking. The problem is the operator forgot to check the separator daily [2][3].

Figure 1 is a population of several types of heavy equipment used in PT XX mining area West Sumatra. It is seen that the population of HD 465-7R is very dominant [4].

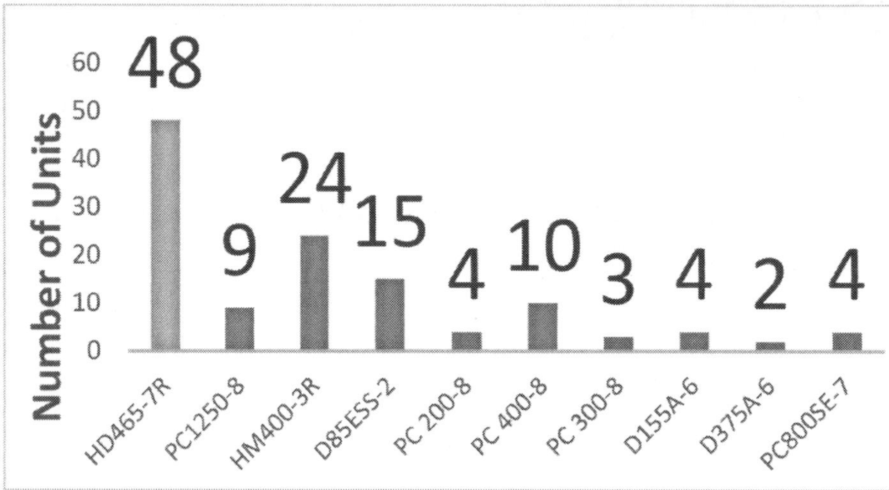

Figure 1. The number of units and types of heavy equipment

Daily monitoring breakdown Data (DMBD) from May 2016 until May 2017 it found that engine breakdown of HD 465-7R around 28% from the whole trouble as figure 2and totally 11232.8 hours. For 11232.88 hours of unit breakdown of the engine system is divided into 5 parts, from the turbo contributed 4675 hours, from low power contributed 2750 hours, the leakage of coolant and oil was 807.88 hours, engine can't start 673 hours and the other 2327 hours [5][6][7].

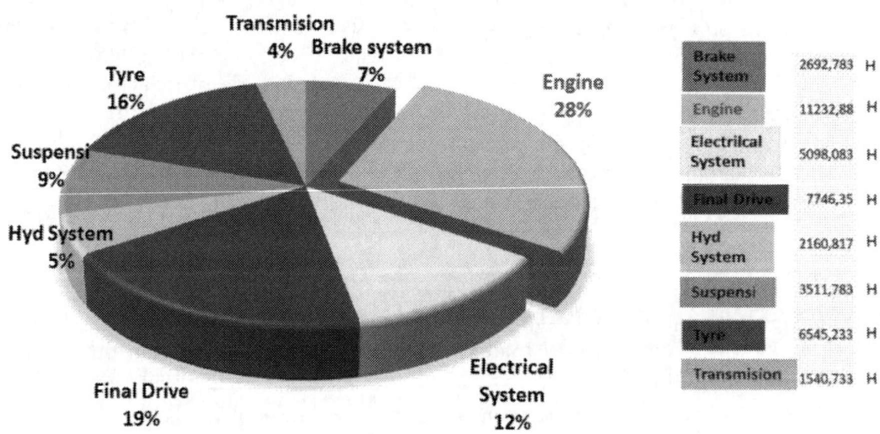

Figure 2. Percentage of downtime unit HD465-7R

2. Low power engine trouble

Damage start from complaining of the operator HD 465-7R that is engine low power causing the unit to be breakdown (unscheduled breakdown), caused by fuel mixed with water (fuel quality is less good). As a result, the unit becomes breakdown and ultimately disrupts the production process. Countermeasures so far are doing the replacement of fuel filters. Table 1 shows the frequency of low power engines during on January 2017 [8][9].

Table 1. Unit HD465-7R Low Power Engine in January 2017

PT. Artamulia Tatapratama
COAL MINE PROJECT MUARO BUNGO - JAMBI
DATE :
System

NO	KODE UNIT	TYPE	MODEL UNIT	HM	PROBLEM
189	AR-36	HD465-7R	RIGID DUMP TRUCK	16230	Turbo Leak (Low Power)
583	AR-16	HD465-7R	RIGID DUMP TRUCK		Material Di Tire And Low Power
740	AR-16	HD465-7R	RIGID DUMP TRUCK	25854	Low Power Engine (Air Cleaner Cloged)
741	AR-30	HD465-7R	RIGID DUMP TRUCK		Low Power
742	AR-16	HD465-7R	RIGID DUMP TRUCK	25789	Low Power
743	AR-01	HD465-7R	RIGID DUMP TRUCK		Low Power
744	AR-16	HD465-7R	RIGID DUMP TRUCK		Low Power
745	AR-18	HD465-7R	RIGID DUMP TRUCK		Low power
746	AR-16	HD465-7R	RIGID DUMP TRUCK		Low power
747	AR-16	HD465-7R	RIGID DUMP TRUCK		Low power
1239	AR-25	HD465-7R	RIGID DUMP TRUCK	22487	Error E-03 (Low power)
1280	AR-18	HD465-7R	RIGID DUMP TRUCK		Engine Low Power
1283	AR-25	HD465-7R	RIGID DUMP TRUCK	22367	E-03 Low Power
1284	AR-25	HD465-7R	RIGID DUMP TRUCK	22368	E-03 Low Power
1325	AR-25	HD465-7R	RIGID DUMP TRUCK	22361	E-03 (Low Power)
1326	AR-25	HD465-7R	RIGID DUMP TRUCK		E-03 (Low Power)
1329	AR-25	HD465-7R	RIGID DUMP TRUCK		E-03 (Low Power)
1330	AR-05	HD465-7R	RIGID DUMP TRUCK		E-03 (Low Power)
1338	AR-25	HD465-7R	RIGID DUMP TRUCK	22413	E/G Low Power
1339	AR-18	HD465-7R	RIGID DUMP TRUCK	19857	E 03 And Low Power
1340	AR-05	HD465-7R	RIGID DUMP TRUCK		E 03 (Low Power)
1341	AR-30	HD465-7R	RIGID DUMP TRUCK		E 03 (Low Power)
1342	AR-05	HD465-7R	RIGID DUMP TRUCK		E 03 (Low Power)
1383	AR-30	HD465-7R	RIGID DUMP TRUCK		E 03 (Low Power)
1734	AR-16	HD465-7R	RIGID DUMP TRUCK		1. Low power OK2. Tyre No 04

3. Solution

By using the fishbone analysis, we got there was water in the separator at maximum level so the fuel mix with water. This condition makes the engine trouble [10]. Solving the problem should made a system that allows water to be automatically drained while the water in separator at the maximum level and also inside the cabin installed the buzzer so that the operator is informed when draining in the engine room. The electrical diagram for discharging water as a figure 3.

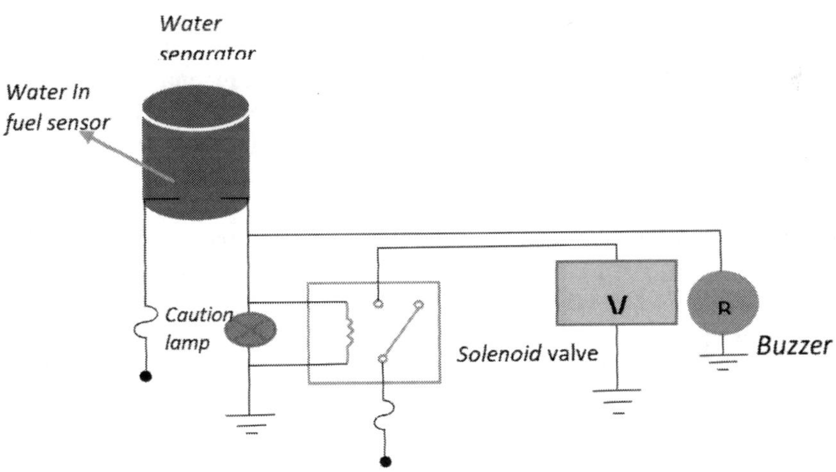

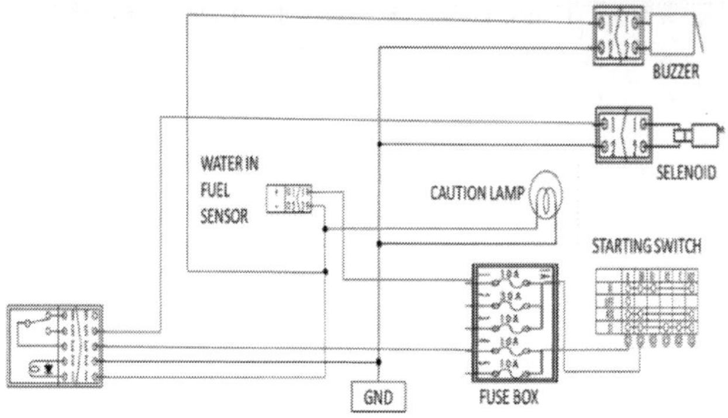

Figure 3. Wiring Diagram of the Automatic Water Draining

The system work as follows:
While in the water separator there is water mixing with fuel, the sensor installed in the water separator will inform the electrical module, and then gives input to actuator and then the output of solenoid is buzzer, and indicator lamp. So, when there is input from the sensor, at the same time the solenoid valve will open and the indicator lamp and buzzer in the cabin will work until the water empty. During that process the operator is prohibited to turn on the engine until the device finished, to avoid air enter to the fuel system. This is done so that no air enters the fuel system.

4. Financial benefit.

Losses due to the unscheduled breakdown of HD 465-7R as follows:

a. One-unit HD465-7R can carry 26 BCM (Bank Cubic Meter) Overburden. So, 1 hour can carry 5 times. 1 BCM = 1.74 USD (≈ IDR 23100), then 1 hour = 5 x 26 BCM = 130 BCM. If one year the total unscheduled breakdown due to low power for all units are 2750 hours, so the company lost is: 130 x 2750 (hours) x (IDR) 23100 = IDR 8,258,250,000
b. Losses due to fuel filter changes. There are 218 times of engine low power troubles. If one-unit use 1 main fuel filter and 2 pre-fuel filters, its mean that there are 218 pieces of main fuel filters change and 436 pieces of pre-fuel filters change. If one main fuel filter ≈ IDR 1789100 and one pre-fuel filter ≈ IDR 253500, the total losses is IDR 1,490,269,800.
c. Losses due to manpower. If 1-hour manpower cost is IDR 350,0000 and one trouble need 2 manpower. For 2750 hours the cost is IDR 700,000 x 2750 (hours) = IDR 1,925,000,000.

5. Conclusion

By using fishbone analysis, the main problem is there was water in the separator over the maximum level that cause the engine was low power. The application of auto drain valve water separator we can reduce losses of units can't operate, spare parts and manpower. All are around IDR 11,673,519,800 can be saved.

References

[1] Shop Manual Komatsu HD 465-7R,Komatsu Ltd. Japan 2010
[2] Shop Manual Engine Komatsu SAA170E-5,Komatsu Ltd. Japan 2010
[3] Technical Training Departement.2008.Basic Electrical System 1.PT United Tractors,Jakarta
[4] UMR 2016 Unit record HD465-7R. PT. Artamulia Tata Pratama
[5] UMR 2017 Unit record HD465-7R. PT. Artamulia Tata Pratama
[6] https://www.google.co.id/search?q=foto+hd+465-7R+komatsu
[7] Huzij, Robert, Angelo Spano, Sean Bennett 2009 *Heavy Equipment System* Delmar
[8] T Gilles 2012 *Automotive Service Inspection, Meintenance*, Repair, 4th Edition, Delmar

[9] S Bennett 2009 Medium/Heavy Duty Truck Engines, Fuel and Computerized Management System, 3th edition Delmar
[10] A J Bell 2007 *Modern Diesel Technology Electricity & Electronics* Delmar

In-Memory Business Intelligence: Concepts and Performance

V P Rantung[1*], O Kembuan[2], P T D Rompas[2], A Mewengkang[2], O E S Liando[2] and J Sumayku[2]
[1] Department of Informatics, Universitas Negeri Manado, Tondano 95618, Sulawesi Utara, Indonesia
[2] Department of Information Technologies and Communication Education, Universitas Negeri Manado, Tondano 95618, Sulawesi Utara, Indonesia

*Corresponding author: vivirantung@unima.ac.id

Abstract.This research aims to discuss in-memory Business Intelligent (BI) and to model the business analysis questions to know the performance of the in-memory BI. By using, the Qlickview application found BI dashboards that easily accessed and modified. The dashboards are developed together using an agile development approach such as pre-study, planning, iterative execution, implementation, and evaluation. At the end, this research helping analyzer in choosing a right implementation for BI solution.

1. Introduction

In-memory business intelligence relies on the technology of In-Memory Database System (IMDS) which is developed precisely to meet demands of performance and resource availability on embedded systems [1]. Memory as temporary storage has functions to supplying data and saving data helps the work of CPU to be faster [2]. Growth in innovation technology influence memory in consequence of data management area. Information system as part of data management area takes opportunity and adopts the concepts of In-Memory Database to In-Memory BI [3]. The impacts of In-Memory Database refer enterprise applications [4]. Business Intelligence is one of them and In-Memory BI adopted the technology framework [5]. If traditional BI loads data from hard disk, In-Memory BI loads data from memory [6]. In recent research review [7], which reports a systematic review of published papers about different data visualization tools that are implemented for Business Intelligence by different organizations for the purpose of business analysis. Past research has paid attention to comparative analysis of in-memory analytic tools and has not created models to evaluate these criteria. Our current research addresses these needs in the field of evaluation of the In-Memory BI performance. The In-Memory BI containing all advantages of In-Memory Database technology. Consideration of In-Memory BI performance raises the following questions: R1. Related to implementing In-Memory BI: How In-Memory BI helps user to get better decisions? ; Does In-Memory BI solution answer the business analysis questions? R2. Related to data model of In-Memory BI: Does user need to consider data model when using In-Memory BI? R3. Related to BI technical requirement [8]: Does In-Memory BI consider about security? ; How about capacity of In-Memory BI? This research was carried out to find answers to the above questions and to provide a case study in order to create BI solution using In-Memory BI application. Although the application type of In-Memory BI now days have many variations, this research focuses on in-memory associative index.

2. Methods

By following agile development approach, we investigate and describe how In-Memory BI works related to the implementation, data model and technical requirement. The model and the user interface are developed together using Qlikview Project Methodology (QPM). This methodology includes the following phases: 1) Pre-study; 2) Planning; 3) Iterative execution; 4) Implementation; 5) Evaluation in Figure 1 [9] [10].

The first stage was Pre-study that includes the following main steps: define initial business objectives and scope, identify initial business requirements and data requirements, prepare enterprise platform. In the second stage, a project planning was conducted with three main stages there are project management planning, prioritized business requirements and defining the data staging requirements (QVD files for larger deployment), define enterprise platform and plan application cycles. The third stage is iterative execution phase that includes the following steps: build, test, user review and refine. Build step includes: build data reload process (configuring of the connections, create incremental load script), data model design, data provisioning, and user interface development. The four and five stages of the research method are implementation and evaluation.

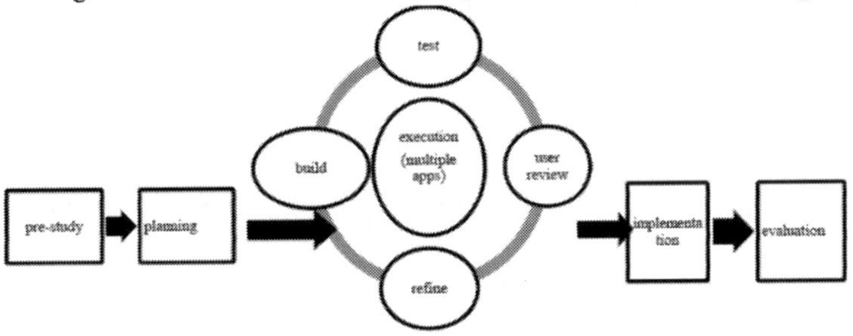

Figure 1. Qlickview Project Methodology [9].

3. Results and discussion

There are many variation of In-Memory BI application. In-Memory application itself divided into 5 main types such us In-Memory OLAP, In-Memory ROLAP, In-memory inverted index, in-memory associative index and in-memory spreadsheet [11][12]. As shown in table 1 the Qlickview application categorized as in-memory associative index and this research using the tool in order to model BI solution to knowing the performance of In-Memory BI. And the data that used for experiment is "NorthWind" database [13].

Before model the BI solution, business analysis questions were made. There are two questions to generate BI solution are: 1) How much of each product category U.S customers order by quarter in 1997? ; 2) Who are top three sales representatives? .Figure 2 shown answer of business question 1 and figure 3 shown answer of business question 2. Figure 2 shown turnovers by time and demography also turnover by products category and suppliers where the dimension used are year, quarter and country. The dimension that used in answering question no.2 are employees name, year order, sheep country, and turnover.

3.1. Implementing In-Memory BI

Implementation of In-Memory BI will help user to get better decision because reporting, querying and analysis become faster since the entire index is in memory. They also treat user in providing an attractive interface. It creates BI solutions easily and considers data complex of details business analysis questions with prepared lots of function for chart, script and dimension. The results above show that answering business analysis questions not a big deal. Moreover Qlickview application challenges user exploring dashboard and find a new analysis fact without re-aggregate.

3.2. Data model of In-Memory BI

According to Evelson that "reporting, querying, analysis can be done without any model constrains, for example any attribute can be instantly reused as fact or as a dimension and every query with an inner join can also show results of an outer join on every column means less modelling required than an OLAP based solution" really happen since implement Nortwind database to Qlickview application [14] [15]. The experiments loads data from excel file, so for more convinced about data modelling need to do more experiment that load data from client server architecture.

Table 1. In-Memory BI Application Variations

In-Memory Applications Type	Description	Vendors
In-memory OLAP	Classic MOLAP cube loaded entirely In-Memory.	IBM Cognos TM1, Actuate BIRT.
In-memory ROLAP	ROLAP metadata loaded entirely in memory.	Micro Strategy
In-memory inverted index	Index (with data) loaded into Memory	SAP Business Objects (BI Accelerator), Endeca.
In-memory associative index	An array/index with every entity/attribute correlated to every other entity/attribute.	QlikView, TIBCO Spotfire, SAS JMP, Advizor Solutions (also OEMd by Information Builders).
In-memory spreadsheet	Spreadsheet like array loaded entirely into Memory	Microsoft (PowerPivot)

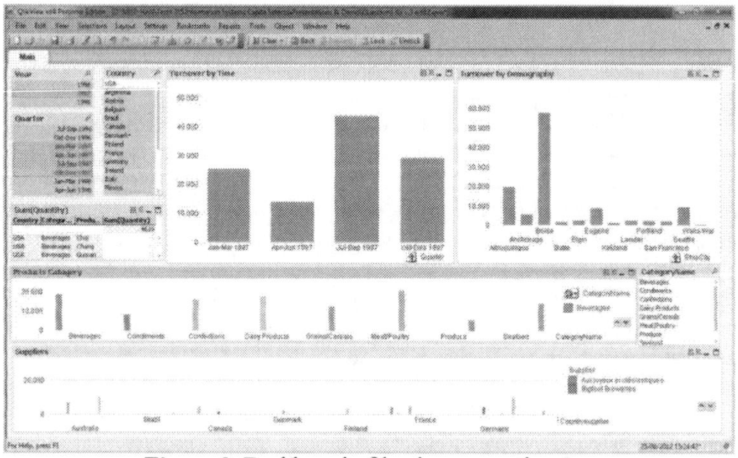

Figure 2. Dashboard of business question 1.

3.3. BI technical requirements

Security: Security is a big issue when deployed BI solutions. The issues regard to how data is accessed, where it is stored and who has access to the data [16]. The Qlickview application provide security mechanism that can be set up in two different ways: It can either be built into document script, or it can be set up through the use of Qlickview publisher. In details, it offers authentication and authorization relation to who can access the data. Unfortunately, they can't meet the completeness of security. They do not provide security in data source level because the best secure location for data is on a centralized server.

Capacity: According to Qlickview research that Qlickview deployment able to handle many thousands of concurrent users and extremely large quantities of data [17]. Nevertheless, we are talking about technology, so all of this depends on memory technology innovation.

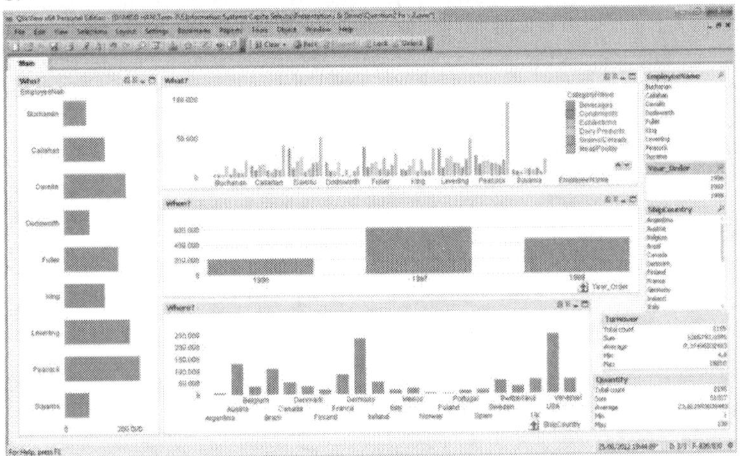

Figure 3. Dashboard of business question 2.

4. Conclusion

The conclusions that achieve are: In-Memory BI helps gaining BI solutions in fast access, storage and manipulation; In-Memory BI application (Qlickview) easy to implement even for complex data and details business analysis questions, eliminates re-aggregate, flexible to zoom in another analysis requirement, has dynamic sources, covers advance analytical, and greatly in data visualization; In-Memory BI application (Qlickview) has highly iterative process development decrease data modelling in BI solution but for client server architecture need to consider; In-Memory BI application (Qlickview) provides security mechanism in offer authorization and authentication; In-Memory BI application (Qlickview) do not provides security in data source level; In-Memory BI depending on amount of RAM.

References

[1] K L Tan, Q Cai, B C Ooi, W F Wong, C Yao and H Zhang 2015 In-memory databases *ACM SIGMOD Rec.* **44** 2 35-40

[2] S Manegold, P Boncz and M Kersten 2002 Optimizing main-memory join on modern hardware *IEEE Trans. Knowl. Data Eng.* **14** 4 709-730

[3] P Loos, J Lechtenbörger, G Vossen, A Zeier, J Krüger and J Müller 2011 In-memory databases in business information systems *Business and Information Systems Engineering* **3** 389-395

[4] J V Brocke, S Debortoli, O Müller and N Reuter 2014 How in-memory technology can create business value: Insights from the hilti case *Commun. Assoc. Inf. Syst.* **34** 1 151-168

[5] H Chen, R H L Chiang and V C Storey 2012 Business intelligence and analytics: from big data to big impact *Mis Q.* **36** 4 1165-1188

[6] H Zhang, G Chen, B C Ooi, K L Tan and M Zhang 2015 In-memory big data management and processing: A survey *IEEE Trans. Knowl. Data Eng.* **27** 7 1920-1948

[7] A Shukla and S Dhir 2016 Tools for data visualization in business intelligence: Case study using the tool qlikview *Information Systems Design and Intelligent Applications* **2** 319-326

[8] S Williams and N Williams 2004 Assessing BI readiness: The key to BI ROI *Bus. Intell. J.* **9** 3 15–23

[9] M Muntean 2014 Toward agile BI by using in-memory analytics *Informatica Economica* **2** 26-40

[10] M Muntean and T Surcel 2013 *Agile BI-the future of BI Inform. Econ.* **17** 114-124

[11] H Plattner and A Zeier 2012 In-memory data management: technology and applications *Springer Science & Business Media*
[12] M Ivan 2014 Characteristics of in-memory business intelligence *Inform. Econ. J.* **18** 3 17-25
[13] Fakas, J Georgios and C Zhi 2008 *Ranking of object summaries* Data Engineering ICDE'09 IEEE 25th International Conference on. IEEE
[14] B Evelson 2010 *I forget whats in memory*. Forrester for application development & delivery professionals [Online] *http://blogs.forrester.com/boris_evelson/10-03-31-i_forget_whats_in_memory*
[15] S Manegold, P Boncz and M Kersten 2002 Optimizing main-memory join on modern hardware *IEEE Trans. Knowl. Data Eng.* **14** 4 709-730
[16] A Cardenas, P K Manadhata and S P Rajan 2013 Big Data Analytics for Security *IEEE Secur. Priv.* **11** 6 74-76
[17] L L Briggs 2013 *In memory Databases Promise Faster Results* Tdwi the premier source for Business Intelligence and Data Warehousing [Online] *http://tdwi.org/Articles/2009/10/07/QA--Inmemory-Databases-Promise-Faster-Results.aspx?sc_lang=en&Page=1*

An Analysis of Website Accessibility in Higher Education in Indonesia Based on WCAG 2.0 Guidelines

W Arasid, A G Abdullah, D Wahyudin, C U Abdullah, I Widiaty, D Zakaria, N Amelia and A Juhana

Departemen Pendidikan Teknik Elektro, Fakultas Pendidikan Teknologi dan Kejuruan, Universitas Pendidikan Indonesia

*wildanarrasyid55@gmail.com

Abstract. Website accessibility is a simple way to access a website by everyone so that information on the website can be easily understood. This study aims to improve the accessibility of universities' website to analyze website accessibility problems based on WCAG 2.0 guidelines. This study analyzed 13 universities' websites in West Java, Indonesia by using TAW as an evaluation tool. The evaluation results were presented in a graph showing the error rate of each university's website. The same errors that occurred in almost all websites were: non-text content, info and relationships, page title, link purpose, language of page, on input, labels and instructions, parsing, and name, role, value criteria. This study was expected to provide information to the university and to perform as guidelines for website accessibility improvements.

1. Introduction

The development of internet-based media has made many organizations aware of the importance of online media to support their business and promote their existence [1]. Likewise, Education cannot be separated from the role of academic websites as information providers and media of promotion for every educational institution.

Some of the main uses of academic websites include promotion of expertise programs, academic research, e-learning support for students, communication to public and vacancy advertisement at schools or at the universities [2]. One effort that can be done to optimize the purpose or usefulness of an academic website is by improving the accessibility of the website to facilitate users in accessing it.

In addition, some research indicates that there are several factors that influence the ranking of websites in webometrics. One of such factors is the accessibility of the website [1].

Academic website accessibility can facilitate everyone to understand the information provided in the website. An accessible website is a website that even the disables can experience, understand, navigate and interact with the website and they can contribute to the website. Accessibility has another benefit for the elders who lost their ability because of aging [3].

Website accessibility is still a problem for many top Universities around the world [3]. Similar to a building, a website should be designed to meet everyone's needs, including people with disabilities. However, many websites ignore the users' need so that it may decrease users' interest to the websites [4].

Website accessibility can be improved by meeting criteria provided in Web Content Accessibility Guidelines (WCAG) [5]. Therefore, College website accessibility is urgently required.

2. Literature Review

2.1. Academic Website

Academic website is one of the most important media for the public to access information and obtain information services [6]. A website is an integral part of formal education. In Taiwan, every school, at any level, has its own website. School websites not only allow students and parents to learn about the activities at school, but also simplify the administrative processes, improve efficiency, and gradually promote teaching with the digital environment [7].

2.2. Website Accessibility

The ease for community in accessing the academic website cannot be separated from the accessibility of the website. The accessibility of a good website is very influential for users of the website to access the available information in it, either for the general public or for the students. The better the accessibility is, the easier the public access the academic website.

Website accessibility means that anyone who uses this type of web browsing technology should be able to visit any site and gain full and complete understanding of the information as well as having full and complete ability to interact with the site [8]. Website accessibility becomes a basic prerequisite to ensure the equal rights for everyone in accessing information [6].

2.3. Web Content Accessibility Guidelines (WCAG) 2.0

To improve the ranking of Higher Education websites, it is necessary to evaluate the content and suitability of the website using a global standard created by the World Wide Web Consortium (W3C), an independent international body that creates Internet standards and programming languages. Since 1999, the W3C has issued several rules and recommendations to improve the accessibility of a website, the rule in question is WCAG guidelines [5].

WCAG is an internationally accepted guideline consisting of guidelines and checkpoints that provide specifications on how to develop the accessibility of a website [9]. The WCAG guidelines can be accessed in the official W3C website (www.w3.org). WCAG plays an important role in establishing a website accessibility standard.

There are several principles that a website must follow to ensure website accessibility:

- Perceivable - Information and user interface components must be presentable to users in ways they can perceive. It means that users must understand the information provided in the website.
- Operable - User interface components and navigation must be operable. This implies that users must be able to operate the interface.
- Understandable - Information and the operation of user interface must be understandable. It indicates that the users should understand the information and user interface operation.
- Robust - Content must be robust enough that it can be interpreted reliably by a wide variety of user agents, including assistive technologies [10].

There are other organizations that focus on accessibility issues, but the W3C guidelines are more sophisticated. Accessibility is supported by developing WCAG guides from version 1-2 [11]. According to the W3C, every website must meet WCAG 2.0 criteria [5].

2.4. Website Accessibility Evaluation

To test the accessibility of a website, there are several different evaluation and improvement tools in using the guidelines. However, most of these tools follow the WCAG guidelines [12].

Website accessibility evaluation tool is an online software or services program that helps determine whether or not a website meets the web accessibility guidelines. These tools can be categorized into two types:

- A general evaluation tool, which evaluates almost any guideline; e.g. TAW and AChecker.
- Specific evaluation tools that evaluate specific topics covered by web accessibility guidelines, such as Color Checker [12].

Accessibility should be evaluated to ensure that it provides the desired destination of the site. Evaluating website accessibility can help designers and developers to improve their website based on evaluation results [13].

2.5. TAW Evaluation Tool

TAW was developed by *Spanish Fundación CTIC.* TAW is a tool used to analyze web accessibility with an overall global analysis of items and pages from a site that aims to check the accessibility of websites and development for access to any individual regardless of the website's characteristics. TAW measurement results are presented in the forms of number of violations or accessibility issues of any WCAG guidelines criteria.

The tool analyzes the site in accordance with the W3C Web accessibility guidelines (WCAG 1.0 and WCAG 2.0) by providing improvements and recommendations. Some errors are generated from TAW regarding Hyper Text Markup Language (HTML), they are not normally generated by other tools [12].

3. Methods

To analyze the website accessibility, this study used content analysis and survey method with qualitative approach. Surveys were conducted by observing the results of website accessibility analysis by using TAW evaluation tool. To get the facts from the problem and to find factual information about the accessibility of the website. The sample of this research is academic websites of universities in West Java under Ministry of Research, Technology and Higher Education.

The evaluation tool used in this research was TAW. TAW is one of the online evaluation tools functioning to measure the accessibility of a website. In the application, it must be connected to the internet network and the result is in the form of number of problems or errors of a website in meeting the criteria of a website according to WCAG 2.0. The data obtained using the TAW evaluation tool was taken on September 23, 2016 [14].

3.1. Research Procedure

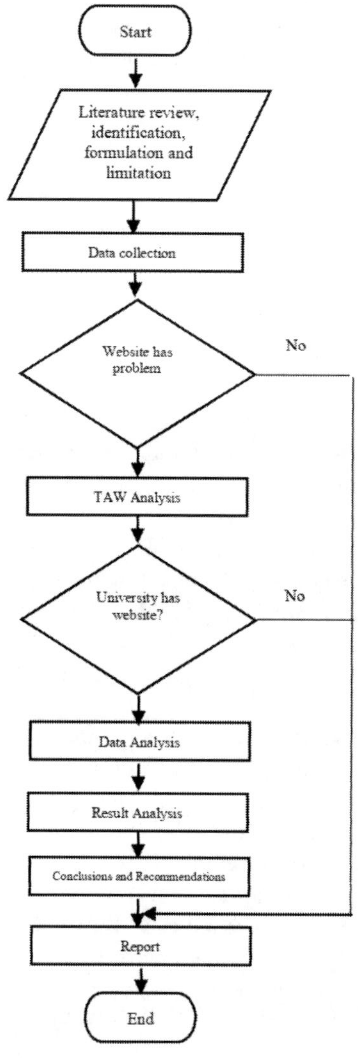

Figure 1. Research Procedure

The first step in this study was to conduct a literature study by reviewing international journals from reliable sources related to website accessibility. This was intended to inspire the author base and knowledge about the importance of accessibility of a website before executing research.

The second step was to collect the data by observing the website in each college which was then analyzed using TAW to get the number of problems or violations from every State University website.

The third step was to do data processing obtained from TAW by using Microsoft Excel. The next step was to make conclusions and recommendations on the accessibility of the website of State Universities in West Java and the final step was making the reports.

3.2. Data Analysis Technique

Once the data was collected, the next step was to analyze the data. Data obtained from TAW were errors or violations on the website. Each data was grouped according to the criteria of the WCAG 2.0 guideline. The data was processed by the error of each success criterion of certain guidelines, and then divided by the number of errors of the success criteria in the guidelines multiplied by 100%.

4. Results and discussion

4.1. TAW Evaluation Tool Result

The following table describes the error criteria on WCAG 2.0 guidelines that become a problem at 13 State Universities in West Java resulting from TAW evaluation tool.

Table 1. TAW Evaluation Tool Result of University in West Java

ERROR CRITERIA	UI	UNPAD	UPI	POLMAN	UNSIL	IPB	ITB	POLBAN	UNSIKA	ISBI	STTT	POLINDRA	STKS
1.1.1 Non-Text Content	17	14	15	0	5	9	0	0	6	1	1	36	0
1.3.1 Info and Relationships	346	14	7	1	17	38	0	1	2	55	10	7	0
2.4.2 Page Title	0	0	0	1	0	0	0	1	0	0	0	0	0
2.4.3. Link Purposes	26	19	23	0	3	20	0	0	12	1	4	14	0
1.1.1 Language of Page	0	1	1	1	1	0	1	1	1	1	1	1	1
3.2.2 On Input	0	0	0	0	0	0	0	0	0	0	0	1	0
3.3.2 Labels or Instructions	17	0	1	0	1	1	0	0	0	1	1	4	0
4.1.1 Parsing	76	8	1	0	16	15	2	1	3	26	14	92	0
4.1.2 Name, Role, Value	19	0	1	0	2	1	0	0	0	1	1	4	2
TOTAL	501	56	46	3	45	84	3	4	24	86	32	159	3

Table 1 shows that the most number of guideline violations found on University of Indonesia website with 501 violations and the smallest number of errors exist in three websites; POLMAN, ITB and STKS with the three error criteria.

4.2. Number of Error Criteria Occurrence

Of the 13 websites of State Universities, there are nine criteria that become errors in accessibility with different number of occurrences. The number of occurrences of each error criterion can be seen in table 2.

Table 2. The number of occurrences

PRINCIPLES	ERROR CRITERIA	OCCURANCES
Understandable	1.1.1 Non-text content	9
	Info and Relationships	11
Operable	Page Title	2
	Link Purpose	9
Understandable	Language of Page	11
	On Input	1
	3.3.2 Labels or Instructions	7
Robust	4.1.1 Parsing	11
	4.1.2 Name, Role, Value	9
	TOTAL	**70**

The occurrence of error criteria in the table above shows that there are three criteria that often arise and violate the website accessibility of State Universities in West Java with 11 occurrences. Those criteria are info and relationships, language on the page, and parsing. From the number of occurrences of the criteria in question, they are presented in the form of diagrams as in Figure 2.

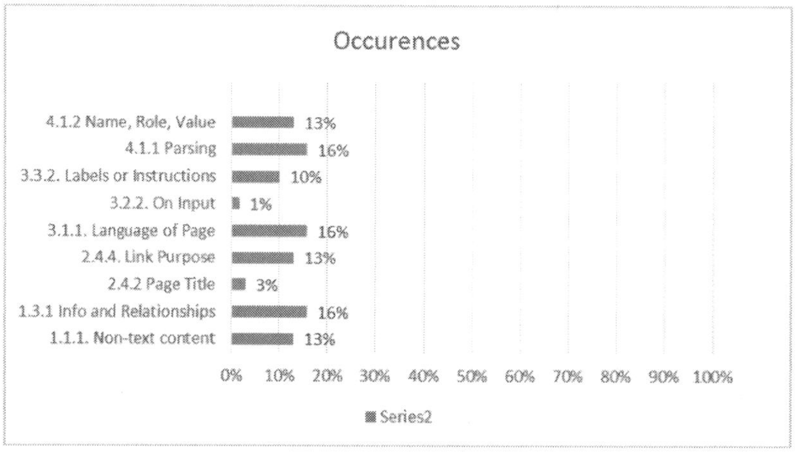

Figure 2. Percentage of Error Criteria Occurrence

4.3. Error Criteria

4.3.1. Non-text Content

All non-text content that is presented to the user has a text alternative that serves the equivalent purpose. These criteria make the information submitted by non-text content accessible through the use of alternative text. Alternative text is the primary way to make information accessible to all users of the website. Providing text alternatives allows information to be input in various ways by the creator of the website to match user needs [10].

4.3.2. Info and Relationships

These criteria encourage website creators to ensure that information and relationships are understood through visual or hearing format. In this criterion the visual presentation formats are:

- The title used larger font size and are bold printed and separated by line space.
- The list of items is preceded by a sign and perhaps indented
- Paragraphs are separated by line spaces
- Different background colors can be used to indicate that some items are related to each other
- Words that have a special status are marked by different fonts, bolded, italicized or underlined.

In addition to the visual format, on this criterion the auditory format can also be used to convey information and relationships. For example, it provides a change of tone that can be used to emphasize important information or to show the quoted text [10].

4.3.3. Page Title

Web pages have titles that describe topic or purpose. This criterion intends to help users find content and adapt within it by ensuring that each web page has a descriptive title. The title identifies the location without requiring the user to read or interpret the page content. When a title appears on a sitemap or a search results list, users can identify the content they need more quickly [10].

4.3.4. Link Purposes

These criteria encourage website designers to help users understand the purpose of each link so they can decide if they want to follow links. Where possible, they provide text on links that identify link destinations without the need for additional context. For example, in a case where links to a document or web application, simply name the link with the name of the document or web application [10].

4.3.5. Language of Page
This criterion encourages website creators to ensure that website creators provide information on their web pages by providing other languages other than the primary language used [10].

4.3.6. On Input
These criteria encourage website makers to ensure that entering data or selecting a form of control has a predictable effect. The benefit of these criteria is to help users who have limitations by making interactive content more predictable. Unexpected changes from the context can be so confusing for users with visual limitations. For example, people with visual limitations may have difficulty knowing when a visual context change has occurred, as new windows pop up. In this case, the user's warning of an interface context change minimizes the confusion when the user finds that the back button is no longer behaving as expected [10].

4.3.7. Labels or Instructions
This criterion encourages the content creators to have the label or instructions of each user interface control used to identify those controls so that the user knows the expected data from the control [10].

4.3.8. Parsing
This criterion encourages the website designer to interpret and parse the content accurately. The content was implemented using markup languages, all elements have complete start and end tags, all elements are nested according to their specifications, all elements do not contain duplicate attributes, and any IDs are unique, except where the specifications allow these features

4.3.9. 4.1.2 Name, Role, Value
According to World Wide Web Consortium (W3C), the purpose of this criterion is to ensure that Assistive Technologies (AT) can collect information, enable set up and keep updating the status of user interface controls in the accessed content.

Assistive Technologies (AT) is a product, equipment, and system that improves learning, working, and daily living for people with disabilities [10].

5. Conclusion
Accessibility of the website is an important thing for the owners to consider in making a university website, because with accessibility, a website can facilitate users in accessing and information submitted. In addition, the accessibility of the website can also affect the ranking of websites in webometrics which is the ranking of university websites in the world. Website accessibility analysis can be done with the help of software. One of which is TAW software.

Website of State Universities in West Java has not fully qualified website accessibility. This is because it still violates some criteria from WCAG 2.0 guidelines in terms of website accessibility. Similar errors occur almost entirely on the entire website: non-text content, info and relationships, page titles, link destinations, languages on the page, when entering data, labels and instructions, parsing, and name, role, value criteria.

References
[1] P Windriyani, R Ferdiana and W Najib 2015 *Accessibility Evaluation Using WCAG 2 . 0 Guidelines Webometrics Based Assessment Criteria* (Case Study : Sebelas Maret University).
[2] D D J Suwawi, E Darwiyanto and M Rochmani 2015 Evaluation of Academic Website Using ISO/IEC 9126. 2015 *3rd International Conference on Information and Communication Technology (ICoICT)* 222–227

[3] S May and A E Q Zhu 2010 *A web accessibility assessment on the Texas public school system* 2010 Univ Access Inf Soc, 9, pp.87–96.

[4] A H Arsyitahadi 2014 *Analisis Usability Website Akademik Perguruan Tinggi Di Indonesia Menggunakan Metode Electree,* Grey Relational Analysis, Dan Weighted Product Model

[5] R Gonçalves, J Martins and F Branco 2014 A Review on the Portuguese Enterprises Web Accessibility Levels – A website accessibility high level improvement proposal *Procedia - Procedia Computer Science* 176–185

[6] Z Sun 2009 On Accessibility of Concept , Principle and Model of Educational Web Sites Design *International Conference on New Trends in Information and Service Science* 730–733

[7] H Sung 2012 Analysis of Vocational High School Websites in Taiwan *International Symposium on Computer, Costumer and Control* 2009–2012

[8] P Zaphiris and R D Ellis 1988 *Website Usability and Content Accessibility of the top USA Universities*

[9] S Hong, P Katerattanakul, H Choi and Y Kang 2007 *Evaluating Government Website Accessibility : A Comparative Study Department of Business Information Systems , Western Michigan University Department of Management Information System , Dong-A University Department of Management Information System , Dong-A. International DSI / Asia and Pacific DSI 2007.*

[10] W3C 2008 *Web Content Accessibility Guidelines (WCAG) 2.0* https://www.w3.org/TR/WCAG20/

[11] M A Aziz, W A R W M Isa and N Nordin 2010 Assessing the accessibility and usability of Malaysia higher education website *Proceedings - 2010 International Conference on User Science and Engineering, i-USEr* **3** 203–208

[12] H S Al-khalifa, M Al-kanhal, H Al-nafisah, N Al-soukaih, E Al-hussain and M Alonzi 2011 A Pilot Study for Evaluating Arabic Websites Using Automated WCAG 2 . 0 Evaluation Tools *International Conference on Innovations in Information Technology* 293–296

[13] A N AbuAli, A Obedidat and H Y Abu-Addose 2013 *Accessibility as an Indicator of Jordanian E-Government 2013 Fourth International Conference on e-Learning* "Best Practices in Management, Design and Development of 2013 Fourth International Conference on e-Learning Best Practices in Management, Design and Development of e-Courses: Standards of Excellence and C 156–160.

[14] Tawdis 2016 *Analisis Aksesibilitas Website* Spanyol: http://www.tawdis.net

IOP Conf. Series: Materials Science and Engineering **306** (2018) 012131 doi:10.1088/1757-899X/306/1/012131

The Effectiveness of Using Interactive Multimedia in Improving the Concept of Fashion Design and Its Application in The Making of Digital Fashion Design

W Wiana

Program Study Fashion Education, Faculty Technology and Vocational Education Universitas Pendidikan Indonesia

*winwinwiana@upi.edu

Abstract. This research is related to the effort to design a more representative learning system to improve the learning result of digital fashion design, through the development of interactive multimedia based on motion graphic. This research is aimed to know the effect of interactive multimedia application based on motion graphic to increase the mastery of the concept and skill of the students to making fashion designing in digital format. The research method used is quasi experiment with research design of Non-equivalent Control Group Design. The lectures are conducted in two different classes, namely class A as the Experimental Class and class B as the Control Class. From the calculation result after interpreted using Normalize Gain, there is an increase of higher learning result in student with interactive learning based on motion graphic, compared with student achievement on conventional learning. In this research, interactive multimedia learning based on motion graphic is effective toward the improvement of student learning in concept mastering indicator and on the aspect of making fashion design in digital format.

1. Introduction

Teachers as one of the agents in the implementing the education system, became a determinant success of education. Teachers as professionals in the transformation system of education, required in the Law of National Education System Number 20-year 2003 CHAPTER XI Article 40 - paragraph (2), to: "a. creating a meaningful, fun, creative, dynamic and dialogical education atmosphere; b. have a professional commitment to improve the quality of education .... ". In relation to the mission that teachers must carry to improve the quality of education, teachers should have competencies that can be transformed to the students so that they have the appropriate competencies required by the employment and ready to enter and compete in the work field. The vital role of teacher as the spearhead in determining the quality of education, implies the ideal expectation of the figure of a teacher with competence that qualified. The professional aspects of teachers related to the competencies he / she must possess as follows [1]: which includes a combination of experiences that empower (1) individual educators, (2) educational teams, and (3) the educational organization to improve (4) curriculum, (5) instruction, and (6) student assessment in order to (7) facilitate student growth and development. However, empirically still encountered teacher weaknesses during the learning process. Exploration of teaching skills is a weakness that is still widely encountered in the learning process. The conventional teaching method by transferring one-way knowledge from teacher to student is still the choice of most teachers in teaching. Another weakness in fulfilling its competence is that many vocational teachers are not technologically oriented in developing their subject matter, so materials that should be facilitated by technological means such as computer devices and fashion design subjects still run the old system in their learning.

Currently, the development of information and communication media, both hardware and software occur rapidly, so that conditions have resulted in shifting the role of teachers, where currently teachers can no longer serve as the only source of information for learning activities of the students. Relate with it, then the teacher needs a media that can be utilized as an alternative source of information. One of the technology products that can be used as an innovative media in the learning process is the computer. The existence of today's computers has helped teachers in various interests related to their duties in designing, implementing and evaluating learning. Computers have been widely used in the teaching and learning process, with one goal that the quality of education going one step ahead with advances in technology. With computer-based learning, it is expected to help learners who have a slower learning speed (slow learner) in order to learn effectively, because with computer to reload the required information, while for faster learners (fast learner) can stimulate learning activities [2].

Fashion Design like other design field, in its making is oriented by using computer technology (Computer Aided Design / CAD), so as to produce characterize fashion design in digital format and effective to be applied in the process of making of fashion product. In this regard, the rational step needs to be taken in order for the learning process to succeed effectively, is to develop a learning program that optimizes all components of the teaching and learning process. An innovation is needed in the learning process that can motivate students to actively construct a knowledge so that the learning process itself becomes more meaningful. Meaningful learning will make students more master the material or concept given and will last longer in the memory of the brain. Such capability will be helpful in facilitating learning the concepts and skills in a lecture, especially in practical class. One component of education that can be developed in the learning process and assumed to have a pretty good influence in efforts to implement the achievement of teaching objectives on the fashion design in digital format, is to develop effective and efficient learning media, so that the student behavior changes in accordance with the objectives to be achieved.

Now Learning media is more diverse, ranging from conventional media such as books or traditional props to modern media audio visual form in cassettes, videos, and another modern visual device. Learning media used to optimize the teaching and learning process, strived to be able to grow creativity and motivation in learning activities to improve the quality of education. One of the media used in learning and believed to be more exciting students interest in lectures is interactive multimedia learning. Learning media is also one of the alternative means that can optimize learning activities based on computer technology. This interactive multimedia application is conditioned to present learning materials with a more interesting and informative display, so it is expected to facilitate and increase student interest to learn.

Efforts are made to obtain maximum results on the learning of making fashion design in digital format needs to be designed and developed by optimizing the use of innovative learning media and related directly in the mechanism of computer-based learning. Efforts are made is expected to improve the quality of learning and can motivate students to actively constructed knowledge and skills in the making of fashion design in digital format, so that the learning process in students becomes more meaningful. The use of multimedia in addition is assumed to increase the mastery of the concept of the fashion design process, is also expected to develop student's skills in making fashion design in digital format.

Information technology in education is applied in the form of interactive multimedia in the form of software (software), which provides facilities to students to learn a material. The use of interactive multimedia applications in learning will improve the efficiency, motivation, and facilitate the active learning, experimental, consistent learning, with student-centered learning [3]. The use of interactive multimedia in learning is also very possible to improve the thinking ability that expected. In general, the benefits that can be gained through the use of interactive multimedia is the learning process can run more interesting, more interactive, the amount of teaching time can be reduced, the quality of student learning can be improved and the learning process can be done anywhere and anytime, and can improve student's logic ability [4].

The improvement of mastering the concept of students who follow the learning using cooperative learning group with help from interactive multimedia is significantly higher than the students who follow the conventional learning [5]. The improvement of mastery of the concept and critical thinking skills of students who follow the magnetic plate learning using online interactive multimedia is significantly higher than the students who follow the conventional learning [4].

This research is specifically done to develop the improvement of the concept and skill of fashion designing in digital format that is closely related to the ability to optimize the use of computer. This research is considered to have a high urgency, related to the demands of the world of fashion industry that is currently beginning to switch to the parameters of digitization in the process of product development, one of them is in the design area. Digital logic is considered richer because it contains programs with various facilities that can help the designers in generating more accurate and complex design ideas.

1.1. Interactive Multimedia

Etymologically multimedia comes from the Latin, the word "Multi" which means a lot; variety and "medium" which means something used to convey or bring something. Some multimedia definitions according to some experts include: 1) Multimedia is various combinations of text, graphics, sound, animation, and video delivered using computers or other electronic devices [6]. 2) Multimedia refers to the various combinations of two or more integrated media formats into the form of information or instruction programs [7] while 3) Multimedia as the presentation of the material using words and images [8]. The words are where the material is presented in verbal form, such as using text or oral text. By drawing, it is intended that the material is presented in pictorial form, such as using static graphics, animation or video. This definition is broad enough to cover every multimedia scenario, ranging from multimedia encyclopedia entries to textbook lessons. Based on these opinions it can be concluded that multimedia is a combination of various media (file format) in the form of text, graphics, audio, and interaction and used to convey messages/information from the sender to the recipient of the message / information.

Multimedia is divided into two categories a) Linear Multimedia and b) Interactive Multimedia. A linear multimedia is a multimedia that is not equipped with any controller that the user can operate. This multimedia runs sequentially. Examples: TV and movies. Interactive multimedia is a multimedia equipped with user-operated controller tools, so users can choose what they want for the next process. Examples of interactive multimedia are: interactive game and CD applications. The description in advance describes an important concept, that is, if the user gets the flexibility in controlling the multimedia, then the device is categorized as interactive multimedia. The most important characteristic of interactive multimedia is the students not only pay attention to media or object, but also required to interact during learning. Interactive multimedia combines and synergizes all media elements consisting of [9] : a) text; b) graphs; c) audio; and d) interactivity. The developing process in learning multimedia based on computer can see in figure 1.

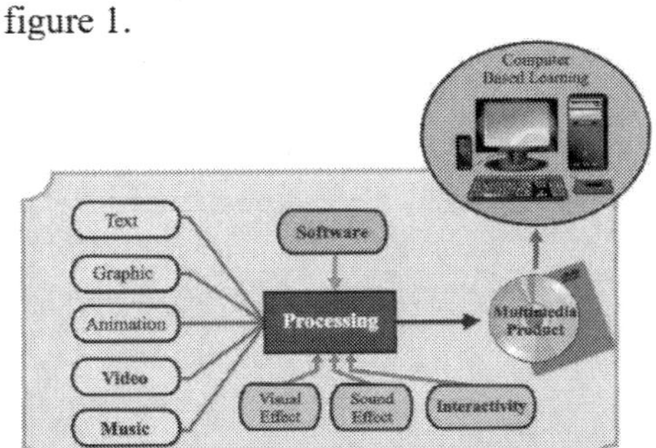

Figure 1. The developing process in learning multimedia based on computer

There are several advantages and disadvantages of interactive multimedia as a medium of learning, such as [10]: (1) Interactive means that multimedia program is programmed or designed to be used by individual students (independent study), (2) Giving individual affective which mean more affective in a more individual way, never forgetting, never bored, very patient in performing instruction, as desired, (3) Increasing motivation to learn (4) Giving feedback (response), and (5) Because interactive multimedia is programmed for self-learning, its utilization controls are entirely within the user. Meanwhile, b) The lack of interactive multimedia are: (1) Development requires a professional team and (2) Development takes a long time.

Utilization of multimedia technology as an interactive learning method, is one of the learning tools for students, has some basic strengths [11], namely: a) Mixed. Media using multimedia technology, various existing conventional media can be integrated into one type of intermediate media, such as text media (whiteboard), audio, video, which if separated will require more media. b) User control. Interactive multimedia implementation technology (IMMI), allows users to browse the teaching materials, according to the ability and background knowledge he/she had, in addition to making users more comfortable in studying media content, repeatedly. c) Simulation and visualization. Simulation and visualization is a special function possessed by interactive multimedia, so with animation technology, simulation and visualization of computer, users will get more real information from an abstract information. In some curriculum requires a complex, abstract, dynamic and microscopic process, so that by simulating and visualizing the learner will be able to develop the mental model in its cognitive aspect. d) Different learning styles. Interactive multimedia has the potential to accommodate users with different learning styles.

1.2. Fashion design in digital format
Design is a design something that can be realized on a real object or human behavior that can be felt, seen, heard and touched [12]. Application of design will be linked with various objects associated with human needs such as architectural design, product design, interior design, fashion design and so on. Fashion design as a form of design to fulfill human needs for products and clothing, today has been transformed from the conventional parameters of the design basis, in digital format parameter. Aside from the fashion design as one of the manifestations of the creative process, it is empirically demonstrated changes format (visual display) a revolutionary. The manual design parameter in the previous decade still widely explored on the various interests of its design, is starting to shift and change the format to the digital format design (made by the process of computerization). Manufacture of computer-based fashion design, providing a visual and tangible image characteristic, so the details that exist on the clothing can be visualized with more expressive.

2. Methodology
The method developed in this research is Educational Research and Development (R & D), is a development of multimedia interactive learning on learning Fashion Design, to improve the understanding of concept and skill of fashion designing in digital format. R & D in education as "a process used to develop and validate education product" , a process used to create, develop and validate research products [13]. Both further states that:... our use of term product include bot only material objects, such a text book, instructional films and so forth, but it also intended to refer to established procedures and processes, such as methods of teaching or methods organizing instruction.

The design of this study applied two groups of respondents, namely the Experimental Group and the Control Group. Both groups were pre-tested to determine the initial state of the respondents, whether there was a difference between the Experimental Group and the Control Group. The design pattern applied in this research refers to Pretest-posttest control group design concept [14], can be see in table 1:

Table 1. Design of Research Class Design Treatment

Class	Design	Treatment	Test	
			Begin	End
Experiment	O_1 X_1 O_2	X_1	Yes	Yes
Control	O_3 X_2 O_4	X_2	Yes	Yes

X_1 = Using Interactive Multimedia
X_2 = Using Conventional Learning

The difference test of two averages of two samples was conducted to determine whether between the Experimental Group and the Control Group there was a difference of N-gain (normalized gain) according to Hake [15], were:

$$(N-gain) = \frac{\%\,\boldsymbol{actual\ gain}}{\%\,\boldsymbol{potential\ gain}} = \frac{\%\,\boldsymbol{skor\ postes} - \%\,\boldsymbol{pretes}}{\boldsymbol{skor\ maksimum} - \%\,\boldsymbol{skor\ pretes}}$$

In this research, N-gain descriptive analysis using N-gain criterion according to Hake [15] is shown in the following table 2:

Table 2. Category level of N-Gain

Limitation	Categori
$g > 0,7$	Hi
$0,3 \leq g \leq 0,7$	Medium
$g < 0,3$	Low

3. Results and Discussion

3.1. Results

Based on the data processing of research results on the initial conditions of mastering the fashion design students showed: in the experimental group the average initial ability is 23.06 while in the control group 23.38.

In the next process, different treatments were given to the Experimental Group and Control Group. The Control Group was guided by the conventional method in the learning process, while in the Experimental Group applied the learning using interactive multimedia based on motion graphic.

Based on the calculation of the increase of learning result using Normalized Gain obtained the average increase of higher gain value in Experimental Class that is 71.31%, compared to average gain of gain value in Control Class equal to 65.24%, as illustrated can see in figure 2:

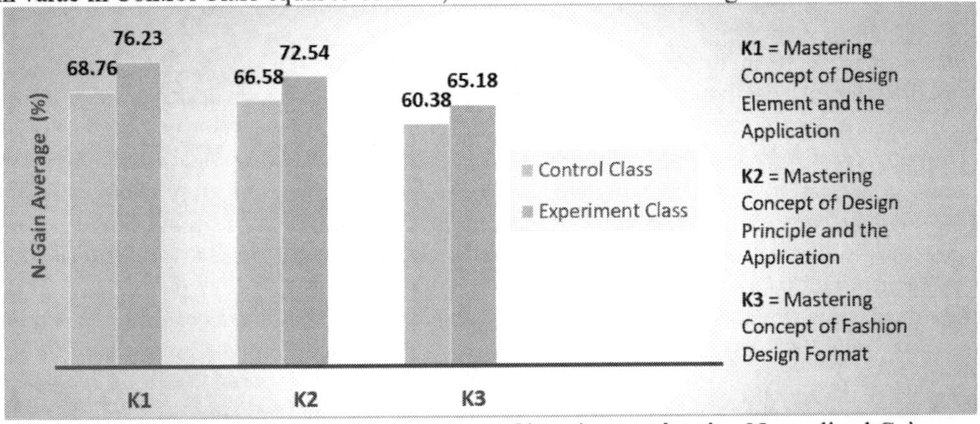

Figure 2. The calculation of the increase of learning result using Normalized Gain

3.2. Discussion

In this study, the data collected is adapted to the needs of analysis, namely the analysis of the concepts of making the fashion design in digital format of the students of Fashion Design Study Program in the subject of Fashion Design that is taught using interactive multimedia in the Experimental Class, and the learning without using interactive multimedia in the Control Class.

The material of the Fashion Design concepts discussed in this research consists of three sub-materials it's Fashion Design Elements, Fashion Design Principles and Fashion Design Format. The average gain of N-gain for the Experimental Class is 71.31% and the Control Class is 65.24%. The average N-gain for the Experimental Class is included in the high category, while in the Control Class included in the moderate category. Based on these data it can be seen that the average N-gain for the Experimental Class is higher than the average N-gain of the Control Class. The comparison result of the average of N-Gain in mastering the concept in the experimental class and control class for each sub material, obtained the largest N-Gain average in Fashion Design Element sub material, while the smallest N-gain of both classes is obtained in Fashion Design Format sub material. This is allegedly due to the characteristics of each sub-material that affects the delivery of the material in the interactive multimedia used. Format Design sub material has a high degree of difficulty in understanding, because the material contained very specific aspects of technical mastery of the fashion design, especially in Design Production Sketching, in the form of an understanding of 1) developing in the format/shape of fashion design, 2) the concept of the forming and accuracy of the parts of fashion products, 3) the material concepts in the manufacturing process of the garment industry, 4) the concept of the standard size of garment products, and 5) the concept of labeling of garment products. Different from the concept mastering of Fashion Design Elements and Principles which tend to be easier to understand because it more relate to daily life in the use of clothing, the concept mastering of fashion design formats need more detailed review and analysis, especially since this material is closely related to the manufacturing process of clothing products in the garment industry. The data acquisition shows that learning using interactive multimedia is able to provide a better learning experience to the learners, especially from the interesting display of material that presented and time to learn that is not limited by the segment of time in the classroom learning.

4. Conclusion

a. Based on the results of processing and data analysis of the results of research that has been done, regarding the use of interactive multimedia based on motion graphic on learning fashion design in digital format, obtained the following conclusion:
   1) Based on the research result, the average of pre-test result showing the students initial ability in the Experimental Class and the Control Class is relatively same. After going through the learning process, it can be seen that there are differences in the average outcomes of students learning using interactive multimedia learning based on motion graphic that is higher than conventional learning
   2) There's an increase in the average of learning outcomes in students who do interactive multimedia learning based on animation compared with students who use conventional learning on learning making the fashion design in digital format
   3) Judging from the mastery of learning indicators, interactive multimedia learning based on motion graphic role in improving student learning outcomes to a lesson material thoroughly better than conventional learning, so that interactive multimedia learning based on motion graphic can be said to be effective against student learning outcomes with greater effectiveness compared with conventional learning

b. The interactive multimedia learning based on motion graphic can be said give a positive response to the students, where the learning of making the fashion design with digital format using interactive multimedia based on motion graphic make the students feel easier to understand the subject matter and feel more confident and satisfied with the learning that has been they do

5. Closing

Based on the results of research and discussion shows that the use of interactive multimedia based on motion graphics significantly influence the improvement of learning outcomes in making fashion design in digital format, the real implementation of the learning process of fashion design becomes urgent to be done immediately. In addition to the conclusion above, to obtain better results expected time planning in learning is one thing that must be carefully set by the next researcher, because many unexpected things that can arise in the learning activities.

References

[1] Gordon S P 2004 *Professional Development for School Improvement: Empowering Learning Communities* (Boston: Pearson Education)
[2] Warsita B 2008 *Teknologi Pembelajaran: Landasan dan Aplikasinya* (Jakarta: Rineka)
[3] Exline 2004 *Workshop: Inquiry-based Learning* (Online) available at: http://www.thirteen.org/edonline/concept222class /inquiry/index_sub2.html. Access Date August 21st 2017
[4] Sutarno 2011 Pengunaan Multimedia Interaktif Pada Pembelajaran Medan Magnet Untuk Meningkatkan Keterampilan Berpikir Generic Sains Mahasiswa *Exacta Journal* **IX** 1 6066
[5] Wiyono K 2012 Peningkatan Keterampilan Berpikir Kritis Mahasiswa Calon Guru Dengan Model Mia-Piza *Forum MIPA (Majalah Ilmiah Jurusan PMIPA FKIP Unsri)* **14** 1 10-16
[6] Vaughan T 1994 *Multimedia: Making it Work (2nd ed.)* (USA: McGraw-Hill)
[7] Heinich R, Molenda M R J, and Smaldino S 2005 *Instructional Media and Technologies for Learning (8th ed.)* (USA: Merrill/Prentice Hall)
[8] Mayer R E 2001 *Multimédia learning* (Cambridge University: Press)
[9] Diartono D A 2008 Media Pembelajaran Desain Grafis Menggunakan Photoshop Berbasis Multimedia, Technology and Information *Journal DINAMIK* **XIII 2** 155-167
[10] Munadi Y 2012 *Media Pembelajaran* (Jakarta: Gaung Persada (GP) press)
[11] Hasrul 2010 Langkah-Langkah Pengembangan Pembelajaran Multimedia Interaktif *MEDTEK Journal* **2**
[12] Mahardhika S and Fathoni C A 2013 Storyboard dalam Pembuatan Motion Graphics *HUMANIORA* **4** 2 1183-1189
[13] Gall M D, Gall J P and Borg W R 2003 *Educational Research, An Introduction* (Boston: Pearson Education. Inc.)
[14] Sugiyono 2012 *Metode Penelitian Kombinasi* (Bandung: Alfabeta)
[15] Widodo W 2010 *Pengembangan Model Pembelajaran "Mikir" pada Perkuliahan Fisiska Dasar untuk Meningkatkan Keterampilan Geberik Sains dan Pemecahan Masalah Calon Guru SMK Program Keahlian Tata Boga* Disertation. Sekolah Pasca Sarjana UPI. Not Published.

IOP Conf. Series: Materials Science and Engineering **306** (2018) 012132 doi:10.1088/1757-899X/306/1/012132

Gender Bias in the Workplace: Should Women be Marginalized in Engineering Job?

Y Kurniawan[1*], I D A Nurhaeni[2], Mugijatna[1] and S K Habsari[1]

[1]English Study Program, Faculty of Cultural Sciences, Universitas Sebelas Maret, Jl. Ir. Sutami 36A Surakarta, Indonesia 57126
[2]Public Administration Study Program, Faculty of Social and Political Sciences, Universitas Sebelas Maret, Jl. Ir. Sutami 36A Surakarta, Indonesia 57126

* yusuf@staff.uns.ac.id

Abstract. Research observing women marginalization in the workplace viewed from the end-users have not been widely conducted. This article discusses about gender bias in the workplace from the perspective of the end-users of vocational higher education in terms of first, the companies' policies in the salary payment and second the availability of companies' policies in fulfilling the gender needs for male and female employees graduating from vocational schools. The research employing gender analysis was conducted in the Ex-Residency of Surakarta, Central Java Indonesia. The data were collected through documentation studies. The result shows that both male and female entrepreneurs have gender-biased view of women's and men's positions within the companies. Consequently, women's salary tends to be lower than that of men for the same job and the companies were still not responsive to the different needs of women and men. Accordingly, local government should supervise the companies to implement gender mainstreaming in the workplace, especially in the employees' career development and give rewards to companies implementing gender equality and otherwise give punishment to companies which have marginalized women in the workplace.

1. Introduction

Although gender-mainstreaming has been echoed by governments of many countries, gender bias is still ubiquitous, not exceptionally in the workplace. In engineering work-field, women merely represent about 15 percent of the employment and nearly 40 percent of female engineers leave engineering. Hiring managers — both men and women — are more likely to prefer male over female candidates for certain STEM (science, technology, engineering and math) jobs [1]. These facts show that gender bias in the workplace does exist. Furthermore, an engineer is 'expected' to be a man, whereas a female engineer is a deviation, both professionally and as a woman. The engineer as a man is also a statistical norm that shapes our stereotypical expectations. We think of a man when we hear the word 'engineer' because most engineers are men [2]. It is frequently claimed that women who enter engineering have to 'fit in' to 'a masculine culture [3].

The research in various countries, including in Indonesia shows that women tend to be marginalized in technical jobs. Gender-stereotype, in which women are positioned in jobs relating to domestic sphere still occurs. There are three types of gender bias in workplace namely "prove-it-again bias", ""tightrope bias" and the "maternal wall" [1]. These mean that women have to look more professional and demonstrate technical prowess at all times to receive the same respect as a male engineer, women must operate within a narrow range of acceptable behavior that is neither too feminine nor too masculine to avoid being branded "emotional", and the ideal worker is always available for work and that mothers do not — or should not — remain committed to their careers.

Another researcher found that gender blindness dominates organizational male cultures in spite of an increasing number of women in management positions [4], and sexual orientation-based and gender identity discrimination is a common occurrence in many workplaces across the country [5].

Men earn more than women at all levels of engineering in New Zealand. Women are offered lower starting salaries and the gap continues throughout their life course, reaching a peak for women between the ages of 45 and 49 [6]. Such practice shows the unequal opportunity to certain positions and career development. Research observing women marginalization in the workplace viewed from the end-users have not been widely done.

This article discusses about gender bias in the workplace based on documentation studies in terms of first, the companies' policies in the salary payment regarding male and female employees graduating from vocational schools, and second the availability of companies' policies in fulfilling the gender needs for male and female employees graduating from vocational schools.

2. Methods

The research was conducted in the Ex-Residency of Surakarta, Central Java Indonesia. It is the second largest city after Semarang, the Capital City of Central Java Province. In Surakarta there are many industries and it is presumed that there are still many gender biases occurring at the companies regarding their workers. Gender bias at the workplace is an iceberg phenomenon and it's difficult to proof. Company managements frequently deny it. Therefore, the data of the research were collected through documentation studies, from previous research regarding gender bias and gender violations taking place at the workplaces.

3. Research Findings

The research findings are dealing with two major problems discussed in this article, i.e. the companies' policies in the salary payment regarding male and female employees graduating from vocational schools and the availability of companies' policies in fulfilling the gender needs for male and female employees graduating from vocational schools.

3.1. Companies' policies on payment

The research reveals that there is no difference in companies' policies regarding salary payment to male and female employees graduating from vocational schools for the same jobs due to the stereotype among the companies assuming that women are more appropriate for certain jobs and occupations and vice versa. Gender stereotype can produce gender bias in the work settings and therefore obstruct women's career and even becomes the basis of biased evaluative judgments and discriminatory treatment of women in work settings [7].

One of the gender-bias problems happening to the employees graduating from vocational schools is the assumption that a woman is (always) single in her status, so that her job health insurance is merely valid to herself, not for her family. This occurs because of the assumption that the one who is responsible in a family is a husband. This occurs because the cultural assumption that men are the breadwinners, while women take care of the household, is still very strong in Indonesian society. As a result, when women are employed, their position is assumed to be merely that of an 'additional' breadwinner. In Indonesia, according to the ILO, women represent approximately 38% of the work force in the civil service. More than a third of the women do domestic works, such as taking care of the households and nurturing. These jobs are paid relatively low and therefore women overall tend to earn less than men [8].

Indonesia in 2015 in the survey conducted by World Economic Forum, Indonesia is positioned on the rank of 114 out of 145 countries in economic participation and opportunity, on the rank of 112 in

IOP Conf. Series: Materials Science and Engineering **306** (2018) 012132 doi:10.1088/1757-899X/306/1/012132

the labor force participation with the score 0.62 (54% female and 86% male). Whereas for wage equality for similar work, Indonesia is on the rank of 67 with the score 0.65 [9].

There are three issues in the project of wage gap of gender in Indonesia, first focuses on collective agreement. In manufacturing sector Cost Benefit Analysis (CBA) makes inventory of housing allowance and secondary facility such as housing allowance and health insurance. For instance, the CBA of the Fish Marine Company, which has already uploaded to *www.gajimu.com*, states that the housing allowance for men is IDR 20,000, for women IDR 10,000 and for widows IDR 15,000. This is the kind of difference that we will try to find in other CBA's and address through the gender pay gap campaign. Second, the gender pay gap in the informal sector such as domestic workers (house maid, driver, gardener, cleaning service, nanny). There is a difference in the income for women and men in domestic workers sector, again with the assumption that men are breadwinners. In addition, many works in informal sector is typical to men or women. For example drivers are mostly men and nannies are mostly women. The campaign will focus on equal pay for equal work. Third, the wage differences between different sectors (e.g. administrators in mining get higher wages than administrators in the garment sector). This has become an option which we can now show, based on the data. In the mining sector, where the majority are men, the wages are higher than that for office workers of whom the majority is female. These gender pay gaps between the sectors can now also be demonstrated, using the data of Wage Indicator [10].

In 2016 the international website *Qerja.com* conducted a survey on wage or salary gap to male and female in Indonesia. A survey by Qerja revealed that male employees generally receive 34% more than female employee in Indonesia. Moreover, the number of male managers in various industries are higher than female managers. Learning the practice that has been going on in the five most favourite industries – banking and financial services, consulting and legal services, media, mining and retail, it is found that all industries value their male employees more than the female.

Below is a table of the range of wage/salary indicating the gap undergone by women in five favorite industries [11].

Table 1. Range of wage/salary in five favorite industries

Industry	Female		Male	
	Average Female Salary	Range of Female Salary	Average Male Salary	Range of Male Salary
Banking & financial service	Rp 9.9 million/month	Rp 5 million-Rp 14 million/month	Rp 12 million/month	Rp 7 million-Rp 15 million/month
Consulting, research & legal services	Rp 15 million/month	Rp 6.4 million-Rp 14 million/month	Rp 21 million/month	Rp 11 million-Rp 25 million/month
Retail & wholesale trading	Rp 9.4 million/month	Rp 4.9 million-Rp 14 million/month	Rp 11 million/month	Rp 6 million-Rp 15 million/month
Mining, oil & gas	Rp 21 million/month	Rp 10 million-Rp 21 million/month	Rp 24 million/month	Rp 13 million-Rp 30 million/month
Media, entertainment & advertising	Rp 8.7 million/month	Rp 4.2 million-Rp 11 million/month	Rp 11 million/month	Rp 7 million-Rp 14 million/month

*Based on the interview and survey from Qerja.com in 2016. The actual salaries can be higher of lower than
Source: Qerja.com/2016

According to the Central Statistics Agency of Indonesia for the period of February 2017, there is a difference between the salary paid to male and female workers. The overall salary/wage of male workers is higher than that of the females. The following is the table of monthly wage/salary payment to male and female workers.

Table 2. Monthly wage/salary payment to male and female workers in Indonesia

Sector	Wage to male workers	Wage to female workers
Agriculture	1,93 million	1,14 million
Mining	4,5 million	3,5 million
Finance	3,7 million	3,6 million
Electricity, Gas, Water	4,4 million	4,7 million
Industry	2,7 million	2 million

Source: BPS, February 2017

3.2. Availability of companies' policies on specific needs

Several companies have attempted to fulfill the gender needs to male and female employees. Some health-examination facilities are provided to female employees when they are having pregnancy for their first and second pregnancies, leaves for giving birth of the first and second children and the insurance for giving birth for the first and second children. However, a number of companies have not provided a breastfeeding room to women who want to give breastfeeding to their children.

Female workers (graduates of senior high schools and vocational high schools) particularly working at textile industries state that they can't possibly ask for a leave for their pregnancy in accordance with the policy set out by the government [12]. The research reveals that one of the companies will even expel the female employees without giving them severance if they get pregnant. Moreover, if the company permits the female employee to get back to work, they should restart their carrier from zero (contract workers with adjusted wages).

Besides the leave entitlement of giving birth, which is not in accordance with the government policy, the right usually not fulfilled by the company is that of female workers for not being permitted to go home for breastfeeding their babies. If they are allowed, they should follow complicated procedure and even can lead to payroll deduction. The company does not provide day care and breastfeeding facilities [12].

Women Research Institute (2016) finds problems frequently encountered by female workers at their workplace, i.e. the company's bathrooms that don't fulfill the hygienic standard, insufficient lighting, unkind health workers, and unavailable dining rooms. Moreover, there are also other problems such as the domination of male workers at the Work-Unit Representative and Labor Union and the low representation of female workers in the organizations of labor union. The problems are still frequently found in the companies having less-good policies [13].

One of the jobs being underestimated is being domestic workers. In fact, the job has become an alternative to those graduating from Senior High Schools and Vocational High Schools having no qualified skill, especially women. In addition to becoming domestic workers, it also covers taking care of children and becoming private nurses. These workers (i.e. housekeepers, nannies, personal care aides, and home health aides) help raise children, care for the elderly and people with disabilities, and free up their employers to engage in other types of productive work. In addition, because arrangements are often informal, domestic workers are subject to coercive tactics on the part of employers, sexual and other types of harassment and threats. Isolated workplaces, where the worker often works in the employer's home with no co-workers, increase the likelihood of domestic workers to withstand abuse and contribute to the poor health outcomes of this population. Domestic work can be hazardous, with potential exposure to toxic chemicals, injuries, and long hours on the job [14].

As a matter of fact, male and female workers deserve proper place and facilities to support their jobs. The lack of legal protections and social vulnerability workers experience increases of the chance that they will be exposed to risky situations that put their health in jeopardy on the job. Workers are described being afraid of speaking up for better conditions because of undocumented status or economic constraints. Repeated experiences of unfair treatment, humiliation, and discrimination within the workplace can lead to chronic stress that can impact the mental health and well-being of workers [14].

4. Conclusion

The research result shows that gender-biased view of the position of women and men within the companies still exists. As a result, women's salary tends to be still lower than that of men for the same job or position. Regarding the companies' policies on specific needs, many companies were still not responsive to the different needs of women and men. In other words, women are still marginalized in engineering job. Even one of the sexes is still harmed in their health during their work due to the presumption that male and female workers are the same. Accordingly, local governments should supervise the companies to implement gender mainstreaming in the workplace, especially in the employees' career development. Moreover, the local governments should also give rewards to companies implementing gender equality and otherwise they should also give punishments to companies which have marginalized women in the workplace.

References

[1] Elejalde-Ruiz A 2016 Women engineers describe unfriendly work environments in study *Chicago Tribune* (online) available at http://www.chicagotribune.com/business/ct-women-engineering-implicit-bias-1028-biz-20161027-story.html
[2] Holth L 2014 Passionate Men and Rational Women: Gender Contradictions in Engineering *International Journal for Masculinity Studies* **9** 97-110
[3] Faulkner W 2009 Doing Gender in Engineering Workplace Cultures. I. Observations from the Field *Journal of Engineering Studies* **1** 3-18
[4] Bergman B 2008 Polarized Gender Thinking-Visions or Norms in Male-Dominated Workplace *International Journal of Qualitative Studies on Health and Well-being* **3** 165-75
[5] Badgett M V L, Lau H, Sears B, and Ho D 2007 Bias in the Workplace: Consistent Evidence of Sexual Orientation and Gender Identity Discrimination
[6] Davies S G, McGrgor J, Pringle J, and Giddings L 2017 Rationalizing Pay Inequity: Women Engineers, Pervasive Patriarchy and the Neoliberal Chimera *Journal of gender Studies* **12** 1-14
[7] Heilman M E 2013 Gender stereotypes and workplace bias *Research in Organizational Behavior* **32** 113-35
[8] Rahmawati I 2017 Indonesian Culture Causes many Gender Pay Gaps (online) available at *Article in wageIndicator.org*, access date September, 28, 2017
[9] World Economic Forum 2015 Global Gender Gap Report 2015- Indonesia's Profile (online) available at *Article in reports.weforum.org,* access date September, 28, 2017
[10] Tijdens K and Van K M 2012 *Frozen in Time: Gender Pay Gap Unchanged for 10 years* (Neterland: Amsterdams Instituut Voor ArbeidsStudies (AIAS))
[11] Marcella B A 2016 Salary Comparison Between Female and Male Employees: Have We Reached Parity? (online) available at *Article in liveolive.com,* access date September, 28, 2017
[12] Subagya S, Indria S, and Budiati A C 2013 Pengarusutamaan Gender dan Optimalisasi Peran Serikat Pekerja Sebagai Upaya Perlindungan Berbasis Gender bagi Perempuan Buruh Pabrik di Kabupaten Karanganyar *Jurnal Forum Ilmu Sosial* **40** 1 44-55
[13] Women Research Institute 2016 *Buruh Perempuan dan Relasi Industrial*
[14] Delgado D, Morris J, Bravo D, Nguyen M, and Gottlieb B 2017 The Health Impacts of Domestic Labor on Women Workers in Massachusetts *Theory in Action* **10** 2 21-40

Reliability Analysis of Differential Relay as Main Protection Transformer Using Fuzzy Logic Algorithm

Y Mulyadi*, T Sucita, Sumarto and M Alpani

Department of Electrical Engineering Education, Universitas Pendidikan Indonesia

*yadielektro@upi.edu

Abstract. Electricity supply demand is increasing every year. It makes PT. PLN (Persero) is required to provide optimal customer service and satisfaction. Optimal service depends on the performance of the equipment of the power system owned, especially the transformer. Power transformer is an electrical equipment that transforms electricity from high voltage to low voltage or vice versa. However, in the electrical power system, is inseparable from interference included in the transformer. But, the disturbance can be minimized by the protection system. The main protection transformer is differential relays. Differential relays working system using Kirchoff law where inflows equal outflows. If there are excessive currents that interfere then the relays will work. But, the relay can also experience decreased performance. Therefore, this final project aims to analyze the reliability of the differential relay on the transformer in three different substations. Referring to the standard applied by the transmission line protection officer, the differential relay shall have slope characteristics of 30% in the first slope and 80% in the second slope when using two slopes and 80% when using one slope with an instant time and the corresponding ratio. So, the results obtained on the Siemens differential release have a reliable slope characteristic with a value of 30 on the fuzzy logic system. In a while, ABB a differential relay is only 80% reliable because two experiments are not reliable. For the time, all the differential relays are instant with a value of 0.06 on the fuzzy logic system. For ratios, the differential relays ABB have a better value than others brand with a value of 151 on the fuzzy logic system.

1. Introduction

In this modern era, the requirement for electrical energy in Indonesia is increasing. This is due to the increased population growth. In addition, electric energy is one of the most influential energy in society [1]. Usually, the electrical energy used by the society is supplied from the power system. The power system itself, composed of several sub-systems [2], It is a generation system, transmission system, and distribution system. These three components need to be maintained in the process by which a general requirement is set for the electric power system [3], among others reliability, quality and stability.

Transmission system is a system used to distribute electrical energy from the generating system with a certain voltage level and then raised the voltage to a higher level before entering the substation. The voltage increase is intended to power the electricity to be supplied to the substation remains in accordance with the needs after experiencing power losses on the transmission line. Generally, the substation has several fixtures, such as power transformers, separators, breakers, busbars and insulators, measuring instruments, releases and safeguards [4].

In the process of distribution, the transmission system, cannot be separated from the appearance of disturbances that interfere with the process. Interference that occurs one of them can be found in the power transformer. Disturbance on the power transformer is avoided, but can still be minimized. In the operation of power transformers, there are 2 kinds of interference that may be experienced, namely internal disturbance and external disturbance [5]. Internal disturbance is a disturbance that occurs in the transformer itself such as short circuit short circuit and core, winding with the tank, as well as insulation (oil) interference on the transformer. While external interference is a disruption that occurs outside the power transformer but can cause interference on the transformer concerned such as short circuit network interference, overload, and lightning surge [6].

Transformer is a major element that is very instrumental in power distribution system because it must be in sync with the amount of electrical energy needed. Because the transformer is the main element and has a high price, the system of protection or protection of a transformer both to the disturbances that occur from the transformer or from outside the transformer is very important [7].

One of the main protective relays on a power transformer is a differential relay. The differential relay is a protection from internal disturbance. The principle of differential relay work is based on Kirchoff's law, where the incoming currents on the primary side will be equal to the amount of outflow on the secondary side [8]. Because as a primary safety, differential relay working time is also made instant so that interference can be overcome. If the relay does not work in instant time, it will be fatal. Therefore, the differential relay must be assured of its reliability. In this case, the cause of the stability problem is focused on analyzing the reliability of the differential relays on the power transformer by analyzing the slope. The determination of slope value is determined by manual calculation and using some software. After getting a slope graph of each method, it will be adjusted to a standard slope setting to determine whether the relay is considered reliable as a primary safety transformer power.

Research data taken from West Bandung Basecamp. In transformer 2 in GIS Cibabat Lama, the differential relays recorded in the substation are still branded GEC type MBCH 12 [9]. GEC type MBCH 12 is a mechanical type of differential relay. However, in the tests performed, the relay on the transformer 2 in GIS Cibabat Lama branded Siemens which is a numeric type of differential relay. From that point, it can be seen that there has been a change of differential relay on transformer 2 in GIS Cibabat Lama. Then there is also a differential ABB relay in transformer 3 GI Cianjur and Toshiba's differential relay in transformer IBT 2 GI Cigelereng which will also be researched.

There are several research methods in analyzing the reliability especially for differential relays. The method used is by direct experiment along with manual calculation [10] and some are using the fuzzy logic method to determine the rule trip [11] and analyze about power transformer [12]. Results from the study will be useful for reference as well as consideration of the reliability of protection applied.

2. Methods

This study was conducted to determine the reliability of a protection in power transformers. To determine the reliability, it will be examined slope characteristic value, work time and its pick-up ratio. In conducting the research, previously required direct testing using the DRTS test tool to determine the values required for the research material. Once obtained these values, then the research can be implemented. In the implementation of research, can be used two methods of calculation manually and using some software. In this study, the software used is Ms. Excel, 87T and Matlab. In the calculation, used Ms. Excel and 87T which have the ability to calculate especially on the 87T software that specifically calculates the differential relays on the transformer. After all values are obtained, it will be concluded on matlab software with fuzzy logic processing. In fuzzy logic, three aspects are given which refer to a reliability of the differential relay. These three aspects are the slope characteristic values, the working time and the pick-up ratio. This research takes the object of three power transformers at Bandung different substations with different types of differential relays.

Table 1 is a power transformer data on three different substations, namely Cibabat Lama, Cianjur and Cigelereng. Transformer studied at GIS Cibabat Lama is a transformer 2 branded Hyundai which has power of 60 MVA, with impedance = 12,50%. Then, the transformer study in GI Cianjur is a transformer 3 branded GEC ALSTHOM which has a power of 60 MVA, with impedance = 12,20%. While the transformer under study in GI Cigereleng is a transformer IBT 2 branded PAUWELS that has a power of 31 MVA, with impedance = 12%. All three have the function of converting high voltage to medium voltage. The third group of the transformer vector is Ynyn0 with NGR resistance varying.

Table 1. Transformator Research Data

Specification	Transformer 2 GIS Cibabat Lama	Transformer 3 GI Cianjur	Transformer IBT 2 GI Cigelereng
Brand	Hyundai	GEC ALSTHOM	PAUWELS
Type	TL-128	-	QRF 31/275
Power	60 MVA	60 MVA	31 MVA
Voltage	150 / 20 kV	150 / 20 kV	150 / 70 kV
Imp. (%Z)	12,50	12,2	12
Vec. Group	YNyn0	YNyn0	YNyn0
R NGR	12 Ohm	12 Ohm	62 Ohm

Table 2 is a differential relay data in each transformer to be studied. In transformer 2 GIS Cibabat Lama used Siemens type differential relay type 7UT61 with Is of 0,3In. Then on transformer 3 GI Cianjur used relay differential brand ABB type RXDSB4 with Is of 0,35In. While the transformer IBT 2 GI Cigelereng used relay differential Toshiba type GRT 100 with Is of 0,3pu. The three differential relays must have their own characteristics.

Table 2. Differential Relay Research Data

Specificatio n	Differential Relay Transform er 2 GIS Cibabat Lama	Differential Relay Transforme r 3 GI Cianjur	Differentia l Relay Transform er IBT 2 GI Cigelereng
Brand	Siemens	ABB	Toshiba
Type	7UT61	RXDSB4	GRT 100
No. Series	BF1009109 727	-	N400082F N001
CT Primary	300/5 A	300/5 A	150/5 A
CT Secondary	2000/5 A	2000/5 A	250/5 A
In	5 A	5 A	5 A
Is	0,3 x In	0,35 x In	0,3 pu

The data will be a research material that will be proven its reliability. This research will be conducted in accordance with the order arranged on the flowchart. For more details, see Figure 1 of the research flowchart.

IOP Conf. Series: Materials Science and Engineering **306** (2018) 012133 doi:10.1088/1757-899X/306/1/012133

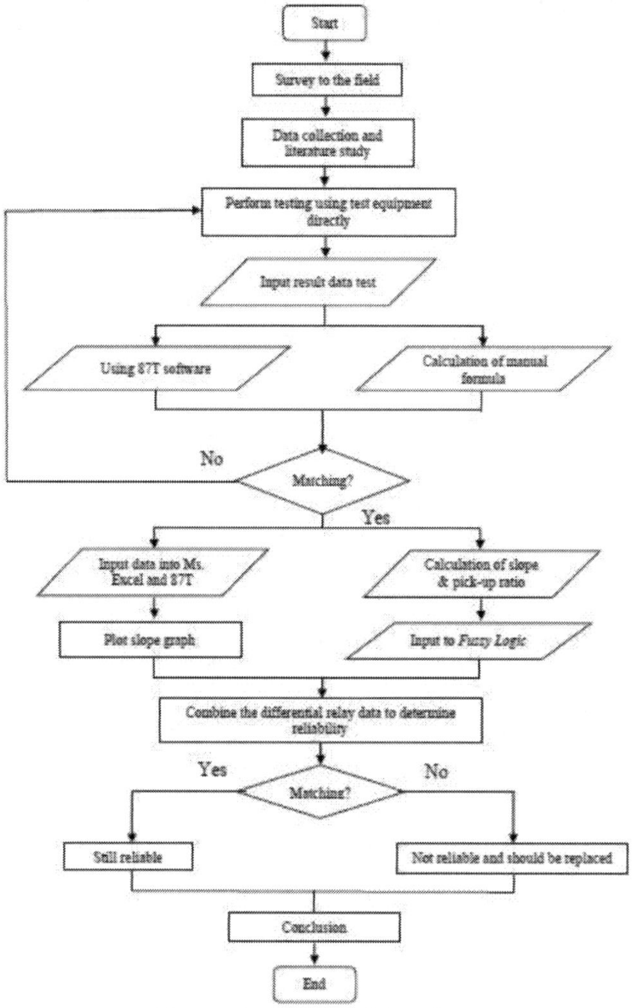

Figure 1. Research Flowchart

3. Results and discussion

Checking of differential relay settings can be seen in Figures 2, 3 and 4. Checking is done by manual calculation or using 87T software. After checking, it goes into the differential testing phase using the DRTS tool. After the differential relay testing, it can be determined the value of restrain currents and differential currents. The result of manual calculation of restrain current and differential current then compared with 87T software calculation to strengthen the calculation result. Comparison of restrain currents can be seen in Figures 5, 6 and 7 while differential current can be seen in Figures 8, 9 and 10. In general, almost all results have the same results. But at the time of testing directly, there are different things, especially on the relay differential brand Toshiba type GRT 100. In the differential relay data listed, the differential relay Toshiba type GRT 100 have nominal current of 5 A. However, the value of the pick-up current at the test is of little value. When a nominal current of 5 A is used, the ratio value will be too far from the standard. But if the nominal current used is 1 A, then the ratio result will still be in standard. Selectable nominal current of 1 A because the nominal current commonly used in the protection system is 1 A and 5 A. Therefore, 1 A is used with the assumption that data on differential relays must be updated. Checking is limited to the current transformer (CT) because there are several differential relays which automatically determine the setting value when given the data. So, to be more efficient enough in doing the check up to CT only.

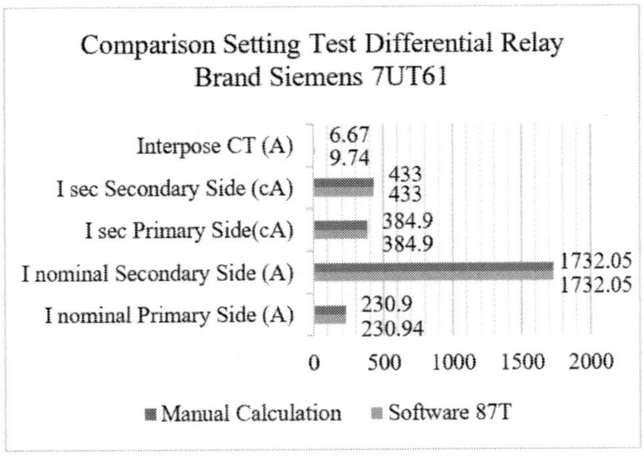

Figure 2. Checking Settings Differential Relay Siemens

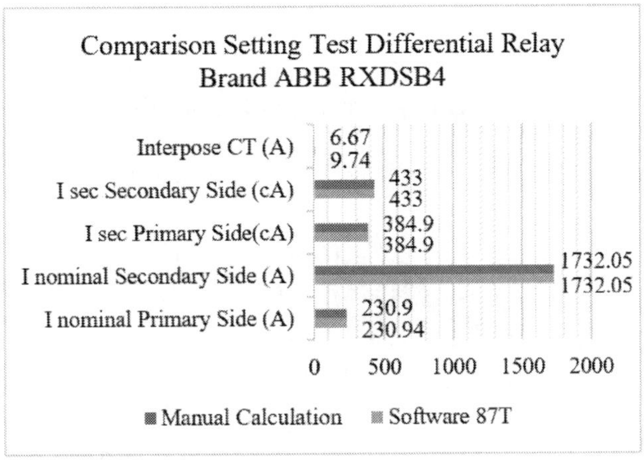

Figure 3. Checking Settings Differential Relay ABB

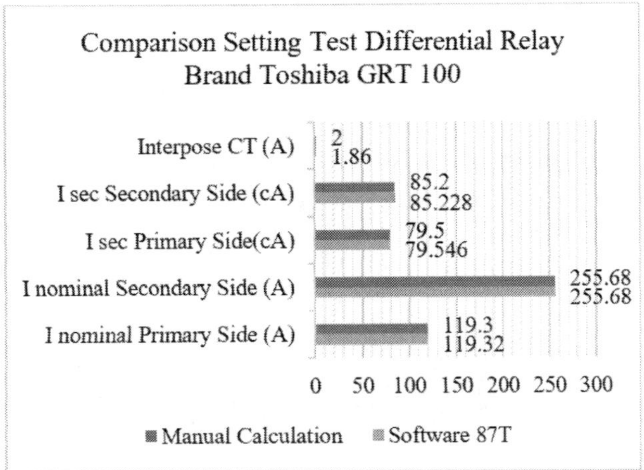

Figure 4. Checking Settings Differential Relay Toshiba

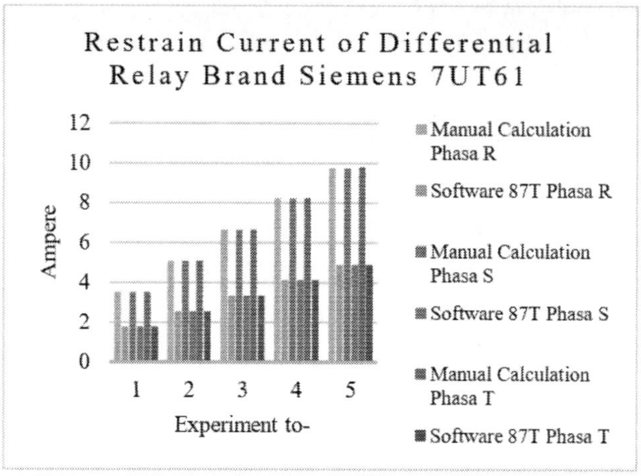

Figure 5. Comparison I_{rest} Differential Relay Siemens

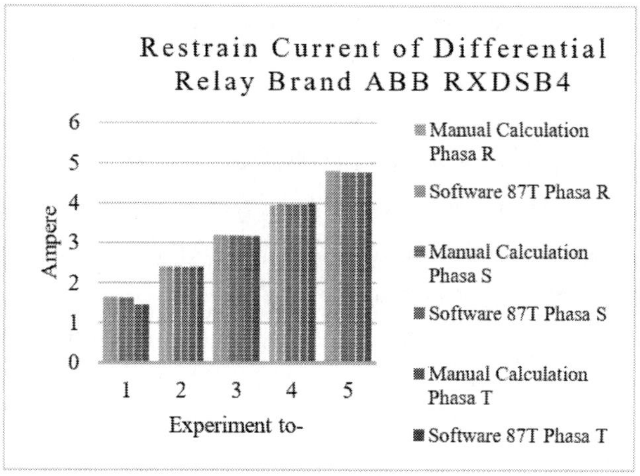

Figure 6. Comparison I_{rest} Differential Relay ABB

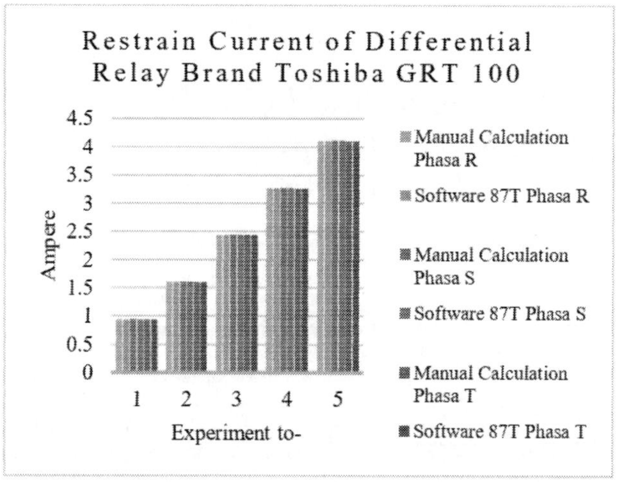

Figure 7. Comparison I_{rest} Differential Relay Toshiba

From the comparison graph of Irest in the above three differential relays there is no difference between manual and 87T software except Siemens. That's because the 87T software is made to calculate the general formula while Siemens different formulas.

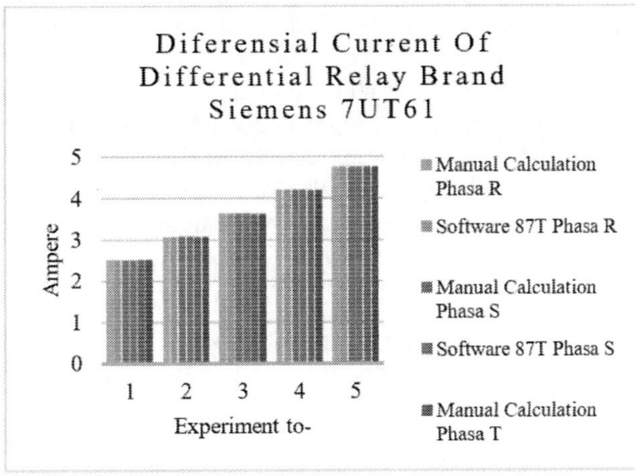

Figure 8. Comparison I_{diff} Differential Relay Siemens

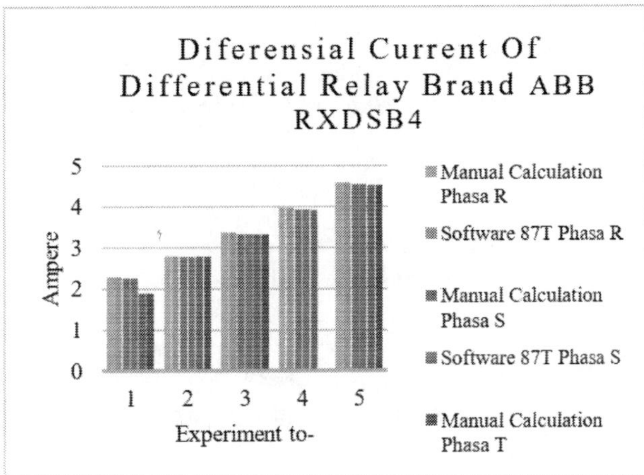

Figure 9. Comparison I_{diff} Differential Relay ABB

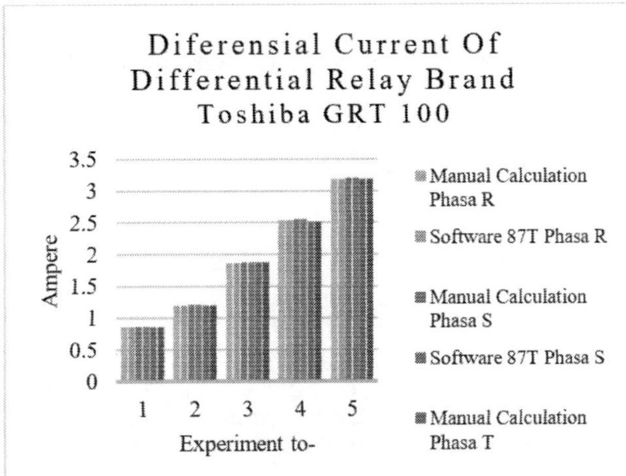

Figure 10. Comparison I_{diff} Differential Relay Toshiba

From Idiff comparison graph in the above three relay relays, there is no difference between manual and 87T software. The values obtained also rise from the first to the fifth experiments, as desired from the differential relay testing.

Having obtained the value of the restrain current and differential current of each differential relay, it can be in the plot of the graphical characteristics of the differential relay slope. The characteristic graph of the differential relay slope refers to the restrain current as the numerator (X) and the differential current as the denominator (Y). From the results graph slope characteristics obtained differential relays form of charts moving up and no decrease, in accordance with the standards of differential relays. To be clearer, the graph of differential relay characteristics can be seen in figures 11, 12 and 13.

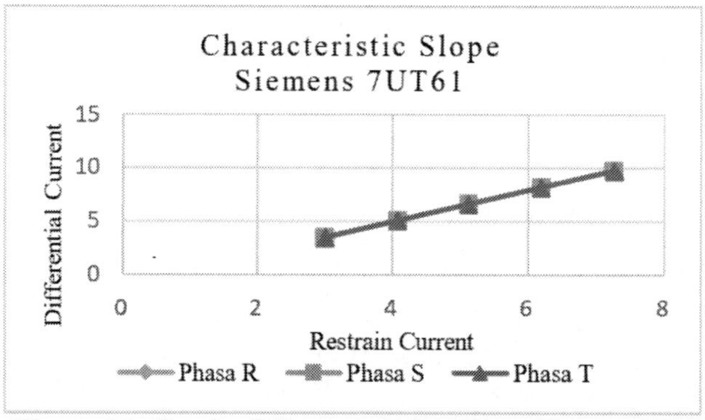

Figure 11. Graph of Characteristics Slope Siemens 7UT61

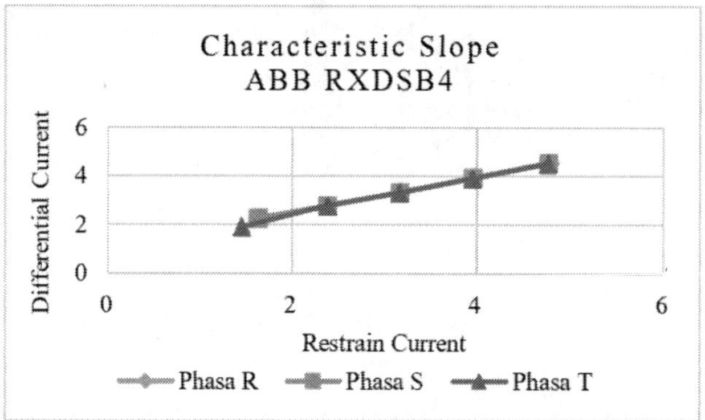

Figure 12. Graph of Characteristics Slope ABB RXDSB4

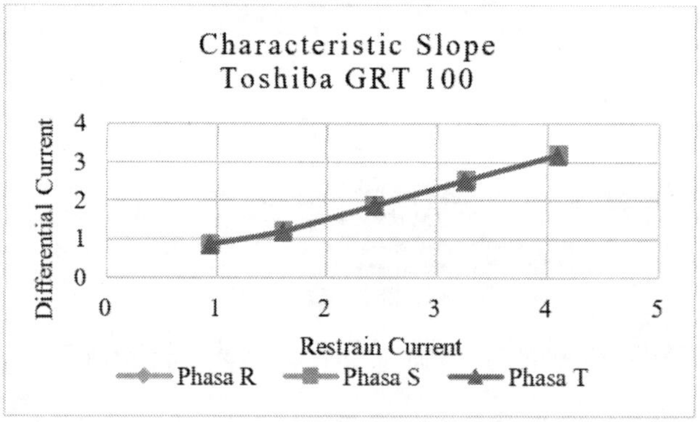

Figure 13. Graph of Characteristics Slope Toshiba GRT 100

Table 3 is the characteristic slope value of the differential relay brand Siemens type 7UT61 on the transformer 2 GIS Cibabat Lama. The value of slope characteristics obtained has values that almost resemble each other in each experiment. This indicates that the performance of the simulated differential relays resembles each other in each phase.

Table 3. Values of Characteristics Slope Siemens 7UT61

Experiment to-	Phasa R	Phasa S	Phasa T
1 & 2	35,8974359	36,30573248	35,8974359
1 & 3	36,1022364	36,10223642	35,8974359
1 & 4	36,0341151	36,03411514	35,8974359
1 & 5	36	36	36
2 & 3	36,3057324	35,8974359	35,8974359
2 & 4	36,1022364	35,8974359	35,8974359
2 & 5	36,0341151	35,8974359	36,0341151
3 & 4	35,8974359	35,8974359	35,8974359
3 & 5	35,8974359	35,8974359	36,1022364
4 & 5	35,8974359	35,8974359	36,3057324

Table 4 is the slope characteristic value of the differential relay branded RBDSB4 type ABB in the transformer 3 GI Cianjur. The value of slope characteristics obtained has various values. Even in two experiments, the slope characteristic values were outside the slope 1 or slope 2 area and found values that exceeded the allowed tolerance limit. Then on each phase, there are some experiments whose values are not similar. This indicates that the performance of the differential relay ABB brand is different from each other in each phase.

Table 4. Values of Characteristics Slope ABB RXDSB4

Experiment to-	Phasa R	Phasa S	Phasa T
1 & 2	67,549668	67,549668	93,617021
1 & 3	69,706840	69,706840	83,040935
1 & 4	71,244635	71,244635	80,239520
1 & 5	72,813990	72,408293	79,154078
2 & 3	71,794871	71,794871	70,129870
2 & 4	73,015873	73,015873	72,204472
2 & 5	74,476987	73,949579	73,417721
3 & 4	74,213836	74,213836	74,213836
3 & 5	75,776397	75	75
4 & 5	77,300613	75,776397	75,776397

Table 5 is the slope characteristic value of the differential relay branded Toshiba type GRT 100 in the transformer IBT 2 GI Cigelereng. The value of slope characteristics obtained has various values. Even in two experiments, the slope characteristic values were outside the slope 1 or slope 2 area and found values that exceeded the allowed tolerance limit. Then on each phase, there are some experiments whose values are not similar. This indicates that Toshiba's differential relay performance is different from each other in each phase.

Table 5. Values of Characteristics Slope Toshiba GRT 100

Experiment to-	Phasa R	Phasa S	Phasa T
1 & 2	49,62406	50,74626	50,74626
1 & 3	66,66666	66,66666	67,10963
1 & 4	71,52034	71,52034	70,96774
1 & 5	73,41772	73,61769	73,41772
2 & 3	80,23952	79,51807	80,23952
2 & 4	80,23952	79,87987	79,15407
2 & 5	79,75951	79,75951	79,51807
3 & 4	80,23952	80,23952	78,04878
3 & 5	79,518072	79,87987	79,15407
4 & 5	78,787878	79,51807	80,23952

In addition to the differential relay slope values, determining a reliability also refers to the work time and the pick-up ratio. Its reason is the character of the differential relay which is a major protection on the transformer, so the working time of the relay must be instantaneous in overcoming the interference. In addition, there is a pick-up ratio where the pick-up current must match with the current setting of the differential relay. That is because, if the pick-up current is too large, the ratio will exceed the current rating setting. That is, the new differential relays work when the current exceeds the setting with a great distance. It is not allowed because of the role of differential relay as the main protection. Smaller pick-up currents are also not allowed because it means the relays work when there is a current that is smaller than the current setting.

Table 6. Work Time and Pick-Up Ratios Differential Relays

Brand	Phasa	Side	Time (s)	Ratio (%)
Siemens 7UT61	R	*Prim*	0,0288	115,3333
		Sec	0,0236	129,3333
	S	*Prim*	0,0263	115,3333
		Sec	0,078	129,3333
	T	*Prim*	0,0332	115,3333
		Sec	0,0388	129,3333
ABB RXDSB4	R	*Prim*	0,0318	101,7143
		Sec	0,0776	101,7143
	S	*Prim*	0,0252	100,5714
		Sec	0,0282	100,5714
	T	*Prim*	0,0659	101,1429
		Sec	0,0179	101,1429
Toshiba GRT 100	R	*Prim*	0,0357	136,6667
		Sec	0,0356	173,3333
	S	*Prim*	0,0354	140
		Sec	0,046	173,3333
	T	*Prim*	0,0532	140
		Sec	0,0638	173,3333

Table 6 is the work time and pick-up ratio differential relay under study. Working time and ratio are reviewed from two sides, namely primary and secondary. From the table above got the value of working time under 0,01 seconds. As for the pick-up ratio has a diverse value. However, the greatest pick-up ratio was in the Toshiba differential relay on the transformer IBT 2 GI Cigelereng with a value on the primary side of phase R is 136,67%, on the primary side of S and T is 140%. Then on the secondary side has the greatest value with a value of 173,33%.

After obtained three aspects to determine the reliability of differential relays, then used Matlab software with fuzzy logic processing. The fuzzy logic used in this study is named Fuzzy_Keandalan with seven inputs and three outputs as needed. For more details, see Figure 14 on fuzzy logic design.

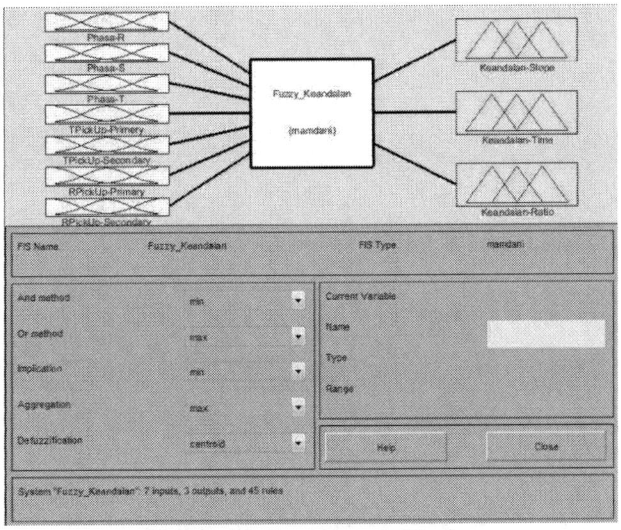

Figure 14. Fuzzy Logic Design

Table 7 is the result of reliability of work time and of pick-up ratio using Fuzzy_keandalan for the three differential relays discussed in this study. The reliability of the differential relay working time is instantaneous with a fuzzy value of 0.06 which is included in the instant set. As for the tolerance of the pick-up ratio is good only on the differential relay ABB. The ABB differential relay is good in all three phases with a fuzzy value of 151. The differential relays of Siemens and Toshiba have a bad pick-up ratio in three phases with a fuzzy value of 49 that belong to a bad set.

Table 7. Reliability of Work Time and Differential Relay Ratio

Name	Phasa	Reliability of Work Time	Reliability of Ratio Pick-up	Explanation
Siemens 7UT61	R	0,06	49	Instant & Bad
	S	0,06	49	Instant & Bad
	T	0,06	49	Instant & Bad
ABB RXDSB4	R	0,06	151	Instant & Good
	S	0,06	151	Instant & Good
	T	0,06	151	Instant & Good
Toshiba GRT 100	R	0,06	49	Instant & Bad
	S	0,06	49	Instant & Bad
	T	0,06	49	Instant & Bad

Table 8 is the reliability result of slope characteristics using Fuzzy_keandalan for the three differential relays discussed in this study. The differential relay which has the best slope characteristic reliability is the differential relay brand Siemens. The Siemens differential relay is reliable in each experiment because it has a fuzzy value of slope reliability of 30 for slope 1 that belongs to a reliable set. The differential relay ABB and Toshiba have two unreliable experiments with a fuzzy value of 55 that fall within the unreliable set. However, in other experiments, they belong to a reliable set with a fuzzy value of 80.

Table 8. Reliability Characteristics Slope of Differential Relay

Name	Phasa	Experiment	Reliability Slope	Explanation
Siemens 7UT61	R, S & T	1 & 2	30	Reliable
		1 & 3	30	Reliable
		1 & 4	30	Reliable
		1 & 5	30	Reliable
		2 & 3	30	Reliable
		2 & 4	30	Reliable
		2 & 5	30	Reliable
		3 & 4	30	Reliable
		3 & 5	30	Reliable
		4 & 5	30	Reliable
ABB RXDSB4	R, S & T	1 & 2	55	Unreliable
		1 & 3	55	Unreliable
		1 & 4	80	Reliable
		1 & 5	80	Reliable
		2 & 3	80	Reliable
		2 & 4	80	Reliable
		2 & 5	80	Reliable
		3 & 4	80	Reliable
		3 & 5	80	Reliable
		4 & 5	80	Reliable
Toshiba GRT 100	R, S & T	1 & 2	55	Unreliable
		1 & 3	55	Unreliable
		1 & 4	80	Reliable
		1 & 5	80	Reliable
		2 & 3	80	Reliable
		2 & 4	80	Reliable
		2 & 5	80	Reliable
		3 & 4	80	Reliable
		3 & 5	80	Reliable
		4 & 5	80	Reliable

From both tables above, it can be determined the final result of research reliability of differential relays on three different substations in the Bandung area. In the Siemens type 7UT61 differential relay in transformer 2 GIS Cibabar Lama has 100% reliable slope characteristics with instant working time. But on the ratio, this relay has a bad value. Then on the differential relay brand ABB type RXDSB4 in transformer 3 GI Cianjur has 80% reliability slope characteristics and 20% slope characteristics are not reliable. However, the working time of this relay is still relatively instant and has the best ratio among other differential relays with a value of 100% on both sides of each phase. Similarly, in the differential relays brand Toshiba type GRT 100 in transformer IBT 2 GI Cigelereng which has 80% reliability slope characteristics and 20% slope characteristics that are not reliable. The working time of this relay is still relatively instant, but the ratio of the differential relay is quite bad on both sides of each phase. To be clearer, it can be seen in Figure 15.

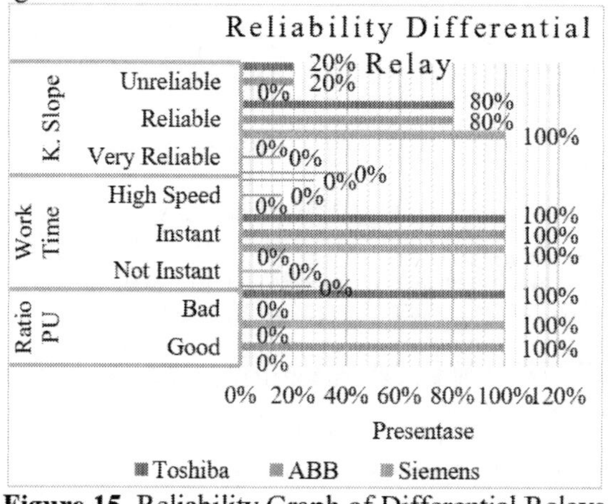

Figure 15. Reliability Graph of Differential Relays

4. Conclusion

First, the calculation result of check setting and testing differential relay with brand Siemens type 7UT61 on transformer 2 150/20 kV 60 MVA in Cibabat Lama, differential relay with brand ABB type RXDSB4 on transformer 3 150/20 kV 60 MVA in Cianjur and differential relay with brand Toshiba type GRT 100 on transformer IBT 2 150/70 kV 31 MVA in Cigelereng has a calculation setting value very close to that applied so that it can be done differential relay testing to find out the values needed to determine reliability. However, in the differential relay testing with brand Toshiba type GRT 100, the pick-up ratio is not suitable if I nominal used for 5 A as in the data transformer. Therefore, 1 A is used for the results to be in accordance with the assumption that the transformer data needs to be updated.

Second, the calculation results slope characteristics, working time and the ratio of the pick-up of third differential relays are different. On differential relays with brand Siemens type 7UT61 on transformers 2 150/20 kV 60 MVA in Cibabat Lama Has a better slope characteristic value than the other two differential relays because the result of the calculation is in accordance with the standard set for the differential relay having two slopes where the value of the Siemens brand differential relay slope characteristic is ± 30% with a tolerance of ± 10% which is classified in the slope 1. While the differential relay brands ABB and Toshiba only have one slope with a standard value of ± 80% with a tolerance of ± 10%. However, for the best working time and pick-up ratios are held by differential relay with brand ABB RXDSB4 type on transformer 3 150/20 kV 60 MVA in Cianjur because it has a pick-up ratio that is still within the limits of tolerance so categorized well. But overall, the working time in all the differential relays is still instant because it works under 0,01 seconds.

Third, the results of reliability decisions through fuzzy logic for differential relays with Siemens brand type 7UT61 on transformers 2 150/20 kV 60 MVA in Cibabat Lama is declared reliable for a while with records to be replaced because it has a ratio that exceeds tolerance. Then on a differential relay with brand ABB type RXDSB4 on transformers 3 150/20 kV 60 MVA in Cianjur declared reliably for a while with the record should be replaced because it has 2 unreliable results from 10 experiments. Finally, the differential relay with brand Toshiba type GRT 100 on the transformer IBT 2 150/70 kV 31 MVA in Cigelereng is not reliable because it has 2 unreliable results from 10 trials and has a bad pick-up ratio value.

References

[1] R Parohon, T Tambunan, and S Handoko 2013 *Analisa Kestabilan Transien dengan Pelepasan Pembangkit dan Beban (Generation / Load Shedding) Pada Sistem Jaringan Distribusi Tragi Sibolga 150 / 20 kV (Studi Kasus Pada Penyulang Tragi Sibolga , Sumut*) 1–11
[2] D P Hariyanto, Tiyono and Sutarno 2009 *Analisis Koordinasi Over Current Relay dan Recloser Di Sistem Proteksi Feeder Gardu Induk Semen Nusantara (SNT 2) Cilacap* **1** 1 1–10
[3] V Aprilia 2016 *Analisis Gangguan 3 Fasa pada Saluran Transmisi terhadap Transient Stability Sistem Multimesin Menggunakan Metode Range-Kutta Orde 5* **5**
[4] N R Fitriani 2017 *Analisis Penggunaan Rele Diferensial sebagai Proteksi pada Transformator Daya 16 MVA di Gardu Induk Jajar*
[5] PT PLN (Persero) PUSDIKLAT 2009 *Pengoprasian Peralatan Gardu Induk*
[6] J C Napitupulu 2013 *Analisis Keandalan Transformator Daya Menggunakan Metode Distribusi Weibull* 5–43
[7] S I Panjaitan 2012 *Studi Pengaruh Beban Lebih Terhadap Kinerja Relai Arus Lebih pada Transformator Daya*
[8] H Wisatawan 2012 *Evaluasi Setting Rele Overall Differential GT 1 . 1 PLTGU Grati dan Rele Jarak GITET Grati pada Bus 500 kV*
[9] PT PLN (Persero) 2006 *Data Setting Relay Proteksi GIS Cibabat Lama*
[10] H Setijasa 2013 *Pengujian Relay Diferensial GI* **9** 2 74–79
[11] K Ramesh and M Sushama 2010 *Power Transformer Protection using Fuzzy Logic Based-Relaying*

[12] M Haris, M Salik, A A Safdar and U Rashid 2013 *Improved Fuzzy Logics Based Differential Protection Scheme* 261–266

Do Technological and Vocational High Schools Differentiate between Male and Female Teachers?

Y Rahayu*, A G Abdullah, E P Asfiyanur, R C Putra

Program Pascasarjana Universitas Pendidikan Indonesia
Jl. Dr. Setiabudi No. 227, Bandung, 40153, Indonesia

*rhyayu08@student.upi.edu

Abstract. High Quality Vocational education is one way to create skilled and professional human resources. In the implementation of teaching and learning process there are many components that are very important one of them is the educator (teacher), where through the quality of teachers materials are expected to be well absorbed by students. Teachers generally consist of male and female teachers where in this era of teacher globalization teachers in vocational schools are not only dominated by male teachers, there are many women who serve as educators in unlimited vocational schools by selected majors. But the polemic is the issue of gender inequality that has been the subject of talks in various countries since 1979. Gender bias in education is the educational reality that benefits certain sexes, leading to gender inequality in which various forms of gender inequality occur in different regions. Female teachers in technology and engineering as a minority are deemed incompetent in understanding vocational materials and are deemed less able in the application of school practices than male teachers. But at this point it can be observed that the large number of female teachers who enter the world of teaching skills in vocational schools in Indonesia. Therefore, this research was conducted to find out the extent of gender differences in the influence of teaching styles on the learning process in SMK with the concentration of technique and technique. This research is planned to be implemented in a vocational high school in Indonesia with concentration of Department of Engineering and Technology which include the use of qualitative research collecting research data by using interview and survey technique

1. Introduction

Gender refer to the economic, social, political and cultural attributes and opportunities associated with female and male. The social definition of what it means to be female or male vary among cultures and changes over time. Gender should be conceptualized as a set of relations, existing in social institutions and reproduced in interpersonal interaction. Gender theory is a social constructionist perspective that simultaneously examines the ideological and the material level of analysis [1].

In this era gender have not been a limitation for career, job, politic, and others, especially in study. A lot of teacher in many vocational schools are dominated by man but its 40% are women. We also can look to senior high school that dominated by women teachers are influence to the interest of student's study. The circumstances on class senior high school more competitive than vocational high school [2]. In this major (technology and engineering) need toughness and masculine of man but it isn't rule out female teacher can do it [3]. In other word nowadays, women can do anything that man can do, so do with teaching in vocational high school. In West Java just a few female teachers who teach the productive subject, but it is not make them feel less confidence. And not less female teacher that teach multiple productive major [4].

Universitas Pendidikan Indonesia is one of the university that graduate female vocational teacher and many university did it too. But some school in West Java put them inside class even they have a good competence about productive major which tough on workshop on reason gender. For this reason, female teacher cannot do improvement on knowledge and competences. And cause of it female teacher feel discriminated and its decrease in interest in become productive teacher [5].

The female teacher can make learning circumstance more attractive one of the cause is student more interested to the female teacher which can adapt in many circumstance. And also, can improve the learning outcomes that will prove by this research. Based on text above the writer take the main problem are: 1) how is the circumstances in learning process when female teacher teaching, 2) how the student receive the learning process, 3) how the improvement of value of learning outcomes

2. Methods

This research will collect data from the teacher and student using interview technique and questionnaire distributed to respondent, using mix method qualitative and quantitative method [6][7]. The population of respondent are the student and teacher of Vocational High School in Bandung area with concentration of Technology and Engineering Department.

The research has following purpose: 1) to establish whether there are differences in terms of vocational study interest caused by teacher gender; 2) to reveal whether there are value differences of study caused by teacher gender.

Hi: Suppose there are difference influenced by women teacher better than man teacher.

Ho: Suppose there are difference influenced by women teacher not better than man teacher.

In order to have trough analysis, researcher formulated the following working hypothesis: there are significant statistical differences value of learning outcomes, investigates the student interest, and the circumstances learning process[8].

3. Results and discussion

Gender Mainstreaming (PUG) is a manifestation of a global commitment to respecting human rights, relating to equal opportunity and treatment for men and women to participate in various areas of life in society. In Indonesia, this commitment is set forth in Presidential Instruction No. 9 of 2000 on Gender Mainstreaming in National Development. Through PUG, among others, it is expected to increase the status, role, and quality of women, to realize gender equality and equity in men and women in gaining access, control, participation, and benefits of each development sector. In the field of education, through permendiknas number 84 of 2008 stated that PUG education is a strategy to integrate gender into an integral dimension of planning, compilation, implementation, monitoring, and evaluation of education policies, programs, and activities. Through PUG, it is expected to create gender responsive education through the arrangement of materials or teaching materials, classroom learning process, and school physical environment, as stated in the permendiknas: "The learning process needs to be improved in order to be fully gender responsive which is shown by (i) the material teaching materials that are generally gender biased; (ii) classroom learning processes that have not fully encouraged equally active participation between male and female students; and (iii) the physical environment of schools that have not addressed the specific needs of boys and girls. In addition, education management also needs to be implemented in a gender-equitable way or provides equal opportunities for men and women to participate in the decision-making process [9].

Deconstructing patriarchal gender values toward egalitarian gender values is not easy. This is not only because the community is still strongly holding on to the traditional essence of gender construction that women are wives and mothers, but also because schools as institutions that are trusted to socialize and "mold" the child themselves are also colored by the nuances of traditional gender values, so that there is a tendency for education (schools) to be "inhibiting" the process of promoting gender equality in society [10].

Men and women are expected to be equal partners in various development activities, so it is appropriate that the gender values being socialized to children today are egalitarian gender values. Awareness of values that uphold the values of equality and anti-discrimination against women or men should be cultivated [11]. The patriarchal gender values that create dichotomies and place men and women in unequal positions must be criticized and deconstructed, since in the traditional gender (patriarchy) position of women and men many are placed imbalanced and unjust [11]. Culture that dominates almost every aspect of life has caused people to simply accept, legitimize, and apply that culture in their lives. Traditional gender values have hegemonized the minds and behaviors of everyone, so the critical power of individuals and society to see the phenomenon of injustice in the value system tends to "die." To break the chain of unfair treatment development, it needs serious and continuous effort to building awareness about equality and relationship equity between men and women. In this context, education, especially schools, has a very strategic role to foster and develop egalitarian gender insight and awareness.

One of the most interesting issues in education is the influence of gender diversity in vocational education. From a gender perspective, there is a stigma that men are considered superior to women. The stigma places women as second class citizens, including in terms of vocational education. Due to the stigma tesebut, then came the view that technical education is a male domain manifested in masculine identity. Until now, people still tend to be skeptical of female teachers. This is reflected in the percentage of female teachers who are still far below male teachers.

3.1. Gender Equality in the Curriculum

The curriculum has a fundamental role in ensuring that boys and girls have equal opportunity in achieving success and entering a higher school level. Nevertheless, gender bias in the curriculum and teaching-learning materials are still commonly found in developing and middle-income countries. Through gender-biased texts and pictures or images, gender stereotypes become more reinforced. The results of gender analysis in textbooks in Indonesia conducted in 2011 also found a considerable gender bias in textbooks in Indonesia.

Good practice in mainstreaming gender equality into curricula and teaching materials including periodic revisions of these materials to include gender-sensitive approaches and gender perspectives. Also included in this effort is the establishment of a formal institution tasked with eliminating gender stereotypes in textbooks and other learning materials [12].

3.2. Gender Equality in Teacher Development

Good practice in teacher development to support gender equality means complementing teachers' understanding of gender equality in the classroom, in the school environment and beyond, and in the community at large [11]. To achieve this, teachers need to have the ability to promote this understanding in the classroom and develop practical strategies and solutions to address the various learning challenges facing boys and girls.

Training in gender equality will be very effective if it is integrated in pre-service and in-service training for teachers [5]. To conduct this training well, teacher training institutes and networks require adequate capacity to teach the pedagogy of active learning [13]. The teaching of the material should have a gender perspective perspective in the development of the training, including the ability and knowledge of the trainer in delivering the training materials. To date, most provincial education quality assurance agencies (LPMP), teacher training institutions, and local teacher professional development networks have never taken part in mainstreaming gender equality.

3.3. Gender Equality in Achieving Learning Outcomes

Gender disparity in achieving learning outcomes is not unavoidable. There is no fundamental difference in the ability of boys and girls in vocational subjects. If conditions support, the performance of boys and girls will be the same in the subjects [14].

In choosing subject specializations at the upper and upper secondary levels, counsel teachers tend not to encourage female students to choose subjects and career possibilities traditionally considered male-dominated [15]. Due to factors such as the perception of the mentor teacher and the gender stereotypes in both men's and women's abilities in vocational subjects, few female students, when compared with boys, take science-related classes or technology in vocational or higher levels . Gender stereotypes encourage women to have careers in administration, child development, education and public health, while men in science, engineering, technology, and law.

External equality is achieved when women and men have equal status in access to goods and resources, in contributing, participating, and utilizing economic, social, cultural and political activities. This has implications for equal and equal opportunities in career and income generation between women and men who share the same qualifications and experience. Although this dimension of equality is outside the education system, the attainment of equality in education serves as a catalyst and contributes to an understanding of equality between women and men in other parts of life, including labor markets and other domestic spaces.

4. Conclusion

The final goal of this research is about the gender concept which is the construction of positioning of female and male teachers by each student in the school. The position of female and male teachers is not fixed according to the culture in each school. Because the positioning of women varies in each gender construction, the injustices they experience are due to different factors. A teacher can teach a lesson well, definitely need a business that requires a mature organizing and all components in teaching situations. Components that include selection: methods, materials, objectives, media, evaluation and learning model. In all teaching and learning activities, the components of the learning model include playing an important role. Because the selection of appropriate teaching strategies in the use of learning models is crucial to student learning outcomes.

Gender equality and gender equity learning needs to be improved because there are still many gender bias symptoms in schools. Men tend to be placed in a more favorable position throughout the entire educational process. The content of textbooks revealing the status and function of women in the family and society is not yet fully gender-sensitive and includes the concept of gender equality.

References

[1] J Campbell and M Gillespie 2017 Gender analysis of spending on the Scottish Modern Apprenticeship programme," 2017.

[2] M T P Carmona and I M García 2013 Influences, values and career aspirations of future professionals in education: a gender perspective *Educ. Rev.* **65** 3 357–371

[3] E Lahelma, S Lappalainen, T Palmu and L Pehkonen 2014 Vocational teachers' gendered reflections on education, teaching and care *Gend. Educ.* **26** 3 293–305

[4] P Fessler and A Schneebaum 2012 Gender and Educational Attainment Across Generations in Austria *Fem. Econ.* **18** 1 161–188

[5] A Braun 2012 Trainee teachers, gender and becoming the 'right' person for the job: Care and authority in the vocational habitus of teaching *Oxford Rev. Educ.* **38** 2 231–246

[6] K B Jensen 2012 *A Handbook of Media and Communication Research: Qualitative and Quantitative Methodologies*

[7] M D Myers 1997 Qualitative research in information systems *Manag. Inf. Syst. Q.* **21** 1–18

[8] A R Smith, J M Colombi and J R Wirthlin 2013 Rapid development: A content analysis comparison of literature and purposive sampling of rapid reaction projects *Procedia Comput. Sci.* **16** 475–482

[9] Menteri Pendidikan Nasional *Permendiknas No. 84 Tahun 2008 Tentang Pedoman Pelaksanaan Pengarusutamaan Gender Bidang Pendidikan*

[10] S W Han 2016 National education systems and gender gaps in STEM occupational expectations *Int. J. Educ. Dev.* **49** 175–187

[11] Editorial 1999 Gender issues in vocational education and training and workplace achievement of 14-19 years olds: an EOC perspective," *Cirriculum J.*, vol. 10, no. 2, pp. 209–229, 1999.

[12] M Owor and H D Musoke 2014 Neglect of gender questions at the vocational stage of judicial education in Uganda *Int. J. Leg. Prof.* **21** 3 281–296

[13] L Darling-Hammond 2017 Teacher education around the world: What can we learn from international practice? *Eur. J. Teach. Educ.* **40** 3 291–309

[14] R Hilal 2012 Vocational Education and Training for women and youth in Palestine: Poverty

reduction and gender equality under occupation *Int. J. Educ. Dev.* **32** 5 686–695
[15] M S Ortega and R G García 2017 Motivations and Decisive Factors in Women's Entrepreneurship. A Gender Perspective in Education and Professional Guidance *Procedia - Soc. Behav. Sci.* **237** 1265–1271

IOP Conf. Series: Materials Science and Engineering **306** (2018) 012135 doi:10.1088/1757-899X/306/1/012135

Personal Computer-less (PC-less) Microcontroller Training Kit

Y Somantri, D Wahyudin*, and I Fushilat

Universitas Pendidikan Indonesia

*deewahyu@upi.edu

Abstract. The need of microcontroller training kit is necessary for practical work of students of electrical engineering education. However, to use available training kit not only costly but also does not meet the need of laboratory requirements. An affordable and portable microcontroller kit could answer such problem. This paper explains the design and development of Personal Computer Less (PC-Less) Microcontroller Training Kit. It was developed based on Lattepanda processor and Arduino microcontroller as target. The training kit equipped with advanced input-output interfaces that adopted the concept of low cost and low power system. The preliminary usability testing proved this device can be used as a tool for microcontroller programming and industrial automation training. By adopting the concept of portability, the device could be operated in the rural area which electricity and computer infrastructure are limited. Furthermore, the training kit is suitable for student of electrical engineering student from university and vocational high school.

1. Introduction

Microcontroller learning generally uses system minimum 89S51 microcontroller type as a training tool then it is developed with AVR 8535 microcontroller. for some years [1]. Student's learning result is quite good from cognitive, affective, and psychomotor domains, yet with modular based microcontroller system student is prefer to used modular system than arranged it independently by themselves, since it is quite complicated. Modular system is considered fun and easy in microcontroller system learning and develop its application system [2]. Then it is designed and developed a modular system with Latepannda central processing unit which features more complete and large enough memory [3]. Lattepanda serves as mini-computer to set up the system, write down a memory program, input-output, serial communication and so on, while as a target (controller) uses Arduino Uno modular that embedded system by central processing unit. This microcontroller training tool's design does not use personal computer or Laptop. It is designed to control input-output system, serial communication, Bluetooth, jack audio, Wi-Fi, USB connector [4].

2. Design of PC-Less microcontroller training kit

The design model in block diagram is shown in figure 1, In general, this training tool consists of 4 process charts, there are: Input (Entry), Process (Compiler & Builder), Target Controller, and Output Interface module [5].

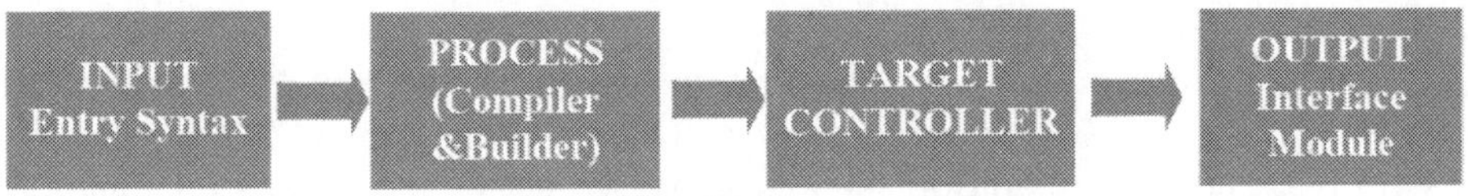

Figure 1. Block Diagram System

Microcontroller Training Kit Personal Computer Less block diagram completely shown in figure 2.

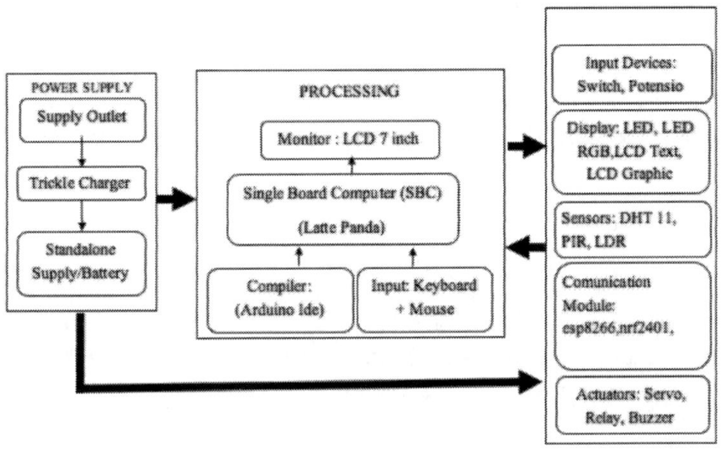

Figure 2. Block Diagram Microcontroller Training Kit Personal Computer Less

2.1. Power Supply

This section is a Direct Current (DC) power supply provider for the whole part and it is in training kit. PSU block's overall specification is as follows: Input: AC Voltage 220V, Current = 1 A, and 50-60 Hz Frequency. Output: DC Voltage 12 V, Current 1,5 A.

There is a section called Trickle Charger in PSU block, this section serves as DC power supply's source yet it also serves for Battery charging. In trickle-charger there is a controller that serves as a power supply direction's regulator, when the battery is fully charged (full Charge) then automatically the supply direction will switch DC supply directly to the circuit in Training kit, and supply to the battery is disconnected. When the voltage from AC Outlet Supply Outlet is absent/unconnected/disconnected the trickle-charger switches the DC voltage supply to the circuit work on the training kit towards the Battery, so that the power source for training kit's work comes from the Battery.

The battery on PSU block is used as backup power source if there is no power in AC Outlet Supply, therefore training kit allows to keep working about 3 hours although there is no power in AC Outlet Supply. Battery's specification: Output: Voltage 12 V, current 2400 m AH.

2.2. Processor

This block is the brain of the training kit which have many tasks i.e., starting from detects and processes learners' input in the form of data or command in training kit's operation, then forwards it to interface block. Single Board Computer (SBC) [6], as a data and command processor use Lattepanda with a Compiler installed, a software named Arduino IDE. This Software that will be used by learners to write syntax/program code in training kit's use, then it changes syntax/program code into understandable machine code by controller training kit, Arduino Mega2560. To control or communicate with interface components, the learner's inputs syntax by available wireless and navigates command by wireless mouse. In this block there is a 7 Inch LCD screen serves as display, learners can view a written program and navigate other program includes debugging and others.

IOP Conf. Series: Materials Science and Engineering **306** (2018) 012135 doi:10.1088/1757-899X/306/1/012135

2.3. Display
This section contains interface component that consists of input: Push Button, Potensio Meter Single, and Display's component: consist of 16-character text LCD, Double Row, LCD Graphic Nokia 5110, LED and RGB LED. Furthermore, sensor component that consists of temperature and humidity sensor (DHT 11), PIR sensor (Passive Infra-Red) and Light Dependent Resistor (LDR) sensor serves as light sensor. This microcontroller training kit is equipped with wireless communication module, there are Bluetooth module, Wi-Fi Module (Esp 8266) and wireless communication module uses 2,4 Ghz frequency (NRF2401), and the last in Interface block is equipped with actuator component such as Motor Servo 1800, Relay 220V 2 A, and Buzzer Active 5V.

3. Usability Testing PC-Less microcontroller training kit
Starting from input process that involves keyboard and mouse, those devices are used as interface between learners and training kit, learners will input syntax program in the form of C++ for Arduino Uno use keyboard, then navigate the command pointer by mouse, which in each input step is displayed on HMI, a 10-inch LCD Display.

The next process is compiling and building, which focusses on training kit "brain"/core in the form of Single Board Computer (SBC) or LattePanda. The SBC is set in such way and equipped with software installed that serves as input-syntax compiler and builder. First SBC will arrange an input syntax program by learners, then translate into Assembly, until this stage SBC acts as Compiler, then syntax Assembly is converted into code in the form of hexadecimal or binary code understood by the Machine in the form of Integrated Circuit (IC) on the target (Arduino).

Next process is downloading, at this stage there is a data transfer from SBC to Arduino in the target, SBC sends machine code that is the result of building syntax process, to IC ATMEGA 2560 in Arduino through USB on SBC as a transmitter through USB Port in Arduino as Receiver. Arduino Uno is a controller that can be determined its performance based on input syntax program, in this Training kit Arduino is on target part, a machine-code form program (*.hex) that is downloaded to Arduino. There is a pin in Arduino that serves as communication and control channel that is connected to the device in Interface part. In this interface, there are several modular interface devices, including: LCD 16 x2, Buzzer, Servo 180°, DHT11, Relay 1, ultrasonic, LED, Potentiometer, Bluetooth, RTC DS 3231, ESP 8266, NRF 24l01, LCD Graphic Nokia 5110, PIR, Infrared Receiver, LED RGB and LDR.

The Arduino interconnection with one of the modules in interfacing is arranged through Dipswitch observation, which every interfacing module has a Dipswitch to communicate with Arduino, the learners must activate communication path by turning Dipswitch on (TURN ON) if they want the Arduino to make contact with one of the interfacing modules in Training Kit. Microcontroller Training Kit Personal Computer Less Physical can see in figure 3.

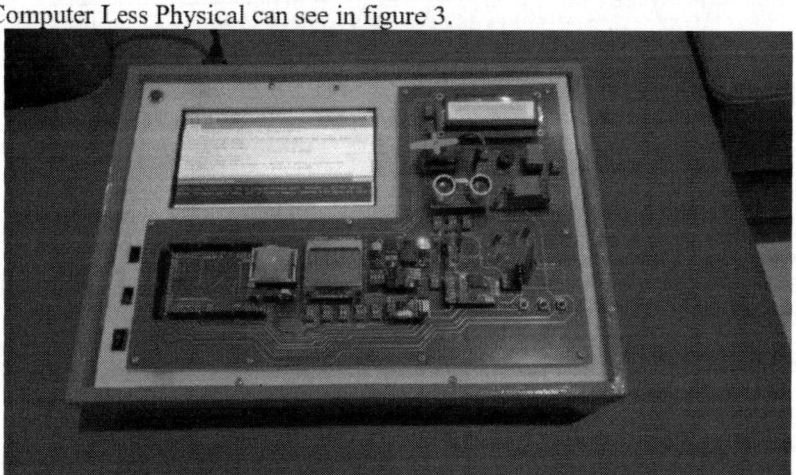

Figure 3. Microcontroller Training Kit Personal Computer Less Physical Form Picture

Interfacing module will show its performance (working) after obtain the right machine code (*hex) derives from syntax program input by learners and has been processed by compiler and builder on SBC. For example, if the learners want to move interfacing module's motor stepper in training kit, he/she should input / create the right syntax program, then perform a code program compiling and building uses software sketch installed on SBC then download a program from SBC to Arduino, then activate the right dipswitch to open communication path between Arduino and interfacing module that is motor stepper module interfacing, by activating dipswitch in that motor stepper module interfacing.

Microcontroller Training Kit Personal Computer Less Hardware's Specification:

1.	Dimension L x W x H	:30 x 30 x10 cm
2.	Display	: TFT LCD 7 inch
3.	Core	: SBC LattePanda
4.	Controller	: Arduino
5.	PCB	: Fiber
6.	Input Device	: Mouse & Keyboard (Wireless)
7.	Power Consumption	: AC 220V ,10 W
8.	Power Backup	: Battery 12V, 1,5 A
9.	Interfacing Module	: LCD16 x 2; Active Buzzer; Motor Servo 180°; Temperature & Humidity Sensor (DHT11); Relay; Bluetooth; Wireless Module NRF 24l01; Wireless Module ESP 8266; LCD Graphic Nokia 5110; PIR Sensor; Infrared Receiver; LED; LED RGB; and LDR

4. Conclusion and Future Work

PC-Less microcontroller kit is an affordable training kit intended to be used for university and vocational high school especially for the school located in rural area. Using Lattepanda, adopted the portability of SBS, the device allows the learners to practice microcontroller programming anywhere with the concept of low-cost and low-power.

The future work of this research is to measure the effectiveness of practical work of microcontroller programming using the device. PC-Less microcontroller kit would be implemented in the course of industrial automation practical work; hence the evaluation will involve some students of electrical engineering education who take this course.

References

[1] Y Somantri and I Fushsilat 2017 Low Cost Advanced Integrated Microcontroller Training Kit *IOP conference Series Materials Science and Engineering* **180** 1 012107
[2] Li Y 2007 Teaching embedded systems using a modular-approach microcontroller training kit *World Transactions on Engineering and Technology Education UICEE* **6** 1
[3] Sheng X, Yin J, and Shang J 2011 An energy-saving control strategy on central air-conditioning system *Electricity* **22** 4 62-66
[4] Yanping Z 2011 Research and practice of course design on microcomputer principle and interface technology *Procedia Engineering* **15** 11 4211-4215
[5] Edi R, Faisal and Fajar 2014 *RaspberryPi, Mikrokontroler Mungil Serba Bisa* (Yogyakarta: Andi Ofset)
[6] http :// www dfrobot. Co. *LattePanda window 10 mini PC (2G/32GB). A Powerful Window.*

IOP Conf. Series: Materials Science and Engineering **306** (2018) 012136 doi:10.1088/1757-899X/306/1/012136

Analysis of Blended Learning Implementation on Waste Treatment Subjects in Agricultural Vocational School

Y Sugiarti*, S Nurmayani and S Mujdalipah

Fakultas Pendidikan Teknologi dan Kejuruan, Universitas Pendidikan Indonesia

*attisugiarti5@upi.edu

Abstract. Waste treatment is one of the productive subjects in vocational high school in programs of Agricultural Processing Technology which is one of the objectives learning has been assigned in graduate competency standards (SKL) of Vocational High School. Based on case studies that have been conducted in SMK Pertanian Pembangunan Negeri Lembang, waste treatment subjects had still use the lecture method or conventional method, and students are less enthusiastic in learning process. Therefore, the implementation of more interactive learning models such as blended learning with *Edmodo* is one of alternative models to resolve the issue. So, the purpose of this study is to formulate the appropriate learning syntax for the implementation of blended learning with *Edmodo* to agree the requirement characteristics of students and waste treatment subject and explain the learning outcome obtained by students in the cognitive aspects on the subjects of waste treatment. This research was conducted by the method of classroom action research (CAR) with a Mc. Tagart model. The result from this research is the implementation of blended learning with *Edmodo* on the subjects of waste treatment can improve student learning outcomes in the cognitive aspects with the maximum increase in the value of N-gain 0.82, as well as student learning completeness criteria reaching 100% on cycle 2. Based on the condition of subject research the formulation of appropriate learning syntax for implementation of blended learning model with *Edmodo* on waste treatment subject are 1) Self-paced learning, 2) Group networking, 3) Live Event- collaboration, 4) Association - communication, 5) Assessment - Performance material support. In summary, implementation of blended learning model with *Edmodo* on waste treatment subject can improve improve student learning outcomes in the cognitive aspects and conducted in five steps on syntax.

1. Introduction

Based on Government Regulation No. 19 of 2005 Article 26 paragraph 3 Vocational High School have the aims to improve intelligence, knowledge, personality, noble character, and skills to live independently and follow further education in accordance with the vocational. The objective of learning SMK is formulated in twenty-three SKL SMK which is an outcome of competence demands to be mastered by students to ajusted the needs of the business and industry field (DU-DI) [1]. One of the existing expertise programs in SMK is Agricultural Product Processing Technology (APPT) that produces graduates according to graduate standards required by the food processing industry. One of the SKLs that must be mastered by students is to utilize the environment productively and responsible. Along with the standard of competence, the purpose of learning subjects of waste treatment is productive utilizing agricultural resources and byproducts and able to process agricultural processing waste.

Based on preliminary observation at SMK Pertanian Pembangunan Negeri Lembang by direct interview with subject teachers, waste treatment subjects are still delivered by using lecture method and learning more dominated by teacher than student. These conditions will affect the quality of learning in achieving learning objectives. In order for learning to take place well and able to achieve learning objectives, then the learning process should refer to the standard process which is one of the standard education. Teacher needs to have the ability to design and implement various learning strategies that are considered to match the interests and talents and in accordance with the stage of student development, including utilizing various sources and instructional media to ensure the effectiveness of learning [2]. On the other hand, SMK Pertanian Pembangunan Negeri Lembang is one of the School that implement one of West Java Province program, it is called as *Jabar Smart School* that is an integration program of information and communication and technology to improve the quality management (on aspects of learning, managerial, supporting system) within the scope of educational units using digital technology In West Java. Learning media used to support the program is Edmodo. Edmodo is a social media platform, such as a *facebook* which is developed specifically for students and teachers in a virtual classroom that can serve to implement interesting and easy-to-use learning [3].

The existence of Edmodo's learning media enables teachers and students to interact more in more direction, one of them is by applying blended learning model. Blended learning is an integration between face to face and online learning to help the classroom experience by developing information and communication technologies [2]. There is an increase in motivation and student learning outcomes due to the application of blended learning [4]. So, with the media *Edmodo* in the program JABAR Smart School followed by SMK Agricultural Development Lembang, blended learning model is very potential to be applied especially efforts to improve the quality of learning on waste treatment subjects. So, this research entitled "Analysis of the implementation of learning blended learning model on waste treatment subjects (case study at SMK Pertanian Pembangunan Negeri Lembang)". The research objectives are to formulate the best of learning strategies in blended learning model with Edmodo as media which agreed with student character and the subject of waste treatment and describe the student learning outcome in cognitive aspect in waste treatment subject. The data analysed with Normalized gain test which is conducted to knows the effectiveness of the improvement of student learning outcomes on the scale of values 1 - 4 and minimal criteria (KKM) in accordance with the Minister of Education and Culture of the Republic of Indonesia No. 104 of 2014 is 2.67. The N-gain value criterion is shown in table 2.1.

2. Observation data Analysis

Data from observation result that is result of student learning on affective and psychomotor aspect is processed with sumative differential analysis and presented in graph.

Table 1. N-gain Value Chriteria

N-gain Score	*N-gain Criteria*
0,70 < *N-gain*	High
0,30 ≤ *N-gain* < 0,70	Middle
N-gain < 0,30	Low

Details:

$$N - Gain = \frac{post\ test\ score - pre\ test\ score}{\text{ideal score} - pre\ test\ score}$$

3. Methods

The research method using classroom action research (CAR) model consisting of 3 cycles with the research design in figures 1 and 2. The subjects of this research are 24 students of APT program. The Data are obtained using data collection technique of objective test and class observation.

IOP Conf. Series: Materials Science and Engineering **306** (2018) 012136 doi:10.1088/1757-899X/306/1/012136

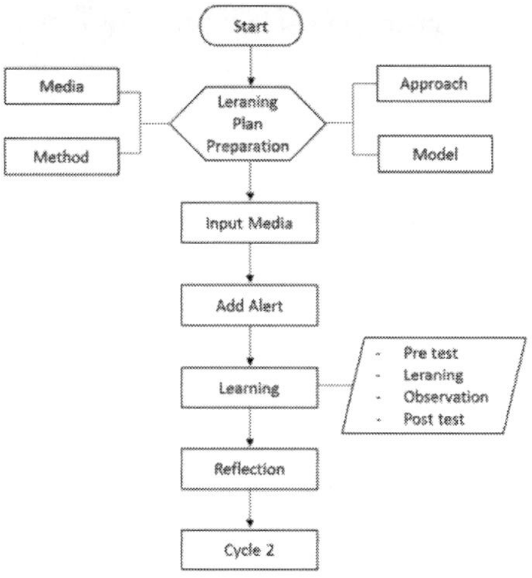

Figure 1. Cycle 1

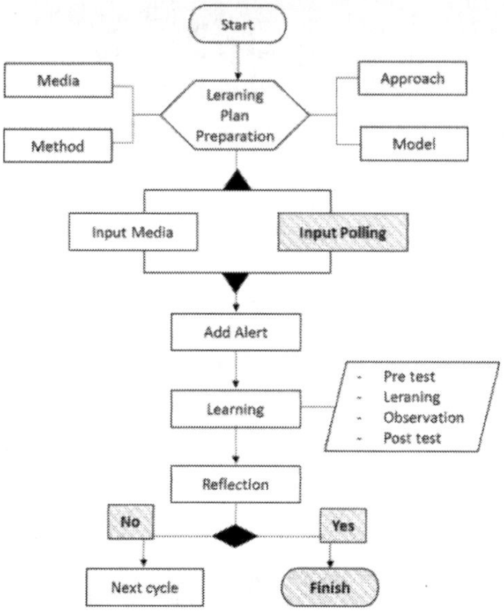

Figure 2. Cycle 2 and 3

3. Results and discussion

3.1. Formulation of learning steps that match the characteristics of students and waste treatment subjects

3.1.1. Planning

Planning prior to learning is to make RPP consistent with blended learning syntax, create a virtual class in Edmodo, upload teaching materials, and ensure that the classroom and student barcode schemes are ready for barcode system tests through the Plickers application.

3.1.2. Implementation

Implementation of learning done for three cycles. Cycle 1 is done according to learning syntax) in Figure 3 [5]:

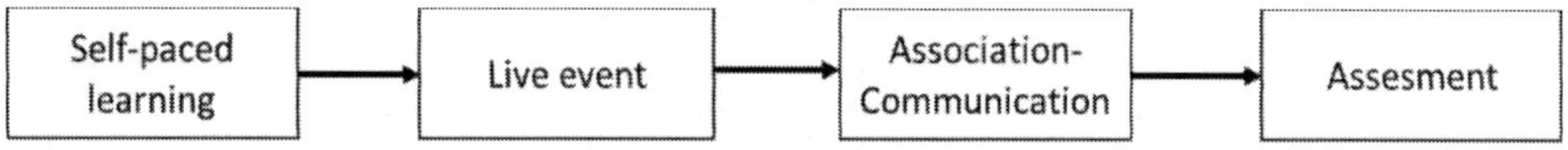

Figure 3. Blended learning Syntax

Based on the result of reflection from cycle 1, it is found that student attendance in virtual class is only 91,67% meaning 8,33% of students are not present in the virtual class and do not download and study the material, so that the self-paced learning stage is less than the maximum and less support stage live event. The impact is 8.33% of students did not get mastery learning with KKM 2.67. Based on that, the syntactic modification needs to be done to better suit the characteristics of the students and the subjects of waste treatment in the next cycle. In cycle 2, the blended learning syntax has been modified and shown in Figure 4.2.

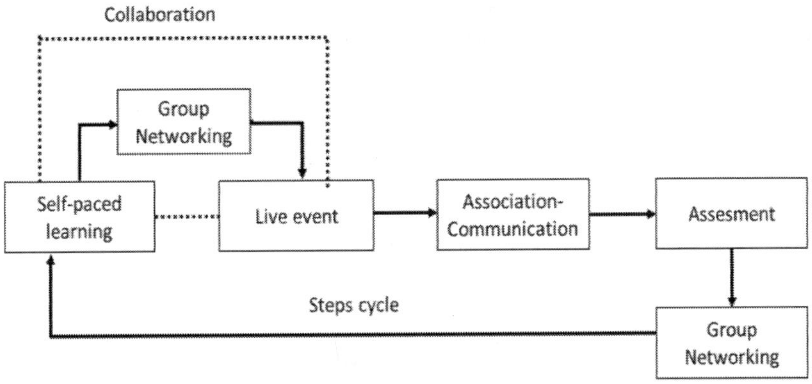

Figure 4. Blended learning syntax based on group networking

The syntax of blended learning based group networking consists of 5 stages as follows:

1. Group networking, the stage students learn the material and do the tasks assigned by *Edmodo*. This stage aims to train the constructive power of student thinking and ensure students can learn the material by exchanging information with members of their group. The result of group networking is a resume that must be individually uploaded by the students. So the risk of students not being present in a virtual class and not studying material before a live event can be suppressed.
2. Self-paced learning, stage of self-learning students before the live event stage in class. Students can directly ask or think according to the material being studied in the virtual classroom and then respond to the teacher and other class members.
3. Live event, face-to-face learning stage in the classroom which is a collaboration between group networking and self-paced learning, in this stage the teacher will explain the material according to the results of the student's constructive to avoid the occurrence of verbalism.
4. Association-Communication, the student stage associates the learning outcomes of group networking, self-paced learning, and live events so that it becomes a meaningful learning experience and is made in a portfolio of notes in a virtual classroom to communicate the learning outcomes.
5. Assessment, the stage students do evaluation of the learning given by teachers either individual or group. This stage can be combined with group networking to start the next learning cycle.

The selection of group networking steps is the application of group learning to students can increase students' motivation and interest up to 89.3% and student achievement cognitively up to 79.825% [6]. Group learning is a process of knowledge acceptance by a group of individuals who perform an activity logically and systematically for the process of behavior change through the improvement of knowledge, skills, attitudes, and abilities [7].

In addition to network grouping actions that can be done to overcome the problems that exist in the cycle one of them, giving punishment to students who do not open teaching materials in *Edmodo* and do not make a resume. The punishment or punishment is a consequence that lowers the probability of something happening [8]. In probability research that may appear is the students download and study the material in *Edmodo* and vice versa, students do not do it, with the application of punishment is expected probability students do not download and do not study the material in *Edmodo* decline. However, in implementation, the application of punishment that is not appropriate will affect the emotional aspects of students. There is a possibility that students cannot accept that they get punishment, but instead do social deviation as a form of emotion [8].

Another treatment that can also be done to anticipate the reflection of cycle 1 is to blend learning model with recitation model. The learning model of recitation is a model of learning that makes students resume their own resume [9]. Through this model students must create a resume sourced from the module in *Edmodo* so that each student must access individually. The downside of this alloy of recitation models is that they cannot address infrastructure problems such as the lack of available internet connections that students have at home. In addition, the results of student recitation are the result of the construction of individual thoughts and materials in the module.

However, based on the material suitability and characteristics of students, the most appropriate action to solve the problems that occurred in cycle 1 is with Group networking. In addition to train student's cohesiveness, the ability to construct students will also be more honed because in addition to source from the module, the material can also be sourced from the opinions of other students in the group. Problems such as internet access facilities of each student can be overcome by providing information to help students access the module.

Based on the results of reflection from cycle 2 obtained information that the attendance of students in the virtual class has 100% and student learning completeness has reached 100% therefore cycle 3 is done according to the syntax of learning blended learning based group networking that has been formulated in cycle 2.

3.2. Results of Student Learning on Waste Treatment Subjects

The results of cognitive learning are one of the indicators to measure the improvement of student learning outcomes, especially in aspects of intelligence, reasoning and analysis power that is needed in waste treatment subject. The development of student learning outcomes on this cognitive aspect was analysed by the N-gain or Normalized gain method of each cycle and presented in Figure 3.3

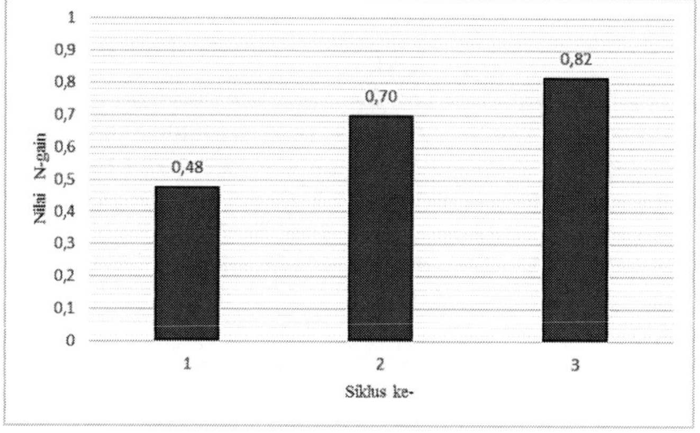

Figure 5. The Increase of N-gain value

Based on Figure 3.3 it can be explained that there is an increase in student learning outcomes from cycle 1 which is initially moderate, then to cycle 2 that has entered a high level, and finally in cycle 3 to achieve higher learning outcomes at the value of 0.82. The students' learning mastery has reached 100% in cycle 2 with the value of mode 3.3. Improving learning outcomes in blended learning may occur because this model helps students to better develop in the learning process, in accordance with learning styles and preferences in learning and provides realistic, practical opportunities for teachers and students for learning Independent, useful, and growing [10].

In addition, blended learning models with *Edmodo* media can also enhance flexible scheduling for students, combining the best aspects of face-to-face and online instruction. The face-to-face classes can be used to engage students in interactive experiences. Online portions provide students with rich, knowledge-rich content at any time, and anywhere as long as students have internet access. Addressing learning problems that require completion through the use of varied learning methods.

4. Conclusion

Based on the result and discussion of the research can be conclude:

1. Formulation learning steps that appropriate with Edmodo and the subject of waste treatment are Self-paced learning, Group networking, Live Event and collaboration, Association and communication, and Assessment.
2. The implementation of blended learning model with Edmodo on waste treatment subject can improve student learning outcomes in cognitive aspects with N-gain maximum increase reaching 0.82, and students learning completeness can reach 100% in cycle 2.

References

[1] P M P Sudira 2006 *Curriculum level of SMK education unit. Jakarta* : Departement of National Education

[2] W Sanjaya 2009 *Learning Strategy Oriented Standards Education Process* Jakarta: Prenada Media Group

[3] Tim Seamolec 2013 *Digital Simulation Volume I* Jakarta: Ministry of Education and Culture

[4] S B Sjukur 2012 Effect of Blended Learning on Student Learning Motivation and Student Result of SMK Level *Journal of Vocational Education* **2** 3 368-378

[5] C J Bonk, R Charles and Graham 2006 *The Handbook of Blended learning* USA : Preiffer

[6] A Setiawan 2015 *Application of Group Learning to Increase Interest and Achievement of Indonesian Language Learning SD Negeri Kepek* [Bachelor thesis] Universitas Negeri Yogyakarta

[7] S D Lonn 2009 *Student use of a learning management system for group projects: a case study investigating interaction, collaboration, and knowledge construction* Michigan University, Michigan.

[8] J W Santrock 2007 P educational psychology (second edition). Jakarta : Kencana.

[9] Y Faliestin 2010 *Improvement of accounting learning achievement through the implementation of problem-based learning model in the students of class XI IPS 2 SMA Negeri 6 Surakarta academic year 2009/2010* (Doctoral dissertation, UNS)

[10] T H Prabowo 2014 *Implementation of Blended Learning Learning Model to Increase Multimedia Student Learning Activity in Class of XI Multimedia 1 SMK Muhammadiyah 2 Klaten Utara Teachings Year 2013/2014* [Bachelor thesis] Faculty of Enginnering Universitas Negeri Yogyakarta

Toolpath Strategy and Optimum Combination of Machining Parameter during Pocket Mill Process of Plastic Mold Steels Material

Y T Wibowo*, S Y Baskoro and V A T Manurung

Politeknik Manufaktur Astra,
Jl. Gaya Motor Raya No. 8, Sunter, Jakarta 14330, Indonesia

*Corresponding author: yohanes.trijoko@polman.astra.ac.id

Abstract. Plastic based products spread all over the world in many aspects of life. The ability to substitute other materials is getting stronger and wider. The use of plastic materials increases and become unavoidable. Plastic based mass production requires injection process as well Mold. The milling process of plastic mold steel material was done using HSS End Mill cutting tool that is widely used in a small and medium enterprise for the reason of its ability to be re sharpened and relatively inexpensive. Study on the effect of the geometry tool states that it has an important effect on the quality improvement. Cutting speed, feed rate, depth of cut and radii are input parameters beside to the tool path strategy. This paper aims to investigate input parameter and cutting tools behaviors within some different tool path strategy. For the reason of experiments efficiency Taguchi method and ANOVA were used. Response studied is surface roughness and cutting behaviors. By achieving the expected quality, no more additional process is required. Finally, the optimal combination of machining parameters will deliver the expected roughness and of course totally reduced cutting time. However actually, SMEs do not optimally use this data for cost reduction.

1. Introduction

Nowadays, in industrial world, the usage of plastic material has been unstoppable anymore. Plastic based good penetrates in so many aspects of life. This condition needs support from plastic injection process as well mold manufacturing. Plastic injection process is kind of mass production process while mold manufacturing process is make to order process. Make to order process is categorized as high cost process so needs good and mature planning, implementation, and quality checking to keep cost component on the right track.

IOP Conf. Series: Materials Science and Engineering **306** (2018) 012137 doi:10.1088/1757-899X/306/1/012137

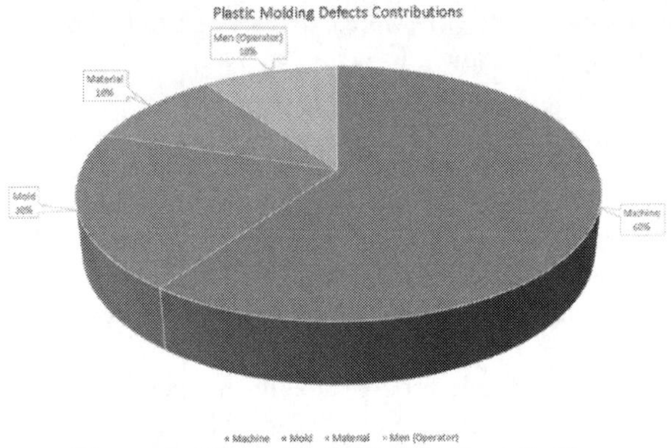

Figure 1. Plastic Mold Defect Contribution
Taken from Texas Plastic Technologies Seminar Material

In the mold manufacturing, as described at figure 1, defect was caused by machining process dominates times contribution. Managing machining process will deliver good quality aspect, beside prevents another unexpected additional process. Simply, additional process can be classified into two mains group which are additional process due to quality mismatch [1] and truly needed additional process. Additional process due to quality mismatch would be the focus of this discussion topic. Surface roughness within mold unit, has certain contribution related to mechanical performance and aesthetics performance. In mechanical performance aspect, it contributes to amount of needed plastic material while in aesthetic performance it is related to how easy the surface visible to the eye or how much the surface is exposed.

Milling machine is machine tools to form any shape which is non-cylindrical forms. In cutting process, milling machine has three main motions that are cutting tool spindle motion (spindle motion), cutting feed motion (feed motion) and depth of cutting motion (depth of cut motion). Combination among these three main motions will deliver expected surface quality, or in simple word would be stated that needed surface roughness would be achieved by combining these three motions in structured combination. The cutting tools selection should be taken to increase the efficiency [2]. The selected cutting tool material is High Speed Steel based because it is widely used and can easily be found in the market and Small Medium Enterprise [3]. The cutting tools are used to change the shape of material into particular desired shapes or certain products should ensure the continuity of the process of cutting/forming with optimum result and should have better properties comparing to the workpiece material [4].

In this paper, examination on surface roughness performed by inserting cutting tools radii aspect and tool path strategy (cutting pass model). Research topic to find out optimum result is still performed. Cutting tools makers have not enough ability to provide parameter formulation convincingly to achieve needed and expected result. The given parameter being provided by the makers are common use parameters such as spindle speed, feed rate, and depth of cut without being completed with value of surface roughness level from machining process as performed on current suggested value. There is information gap between supplied information from the maker and demand information from user.

The current condition drives some research to investigate the relationship among cutting tool spindle motion, feed rate, depth of cut, radii and cutting tool strategy (cutting pass) model. The most influencing factor in surface roughness machining is spindle rotation, cutting speed and then followed by depth of cut [5]. Other side informs that generally, by increasing cutting speed, the tool chip interface temperature will increase. It will affect surface roughness decreasing [6]. The cutting pass model especially Contour-in has simple path compared to zigzag strategy but has competitive point in cutting time reducing. By simplifying the tool path, the number of turning point and corner regions can be reduced. In turn, the effect of them such as high cutting temperature and excessive cutting force can be eliminated [7]. Contour-in cutting strategy is known also as helix cutting pass model. Within this research, the effects of listed input parameters were expected to be known, so the result should be used as reference to achieve needed quality standard.

2. Methods

The method used in this research consists of some phases which are field observation, study of literature, and data collecting. Within this observation, some discussions are conducted to dig related information with this research. On the study of literature, searching and reviewing references are carried out to have broaden view from theoretical aspect and data collecting of research testing was conducted at Manufacturing 2 Laboratory of Politeknik Manufaktur Astra Jakarta with the supporting of industrial partner.

Taguchi method is method to improve product quality as well as the process with capability to suppress cost. Taguchi which has another name as Design of Experiments showed to be very robust method and allowed to improve the surface roughness, and for a specific range of parameters, the experimental result could identify the best cutting parameters [8]. This experiment was taken using method of Design of Experiments to not expend extra cost for the optimized result [9][10] and for confirming best possible of optimized result, ANOVA was chosen [11]. In this experiment, there are 3 parameters, and will be 3 levels for each parameter. Considering data adequacy and best possible result, Taguchi array that using is L_{27} (3^{13}) orthogonal array. There will be 27 runs of experiments.

This research examination was taken by using computerized numerical controlled based machine which has certain reliability. This examination purpose is to produce surface roughness level from plastic mold steel material using specific machining process named pocket milling in certain area. The machine used for this research is Makino CNC Milling Machine S-33 series, as figured at figure 2.

Figure 2. Makino CNC Milling Machine Series S-33

The data collection was retrieved from surface roughness of workpiece. The used measuring tool was surface roughness measuring tools made by Tesa named Rugosurf 10. The displayed measured result is roughness average (Ra) because Ra is the most used common result representing surface roughness [14] as figured at figure 3.

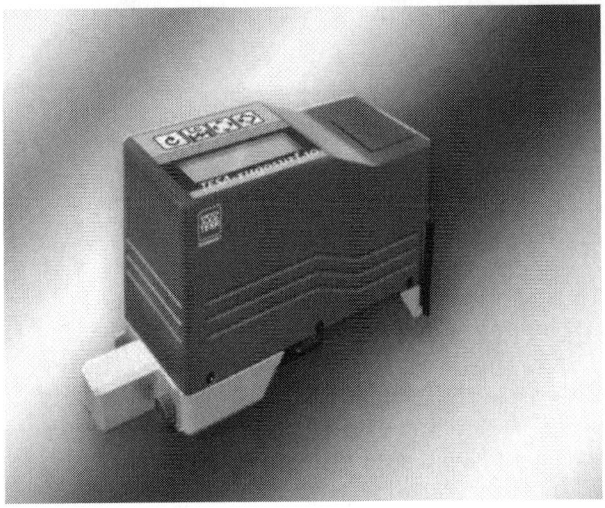

Figure 3. Tesa Rugosurf 10
Taken from http://www.tesatechnology.com

In this research, there are three cutting pass models used. These models are three most common used by CADCAM user [12]. The first cutting pass model is zig or one way cutting pass as figured on figure 4. The grey curve represents shape of workpiece. The black curve simply represents basic shape of contour. The green line represents cutting motion and the red line represents non-cutting motion.

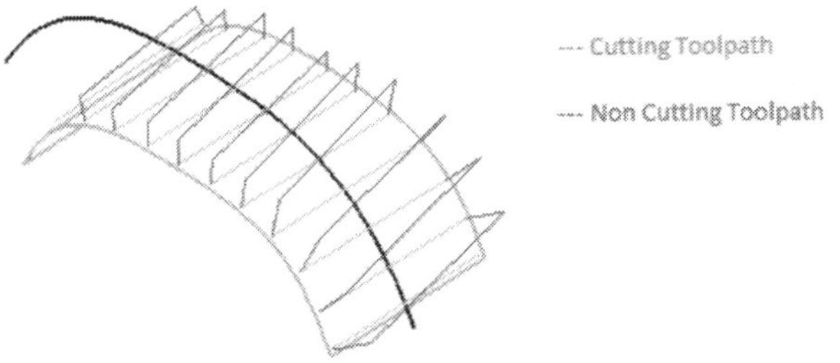

Figure 4. Zig or One Way Cutting Pass Model

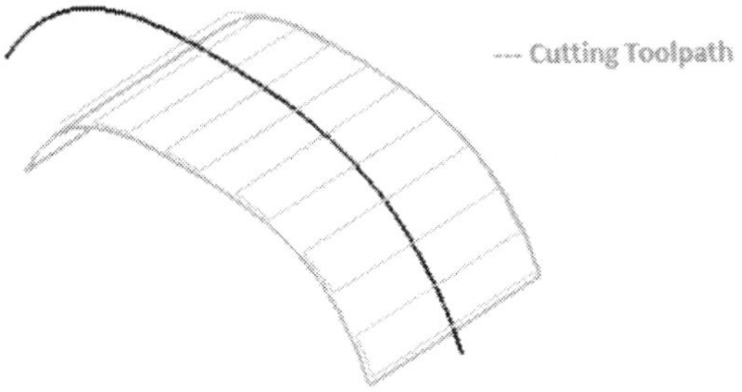

Figure 5. Zig zag or Two Way Cutting Pass Model

IOP Conf. Series: Materials Science and Engineering **306** (2018) 012137 doi:10.1088/1757-899X/306/1/012137

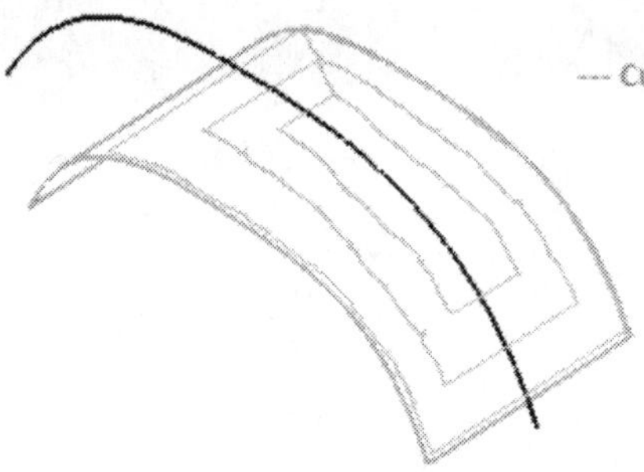

Figure 6. Helix Cutting Pass Model

The next two cutting passes are zig zag and Helix as figured on figured 5 and figure 6. On the both of next two cutting pass models, non-cutting toolpath (red line) is mostly eliminated for efficiency reason.

The data from this research and testing were summarised and presented in this paper. The data were discussed based on achieved result.

3. Results and discussion

3.1. Machinery Testing

Testing process initiated by specimen hardness testing to identify the homogeneity of testing specimen hardness. This testing was taken on four sides with specified two testing points on every side as listed on table 1.

Table 1. Specimen Homogeneity Hardness

Surface	Testing Point	Hardness - HRc
1	1	36
	2	37
2	1	36
	2	36
3	1	37
	2	36
4	1	36
	2	36

Machining process that will produce surface roughness quality was carried out by implementing input parameters below such as using defined spindle speed, defined feed rate, defined depth of cut, and as well-defined cutting pass model. By implementing defined listed parameters, expected result should be achieved. The listed parameters are as informed on table 2.

Table 2. Machining Process Parameter

No	Feed Rate (mm/min)	Depth of Cut (mm)	Cutting Model
1	35	0.07	Zig
2	35	0.07	Zigzag
3	35	0.07	Helix
4	35	0.14	Zig
5	35	0.14	Zigzag
6	35	0.14	Helix
7	35	0.20	Zig
8	35	0.20	Zigzag
9	35	0.20	Helix
10	121	0.07	Zig
11	121	0.07	Zigzag
12	121	0.07	Helix
13	121	0.14	Zig
14	121	0.14	Zigzag
15	121	0.14	Helix
16	121	0.20	Zig
17	121	0.20	Zigzag
18	121	0.20	Helix
19	277	0.07	Zig
20	277	0.07	Zigzag
21	277	0.07	Helix
22	277	0.14	Zig
23	277	0.14	Zigzag
24	277	0.14	Helix
25	277	0.20	Zig
26	277	0.20	Zigzag
27	277	0.20	Helix

3.2. Examination Using Machining Process

In inform earlier, machining process with defined input parameters will generate surface roughness at work piece surface. The surface roughness was measured using Tesa Rugosurf 10. To prevent mistake on data collecting as well as make sure the result, every work piece would be measured three times. The result then logged and was compared with two other measurements. The maximum allowable disparity is 0.1 micron. In case of the range was exceeded, measuring process would be re take for conforming the result within disparity range.

Table 3 describes surface roughness result recorded by Tesa Rugosurf 10. Every workpiece was checked for 3 times. The displayed value was using micron unit. The specimen numbers represented no of testing workpiece.

Table 3. Surface Roughness Average (Ra) Measurement Result (in um)

Specimen No	Check Poin 1	Check Poin 2	Check Poin 3
1	1.67	1.64	1.63
2	1.35	1.40	1.37
3	1.90	1.92	1.96
4	1.56	1.62	1.55
5	2.03	2.01	1.98
6	1.41	1.36	1.44
7	1.31	1.35	1.35
8	1.26	1.32	1.27
9	1.37	1.37	1.37
10	1.19	1.22	1.22
11	1.80	1.78	1.81
12	1.98	1.95	1.90
13	1.53	1.55	1.55
14	1.72	1.77	1.69
15	1.35	1.35	1.34
16	1.68	1.70	1.65
17	1.42	1.40	1.38
18	1.35	1.35	1.36
19	1.60	1.62	1.64
20	1.60	1.65	1.66
21	1.92	1.87	1.91
22	1.99	1.97	1.95
23	1.78	1.80	1.75
24	1.82	1.80	1.80
25	1.48	1.50	1.48
26	2.01	2.00	1.92
27	1.66	1.70	1.65

Table 4. Input Parameters and Average Ra Measurement Result (in um)

Ra (um)		Cutting Pass Model								
		Zig			ZigZag			Helix		
		Depth Of Cut (mm)			Depth Of Cut (mm)			Depth Of Cut (mm)		
		0.07	0.14	0.2	0.07	0.14	0.2	0.07	0.14	0.2
Feed Rate (mm/min)	35	1.65	1.37	1.93	1.58	2.01**	1.40	1.34	1.28	1.37
	121	1.21*	1.80	1.94	1.54	1.73	1.35	1.68	1.40	1.35
	277	1.62	1.64	1.90	1.97	1.78	1.81	1.49	1.98	1.67

Table 4 describes input parameter used for machining process, and the roughness level result as measured by Rugosurf 10. It also can be seen both of the best surface roughness level and the worst. The best had been achieved using the combination of input parameters as Zig cutting pass model, 121 mm/min as feed rate and 0.07 mm as depth of cut, while the worst is the combination of input parameters as Zigzag cutting pass, 35 mm/min as feed rate and 0.14 mm as depth of cut. The best surface roughness is 1.21 um (as indicated using *) and the worst is 2.01 um (as indicated using **).

4. Conclusion

Based on the results of testing that were completed with some mathematic calculations, it can be concluded that feed rate and depth of cut affect the level of surface roughness. Combined parameter of feed rate and depth of cut affects the level of surface roughness as well combined parameter of depth of cut and cutting pass model. Finally, interaction among some listed parameters, the lowest (finest) surface roughness level was achieved by combining of feed rate parameter at 121 mm/minutes, depth of cut parameter at 0.07 mm, and the cutting pass model is Zig even though on previous conclusion clearly informed that cutting pass model does not affect the level of surface roughness. The best (finest) achieved surface roughness level is that Roughness Average (Ra) 1.21 um which is equivalent to surface roughness level N7. On this experiment, optimum combination in producing the best surface roughness is described above that implemented on cutting process using zig toolpath strategy.

Acknowledgements

For great work, suggestions and the support from the civitas academic of Program Studi Pembuatan Peralatan dan Perkakas Produksi (Tool Making), Politeknik Manufaktur Astra, and industrial partner from Engineering Development Centre of PT. Astra Otoparts Tbk, the author are very grateful and think that it is worth to be mentioned.

References

[1] P Zgorniak and A Grdulska 2012 Investigation of Temperature Distribution during Milling Process of Az91hp Magnesium Alloys *Mechanics and Mechanical Engineering* **16** 1 33-40
[2] P Kayal 2007 Optimal Ball-end and Filletend Mills Selection for 3 Axis Finish Machining of Point-based Surfaces *International Journal of CAD/CAM* **7** 1
[3] D G E Paul 1984 *Material and Processes in Manufacturing* MacMillan Publishing Company.
[4] T Rochim 1993 *Teori & Teknologi Proses Pemesinan* Proyek HEDS
[5] Y T Wibowo Analisis Tingkat Kekasaran Permukaan Hasil Proses Mill Slot pada Baja Permesianan SCM 440 *Prosiding SMNTK 2015* – UNJ Jakarta, 100-105
[6] B Jabbaripour, M H Sadeghi and Sh Faridvand 2010 A study of the effects of cutter path strategies and cutting speed variations in milling of thin walled parts *The 7th Jordanian International Mechanical Engineering Conference (JIMEC 7)* 27-29 September 2010 Amman – Jordan.
[7] E Y T Adesta, M Riza, M Y Suprianto and R Hamidon 2016 The effect of tool path strategies on cutting temperature and cutting force during pocket milling of AISI H13 *ARPN Journal of Engineering and Applied Sciences* **11** 1 337-344
[8] J Riberuro, H Lopes, L Queijo and D Figueiredo 2017 Optimization of Cutting Parameters to Minimize the Surface Roughness in the End illing Process Using the Taguchi Method. *Periodica Polytechnica Mechanical Engineering* **61** 1 30-35
[9] C M Douglas 2013 *Design and Analysis of Experiments* 8th Edition, Wiley
[10] A Yadav, A Bangar, R Sharma and D Pal 2012 Optimization of Turning Process Parameters for Their Effect on En 8 Material Work Piece Hardness by Using Taguchi Parametric Optimization Method *International Journal of Mechanical and Industrial Engineering* **1** 2231-6477
[11] G B Murugan, M Biswanath and G Sukamal 2009 Taguchi Method and ANOVA: An Approach for Process Parameters Optimization of Hard Machining while Machining Hardened Steel *Journal of Scientific and Industrial Research (JSIR)* **68**
[12] D A Stephenson and J S Agapiou 2016 *Metal Cutting Theory and Practice Third Edition* CRC Press
[13] P G Mikell 2010 *Fundamental of Modern Manufacturing, 4th Edition* Wiley
[14] S B Jit 2014 *RSM: A Key to Optimize Machining, Anchor Academic Publishing*

ICT Literacy of Vocational High School Students

Y Z Miraj*, D Rohendi, Yannuar, Nurhabibah and H F Wendi

Pendidikan Teknologi dan Kejuruan, Universitas Pendidikan Indonesia, Bandung

*yusepzaenimiraj@gmail.com

Abstract. The development of information and communication technology (ICT) has brought about change in education in Indonesia. With ICT students can easily access a variety of information in supporting the learning process. But there was a problem where ICT infrastructure across Indonesia including in the uneven Siak Regency, this can lead to gaps in technology and also weak ICT literacy. This research aims to know the ICT literacy of the students of SMK in Siak Regency. The research method used was survey with the quantitative approach. The results of this study showed that the use of ICT (computers, Internet) has been used by students of SMK, but usability of ICT has not been fullest support the learning process.

1. Introduction

The development of information and communication technology (ICT) in a relatively short period of time has given the changes in the various joints of the life, is no exception in the world of education. World Education is believed to be the most dominant factor in developing the quality of human resources required generate human able to compete in the competition on the current globalization. Various efforts undertaken to improve the quality of education, quality and ability of students at all levels of education, including in Vocational high school (SMK). [1,2]

The presence of ICT (internet) for example in the development so far, it has brought change to the world of education. It's just that, the digital divide, especially in developing countries like Indonesia is still a problem to look for a solution. One of the factors that influence the digital divide, including ICT literacy, is a factor of economic and social geography. Both of these factors are very influential in computer usage and retain the opportunity to access the internet. The results of this research also suggest that the use of computers and the internet able to improve knowledge and skills of the informant. [3]

Changes to the curriculum of the curriculum unit level education (KTSP) into 2013 curriculum where subject's computer skills and information management (KKPI) was replaced with service information and communication technology (ICT), [4], greatly affect students in using and leveraging technology in a positive and responsible.

Approximately 7,127 SMK students in Siak District [5]. Issues faced by students, in General is the lack of access to information, economic levels and weak old man. Such condition is the cause and effect between the two, meaning low economic level result against low levels of education, and low levels of access to information may also result in low levels of economy.

Based on the above description, then the main problem to be traced in this research is the extent of information technology literacy and communication of vocational students in Siak District. The purpose of this study is to determine the extent of information technology literacy and vocational student communication in learning in Siak District.

2. Literacy ICT

The ability to effectively and efficiently comprehend and utilize mass media content. Suggests some elements of media literacy : (1) An awareness of the impact of media; (2) An understanding of the process of mass communication; (3) Strategies for analyzing and discussing media messages; (4) An understanding of media content as a text that provides insight into our culture and our lives; (5) The ability to enjoy, understand, and appreciate media content; (6) An understanding of the ethical and moral obligations of media practitioners; (7) Development of appropriate and effective production skills [6].

One model for measuring the level of ICT literacy in the education world is usually used Personal Capability Maturity Model (P-CMM). In this case, ICT literacy is categorized into five levels, namely: Zero rate = if an individual is completely ignorant and does not care about the importance of information and technology for everyday life. Level one = if an individual has had one or two experiences, where information is an essential component of achieving desired and problem solving and has involved information technology to seek it. Level two = if an individual has repeatedly used technology to aid in daily activities and has had a pattern of repetition in its use. Level three = if an individual has a standard of mastery and understanding of the information and technology he enacts and consistently uses the standard as a reference for the organization of daily activities. Level four = if an individual has been able to significantly increase (quantitatively) quantitatively the performance of his or her daily life activities through the use of information and technology. Level five = if an individual has regarded information and technology as an integral part of everyday activities and directly or indirectly colored his or her behavior and culture [7,8].

To understand the literacy of students to ICT should be known first various elements of knowledge, literacy experience, and other skills that must be owned by students. To measure the level of student literacy on information and communication technology, it can be done from the level of recognition, use, and purpose of using information and communication technology (ICT) and its various elements. This research will look at the level of ICT literacy, in this case, computer and the internet. The literacy used is the use of both media by the respondent, the location is often used when utilization and frequency using ICT media.

The level of understanding of ICT literacy is measured using the Personal Capability Maturity Model (P-CMM). From this, it can be seen that the respondent's ICT literacy is at what stage.

3. Methods

This research uses survey method with quantitative approach. [9] Data collection was conducted through a survey of the students of SMK in addition to the field of study of information and communication technology in Siak Regency. The principle of choosing the study site is based on two considerations, firstly, consideration of the representation of location and population. Second, consideration of the effectiveness of the research implementation. Related to these considerations. In order to produce a sample that can represent the population in all research sites, the sampling technique uses purposive random sampling method. While the ICT media discussed in this research is limited to two products of ICT namely computer and internet. The selection of this media is as the most familiar media and widely used in education. To obtain the validity and reliability of the instruments used, a pre-test is first performed, to ascertain whether the instrument is an accurate and reliable measuring instrument. Validity indicates the extent to which a measuring instrument measures what it wants to measure.

While the reliability shows the extent to which a measurement result is relatively consistent if the measurement of the same aspect of the same measuring instrument or also called internal consistency reliability [10].

Initial research took 20 informants as a sample. The informant identity of the research result shows the sex informant, the most dominant man in this research, that is 14 informants following the women as many as 6 informants. While the age of informants between 16 to 18 years. These results indicate that students or students can master the ICT device more quickly because in accordance with technological developments and tasks provided by teachers. This is evident from the results of research where most students have access to ICT. In addition, the age of the majority of informants is still young, between

16 to 18 indicate that the penetration of ICT more quickly penetrated among students. This shows that young people are more likely to access ICT.

4. Use of Computer

This section is an analysis of the use / use of computers by informants. Majority of the informants, 16 respondents have their own computers and can take advantage of computers, while 4 informants said they cannot use the computer. This is reflected in the age of informants and of course also related to informant education, young age, and educated tend to use and utilize the computer as a necessity and its relation to the lesson.

From 20 informants, as many as 16 informants stated using computers in the last month. While as many as 4 informants stated not to use computer during last one month. Therefore, it can be said that the computer becomes an important thing for the informant because only about 4 informants who did not declare use in the last month.

For frequency of computer usage in a week, as many as 12 informants stated using computer more than four times a week, 2 informants stated using computer 3 times a week, and 2 informants use computer twice a week.

Place informants use the most computers are in their own homes as many as 12 informants. Following at the school of 6 informants, and at the place of two informants.

The purpose of informants using the computer is to play this game is the most dominant goal of as many as 10 informants, while 7 informants is to process words, process data, and process numbers. Meanwhile, as many as 3 respondents admitted to spending less than two hours a day to have fun using the computer.

5. Use of the Internet

Section discusses internet usage by informants. All informants claimed to have used the internet. This means that informants are already using the internet in their daily lives.

As for the utilization of the internet, the results showed that, of the 20 informants who are utilizing the internet, 17 of whom the informant who claimed to use internet in one this last week. While the three informants stated not to use the internet in a week. Next is the most predominant frequencies used by informants in the use of the internet was four times in a week. Recorded 16 informants chose the answer. 4 while the informant that States use the internet twice a week.

The place of informants when accessing or using the internet based on the results of research shows that, most informants use the internet in the free internet area that is 10 informants, followed in school as many as 5 informants, 4 informants at home and using internet cafe as much as 1 informant. free internet area becomes a favorite place for informants to access the internet.

Most informants 18 informants in a day spent two to four hours / day while using the internet, and only 2 informants who used six to eight hours a day while using the internet.

Meanwhile, internet facilities owned by informants, based on the results of research that all informants have an email address, then followed by ownership of personal accounts on social networking sites such as Facebook, twitter, and others as many as 18 informants.

The results showed that the purpose of internet use by informants dominated to access social networking (Facebook, Friendster, twitter, etc.) as many as 10 informants, then access data and information as much as 7 informants, to play (game) as much as 2 informants, as many as 1 informant.

Cannot be denied that today the internet has become part of the everyday life of students. They recognize that the internet is very beneficial both to support the school, information sharing, and others. The results showed that, as much as the 18 informants stated that the internet greatly helped them in carrying out home work (tasks). Only 2 informants stated that the internet does not help homework (tasks).

6. Results and discussion

The level of literacy of information and communication technology (ICT) in this research is a tool used by someone to communicate and or exchange data and information digitally (online) covering computer and internet. From the data obtained can be concluded that the use of computers already used by the majority of students or more than 60% of students use computers at most four times a week for two to four hours a day at home and at school. Students aim to use majority computer to play games, and process words, process data and process numbers.

Most students are already using and using the internet. On average four times a week they use the internet in the area of free internet and at school for two to four hours a day. The average informer has a personal email and social networking site account. The purpose of using the internet is to access data and information, open accounts social networking sites, download, email, chat, and gaming. The Internet also used informants to look for schoolwork for one to two hours a day. Informants also use the internet to play online games (fun) less than two hours a day.

Based on the analysis of research data above shows that in the development so far, the presence of ICT especially computers and the Internet is no longer a rare and expensive goods. The results of this research also show that students are already entered in the category of ICT literacy in which they as individuals have been able to use ICT devices and mastery of the management has not been fullest support the learning process.

Referring to the theory of Personal Capabilty Maturity Model (P-CMM), the results of this study indicate that the average ICT literacy of students on computer ICT media can be categorized at level four, where the majority of individual informants have considered technology as an integral part of daily activities and directly or indirectly color the behavior and culture of his life. As for other ICT media literacy, the average internet informant is still at level three, the informant has a standard of mastery and understanding of the information and technology needed, and consistently use the standard as a reference for the organization of daily activities. It is shown that the informant has been utilizing the internet in everyday life.

7. Conclusion

The literacy information and communication technology (ICT) students in Siak Regency is already adequate. The Internet is a medium that is already widely used by informant following the computer. And for the utilization of these two media, respondent majority already up on stage four, where the majority of individual informants have considered information and technology as part of daily activities and directly or indirect has been coloring his life and cultural behavior or part of the information society or human cultured information.

The use of internet media, informants have had standards mastery and understanding of the information and technology needs, and consistently apply such standards as a reference for conducting daily activities. This indicates that the informant has been utilizing the internet in daily life but is not used to its full potential in supporting the learning process.

8. Suggestion

Still required the involvement of teachers in an effort to increase ICT skills of students. And students need to be redirected to sites that are positive and carried out socialization about how to take advantage of ICT is good but just simply access the social media and online gaming.

References

[1] E Nugroho 2013 *Skripsi "Pengaruh Persepsi Guru Tentang TIK dan Motivasi Kerja Guru Terhadap Pemanfaatan TIK dalam Pembelajaran SMK RSBI Bidang Keahlian Teknik Elektro di DIY* Universitas Negeri Yogyakarta

[2] An Ibrahim 2016 *School Factors Influencing Adoption of Information Communication Technology in Teaching and Learning in Secondary Schools in Westlands, Kenya* PhD Thesis. University of Nairobi

[3] H Hanif and B Saleh 2009 Penggunaan Komputer dan Internet di Indonesia *Jurnal Pekomnas Penelitian Komunikasi dan Media Massa Makassar* **12** 1529
[4] Permendikbud No 68 tahun 2014
[5] data.kemdikbud.go.id http://referensi.data.kemdikbud.go.id/pd_index.php?kode=091100&level=2
[6] S Baran and D Davis 2010 *Mass Communication Teory: Foundations, Frement and Future. (Terj. Alfrianto Daud dan Putri)* Jakarta: Salemba Humanika
[7] Badan Litbang SDM Kemkominfo 2013 *Dinamika Perkembangan Pemanfaatan TIK serta Implikasinya di Masyarakat* Jakarta: Media Bangsa.
[8] Chen, Yin-Che, Wang and Yen-Jung 2016 Application and development of the people capability maturity model level of an organisation *Total Quality Management & Business Excellence* 1-17
[9] J Brannen 2017 *Mixing methods: Qualitative and quantitative research* Routledge
[10] S Masri and S Effendi 1987 *Metode Penelitian Survei* Jakarta: LP3ES

Vocational Teacher Perceptions on the use of ICT in Learning Computer Network

Yannuar*, **D Rohendi, H Yanti, Nurhabibah and Y Z Mi'raj**

Technology and Vocational Education, Universitas Pendidikan Indonesia (UPI) Jl. Dr. Setiabudhi 207 Bandung 40154 Indonesia

*yannuar1605713@student.upi.edu

Abstract. ICT has been widely used in primary education to vocational schools, but has not been so clearly integrate ICT in the learning process. While the teacher is the key to the effective use of ICT processed. This paper reports a study of surveys that examine the perspective of vocational school teachers. Current research aims to examine a vocational school teacher knowledge about ICT and support for computer use for learning. The sample in this research group consists of 25 teachers of vocational schools. The findings of this research use descriptive method with engineering survey with sampling purposes. Resources in research is journals and book report research results. The results showed teachers have a positive outlook towards the use of ICT in learning. The conclusions resulting from this research is the use of ICT to help teachers be more effective in teaching in the classroom and can improve student learning.

1. Introduction

Teacher holding a major role and very important in achieving the learning objectives by applying the right learning model. Further integration of ICT-based learning models into is one of the new innovations in the learning in the classroom [1] [2] [3]. 21st century learning is a learning experience that helps students develop the potential of the socio cultural, cognitive, Metacognition, productive, and technology to work [4] [5]. Digital technology is often presented as the driving force of the transformation of education and ICT will contribute positively toward this transformation [6] [7] [8]. Teachers who have experience and are accustomed to utilizing ICT in learning such as accessing the internet, using a communication application, using word processing applications and educational CDs, still requires a basic knowledge of How to develop a web page and multimedia applications, how to make the application of simulation learning and how to make computer programs and web pages that require the capability of programming, graphics processing and editing sounds like the ability of an already professional [9]. Most of the teachers are not capable of using the hardware in the learning process because of lack of resources mainly teaching and learning. This shows that most teachers in college does not integrate ICT in their teaching courses [10]. Teachers are developing and testing a new strategy that is specific to mediating learning ICT-supported. In particular, some form of ICT by focusing the attention of students to the learning objectives of the underlying [11].

Gap with previous research i.e. research analyzing the relationship with coordinating demographics as independent variable, such as the location of the school, the type of school and school accreditation status at present, and to give preference a more thorough information about the attitude, and the willingness of teachers to learn more skills in ICT.

IOP Conf. Series: Materials Science and Engineering **306** (2018) 012139 doi:10.1088/1757-899X/306/1/012139

Research on factors influencing teachers ' use of ICT in education one of them comparing the two theories about the factors that affect teachers in implementing ICT in learning. The results showed that the theory of self-efficacy and attitudes have been widely reported in the U.S. research symbiotically related to ICT and its use in classrooms [12].

A review of research through interviews to teachers who teach the subjects the basic computer and network shows still awaited the teacher who does not make use of ICT in learning. Teachers who have yet to maximize the use of ICT in the lesson due to lack of creativity, not innovative and are still using the old traditions so that the decrease in the level of professionalism and adaptability of teachers towards new technologies. The main problem arises from some discussion of how the perception of teacher research against the use of ICT and the factors affecting the use of ICT?

Based on the problems that are found, then do some research on how the perception of teachers in the use of ICT in learning computer networking, so that the formulation of the problem can be formulated into several research questions as the following:

1. How the perception of teachers of subjects the basic computer and network against the use of ICT?
2. What are the factors affecting the use of ICT in learning basic computer and network?

2. Experimental Methods

This research uses descriptive techniques with engineering literature and documentation. The descriptive method was used because it is only aimed to know the perceptions of teachers on the use of ICT in learning computer networking. The design of the research can be seen in Figure 1 as follows.

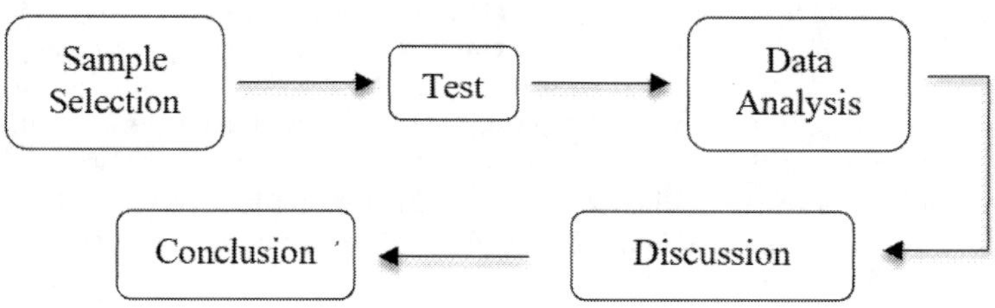

Figure 1. Research Design

Review of the research conducted by obtaining background information of the respondents, the response and the interest of the respondent against the ICT in learning, increase proficiency level of respondents in using ICT for learning purposes, passed the level support for ICT and the problems encountered in using ICT.

The population of the research was obtained from all subjects. They are teacher computer and basic network, totaling 45 persons from 16 Vocational high school in Dumai. As for the sample of the study amounted to 25 teachers who have been teaching for more than 5 years. Sampling technique was purposive sampling. This research defines a purposive sampling as the unit of random selection from the population that is believed to produce the desired estimate of the parameters of the wanted population [13] [14]. Research conducted for one week to answer the question in the form of a detailed questionnaire.

IOP Conf. Series: Materials Science and Engineering **306** (2018) 012139 doi:10.1088/1757-899X/306/1/012139

Table 1. Respondents by Location and School Status.

No	School Name	Status	Accreditation	*Distance from city (km)*	Number of teachers
1	SMKN 1 Dumai	Public	A	5	2
2	SMKN 2 Dumai	Public	A	3	4
3	SMKN 3 Dumai	Public	B	45	1
4	SMKN 4 Dumai	Public	A	28	3
5	SMKN 5 Dumai	Public	A	9	3
6	SMKN 6 Dumai	Public	-	20	1
7	SMK Taruna Persada	Private	A	6	3
8	SMK Al Munawarah	Private	B	4	1
9	SMK Erna	Private	A	1.5	2
10	SMK Budi Dharma	Private	B	1	2
11	SMK Ikasari	Private	-	7	1
12	SMK PGRI	Private	-	4	1
13	SMK Pelayaran	Private	B	1	1

Based on table 1, the sample used is a teacher who has taught the subjects of the basic computer and network at his school. Sampling techniques in the study using a Purposive Sampling with a particular purpose after the test then the data analysis and discussion then made the conclusions of the study results. 25 questionnaires were randomly distributed to teachers of subjects to computers and computer networks from 16 vocational school located in Dumai, and 25 questionnaires filled out. It gives 100% response rate.

3. Results and discussion

Based on table 2, researchers looked at the level of teacher interest in ICT in computer use in various work and learning, use of e-mail / messenger, internet and ICT utilization in learning process.

Table 2. Level of Teacher Interests ICT

No	Level of teacher's perception in ICT	High (%)	Medium (%)	Low (%)
1	The use of computers for various jobs and learning	23 92%	2 8%	0 0%
2	Use of e-mail / messenger to communicate	22 88%	2 8%	1 4%
3	Finding interesting graphic material	17 68%	6 24%	2 8%
4	Efforts to improve ICT skills for learning in various ways	18 72%	5 20%	2 8%
5	Learn the use of ICT from other teachers	19 76%	5 20%	1 4%
6	Getting the latest information for learning via the internet	20 80%	4 16%	1 4%
7	Proficient using ICT tools in learning	21 84%	3 12%	1 4%
8	Attend government-run ICT training	25 100%	0 0%	0 0%

Based on table 2, researchers looking at the level of interest in using ICT in teacher against the computer in a variety of work and learning, the use of mail/messenger, internet and utilization of ICT in the learning process.

Through this research question, the items generally respondents have a very high level of interest towards ICT i.e. average 82.5% to 4% of the overall respondents. While 13.5% have a medium level of interest towards ICT. Use of computers in a variety of work and learning, the second highest position i.e. 92% interest teachers to training after ICT organized government that is 100%. The difficulty in finding teachers learning materials an attractive graphic shape is lowest with the results of question items that is 68%.

Through the exposure data obtained from this research indicate a very high teacher's interest in using computers and ICT in learning. The Internet application and the messenger is one of the media that are effective in improving teachers ' professional knowledge and abilities.

Some of the factors affecting the use of ICT in learning basic computer and network. In research presented i.e. theory of self-efficacy [15] and the other is the behavioral relations theory [16]. The second theory is used on research learning in ICT utilization.

A second theory suggests that a positive attitude that is associated specifically with ICT in learning and feelings of self-efficacy in the use of computers in education seems to affect the use of ICT more. A positive attitude toward ICT generally advised not to contribute to the use of ICT in the classroom. The difference between the importance of specific and general attitudes towards the use of ICT and the emphasis on self-efficacy contributes to the research. Self-efficacy and attitudes associated with each other are advised to use ICT.

4. Conclusion

Perceptions of teachers is an affective domain which serves to encourage teachers to incorporate the use of ICT in particular direction with regard to the task of teaching. Perception towards ICT will also increase the willingness of teachers to advance yourself and change your attitude toward more positive. Research results through the analysis that has been performed, clearly states the teachers the basic computer and network has high interest i.e. 82.5% in the use of ICT as a medium of instruction. ICT is no longer something a new phenomenon in the world of education. Master teacher will also show an increased knowledge and skills related to ICT viewing from the willingness of teachers to follow a course that is 100%. Teachers also strive themselves through fellow colleagues about how the use of ICT equipment for the purpose of teaching and learning.

Acknowledgement

The author wishes to thanks for Mr. Dedi Rohendi, Nurhabibah, Halida Yanti and Yusep Zaini Mi'raj for the journal reviewers and editors who helped to improve the paper.

References

[1] R Schibeci, J Maccallum, W C Potvin, C Durrant, B Kissane, E J Miller 2008 Teachers ' journeys towards critical use of ICT *Learn Media Technol* **33** 4 313–327

[2] J McKenzie 2002 Beyond toolishness *Multimed Sch.* **4**

[3] M E Webb 2005 Affordances of ICT in science learning: Implications for an integrated pedagogy *Int J Sci Educ.* **27** 705–735

[4] J Hwee, L Koh, C Sing, C Wong and B H Hong 2015 Technological Pedagogical Content Knowledge (TPACK) and Design Thinking : A Framework to Support ICT Lesson Design for 21st Century Learning *Asia-Pacific Educ Res.*

[5] C Durrant and B Green 2000 Literacy and the new technologies in school education: Meeting the l(IT)eracy challenge? *Aust J Lang Lit.* **23** 2 89–108

[6] T Fisher 2006 Educational transformation: Is it, like "beauty", in the eye of the beholder, or will we know it when we see it? *Educ Inf Technol.* **11** (3–4) 293–303

[7] M Nivala 2009 Simple answers for complex problems: Education and ICT in Finnish information society strategies *Media, Cult Soc.* **31** 3 433–448

[8] G Ottestad 2010 Innovative pedagogical practice with ICT in three Nordic countries – Differences and similarities *J Comput Assist Learn.* **26** 478–491

[9] S Aydin 2013 Teachers' perceptions about the use of computers in EFL teaching and learning : the case of Turkey *Comput Assist Lang Learn* **26** 3 214–233
[10] M A Gebremedhin Fenta AA. Assessing Teachers ' Perception on Integrating ICT in Teaching-Learning Process : The Case of Adwa College. J Educ Pract. 2015;6(4):114–25.
[11] Hennessy S, Ruthven K, Brindley S. Teacher perspectives on integrating ICT into subject teaching : commitment , constraints , caution , and change Teacher perspectives on integrating ICT into subject teaching : commitment , constraints , caution , and change. J Curric Stud. 2005;37(2):155–92.
[12] Player-koro C 2012 Factors Influencing Teachers' Use of ICT in Education *Educ Inq.* **3** 1 93–108
[13] J M Guarte and E B Barrios 2006 Estimation under purposive sampling *Commun Stat Simul Comput.* **35** 2 277–284
[14] J W Creswell 2013 *Research Design: Qualitative, Quantitative, and Mixed Methods Approaches. Research design Qualitative quantitative and mixed methods approaches*. **398**
[15] A Bandura, C Argyris, Bandura and S S Stevens 1977 *Self-efficacy mechanism in human agency. Psychol Rev* [Internet] **55** 2 191–215 Available from: http://www.westernsnowandice.com/09-Presos/DoubleLoop.pdf
[16] R H Fazio and C J Williams 1986 Attitude accessibility as a moderator of the attitude–perception and attitude–behavior relations: An investigation of the 1984 presidential election. *J Pers Soc Psychol* **51** 3 505–514

The Application of Problem-Based Learning in Mechanical Engineering

Z A Putra[1*] and M Dewi[2]

[1]Universitas Negeri Padang, Indonesia
[2]Universitas Putra Indonesia YPTK Padang, Indonesia

*zonnyap@gmail.com

Abstract. The course of Technology and Material Testing prepare students with the ability to do a variety of material testing in the study of mechanical engineering. Students find it difficult to understand the materials to make them unable to carry out the material testing in accordance with the purpose of study. This happens because they knowledge is not adequately supported by the competence to find and construct learning experience. In this study, quasy experiment research method with pre-post-test with control group design was used. The subjects of the study were students divided in two groups; control and experiment with twenty-two students in each group. Study result: their grades showed no difference in between the pre-test or post-test in control group, but the difference in grade existed between the pre-test and post-test in experiment group. Yet, there is no significant difference in the study result on both groups. The researcher recommend that it is necessary to develop Problem-Based Learning that suits need analysis on D3 Program for Mechanical Engineering Department at the State University of Padang, to ensure the compatibility between Model of Study and problems and need. This study aims to analyze how Problem-Based Learning effects on the course of Technology and Material Testing for the students of D3 Program of Mechanical Engineering of the State University of Padang.

1. Introduction

National Education System is aimed to yield brilliant and competitive Indonesians with excellent personality and fond of quality, high-spirited, strong-willed, independent, able to build up networking, friendly with changes, innovative, willing to be the agent of change, productive, aware of quality, global oriented, and study for the whole life time [1]. Brilliant in this term involves spiritual, emotional, social, intellectual and kinetic intelligence. He becomes a person with competitive competence is capable to compete and fight for his life and becomes a person who can give meaning to his own life and the lives of others which is specified by UNESCO as 'How to Live Together". From this point of view, it can be seen that the process of producing intelligent and competitive human resources is standardized on education. Since the development of quantity triggers high competition among universities, it is necessary for the university to present excellent quality graduates who can stand out in competition [2].

The Indonesian Law No. 20 2003 states that education is an effort intentionally made and well-planned planning to provide a good environment for study to enable students to actively develop their potentials, to have spiritual power, self-control, intelligence, good-manner, and as well as skills necessary for himself and other people. The vision of national education of 2025 is to produce intelligent and competitive Indonesians well-prepared to face anything to come. One of those efforts made to achieve these targets of national education is to provide people with vocational education. The basic philosophy of vocational education is the interaction of students and business-like environment [3]. As a result, practitioners in vocational education have to pay a closed attention to students through educational programs.

As one of the educational programs at the Vocational pathways, the students of D3 Mechanical Engineering are educated through the learning program integrated with the needs of graduates' provision used in industry. Generally Technical Vocational Education plays an important role in preparing its graduates to face globalization and technological changes that affect the life of the social, political, economic and economy of the society. The challenge of the business world with job competence that increases with the advancement of technology requires that a higher education institution should be able to anticipate and address changes that occur by utilizing a variety of capacities by preparing the students with knowledge during their studies [4].

Problems that occur in the field associated with the process of learning in the course of Technology and Materials Testing which is a must for students of D3 Mechanical Engineering of the State University of Padang that the learning process is still carried out with conventional ways. Learning models that are less-triggering passion and creativity of students, learning activities are monotonous and cause boredom, the lack of activity of thinking which allows students to not gain the ability to solve problems that are factually and contextually appropriate to the reality in the field of employment. As a result, less innovative learning process which has been carried out causes the students to have low abilities in the subject of Technology and Testing of Materials. Students who are expected to develop reasoning and critical thinking in testing tend to have low ability.

Based on their motor skills students in practical coursework met the fact that they were not able to carry out the testing of materials in accordance with the standards of competence expected, although the course has been equipped with worksheet and teaching materials, as well as the application of the teaching method of discussion, lectures, a FAQ, cooperative learning, assignments and practices but the results of student learning is still inadequate. All of these result in low learning outcome, and many students get inadequate grade and thus have to retake the course in the next semester. The researchers face that the problem tends to re-occur every year so that it triggers polemics and thought that the material that is given to students may be too difficult for them to understand. If it is not based on the ability of students to find their own knowledge and construct a learning experience to solve problems through learning undertaken related with material Technology and Materials Testing. Departing from the above problems, the researcher has the initiative of researchers to apply Model-Based Problem learning so that the phenomenon of failure in learning Materials Testing can be overcome.

2. Theoretical study

2.1. The Course of Technology and Material Testing

The course of Technology and Materials Testing is the practice course through which students are equipped with knowledge about the various tests such as tensile testing, impact testing, and the testing of microstructure hardness of a material. This subject weigh 3 credits (2 credits 1 credits theory and practice). The contents of the course to the theory include: classification of engineering materials, the properties of engineering materials, phase diagrams, standardization of engineering materials, ferrous metals, nonferrous metals, non-metals, materials selection criteria, and corrosion. As for the practice include: testing of hardness, tensile testing, deflection testing, torsion testing, testing at the notch, the macro and micro structure.

Materials provided on this course was developed in accordance with the competencies to be developed concerning: 1) Describing the classification of engineering materials, 2) Describing the properties of engineering materials, 3) Explaining the diagram phase, 4) Describing the standardization of materials engineering, 5) Describing Miscellaneous uses of ferrous metals, 6) Describing the kinds and applications of nonferrous, 7) Describing the kinds and uses of non-metallic materials engineering, 8) Describing the criteria for the selection of technical material in the application, 9) Describing the corrosion and prevention, 10) Having the ability to perform tests and analyze the results of testing of hardness, 11) Being able to perform testing and to analyze the results of tensile test, 12) Being able to perform testing and to analyze the results of torsion test, 13) Being able to perform testing and analyzing the test results at a notch, 14) Being able to perform the testing and analyzing the test results deflections, 15) Being able to do a test and analyze the test results of microstructure.

The Interest of the course of Technology and Materials Testing is that students are able to classify materials engineering, properties of engineering materials, diagrams phase, standardization of engineering materials, ferrous metals, nonferrous, non-metallic, criteria for the selection of materials and corrosion and practically able to do testing hardness, tensile testing, testing a deflection, torsion testing, testing at the notch, the macro and micro structure. From the learning objectives developed by the Syllabus, educators have to make some efforts to implement a learning model that can develop the ability of students to solve problems and become independent learners. Because it is believed that learners will never gain the ability if they do not have the experience that encourages their development [5]. Therefore, to achieve this goal of study, the researcher as an educator implements Problem Based Learning Model that is carried out in accordance with the lesson plan with Syntac expressed by the experts who developed this learning model.

2.2. Problem Based Learning Model

In achieving national educational goals that have been implemented, the quality of the defense-distance must be considered. The job of an educator to create learning that can help students understand the lessons given and realize the change of behavior from not understanding to understanding with a fun process, continuously and continues to change as thoughts and feelings that continue to be built and developed on an ongoing basis [6]. One effort in order to realize the quality learning is to use a learning model which has proven its accuracy and usefulness in achieving the learning objectives.

Teaching models that have been designed by experts through research that that has been done are considered to be a way to help learners acquire information, ideas, skills, values the way of thinking and purpose of expressing self-learners, so the capability of learners in the learning becomes easier and more effectively in the future. Embodiments of learning success through appropriate learning model obtained for students to better cope with the learning process. Through the PBL students can actively build a mental model of the problem and its solution, PBL encourages self-direction, so that students can set goals of study by themselves, identifying barriers and arbitrary been awarded by his own efforts in learning [7].

PBL allows students to be improve their ability to analyze a material, PBL is an instructional method that challenges students to learn, to work together in groups, and to find solutions to real problems that these problems can be used to remind curiosity as well as analytical and initiatives on the subject matter. Their curiosity can be enhanced through the model PBL, to attract students to learn is by presenting a problem that can be attractive so that the learning process is not monotonous and boring begins with providing an appropriate issue to the knowledge base of students that will foster the students' enthusiasm [8].

Thus, the models of educational practices that are considered beneficial in the century industrial, such as learning the facts, drill and practices, rules and procedures were replaced by learning in the context of real-world, authentic through problems and projects, inquiry, discovery, and invention into practice knowledge age. One model that is aligned with the development of information technology is a Model Problem Based Leaning.

PBL is a set of models of teaching that makes the problem the focus on developing problem-solving skills, materials and self-regulation [9]. Model Problem-based learning is an innovation in the creation-learning because this model is the ability to think Learners truly optimized through the process of the work group or team is a systematic, so that learners can empower, sharpen, test and develop the capacity to think continuously [10].

PBL model is the most significant innovation in education [11]. Model PBL helps to enhance the development of skills for lifelong learning in a mindset that is open, reflective, critical, and active learning, as well as facilitating the success of problem solving, communication, teamwork and interpersonal skills better than other models.

PBL is an instructional model that involves learners to solve problems through the stages of the scientific method so that learners can learn the knowledge related to these issues and also have the skills to solve problems [12]. PBL is a learning model that challenges learners to "learn how to learn", to work in groups to seek solutions to real-world problems. This problem used to bind Learners curiosity on learning in question. PBL is an approach to learning in which learners are faced with the problem of authentic (real) so hopefully they can draw up his own knowledge, to develop high-level skills and inquiry, to help learners be independent, and to increase their confidence [13].

It can be concluded that the model PBL is an instructional model that is designed where learners actively confronted with complex issues in a real-life situation learning model is centered on students, and the curriculum is presented in the form of the existing problems so that learners have a curiosity out, which in turn will solve the problem. Based on the inferred meaning of understanding the problem-based learning is considered a good learning model used in practical learning for students Materials Testing D3 Mechanical Engineering.

2.3. Application Procedures Learning Model

PBL characteristics is learning focuses on problem-solving, 2) responsibility in resolving the problem rests on the students, 3) teachers support the process when students are working on the problem. Based on these studies, the researchers designing and implementing problem-based learning is based Characteristics presented with descriptions syntac as follows [5]:

2.3.1. Reviewing and Presenting Problems. In this phase, the researchers begin by reviewing the initial knowledge needed to solve the problem and then presents a problem themselves. Researchers carefully consider the progress of the students in identifying the problems encountered in the material presented, to be more effective, the researchers in this phase determines the issues to be discussed by the students, it is for effective learning. The researcher also provides training in identifying problems associated with testing material. In this phase also formed group of study consisting of 4 to 5 students in one group.

2.3.2. Developing Strategies. In phase 2 students are directed to strategize in solving problems that have been identified. Researchers provide guidance in order to avoid waste of time for students in determining the strategy in solving the problem. In the concept of giving the guidance, the researcher simply directed students and still provided decision-making in the hands of student problem-solving strategies

2.3.3. Implementing Strategies Researchers apply the learning support to help students complete the tasks that are not able to be completed by themselves in applying problem-solving strategies.
The researcher studied carefully when he was able to implement troubleshooting strategy. So that students do not spend time in guessing the correct answer from the strategy they charge.

2.3.4. Discussing and evaluating Results. In this last phase the researchers asked students to discuss and display the problem-solving strategies that they have done in front of the class. Ask students to explain and evaluate its own strategy of testing materials that have been done assessing the validity of the strategy used. Other class members would ask questions and members of the group will maintain the strategy used while researchers are directing the address the complexities and difficulties that may be encountered disclosure of the results of students in solving problems that they have done. The final result is a draw the conclusion of solving problems that do and direct class members to draw the conclusion together.

3. Methods

3.1. Research procedure
This study uses a quasi-experimental study (quasi-experimental) design with pre-post-test with control group design ", with the treatment of PBL Model. This study was conducted to determine the change in the form of students' ability to learn the results before and after treatment of the application of learning models, then researchers compared two groups of students, namely the treatment group and the control group. The initial stage is to divide the two groups of students in the classroom and then take the initial value as a data protest.

3.2. Data analysis technique
Subjects in this study consisted of two groups of classes, namely class control group (Q1) and an Experiment Group (Q2). The number of subjects of each class is 22 students. The procedures performed in the study were (1) Perform the protest for each class at the beginning of the semester courses, (2) treatment group taught by the application of the Model Problem Based Learning (3) The control group were still taught by conventional Model, for example the method of lecture and using the practice of using job sheet without using the problem Based Learning Model. (4) At the end of the semester was done post-test. Data were obtained using an assessment rubric for the practice, and essay questions to measure the ability of theory. The data analysis was done to describe the data of each group by revealing Central Tendency data and formulas Percentage Rate of Achieved Respondents were categorized. Whereas, to measure the difference in the effect of the use of Model-Based Problem using Statistical Analysis The test (Paired sample t test) to find whether there is any difference in effect in the current group protest and post-test and use statistic test (independent sample t test) to find whether there is any difference in effect treatment control group and experimental group.

4. Results and discussion

4.1. Initial Capabilities Equality Group Control and Experiment
Based on data analysis that has been done it can be explained description of the results of research on each group during the protest associated with early ability students, a more complete can be seen in Table 1:

Table 1. Description of Study Result of Pretest

Group	N	Mean	Median	SD	Sum	Min-Max
Control (Q0)	22	68.41	68.00	5.903	1505	60-82
Treatment(Q1)	22	67.45	66.50	7.513	1484	55-80

Table 1. Based on the table above it is seen that the prior knowledge of students in the control group (Q0) and experimental (Q1) have on average relatively similar (Q0) = 68.41, and (Q1) = 67.45. While based on the analysis of the t test (independent Sample T Test) to see the equality of both groups note that the significance (Sig.) is placed on 0082 scores. This score means that the two groups have no significant difference in the results. It means that the two groups have in common tendency capability significantly.

4.2. Differences Results Pretest and Posttest Experimental Group and Control
To determine differences in learning outcomes at protest and post-test in the experimental and control groups, it can be seen in the description following studies:

Table 2. The result difference between Pretest and Post-test

Group	Mean	95% (CI)	t hit	t tab	P value
		Control (Q0)			
Before	68.41	2.622 - 1.349	.667	1.717	0.512
After	69.05				
Diff	0.64				
		Experiment (Q1)			
Before	67.45	12.877- 4.941	4.669	1.717	0.000
After	76.36				
Diff	8.91				

Based on the data analysis, it shows that the learning outcomes of students in the control group had an average difference of 0.64. Based on t test known that score t <t table which means that there are no differences in learning outcomes and initial ability of students between protest and post-test, boosted with p value = 0512 stating that no tar-can be a real difference at 95% significance level. In the experiment group it was discovered that the average difference was 8.91. Based on t test known that score t> t table which means that there are differences in learning outcomes and initial ability of students in the subject of Materials Testing between protest and post-test, boosted by the significance of the score of 0.000 in the sense that there are real differences were significant at the significance level 95%.

4.3. Difference between Learning Outcomes Group Control and Experiment

Based on data analysis that has been done it can be explained description of the results of research in each group at the time associated with the post-test of Student Learning Outcomes in Subjects Testing Materials, more complete can be seen in Table 3 below:

Table 3. Description of study result of Post-test

Group	N	Mean	Median	SD	Sum	Min-Max
Control (Q0)	22	69.05	70.00	7.377	1519	55-80
Treatment(Q1)	22	76.36	76.50	8.561	1680	55-90

Based on the above table it is seen that student learning outcomes in the control group (Q0) and experimental (Q1) have on average relatively distinct (Q0) = 69.05, and (Q1) = 76.36. While based on the analysis of the t test (independent Sample T Test) to see the difference in the results of both groups note that the significance (Sig.) on the score of 0973. This score means that the two groups have no significant difference in the results. It means that the two groups have in common tendency capability significantly. This shows that with the implementation of Problem Based Learning Model protest and post-test although there are differences in the results of the experiment group, but there are differences in learning outcomes significantly in experimental and control groups.

5. Conclusion

The study aimed to analyze the effect of the application of the Model Problem Based Learning (PBL) on the subjects of Technology and Materials Testing showed that there was no difference in the value protest and post-test in the group taught by using conventional means that there is no increase in learning outcomes in this group. There are differences in the value protest and post-test student groups taught by PBL models, which means that there is an increase learning outcome through the implementation of

PBL Model. But the results of the study to analyze the differences in learning outcomes significantly in both groups showed that there were no differences in learning outcomes in both groups significantly and note the difference in the success of studying the treatment group was not too big (protest 67.45 increased to 76.36 with a difference of 8.91 when post-test). This shows that with the application of PBL model although there are differences in protest and post-test results in the group being taught by PBL models but there are differences in learning outcomes significantly in experimental and control groups at the end of the study. This shows that there are still many things that need to be considered in applying the model PBL because based on this study both groups showed significantly different results. Related to the results of this study, researchers recommend that there are things that are associated with the process of learning PBL to note, considering learning process is a system that is interconnected between one component to another, if educators do not pay attention to a component, the possibility of success in study will be hard to achieve. Rusman (2012: 93) argues that learning is seen as a system consisting of various components that relate to one another. The components cover objectives, materials, methods of study, strategies and evaluation. The method is an attempt to implement a plan that has been prepared in real activities to achieve the goal [14]. The method is a learning procedure chosen by educators to assist students in achieving the learning objectives with maximum outcome. The learning method is a series of activities deliberated with designing, developing, aligning implementation, and evaluating a particular method for backing facilitate learners with the goal of achieving a competency. As educators need me-develop a model that is closer with the needs and circumstances of the problem of system in place to teach, therefore it is necessary to develop PBL learning model that works for student learning needs D3 Mechanical Engineering University of Padang to overcome existing deficiencies in the PBL model that has been proposed by experts.

References

[1] S Munadi 2008 *Transformasi Teknologi pada Pendidikan Kejuruan* FT UNP: Temu Karya Konvensi Nasional Aptekindo V

[2] E Rosalin 2010 Membangun Kompetitif Advantage Perguruan Tinggi dalam Menghadapi Tantaangan dan Perubahan Abad 21 *Jurnal Manajemen Pendidikan* Nomor 02/Tahun VI/Oktober 010

[3] B C Miller and E M Duvall 1985 *Marriage and Family Development* (9th Ed)

[4] A M Taufiq 2009 *Inovasi pendidikan melalui problem based learning. Bagaimana pendidik memberdayakan pemelajar di era pengetahuan* Jakarta: Kencana Prenada Media Group

[5] Eggen and P D Kouchak 2012 *Strategi dan Model Pembelajaran*. Jakarta: PT. Indeks

[6] Joyce, M Weil and E Calhoun 2009 *Models of Teaching* Edisi Delapan. Jogyakarta. Pustaka Pelajar

[7] Williams, C Judith and D J Paltridge 2016 *What We Think We Know About the Tutor in Problem-Based Learning. Health Professions Education* Available online at www.sciencedirect.com.

[8] S Maggi and H M Claire 2000 *Founda-tions of problem-based learning* New York: Open University Press

[9] E P Sarafino 1998 *Health Psychology: Biopsychosocial Interactions* Third

[10] Tan Oon-Seng 2004 *Enhancing thingking through problem-based learning approaches* Cengage Learning

[11] Bound dan Felleti 1997 *Model Pembelajaran Berbasis Masalah* Online. http://duniapembelajaran.com.2012

[12] W Kamdi 2007 *Model- Model Pembelajaran Inovatif*. Universitas Negeri Yogyakarta

[13] Arends 2007 *Learning to Teach (belajar untuk mengajar)* Yogyakarta: Pustaka Pelajar

[14] R Heinich, M Molenda and J D Russel 1989 *Instructional and new technologies of instruction* New York: Macmillan, Inc.

AUTHOR INDEX

AUTHOR INDEX

AUTHOR INDEX